ISSAC 2013

Boston, USA, June 26–29, 2013

Proceedings of the
38th International Symposium on
Symbolic and Algebraic Computation

Manuel Kauers, editor

Association for Computing Machinery

Advancing Computing as a Science & Profession

Association for
Computing Machinery

Advancing Computing as a Science & Profession

The Association for Computing Machinery
2 Penn Plaza, Suite 701
New York, New York 10121-0701

ISBN: 978-1-4503-2059-7

Additional copies may be ordered prepaid from:

ACM Order Department
PO Box 30777
New York, NY 10087-0777, USA

Phone: 1-800-342-6626 (USA and Canada)
+1-212-626-0500 (Global)
Fax: +1-212-944-1318
E-mail: acmhelp@acm.org
Hours of Operation: 8:30 am – 4:30 pm ET

ACM Order Number: 505130

Printed in the USA

Foreword

The 2013 International Symposium on Symbolic and Algebraic Computation (ISSAC) is the premier conference for research in symbolic computation and computer algebra. ISSAC 2013, held at Northeastern University in Boston, USA, is the 38th meeting in the series, which began in 1966 with the seminal ACM Symposium on Symbolic and Algebraic Manipulation. ISSAC 2013 is fully sponsored by the Association for Computing Machinery and its Special Interest Group on Symbolic Manipulation (ACM SIGSAM).

The ISSAC meeting is a showcase for original research contributions on all aspects of computer algebra and symbolic mathematical computation, including:

Algorithmic aspects:

- Exact and symbolic linear, polynomial and differential algebra
- Symbolic-numeric, homotopy, perturbation and series methods
- Computational algebraic geometry, group theory and number theory
- Computer arithmetic
- Summation, recurrence equations, integration, solution of ODEs and PDEs
- Symbolic methods in other areas of pure and applied mathematics
- Complexity of algebraic algorithms and algebraic complexity

Software aspects:

- Design of symbolic computation packages and systems
- Language design and type systems for symbolic computation
- Data representation
- Considerations for modern hardware
- Algorithm implementation and performance tuning
- Mathematical user interfaces

Application aspects:

- Applications that stretch the current limits of computer algebra algorithms or systems, use computer algebra in new areas or new ways, or apply it in situations with broad impact.

The ISSAC Program Committee adhered to the highest standards and practices in the evaluation of submitted papers, and we are very pleased with the quality of the papers appearing at the conference. All papers submitted to ISSAC were judged, and accepted or rejected, solely according to their scientific novelty and excellence. An average of more than 3.3 referee reports were obtained for each submission, and 47 papers were ultimately accepted for publication. Each submitted paper was assigned to two Program Committee members, but all members could and did actively participate in the evaluation of other papers. Strict conflict of interest rules were enforced, disallowing any access to the evaluation process of papers by institutional colleagues, recent co-authors, supervisors, students or under any other biasing circumstance. All papers for which broad agreement was not obtained were voted upon in a final and binding ballot. The Program Committee thanks all the authors of all submitted papers for considering ISSAC for their best work, and hopes that the high quality of the accepted papers validates their choice of venues and encourages submission to ISSAC in the future.

ACM SIGSAM sponsors awards for Distinguished Papers and Distinguished Student Authors at every ISSAC, and these will be selected by a vote of the Program Committee. While the winners are not known at the time of this writing, the quality of the candidates ensures that these papers should have great merit and impact on our field.

ISSAC features invited talks, contributed papers, tutorials, posters and software demonstrations. These Proceedings contain all accepted contributed papers, and abstracts of the invited talks and tutorials. Abstracts of posters and software demonstrations will appear in an upcoming issue of the ACM SIGSAM Communications in Computer Algebra. We are thrilled with the exceptional scientific stature of our invited presenters, and the high quality of all the contributed works, and thank everyone for their investment in ISSAC 2013.

Running a large conference such as ISSAC 2013 depends upon many volunteers, and hopefully all of these people are listed on the following pages. Without their contributions, this conference would not have been possible. We gratefully acknowledge the financial sponsors of ISSAC 2013; their generous and continued support enhances the quality and accessibility of the conference. Finally, we write this Foreword in anticipation of a scientifically rewarding, educational and exciting ISSAC 2013 for all participants!

Gene Cooperman *(Co-General Chair)*
Mark Giesbrecht *(Program Committee Chair)*
Manuel Kauers *(Proceedings Editor)*
Michael Monagan *(Co-General Chair)*

Table of Contents

ISSAC 2013 Conference Organization

General Chairs: Michael Monagan *(Simon Fraser University, Burnaby, Canada)*
Gene Cooperman *(Northeastern University, Boston, USA)*

Program Chair: Mark Giesbrecht *(University of Waterloo, Waterloo, Canada)*

Proceedings Editor: Manuel Kauers *(RISC, Johannes Kepler University, Linz, Austria)*

Local Arrangements Chair: Gene Cooperman *(Northeastern University, Boston, USA)*

Tutorial Chair: David Saunders *(University of Delaware, Newark, USA)*

Treasurer: Daniel Roche *(US Naval Academy, Annapolis, USA)*

Publicity Chair: Jeremy Johnson *(Drexel University, Philadelphia, USA)*

Webmaster: Andy Novocin *(University of Waterloo, Waterloo, Canada)*

Program Committee: Hirokazu Anai *(Fujitsu Laboratories, Kanagawa, Japan)*
Xavier Dahan *(Kyushu University, Fukuoka, Japan)*
James Davenport *(University of Bath, Bath, UK)*
Joachim von zur Gathen *(B-IT/University of Bonn, Bonn, Germany)*
Mark Giesbrecht *(University of Waterloo, Waterloo, Canada)* – Chair
Mark van Hoeij *(Florida State University, Tallahassee, USA)*
Erich Kaltofen *(North Carolina State University, Raleigh, USA)*
Manuel Kauers *(RISC, Johannes Kepler University, Linz, Austria)*
Anton Leykin *(Georgia Tech, Atlanta, USA)*
Steve Linton *(University of St. Andrews, St. Andrews, UK)*
Elisabeth Mansfield *(University of Kent, Canterbury, UK)*
Tomas Recio *(Universidad de Cantabria, Santander, Spain)*
Bruno Salvy *(INRIA, Lyon, France)*
Hans Schönemann *(Univ. of Kaiserslautern, Kaiserslautern, Germany)*
Éric Schost *(Western University, London, Canada)*
Allan Steel *(University of Sydney, Sydney, Canada)*
Barry Trager *(IBM Research, Yorktown, USA)*
Lihong Zhi *(Academia Sinica, Beijing, China)*
Paul Zimmermann *(INRIA, Nancy, France)*

Poster Committee: Alin Bostan *(INRIA Saclay Ile-de-France, Palaiseau, France)* – Chair
Shaoshi Chen *(North Carolina State University, Raleigh, USA)*
Viktor Levandovskyy *(RWTH, Aachen, Germany)*
Guénaël Renault *(Université Pierre et Marie Curie, Paris, France)*
Agnes Szanto *(North Carolina State University, Raleigh, USA)*

Software Demos Committee: Joris van der Hoeven *(Ecole Polytechnique, Paris, France)* – Chair
Roman Pearce *(Simon Fraser University, Burnaby, Canada)*

External reviewers:

John Abbott
Sergei Abramov
Andrew Arnold
Roberto Barrio
Dan Bates
Carlos Beltran
Frits Beukers
Greg Blekherman
Johannes Bluemlein
Marco Bodrato
Janko Boehm
Alin Bostan
François Boulier
Sylvain Boulmé
Andrew Bremner
Richard Brent
John Brzozowski
Laurent Busé
Massimo Caboara
Jacques Carette
Francisco-Jesús Castro-Jimenez
Changbo Chen
Shaoshi Chen
Jin-San Cheng
John Chinneck
Frédéric Chyzak
Thomas Cluzeau
Michel Coste
Brendan Creutz
James Davenport
Bart De Bruyn
Thomas Dreyfus
Christian Eder
Matthew England
Claus Fieker
Shuhong Gao
Pierrick Gaudry
Jürgen Gerhard
Stefan Gerhold
Patrizia Gianni
Pascal Giorgi
Laureano Gonzalez Vega
Sergey Gorchinsky
Dima Grigoriev
Bjarke Hammersholt Roune
Jonathan Hauenstein
Albert Heinle
Joos Heintz
Christopher Hillar
Eckhard Hitzer
Milan Hladik
Michael Hoffman
Hoon Hong
Hidenao Iwane
Claude-Pierre Jeannerod
David Jeffrey
Xiaohong Jia
Dejan Jovanovic
Bettina Just
Hiroshi Kai
Kinji Kimura
Simon King
Thorsten Kleinjung
David Kohel
Pascal Koiran
Risi Kondor
Marek Kosta
Ilias Kotsireas
Christoph Koutschan
Martin Kreuzer
George Labahn
Daniel Lazard
Sylvain Lazard
Francois Le Gall
Gregoire Lecerf
Viktor Levandovskyy
Ziming Li
Daniel Loebenberger
Ulrich Loup
Kamal Khuri Makdisi
Gregorio Malajovich
Angelos Mantzaflaris
Jose Martin-Garcia
Scott McCallum
Kurt Mehlhorn
Niels Moeller
Michael Moller
Antonio Montes
Teo Mora
Marc Moreno-Maza
Jiawang Nie
Michael Nüsken
Alexey Ovchinnikov
Senshan Pan
Dima Pasechnik
Grant Passmore
Tim Penttila
Clement Pernet
Chris Peterson
Veronika Pillwein
Clemens G. Raab
Alyson Reeves
Greg Reid
Daniel Roche
Fabrice Rouillier
Bjarke Roune
Sonia Rueda
Michael Sagraloff
David Saunders
Carsten Schneider
Mathias Schulze
Markus Schweighofer
Hiroshi Sekigawa
Jason Selby
Tony Shaska
Ekaterina Shemyakova
Kiyoshi Shirayanagi
Ilie Silvana
Michael Singer
Jonathan Sorenson
Stefan Steidel
Arne Storjohann
Adam Strzebonski
Akira Terui
Laurent Théry
Will Traves
William Unger
Joris van der Hoeven
Sven Verdoolaege
Jan Verschelde
Raimundas Vidunas
Gilles Villard
Paul Vrbik
Dingkang Wang
Stephen Watt
Jacques-Arthur Weil
Michael Wibmer
David Wilson
Franz Winkler
Jean Claude Yakoubsohn
Hitoshi Yanami
Kazuhiro Yokoyama
Josephine Yu
Konstantin Ziegler

ISSAC 2013 Sponsors & Supporters

ISSAC acknowledges the generous support of the following institutions.

Sponsor: ACM Special Interest Group on Symbolic and Algebraic Computation

Supporters: Maplesoft

Microsoft Research

Northeastern University

Wolfram Research

Solving Equations with Size Constraints for the Solutions

[Invited Talk]

Henry Cohn
Microsoft Research New England
One Memorial Drive
Cambridge, MA 02142
cohn@microsoft.com

ABSTRACT

In this talk, I'll survey some key applications within coding theory and cryptography for solving polynomial equations with size constraints on the solutions. Specifically, we seek solutions that are small integers or low-degree polynomials. In the single-variable case this is relatively well understood, thanks to important work in the late 1990's by Guruswami and Sudan and by Coppersmith, but higher dimensions hold many mysteries. I'll highlight connections with hot topics such as fully homomorphic encryption, as well as some problems for which progress should be possible.

Categories and Subject Descriptors

I.1.2 [**Computing Methodologies**]: Symbolic and Algebraic Manipulation—*Algorithms*

Keywords

coding theory; polynomial systems

ISSAC'13, June 26–29, 2013, Boston, Massachusetts, USA.
ACM 978-1-4503-2059-7/13/06 .

Lattices with Symmetry

[Invited Talk]

Hendrik Lenstra
Mathematisch Instituut
Universiteit Leiden
Postbus 9512
2300 RA Leiden, The Netherlands
hwl@math.leidenuniv.nl

ABSTRACT

It is a notoriously difficult algorithmic problem to decide whether a given lattice admits an orthonormal basis. However, this problem becomes doable if the lattice is given along with a suitably large abelian group of symmetries. The lecture is devoted to a precise formulation of this result and to an outline of the algorithm that underlies its proof. One of the ingredients is an elegant algorithmic technique that C. Gentry and M. Szydlo introduced several years ago in the context of cryptography, but that can be recast in algebraic language. Other ingredients are taken from analytic number theory and from commutative algebra. Part of the work reported on was done jointly with Alice Silverberg and René Schoof.

Categories and Subject Descriptors

I.1.2 [**Computing Methodologies**]: Symbolic and Algebraic Manipulation—*Algorithms*

Keyword

lattices

ISSAC'13, June 26–29, 2013, Boston, Massachusetts, USA.
ACM 978-1-4503-2059-7/13/06 .

Critical Point Methods and Effective Real Algebraic Geometry: New Results and Trends

Mohab Safey El Din
Université Pierre and Marie Curie
Institut Universitaire de France
INRIA Paris-Rocquencourt
CNRS – LIP6 UMR 7606
Mohab.Safey@lip6.fr

ABSTRACT

Critical point methods are at the core of the interplay between polynomial optimization and polynomial system solving over the reals. These methods are used in algorithms for solving various problems such as deciding the existence of real solutions of polynomial systems, performing one-block real quantifier elimination, computing the real dimension of the solution set, etc.

The input consists of s polynomials in n variables of degree at most D. Usually, the complexity of the algorithms is $(s\,D)^{O(n^\alpha)}$ where α is a constant. In the past decade, tremendous efforts have been deployed to improve the exponents in the complexity bounds. This led to efficient implementations and new geometric procedures for solving polynomial systems over the reals that exploit properties of critical points. In this talk, we present an overview of these techniques and their impact on practical algorithms. Also, we show how we can tune them to exploit algebraic and geometric structures in two fundamental problems.

The first one is real root finding of determinants of n-variate linear matrices of size $k \times k$. We introduce an algorithm whose complexity is polynomial in $\binom{n+k}{k}$ (joint work with S. Naldi and D. Henrion). This improves the previously known $k^{O(n)}$ bound. The second one is about computing the real dimension of a semi-algebraic set. We present a probabilistic algorithm with complexity $(s\,D)^{O(n)}$, that improves the long-standing $(s\,D)^{O(n^2)}$ bound obtained by Koiran (joint work with E. Tsigaridas).

Categories and Subject Descriptors

I.1.2 [**Symbolic and Algebraic Manipulation**]: Algorithms—*algebraic algorithms*; F.2.2 [**Analysis of Algorithms and Problem Complexity**]: Non numerical algorithms and problems—*complexity of proof procedures*

General Terms

Theory, algorithms

Keywords

polynomial system solving, real roots, effective real algebraic geometry

Introduction. Many important results in combinatorial and computational geometry (see *e.g.* [12, 27]), in theoretical computer science (see *e.g.* results on non-negative matrix factorization [1] or game theory [18]) rely on effective real algebraic geometry. Polynomial system solving over the reals has also many applications in engineering sciences, *e.g.* in robotics [2], and control theory [11] among other areas.

Typical computational challenges in real algebraic geometry are: deciding the emptiness of semi-algebraic sets, performing geometric operations such as projection (quantifier elimination), answering connectivity queries (roadmaps), computing the real dimension or computing the Euler-Poincaré characteristic, Betti numbers, etc.

Huge efforts have been invested during the last 25 years to derive algorithms that improve the doubly exponential complexity in n of Cylindrical Algebraic Decomposition [13]. This has led to algorithms for deciding the emptiness of semi-algebraic sets (in time $(s\,D)^{O(n)}$) [7], performing one-block quantifier elimination [6], computing the real dimension [22], answering connectivity queries (in time $(s\,D)^{O(n^2)}$) [8, 12]; see [9] for a self-contained overview.

Critical point methods are at the heart of these results. They consist in extracting important properties of semi-algebraic sets from the *critical points* of a well-chosen map. These are points at which the differential (of the map) is not surjective; local extrema of the map are reached at its critical points. These methods were used in combination with the introduction of infinitesimals that deform the input. This allows us to obtain cheap reductions to smooth and bounded semi-algebraic sets but affects the cost of arithmetic operations and hence practical performance.

It has been a long-standing problem to obtain efficient implementations for real-world problems based on critical point methods. Indeed, it requires to improve the exponents in the complexity bounds by introducing new algebraic and geometric techniques to avoid the use of infinitesimals. One successful research direction is to identify properties of critical points or polar varieties and to exploit them computationally using algorithms of elimination theory.

This trend started with [3] and has been developed for a decade (*e.g.* [4, 15, 19, 23, 24] and references therein) to understand the properties of these sets of points. We refer to [5] for an exposition of properties of polar varieties. Once these properties are understood, they can be exploited to design geometric procedures for solving. For instance, the first improvement of the long-standing $O(n^2)$ exponent in the complexity of Canny's probabilistic algorithm [12] to $O(n^{3/2})$ is based on a new geometric connectivity result obtained by investigating properties of polar varieties in [25] (see also [10] for a further generalization to general algebraic sets).

This talk presents an overview of critical point methods. We highlight recent advances that lead to practically fast algorithms for deciding the existence of real solutions of polynomial systems. We describe new ways of exploiting structural properties of critical points and we introduce new geometric procedures to improve the complexity bounds for solving two important problems: *(i)* real root finding of determinants of matrices whose entries are linear

ISSAC'13, June 26–29, 2013, Boston, Massachusetts, USA.
ACM 978-1-4503-2059-7/13/06.

forms (linear matrices) and *(ii)* computing the real dimension of a semi-algebraic set.

Real root finding of determinants of linear matrices. Let $\mathsf{M}_0, \ldots, \mathsf{M}_n$ be matrices of size $k \times k$ with rational entries, $X_1, \ldots, X_n$ be variables and $\mathsf{M} = \mathsf{M}_0 + X_1\mathsf{M}_1 + \cdots + X_n\mathsf{M}_n$. We consider the problem of finding real roots of the determinant of M. This is a generalization of the eigenvalue problem. It is also related to simultaneous stabilization problems in control theory (*e.g.* [11]). Moreover, if M is symmetric and of full rank, then $\det(\mathsf{M})$ shapes the boundary of the feasible solution set of the Linear Matrix Inequality $\mathsf{M} \succeq 0$. In this case, exact algorithms for finding real points in the feasible set start by computing points on its boundary.

To compute real roots of the determinant of M, the traditional procedure consists in applying algorithms for deciding the emptiness of the real solution set of the equation $\det(\mathsf{M}) = 0$. Using this strategy, the cost is $k^{O(n)}$ arithmetic operations. Additionally, the equation $\det(\mathsf{M}) = 0$ defines a hypersurface with *generic* singularities (corresponding to rank deficiencies greater than 1).

Our approach consists in studying the variety defined by the bilinear system $\mathsf{M}.\mathbf{Y} = 0$ where $\mathbf{Y}$ is a vector of new homogeneous variables. Under some genericity assumptions on the entries of M, this new set of bi-linear equations defines a smooth algebraic set. We show how to reduce our problem to global optimization problems that preserve the bi-linear structure of the system. We model the global optimization problems using Lagrange multipliers that in turn leads to solve multi-linear polynomial systems. The multi-homogeneous bound associated to these systems is dominated by $\binom{n+k}{k}^2$. Using algebraic elimination routines which take advantage of multi-linear structures (*e.g.* [14, 17]), we obtain algorithms whose complexity is polynomial in this quantity. Hence, families of problems where k remains constant can be solved in polynomial time. Preliminary implementations allow to handle problems involving 20 variables (for 6×6 matrices).

Computation of the dimension of semi-algebraic sets. For computing the real dimension of a semi-algebraic set, the best previously known complexity bound, due to Koiran, was $(s\,D)^{O(n^2)}$ [22] in the worst case (see also [28] for a partial improvement). It is based on quantifier elimination techniques, see [9, Alg. 14.10] and references therein. On the other hand, in the complex case, it is well understood that we can compute the (Krull) dimension of an algebraic variety over algebraically closed fields in time $D^{O(n)}$ [16], see also [21].

It is of great interest to know if the problem of computing the dimension admits the same complexity bound in the real case and in the algebraically closed case. This problem also finds applications in geometric modeling and mechanics (see *e.g.* [20]).

We present a probabilistic algorithm for computing the real dimension of a semi-algebraic set $\mathscr{S} \subset \mathbb{R}^n$ in time $(s\,D)^{O(n)}$ [26]. First, we perform a classical reduction to the case of bounded semi-algebraic sets. Next, we consider the critical loci W_i of the restrictions of projections $(x_1, \ldots, x_n) \to (x_1, \ldots, x_i)$ to a smooth deformation of $\mathscr{S}$. It turns out that these critical loci coincide with $\mathscr{S}$ when $i \geq \dim(\mathscr{S})+1$. The algorithm exploits this structural property. It uses a subroutine that finds the largest integer i such that $W_i \neq W_{i-1}$ in time $(s\,D)^{O(n)}$; this integer equals $\dim(\mathscr{S}) + 1$. Finally, we obtain an algorithm which computes the real dimension of a semi-algebraic set in time $(s\,D)^{O(n)}$.

References

[1] S. Arora, R. Ge, R. Kannan, and A. Moitra. Computing a nonnegative matrix factorization–provably. In *Proc. STOC'12*, pp. 145–162. ACM, 2012.

[2] C. Bajaj. *Some applications of constructive real algebraic geometry*. Springer, 1994.

[3] B. Bank, M. Giusti, J. Heintz, and G.-M. Mbakop. Polar varieties and efficient real equation solving: the hypersurface case. *J. Complexity*, 13(1):5–27, 1997.

[4] B. Bank, M. Giusti, J. Heintz, and L.-M. Pardo. Generalized polar varieties: geometry and algorithms. *J. Complexity*, 21(4):377–412, 2005.

[5] B. Bank, M. Giusti, J. Heintz, M. Safey El Din, and E. Schost. On the geometry of polar varieties. *Applicable Algebra in Engineering, Communication and Computing*, 2010.

[6] S. Basu, R. Pollack, and M.-F. Roy. On the combinatorial and algebraic complexity of quantifier elimination. *J. ACM*, 43(6):1002–1045, Nov. 1996.

[7] S. Basu, R. Pollack, and M.-F. Roy. A new algorithm to find a point in every cell defined by a family of polynomials. In *Quantifier elimination and cylindrical algebraic decomposition*. Springer-Verlag, 1998.

[8] S. Basu, R. Pollack, and M.-F. Roy. Computing roadmaps of semi-algebraic sets on a variety. *J. AMS*, 3(1):55–82, 1999.

[9] S. Basu, R. Pollack, and M.-F. Roy. *Algorithms in real algebraic geometry*, volume 10 of *Algorithms and Computation in Mathematics*. Springer-Verlag, second edition, 2006.

[10] S. Basu, M.-F. Roy, M. Safey El Din, and E. Schost. A baby-step giant-step roadmap algorithm for general real algebraic sets. *submitted to Foundations of Computational Mathematics*, 2012.

[11] V. D. Blondel. Simultaneous stabilization of linear systems and interpolation with rational functions. In *Open problems in mathematical systems and control theory*, pp. 53–59. Springer, 1999.

[12] J. Canny. *The complexity of robot motion planning*. PhD thesis, MIT, 1987.

[13] G. Collins. Quantifier elimination for real closed fields by cylindrical algebraic decompostion. In *Automata Theory and Formal Languages*, pp. 134–183. Springer, 1975.

[14] J.-C. Faugère, M. Safey El Din, and P.-J. Spaenlehauer. Gröbner bases of bihomogeneous ideals generated by polynomials of bidegree (1, 1): Algorithms and complexity. *J. Symb. Comput.*, 46(4):406–437, 2011.

[15] J.-C. Faugère, M. Safey El Din, and P.-J. Spaenlehauer. Critical points and Gröbner bases: the unmixed case. *ISSAC'12*, pp. 162–169. ACM, 2012.

[16] M. Giusti and J. Heintz. La détermination de la dimension et des points isolées d'une variété algébrique peuvent s'effectuer en temps polynomial. In *Computational Algebraic Geometry and Commutative Algebra, Cortona*, volume 34, pp. 216–256, 1991.

[17] M. Giusti, G. Lecerf, and B. Salvy. A Gröbner-free alternative for polynomial system solving. *J. Complexity*, 17(1):154–211, 2001.

[18] K. A. Hansen, M. Koucky, N. Lauritzen, P. B. Miltersen, and E. P. Tsigaridas. Exact algorithms for solving stochastic games. In *Proc. STOC'11*, pp. 205–214. ACM, 2011.

[19] G. Jeronimo, D. Perrucci, and J. Sabia. On sign conditions over real multivariate polynomials. *Discrete Comput. Geom.*, 44(1):195–222, 2010.

[20] Q. Jin and T. Yang. Overconstraint analysis on spatial 6-link loops. *Mechanism and machine theory*, 37(3):267–278, 2002.

[21] P. Koiran. Randomized and deterministic algorithms for the dimension of algebraic varieties. In *Proc. 38th Annual Symposium on Foundations of Computer Science.*, pp. 36–45. IEEE, 1997.

[22] P. Koiran. The Real Dimension Problem is $\mathrm{NP}_{\mathbb{R}}$-complete. *J. Complexity*, 15(2):227–238, 1999.

[23] M. Safey El Din. Testing sign conditions on a multivariate polynomial and applications. *Mathematics in Computer Science*, 1(1):177–207, December 2007.

[24] M. Safey El Din and E. Schost. Polar varieties and computation of one point in each connected component of a smooth real algebraic set. In *ISSAC'03*, pp. 224–231. ACM, 2003.

[25] M. Safey El Din and E. Schost. A baby steps/giant steps probabilistic algorithm for computing roadmaps in smooth bounded real hypersurface. *Discrete Comput. Geom.*, 45(1):181–220, 2011.

[26] M. Safey El Din and E. Tsigaridas. A probabilistic algorithm to compute the real dimension of a semi-algebraic set. *ArXiv 1304.1928*, Apr. 2013.

[27] J. Solymosi and T. Tao. An incidence theorem in higher dimensions. *Discrete Comput. Geom.*, pp. 1–26, 2012.

[28] N. Vorobjov. Complexity of computing the local dimension of a semi-algebraic set. *J. Symb. Comput.*, 27(6):565–579, 1999.

Computer Algebra: A 32-Year Update

[Invited Software Talk]

Stephen Wolfram
Wolfram Research, Inc.
100 Trade Center Dr.
Champaign, IL 61820
sw-staff@wolfram.com

ABSTRACT

I last spoke at a computer algebra conference in August 1981. Since that time I created Mathematica (launched 1988) and Wolfram|Alpha (launched 2009). This talk will survey perspectives on computer algebra gained through these activities, as well as through my work in basic science. I will also describe what I see as being key future directions and aspirations for computer algebra.

Categories and Subject Descriptors

I.1.3 [**Computing Methodologies**]: Symbolic and Algebraic Manipulation—*Languages and Systems*

Keywords

computer algebra

ISSAC'13, June 26–29, 2013, Boston, Massachusetts, USA.
ACM 978-1-4503-2059-7/13/06.

Convex Algebraic Geometry and Semidefinite Optimization

[Extended Abstract]

Pablo A. Parrilo
Laboratory for Information and Decision Systems
Massachusetts Institute of Technology
77 Massachusetts Avenue, Room 32D-726
Cambridge, MA 02139, USA
parrilo@mit.edu
http://www.mit.edu/~parrilo

ABSTRACT

In the past decade there has been a surge of interest in algebraic approaches to optimization problems defined by multivariate polynomials. Fundamental mathematical challenges that arise in this area include understanding the structure of nonnegative polynomials, the interplay between efficiency and complexity of different representations of algebraic sets, and the development of effective algorithms. Remarkably, and perhaps unexpectedly, convexity provides a new viewpoint and a powerful framework for addressing these questions. This naturally brings us to the intersection of *algebraic geometry*, *optimization*, and *convex geometry*, with an emphasis on algorithms and computation. This emerging area has become known as *convex algebraic geometry* [1].

This tutorial will focus on basic and recent developments in convex algebraic geometry, and the associated computational methods based on semidefinite programming for optimization problems involving polynomial equations and inequalities; see e.g. [2, 3, 4, 5, 6, 7, 8, 9, 10, 11]. There has been much recent progress, by combining theoretical results in real algebraic geometry with semidefinite programming to develop effective computational approaches to these problems. We will make particular emphasis on sum of squares decompositions, general duality properties, infeasibility certificates, approximation/inapproximability results, as well as survey the many exciting developments that have taken place in the last few years.

Categories and Subject Descriptors

I.1.2 [**Symbolic and algebraic manipulation**]: Algebraic algorithms; G.1.6 [**Numerical Analysis**]: Optimization—*Convex programming*

General Terms

Theory, Algorithms

Keywords

convex algebraic geometry; multivariate polynomials; semidefinite programming; convex optimization

ISSAC'13, June 26–29, 2013, Boston, Massachusetts, USA.
ACM 978-1-4503-2059-7/13/06.

1. ACKNOWLEDGMENTS

This work was partially supported by NSF grant DMS-0757207 and a Finmeccanica Career Development Chair.

2. REFERENCES

[1] G. Blekherman, P. A. Parrilo, and R. Thomas, editors. *Semidefinite optimization and convex algebraic geometry*, volume 13 of *MOS-SIAM Series on Optimization*. SIAM, 2012.

[2] K. Gatermann and P. A. Parrilo. Symmetry groups, semidefinite programs, and sums of squares. *Journal of Pure and Applied Algebra*, 192(1-3):95–128, 2004.

[3] M. Kojima, S. Kim, and H. Waki. Sparsity in sums of squares of polynomials. *Mathematical Programming*, 103(1):45–62, May 2005.

[4] J. B. Lasserre. Global optimization with polynomials and the problem of moments. *SIAM J. Optim.*, 11(3):796–817, 2001.

[5] J. Löfberg. Pre-and post-processing sum-of-squares programs in practice. *IEEE Transactions on Automatic Control*, 54(5):1007–1011, 2009.

[6] Y. Nesterov. Squared functional systems and optimization problems. In J. Frenk, C. Roos, T. Terlaky, and S. Zhang, editors, *High Performance Optimization*, pages 405–440. Kluwer Academic Publishers, 2000.

[7] P. A. Parrilo. *Structured semidefinite programs and semialgebraic geometry methods in robustness and optimization*. PhD thesis, California Institute of Technology, May 2000.

[8] P. A. Parrilo. Semidefinite programming relaxations for semialgebraic problems. *Math. Prog.*, 96(2, Ser. B):293–320, 2003.

[9] H. Peyrl and P. Parrilo. Computing sum of squares decompositions with rational coefficients. *Theoretical Computer Science*, 409(2):269–281, 2008.

[10] S. Prajna, A. Papachristodoulou, and P. A. Parrilo. *SOSTOOLS: Sum of squares optimization toolbox for MATLAB*, 2002-05. Available from `http://www.cds.caltech.edu/sostools` and `http://www.mit.edu/~parrilo/sostools`.

[11] N. Z. Shor. Class of global minimum bounds of polynomial functions. *Cybernetics*, 23(6):731–734, 1987. (Russian orig.: Kibernetika, No. 6, (1987), 9–11).

Exact Linear and Integer Programming

Tutorial Abstract

Daniel E. Steffy
Department of Mathematics and Statistics
Oakland University
2200 N. Squirrel Rd.
Rochester, Michigan 48309
steffy@oakland.edu

ABSTRACT

This tutorial surveys state-of-the-art algorithms and computational methods for computing exact solutions to linear and mixed-integer programming problems.

Categories and Subject Descriptors

G.1.6 [**Mathematics of Computing**]: Numerical Analysis—*Optimization*; I.1.2 [**Computing methodologies**]: Symbolic and algebraic manipulation—*Algorithms*

Keywords

Linear programming; Integer programming; Hybrid symbolic-numeric computation

TUTORIAL OVERVIEW

A *linear program* (LP) is an optimization problem of minimizing a linear function subject to a finite set of linear inequalities and can be expressed as: $\min_{x\in\mathbb{R}^n}\{c^T x : Ax \geq b\}$, where x is a vector of decision variables. A *mixed-integer program* (MIP) additionally restricts a subset of the decision variables to take integer values, if all decision variables are required to be integers it is called an *integer program* (IP). Many decision problems can be modeled in these formats and a steady line of research has focused on developing and improving algorithms and numerical methods to solve them. Over the past 20 years some commercial software packages have increased their speed by a factor of more than ten thousand due to algorithmic improvements alone.

Most high-performance optimization software used today relies on inexact floating-point arithmetic. Although numerical solutions are adequate for many applications, exact conclusions are sometimes necessary. For example, one component of Thomas Hales' proof of the Kepler Conjecture relies on bounding the value of certain LPs and certain industrial applications, including design verification, also require exact solutions. A number of recent efforts have focused on developing efficient techniques to compute exact solutions and/or rigorous bounds; the primary goal of this tutorial is to survey these efforts, many of which use a hybrid of numerical and exact computation.

Of the aforementioned classes of problems, LPs are considered the easiest. One widely used algorithm, the simplex method, works by pivoting between vertices of the feasible region, a convex polyhedron, until an optimal vertex is found. It computes not only a solution, but a structural description of that solution. One way to solve linear programs exactly is thus to apply numerical computation to identify candidate optimal vertices of the polyhedron and recompute the solution by solving a system of equations exactly. Even if the vertex returned by the numerical solver is not feasible or optimal, it often provides useful information which can be used to warm start additional computations.

Algorithms for mixed-integer programming often rely on solving LPs as a subroutine. The *branch-and-bound* method starts with the LP relaxation of the problem, dropping the integrality constraints, and successively subdivides the search space, solving a sequence of related LPs to arrive at the solution. Computation of LP dual bounds often allows large portions of the search space to be eliminated based on the objective value of known feasible solutions. The *cutting plane* method also starts with the LP relaxation; instead of subdividing the feasible region, it iteratively adds additional inequality constraints in an effort to approximate the convex hull of the feasible solutions, always working with a relaxation of the original feasible region. It continues solving LPs and adding new inequalities until a solution satisfying the integrality constraints is discovered. In practice these methods are integrated together as a *branch-and-cut* method, also incorporating a number of other modules such as preprocessing and heuristics for finding feasible solutions. Due to the complex nature of the overall solution algorithm, there are many different ways in which numerical errors can creep in and invalidate the result.

Avoiding errors by calling an exact LP solver at every step of the branch-and-bound method is a natural strategy. However, the computation of safe dual bounds using interval arithmetic or other fast methods can allow many portions of the search space to be safely eliminated, only leaving a relatively small number of exact LPs to be solved. For the cutting plane method, rounding errors must be avoided when computing the newly added inequalities, otherwise they may cut off feasible solutions. Some recent studies have applied directed rounding to safely generate valid cutting planes using floating-point arithmetic. In many cases exact solutions and safe bounds can be computed with a surprisingly low overhead beyond the cost of the existing numerical methods. Still, there are many open questions and further areas for improvement that will also be discussed in this tutorial.

Little prior experience with optimization is assumed and,

ISSAC'13, June 26–29, 2013, Boston, Massachusetts, USA.
ACM 978-1-4503-2059-7/13/06.

in addition to describing exact methods, this tutorial also serves as a self-contained introduction to linear and integer programming algorithms. With this goal in mind, part of the tutorial discusses ideas widely applied to tailor the general algorithms into highly efficient problem specific algorithms, including the dynamic generation of variables and constraints and methods for reformulation and decomposition.

The Complexity of Factoring Univariate Polynomials over the Rationals

Tutorial Abstract

Mark van Hoeij
Dept. of Mathematics, Florida State University
Tallahassee, Florida 32306-3027, USA
hoeij@math.fsu.edu

ABSTRACT

This tutorial will explain the algorithm behind the currently fastest implementations for univariate factorization over the rationals. The complexity will be analyzed; it turns out that modifications were needed in order to prove a polynomial time complexity while preserving the best practical performance.

The complexity analysis leads to two results: (1) it shows that the practical performance on common inputs can be improved without harming the worst case performance, and (2) it leads to an improved complexity, not only for factoring, but for LLL reduction as well.

Categories and Subject Descriptors

I.1.2 [**Symbolic and Algebraic Manipulation**]: Algorithms; G.4 [**Mathematics of Computing**]: Mathematical Software

General Terms

Algorithms

Keywords

Symbolic Computation, Polynomial Factorization

Most practical factoring algorithms in $\mathbb{Q}[x]$ use a structure similar to Zassenhaus [10]: factor f modulo a small prime p, Hensel lift this factorization, and recombine these local factors into factors in $\mathbb{Q}[x]$. To reduce the combinatorial part, one first tries several primes, and then selects a p for which the number (denoted r) of factors mod p is smallest.

Polynomial time algorithms, based on lattice reduction, were given in [7, 9]. For a polynomial f of degree N, and entries bounded in absolute value by 2^h, these algorithms perform $\mathcal{O}(N^2(N+h))$ LLL switches. Schönhage [9] gave a modification to LLL reduction, and obtained the following complexity for factoring: $\tilde{\mathcal{O}}(N^4(N+h)^2)$, where $\sim$ indicates that logarithmic factors are ignored.

ISSAC'13, June 26–29, 2013, Boston, Massachusetts, USA.
ACM 978-1-4503-2059-7/13/06.

When comparing an algorithm with other algorithms, it is useful to know

Q.1 What is the worst-case asymptotic behavior?

Q.2 What is the best proven upper bound?

Q.3 Is the bound a good predictor for the running time on worst-case inputs?

Q.4 What about the performance on typical inputs?

Most attention is usually spent on Q.2 even though the other questions are more interesting.

Despite its exponential complexity, Zassenhaus [10] is usually faster than the polynomial time algorithms in [7, 9]. To explain this, note that [10] is not exponential in terms of N, it is exponential in r, which is usually $\ll N$. However, it is possible for r and N to be comparable in size. The best-known examples are Swinnerton-Dyer polynomials. They have $r = N/2$, which is worst-case ($r > N/2$ indicates the presence of factors of low degree).

The algorithm with the best complexity is described in [4]. It is based on the strategy in [5, 8]. It is efficient in practice and improves Schönhage's complexity, except if one makes the highly restrictive assumption that $N + h = \mathcal{O}(r)$, in which case [4] is only faster by a large constant.

We will denote the complexity of the algorithm in [4] as $C_H + C_L + C_O$ where C_H is the time spent on Hensel lifting, C_L is the time spent inside a variation of the LLL algorithm, and C_O are other costs (factoring f modulo several primes, preparing the input for LLL, reconstructing factors in $\mathbb{Z}[x]$ by multiplying p-adic factors, etc.).

For the LLL cost C_L, the answer to question Q.2 is $\tilde{\mathcal{O}}(r^6)$. The main ingredient is to prove that the number of LLL switches is bounded by $\mathcal{O}(r^3)$ (no logarithmic factors). Note that the bound for C_L is independent of both N and h! Regarding question Q.1, experiments suggest that C_L grows less than r^6, but this does not imply that the asymptotic complexity has degree < 6 (perhaps r^6 still manifests itself for impractically large inputs?). Regarding question Q.3 for C_L, it seems that r^6 is roughly in the right ball-park.

C_H has lower degree than C_L, but this does not imply that C_H is asymptotically smaller than C_L. After all, C_H depends on both N and h which are generally much larger than r. Indeed, for typical inputs, C_H dominates the CPU time. Ideally, one would have a bound for C_H that (a) can be proven, and (b) reflects the actual behavior of the algorithm.

Hensel lifting is highly optimized [3]. In order to know how much time [4] will spend on Hensel lifting, we need to know the p-adic precision that [4] will lift to. That turns out to be difficult to predict, it can vary considerably, and our bound appears to be a factor N higher than what can actually occur.

In practice, [4] lifts far less than [7, 9]. Regarding question Q.1, we conjecture that (just like [10]) it never lifts further than p^a with $\log(p^a) = \tilde{O}(N + h)$. This can be translated into a number theoretical problem. This problem has some resemblance to the abc-conjecture and looks plausible, but unfortunately, we could not prove it. Regarding question Q.2, our bound from [2] is $\log(p^a) = \mathcal{O}(N(N+h))$, which is the same as in [7, 9].

Even though we could not prove an upper bound with $\log(p^a) = \tilde{O}(N + h)$, the actual amount of Hensel lifting in [4] is often smaller still, through a technique called early-termination. Suppose for example that the input f is irreducible (that does not mean that factoring is an empty task; one still has to compute an irreducibility proof). Though counter intuitive, computing mod p^a often suffices to prove irreducibility of a polynomial with coefficients $\gg p^a$.

It would not harm the asymptotic worst-case complexity if the algorithm starts by Hensel lifting the p-adic factors mod p^a with $\log(p^a) = \mathcal{O}(N + h)$. But it would not be a good design choice. For instance, suppose f has two factors, $f = f_1 f_2 \in \mathbb{Z}[x]$, and that f_1 has 10-digit while f_2 has 1000-digit coefficients. Suppose also that the combinatorial problem is solved once $\log_{10}(p^a)$ reaches 50. Then f_1 (but not f_2) can be reconstructed from its image mod p^a. But f_2 can be constructed as f/f_1. So even though f had coefficients with $\approx$ 1000 digits, a far smaller p-adic precision could be sufficient to factor it.

Inputs where where one of the irreducible factors is large, and all the others are small, are common (e.g. irreducible inputs). For questions Q.3 and Q.4, this means that for polynomials with large coefficients, a good worst-case predictor for C_H can overestimate the CPU time on typical inputs.

The goal in [4] was to set up the algorithm in such a way that we make no compromises on the practical performance (typical inputs and worst-case inputs) while simultaneously designing it in such a way that it satisfies the best upper bound that we are able to prove for a factoring algorithm.

The bound for C_L is quite good, but the bound for C_H is not. It is highly implausible that $\log(p^a) = \mathcal{O}(N(N + h))$ is close to sharp in the worst case. But it is also difficult to prove a better bound. Settling this issue would lead to a much better (Q.1 instead of Q.2) description of the complexity of factoring.

I used to think that complexity analysis is not useful in general, that it was useful if and only if it gives a good description for the actual CPU time. If known bounds are far from the actual performance, then, in order to avoid confusion, I thought it is better to give no bound (i.e. complexity = unknown). My view has become more nuanced since then. The complexity analysis of C_L, the LLL cost during factoring, has lead to improvements ("gradual feeding", discussed at [1] and analyzed in [8, 6]) for other LLL applications as well. Moreover, the complexity analysis also showed how a strategy called "early termination" (to improve CPU timings on typical inputs) could be set up in such a way that the CPU time on worst-case inputs will not be adversely affected. So even if only CPU timings matter, as I used to think, complexity analysis still turned out to be useful.

1. REFERENCES

[1] *Open Questions From AIM Workshop*, 2006. www.aimath.org/WWN/polyfactor/polyfactor.pdf

[2] K. Belabas, M. van Hoeij, J. Klüners, and A. Steel. *Factoring polynomials over global fields. J. de Théorie des Nombres de Bordeaux*, **21**, 15–39, 2009.

[3] A. Bostan, G. Lecerf, B. Salvy, É. Schost, and B. Wiebelt. *Complexity issues in bivariate polynomial factorization*, Proceedings of ISSAC, 42–49, 2004.

[4] W. Hart, M. van Hoeij, A. Novocin, *Practical Polynomial Factoring in Polynomial Time*, ISSAC'2011 Proceedings, 163–170, 2011.

[5] M. van Hoeij, *Factoring polynomials and the knapsack problem, J. of Number Theory*, **95**, 167–189, 2002.

[6] Mark van Hoeij and Andrew Novocin. *Gradual sub-lattice reduction and a new complexity for factoring polynomials.* LATIN, 539–553, 2010.

[7] A. K. Lenstra, H. W. Lenstra, Jr., and L. Lovász. *Factoring polynomials with rational coefficients, Math Ann.*, **261**, 515–534, 1982.

[8] A. Novocin. *Factoring Univariate Polynomials over the Rationals.* PhD thesis, Florida State University, 2008.

[9] A. Schönhage. *Factorization of univariate integer polynomials by Diophantine approximation and improved basis reduction algorithm.* Proceedings of the 1984 International Colloquium on Automata, Languages and Programming (ICALP 1984), **172** LNCS, 436–447, 1984.

[10] H. Zassenhaus. *On Hensel Factorization I.* J. Number Theory, **1, 291–311, 1969.**

A New Truncated Fourier Transform Algorithm

Andrew Arnold
Symbolic Computation Group
University of Waterloo
Waterloo, Ontario, Canada
a.arnold55@gmail.com
http://cs.uwaterloo.ca/~a4arnold

ABSTRACT

Truncated Fourier Transforms (TFTs), first introduced by van der Hoeven, refer to a family of algorithms that attempt to smooth "jumps" in complexity exhibited by FFT algorithms. We present an in-place TFT whose time complexity, measured in terms of ring operations, is asymptotically equivalent to existing not-in-place TFT methods. We also describe a transformation that maps between two families of TFT algorithms that use different sets of evaluation points.

Categories and Subject Descriptors

F.2.1 [**Analysis of Algorithms and Problem Complexity**]: Numerical Algorithms and Problems—*Computation of transforms*; G.4 [**Mathematical Software**]: Algorithm design and analysis

Keywords

Truncated Fourier Transform; in-place algorithms

1. INTRODUCTION

Let $\mathcal{R}$ be a ring containing an N-th principal root of unity ω. Given two polynomials $f, g \in \mathcal{R}[z]$, $\deg(fg) < N$, we can compute fg by way of the Discrete Fourier Transform (DFT): a linear, invertible map which evaluates a given polynomial at the powers of ω.

Computing the DFT naively is quadratic-time. However, if N is strictly comprised of small prime factors, one can compute a DFT using $\mathcal{O}(N \log N)$ arithmetic operations by way of the Fast Fourier Transform (FFT). The most widely-known FFT, the *radix-2* FFT, requires that N is a power of two. To compute the DFT of an input of arbitrary size, one typically appends zeros to the input to give it power-of-two length, and then applies a radix-2 FFT. This method exhibits significant jumps in its time and memory costs.

Truncated Fourier Transforms (TFTs) flatten these jumps in complexity. A TFT takes an input of length $n \le N$ and returns a size-n subset of its length-N DFT, with time complexity that grows comparatively smoothly with $n \log n$. Typically one chooses the first n entries of the DFT, with the DFT sorted in bit-reversed order. This is a natural choice as it comprises the first n entries of the output of an in-place FFT. We will call such a TFT the *bit-reversed* TFT.

Van der Hoeven [11] showed how one can obtain a polynomial $f(z)$ from its bit-reversed TFT, provided one knows the terms of $f(z)$ with degree at least n. This allows for faster FFT-based polynomial multiplication, particularly for products whose degree is a power of two or slightly larger. Harvey and Roche showed further in [5] how the bit-reversed TFT can be computed in-place, at the cost of a constant factor additional ring multiplications.

Mateer [7] devised a TFT algorithm based on a series of modular reductions, that acts as a preprocessor to the FFT. Mateer's TFT algorithm, which we discuss in section 3, reduces $f(z)$, $\deg(f) < n$, modulo cyclotomic polynomials of the form $z^k + 1$, k a power-of-two. We will call this TFT the *cyclotomic* TFT.

In [9], Sergeev shows how the cyclotomic TFT can be made in-place with cost asymptotically equivalent to not-in-place TFT algorithms. In section 4, we describe Sergeev's algorithm. In section 5, we give an in-place algorithm of similar cost, related to Sergeev's, for computing the cyclotomic TFT. One incremental improvement of this new TFT algorithm is that, for input sizes n of low Hamming weight, it requires fewer passes through our input array than Sergeev's algorithm in order to produce the polynomial images.

One caveat of the cyclotomic TFT is that different-sized inputs may use entirely different sets of evaluation points. This is problematic in applications to multivariate polynomial multiplication. In section 6, we show how an algorithm that computes the cyclotomic TFT can be modified to compute a bit-reversed TFT by way of an affine transformation.

As a proof of concept we implemented the algorithms introduced in this paper in Python. These implementations can be found at `http://cs.uwaterloo.ca/~a4arnold/tft`.

2. PRELIMINARIES

2.1 The Discrete Fourier Transform

The *Discrete Fourier Transform* (DFT) [4] of a polynomial $f(z)$ is its vector of evaluations at the distinct powers of a root of unity. Specifically, if $f(z) = \sum_{i=0}^{N-1} a_i z^i$ is a polynomial over a ring $\mathcal{R}$ containing an order-N root of unity ω, then we define the Discrete Fourier Transform of $f(z)$ as

$$\mathrm{DFT}_\omega(f) = \left(f(\omega^0), \ldots, f(\omega^{N-1})\right).$$

We treat the polynomial f and its vector of coefficients $a = (a_0, a_1, \ldots, a_{N-1})$ as equivalent and use the notation $\mathrm{DFT}_\omega(a)$ and $\mathrm{DFT}_\omega(f)$ interchangeably. If we take addition and multiplication component-wise in $\mathcal{R}^N$, then the map $\mathrm{DFT}_\omega \colon \mathcal{R}[z]/(z^N-1) \to \mathcal{R}^N$ forms a ring homomorphism. Throughout this paper we will assume ω is a *principal* root of unity, that is, for j not divisible by N, the order of ω, $\sum_{i=0}^{N-1} \omega^{ij} = 0$. In such case DFT_ω has an inverse map $\mathrm{IDFT}_\omega \colon \mathcal{R}^N \to \mathcal{R}[z]/(z^N-1)$, defined by

$$\mathrm{IDFT}_\omega(\hat{a}) = \tfrac{1}{N}\mathrm{DFT}_{\omega^{-1}}(\hat{a}),$$

where $\hat{a} \in \mathcal{R}^N$ and we again treat a polynomial as equivalent to its vector of coefficients. This suggests a multiplication algorithm for $f, g \in \mathcal{R}[z]$.

THEOREM 1 (THE CONVOLUTION THEOREM). *Let f, g be polynomials over a ring $\mathcal{R}$ containing an N-th principal root of unity ω. Then*

$$fg \bmod (z^N - 1) = \mathrm{IDFT}_\omega\left(\mathrm{DFT}_\omega(f) \cdot \mathrm{DFT}_\omega(g)\right),$$

where "$\cdot$" is the vector component-wise product, and, given two polynomials $s(z), t(z) \in \mathcal{R}[z]$, $s(z) \bmod t(z)$ denotes the unique polynomial $r(z)$ such that $t(z)$ divides $r(z) - s(z)$ and $\deg(r) < \deg(t)$ throughout.

Thus to multiply f and g, we can choose $N > \deg(fg)$ and an N-th principal root of unity $\omega \in \mathcal{R}$, compute the length-N DFTs of f and g, take their component-wise product, and take the inverse DFT of that product.

2.2 The Fast Fourier Transform

We can compute the Discrete Fourier Transform of $f(z)$, f reduced modulo $(z^N - 1)$, by way of a *Fast Fourier Transform* (FFT). The FFT is believed to have been first discovered by Gauss, but did not become well known until it was famously rediscovered by Cooley and Tukey [1]. For a detailed history of the FFT we refer the reader to [6].

The simplest FFT, the *radix-2* FFT, assumes $N = 2^p$ for some $p \in \mathbb{Z}_{\geq 0}$. We describe the radix-2 FFT in terms of modular reductions. We break f into images modulo polynomials of decreasing degree until we have the images $f \bmod (z - \omega^i) = f(\omega^i)$, $0 \leq i < N$. At the start of the first iteration we have f reduced modulo $z^N - 1$. At the start of the i-th iteration, we will have the 2^{i-1} images

$$f \bmod (z^{2u} - \omega^{2uj}) \quad \text{for} \quad 0 \leq j < 2^{i-1},$$

where $u = N/2^i$. If $i = p+1$ this gives us the DFT of f. Consider then the image $f' = f \bmod (z^{2u} - \omega^{2uj}) = \sum_{k=0}^{2u-1} b_k z^k$, for some j, $0 \leq j < 2^{i-1}$. We break this image into two images f_0 and f_1, where

$$f_0 = f' \bmod (z^u - \omega^{uj}), \text{ and}$$

$$f_1 = f' \bmod (z^u + \omega^{uj}) = f' \bmod (z^u - \omega^{uj+N/2}).$$

We can write f_0 and f_1 in terms of the coefficients b_k:

$$f_0 = \sum_{k=0}^{u-1}(b_k + \omega^{uj} b_{k+u}) z^k, \quad f_1 = \sum_{k=0}^{u-1}(b_k - \omega^{uj} b_{k+u}) z^k.$$

Thus, given an array containing the coefficients b_k of f', we can write f_0 and f_1 in place of f' by way of operations

$$\begin{bmatrix} b_k \\ b_{k+u} \end{bmatrix} \longleftarrow \begin{bmatrix} 1 & \omega^{uj} \\ 1 & -\omega^{uj} \end{bmatrix} \begin{bmatrix} b_k \\ b_{k+u} \end{bmatrix}, \quad 0 \leq i < u. \qquad (1)$$

The pair of assignments (1) are known as a *butterfly operation*, and can be performed with a ring multiplication by the *twiddle factor* ω^{uj}, and two additions. Note f_0 and f_1 are in a similar form as f', and if $u > 1$ we can break those images into smaller images in the same fashion. Starting this method with input $f \bmod (z^N - 1)$, will give us $f \bmod (z - \omega^j) = f(\omega^j)$, for $0 \leq j < N$.

If the butterfly operations are performed in place, the resulting evaluations $f(\omega^j)$ will be written in *bit-reversed* order. More precisely, if we let $[j]_p$ denote the integer resulting from reversing the p bits of j, $0 \leq j < 2^p$, we have that $f(\omega^j)$ will be written in place of a_k, where $k = [j]_p$. As an example,

$$[13]_5 = [01101_2]_5 = 10110_2 = 16 + 4 + 2 = 22.$$

We can make the FFT entirely in-place by computing the powers of ω^u sequentially at every iteration. This entails traversing the array in a non-sequential order. Procedure FFT describes such an implementation.

If we observe that

$$\begin{bmatrix} 1 & \omega^{uj} \\ 1 & -\omega^{uj} \end{bmatrix}^{-1} = \frac{1}{2}\begin{bmatrix} 1 & 1 \\ \omega^{-uj} & -\omega^{-uj} \end{bmatrix},$$

then we can implement an inverse FFT by inverting the butterfly operations in reversed order. We can, moreover, delay multiplications by powers of $\frac{1}{2}$ until the end of the inverse FFT computation. This entails multiplying the result by $\frac{1}{N}$.

Procedure FFT($\mathbf{a}$, N), an in-place implementation of the radix-2 FFT

Input: $\mathbf{a}$, a length $N = 2^p$ array containing $f \in \mathcal{R}[z]$.
Result: $\mathrm{DFT}_\omega(f)$ is written to $\mathbf{a}$ in bit-reversed order.

for $i \longleftarrow 1$ **to** p **do**
  $u \longleftarrow N/2^i$
  for $j \longleftarrow 0$ **to** $\frac{N}{2u} - 1$ **do**
    $t \longleftarrow 2[j]_i u$
    for $k \longleftarrow 0$ **to** $u - 1$ **do**
      // Butterfly operations (1)
      $\begin{bmatrix} \mathbf{a}_{t+k} \\ \mathbf{a}_{t+k+u} \end{bmatrix} \longleftarrow \begin{bmatrix} 1 & \omega^{uj} \\ 1 & -\omega^{uj} \end{bmatrix} \begin{bmatrix} \mathbf{a}_{t+k} \\ \mathbf{a}_{t+k+u} \end{bmatrix}$

THEOREM 2. *Let N be a power of two and ω an N-th root of unity. Then* FFT *computes a length-N DFT in place using no more than $\frac{1}{2}N \log N + \mathcal{O}(N)$ ring multiplications and $N \log N + \mathcal{O}(N)$ ring additions, where* log *is taken to be base-2 throughout. The inverse FFT can be computed in-place with asymptotically equivalent cost.*

Using the radix-2 FFT, if $d = \deg(fg)$, we choose N to be the least power of 2 exceeding d. This entails appending zeros to the input arrays containing the coefficients of f and g respectively. By this method, computing a product of degree 2^p costs at least double that a product of degree less than 2^p. Crandall's "devil's convolution" algorithm [2] somewhat flattens these jumps in complexity, though not entirely. It works by reducing a discrete convolution of arbitrary length into more easily computable convolutions. More recently, Truncated Fourier Transform (TFT) algorithms, described hereafter, have addressed this issue.

2.3 Truncated Fourier Transforms

In many applications, it is useful to compute a *pruned* DFT, a subset of a length-N DFT, at a cost less than that to compute a complete DFT. In 2004, van der Hoeven [10] showed that, given some knowledge of the form of the input, one can invert some pruned DFTs. The inverse transform relies on the observation that, given any two of the inputs/outputs to a butterfly operation (1), one can compute the other two values. Suppose $n \in \mathbb{Z}_{>0}$ is arbitrary and N is the least power of two at least n. For ω, a primitive root of unity of order $N = 2^p$, van der Hoeven showed how to invert the length-n Truncated Fourier Transform,

$$\mathrm{TFT}_{\omega,n}(f) = \left(f(\omega^{[i]_p})\right)_{0 \le i < n},$$

when we know the terms of $f(z)$ of degree at least n (e.g., when $\deg(f) < n$). To distinguish this particular TFT we will call it the *bit-reversed* TFT.

THEOREM 3 (VAN DER HOEVEN, [11]). *Suppose f is a polynomial over $\mathcal{R}$ of degree less than n. $\mathrm{TFT}_{\omega,n}(f)$ can be computed using $n \log n + \mathcal{O}(n)$ ring additions and $\frac{1}{2} n \log n + \mathcal{O}(n)$ multiplications by powers of ω.*

$f(z)$ can be recovered from $\mathrm{TFT}_{\omega,n}(f)$ using $n \log n + \mathcal{O}(n)$ shifted *ring additions and $\frac{1}{2} n \log n + \mathcal{O}(n)$ multiplications.*

A shifted ring addition in this context merely means an addition plus a multiplication by $2^{\pm 1}$. Van der Hoeven's algorithm generalizes to allow us to compute arbitrary subsets of the DFT. Given subsets $S, T \subseteq \{0, 1, \ldots, N-1\}$, we define

$$\mathrm{TFT}_{\omega,S,T}(f) = \left(f(\omega^{[i]_p})\right)_{i \in T},$$

where we now assume f is of the form $f(z) = \sum_{i \in S} a_i z^i$. However, such a transform may have a greater complexity than stated in theorem 3, taking $n = \#S$. Moreover, such a map is not necessarily invertible, even in the case $S = T$.

EXAMPLE 4. *Let $N = 8$. Taking an 8-th principal root of unity ω with $S = T = \{0, 3, 4, 5\}$ gives an uninvertible map. To see that the map $\mathrm{TFT}_{\omega,S,T}$ is not invertible, one can check that the polynomial $f(z) = (1+\omega^2)z^5 - z^4 + (1-\omega^2)z^3 - 1$ evaluates to 0 for $z = \omega^k$, $k \in \{[0]_3, [3]_3, [4]_3, [5]_3\} = \{0, 6, 1, 5\}$.*

Van der Hoeven's method still exhibits significant jumps in space complexity, as it requires space for N ring elements regardless of n. In 2010, Harvey and Roche [5] introduced an in-place TFT algorithm, requiring $n + \mathcal{O}(1)$ ring elements plus an additional $\mathcal{O}(1)$ bounded-precision integers to compute $\mathrm{TFT}_{\omega,n}(f)$. Their method potentially requires evaluating polynomials using linear-time methods. This adds an additional constant factor to the algorithm's worst-case cost.

THEOREM 5 (ROCHE, THEOREM 3.5, [8]). *Let N, n, f and ω be as in theorem 3. Then $\mathrm{TFT}_{\omega,n}(f)$ can be computed in-place using at most $\frac{5}{6} n \log n + \mathcal{O}(n)$ ring multiplications and $\mathcal{O}(n \log n)$ ring additions.*

The inverse in-place transform entails similarly many ring multiplications and $\mathcal{O}(n \log n)$ shifted ring additions. As an application, Harvey and Roche used this transform towards asymptotically fast in-place polynomial multiplication.

3. THE CYCLOTOMIC TFT

3.1 Notation

We use the following notation throughout section 3 and thereafter. Suppose now that we have a polynomial $f(z) = \sum_{j=0}^{n-1} a_j z^j \in \mathcal{R}[z]$, where n is not necessarily a power of two. For the remainder of this paper, we write n as $n = \sum_{i=1}^{s} n_i$, where $n_i = 2^{n(i)}$, $n(i) \in \mathbb{Z}_{\ge 0}$ and $n_i > n_j$ for $1 \le i < j \le s$. Here we let N be the smallest power of two exceeding n. For $1 \le i \le s$, we let

$$\Phi_i = z^{n_i} + 1.$$

The TFT algorithms of sections 3 and thereafter will compute the evaluations of $f(z)$ at the roots of Φ_i. Namely, if we fix a canonical root $\omega = \omega_1$ of Φ_1, and then let, for $2 \le i \le s$, $\omega_i = \omega_1^{n_1/n_i}$, a root of Φ_i, these algorithms will compute

$$f(\omega_i^{2j+1}) \quad \text{for} \quad 0 \le j < n_i, \qquad 1 \le i \le s, \tag{2}$$

the evaluation of $f(z)$ at the roots of the cyclotomic polynomials Φ_i. As such, we will call it here the *cyclotomic* TFT and denote it by $\mathrm{TFT}'_{\omega,n}(f)$. The choice of our order-N root ω only affects the ordering of the elements of the cyclotomic TFT. If we let

$$T(n) = \{k : n_i \le k < 2n_i \text{ for some } i,\ 1 \le i \le s\},$$

then we have that $\mathrm{TFT}'_{\omega,n}$ uses the same set of evaluation points as $\mathrm{TFT}_{\omega,S,T(n)}(f)$.

We define, for $1 \le i \le s$, the images

$$f_i = f \bmod \Phi_i.$$

The algorithms for computing a cyclotomic TFT all follow a similar template: we will produce the images f_i sequentially, and then evaluate f at the roots of Φ_i in place of each image f_i.

3.2 Discrete Weighted Transforms

Given the images $f_i = f \bmod \Phi_i$, one can evaluate f at the roots of Φ_i by way of a *Discrete Weighted Transform* (DWT), which comprises an affine transformation followed by an FFT [3].

In a more general setting, suppose we have an image $f^* = f \bmod (z^K - c)$, where K is a power-of-two. Assuming c has an K-th root over our ring $\mathcal{R}$, the roots of $z^K - c$ are all of the form $c^{1/K}\gamma^i$, $0 \le i < K$, where γ is an order-K root of unity. Thus to evaluate f at the roots of $(z^K - c)$ one can replace $f^*(z)$ with $f^*(c^{1/K}z)$, and then compute $\mathrm{DFT}_\gamma(f^*(c^{1/K}z))$. Replacing $f^*(z)$ with $f^*(c^{1/K}z)$ iteratively term-by-term entails fewer than $2K$ ring multiplications.

To evaluate f at the roots of $\Phi_i(z) = z^{n_i} - \omega_i^{n_i}$, one would write $f_i(\omega_i z)$ in place of $f_i(z)$, then compute $\mathrm{DFT}_{\omega_i^2}(f_i(\omega_i z))$ by way of the FFT. As both the FFT and the affine transformation are invertible, a Discrete Weighted Transform is easily invertible as well.

Procedure DWT($\mathbf{a}$, K, v), the Discrete Weighted Transform

Input: $\mathbf{a}$, a length-K array; $v \in \mathcal{R}$, a weight.
for $i \longleftarrow 0$ **to** $K-1$ **do** $\mathbf{a}_i \longleftarrow v^i \mathbf{a}_i$
FFT($\mathbf{a}, K$)

3.3 Mateer's TFT algorithm

Procedure MateerTFT below gives a short description of Mateer's algorithm [7] for computing $\mathrm{TFT}'_{\omega,n}(f)$. Given array $\mathbf{a}$ and integer k we let the array $\mathbf{a}+k$ denote the array $\mathbf{b}$ such that $\mathbf{a}_{\ell+k}$ and $\mathbf{b}_\ell$ refer to the same element for all ℓ. MateerTFT will write f_i to $\mathbf{a}+n_i$.

Mateer's algorithm computes the images $f \bmod (z^K+1)$ for $K=2^\ell$, $n(s) \le \ell \le n(1)$, and the image $f \bmod (z^{n_s}-1)$. For the images $f \bmod (z^K+1)$ where $K=n_i$, $1 \le i \le s$, we perform a DWT to evaluate f at the roots of Φ_i. The remaining images we can simply discard.

Procedure MateerTFT($\mathbf{a}, n$)

```
Input: a, a length N = 2^(⌊log n⌋+1) array containing
       f ∈ R[z].
Result: TFT'_{ω,n}(f) is written to a.
K ⟵ N/2
while K > n_s do
    for e ⟵ 0 to K do
        (a_e, a_{e+K}) ⟵ (a_e + a_{e+K}, a_e − a_{e+K})
    K ⟵ K/2
for i ⟵ 1 to s do DWT(a + n_i, n_i, ω_i)
```

On input we are given $f(z)$ reduced modulo z^N-1. Given $f \bmod (z^{2K}-1) = \sum_{d=0}^{2K-1} b_d z^d$, Mateer's TFT breaks $f \bmod (z^{2K}-1)$ into the images

$$f \bmod (z^K-1) = \sum_{d=0}^{K-1} (b_d + b_{d+K}) z^d, \text{ and} \tag{3}$$

$$f \bmod (z^K+1) = \sum_{d=0}^{K-1} (b_d - b_{d+K}) z^d. \tag{4}$$

Performing the aforementioned modular reductions for $K=N/2,\ldots,n_s$ amounts to $\mathcal{O}(n)$ additions. The number of multiplications required to perform the DWTs is bounded by

$$\sum_{i=1}^{s} \tfrac{1}{2} n_i \log n_i + cn_i \le \tfrac{1}{2} n \log n + \mathcal{O}(n), \tag{5}$$

for some constant c. A similar analysis for the ring additions due to the DWTs gives us the following complexity:

Lemma 6 (Mateer). *Procedure* MateerTFT *computes* $\mathrm{TFT}'_{\omega,n}(f)$ *from* f *using* $\frac{1}{2} n\log n + \mathcal{O}(n)$ *ring multiplications and* $n \log n + \mathcal{O}(n)$ *ring additions.*

Mateer gives a method of inverting the cyclotomic TFT with asymptotically equivalent cost, the details of which we omit here. MateerTFT is not in-place as the images $f \bmod (z^{N/2}-1)$ and $f \bmod (z^{N/2}+1)$ may have maximal degree.

4. COMPUTING THE CYCLOTOMIC TFT WITHOUT ADDITIONAL MEMORY

In order to compute the cyclotomic TFT in-place, it appears, unlike the Mateer TFT, that we need to use some of the information from the images $f_1,\ldots,f_i$ towards producing the image f_{i+1}. Both Sergeev's TFT and the new algorithm presented thereafter work in this manner.

For $1 \le i \le s$, let

$$\Gamma_i(z) = \prod_{j=1}^{i} \Phi_j(z) \quad \text{and} \quad C_i = f \bmod \Gamma_i.$$

We call C_i the *combined* image of f, as it is the result of Chinese remaindering on the images $f_1,\ldots,f_i$. We also define

$$q_i = \begin{cases} f & \text{if } i=0, \\ f \text{ quo } \Gamma_i & \text{if } 1 \le i \le s, \end{cases}$$

the quotient produced dividing f by Γ_i, as well as

$$n_i^* = \begin{cases} n & \text{if } i=0, \\ n \bmod n_i & \text{if } 1 \le i \le s. \end{cases}$$

Note that, as $\Phi_i \bmod \Phi_j = 2$ for $j>i$, we also have $\Gamma_i \bmod \Phi_j = 2^i$ for $j>i$. Similarly, $\Gamma_i \bmod (z^K-1) = 2^i$ for K, a power of two at most n_i.

For any choice of $1 \le i \le s$, we have

$$f(z) = C_i + \Gamma_i q_i.$$

It is straightforward to obtain q_i, given f. Note that the degrees of any two distinct terms of Γ_i differ by at least n_i, and that $\deg(q_i) < n_i$. Thus, as Γ_i is monic, we have that the coefficients of q_i merely comprise the coefficients of the higher-degree terms of f. More precisely,

$$q_i = \sum_{e=0}^{n_i^*-1} a_{n-n_i^*+e} z^e.$$

By a similar argument, we also have that, for $1 \le i \le s$,

$$q_i = q_{i-1} \text{ quo } \Phi_i.$$

We note that $q_s = 0$.

4.1 Computing images of C_i without explicitly computing C_i

We will express the combined image C_i, C_i reduced modulo $z^m \pm 1$, m a power of two, in terms of the coefficients of the images $f_1,\ldots,f_i$. To this end we introduce the following notation. Given an integer e, we will let $e[i]$ refer to the i-th bit of e, i.e.,

$$e = \sum_{i=0}^{\lfloor \log(e) \rfloor} e[i] 2^i, \qquad e[i] \in \{0,1\}.$$

Sergeev's TFT relies on the following lemma, albeit stated differently here than in [9].

Lemma 7. *Fix i and j such that $1 \le j \le i \le s$. Suppose that $f_j = z^e$ and that $f_\ell = 0$ for all $\ell \ne j$, $1 \le \ell \le i$. Let m be a power of two at most n_i. Then $C_i \bmod (z^m-1)$ is non-zero only if $e[n(\ell)] = 1$ for all $\ell \in \{j+1, j+2, \ldots, i\}$, in which case*

$$\begin{aligned} C_i \bmod (z^m-1) &= 2^{i-j} z^e \bmod (z^m-1), \\ &= 2^{i-j} z^{e \bmod m}. \end{aligned} \tag{6}$$

Lemma 7 can be derived from lemma 1 in [9]. As Φ_k, $k>i$, divides $z^{n_i}-1$, lemma 7 gives us the following corollary.

Corollary 8. *Fix i, j and k such that $1 \le j \le i < k \le s$. Suppose that $f_j = z^e$ and that $f_\ell = 0$ for all $\ell \ne j$,*

$1 \le \ell \le i$. *Then* $C_i \bmod \Phi_k$ *is non-zero only if* $e[n(\ell)] = 1$ *for all* $\ell \in \{j+1, j+2, \ldots, i\}$, *in which case*

$$C_i \bmod \Phi_k = 2^{i-j} z^e \bmod \Phi_k.$$

Given that $z^{n_k} \bmod \Phi_k = -1$, we have that

$$2^{i-j} z^e \bmod \Phi_k = (-1)^{e[n(k)]} 2^{i-j} z^{(e \bmod n_k)},$$

where $e \bmod n_k$ is the integer e^* such that $n_k \mid (e - e^*)$ and $0 \le e^* < n_k$. The values $e[n(\ell)]$ can be determined from $n_\ell = 2^{n(\ell)}$ and e by way of a bitwise "and" operation.

Example 9. *Suppose the input size is* $n = 86$ $(n = 64 + 16 + 4 + 2)$, *and suppose that*

$$f_1 = f \bmod (z^{64} + 1) = z^e, \qquad f_2 = 0, \qquad f_3 = 0.$$

In this example,

$$C_3 = f \bmod \left[(z^{64}+1)(z^{16}+1)(z^4+1)\right].$$

Let $g = C_3 \bmod (z^2 + 1)$. *Then by corollary 8,*

$$g = \begin{cases} 0 & \text{if } e \in [0,20) \cup [24,28) \cup [32,52) \cup [56,60), \\ 4 & \text{if } e = 20, 28, 52, \text{ or } 56, \\ 4z & \text{if } e = 21, 29, 53, \text{ or } 57, \\ -4 & \text{if } e = 22, 30, 54, \text{ or } 58, \\ -4z & \text{if } e = 23, 31, 55, \text{ or } 59. \end{cases}$$

Remark 10. *Let* $1 \le j \le i \le s$. *A proportion of* 2^{j-i} *terms of* f_j *have an exponent satisfying the non-zero criterion of lemma 7 and corollary 8.*

Proof of lemma 7. We fix e and j and prove the lemma by induction on i.

Base case: Suppose $i = j$, in which case the non-zero criterion of the lemma always holds and we need only to show (6). We have that $C_i \bmod \Phi_i = z^e$ and $C_i \bmod \Phi_\ell = 0$ for $\ell < i$. Chinese remaindering gives us

$$C_i = C_{i-1} + \Gamma_{i-1}\left(\Gamma_{i-1}^{-1}(z^e - C_{i-1}) \bmod \Phi_i\right).$$

Furthermore, as $f_\ell = 0$ for $1 \le \ell < i$, we have $C_{i-1} = 0$, and $C_i = 2^{1-i}\Gamma_{i-1} z^e$. Reducing this modulo $z^m - 1$, we have $C_i = z^{e \bmod m}$ as desired.

Inductive step: Suppose now that the lemma holds for a fixed $i \ge j$, and consider $C_{i+1} \bmod (z^m - 1)$, m a power of two dividing n_{i+1}. We suppose that $f_\ell = 0$ for $1 \le \ell \ne j \le i+1$ and $f_j = z^e$. By Chinese remaindering,

$$C_{i+1} = C_i - \Gamma_i\left(\Gamma_i^{-1} C_i \bmod \Phi_{i+1}\right). \qquad (7)$$

We prove the inductive step by cases:

Case 1: If $e[n(\ell)] = 0$ for some ℓ, $j < \ell \le i$, then by the induction hypothesis, $C_i \bmod (z^{n_i} - 1) = 0$. As Φ_{i+1} and $z^m - 1$ both divide $z^{n_i} - 1$, the images $C_i \bmod \Phi_{i+1}$ and $C_i \bmod (z^m - 1)$ are necessarily zero as well. It follows from (7) that $C_{i+1} \bmod (z^m - 1)$ is also zero.

Case 2: If $e[n(\ell)] = 1$ for all ℓ, $j < \ell \le i$, then by the induction hypothesis,

$$C_i \bmod (z^{n_i} - 1) = 2^{i-j} z^e \bmod (z^{n_i} - 1).$$

Reducing (7) modulo $z^m - 1$, and again using that $z^m - 1$ and Φ_{i+1} divide $z^{n_i} - 1$, we have

$$\begin{aligned} & C_{i+1} \bmod (z^m - 1), \\ & = 2^{i-j} z^e - \Gamma_i\left(\Gamma_i^{-1} 2^{i-j} z^e \bmod \Phi_{i+1}\right) \bmod (z^m - 1), \\ & = 2^{i-j} z^{e \bmod m} - 2^{i-j}\left(z^e \bmod \Phi_{i+1}\right) \bmod (z^m - 1), \\ & = 2^{i-j}\left(1 - (-1)^{e[n(i+1)]}\right) z^{e \bmod m}. \end{aligned}$$

Since $1 - (-1)^{e[n(i+1)]}$ evaluates to 2 if $e[n(i+1)] = 1$ and 0 otherwise, this completes the proof. $\square$

4.2 Sergeev's in-place cyclotomic TFT

In [9], Sergeev describes an algorithm for computing a length-n cyclotomic TFT requiring space for $n + \mathcal{O}(1)$ ring elements, with cost asymptotically equivalent to van der Hoeven's algorithm for the bit-reversed TFT. This algorithm, like Mateer's, breaks f into the images f_i with linear cost, and then applies a DWT on each image. Procedure SergeevBreakIntoImages gives Sergeev's algorithm for computing the images f_i in place of f. Using our array notation, we let $\mathbf{a}^{(i)} = \mathbf{a} + n_1 + \cdots + n_{i-1}$ and write f_i to $\mathbf{a}_i$.

Procedure SergeevBreakIntoImages($\mathbf{a}, n$)

Input: $\mathbf{a}$, a length-n array containing $f \in \mathcal{R}[z]$.
Result: $f_1, \ldots, f_s$ is written in place of f.

```
a^(1) ⟵ a
for i ⟵ 1 to s − 1 do a^(i+1) ⟵ a^(i) + n_i

N ⟵ 2^(⌊log n⌋+1)
(K, i) ⟵ (N/2, 0)
while K ≥ n_s do
    if K > n_{i+1} then
        for d ⟵ 0 to n_i^* − 1 do
            a_d^(i+1) += Σ_{j=1}^{i} 2^(i−j) Σ_{e=0, e=d+K mod 2K, e[n(ℓ)]=1, j<ℓ≤i}^{n_j−1} a_e^(j)
    else
        for d ⟵ n_{i+1}^* to n_{i+1} − 1 do
            a_d^(i+1) −= Σ_{j=1}^{i} 2^(i−j) Σ_{e=0, e=d+K mod 2K, e[n(ℓ)]=1, j<ℓ≤i}^{n_j−1} a_e^(j)
        for d ⟵ 0 to n_{i+1}^* − 1 do
            [a_d^(i+1), a_{d+K}^(i+1)] ⟵ [a_d^(i+1) − a_{d+K}^(i+1), a_d^(i+1) + a_{d+K}^(i+1)]
        i ⟵ i + 1
    K ⟵ K/2
```

At the start of an iteration of the while loop of Sergeev's TFT, we have $\mathbf{a}^{(j)}$ containing f_j for $1 \le j \le i$, and $\mathbf{a}^{(i+1)}$ containing the first n_i^* coefficients of $f \bmod (z^{2K} - 1)$, for a power of two $K \in [n_{i+1}, n_i)$. If $K > n_{i+1}$, the algorithm writes the first n_i^* coefficients of $f \bmod (z^K - 1)$ to $\mathbf{a}^{(i+1)}$. If $K = n_{i+1}$, the algorithm writes f_{i+1} to $\mathbf{a}^{(i+1)}$ and the first n_{i+1}^* coefficients of $f \bmod (z^K - 1)$ to $\mathbf{a}^{(i+2)}$.

Consider then the case $K > n_i$. Write $f \bmod (z^{2K} - 1) = \sum_{d=0}^{2K-1} b_d z^d$ and $f_j = \sum_{e=0}^{n_j} a_e^{(j)} z^e$. The coefficients of

$f \bmod (z^K - 1)$ are given by (3), as is used in Mateer's algorithm. To write the length-n_i^* truncation of $f \bmod (z^K - 1)$ in place of $f \bmod (z^{2K} - 1)$ we merely add b_{d+K} to b_d for $0 \le d < n_i^*$. Since

$$f \bmod (z^{2K} - 1) = C_i + 2^i q_i \bmod (z^{2K} - 1)$$

and $\deg(q_i) < n_i^* < K$, it follows that b_{d+K} depends strictly on $C_i \bmod (z^{2K} - 1)$. Thus by lemma 7,

$$b_{d+K} = \sum_{j=1}^{i} 2^{i-j} \sum_{e \in S_{j,i}^d} a_e^{(j)}, \qquad (8)$$

where

$$S_{j,i}^d = \left\{ e \in [0, n_j) : 2K \mid (e - d - K), \prod_{\ell=j+1}^{i} e[n(\ell)] = 1 \right\}.$$

We call the sum (8) the *contributions* of $f_1, \ldots, f_i$ towards b_{d+K}.

For the case $K = n_{i+1}$, the coefficients of $f_{i+1} = f \bmod (z^K + 1)$ are given by (4). We compute the first n_{i+1}^* coefficients of $f \bmod (z^K \pm 1)$, $b_d \mp b_{d+K}$, in place of the stored values b_d and b_{d+K}, for $0 \le d < n_{i+1}^*$. For the remaining coefficients of f_{i+1} we compute b_{d+K} using (8) as in the first case.

If we compute (8) in an intelligent order, then the cost of Sergeev's algorithm amounts to linearly many additions and multiplications by powers of 2, plus the cost of the Discrete Weighted Transforms. Sergeev measured the cost of his approach for computing the images f_i in terms of ring additions and multiplications by powers of 2:

THEOREM 11. *Let $f \in \mathcal{R}[z]$ have degree less than n. Then one can compute $f_1, \ldots, f_s$ in place of f using $4n - 6n_1$ ring multiplications by powers of 2 and $4n - 4n_1$ ring additions.*

This linear cost is absorbed into the $\mathcal{O}(n)$ factor appearing in the cost bound (5) of the Discrete Weighted Transforms. Thus Sergeev's algorithm is asymptotically equivalent to Mateer's and van der Hoeven's TFT algorithms.

In order to compute a new coefficient of an image of f in Sergeev's algorithm, one effectively has to make a pass through the array in order to sum the contributions (8), for every coefficient computed in this manner. This is because, in order to keep things in-place without increasing the cost, one has to compute the entire sum (8) before adding it back into the array. One improvement is to instead work with the *weighted* images

$$f_j^* = 2^{-j+1} f_j$$

instead of f_j, and $2^{-i-1} f \bmod (z^{2K} - 1)$ instead of $f \bmod (z^{2K} - 1)$, where $K \in (n_{i+1}, n_i]$. If we now let $a_d^{(j)}$ and b_d be the coefficients of these weighted images, our equation (8) for b_{d+K} becomes

$$b_{d+K} = \sum_{j=1}^{i} \sum_{e \in S_{j,i}^d} a_e^{(j)}.$$

This allows us to progressively add b_{d+K} to another value in our array without having to compute b_{d+K} in entirety first. We use such weighted images in our implementation of Sergeev's algorithm, as well as in the algorithm we describe in the next section, where for that method we describe the reweighting in greater detail.

5. A NEW IN-PLACE CYCLOTOMIC TFT ALGORITHM

We present an algorithm related to Sergeev's algorithm, also in-place, with asymptotically equivalent cost. Unlike Sergeev's algorithm, we will forego producing part of the images $f \bmod (z^K - 1)$, K a power of two. We will compute f_{i+1} immediately after producing f_i. An advantage of such an approach is, we require fewer passes through our array for input sizes n of low Hamming weight, i.e., n containing few non-zero bits.

We write $f_1, \ldots, f_s$ in place of f in three steps: we first compute the remainders produced by dividing q_{i-1} by Φ_i,

$$r_i = q_{i-1} \bmod \Phi_i, \quad 1 \le i \le s,$$

in place of f; we then iteratively write f_i^* in place of r_i for $i = 1, 2, \ldots, s$; lastly we reweight f_i^* to get f_i. After we have the images f_i we again compute a DWT of each image f_i separately to give us the weighted evaluation points.

5.1 Breaking f into the remainders r_i

We first break $f = q_0$ into its quotient and remainder dividing by $z^{n_1} + 1$,

$$r_1 = f \bmod (z^{n_1} + 1) = \sum_{i=0}^{n_1 - 1} (a_i - a_{i+n_1}) z^i,$$

$$q_1 = f \text{ quo } (z^{n_1} + 1) = \sum_{i=0}^{n_1^* - 1} a_{i+n_1} z^i,$$

where $a_i = 0$ for $i \ge n$. This can be done in place with n_1^* subtractions in $\mathcal{R}$. We then similarly break q_1 into r_2 and q_2, then q_2 into r_3 and q_3, and continue until we have $r_1, \ldots, r_{s-1}$ and q_{s-1}. Since $\deg(q_{s-1}) < n_s = \deg(\Phi_s)$, r_s is exactly q_{s-1}.

For the purposes of the inverse transform, computing f from the remainders r_i is equally uncomplicated. Given q_{i+1} and r_{i+1}, we recompute q_i in place of q_{i+1} and r_{i+1} as $q_i = q_{i+1}(z^{n_{i+1}} + 1) + r_{i+1}$.

5.2 Computing f_i^* in place of r_i

We first note that f_1^* is precisely r_1. We will iteratively produce the remaining weighted images. Suppose, at the start of the i-th iteration, we have f_j^*, and r_k, for $1 \le j \le i < k \le s$. We want to write f_{i+1}^* in place of r_{i+1}. We have

$$\begin{aligned} f_{i+1}^* &= 2^{-i} f \bmod \Phi_{i+1}, \\ &= 2^{-i} (\Gamma_i q_i + C_i) \bmod \Phi_{i+1}, \\ &= \left(q_i + 2^{-i} C_i\right) \bmod \Phi_{i+1}, \\ &= r_{i+1} + \left(2^{-i} C_i \bmod \Phi_{i+1}\right). \end{aligned} \qquad (9)$$

Unfortunately, we do not have the combined image C_i, but rather the weighted images f_j^*, $1 \le j \le i$, from which we can reconstruct C_i. We would like to be able to compute the sum (9) in place from the remainder r_{i+1} and the weighted images f_j^*.

Corollary 8 tells us the contribution of f_i^* towards subsequent images. If e satisfies the non-zero criterion of the corollary, then by (9), a term $c_{j,e} z^e$ of f_j^*, $j \le i$ will contribute $\frac{1}{2}(-1)^{e[n(i+1)]} z^{(e \bmod n_{i+1})}$ to f_{i+1}^*. In order to make the contributions have weight ± 1, we instead first reweight r_i by 2 and compute $2f_i^*$, and then divide by 2 thereafter.

We note that this cost of reweighting by $2^{\pm 1}$ requires fewer multiplications than instead introducing a factor $\frac{1}{2}$ into all the contributions. AddContributions describes how we add the contributions of $f_1^*, \ldots, f_i^*$ to f_{i+1}^*.

Procedure AddContributions($\mathbf{a}$, n, i)

Input: $\mathbf{a}$, a length-n array containing $f_1^*, \ldots, f_i^*$ and $2r_{i+1}$, in that order.
Result: The contributions of $f_1^*, \ldots, f_i^*$ towards $2f_{i+1}^*$ are added to $2r_{i+1}$. As a result we will have $2f_{i+1}^*$ in place of $2r_{i+1}$.

```
a^(1) ⟵ a
for j ⟵ 1 to i do a^(j+1) ⟵ a^(j) + n_j

for j ⟵ 1 to i do
    // Add contribution of f_j^* to f_i^*
    for e ⟵ 0 to n_j − 1 do
        if e[n(ℓ)] = 0 for all ℓ, j < ℓ ≤ i then
            a^(i+1)_{e mod n_{i+1}} += (−1)^{e[n(i+1)]} a^(j)_e
```

According to corollary 8, only a proportion of 2^{j-i} of the terms of f_j^* will have a non-zero contribution to f_{i+1}^*. Thus the total cost of adding contributions of f_j^* towards f_ℓ^*, for all $\ell > j$, is less than $2\#f_j^* = 2n_j$. It follows that the total additions and subtractions in $\mathcal{R}$ required to add all these contributions are bounded by $2n$. Since AddContributions only scales array ring elements by ± 1, we have the following complexity:

LEMMA 12. *Calling* AddContributions($\mathbf{a}, n, i$) *for* $1 \leq i < s$ *entails no more than* $2n$ *ring additions and no ring multiplications.*

In the manner we have chosen to add these contributions, we will have to make $s-1$ passes through our array to add them all. One way we could avoid this is to instead add all the contributions from f_1^*, and then all the contributions from f_2^*, and so forth, adding up all the contributions from a single term at once. We could use that a term $c_{j,e}z^e$ of f_j^* that does not contribute towards f_{i+1}^* will not contribute to f_k^* for any $k > i$. This would reduce the number of passes we make through the larger portion of the array, though the cache performance of potentially writing to $s-1$ images at once raises questions.

When adding contributions to f_{i+1}^*, we need only inspect the non-zero criterion of one exponent e in a block of exponents $kn_i \leq e < (k+1)n_i$. Similarly, we need only inspect one exponent in a block of n_{i+1} consecutive exponents in order to determine their shared value of $(-1)^{e[n(i+1)]}$. There may appear long segments of consecutive exponents not satisfying the non-zero criterion. To avoiding traversing such segments unnecessarily, our implementation computes the next exponent that satisfies the non-zero criterion by way of bit operations.

We note here that, for the purposes of the inverse transform, we can as easily reobtain r_i from f_i^*. We merely multiply f_i^* by 2, instead subtract the contributions to get $2r_i$, and then multiply by $\frac{1}{2}$ to get r_i.

Procedure BreakIntoImages breaks f into the images f_i, after which we can again use Discrete Weighted Transforms to compute $\mathrm{TFT}'_{n,\omega}(f)$.

Procedure BreakIntoImages($\mathbf{a}$, n)

Input: $\mathbf{a}$, a length-n array containing $f \in \mathcal{R}[z]$.
Result: The images $f_1, \ldots, f_s$ are written in place of f.

```
a^(1) ⟵ a
for i ⟵ 1 to s − 1 do a^(i+1) ⟵ a^(i) + n_i

// 1: Write r_1, ..., r_s in place of f
for i ⟵ 1 to s − 1 do
    for e ⟵ 0 to n_i^* − 1 do
        a^(i)_e ⟵ a^(i)_e − a^(i)_{e+n_i}

// 2: Write f_{i+1}^* in place of r_{i+1}
for i ⟵ 1 to s − 1 do
    for e ⟵ 0 to n_{i+1} − 1 do a^(i+1)_e ⟵ 2a^(i+1)_e
    AddContributions(a, n, i)
    for e ⟵ 0 to n_{i+1} − 1 do a^(i+1)_e ⟵ ½a^(i+1)_e

// 3: Reweight f_i^* to get f_i
for i ⟵ 1 to s − 1 do
    for e ⟵ 0 to n_i^* − 1 do a^(i+1)_e ⟵ 2a^(i+1)_e
```

5.3 Cost analysis

Procedure BreakIntoImages effectively has three parts. In the first part of the algorithm, we break f into the remainders r_i. Producing r_i entails $n_i^* < n_i$ additions, and so producing all the r_i entails less than $\sum_{i=1}^{s} n_i = n$ ring additions.

In the second part we write the weighted images f_i^* in place of the r_i. Adding all the contributions, per lemma 12, requires $2n$ additions. We reweight the last $n - n_1$ coefficients of f by 2, then by $\frac{1}{2}$. This constitutes less than n multiplications in total.

In the third part we reweight the weighted images f_i^* to get the images f_i. This entails less than n multiplications by 2. This gives us the following complexity:

LEMMA 13. *Procedure* BreakIntoImages($\mathbf{a}, n$) *requires at most* $3n$ *ring additions and* $2n$ *ring multiplications by* $2^{\pm 1}$.

Thus, like Mateer's and Sergeev's algorithms, the TFT algorithm suggested in this section requires a linear cost to separate f into the images f_i, and again has cost asymptotically equivalent to van der Hoeven's bit-reversed TFT.

We briefly mention that it is reasonably straightforward to invert BreakIntoImages. We have described in sections 5.1 and 5.2 how to reverse the first two parts of the algorithm. Reversing the third part is trivial. Without giving a detailed analysis of the inverse transform, we remark that computing f from $f_1, \ldots, f_s$ also has a linear cost.

6. A LINK BETWEEN BIT-REVERSED & CYCLOTOMIC TFT METHODS

The bit-reversed TFT has the property that, for $m < n$, $\mathrm{TFT}_{\omega,m}(f)$ is merely the first m entries of $\mathrm{TFT}_{\omega,n}(f)$, whereas the cyclotomic TFT does not, in general, have this property. Hence the bit-reversed TFT lends itself more readily to multivariate polynomial arithmetic. We show how an algorithm for computing the cyclotomic TFT may be modified to compute a bit-reversed TFT.

As before, let $n = \sum_{i=1}^{s} n_i = \sum_{i=1}^{s} 2^{n(i)}$ and let $N = 2^p$ be the least power of two exceeding n. Let $\omega = \omega_1$ be a root of Φ_1 and $\omega_i = \omega^{n_1/n_i}$ be a root of Φ_i. Define

$$\Omega_i = \prod_{\ell=1}^{i} \omega_\ell \qquad \text{and} \qquad \Psi_j(z) = z^{n_j} - \Omega_{j-1}^{n_j},$$

for $0 \le i \le s$ and $1 \le j \le s$. $\mathrm{TFT}_{\omega,n}(f)$ is comprised of the evaluation of $f(z)$ at the roots of the polynomials $\Psi_j(z)$. If $n_1 + \cdots + n_{j-1} \le \ell < n_1 + \cdots + n_j$, we have that $\omega^{[\ell]_p}$ is a root of Ψ_j. To see this, write $\ell = n_1 + \cdots + n_{j-1} + \ell'$, $0 \le \ell' < n_j$, and observe that

$$\begin{aligned} \omega^{[\ell]_p n_j} &= \omega^{\left(N/(2n_1)+\cdots+N/(2n_{j-1})\right)n_j} \omega^{[\ell']_p n_j}, \\ &= (\omega_1 \cdots \omega_{j-1})^{n_j} \omega^{[\ell']_p n_j}, \\ &= \Omega_{j-1}^{n_j} \omega^{[\ell']_p n_j}. \end{aligned} \tag{10}$$

Every n_j-th root of unity in $\mathcal{R}$ is of the form $\omega^{[k]_p}$, where $0 \le k < n_j$. In particular, $\omega^{[\ell']_p}$ is an n_j-th root of unity. Thus (10) is precisely $(\Omega_{j-1})^{n_j}$ and $\omega^{[\ell]_p}$ is a root of Ψ_j.

Consider the affine transformation $z \mapsto \Omega_s z$. Then

$$\begin{aligned} \Psi_i(\Omega_s z) &= \Omega_s^{n_i} z^{n_i} - \Omega_{i-1}^{n_i}, \\ &= \Omega_{i-1}^{n_i} \left[\left(\prod_{j=i}^{s} \omega_j^{n_i} \right) z^{n_i} - 1 \right], \\ &= -\Omega_{i-1}^{n_i}(z^{n_i} + 1) = -\Omega_{i-1}^{n_i} \Phi_i. \end{aligned}$$

Thus, for a polynomial $f(z)$, $f(z) \bmod \Phi_i(z) = f(z) \bmod \Psi_i(\Omega_s z)$. We can break $f(z)$ into its images modulo the polynomials Φ_i as follows:

1. Replace $f(z)$ with $f(\Omega_s z)$.
2. Break $f(\Omega_s z)$ into its images modulo $\Phi_i(z)$ per a cyclotomic TFT method. Equivalently, this gives us the images $f(\Omega_s z) \bmod \Psi_i(\Omega_s z)$.
3. Apply transformation $z \mapsto \Omega_s^{-1} z$ to every image to give us $f(z) \bmod \Psi_i(z)$ in place of $f(\Omega_s z) \bmod \Psi_i(\Omega_s z)$.

Note that the affine transformations of steps 1 and 3 both have complexity $\mathcal{O}(n)$. We can get $\mathrm{TFT}_{\omega,n}(f)$ from the images $f \bmod \Psi_i$ by applying a DWT with weight Ω_{i-1} to each image $f \bmod \Psi_i$. As each step here is invertible, we can invert a bit-reversed TFT using an inverse cyclotomic TFT algorithm. We can similarly use a bit-reversed TFT algorithm to compute a cyclotomic TFT.

7. CONCLUSION

We have presented a method of computing a cyclotomic TFT in-place, with cost, in terms of ring multiplications, asymptotically equivalent to out-of-place TFT methods. We have also shown a means of using a cyclotomic TFT algorithm to compute a bit-reversed TFT.

As future work, we would like to make fine-tuned implementations of Sergeev's TFT algorithm and the TFT method presented here, to gauge how well they perform compared to existing competitive TFT implementations.

We also would like to better understand exactly which families of Truncated Fourier Transforms are equivalent by way of transformations as in section 6.

8. ACKNOWLEDGMENTS

Thanks to: Igor Sergeev for bringing [7] and [9] to my attention, and for his assistance in helping me understand his TFT algorithm; Joris van der Hoeven for posing the question of non-invertible TFTs; Dan Roche for providing his implementation of bit-reversed TFT algorithms; and Curtis Bright, Reinhold Burger, Mustafa Elsheikh, Mark Giesbrecht, and Colton Pauderis for their comments and feedback.

The author would also like to thank the referees for their careful reading of the original draft of this paper and for their many suggestions.

9. REFERENCES

[1] J. W. Cooley and J. W. Tukey. An algorithm for the machine calculation of complex Fourier series. *Math. Comp.*, 19:297–301, 1965.

[2] R. Crandall. *Advanced Topics in Scientific Computation.* Springer-Verlag, 1996. ISBN 0-387-94473-7.

[3] R. Crandall and B. Fagin. Discrete weighted transforms and large-integer arithmetic. *Math. Comp.*, 62(205):305–324, 1994.

[4] J. V. Z. Gathen and J. Gerhard. *Modern Computer Algebra.* Cambridge University Press, New York, NY, USA, 2-nd edition, 2003.

[5] D. Harvey and D. S. Roche. An in-place truncated Fourier transform and applications to polynomial multiplication. In *Proceedings of the 2010 International Symposium on Symbolic and Algebraic Computation*, ISSAC '10, pages 325–329, New York, NY, USA, 2010. ACM.

[6] M. Heideman, D. Johnson, and C. Burrus. Gauss and the history of the fast Fourier transform. *ASSP Magazine, IEEE*, 1(4):14–21, 1984.

[7] T. Mateer. Fast Fourier Transform Algorithms with Applications. Master's thesis, Clemson University, 2008.

[8] D. Roche. *Efficient Computation With Sparse and Dense Polynomials.* PhD thesis, University of Waterloo, 2011.

[9] I. S. Sergeev. Regular estimates for the complexity of polynomial multiplication and truncated Fourier transform. *Prikl. Diskr. Mat.*, pages 72–88, 2011. (Russian) `http://mi.mathnet.ru/eng/pdm347`.

[10] J. van der Hoeven. The Truncated Fourier Transform and Applications. In J. Gutierrez, editor, *Proc. ISSAC 2004*, pages 290–296, Univ. of Cantabria, Santander, Spain, July 4–7 2004.

[11] J. van der Hoeven. Notes on the Truncated Fourier Transform. Technical Report 2005-5, Université Paris-Sud, Orsay, France, 2005.

Approximately Counting Semismooth Integers

Eric Bach
Computer Sciences Department
University of Wisconsin-Madison
bach@cs.wisc.edu

Jonathan Sorenson
Computer Science and Software Engineering Department
Butler University
sorenson@butler.edu

ABSTRACT

An integer n is (y,z)-semismooth if $n = pm$ where m is an integer with all prime divisors $\leq y$ and p is 1 or a prime $\leq z$. Large quantities of semismooth integers are utilized in modern integer factoring algorithms, such as the number field sieve, that incorporate the so-called *large prime* variant. Thus, it is useful for factoring practitioners to be able to estimate the value of $\Psi(x,y,z)$, the number of (y,z)-semismooth integers up to x, so that they can better set algorithm parameters and minimize running times, which could be weeks or months on a cluster supercomputer. In this paper, we explore several algorithms to approximate $\Psi(x,y,z)$ using a generalization of Buchstab's identity with numeric integration.

Categories and Subject Descriptors

F.2.1 [**Analysis of Algorithms and Problem Complexity**]: Numerical Algorithms and Problems—*Number-theoretic computations*; G.1.1 [**Numerical Analysis**]: Interpolation; I.1.2 [**Symbolic and Algebraic Manipulation**]: Algorithms—*Algebraic algorithms*; G.4 [**Mathematical Software**]: Algorithm Design and Analysis

Keywords

Semismooth Numbers, Integer Factoring, Smooth Numbers, Number Theoretic Algorithms

1. INTRODUCTION

The security of the public-key cryptosystem RSA [18, 22] is based on the practical difficulty of integer factoring.

The fastest current general-purpose integer factoring algorithm is the number field sieve [9, 17], which in its basic form makes use of *smooth numbers*, integers with only small prime divisors. This has inspired research into algorithms to approximately count smooth numbers [5, 6, 15, 20, 24, 26, 27]. However, most implementations of the number field sieve make use of the so-called *large prime* variant [9, §6.1.4].

ISSAC'13, June 26–29, 2013, Boston, Massachusetts, USA.

So we want, in fact, to count smooth integers that admit at most one slightly larger prime divisor, or *semismooth* numbers. (See, for example, the details on the factorization of a 768-bit RSA modulus [16] where smoothness bounds are discussed near the end of §2.2.)

The principal contribution of this paper is twofold:

1. We present data showing that the key to estimating $\Psi(x,y,z)$ accurately is an algorithm to estimate $\Psi(x,y)$ accurately, and

2. We present head-to-head comparisons of five algorithms for estimating $\Psi(x,y,z)$.

Previous work was done by Bach and Peralta [3] and generalized by Zhang and Ekkelkamp [10, 29]; we discuss this below.

This paper is organized as follows. We begin with some definitions, and briefly discuss computing exact counts of semismooth integers. We then give our main theoretical result, a generalized Buchstab identity, which together with numerical integration, is the basis of all our algorithms. We then present five different algorithms in some detail, two based on the Dickman ρ function and three based on the saddle point methods of Hildebrand and Tenenbaum [14], along with empirical results for each algorithm. As one might expect, we discover a tradeoff in algorithm choice between speed and accuracy. We follow this up with an elaboration on some numerical details.

2. DEFINITIONS

Let $P(n)$ denote the largest prime divisor of the positive integer n, with $P(1) = 1$. An integer n is y-*smooth* if $P(n) \leq y$, and $\Psi(x,y)$ counts the integers $n \leq x$ that are y-smooth.

An integer n is (y,z)-*semismooth* if we can write $n = mp$ where m is y-smooth and $p \leq z$ is a prime or 1. $\Psi(x,y,z)$ counts the integers $n \leq x$ that are (y,z)-semismooth. (Generalizations to more than one exceptional prime have been defined by Zhang and Ekkelkamp [10, 29].) Observe that $\Psi(x,y,y) = \Psi(x,y)$, the function $\Psi(x,y,x)$ counts integers whose second-largest prime divisor is bounded by y, and $\Psi(x,1,z) = \min\{\pi(x), \pi(z)\}$, where $\pi(x)$ is the number of primes up to x.

Our basic unit of work is the floating point operation. Along with the four basic arithmetic operations $(+, -, \times, \div)$ we include square roots, logarithms, and exponentials, since their complexity is close to that of multiplication (see for example [7]).

Table 1: Exact Values of $\Psi(x, y, z)$, $x = 2^{40}$

y	$z = 2^{10}$	$z = 2^{12}$	$z = 2^{14}$	$z = 2^{16}$	$z = 2^{18}$	$z = 2^{20}$
2^2	58916	170906	503392	1500366	4513650	13597105
2^4	6132454	15111450	36766896	88920834	213965871	508848834
2^6	323105012	678707129	1326493628	2499496319	4603776946	8298713253
2^8	3157707079	6694272918	11837179134	19296840890	30059136386	45290571262
2^{10}	7138986245	21494669620	39400743040	61719198990	89501569374	123782024151
2^{12}	—	30641713551	68600140477	111769092210	160884758713	215725604647
2^{14}	—	—	80324574755	145583683889	214469637137	286977146180
2^{16}	—	—	—	155283653287	241316058768	329068435579
2^{18}	—	—	—	—	248857736183	349745847766
2^{20}	—	—	—	—	—	354983289990

3. EXACT COUNTS

Using a prime number sieve, such as the sieve of Eratosthenes, we can completely factor all integers up to x in $O(x \log \log x)$ arithmetic operations, and thereby compute exact values of $\Psi(x, y, z)$. Of course this is not a practical approach for large x, but it is useful for evaluating the accuracy of approximation algorithms, which is what we do here. So we wrote a program to do this, based on a segmented sieve of Eratosthenes (see [25] for prime number sieve references), and we ran our program up to $x = 1099511627776 = 2^{40}$. Our results for this largest value for x appear in Table 1, which took just over 100 CPU hours to compute.

4. A GENERALIZED BUCHSTAB IDENTITY

We have the following version of Buchstab's identity (see for example [28, p. 365]):

$$\Psi(x, y) = \Psi(x, 2) + \sum_{2<p\le y} \Psi(x/p, p), \tag{1}$$

which is obtained by summing over the largest prime divisor of y-smooth integers $n \le x$. Using this same idea gives us the following:

$$\Psi(x, y, z) = \Psi(x, y) + \sum_{y<p\le z} \Psi(x/p, y). \tag{2}$$

As one can see, the identity is obtained by summing over the largest prime divisor.

5. APPROXIMATE COUNTS

As mentioned in the Introduction, there are many algorithms to estimate values of $\Psi(x, y)$. We could choose one of them, compute a list of primes up to z, and then apply (2) to approximate $\Psi(x, y, z)$. We found that this does, in fact, give fairly accurate estimates, but the resulting algorithms are quite slow since roughly $O(z/\log z)$ evaluations of $\Psi(x/p, y)$ (one for each p, $y < p \le z$) are needed.

Our approach, then, is to replace the sum in (2) with an integral, and then use numeric integration to evaluate it [8, §7.2]; we used Simpson's rule. We found that in practice, the relative error introduced by replacing the sum with an integral that was then estimated, was less than the relative error introduced by the approximation algorithms for $\Psi(x, y)$.

Let us define $\mathrm{li}(x) := \int_2^x dt/\log t$, and let $e(x) := \pi(x) - \mathrm{li}(x)$. By the prime number theorem, $e(x) = x/\exp[\Omega(\sqrt{\log x})]$; if we assume the Riemann Hypothesis, $e(x) = O(\sqrt{x} \log x)$ [23].

We have the following (see [4, §2.7]):

LEMMA 1. *Let f be a continuously differentiable function on an open interval containing $[2, z]$, and let $2 \le y \le z$. Then*

$$\sum_{y<p\le z} f(p) = \int_y^z \frac{f(t)}{\log t} dt + f(z)e(z) - f(y)e(y) - \int_y^z e(t) f'(t) dt.$$

Of course we cannot apply this lemma to $\Psi(x, y)$ directly, so we use an estimate instead. Define $\rho(u)$ as the unique continuous solution to

$$\rho(u) = 1 \quad (0 \le u \le 1)$$
$$\rho(u-1) + u\rho'(u) = 0 \quad (u > 1).$$

Note that $\rho \in C^1$ for $u > 1$. Hildebrand [12] proved that for $\epsilon > 0$ we have

$$\Psi(x, y) = x\rho(u) \left(1 + O_\epsilon \left(\frac{\log(u+1)}{\log y}\right)\right), \tag{3}$$

uniformly on the set defined by $1 \le u \le \exp((\log y)^{3/5-\epsilon})$ and $y \ge 2$. Here, $u := u(x, y) = \log x/\log y$.

THEOREM 1. *Given $2 \le y \le z \le x$, let $\epsilon > 0$, and assume that $1 \le \log(x/z)/\log y \le \log x/\log y \le \exp((\log y)^{3/5-\epsilon})$. Then*

$$\Psi(x, y, z) = \left(\Psi(x, y) + \int_y^z \frac{\Psi(x/t, y)}{\log t} dt\right) (1 + o(1)). \tag{4}$$

For asymptotic notation, we are assuming y is large.

PROOF. Define $f(t) := (x/t)\rho(\log(x/t)/\log y)$. By (3) we have $f(p) = \Psi(x/p, y)(1 + o(1))$ for all primes $y < p \le z$. f is differentiable and continuous, with $f'(t) \sim -f(t)/t$ (for large y). It is then straightforward to show that $e(t)f'(t) = o(f(t)/\log t)$. Also since f is decreasing we can show $e(z)f(z) - e(y)f(y) = o(\pi(z) - \pi(y))f(z) + O(|e(y)|f(y))$ and then $(\pi(z) - \pi(y))f(z) \le \sum_{y<p\le z} f(p)$ and $|e(y)|f(y) = o(\Psi(x, y))$. In the case when u is large, we make use of Lemma 8.1 and (61) from [28, §5.4]. We then apply Lemma 1 and use (2), and finally substitute $\Psi(x/t, y)$ back in for $f(t)$ to complete the proof. □

It would be nice to have some function $g(y, z)$ where

$$\Psi(x, y, z) \approx \Psi(x, y) \cdot g(y, z) \tag{5}$$

Table 2: $x \cdot \sigma(u, v)/\Psi(x, y, z)$, $x = 2^{40}$

y	$z = 2^{10}$	$z = 2^{12}$	$z = 2^{14}$	$z = 2^{16}$	$z = 2^{18}$	$z = 2^{20}$
2^2	7.0984e-14	1.2946e-12	2.1963e-11	3.4242e-10	4.8536e-09	6.2227e-08
2^4	0.0042182	0.007326	0.012701	0.021657	0.035951	0.058
2^6	0.26627	0.2956	0.32972	0.36872	0.41167	0.4585
2^8	0.64392	0.66898	0.68836	0.70731	0.72686	0.74754
2^{10}	0.75636	0.80963	0.82644	0.83778	0.84679	0.85628
2^{12}	—	0.84863	0.87886	0.89096	0.89979	0.90729
2^{14}	—	—	0.89495	0.9169	0.92614	0.93196
2^{16}	—	—	—	0.92275	0.93539	0.9421
2^{18}	—	—	—	—	0.93769	0.94722
2^{20}	—	—	—	—	—	0.95043

Table 3: $E(x, y, z)/\Psi(x, y, z)$, $x = 2^{40}$

y	$z = 2^{10}$	$z = 2^{12}$	$z = 2^{14}$	$z = 2^{16}$	$z = 2^{18}$	$z = 2^{20}$
2^2	1.6195e-13	2.9202e-12	4.8929e-11	7.5256e-10	1.0508e-08	1.325e-07
2^4	0.0063754	0.010979	0.018858	0.031834	0.052265	0.0833
2^6	0.34328	0.37895	0.42001	0.46628	0.51656	0.57018
2^8	0.76655	0.79204	0.81087	0.82903	0.84748	0.86587
2^{10}	0.87052	0.91765	0.93094	0.93899	0.94491	0.95185
2^{12}	—	0.94459	0.96506	0.97334	0.97836	0.98148
2^{14}	—	—	0.97446	0.98537	0.98863	0.99038
2^{16}	—	—	—	0.98694	0.99092	0.99305
2^{18}	—	—	—	—	0.99154	0.99516
2^{20}	—	—	—	—	—	0.99766

if this is possible. Rewriting (2) we have

$$\Psi(x, y, z) = \Psi(x, y) \cdot \left(1 + \sum_{y<p\leq z} \frac{\Psi(x/p, y)}{\Psi(x, y)}\right). \quad (6)$$

A very crude estimate of $\Psi(x/p, y)/\Psi(x, y) \approx 1/p$ leads to

$$\Psi(x, y, z) \approx \Psi(x, y) \cdot (1 + \log(\log z/\log y))). \quad (7)$$

In practice, this is too crude to be useful. However, this estimate can certainly be improved using, for example, Theorem 11 from [28, §5.5]. This is a possible direction for future work.

5.1 The Method of Bach and Peralta

The first algorithm to try would be to use the estimate $\Psi(x, y) \approx x\rho(u)$ from (3) and plug it into Theorem 1. This, in fact, is simply another way to derive the algorithm of Bach and Peralta [3]. They define $u := \log x/\log y$, $v := \log x/\log z$ and

$$\sigma(u, v) := \rho(u) + \int_v^u (\rho(u - u/w)/w)dw. \quad (8)$$

They then prove that for fixed u, v and $x \to \infty$,

$$\Psi(x, y, z) \approx x\sigma(u, v). \quad (9)$$

We can use (2) to obtain the same approximation, as follows:

$$\begin{aligned}\Psi(x, y, z) &\approx \Psi(x, y) + \int_y^z (\Psi(x/t, y)/\log t)dt \\ &\approx x \cdot \rho(\log x/\log y) \\ &\quad + \int_y^z (x/t\log t)\rho(\log(x/t)/\log y)dt \\ &= x \cdot \left(\rho(u) + \int_v^u (\rho(u - u/w)/w)dw\right) \\ &= x \cdot \sigma(u, v).\end{aligned}$$

We adapted code written by Peralta to compute values of the Dickman ρ function (accurate to roughly 8 decimal digits) to compute σ. Bach and Peralta discuss methods to compute both ρ and σ in some detail in [3]. We present the results from this algorithm in Table 2. As for all our algorithms, we show the ratio of what the algorithm produces as an estimate divided by the actual values we computed earlier. The closer we are to 1 the better the estimates.

This algorithm is fast, but the results are not as good as we might desire. As one might expect, they are about as good as what was found for estimating counts of smooth numbers using (3) in [15].

5.2 Ekkelkamp's Improvement

Ekkelkamp [10] pointed out that $\sigma(u, v)$ could be made more accurate by adding a quantity which, in our notation, is

$$\frac{(1-\gamma)x}{\log x}\left[\rho(u-1) + \int_v^u \frac{\rho(u - u/w - 1)}{w-1}dw\right].$$

Table 4: $HT(x,y,z)/\Psi(x,y,z)$, $x = 2^{40}$

y	$z = 2^{10}$	$z = 2^{12}$	$z = 2^{14}$	$z = 2^{16}$	$z = 2^{18}$	$z = 2^{20}$
2^2	1.0969	1.0712	1.0641	1.067	1.0728	1.0833
2^4	1.0592	1.039	1.031	1.0353	1.0507	1.0829
2^6	1.0414	1.0281	1.0205	1.024	1.0566	1.1582
2^8	1.0194	1.0165	1.0135	1.0127	1.0296	1.1338
2^{10}	1.0024	1.0104	1.0101	1.01	1.0112	1.0437
2^{12}	—	1.0043	1.0104	1.01	1.0078	1.0107
2^{14}	—	—	1.0046	1.0084	1.0115	1.0143
2^{16}	—	—	—	1.0101	1.016	1.017
2^{18}	—	—	—	—	1.0142	1.0116
2^{20}	—	—	—	—	—	1.0083

This can be derived by using the better approximation

$$\Psi(x,y) \approx x\rho(u) + \frac{(1-\gamma)x}{\log x}\rho(u-1),$$

due to Ramaswami [21, Theorem 1]. To get her formula, substitute $w = 1/\lambda$ in the integral; note that we need the first term inside the brackets, since our definition of semismoothness differs from hers. (Our "large prime" can be 1.) This correction does not require much additional effort; essentially just one more numerical integration. Let $E(x,y,z)$ denote this approximation.

Our experiments with this indicate that the additional term enhances accuracy significantly when y and z are large. This is roughly the bottom corner of Table 2. However, the accuracy still drops off dramatically for smaller values of y. See Table 3.

5.3 A Saddlepoint Method

Our third algorithm is based on Algorithm HT for estimating $\Psi(x,y)$ presented in [15]. Define

$$\begin{aligned}\zeta(s,y) &:= \prod_{p\le y}(1-p^{-s})^{-1};\\ \phi(s,y) &:= \log\zeta(s,y);\\ \phi_k(s,y) &:= \frac{d^k}{ds^k}\phi(s,y) \qquad (k\ge 1).\end{aligned}$$

The functions ϕ_k can be expressed as sums over primes. Indeed, we have

$$\begin{aligned}\phi(s,y) &= -\sum_{p\le y}\log(1-p^{-s});\\ \phi_1(s,y) &= -\sum_{p\le y}\frac{\log p}{p^s-1};\\ \phi_2(s,y) &= \sum_{p\le y}\frac{p^s(\log p)^2}{(p^s-1)^2}.\end{aligned}$$

Thus, with a list of primes up to y, the quantities $\zeta(s,y)$, $\phi_1(s,y)$, and $\phi_2(s,y)$ can be computed in $O(y/\log y)$ floating point operations.

Define

$$HT(x,y,s) := \frac{x^s\zeta(s,y)}{s\sqrt{2\pi\phi_2(s,y)}},$$

and let α be the unique solution to $\phi_1(\alpha,y) + \log x = 0$. Hildebrand and Tenenbaum proved the following [13]:

THEOREM 2.

$$\Psi(x,y) = HT(x,y,\alpha)\cdot\left(1 + O\left(\frac{1}{u} + \frac{(\log y)}{y}\right)\right) \qquad (10)$$

uniformly for $2 \le y \le x$.

This gives us Algorithm HT [15]:

1. Find the primes up to y.
2. Compute an approximation α' to α using binary search and Newton's method. Make sure that $|\alpha - \alpha'| = O(1/(u\log x))$.
3. Output $HT(x,y,\alpha')$.

Write $HT(x,y)$ for the value output in the last step. The running time is $O\left(\frac{y\log\log x}{\log y} + \frac{y}{\log\log y}\right)$ floating point operations.

We simply plugged Algorithm HT into Theorem 1 to estimate $\Psi(x,y,z)$ using the saddle point method as follows:

$$HT(x,y,z) := HT(x,y) + \int_y^z (HT(x/t,y)/\log t)dt. \qquad (11)$$

In Table 4 we give the results for this algorithm, which are quite good. The method is, however, a bit slow.

Using the summation algorithms described in [2], we can lower the exponent of y in the running time from 1 to 2/3. We give some details for this in §6.3. We did not implement this improvement, however, because it would not change the computed results, and the method of the section below is faster.

5.4 Assuming Riemann's Hypothesis

This is the same as Algorithm HT, only sums over primes (ζ,ϕ_1,ϕ_2) above roughly $\sqrt{y}$ are estimated using the prime number theorem plus the Riemann Hypothesis [24, 23]. It is *much* faster than Algorithm HT and nearly as accurate; its running time is roughly $\sqrt{y}$. Let $HT_f(x,y)$ denote the estimate this algorithm computes for $\Psi(x,y)$, and $HT_f(x,y,z)$ the estimate after using $HT_f(x,y)$ with Theorem 1. We present our results in Table 5.

We recommend this method.

5.5 Suzuki's Algorithm

In successive papers, Suzuki [26, 27] develops a very fast algorithm, with cost $O(\sqrt{\log x \log y})$ operations, using the saddle point method to estimate $\Psi(x,y)$. This is based on

Table 5: $HT_f(x,y,z)/\Psi(x,y,z)$, $x = 2^{40}$

y	$z=2^{10}$	$z=2^{12}$	$z=2^{14}$	$z=2^{16}$	$z=2^{18}$	$z=2^{20}$
2^2	1.0969	1.0712	1.0641	1.067	1.0728	1.0833
2^4	1.0592	1.039	1.031	1.0353	1.0507	1.0829
2^6	1.0414	1.0281	1.0205	1.024	1.0566	1.1582
2^8	1.0194	1.0165	1.0135	1.0127	1.0296	1.1338
2^{10}	0.99773	1.0069	1.007	1.0073	1.0088	1.0413
2^{12}	—	1.0152	1.0186	1.0171	1.0141	1.0164
2^{14}	—	—	1.0177	1.0186	1.0203	1.0222
2^{16}	—	—	—	1.021	1.0247	1.0246
2^{18}	—	—	—	—	1.021	1.0172
2^{20}	—	—	—	—	—	1.013

Table 6: $S(x,y,z)/\Psi(x,y,z)$, $x = 2^{40}$

y	$z=2^{10}$	$z=2^{12}$	$z=2^{14}$	$z=2^{16}$	$z=2^{18}$	$z=2^{20}$
2^2	0	0	0	0	0	0
2^4	0	0	0	0	0	2.3032
2^6	0.57462	0.60753	0.64352	0.68488	0.7388	0.82405
2^8	0.90295	0.90963	0.91195	0.91382	0.92929	1.0201
2^{10}	0.94404	0.93337	0.92495	0.9175	0.9108	0.93201
2^{12}	—	0.93348	0.90717	0.89212	0.87883	0.87083
2^{14}	—	—	0.90403	0.87455	0.85967	0.84873
2^{16}	—	—	—	0.88064	0.85461	0.83658
2^{18}	—	—	—	—	0.85917	0.82827
2^{20}	—	—	—	—	—	0.83353

good approximations for α and the prime sums ζ, ϕ_1, ϕ_2 using the prime number theorem.

For $u > 1$, let ξ be the positive solution to the equation $e^{\xi} = 1 + u\xi$, or equivalently, $\xi = \log(1 + u\xi)$. This last equation implies that $\xi \approx \log(u \log u)$, and can be used iteratively to evaluate ξ. (See §6.2 for more information on this point.)

Let $\gamma = 0.57721...$ be Euler's constant. We now define

$$\alpha_s := 1 - \frac{\xi}{\log y} \tag{12}$$

and

$$S(x,y) := \frac{x^{\alpha_s} e^{\gamma + \int_0^{\xi} t^{-1}(e^t - 1)dt}}{\alpha_s \sqrt{2\pi u (1 + (\log x)/y)}}. \tag{13}$$

Suzuki proves the following [27, Theorem 1.1]:

THEOREM 3. *Let $\epsilon \leq 1/2$. If $(\log\log x)^{5/3-\epsilon} < \log y < e^{-1}(1-\epsilon)\log x$, then*

$$\Psi(x,y) = S(x,y)(1 + o(1)). \tag{14}$$

Suzuki proposed using the midpoint method to evaluate the integral $\int_0^{\xi} t^{-1}(e^t - 1)dt$. We used its Maclaurin series $\sum_{n \geq 1} \xi^n/(n \cdot n!)$ [1, formulas 5.1.10 and 5.1.40]. We write $S(x,y,z)$ for the function to estimate $\Psi(x,y,z)$ using $S(x,y)$ to estimate $\Psi(x,y)$ in Theorem 1.

Our results are presented in Table 6. We found the algorithm to be extremely fast, but not accurate for small y, which is not surprising given the approximations used.

Earlier, in [26], Suzuki discussed $\mathrm{HT}(x,y,\alpha_s)$ as an approximation to $\Psi(x,y)$. Although this is faster than Algorithm HT by a factor of $\log\log x$, it is not as accurate, and so we chose not to test its use.

5.6 Speed

Below we give timing results for the algorithms presented above. This is the total time it took to compute the estimates given in the tables in this section.

Algorithm	*Time in Seconds*
Suzuki	0.09
Bach and Peralta	0.40
Ekkelkamp	0.78
HT_f	1.90
HT	20.0

We used the Gnu **g++** compiler on a unix server with an intel CPU.

6. NUMERICAL DETAILS

We elaborate on some details from the algorithms presented above.

6.1 Estimates for α

In this subsection, we give more information about the function $\alpha(x,y)$, defined implicitly by the equation $\phi_1(\alpha, y) + \log x = 0$ (here $2 \leq y \leq x$). In particular we will prove that

$$\frac{1}{2\log x} \leq \alpha \leq 2. \tag{15}$$

We prove the lower bound first. From the prime number sum for ϕ_1, we see that for $s > 0$,

$$-\phi_1 \geq \frac{\log 2}{2^s - 1}.$$

So α, the solution to $\phi_1(\alpha, y) + \log x = 0$, is lower bounded by the solution β to

$$\frac{\log 2}{2^\beta - 1} = \log x.$$

Solving, we get

$$\alpha \geq \beta = \frac{\log(1 + (\log 2)/(\log x))}{\log 2} \geq \frac{1}{2\log x},$$

the last inequality holding whenever $x \geq 2$.

Next we show the upper bound. Let ζ denote the Riemann zeta function. By examining their Dirichlet series, we can see that $-\phi_1(s, y) \leq -\zeta'/\zeta(s)$. Both sides are decreasing smooth functions of s on $(0, \infty)$. It follows that α is upper bounded by the solution to $\zeta'(s)/\zeta(s) = \log 2$, which is less than 2.

More precise information can be found in [13]. In particular,

$$\alpha = \frac{\log(1 + y/\log x)}{\log y}\left\{1 + O\left(\frac{\log\log(1+y)}{\log y}\right)\right\}$$

holds uniformly for $x \geq y \geq 2$, with the explicit lower bound

$$\alpha \geq \frac{\log(1 + y/(5\log x))}{\log y}.$$

The strength of this is similar to (15) if y is fixed and $x \to \infty$.

6.2 Computing ξ

Here we discuss some numerical methods for solving $e^\xi = 1 + u\xi$, when $u > 1$.

Let $f(x) = x - \log(1 + ux)$. Then f is convex on $(0, \infty)$ with a minimum at $x = 1 - 1/u$. This gives the lower bound $\xi \geq 1 - 1/u$. To get an upper bound, we observe that $e^x > 1 + x + x^2/2$, so ξ is no larger than the positive solution to $1 + \xi + \xi^2/2 = 1 + u\xi$, which is $2(u-1)$. Using binary search between these bounds, we can get $\lfloor \xi \rfloor$ plus d bits of its fraction, with $O(\log u + d)$ evaluations of f.

Starting from the defining equation and taking logarithms, we get

$$\xi = \log(u\xi + 1).$$

Suzuki [26, Lemma 2.2] proves that this iteration, starting from $\log u$, is linearly convergent to ξ. (Here $u > e$ is fixed.)

In practice, we can use the Newton iteration

$$\xi := \xi - \frac{(\xi - \log(1 + u\xi))(1 + u\xi)}{1 + u(\xi - 1)},$$

starting with the upper bound $2(u-1)$. By convexity, the iterates decrease toward the root. We tested values of u from 2 to 1000, and for these u, about 5 iterations were enough to get machine accuracy (about 15D). (Note that when x is large, $f(x) \sim x$, so even if we start with a large upper bound, the second iterate will be much closer to the root.)

Newton iteration does not work well when u is close to 1, for the following reason. Suppose that $u = 1 + \epsilon$. Then, we have $\epsilon + O(\epsilon^2) < \xi < 2\epsilon$, making $\xi - \log(1 + u\xi)$ vanish to first order in ϵ. Thus, when computing this factor, we will lose precision due to cancellation.

If special function software is available, ξ can be expressed using the Lambert W function. For example, in the notation of a well known computer algebra system of Canadian origin, $\xi = -1/u - \mathrm{LambertW}(-1, -e^{-1/u}/u)$. (The argument -1 indicates which branch should be used.)

6.3 ATM Summation

The purpose of the next two subsections is to justify the remark made earlier that the cost of evaluating the formula HT can be lowered to $y^{2/3+o(1)}$. Here, the "$o(1)$" term includes factors of order $\log x$, so we are implicitly assuming that x is not outrageously large.

If f is a function defined on the positive integers, it is *multiplicative* if $f(mn) = f(m)f(n)$. This is a stronger requirement than is usual in number theory, where m and n need only be coprime. The concept of an *additive* function is defined similarly; we require $f(mn) = f(m) + f(n)$. We will call f an *ATM function* (additive times multiplicative) if $f = gh$, where g is additive and h is multiplicative.

The paper [2] gave an algorithm that evaluates the prime sum $\sum_{p \leq y} f(p)$, with f an ATM function, in $y^{2/3+o(1)}$ steps. This generalized a previously known result, also explained in [2], in which f could be multiplicative.

We now explain how these summation algorithms can be used to evaluate ϕ_1 and $\log\zeta$. The basic idea is to approximate each of these by a "small" number of ATM or multiplicative prime sums. With $\log\zeta$ in hand, we can exponentiate to get ζ.

We first assume $s > 0$.

Let us consider ϕ_1 first. By summing geometric series, we see that

$$-\phi_1 = \sum_{p \leq y} \frac{\log p}{p^s - 1} = \sum_{k \geq 1} \sum_{p \leq y} \frac{\log p}{p^{ks}}.$$

Note that each inner sum involves an ATM function. We will restrict the outer sum to $1 \leq k \leq N$, and choose N to make the truncation error,

$$\sum_{k \geq N+1} \sum_{p \leq y} \frac{\log p}{p^{ks}}.$$

small.

If we interchange the order of summation, allow all $p \geq 2$, and sum geometric series, we can express the truncation error as

$$\sum_{p \geq 2} \frac{\log p}{p^{Ns}(p^s - 1)}.$$

Using the globally convergent Maclaurin series for p^s, we see that $1/(p^s - 1) \leq 1/(s \log p)$. If we plug this in, the $\log p$ factors cancel and we get the upper bound

$$\frac{1}{s}\sum_{p \geq 2} \frac{1}{p^{Ns}} \leq \frac{1}{s}\left(\frac{1}{2^{Ns}} + \int_2^\infty \frac{dt}{t^{Ns}}\right) \leq \frac{3}{s 2^{Ns}},$$

provided that $Ns \geq 2$. For us, $s \geq 1/(2\log x)$, and with this additional assumption we get

$$[\text{truncation error}] \leq \frac{6\log x}{2^{Ns}}.$$

Thus, to achieve truncation error less than 2^{-d}, we can use $N = \Theta((\log x)(\log\log x + d))$.

Similarly, we can use

$$\log\zeta(s, y) = -\sum_{p \leq y} \log(1 - p^{-s}) = \sum_{k \geq 1} \frac{1}{k} \sum_{p \leq y} \frac{1}{p^{ks}}.$$

Now each inner sum involves a multiplicative function. If we use only the inner sums with $k \leq N$, a similar analysis shows that choosing $N = \Theta((\log x)(\log\log x + d))$ will keep the truncation error below 2^{-d}.

6.4 Numerical Differentiation

In this subsection, we explain how to evaluate

$$\phi_2(s,y) = \sum_{p\le y} \frac{\log^2 p \cdot p^s}{(p^s-1)^2},$$

for use in Algorithm HT. There is no obvious way to reduce this to the kind of sums treated in [2], so we will approximate it by a difference.

Using balanced numerical differentiation [8, p. 297], we have

$$\phi_2(s,y) = \frac{\phi_1(s+h,y) - \phi_1(s-h,y)}{2h} + \epsilon, \qquad \epsilon = \frac{h^2}{6}\phi_4(\eta, y)$$

for some $\eta \in [s-h, s+h]$. Let us determine how much precision will be necessary to deliver d bits of $\phi_2(s,y)$ accurately, when we use this formula.

After differentiating the sum for ϕ_2 twice, we see that

$$\phi_4(s,y) = \sum_{p\le y} \frac{p^s \log^2 p}{(p^s-1)^2} \times \frac{(p^{2s}+4p^s+1)\log^2 p}{(p^s-1)^2}.$$

Observing that $(t^2+4t+1)/(t-1)^2$ is decreasing for $t>1$, we get the estimate

$$\frac{h^2}{6}\phi_4(s,y) \le \frac{h^2\log^2 y}{6}\left(\frac{2^{2s}+4\cdot 2^s+1}{(2^s-1)^2}\right)\phi_2(s,y).$$

If $0 \le s \le 2$, the factor in parentheses is bounded by $15/s^2$. (Numerically, anyway.) So, if we could use exact arithmetic, numerical differentiation would give us

$$[\text{relative error}] \le \frac{5}{2}\,\frac{h^2\log^2 y}{s^2} \le 10h^2\log^2 x \log^2 y.$$

However, we don't have exact arithmetic, so we must also analyze the loss of precision due to cancellation. For this, we use an ad hoc theory. When h is small, the number of bits lost, when using the balanced difference formula to compute $\phi_1'(s)$, is about

$$-\log_2\left|\frac{\phi_1(s+h)-\phi_1(s-h)}{\phi_1(s)}\right|,$$

since dividing by h causes no loss of precision. Note that this is a centered version of the usual relative error formula. Since $\phi_1(s+h)-\phi_1(s-h) \sim 2h\phi_1'(s)$, we must bound the logarithmic derivative of ϕ_1, or, what is the same thing, relate $\phi_2 = \phi_1'$ to ϕ_1.

In our case, we require a lower bound for ϕ_2. Then since $t/(t-1)$ is decreasing on $(1,\infty)$,

$$\phi_2(s,y) = \sum_{p\le y}\frac{\log p}{p^s-1} \times \frac{p^s \log p}{p^s-1}$$

$$\ge \sum_{p\le y}\frac{\log p}{(p^s-1)}\log p \ge -\phi_1(s,y)\log 2.$$

Therefore,

$$\phi_1(s+h,y)-\phi_1(s-h,y) \sim 2h\phi_2(s,y) \ge -\phi_1(s,y)\cdot 2h\log 2,$$

as $h\to 0$,

Let us now translate these results into practical advice. Suppose our goal is to obtain ϕ_2 to d bits of precision, in the sense of relative error. By the exact arithmetic formula, we should choose $h \le 2^{-d/2}/(\sqrt{10}\log^2 x)$. Then we need to use $\log_2 h^{-1} + O(1)$ guard bits in our computation. Put more crudely, unless x is very large, doubling the working precision should be enough, if we select h properly.

The following example indicates that the theory above is roughly correct. Suppose we want 10 digits of ϕ_2, for $x = 10^6$, $y = 10^3$, and $s = 1/(2\log x) = 0.03619...$. Our recipe allows us to take $h = 10^{-7}$. With 17 digit arithmetic, we obtained

$$\begin{aligned} \text{numerical derivative} &= 127790.77386350000 \\ \text{summation for } \phi_2 &= 127790.77386041727 \end{aligned}$$

which agree to 11 figures.

7. CONCLUSION AND FUTURE WORK

In summary, we recommend the estimate $HT_f(x,y,z)$ of §5.4 for approximating $\Psi(x,y,z)$. We feel it gives high accuracy while retaining sufficient speed to be very practical.

For future work, we hope to generalize our results to 2 or more large primes. We also hope to further examine estimates of the form of (5).

8. ACKNOWLEDGMENTS

We want to thank the referees, whose comments helped improve this paper.

Abstract presented at the AMS-MAA Joint Mathematics Meetings, January 2012, Boston MA, and at the CMS Summer Meeting, June 2013, Halifax Nova Scotia.

Supported in part by grants from the Holcomb Awards Committee, NSF (CCF-635355), and ARO (W911NF-09-1-0439).

9. REFERENCES

[1] M. Abramowitz and I. A. Stegun. *Handbook of Mathematical Functions.* Dover, 1970.

[2] E. Bach. Sums over primes. In A. M. Ernvall-Hytönen, M. Jutila, J. Karhumäki, and A. Lepistö, editors, *Proceedings of Conference on Algorithmic Number Theory 2007*, number 46 in Turku Centre for Computer Science, pages 40–44, 2007.

[3] E. Bach and R. Peralta. Asymptotic semismoothness probabilities. *Math. Comp.*, 65(216):1701–1715, 1996.

[4] E. Bach and J. O. Shallit. *Algorithmic Number Theory*, volume 1. MIT Press, 1996.

[5] D. J. Bernstein. Bounding smooth integers. In J. P. Buhler, editor, *Third International Algorithmic Number Theory Symposium*, pages 128–130, Portland, Oregon, June 1998. Springer. LNCS 1423.

[6] D. J. Bernstein. Arbitrarily tight bounds on the distribution of smooth integers. In Bennett, Berndt, Boston, Diamond, Hildebrand, and Philipp, editors, *Proceedings of the Millennial Conference on Number Theory*, volume 1, pages 49–66. A. K. Peters, 2002.

[7] R. P. Brent. Multiple precision zero-finding methods and the complexity of elementary function evaluation. In J. F. Traub, editor, *Analytic Computational Complexity*, pages 151–176. Academic Press, 1976.

[8] S. D. Conte and C. W. D. Boor. *Elementary Numerical Analysis: An Algorithmic Approach.* McGraw-Hill Higher Education, 3rd edition, 1980.

[9] R. Crandall and C. Pomerance. *Prime Numbers, a Computational Perspective.* Springer, 2001.

[10] W. H. Ekkelkamp. The role of semismooth numbers in factoring large numbers. In A.-M. Ernvall-Hytönen, M. Jutila, J. Karhumäki, and A. Lepistö, editors, *Proceedings of Conference on Algorithmic Number Theory 2007*, number 46 in TUCS General Publication, pages 40–44. Turku Centre for Computer Science, 2007.

[11] A. Granville. Smooth numbers: computational number theory and beyond. In *Algorithmic number theory: lattices, number fields, curves and cryptography*, volume 44 of *Math. Sci. Res. Inst. Publ.*, pages 267–323. Cambridge Univ. Press, Cambridge, 2008.

[12] A. Hildebrand. On the number of positive integers $\leq x$ and free of prime factors $> y$. *Journal of Number Theory*, 22:289–307, 1986.

[13] A. Hildebrand and G. Tenenbaum. On integers free of large prime factors. *Trans. AMS*, 296(1):265–290, 1986.

[14] A. Hildebrand and G. Tenenbaum. Integers without large prime factors. *Journal de Théorie des Nombres de Bordeaux*, 5:411–484, 1993.

[15] S. Hunter and J. P. Sorenson. Approximating the number of integers free of large prime factors. *Mathematics of Computation*, 66(220):1729–1741, 1997.

[16] T. Kleinjung, K. Aoki, J. Franke, A. K. Lenstra, E. Thomé, P. Gaudry, P. L. Montgomery, D. A. Osvik, H. T. Riele, A. Timofeev, P. Zimmermann, and et al. Factorization of a 768-bit rsa modulus. `http://eprint.iacr.org/2010/006.pdf`, 2010.

[17] A. K. Lenstra and H. W. L. Jr., editors. *The Development of the Number Field Sieve*, volume 1554 of *Lecture Notes in Mathematics*. Springer-Verlag, Berlin and Heidelberg, Germany, 1993.

[18] A. J. Menezes, P. C. van Oorschot, and S. A. Vanstone. *Handbook of Applied Cryptography*. CRC Press, Boca Raton, 1997.

[19] P. Moree. Nicolaas Govert de Bruijn, the enchanter of friable integers. *Indag. Math.*, 2013. to appear; available from arxiv.org:1212.1579.

[20] S. Parsell and J. P. Sorenson. Fast bounds on the distribution of smooth numbers. In F. Hess, S. Pauli, and M. Pohst, editors, *Proceedings of the 7th International Symposium on Algorithmic Number Theory (ANTS-VII)*, pages 168–181, Berlin, Germany, July 2006. Springer. LNCS 4076, ISBN 3-540-36075-1.

[21] V. Ramaswami. The number of positive integers $\leq x$ and free of prime divisors x^c, and a problem of S. S. Pillai. *Duke Mathematics Journal*, 16(1):99–109, 1949.

[22] R. Rivest, A. Shamir, and L. Adleman. A method for obtaining digital signatures and public-key cryptosystems. *Communications of the ACM*, 21(2):120–126, 1978.

[23] L. Schoenfeld. Sharper bounds for the Chebyshev functions $\theta(x)$ and $\psi(x)$. II. *Mathematics of Computation*, 30(134):337–360, 1976.

[24] J. P. Sorenson. A fast algorithm for approximately counting smooth numbers. In W. Bosma, editor, *Proceedings of the Fourth International Algorithmic Number Theory Symposium (ANTS IV)*, pages 539–549, Leiden, The Netherlands, 2000. LNCS 1838.

[25] J. P. Sorenson. The pseudosquares prime sieve. In F. Hess, S. Pauli, and M. Pohst, editors, *Proceedings of the 7th International Symposium on Algorithmic Number Theory (ANTS-VII)*, pages 193–207, Berlin, Germany, July 2006. Springer. LNCS 4076, ISBN 3-540-36075-1.

[26] K. Suzuki. An estimate for the number of integers without large prime factors. *Mathematics of Computation*, 73:1013–1022, 2004. MR 2031422 (2005a:11142).

[27] K. Suzuki. Approximating the number of integers without large prime factors. *Mathematics of Computation*, 75:1015–1024, 2006.

[28] G. Tenenbaum. *Introduction to Analytic and Probabilistic Number Theory*, volume 46 of *Cambridge Studies in Advanced Mathematics*. Cambridge University Press, english edition, 1995.

[29] C. Zhang. *An extension of the Dickman function and its application*. PhD thesis, Purdue University, 2002.

An Explicit Expression of the Lüroth Invariant

Romain Basson
IRMAR
Université de Rennes 1,
Campus de Beaulieu
35042 Rennes
France
romain.basson@univ-rennes1.fr

Reynald Lercier
DGA & IRMAR
Université de Rennes 1,
Campus de Beaulieu
35042 Rennes
France
reynald.lercier@m4x.org

Christophe Ritzenthaler
Institut de Mathématiques de Luminy, UMR 6206 du CNRS,
Luminy, Case 907
13288 Marseille
France
ritzenth@iml.univ-mrs.fr

Jeroen Sijsling
Mathematics Institute,
Zeeman Building, University of Warwick
Coventry CV4 7AL
United Kingdom
sijsling@gmail.com

ABSTRACT

In this short note, we give an algorithm that returns an explicit expression of the Lüroth invariant in terms of the Dixmier-Ohno invariants of plane quartic curves. We also obtain an explicit factorized expression on the locus of Ciani quartics in terms of the coefficients. After this calculation, we extend our methods to answer two open theoretical questions concerning the sub-locus of singular Lüroth quartics.

Categories and Subject Descriptors

J.2 [**Computer applications**]: Mathematics and statistics

Keywords

Invariant ; ternary quartic ; genus 3 ; Ciani quartic ; algorithm ; Dixmier-Ohno invariants.

1. LÜROTH QUARTICS

This note considers *Lüroth quartics*, which are plane quartics containing the ten vertices of a non-degenerate pentalateral. To make these notions precise, we give the following definitions. Let V be a three-dimensional vector space over the complex field $\mathbb{C}$.

Definition 1. *A* complete pentalateral *in the projective plane* $\mathbb{P}V$ *is a curve* $C \subset \mathbb{P}V$ *consisting of the union of five lines* $\ell_1, \ldots, \ell_5$ *that are three by three linearly independent (which is to say that the pairwise intersections of the lines* ℓ_i *yield exactly* 10 *distinct points).*

The vertices *of a complete pentalateral* C *are the double points of* C*, that is, the* 10 *points* $\bigcup_{i \neq j}((\ell_i = 0) \cap (\ell_j = 0))$.

Definition 2. *Let* $Q \subset \mathbb{P}V$ *be a non-singular quartic. Then* Q *is called a* non-singular Lüroth quartic *if it contains the vertices of a complete pentalateral in* $\mathbb{P}V$.

The set of plane quartics in $\mathbb{P}V$ can be identified with the projective space $\mathbb{P}\,\mathrm{Sym}^4(V^*)$ over the fourth symmetric power of the dual vector space V^* of V. This inherits an action of the group $\mathrm{GL}(V)$ and its derived subgroup $\mathrm{SL}(V)$, which act canonically on V. Choosing a basis, as we will do during our calculations, identifies V with $\mathbb{C}^3$ and the set of quartics $\mathbb{P}\,\mathrm{Sym}^4(V^*)$ with the projectivization of the vector space on the homogeneous degree 4 monomials in the canonical basis x, y, z of the dual space $(\mathbb{C}^3)^*$. We will in turn identify this projective space with $\mathbb{P}^{14}$ by choosing some ordering of these 15 monomials. In this way, $\mathbb{P}^{14}$ inherits an action of the groups $\mathrm{GL}_3(\mathbb{C})$ and $\mathrm{SL}_3(\mathbb{C})$.

The classical study of Lüroth quartics culminated in 1919 with the work of Morley [9]. This showed that the Zariski closure of the locus of non-singular Lüroth quartics in the projective space $\mathbb{P}\,\mathrm{Sym}^4(V^*)$ is an irreducible hypersurface described as the vanishing locus of a single homogeneous polynomial function L on the projective space of quartics $\mathbb{P}\,\mathrm{Sym}^4(V^*)$, well-defined up to scalars. We shall call this L the *Lüroth invariant.* Morley showed that L is of degree 54.

Definition 3. *Let* $Q \subset \mathbb{P}V$ *be a quartic. Then* Q *is called a* Lüroth quartic *if* $L(Q) = 0$.

In recent years, after the seminal work of [1], several authors have revived this subject in [2, 11, 12, 13] (see also [14] on the undulation invariant).

However, an explicit expression of L was still missing. In the following section we explain how to compute such an expression. Our main new technique lies in an effective use of [10], an unfortunately unpublished article in which Ohno gives a complete set of generators for the invariants of ternary quartics under the action of $\mathrm{SL}_3(\mathbb{C})$, completing the set of primary invariants found in [3]. These invariants were

ISSAC'13, June 26–29, 2013, Boston, Massachusetts, USA.

also used in [5], and new effective methods to verify their correctness can be found in [4]. Our calculations use the implementation of these invariants in Magma, which is due to Kohel.

2. THE ALGORITHM

The key point is the following observation:

PROPOSITION 1. *The homogeneous polynomial function L on $\mathbb{P}\operatorname{Sym}^4(V^*)$ is $\mathrm{GL}(V)$-invariant up to scalars. In particular, L is $\mathrm{SL}(V)$-invariant.*

PROOF. Since we are working over an algebraically closed field, this is obvious from the obvious fact that any $\mathrm{GL}(V)$-transform of a Lüroth quartic is again a Lüroth quartic. □

Let

$$R = S[\mathbb{P}\operatorname{Sym}^4(V^*)]^{\mathrm{SL}(V)}$$

be the ring of $\mathrm{SL}(V)$-invariant homogeneous polynomial functions on $\mathbb{P}\operatorname{Sym}^4(V^*)$, which coincides with those functions that are $\mathrm{GL}(V)$-invariant up to scalars. The structure of the ring R is known. Indeed, let $I = (I_3, I_6, I_9, I_{12}, I_{15}, I_{18}, I_{27})$ be the primary invariants for ternary quartics under the action of $\mathrm{SL}_3(\mathbb{C})$ found by Dixmier in [3], and let $J = (J_9, J_{12}, J_{15}, J_{18}, J_{21}, J_{21})$ be the secondary invariants found by Ohno [10]. In both cases, the index specifies the degree of the invariant as a homogeneous function. Then we have:

THEOREM 1 (DIXMIER-OHNO). *We have $R = \mathbb{C}[I, J]$.*

We now choose a basis for V and a corresponding coordinatization of $\mathbb{P}\operatorname{Sym}^4(V^*) \cong \mathbb{P}^{14}$ as the projective space over the degree 4 monomials in x, y, z. The function L then becomes a homogeneous expression in the coefficients of these monomials. It is unlikely that this function can be written down in any reasonable way (see also the final remark in Section 5). However, by Proposition 1 and Theorem 1, we can express L as a polynomial in the invariant functions in I and J. This expression is not unique, since as we shall see, there are relations between these invariant monomials in degree 54.

To obtain an expression for L, we apply the method of evaluation-interpolation. This is based on the following observation:

PROPOSITION 2. *Let $S = \mathbb{C}[x_3, \ldots, x_{27}, y_9, \ldots, y_{21}, y_{21}']$ be a graded polynomial algebra in 13 variables, weighted by indices, and consider the surjection q given by*

$$\begin{aligned} q : S &\longrightarrow R \\ x_k &\longmapsto I_k \\ y_\ell &\longmapsto J_\ell . \end{aligned}$$

Let $R_{54} \subset R$ be the set of homogeneous functions of degree 54, and define $S_{54} \subset S$ analogously. Let K be the kernel of the map $S_{54} \to R_{54}$. Then the $\dim(K) = 215$.

Let X be a finite set of Lüroth quartics. Consider the linear map $q' : S_{54} \to \mathbb{C}^X$ given by evaluating at the polynomials in X. Let K' be the kernel of q', and suppose that $\dim(K') = 216$. Let $\widetilde{L}$ be an element of $K' \backslash K$. Then the image $L = q(\widetilde{L})$ equals the Lüroth invariant.

PROOF. One calculates that the $\dim(S_{54}) = 1380$. Since calculating the Hilbert polynomial of R as in [15, p.1045] yields $\dim(S_{54}) = 1165$, we indeed find that $\dim(K) = 215$. The rest is straightforward considering the uniqueness of L up to scalars. □

The details of the calculation are therefore as follows.

1. Construct the 1380 monomials
$$\mathcal{I} = \{I_3^{18}, I_3^{16} I_6, I_3^{15} I_9, I_3^{15} J_9, \ldots, J_{18}^3, I_{27}^2\}$$
of degree 54 that generate the $\mathbb{C}$-vector space of invariants of degree 54.
2. Generate a sufficiently large finite set $\mathcal{Q}$ of cardinality q of random plane quartics with rational coefficients.
3. Generate a sufficiently large finite set $\mathcal{L}$ of cardinality l of random Lüroth quartics of the form
$$\ell_1\ell_2\ell_3\ell_4 + c_1 \cdot \ell_2\ell_3\ell_4\ell_5 + c_2 \cdot \ell_1\ell_3\ell_4\ell_5 + c_3 \cdot \ell_1\ell_2\ell_4\ell_5 + c_4 \cdot \ell_1\ell_2\ell_3\ell_5$$
where $\ell_1 = x, \ell_2 = y, \ell_3 = z, \ell_4 = x + y + z$, ℓ_5 is a line with random rational coefficients and c_i are rational coefficients.
4. Compute the matrix $M_1 = (I(q))_{I \in \mathcal{I}, q \in \mathcal{Q}}$, evaluating the monomials in $\mathcal{I}$ at the quartics in $\mathcal{Q}$.
5. Compute the matrix $M_2 = (I(q))_{I \in \mathcal{I}, q \in \mathcal{L}}$, evaluating the monomials in $\mathcal{I}$ at the Lüroth quartics in $\mathcal{L}$.
6. Compute the 215-dimensional kernel N_1 of M_1. This gives a basis of the homogeneous relations of degree 54 that are satisfied by the invariants of all ternary quartics.
7. Compute the 216-dimensional kernel N_2 of M_2. This gives a basis of the homogeneous relations of degree 54 that are satisfied by all Lüroth quartics.
8. A non-zero element in the complement of N_2 in N_1 is an expression for L in terms of the Dixmier-Ohno invariants.

All these computations were done with Magma software. Over finite fields $\mathbb{F}_p$ with prime cardinality $p = 2017$, 10007, 100003 or even 1000003, computations can be done in less than a minute. However, getting the result over the rationals is more challenging. The main concern is to deal with matrices M_1 and M_2 whose coefficients are as small as possible. So, at Step (2) of the algorithm, we generate plane quartics with random integer coefficients only equal to -1, 0, or 1. Similarly, we restrict Step (3) to Lüroth forms defined by integer coefficients c_i bounded in absolute value by 4.

We can estimate the size of the computations involved in this run of the algorithm by using the Hadamard bounds for our matrices M_1 and M_2; the quartics under consideration yield bounds slightly smaller than $2^{200\,000}$ for M_1 and $2^{350\,000}$ for M_2. As a sanity check before running the code over the rationals, we verify that this subset of quartics yields a valid result modulo small primes.

Most of the time is spent at Step (6) and Step (7) of the algorithm, precisely 5 and 9 hours on our laptop (based on a INTEL CORE I7 M620 2.67GHz processor).

A program to get the result is available on the web page of the authors[1]. It uses the implementation of the Dixmier-Ohno invariants in Magma by Kohel[2]. The 1.4Mb result is also available online[3]. It is given by 1164 monomials with rational coefficients, the largest of which is a quotient of a 680-digit integer by a coprime 671-digit integer. Modulo 1000003, the expression starts as

$$\begin{aligned}
& I_3^{18} + 469313 I_3^2 I_6^8 + 710780 I_6^9 + 969230 I_3^3 I_6^6 I_9 \\
& + 374233 I_3 I_6^7 I_9 + 276144 I_3^2 I_6^5 I_9^2 \\
& + 602674 I_6^6 I_9^2 + 527614 I_3^3 I_6^3 I_9^3 \\
& + 538637 I_3 I_6^4 I_9^3 + 392526 I_3^4 I_6 I_9^4 \\
& + 645841 I_3^2 I_6^2 I_9^4 + 914224 I_6^3 I_9^4 + 207808 I_3^3 I_9^5 \\
& + 31577 I_3 I_6 I_9^5 + 635768 I_9^6 + 668878 I_3^{15} J_9 \\
& + 507293 I_3^3 I_6^6 J_9 + 318476 I_3 I_6^7 J_9 \\
& + 59775 I_3^2 I_6^5 I_9 J_9 + 581086 I_6^6 I_9 J_9 \\
& + 830307 I_3^3 I_6^3 I_9^2 J_9 + 804817 I_3 I_6^4 I_9^2 J_9 \\
& + 6418 I_3^6 I_9^3 J_9 + 578316 I_3^4 I_6 I_9^3 J_9 \\
& + 741618 I_3^2 I_6^2 I_9^3 J_9 + 452974 I_6^3 I_9^3 J_9 \\
& + 36214 I_3^3 I_9^4 J_9 + 522408 I_3 I_6 I_9^4 J_9 \\
& + 253043 I_9^5 J_9 + 469299 I_3^2 I_6^5 J_9^2 + \dots
\end{aligned}$$

3. CIANI QUARTICS

We call a *Ciani quartic* a plane quartic of the form

$$ax^4 + bx^2y^2 + cx^2z^2 + dy^4 + ey^2z^2 + fz^4.$$

A generic Ciani quartic has automorphism group isomorphic with $\mathbb{Z}/2\mathbb{Z} \times \mathbb{Z}/2\mathbb{Z}$, and conversely every quartic with this property is $\mathbb{C}$-isomorphic to a Ciani quartic. The dimension of the substratum of Ciani quartics in the full dimension 6 moduli space of plane quartics equals 3.

In [7, Sec.5], using different techniques, Hauenstein and Sottile obtained the factorization on Ciani quartics of the Lüroth invariant as

$$G^4 H^2 J$$

with $G, H, J \in \mathbb{C}[a, b, c, d, e, f]$ homogeneous of respective degree 6, 9 and 12. Using our expression, it is easy to confirm their decomposition. We give a slightly different version of the result, which is due to the fact the coefficients b, c, e are replaced by $2b, 2c, 2e$ in [7, Sec.5]:

$$G = a \cdot d \cdot f \cdot (adf - (1/4)ae^2 - (1/4)b^2 f - (1/4)bce - (1/4)c^2 d),$$

$$\begin{aligned}
H = & (adf - (1/4)ae^2 - (1/4)b^2 f + (1/4)bce + (3/4)c^2 d) \\
& \cdot (adf - (1/4)ae^2 + (3/4)b^2 f + (1/4)bce - (1/4)c^2 d) \\
& \cdot (adf + (3/4)ae^2 - (1/4)b^2 f + (1/4)bce - (1/4)c^2 d),
\end{aligned}$$

$$\begin{aligned}
J = & a^4d^4f^4 - (1/49)a^4d^3e^2f^3 + (51/19208)a^4d^2e^4f^2 \\
& - (1/38416)a^4de^6f + (1/614656)a^4e^8 - (1/49)a^3b^2d^3f^4 \\
& - (205/9604)a^3b^2d^2e^2f^3 - (3/38416)a^3b^2de^4f^2 \\
& + (1/153664)a^3b^2e^6f + (15/343)a^3bcd^3ef^3 \\
& + (29/9604)a^3bcd^2e^3f^2 - (5/38416)a^3bcde^5f \\
& - (1/153664)a^3bce^7 - (1/49)a^3c^2d^4f^3 \\
& - (205/9604)a^3c^2d^3e^2f^2 - (3/38416)a^3c^2d^2e^4f \\
& + (1/153664)a^3c^2de^6 + (51/19208)a^2b^4d^2f^4 \\
& - (3/38416)a^2b^4de^2f^3 + (3/307328)a^2b^4e^4f^2 \\
& + (29/9604)a^2b^3cd^2ef^3 - (5/19208)a^2b^3cde^3f^2 \\
& - (3/153664)a^2b^3ce^5f - (205/9604)a^2b^2c^2d^3f^3 \\
& + (2/2401)a^2b^2c^2d^2e^2f^2 + (55/153664)a^2b^2c^2de^4f \\
& + (3/307328)a^2b^2c^2e^6 + (29/9604)a^2bc^3d^3ef^2 \\
& - (5/19208)a^2bc^3d^2e^3f - (3/153664)a^2bc^3de^5 \\
& + (51/19208)a^2c^4d^4f^2 - (3/38416)a^2c^4d^3e^2f \\
& + (3/307328)a^2c^4d^2e^4 - (1/38416)ab^6df^4 \\
& + (1/153664)ab^6e^2f^3 - (5/38416)ab^5cdef^3 \\
& - (3/153664)ab^5ce^3f^2 - (3/38416)ab^4c^2d^2f^3 \\
& + (55/153664)ab^4c^2de^2f^2 + (3/153664)ab^4c^2e^4f \\
& - (5/19208)ab^3c^3d^2ef^2 - (17/76832)ab^3c^3de^3f \\
& - (1/153664)ab^3c^3e^5 - (3/38416)ab^2c^4d^3f^2 \\
& + (55/153664)ab^2c^4d^2e^2f + (3/153664)ab^2c^4de^4 \\
& - (5/38416)abc^5d^3ef - (3/153664)abc^5d^2e^3 \\
& - (1/38416)ac^6d^4f + (1/153664)ac^6d^3e^2 + (1/614656)b^8f^4 \\
& - (1/153664)b^7cef^3 + (1/153664)b^6c^2df^3 \\
& + (3/307328)b^6c^2e^2f^2 - (3/153664)b^5c^3def^2 \\
& - (1/153664)b^5c^3e^3f + (3/307328)b^4c^4d^2f^2 \\
& + (3/153664)b^4c^4de^2f + (1/614656)b^4c^4e^4 \\
& - (3/153664)b^3c^5d^2ef - (1/153664)b^3c^5de^3 \\
& + (1/153664)b^2c^6d^3f + (3/307328)b^2c^6d^2e^2 \\
& - (1/153664)bc^7d^3e + (1/614656)c^8d^4.
\end{aligned}$$

The product G^4H^2J has 1695 monomials. Note that the total amount of weighted monomials in a, b, c, d, e and f in a generic degree 54 invariant is 3439.

4. SINGULAR LÜROTH QUARTICS

Let $\mathcal{L} \subset \mathbb{P}\,\mathrm{Sym}^4(V^*)$ be the locus of Lüroth quartics, and let $\mathcal{D} \subset \mathbb{P}\,\mathrm{Sym}^4(V^*)$ be the discriminantal hypersurface defined by the equation $I_{27} = 0$. We will now obtain new results on the geometry of the locus $\mathcal{L} \cap \mathcal{D}$ of *singular Lüroth quartics*. Work by Le Potier and Tikhomirov [8] shows that

$$\mathcal{L} \cap \mathcal{D} = \mathcal{L}_1 \cup \mathcal{L}_2,$$

where $\mathcal{L}_1$ and $\mathcal{L}_2$ are irreducible subschemes of $\mathbb{P}\,\mathrm{Sym}^4(V^*)$ of codimension 2 whose respective degrees as subschemes of $\mathcal{D}$ equal 24 and 30 respectively. Moreover, while $\mathcal{L}_1$ is reduced, the reduced subscheme $(\mathcal{L}_2)_{\mathrm{red}}$ of $\mathcal{L}_2$ is of degree 15.

In [12], Ottaviani and Sernesi showed that no new degree 15 invariant vanishes on $(\mathcal{L}_2)_{\mathrm{red}}$, which implies that this scheme is not a principal hypersurface in $\mathcal{L}$. We will prove a stronger result, namely that none of $\mathcal{L}_1$, $\mathcal{L}_2$, $(\mathcal{L}_2)_{\mathrm{red}}$

[1] http://iml.univ-mrs.fr/~ritzenth/programme/luroth/luroth.m

[2] http://echidna.maths.usyd.edu.au/kohel/alg/index.html

[3] http://iml.univ-mrs.fr/~ritzenth/programme/luroth/LurothInvF.m

is a complete intersection. We apply the same methods as in Section 2. The main problem now is to generate quartics in $\mathcal{L}_1$ and $\mathcal{L}_2$.

For $\mathcal{L}_2$, we can proceed by using Remark 3.3 in [12]: we now choose the lines in Step (3) above such that three of them have a common point of intersection.

For $\mathcal{L}_1$, we have to perform the constructions and results in [12, p. 1759]. The procedure is as follows.

1. Construct a cubic surface S with two skew lines l, m.
2. Calculate the double cover $f : l \to m$ sending $p \in l$ to the intersection $T_pS \cap m$ of the tangent plane T_pS to p at S with the line m, and construct $g : m \to l$ analogously.
3. Let $B_f \subset m$ (resp. $B_g \subset l$) be the branch divisor of f (resp. g). Construct morphisms $f' : l \to \mathbb{P}^1$ (resp. $g' : m \to \mathbb{P}^1$) ramifying over B_g (resp. B_f).
4. Construct $Q \in m \subset S$ such that $f^{-1}(Q)$ is also a fiber of f'.
5. Construct the ramification locus of the degree 2 projection $S \to \mathbb{P}(T_QS)$ from the point $Q \in S$. Then by Proposition 3.1(i) of [12], we obtain a quartic in $\mathcal{L}_1$.

The following string of propositions and remarks show how these steps can be implemented without extending the base field (we will take $k = \mathbb{Q}$ throughout).

PROPOSITION 3. *Let k be a field, and suppose that we are given six rational points $p_1, \ldots, p_6 \in \mathbb{P}V(k)$ in the projective plane over k. Suppose additionally that this set of points is sufficiently general in the sense that the complete linear system C of cubics passing through them has dimension 4. Construct the Clebsch rational map $c : \mathbb{P}V \to \mathbb{P}C$, and let $S \subset \mathbb{P}C$ be the Zariski closure of $c(\mathbb{P}V)$. Then S is the vanishing locus of a quaternary cubic form $F \in \mathrm{Sym}^3(C^*)$ over k.*

The rational map c restricts to a birational map between $\mathbb{P}V$ and S. Let $l_0 \subset \mathbb{P}V$ be the rational line containing p_1 and p_2, and let $m_0 \subset \mathbb{P}V$ be the rational line containing p_1 and p_3. Let x and x' be two points in l not equal to p_1 or p_2, and let y and y' be two points in l not equal to p_1 or p_3. Then the images $c(x)$ and $c(x')$ are well-defined elements of S, and the line l through them is defined over k and included in S. Analogously, one obtains a line m through $c(y)$ and $c(y')$. The lines l and m are skew.

PROOF. This is a standard result from the theory of cubic surfaces, see [6, Section V.4]. □

This deals with part 1. To perform the calculations in point 2 explicitly, choose coordinates on l by taking two points l_1, l_2 on l and sending $(x : y) \in \mathbb{P}^1$ to $xl_1 + yl_2$, and similarly on m by choosing $m_1, m_2 \in m$. To determine the morphism $f : l \to m$ explicitly in these in coordinates, we choose two equations $M_1 = M_2 = 0$ defining m. Given $p \in l$ with coordinates $(x : y) \in \mathbb{P}^1$, the point $f(p) = T_pS \cap m$ corresponds to the vector space that is the kernel of the matrix whose rows are given by M_1, M_2 and the partial derivatives of F. A generating vector for this space will be a combination of m_1 and m_2 with homogeneous quadratic coefficients $f_1(x,y), f_2(x,y)$ in (x,y). The morphism f now corresponds to the map $\mathbb{P}^1 \to \mathbb{P}^1$ given by (f_1, f_2). Similarly, one determines g. For point 3, we use the following result.

PROPOSITION 4. *Let l be a projective line over k, with homogeneous coordinates x and y, and let D be a k-rational divisor of degree 2 on l. Then the following formulae determine a degree 2 morphism $f' : l \to \mathbb{P}^1$ over k whose ramification locus equals D.*

- *If D consists of two points $(x_1 : y_1)$ and $(x_2 : y_2)$ that are rational over k, then one can take*
$$f'(x:y) = ((y_1x - x_1y)^2 : (y_2x - x_2y)^2).$$
- *If D is defined by an equation $rx^2 + ty^2 = 0$, then one can take*
$$f'(x:y) = (rx^2 - 2txy - ty^2 : rx^2 + 2txy - ty^2);$$
- *If D is defined by an equation $rx^2 + sxy + ty^2 = 0$ with $s \neq 0$, then one can take*
$$\begin{aligned} f'(x:y) =& (r^2sx^2 + 2r(s^2 - 2rt)xy + (s^3 - 3rst)y^2 : \\ & r(rsx^2 + 4rtxy + sty^2)). \end{aligned}$$

PROOF. Once the answer is given, the verification is trivial. But let us illustrate how to find these expressions by treating the thir case, where the points of D are defined over a proper quadratic extension of k. We use the affine coordinate $t = x/y$ on l. Suppose that using this coordinate, the divisor $D = [d] + [\bar{d}]$ consists of two conjugate points, not summing to zero because we are in case (iii). Then $f' = ((t-d)/(t-\bar{d}))^2$ of above now has the property that $\overline{f'} = 1/f'$. But then one verifies that $(df' + \bar{d})/(\bar{d}f' + d)$ is a fractional linear transformation of f' that is stable under conjugation and hence defines a morphism over the ground field. Homogenizing, one obtains the more elegant expression given in the statement of the proposition. □

We now treat point 4.

PROPOSITION 5. *Let $f, f' : \mathbb{P}^1 \to \mathbb{P}^1$ be two degree 2 morphisms, neither of which can be obtained from the other by postcomposing with an automorphism of $\mathbb{P}^1$. Then the q in $\mathbb{P}^1$ such that the fiber of f over q is also a fiber of f' over a point q' can be obtained as follows.*

Write
$$f(x:y) = (a_1x^2 + b_1xy + c_1y^2 : a_2x^2 + b_2xy + c_2y^2)$$
and
$$f'(x:y) = (a_1'x^2 + b_1'xy + c_1'y^2 : a_2'x^2 + b_2'xy + c_2'y^2).$$
Then $q = (\lambda_1 : \lambda_2)$, where $(\lambda_1, \lambda_2, \lambda_1', \lambda_2')$ generates the kernel of the matrix
$$\begin{pmatrix} a_1 & -a_2 & -a_1' & a_2' \\ b_1 & -b_2 & -b_1' & b_2' \\ c_1 & -c_2 & -c_1' & c_2' \end{pmatrix}.$$

PROOF. If we let $q = (\lambda_1 : \lambda_2)$ and $q' = (\lambda_1' : \lambda_2')$, then finding q and q' comes down to solving the equation
$$\begin{aligned} &\lambda_2(a_1x^2 + b_1xy + c_1y^2) - \lambda_1(a_2x^2 + b_2xy + c_2y^2) \\ =& \lambda_2'(a_1'x^2 + b_1'xy + c_1'y^2) - \lambda_1'(a_2'x^2 + b_2'xy + c_2'y^2), \end{aligned}$$
which evidently corresponds to the determination of the kernel of the matrix in question. □

To calculate point 5, we choose an isomorphism $\mathbb{P}C \cong \mathbb{P}^3$ mapping the point Q to $(1:0:0:0)$ and apply the following elementary result.

PROPOSITION 6. *Let $S \subset \mathbb{P}^3$ be a cubic surface containing $Q = (1:0:0:0)$ that is defined by a quaternary cubic form $F \in k[w,x,y,z]$. Let q be the projection $S \to \mathbb{P}(T_QS)$ from the point $Q \in S$. Then the ramification locus of q is isomorphic to the quartic curve in $\mathbb{P}^2$ determined by the vanishing of the discriminant of the quadratic polynomial $F(1,xt,yt,zt)/t$.*

PROOF. Since S is a subscheme of $\mathbb{P}^3$, we get an induced coordinatization of T_QS by sending $(x:y:z)$ to the tangent direction given by the line through the points $(1:0:0:0)$ and $(1:x:y:z)$. Then $(x:y:z)$ is a ramification point of the projection $S \to \mathbb{P}(T_QS)$ if and only if the equation $F(1,xt,yt,zt) = 0$ has a double root outside 0, or in other words if the discriminant of the quadratic polynomial $F(1,xt,yt,zt)/t$ vanishes. This discriminant is a homogeneous quartic form in the variables x,y,z, which defines the plane quartic in $\mathcal{L}_1$ that we were looking for. □

A program to generate quartics in $\mathcal{L}_1$ by using the steps above is available online[4].

REMARK 1. *We also tried to generate quartics in $\mathcal{L}_1$ by using Remark 3.4 of [12]. We take cubics S of the form*

$$t^2x + t(ax^2 + 2bxy + 2cxz + dy^2 + 2eyz + fz^2) + g(x,y,z)$$

with g a random degree 3 homogeneous polynomial, such that S is non singular and $e^2 = df$. The last condition ensures that $p = (0:0:0:1)$ belongs to the Hessian H of S and, after checking that p is non singular, we take quartics which are tangent plane sections of H at p. Unfortunately, it seems that these quartics are special in L_1, since there are degree 24 relations between their invariants (there is a 27 dimensional space of relations in degree 24 between randomly generated quartics of this form).

Having generated a sufficiently large database[5] of curves in $\mathcal{L}_1$ by choosing random 6-tuples $\{p_1,\ldots,p_6\}$, we can again proceed as in Section 2. Up to degree 30, all invariants vanishing on the quartics in this databases for $\mathcal{L}_1$ and $\mathcal{L}_2$ are multiples of I_{27}. Since the codimension 2 components $\mathcal{L}_1$, $\mathcal{L}_2$, $(\mathcal{L}_2)_{\text{red}}$ of $\mathcal{L}\cap\mathcal{D}$ have degree at most $24\cdot27, 24\cdot30, 24\cdot15$, which are all smaller than $(30)^2$, we have the following result.

THEOREM 2. *The subschemes $\mathcal{L}_1$, $\mathcal{L}_2$, $(\mathcal{L}_2)_{\text{red}}$ of the projective space of quartic curves $\mathbb{P}\,\mathrm{Sym}^4(V^*)$ (and hence their images in the coarse moduli space of plane quartic curves) are not complete intersections. In particular, they are not principal hypersurfaces in the discriminant locus $\mathcal{D}$.*

As there is no degree 24 invariant vanishing on $\mathcal{L}_1$, Morley's putative construction of such an invariant I_{24} in [9, p.282] is incorrect. On the authors' webpage, a Magma program[6] is available to check all steps on the way to Theorem 2.

5. OPEN QUESTIONS

The expression L of the Lüroth invariant that we found depends on several arbitrary choices that may explain its cumbersomeness. First, there is the choice of the basis of invariants. Though some of the Dixmier invariants have geometrical interpretations that are 'natural', the same is far from evident for the new Ohno invariants. Secondly, our choice can be modified by any element of the kernel N_1. Beyond a cancellation of the coefficients of 215 of these monomials that we have already accomplished by simple linear algebra, further minimization of the number of monomials in the expression for L could in theory be achieved by techniques based on coding theory. Still, the parameters seem too large to make this feasible in practice.
The negative answers concerning the existence of degree 24 and 30 invariants in Section 4 exclude the decomposition from [12, p.1764]. The geometry of the situation does not seem to give a clue for the existence of another such decomposition.
An expression in terms of the 15 coefficients of the generic quartic would of course be useful. However, it is not even practically achievable to formally express the fundamental Dixmier-Ohno invariants in this way, since these expressions contain far too many monomials, as is for instance the case for the discriminant I_{27}. A count of weighted monomials in 15 variables for degree 54 invariants leads to a total of 62 422 531 333. Of course only a fraction of these monomials may occur in the final expression of L, but we could not figure out their number, let alone the Newton polytope of L.

Acknowledgments. This note arose from Giorgio Ottaviani's great lectures at a workshop on invariant theory and projective geometry held in Trento in September 2012, in which the main problems treated in it were mentioned. The authors wish cordially to thank Ottaviani for his further elaborations on the subject, as well as his careful reading of a first draft of this article.

The authors acknowledge support by the grants ANR-09-BLAN-0020-01. The fourth author was additionally supported by the Marie Curie Fellowship IEF-GA-2011-299887.

6. REFERENCES

[1] W. Barth. Moduli of vector bundles on the projective plane. *Invent. Math.*, 42:63–91, 1977.

[2] C. Böhning and H.-C. G. von Bothmer. On the rationality of the moduli space of Lüroth quartics. *Math. Ann.*, 353(4):1273–1281, 2012.

[3] J. Dixmier. On the projective invariants of quartic plane curves. *Adv. in Math.*, 64:279–304, 1987.

[4] A.-S. Elsenhans. Explicit computations of invariants of plane quartic curves. To appear in the proceedings of MEGA 2013.

[5] M. Girard and D. R. Kohel. Classification of genus 3 curves in special strata of the moduli space. Hess, Florian (ed.) et al., Algorithmic number theory. 7th international symposium, ANTS-VII, Berlin, Germany, July 23–28, 2006. Proceedings. Berlin: Springer. Lecture Notes in Computer Science 4076, 346-360, 2006.

[6] R. Hartshorne. Algebraic Geometry. Springer, 1977.

[7] J. Hauenstein and F. Sottile. Newton polytopes and witness sets, 2012. Preprint available at `arxiv.org/abs/1210.2726`.

[8] J. Le Potier and A. Tikhomirov. Sur le morphisme de Barth. *Annales scientifiques de l'École Normale Supérieure*, 34(4):573–629, 2001.

[4] `http://iml.univ-mrs.fr/~ritzenth/programme/luroth/GenerateL1.m`

[5] 1024 curves over $\mathbb{Q}$ are available at **`http://iml.univ-mrs.fr/~ritzenth/programme/luroth/L1Database.m`**

[6] `http://iml.univ-mrs.fr/~ritzenth/programme/luroth/SingularLurothInv.m`

[9] F. Morley. On the Lüroth Quartic Curve. *Amer. J. Math.*, 41(4):279–282, 1919.

[10] T. Ohno. The graded ring of invariants of ternary quartics I, 2005? Unpublished.

[11] G. Ottaviani and E. Sernesi. On the hypersurface of Lüroth quartics. *Michigan Math. J.*, 59(2):365–394, 2010.

[12] G. Ottaviani and E. Sernesi. On singular Lüroth quartics. *Sci. China Math.*, 54(8):1757–1766, 2011.

[13] G. Ottaviani. A computational approach to Lüroth quartics, 2012. *Rendiconti del Circolo Matematico di Palermo* (2)62, no.1, 165–177, 2013.

[14] A. Popolitov and S. Shakirov. On undulation invariants of plane curves, 2012. Preprint available at `http://arxiv.org/abs/1208.5775`.

[15] T. Shioda. On the graded ring of invariants of binary octavics. *American J. of Math.*, 89(4):1022–1046, 1967.

Multiple GCDs. Probabilistic Analysis of the Plain Algorithm

Valérie Berthé
LIAFA (CNRS and Université Paris Diderot), France
berthe@liafa.univ-paris-diderot.fr

Jean Creusefond
INSA Rouen, France
jean.creusefond@insa-rouen.fr

Loïck Lhote
GREYC (CNRS and ENSICAEN), France
loick.lhote@ensicaen.fr

Brigitte Vallée
GREYC (CNRS, and Université of Caen), France
brigitte.vallee@unicaen.fr

ABSTRACT

This paper provides a probabilistic analysis of an algorithm which computes the gcd of ℓ inputs (with $\ell \geq 2$), with a succession of $\ell - 1$ phases, each of them being the Euclid algorithm on two entries. This algorithm is both basic and natural, and two kinds of inputs are studied: polynomials over the finite field $\mathbb{F}_q$ and integers. The analysis exhibits the precise probabilistic behaviour of the main parameters, namely the number of iterations in each phase and the evolution of the length of the current gcd along the execution. We first provide an average-case analysis. Then we make it even more precise by a distributional analysis. Our results rigorously exhibit two phenomena: (i) there is a strong difference between the first phase, where most of the computations are done and the remaining phases; (ii) there is a strong similarity between the polynomial and integer cases.

Categories and Subject Descriptors

F.2.1 [**Numerical Algorithms and Problems**]: Computation on polynomials; G.2.1 [**Combinatorics**]: Generating functions

Keywords

GCD algorithm; average-case analysis; analytic combinatorics; generating functions; dynamical analysis

1. INTRODUCTION

Computing gcds is a common operation, perhaps the fifth main arithmetic operation. Indeed, in many symbolic computation systems, a large proportion of the time is devoted to computing gcds on numbers or polynomials in order to keep fractions under an irreducible form (see [19]). Here, we deal with the case when inputs are either integers or polynomials over a finite field $\mathbb{F}_q$. When there are only two inputs (polynomials or integers), various methods have been designed to compute gcds. The Euclid Algorithm or its variants play a central role here, and Knuth writes in [14] that the Euclid algorithm can be called "the granddaddy of all algorithms, because it is the oldest nontrivial algorithm that has survived to the present day".

In this paper, we are concerned by the case when the gcd of ℓ inputs $x_1, \ldots, x_\ell$, with $\ell \geq 2$ has to be computed. The most straightforward algorithm, described in Knuth's book [14], is a sequence of $\ell - 1$ gcd computations on two entries: one lets $y_1 := x_1$, then, for $k \in [2..\ell]$, one successively computes $y_k := \gcd(x_k, y_{k-1}) = \gcd(x_1, x_2, \ldots, x_k)$. The "total" gcd is $y_\ell := \gcd(x_1, x_2, \ldots, x_\ell)$, and it is obtained after $\ell - 1$ phases. We call it the plain ℓ-Euclid algorithm.

This algorithm cannot be easily extended for computing small Bézout coefficients and we do not claim it is efficient. However, a first step in analysis of algorithms is to understand and precisely analyze even the simplest algorithms; such an analysis then provides a basis of comparison for other algorithms of the same class.

To the best of our knowledge, the plain ℓ–Euclid algorithm has not been yet analyzed[1]. The situation contrasts with the case $\ell = 2$, where the classical Euclid algorithm and its main variants running on integers or on polynomials are now precisely analyzed. See [2, 10, 13, 16] for analyses on polynomials, [11, 5] for the first analyses on integers and [12, 18, 1, 15] for more recent ones. As usual, the results are similar in the two cases. The length is the degree of the polynomial, or the logarithm of the integer, and the length of a pair is the maximum of its components. With respect to the length, the mean number of iterations is linear, the arithmetic complexity is quadratic. There are also precise results on the distribution of the number of iterations, which is asymptotically gaussian. Observe that the analysis is much more difficult in the integer case. In the polynomial case, classical tools of analytic combinatorics,

ISSAC'13, June 26–29, 2013, Boston, Massachusetts, USA.

[1]The analysis of the plain algorithm was proposed as an exercise in Knuth's book (second edition) [14], and quoted as a difficult one (HM48). However, for reasons we do not understand, the exercise disappears in the third edition....Too bad!

described in [8], may be used, whereas the analysis in the integer case is based on a methodology which mixes analytic combinatorics and dynamical systems [18].

For the ℓ-Euclid algorithm, it is natural to choose, as the length of the input, the sum of the length of its components. Changing the input length yields a different probabilistic behavior, even when $\ell = 2$. We show the following: in the case $\ell = 2$, the mean number of iterations remains linear with respect to the (new) length but the distribution of number of iterations is now asymptotically uniform. In the case $\ell \geq 3$, our analysis exhibits a strong difference between the first phase and the following phases. In the first phase, the number of iterations has a linear mean and follows a beta law (the same as the law followed by the minimum of $\ell - 1$ reals i.i.d. in the unit interval). In the following phases, the number of iterations is constant on the average and follows a geometric law. These results can be expected, as Knuth wrote: "In most cases, the length of the partial gcd decreases rapidly during the first few phases of the calculation, and this will make the remainder of the computation quite fast" [14]. Our analysis shows that, in most cases, "almost all the calculation" is done during the first phase. This will make possible to compare (in the extended version of this paper) the ℓ-Euclid algorithm to another strategy proposed in [20].

Plan of the paper. Sections 2, 3, 4 are devoted to the polynomial case, and Section 5 to the integer case. Section 2 describes the algorithm and states the main results, then, Sections 3 and 4 are devoted to the proofs, mainly based on analytical combinatorics: we use generating functions, built in Section 3, and used in Section 4. Section 5 explains the strong similarity between the two cases (polynomial and integer); it presents the useful generating functions (here of Dirichlet type), states the main results (only for the average-case) and briefly describes the main steps of the proof. We also explain why distributional results may be expected.

2. MAIN RESULTS FOR POLYNOMIALS.

We consider the ring $\mathbb{F}_q[X]$ of polynomials over the finite field $\mathbb{F}_q$ with q elements, and the degree of a non-zero polynomial x is denoted by $\mathrm{d}(x)$.

The plain ℓ-Euclid algorithm on $\mathbb{F}_q[X]$ deals with a sequence of ℓ nonzero polynomials and computes their greatest common divisor y via a sequence of $\ell - 1$ gcd computations between two polynomials, as we previously explained.

Each phase is a 2–Euclid Algorithm which performs a gcd computation between two polynomials, with a sequence of Euclidean divisions, as recalled as follows:

The Euclid Algorithm on (a_1, a_2) with $\mathrm{d}(a_1) > \mathrm{d}(a_2)$

$$\begin{array}{lllll} a_1 & = & m_1\, a_2 & + & a_3 & 0 < \mathrm{d}(a_3) < \mathrm{d}(a_2) \\ a_2 & = & m_2\, a_3 & + & a_4 & 0 < \mathrm{d}(a_4) < \mathrm{d}(a_3) \\ \dots & = & \dots & + & & \\ a_{r-1} & = & m_{r-1}\, a_r & + & a_{r+1} & 0 < \mathrm{d}(a_{r+1}) < \mathrm{d}(a_r) \\ a_r & = & m_r\, a_{r+1} & + & 0 & \end{array}$$

Then, $\gcd(a_1, a_2)$ is the last non-zero remainder a_{r+1}. It can be chosen monic. The number of steps (here equal to r) is one of the main parameters of interest.

This paper aims at precisely understanding the random behavior of the plain algorithm. Since the algorithm is a succession of phases, it is important to describe each phase of index k ($k \in [1..\ell - 1]$), with the following parameters:

(i) the number L_k of divisions performed during the k-th phase,

(ii) the degree D_k of the gcd y_k at the beginning of the k-th phase.

We are also interested in the analysis of global parameters.

(iii) The algorithm may be interrupted as soon as $D_k = 0$, and Π is the number of useful phases. Namely:

$\Pi = 0$ if $D_1 = 0$ and $\Pi := \max\{k | D_k > 0\}$ if $D_1 > 0$.

(iv) The total number of divisions of the interrupted algorithm is defined as

$L = 0$ if $\Pi = 0$ and $L = L_1 + L_2 + \ldots + L_\Pi$ if $\Pi > 0$.

The possible inputs are all the sequences $\underline{x}$ formed of ℓ polynomials, and we limit ourselves to monic polynomials, without loss of generality. Then the set of inputs is $\Omega = \mathcal{U}^\ell$, where $\mathcal{U}$ is the set of monic polynomials. Here, the size of the input $\underline{x}$ is the total degree of the sequence (and not the maximum, as it is often the case), and we let

$$\mathrm{d}(\underline{x}) = \mathrm{d}(x_1, x_2, \ldots, x_\ell) := \mathrm{d}(x_1) + \mathrm{d}(x_2) + \ldots + \mathrm{d}(x_\ell).$$

The subset Ω_n formed with the inputs of size n is a finite set, and it is endowed with the uniform probability. Now, the parameters of interest become random variables and we are interested in their probabilistic features.

With analytic combinatorics methodology, we prove the following two results. The first one deals with the expected values, whereas the second one describes asymptotic limit laws.

2.1 Average-case analysis.

Theorem 1 below exhibits a strong difference between the first phase and the following ones. It shows that, on average, the first phase performs a linear number of iterations which involves the entropy $2q/(q-1)$ of the Euclid dynamical system; then, even if the degree of the first gcd is linear with respect to the input size, the degree of the gcd is proven to be of constant order after the first phase on the average. Then, the number of divisions L_k which will be performed in the following phases, together with the degrees D_k of the following gcds will be of constant order.

THEOREM 1. *[Expectations]. When the set Ω_n is endowed with the uniform distribution, the following holds:*

(a) *The expectation of the number of iterations L_1 during the first phase is linear with respect to the size n and satisfies*

$$\mathbb{E}_n[L_1] = \frac{q-1}{2q}\frac{n}{\ell} + \frac{3q+1}{4q} + O\left(\frac{1}{n}\right).$$

(b) *For any $k \in [2..\ell - 1]$, the expectation of the number of iterations L_k during the k-th phase is asymptotic to a constant, and satisfies*

$$\mathbb{E}_n[L_k] = \frac{q^k - 1}{q^k - q}\left[1 + O\left(\frac{1}{n}\right)\right].$$

(c) *The expectation of the degree of the first polynomial x_1 is linear with respect to the size n and satisfies*

$$\mathbb{E}_n[D_1] = \frac{n}{\ell}.$$

(d) *For any $k \in [2..\ell - 1]$, the expectation of the degree D_k of the gcd y_k at the beginning of the k-th phase is asymptotic to a constant, and satisfies*

$$\mathbb{E}_n[D_k] = \frac{1}{q^{k-1} - 1}\left[1 + O\left(\frac{1}{n}\right)\right].$$

2.2 Limit Laws.

The following result makes more precise the results obtained in Theorem 1, and explains more deeply the difference between the first phase ($k = 1$) and the following phases. For $k = 1$, the expected degrees of the first two polynomials x_1 and x_2 are linear, and the number of divisions is closely related to $\min(\mathrm{d}(x_1), \mathrm{d}(x_2))$. Then, it is natural to expect beta laws for the first phase, more precisely a beta law of parameter $(1, \ell - 1)$, since it is the law of the minimum Y of $\ell - 1$ random variables i.i.d. on the unit interval, which satisfies $\mathbb{P}[Y \geq x] = (1 - x)^{\ell - 1}$. Such a law has a density equal to $(\ell - 1)(1 - x)^{\ell - 2}$. For $\ell = 2$, this is the uniform law. For the following phases, geometric laws are expected, since the means of the gcd degrees are of constant order.

The next result describes the asymptotic distribution of L_k [Assertion (a)] and D_k [Assertion (b)]. At first glance, the results do not seem to heavily depend on the index k of the phase. However, it is not true, since the two ratios $p_k := (q-1)/(q^k - 1)$ (in case L), $r_k := q^{1-k}$ (in case D), are equal to 1 for $k = 1$ and strictly less than 1 for $k \geq 2$.

THEOREM 2. *[Limit laws.] When the set Ω_n is endowed with the uniform distribution, the following holds:*

(a) The number of iterations L_1 during the first phase asymptotically follows a beta law of parameter $(1, \ell - 1)$ on the interval $[0, (q-1)/(2q)]$ whereas, the number of iterations L_k during each following phase asymptotically follows a geometric law with ratio $p_k := (q-1)/(q^k - 1)$. One has:

$$\mathbb{P}_n\left[L_k > n/(k+1)\right] = 0, \quad \textit{for any } k.$$

For any k, the distribution of L_k satisfies when $n \to \infty$,

$$\mathbb{P}_n[L_k > m] = \left(\frac{q-1}{q^k - 1}\right)^m \left[1 + O\left(\frac{m}{n}\right)\right] \quad \textit{for } m = o(n),$$

and for $m/n \to c$ with $c \in]0, \frac{1}{k+1}\frac{q^k-1}{q^k}[$

$$\mathbb{P}_n[L_k > m] = \left(\frac{q-1}{q^k - 1}\right)^m \left(1 - \frac{(k+1)q^k}{q^k - 1}c\right)^{\ell - 1} \left[1 + O\left(\frac{1}{n}\right)\right].$$

(b) The degree D_1 of the first polynomial x_1 asymptotically follows a beta law of parameter $(1, \ell - 1)$ on the interval $[0, 1]$, whereas the degree D_k of the gcd y_k at the beginning of each following phase asymptotically follows a geometric law with ratio $r_k := q^{1-k}$. One has: $\mathbb{P}_n\left[D_k > n/k\right] = 0$ for any k. For any k, the distribution of D_k satisfies when $n \to \infty$,

$$\mathbb{P}_n[D_k \geq m] = q^{(1-k)m} \left[1 + O\left(\frac{m}{n}\right)\right] \quad \textit{for } m = o(n),$$

and for $m/n \to c$ with $c \in]0, 1/k[$

$$\mathbb{P}_n[D_k \geq m] = q^{(1-k)m} (1 - kc)^{\ell - 1} \left[1 + O\left(\frac{1}{n}\right)\right].$$

2.3 Global parameters.

The interrupted algorithm stops as soon as the gcd y_k is of degree 0. Let $\Pi(x_1, \ldots, x_\ell)$ be the number of useful phases. Since the event $[\Pi \geq k]$ coincides with the event $[D_k \geq 1]$ for $k \in [1..\ell - 1]$, Theorem 2 leads to the following:

THEOREM 3. *When the set Ω_n is endowed with the uniform distribution, the distribution of the number Π of useful phases satisfies $\mathbb{P}_n[\Pi \geq 0] = 1$, $\mathbb{P}_n[\Pi \geq \ell] = 0$,*

$$\mathbb{P}_n[\Pi \geq k] = q^{1-k} \left[1 + O\left(\frac{1}{n}\right)\right] \quad \textit{for } k \in [1..\ell - 1].$$

The following result studies the total number L of divisions performed by the interrupted version. With Theorems 1 and 3, the mean is easy to study. Moreover, Theorems 1 and 2 prove that the variance of L_1 is quadratic whereas the variance of L_k (for $k \geq 2$) is of constant order. Then, the linearity of the mean with the Markov inequality entail that L has the same asymptotic distribution as L_1.

THEOREM 4. *When the set Ω_n is endowed with the uniform distribution, the total number of divisions L performed by the interrupted version of the ℓ-Euclid algorithm has an expected value $\mathbb{E}_n[L]$ equal to*

$$\frac{q-1}{2q}\frac{n}{\ell} + \frac{3q+1}{4q} + \sum_{k=2}^{\ell-1}\left[\frac{q}{q^k} + \frac{q-1}{q^k - 1}\right] + O\left(\frac{\ell^2}{n}\right).$$

Moreover, the number L asymptotically follows a beta distribution of parameter $(1, \ell - 1)$ on the interval $[0, (q-1)/(2q)]$.

3. GENERATING FUNCTIONS

3.1 General setting

We use the analytic combinatorics methodology, and deal with its main tool, the generating functions. We use a variable z_i to mark the degree of the i-th polynomial x_i, and the generating function $F(z_1, z_2, \ldots z_\ell)$ of the set $\Omega = \mathcal{U}^\ell$, relative to the size d, is defined as

$$F(z_1, z_2, \ldots z_\ell) := \sum_{\underline{x} \in \mathcal{U}^\ell} z_1^{\mathrm{d}(x_1)} z_2^{\mathrm{d}(x_2)} \ldots z_\ell^{\mathrm{d}(x_\ell)}.$$

It is equal to the product $U(z_1)U(z_2)\ldots U(z_\ell)$, where $U(z)$ is the generating function of the set $\mathcal{U}$ of the monic polynomials relative to the size d, namely

$$U(z) = \sum_{x \in \mathcal{U}} z^{\mathrm{d}(x)} = \sum_{n \geq 0} q^n z^n = \frac{1}{1 - qz}.$$

Most of the time, we limit ourselves to the case when all the variables z_i are equal, and we write $F(z)$ instead of $F(z, \cdots, z)$. For studying a parameter (or a cost C) on $\Omega = \mathcal{U}^\ell$, a main tool is the bivariate generating function relative to some cost C, obtained by introducing a further variable u to mark the cost, and defined as

$$F(z, u) := \sum_{\underline{x} \in \mathcal{U}^\ell} z^{\mathrm{d}(\underline{x})} u^{C(\underline{x})}.$$

We are first interested in the mean value of parameter C, and we deal with the cumulative generating function

$$\widehat{F}(z) := \frac{\partial F}{\partial u}(z, u)|_{u=1}, \quad \text{and} \quad \mathbb{E}_n[C] = \frac{[z^n]\widehat{F}(z)}{[z^n]F(z)}. \tag{1}$$

The probability distribution of the cost C can be studied with the generating function $F(z, u)$, via the relation

$$\mathbb{P}_n[C = i] = \frac{[z^n\, u^i]F(z, u)}{[z^n]F(z)}.$$

Then, the probabilities $\mathbb{P}_n[C \geq m]$ are expressed with the "cumulative bivariate generating functions" defined as

$$\widehat{F}^{[m]}(z) := \sum_{i \geq m} [u^i]F(z, u), \quad \text{and} \quad \mathbb{P}_n[C \geq m] = \frac{[z^n]\widehat{F}^{[m]}(z)}{[z^n]F(z)}. \tag{2}$$

3.2 Another expression for the generating function

We first obtain an alternative expression for the generating function $F(z)$ with a product of $\ell - 1$ factors, each of them describing a phase of the algorithm.

PROPOSITION 1. *The generating function of the set $\Omega = \mathcal{U}^\ell$ with the size equal to the total degree decomposes as*

$$F(z) = U(z)^\ell = U(z^\ell) \cdot \prod_{k=1}^{\ell-1} T(z, z^k)$$

and involves the phase-function T defined as

$$T(z,t) = \frac{U(z) + U(t) - 1}{1 - G(zt)}, \tag{3}$$

the generating function $U(z)$ of monic polynomials, and the generating function $G(z)$ of general polynomials with strictly positive degree, i.e.,

$$U(z) = \frac{1}{1-qz}, \qquad G(z) = \frac{(q-1)qz}{1-qz}.$$

PROOF. The Euclid algorithm first compares the degrees of x_1 and x_2. There are three cases:

$$\mathrm{d}(x_1) = \mathrm{d}(x_2), \qquad \mathrm{d}(x_1) > \mathrm{d}(x_2), \qquad \mathrm{d}(x_1) < \mathrm{d}(x_2).$$

In the first case, there is an extra subtraction, which can be viewed as a division with a quotient equal to 1.

In all the cases, the gcd $y := \gcd(x_1, x_2)$ together with the sequence of quotients $(m_1, m_2, \ldots m_r)$ completely determines the input pair (x_1, x_2). More precisely, one writes $(x_1, x_2) = (y\,\widehat{x}_1, y\,\widehat{x}_2)$ with a coprime pair $(\widehat{x}_1, \widehat{x}_2)$ and the execution of the Euclid algorithm on the pair $(\widehat{x}_1, \widehat{x}_2)$ produces the same sequence $(m_1, m_2, \ldots m_r)$ as the pair (x_1, x_2). The first quotient m_1 is monic (this is due to the fact that x_1 and x_2 are monic) and the remainder of the sequence $\Sigma = (m_2, \ldots m_r)$ is formed with general polynomials m_i (no longer monic) with $\mathrm{d}(m_i) \geq 1$. As previously, the total degree of Σ is $\mathrm{d}(\Sigma) = \mathrm{d}(m_2) + \ldots + \mathrm{d}(m_r)$.

We now focus on the first quotient m_1, and we consider the three following possible cases:

(*a*) If $\mathrm{d}(x_1) = \mathrm{d}(x_2)$, then $m_1 = 1$.

(*b*) If $\mathrm{d}(x_1) > \mathrm{d}(x_2)$ then
$\mathrm{d}(m_1) \geq 1, \quad \mathrm{d}(\widehat{x}_2) = \mathrm{d}(\Sigma), \quad \mathrm{d}(\widehat{x}_1) = \mathrm{d}(m_1) + \mathrm{d}(\Sigma).$

(*c*) If $\mathrm{d}(x_1) < \mathrm{d}(x_2)$ then
$\mathrm{d}(m_1) \geq 1, \quad \mathrm{d}(\widehat{x}_1) = \mathrm{d}(\Sigma), \quad \mathrm{d}(\widehat{x}_2) = \mathrm{d}(m_1) + \mathrm{d}(\Sigma).$

All these remarks provide an alternative expression of the product $U(z_1)U(z_2)$. Indeed, we use the two relations

$$z_1^{\mathrm{d}(x_1)} z_2^{\mathrm{d}(x_2)} = (z_1 z_2)^{\mathrm{d}(y)} \cdot z_1^{\mathrm{d}(\widehat{x}_1)} z_2^{\mathrm{d}(\widehat{x}_2)},$$

$$\sum_{\widehat{x}_1, \widehat{x}_2} z_1^{\mathrm{d}(\widehat{x}_1)} z_2^{\mathrm{d}(\widehat{x}_2)} = \left[1 + \sum_{m_1} z_1^{\mathrm{d}(m_1)} + z_2^{\mathrm{d}(m_1)}\right]\left[\sum_{\Sigma} (z_1 z_2)^{\mathrm{d}(\Sigma)}\right],$$

together with the conditions previously described on the first quotient m_1, the sequence Σ and the gcd y. The first factor gives rise to the generating function

$$1 + (U(z_1) - 1) + (U(z_2) - 1) = U(z_1) + U(z_2) - 1$$

which involves the generating function $U(z)$ of monic polynomials, whereas the second factor is expressed with the generating function $G(z)$ of general polynomials with a strictly positive degree, under the form $1/(1 - G(z_1 z_2))$. We have then proven the following alternative form for the product

$$U(z_1)U(z_2) = U(z_1 z_2) \cdot T(z_1, z_2).$$

When we replace this expression into the total product

$$U(z_1)U(z_2) \ldots U(z_\ell) = F(z_1, z_2, \ldots, z_\ell)$$

and iterate the transformation, we obtain an alternative expression for the generating function $F(z_1, z_2, \ldots, z_\ell)$ with a product of $\ell - 1$ factors, each of them involving the phase-function T at points z_k and $t_k = z_1 \ldots z_k$

$$F(z_1, z_2, \ldots z_\ell) = U(t_\ell) \cdot \prod_{k=1}^{\ell-1} T(z_{k+1}, t_k).$$

It can be useful in some studies to keep all the variables z_i, but here, we let $z_1 = z_2 = \ldots = z_\ell$, and we obtain an expression of the generating function $F(z)$. □

3.3 Generating functions for parameters of interest

When studying the parameter L_k (number of steps in the k-th phase), we use an extra variable u which marks each step of the k-th iteration, and we deal with the generating function

$$T(z,t,u) = u\frac{U(z) + U(t) - 1}{1 - uG(zt)} \qquad \text{with } t = z^k,$$

which replaces $T(z, z^k)$ inside $F(z)$.

When studying the parameter D_k (degree of the gcd at the beginning of the k-th phase), we use an extra variable u which marks the degree of the gcd y_k, and we deal with the generating function

$$U(t,u) = \frac{1}{1 - qut} \qquad \text{with } t = z^k,$$

which replaces $U(z^k)$ inside $F(z)$. Finally:

PROPOSITION 2. *For any $k \in [1..\ell - 1]$, the bivariate generating function $L_k(z,u)$ relative to the number of divisions during the k-th phase satisfies*

$$L_k(z,u) = U(z)^\ell \cdot \frac{T(z, z^k, u)}{T(z, z^k)}.$$

For any $k \in [1..\ell - 1]$, the bivariate generating function $D_k(z,u)$ relative to the degree of the k-th gcd y_k at the beginning of the k-th phase satisfies

$$D_k(z,u) = U(z)^\ell \cdot \frac{U(z^k, u)}{U(z^k)}.$$

With Proposition 1, the functions $T(z, z^k, u)$ and $U(z^k, u)$ admit precise expressions, and Proposition 2 leads to the expression of the bivariate generating functions $L_k(z,u)$ and $D_k(z,u)$

$$\frac{L_k(z,u)}{U(z)^\ell} = u\frac{1 - G(z^{k+1})}{1 - uG(z^{k+1})}, \qquad \frac{D_k(z,u)}{U(z)^\ell} = \frac{1 - qz^k}{1 - quz^k}. \tag{4}$$

Finally, taking the derivative with respect to u, we obtain the cumulative generating functions

$$\frac{\widehat{L}_k(z)}{U(z)^\ell} = \frac{1 - qz^{k+1}}{1 - q^2 z^{k+1}}, \qquad \frac{\widehat{D}_k(z)}{U(z)^\ell} = \frac{qz^k}{1 - qz^k}. \tag{5}$$

Extracting in (4) the coefficient of $[u^i]$ in the bivariate generating functions and taking the sum over $i \geq m$ gives

$$\frac{\widehat{L}_k^{[m]}(z)}{U(z)^\ell} = G(z^{k+1})^{m-1}, \qquad \frac{\widehat{D}_k^{[m]}(z)}{U(z)^\ell} = (qz^k)^m. \tag{6}$$

4. FINAL STEPS FOR THE PROOFS.

We have obtained in (5) and (6) the expressions of useful generating functions, that are always here fractional functions. It is then possible to use (1) and (2) to obtain an exact expression of the expectation and probability distribution of the parameters D_k and L_k. However, we are mainly interested in the asymptotic behaviour (as $n \to \infty$) of these probabilistic features. Singularity analysis describes the possible transfer between the behavior of a generating function, viewed as an analytic function, near its dominant singularity (the singularity closest to 0) and the asymptotic behavior of its coefficients. More precisely, the position and the nature of the dominant singularity play a fundamental role.

4.1 Average-case analysis.

Here, the main tools are the cumulative generating functions (5). They both admit a dominant pole in $z = 1/q$ but the order of the pole is different according to the phase. For the first phase ($k = 1$), the pole $z = 1/q$ is of order $\ell + 1$ whereas for the other phases ($k \geq 2$), this pole remains of order ℓ.

We now use the following result, here quite trivial since we deal with rational fractions.

LEMMA 1. *Consider a function $f(z) = g(z)/(1-qz)^j$ with $j \geq 2$ which involves a function $g(z)$ which is analytic in the disk $|z| > 1/q$ and satisfies $g(1/q) \neq 0$. Then,*

$$[z^n]f(z) = g\left(\frac{1}{q}\right)\binom{n+j-1}{j-1}q^n - g'\left(\frac{1}{q}\right)\binom{n+j-2}{j-2}q^{n-1},$$

with a remainder term in $O\left(n^{j-3}q^n\right)$. The previous estimate also holds if g admits simple isolated poles on the punctured circle $\{|z| = 1/q, z \neq 1/q\}$, as soon as $j \geq 3$.

We first consider the case when the phase index k is at least equal to 2. In this case, with the expressions of the cumulative generating functions $\widehat{D}_k(z)$ and $\widehat{L}_k(z)$ given in (5), the lemma applies with $j = \ell$ and

$$g(z) = \frac{1 - qz^{k+1}}{1 - q^2 z^{k+1}}, \quad g\left(\frac{1}{q}\right) = \frac{q^k - 1}{q^k - q} \quad \text{(case } L\text{)},$$

$$g(z) = \frac{qz^k}{1 - qz^k} \quad g\left(\frac{1}{q}\right) = \frac{1}{q^{k-1} - 1} \quad \text{(case } D\text{)}.$$

Consider now the case when the phase index k equals 1. In this case, the integer j equals $\ell + 1$, and

$$g(z) = qz \quad \text{(case } D\text{)}, \qquad g(z) = \frac{1 - qz^2}{1 + qz} \quad \text{(case } L\text{)}.$$

In both cases L and D, the lemma applies. In case D, we obtain:

$$[z^n]\widehat{D}_1(z) = q^n\binom{n+\ell-1}{\ell}.$$

We remark that, in case L, the function g admits $z = -1/q$ as a simple pole and we obtain

$$[z^n]\widehat{L}_1(z) =$$

$$= \frac{q-1}{2q}q^n\binom{n+\ell}{\ell} + \frac{q+3}{4}q^{n-1}\binom{n+\ell-1}{\ell-1} + O\left(n^{\ell-2}q^n\right).$$

In all the cases, the normalization by $\text{card}(\Omega_n) = \binom{n+\ell-1}{\ell-1}q^n$ proves Theorem 1.

4.2 A general framework for limit laws.

The "cumulative bivariate generating functions" which are useful for the study of the distributions at the k-th phase are, due to (6), of the form

$$U(z)^\ell \cdot A_k(z)^m,$$

where the function $A_k(z)$ depends on the index of the phase, and the type of parameter. One has

$$A_k(z) = qz^k \quad \text{(type } D\text{)}, \tag{7}$$

$$A_k(z) = \frac{(q-1)qz^{k+1}}{1 - qz^{k+1}} = G(z^{k+1}) \quad \text{(type } L\text{)}. \tag{8}$$

In all the cases, the generating functions are expressed as a product of the function $U(z)^\ell$ which has a pole of order ℓ at $z = 1/q$, with a "large power" of a function $A(z)$. The term "large power" is used since the exponent m may depend on the size n. As the following proposition shows, there are two main cases, according as the value of A at $z = 1/q$ is equal to 1 or not. The first case happens for $k = 1$ and leads to beta distribution, whereas the second case happens for $k \geq 2$ and leads to geometric distributions.

PROPOSITION 3. *Consider the function*

$$F^{[m]}(z) = \frac{1}{(1-z)^\ell} \cdot A(z)^m,$$

where $A(z)$ is analytic on the disk $|z| \leq \rho$ with $\rho > 1$ and satisfies $a := A(1) \neq 0$, $b := A'(1) > 0$ and for $|z|$ close enough to 1, $|A(z)| \leq A(|z|)$.
Then, the coefficient of z^n in $F^{[m]}(z)$ satisfies the following:

(a) *When $m/n \to 0$, then*

$$[z^n]F^{[m]}(z) = \frac{n^{\ell-1}}{(\ell-1)!} \cdot a^m \left[1 + O\left(\frac{m}{n}\right)\right].$$

(b) *When $m/n \to c$ for some $c \in]0, a/b[$ then*

$$[z^n]F^{[m]}(z) = \frac{n^{\ell-1}}{(\ell-1)!} a^m \left(1 - \frac{b}{a}c\right)^{\ell-1}\left[1 + O\left(\frac{1}{n}\right)\right].$$

4.3 End of the proof of Theorem 2.

The probabilities $\mathbb{P}_n[L_k > m]$, $\mathbb{P}_n[D_k > m]$ are related to the coefficient of z^n in the generating functions described in (6), and the functions $A_k(z)$ of interest are given in (7) and (8). Since, in case D, $A_k(z)$ is multiple of z^k, and, in case L, $A_k(z)$ is multiple of z^{k+1}, we first deduce the equalities

$$\mathbb{P}_n\left[D_k > n/k\right] = 0, \qquad \mathbb{P}_n\left[L_k > n/(k+1)\right] = 0.$$

After a change of variable $z \to z/q$, the hypotheses of Proposition 3 are fulfilled for $z \mapsto A_k(z/q)$, and one has,

$$a_k = q^{1-k}, \quad \frac{b_k}{a_k} = k \qquad \text{(case } D\text{)},$$

$$a_k = \frac{q-1}{q^k - 1}, \quad \frac{b_k}{a_k} = (k+1)\frac{q^k}{q^k - 1}, \qquad \text{(case } L\text{)}.$$

For $k = 1$, the constants a_k are equal to 1, whereas they are strictly less than 1 for $k \geq 2$. Finally, the normalization by

$$\text{card}(\Omega_n) = \frac{n^{\ell-1}}{(\ell-1)!}q^n\left[1 + O\left(\frac{1}{n}\right)\right],$$

proves Assertions (a) and (b) of Theorem 2.

5. THE INTEGER CASE.

We now briefly explain how a similar study can be performed in the case of the Euclid algorithm on integers. As usual (see for instance [15, 18]), the polynomial study shows the road, and similar results are expected in the integer study, even if they are often more difficult to obtain and less precise.

5.1 The plain algorithm on integers.

In the integer case, the ℓ–plain Euclid algorithm has exactly the same structure as in the polynomial case. It is composed with $\ell - 1$ phases, each of them being the Euclid algorithm which performs the gcd computation between two integers. Its execution is described as in the figure of Section 2, and the degree is just replaced by the integers themselves. The main parameters of interest are the same, and D_k is now the length of the gcd y_k, namely its logarithm.

The possible inputs are all the sequences $\underline{x}$ formed of ℓ integers, and we limit ourselves to positive integers, without loss of generality. Then the set of inputs is $\Omega = \mathbb{N}^\ell$, where $\mathbb{N}$ is the set of positive integers. Here, the size of the input $\underline{x}$ equals the product of its components $x_1 \cdot x_2 \cdot \ldots \cdot x_\ell$, and the length of the input is the sum of the lengths of its components.

Here again, the subset Ω_N formed with the inputs of size at most N is a finite set, and it is endowed with the uniform probability.

5.2 Dirichlet generating functions.

The generating functions are now of Dirichlet type, and the basic one is the generating function of the set $\mathbb{N}^\ell$. We deal with ℓ–uples $\underline{x}$ of integers $\underline{x} = (x_1, x_2, \ldots x_\ell)$ and consider the generating function

$$F(s_1, s_2, \ldots s_\ell) = \sum_{\underline{x} \in \mathbb{N}^\ell} \frac{1}{x_1^{s_1}} \frac{1}{x_2^{s_2}} \cdots \frac{1}{x_\ell^{s_\ell}} = \zeta(s_1) \ldots \zeta(s_\ell),$$

where the ζ function is the generating function of $\mathbb{N}$,

$$\zeta(s) := \sum_{n \geq 1} \frac{1}{n^s}.$$

The main case of interest $s_1 = s_2 = \ldots s_\ell = s$ gives rise to

$$F(s) := F(s, \ldots, s) = \zeta(s)^\ell.$$

However, we begin with the generic case when the ℓ–uple $s = (s_1, \ldots, s_\ell)$ is general.

The first step derives, as previously, a decomposition for the product $\zeta(s_1)\,\zeta(s_2)$, of the form

$$\zeta(s_1)\,\zeta(s_2) = \zeta(s_1 + s_2) \cdot T(s_1, s_2),$$

where $T(s_1, s_2)$ is the phase generating function which describes the Euclid algorithm on two integers. In the integer case, the 2-Euclid algorithm is described with the transfer operator, related to the underlying dynamical system, and introduced by Ruelle in [17] in a general setting. This operator deals here with the Gauss map S defined by $S(x) := 1/x - [1/x]$, and the set $\mathcal{G}$ of its inverse branches, namely

$$\mathcal{G} := \left\{ h_m(t) : t \mapsto \frac{1}{m+t};\quad m \geq 1 \right\}.$$

It depends on a complex parameter s, and acts for $\Re s > 1$ on functions defined on the unit interval as

$$\mathbf{G}_s[f](t) = \sum_{h \in \mathcal{G}} |h'(t)|^{s/2} f \circ h(t). \tag{9}$$

The following result provides an analog of Proposition 1.

PROPOSITION 4. *The generating function of $\Omega = \mathbb{N}^\ell$ with the size "product of inputs" can be written as*

$$F(s) = \zeta(s)^\ell = \zeta(\ell s) \cdot \prod_{k=1}^{\ell-1} T(s, ks),$$

and involves the phase-function T defined as

$$T(s,t) = \frac{1}{2}\left[(I - \mathbf{G}_{s+t})^{-1} \circ (\mathbf{G}_s + \mathbf{G}_t)[1](0)\right], \tag{10}$$

where $\mathbf{G}_s$ is the transfer operator relative to the Euclid dynamical system, defined in (9).

PROOF. The Euclid algorithm first compares the two integers x_1 and x_2. There are three cases:

$$x_1 = x_2, \qquad x_1 > x_2, \qquad x_1 < x_2.$$

In all the cases, the gcd $y := \gcd(x_1, x_2)$ together with the sequence of quotients $(m_1, m_2, \ldots m_r)$ completely determines the input pair (x_1, x_2). More precisely, one writes $(x_1, x_2) = (y\,\widehat{x}_1, y\,\widehat{x}_2)$ with a coprime pair $(\widehat{x}_1, \widehat{x}_2)$ and the execution of the Euclid algorithm on the pair $(\widehat{x}_1, \widehat{x}_2)$ produces the same sequence $(m_1, m_2, \ldots m_r)$ as the pair (x_1, x_2), with now remainders $\widehat{x}_i$ which satisfy $x_i = y\widehat{x}_i$.

The execution of the Euclid algorithm on the pair $(\widehat{x}_1, \widehat{x}_2)$ with $\widehat{x}_1 > \widehat{x}_2$ builds continued fraction expansions

$$\frac{\widehat{x}_2}{\widehat{x}_1} = h \circ g(0), \qquad \frac{\widehat{x}_3}{\widehat{x}_2} = g(0).$$

Here, $h := h_{m_1}$ is related to the first quotient and $g = h_{m_2} \circ h_{m_3} \circ \ldots \circ h_{m_r}$ is related to the sequence $(m_2, m_3, \ldots, m_r)$. Since the two pairs $(\widehat{x}_1, \widehat{x}_2)$ and $(\widehat{x}_2, \widehat{x}_3)$ are coprime, the denominators of each rational are expressed with derivatives,

$$\frac{1}{\widehat{x}_1^2} = |(h \circ g)'(0)| = |h'(g(0))| \cdot |g'(0)|, \qquad \frac{1}{\widehat{x}_2^2} = |g'(0)|.$$

Then, in the case when $\widehat{x}_1 \geq \widehat{x}_2$, the sum

$$\sum_{\widehat{x}_1 \geq \widehat{x}_2} \frac{1}{\widehat{x}_1^{s_1}} \frac{1}{\widehat{x}_2^{s_2}} = 1 + \sum_{h,g} |h'(g(0))|^{s_1/2} \cdot |g'(0)|^{(s_1+s_2)/2},$$

can be expressed with the transfer operator $\mathbf{G}_s$ as[2]

$$\frac{1}{2}\left[(I - \mathbf{G}_{s_1+s_2})^{-1} \circ \mathbf{G}_{s_1}[1](0) + 1\right].$$

The case $\widehat{x}_2 \geq \widehat{x}_1$ can be dealt with exchanging the roles of $\widehat{x}_1$ and $\widehat{x}_2$. Finally, the relations

$$\sum_{x_1, x_2} \frac{1}{x_1^{s_1}} \frac{1}{x_2^{s_2}} = \sum_{v} \frac{1}{y^{s_1+s_2}} \sum_{\widehat{x}_1, \widehat{x}_2} \frac{1}{\widehat{x}_1^{s_1}} \frac{1}{\widehat{x}_2^{s_2}}$$

entail the equality $\zeta(s_1)\zeta(s_2) = \zeta(s_1 + s_2) \cdot T(s_1, s_2)$, where T is defined in (10). When we replace this expression into the total product

$$\zeta(s_1)\,\zeta(s_2) \ldots \zeta(s_\ell) = F(s_1, s_2, \ldots, s_\ell),$$

and iterate the transformation, we obtain an alternative expression for the generating function $F(s_1, s_2, \ldots, s_\ell)$ with a product of $\ell - 1$ factors, each of them involving the phase-function T at points s_k and $t_k = s_1 + \ldots + s_k$,

$$F(s_1, s_2, \ldots s_\ell) = \zeta(t_\ell) \cdot \prod_{k=1}^{\ell-1} T(s_{k+1}, t_k).$$

[2] The factor $(1/2)$ is here to take into account the fact that any rational of $]0,1]$ admits two continued fraction expansions: the proper one and the improper one.

It may be useful in some studies to keep all the variables s_i, but, here again, we let $s_1 = s_2 = \ldots = s_\ell = s$, and we obtain the expression of the generating function $F(s)$. □

5.3 Dirichlet generating functions for parameters of interest

When studying the parameter L_k (number of steps in the k-th phase), we use an extra variable u which marks each step of the k-th iteration, and we deal with the generating function

$$2T(s,t,u) = u(1 - u\mathbf{G}_{s+t})^{-1} \circ (\mathbf{G}_s + \mathbf{G}_t)[1](0),$$

with $t = ks$, which replaces $2T(s, ks)$ inside $F(s)$.
When studying the parameter D_k (length of the gcd at the beginning of the k-th phase), we use an extra variable u which marks the length of the gcd y_k, and we deal with the generating function

$$\zeta(t,u) = \sum_{n\geq 1} \frac{u^{\log n}}{n^t} = \zeta(t - \log u) \qquad \text{with } t = ks,$$

which replaces $\zeta(ks)$ inside $F(s)$. Finally, we obtain an analog of Proposition 2.

PROPOSITION 5. *For any $k \in [1..\ell-1]$, the bivariate generating function $L_k(s,u)$ relative to the number of divisions during the k-th phase satisfies*

$$L_k(s,u) = \zeta(s)^\ell \cdot \frac{T(s,ks,u)}{T(s,ks)}.$$

For any $k \in [1..\ell-1]$, the bivariate generating function $D_k(z,u)$ relative to the size of the k-th gcd y_k at the beginning of the k-th phase satisfies

$$D_k(s,u) = \zeta(s)^\ell \cdot \frac{\zeta(ks - \log u)}{\zeta(ks)}.$$

With Proposition 4, the function $T(s, ks, u)$ admits a precise expression. Using Proposition 5, and taking the derivative with respect to u, we obtain the cumulative generating functions

$$\frac{\widehat{D}_k(s)}{\zeta(s)^\ell} = \frac{\zeta'(ks)}{\zeta(ks)}, \qquad \frac{\widehat{L}_k(s)}{\zeta(s)^\ell} = \frac{\widehat{T}(s,ks)}{T(s,ks)}. \tag{11}$$

Remark that the Dirichlet series

$$\widehat{T}(s,ks) := \frac{\partial T}{\partial u}(s,ks,u)|_{u=1}$$

involves two occurrences of the quasi-inverse $(I - \mathbf{G}_{(k+1)s})^{-1}$.

5.4 Average-case analysis results.

The following result is an exact analog of Theorem 1. In particular, in Assertion (a), the entropy $\pi^2/(6\log 2)$ of the integer Euclidean system replaces its polynomial analog $(2q)/(q-1)$ on $\mathbb{F}_q[X]$.

THEOREM 5. *[Expectations]. When the set Ω_N is endowed with the uniform distribution, the following holds:*

(a) The expectation of the number of iterations L_1 during the first phase is linear with respect to the length $\log N$ and satisfies

$$\mathbb{E}_N[L_1] = \frac{6\log 2}{\pi^2}\left(\frac{1}{\ell}\log N\right)\left[1 + O\left(\frac{1}{\log N}\right)\right].$$

(b) For any $k \in [2..\ell-1]$, the expectation of the number of iterations L_k during the k-th phase is asymptotic to a constant a_k which is expressed with the operator $\mathbf{G}_s$ at $s = k+1$,

$$\mathbb{E}_N[L_k] = a_k\left[1 + O\left(\frac{1}{\log N}\right)\right].$$

(c) The expectation of the length of the first integer u_1 is linear with respect to the length $\log N$ and satisfies

$$\mathbb{E}_N[D_1] = \frac{1}{\ell}\log N.$$

(d) For any $k \in [2..\ell-1]$, the expectation of the length of the gcd y_k at the beginning of the k-th phase is asymptotic to a constant, and satisfies

$$\mathbb{E}_N[D_k] = \frac{\zeta'(k)}{\zeta(k)}\left[1 + O\left(\frac{1}{\log N}\right)\right].$$

5.5 Main principles for the analysis.

We have obtained in (11) the expressions of the cumulative generating functions. It is now possible to "extract" coefficients of these Dirichlet series, with an analog of (1).

However, for Dirichlet generating functions, it is (very often) only possible to study the sum of the coefficients for $n \leq N$, and this is why we deal with the set Ω_N of inputs with size at most N. Then the mean value of cost C on Ω_N is obtained from the cumulative generating function as

$$\mathbb{E}_N[C] = \frac{\sum_{n\leq N}[n^{-s}]\widehat{F}(s)}{\sum_{n\leq N}[n^{-s}]F(s)}.$$

As previously, singularity analysis performs a transfer between the behavior of a Dirichlet generating function, viewed as an analytic function, near its dominant singularity (here, the singularity with the largest real part) and the asymptotic behavior of its coefficients. The position and the nature of the dominant singularity play a fundamental role.

However, this transfer is more difficult for Dirichlet series. As previously, the basic tool is the Cauchy formula, but, here, the circles centered at 0 are replaced by vertical lines, which are not compact. For this short version, we use here the following Tauberian theorem, due to Delange [4], which deals with Dirichlet series with positive coefficients, but does not provide any remainder term. This is why we only give, in this short version, a sketch of the proof of a weak version of Theorem 5, without remainder terms.

THEOREM 6. *Let $F(s) = \sum_{n\geq 1} a_n n^{-s}$ be a Dirichlet series with non-negative coefficients such that $F(s)$ converges for $\Re(s) > \sigma > 0$. Assume the following:*

(i) $F(s)$ is analytic on $\Re(s) = \sigma$, $s \neq \sigma$,

(ii) for some $\gamma \geq 0$, one has

$$F(s) = A(s)(s-\sigma)^{-\gamma-1} + C(s),$$

where A, C are analytic at σ, with $A(\sigma) \neq 0$.
Then, as $N \to \infty$,

$$\sum_{n\leq N} a_n = \frac{A(\sigma)}{\sigma\Gamma(\gamma+1)} N^\sigma \log^\gamma N \quad [1 + \varepsilon(N)], \qquad \varepsilon(N) \to 0.$$

5.6 Sketch of proof for Theorem 5.

We apply Theorem 6 with $\sigma = 1$ to the cumulative functions defined in (11). We then study the series $\zeta(s)^\ell, \zeta'(s)$, $\widehat{T}(s,ks), T(s,ks)$. For the first two, enough is known for an easy application of Theorem 6 in case D (see [7]).

We focus now on case L, where we use precise results on the operator $\mathbf{G}_s$, when it acts on the Banach space $\mathcal{B} := \mathcal{C}^1([0,1])$. First, the function $\widehat{\zeta}_s = \mathbf{G}_s[1]$ has a pole of order 1 at $s=1$, and near $s=1$, one has $\widehat{\zeta}_s \sim C/(s-1)$, where C belongs to $\mathcal{B}$. Then, near $s=1$,

$$2^{\delta(k,1)}\, T(s,ks) \sim 1/(s-1)(I - \mathbf{G}_{(k+1)s})^{-1}[C](0),$$

for any k, ($\delta(i,j)$ is the Kronecker symbol).

Now, the following is known: for $\Re t > 2$, the operator $\mathbf{G}_t$ has a spectral radius strictly less than 1, and the quasi-inverse $(I - \mathbf{G}_t)^{-1}$ is analytic there. On the line $\Re t = 2$, the quasi-inverse $(I - \mathbf{G}_t)^{-1}$ is analytic except at $t=2$ where it admits a simple pole with a residue which involves the entropy. As $T(s,ks)$ (resp. $\widehat{T}(s,ks)$) contains one (resp. two) occurrence(s) of the quasi-inverse, the following holds:
- for $k \geq 2$, the series $T(s,ks)$ and $\widehat{T}(s,ks)$ have a simple pole at $s=1$,
- for $k=1$, the series $T(s,s)$ has a pole of order 2 at $s=1$, and the series $\widehat{T}(s,s)$ has a pole of order 3 at $s=1$.

Finally, the function $\widehat{L}_k(s)$ satisfies the hypotheses of Theorem 6, with $\sigma = 1$; the exponent γ equals ℓ for $k \geq 2$ and $\ell+1$ for $k=1$. This concludes the "proof" in case L.

5.7 What next ?

We can also study the number of useful phases, and recover the classical result $\mathbb{P}_N[D_k = 0] \sim 1/\zeta(k)$. If we wish to obtain remainder terms in Theorem 5, and an analog of distributional results obtained in Theorem 2, we need the analog of Proposition 3, and have to deal with the Perron Formula, as for previous distributional analyses.

The Perron Formula provides remainder terms as soon as the Dirichlet series of interest possess a vertical strip on the left of the vertical line $\Re s = 1$, where $s=1$ is their only pole and they are of polynomial growth for $|\Im s| \to \infty$. Classical results [7] entail such properties for the zeta function and its derivatives, and results due to Dolgopyat-Baladi-Vallée [6, 1] prove that they also hold for the transfer operator $\mathbf{G}_{2s}$. Then, it would be possible to obtain the analog of Theorem 2 where n, m are replaced by $\log N, \log M$ and the ratios p_k, r_k of Theorem 2 (a) and (b) are replaced by

$$\widehat{p}_k := \lambda\left((k+1)/2\right) \ (\text{case } L), \quad \widehat{r}_k := \exp(1-k) \ (\text{case } D).$$

Here, $\lambda(s)$ is the dominant eigenvalue of the operator $\mathbf{G}_{2s}$ (when acting on the functional space $\mathcal{B}$), which plays a central role in many analyses of Euclidean type. It satisfies $\lambda(s) = 1$, and, for $k=3$, we recover the constant $\lambda(2)$ which plays a central role in Euclidean dynamics, notably in the analysis of the Gauss lattice reduction algorithm [3, 9].

5.8 Some experiments

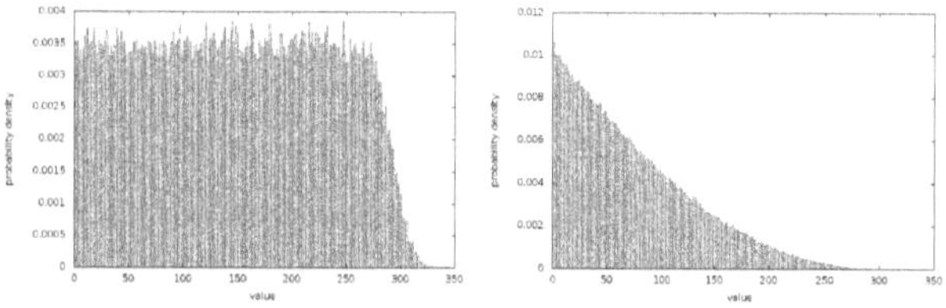

The figure shows the experimental densities of L_1 for $\ell = 2$ (left) and $\ell = 4$ (right) and exhibits the uniform limit law (left) and the beta limit law (right). These experiments were obtained with $5 \cdot 10^5$ executions on random integer inputs $(x_1, \ldots, x_\ell)$ whose (total) length satisfied $\log_2 N = 10^4$.

Acknowledgements. Thanks to the two ANR Projects:
- ANR BOOLE (ANR 2009 BLAN 0011)
- ANR MAGNUM (ANR 2010 BLAN 0204)

6. REFERENCES

[1] V. Baladi and B. Vallée. Euclidean algorithms are Gaussian. *Journal of Number Theory*, 110:331–386, 2006.

[2] V. Berthé and H. Nakada. On continued fraction expansions in positive characteristic: Equivalence relations and some metric properties. *Expositiones Mathematicae*, 18:257–284, 2000.

[3] H. Daudé, P. Flajolet, and B. Vallée. An average-case analysis of the Gaussian Algorithm for Lattice Reduction. *Combinatorics, Probability & Computing*, 6(4):397–433, 1997.

[4] H. Delange. Généralisation du théorème d'Ikehara. *Ann. Sc. ENS*, 71:213–422, 1954.

[5] J. D. Dixon. The number of steps in the Euclidean algorithm. *Journal of Number Theory*, 2:414–422, 1970.

[6] D. Dolgopyat. On decay of correlations in Anosov flows. *Ann. of Math.*, 147(2):357–390, 1998.

[7] H. Edwards. *Riemann's Zeta Function.* Dover Publications, 2001.

[8] P. Flajolet and R. Sedgewick. *Analytic Combinatorics.* Cambridge University Press, 2009.

[9] P. Flajolet and B. Vallée. Continued fraction algorithms, functional operators, and structure constants. *Theor. Comput. Sci.*, 194(1-2):1–34, 1998.

[10] C. Friesen and D. Hensley. The statistics of continued fractions for polynomials over a finite field. *Proc. Amer. Math. Soc*, 124:2661–2673, 1996.

[11] H. Heilbronn. On the average length of a class of continued fractions. In P. Turan, editor, *Number Theory and Analysis*, pages 87–96, 1969.

[12] D. Hensley. The number of steps in the Euclidean algorithm. *Journal of Number Theory*, 2(49):149–182, 1994.

[13] A. Knopfmacher and J. Knopfmacher. The exact length of the Euclidean algorithm in $F_q[X]$. *Mathematika*, 35:297–304, 1988.

[14] D. E. Knuth. *Seminumerical Algorithms*, volume 2 of *The Art of Computer Programming.* Addison-Wesley, 3rd edition, 1998.

[15] L. Lhote and B. Vallée. Gaussian laws for the main parameters of the Euclid algorithms. *Algorithmica*, 50(4):497–554, 2008.

[16] K. Ma and J. von zur Gathen. Analysis of Euclidean algorithms for polynomials over finite fields. *Journal of Symbolic Computation*, 9(4):429 – 455, 1990.

[17] D. Ruelle. *Thermodynamic Formalism.* Cambridge University Press, 2004.

[18] B. Vallée. Euclidean dynamics. *Discrete and Continuous Dynamical Systems*, 1(15):281–352, 2006.

[19] J. von zur Gathen and J. Gerhard. *Modern Computer Algebra.* Cambridge University Press, 2003.

[20] J. von zur Gathen, M. Mignotte, and I. Shparlinski. Approximate polynomial gcd: Small degree and small height perturbations. *Journal of Symbolic Computation*, 45(8):879–886, 2010.

Integrability Conditions for Parameterized Linear Difference Equations*

Mariya Bessonov
Cornell University
Department of Mathematics
310 Malott Hall
Ithaca, NY 14853, USA
myb@math.cornell.edu

Alexey Ovchinnikov
CUNY Queens College
Department of Mathematics
65-30 Kissena Blvd
Queens, NY 11367, USA
aovchinnikov@qc.cuny.edu

Maxwell Shapiro
CUNY Graduate Center
Department of Mathematics
365 Fifth Avenue
New York, NY 10016, USA
mgshapiro100@yahoo.com

ABSTRACT

We study integrability conditions for systems of parameterized linear difference equations and related properties of linear differential algebraic groups. We show that isomonodromicity of such a system is equivalent to isomonodromicity with respect to each parameter separately under a linearly differentially closed assumption on the field of differential parameters. Due to our result, it is no longer necessary to solve non-linear differential equations to verify isomonodromicity, which will improve efficiency of computation with these systems. Moreover, it is not possible to further strengthen this result by removing the requirement on the parameters, as we show by giving a counterexample. We also discuss the relation between isomonodromicity and the properties of the associated parameterized difference Galois group.

Categories and Subject Descriptors

I.1.2 [**Computing methodologies**]: Symbolic and algebraic manipulation—*Algebraic algorithms*

General Terms

Algorithms, Theory

Keywords

Differential algebra, difference algebra, integrability conditions, difference Galois theory, differential algebraic groups

1. INTRODUCTION

In this paper, we improve the algorithm that verifies if a system of linear difference equations with differential parameters is isomonodromic (Definition 1). Given a system of difference equations, it is natural to ask if its solutions satisfy extra differential equations. For example, it is a famous result of Hölder [19] that the gamma function Γ satisfying the linear difference equation

$$y(x+1) = xy(x)$$

satisfies no polynomial differential equations with rational coefficients. This was also recently shown in [13] using parameterized difference Galois theory (see also [12, 15, 17, 16, 18, 14, 38, 11, 31, 36]). To explain the results of our paper, consider a decision procedure whose input consists of a field $\mathbf{K}$ with commuting automorphisms $\phi_1, \ldots, \phi_q$ and derivations $\partial_1, \ldots, \partial_m$ and a system of difference equations

$$\phi_1(Y) = A_1 Y, \ldots, \phi_q(Y) = A_q Y, \tag{1}$$

where $A_i \in \mathbf{GL}_n(\mathbf{K})$. Its output consists of such additional linear differential equations

$$\partial_1 Y = B_1 Y, \ldots, \partial_m Y = B_m Y, \tag{2}$$

where $B_i \in \mathbf{M}_n(\mathbf{K})$, if they exist, for which there exists an invertible matrix solution of (1) that satisfies (2) as well. For solutions of (1) to possibly exist, the A_i's must satisfy the integrability conditions (4). Moreover, existence of the B_i's above is equivalent to the existence of B_i's satisfying another large collection of integrability conditions (6) and (7). Such systems (1) are called isomonodromic in analogy with differential equations [33, 34]. Isomonodromy problems for q-difference equations and their relations with the q-difference Painlevé equations were studied in [22, 23].

In our main result, Theorem 1, we show that (7), which are non-linear differential equations, do not have to be verified to check the existence of the B_i's. More precisely, we prove that the existence of B_i's satisfying (6) implies the existence of new matrices that satisfy both (6) and (7).

Since, due to our result, we only need to check the existence of solutions (that have entries in the ground field $\mathbf{K}$) of a system of linear difference equations, a complexity estimate for verifying whether a system of difference equations is isomonodromic becomes possible. For this, a full complexity analysis for finding a rational solution of an inhomogeneous linear difference equation is sufficient, which we expect to appear in the near future. For linear differential equations there are already such results [2].

Our main result, Theorem 1, has a restriction on the field of constants. Namely, we require that this field be linearly differentially closed with respect to all but, possibly, one of the derivations (Definition 2). However, if this restriction

*This work has been partially supported by the NSF grants CCF-0952591 and DMS-0739164

ISSAC'13, June 26–29, 2013, Boston, Massachusetts, USA.

were not imposed, then the conclusion of Theorem 1 would not hold, as our Example 3 shows. Moreover, we also conjecture that this restriction is necessary and leave the discovery of a proof for future research.

A similar problem but for systems of differential equations was considered in [10], motivated by the classical results [20, 21]. Differential categories developed in [9] formed the main technical tool in [10]. On the contrary, our proofs are written in elementary terms, which makes them more accessible. Moreover, from our proof, one can produce an algorithm that, given a common solution of (6), computes a common solution of both (6) and (7).

Given a system of linear difference equations with parameters, parameterized difference Galois theory [13] associates a linear differential algebraic group [3, 4, 32, 6, 35, 27, 28], which is called the parameterized difference Galois group and is a group of matrices whose entries satisfy a system of polynomial differential equations. In addition to our main result, we also show in Proposition 2 how isomonodromicity can be characterized using the Galois group. This extends the corresponding result of [13], as our version does not require the constants to be differentially closed.

Our main result has further applications. In parameterized differential Galois theory [5], there are several algorithms for computing the Galois groups. For 2×2 systems, they are given in [1, 7]. An algorithm, more general in terms of the order of the system, [30, 29] uses the differential analogues [10] of our results to make the computation more efficient. Our results may prove useful in the design of an algorithm for computing parameterized difference Galois groups, as it has happened in the differential case.

Also note that the algorithm in [30] performs simultaneous prolongations with respect to all derivations, which gives a linear growth of algebraic indeterminates to be dealt with. On the other hand, if one performs prolongations with respect to each derivation separately, there is no such growth. The results of our paper, as well as [10], describe a situation in which one can deal with each derivation separately. We hope that more such situations will be discovered in the future to speed up the algorithms that compute parameterized differential and difference Galois groups.

The paper is organized as follows. We introduce our notation and review the basic notions of differential and difference algebra in §2. The integrability conditions are introduced in §3.1. Our main result is described in §3.2. In §3.3, we demonstrate how isomonodromicity is reflected in the associated parameterized difference Galois group. An example showing that assumption 1 cannot be removed from our main result, Theorem 1, is given in §3.4.

2. BASIC DEFINITIONS

We will start with the basic definitions and notation of differential and difference algebra. A Δ-ring is a commutative associative ring with unit 1 together with a set $\Delta=\{\partial_1,\ldots,\partial_m\}$ of commuting derivations $\partial_i\colon R\to R$ such that

$$\partial_i(a+b)=\partial_i(a)+\partial_i(b),\quad \partial_i(ab)=\partial_i(a)b+a\partial_i(b)$$

for all $a,b\in R$. For example, $\mathbb{Q}$ is a $\{\partial\}$-field with the unique zero derivation. For every $f\in\mathbb{C}(x)$, there exists a unique derivation $\partial:\mathbb{C}(x)\to\mathbb{C}(x)$ with $\partial(x)=f$, turning $\mathbb{C}(x)$ into a $\{\partial\}$-field. Let

$$\Theta=\{\partial_1^{i_1}\cdot\ldots\cdot\partial_m^{i_m}\mid i_j\geqslant 0\}.$$

Since ∂_i acts on R, there is a natural action of Θ on R. Let R and B be Δ-rings. If $B\supset R$, then B is a Δ-R-algebra if, for all i, $1\leqslant i\leqslant m$, the action of ∂_i on B extends the action of ∂_i on R. Let $Y=\{y_1,\ldots,y_n\}$ be a set of variables and

$$\Theta Y:=\{\theta y_j \mid \theta\in\Theta,\ 1\leqslant j\leqslant n\}.$$

The ring of differential polynomials $R\{Y\}_\Delta$ in differential indeterminates Y over R is $R[\Theta Y]$ with the derivations ∂_i that extends the ∂_i-action on R as follows:

$$\partial_i(\theta y_j):=(\partial_i\cdot\theta)y_j,\quad 1\leqslant j\leqslant n.$$

An ideal I in a Δ-ring R is called a differential ideal if $\partial_i(a)\in I$ for all $a\in I$, $1\leqslant i\leqslant m$.

Let $\mathbf{k}$ be a Δ-field of characteristic zero. In what follows, $\mathbf{M}_n(\mathbf{k})$ denotes the set of $n\times n$ matrices with entries in $\mathbf{k}$ and $\mathbf{GL}_n(\mathbf{k})$ are the invertible matrices in $\mathbf{M}_n(\mathbf{k})$.

A Φ-ring R is a commutative associative ring with unit 1 and a set $\Phi=\{\phi_1,\ldots,\phi_q\}$ of commuting automorphisms $\phi_i:R\to R$. A $\{\Phi,\Delta\}$-ring R is a Φ-ring and a Δ-ring such that, for all $\phi\in\Phi$ and $\partial\in\Delta$, $\phi\partial=\partial\phi$.

Example 1. We will list a few examples of $\{\Phi,\Delta\}$-rings:

1. $R=\mathbb{Q}(x_1,\ldots,x_m)$ with $\Phi=\{\phi_1,\ldots,\phi_m\}$ and $\Delta=\{\partial_1,\ldots,\partial_m\}$ defined by

$$\phi_i(f)(x_1,\ldots,x_m)=f(x_1,\ldots,x_i+1,\ldots,x_m),\ f\in R,$$
$$\partial_i=\partial/\partial x_i,\quad 1\leqslant i\leqslant m.$$

2. $R=\mathbb{Q}(x,y)$ with $\Phi=\{\phi\}$ and $\Delta=\{\partial_1,\partial_2\}$ defined by

$$\phi(f)(x,y)=f(xy,y),\ \partial_1=x\partial/\partial_x,\ \partial_2=\partial/\partial_y.$$

3. MAIN RESULT

We will start by introducing integrability conditions in §3.1 and then show our main result in §3.2. Applications of the parameterized difference Picard–Vessiot theory to this will be discussed in §3.3. §3.4 contains an example showing that the conclusion of Theorem 1 does not hold if one drops the linearly differentially closed assumption 1.

3.1 Integrability conditions

Let $\mathbf{K}$ be a $\{\Phi,\Delta\}$-field of characteristic zero and

$$A_1,\ldots,A_q\in\mathbf{GL}_n(\mathbf{K}).$$

Consider the system of difference equations

$$\phi_1(Y)=A_1Y,\ldots,\phi_q(Y)=A_qY. \tag{3}$$

If $L\supset\mathbf{K}$ is a Φ-field extension and $Z\in\mathbf{GL}_n(L)$ satisfies (3) then, for all i, j, $1\leqslant i,j\leqslant q$, we have

$$\phi_j(A_i)A_jZ=\phi_j(\phi_i(Z))=\phi_i(\phi_j(Z))=\phi_i(A_j)A_iZ.$$

Therefore, we obtain

$$\phi_j(A_i)A_j=\phi_i(A_j)A_i \tag{4}$$

Moreover, if, in addition, L is a $\{\Phi,\Delta\}$-field extension of $\mathbf{K}$ and there exist $B_1,\ldots,B_m\in\mathbf{M}_n(\mathbf{K})$ such that

$$\partial_1(Z)=B_1Z,\ldots,\partial_m(Z)=B_mZ, \tag{5}$$

then

$$\phi_j(B_i)A_jZ = \phi_j(\partial_i(Z)) = \\ = \partial_i(\phi_j(Z)) = \partial_i(A_j)Z + A_jB_iZ$$

and

$$\partial_j(B_i)Z + B_iB_jZ = \partial_j(\partial_i(Z)) = \\ = \partial_i(\partial_j(Z)) = \partial_i(B_j)Z + B_jB_iZ,$$

which imply that

$$\phi_j(B_i) = \partial_i(A_j)A_j^{-1} + A_jB_iA_j^{-1}, 1 \leqslant i \leqslant m, 1 \leqslant j \leqslant q, \quad (6)$$

and

$$\partial_j(B_i) - \partial_i(B_j) = [B_j, B_i], \quad 1 \leqslant i, j \leqslant m. \quad (7)$$

Therefore, (4), (6), and (7) are necessary conditions for the existence of a common invertible matrix solution of (3) and (5) with entries in a field L (see Definition 6.6(d) in [13]). They are also sufficient. Indeed, let

$$L = \mathbf{K}(x_{11}, \ldots, x_{nn}),$$

$$\partial_i((x_{rs})) = B_i(x_{rs}), \quad 1 \leqslant i \leqslant m,$$

and

$$\phi_j((x_{rs})) = A_j(x_{rs}), \quad 1 \leqslant j \leqslant q.$$

Then (4), (6), and (7) imply that L is a $\{\Phi, \Delta\}$-field.

Definition 1. The system of linear difference equations

$$\phi_1(Y) = A_1Y, \ldots, \phi_q(Y) = A_qY$$

with $A_1, \ldots, A_q \in \mathbf{GL}_n(\mathbf{K})$ satisfying (4) is called *isomonodromic* if there exist $B_1, \ldots, B_m \in \mathbf{M}_n(\mathbf{K})$ satisfying (6) and (7).

Connections to analytic interpretations of isomonodromy can be found, e.g., in §5 of [5] and §6.2 of [10]. We will need the following definition in §3.2 (which is a weaker restriction than differentially closed) to state the main result.

Definition 2. For $\Delta' = \{\delta_1, \ldots, \delta_s\} \subset \Delta$, a Δ-field $\mathbf{k}$ is called *linearly Δ'-closed* if, for all $n \geqslant 1$ and $B_1, \ldots, B_s \in \mathbf{M}_n(\mathbf{k})$ such that

$$\delta_i(B_j) - \delta_j(B_i) = [B_i, B_j], \quad 1 \leqslant i, j \leqslant s,$$

there exists $Z \in \mathbf{GL}_n(\mathbf{k})$ such that

$$\delta_1(Z) = B_1Z, \ldots, \delta_s(Z) = B_sZ.$$

3.2 Statement and proof of the main result

In this section, we will state and prove our main result, Theorem 1, which allows us to reduce the number of integrability conditions to be tested. For this, we need to impose a restriction on the fields we consider, which we will describe now. Note that, if this restriction were not added, the conclusion of Theorem 1 would no longer be true (see §3.4).

Assumption 1. Let $\mathbf{K}$ be a $\{\Phi, \Delta\}$-field such that, after some renumbering of $\partial_1, \ldots, \partial_m$, for all k, $1 \leqslant k \leqslant m-1$,

$$\mathbf{K}^\Phi := \{a \in \mathbf{K} \mid \phi(a) = a, \ \phi \in \Phi\}$$

is linearly $\{\partial_1, \ldots, \partial_k\}$-closed.

Example 2. Let $\mathcal{U}$ be a linearly closed ∂_1-field and $\mathbf{K} := \mathcal{U}(x)$, the field of rational functions in x, with $\Phi = \{\phi\}$ and $\Delta = \{\partial/\partial x, \partial_1\}$, where $\phi(f)(x) = f(x+1)$, $f \in \mathbf{K}$, and $\partial_1(x) = 0$. Then $\mathbf{K}^\Phi = \mathcal{U}$ with $\{\partial/\partial x, \partial_1\}$, and $\mathbf{K}$ satisfies assumption 1 with the renumbering $\{\partial_1, \partial/\partial x\}$.

Theorem 1. *Let $\mathbf{K}$ satisfy assumption 1. Then, for all $A_1, \ldots, A_q \in \mathbf{GL}_n(\mathbf{K})$ satisfying (4), if there exist $B_1, \ldots, B_m \in \mathbf{M}_n(\mathbf{K})$ satisfying (6), then there exist $D_1, \ldots, D_m \in \mathbf{M}_n(\mathbf{K})$ such that all integrability conditions are satisfied, that is,*

$$\phi_j(D_i) = \partial_i(A_j)A_j^{-1} + A_jD_iA_j^{-1}, \ 1 \leqslant i \leqslant m, 1 \leqslant j \leqslant q, \\ \partial_i(D_j) - \partial_j(D_i) = [D_i, D_j], \quad 1 \leqslant i, j \leqslant m. \quad (8)$$

Proof. This will be done by induction. Suppose that a renumbering from assumption 1 has been performed. Let there exist $B_1, \ldots, B_m$ such that (6) is satisfied and let $k \leqslant m$. Suppose that there exist

$$D_1, \ldots, D_{k-1}$$

that satisfy both (6) and (7) ($k-1$ is substituted for m and the D_i's are substituted for the B_i's). We claim that there exists $E \in \mathbf{M}_n(\mathbf{K})$ such that

$$(D_1, \ldots, D_{k-1}, B_k + E)$$

satisfies both (6) and (7) (under the substitution as above), so that we can take

$$(D_1, \ldots, D_k) := (D_1, \ldots, D_{k-1}, B_k + E)$$

to satisfy (8) by induction. In order to show that $B_k + E$ satisfies (6) and (7), we need to show that

$$\phi_j(B_k + E) = \partial_k(A_j)A_j^{-1} + A_j(B_k + E)A_j^{-1} \quad (9)$$

and

$$\partial_i(B_k + E) - \partial_k(D_i) = [D_i, B_k + E] \quad (10)$$

for all i and j, $1 \leqslant j \leqslant q$, $1 \leqslant i < k$. However, expanding the left-hand side of (9) using (6), we see that

$$\phi_j(B_k + E) = \phi_j(B_k) + \phi_j(E) = \\ = \partial_k(A_j)A_j^{-1} + A_jB_kA_j^{-1} + \phi_j(E).$$

Moreover, rearranging the terms in (10), we see that we need to find $E \in \mathbf{M}_n(\mathbf{K})$ such that the following two conditions are satisfied:

$$\phi_j(E) = A_jEA_j^{-1} \quad (11)$$

and

$$\partial_i(E) + [E, D_i] = \partial_k(D_i) - \partial_i(B_k) + [D_i, B_k] \quad (12)$$

for all i and j, $1 \leqslant j \leqslant q$, $1 \leqslant i < k$. We now show that

$$\partial_k(D_i) - \partial_i(B_k) + [D_i, B_k]$$

satisfies (11) as an equation in E. We have:

$$\begin{aligned} &\phi_j(\partial_k(D_i) - \partial_i(B_k) + [D_i, B_k]) = \\ &= \phi_j(\partial_k(D_i)) - \phi_j(\partial_i(B_k)) + \phi_j(D_iB_k) - \phi_j(B_kD_i) = \\ &= \partial_k(\phi_j(D_i)) - \partial_i(\phi_j(B_k)) + \\ &\quad + \phi_j(D_i)\phi_j(B_k) - \phi_j(B_k)\phi_j(D_i) = \\ &= \partial_k\left(\partial_i(A_j)A_j^{-1}\right) + \partial_k\left(A_jD_iA_j^{-1}\right) - \\ &\quad - \partial_i\left(\partial_k(A_j)A_j^{-1}\right) - \partial_i\left(A_jB_kA_j^{-1}\right) + \\ &\quad + \left(\partial_i(A_j)A_j^{-1} + A_jD_iA_j^{-1}\right)\left(\partial_k(A_j)A_j^{-1} + A_jB_kA_j^{-1}\right) - \\ &\quad - \left(\partial_k(A_j)A_j^{-1} + A_jB_kA_j^{-1}\right)\left(\partial_i(A_j)A_j^{-1} + A_jD_iA_j^{-1}\right). \end{aligned}$$

Using the relation

$$\partial_d(A_j^{-1}) = -A_j^{-1}\partial_d(A_j)A_j^{-1},$$

and after we expand and cancel out terms, we are left with

$$\begin{aligned} &A_j\partial_k(D_i)A_j^{-1} - A_j\partial_i(B_k)A_j^{-1} + \\ &\quad + A_jD_iB_kA_j^{-1} - A_jB_kD_iA_j^{-1} = \\ &= A_j(\partial_k(D_i) - \partial_i(B_k) + (D_iB_k - B_kD_i))A_j^{-1} \end{aligned}$$

as desired.

For any $Z \in \mathbf{M}_n(\mathbf{K})$ satisfying (11), we now show that

$$\partial_i(Z) + [Z, D_i]$$

also satisfies (11) for $1 \leqslant j \leqslant q$, $1 \leqslant i < k$. Indeed,

$$\begin{aligned} &\phi_j(\partial_i(Z) + [Z, D_i]) = \\ &= \partial_i(\phi_j(Z)) + \phi_j(Z)\phi_j(D_i) - \phi_j(D_i)\phi_j(Z) = \\ &= \partial_i(A_jZA_j^{-1}) + (A_jZA_j^{-1})(\partial_i(A_j)A_j^{-1} + A_jD_iA_j^{-1}) - \\ &\quad - (\partial_i(A_j)A_j^{-1} + A_jD_iA_j^{-1})(A_jZA_j^{-1}) = \\ &= \partial_i(A_j)ZA_j^{-1} + A_j\partial_i(Z)A_j^{-1} + A_jZ\partial_i(A_j^{-1}) + \\ &\quad + A_jZA_j^{-1}\partial_i(A_j)A_j^{-1} + A_jZD_iA_j^{-1} - \\ &\quad - \partial_i(A_j)ZA_j^{-1} - A_jD_iZA_j^{-1} = \\ &= A_j\partial_i(Z)A_j^{-1} + A_jZD_iA_j^{-1} - A_jD_iZA_j^{-1} = \\ &= A(\partial_i(Z) + [Z, D_i])A_j^{-1}. \end{aligned}$$

Since the set S of solutions of (11) inside $\mathbf{M}_n(\mathbf{K})$ is a finite-dimensional vector space over $\mathbf{K}^\Phi$, say, $\dim_{\mathbf{K}^\Phi} S = p$, there exists a $\mathbf{K}^\Phi$-basis of S and $C_{i,s} \in \mathbf{M}_p(\mathbf{K}^\Phi)$ for $s = 1, 2$, such that C_{i2} represents

$$\partial_k(D_i) - \partial_i(B_k) + [D_i, B_k]$$

in this basis and C_{i1} is such that the differential operator defined by

$$\partial_i(Z) + [Z, D_i],$$

when restricted to the matrices from S, has the form

$$\partial_i(Y) + C_{i1}Y = C_{i2} \tag{13}$$

for all i, $1 \leqslant i < k$. We will now show that these equations are consistent, that is, they have a common solution with entries in some Δ-field extension of $\mathbf{K}^\Phi$. For this, it is sufficient to show that (12) are consistent. By Proposition IV.6.3 of [25], to prove this, it is sufficient to show that (12) and the inductive hypothesis for (8) imply

$$\partial_i(\partial_j(E)) = \partial_j(\partial_i(E)), \quad 1 \leqslant i, j < k. \tag{14}$$

To do this, first note that (12) implies that

$$\partial_j\partial_i(E) = \partial_j\partial_k(D_i) - \partial_j\partial_i(B_k) + \partial_j[D_i, B_k + E]$$

and

$$\partial_i\partial_j(E) = \partial_i\partial_k(D_j) - \partial_i\partial_j(B_k) + \partial_i[D_j, B_k + E]$$

Since $\partial_j\partial_i(B_k) = \partial_i\partial_j(B_k)$ and $\partial_j\partial_k(D_i) = \partial_k\partial_j(D_i)$ and, by the inductive hypothesis for (8),

$$\begin{aligned} &\partial_j\partial_i(E) - \partial_i\partial_j(E) = \partial_k(\partial_j(D_i) - \partial_i(D_j)) + \\ &\quad + \partial_j([D_i, B_k + E]) - \partial_i([D_j, B_k + E]) = \\ &= \partial_k([D_j, D_i]) + \partial_j([D_i, B_k + E]) - \partial_i([D_j, B_k + E]) = \\ &= \partial_k([D_j, D_i]) + [[D_j, D_i], B_k + E] + \\ &\quad + [D_i, \partial_j(B_k + E)] - [D_j, \partial_i(B_k + E)]. \end{aligned}$$

Now, substituting the expression for $\partial_i(E)$ from (12), and similarly for $\partial_j(E)$,

$$\begin{aligned} &\partial_j\partial_i(E) - \partial_i\partial_j(E) = \partial_k([D_j, D_i]) + [[D_j, D_i], B_k + E] + \\ &\quad + [D_i, \partial_j(B_k)] - [D_j, \partial_i(B_k)] + \\ &\quad + [D_i, \partial_k(D_j) - \partial_j(B_k) + [D_j, B_k + E]] - \\ &\quad - [D_j, \partial_k(D_i) - \partial_i(B_k) + [D_i, B_k + E]]. \end{aligned}$$

Rearranging terms,

$$\begin{aligned} &\partial_j\partial_i(E) - \partial_i\partial_j(E) = \partial_k([D_j, D_i]) + [D_i, \partial_k(D_j)] - \\ &\quad - [D_j, \partial_k(D_i)] + [[D_j, D_i], B_k + E] + \\ &\quad + [D_i, [D_j, B_k + E]] - [D_j, [D_i, B_k + E]] + \\ &\quad + [D_i, \partial_j(B_k)] - [D_j, \partial_i(B_k)] + \\ &\quad + [D_j, \partial_i(B_k)] - [D_i, \partial_j(B_k)] = \\ &= 0 + [[D_j, D_i], B_k + E] + [D_i, [D_j, B_k + E]] - \\ &\quad - [D_j, [D_i, B_k + E]] + 0 = 0 \end{aligned}$$

by the Jacobi identity. Therefore, (14) holds and, thus, (13) is consistent.

Since $\mathbf{K}^\Phi$ is linearly $\{\partial_1, \ldots, \partial_{k-1}\}$-closed, there exists a solution $F \in \mathbf{M}_p(\mathbf{K}^\Phi)$ to (13), which implies the existence of $E \in \mathbf{M}_n(\mathbf{K})$ satisfying (11) and (12). Thus,

$$(D_1, \ldots, B_k + E)$$

satisfies both (6) and (7). □

3.3 Using difference Picard–Vessiot theory

In this section, we will see in Proposition 2 how isomonodromicity can be detected using the methods of parameterized difference Picard–Vessiot (PPV) theory [13] in light of Theorem 1.

Recall that a Δ-field $\mathcal{U}$ is called differentially closed if any system of polynomial differential equations with coefficients in $\mathcal{U}$ having a solution with entries in some Δ-extension of $\mathcal{U}$ already has a solution with entries in $\mathcal{U}$ (for various equivalent versions of this definition, see Definition 3.2 in [5], Definition 4 in [37], and the references given there). Recall also that every Δ-field $\mathbf{k}$ is contained in some differentially closed Δ-field $\mathcal{U}$.

Definition 3. A *Kolchin-closed* subset W of $\mathcal{U}^n$ defined over $\mathbf{k}$ is the set of common zeroes of a system of differential algebraic equations with coefficients in $\mathbf{k}$, that is, there exist $f_1, \ldots, f_r \in \mathbf{k}\{Y\}$ such that

$$W = \{a \in \mathcal{U}^n \mid f_1(a) = \ldots = f_r(a) = 0\}.$$

Its coordinate ring is defined to be

$$\mathbf{k}\{W\} := \mathbf{k}\{Y\}/\{f_1, \ldots, f_r\},$$

where $\{f_1, \ldots, f_r\}$ denotes the radical differential ideal generated by $f_1, \ldots, f_r$.

Definition 4. (Ch. II, §1, p. 905 of [3]) A *linear differential algebraic group* (LDAG) defined over $\mathbf{k}$ is a Kolchin-closed subgroup G of $\mathbf{GL}_n(\mathcal{U})$, over $\mathbf{k}$ that is, an intersection of a Kolchin-closed subset of $\mathcal{U}^{n^2}$ over $\mathbf{k}$ with $\mathbf{GL}_n(\mathcal{U})$, which is closed under the group operations.

We will now briefly recall difference PPV theory. Let $\mathbf{K}$ be a $\{\Phi, \Delta\}$-field and $A_1, \ldots, A_q \in \mathbf{M}_n(\mathbf{K})$ be given. Contrary to [13], we do not assume $\mathbf{K}^\Phi$ to be differentially closed. A *parameterized Picard-Vessiot ring (PPV-ring)* R of $\mathbf{K}$ associated with (3) is a $\{\Phi,\Delta\}$-$\mathbf{K}$-algebra R with no $\{\Phi,\Delta\}$-ideals such that there exists a $Z \in \mathbf{GL}_n(R)$ satisfying (3), $(\mathrm{Quot}(R))^{\Phi} = \mathbf{K}^{\Phi}$, and R is Δ-generated over $\mathbf{K}$ by the entries of Z and $1/\det Z$ (that is, $R = \mathbf{K}\{Z, 1/\det Z\}$).

$\mathbf{K}^{\Phi}$ is a Δ-field and, if it is differentially closed, a PPV-ring associated with (3) always exists and is unique up to Δ-$\mathbf{K}$-isomorphism (Propositions 6.14 and 6.16 of [13]). If $R = \mathbf{K}\{Z, 1/\det Z\}$ is a PPV-ring of $\mathbf{K}$, one defines the *parameterized Picard–Vessiot group (PPV-group)*, denoted by $\mathbf{Gal}(R/\mathbf{K})$, of R over $\mathbf{K}$ to be

$$\begin{aligned} G := \{\sigma : R \to R \mid &\ \sigma \text{ is an automorphism},\\ &\sigma\delta = \delta\sigma \text{ for all } \delta \in \{\Phi,\Delta\}, \text{ and}\\ &\sigma(a) = a,\ a \in \mathbf{K}\}. \end{aligned}$$

For any $g \in G$, one can show that there exists a matrix $C_g \in \mathbf{GL}_n(\mathbf{K}^{\Phi})$ such that

$$g(Z) = ZC_g,$$

and the map $\sigma \mapsto C_g$ is an isomorphism of G onto a differential algebraic subgroup of $\mathbf{GL}_n(\mathbf{K}^{\Phi})$. One can also give a functorial definition of $\mathbf{Gal}$ (as on pages 368–369 of [13]) turning $\mathbf{Gal}$ into a functor from $\mathbf{K}^{\Phi}$-Δ-algebras to groups: $A \mapsto \mathbf{Gal}(A)$ (each $g \in \mathbf{Gal}(A)$ becomes $\{\Phi,\Delta\}$-automorphism $R \otimes_{\mathbf{K}^{\Phi}} A \to R \otimes_{\mathbf{K}^{\Phi}} A$), that is, the A-points of $\mathbf{Gal}$ (see also page 420 of [32]). We will need the functorial definition of $\mathbf{Gal}$ in the proof of Proposition 2.

Proposition 2.9 of [13] gives a criterion for isomonodromicity via PPV theory, but requires $\mathbf{K}^{\Phi}$ to be differentially closed. This is an obstacle for potential applications, and our version of this result, Proposition 3, avoids it. Moreover, several recent results [9, 39] on differential PPV theory, in which the constants are not required to be differentially closed (weaker assumptions are made there), indicate that their difference analogues could be constructed using [24]. This encourages us to extend the result of [13] by not requiring that $\mathbf{K}^{\Phi}$ be differentially closed.

PROPOSITION 2. *Let the system of difference equations (3) be such that difference integrability conditions (4) are satisfied. Assume also that there exists a PPV-ring of $\mathbf{K}$ for (3) and G is its PPV Galois group.*

Then (3) is isomonodromic if and only if, for all i, $1 \leqslant i \leqslant m$, there exists $D_i \in \mathbf{M}_n(\mathbf{K}^{\Phi})$ satisfying

$$\partial_u(D_v) - \partial_v(D_u) = [D_u, D_v], \quad 1 \leqslant u, v \leqslant m \tag{15}$$

such that, for all r and s, $1 \leqslant r, s \leqslant n$, the following equation is in the defining ideal of G:

$$\partial_i(x_{rs}) + [(x_{rs}), D_i] = 0. \tag{16}$$

Moreover, if (16) is in the defining ideal of G, then there exists a finitely generated Δ-field extension F of $\mathbf{K}^{\Phi}$ and $C_i \in \mathbf{GL}_n(F)$ such that, for all r and s, $1 \leqslant r, s \leqslant n$,

$$\partial_i(C_i^{-1}(x_{rs})C_i) = 0 \tag{17}$$

is in the defining ideal of G, that is, G is conjugate over F to a group of matrices with ∂_i-constant entries.

PROOF. Let R be a PPV-ring of $\mathbf{K}$ for (3) and $Z \in \mathbf{GL}_n(R)$ be a fundamental solution matrix.

Let there exist $B_i \in \mathbf{GL}_n(\mathbf{K})$ such that (6) and (7) are satisfied. For all j, $1 \leqslant j \leqslant q$, we have

$$\begin{aligned} &\phi_j(\partial_i(Z) - B_iZ) = \partial_i(\phi_j(Z)) - \phi_j(B_i)\phi_j(Z) =\\ &= \partial_i(AZ) - (\partial_i(A)A^{-1} + AB_iA^{-1})AZ = A(\partial_i(Z) - B_iZ). \end{aligned}$$

Therefore, since $R^{\Phi} = \mathbf{K}^{\Phi}$, there exists $D_i \in \mathbf{M}_n(\mathbf{K}^{\Phi})$ such that

$$\partial_i(Z) = B_iZ - ZD_i.$$

Therefore,

$$\begin{aligned} \partial_j(\partial_i(Z)) &= \partial_j(B_i)Z + B_i(B_jZ - ZD_j) +\\ &\quad + (B_jZ - ZD_j)D_i + Z\partial_j(D_i), \qquad (18)\\ \partial_i(\partial_j(Z)) &= \partial_i(B_j)Z + B_j(B_iZ - ZD_i) +\\ &\quad + (B_iZ - ZD_i)D_j + Z\partial_i(D_j). \qquad (19) \end{aligned}$$

Hence, by (7), we have

$$Z\partial_j(D_i) - ZD_jD_i = Z\partial_i(D_j) - ZD_iD_j,$$

which implies (15) since Z is invertible. Now, for every $\mathbf{K}^{\Phi}$-Δ-algebra L and $g \in G(L)$, let $C_g \in \mathbf{GL}_n(L)$ be such that $g(Z) = ZC_g$. Then, on the one hand,

$$g(\partial_i(Z)) = g(BZ - ZD_i) = BZC_g - ZC_gD_i.$$

On the other hand,

$$g(\partial_i(Z)) = \partial_i(g(Z)) = \partial_i(ZC_g) = BZC_g - ZD_iC_g + Z\partial_i(C_g).$$

Therefore, for all $g \in G$, we have

$$\partial_i(C_g) + [C_g, D_i] = 0, \quad 1 \leqslant i \leqslant m,$$

showing (16). To show that (17) is in the defining ideal of G, let F be the Δ-field generated over $\mathbf{K}^{\Phi}$ by the entries of an invertible solution C of

$$\partial_i(Y) = D_iY, \quad 1 \leqslant i \leqslant m,$$

inside of a differential closure of $\mathbf{K}^{\Phi}$ ((15) is satisfied). Then,

$$\begin{aligned} \partial_i(C^{-1}C_gC) &= -C^{-1}D_iCC^{-1}C_gC -\\ &\quad - C^{-1}[C_g, D_i]C + C^{-1}C_gD_iC = 0. \end{aligned}$$

Suppose now that, for all $\mathbf{K}^{\Phi}$-∂-algebras L and $g \in G(L)$,

$$\partial_i(C_g) + [C_g, D_i] = 0, \quad 1 \leqslant i \leqslant m,$$

where $C_g := Z^{-1}g(Z)$, and (15) is satisfied. Let

$$B_i := \partial(Z)Z^{-1} + ZD_iZ^{-1}.$$

Then, for all $g \in G(L)$,

$$\begin{aligned} &g(B_i) = \partial_i(ZC_g)C_g^{-1}Z^{-1} + ZC_gD_iC_g^{-1}Z^{-1} =\\ &= \partial_i(Z)Z^{-1} - Z[C_g, D_i]C_g^{-1}Z^{-1} + ZC_gD_iC_g^{-1}Z^{-1} = B_i. \end{aligned}$$

Hence, by Lemma 6.19 of [13], $B_i \in \mathbf{GL}_n(\mathbf{K})$. Moreover, for all j, $1 \leqslant j \leqslant q$, we have

$$\begin{aligned} &\phi_j(B_i) = \partial_i(\phi_j(Z))(\phi_j(Z))^{-1} + \phi_j(Z)D_i(\phi_j(Z))^{-1} =\\ &= \partial_i(A_jZ)(A_jZ)^{-1} + A_jZD_i(A_jZ)^{-1} =\\ &= \partial_i(A_j)A_j^{-1} + A_j\partial_i(Z)Z^{-1}A_j^{-1} + A_jZD_iZ^{-1}A_j^{-1} =\\ &= \partial_i(A_j)A_j^{-1} + A_jB_iA_j^{-1}, \end{aligned}$$

showing (6). To show (7), we proceed as in (18) and (19). □

Remark 1. It follows from Theorem 1 that, if $\mathbf{K}$ satisfies assumption 1, then one can remove requirement (15) from the statement of Proposition 2.

3.4 Example

We will now give an example that shows that the linearly differentially closed restriction in the statement of Theorem 1 cannot be waived.

Example 3. This example is based on Example 6.7 from [10], but requires a modification described below. Starting with $\mathbf{K} := \overline{\mathbb{Q}(t_1,t_2)}$ (the algebraic closure), $\Delta := \{\partial_{t_1}, \partial_{t_2}\}$, and $\Phi = \{\phi\}$ with ϕ acting as identity on $\mathbf{K}$, we let

$$F := \mathbf{K}\left(\phi^i\partial_{t_1}^{j_1}\partial_{t_2}^{j_2}I_k,\ j_1,j_2 \geqslant 0,\ i\in\mathbb{Z},\ k=1,2\right),$$

be the field of $\{\phi,\partial_{t_1},\partial_{t_2}\}$-rational functions in the difference-differential indeterminates I_1 and I_2 over $\mathbf{K}$. Notice that $\mathbf{K}$ is neither ∂_{t_1}- nor ∂_{t_2}-linearly closed as $\partial_{t_i}(y) = y$ has no non-zero solutions in $\mathbf{K}$, $i=1,2$. Let S be a PPV-ring of F for the difference equation

$$\phi(y) - y = (\phi(I_1) - I_1)\cdot I_2, \qquad (20)$$

$L := \operatorname{Quot}(S)$, and $I \in L$ be a solution of (20) (the existence of S is shown as in Proposition 5.4 of [10], in particular,

$$L = \operatorname{Quot}\left(F\{x\}_\Delta/[D_1(x)-a_1,\ldots,D_r(x)-a_r]\right),$$

where

$$\phi(x) := x + (\phi(I_1) - I_1)\cdot I_2,$$

$a_1,\ldots,a_r \in F$, and the D_i's are some homogeneous linear Δ-operators, possibly, all equal to zero). We now show that there are no $a\in F$ and non-zero homogeneous linear Δ-operator D with coefficients in $\mathbf{k}$ such that

$$\phi(a) - a = D((\phi(I_1)-I_1)\cdot I_2). \qquad (21)$$

For this, let $\frac{P}{Q} \in F$ be such that P and Q are relatively prime polynomials. Then $\frac{\phi(P)}{\phi(Q)} - \frac{P}{Q}$ is a polynomial if and only if Q is a non-zero constant. Indeed, suppose

$$\frac{\phi(P)}{\phi(Q)} - \frac{P}{Q} = R,$$

with $R\in F$ is a polynomial. Rearranging the terms, we have

$$\phi(P) = \phi(Q)R + \frac{\phi(Q)P}{Q}.$$

Since $\phi(P)$ is a polynomial and P and Q are relatively prime, $\frac{\phi(Q)}{Q}$ must be a polynomial. Let $\phi(Q) = Q\tilde{R}$, $\tilde{R}\in F$ a polynomial. Suppose that Q is nonconstant and take a term of maximal total degree in Q. The corresponding term of $\phi(Q)$ is of the same degree and is a term of maximal degree in $\phi(Q)$. Thus, $\tilde{R}$ must be constant and, therefore, Q must be a constant multiple of $\phi(Q)$. But this is only possible if Q is a constant. The other direction is automatic.

Now we prove the claim. If $a\in F$ and D satisfy (21), since the right-hand side of (21) is a polynomial, by the above, a is also a polynomial. Then,

$$\phi(a) - a = D\left((\phi(I_1)-I_1)\cdot I_2\right) = \tilde{D}\left(\phi(I_1)-I_1\right)\cdot I_2 + R,$$

where $\tilde{D}$ is another Δ-operator with coefficients in $\mathbf{K}$ and $R\in F$ a polynomial with no terms containing $\phi^i(I_2)$, $i\in\mathbb{Z}$ (including I_2 itself). Thus, $\phi(a)$ or a has a term of the form $c\cdot I_2$, with c a polynomial in F not containing I_2. Suppose this term is in a (the other case can be treated similarly). Then it can be seen by induction that $\phi(a)$ must have a term $\phi^n(cI_2)$ for all $n\in\mathbb{N}$, which contradicts $\phi(a)\in F$.

Since $\mathbf{K}$ is algebraically closed, Proposition 3.1 and the descent argument from the proof of Corollary 3.2 of [13] imply that the elements

$$\partial_{t_1}^{i_1}\partial_{t_2}^{i_2}I \in L,\quad i_1, i_2 \geqslant 0,$$

are algebraically independent over F. Let $\mathbf{K}\subset L$ be the $\{\Phi,\Delta\}$-subfield generated by

$$\phi(I_k) - I_k,\ \partial_{t_1}I_m,\ \partial_{t_2}I_m,\quad k=1,2,$$
$$J_1 := \partial_{t_1}I - \partial_{t_1}I_1\cdot I_2 - I_2/t_1,$$
$$J_2 := \partial_{t_2}I - \partial_{t_2}I_1\cdot I_2 + I_1/t_2.$$

Since I satisfies (20) and $J_1,J_2\in\mathbf{K}$, for all $(i,j_1,j_2)\neq(0,0,0)$, we have

$$(\phi-\mathrm{id})^i\partial_{t_1}^{j_1}\partial_{t_2}^{j_2}(I)\in\mathbf{K}(I_1,I_2).$$

Indeed, we will show this by induction on the triples (i,j_1,j_2), $i,j_1,j_2\geqslant 0$ ordered degree-lexicographically. We denote the nth triple by a_n, with $a_1 = (0,0,1)$. For $n=1$,

$$(\phi-\mathrm{id})^0\partial_{t_1}^0\partial_{t_2}(I) = \partial_{t_2}(I) = $$
$$= J_2 + \partial_{t_2}(I_1)\cdot I_2 - I_1/t_2 \in \mathbf{K}(I_1,I_2).$$

Supposing that the result holds for a_{n-1}, and letting $a_n = (i,j_1,j_2)$, first note that, for $n\geqslant 2$,

$$a_{n-1} = \begin{cases} (i,j_1,j_2-1) & \text{if} \quad j_2\neq 0\\ (i,j_1-1,j_2) & \text{if} \quad j_2 = 0 \text{ and } j_1\neq 0\\ (i-1,j_1,j_2) & \text{if} \quad j_1=j_2=0 \text{ and } i\neq 0\end{cases}$$

In the first case, if $a_{n-1} = (i,j_1,j_2-1)$, let

$$(\phi-\mathrm{id})^i\partial_{t_1}^{j_1}\partial_{t_2}^{j_2-1}(I) = \frac{P(I_1,I_2)}{Q(I_1,I_2)} \in K(I_1,I_2),$$

where $P,Q\in\mathbf{K}[I_1,I_2]$ and $Q\neq 0$. Then

$$(\phi-\mathrm{id})^i\partial_{t_1}^{j_1}\partial_{t_2}^{j_2}(I) = \partial_{t_2}\left(\frac{P(I_1,I_2)}{Q(I_1,I_2)}\right) =$$
$$= \frac{\partial_{t_2}(P(I_1,I_2))\cdot Q(I_1,I_2) - P(I_1,I_2)\cdot\partial_{t_2}(Q(I_1,I_2))}{Q(I_1,I_2)^2},$$

which is in $\mathbf{K}(I_1,I_2)$ since $\partial_{t_2}(I_k)\in\mathbf{K}$, $k=1,2$. In the second case, when $a_{n-1} = (i,j_1-1,j_2)$, it is similarly shown that

$$(\phi-\mathrm{id})^i\partial_{t_1}^{j_1}\partial_{t_2}^{j_2}(I)\in\mathbf{K}(I_1,I_2).$$

In the third case, when $a_{n-1} = (i-1,j_1,j_2)$,

$$(\phi-\mathrm{id})^i\partial_{t_1}^{j_1}\partial_{t_2}^{j_2}(I) = (\phi-\mathrm{id})\left(\frac{P(I_1,I_2)}{Q(I_1,I_2)}\right) =$$
$$= \frac{P(\phi(I_1),\phi(I_2))}{Q(\phi(I_1),\phi(I_2))} - \frac{P(I_1,I_2)}{Q(I_1,I_2)} \in\mathbf{K}(I_1,I_2),$$

since $\phi(I_k) - I_k\in\mathbf{K}$, $k=1,2$. When $i<0$, a similar argument applies. Thus,

$$L = \mathbf{K}(I_1,I_2,I).$$

It follows from Lemma 3.6 of [8] (see also [26, 25]) that I_1,I_2,I are algebraically independent over $\mathbf{K}$ using a characteristic set argument with respect to any difference-differential ranking with $I>I_1>I_2$ and $\phi>\partial_{t_1}>\partial_{t_2}$ (the corresponding characteristic set C will have leaders that are strictly greater than I in ranking and, therefore, no polynomial in I,I_1,I_2 can be reduced to zero with respect to C). Put

$$f_i := \phi(I_i) - I_i\in\mathbf{K},\quad i=1,2$$

and consider the equation

$$\phi(Y) = AY, \quad A := \begin{pmatrix} 1 & f_1 & 0 \\ 0 & 1 & f_2 \\ 0 & 0 & 1 \end{pmatrix}. \tag{22}$$

Then

$$S := \begin{pmatrix} 1 & I_1 & I \\ 0 & 1 & I_2 \\ 0 & 0 & 1 \end{pmatrix}$$

is a fundamental matrix for equation (22). Hence, L is a PPV extension of $\mathbf{K}$ for equation (22). Let U be the differential algebraic group of matrices of the form

$$g(u_1, u_2, v) = \begin{pmatrix} 1 & u_1 & v \\ 0 & 1 & u_2 \\ 0 & 0 & 1 \end{pmatrix}$$

and G be its differential algebraic subgroup defined by the equations:

$$\partial_{t_i} u_j = 0,\ i = 1, 2,\ j = 1, 2,\ \partial_{t_1} v = \frac{1}{t_1} u_2,\ \partial_{t_2} v = -\frac{1}{t_2} u_1.$$

If we let

$$D_1 := \begin{pmatrix} 0 & 1/t_1 & 0 \\ 0 & 0 & 0 \\ 0 & 0 & 0 \end{pmatrix} \quad \text{and} \quad D_2 := \begin{pmatrix} 0 & 0 & 0 \\ 0 & 0 & 1/t_2 \\ 0 & 0 & 0 \end{pmatrix},$$

then the defining equations for G turn into

$$\partial_{t_i}(x_{rs}) + [(x_{rs}), D_i], \quad i = 1, 2. \tag{23}$$

Note that $\mathbf{Gal}(L/\mathbf{K})$ is a differential algebraic subgroup in U, where U acts on S by multiplication from the right. For all $g(u_1, u_2, v) \in G$, we have

$$\begin{aligned} &g(u_1, u_2, v)(\partial_{t_i}(I_i)) = \partial_{t_i}(g(u_1, u_2, v)(I_i)) = \\ &= \partial_{t_i}(I_i + u_i) = \partial_{t_i}(I_i), \\ &g(u_1, u_2, v)(\phi(I_i) - I_i) = \phi(I_i + u_i) - I_i - u_i = \\ &= \phi(I_i) - I_i,\ i = 1, 2, \\ &g(u_1, u_2, v)(J_1) = \partial_{t_1}(I + I_1 u_2 + v) - \\ &\quad - \partial_{t_1}(I_1 + u_1)(I_2 + u_2) - (I_2 + u_2)/t_1 = J_1, \\ &g(u_1, u_2, v)(J_2) = \partial_{t_2}(I + I_1 u_2 + v) - \\ &\quad - \partial_{t_2}(I_1 + u_1)(I_2 + u_2) + (I_1 + u_1)/t_2 = J_2. \end{aligned}$$

Therefore, $K \subset L^G$. Since

$$\mathbf{k}\{G\} \cong \mathbf{k}[u_1, u_2, v] \quad \text{and} \quad \operatorname{tr.deg}_{\mathbf{K}} L = 3$$

by the above, we have $\mathbf{Gal}(L/\mathbf{K}) = G$ (see Lemma 6.19 and Proposition 6.26 in [13]). Therefore, by Example 4.7 of [10] and Proposition 2, equation (22) is not isomonodromic. On the other hand, by (23) and Proposition 2, equation (22) is isomonodromic with respect to each ∂_{t_i}, $i = 1, 2$, separately with the corresponding matrices given by

$$B_i := S \cdot D_i \cdot S^{-1} + \partial_{t_i} S \cdot S^{-1}, \quad i = 1, 2,$$

that is,

$$B_1 = \begin{pmatrix} 0 & 1/t_1 + \partial_{t_1} I_1 & J_1 \\ 0 & 0 & \partial_{t_1} I_2 \\ 0 & 0 & 0 \end{pmatrix},$$

$$B_2 = \begin{pmatrix} 0 & \partial_{t_2} I_1 & J_2 \\ 0 & 0 & 1/t_2 + \partial_{t_2} I_2 \\ 0 & 0 & 0 \end{pmatrix}.$$

4. ACKNOWLEDGMENTS

We are highly grateful to Sergey Gorchinskiy and the referees their very helpful suggestions.

5. REFERENCES

[1] Arreche, C.: Computing the differential Galois group of a one-parameter family of second order linear differential equations (2012). URL http://arxiv.org/abs/1208.2226 1

[2] Barkatou, M., Cluzeau, T., El Bacha, C., Weil, J.A.: Computing closed form solutions of integrable connections. In: Proceedings of the 36th international symposium on symbolic and algebraic computation, ISSAC '12. ACM, New York, NY, USA (2012) 1

[3] Cassidy, P.: Differential algebraic groups. American Journal of Mathematics **94**, 891–954 (1972). URL http://www.jstor.org/stable/2373764 1, 4

[4] Cassidy, P.: The differential rational representation algebra on a linear differential algebraic group. Journal of Algebra **37**(2), 223–238 (1975). URL http://dx.doi.org/10.1016/0021-8693(75)90075-7 1

[5] Cassidy, P., Singer, M.: Galois theory of parametrized differential equations and linear differential algebraic group. IRMA Lectures in Mathematics and Theoretical Physics **9**, 113–157 (2007). URL http://dx.doi.org/10.4171/020-1/7 1, 3.1, 3.3

[6] Cassidy, P., Singer, M.: A Jordan–Hölder theorem for differential algebraic groups. Journal of Algebra **328**(1), 190–217 (2011). URL http://dx.doi.org/10.1016/j.jalgebra.2010.08.019 1

[7] Dreyfus, T.: Computing the Galois group of some parameterized linear differential equation of order two (2012). URL http://arxiv.org/abs/1110.1053. To appear in the Proceedings of the AMS 1

[8] Gao, X.S., Van der Hoeven, J., Yuana, C.M., Zhanga, G.L.: Characteristic set method for differential–difference polynomial systems. Journal of Symbolic Computation **44**(9) (2009). URL http://dx.doi.org/10.1016/j.jsc.2008.02.010 3

[9] Gillet, H., Gorchinskiy, S., Ovchinnikov, A.: Parameterized Picard–Vessiot extensions and Atiyah extensions. Advances in Mathematics **238**, 322–411 (2013). URL http://dx.doi.org/10.1016/j.aim.2013.02.006 1, 3.3

[10] Gorchinskiy, S., Ovchinnikov, A.: Isomonodromic differential equations and differential categories (2012). URL http://arxiv.org/abs/1202.0927 1, 3.1, 3, 3, 3

[11] Granier, A.: A Galois D-groupoid for q-difference equations. Annales de l'Institut Fourier **61**(4), 1493–1516 (2011) 1

[12] Hardouin, C.: Hypertranscendance des systèmes aux différences diagonaux. Compositio Mathematica **144**(3), 565–581 (2008) 1

[13] Hardouin, C., Singer, M.: Differential Galois theory of linear difference equations. Mathematische Annalen **342**(2), 333–377 (2008). URL http://dx.doi.org/10.1007/s00208-008-0238-z 1, 1, 3.1, 3.3, 3.3, 3.3, 3, 3

[14] Hardouin, C., di Vizio, L.: Courbures, groupes de Galois génériques et D-groupoïdes de Galois d'un

système aux D-différences. C. R. Math. Acad. Sci. Paris **348**(17–18), 951–954 (2010). URL http://dx.doi.org/10.1016/j.crma.2010.08.001 1

[15] Hardouin, C., di Vizio, L.: Galois theories for q-difference equations: comparison theorems (2011). URL http://arxiv.org/abs/1002.4839 1

[16] Hardouin, C., di Vizio, L.: On the Grothendieck conjecture on p-curvatures for q-difference equations (2011). URL http://arxiv.org/abs/1002.4839 1

[17] Hardouin, C., di Vizio, L.: Parameterized generic Galois groups for q-difference equations, followed by the appendix "The Galois D-groupoid of a q-difference system" by Anne Granier (2011). URL http://arxiv.org/abs/1002.4839 1

[18] Hardouin, C., di Vizio, L.: Descent for differential Galois theory of difference equations. Confluence and q-dependency. Pacific Journal of Mathematics **256**(1), 79–104 (2012). URL http://dx.doi.org/10.2140/pjm.2012.256.79 1

[19] Hölder, O.: Über die Eigenschaft der Gamma Funktion keiner algebraische Differentialgleichung zu genügen. Mathematische Annalen **28**, 248–251 (1887). URL http://resolver.sub.uni-goettingen.de/purl?GDZPPN002249537 1

[20] Jimbo, M., Miwa, T.: Deformation of linear ordinary differential equations. I. Japan Academy. Proceedings. Series A. Mathematical Sciences **56**(4), 143–148 (1980). URL http://dx.doi.org/10.3792/pjaa.56.143 1

[21] Jimbo, M., Miwa, T., Ueno, K.: Monodromy preserving deformation of linear ordinary differential equations with rational coefficients. I. General theory and τ-function. Phyica D: Nonlinear Phenomena **2**(2), 306–352 (1981). URL http://dx.doi.org/10.1016/0167-2789(81)90013-0 1

[22] Joshi, N., Shi, Y.: Exact solutions of a q-discrete second Painlevé equation from its iso-monodromy deformation problem: I. Rational solutions. Proceedings of the Royal Society A: Mathematical, Physical, and Engineering Sciences **467**(2136), 3443–3468 (2011). URL http://dx.doi.org/10.1098/rspa.2011.0167 1

[23] Joshi, N., Shi, Y.: Exact solutions of a q-discrete second Painlevé equation from its iso-monodromy deformation problem. II. Hypergeometric solutions. Proceedings of the Royal Society A: Mathematical, Physical, and Engineering Sciences **468**(2146), 3247–3264 (2012). URL http://dx.doi.org/10.1098/rspa.2012.0224 1

[24] Kamensky, M.: Tannakian formalism over fields with operators. International Mathematics Research Notices **361**, 163–171 (2012). URL http://dx.doi.org/10.1093/imrn/rns190 3.3

[25] Kolchin, E.: Differential Algebra and Algebraic Groups. Academic Press, New York (1973) 3.2, 3

[26] Levin, A.: Difference Algebra. Springer (2008). URL http://dx.doi.org/10.1007/978-1-4020-6947-5 3

[27] Minchenko, A., Ovchinnikov, A.: Zariski closures of reductive linear differential algebraic groups. Advances in Mathematics **227**(3), 1195–1224 (2011). URL http://dx.doi.org/10.1016/j.aim.2011.03.002 1

[28] Minchenko, A., Ovchinnikov, A.: Extensions of differential representations of $\mathbf{SL}_2$ and tori. Journal of the Institute of Mathematics of Jussieu **12**(1), 199–224 (2013). URL http://dx.doi.org/10.1017/S1474748012000692 1

[29] Minchenko, A., Ovchinnikov, A., Singer, M.F.: Reductive linear differential algebraic groups and the Galois groups of parameterized linear differential equations (2013). URL http://arxiv.org/abs/1304.2693 1

[30] Minchenko, A., Ovchinnikov, A., Singer, M.F.: Unipotent differential algebraic groups as parameterized differential Galois groups (2013). URL http://arxiv.org/abs/1301.0092 1

[31] Morikawa, S.: On a general difference Galois theory. I. Annales de l'Institut Fourier **59**(7), 2709–2732 (2009) 1

[32] Ovchinnikov, A.: Tannakian approach to linear differential algebraic groups. Transformation Groups **13**(2), 413–446 (2008). URL http://dx.doi.org/10.1007/s00031-008-9010-4 1, 3.3

[33] Sabbah, C.: The work of Andrey Bolibrukh on isomonodromic deformations. IRMA Lectures in Mathematics and Theoretical Physics **9**, 9–25 (2007). URL http://dx.doi.org/10.4171/020-1/2 1

[34] Sibuya, Y.: Linear differential equations in the complex domain: problems of analytic continuation, vol. 82. American Mathematical Society, Providence, RI (1990) 1

[35] Singer, M.: Linear algebraic groups as parameterized Picard–Vessiot Galois groups. Journal of Algebra **373**(1), 153–161 (2013). URL http://dx.doi.org/10.1016/j.jalgebra.2012.09.037 1

[36] Takano, K.: On the hypertranscendency of solutions of a difference equation of Kimura. Funkcialaj Ekvacioj. Serio Internacia **16**, 241–254 (1973). URL http://fe.math.kobe-u.ac.jp/FE/FE_pdf_with_bookmark/FE11-20-en_KML/fe16-241-254/fe16-241-254.pdf 1

[37] Trushin, D.: Splitting fields and general differential Galois theory. Sbornik: Mathematics **201**(9), 1323–1353 (2010). URL http://dx.doi.org/10.1070/SM2010v201n09ABEH004114 3.3

[38] di Vizio, L.: Approche galoisienne de la transcendance différentielle (2012). Journée annuelle 2012 de la SMF URL http://divizio.perso.math.cnrs.fr/PREPRINTS/16-JourneeAnnuelleSMF/difftransc.pdf. 1

[39] Wibmer, M.: Existence of ∂-parameterized Picard–Vessiot extensions over fields with algebraically closed constants. Journal of Algebra **361**, 163–171 (2012). URL http://dx.doi.org/10.1016/j.jalgebra.2012.03.035 3.3

Rainbow Cliques and the Classification of Small BLT-Sets

Anton Betten
Department of Mathematics
Colorado State University
Fort Collins, CO, U.S.A.
betten@math.colostate.edu

ABSTRACT

In Finite Geometry, a class of objects known as BLT-sets play an important role. They are points on the $Q(4,q)$ quadric satisfying a condition on triples. This paper is a contribution to the difficult problem of classifying these sets up to isomorphism, i.e., up to the action of the automorphism group of the quadric. We reduce the classification problem of these sets to the problem of classifying rainbow cliques in graphs. This allows us to classify BLT-sets for all orders q in the range 31 to 67.

Categories and Subject Descriptors

G.2.1 [**Mathematics of Computing**]: Discrete Mathematics—*Combinatorics*; G.2.2 [**Mathematics of Computing**]: Discrete Mathematics—*Graph Theory*; I.1.2 [**Computing Methodologies**]: Symbolic and Algebraic Manipulation—*Algorithms*

Keywords

Isomorphism, Classification, Finite Geometry

1. INTRODUCTION

BLT-sets have been introduced in [1] in relation with the study of flocks of a quadratic cone in projective 3-space (cf. [15, 18]). A BLT-set of order q is a set S of $q+1$ points on the parabolic quadric $Q(4,q)$ such that $P^\perp \cap Q^\perp \cap R^\perp$ is empty for any three points P, Q, R in S. BLT-sets are important in finite geometry, due to the connections to translation planes and to generalized quadrangles. We refer to the introduction of [16] for more details. In a curious twist, BLT-sets predate themselves by six years, as they first arose in [7] under the name $(0,2)$-sets.

It is known that BLT-sets of order q exist if and only if q is odd. Two BLT-sets S and T of order q are equivalent if there is a group element g in the automorphism group of $Q(4,q)$ such that $S^g = T$. It is an important problem to classify all BLT-sets of $Q(4,q)$ up to isomorphism, and the problem is open for $q \geq 31$. It is known that the quadric $Q(4,q)$ has $(q+1)(q^2+1)$ points and that its automorphism group is the orthogonal semilinear group $\mathrm{P\Gamma O}(5,q)$, of order $h(q^4-1)(q^2-1)q^4$, where $q = p^h$ for some prime p and some integer h. We let G denote the group $\mathrm{P\Gamma O}(5,q)$. In this paper, we classify all BLT-sets of orders $q = 31, 37, 41, 43, 47, 49, 53, 59, 61, 67$. The reason for stopping at $q = 67$ is that we have reached the limits of what can be computed using the resources available to us.

q	BLT	F	q	BLT	F
3	1	1	29	9	28
5	2	2	31	8	33
7	2	2	37	7	37
9	3	3	41	10	51
11	4	4	43	6	50
13	3	4	47	10	51
17	6	9	49	8	24
19	5	8	53	8	39
23	9	18	59	9	48
25	6	12	61	5	36
27	6	14	67	6	39

Table 1: The Number of Isomorphism Classes of BLT-Sets (BLT) and Flocks (F) of Order q (The Numbers for $31 \leq q \leq 67$ Are New)

THEOREM 1. *The number of isomorphism types of BLT-sets and of flocks of the quadratic cone in* $\mathrm{PG}(3,q)$ *for any given order* $q \leq 67$ *that is relevant is given in Table 1.*

PROOF. By computer, using the classification algorithm decribed in this paper. In Sections 2,3,4, and 6, we describe our algorithm to search for BLT-sets. In Section 5 we present our algorithm to solve the isomorphism problem for BLT-sets. In Section 7 we present the results from the search. This constitutes our proof of the theorem. A few remarks on the correctness of the result are contained in Section 8. □

A brief description of all BLT-sets in this range is given in Appendix A. In Section 9, we describe an invariant, called the plane type, that can distinguish between all isomorphism types of BLT-sets of order at most 67.

2. THE ALGORITHM

Let us now describe our search algorithm and the results of the search. This will constitute the basis for Theorem 1.

The classification of BLT-sets up to isomorphism suffers from the fact that there is a large number of nonisomorphic partial objects, many of which either do not extend or are embedded in BLT-set in different ways. Thus, classifying partial objects becomes infeasible. On the other hand, doing a search without isomorph rejection is almost sure to fail, since the number of BLT-sets of $Q(4, q)$ (not considering isomorphism) is simply too large. In addition, checking BLT-sets for isomorphism is difficult.

Our approach is a combination of techniques from geometry, group theory and combinatorics. We reduce the problem to a search for rainbow cliques in colored graphs. While this is still NP-hard, it allows us to find and classify all BLT-sets for the orders relevant for Theorem 1.

Our main method is a technique called "breaking the symmetry." This is a strategy to attack classification problems involving symmetry and the resulting issues of isomorphism. The terminology seems to be due to [5], but the methodology is part of the folklore. It is the idea behind the "we may assume" term that is frequently found in proofs.

We separate the search into different stages. The goal is to find suitable *starter configurations* such that every BLT-set can be obtained (in the sense of a practical computation) from at least one of these starter configurations. A good choice for these starter configurations are partial BLT-sets of a certain size. A partial BLT-set is simply a set S of points on $Q(4, q)$ such that $P^\perp \cap Q^\perp \cap R^\perp$ is empty for any three points P, Q, R in S. Thus, a BLT-set is a partial BLT-set of size $q + 1$ and every subset of a BLT-set is a partial BLT-set. We take as starter configurations the partial BLT-sets of size s, for some reasonably chosen integer s. In Section 3, the starter configurations are classified up to isomorphism. The orbits of starter configurations are called *starters,* and the chosen representatives of these orbits are called *starter sets.* The integer s is chosen so that the average starter configuration has only a very small automorphism group, but at the same time the number of starters is reasonably small ($s = 5$ seems to work well). The fact that starter configurations have small average stabilizers explains the terminology of breaking the symmetry.

In a second step, described in Section 4, each starter set is considered in turn and all BLT-sets containing this set are constructed using a technique from graph theory, called rainbow cliques. Computationally, this is the dominant part of the algorithm. To this end, a graph Γ_S is defined in such a way that all BLT-sets B containing S correspond to cliques of a certain size in Γ_S. In fact, we can find a colored graph $\Gamma_{S,\ell}$ such that the BLT-sets B containing S correspond to the rainbow cliques in $\Gamma_{S,\ell}$.

In Section 5 we perform the isomorphism testing to classify the BLT-sets that arise by means of rainbow cliques. In Section 6 we describe a technique to speed up the search, using the lexicographical ordering of subsets.

3. STARTER CONFIGURATIONS

In order to compute the orbits of partial BLT-sets of a fixed size s under the symmetry group of $Q(4, q)$, we employ the orbit algorithm "Snakes and Ladders," due to Schmalz [22]. The algorithm builds up a data structure that stores all orbit representatives and the associated automorphism groups. We choose the orbit representative to be the

q	$\lvert Q(4,q)\rvert$	$\lvert P\Gamma O(5,q)\rvert$	5-Orbits	Time
31	30,784	8.1×10^{14}	2,693	52 sec
37	52,060	4.8×10^{15}	6,739	2 min 52 sec
41	70,644	1.3×10^{16}	11,478	6 min 24 sec
43	81,400	2.1×10^{16}	14,693	8 min 33 sec
47	106,080	5.2×10^{16}	23,312	17 min 31 sec
49	120,100	1.5×10^{17}	14,542	14 min 10 sec
53	151,740	1.7×10^{17}	43,465	44 min 36 sec
59	208,920	5.1×10^{17}	75,707	1 hr 46 min
61	230,764	7.1×10^{17}	89,954	2 hrs 14 min
67	305,320	1.8×10^{18}	146,009	4 hrs 50 min

Table 2: Summary of the Classification of Starters

lexicographically least sets in their respective orbits, relative to the ordering induced from some total ordering of the points of $Q(4, q)$. The algorithm also stores additional data that can be used to identify the representative R of any partial BLT-set S of size at most s quickly and to provide a group element $g \in G$ such that $S^g = R$.

Table 2 gives information about the classification of partial BLT-sets. In the table, we list the number of 5-orbits in each of the cases $q = 31, \ldots, 67$. These are the orbits on partial BLT-sets of size 5.

4. RAINBOW CLIQUES

Given a partial BLT-set S, we employ techniques from graph theory to find all BLT-sets containing S. For each partial BLT-set S, we define a graph Γ_S. The vertices of Γ_S are the points of $Q(4, q)$ that are admissible. A point P is admissible if $S \cup \{P\}$ is a partial BLT-set also. The edges in Γ_S are between points P and Q such that $S \cup \{P, Q\}$ is partial BLT-set. It is clear that cliques of size $q + 1 - s$ in Γ_S are necessary for the existence of a BLT-set T containing the starter S. Interestingly, the existence of such cliques is also sufficient for the existence of a BLT-set T containing S:

LEMMA 1. *Let S be a partial BLT-set of size s in the $Q(4, q)$ quadric, and let Γ_S be the associated graph as defined above. Then*

1. *The stabilizer of S induces a group of automorphisms of Γ_S.*
2. *The BLT-sets T containing S are in one-to-one correspondence to the cliques of Γ_S of size $q + 1 - s$.*

PROOF. The first statement is clear, so we look at the second. It follows from the definition of the graph Γ_S that a BLT-set T containing S gives rise to a clique of size $q+1-s$ in Γ_S. For the converse, we need to show that a clique C in Γ_S of size $q + 1 - s$ defines a BLT-set $T := S \cup C$. For this purpose, we need to look at all triples $P, Q, R \in T$. The only interesting case is when $P, Q, R \in C$. In this case, pick a point $P_0 \in S$. The presence of the three edges PQ, PR, and QR in Γ_S implies that the triples P_0PQ, P_0PR and P_0QR are partial BLT-sets. By [2, Lemma 4.3], the triple PQR is partial BLT, too. □

Thus we have reduced the problem of finding BLT-sets containing a given starter into the problem of finding certain

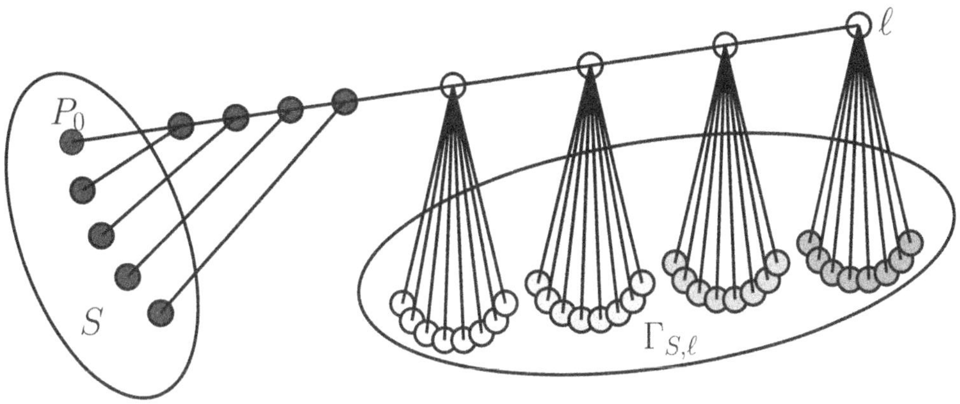

Figure 1: The Colored Graph $\Gamma_{S,\ell}$ (Edges Not Shown)

cliques in a certain graph. We will now consider a refinement of this technique, based on a coloring of the graph Γ_S. Since $Q(4,q)$ is a generalized quadrangle [19], a point not on a line is collinear with exactly one point on that line (this is true for any non-incident point/line pair). We can define a colored graph $\Gamma_{S,\ell}$ as follows. Suppose we fix a point $P_0 \in S$, together with a line ℓ of $Q(4,q)$ that passes through P_0. Let $\mathcal{C}$ be the points on ℓ that are not collinear to any of the points of S other than P_0. Any point P in Γ_S is collinear to exactly one point $Q \in \mathcal{C}$. So, by labeling the point P with the color Q, we find that Γ_S can be colored with the points in $\mathcal{C}$. This gives a new graph $\Gamma_{S,\ell}$ (cf. Fig. 1). The next result shows that the cliques T containing S are in one-to-one correspondence to the rainbow cliques in $\Gamma_{S,\ell}$. Here, a rainbow clique is a clique that intersects each color class in exactly one element.

LEMMA 2. *Let S be a partial BLT-set of size s in the $Q(4,q)$ quadric, and let Γ_S be the associated graph as defined above. Let $P_0 \in S$ and let ℓ be a line of $Q(4,q)$ containing P_0. Let $\Gamma_{S,\ell}$ be the graph Γ_S after coloring the vertices according to the line ℓ as described above. Then*

1. *The stabilizer of S induces a group of automorphisms of $\Gamma_{S,\ell}$ that permutes the color classes among themselves.*
2. *The BLT-sets T containing S are in one-to-one correspondence to the cliques of $\Gamma_{S,\ell}$ of size $q+1-s$ with one vertex from each color class.*

PROOF. The first statement is clear. Regarding the second, we observe that the collinearity relation establishes a bijection between the $s-1$ points of $S\setminus\{P_0\}$ and some $s-1$ points on $\ell\setminus\{P_0\}$. For this, observe that S is partial BLT, and hence no two points P_i, P_j of $S\setminus\{P_0\}$ are collinear to a point $Q \in \ell\setminus\{P_0\}$, for otherwise $Q \in P_i^\perp \cap P_j^\perp \cap P_0^\perp$. Thus, the set $\mathcal{C}$ of points on $\ell\setminus\{P_0\}$ *not* collinear with S has size $q+1-s$. In any BLT-set T containing S, the relation of collinearity induces a bijection between the points on the line ℓ different from P_0 and the points in $T\setminus\{P_0\}$. The points in $S\setminus\{P_0\}$ are paired to some $s-1$ points on $\ell\setminus\{P_0\}$. Thus, the points in $T\setminus S$ are paired to the points in $\mathcal{C}$. This shows that $T\setminus S$ (thought of as a set of vertices in the graph $\Gamma_{S,\ell}$) intersects each color class in exactly one element. By Lemma 1, we know that T is a clique. Therefore, T is a rainbow clique of size $q+1-s$ in $\Gamma_{S,\ell}$. □

Rainbow cliques are easier to search for than ordinary cliques. One reason is that the size of the clique is known beforehand, and thus a condition on the number of neighbors of any live point in the search can be facilitated. The second reason is that the color classes can be used to cut down the size of the choice set encountered at each stage of the backtracking. In Section 6 below, we will use the lexicographic ordering to improve the search algorithm yet again.

We observe:

COROLLARY 1. *There is no BLT-set T containing the partial BLT-set S if, for some line ℓ with $|\ell \cap S| = 1$, the graph $\Gamma_{S,\ell}$ has an empty color class.*

5. ISOMORPH TESTING

Let G be a group acting on finite sets $\mathcal{X}$ and $\mathcal{Y}$, and let $\mathcal{R}$ be a relation between $\mathcal{X}$ and $\mathcal{Y}$ (i.e., a subset of $\mathcal{X}\times\mathcal{Y}$). Let $\mathcal{P}_1,\ldots,\mathcal{P}_m$ be the orbits of G on $\mathcal{X}$ and let $\mathcal{Q}_1,\ldots,\mathcal{Q}_n$ be the orbits of G on $\mathcal{Y}$. Let P_i be a representative of orbit $\mathcal{P}_i$ $(i=1,\ldots,m)$ and let Q_j be a represenative of orbit $\mathcal{Q}_j$ $(j=1,\ldots,n)$. Let $\mathcal{T}_{i,1},\ldots,\mathcal{T}_{i,u_i}$ be the orbits of $\mathrm{Stab}(P_i)$ on the set $\{Y\in\mathcal{Y}\mid (P_i,Y)\in\mathcal{R}\}$, with representatives $T_{i,j}$ for $1\le i\le m$ and $1\le j\le u_i$. Let $\mathcal{S}_{j,1},\ldots,\mathcal{S}_{j,v_j}$ be the orbits of $\mathrm{Stab}(Q_j)$ on the set $\{X\in\mathcal{X}\mid (X,Q_j)\in\mathcal{R}\}$, with representatives $S_{j,i}$ for $1\le j\le n$ and $1\le i\le v_j$.

LEMMA 3. *There is a bijection between the orbits*

$$\{\mathcal{S}_{j,i}\mid j=1,\ldots,n,\ i=1,\ldots,v_j\}$$

and the orbits

$$\{\mathcal{T}_{i,j}\mid i=1,\ldots,m,\ j=1,\ldots,u_i\}.$$

PROOF. Consider the action of G on $\mathcal{R}$. Both sets are in one-to-one correspondence to the orbits of G on pairs $(X,Y)\in\mathcal{X}\times\mathcal{Y}$ with $(X,Y)\in\mathcal{R}$. □

The *decomposition matrix* is the $m\times n$ integer matrix whose (i,j) entry counts the number of h such that $\mathcal{T}_{i,h}$ is paired with an orbit $\mathcal{S}_{j,k}$ for some k.

For the purposes of this paper, *two relations* are important. The first one, denoted as $\mathcal{R}$, is between partial BLT-sets of size s and partial BLT-sets of size $q+1$. The relation is inclusion of sets. The group $G = \mathrm{P\Gamma O}(5,q)$ from above acts on $\mathcal{R}$ component-wise. The second relation will be introduced in Section 6.

Suppose that, for a fixed order q, we have classified the G-orbits $\mathcal{P}_1,\ldots,\mathcal{P}_m$ of partial BLT-sets of $Q(4,q)$ of size s (with representatives $P_i \in \mathcal{P}_i$). Suppose further that we have computed for each i with $1 \le i \le m$ the orbits $\mathcal{T}_{i,1},\ldots,\mathcal{T}_{i,u_i}$ of $\mathrm{Aut}(P_i)$ on the BLT-sets containing P_i, with representatives $T_{i,j} \in \mathcal{T}_{i,j}$. The following algorithm computes the orbits $\mathcal{Q}_1,\ldots,\mathcal{Q}_n$ of BLT-sets, with representatives $Q_i \in \mathcal{Q}_i$ and stabilizers $\mathrm{Aut}(Q_i)$.

ALGORITHM 1. **(Classification)**
Input: *Orbits $\mathcal{P}_1,\ldots,\mathcal{P}_m$ of partial BLT-sets of size s, with representatives P_i and automorphism groups $\mathrm{Aut}(P_i)$.*
For each $i = 1,\ldots,m$, the set of orbits $\mathcal{T}_{i,j}$ $(1 \le j \le u_i)$ of BLT-sets containing P_i under the group $\mathrm{Aut}(P_i)$. Representatives $T_{i,j}$ of the orbit $\mathcal{T}_{i,j}$ and groups $\mathrm{Stab}_{\mathrm{Aut}(P_i)}(T_{i,j})$ for $j = 1,\ldots,u_i$.
Output: *Representatives $Q_1,\ldots,Q_n$ for the G-orbits of BLT-sets of $Q(4,q)$, together with their stabilizers.*

1. *Initialize by marking all orbits $\mathcal{T}_{i,j}$ as unprocessed.*
2. *Consider the first/next unprocessed $\mathcal{T}_{i,j}$.*
3. *Mark $\mathcal{T}_{i,j}$ as processed.*
4. *Define a new isomorphism type $\mathcal{Q}$ of BLT-sets represented by $Q = T_{i,j}$. Record $\mathrm{Stab}_{\mathrm{Aut}(p_i)}(T_{i,j})$ as subgroup of $\mathrm{Aut}(Q)$.*
5. *Loop over all s-subsets of $T_{i,j}$*
6. *Let S be the first/next unprocessed s-subset of $T_{i,j}$.*
7. *Determine the index a such that the partial BLT-set S lies in the orbit $\mathcal{P}_a$.*
8. *Determine a group element $g_1 \in G$ such that $S^{g_1} = P_a$.*
9. *Determine the index $b \le u_a$ such that $T_{i,j}^{g_1}$ is contained in the orbit $\mathcal{T}_{a,b}$ of BLT-sets containing P_a.*
10. *Determine a group element $g_2 \in \mathrm{Aut}(P_a)$ such that $T_{i,j}^{g_1 g_2} = T_{a,b}$.*
11. *If $a = i$ and $b = j$, record g_1g_2 as generator for $\mathrm{Aut}(Q)$.*
12. *Otherwise, mark $T_{a,b}$ as processed. Store the index pair (i,j) and the group element $\theta_{a,b} := g_1g_2$ with it.*
13. *Continue with the next s-subset of $T_{i,j}$ in step 6.*
14. *Continue with the next unprocessed orbit $\mathcal{T}_{i,j}$ in step 2.*

In practice, Steps 7 and 8 will be performed in parallel, as well as Steps 9 and 10. After the algorithm terminates, the $Q := T_{i,j}$ associated to $\mathcal{T}_{i,j}$ considered in Step 2 form a system of representatives for the G-orbits on BLT-sets. The automorphismgroup $\mathrm{Aut}(Q)$ is obtained by extending $\mathrm{Stab}_{\mathrm{Aut}(P_i)}(T_{i,j})$ from Step 4 with all elements $\theta_{a,b}$ encountered in Step 11.

In Step 5, we loop over all subsets of the new BLT-set $Q = T_{i,j}$. This can be improved as follows. Consider the orbits of the group $\mathrm{Stab}_{\mathrm{Aut}(P_i)}(T_{i,j})$ as computed in Step 4 on s-subsets of Q. The loop in Step 5 is over the orbit representatives under this group. In fact, as this group gets extended by automorphisms g_1g_2 in Step 11, we take orbit representatives under that larger group. This will help reduce the number of s-subsets that need to be considered, in particular for BLT-sets with a large automorphism group.

6. LEX-REDUCTION

We can use the lexicographic ordering of points of the $Q(4,q)$ quadric to reduce the number of times that representatives from the same isomorphism type of BLT-sets are constructed. Choose any total ordering of points of the $Q(4,q)$ quadric. Consider the lexicographic ordering of subsets induced from this total ordering of points.

Let us introduce the following two notations.

Let G act on the finite, totally ordered set X. Let S and T be subsets of X with $S \subseteq T$. Let $\max S$ be the largest element of S. The pair (S,T) is *admissible* if $T \setminus S$ intersects trivially the $\mathrm{Aut}(S)$-orbits of the points $1,\ldots,\max S$. Also, if A is a subset of X, the *prefix of size s* of A is the lex-least s-subset of A. Thus, if $A = \{a_1,\ldots,a_k\}$ with $a_1 < a_2 < \cdots < a_k$, then the prefix of size s of A is $\{a_1,\ldots,a_s\}$.

Observe that (S,T) is not admissible if and only if there is an element $x \in T \setminus S$ such that $x^g < \max S$ for some $g \in \mathrm{Stab}(S)$.

We consider the relation $\mathcal{R}_2$ consisting of pairs (S,T) with S and T partial BLT-sets of size s and $q+1$ respectively, and (S,T) admissible. The relation $\mathcal{R}_2$ is obtained by restricting $\mathcal{R}$ to admissible pairs. However, $\mathcal{R}_2$ is no longer G-invariant. We will see that we can modify Algorithm 1 to classify orbits on BLT-sets of size $q+1$.

Let Γ_S^* be the graph whose vertices are the vertices $x \in \Gamma_S$ such that for no group element $g \in \mathrm{Aut}(S)$ we have $x^g < \max S$. In the same way, let $\Gamma_{S,\ell}^*$ be the graph obtained from $\Gamma_{S,\ell}$. We say that Γ_S^* ($\Gamma_{S,\ell}^*$, resp.) is obtained from Γ_S ($\Gamma_{S,\ell}$, resp.) by *lex-reduction*.

If lex-reduction is used, not all clique orbits associated to a given starter are present. Therefore, it is necessary to modify Algorithm 1. We replace Steps 7-10 of Algorithm 1 by the following steps. The idea is the following. Given a BLT-set T, determine the prefix S of T (assume that S is a representative, otherwise move T using a group element $g \in G$). If T cannot be found among the stored BLT-sets associated to S, then (S,T) is not admissible, and we can find a group element $g \in \mathrm{Aut}(S)$ such that T^g has a lexicographically smaller prefix S' and we repeat (with Step 3). This procedure must terminate because the number of sets that preceed S in the lexicographic order is finite.

1. Let T be the set $T_{i,j}$. Let $h := id$.
2. Repeat the following loop:
3. Let S be the prefix of size s of T.
4. Determine the index a such that $S \in \mathcal{P}_a$. Let g_1 be a group element such that $S^{g_1} = P_a$.

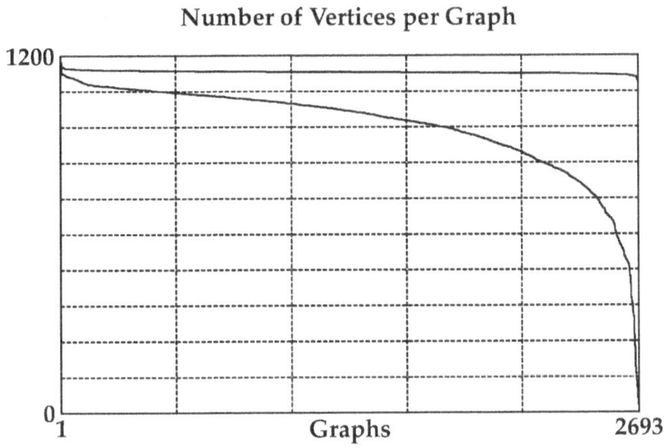

Figure 2: Distribution of the Number of Vertices per Graph when $q = 31$: Top Curve is Without Lex-Reduction, Bottom Curve is With Lex-Reduction

5. Try to find the orbit $\mathcal{T}_{a,b}$ containing T with $1 \leq b \leq u_a$.
6. If that orbit exists, break out of the loop and go to step 8.
7. If no such orbit exists, T must contain an element x such that $x^g < \max S$ for some $g \in \mathrm{Aut}(P_a)$. Replace T by T^g and h by hg and repeat with Step 3.
8. Replace g_1 by hg_1 and find a group element g_2 such that $T_{i,j}^{g_1 g_2}$ is the orbit representative $\mathcal{T}_{a,b}$.

We illustrate the effect of lex-reduction by an example. We consider the partial BLT-sets of size $s = 5$ of $Q(4,31)$. Recall from Table 2 that we have $2,693$ starters in this case, each associated with one graph. In Figure 2, we plot the distribution of the number of vertices in these graphs (we arrange the graphs in order of increasing number of vertices). The top curves shows the number of vertices in the graphs $\Gamma_{S,\ell}$. The bottom curve shows the number of vertices in the graphs $\Gamma^*_{S,\ell}$, i.e., after the lex-reduction has been applied to the vertex set. In Figure 3, we plot the distribution of the number of rainbow cliques in these graphs (we arrange the graphs in order of increasing number of cliques). Again, the top curve corresponds to the graphs $\Gamma_{S,\ell}$ while the bottom curve corresponds to the graphs $\Gamma^*_{S,\ell}$ obtained from lex-reduction. The total number of cliques is $180,816$ in the first case and $19,989$ in the latter. These numbers can be thought of as the area under the curves. Without lex-reduction, all but one graph have cliques. With lex-reduction, 960 graphs have no cliques. In each of the two cases, one graph has a large number of cliques. Without lex-reduction, this graph has 485 cliques. After lex-reduction, 416 cliques remain.

Since the lex-reduction removes whole orbits under $\mathrm{Aut}(S)$, the stabilizer of S, the group $\mathrm{Aut}(S)$ acts as a group of symmetries on the resulting graphs Γ^*_S and $\Gamma^*_{S,\ell}$. Thus, $\mathrm{Aut}(S)$ also acts on the set of cliques (rainbow cliques, respectively) of these graphs. Considering the case $q = 31$ once again, Figure 4 shows the distributions of the number of orbits under $\mathrm{Aut}(S)$ on rainbow cliques per graph over all starter sets S. The top curve is without lex-reduction, while the bottom

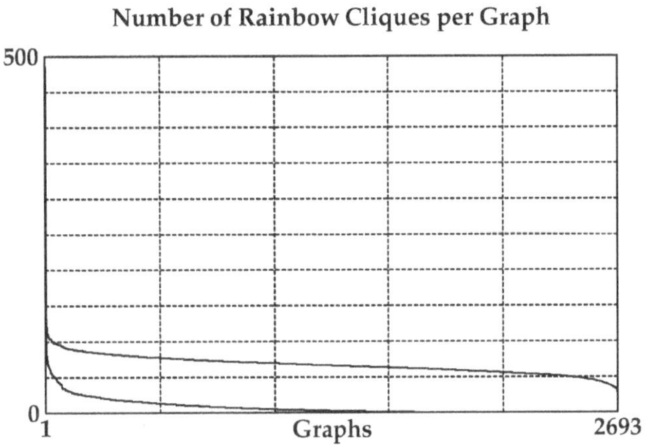

Figure 3: Distribution of the Number of Rainbow Cliques per Graph when $q = 31$: Top Curve is Without Lex-Reduction, Bottom Curve is With Lex-Reduction

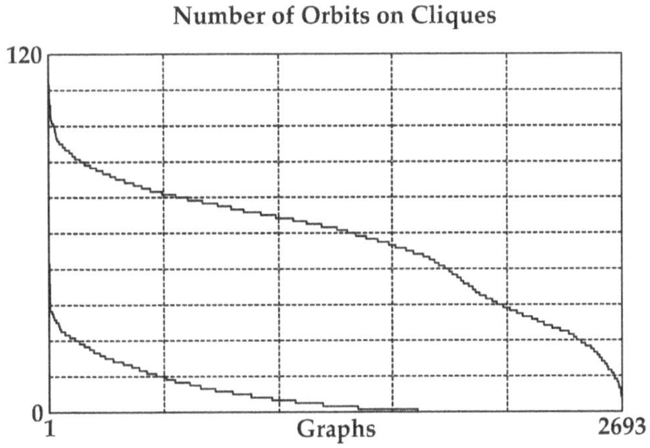

Figure 4: Distribution of the Number of Orbits on Rainbow Cliques per Graph when $q = 31$: Top Curve is Without Lex-Reduction, Bottom Curve is With Lex-Reduction

q	Cliques	Time
31	19,989	9 min 8 sec
37	39,969	1 hr 13 min
41	82,156	4 hrs 58 min
43	94,797	10 hrs 5 min
47	154,377	13 days
49	62,618	11 days
53	126,824	85 days
59	249,466	473 days
61	208,710	921 days
67	304,604	16 years

Table 3: Cumulative Number of Rainbow Cliques Obtained by Exhaustive Search Over All Graphs $\Gamma^*_{S,\ell}$ Associated to Starter Sets S (i.e., with Lex-Reduction)

q	Orbits on Rainbow Cliques
31	15,893
37	32,743
41	68,078
43	79,746
47	131,728
49	51,565
53	107,409
59	216,140
61	181,460
67	265,461

Table 4: Cumulative Number of Stabilizer Orbits on Rainbow Cliques with Lex-Reduction

curve is with lex-reduction. In total, there are $152,402$ orbits without lex-reduction and $15,893$ orbits with lex-reduction.

7. RESULTS OF THE SEARCH

Table 3 displays the results of the search for rainbow cliques from the starters computed in Section 3, using the lexicographic condition. For each order q, the table shows the total number of cliques that arise from the graphs $\Gamma^*_{S,\ell}$, where S runs over a system of representatives of starters and ℓ is a line on a point $P_0 \in S$. The table also shows the time to compute these cliques.

Table 4 shows the cumulative number of orbits of the groups $\mathrm{Aut}(S)$ on the cliques associated with $\Gamma^*_{S,\ell}$, where S ranges over all starter sets.

8. CORRECTNESS OF THE RESULTS

Lemma 3 allows for an easy check of correctness of the classification when lex-reduction is *not* used. In terms of the decomposition matrix (see Section 5), we sum up the entries in two different ways (by rows and by columns). We simply check if the cumulative number of $\mathrm{Aut}(S)$-orbits on cliques over all starter sets S of size s equals the cumulative number of orbits of $\mathrm{Aut}(T)$ on s-subsets of T where T runs through a set of representatives of the BLT-sets obtained from the classification in Section 7. For instance, when $q = 31$, the number of 5-orbits of $\mathrm{Aut}(T)$ for each of the 8 BLT-sets T is shown in the middle column of Table 5. The fact that we

BLT-set	5-Orbits	Special 5-Orbits
31#1	11	1
31#2	387	70
31#3	2,490	269
31#4	50,449	5343
31#5	25,277	2573
31#6	50,344	5114
31#7	20,245	2111
31#8	3,199	412
Total:	152,402	15,893

Table 5: Orbits on 5-Subsets for each of the BLT-sets of $Q(4, 31)$ (Labeling of BLT-sets according to Appendix A)

have $152,402$ orbits in total, which is the same as the number of orbits on rainbow-cliques as described in Section 6 (when lex-reduction is not used) corroborates the correctness of our classification in this case.

If lex-reduction is used, we must proceed differently. An s-orbit represented by an s-subset O of a BLT-set T is called *special* if it has the following property: Let $g \in G$ be such that $O^g = S$ is one of the chosen starter sets (such an element g exists since O is a partial BLT-set of size s and since the starters provide an exhaustive list of all partial BLT-sets of size s). The s-orbit O is special if $(S, (T \setminus O)^g)$ is admissible (in the sense of Section 6). In the third column of Table 5, we list the number of special 5-orbits for each of the 8 BLT-sets of order 31. The total number of these special orbits is $15,893$, which equals the number of orbits on rainbow cliques when lex-reduction is used (as shown in Section 6). This is evidence that the classification algorithm is correct in the case that lex-reduction is used.

9. THE PLANE TYPE

It is helpful to have invariants to distinguish between non-isomorphic BLT-sets. Let B be a BLT-set of $Q(4, q)$ inside $\mathrm{PG}(4, q)$. Let a_i be the number of planes π of $\mathrm{PG}(4, q)$ such that $|\pi \cap B| = i$. The plane type of B is the vector $(a_0, \ldots, a_{q+1})$. Counting all planes yields $\sum_{i=0}^{q+1} a_i = (q^4+q^3+q^2+q+1)(q^2+1)$. For this reason, we may omit the coefficient a_0. In addition, it suffices to list only the non-zero coefficients among the $(a_1, \ldots, a_{q+1})$. We list these coefficients in exponential notation i^{a_i}. For instance (a^m, b^n, c^o) means that there are m planes intersecting in a points, n planes intersecting in b points and o planes intersecting in c points. Several families of BLT-sets can be identified by their plane type. For instance, the linear BLT-set consists of a conic on a plane. The Fisher BLT-set consists of two halved conics on two planes. The FTWKB BLT-set has no 4 points coplanar.

10. FINAL REMARKS

It is possible to formulate the problem of finding all BLT-sets above a starter set as an exact cover problem, and then hand it over to Knuth's dancing links (DLX) algorithm [14]. It should be observed though that the size of the system tends to get large. The rows correspond to the points on all lines through points of the starter sets that are not collinear to a point of the starter set. The columns correspond to the vertices in the graph Γ_S. It seems that the large input size of this system makes this approach impractical. Also, if only one line would be used, then the covering might not be a BLT-set, since the system would not have sufficient information to tell whether two points can be added simultaneously. We have not yet discussed how to choose the line ℓ that gave us the colored graphs. We did not try very hard: We chose the first line on the first point in the starter set, using a certain total order on the lines of $Q(4, q)$. We do not know if a clever choice of line would make the algorithm faster.

11. ACKNOWLEDGEMENTS

This research was done using resources provided by the Teragrid and the Open Science Grid, which are both supported by the National Science Foundation. The Open Science Grid is also supported by the U.S. Department of Energy's Office of Science. In addition, the author thanks Tim Penttila for introducing him to the idea of using cliques to classify BLT-sets.

APPENDIX

A. TABLES OF BLT-SETS

In Tables 5-6, we list all BLT-sets of orders 31-67. For each BLT-set $\mathcal{B}$, we indicate: 1. A unique identifier of the form $q\#i$, with q the order and i a number that we assign to distinguish BLT-sets of order q. 2. The order of the automorphism group $A = \mathrm{Aut}(\mathcal{B})$ of $\mathcal{B}$. 3. The orbit structure of A on $\mathcal{B}$. 4. The plane type of $\mathcal{B}$ (omitting 3-planes, 2-planes and 1-planes, as they can be computed). 5. The common name of $\mathcal{B}$ (if there is one). We use the following shortcuts: Orb = Orbit structure on points, Ago = Automorphism group order, PT = Plane type, L = Linear, Fi = [8], DCH = [6], PR = [21], K1 = [12], K2 = [13], K3 = GJT / [12], FTWKB = Fisher, Thas [8] / Kantor [11] / Walker [23] / Betten [4], LP = [17], KS = Kantor semifield, G = [9], GJT = [10], DCP = De Clerck, Penttila, unpublished result.

ID	Ago	Orb	PT	Ref
31#1	1904640	(32)	(32^1)	L
31#2	2048	(32)	$(4^{64}, 16^2)$	Fi
31#3	96	$(24, 6, 2)$	$(4^{342}, 5^8, 8^1)$	LP96
31#4	4	$(4^7, 2^2)$	$(4^{142}, 5^4)$	LP4b
31#5	8	$(8^3, 4^2)$	$(4^{117}, 5^8)$	LP8
31#6	4	(4^8)	$(4^{136}, 5^4)$	LP4a
31#7	10	$(10^2, 5^2, 2)$	(4^{128})	LP10
31#8	64	(32)	(4^{145})	[20]
37#1	3846816	(38)	(38^1)	L
37#2	2888	(38)	(19^2)	Fi
37#3	4	$(4^9, 2)$	$(4^{190}, 5^8)$	LP4a
37#4	4	$(4^9, 1^2)$	(4^{230})	LP4b
37#5	4	$(4^9, 2)$	(4^{210})	New
37#6	72	$(36, 2)$	(4^{216})	K2
37#7	72	$(36, 2)$	$(4^{270}, 5^{12})$	LP72
41#1	5785920	(42)	(42^1)	L
41#2	3528	(42)	(21^2)	Fi
41#3	2	(2^{21})	$(4^{258}, 5^6)$	New
41#4	3	(3^{14})	$(4^{210}, 5^9)$	New
41#5	8	$(8^5, 2)$	$(4^{276}, 5^{16}, 6^2)$	LP
41#6	24	$(24, 12, 6)$	$(4^{228}, 5^{12}, 6^1)$	LP
41#7	60	$(30, 12)$	$(4^{210}, 5^{30})$	LP
41#8	84	(42)	$(4^{126}, 7^6)$	[3]
41#9	84	(42)	(4^{147})	[20]
41#10	68880	(42)	$()$	FTWKB
43#1	6992832	(44)	(44^1)	L
43#2	3872	(44)	$(4^{121}, 22^2)$	Fi
43#3	2	$(2^{21}, 1^2)$	$(4^{297}, 5^9)$	New
43#4	4	$(4^{10}, 2^2)$	(4^{275})	New
43#5	4	(4^{11})	(4^{306})	LP
43#6	84	$(42, 2)$	(4^{210})	K
47#1	9962496	(48)	(48^1)	L
47#2	4608	(48)	$(4^{144}, 24^2)$	Fi
47#3	2304	(48)	$(4^{1656}, 6^{32}, 8^{18})$	DCP
47#4	2	$(2^{23}, 1^2)$	$(4^{371}, 5^{10})$	New
47#5	3	(3^{16})	$(4^{276}, 5^{15})$	LP
47#6	8	$(8^5, 4^2)$	$(4^{327}, 5^4)$	New
47#7	12	$(12^3, 6^2)$	$(4^{267}, 5^6, 6^2)$	New
47#8	24	(24^2)	(4^{384})	LP
47#9	92	$(46, 2)$	(4^{207})	K
47#10	103776	(48)	$()$	FTWKB
49#1	23520000	(50)	(50^1)	L
49#2	10000	(50)	(25^2)	Fi
49#3	940800	(50)	(8^{350})	KS
49#4	20	$(20^2, 5^2)$	(4^{420})	LP
49#5	40	$(40, 10)$	(4^{450})	LP
49#6	8	$(8^6, 2)$	$(4^{356}, 5^8)$	New
49#7	8	$(8^6, 2)$	(4^{408})	New
49#8	200	(50)	$(4^{350}, 5^{10})$	[20]

Figure 5: The BLT-sets, Part I

B. REFERENCES

[1] Laura Bader, Guglielmo Lunardon, and Joseph A. Thas. Derivation of flocks of quadratic cones. *Forum Math.*, 2(2):163–174, 1990.

[2] Laura Bader, Christine M. O'Keefe, and Tim Penttila. Some remarks on flocks. *J. Aust. Math. Soc.*, 76(3):329–343, 2004.

[3] Anton Betten. A class of transitive BLT-sets. *Note di Matematica* 28 (2010), 2-10.

[4] Dieter Betten. 4-dimensionale Translationsebenen mit 8-dimensionaler Kollineationsgruppe. *Geometriae Dedicata*, 2:327–339, 1973.

[5] Cynthia A. Brown, Larry Finkelstein, and Paul Walton Purdom, Jr. Backtrack searching in the presence of symmetry. In *Applied algebra, algebraic algorithms and error-correcting codes (Rome, 1988)*, volume 357 of *Lecture Notes in Comput. Sci.*, pages 99–110. Springer, Berlin, 1989.

[6] F. De Clerck and C. Herssens. Flocks of the quadratic cone in PG(3, q), for q small. The CAGe reports 8, Computer Algebra Group, The University of Ghent, Ghent, Belgium, 1992.

ID	Ago	Orb	PT	Ref
53#1	16072992	(54)	(54^1)	L
53#2	5832	(54)	(27^2)	Fi
53#3	3	(3^{18})	$(4^{387}, 5^{12})$	New
53#4	12	$(12^4, 6)$	$(4^{381}, 5^{24})$	LP
53#5	24	$(24^2, 6)$	$(4^{540}, 6^1)$	LP
53#6	8	$(8^6, 4, 2)$	$(4^{414}, 5^{20})$	New
53#7	104	$(52, 2)$	(4^{390})	K2
53#8	148824	(54)	()	FTWKB
59#1	24638400	(60)	(60^1)	L
59#2	7200	(60)	$(4^{225}, 30^2)$	Fi
59#3	8	$(8^7, 4)$	$(4^{537}, 5^8, 6^2)$	New
59#4	3	(3^{20})	$(4^{453}, 5^{24})$	New
59#5	120	(60)	$(4^{240}, 6^{30})$	LP
59#6	120	(60)	$(4^{390}, 5^{12})$	[20]
59#7	5	(5^{12})	(4^{600})	LP
59#8	24	$(24^2, 12)$	(4^{564})	LP
59#9	205320	(60)	()	FTWKB
61#1	28138080	(62)	(62^1)	L
61#2	7688	(62)	(31^2)	Fi
61#3	4	$(4^{15}, 2)$	(4^{582})	New
61#4	4	$(4^{15}, 1^2)$	$(4^{572}, 5^{12})$	New
61#5	124	(62)	(4^{496})	[20]
67#1	40894656	(68)	(68^1)	L
67#2	9248	(68)	$(4^{289}, 34^2)$	Fi
67#3	4	$(4^{16}, 2)$	(4^{761})	New
67#4	4	(4^{17})	(4^{686})	New
67#5	68	(68)	(4^{1122})	New
67#6	132	$(66, 2)$	(4^{462})	K2

Figure 6: The BLT-sets, Part II

[7] M. De Soete and J. A. Thas. A characterization theorem for the generalized quadrangle $T_2^*(O)$ of order $(s, s+2)$. *Ars Combin.*, 17:225–242, 1984.

[8] J. Chris Fisher and Joseph A. Thas. Flocks in PG(3, q). *Math. Z.*, 169(1):1–11, 1979.

[9] Michael J. Ganley. Central weak nucleus semifields. *European J. Combin.*, 2(4):339–347, 1981.

[10] H. Gevaert, N. L. Johnson, and J. A. Thas. Spreads covered by reguli. *Simon Stevin*, 62(1):51–62, 1988.

[11] William M. Kantor. Generalized quadrangles associated with $G_2(q)$. *J. Combin. Theory Ser. A*, 29(2):212–219, 1980.

[12] William M. Kantor. On point-transitive affine planes. *Israel J. Math.*, 42(3):227–234, 1982.

[13] William M. Kantor. Some generalized quadrangles with parameters q^2, q. *Math. Z.*, 192(1):45–50, 1986.

[14] D. E. Knuth. Dancing links. *eprint arXiv:cs/0011047*, November 2000. in Davies, Jim; Roscoe, Bill; Woodcock, Jim, Millennial Perspectives in Computer Science: Proceedings of the 1999 Oxford-Microsoft Symposium in Honour of Sir Tony Hoare, Palgrave, pp. 187-214.

[15] Maska Law. Flocks, generalised quadrangles and translation planes from BLT-sets. Thesis presented to the Department of Mathematics and Statistics, The University of Western Australia, March 2003.

[16] Maska Law and Tim Penttila. Classification of flocks of the quadratic cone over fields of order at most 29. *Adv. Geom.*, (suppl.):S232–S244, 2003. Special issue dedicated to Adriano Barlotti.

[17] Maska Law and Tim Penttila. Construction of BLT-sets over small fields. *European J. Combin.*, 25(1):1–22, 2004.

[18] Maska Law and Tim Penttila: Flocks, ovals and generalized quadrangles (Four lectures in Napoli, June 2000).

[19] Stanley E. Payne and Joseph A. Thas. *Finite generalized quadrangles*. EMS Series of Lectures in Mathematics. European Mathematical Society (EMS), Zürich, second edition, 2009.

[20] T. Penttila. Regular cyclic BLT-sets. *Rend. Circ. Mat. Palermo (2) Suppl.*, (53):167–172, 1998. Combinatorics '98 (Mondello).

[21] Tim Penttila and Gordon F. Royle. BLT-sets over small fields. *Australas. J. Combin.*, 17:295–307, 1998.

[22] B. Schmalz. t-Designs zu vorgegebener Automorphismengruppe. *Bayreuth. Math. Schr.*, 41:1–164, 1992. Dissertation, Universität Bayreuth, Bayreuth, 1992.

[23] Michael Walker. A class of translation planes. *Geometriae Dedicata*, 5(2):135–146, 1976.

Sub-Linear Root Detection, and New Hardness Results, for Sparse Polynomials Over Finite Fields

Jingguo Bi[*]
Institute for Advanced Study
Tsinghua University
Beijing, China
jingguobi@mail.tsinghua.edu.cn

Qi Cheng[†]
School of Computer Science
University of Oklahoma,
Norman, Oklahoma 73019,
USA.
qcheng@cs.ou.edu

J. Maurice Rojas[‡]
Department of Mathematics
Texas A&M University College
Station, Texas 77843-3368,
USA.
rojas@math.tamu.edu

ABSTRACT

We present a deterministic $2^{O(t)}q^{\frac{t-2}{t-1}+o(1)}$ algorithm to decide whether a univariate polynomial f, with exactly t monomial terms and degree $< q$, has a root in $\mathbb{F}_q$. Our method is the first with complexity *sub-linear* in q when t is fixed. We also prove a structural property for the nonzero roots in $\mathbb{F}_q$ of any t-nomial: the nonzero roots always admit a partition into no more than $2\sqrt{t-1}(q-1)^{\frac{t-2}{t-1}}$ cosets of two subgroups $S_1 \subseteq S_2$ of $\mathbb{F}_q^*$. This can be thought of as a finite field analogue of Descartes' Rule. A corollary of our results is the first deterministic sub-linear algorithm for detecting common degree one factors of k-tuples of t-nomials in $\mathbb{F}_q[x]$ when k and t are fixed.

When t is not fixed we show that, for p prime, detecting roots in $\mathbb{F}_p$ for f is **NP**-hard with respect to **BPP**-reductions. Finally, we prove that if the complexity of root detection is sub-linear (in a refined sense), relative to the *straight-line program encoding*, then $\mathbf{NEXP} \not\subseteq \mathbf{P/poly}$.

Categories and Subject Descriptors

F.2.1 [**Numerical Algorithms and Problems**]: Computations in finite fields; F.1.3 [**Complexity Measures and Classes**]: Reducibility and completeness

Keywords

Solvability, sparse polynomial, finite fields, **NP**-hardness, gcd, square-free, discriminant, resultant

*J.B. was partially supported by: 973 Program (Grant 2013CB834205) and NSF of China under grants No.61133013 & 61272035.

†Q.C. was partially supported by 973 Program (Grant 2013CB834201) and NSF under grants CCF-0830522 and CCF-0830524.

‡J.M.R. was partially supported by NSF MCS grant DMS-0915245, DOE ASCR grant DE-SC0002505, and Sandia National Laboratories.

1. INTRODUCTION

The solvability of univariate sparse polynomials is a fundamental problem in computer algebra, and an important precursor to deep questions in polynomial system solving and circuit complexity. Cucker, Koiran, and Smale [CKS99] found a polynomial-time algorithm to find all integer roots of a univariate polynomial f in $\mathbb{Z}[x]$ with exactly t terms, i.e., a *univariate t-nomial.* Shortly afterward, H. W. Lenstra, Jr. [Len99] gave a polynomial-time algorithm to compute all factors of fixed degree over an algebraic extension of $\mathbb{Q}$ of fixed degree (and thereby all rational roots). Independently, Kaltofen and Koiran [KK05] and Avendano, Krick, and Sombra [AKS07] extended this to finding bounded-degree factors of sparse polynomials in $\mathbb{Q}[x, y]$ in polynomial-time. Unlike the famous LLL factoring algorithm [LLL82], the complexity of the algorithms from [CKS99, Len99, KK05, AKS07] was relative to the *sparse encoding* (cf. Definition 2.1 of Section 2 below) and thus polynomial in $t + \log \deg f$.

Changing the ground field dramatically changes the complexity. For instance, while polynomial-time algorithms are now known for detecting real roots for trinomials in $\mathbb{Z}[x]$ [RY05, BRS09], no polynomial-time algorithm is known for tetranomials [BHPR11]. Also, detecting p-adic rational roots for trinomials in $\mathbb{Z}[x]$ was only recently shown to lie in **NP** (for a fixed prime p), as was **NP**-hardness with respect to **ZPP**-reductions for t-nomials when neither t nor p are fixed [AIRR12, Thm. 1.4 & Cor. 1.5].

Here, we focus on the complexity of detecting solutions of univariate t-nomials over finite fields.

1.1 Main Results and Related Work

While deciding the existence of a $d^{\underline{\text{th}}}$ root of an element of the q-element field $\mathbb{F}_q$ is doable in time polynomial in $\log(d) + \log q$ (see, e.g., [BS96, Thms. 5.6.2 & 5.7.2, pg. 109]), detecting roots for a *trinomial* equation $a + bx^{d_0} + cx^d = 0$ with $d > d_0 > 0$ within time sub-linear in d and q is already a mystery. (Erich Kaltofen and David A. Cox independently asked about such polynomial-time algorithms around 2003 [Kal03, Cox04].) We make progress on a natural extension of this question. In what follows, we use $|S|$ for the cardinality of a set S.

THEOREM 1.1. *Given any univariate t-nomial*

$$f(x) := c_1 + c_2x^{a_2} + c_3x^{a_3} + \cdots + c_tx^{a_t} \in \mathbb{F}_q[x]$$

with degree $< q$, we can decide, within $4^{t+o(1)}q^{\frac{t-2}{t-1}+o(1)}$ deterministic bit operations, whether f has a root in $\mathbb{F}_q$. Moreover, letting $\delta := \gcd(q-1, a_2, \ldots, a_t)$ and $\eta := \sqrt{t-1}\left(\frac{q-1}{\delta}\right)^{\frac{t-2}{t-1}}$, *the entire set of nonzero roots of f in $\mathbb{F}_q$ is a union of at most 2η cosets of two subgroups $S_1 \subseteq S_2$ of $\mathbb{F}_q^*$, where $|S_1| = \delta$, $|S_2|$ can be determined in time $4^{t+o(1)}(\log q)^{O(1)}$ and* $|S_2| \geq \frac{\delta^{\frac{t-2}{t-1}}(q-1)^{\frac{1}{t-1}}}{\sqrt{t-1}}$.

The degree assumption is natural since $x^q = x$ in $\mathbb{F}_q[x]$. Note also that deciding whether an f as above has a root in $\mathbb{F}_q$ via brute-force search takes $q^{1+o(1)}$ bit operations, assuming t is fixed.

The classic *Descartes' Rule* [SL54] implies that the number of distinct real roots of a real univariate t-nomial is at most $2t-1$, regardless of the degree. At first glance, one would think that the polynomial $x^{q-1}-1 \in \mathbb{F}_q[x]$ immediately renders a finite field analogue impossible. On the other hand, note that the nonzero roots of any binomial form a coset of a subgroup of $\mathbb{F}_q^*$. Our first main result indicates that, over a finite field, the number of *cosets needed to cover* the set of nonzero roots of a sparse polynomial f is much smaller than the degree of f. We thus obtain a finite field analogue of Descartes' Rule. We consider the new idea of counting by cosets as one of the main contributions of this paper. More to the point, Theorem 1.1 provides new structural and algorithmic information, complementing an earlier finite field analogue of Descartes' Rule [CFKLLS00, Lemma 7]. Theorem 1.1 can also be thought of as a refined, positive characteristic analogue of results of Tao and Meshulam [Tao05, Mes06] bounding the number of complex roots of unity at which a sparse polynomial can vanish (a.k.a. uncertainty inequalities over finite groups).

Note that if we pick $a_2, \ldots, a_t$ uniformly randomly in $\{-M, \ldots, M\}$ then, as $M \longrightarrow \infty$, the probability that $\gcd(a_2, \cdots, a_t) = 1$ approaches $1/\zeta(t-1)$ (see, e.g., [Chr56]). The latter quantity increases from $\frac{6}{\pi^2} \approx 0.6079$ to 1 as t goes from 3 to ∞. Our theorem thus implies that, with "high" probability, the nonzero roots of a sparse polynomial over a finite field can be divided into two components: one component consisting of no more than q^c (for some $c<1$) isolated roots, and the other component consisting of q^c cosets of a (potentially large) subgroup of $\mathbb{F}_q^*$. Put another way, if the number of roots of a univariate t-nomial in $\mathbb{F}_q^*$ is much larger than $q^{\frac{t-2}{t-1}}$, then the roots must exhibit a strong multiplicative structure.

Since detecting roots over $\mathbb{F}_q$ is the same as detecting linear factors of polynomials in $\mathbb{F}_q[x]$, it is natural to ask about the complexity of factoring sparse polynomials over $\mathbb{F}_q[x]$. The asymptotically fastest randomized algorithm for factoring arbitrary $f \in \mathbb{F}_q[x]$ of degree d uses $O(d^{1.5} + d^{1+o(1)} \log q)$ arithmetic operations in $\mathbb{F}_q$ [KU11], but no complexity bound polynomial in $t + \log(d) + \log q$ is known. (See [Ber70, CZ81, GS92, KS98, Uma08] for some important milestones, and [GP01, Kal03, vzGat06] for an extensive survey on factoring.) However, to detect roots in $\mathbb{F}_q$, we don't need the full power of factoring: we need only decide whether $\gcd(x^q - x, f(x))$ has positive degree. Indeed, a consequence of our first main result is a speed-up for a variant of the latter decision problem.

COROLLARY 1.2. *Given any univariate t-nomials $f_1, \ldots, f_k \in \mathbb{F}_q[x]$, we can decide if $f_1, \ldots, f_k$ have a common degree one factor in $\mathbb{F}_q[x]$ via a deterministic algorithm with complexity* $4^{kt-k+o(1)} q^{\frac{kt-k-1}{kt-k}+o(1)}$.

Corollary 1.2 (proved in Section 3.2) appears to give the first sub-linear algorithm for detecting roots of k-tuples of univariate t-nomials for k and t fixed.

REMARK 1.3. *It is important to note that the $k=2$ case is* not *the same as deciding whether the gcd of two* general *polynomials has positive degree: the latter problem is the same as detecting common factors of* arbitrary *degree, or degree one factors over an extension field. Finding an algorithm for the latter problem with complexity sub-linear in q is already an open problem for $k=2$ and $t \geq 3$: see [EP05], and Theorem 1.5 and Remark 1.7 below.* ◇

One reason why it is challenging to attain complexity sub-linear in q is that detecting roots in $\mathbb{F}_q$ for t-nomials is **NP**-hard when t is not fixed, even restricting to one variable and prime q.

THEOREM 1.4. *Suppose that, for any input (f,p) with p a prime and $f \in \mathbb{F}_p[x]$ a t-nomial of degree $< p$, one could decide whether f has a root in $\mathbb{F}_p$ within* **BPP**, *using $t + \log p$ as the underlying input size. Then* **NP** $\subseteq$ **BPP**.

The least n making root detection in $\mathbb{F}_p^n$ be **NP**-hard for polynomials in $\mathbb{F}_p[x_1, \ldots, x_n]$ (for p prime, and relative to the sparse encoding) appears to have been unknown. Theorem 1.4 thus comes close to settling this problem. Theorem 1.4 also complements an earlier result of Kipnis and Shamir proving **NP**-hardness for detecting roots of univariate sparse polynomials over fields of the form $\mathbb{F}_{2^\ell}$ [KiSha99]. Furthermore, Theorem 1.4 improves another recent **NP**-hardness result where the underlying input size was instead the (smaller) straight-line program complexity [CHW11].

Let $\overline{\mathbb{F}}_q$ denote the algebraic closure of $\mathbb{F}_q$. A consequence of our last complexity lower bound is the hardness of detecting *degenerate* roots over $\mathbb{F}_p$ and $\overline{\mathbb{F}}_q$:

THEOREM 1.5. *Consider the following two problems, each with input (f,p) where p is a prime and $f \in \mathbb{F}_p[x]$ is a t-nomial of degree $< p$.*

1. *Decide whether f is divisible by the square of a degree one polynomial in $\mathbb{F}_p[x]$.*
2. *Decide whether f is divisible by the square of a degree one polynomial in $\overline{\mathbb{F}}_p[x]$.*

Then, using $t + \log p$ as the underlying input size, each of these problems is **NP**-*hard with respect to* **BPP**-*reductions.*

The **NP**-hardness of both problems had been previously unknown. Theorem 1.5 thus improves [KaShp99, Cor. 2] where **NP**-hardness (with respect to **BPP**-reductions) was proved for the *harder* variant of Problem (2) where one allows f in the larger ring $\overline{\mathbb{F}}_p[x]$.

REMARK 1.6. *Note that detecting a degenerate root for f is the same as detecting a common degree one factor of f and $\frac{\partial f}{\partial x}$, at least when $\deg f$ is less than the characteristic of the field. So an immediate consequence of Theorem 1.5 is that detecting common degree one factors in $\mathbb{F}_p[x]$ (resp. $\overline{\mathbb{F}}_p[x]$) for pairs of polynomials in $\mathbb{F}_p[x]$ is* **NP**-*hard with respect to* **BPP**-*reductions. We thus also strengthen earlier work proving similar complexity lower bounds for detecting common degree one factors in $\mathbb{F}_q[x]$ (resp. $\overline{\mathbb{F}}_q[x]$) [vzGKS96, Thm. 4.11].* ◇

REMARK 1.7. *It should be noted that Problem (2) is equivalent to deciding the vanishing of univariate $\mathcal{A}$-discriminants (see [GKZ94, Ch. 12, pp. 403–408] and Definitions 2.6 and 2.8 of Section 2.2 below). While the trinomial case of Problem (2) can be done in* **P** *(see [AIRR12, Lemma 5.3]), we are unaware of any other speed-ups for fixed t. In particular, it follows immediately from Theorem 1.5 that deciding the vanishing of univariate* resultants *(see, e.g., [GKZ94, Ch. 12, Sec. 1, pp. 397–402] and Definition 2.6 of Section 2.2 below), of polynomials in $\mathbb{F}_p[x]$, is also* **NP**-*hard with respect to* **BPP**-*reductions.* ◇

Our final result is a complexity separation depending on a weak tractability assumption for detecting roots of univariate polynomials given as *straight-line programs (SLPs)*.

THEOREM 1.8. *Suppose that, given any straight-line program of size L representing a polynomial $f\in\mathbb{F}_{2^\ell}[x]$, we could decide if f has a root in $\mathbb{F}_{2^\ell}$ within time $L^{O(1)}2^{\ell-\omega(\log\ell)}$. Then* **NEXP** $\not\subseteq$ **P/poly**.

One should recall that **NEXP** $\subseteq$ **P/poly** $\Longleftrightarrow$ **NEXP** = **MA** [IKW01]. So the conditional assertion of our last theorem indeed implies a new separation of complexity classes. It may actually be the case that there is no algorithm for detecting roots in $\mathbb{F}_{2^\ell}$ better than brute-force search. Such a result would be in line with the Exponential Time Hypothesis [IP01] and the widely-held belief in the cryptographic community that the only way to break a well-designed block cipher is by exhaustive search.

1.2 Highlights of Main Techniques

Let e be a positive integer such that $gcd(e,q-1)=1$. If we replace x by x^e in

$$f(x)=c_1+c_2x^{a_2}+c_3x^{a_3}+\cdots+c_tx^{a_t}\in\mathbb{F}_q[x],$$

then we obtain

$$f(x^e)=c_1+c_2x^{ea_2}+c_3x^{ea_3}+\cdots+c_tx^{ea_t}.$$

These two polynomials have the same number of roots in $\mathbb{F}_q$ since the map from $\mathbb{F}_q$ to $\mathbb{F}_q$ given by $x\mapsto x^e$ is one-to-one. Now suppose that $(m_2,m_3,\cdots,m_t)\in\mathbb{Z}^{t-1}$ satisfies

$$m_2\equiv ea_2 \bmod q-1,\ldots,m_t\equiv ea_t \bmod q-1.$$

Then f has a root in $\mathbb{F}_q$ iff the polynomial

$$c_1+c_2x^{m_2}+c_3x^{m_3}+\cdots+c_tx^{m_t}$$

has a root in $\mathbb{F}_q$. The key new advance needed to attain our speed-ups is a method employing recent fast algorithms for the *Shortest Vector Problem (SVP)* (see [MV10] and Section 2.1). In particular, our method finds a suitable e that lowers the degree of any sparse polynomial in $\mathbb{F}_q[x]$ to a power of q strictly less than 1 while still preserving solvability over $\mathbb{F}_q$.

LEMMA 1.9. *Given integers $a_1,\cdots,a_t,N$ satisfying $0<a_1<\cdots<a_t<N$ and $\gcd(N,a_1,\cdots,a_t)=1$, one can find, within $4^t(t\log N)^{O(1)}$ bit operations, an integer e with the following property for all $i\in\{1,\ldots,t\}$: if $m_i\in\{-\lfloor N/2\rfloor,\ldots,\lceil N/2\rceil\}$ is the unique integer congruent to ea_i mod N then $|m_i|\leq\sqrt{t}N^{1-t^{-1}}$.*

We prove this lemma in Section 2.1. The lemma can be applied to the exponents of a general sparse polynomial to yield Theorem 1.1 in Section 3.1, after overcoming two potential difficulties: one can sometimes have $\gcd(q-1,a_1,\cdots,a_t)>1$ or $\gcd(e,q-1)>1$.

Recall that any Boolean expression of one of the following forms:

(◇) $y_i\vee y_j\vee y_k,\ \neg y_i\vee y_j\vee y_k,\ \neg y_i\vee\neg y_j\vee y_k,\ \neg y_i\vee\neg y_j\vee\neg y_k$,
with $i,j,k\in[3n]$, is a `3CNFSAT` *clause*. A *satisfying assigment* for an arbitrary Boolean formula $B(y_1,\ldots,y_n)$ is an assignment of values from $\{0,1\}$ to the variables $y_1,\ldots,y_n$ which makes the equality $B(y_1,\ldots,y_n)=1$ true.[1]

A key construction behind the proofs of Theorems 1.4 and 1.5 in Section 4 is a highly structured randomized reduction from `3CNFSAT` to detecting roots of univariate polynomial systems over finite fields. In particular, the finite fields arising in this reduction have cardinality coming from a very particular family of prime numbers. (See Definition 2.1 from Section 2 for our definition of input size.)

THEOREM 1.10. *Given any* `3CNFSAT` *instance $B(y_1,\ldots,y_n)$ in $n\geq 4$ variables with k clauses, there is a (Las Vegas) randomized polynomial-time algorithm that produces positive integers $c,p_1,\ldots,p_n$ and a k-tuple of polynomials $(f_1,\ldots,f_k)\in(\mathbb{Z}[x])^k$ with the following properties:*

1. *$c\geq 11$ and $\log(cp_1\cdots p_n)=n^{O(1)}$.*
2. *$p_1,\ldots,p_n$ is an increasing sequence of primes and $p:=1+cp_1\cdots p_n$ is prime.*
3. *For all i, f_i is monic, $f_i(0)\neq 0$, $\deg f_i<p_1\cdots p_n$, and $\text{size}(f_i)=n^{O(1)}$.*
4. *For all i, the mod p reduction of f_i has exactly $\deg f_i$ distinct roots in $\mathbb{F}_p$.*
5. *B has a satisfying assignment if and only if the mod p reduction of $(f_1,\ldots,f_k)$ has a root in $\mathbb{F}_p$.* ■

Theorem 1.10 is based on an earlier reduction of Plaisted involving complex roots of unity [Pla84, Sec. 3, pp. 127–129] and was refined into the form below in [AIRR12, Secs. 6.2–6.3].[2]

We now review some additional background necessary for our proofs.

2. BACKGROUND

Our main notion of input size essentially reduces to how long it takes to write down monomial term expansions, a.k.a. the *sparse encoding*.

DEFINITION 2.1. *For any polynomial $f\in\mathbb{Z}[x_1,\ldots,x_n]$ written $f(x)=\sum_{i=1}^t c_ix_1^{a_{1,i}}\cdots x_n^{a_{n,i}}$, we define*

$$\text{size}(f):=\sum_{i=1}^t\log_2\left[(2+|c_i|)(2+|a_{1,i}|)\cdots(2+|a_{n,i}|)\right].$$

Also, when $F:=(f_1,\ldots,f_k)$, we define $\text{size}(F):=\sum_{i=1}^k\text{size}(f_i)$. ◇

The definition above is also sometimes known as the *sparse* size of a polynomial. Note that $\text{size}(c)=O(\log|c|)$ for any integer c.

[1]We respectively identify 0 and 1 with "False" and "True".
[2][AIRR12] in fact contains a version of Theorem 1.10 with $c\geq 2$, but $c\geq 11$ can be attained by a trivial modification of the proof there.

A useful fact, easily obtainable from the famous *Schwartz-Zippel Lemma* is that systems of univariate polynomial equations can, at the expense of some randomization, be reduced to *pairs* of univariate equations. (See [GH93] for a multivariate version.)

LEMMA 2.2. *Given any prime power q and $f_1, \ldots, f_k \in \mathbb{F}_q[x]$, let $Z(f_1, \ldots, f_k)$ denote the set of solutions of $f_1 = \cdots = f_k = 0$ in $\overline{\mathbb{F}}_q$. Also let $d := \max_i \deg f_i$. Then at least a fraction of $1 - \frac{d}{q}$ of the $(u_2, \ldots, u_k) \in \mathbb{F}_q^{k-1}$ satisfy $Z(f_1, \ldots, f_k) = Z(f_1, u_2 f_2 + \cdots + u_k f_k)$.* ■

REMARK 2.3. *For this lemma to yield a high-probability reduction from $k \times 1$ systems to 2×1 systems, we will of course need to assume that d is a small constant fraction of q. This will indeed be the case in our upcoming applications.* ◇

Let us now observe the following complexity bound for root detection for (not necessarily sparse) polynomials over finite fields.

PROPOSITION 2.4. *Given any polynomial $f \in \mathbb{F}_q[x]$ of degree d and $N|(q-1)$, we can decide within $d^{1+o(1)}(\log q)^{2+o(1)}$ deterministic bit operations whether f has a root in the order N subgroup of $\mathbb{F}_q^*$.* ■

Since detecting roots for f as above is the same as deciding whether $\gcd(x^N - 1, f(x))$ has positive degree, the complexity bound above can be attained as follows: compute $r(x) := x^N \bmod f(x)$ via recursive squaring [BS96, Thm. 5.4.1, pg. 103], and then compute $\gcd(r(x) - 1, f(x))$ in time $d^{1+o(1)}(\log q)^{1+o(1)}$ via the Knuth-Schönhage algorithm [BCS97, Ch. 3].

2.1 Geometry of Numbers for Speed-Ups

Recall that a *lattice* in $\mathbb{R}^m$ is the set $\mathcal{L}(\mathbf{b}_1, \ldots, \mathbf{b}_d) = \left\{ \sum_{i=1}^{d} x_i \mathbf{b}_i \;\middle|\; x_i \in \mathbb{Z} \right\}$ of all integral combinations of d linearly independent vectors $\mathbf{b}_1, \ldots, \mathbf{b}_d \in \mathbb{R}^m$. The integers d and m are respectively called the *rank* and *dimension* of the lattice. The determinant $\det(\mathcal{L})$ of the lattice $\mathcal{L}$ is the volume of the d-dimensional parallelepiped spanned by the origin and the vectors of any $\mathbb{Z}$-basis for $\mathcal{L}$. Any lattice can be conveniently represented by a $d \times m$ matrix $\mathbf{B}$ with rows $\mathbf{b}_1, \ldots, \mathbf{b}_d$. The determinant of the lattice $\mathcal{L}$ can then be computed as $\det(\mathcal{L}(\mathbf{B})) = \sqrt{\det(\mathbf{B}\mathbf{B}^\top)}$.

Let $\|\cdot\|$ denote the Euclidean norm on $\mathbb{R}^n$ for any n. Perhaps the most famous computational problem on lattices is the *(exact) Shortest Vector Problem (SVP)*: Given a basis of a lattice $\mathcal{L}$, find a non-zero vector $\mathbf{u} \in \mathcal{L}$, such that $\|\mathbf{v}\| \geq \|\mathbf{u}\|$ for any vector $\mathbf{v} \in \mathcal{L} \setminus \mathbf{0}$. The following is a well-known upper bound on the shortest vector length in lattice $\mathcal{L}$.

MINKOWSKI'S THEOREM. *Any lattice $\mathcal{L}$ of rank d contains a non-zero vector $\mathbf{v}$ with $\|\mathbf{v}\| \leq \sqrt{d} \det(\mathcal{L})^{1/d}$.* ■

Given a lattice with rank d, the celebrated *LLL algorithm* [LLL82] can find, in time polynomial in the bit-size of a given basis for $\mathcal{L}$, a vector whose length is at most $2^{\frac{d}{2}}$ times the length of the shortest nonzero vector in $\mathcal{L}$. An algorithm with arithmetic complexity $d^{O(1)}4^d$, proposed in [MV10, Sec. 5] by Micciancio and Voulgaris, is currently the fastest deterministic algorithm for solving SVP. (See [Ngu11] for a survey of other SVP algorithms.)

Let us now prepare for our degree-lowering tricks. First, we construct the lattice $\mathcal{L}$ spanned by the rows of matrix $\mathbf{B}$, where

$$(\star\star) \qquad \mathbf{B} = \begin{bmatrix} a_1 & a_2 & \cdots & a_t \\ N & 0 & \cdots & 0 \\ 0 & N & \cdots & 0 \\ \vdots & \vdots & \ddots & 0 \\ 0 & 0 & \cdots & N \end{bmatrix}$$

Letting $\mathbf{v} := (m_1, m_2, \cdots, m_t)$ be the shortest vector of lattice $\mathcal{L}$, there then clearly exists an integer e such that $ea_1 \equiv m_1, \ldots, ea_t \equiv m_t \bmod N$. (In fact, e is merely the coefficient of $(a_1, \ldots, a_t)$ in the underlying linear combination defining $\mathbf{v}$.) Most importantly, the factorization of $\det(\mathcal{L})$ is rather restricted when the a_i are relatively prime.

LEMMA 2.5. *If* $\gcd(N, a_1, \ldots, a_t) = 1$ *then* $\det(\mathcal{L}) | N^{t-1}$.

Proof: Since the rows of $\mathbf{B}$ do not form a basis for the lattice $\mathcal{L}$, one can not use the formula $\det(\mathcal{L}(\mathbf{B})) = \sqrt{\det(\mathbf{B}\mathbf{B}^\top)}$ to calculate the determinant of $\mathcal{L}$. Let $\mathcal{L}_i$ denote the sublattice of $\mathcal{L}$ generated by all rows of $\mathbf{B}$ save the ith row. Clearly then, $\det(\mathcal{L}) | \det(\mathcal{L}_i)$ for all i. Moreover, we have $\det(\mathcal{L}_1) = N^t$ and, via minor expansion from the ith column of $\mathbf{B}$, we have $\det(\mathcal{L}_{i+1}) = a_i N^{t-1}$ for all $i \in \{1, \ldots, t\}$. So $\det(\mathcal{L})$ divides $a_1 N^{t-1}, \ldots, a_t N^{t-1}$ and we are done. ■

We are now ready to prove Lemma 1.9.

Proof of Lemma 1.9: From Lemma 2.5 and Minkowski's theorem, there exists a shortest vector $\mathbf{v}$ of $\mathcal{L}$ satisfying $\|\mathbf{v}\| \leq \sqrt{t} N^{1-t^{-1}}$. By invoking the exact SVP algorithm from [MV10] we can then find the shortest vector $\mathbf{v}$ in time $4^t (t \log N)^{O(1)}$. Let $\mathbf{v} := (m_1, \ldots, m_t)$. Clearly, by shortness, we may assume $|m_i| \leq N/2$ for all $i \in \{1, \ldots, t\}$. (Otherwise, we would be able to reduce m_i in absolute value by subtracting a suitable row of the matrix $\mathbf{B}$ from $\mathbf{v}$.) Also, by construction, there is an e such that $ea_i \equiv m_i \bmod N$ for all $i \in \{1, \ldots, t\}$. ■

2.2 Resultants, $\mathcal{A}$-discriminants, and Square-Freeness

Let us first recall the classical univariate resultant.

DEFINITION 2.6. *(See, e.g., [GKZ94, Ch. 12, Sec. 1, pp. 397–402].) Suppose $f(x) = a_0 + \cdots + a_d x^d$ and $g(x) = b_0 + \cdots + b_{d'} x^{d'}$ are polynomials with indeterminate coefficients. We define their* Sylvester matrix *to be the $(d+d') \times (d+d')$ matrix*

$$\mathcal{S}_{(d,d')}(f,g) := \begin{bmatrix} a_0 & \cdots & a_d & 0 & \cdots & 0 \\ & \ddots & & & \ddots & \\ 0 & \cdots & 0 & a_0 & \cdots & a_d \\ b_0 & \cdots & b_{d'} & 0 & \cdots & 0 \\ & \ddots & & & \ddots & \\ 0 & \cdots & 0 & b_0 & \cdots & b_{d'} \end{bmatrix} \begin{matrix} \left.\vphantom{\begin{matrix}a\\a\\a\end{matrix}}\right\} d' \text{ rows} \\ \left.\vphantom{\begin{matrix}a\\a\\a\end{matrix}}\right\} d \text{ rows} \end{matrix}$$

and their Sylvester resultant *to be*

$$\mathrm{Res}_{(d,d')}(f,g) := \det \mathcal{S}_{(d,d')}(f,g). \quad \diamond$$

LEMMA 2.7. *Following the notation of Definition 2.6, assume $f, g \in K[x]$ for some field K, and that a_d and $b_{d'}$ are not both 0. Then $f = g = 0$ has a root in the* *algebraic closure of K if and only if* $\mathrm{Res}_{(d,d')}(f,g)=0$. *More precisely, we have* $\mathrm{Res}_{(d,d')}(f,g)=a_d^{d'}\prod_{f(\zeta)=0} g(\zeta)$, *where the product counts multiplicity.* ■

The lemma is classical: see, e.g., [GKZ94, Ch. 12, Sec. 1, pp. 397–402], [RS02, pg. 9], and [BPR06, Thm. 4.16, pg. 107] for a more modern treatment.

We may now define a refinement of the classical *discriminant.*

DEFINITION 2.8. *(See also [GKZ94, Ch. 12, pp. 403–408].) Let* $\mathcal{A}:=\{a_1,\ldots,a_t\}\subset\mathbb{N}\cup\{0\}$ *and* $f(x):=\sum_{i=1}^t c_ix^{a_i}$, *where* $0\leq a_1<\cdots<a_t$ *and the* c_i *are indeterminates. We then define the* $\mathcal{A}$-discriminant *of* f, $\Delta_{\mathcal{A}}(f)$, *to be*

$$\mathrm{Res}_{(\bar{a}_t,\bar{a}_t-\bar{a}_2)}\left(\bar{f},\frac{\partial\bar{f}}{\partial x}\Big/x^{\bar{a}_2-1}\right)\Big/c_t^{\bar{a}_t-\bar{a}_{t-1}},$$

where $\bar{a}_i:=(a_i-a_1)/g$ *for all* i, $\bar{f}(x):=\sum_{i=1}^t c_ix^{\bar{a}_i}$, *and* $g:=\gcd(a_2-a_1,\ldots,a_t-a_1)$. ◇

REMARK 2.9. *Note that when* $\mathcal{A}=\{0,\ldots,d\}$ *we have* $\Delta_{\mathcal{A}}(f)=\mathrm{Res}_{(d,d-1)}(f,f')/c_d$, *i.e., for dense polynomials, the* $\mathcal{A}$*-discriminant agrees with the classical discriminant.* ◇

LEMMA 2.10. *Suppose* p *is any prime and* $f,g\in\mathbb{F}_p[x]$ *are relatively prime polynomials satisfying* $f(0)g(0)\neq 0$, $d:=\deg g\geq\deg f$, *and* $p>d$. *Then the polynomial* $f+ag$ *is square-free for at least a fraction of* $1-\frac{2d-1}{p}$ *of the* $a\in\mathbb{F}_p$.

Proof: For $2d-1\geq p$ the lemma is vacuous, so let us assume $2d-1<p$. Note also that the polynomial $f+ag$ is irreducible in $\mathbb{F}_p[x,a]$, since f and g have no common factors in $\mathbb{F}_p[x]$. The splitting field $L\subsetneq\overline{\mathbb{F}_p(a)}$ of $f(x)+ag(x)$ must have degree $[L:\mathbb{F}_p(a)]$ dividing $(\deg f)!$. Since $\deg f\leq d<p$, p can not divide $[L:\mathbb{F}_p(a)]$ and thus L is a separable extension of $\mathbb{F}_p(a)$, i.e., $f+ag$ has no degenerate roots in $\overline{\mathbb{F}_p(a)}$. So the classical discriminant of $f+ag$ (where the coefficients are considered as polynomials in a) is a polynomial in a that is not identically zero. Furthermore, from Definition 2.6, $\mathrm{Res}_{(d,d-1)}(f+ag,f'+ag')\in\mathbb{F}_p[a]$ has degree at most $d+d-1=2d-1$. So by Lemma 2.2, the classical discriminant of $f+ag$ is non-zero for at least $1-\frac{2d-1}{p}$ of the $a\in\mathbb{F}_p$. Thanks to Lemma 2.7, we thus obtain that $f+ag$ is square-free for at least a fraction of $1-\frac{2d-1}{p}$ of the $a\in\mathbb{F}_p$. ■

REMARK 2.11. *Just as for Lemma 2.2, we will need to assume that* d *is a small constant fraction of* q *for Lemma 2.10 to be useful. This will indeed be the case in our upcoming applications.* ◇

A stronger assertion, satisfied on a much smaller set of a, was observed earlier in the proof of Theorem 1 of [KaShp99]. For our purposes, easily finding an a with $f+ag$ square-free will be crucial.

3. FASTER ROOT DETECTION: PROVING THEOREM 1.1 AND COROLLARY 1.2

3.1 Proving Theorem 1.1

Before proving Theorem 1.1, let us first prove a result that will in fact enable sub-linear root detection in *arbitrary* subgroups of $\mathbb{F}_q^*$.

LEMMA 3.1. *Given a finite field* $\mathbb{F}_q$ *and the polynomials*

$(\star\star\star)$ $\qquad x^N-1$ *and* $c_1+c_2x^{a_2}+\cdots+c_tx^{a_t}$,

in $\mathbb{F}_q[x]$ *with* $0<a_2<\cdots<a_t<N$, $\gcd(N,a_2,\cdots,a_t)=1$, $c_i\neq 0$ *for all* i, *and* $N|(q-1)$, *there exists a deterministic* $q^{1/4}(\log q)^{O(1)}+4^t(t\log N)^{O(1)}+t^{\frac{1}{2}+o(1)}N^{\frac{t-2}{t-1}+o(1)}(\log q)^{2+o(1)}$ *algorithm to decide whether these two polynomials share a root in* $\mathbb{F}_q$. *Furthermore, for some* $\delta'|N$ *with* $\delta'\leq\sqrt{t-1}N^{\frac{t-2}{t-1}}$ *and* $\gamma\in\{1,\ldots,\delta'\}$, *the set of roots of* $(\star\star\star)$ *is equal to the union of a set of cardinality at most* $2\gamma\sqrt{t-1}N^{\frac{t-2}{t-1}}/\delta'$ *and the union of* $\delta'-\gamma$ *cosets of a subgroup of* $\mathbb{F}_q^*$ *of order* N/δ'.

Proof of Lemma 3.1: By Lemma 1.9 we can find an integer e such that, if $m_2,\ldots,m_t$ are the unique integers in the range $[-\lfloor N/2\rfloor,\lceil N/2\rceil]$ respectively congruent to $ea_2,\ldots,ea_t$, then $|m_i|<\sqrt{t-1}N^{\frac{t-2}{t-1}}$ for each $i\in\{2,\ldots,t\}$. Thanks to [MV10], this takes $4^t(t\log N)^{O(1)}$ deterministic bit operations. By [Shp96], we can then find a generator σ of $\mathbb{F}_q^*$ within $q^{1/4}(\log q)^{O(1)}$ bit operations. For any $\tau\in\mathbb{F}_q^*$, let $\langle\tau\rangle$ denote the multiplicative subgroup of $\mathbb{F}_q^*$ generated by τ.

Now, x^N-1 vanishing is the same as $x\in\langle\sigma^{\frac{q-1}{N}}\rangle$ since $N|(q-1)$. Let $\zeta_N:=\sigma^{\frac{q-1}{N}}$ and define $\delta':=\gcd(e,N)$. If $\delta'=1$ then the map from $\langle\zeta_N\rangle$ to $\langle\zeta_N\rangle$ given by $x\mapsto x^e$ is one-to-one. So finding a solution for $(\star\star\star)$ is equivalent to finding $x\in\langle\zeta_N\rangle$ such that $c_1+c_2x^{ea_2}+\cdots+c_tx^{ea_t}=0$. Thanks to Lemma 1.9, the last equation can be rewritten as the lower degree equation $c_1+c_2x^{m_2}+\cdots+c_tx^{m_t}=0$, and we may conclude our proof by applying Proposition 2.4.

However, we may have $\delta'>1$. In which case, the map from $\langle\zeta_N\rangle$ to $\langle\zeta_N\rangle$ given by $x\mapsto x^e$ is no longer one-to-one. Instead, it sends $\langle\zeta_N\rangle$ to a smaller subgroup $\langle\zeta_N^{\delta'}\rangle$ of order N/δ'. We first bound δ': re-ordering monomials if necessary, we may assume that $m_2\neq 0$. We then obtain

$\delta'=\gcd(e,N)\leq\gcd(ea_2,N)=\gcd(m_2,N)\leq|m_2|\leq\sqrt{t-1}N^{\frac{t-2}{t-1}}$.

Any element $x\in\langle\zeta_N\rangle$ can be written as ζ_N^iz for some $i\in\{0,\ldots,\delta'-1\}$ and $z\in\langle\zeta_N^{\delta'}\rangle$. It is then clear that $x^N-1=c_1+c_2x^{a_2}+\cdots+c_tx^{a_t}=0$ has a root in $\mathbb{F}_q^*$ if and only if there is an $i\in\{0,\ldots,\delta'-1\}$ and a $z\in\langle\zeta_N^{\delta'}\rangle$ with $c_1+c_2(\zeta_N^iz)^{a_2}+\cdots+c_t(\zeta_N^iz)^{a_t}=0$. Now, $\gcd(e/\delta',N/\delta')=1$. So the map from $\langle\zeta_N^{\delta'}\rangle$ to $\langle\zeta_N^{\delta'}\rangle$ given by $x\mapsto x^{e/\delta'}$ is one-to-one. By the definition of the m_i, $(\star\star\star)$ having a solution is thus equivalent to there being an $i\in\{0,\ldots,\delta'-1\}$ and a $z\in\langle\zeta_N^{\delta'}\rangle$ with $c_1+c_2\zeta_N^{a_2i}z^{m_t/\delta'}+\cdots+c_t\zeta_N^{a_ti}z^{m_t/\delta'}=0$. So define the Laurent polynomial

$$f_i(z):=c_1+c_2(\zeta_N^i)^{a_2}z^{m_2/\delta'}+\cdots+c_t(\zeta_N^i)^{a_t}z^{m_t/\delta'}$$

If f_i is identically zero then we have found a whole set of solutions for $(\star\star\star)$: the coset $\zeta_N^i\langle\zeta_N^{\delta'}\rangle$. If f_i is not identically zero then let $\ell:=\min_i\min\{m_i/\delta',0\}$. The polynomial $z^{-\ell}f_i(z)$ then has degree bounded from above by $2\sqrt{t-1}N^{\frac{t-2}{t-1}}/\delta'$. Deciding whether the pair of equations $z^{N/\delta'}-1=z^{-\ell}f_i(z)=0$ has a solution for some i takes deterministic time

$$\delta'\left(\sqrt{t-1}N^{\frac{t-2}{t-1}}/\delta'\right)^{1+o(1)}(\log q)^{2+o(1)},$$

applying Proposition 2.4 δ' times.

The final statement characterizing the set of solutions to $(\star\star\star)$ then follows immediately upon defining γ to be the number of $i\in\{0,\ldots,\delta'-1\}$ such that f_i is not identically zero. In particular, $\gamma\geq 1$ since $\deg f<N$ and thus f is not identically zero on the order N subgroup of $\mathbb{F}_q^*$. ■

EXAMPLE 3.2. *Consider any polynomial of the form*

$$f(x)=c_1+c_2x+c_3x^{2^{200}+26}+c_4x^{2^{200}+27}\in\mathbb{F}_q[x]$$

where $q := 6(2^{200}+26)+1$ *(which is a 61-digit prime) and* $c_1c_4 \neq 0$. *Considering the lattice generated by the vectors* $(1, 2^{200}+26, 2^{200}+27), (q-1, 0, 0), (0, q-1, 0), (0, 0, q-1)$, *it is not hard to see that* $(6, 0, 6)$ *is a minimal length vector in this lattice. Moreover,* $6 \cdot 1 \equiv 6$, $6(2^{200}+26) \equiv 0$, $6(2^{200}+27) \equiv 6$ *mod* $q-1$. *Letting* σ *be any generator of* $\mathbb{F}_q^*$ *it is clear that any* $x \in \mathbb{F}_q^*$ *can be written as* $x = \sigma^i z$ *for some* $i \in \{0, \ldots, 5\}$ *and* $z \in \mathbb{F}_q^*$ *satisfying* $z^{\frac{q-1}{6}} = 1$. *So then, we see that solving* $f(x) = 0$ *is equivalent to finding an* $i \in \{0, \ldots, 5\}$ *and a* $z \in \mathbb{F}_q^*$ *with*

$$\left(c_1 + c_3\sigma^{(2^{200}+26)i}\right) + \left(c_2\sigma^i + c_4\sigma^{(2^{200}+27)i}\right)z^6 = z^{\frac{q-1}{6}} - 1 = 0.$$

$\diamond$

REMARK 3.3. *Via fast randomized factoring, we can also pick out a representative from each coset of roots within essentially the same time bound. Note also that it is possible for some of the Laurent polynomials* f_i *to vanish identically: the polynomial* $1 + x - x^2 - x^3$ *and the prime* $q = 13$, *obtained by mimicking Example 3.2, provide one such example (with* $\delta' = 6$ *and* $\gamma = 1$*).* $\diamond$

We are now ready to prove our first main theorem.

Proof of Theorem 1.1: Let $\delta := \gcd(q-1, a_2, \ldots, a_t)$ and $y = x^\delta$. Then the solvability of f is equivalent to the solvability of the following system of equations:

$$c_1 + c_2y^{a_2/\delta} + \cdots + c_ty^{a_t/\delta} = 0$$
$$y^{\frac{q-1}{\delta}} = 1$$

Since $\gcd(\frac{a_1}{\delta}, \ldots, \frac{a_t}{\delta}, \frac{q-1}{\delta}) = 1$, we can solve this problem via Lemma 3.1 (with $N = \frac{q-1}{\delta}$), within the stated time bound. (Note that $q^{1/4} \leq q^{\frac{t-2}{t-1}}$ for all $t \geq 3$. Also, the computation of $\gcd(q-1, a_2, \ldots, a_t)$ is dominated by the other steps of the algorithm underlying Lemma 3.1.) Also, since $y^{\frac{q-1}{\delta}} = 1$, each solution y of the preceding 2×1 system induces exactly δ roots of f in $\mathbb{F}_q$. So we can indeed efficiently detect roots of f, and the second assertion of Lemma 3.1 gives us the stated characterization of the roots of f. In particular, S_1 is the unique order δ subgroup of $\mathbb{F}_q^*$, and S_2 is the unique order $\frac{q-1}{\delta'}$ subgroup of $\mathbb{F}_q^*$ (following the notation of the proof of Lemma 3.1).

3.2 The Proof of Corollary 1.2

Deciding whether 0 is a root of all the f_i is trivial, so let us divide all the f_i by a suitable power of x so that all the f_i have a nonzero constant term. Next, concatenate all the nonzero exponents of the f_i into a single vector of length $T \leq k(t-1)$. Applying Lemma 1.9, and repeating our power substitution trick from our proof of Theorem 1.1, we can then reduce to the case where each f_i has degree at most $2\sqrt{T}q^{1-T^{-1}}$, at the expense of $4^T(T\log q)^{O(1)}$ deterministic bit operations.

At this stage, we then simply compute

$$g(x) := ((\cdots(\gcd(\gcd(f_1, f_2), f_3), \ldots), f_k)$$

via $k-1$ applications of the Knuth-Schönhage algorithm [BCS97, Ch. 3]. This takes

$$(k-1)\left(2\sqrt{T}q^{1-T^{-1}}\right)^{1+o(1)}(\log q)^{1+o(1)}$$

deterministic bit operations. We then conclude via Proposition 2.4, at a cost of $\left(2\sqrt{T}q^{1-T^{-1}}\right)^{1+o(1)}(\log q)^{2+o(1)}$ bit operations.

Summing the complexities of our steps, we arrive at our stated complexity bound. ■

4. HARDNESS IN ONE VARIABLE: PROVING THEOREMS 1.4, 1.5, AND 1.8

4.1 The Proof of Theorem 1.4

Thanks to Theorem 1.10 we obtain an immediate ZPP-reduction from 3CNFSAT to the detection of roots in $\mathbb{F}_p$ for systems of univariate polynomials in $\mathbb{F}_p[x]$. By Lemma 2.2 and Remark 2.3 we then obtain a BPP-reduction to 2×1 systems. Let us now describe a **ZPP**-reduction from 2×1 systems to 1×1 systems.

Suppose $\chi \in \mathbb{F}_q$ is a quadratic non-residue. Clearly, the only root in $\mathbb{F}_q^2$ of the quadratic form $x^2 - \chi y^2$ is $(0, 0)$. So we can decide the solvability of $f_1(x) = f_2(x) = 0$ over $\mathbb{F}_q$ by deciding the solvability of $f_1^2 - \chi f_2^2$ over $\mathbb{F}_q$. Finding a usable χ is easily done in **ZPP** via random-sampling and polynomial-time Jacobi symbol calculation (see, e.g., [BS96, Cor. 5.7.5 & Thm. 5.9.3, pg. 110 & 113]).

So there is indeed a **BPP**-reduction from 3CNFSAT to our main problem, and we are done. ■

4.2 The Proof of Theorem 1.5

First note that the hardness of detecting common degree one factors in $\mathbb{F}_p[x]$ (or $\overline{\mathbb{F}}_p[x]$) for *pairs* of polynomials in $\mathbb{F}_p[x]$ follows immediately from Theorem 1.10 and Lemma 2.2: the proof of Theorem 1.4 above already tells us that there is a **BPP**-reduction from 3CNFSAT to detecting common roots in $\overline{\mathbb{F}}_p$ of pairs of polynomials in $\mathbb{F}_p[x]$. Thanks to Assertion (4) of Theorem 1.10, we also obtain a **BPP**-reduction to detecting common roots, in $\mathbb{F}_p$ instead, for pairs of polynomials in $\mathbb{F}_p[x]$.

So why does this imply hardness for deciding divisibility by the square of a degree one polynomial in $\overline{\mathbb{F}}_p[x]$ (or $\mathbb{F}_p[x]$)? Assume temporarily that Problem (2) is doable in **BPP**. Consider then, for any $f, g \in \mathbb{F}_p[x]$, the polynomial $H := (f + ag)(f + bg)$ where $\{a, b\} \subset \mathbb{F}_p$ is a uniformly random subset of cardinality 2. Note that should f and g have a common factor in $\overline{\mathbb{F}}_p[x]$, then H has a repeated factor in $\overline{\mathbb{F}}_p[x]$.

On the other hand, if f and g have no common factor, then $f + ag$ and $f + bg$ clearly have no common factors. Moreover, thanks to Lemma 2.10 and Remark 2.11, the probability that $f + ag$ and $f + bg$ are both square-free — and thus H is square-free — is at least $\left(1 - \frac{2d-1}{q}\right)\left(1 - \frac{2d-2}{q}\right)$, assuming f and g satisfy the hypothesis of the lemma.

In other words, to test f and g for common factors, it's enough to check square-freeness of H for random (a, b).

To conclude, thanks to Theorem 1.10, the pairs of polynomials arising from our **BPP**-reduction from 3CNFSAT satisfy the hypothesis of Lemma 2.10. Furthermore, thanks to Assertion (1) of Theorem 1.10, our success probability is at least $(1 - \frac{2}{11})^2 \geq \frac{2}{3}$, so we are done. ■

4.3 Proving Theorem 1.8

We will need the following proposition, due to Ryan Williams.

PROPOSITION 4.1. *[Wil10] Assume, for any Boolean circuit with* n *inputs and size polynomial in* n, *that the Circuit Satisfiability Problem can be solved in time* $2^{n-\omega(\log n)}$. *Then* $\mathbf{NEXP} \not\subseteq \mathbf{P/poly}$. ■

We will also need the following lemma, which is implicit in [KiSha99]. For completeness, we supply a proof below.

LEMMA 4.2. *Given a Boolean circuit with d inputs and L gates, we can find a straight-line program of size $L^{O(1)}$ for a polynomial $f \in \mathbb{F}_{2^d}[x]$ such that the circuit is satisfied if and only if f has a root in $\mathbb{F}_{2^d}$.*

Proof: A Boolean circuit can be viewed as a straight-line program using Boolean variables and Boolean operations. One can replace the Boolean operations by polynomials over $\mathbb{F}_2$:

$$x_1 \wedge x_2 = x_1 x_2$$
$$x_1 \vee x_2 = x_1 + x_2 + x_1 x_2$$
$$\neg x_1 = 1 - x_1$$

Hence a straight-line program for a Boolean function of size L with d inputs can be converted into a straight-line program for a polynomial $f(x_0, x_1, \cdots, x_{d-1}) \in \mathbb{F}_2[x_0, x_1, \cdots, x_{d-1}]$ of size $O(L)$.

Let $b(x)$ be an irreducible polynomial of degree d over $\mathbb{F}_2$. Let α be one root of $b(x)$. Then $\{1, \alpha, \alpha^2, \ldots, \alpha^{d-1}\}$ is a basis for $\mathbb{F}_{2^d}$ over $\mathbb{F}_2$. Then any element $x \in \mathbb{F}_{2^d}$ can be written uniquely as $x = x_0 + x_1\alpha + \cdots + x_{d-1}\alpha^{d-1}$, where $x_i \in \mathbb{F}_2$ for all i. So we obtain the system of linear equations

$$\begin{bmatrix} 1 & \alpha & \cdots & \alpha^{d-1} \\ 1 & \alpha^2 & \cdots & \alpha^{2(d-1)} \\ 1 & \alpha^4 & \cdots & \alpha^{4(d-1)} \\ \vdots & & & \\ 1 & \alpha^{2^{d-1}} & \cdots & \alpha^{2^{d-1}(d-1)} \end{bmatrix} \begin{bmatrix} x_0 \\ x_1 \\ x_2 \\ \vdots \\ x_{d-1} \end{bmatrix} = \begin{bmatrix} x \\ x^2 \\ x^4 \\ \vdots \\ x^{2^{d-1}} \end{bmatrix}.$$

The underlying matrix is Vandermonde and thus non-singular. So we can represent each x_i as a linear combination of $x, x^{2^1}, x^{2^2}, \ldots, x^{2^{d-1}}$ over $\mathbb{F}_{2^d}$. Replacing each x_i by the appropriate linear combination of high powers of x, in the SLP for f, we obtain our lemma. ■

Proof of Theorem 1.8: From Lemma 4.2, an algorithm as hypothesized in Theorem 1.8 would imply a $2^{\ell-\omega(\log \ell)}$ algorithm for any instance of the Circuit Satisfiability Problem of ℓ inputs and size polynomial in ℓ. By Proposition 4.1, we would then obtain $\mathbf{NEXP} \not\subseteq \mathbf{P/poly}$. ■

Acknowledgements

We would like to thank Igor Shparlinski for insightful comments on an earlier draft of this paper. We also thank the referees for their comments.

5. REFERENCES

[AIRR12] Avendaño, Martíin; Ibrahim, Ashraf, Rojas, J. Maurice; and Rusek, Korben, *"Faster p-adic Feasibility for Certain Multivariate Sparse Polynomials,"* Journal of Symbolic Computation, special issue in honor of 60th birthday of Joachim von zur Gathen, vol. 47, no. 4, pp. 454–479 (April 2012).

[AKS07] Avendaño, Martín; Krick, Teresa; and Sombra, Martín, *"Factoring bivariate sparse (lacunary) polynomials,"* J. Complexity, vol. 23 (2007), pp. 193–216.

[BS96] Bach, Eric and Shallit, Jeff, *Algorithmic Number Theory, Vol. I: Efficient Algorithms,* MIT Press, Cambridge, MA, 1996.

[BHPR11] Bastani, Osbert; C. Hillar, D. Popov, and J. M. Rojas, *"Randomization, Sums of Squares, and Faster Real Root Counting for Tetranomials and Beyond,"* Randomization, Relaxation, and Complexity in Polynomial Equation Solving, Contemporary Mathematics, vol. 556, pp. 145–166, AMS Press, 2011.

[BPR06] Basu, Saugata; Pollack, Ricky; and Roy, Marie-Francoise, *Algorithms in Real Algebraic Geometry,* Algorithms and Computation in Mathematics, vol. 10, Springer-Verlag, 2006.

[Ber70] Berlekamp, Elwyn R., *"Factoring polynomials over large finite fields,"* Math. Comp. 24, pp. 713–735 (1970).

[BRS09] Bihan, Frederic; Rojas, J. Maurice; Stella, Case E., *"Faster Real Feasibility via Circuit Discriminants,"* proceedings of International Symposium on Symbolic and Algebraic Computation (ISSAC 2009, July 28–31, Seoul, Korea), pp. 39–46, ACM Press, 2009.

[BCS97] Bürgisser, Peter; Clausen, Michael; and Shokrollahi, M. Amin, *Algebraic complexity theory,* with the collaboration of Thomas Lickteig, Grundlehren der Mathematischen Wissenschaften [Fundamental Principles of Mathematical Sciences], 315, Springer-Verlag, Berlin, 1997.

[CFKLLS00] Canetti, Ran; Friedlander, John B.; Konyagin, Sergei; Larsen, Michael; Lieman, Daniel; and Shparlinski, Igor E., *"On the statistical properties of Diffie-Hellman distributions,"* Israel J. Math. 120 (2000), pp. 23–46.

[CZ81] Cantor, David G. and Zassenhaus, Hans, *"A new algorithm for factoring polynomials over finite fields,"* Math. Comp. 36 (1981), no. 154, pp. 587–592.

[CHW11] Cheng, Qi; Hill, Joshua E.; and Wan, Daqing, *"Counting Value Sets: Algorithm and Complexity,"* Math ArXiV preprint `1111.1224` .

[Chr56] Christopher, John, *"The Asymptotic Density of Some k-Dimensional Sets,"* the American Mathematical Monthly, vol. 63, no. 6 (Jun.–Jul., 1956), pp. 399-401.

[Cox04] Cox, David A., *personal communication,* August, 2004.

[CKS99] Cucker, Felipe; Koiran, Pascal; and Smale, Steve, *"A polynomial time algorithm for Diophantine equations in one variable,"* J. Symbolic Comput. 27 (1999), pp. 21–29.

[EP05] Emiris, Ioannis Z. and Pan, Victor, *"Improved algorithms for computing determinants and resultants,"* J. Complexity (FOCM 2002 special issue), Vol. 21, no. 1, February 2005, pp. 43–71.

[vzGat06] von zur Gathen, Joachim, *"Who was who in polynomial factorization,"* Proceedings of ISSAC 2006 (B. M. Trager, ed.), pp. 2–3, ACM Press, 2006.

[vzGKS96] von zur Gathen, Joachim; Karpinski, Marek; and Shparlinski, Igor E., *"Counting curves and their projections,"* Computational Complexity 6, no. 1 (1996/1997), pp. 64–99.

[GP01] von zur Gathen, Joachim and Panario, Daniel, *"Factoring polynomials over finite fields: A survey,"* J. Symb. Comput., 31(1/2):3–17, 2001.

[GS92] von zur Gathen, Joachim and Shoup, Victor, *"Computing Frobenius maps and factoring polynomials,"* Computational Complexity 2:187–224, 1992.

[GKZ94] Gel'fand, Israel Moseyevitch; Kapranov, Misha M.; and Zelevinsky, Andrei V.; *Discriminants, Resultants and Multidimensional Determinants,* Birkhäuser, Boston, 1994.

[GH93] Giusti, Marc and Heintz, Joos, *"La détermination des points isolés et la dimension d'une variété algébrique peut se faire en temps polynomial,"* Computational Algebraic Geometry and Commutative Algebra (Cortona, 1991), Sympos. Math. XXXIV, pp. 216–256, Cambridge University

[IKW01] Impagliazzo, Russell; Kabanets, Valentine; and Wigderson, Avi, *"In Search of an Easy Witness: Exponential Time vs. Probabilistic Polynomial Time,"* Journal of Computer and System Sciences, 65(4), pp. 672–694, 2002.

[IP01] Impagliazzo, Russell and Paturi, Ramamohan, *"The Complexity of k-SAT,"* Journal of Computer and System Sciences, Volume 62, Issue 2, March 2001, pp. 367–375.

[JS07] Jeronimo, Gabriela and Sabia, Juan, *"Computing multihomogeneous resultants using straight-line programs,"* J. Symbolic Comput. 42 (2007), no. 1–2, pp. 218–235.

[Kal03] Kaltofen, Erich, *"Polynomial factorization: a success story,"* In ISSAC 2003 Proc. 2003 Internat. Symp. Symbolic Algebraic Comput. (New York, N.Y., 2003), J. R. Sendra, Ed., ACM Press, pp. 3–4.

[KK05] Kaltofen, Erich and Koiran, Pascal, *"On the complexity of factoring bivariate supersparse (lacunary) polynomials,"* ISSAC05, Proceedings of 2005 International Symposium Symbolic Algebraic Computation, ACM Press, New York, 2005.

[KS98] Kaltofen, Erich and Shoup, Victor, *"Subquadratic-time factoring of polynomials over finite fields,"* Math. Comp. 67 (1998), no. 223, pp. 1179–1197.

[KaShp99] Karpinski, Marek and Shparlinski, Igor E., *"On the computational hardness of testing square-freeness of sparse polynomials,"* Applied algebra, algebraic algorithms and error-correcting codes (Honolulu, HI, 1999), pp. 492–497, Lecture Notes in Comput. Sci., 1719, Springer, Berlin, 1999.

[KU11] Kedlaya, Kiran and Umans, C., *"Fast polynomial factorization and modular composition,"* SIAM Journal on Computing, Vol. 40, No. 6, pp. 1767–1802, 2011.

[KiSha99] Kipnis, Aviad and Shamir, Adi, *"Cryptanalysis of the HFE public key cryptosystem by relinearization,"* Advances in cryptology — CRYPTO '99 (Santa Barbara, CA), pp. 19–30, Lecture Notes in Comput. Sci. 1666, Springer, Berlin, 1999.

[LLL82] Lenstra, Arjen K.; Lenstra, Hendrik W., Jr.; Lovász, L., *"Factoring polynomials with rational coefficients,"* Math. Ann. 261 (1982), no. 4, pp. 515–534.

[Len99] Lenstra (Jr.), Hendrik W., *"Finding Small Degree Factors of Lacunary Polynomials,"* Number Theory in Progress, Vol. 1 (Zakopane-Kóscielisko, 1997), pp. 267–276, de Gruyter, Berlin, 1999.

[Mes06] Meshulam, Roy, *"An uncertainty inequality for finite abelian groups,"* European J. of Combinatorics, 27 (2006), pp. 63–67.

[MV10] Micciancio, D. and Voulgaris, P., *"A deterministic single exponential time algorithm for most lattice problems based on voronoi cell computations,"* SIAM J. Computing, special issue, to appear.

[Ngu11] Nguyen, Phong Q., *"Lattice Reduction Algorithms: Theory and Practice,"* K.G. Paterson (ed.): Eurocrypt 2011, LNCS 6632, pp. 2–6, 2011.

[Pla84] Plaisted, David A., *"New NP-Hard and NP-Complete Polynomial and Integer Divisibility Problems,"* Theoret. Comput. Sci. 31 (1984), no. 1–2, 125–138.

[RS02] Rahman, Qazi Ibadur; and Schmeisser, Gerhard, *Analytic Theory of Polynomials,* Clarendon Press, London Mathematical Society Monographs 26, 2002.

[RY05] Rojas, J. Maurice and Ye, Yinyu, *"On Solving Sparse Polynomials in Logarithmic Time,"* Journal of Complexity, special issue for the 2002 Foundations of Computation Mathematics (FOCM) meeting, February 2005, pp. 87–110.

[Sch80] Schwartz, Jacob T., *"Fast Probabilistic Algorithms for Verification of Polynomial Identities,"* J. of the ACM 27, 701–717, 1980.

[Shp96] Shparlinski, Igor E., *"On finding primitive roots in finite fields,"* Theoretical Computer Science, Vol. 157, Issue 2, 5 May 1996, pp. 273–275.

[SL54] Smith, David Eugene and Latham, Marcia L., *The Geometry of René Descartes,* translated from the French and Latin (with a facsimile of Descartes' 1637 French edition), Dover Publications Inc., New York (1954).

[Tao05] Tao, Terence, *"An Uncertainty Principle for Cyclic Groups of Prime Order,"* Math. Res. Lett. 12 (1) (2005), pp. 121–127.

[Uma08] Umans, Christopher, *"Fast polynomial factorization and modular composition in small characteristic,"* STOC'08, pp. 481–490, ACM, New York, 2008.

[Wil10] Williams, Ryan, *"Improving exhaustive search implies superpolynomial lower bounds,"* in STOC, 231–240, 2010.

[Zip89] Zippel, Richard, *"An explicit separation of relativised random polynomial time and relativised deterministic polynomial time,"* Technical report #965, Department of Computer Science, Cornell University, 1989.

A Term Rewriting System for the Calculus of Moving Surfaces

Mark Boady
Department of Computer Science
Drexel University
3141 Chestnut St.
Philadelphia, PA 19104, USA
mwb33@drexel.edu

Pavel Grinfeld
Department of Mathematics
Drexel University
Korman Center
33rd and Market Streets
Philadelphia, PA 19104, USA
pg77@drexel.edu

Jeremy Johnson
Department of Computer Science
Drexel University
3141 Chestnut St.
Philadelphia, PA 19104, USA
jjohnson@cs.drexel.edu

ABSTRACT

The calculus of moving surfaces (CMS) is an analytic framework that extends the tensor calculus to deforming manifolds. We have applied the CMS to a number of boundary variation problems using a Term Rewrite System (TRS). The TRS is used to convert the initial CMS expression into a form that can be evaluated. The CMS produces expressions that are true for all coordinate spaces. This makes it very powerful but applications remain limited by a rapid growth in the size of expressions. We have extended results on existing problems to orders that had been previously intractable. In this paper, we describe our TRS and our method for evaluating CMS expressions on a specific coordinate system. Our work has already provided new insight into problems of current interest to researchers in the CMS.

Categories and Subject Descriptors

I.1 [SYMBOLIC AND ALGEBRAIC MANIPULATION]: Simplification of expressions; I.1.3 [SYMBOLIC AND ALGEBRAIC MANIPULATION]: Languages and Systems—*Special-purpose algebraic systems*

Keywords

Calculus of Moving Surfaces, Tensor Analysis, Term Rewrite Systems

1. INTRODUCTION

In recent papers, [3] and [4], we have examined boundary variation problems using the calculus of moving surfaces (CMS). The CMS provided valuable insight into these problems. The results provided in our previous publications, [3] and [4], were derived by using a custom Term Rewrite System (TRS) and evaluated in Maple [24]. The TRS was treated as a black box in those papers. Here we detail its implementation.

In 2004, Grinfeld and Strang [10] posed the following boundary problem. *What is the series in $1/N$ for the simple Laplace eigenvalues λ_N on a regular polygon with N sides under Dirichlet boundary conditions?* More precisely, consider a regular polygon Ω_N with N sides inscribed into a unit circle Ω_∞. We view Ω_N as a boundary perturbation of Ω_∞ and ask the question of the relationship between the radial eigenvalues $\lambda_{\infty,n}$ on the unit circle and the corresponding eigenvalues $\lambda_{N,n}$ on the polygon. In [10], the idea of expressing $\lambda_{N,n}$ as a series in $1/N$ was put forth and in [11] the first several terms were computed using the calculus of moving surfaces.

Determining additional terms in this series was restricted by the rapid growth in complexity. A recipe for calculating terms was known, but the process is error-prone when done by hand and quickly becomes intractable. This problem is one of the motivations for our TRS. The TRS automates the derivation and assists the final evaluation. Our automated tools found an error in the fourth term of the hand calculation in the series expansion in [11] and was used to compute additional terms.

A model CMS problem was examined in [4]. We analyzed the evolution of Poisson's equation under arbitrary smooth deformations of the domain. Specifically, an approximate solution to $\Delta u = 1$ on a regular N-sided polygon was found. We also derived a partial series in $1/N$ for the Poisson energy E_N. This series was used to study the asymptotic behavior as $N \to \infty$.

In [3], we examined a classical question in boundary variations [17] that had been established correctly to second order. The seminal work [17] gives a third order expression. We obtained a partial series for the Laplace eigenvalues on an ellipse. The deforming surface was an expanding ellipse $\Omega(t)$ with semi-axes $A = 1$ and $B = 1 + t$. We determined the first seven Taylor terms of the series. The sixth term required calculating the sum of 11,024 terms and the seventh included 115,249 terms. This rapid increase in the number of terms is a consistent feature of high order perturbation problems. Many problems can be described by the CMS, but their calculation quickly becomes intractable.

Equations from the CMS were translated into directional rewrite rules. The direction of each rule was selected to provide results that can be evaluated, most importantly removal of the $\frac{\delta}{\delta t}$-derivative, the central differential operator in the CMS. Our rule set leads to normal forms that can be evaluated. The normal form expression is true for any coordinate system. The evaluation is performed by generating Maple code in a specific coordinate system. We demo how these rules are applied to an example problem in Section 7. Given a circle being stretched into an ellipse at rate c, determine an expression for the contour length of the shape in terms of c.

We begin with background information in Section 2. The formal language of the CMS and the signature of our TRS are given in Section 3. After listing our reduction rules in Section 4 and showing that they produce a unique normal form in Section 5, we describe the implementation in Section 6. Finally, we detail the solution to our example problem in Section 7 and describe our applications in Section 8.

2. BACKGROUND

The CMS is an extension of tensor calculus to support deforming manifolds. To our knowledge, existing algebraic packages do not address moving surfaces. A key feature of the tensor calculus is that its expressions can be evaluated in any coordinate system. A general purpose computer algebra system can be used if a problem is restricted to a specific coordinate system.

A TRS is a computational model for equational reasoning. The TRS has two components. A signature describes the language of the TRS. The signature contains all the function symbols that are used to generate terms. The TRS also contains a set of directional rewrite rules. Each rewrite rule follows the form $l \rightarrow r$. When a term is matched against the l pattern it is rewritten or reduced to the r pattern. Reduction continues until the term can no longer be matched to the left side of any rules. A term that does not match any rules is called a normal form. A normal form is the result of applying the TRS to an input term. The TRS provides an important method for algebraic simplification [1]. It can also automate mathematical formalism outside of general purpose computer algebra systems. A survey on the theory of TRSs can be found in [16].

Although no computer algebra systems have been developed for the CMS, software exists for tensor calculus. These packages originated in the theory of relativity and do not implement stationary manifolds, let alone the CMS. The two packages which closely resemble our TRS are MathTensor [26] and Cadabra [28]. In addition to rewriting, the symbolic manipulation of tensor indexes is also a challenging problem [25]. Other symbolic manipulation packages focusing on relativity have been developed.

These packages have proven quite successful in their fields, typically relativity. MathTensor has supported research in general relativity, such as [29] and [19]. It has also been successful in quantum field theory [13]. Cadabra is also used in general relativity [5, 23].

These packages could be modified to solve our example problem in Section 7, but the extensions needed for [3] and [4] would require fundamental changes to the basic data structures. This provided a motivation to begin with a new TRS instead of extending existing ones.

3. CALCULUS OF MOVING SURFACES

The signature of our TRS is derived from the formalism of the CMS. In this section, we describe the language of the CMS. We restrict our view to those objects and functions required for our model problems. The CMS has been described in great detail in earlier publications [12, 8]. The CMS, deeply rooted in tensor calculus [21], [27], [31], was originated by Jacques Hadamard. A historical review of the CMS can be found in [7].

In Section 7, we will use the CMS to examine a circle being stretched into a ellipse. At time zero, the unit circle has a contour length of 2π. We stretch the horizontal axis of the circle to $1+ct$ to create an ellipse. Our goal is to create a Taylor series, in terms of c, for the contour length at time $t = 1$. For this simple problem, a series can be calculated without the CMS. The beauty of the CMS is its generality. The CMS expressions we derive are correct for the contour length of any deforming manifold.

The fundamental object of the CMS is the tensor. A tensor is a geometric field defining a linear and homogeneous transformation [30]. In the TRS, a tensor is a named value and has a set of properties. The tensor can be a spacial tensor, existing in the entire space, or a surface tensor, restricted to a manifold. The tensor also has an ordered list of named indexes. Each index can be a spacial or surface index. The index is either a contravariant or covariant index. This property describes how the tensor changes with respect to a change in coordinate systems [30].

We give two example tensors to clarify this description $X^{i}_{.j}$ and Y_α. The symbol $X^{i}_{.j}$ is a spacial tensor named X. It has two indexes. The first index is named i and is a contravariant space index. The second index, j, is a covariant space index. Lower case Roman letters are used for space indexes and lower case Greek letters for surface indexes. Contravariant indexes are upper indexes. Covariant indexes are lower indexes. The indexes are a single ordered list. A dot is used to show that j is after i in $X^{i}_{.j}$. This will be done when the position of indexes is not clear. The second tensor, Y_α, is a surface tensor with one covariant surface index named α.

The indexes of the tensor are ordered. Returning to the previous examples, $X^{i}_{.j}$ is a two dimensional system and Y_α is a one dimensional system. To give explicit values, a coordinate system must be selected. The number of elements in each dimension is determined by the dimensions of the space.

We now give the tensor objects needed to describe our TRS. The covariant metric tensor, Z_{ij}, describes the ambient coordinate space.

The surface is described by its own coordinate system. The shift tensor $Z^{i}_{.\alpha}$ selects the surface part of a space tensor. Multiplication is the outer product. Contraction is a summation over a pair of indexes, shown by a repeated index. The covariant or contravariant property of an index can be changed by multiplication and contraction with the metric tensor. Raising or lowering an index is called index juggling. When contracting an index, the contracted name must appear exactly twice, once as a covariant index and once as a contravariant index. Both indexes must be of the same type, space or surface. This notation provides a shorthand for the summation $Z_{ki}Z^{k}_{.\alpha} = \sum_{k=\dots}\left(Z_{ki}Z^{k}_{.\alpha}\right)$.

The surface metric tensors are defined by the shift tensor $S_{\alpha\beta} = Z^{i\alpha}Z_{i\beta}$ and it's inverse $S^{\alpha\beta}$.The surface normal is given by N^i. The curvature tensor is B^{α}_{β} and its trace B^{α}_{α} is mean curvature.

We can take derivatives with respect to the space and surface coordinate. The covariant surface derivative is a function of one input ∇_α. Contravariant and spacial derivatives

Symbol	Description
C	Surface Velocity
N^i	Surface Normal
$B^{\alpha}_{.\beta}$	Curvature
$Z^i_{.\alpha}$	Shift Tensor
$R^{\gamma}_{.\alpha\beta}$	Commutator
$+$	Addition
Juxtaposition	Multiplication
Repeated Indexes	Contraction
Integer Superscript	Exponent
∇_i	Covariant Space Derivative
∇_α	Covariant Surface Derivative
$\frac{\delta}{\delta t}$	$\delta/\delta t$-derivative
$\frac{\partial}{\partial t}$	Partial Time Derivative

Table 1: TRS Signature

all exist and are defined by the index of the ∇ symbol. The mechanics can be found in [30].

The CMS describes surfaces that deform over time. The surface velocity C is the rate of deformation of the surface in the normal direction. Tensors on the surface now changes as the surface changes. We can study how these fields change over time by using the $\frac{\delta}{\delta t}$-derivative. These functions are described in [7]. The mean curvature of an ellipse can be described with respect to time, and the derivative taken directly. The curvature of a red blood cell is more difficult to describe and evaluate [22].

We introduce the commutator tensor to facilitate switching the order of $\frac{\delta}{\delta t}$ and ∇_α [7]. It is a shorthand for the following:

$$R^{\gamma}_{.\alpha\beta} = \nabla^{\gamma}(CB_{\alpha\beta}) - \nabla^{\alpha}(CB^{\gamma}_{\beta}) - \nabla_{\beta}(CB^{\gamma}_{\alpha}) \quad (1)$$

The signature of our TRS is summarized in Table 1. Tensors that are only required for the evaluation of expressions are not included. Transformations given by index juggling are implemented but not explicitly listed. The integers and rationals are implemented but not listed in the table.

4. REWRITE RULES

The majority of our rewrite rules are related to derivative calculations. For the reduction rules, index names are always considered variables. Unless explicitly noted, all other properties of the index must match exactly. F and G are variables for general tensors. In the CMS any valid expression is a tensor. If no indexes are attached to the variables F and G, then any combination of indexes is valid. Constant integers and rationals are given by $x_1, x_2, \cdots, x_n$.

The covariant and contravariant derivatives are defined by rules (2) to (5). These rules are true for any index of the derivative. Only the rules for ∇_α are shown. These rules are repeated for the other indexes of the derivative. The summation symbol is used to show a contraction summation in rule (5).

$$\nabla_\alpha(FG) \to G\nabla_\alpha(F) + F\nabla_\alpha(G) \quad (2)$$
$$\nabla_\alpha(x_1) \to 0 \quad (3)$$
$$\nabla_\alpha(F+G) \to \nabla_\alpha(F) + \nabla_\alpha(G) \quad (4)$$
$$\nabla_\alpha(\sum_i F^{\cdots i \cdots}_{\cdots i \cdots}) \to \sum_i \nabla_\alpha(F^{\cdots i \cdots}_{\cdots i \cdots}) \quad (5)$$

The $\frac{\delta}{\delta t}$-derivative is at the heart of the CMS. Calculating expressions using this derivative was the original motivation for our TRS. Although a mechanism for evaluating this derivative exists, it is challenging to evaluate. In [22], the Helfrich energy governing the shape of a red blood cell is examined. The fourth order derivative is essential to its analysis, but remains an open problem for a distorted torus. The application of the $\frac{\delta}{\delta t}$-derivative to the problem in [10] becomes intractable because some fields are described by infinite Fourier series. Our target normal form is an expression where $\frac{\delta}{\delta t}$-derivatives can be evaluated. The only appearance of the $\frac{\delta}{\delta t}$-derivative is when it is applied to the Surface Velocity or Commutator tensors.

First, the differentiation table for specific tensor objects is given.

$$\frac{\delta B^{\alpha}_{.\beta}}{\delta t} \to \nabla^{\alpha}\nabla_{\beta}C + CB^{\alpha}_{.\gamma}B^{\gamma}_{.\beta} \quad (6)$$
$$\frac{\delta B^{\alpha\beta}}{\delta t} \to \nabla^{\alpha}\nabla^{\beta}C + 3CB^{\alpha}_{.\gamma}B^{\gamma\beta} \quad (7)$$
$$\frac{\delta B_{\alpha\beta}}{\delta t} \to \nabla_{\alpha}\nabla_{\beta}C - CB_{\alpha\gamma}B^{\gamma}_{.\beta} \quad (8)$$
$$\frac{\delta N^i}{\delta t} \to -Z^i_{.\alpha}\nabla^{\alpha}C \quad (9)$$
$$\frac{\delta N_i}{\delta t} \to -Z_{i\alpha}\nabla^{\alpha}C \quad (10)$$
$$\frac{\delta Z^i_{.\alpha}}{\delta t} \to N^i\nabla_{\alpha}C - CZ^i_{.\beta}B^{\beta}_{.\alpha} \quad (11)$$
$$\frac{\delta Z_{i\alpha}}{\delta t} \to N_i\nabla_{\alpha}C - CZ_{i\beta}B^{\beta}_{.\alpha} \quad (12)$$
$$\frac{\delta Z^{i\alpha}}{\delta t} \to N^i\nabla^{\alpha}C - CZ^i_{.\beta}B^{\beta\alpha} \quad (13)$$
$$\frac{\delta Z_i^{.\alpha}}{\delta t} \to N_i\nabla^{\alpha}C - CZ_{i\beta}B^{\beta\alpha} \quad (14)$$
$$\frac{\delta C^{x_1}}{\delta t} \to x_1 C^{x_1-1}\frac{\delta C}{\delta t} \quad (15)$$
$$\frac{\delta x_1}{\delta t} \to 0 \quad (16)$$

The $\frac{\delta}{\delta t}$-derivative commutes with contraction and satisfies the sum and the product rules. The summation symbol is again used to show a contraction.

$$\frac{\delta \sum_i F^{\cdots i \cdots}_{\cdots i \cdots}}{\delta t} \to \sum_i \frac{\delta F^{\cdots i \cdots}_{\cdots i \cdots}}{\delta t} \quad (17)$$
$$\frac{\delta FG}{\delta t} \to G\frac{\delta F}{\delta t} + F\frac{\delta G}{\delta t} \quad (18)$$
$$\frac{\delta(F+G)}{\delta t} \to \frac{\delta F}{\delta t} + \frac{\delta G}{\delta t} \quad (19)$$

Reordering the $\frac{\delta}{\delta t}$-derivative and surface derivative introduces a collection of rules. These rules are given for the variable tensor with no indexes A. For each index in A, an additional term is added to the sum. Examples for all variations of A with one index are shown.

$$\frac{\delta \nabla_\alpha A}{\delta t} \to \nabla_\alpha \frac{\delta A}{\delta t} \tag{20}$$

$$\frac{\delta \nabla_\alpha A^\beta}{\delta t} \to\to \nabla_\alpha \frac{\delta A^\beta}{\delta t} + R^\beta_{.\alpha\gamma} A^\gamma \tag{21}$$

$$\frac{\delta \nabla_\alpha A_\beta}{\delta t} \to\to \nabla_\alpha \frac{\delta A_\beta}{\delta t} - R^\gamma_{.\alpha\beta} A_\gamma \tag{22}$$

$$\frac{\delta \nabla^\alpha A}{\delta t} \to \nabla^\alpha \frac{\delta A}{\delta t} + 2CB^{\alpha\gamma} \nabla_\gamma A \tag{23}$$

$$\frac{\delta \nabla^\alpha A^\beta}{\delta t} \to \nabla^\alpha \frac{\delta A^\beta}{\delta t} + 2CB^{\alpha\gamma} \nabla_\gamma A^\beta + R^{\beta\alpha}_{..\gamma} A^\gamma \tag{24}$$

$$\frac{\delta \nabla^\alpha A_\beta}{\delta t} \to \nabla^\alpha \frac{\delta A_\beta}{\delta t} + 2CB^{\alpha\gamma} \nabla_\gamma A_\beta - R^{\gamma\alpha}_{..\beta} A_\gamma \tag{25}$$

For each index in the A tensor, a new $R^{:::}_{:::}A^{:::}_{:::}$ term is added. The form and contraction are defined by the index. The four possible terms are shown. The sign of the term is determined by the contracted index's status as covariant or contravariant.

The partial derivative $\frac{\partial}{\partial t}$ is defined by the following rules.

$$\frac{\partial FG}{\partial t} \to F\frac{\partial G}{\partial t} + G\frac{\partial F}{\partial t} \tag{26}$$

$$\frac{\partial x_1}{\partial t} \to 0 \tag{27}$$

$$\frac{\partial (F+G)}{\partial t} \to \frac{\partial F}{\partial t} + \frac{\partial G}{\partial t} \tag{28}$$

$$\frac{\partial \sum_i F^{\cdots i \cdots}_{\cdots i \cdots}}{\partial t} \to \sum_i \frac{\partial F^{\cdots i \cdots}_{\cdots i \cdots}}{\partial t} \tag{29}$$

$$\frac{\partial \nabla_\alpha F}{\partial t} \to \nabla_\alpha \frac{\partial F}{\partial t} \tag{30}$$

Rule (30) is true for all index variations of the derivative.

To complete the TRS, we add a few additional rules for simplification.

$$A^{x_1} A^{x_2} \to A^{x_1+x_2} \tag{31}$$

$$A + 0 \to A \tag{32}$$

$$A(F+G) \to AF + AG \tag{33}$$

$$0\,A \to 0 \tag{34}$$

Expressions with rationals are calculated immediately. After reaching a normal form, like terms are combined to decrease the size of the result term. This is handled by a separate routine outside the TRS. We store the terms as an expression tree and implement tree matching algorithms to combine terms.

5. CONFLUENCE AND TERMINATION

For successful application of our TRS, it is crucial that expressions only contain symbols that can be evaluated. The TRS terminates when it reaches a normal form, a term that does not match any reduction rules. We oriented our rules so that if a normal form is reached it can be evaluated. Additionally, we required that the TRS provides a consistent output given a consistent input.

If a term matches more than one rule, the decision of which rule to apply could result in multiple normal forms. If a TRS is confluent, all paths that diverge will eventually converge. This property ensures that normal forms are unique. We will briefly justify that our system is confluent and terminating. A more comprehensive examination of these concepts is presented in [16].

Our TRS includes associative and commutative operators, but both these properties can be extended to TRSs modulo sets of equations as shown in [18] and [2]. This extensions means that normal forms will be unique in the equivalence class defined by the associative and commutative properties. Index juggling and renaming of contracted indexes follow the same logic as these properties, allowing them to be used as an equivalence class.

Critical pairs can be used to prove confluence in a TRS as shown in the classic work by Knuth & Bendix [20]. A terminating TRS is confluent if and only if no critical pair diverges [20, 15]. This can be extended to TRS modulo equational theories [18].

A critical pair (s, t), is a pair of terms created by a pattern that matches multiple rules. Using the tensor variables F and G we create a pattern that matches reduction rules (32) and (33). The pattern $F(G+0)$ is matched to both rules. A critical pair is created by following both possible reduction paths. First, $F(G+0) \to_{32} FG$. The second possible reduction is $F(G+0) \to_{33} FG + F0$. The critical pair is the set $(FG, FG + F0)$. The critical pair (s,t) converges if $s \twoheadrightarrow s'$, $t \twoheadrightarrow t'$, and $s' = t'$. The $\twoheadrightarrow$ notation is used to show a sequence of reductions. In this case, equals is defined specifically as syntactically equal with respect to associativity, commutativity, and index juggling. For our example critical pair, FG has no remaining reductions and $FG + F0 \to_{32} FG$. This shows that the critical pair converges to the term FG. It is easy to show that there exist no diverging critical pairs in our TRS, therefore it is confluent if it is terminating.

A TRS is terminating if there exists a well-founded ordering on the terms. Again referring to [20], a TRS is terminating if there exists a well-founded ordering $>$ such that for each reduction rule $l \to r$, $l > r$. An ordering is well-founded if it contains a minimum element. We define the ordering on our reduction rules by giving an ordering on the signature and referring to the process of Associative Path Ordering described in [2]. A termination ordering on the function symbols is given in equation (35). For readability, we have introduced symbols: x^y for exponents, $\sum$ for contraction, and * for multiplication. A constant is any function symbol that takes no inputs, therefore both 10 and B^α_β would be considered constants for this ordering.

$$\frac{\delta}{\delta t} > \frac{\partial}{\partial t} > \nabla > * > x^y > + > \sum > Constant \tag{35}$$

We now compare the left and right hand sides of our rules using this ordering. Rules that select elements, such as rule (32), or reduce to a constant, rule (16), are trivially ordered.

The remaining rules are ordered using a recursive examination of the function symbols. As an example, we examine rule (20). The outermost function symbol for this rule is decreasing by the ordering from equation (35), $\frac{\delta}{\delta t} > \nabla$. In addition, we must recursively compare the inputs to the function on the right hand side. The rule is not decreasing if a larger element has been created as a subterm of the right hand side. This means that the next test is between $\frac{\delta \nabla_\alpha A}{\delta t}$ and $\frac{\delta A}{\delta t}$. In this recursive test, the outermost function symbols are equal and $\nabla_\alpha A > A$. This proves that the entire reduction rule is decreasing with respect to our ordering.

In rule (21), the same process is used. The outermost level is decreasing, $\frac{\delta}{\delta t} > +$. The first input to the addition function is less then the left side by the same process we used for rule (20). It remains to be shown that $\frac{\delta \nabla_\alpha A^\beta}{\delta t} > R^\beta_{\alpha\gamma} A^\gamma$. A continuation of the recursive process shows that this ordering is correct and the reduction is decreasing. By extension, the reduction patterns given by rules (21) to (25) are decreasing.

The process of critical pair convergence and path ordering show that our TRS is both confluent and terminating. This means that any input will lead to a unique normal form. These properties ensure our TRS produces expressions that we can evaluate.

This is true for the CMS objects for which we have defined rules. Additional objects, such as ϵ_{ijk}, a Levi-Civita Symbol, have rules that do not lead to obvious termination orderings. We will attempt the Knuth-Bendix Algorithm [20] when extending our rule set for these objects.

6. IMPLEMENTATION

We implemented our TRS in the Java programming language. In addition to providing basic TRS functionality, we also implemented automatic code generation to Maple. This allowed for the automated testing of results without an intermediate program. The TRS implemented class objects for the signature described in Section 3. Expressions were then stored in a tree structure. The rules from Section 4 were implemented to match and transform the expression tree. The associative and commutative properties of functions were handled by allowing these objects children to have an unordered set of children and flattening repeated function applications. Automating the solution to a problem such as that in Section 7 required creating a driver program to generate terms, call the TRS for reduction to normal form, and then generate Maple code. The expression tree resulting from one iteration was used to create the starting term of the next iteration.

There were a number of reasons we decided to implement a custom TRS instead of building on an existing TRS such as Maude [6] or Mathematica [32]. The symbols of the CMS are inherently objects. This meant that implementing them as objects proved more efficient then using a flat functional notation. The language of object oriented programming closely matches the native language of the CMS. The ability to use inheritance also decreased the total number of rules that needed to be implemented. For example, rules (2) to (5) were easily implemented using an inheritance model on derivatives. The sequence of rules starting with rule (20), was also simple to implement using loops and objects. These problems could all be overcome in a standard TRS framework, but we felt that these gains were significant. Additionally, using a custom TRS allowed for two additional components to be built in the same framework. To minimize the number of terms that need to be evaluated, it is convenient to combine terms using associativity, commutativity, and index juggling. The most important of these equivalence classes is the ability to exchange the contravariant and covariant property of contracted indexes. While we could implemented a secondary module for compressing the normal form by combining terms, it was helpful to do this as part of the TRS. Since the result of the TRS was already a tree structure, matching subterms was reduced to a graph matching problem. A final motivation for creating our own TRS was code generation.

To evaluate expressions, they were translated to a computer algebra system. The object oriented nature of our TRS allowed this code generation to be encapsulated by the object classes. Maple code for the expression could then be generated using a simple tree walk. We created a custom Maple library for working with CMS expressions. Maple provides a library for tensor calculus, but it does not provide support for embedded manifolds, which required more general data structures and operations. Our libraries directly implemented the textbook definitions of the objects and functions from [30] and [7].

We initialized our Maple worksheet by defining the surface and spacial coordinate system. We provided function calls to create the primary objects which are generated from the coordinate system. Using function calls, we built up the remaining objects. For example, B^α_α is created by evaluating $A^{\alpha\beta}B_{\alpha\beta}$. The Maple code for this using our library is `BAb:=contract(prod(SAB,Bab),[2,3]):`. Our Maple code can generate all the objects in our signature. We can describe our expressions using these objects and functions. The generated code allows us to evaluate our CMS expressions for a defined coordinate system.

7. CONTOUR LENGTH

In this section, we show an example of calculations performed using our TRS. We have selected a simple contour length problem. This problem is based on [4] but can be easily calculated outside the CMS.

Consider the problem of evaluating the contour length of an ellipse with semiaxes $1+c$ and 1. It is a simple problem for which the answer is known by other means. This makes it a good test problem. The CMS approaches this problem by considering a smooth evolution from the unit circle at time $t=0$ to the ellipse at time $t=1$. Such an evolution may be parameterized as follows:

$$x(t,\theta) = (1+ct)\cos\theta \tag{36}$$
$$y(t,\theta) = \sin\theta \tag{37}$$

Let L(t) denote the contour length at time t:

$$L(t) = \int_{S(t)} 1dS \tag{38}$$

The contour length of the ellipse with semiaxes 1+c and 1 is given by L(1). It is estimated by the Taylor series

$$L(1) = L(0) + L'(0) + \frac{1}{2}L^2(0) + \frac{1}{6}L^3(0) + \cdots \tag{39}$$

where the expressions for the derivatives, $L^n(0)$, are derived by evaluating expressions generated by our TRS. We calculate repeated derivatives and evaluate them at $t=0$. The first term in the series is $L(0) = \int_S 1dS = 2\pi$.

The CMS is used to determine the higher order derivatives of $L(t)$. The first derivative is

$$L'(t) = \frac{\delta}{\delta t}\int_S 1dS$$
$$= \int_S (\frac{\delta 1}{\delta t} - CB^{\alpha}_{\alpha})dS$$
$$= -\int_S CB^{\alpha}_{\alpha}dS \quad (40)$$

The CMS provides a means for taking the $\frac{\delta}{\delta t}$ derivatives of an integral $\int_S dS$. In addition to the base derivative $\frac{\delta 1}{\delta t} = 0$, new terms are added to account for the changing surface.

The integral range will be the same for each order derivative. Our TRS calculates the series of integrands, which are then evaluated and integrated in Maple. To calculate the nth order derivative, we determine

$$M_n = \frac{\delta M_{n-1}}{\delta t} - CB^{\alpha}_{\alpha}M_{n-1} \quad (41)$$
$$L^n(0) = \int_S M_n|_{t=0} dS \quad (42)$$

Having already determined the first integrand $M_1 = -CB^{\alpha}_{\alpha}$, we apply this recursive definition for M_2.

$$M_2 = -\frac{\delta(CB^{\alpha}_{\alpha})}{\delta t} + C^2 B^{\alpha}_{\alpha} B^{\beta}_{\beta} \quad (43)$$

Equation (43) is the first expression that requires rewriting by our TRS. This expression is invariant; it is valid for the contour length of any surface deformation. In our specific case, the value $\frac{\delta B^{\alpha}_{\alpha}}{\delta t}$ can be evaluated directly. For many surfaces, evaluating the derivative of mean curvature is extremely complex. Our TRS produces a normal form where all terms can be evaluated. This means the $\frac{\delta}{\delta t}$-derivatives will only be on the Surface Velocity and Commutator. These are defined by the change in surface and their $\frac{\delta}{\delta t}$-derivatives can be evaluated.

The TRS reduces this expression to a normal form. The total reduction takes 31 rewrites including structure changes like rule (17) and simple reductions like rule (32). We highlight some of key reductions below. Subscripts are attached to the arrow symbol referencing the rule list in Section 4.

$$\frac{\delta(-CB^{\alpha}_{\alpha})}{\delta t} + C^2 B^{\alpha}_{\alpha} B^{\beta}_{\beta} \to_{18} -B^{\alpha}_{\alpha}\frac{\delta C}{\delta t} - C\frac{\delta B^{\alpha}_{\alpha}}{\delta t} + C^2 B^{\alpha}_{\alpha} B^{\beta}_{\beta}$$

$$\to_6 -B^{\alpha}_{\alpha}\frac{\delta C}{\delta t} - C(\nabla^{\alpha}\nabla_{\alpha}C + CB^{\alpha}_{\gamma}B^{\gamma}_{\alpha}) + C^2 B^{\alpha}_{\alpha} B^{\beta}_{\beta}$$
$$\to_{33} -B^{\alpha}_{\alpha}\frac{\delta C}{\delta t} - C\nabla^{\alpha}\nabla_{\alpha}C - C^2 B^{\alpha}_{\gamma}B^{\gamma}_{\alpha} + C^2 B^{\alpha}_{\alpha} B^{\beta}_{\beta} \quad (44)$$

The normal form for M_2 is given by equation (44). This expression is true for the contour length of any deforming manifold. We now generate code to evaluate the expression for our realization of the problem. Given a specific coordinate system, our Maple library calculates the value of the symbols. We precalculate the derivatives of C as $C0, C1, C2, \cdots$. Tensors are stored as multidimensional Maple Arrays. The Maple tensor library does not support rectangular tensors, so we created our own functions using the existing libraries. The Maple code used for evaluation is shown below.

```
Temp0:=contract(prodlist(intTensor(-1/1), BAb, C1),
[1,2]):
Temp0:=Temp0:-apply([θ,0]):
Temp1:=contract(prodlist(intTensor(-1/1), C0, ddSA(
ddSa( C0 ))), [1,2]):
Temp1:=Temp1:-apply([θ,0]):
Temp2:=contract(prodlist(intTensor(-1/1), BAb, BAb,
TensorExp(C0, 2)), [1,4 , 2,3]):
Temp2:=Temp2:-apply([θ,0]):
Temp3:=contract(prodlist( BAb, BAb, TensorExp( C0,
2)), [3,4 , 1,2]):
Temp3:=Temp3:-apply([θ,0]):
solution:=lin_com(Temp0, Temp1, Temp2, Temp3);
```

Each term in the sum is calculated, evaluated at $t = 0$, and then the sum is evaluated. The result of this expression in our coordinate system, equation (48), is

$$M_2|_{t=0} = c^2(7\cos^2\theta - 5)\cos^2\theta \quad (45)$$

Taking the integral gives the second term in the series.

$$\int_0^{2\pi} (M_2|_{t=0})\, d\theta = \frac{1}{4}c^2\pi \quad (46)$$

The TRS repeats this process and determines the normal form of M_3 which requires 118 rewrites. In addition, we have combined terms to shorten the expression.

$$M_3 = -C^3 B^{\alpha}_{\alpha}B^{\beta}_{\beta}B^{\gamma}_{\gamma} + 3C^3 B^{\alpha}_{\beta}B^{\beta}_{\alpha}B^{\gamma}_{\gamma} - 2C^3 B^{\alpha}_{\beta}B^{\gamma}_{\alpha}B^{\beta}_{\gamma}$$
$$+ 3C^2 B^{\alpha}_{\alpha}\nabla^{\beta}\nabla_{\beta}C - 4C^2 B^{\alpha\beta}\nabla_{\beta}\nabla_{\alpha}C$$
$$+ 3C\frac{\delta C}{\delta t}B^{\alpha}_{\alpha}B^{\beta}_{\beta} - 3C\frac{\delta C}{\delta t}B^{\alpha}_{\beta}B^{\beta}_{\alpha} - 2\frac{\delta C}{\delta t}\nabla^{\alpha}\nabla_{\alpha}C$$
$$- \frac{\delta^2 C}{\delta^2 t}B^{\alpha}_{\alpha} - C\nabla^{\alpha}\nabla_{\alpha}\frac{\delta C}{\delta t} + CR^{\alpha\beta}_{\beta}\nabla_{\alpha}C \quad (47)$$

Equation (47) already shows the rapid growth of expressions in the CMS. The challenge of calculating M_4 without an automated system is obvious. M_4 is the sum of 94 products and requires 595 rewrites. An important feature of the CMS remains in M_3, this expression is valid for any surface deformation.

Although these expressions are true for any surface, the values for each tensor can only be defined after selecting surface and space coordinate systems. In Maple, these are defined as global variables and the objects are calculated by library calls. We will now evaluate equation (40) to show the process. For the ambient space, we will use Cartesian coordinates $[x, y]$. For the surface, we will define a circle that is being stretched into an ellipse. The function $S(\theta, t)$ maps a point in the surface coordinates to the ambient coordinates. The constant c controls the rate of change over time.

$$S(\theta, t) = [(1 + ct)\cos\theta, \sin\theta] \quad (48)$$

First, we generate the tensors needed to determine B^{α}_{α} and C. We evaluate all tensors at $t = 0$. The shift tensor is defined by the surface, it is used to restrict a spacial field to the surface.

$$Z^i_{\alpha} = \begin{bmatrix} [-(1+ct)\sin\theta] \\ [\cos\theta] \end{bmatrix} \quad (49)$$

The surface normal, N_i, surface velocity, C, and surface metric $S^{\alpha\beta}$, are also derived from the surface.

$$N_i|_{t=0} = \begin{bmatrix}\cos\theta \\ \sin\theta\end{bmatrix}$$
$$C|_{t=0} = c\cos^2\theta$$
$$S^{\alpha\beta}|_{t=0} = [[1]]$$

Having defined these symbols, we now evaluate the mean curvature B^α_α.

$$\nabla_\beta N_i|_{t=0} = \left[\begin{bmatrix}-\sin\theta \\ \cos\theta\end{bmatrix}\right]$$
$$B_{\alpha\beta}|_{t=0} = \left(-Z^i_\alpha \nabla_\beta N_i\right)|_{t=0} = [[-\sin^2\theta - \cos^2\theta]]$$
$$B^\alpha_\beta|_{t=0} = (S^{\alpha\gamma}B_{\gamma\beta})|_{t=0} = [[-\sin^2\theta - \cos^2\theta]]$$
$$B^\alpha_\alpha|_{t=0} = -\sin^2\theta - \cos^2\theta = -1$$

Now that the values of C and B^α_α are known, we evaluate

$$M_1|_{t=0} = -CB^\alpha_\alpha = c\cos^2\theta \tag{50}$$

Next we determine the integral of equation (50) to find $L'(0)$.

$$L'(0) = \int_0^{2\pi} c\cos^2\theta d\theta = c\pi \tag{51}$$

The series for $L(1)$ can be calculated independently of the CMS to determine if this result is correct.

$$\begin{aligned} L(1) &= \int_0^{2\pi}\left(\sqrt{(1+c)^2\cos^2(\theta)+\sin^2(\theta)}\right)d\theta \\ &= \left(2 + c + \frac{1}{8}c^2 - \frac{1}{16}c^3 + \frac{17}{512}c^4 - \frac{19}{1024}c^5 + \cdots\right)\pi \end{aligned} \tag{52}$$

We calculate the next two terms in the series using equations (44) and (47).

$$\frac{1}{2!}L^2(0) = \frac{1}{2}\left(\frac{1}{4}c^2\pi\right) = \frac{1}{8}c^2\pi \tag{53}$$
$$\frac{1}{3!}L^3(0) = \frac{1}{6}\left(-\frac{3}{8}\pi c^3\right) = -\frac{1}{16}c^3\pi \tag{54}$$

Using our TRS and evaluating in Maple, we were able to confirm that the first 7 terms in the series are correct.

8. APPLICATIONS

We have applied our TRS to problems that are of current interest to researchers in the CMS. In [4], our implementation was used to analyze the evolution of Poisson's equation under arbitrary smooth deformations of the domain. We minimized the energy E

$$E = \int_\Omega\left(\frac{1}{2}\nabla_i u\nabla^i u + fu\right)d\Omega \tag{55}$$

on the domain Ω with the zero Dirichlet boundary condition

$$u|_S = 0 \tag{56}$$

Order	Products	TRS	Maple
2	5	85 millisec	1 second
3	25	149 millisec	1 second
4	152	746 millisec	6 seconds
5	1,138	7,169 millisec	2.6 minutes

Table 2: Number of terms and time to evaluate the Poisson series.

Order	Products	TRS	Maple
2	4	1 sec	1 sec
3	24	1 sec	1 sec
4	155	5 sec	4 min
5	1,221	42 sec	6 min
6	11,024	11 min	1.5 hours
7	115,249	3 hours	18 hours

Table 3: Number of terms and time to evaluate the $\lambda(c)$ series.

along the boundary S. Minimizing u is governed by the Poisson equation

$$\nabla_i u \nabla^i u = f. \tag{57}$$

We have calculated the first 5 derivatives and evaluated them at time $t = 0$, providing the terms in a series for the energy. The number of CMS products that needed to be summed and the time to calculate values are given in Table 2.[1] We give the initial terms in the series:

$$E_N = -\frac{\pi}{16}\left(1 - \frac{8\zeta(2)}{N^2} - \frac{8\zeta(3)}{N^3} + \cdots\right). \tag{58}$$

We have also used our system to calculate the Laplace-Dirichlet eigenvalues on an ellipse in [3]. This is a classical question in boundary variations [17] and has been established correctly to second order. We have used our system to determine the first 7 terms. We give the first four terms in the series in eccentricity e for the lowest eigenvalue $\lambda(e)$ below

$$\begin{aligned} \lambda(e) =& \lambda - \frac{\lambda}{2}e^2 + \frac{\lambda - 6}{32}e^4\lambda + \frac{\lambda - 6}{64}e^6\lambda \\ & - \frac{7\lambda^3 - 58\lambda^2 - 192\lambda + 1792}{32768}e^8\lambda + \cdots \end{aligned} \tag{59}$$

In this equation, λ is the corresponding eigenvalue on the unit circle. The number of products that needed to be summed and time to evaluate each expression is given in Table 3. The first 7 terms are presented in [3].

9. CONCLUSIONS

Although we have limited the objects defined in our initial implementation, the CMS has a wide range of uses. In high order stability analysis, the CMS can be used for shape optimization [14]. Biological models, such as blood cell membranes, can also be modeled [8]. The CMS introduces a great deal of analytical order into fluid film dynamics [9]. Having shown the success of a TRS for the CMS on boundary variation problems, we plan to expand the rule set to include additional rules and objects from the CMS. This will make proving confluence and termination more difficult problems.

[1]All timings run on Mac OS 10.7 3.06 GHz Intel 4GB ram with Maple 16 and Java 1.7.

The CMS has been hindered by a lack of automated computation and a rapid growth in expression length. We have proposed a TRS to solve boundary variation problems in the CMS. This system has already shown the power of applying a TRS to the CMS. It was used to support current research in the CMS, with results presented in [3] and [4].

Although we have only implemented a subset of the CMS, these results illustrate the power of applying a TRS to the problems of the CMS. We have found high order variations in problems that had previously been intractable. We view this TRS as a first step towards comprehensive implementation of the CMS.

10. REFERENCES

[1] L. B. Buchberger and K. R. Loos. Algebraic simplification. *Computing, Suppl.*, 4:11–43, 1982.

[2] L. Bachmair and D. A. Plaisted. Termination orderings for associative-commutative rewrite systems. *J. Symbolic Computation*, 1:329–349, 1985.

[3] M. Boady, P. Grinfeld, and J. Johnson. Laplace eigenvalues on the ellipse and the symbolic calculus of moving surfaces. *In preparation.*

[4] M. Boady, P. Grinfeld, and J. Johnson. Boundary variation of poisson's equation: a model problem for symbolic calculus of moving surfaces. *Int. J. Math. Comp. Sci.*, 6(2), 2011.

[5] A. J. Christopherson, K. A. Malik, D. R. Matravers, and K. Nakamura. Comparing two different formulations of metric cosmological perturbation theory. *Classical and Quantum Gravity*, 28(22):225024, 2011.

[6] M. Clavel, F. Durán, S. Eker, P. Lincoln, N. Martí-Oliet, J. Meseguer, and C. Talcott. The maude 2.0 system. *Proc. Rewriting Techniques and Applications*, pages 76–87, June 2003.

[7] P. Grinfeld. Hadamard's formula inside and out. *J. Opt. Theory and Appl.*, 146(3):654–690, 2009.

[8] P. Grinfeld. Hamiltonian dynamic equations for fluid films. *Stud. Appl. Math.*, 125:223–264, 2010.

[9] P. Grinfeld. A variable thickness model for fluid films under large displacements. *Phys. Rev. Lett.*, 105:137802, 2010.

[10] P. Grinfeld and G. Strang. Laplace eigenvalues on polygons. *Computers and Mathematics with Applications*, 48:1121–1133, 2004.

[11] P. Grinfeld and G. Strang. Laplace eigenvalues on regular polygons: A series in $1/N$. *Journal of Mathematical Analysis and Applications*, 385(1):135 – 149, 2012.

[12] P. Grinfeld and J. Wisdom. A way to compute the gravitational potential for near-spherical geometries. *Quart. Appl. Math.*, 64(2):229–252, 2006.

[13] Y. V. Gusev. Heat kernel expansion in the covariant perturbation theory. *Nuclear Physics B*, 807(3):566 – 590, 2009.

[14] H. Howards, M. Hutchings, and F. Morgan. The isoperimetric problem on surfaces. *The American Mathematical Monthly*, 106(5):pp. 430–439, 1999.

[15] G. Huet. Confluent reductions: Abstract properties and applications to term rewriting systems. *J. Assoc. Comp. Mach.*, 27:797–821, 1980.

[16] G. Huet and D. C. Oppen. Equations and rewrite rules – a survey. Technical report, Stanford University, Jan 1980.

[17] D. Joseph. Parameter and domain dependence of eigenvalues of elliptic partial differential equations. 24(5):325–361, 1967.

[18] J.-P. Jouannaud and H. Kirchner. Completion of a set of rules modulo a set of equations. In *Proceedings of the 11th ACM SIGACT-SIGPLAN symposium on Principles of programming languages*, POPL '84, pages 83–92, New York, NY, USA, 1984. ACM.

[19] J. Katz and G. I. Livshits. Superpotentials from variational derivatives rather than lagrangians in relativistic theories of gravity. *Classical and Quantum Gravity*, 25(17):175024, 2008.

[20] D. Knuth and P. Bendix. Simple word problems in universal algebras. *Computational Problems in Abstract Algebra*, pages 263–297, 1970.

[21] T. Levi-Civita. *The Absolute Differential Calculus (Calculus of Tensors)*. Dover Publications, 1977.

[22] C.-H. L. Lin, M.-H. Lo, and Y.-C. Tsai. Shape Energy of Fluid Membranes —Analytic Expressions for the Fourth-Order Variation of the Bending Energy—. *Progress of Theoretical Physics*, 109:591–618, Apr. 2003.

[23] T. Málek and V. Pravda. Kerra-schild spacetimes with an (a)ds background. *Classical and Quantum Gravity*, 28(12):125011, 2011.

[24] Maplesoft, a division of Waterloo Maple Inc., Waterloo, Ontario, Canada. *Maple User Manual*, 2012.

[25] J. M. Maran-Garca. xperm: fast index canonicalization for tensor computer algebra. *Computer Physics Communications*, 179(8):597 – 603, 2008.

[26] MathTensor Inc. Mathtensor - tensor analysis for mathematica. http://smc.vnet.net/MathTensor.html.

[27] A. McConnell. *Applications of Tensor Analysis*. Dover Publications, New York, 1957.

[28] K. Peeters. *Cadabra: reference guide and tutorial.* http://cadabra.phi-sci.com/cadabra.pdf, June 2008.

[29] C. F. Steinwachs and A. Y. Kamenshchik. One-loop divergences for gravity nonminimally coupled to a multiplet of scalar fields: Calculation in the jordan frame. i. the main results. *Physical Review D*, 84(2):024026, July 2011.

[30] J. Synge and A. Schild. *Tensor Calculus*. Dover Publications, Inc., 1949.

[31] T. Thomas. *Concepts from Tensor Analysis And Differential Geometry*. Academic Press, New York, 1965.

[32] Wolfram Research, Champaign, IL. *Wolfram Mathematica 9 Documentation Center*, 2012.

Hermite Reduction and Creative Telescoping for Hyperexponential Functions*

Alin Bostan[1], Shaoshi Chen[2], Frédéric Chyzak[1], Ziming Li[3], Guoce Xin[4]
[1]INRIA, Palaiseau, 91120, (France)
[2]Department of Mathematics, NCSU, Raleigh, 27695-8025, (USA)
[3]KLMM, AMSS, Chinese Academy of Sciences, Beijing 100190, (China)
[4]Department of Mathematics, Capital Normal University, Beijing 100048, (China)
schen@amss.ac.cn, {alin.bostan, frederic.chyzak}@inria.fr
zmli@mmrc.iss.ac.cn, guoce_xin@163.com

ABSTRACT

We present a new reduction algorithm that simultaneously extends Hermite's reduction for rational functions and the Hermite-like reduction for hyperexponential functions. It yields a unique additive decomposition that allows to decide hyperexponential integrability. Based on this reduction algorithm, we design a new algorithm to compute minimal telescopers for bivariate hyperexponential functions. One of its main features is that it can avoid the costly computation of certificates. Its implementation outperforms Maple's function DEtools[Zeilberger]. We also derive an order bound on minimal telescopers that is tighter than the known ones.

Categories and Subject Descriptors

I.1.2 [**Computing Methodologies**]: Symbolic and Algebraic Manipulation—*Algebraic Algorithms*

General Terms

Algorithms, Theory

Keywords

Hermite reduction, Hyperexponential function, Telescoper

1. INTRODUCTION

Given a univariate rational function r, Hermite reduction in [17, 14, 6] finds rational functions r_1 and r_2 s.t. (i) $r = r_1 + r_2$, (ii) r_1 is rational integrable, (iii) r_2 is a proper fraction with a squarefree denominator. The additive decomposition is unique, and r is rational integrable if and only if $r_2 = 0$.

A univariate function is hyperexponential if its logarithmic derivative is rational. Exponential, radical and rational functions are hyperexponential. Rational Hermite reduction was extended to hyperexponential functions by Davenport in [11] and by Geddes, Le and Li in [12]. The former aims at solving Risch's equation; the latter is a differential analogue of the reduction algorithm for hypergeometric terms in [2]. For a given hyperexponential function H, the reduction algorithms in [11, 12] compute two hyperexponential functions H_1 and H_2 s.t. (i) $H = H_1 + H_2$, (ii) H_1 is hyperexponential integrable, (iii) H_2 is minimal in some sense. However, H_2 is not unique in general and it may be nonzero even when H is hyperexponential integrable. To decide the integrability of H, one additionally needs to compute polynomial solutions of a first-order linear differential equation.

The method of creative telescoping, developed initially for hyperexponential functions by Almkvist and Zeilberger [3], then extended by Chyzak to general holonomic functions [10], is nowadays an important automatic tool for computing definite integrals. Recently, it has also played an important role in the resolution of intriguing problems in enumerative combinatorics [15, 16]. For a bivariate hyperexponential function $H(x, y)$, the problem of creative telescoping is to find a nonzero operator $L(x, D_x)$ in $\mathbb{F}(x)\langle D_x\rangle$, the ring of linear differential operators over the rational-function field $\mathbb{F}(x)$, s.t.

$$L(x, D_x)(H) = D_y(G) \tag{1}$$

for some hyperexponential function G, where $D_x = \partial/\partial x$ and $D_y = \partial/\partial y$. The operator L above is called a *telescoper* for H, and G is the corresponding *certificate*. An algorithm for solving (1) is given in [3], which is based on a differential version of Gosper's algorithm. An algorithm for rational-function telescoping is given in [5], which is based on Hermite reduction. The latter separates the computation of telescopers from that of certificates, and has a lower complexity than the former for rational functions.

In the present paper, we develop a reduction algorithm which, given a univariate hyperexponential function H, constructs two hyperexponential functions H_1 and H_2 s.t. (i) $H = H_1 + H_2$, (ii) H_1 is hyperexponential integrable, and (iii) H_2 is either zero or not hyperexponential integrable. We

*S.C. was supported by the National Science Foundation (NSF) grant CCF-1017217, A.B. and F.C. were supported in part by the MSR-INRIA Joint Centre, Z.L. by two NSFC grants (91118001, 60821002/F02) and a 973 project (2011CB302401), and G.X. by NSFC grant (11171231).

ISSAC'13, June 26–29, 2013, Boston, Massachusetts, USA.

show that H_2 in the above additive decomposition is unique in a certain technical sense and can be obtained without computing polynomial solutions of any differential equation. Our algorithm is based on the Hermite-like reduction in [12], a differential variant of the polynomial reduction in [2] and on the idea for reducing simple radicals in [18, Proposition 7]. The main new ingredient is property (iii), which is crucial in many applications. Using the reduction algorithm, we extend the rational telescoping algorithm in [5] to the hyperexponential case, and derive an order bound on the telescopers. The new telescoping algorithm avoids the costly computation of certificates, and the order bound is tighter than that obtained in [4] and [8].

The rest of the paper is organized as follows. We review the notion of hyperexponential functions and Hermite-like reduction in Sections 2 and 3, respectively. A new reduction algorithm is developed for hyperexponential functions in Section 4. After introducing kernel reduction in Section 5, we present a reduction-based telescoping algorithm for bivariate hyperexponential functions, and derive an upper bound on the order of minimal telescopers in Section 6. We briefly describe an implementation of the new telescoping algorithm, and present some experimental results in Section 7, which validate its practical relevance.

As a matter of notation, we let $\mathbb{F}$ be a field of characteristic zero and $\mathbb{F}(y)$ be the field of rational functions in y over $\mathbb{F}$. For a polynomial $p \in \mathbb{F}[y]$, we denote by $\deg(p)$ and $\operatorname{lc}(p)$ the degree and leading coefficient of p, respectively. Let D_y denote the usual derivation d/dy on $\mathbb{F}(y)$. Then $(\mathbb{F}(y), D_y)$ is a differential field.

2. HYPEREXPONENTIAL FUNCTIONS

Hyperexponential functions share the common properties of rational functions, simple radicals, and exponential functions. Together with hypergeometric terms, they are frequently viewed as a special and important class of "closed-form" solutions of linear differential and difference equations with polynomial coefficients.

Definition 1. *Let Φ be a differential field extension of $\mathbb{F}(y)$. A nonzero element $H \in \Phi$ is said to be* hyperexponential *over $\mathbb{F}(y)$ if its logarithmic derivative $D_y(H)/H$ is in $\mathbb{F}(y)$.*

The product of hyperexponential functions is also hyperexponential. Two hyperexponential functions H_1, H_2 are said to be *similar* if there exists $r \in \mathbb{F}(y)$ s.t. $H_1 = rH_2$. The sum of similar hyperexponential functions is still hyperexponential, provided that it is nonzero.

For brevity, we use the notation $\exp(\int f dy)$ to indicate a hyperexponential function whose logarithmic derivative is f. For a rational function $r \in \mathbb{F}(y)$, we have

$$r \exp\left(\int f\, dy\right) = \exp\left(\int \left(f + D_y(r)/r\right) dy\right).$$

A univariate hyperexponential function H is said to be *hyperexponential integrable* if it is the derivative of another hyperexponential function. For brevity, we say "integrable" instead of "hyperexponential integrable" in the sequel.

A hyperexponential function H can be expressed as a product $r \exp\left(\int f dy\right)$ for some $r, f \in \mathbb{F}(y)$. Assume that H is integrable. Then it is equal to $D_y(G)$ for some hyperexponential function G. A straightforward calculation shows that G is similar to H. In other words, $G = s \exp\left(\int f dy\right)$ for some $s \in \mathbb{F}(y)$. It follows that $H = D_y(G)$ if and only if

$$r = D_y(s) + f\, s. \tag{2}$$

Deciding the integrability of H amounts to finding a rational solution s s.t. the above equation holds.

3. HERMITE-LIKE REDUCTION

Reduction algorithms have been developed for computing additive decompositions of rational functions [17, 14], hypergeometric terms [1, 2], and hyperexponential functions [11, 12]. Those algorithms can be viewed as generalizations of Gosper's algorithm [13] and its differential analogue [3, §5].

For a hyperexponential function H, a reduction algorithm computes two hyperexponential functions H_1, H_2 s.t.

$$H = D_y(H_1) + H_2. \tag{3}$$

This implies that H, H_1 and H_2 are similar. So we may write $H = r \exp\left(\int f dy\right)$ and $H_i = r_i \exp\left(\int f dy\right)$, where r, r_i, f belong to $\mathbb{F}(y)$ and $i = 1, 2$. Then (3) translates into

$$r = D_y(r_1) + f\, r_1 + r_2. \tag{4}$$

A reduction algorithm for computing (3) amounts to choosing rational functions r, f and r_1 so that r_2 satisfies properties similar to those obtained in Hermite reduction for rational functions. There are at least two approaches to this end. One is given in [11], and the other in [12]. We review the latter, because the notion of differential-reduced rational functions plays a key role in Lemma 6 in Section 4.

Recall [12, §2] that a rational function $r = a/b \in \mathbb{F}(y)$ is said to be *differential-reduced* w.r.t. y if

$$\gcd\left(b, a - i\, D_y(b)\right) = 1 \quad \text{for all } i \in \mathbb{Z}.$$

By Lemma 2 in [12], r is differential-reduced if and only if none of its residues is an integer. The *differential rational canonical form* of a rational function f in $\mathbb{F}(y)$ is a pair (K, S) in $\mathbb{F}(y) \times \mathbb{F}(y)$ s.t. (i) K is differential-reduced; (ii) the denominator of S is coprime with that of K; and (iii) f is equal to $K + D_y(S)/S$. Every rational function has a unique canonical form in the sense that K is unique and S is unique up to a multiplicative constant in $\mathbb{F}$ [12, §3]. We call K and S the *kernel* and *shell* of f, respectively. They can be constructed by the method described in [12, §3].

Let H be a univariate hyperexponential function over $\mathbb{F}(y)$, in the form $\exp(\int f dy)$. Assume that K and S are the kernel and shell of f, respectively. Then $H = S \exp\left(\int K\, dy\right)$. Note that $K = 0$ if and only if H is a rational function, which is then equal to cS for some $c \in \mathbb{F}$.

Example 2. *Let $H = \sqrt{y^2+1}/(y-1)^2$. The logarithmic derivative of H is*

$$\frac{D_y H}{H} = \frac{D_y(1/(y-1)^2)}{1/(y-1)^2} + \frac{y}{y^2+1},$$

where $y/(y^2+1)$ is differential-reduced. The kernel and shell of $D_y(H)/H$ are $y/(y^2+1)$ and $1/(y-1)^2$, respectively. So $H = \exp\left(\int y/(y^2+1)\, dy\right)/(y-1)^2$.

For brevity, we make a notational convention.

Convention 3. *Let H denote a hyperexponential function whose logarithmic derivative has kernel K and shell S. Assume that K is nonzero, that is, H is not a rational function. Set $T = \exp\left(\int K\, dy\right)$. Moreover, write $K = k_1/k_2$, where k_1, k_2 are polynomials in $\mathbb{F}[y]$ with $\gcd(k_1, k_2) = 1$.*

The algorithm **ReduceCert** in [12] computes a rational function S_1 s.t.

$$S = D_y(S_1) + S_1 K + \frac{a}{bk_2}, \quad (5)$$

where $a \in \mathbb{F}[y]$ and b is the squarefree part of the denominator of S. Thus, $\gcd(b, k_2) = 1$ by the definition of canonical forms. Note that a is not necessarily coprime with bk_2. As the algorithm **ReduceCert** only reduces the shell S, it is referred to as *shell reduction*. It follows from (5) that

$$H = D_y(S_1 T) + \frac{a}{bk_2} T. \quad (6)$$

By Theorem 4 in [12], a/b belongs to $\mathbb{F}[y]$ if H is integrable.

Example 4. *Let H be the same hyperexponential function as in Example 2. Then $D_y(H)/H$ has kernel $K = y/(y^2+1)$ and shell $S = 1/(y-1)^2$. Shell reduction yields*

$$S = D_y(S_1) + S_1 K + \frac{y}{(y-1)k_2},$$

where $S_1 = -1/(y-1)$ and $k_2 = y^2 + 1$. Then

$$H = D_y(S_1 T) + \frac{yT}{(y-1)k_2}, \quad \text{where } T = \sqrt{y^2+1}.$$

By Theorem 4 in [12], H is not integrable.

Remark that a in (6) can be nonzero for an integrable H:

Example 5. *Let $H = y\exp(y)$ whose logarithmic derivative has kernel 1 and shell y, i.e., $H = y\exp\left(\int 1 dy\right)$, for $S_1 = 0$. But H is integrable as it is equal to $D_y\left(y\exp(y) - \exp(y)\right)$.*

Thus, shell reduction cannot be directly used to decide hyperexponential integrability, which is a difference to the rational case. To amend this, the solution proposed in [12, Algorithm **ReduceHyperexp**] was to find the polynomial solutions of an auxiliary first-order linear differential equation. In the following section, we show how this can be avoided and improved.

4. HERMITE REDUCTION FOR HYPEREXPONENTIAL FUNCTIONS

After the shell reduction described in (6), it remains to decide the integrability of $(a/bk_2)T$. In the rational case, i.e., when the kernel K is equal to zero, a in (6) can be chosen s.t. $\deg(a) < \deg(b)$, because all polynomials are rational integrable. But a hyperexponential function with a polynomial shell is not necessarily integrable. For example, $H = \exp\left(y^2\right)$.

We present a differential variant of [2, Theorem 7] to bound the degree of a in (5). The variant leads not only to a canonical additive decomposition of hyperexponential functions, but also a direct way to decide their integrability.

4.1 Polynomial reduction

With Convention 3, we define

$$\mathcal{M}_K = \{k_2 D_y(p) + k_1 p \mid p \in \mathbb{F}[y]\}.$$

It is an $\mathbb{F}$-linear subspace in $\mathbb{F}[y]$. We call $\mathcal{M}_K$ the *subspace for polynomial reduction w.r.t. K*. Moreover, let ϕ_K be the $\mathbb{F}$-linear map from $\mathbb{F}[y]$ to $\mathcal{M}_K$ that sends p to $k_2 D_y(p) + k_1 p$ for every $p \in \mathbb{F}[y]$. We call ϕ_K the *map for polynomial reduction w.r.t. K*.

Concerning the subspace $\mathcal{M}_K$ and the map ϕ_K, we have:

Lemma 6. *(i) If $k_2 D_y(g) + k_1 g \in \mathbb{F}[y]$ for some $g \in \mathbb{F}(y)$, then $g \in \mathbb{F}[y]$. (ii) The map ϕ_K is bijective.*

Proof. Assume that g has a pole. Without loss of generality, we assume that the pole is $y = 0$ and has order m, because the following argument is also applicable over the algebraic closure of $\mathbb{F}$. Expanding g around the origin yields

$$g = \frac{r}{y^m} + \text{terms of higher orders in } y,$$

where $r \in \mathbb{F} \setminus \{0\}$. It follows from $k_2 D_y(g) + k_1 g \in \mathbb{F}[y]$ that $y = 0$ is a pole of

$$\left(-\frac{mr}{y^{m+1}} + \text{higher terms}\right) + K\left(\frac{r}{y^m} + \text{higher terms}\right)$$

with order no more than that of K. This implies that $y{=}0$ is a simple pole of K with residue m, which is incompatible with K being differential-reduced. The first assertion holds.

The map ϕ_K is surjective by definition. If $\phi_K(p) = 0$ for some nonzero polynomial $p \in \mathbb{F}[y]$, then K equals $-D_y(p)/p$, which is nonzero since $K \neq 0$. So K is not differential-reduced, a contradiction. The second assertion holds. ∎

An $\mathbb{F}$-basis of $\mathcal{M}_K$ is called an *echelon basis* if distinct elements in the basis have distinct degrees. Echelon bases always exist and their degrees form a unique subset of $\mathbb{N}$. Let $\mathcal{B}$ be an echelon basis of $\mathcal{M}_K$. Define

$$\mathcal{N}_K = \operatorname{span}_{\mathbb{F}}\left\{y^\ell \mid \ell \in \mathbb{N} \text{ and } \ell \neq \deg(f) \text{ for all } f \in \mathcal{B}\right\}.$$

Then $\mathbb{F}[y] = \mathcal{M}_K \oplus \mathcal{N}_K$. We call $\mathcal{N}_K$ the *standard complement of $\mathcal{M}_K$*. Using an echelon basis of $\mathcal{M}_K$, one can reduce a polynomial p to a unique polynomial $\tilde{p} \in \mathcal{N}_K$ s.t. $p - \tilde{p} \in \mathcal{M}_K$.

In order to find an echelon basis of $\mathcal{M}_K$, we set $d_1 = \deg k_1$, $d_2 = \deg k_2$, $\tau_K = -\operatorname{lc}(k_1)/\operatorname{lc}(k_2)$, and $\mathcal{B} = \{\phi_K(y^n) \mid n \in \mathbb{N}\}$. By Lemma 6 (ii), $\mathcal{B}$ is an $\mathbb{F}$-basis of $\mathcal{M}_K$. Let p be a nonzero polynomial in $\mathbb{F}[y]$. We make the following case distinction.

Case 1. $d_1 \geq d_2$. Then

$$\phi_K(p) = \operatorname{lc}(k_1)\operatorname{lc}(p) y^{d_1 + \deg p} + \text{lower terms}.$$

So $\mathcal{B}$ is an echelon basis, in which $\deg \phi_K(y^n) = d_1 + n$ for all $n \in \mathbb{N}$. Accordingly, $\mathcal{N}_K$ is spanned by $1, y, \ldots, y^{d_1 - 1}$. It follows that $p \equiv q \bmod \mathcal{M}_K$ for some $q \in \mathcal{N}_K$ with $\deg q < d_1$.

Case 2. $d_1 = d_2 - 1$ and τ_K is not a positive integer. Then

$$\phi_K(p) = \left(\deg(p)\operatorname{lc}(k_2) + \operatorname{lc}(k_1)\right)\operatorname{lc}(p) y^{d_1 + \deg p} + \text{lower terms}. \quad (7)$$

Since τ_K is not a positive integer, $\deg \phi_K(y^n) = d_1 + n$. Thus, $\mathcal{M}_K$ and $\mathcal{N}_K$ have the same bases as in Case 1. Furthermore, $p \equiv q \bmod \mathcal{M}_K$ for some $q \in \mathcal{N}_K$ with $\deg q < d_1$.

Case 3. $d_1 < d_2 - 1$. If $\deg(p) > 0$, then

$$\phi_K(p) = \deg(p)\operatorname{lc}(k_2)\operatorname{lc}(p) y^{d_2 + \deg(p) - 1} + \text{lower terms}.$$

Otherwise, $\deg(p) = 0$ and $\phi_K(p) = k_1 p$. Therefore, $\mathcal{B}$ is again an echelon basis, in which

$$\deg \phi_K(1) = d_1 \text{ and } \deg \phi_K(y^n) = d_2 + n - 1 \text{ for all } n \geq 1.$$

Accordingly, $\mathcal{N}_K$ has a basis $1, \ldots, y^{d_1 - 1}, y^{d_1 + 1}, \ldots, y^{d_2 - 1}$. It follows that there is $q \in \mathcal{N}_K$ s.t. $p \equiv q \bmod \mathcal{M}_K$, $\deg q < d_2$ and the coefficient of y^{d_1} in q is equal to zero.

Case 4. $d_1 = d_2 - 1$ and τ_K is a positive integer. It follows from (7) that $\deg \phi(y^n) = d_1 + n$ if $n \neq \tau_K$. Furthermore, for every polynomial p of degree τ_K, $\phi_K(p)$ is of degree less than $d_1+\tau_K$. So any echelon basis of $\mathcal{M}_K$ does not contain a polynomial of degree $d_1+\tau_K$. Set

$$\mathcal{B}' = \{\phi_K(y^n) \,|\, n \in \mathbb{N}, n \neq \tau_K\}.$$

Reducing $\phi_K(y^{\tau_K})$ by the polynomials in $\mathcal{B}'$, we obtain a polynomial a of degree less than d_1. Note that a is nonzero, because $\mathcal{B}$ is an $\mathbb{F}$-linearly independent set. Hence, $\mathcal{B}'\cup\{a\}$ is an echelon basis of $\mathcal{M}_K$. As a consequence, $\mathcal{N}_K$ has an $\mathbb{F}$-basis $\left\{1, y, \ldots, y^{\deg(a)-1}, y^{\deg(a)+1}, \ldots, y^{d_1-1}, y^{d_1+\tau_K}\right\}$. It follows that there exists $r \in \mathbb{F}[y]$ of degree less than d_1 s.t.

$$p \equiv s y^{d_1+\tau_K} + r \mod \mathcal{M}_K \quad \text{for some } s \in \mathbb{F}.$$

Moreover, $sy^{d_1+\tau_K}+r \in \mathcal{N}_K$, and r has at most d_1-1 terms.

Example 7. *Let $K = -6y^3/(y^4+1)$, which is differential-reduced. Then $\tau_K = 6$. According to Case 4, $\mathcal{M}_K$ has an echelon basis $\{y\} \cup \{(n-6)y^{n+3}+ny^{n-1} | n \in \mathbb{N}, n \neq 6\}$. Moreover, $\mathcal{N}_K$ has a basis $\{1, y^2, y^9\}$.*

The next lemma enables us to derive an order bound on telescopers for hyperexponential functions.

Lemma 8. *With Convention 3, we further let $d_1 = \deg(k_1)$ and $d_2 = \deg(k_2)$. Then there exists $\mathcal{P} \subset \{y^n | n \in \mathbb{N}\}$ with*

$$|\mathcal{P}| \leq \max(d_1, d_2-1)$$

s.t. every polynomial in $\mathbb{F}[y]$ can be reduced modulo $\mathcal{M}_K$ to an $\mathbb{F}$-linear combination of the elements in $\mathcal{P}$.

Proof. By the above case distinction, the dimension of $\mathcal{N}_K$ over $\mathbb{F}$ is at most $\max(d_1, d_2-1)$. The lemma follows. ∎

4.2 Hyperexponential integrability

With Convention 3, we further assume that the polynomials a and b are obtained by shell reduction in (6). So the decomposition (6) holds for the present section. Moreover, let $\mathcal{M}_K$ be the subspace for polynomial reduction w.r.t. K, and let $\mathcal{N}_K$ be its standard complement.

We are going to determine necessary and sufficient conditions on hyperexponential integrability. Since $\gcd(b, k_2)=1$,

$$\frac{a}{bk_2} = \frac{q}{b} + \frac{r}{k_2}, \tag{8}$$

where $q, r \in \mathbb{F}[y]$ and $\deg(q) < \deg(b)$. Using an echelon basis of $\mathcal{M}_K$, we compute two polynomials $u \in \mathcal{M}_K$ and $v \in \mathcal{N}_K$ s.t. $r = u+v$. By the definition of $\mathcal{M}_K$, there exists w in $\mathbb{F}[y]$ s.t. $u = k_2 D_y(w) + k_1 w$. By (8), we get

$$\frac{a}{bk_2} = \frac{q}{b} + \frac{k_2 D_y(w) + k_1 w + v}{k_2} = D_y(w) + Kw + \frac{q}{b} + \frac{v}{k_2}.$$

It follows from the equivalence of (4) and (3) that

$$\frac{a}{bk_2} T = D_y(wT) + \left(\frac{q}{b} + \frac{v}{k_2}\right) T. \tag{9}$$

The previous process for obtaining (9) is referred to as the *polynomial reduction for $(a/(bk_2))T$ w.r.t. K*, as it makes essential use of the subspaces $\mathcal{M}_K$ and $\mathcal{N}_K$. By (9) and (6),

$$H = D_y((S_1+w)T) + \left(\frac{q}{b} + \frac{v}{k_2}\right) T, \tag{10}$$

which motivates us to introduce a notion of *residual forms.*

Definition 9. *With Convention 3, we further let f be a rational function in $\mathbb{F}(y)$. Another rational function $r \in \mathbb{F}(y)$ is said to be a* residual form *of f w.r.t. K if there exist g in $\mathbb{F}(y)$ and q, b, v in $\mathbb{F}[y]$ s.t.*

$$f = D_y(g) + Kg + r \quad \text{and} \quad r = \frac{q}{b} + \frac{v}{k_2},$$

where b is squarefree, $\gcd(b, k_2) = 1$, $\deg(q) < \deg(b)$, and v is in the standard complement $\mathcal{N}_K$ of the subspace of polynomial reduction w.r.t. K. For brevity, we say that r is a residual form *w.r.t K if f is clear from the context.*

Remark 10. *The set of residual forms w.r.t. K is an $\mathbb{F}$-linear subspace of $\mathbb{F}(y)$ by the four conditions on b, k_2, q and v in the above definition.*

Residual forms are closely related to the integrability of hyperexponential functions.

Lemma 11. *With Convention 3, we further assume that r is a nonzero residual form w.r.t. K. Then the hyperexponential function rT is not integrable.*

Proof. Suppose on the contrary that rT is integrable. We let $\mathcal{M}_K$ be the subspace for polynomial reduction, and $\mathcal{N}_K$ its standard complement w.r.t. K. By the definition of residual forms, there exist $b, q \in \mathbb{F}[y]$ with b being squarefree and $v \in \mathcal{N}_K$ s.t.

$$\deg(q) < \deg(b), \ \gcd(b, k_2) = 1, \text{ and } r = \frac{q}{b} + \frac{v}{k_2}. \tag{11}$$

Thus, r can be rewritten as $(k_2 q + bv)/(bk_2)$. It follows that

$$rT = \frac{k_2 q + bv}{b} \exp\left(\int \frac{k_1 - D_y(k_2)}{k_2}\, dy\right).$$

The pair $((k_2 q + bv)/b, (k_1 - D_y(k_2))/k_2)$ is an indecomposable pair according to Definition 2 in [12], since the rational function $(k_1 - D_y(k_2))/k_2$ is differential-reduced, k_2 and b are coprime, and b is squarefree. By Theorem 4 in [12], $(k_2 q + bv)/b$ is a polynomial in $\mathbb{F}[y]$. So $q = 0$ because $\gcd(b, k_2)=1$. It follows from the last equality in (11) that $(v/k_2)T$ is integrable. By (2), $v = k_2 D_y(s) + k_1 s$ for some $s \in \mathbb{F}(y)$. Since $v \in \mathbb{F}[y]$, $s \in \mathbb{F}[y]$ by Lemma 6 (i), and, thus, $v \in \mathcal{M}_K$ by the definition of $\mathcal{M}_K$ at the beginning of Section 4.1, which, together with $v \in \mathcal{N}_K$, implies that $v = 0$. Consequently, $r = 0$, a contradiction to the assumption that $r \neq 0$. ∎

The existence and uniqueness of residual forms are described below.

Lemma 12. *With Convention 3, we have that the shell S has a residual form w.r.t. the kernel K. If a rational function has two residual forms w.r.t. K, then they are equal.*

Proof. By (10), $S = D_y(S_1+w) + (S_1+w)K + q/b + v/k_2$. So $q/b + v/k_2$ is a required form.

Let r and r' be two residual forms of a rational function w.r.t. K. Then $D_y(f) + fK + r = D_y(f') + f'K + r'$ for some $f, f' \in \mathbb{F}(y)$. So $D_y(f-f') + (f-f')K + r - r' = 0$. Consequently, $(r-r')T$ is integrable by (2). We conclude that $r = r'$ by Remark 10 and Lemma 11. ∎

Below is the main result of the present section.

Theorem 13. *Let H be a hyperexponential function whose logarithmic derivative has kernel K and shell S. Then there is an algorithm for computing a rational function h in $\mathbb{F}(y)$ and a unique residual form r w.r.t. K s.t.*

$$H = D_y\left(h \exp\left(\int K\, dy\right)\right) + r \exp\left(\int K\, dy\right). \quad (12)$$

Moreover, H is integrable if and only if $r = 0$.

Proof. Let $T = \exp\left(\int K\, dy\right)$. Applying shell reduction to H w.r.t. K, we find a rational function S_1, and two polynomials a, b s.t. (5) holds. Then we apply polynomial reduction to $a/(bk_2)T$ to obtain the residual form $r = q/b + v/k_2$ s.t. (12) holds.

Suppose that there exists another decomposition

$$H = D_y\left(h'T\right) + r'T \quad (13)$$

for some $h' \in \mathbb{F}(y)$ and some residual form r' w.r.t. K. Then both r and r' are residual forms of S by (12), (13) and the fact $H = ST$. So $r = r'$ by Lemma 12.

If $r = 0$, then H is obviously integrable. Conversely, assume that H is integrable. Then rT is also integrable by (12). So $r = 0$ by Lemma 11. ∎

The reduction algorithm described in the proof of Theorem 13 decomposes a hyperexponential function into a sum of an integrable one and a non-integrable one in a canonical way. The given function is integrable if and only if the non-integrable part is trivial. As a byproduct, it decides hyperexponential integrability without computing a polynomial solution of any first-order linear differential equation, which enables us to construct telescopers for hyperexponential functions using merely linear algebra in Section 6. The algorithm will be referred to as *Hermite reduction for hyperexponential functions* in the sequel, because it extends all important features in Hermite reduction for rational functions to hyperexponential ones.

Example 14. *Let H be the same hyperexponential function as in Example 2. Then $K = y/(y^2+1)$ and $S = 1/(y-1)^2$. Set $T = \sqrt{y^2+1}$. By the shell reduction in Example 4,*

$$H = D_y\left(\frac{-1}{y-1}T\right) + \frac{y}{bk_2}T,$$

where $b = y - 1$ and $k_2 = y^2 + 1$. The polynomial reduction yields $(y/(bk_2))T = D_y\left(-T/2\right) + (1/(2b) + 1/(2k_2))\,T$. Combining the above equations, we decompose H as

$$H = D_y\left(\frac{-(y+1)}{2(y-1)}T\right) + \left(\frac{1}{2b} + \frac{1}{2k_2}\right)T.$$

Example 15. *Consider $H = y\exp(y)$ as given in Example 5. Since its logarithmic derivative has kernel $K=1$, the subspace $\mathcal{M}_K$ for polynomial reduction is equal to $\mathbb{F}[y]$. Thus, y is in $\mathcal{M}_K$ and H is integrable. More generally, $\mathcal{M}_K = \mathbb{F}[y]$ corresponds to the well-known fact that $p(y)\exp(y)$ is integrable for all $p \in \mathbb{F}[y] \setminus \{0\}$.*

5. KERNEL REDUCTION

Let $K = k_1/k_2$ be a nonzero differential-reduced rational function in $\mathbb{F}(y)$ with $\gcd(k_1, k_2) = 1$. We may want to reduce a hyperexponential function in the form

$$\frac{p}{k_2^m}\exp\left(\int K\, dy\right) \quad \text{for some } p \in \mathbb{F}[y] \text{ and } m \in \mathbb{N}.$$

One way would be to rewrite the above function as

$$p \exp\left(\int \frac{k_1 - mD_y(k_2)}{k_2}\, dy\right),$$

and proceed by polynomial reduction w.r.t. the new kernel $(k_1 - mD_y(k_2))/k_2$. However, it will prove to be more convenient in Section 6 to reduce the given function w.r.t. the initial kernel K. To this end, we introduce another type of reduction, based on ideas in [11, 18].

Lemma 16. *With Convention 3, we let $p \in \mathbb{F}[y]$ and $m \geq 1$. Then there exist $p_1, p_2 \in \mathbb{F}[y]$ s.t.*

$$\frac{p}{k_2^m} = D_y\left(\frac{p_1}{k_2^{m-1}}\right) + \frac{p_1}{k_2^{m-1}}K + \frac{p_2}{k_2}. \quad (14)$$

Proof. We proceed by induction on m. If $m = 1$, then taking $p_1 = 0$ and $p_2 = p$ yields the claimed form. Assume that $m > 1$. We first show that there exist $\tilde{p}_1, \tilde{p}_2 \in \mathbb{F}[y]$ s.t.

$$\frac{p}{k_2^m} = D_y\left(\frac{\tilde{p}_1}{k_2^{m-1}}\right) + \frac{\tilde{p}_1}{k_2^{m-1}}K + \frac{\tilde{p}_2}{k_2^{m-1}},$$

which is equivalent to

$$p = \tilde{p}_1(k_1 - (m-1)D_y(k_2)) + (D_y(\tilde{p}_1) + \tilde{p}_2)k_2.$$

Since k_1/k_2 is differential-reduced, there exist $u, v \in \mathbb{F}[y]$ s.t. $p = u(k_1 - (m-1)D_y(k_2)) + vk_2$ by the extended Euclidean algorithm. So we can take $\tilde{p}_1 = u$ and $\tilde{p}_2 = v - D_y(u)$. By the induction hypothesis, there exist $p_1', p_2' \in \mathbb{F}[y]$ s.t.

$$\frac{\tilde{p}_2}{k_2^{m-1}} = D_y\left(\frac{p_1'}{k_2^{m-2}}\right) + \frac{p_1'}{k_2^{m-2}}K + \frac{p_2'}{k_2}.$$

Setting $p_1 = p_1'k_2 + \tilde{p}_1$ and $p_2 = p_2'$ completes the proof. ∎

With Convention 3, we have

$$\frac{p}{k_2^m}T = D_y\left(\frac{p_1}{k_2^{m-1}}T\right) + \frac{p_2}{k_2}T$$

by Lemma 16. This reduction will be referred to as the *kernel reduction for $(p/k_2^m)T$ w.r.t. K*.

6. TELESCOPING VIA REDUCTIONS

Hermite reduction has been used to construct telescopers for bivariate rational functions in [5]. We extend the idea in [5] and apply Theorem 13 to develop a reduction-based telescoping method for bivariate hyperexponential functions. The method also yields an order bound on minimal telescopers, which is tighter than those given in [4, 8]

6.1 Creative telescoping for bivariate hyperexponential functions

A nonzero element H in some differential field extension of $\mathbb{F}(x, y)$ is said to be *hyperexponential* over $\mathbb{F}(x, y)$ if its logarithmic derivatives $D_x(H)/H$ and $D_y(H)/H$ are in $\mathbb{F}(x, y)$.

Set $f = D_x(H)/H$ and $g = D_y(H)/H$. Then $D_y(f) = D_x(g)$ because D_x and D_y commute. Therefore, it is legitimate to denote H by $\exp(\int f\, dx + g\, dy)$. For every nonzero rational function $r \in \mathbb{F}(x, y)$,

$$rH = \exp\left(\int (f + D_x(r)/r)\, dx + (g + D_y(r)/r)\, dy\right).$$

The following fact is immediate from [12, Lemma 8].

Fact 17. *Let f and g be rational functions in $\mathbb{F}(x,y)$ satisfying $D_y(f) = D_x(g)$. Then the denominator of f divides that of g in $\mathbb{F}(x)[y]$.*

For a hyperexponential function H over $\mathbb{F}(x,y)$, the telescoping problem is to construct a linear ordinary differential operator $L(x, D_x)$ in $\mathbb{F}(x)\langle D_x\rangle$ s.t.

$$L(x, D_x)(H) = D_y(G)$$

for some hyperexponential function G over $\mathbb{F}(x,y)$. As in the rational case [5], we apply the Hermite reduction for univariate hyperexponential functions w.r.t. y to the derivatives $D_x^i(H)$ iteratively, and then find a linear dependency among the residual forms over $\mathbb{F}(x)$.

Lemma 18. *Let $H = \exp(\int f\,dx + g\,dy)$ be a hyperexponential function over $\mathbb{F}(x,y)$. Let K be the kernel and S the shell of g w.r.t. y. Then, for every $i \in \mathbb{N}$, the i-th derivative $D_x^i(H)$ can be decomposed into*

$$D_x^i(H) = D_y(u_i T) + r_i T, \tag{15}$$

where $u_i \in \mathbb{F}(x,y)$, $T = \exp(\int (f - D_x(S)/S)\,dx + K\,dy)$ and $r_i \in \mathbb{F}(x,y)$ is a residual form w.r.t. K. Moreover, let k_2 be the denominator of K, b the squarefree part of the denominator of S, and $\mathcal{N}_K$ the standard complement of the subspace for polynomial reduction w.r.t. K. Then

$$r_i = \frac{q_i}{b} + \frac{v_i}{k_2} \tag{16}$$

for some $q_i \in \mathbb{F}(x)[y]$ with $\deg_y q_i < \deg_y b$ and $v_i \in \mathcal{N}_K$.

Proof. We proceed by induction on i. If $i = 0$, then the assertion holds by Theorem 13.

Assume that $D_x^i(H)$ can be decomposed into (15) and assume that (16) holds. Moreover, let $\tilde{f} = f - D_x(S)/S$. Consider the $(i+1)$-th derivative $D_x^{i+1}(H)$. There exists a polynomial a in $\mathbb{F}(x)[y]$ s.t. $\tilde{f} = a/k_2$ by $D_y\left(\tilde{f}\right) = D_x(K)$ and Fact 17. A direct calculation leads to

$$\begin{aligned} D_x^{i+1}(H) =& D_y(D_x(u_i T)) + \left(\frac{q_i a}{b k_2} + \frac{D_x(q_i)}{b} + \frac{D_x(v_i)}{k_2}\right) T \\ &+ \left(\frac{-q_i D_x(b)}{b^2} + \frac{(a - D_x(k_2))v_i}{k_2^2}\right) T. \end{aligned}$$

Applying shell reduction to $\left(-q_i D_x(b)/b^2\right) T$ and kernel reduction to $\left((a - D_x(k_2))v_i/k_2^2\right) T$ w.r.t. y, we get

$$\begin{aligned} \frac{-q_i D_x(b)}{b^2} &= D_y\left(\frac{w_1}{b}\right) + \frac{w_1}{b}K + \frac{w_2}{b k_2}, \\ \frac{(a - D_x(k_2))v_i}{k_2^2} &= D_y\left(\frac{p_1}{k_2}\right) + \frac{p_1}{k_2}K + \frac{p_2}{k_2}, \end{aligned}$$

where w_1, w_2, p_1 and p_2 are in $\mathbb{F}(x)[y]$. We then apply polynomial reduction to $\tilde{S}T$ w.r.t. K, where

$$\tilde{S} = \frac{w_2}{b k_2} + \frac{p_2}{k_2} + \frac{a q_i}{b k_2} + \frac{D_x(q_i)}{b} + \frac{D_x(v_i)}{k_2},$$

which leads to $\tilde{S} = D_y(w) + wK + (q_{i+1}/b + v_{i+1}/k_2)$, where w is in $\mathbb{F}(x,y)$ and $q_{i+1}/b + v_{i+1}/k_2$ is the residual form of $\tilde{S}$ w.r.t. K. It follows from a direct calculation that

$$D_x^{i+1}(H) = D_y(u_{i+1}T) + \left(\frac{q_{i+1}}{b} + \frac{v_{i+1}}{k_2}\right) T,$$

where $u_{i+1} = D_x(u_i) + u_i \tilde{f} + w_1/b + p_1/k_2 + w$. ∎

The main results in the present section are given below.

Theorem 19. *With the notation introduced in Lemma 18, we let $L = \sum_{i=0}^{\rho} e_i D_x^i$ with $e_0, \ldots, e_\rho \in \mathbb{F}(x)$, not all zero.*

(i) L is a telescoper for H if and only if $\sum_{i=0}^{\rho} e_i r_i = 0$.

(ii) The order of a minimal telescoper for H is no more than $\deg_y(b) + \max(\deg_y(k_1), \deg_y(k_2) - 1)$.

Proof. We regard hyperexponential functions involved in the proof as univariate ones in y. Moreover, let $u = \sum_{i=0}^{\rho} e_i u_i$ and $r = \sum_{i=0}^{\rho} e_i r_i$. By (15), we have

$$L(H) = D_y(uT) + rT. \tag{17}$$

If $r = 0$, then L is a telescoper by (17). Conversely, assume that L is a telescoper of H. Then rT is integrable w.r.t. y by (17). Since r is a residual form by Remark 10, it is equal to zero by Lemma 11. The first assertion holds.

Set $\lambda = \max(\deg_y(k_1), \deg_y(k_2) - 1)$. Let the residual form $r_i = q_i/b + v_i/k_2$ be as defined in (15) and (16). By Lemma 8, the v_i's have a common set $\mathcal{P}$ of supporting monomials with $|\mathcal{P}| \le \lambda$. Moreover, $\deg_y(q_i) < \deg_y(b)$ and $\gcd(b, k_2) = 1$. Therefore, the residual forms $r_0, \ldots, r_\rho$ are linearly dependent over $\mathbb{F}(x)$ if $\rho \ge \deg_y(b) + \lambda$. The second assertion holds ∎

Remark 20. *By Theorem 19, a linear dependency among the residual forms r_0, ..., r_σ, for minimal σ, gives rise to a minimal telescoper of H.*

With the notation introduced in Lemma 18, we outline a reduction-based telescoping algorithm for bivariate hyperexponential functions.

Algorithm. HermiteTelescoping: Given a bivariate hyperexponential function $H = \exp(\int f\,dx + g\,dy)$ over $\mathbb{F}(x,y)$, compute a minimal telescoper L and its certificate w.r.t. y.

1. Find the kernel K and shell S of $D_y(H)/H$ w.r.t. y. Set b to be the squarefree part of the denominator of S.

2. Decompose H into $H = D_y(u_0 T) + r_0 T$ using the Hermite reduction for hyperexponential functions given in Theorem 13. If $r_0 = 0$, return $(1, u_0 T)$.

3. Set $\rho := \deg_y(b) + \max(\deg_y(k_1), \deg_y(k_2) - 1)$.

4. For i from 1 to ρ do

   4.1. Compute (u_i, r_i) incrementally s.t.

   $$D_x^i(H) = D_y(u_i T) + r_i T$$

   by the shell, kernel and polynomial reductions described in Lemma 18.

   4.2. Find $\eta_j \in \mathbb{F}(x)$ s.t. $\sum_{j=0}^{i} \eta_j r_j = 0$ by solving a linear system over $\mathbb{F}(x)$. If there is a nontrivial solution, return $\left(\sum_{j=0}^{i} \eta_j D_x^j, \sum_{j=0}^{i} \eta_j u_j T\right)$.

Example 21. *Let $H = \sqrt{x - 2y}\exp(x^2 y)$. Then $D_x(H)/H$ and $D_y(H)/H$ are, respectively,*

$$f = \frac{1 + 4x^2 y - 8xy^2}{2(x - 2y)} \quad \textit{and} \quad g = \frac{-1 + x^3 - 2x^2 y}{x - 2y}.$$

Since g is differential-reduced w.r.t. y, g is the kernel and 1 is the shell of $D_y(H)/H$ w.r.t. y. By Hermite reduction,

$$H = D_y\left(\frac{1}{x^2}H\right) + \frac{1}{x^2k_2}H. \qquad (18)$$

Applying D_x to the above equation yields

$$D_x(H) = D_y\left(\frac{-3x+8y+4x^3y-8x^2y^2}{2x^3(x-2y)}H\right) + rH,$$

where $r = (-5x+8y+4x^3y-8x^2y^2)/(2x^3k_2^2)$. The shell, kernel and polynomial reduction given in Lemma 18 yields

$$D_x(H) = D_y\left(\frac{2x^2y-3}{x^3}\cdot H\right) + \frac{3x^3-6}{2x^3k_2}H \qquad (19)$$

Combining (18) and (19), we get $L = (6-3x^3) + 2xD_x$ is a minimal telescoper for H and $G = (4y-3x)H$ is the corresponding certificate.

The algorithm above separates the computation of minimal telescopers from that of certificates. One may neglect the computation for certificates in the algorithm when they are irrelevant in applications. Moreover, one may opt for unnormalized certificates in the form wT, where $w = \sum_j w_j$ with $w_j \in \mathbb{F}(x,y)$ as described in step 4.2. Experiments carried out in Section 7 reveal that it is time-consuming to normalize w as a fraction p/q with $p,q \in \mathbb{F}[x,y]$ and $\gcd(p,q)=1$. In fact, unnormalized certificates are sufficient for many applications. For instance, we may want to compute $w(x,s)$ for $s \in F$ with $q(x,s) \neq 0$ when evaluating definite integrals. This can be achieved by unnormmalized certificates, because $w(x,s)$ equals the sum of all residues of $w_j/(y-s)$ at $y=s$.

Remark 22. *Another idea for computing a minimal telescoper of H is the following: We first compute a nonzero operator $L_1 \in \mathbb{F}(x)\langle D_x\rangle$ of minimal order s.t.*

$$L_1(H) = D_y(G_1) + (p/k_2)T$$

for some hyperexponential function G_1 and polynomial p. Note that such operators always exist, because $\deg_y q_i$ in (16) is less than $\deg_y b$. Then we apply the algorithm HermiteTelescoping to get a minimal telescoper L_2 for $(p/k_2)T$. One can show that L_2L_1 is a minimal telescoper of H.

Let $\ell_1 = \deg_y b$ and $\ell_2 = \max(\deg_y(k_1), \deg_y(k_2)-1)$, where b, k_1 and k_2 are given in Theorem 19. The algorithm HermiteTelescoping solves linear systems of at most $\ell_1+\ell_2$ equations over $\mathbb{F}(x)$ to obtain the minimal telescoper L, while an algorithm based on the idea given above solves linear systems of at most ℓ_1 equations to obtain L_1, and then solves linear systems of at most ℓ_2 equations to obtain L_2. However, the linear systems over $\mathbb{F}(x)$ corresponding to L_2 have coefficients of high degrees in x. In addition, it takes time to expand the product of L_2L_1. Preliminary experiments reveal that such an algorithm may outperform HermiteTelescoping in practice only when ℓ_2 is no more than three.

6.2 Comparison with the Apagodu-Zeilberger bound

Assume that H is a nonzero hyperexponential fundtion over $\mathbb{F}(x,y)$ of the form

$$H = u\exp\left(\frac{r_1}{r_2}\right)\prod_{i=1}^{m} p_i(x,y)^{c_i}, \qquad (20)$$

where $u, r_1, r_2, p_1, \ldots, p_m$ are nonzero polynomials in $\mathbb{F}[x,y]$ and $c_1, \ldots, c_m$ are *distinct indeterminates*.

Theorem cAZ in [4] asserts that the order of minimal telescopers for H is bounded by

$$\alpha := \deg_y(r_2) + \max\left(\deg_y(r_1), \deg_y(r_2)\right) + \sum_{i=1}^{m}\deg_y(p_i) - 1.$$

Note that H can be viewed as a hyperexponential function over $\mathbb{F}(c_1,\ldots,c_m)(x,y)$. We now show that α given above is at least the order bound on minimal telescopers for H obtained from Theorem 19 (ii). The kernel and shell of the logarithmic derivative $D_y(H)/H$ are

$$K := D_y\left(\frac{r_1}{r_2}\right) + \sum_{i=1}^{m} c_i\frac{D_y(p_i)}{p_i} \quad \text{and} \quad S := u,$$

respectively, because K has no integral residue at any simple pole, S is a polynomial in $\mathbb{F}[x,y]$, and $D_y(H)/H$ is equal to $K + D_y(S)/S$. Let $K = k_1/k_2$ with $\gcd(k_1,k_2)=1$. A direct calculation leads to

$$\deg_y(k_1) \leq \deg_y(r_1) + \deg_y(r_2) + \sum_{i=1}^{m}\deg_y(p_i) - 1,$$

and $\deg_y(k_2) \leq 2\deg_y(r_2) + \sum_{i=1}^{m}\deg_y(p_i)$. By Theorem 19, the order of minimal telescopers for H is no more than $\max\left(\deg_y(k_1), \deg_y(k_2)-1\right)$, which is no more than α by the above two inequalities.

Indeed, the order bound in Theorem 19 (ii) may be smaller than that in Theorem cAZ.

Example 23. *Let $H = q^c\exp(a/q)$, where a, q are irreducible polynomials in $\mathbb{F}[x,y]$ with $\deg_y(a) < \deg_y(q)$, and c is a transcendental constant over $\mathbb{F}$. By Theorem cAZ, a minimal telescoper for H has order no more than $3\deg_y q - 1$. On the other hand, the kernel and shell of $D_y(H)/H$ are equal to $\left(D_y(a)q - aD_y(q) + cqD_y(q)\right)/q^2$ and 1, respectively. A minimal telescoper has order no more than $2\deg_y q - 1$ by Theorem 19 (ii).*

Without assuming that the exponents $c_1, \ldots, c_m$ in (20) are distinct indeterminates, Theorem 14 in [8] derives order and degree bounds for minimal telescopers, in which the order bound is the same as that in Theorem cAZ. Furthermore, Christopher's Theorem in [9, 7] states that a general hyperexponential function over $\mathbb{F}(x,y)$ can be written as:

$$\frac{u}{v}\exp\left(\frac{r_1}{r_2}\right)\prod_{i=1}^{m} p_i(x,y)^{c_i}, \qquad (21)$$

where $u, v, r_1, r_2 \in \mathbb{F}[x,y]$, c_i is algebraic over $\mathbb{F}$, and p_i is in $\mathbb{F}(c_i)[x,y]$, $i = 1, \ldots, m$. So H given in (20) is a special instance of hyperexponential functions. In addition, it is easier to compute the kernel and shell w.r.t. y than to compute the decompositions (20) and (21) when a hyperexponential function is given by its logarithmic derivatives.

7. IMPLEMENTATION AND TIMINGS

We have implemented the algorithm HermiteTelescoping in the computer algebra system Maple 16. Our Maple code is available from

```
http://www.mmrc.iss.ac.cn/~zmli/HermiteCT.html
```

We now compare the performance of our algorithm with the Maple implementation DEtools[Zeilberger] of the telescoping algorithm in [3]. We take examples of the form

$$\frac{p}{q^m} \cdot \sqrt{\frac{a}{b}} \cdot \exp\left(\frac{u}{v}\right),$$

where $m \in \mathbb{N}$, $p, q, a, b, u, v \in \mathbb{Z}[x, y]$ are irreducible and their coefficients are randomly chosen. For simplicity, we choose $\lambda = \deg_y(p) = \deg_y(q)$, $\mu = \deg_y(a) = \deg_y(b)$, and $\nu = \deg_y(u) = \deg_y(v)$. The runtime comparison (in seconds) for different examples is shown in Table 1, in which

- ZT: the Maple function DEtools[Zeilberger].
- HT_un: the algorithm HermiteTelescoping, which returns telescopers and unnormalized certificates.
- HT_n: the algorithm HermiteTelescoping, which returns telescopers and normalized certificates.
- order: the order of the computed minimal telescoper.
- OOM: Maple runs out of memory.

(λ, μ, ν, m)	ZT	HT_un	HT_n	order
(2, 0, 2, 1)	2.16	2.01	3.80	5
(2, 0, 2, 2)	2.06	1.98	2.59	5
(3, 0, 2, 1)	8.68	6.54	14.01	6
(3, 0, 2, 2)	9.23	6.06	13.72	6
(6, 0, 1, 1)	44.04	24.39	70.49	7
(6, 0, 1, 2)	41.85	22.74	59.50	7
(2, 2, 2, 1)	1399.2	155.54	570.40	9
(2, 2, 2, 2)	1397.7	142.34	510.11	9
(3, 0, 3, 1)	151.84	44.07	120.44	8
(3, 0, 3, 2)	150.14	43.46	122.36	8
(3, 3, 0, 1)	206.90	46.15	165.67	8
(3, 3, 0, 2)	217.81	44.95	161.25	8
(3, 2, 1, 1)	300.93	60.33	184.71	8
(3, 2, 1, 2)	333.75	55.86	176.78	8
(3, 1, 3, 1)	OOM	361.79	1556.1	10
(3, 1, 3, 2)	OOM	370.18	1535.7	10

Table 1: Timings (in seconds) measured on a Mac OS X computer, 4Gb RAM, 3.06 GHz Core 2 Duo processor.

Our empirical results in the above table illustrate that HermiteTelescoping is markedly superior to Maple's function DEtools[Zeilberger] if it computes minimal telescopers and unnormalized certificates, and that it is either comparable to or faster than DEtools[Zeilberger] when it computes telescopers and normalized certificates.

Remark 24. *The orders of the computed minimal telescopers in our experiments equal the predicted order bounds given in Theorem 19.*

8. REFERENCES

[1] S.A. Abramov. The rational component of the solution of a first order linear recurrence relation with rational right hand-side. *Ž. Vyčisl. Mat. i Mat. Fiz.*, 15(4):1035–1039, 1090, 1975.

[2] S.A. Abramov and M. Petkovšek. Minimal decomposition of indefinite hypergeometric sums. In *ISSAC'01: Proceedings of the 2001 International Symposium on Symbolic and Algebraic Computation*, pages 7–14, New York, 2001. ACM.

[3] G. Almkvist and D. Zeilberger. The method of differentiating under the integral sign. *J. Symbolic Comput.*, 10:571–591, 1990.

[4] M. Apagodu and D. Zeilberger. Multi-variable Zeilberger and Almkvist-Zeilberger algorithms and the sharpening of Wilf- Zeilberger theory. *Adv. in Appl. Math.*, 37(2):139–152, 2006.

[5] A. Bostan, S. Chen, F. Chyzak, and Z. Li. Complexity of creative telescoping for bivariate rational functions. In *ISSAC'10: Proceedings of the 2010 International Symposium on Symbolic and Algebraic Computation*, pages 203–210, New York, NY, USA, 2010. ACM.

[6] M. Bronstein. *Symbolic Integration I: Transcendental Functions*, volume 1 of *Algorithms and Computation in Mathematics*. Springer-Verlag, Berlin, second edition, 2005.

[7] S. Chen. *Some Applications of Differential-Difference Algebra to Creative Telescoping*. PhD thesis, École Polytechnique (Palaiseau, France), February 2011.

[8] S. Chen and M. Kauers. Trading order for degree in creative telescoping. *J. of Symbolic Computation*, 47:968–995, 2012.

[9] C. Christopher. Liouvillian first integrals of second order polynomial differential equations. *Electron. J. Differential Equations*, 49:1–7, 1999.

[10] F. Chyzak. An extension of Zeilberger's fast algorithm to general holonomic functions. *Disc. Math.*, 217(1-3):115–134, 2000.

[11] J.H. Davenport. The Risch differential equation problem. *SIAM J. Comput.*, 15(4):903–918, 1986.

[12] K.O. Geddes, H.Q. Le, and Z. Li. Differential rational normal forms and a reduction algorithm for hyperexponential functions. In *ISSAC'04: Proceedings of the 2004 International Symposium on Symbolic and Algebraic Computation*, pages 183–190, New York, USA, 2004. ACM.

[13] R.W. Gosper, Jr. Decision procedure for indefinite hypergeometric summation. *Proc. Nat. Acad. Sci. U.S.A.*, 75(1):40–42, 1978.

[14] C. Hermite. Sur l'intégration des fractions rationnelles. *Ann. Sci. École Norm. Sup. (2)*, 1:215–218, 1872.

[15] M. Kauers, C. Koutschan, and D. Zeilberger. Proof of Ira Gessel's lattice path conjecture. *Proc. Natl. Acad. Sci. USA*, 106(28):11502–11505, 2009.

[16] C. Koutschan, M. Kauers, and D. Zeilberger. Proof of George Andrews's and David Robbins's q-TSPP conjecture. *Proc. Natl. Acad. Sci. USA*, 108(6):2196–2199, 2011.

[17] M.V. Ostrogradskiĭ. De l'intégration des fractions rationnelles. *Bull. de la classe physico-mathématique de l'Acad. Impériale des Sciences de Saint-Pétersbourg*, 4:145–167, 286–300, 1845.

[18] G. Xin and T.Y.J. Zhang. Enumeration of bilaterally symmetric 3-noncrossing partitions. *Discrete Math.*, 309(8):2497–2509, 2009.

Complexity Estimates for Two Uncoupling Algorithms*

Alin Bostan
INRIA (France)
alin.bostan@inria.fr

Frédéric Chyzak
INRIA (France)
frederic.chyzak@inria.fr

Élie de Panafieu
LIAFA (France)
depanafieuelie@gmail.com

ABSTRACT

Uncoupling algorithms transform a linear differential system of first order into one or several scalar differential equations. We examine two approaches to uncoupling: the cyclic-vector method (CVM) and the Danilevski-Barkatou-Zürcher algorithm (DBZ). We give tight size bounds on the scalar equations produced by CVM, and design a fast variant of CVM whose complexity is quasi-optimal with respect to the output size. We exhibit a strong structural link between CVM and DBZ enabling to show that, in the generic case, DBZ has polynomial complexity and that it produces a single equation, strongly related to the output of CVM. We prove that algorithm CVM is faster than DBZ by almost two orders of magnitude, and provide experimental results that validate the theoretical complexity analyses.

Categories and Subject Descriptors:
I.1.2 [**Computing Methodologies**]: Symbolic and Algebraic Manipulations — *Algebraic Algorithms*

General Terms: Algorithms, Theory.

Keywords: Danilevski-Barkatou-Zürcher algorithm, gauge equivalence, uncoupling, cyclic-vector method, complexity.

1. INTRODUCTION

1.1 Motivation

Uncoupling is the transformation of a linear differential system of first order, $Y' = MY$, for a square matrix M with coefficients in a rational-function field $\mathbb{K}(X)$ of characteristic zero, into one or several scalar differential equations $y^{(n)} = c_{n-1}y^{(n-1)} + \cdots + c_0 y$, with coefficients c_i in $\mathbb{K}(X)$. This change of representation makes it possible to apply algorithms that input scalar differential equations to systems.

In the present article, we examine two uncoupling algorithms: the cyclic-vector method (CVM) [24, 10, 8] and the Danilevski-Barkatou-Zürcher algorithm (DBZ) [12, 5, 30, 6]. While CVM always outputs only one equivalent differential equation, DBZ can decompose the system into several differential equations. This makes us consider two scenarios: the *generic case* corresponds to situations in which DBZ does not split the system into uncoupled equations, whereas in the *general case*, several equations can be output.

For some applications, getting several differential equations is more desirable than a single one. Besides, although the complexity of CVM is rarely discussed, its output is said to be "very complicated" in comparison to other uncoupling methods [18, 5, 30, 2, 16]. For these reasons, CVM has had bad reputation. Because of this general belief, uncoupling algorithms have not yet been studied from the complexity viewpoint. The lack and need of such an analysis is however striking when one considers, for instance, statements like *We tried to avoid [...] cyclic vectors, because we do not have sufficiently strong statements about their size or complexity* in a recent work on Jacobson forms [17]. One of our goals is a first step towards filling this gap and rehabilitating CVM.

1.2 Contribution

In relation to the differential system $Y' = MY$, a classical tool of differential algebra is the map δ, defined at a matrix or a row vector u by $\delta(u) = uM + u'$. A common objective of both CVM and DBZ, explicitly for the former and implicitly for the latter, as we shall prove in the present article, is to discover a basis P of $\mathbb{K}(X)^n$ with respect to which the matrix of the application δ is very simple. This matrix is the matrix $P[M]$ defined in §1.5. In contrast with CVM, which operates only on P, DBZ operates only on $P[M]$ by performing pivot manipulations without considering P. An important part of our contribution is to provide an algebraic interpretation of the operations in DBZ in terms of transformations of the basis P (§3.1 and §3.2).

More specifically, we analyse the degree of the outputs from CVM and DBZ, so as to compare them. Interestingly enough, an immediate degree analysis of (the first, generically dominating part of) DBZ provides us with a pessimistic exponential growth (§3.3). We prove that this estimate is far from tight: the degree growth is in fact only quadratic (Theorem 12). Surprising simplifications between numerators and denominators explain the result. This leads to the first complexity analysis of DBZ. It appears that, in contradiction to the well-established belief, DBZ and CVM have the same output on generic input systems (Theorem 13).

With respect to complexity, another surprising contribution of our work is that both CVM and DBZ have polynomial complexity in the generic case. Combining results, we design a fast variant of CVM (Theorem 8). Even more surprisingly, it turns out that this fast CVM has better com-

*Supported in part by the MSR-INRIA Joint Centre.

ISSAC'13, June 26–29, 2013, Boston, Massachusetts, USA.

plexity ($\approx n^{\theta+1}d$, for $2 \le \theta \le 3$) than DBZ ($\approx n^5 d$), when applied to systems of size n and degree d. As the output size is proved to be generically asymptotically proportional to $n^3 d$, our improvement of CVM is quasi-optimal with respect to the output size.

Another uncoupling algorithm is part of the work in [1]. We briefly analyse its complexity in §3.5 and obtain the same complexity bound as for DBZ (Theorem 17).

Our results remain valid for large positive characteristic. They are experimentally well corroborated in §4.

1.3 Previous work

Uncoupling techniques have been known for a long time; CVM can be traced back at least to Schlesinger [24, p. 156–160]. Loewy [20, 21] was seemingly the first to prove that every square matrix over an ordinary differential field of characteristic zero is gauge-similar to a companion matrix.

That a linear system of first order is equivalent to *several* scalar linear equations is a consequence of Dickson's [14, p. 173–174] and Wedderburn's [29, Th. 10.1] algorithmic results on the Smith-like form of matrices with entries in non-commutative domains; see also [23, Chap III, Sec. 11]. Its refinement nowadays called *Jacobson form* [19, Th. 3 & 4] implies equivalence to a *single* scalar equation. This approach was further explored in [15, §6] and [11, 3].

Cope [10, §6] rediscovered Schlesinger's CVM, and additionally showed that for a system of n equations over $\mathbb{K}(X)$, one can always choose a polynomial cyclic vector of degree less than n in X. A generalisation to arbitrary differential fields was given in [8, §7]. The subject of differential cyclic vectors gained a renewed interest in the 1970's, starting from Deligne's non-effective proof [13, Ch. II, §1]. An effective version, with no restriction on the characteristic, was given in [8, §3]. CVM has bad reputation, its output being said to have *very complicated coefficients* [18, 5, 2]. However, adhoc examples apart, very few degree bounds and complexity analyses are available [9, §2]. The few complexity results we are aware of [8, 16] only measure operations in $\mathbb{K}(X)$, and do not take degrees in X into account.

Barkatou [5] proposed an alternative uncoupling method reminiscent of Danilevski's algorithm for computing weak Frobenius forms [12]. Danilevski's algorithm was also generalised by Zürcher [30]; see also [6, §5]. This is what we call DBZ, for Danilevski, Barkatou, and Zürcher. Various uncoupling algorithms are described and analysed in [16, 22], including the already mentioned algorithm from [1].

1.4 Notation and Conventions

Algebraic Structures and Complexity. Let $\mathbb{K}$ denote a field of characteristic zero, $\mathbb{K}_d[X]$ the set of polynomials of degree at most d, and $\mathcal{M}_n(S)$ and $\mathcal{M}_{1,n}(S)$, respectively, the sets of square matrices of dimension $n \times n$ and of row vectors of dimension n, each with coefficients in some set S. The *arithmetic size* of a matrix in $\mathcal{M}_n(\mathbb{K}(X))$ is the number of elements of $\mathbb{K}$ in its dense representation.

We consider the *arithmetic complexity* of algorithms, that is, the number of operations they require in the base field $\mathbb{K}$. For asymptotic complexity, we employ the classical notation $\mathcal{O}(\cdot)$, $\Omega(\cdot)$, and $\Theta(\cdot)$, as well as the notation $g = \tilde{\mathcal{O}}(f)$ if there exists k such that $g/f = \mathcal{O}(\log^k n)$. We denote by $\mathsf{M}(d)$ (resp. $\mathsf{MM}(n,d)$) the arithmetic complexity of the multiplication in $\mathbb{K}_d[X]$ (resp. in $\mathcal{M}_n(\mathbb{K}_d[X])$). When the complexity of an algorithm is expressed in terms of M and MM, it means that the algorithm can use multiplication (in $\mathbb{K}[X]$ and $\mathcal{M}_n(\mathbb{K}[X])$) as "black boxes". Estimates for $\mathsf{M}(d)$ and $\mathsf{MM}(n,d)$ are summarised in the following table, where θ is a constant between 2 and 3 that depends on the matrix-multiplication algorithm used. For instance, $\theta = 3$ for the schoolbook algorithm, and $\theta = \log_2(7) \approx 2.807$ for Strassen's algorithm [27]. The current tightest upper bound, due to Vassilevska Williams [28], is $\theta < 2.3727$.

Structure	Notation	Naive	Fast	Size
$\mathbb{K}$	–	1	1	1
$\mathbb{K}_d[X]$	$\mathsf{M}(d)$	$\mathcal{O}(d^2)$	$\tilde{\mathcal{O}}(d)$ [25]	$\Theta(d)$
$\mathcal{M}_n(\mathbb{K}_d[X])$	$\mathsf{MM}(n,d)$	$\mathcal{O}(n^3d^2)$	$\tilde{\mathcal{O}}(n^\theta d)$ [7]	$\Theta(n^2d)$

The complexity of an algorithm is said to be *quasi-optimal* when, assuming $\theta = 2$, it matches the arithmetic size of its output up to logarithmic factors. For instance, the algorithms in the table above have quasi-optimal complexity.

Generic Matrices. The notion of *genericity* is useful to analyse algorithms on inputs that are not "particular". For parameters in some $\mathbb{K}^r$, a property is classically said to be *generic* if it holds out of the zero set of a non-zero r-variate polynomial. To define the notion of *generic matrices*, we identify $M \in \mathcal{M}_n(\mathbb{K}_d[X])$ with the family $\{m_{i,j,k}\}$, indexed by $1 \le i,j \le n$, $0 \le k \le d$, of its coefficients in $\mathbb{K}$.

Conventions. In this text, M is always the input of the uncoupling algorithms. It is assumed to be a matrix in $q(X)^{-1}\mathcal{M}_n(\mathbb{K}_d[X])$ with $q(X) \in \mathbb{K}_d[X]$. It defines δ on a matrix or a row vector u by $\delta(u) = uM + u'$.

For a matrix A, we respectively denote its ith row and jth column by $A_{i,*}$ and $A_{*,j}$. We write $\mathsf{VJoin}(r^{(1)}, \dots, r^{(n)})$ for the matrix A such that for all i, $A_{i,*} = r^{(i)}$. We define the rows e_i by $I_n = \mathsf{VJoin}(e_1, \dots, e_n)$.

A square matrix C is said to be *companion* if beside zero coefficients, it has 1s on its upper diagonal and arbitrary coefficients on its last row, $c = (c_0, \dots, c_{n-1})$. Thus:

$$C = \mathsf{VJoin}(e_2, \dots, e_{n-1}, c). \tag{1}$$

We say A has its ith row *in companion shape* if $A_{i,*} = e_{i+1}$.

We write $\operatorname{diag}(B^{(1)}, \dots, B^{(n)})$ for a diagonal block matrix given by square blocks $B^{(i)}$.

Degrees of rational functions and matrices. In the present paper, the degree of a rational function is the maximum of the degrees of its numerator and denominator. The degree of a vector or a matrix with rational-function coefficients is the maximum of the degrees of its coefficients. The following lemma expresses the generic degrees encountered when solving a generic matrix.

Lemma 1 *Let A be a matrix in $\mathcal{M}_n(\mathbb{K}[X])$. Define a_i as $\deg(A_{i,*})$ and D as $\sum_i a_i$. Then, $\deg(\det(A)) \le D$ and, for all i, $\deg(\det(A)(A^{-1})_{*,i}) \le D - a_i$. When A is generic with $\deg(A_{i,*}) = a_i$ for all i, those bounds are reached.*

PROOF. Proofs use classical techniques and are omitted. We simply observe that $\operatorname{diag}(x^{a_1}, \dots, x^{a_n})\,N$ reaches the announced bounds when $N \in \mathcal{M}_n(\mathbb{K} \setminus \{0\})$ and $\det N \ne 0$. □

1.5 Companion matrices and uncoupling

For an invertible matrix P, let us perform the change of unknowns $Z = PY$ in a system $Y' = MY$. Then, $Z' = PY' + P'Y = (PM + P')P^{-1}Z$. The system is therefore *equivalent* to $Z' = P[M]\,Z$ where $P[M]$ denotes $(PM +$

$P')P^{-1}$, in the sense that the solutions of both systems, whether meromorphic or rational, are in bijection under P.

We call *gauge transformation* of a matrix A by an invertible matrix P the matrix $P[A] = (PA+P')P^{-1}$. When $B = P[A]$, we say that A and B are *gauge-similar*. The gauge-similarity relation is transitive since $P[Q[A]] = (PQ)[A]$. With the notation introduced above, $P[M] = \delta(P)\,P^{-1}$.

The following folklore theorem relates the solutions of a system with the solutions of the uncoupled equations obtained from a suitable gauge-similar system: it states that, to uncouple the system $Y' = MY$, it suffices to find an invertible matrix P such that $P[M]$ is in diagonal companion block form. This is the main motivation for uncoupling. We omit its proof, as it has similarity with the proofs in §2.1, and because we use no consequence of it later in this article. (We write ∂f instead of f' for derivations.)

Theorem 2 *Let P be an invertible matrix such that*

$$P[M] = \operatorname{diag}(C^{(1)}, \ldots, C^{(t)}) \tag{2}$$

with $C^{(i)}$ companion of dimension k_i. Denote the last row of $C^{(i)}$ by $\big(c_0^{(i)}, \ldots, c_{k_i-1}^{(i)}\big)$. Then, $\partial Y = MY$ if and only if

$$PY = \mathsf{VJoin}(Z^{(1)}, \ldots, Z^{(t)})$$

where $Z^{(i)} = \big(z^{(i)}, \partial z^{(i)}, \ldots, \partial^{k_i-1} z^{(i)}\big)^T$ and

$$\partial^{k_i} z^{(i)} = c_{k_i-1}^{(i)} \partial^{k_i-1} z^{(i)} + \cdots + c_0^{(i)} z^{(i)}.$$

2. CYCLIC-VECTOR METHOD

Two versions of CVM are available, depending on how the first row of P is obtained: a version ProbCV picks this first row at random, and thus potentially fails, but with tiny probability; a deterministic version DetCV computes a first row in such a way that the subsequent process provably cannot fail. In both cases, CVM produces no non-trivial diagonal companion block decomposition but only one block.

We present the randomised CVM only, before analysing the degree of its output and giving a fast variant.

2.1 Structure theorems

Let $\Delta^k(u)$ denote the matrix $\mathsf{VJoin}(u, \delta(u), \ldots, \delta^{k-1}(u))$ of dimension $k \times n$. The diagonal companion block decomposition (2) is based on the following folklore result.

Lemma 3 *Let P be an invertible matrix. Then, there exists a companion matrix C of dimension k such that*

$$P[M] = \left(\begin{smallmatrix} C & 0 \\ * & * \end{smallmatrix}\right) \tag{3}$$

if and only if there exists a vector u such that $P = \left(\begin{smallmatrix} \Delta^k(u) \\ * \end{smallmatrix}\right)$ and $\delta^k(u) \in \operatorname{span}\big(u, \delta(u), \ldots, \delta^{k-1}(u)\big)$.

PROOF. Set $\left(\begin{smallmatrix} U \\ R \end{smallmatrix}\right) := P$ where U has k rows. Equality (3) is equivalent to $\delta\left(\begin{smallmatrix} U \\ R \end{smallmatrix}\right) = \left(\begin{smallmatrix} C & 0 \\ * & * \end{smallmatrix}\right)\left(\begin{smallmatrix} U \\ R \end{smallmatrix}\right)$, then with $\delta(U) = CU$. This can be rewritten:

$$\mathsf{VJoin}(\delta(U_{1,*}), \ldots, \delta(U_{k,*})) = \mathsf{VJoin}(U_{2,*}, \ldots, U_{k,*}, C_{k,*}U).$$

Set u to the first row of U. This equation is satisfied if and only if $U = \Delta^k(u)$ and $\delta^k(u) \in \operatorname{span}(\Delta^k(u))$. □

The following corollaries for partial companion decomposition and diagonal companion block decomposition are proved in a very similar fashion to the preceding lemma. They will be used for the analysis of CVM and DBZ.

Corollary 4 *Let P be an invertible matrix. Then, $P[M]$ has its first $k-1$ rows in companion shape if and only if there exists a row vector u such that $P = \left(\begin{smallmatrix} \Delta^k(u) \\ * \end{smallmatrix}\right)$.*

Corollary 5 *Let P be an invertible matrix and $\{C^{(i)}\}_{1 \le i \le t}$ a family of companion matrices of dimension k_i. Then,*

$$P[M] = \operatorname{diag}(C^{(1)}, \ldots, C^{(t)})$$

if and only if there exist t row vectors $\{u^{(i)}\}_{1 \le i \le t}$ such that $P = \mathsf{VJoin}(\Delta^{k_1}(u^{(1)}), \ldots, \Delta^{k_t}(u^{(t)}))$ and for all i, $\delta^{k_i}(u^{(i)})$ is in $\operatorname{span}(\Delta^{k_i}(u^{(i)}))$.

2.2 Classical algorithms for cyclic vectors

The name CVM comes from the following notion.

Definition *A* cyclic vector *is a row vector $u \in \mathbb{K}(X)^n$ for which the matrix $\Delta^n(u)$ is invertible, or, equivalently, such that the cyclic module generated by u over $\mathbb{K}(X)\langle\delta\rangle$ is the full vector space $\mathcal{M}_{1,n}\big(\mathbb{K}(X)\big)$ of row vectors.*

The next folklore method [5, 8] is justified by Theorem 6 below, which means that CVTrial will not fail too often.

Algorithm CVTrial: Testing if a vector is a cyclic vector

Input: $M \in q(X)^{-1}\mathcal{M}_n(\mathbb{K}_d[X])$ and $u \in \mathcal{M}_{1,n}(\mathbb{K}_{n-1}[X])$
Output: P, C with C companion and $\delta(P)P^{-1} = C$

1: set P to the square zero matrix of dimension n
2: $P_{1,*} := u$
3: for $i = 1$ to $n-1$, do $P_{i+1,*} := \delta(P_{i,*})$
4: if P is not invertible, return "Not a cyclic vector"
5: $C := \delta(P)P^{-1}$
6: return (P, C)

Theorem 6 *[10, 8] When u is generic of degree less than n, the matrix $P = \Delta^n(u)$ is invertible.*

PROOF. It is proved in [10, 8] that every matrix M admits a cyclic vector u of degree less than n. Then, $\det(\cdot)$ is a non-zero polynomial function of the matrix coefficients. □

In ProbCV, u is chosen randomly, leading to a Las Vegas algorithm for finding a cyclic vector. The proof above refers to the theorem that every matrix M admits a cyclic vector. Churchill and Kovacic give in [8] a good survey of this subject. They also provide an algorithm that we denote DetCV that takes as input a square matrix M and deterministically outputs a cyclic vector u. The arithmetic complexity of this algorithm is polynomial, but worse than that of ProbCV.

2.3 Degree analysis and fast algorithm

CVTrial computes two matrices, P and C, whose sizes we now analyse. We shall find the common bound $\mathcal{O}(n^3 d)$. When $u \in \mathcal{M}_{1,n}(\mathbb{K}_{n-1}[X])$ is generic, this bound is reached. The size $\Theta(n^3 d)$ of the output of CVTrial is then a lower bound on the complexity of any algorithm specifying CVM. After the remark that the complexity of the simple algorithm is above this bound, we give a fast algorithm.

We start by bounding the degree of the matrix $\Delta^n(u)$. Following the result of Churchill and Kovacic, we make the assumption that $\deg(u)$ is less than n.

Lemma 7 *The row vector $q^k \delta^k(u)$ consists of polynomials of degree at most $\deg(u) + kd$.*

PROOF. The proof proceeds by induction after noting that $q^{k+1}\delta^{k+1} = q(\delta q^k - kq^{k-1}q')\delta^k = (q\delta - kq')q^k\delta^k$. □

We list further bounds on degrees and arithmetic sizes, some of which are already in [9, §2]:

	Degree	Arithmetic size
P	$\deg(u) + (n-1)d$	$\mathcal{O}(n^3 d + n^2 \deg(u))$
P^{-1}	$n\deg(u) + \frac{n(n-1)}{2} d$	$\mathcal{O}(n^4 d + n^3 \deg(u))$
$\delta^n(u)$	$\deg(u) + nd$	$\mathcal{O}(n^2 d + n \deg(u))$
C	$n\deg(u) + \frac{n(n+1)}{2} d$	$\mathcal{O}(n^3 d + n^2 \deg(u))$

Theorem 8 *Algorithm CVTrial implements CVM in quasi-optimal complexity $\tilde{\mathcal{O}}(n^{\theta+1}d)$.*

PROOF. At Step 3, CVTrial computes $\delta(P_{i,*}) = P_{i,*}M + P'_{i,*}$ for successive i's. Addition and derivation have softly linear complexity, so we focus on the product $P_{i,*}M$. Computing it by a vector-matrix product would have complexity $\Omega(n^2 id)$ and the complexity of the loop at Step 3 would then be $\Omega(n^4 d)$. The row $P_{i,*}$ has higher degree than M, so the classical idea to make it a matrix to balance the product applies. Let $A_k \in \mathbb{K}_d[X]^n$ be the rows defined by $P_{i,*} = \sum_{k=0}^{n-1} A_k X^{kd}$, and $A := \mathsf{VJoin}(A_0, \ldots, A_{n-1})$. The product AM is computed in complexity $\mathcal{O}(\mathsf{MM}(n,d))$, and $P_{i,*}M$ is reconstructed in linear complexity, thus performing the whole loop in complexity $\mathcal{O}(n\,\mathsf{MM}(n,d))$.

Only the last row $\delta^n(u)P^{-1}$ of C needs to be computed at Step 5, and the size $\Theta(n^4 d)$ of P^{-1} bans the computation of P^{-1} from any low complexity algorithm. C can be computed in complexity $\tilde{\mathcal{O}}(\mathsf{MM}(n, nd))$. This is achieved by solving $PY = \delta^n(u)$ by Storjohann's algorithm [26], which inputs $\delta^n(u)$ and P, of degree $\Theta(nd)$, and outputs the last row $\delta^n(u)P^{-1}$ of C in $\mathcal{O}(\mathsf{MM}(n, nd)\log(nd))$ operations. □

3. THE DANILEVSKI-BARKATOU-ZÜRCHER ALGORITHM

We begin this section with a description of algorithm DBZ, before a naive analysis that gives exponential bounds. Experiments show a polynomial practical complexity, whence the need for a finer analysis. To obtain it, we develop an algebraic interpretation of the algorithm.

3.1 Description of the algorithm

The input to DBZ is a matrix $M \in \mathcal{M}_n(\mathbb{K}(X))$; its output $(P, C^{(1)}, \ldots, C^{(t)})$ satisfies $P[M] = \mathrm{diag}(C^{(1)}, \ldots, C^{(t)})$ for companion matrices $C^{(i)}$. DBZ iterates over $P[M]$ to make it progressively diagonal block-companion. To do so, three sub-algorithms are used, $\mathsf{DBZ}^{\mathrm{I}}$, $\mathsf{DBZ}^{\mathrm{II}}$, and $\mathsf{DBZ}^{\mathrm{III}}$, in order to achieve special intermediate forms for $P[M]$. These forms, respectively Shape (I), (II), and (III), are:

$$\begin{pmatrix} C & 0 \\ \alpha & \beta \end{pmatrix}, \quad \begin{pmatrix} C & \begin{matrix} 0 & \cdots & 0 \\ \vdots & & \vdots \\ 0 & \cdots & 0 \end{matrix} \\ \begin{matrix} & 0 & \cdots & 0 \\ v & \vdots & & \vdots \\ & 0 & \cdots & 0 \end{matrix} & \beta \end{pmatrix}, \quad \begin{pmatrix} * & 1\ 0 \cdots 0 & * \cdots * \\ 0 & & 0 \cdots 0 \\ \vdots & C & \vdots \quad \vdots \\ 0 & & 0 \cdots 0 \\ * & 0 \cdots \cdots 0 & * \cdots * \\ \vdots & \vdots \quad\quad \vdots & \vdots \quad \vdots \\ * & 0 \cdots \cdots 0 & * \cdots * \end{pmatrix},$$

where in each case C denotes a companion matrix, v is a column vector, α, β are general matrices, and β is square.

In the course of DBZ, first, $\mathsf{DBZ}^{\mathrm{I}}$ computes P^{I} and M^{I} such that $P^{\mathrm{I}}[M] = M^{\mathrm{I}}$ has Shape (I). If $M^{\mathrm{I}} = C$ is companion — that is, if α and β do not occur — then DBZ returns $(P^{\mathrm{I}}, M^{\mathrm{I}})$. If not, at this point, DBZ has obtained a first companion block C and we would hope that α is zero to apply DBZ recursively to β. So, in general, DBZ tries to cancel α, by appealing to $\mathsf{DBZ}^{\mathrm{II}}$ to compute P^{II} and M^{II} such that $P^{\mathrm{II}}[M^{\mathrm{I}}] = M^{\mathrm{II}}$ has Shape (II). If the obtained v is zero, then DBZ can go recursively to β.

If not, DBZ seems to have failed and restarts on a matrix $P^{\mathrm{III}}[M^{\mathrm{II}}] = M^{\mathrm{III}}$ with Shape (III), which ensures $\mathsf{DBZ}^{\mathrm{I}}$ can treat at least one more row than previously (as proved in [6]). Therefore, the algorithm does not loop forever. The matrix M^{III} on which DBZ starts over and the differential change of basis P^{III} associated are computed by $\mathsf{DBZ}^{\mathrm{III}}$.

Algorithm DBZ (Danilevski-Barkatou-Zürcher)

Input: $M \in \mathcal{M}_n(\mathbb{K}(X))$
Output: $(P, C^{(1)}, \ldots, C^{(t)})$ with $C^{(i)}$ companion matrices and $P[M] = \mathrm{diag}(C^{(1)}, \ldots, C^{(t)})$

1: $(P^{\mathrm{I}}, M^{\mathrm{I}}) := \mathsf{DBZ}^{\mathrm{I}}(M)$
2: if M^{I} is companion then return$(P^{\mathrm{I}}, M^{\mathrm{I}})$
3: $(P^{\mathrm{II}}, M^{\mathrm{II}}) := \mathsf{DBZ}^{\mathrm{II}}(M^{\mathrm{I}})$
4: $\left(\begin{smallmatrix} C & & 0 \\ v & 0 & \beta \end{smallmatrix}\right) := M^{\mathrm{II}}$ where $v \in \mathcal{M}_{n-k,1}(\mathbb{K}(X))$
5: if $v = 0$ then
6: $\quad (P, C^{(1)}, \ldots, C^{(t)}) := \mathsf{DBZ}(\beta)$
7: $\quad$ return $(\mathrm{diag}(I_k, P)P^{\mathrm{II}}P^{\mathrm{I}}, C, C^{(1)}, \ldots, C^{(t)})$
8: $(P^{\mathrm{III}}, M^{\mathrm{III}}) := \mathsf{DBZ}^{\mathrm{III}}(M^{\mathrm{II}})$
9: $(P, C^{(1)}, \ldots, C^{(t)}) := \mathsf{DBZ}(M^{\mathrm{III}})$
10: return $(PP^{\mathrm{III}}P^{\mathrm{II}}P^{\mathrm{I}}, C^{(1)}, \ldots, C^{(t)})$

3.2 Description of the sub-algorithms

We now describe $\mathsf{DBZ}^{\mathrm{I}}$, $\mathsf{DBZ}^{\mathrm{II}}$, and $\mathsf{DBZ}^{\mathrm{III}}$ in more details. By $E_{i,j}(t)$, we denote the matrix obtained after replacing by t the (i,j) coefficient in the identity matrix I_n, and by $E_i(u)$ the matrix obtained after replacing the ith row by the row vector u in I_n. Let $\mathsf{Inv}^{(j,n)}$ denote the matrix obtained from I_n after exchanging the jth and nth rows. Set $\mathsf{Rot} = \mathsf{VJoin}(e_n, e_1, \ldots, e_{n-1})$.

The algorithms developed below rely on a common Pivot subtask, which inputs $(M, P, T) \in \mathcal{M}_n(\mathbb{K}(X))^3$ with invertible P and T, and outputs the update of (M, P) under T. This really behaves like a Gauge transformation, changing (M, P) to $\big(T[M], TP\big)$. This modification of M and P only ensures the invariant $M = P[M_{\mathrm{initial}}]$.

$\mathsf{DBZ}^{\mathrm{I}}$ inputs M and outputs the tuple of matrices $(P^{\mathrm{I}}, M^{\mathrm{I}})$ with M^{I} in Shape (I). It starts with $M^{\mathrm{I}} = M$ and modifies its rows one by one. At the ith iteration of the loop (Step 2), the matrix M^{I} has its first $i-1$ rows in companion form. To put the ith row in companion form, $\mathsf{DBZ}^{\mathrm{I}}$ sets $M^{\mathrm{I}}_{i+1,i+2}$ to 1 (Step 6), and uses it as a pivot to cancel the other coefficients of the row (loop at Step 7).

Algorithm $\mathsf{DBZ}^{\mathrm{I}}$

Input: $M \in \mathcal{M}_n(\mathbb{K}(X))$
Output: $(P^{\mathrm{I}}, M^{\mathrm{I}})$ with M^{I} in Shape (I) and $P^{\mathrm{I}}[M] = M^{\mathrm{I}}$

1: $(P^{\mathrm{I}}, M^{\mathrm{I}}) := (I, M)$
2: for $i = 1$ to $n-1$ do
3: $\quad r := \min\big(\{j \mid M^{\mathrm{I}}_{i,j} \neq 0 \text{ and } j > i\} \cup \{n+1\}\big)$
4: $\quad$ if $r = n+1$ then return $(P^{\mathrm{I}}, M^{\mathrm{I}})$
5: $\quad (M^{\mathrm{I}}, P^{\mathrm{I}}) := \mathsf{Pivot}(M^{\mathrm{I}}, P^{\mathrm{I}}, \mathsf{Inv}^{(i+1,r)})$
6: $\quad (M^{\mathrm{I}}, P^{\mathrm{I}}) := \mathsf{Pivot}(M^{\mathrm{I}}, P^{\mathrm{I}}, E_{i+1,i+1}(M^{\mathrm{I}}_{i,i+1})^{-1})$
7: $\quad$ for $j = 1$ to n with $j \neq i+1$ do
8: $\quad\quad (M^{\mathrm{I}}, P^{\mathrm{I}}) := \mathsf{Pivot}(M^{\mathrm{I}}, P^{\mathrm{I}}, E_{i+1,j}(-M^{\mathrm{I}}_{i,j}))$
9: return $(P^{\mathrm{I}}, M^{\mathrm{I}})$

If $M^{\mathrm{I}}_{i+1,i+2} = 0$ and there is a non-zero coefficient farther on the row, then the corresponding columns are inverted at Step 5 and $\mathsf{DBZ}^{\mathrm{I}}$ goes on. If there is no such coefficient, the matrix has reached Shape (I) and is returned at Step 4.

$\mathsf{DBZ}^{\mathrm{II}}$ inputs M^{I} and outputs a tuple $(P^{\mathrm{II}}, M^{\mathrm{II}})$ with M^{II} in Shape (II). At Step 3, it cancels the columns of the lower-left block of M^{I} one by one, from the last one to the second one, using the 1's of C as pivots. At the ℓth iteration, the lower-left block of M^{I} ends with ℓ zero columns.

Algorithm $\mathsf{DBZ}^{\mathrm{II}}$

Input: M^{I} in Shape (I)
Output: $(P^{\mathrm{II}}, M^{\mathrm{II}})$ in Shape (II) such that $P^{\mathrm{II}}[M^{\mathrm{I}}] = M^{\mathrm{II}}$

1: $k :=$ size of the companion block of M^{I}
2: $(P^{\mathrm{II}}, M^{\mathrm{II}}) := (I, M^{\mathrm{I}})$
3: for $j = k$ down to 2 do
4:     for $i = k+1$ to n do
5:         $(M^{\mathrm{II}}, P^{\mathrm{II}}) := \mathsf{Pivot}(M^{\mathrm{II}}, P^{\mathrm{II}}, E_{i,j-1}(-M^{\mathrm{II}}_{i,j}))$
6: return $(P^{\mathrm{II}}, M^{\mathrm{II}})$

$\mathsf{DBZ}^{\mathrm{III}}$ inputs M^{II} with v non-zero and outputs the tuple $(P^{\mathrm{III}}, M^{\mathrm{III}})$ with M^{III} in Shape (III). The transformation of M at Step 3 reverses v to put a non-zero coefficient on the last row of M^{II}. Then it sets it to 1 at Step 4 and uses it as a pivot to cancel the other v_j's (Step 6). Finally, at Step 7, a cyclic permutation is applied to the rows and columns: last row becomes first, last column becomes first.

Algorithm $\mathsf{DBZ}^{\mathrm{III}}$

Input: M^{II} in Shape (II), under the constraints C has dimension k and $v \neq 0$
Output: $(P^{\mathrm{III}}, M^{\mathrm{III}})$ in Shape (III) where $P^{\mathrm{III}}[M^{\mathrm{II}}] = M^{\mathrm{III}}$

1: $M^{\mathrm{III}} := M^{\mathrm{II}}$
2: $h := \max\{i \mid M^{\mathrm{III}}_{i,1} \neq 0\}$
3: $(M^{\mathrm{III}}, P^{\mathrm{III}}) := \mathsf{Pivot}(M^{\mathrm{III}}, P^{\mathrm{III}}, \mathsf{Inv}^{(h,n)})$
4: $(M^{\mathrm{III}}, P^{\mathrm{III}}) := \mathsf{Pivot}(M^{\mathrm{III}}, P^{\mathrm{III}}, E_{n,n}(1/M^{\mathrm{III}}_{n,1}))$
5: for $i = k+1$ to $n-1$ do
6:     $(M^{\mathrm{III}}, P^{\mathrm{III}}) := \mathsf{Pivot}(M^{\mathrm{III}}, P^{\mathrm{III}}, E_{i,n}(-M^{\mathrm{III}}_{i,1}))$
7: $(M^{\mathrm{III}}, P^{\mathrm{III}}) := \mathsf{Pivot}(M^{\mathrm{III}}, P^{\mathrm{III}}, \mathsf{Rot})$
8: return $(P^{\mathrm{III}}, M^{\mathrm{III}})$

3.3 A naive degree analysis of the generic case

When M is generic, $\mathsf{DBZ}^{\mathrm{I}}$ outputs a companion matrix, so DBZ terminates at Step 2 in the generic case with only one companion matrix in the diagonal companion block decomposition. The proof of this fact will be given in §3.4. Therefore, the complexity of $\mathsf{DBZ}^{\mathrm{I}}$ is interesting in itself.

A lower bound on this complexity is the degree of its output. We explain here why a naive analysis of $\mathsf{DBZ}^{\mathrm{I}}$ only gives an exponential upper bound on this degree.

Let $M^{(i)}$ be the value of M^{I} just before the ith iteration of the loop at Step 2 (in particular, $M^{(1)} = M$), and $M^{(n)}$ the output value. Remark that the matrices involved in the gauge transformations at Steps 6 and 8 commute with one another. Their product is equal to $E_{i+1}(\mathsf{Inv}^{(i+1,r)}[M^{(i)}]_{i,*})$.

Lemma 9 *If $A \in \mathcal{M}_n(\mathbb{K}_d[X])$ is a generic matrix with its first $i-1$ rows in companion form and $T = E_{i+1}(A_{i,*})$, then $T[A]$ has degree $3d$.*

An exponential bound on the output of $\mathsf{DBZ}^{\mathrm{I}}$ is easily deduced: $\deg(M^{(n)}) \leq 3^{n-1}\deg(M)$. We will dramatically improve this bound in the following section.

3.4 Algebraic interpretation and better bounds

To prove the announced tight bound, it could in principle be possible to follow the same pattern as in Bareiss' method [4]: give an explicit form for the coefficients of the transformed matrices M, from which the degree analysis becomes obvious. But it proves more fruitful to find a link between CVM and DBZ, and our approach involves almost no computation.

Algorithm DBZ reshapes the input matrix by successive elementary gauge transformations. It completely relies on the shape of M, M^{I}, M^{II}, and M^{III}, while the construction of the matrices P is only a side-effect. As illustrated in §3.3, this approach is not well suited for degree analysis.

In this section, we focus on P, P^{I}, P^{II}, and P^{III}. It turns out that these matrices allow nice algebraic formulations, leading to sharp degree and complexity analyses of DBZ.

The following lemma, whose omitted proof is immediate from the design of DBZ, provides the complexity of the computation of an elementary gauge transformation.

Lemma 10 *If $t \in \mathbb{K}_d[X]$ and $M \in \mathcal{M}_n(\mathbb{K}_d[X])$, then the gauge transformation of M by $E_{i,j}(t)$ can be computed in $\mathcal{O}(n\,\mathsf{M}(d)) = \tilde{\mathcal{O}}(nd)$ operations in $\mathbb{K}$.*

3.4.1 *Analysis of* $\mathsf{DBZ}^{\mathrm{I}}$

We consider an execution of Algorithm $\mathsf{DBZ}^{\mathrm{I}}$ on a matrix M of denominator q, where $\deg(q)$ and $\deg(qM)$ are equal to d. Let $P^{(i)}$ and $M^{(i)}$ be the values of the matrices P^{I} and M^{I} when entering the ith iteration of the loop at Step 2, and $P^{(n)}$ and $M^{(n)}$ the values they have at Step 9 if this step is reached. Set k to either the last value of i before returning at Step 4 or n if Step 9 is reached. Consequently, the companion block C of M^{I} has dimension k. The use of Algorithm Pivot ensures the invariant $M^{(i)} = P^{(i)}[M]$ for all $i \leq k$.

The following lemma gives the shape of the matrices $P^{(i)}$.

Lemma 11 *For each i, $P^{(i)} = \mathsf{VJoin}(\Delta^i(e_1), Q^{(i)})$ for some $Q^{(i)}$ whose rows are in $\{e_2, \ldots, e_n\}$.*

Proof. The first $i-1$ rows of $P^{(i)}[M]$ have companion shape by design of $\mathsf{DBZ}^{\mathrm{I}}$. So, by Corollary 4, there exist a vector u and a matrix $Q^{(i)}$ with $P^{(i)} = \mathsf{VJoin}(\Delta^i(u), Q^{(i)})$.

We now prove by induction that $P^{(i)}_{1,*} = e_1$ and that for all $a > i$, $P^{(i)}_{a,*} \in \{e_2, \ldots, e_n\}$. First for $i = 1$, $P^{(1)} = I$, so the property holds. Now, we assume the property for $P^{(i)}$ and consider the ith iteration of the loop at Step 2. For $j \neq i+1$, let $T^{(j)}$ denote the value of the matrix involved in the gauge transformation at Step 8 during the jth iteration of the loop at Step 7, and let $T^{(i+1)}$ denote the matrix used at Step 6. Let r denote the integer defined at Step 3.

The transformations of P at Steps 5, 6, and 8 imply $P^{(i+1)} = T^{(n)} \cdots T^{(i+2)} T^{(i)} \cdots T^{(1)} T^{(i+1)} \mathsf{Inv}^{(i+1,r)} P^{(i)}$. The first row of each matrix $T^{(j)}$ and each matrix $\mathsf{Inv}^{(i+1,r)}$ is e_1, so $P^{(i+1)}_{1,*} = P^{(i)}_{1,*}$ which is, by induction, equal to e_1.

For each integer $a > i+1$ and each j, by definition $T^{(j)}_{a,*} = e_a$. Therefore, $P^{(i+1)}_{a,*} = \mathsf{Inv}^{(i+1,r)}_{a,*} P^{(i)}$. Moreover, if $a = r$, then $\mathsf{Inv}_{a,*}^{(i+1,r)}$ is equal to e_{i+1}; if not, it is equal to e_a. In both cases, by induction, $\mathsf{Inv}_{a,*}^{(i+1,r)} P^{(i)} \in \{e_2, \dots, e_n\}$. □

We are now able to give precise bounds on the degree of the output and the complexity of $\mathsf{DBZ}^{\mathrm{I}}$, using Lemma 1.

Theorem 12 *Let k be the dimension of the companion block output from DBZ^{I}. The degree of $q^{k-1}P^{(k)}$ is $\mathcal{O}(kd)$ and the degree of $\left(q^{k(k+1)/2}\det(P^{(k)})\right)M^{(k)}$ is $\mathcal{O}(k^2 d)$. DBZ^{I} has complexity $\mathcal{O}(n^2 k\,\mathsf{M}(k^2 d)) = \tilde{\mathcal{O}}(n^2 k^3 d)$.*

It is possible to give more precise bounds on the degrees and even to prove that they are reached in the generic case.

Proof. By Lemmas 11 and 7, the degrees of the rows of $\operatorname{diag}(1, q, \dots, q^{i-1}, 1, \dots, 1)P^{(i)}$ are upper bounded by $(0, d, \dots, (i-1)d, 0, \dots, 0)$. Now Lemma 1 implies that the degree of $q^{i(i-1)/2}\det(P^{(i)}){P^{(i)}}^{-1}$ is $\mathcal{O}(i^2 d)$. By the invariant of Pivot and $P[M] = \delta(P)\,P^{-1}$, $M^{(i)} = \delta(P^{(i)}){P^{(i)}}^{-1}$, we deduce that the lcm of the denominators in $M^{(i)}$ divides $L_i := q^{i(i+1)/2}\det(P^{(i)})$ and that $\deg(L_i M^{(i)})$ is in $\mathcal{O}(i^2 d)$. The degrees of the theorem follow for $i = k$.

The computation of $M^{(i+1)}$ from $M^{(i)}$ uses n elementary gauge transformations on $M^{(i)}$, leading by Lemma 10 to a complexity $\mathcal{O}(n^2\mathsf{M}(i^2 d))$. The announced complexity for $\mathsf{DBZ}^{\mathrm{I}}$ is obtained upon summation over i from 1 to k. □

The output matrix $P^{\mathrm{I}} = P^{(k)}$ is invertible, so $i < k$ implies $\delta^i(e_1) \notin \operatorname{span}(\Delta^i(e_1))$. Therefore, k is characterised as the least $i \in \mathbb{N} \setminus \{0\}$ such that $\delta^i(e_1) \in \operatorname{span}(\Delta^i(e_1))$.

Informally, for random M, the $\delta^i(e_i)$ are random, so most probably k is n. Indeed, when we experiment DBZ on random matrices, it always computes only one call to $\mathsf{DBZ}^{\mathrm{I}}$ and outputs a single companion matrix. We make this rigourous.

Theorem 13 *When M is generic, then DBZ has the same output as CVM with initial vector e_1.*

Proof. For indeterminates q_k and $m_{i,j,k}$, let $\hat{M}$ be the $n \times n$ matrix whose (i,j)-coefficient $\hat{m}_{i,j}/\hat{q}$ has numerator $\hat{m}_{i,j} = \sum_{k=0}^{d}\hat{m}_{i,j,k}X^k$ and denominator $\hat{q} = \sum_{k=0}^{d}\hat{q}_k X^k$. Replacing M by $\hat{M}$ formally in $\det(\Delta^n(e_1))$, we obtain a polynomial in the $\hat{q}_k$'s and the $\hat{m}_{i,j,k}$'s. This polynomial is non-zero since for $M = \mathsf{VJoin}(e_2, \dots, e_n, e_1)$, $\Delta^n(e_1) = I$. This proves that when M is generic, e_1 is a cyclic vector for M, so Shape (I) is reached with empty α and β, and DBZ behaves as $\mathsf{DBZ}^{\mathrm{I}}$ and as CVM with initial vector e_1. □

3.4.2 Analysis of $\mathsf{DBZ}^{\mathrm{II}}$

As for $\mathsf{DBZ}^{\mathrm{I}}$, a naive analysis would lead to the conclusion that the degrees in P^{II} increase exponentially during execution of $\mathsf{DBZ}^{\mathrm{II}}$. We give an algebraic interpretation of P^{II} that permits a tighter degree and complexity analysis of $\mathsf{DBZ}^{\mathrm{II}}$.

In this section, we consider the computation of $\mathsf{DBZ}^{\mathrm{II}}$ on a matrix M^{I} in Shape (I) whose block C has dimension k. Let q_{I} denote the denominator of M^{I}, and d_{I} be a common bound on $\deg(q_{\mathrm{I}})$ and $\deg(q_{\mathrm{I}}M^{\mathrm{I}})$.

Let Γ (or Γ_β) denote the operator on vectors or matrices with $n-k$ rows defined by $\Gamma(v) = \beta v - v'$. Observe that, as in Lemma 7, $\deg(q_{\mathrm{I}}^k\Gamma^k(v))$ is bounded by $\deg(v) + kd_{\mathrm{I}}$.

The loop at Step 3 processes the j's in decreasing order. Let $P^{(j)}$ and $M^{(j)}$ be the values of the matrices P^{II} and M^{II} just before executing the loop at Step 4 in $\mathsf{DBZ}^{\mathrm{II}}$.

Lemma 14 *For each j, the matrix $P^{(j)}$ has the shape*

$$P^{(j)} = \begin{pmatrix} I & 0 \\ A^{(j)} & I \end{pmatrix} \tag{4}$$

where $A^{(j)}$ is a matrix of dimension $(n-k)\times k$. Furthermore, for all $a < j$ and for $a = k$, $A^{(j)}_{*,a} = 0$ and for $j \le a < k$,

$$A^{(j)}_{*,a} = \Gamma(A^{(j)}_{*,a+1}) - \alpha_{*,a+1}. \tag{5}$$

Proof. The matrix $P^{(k)}$ is the identity, owing to Step 2; for $k < j$, $P^{(j)}$ is equal to the product of all the matrices T previously introduced for the gauge transformations at Step 5 for greater values of j. Each of those matrices has a block decomposition of the form $\left(\begin{smallmatrix} I & 0 \\ B & I \end{smallmatrix}\right)$. Therefore, their product $P^{(j)}$ has shape (4), where $A^{(j)}$ is the sum of the blocks B's. Whether $j = k$ or $j < k$, for $a < j$ and for $a = k$, and for each T, $B_{*,a} = 0$; therefore, $A^{(j)}_{*,a} = 0$.

Since $P^{(j)} = \left(\begin{smallmatrix} I & 0 \\ A^{(j)} & I \end{smallmatrix}\right)$, its inverse is $\left(\begin{smallmatrix} I & 0 \\ -A^{(j)} & I \end{smallmatrix}\right)$

$$M^{(j)} = P^{(j)}[M^{\mathrm{I}}] = \begin{pmatrix} C & 0 \\ A^{(j)}C + \alpha - \Gamma(A^{(j)}) & \beta \end{pmatrix}. \tag{6}$$

By the design of $\mathsf{DBZ}^{\mathrm{II}}$, $A^{(j)}C + \alpha - \Gamma(A^{(j)})$ ends with $k-j$ zero columns. We consider the $(a+1)$th column of (6) and use the fact that $A^{(j)}_{*,k} = 0$, to obtain (5). □

This leads to the degree and complexity analysis of $\mathsf{DBZ}^{\mathrm{II}}$.

Proposition 15 *Both $\deg\left(q_I^{k-1}P^{II}\right)$ and $\deg\left(q_I^k M^{II}\right)$ are in $\mathcal{O}(kd_I)$. The complexity of DBZ^{II} is $\mathcal{O}((n-k)^2 k^2\mathsf{M}(d_I)) = \tilde{\mathcal{O}}((n-k)^2 k^2 d_I)$.*

Proof. The degree of $P^{(j)}$ is equal to the degree of $A^{(j)}$. Equation (5) implies, after using $A^{(k)}_{*,a} = 0$, that for each a,

$$A^{(j)}_{*,a} = -\sum_{i=1}^{k-a}\Gamma^{i-1}(\alpha_{*,a+i}).$$

Therefore, $\deg(q_{\mathrm{I}}^{k-j+1}A^{(j)})$ is $\mathcal{O}((k-j)d_{\mathrm{I}})$. From Equation (6), it follows that the degree of $q_{\mathrm{I}}^{k-j+2}M^{(j)}$ is also $\mathcal{O}((k-j)d_{\mathrm{I}})$. For $j = 2$, we conclude that both $\deg(q_{\mathrm{I}}^{k-1}P^{\mathrm{II}})$ and $\deg(q_{\mathrm{I}}^k M^{\mathrm{II}})$ are $\mathcal{O}(kd_{\mathrm{I}})$.

The computation of $M^{(j+1)}$ from $M^{(j)}$ by the loop at Step 4 involves $n-k$ elementary gauge transformations. Each one computes $n-k$ (unbalanced) multiplications of elements of β with elements of $M^{(j)}$. The cost is then $\mathcal{O}((n-k)^2(k-j)\mathsf{M}(d_{\mathrm{I}}))$. We obtain the complexity of $\mathsf{DBZ}^{\mathrm{II}}$ by summation over j from 2 to k. □

3.4.3 Analysis of $\mathsf{DBZ}^{\mathrm{III}}$ and DBZ

Let $(P, C^{(1)}, \dots, C^{(t)})$ be the output of DBZ on M. Corollary 5 states that $P = \mathsf{VJoin}(\Delta^{k_1}(u^{(1)}), \dots, \Delta^{k_t}(u^{(t)}))$. The degrees of the matrices transformed by DBZ, and thus its complexity, are obviously linked to the degrees of the vectors $u^{(i)}$. Focusing the analysis on the degree of $u^{(1)}$ will result in the exponential degree bound $\mathcal{O}\left(n^{\mathcal{O}(n)}d\right)$, which we believe is not pessimistic. In turn, this seems to be a lower bound on the complexity of DBZ.

Conjecture *The complexity of DBZ is more than exponential in the worst case.*

We shall show that this explosion originates in the recursive calls at Step 10. Unfortunately, we have been unable to exhibit a matrix M leading to an execution with more than one recursive call, such cases being *very* degenerate.

We now drop the exponent and write u for $u^{(1)}$. As u can only be modified at Step 10, we consider the initial flow of an execution, as long as the M^{I}'s are not companion and the v's are non-zero; this excludes any return at Step 2 or 7.

Set $P^{(\mathrm{I},r)}$, $M^{(\mathrm{I},r)}$, $P^{(\mathrm{II},r)}$, and $P^{(\mathrm{III},r)}$ to the values of P^{I}, M^{I}, P^{II}, and P^{III} just before the rth call at Step 10. The matrix $M^{(\mathrm{I},r)}$ has Shape (I) and is by construction gauge-similar to M: for some invertible $P^{(r)}$, $P^{(r)}[M] = M^{(\mathrm{I},r)}$, and, by Lemma 3, there exist $u^{(r)}$, $Q^{(r)}$, and k_r such that

$$P^{(r)} = \mathsf{VJoin}\big(\Delta^{k_r}(u^{(r)}), Q^{(r)}\big).$$

This leads to a new interpretation of DBZ: it tests several vectors $u^{(r)}$ and iterates δ on them to construct the matrices $P^{(r)}$, until $P^{(r)}[M]$ is companion (Step 2) or allows a block decomposition (Step 7).

Proposition 16 *Write $P^{(II,r)}$ by blocks as $\left(\begin{smallmatrix} I & 0 \\ A^{(r)} & I \end{smallmatrix}\right)$. There exist an integer h and a rational function w such that*

$$u^{(r+1)} = w(A^{(r)}\, \Delta^{k_r}(u^{(r)}) + Q^{(r)})_{h,*}. \qquad (7)$$

Proof. By definition, $u^{(r+1)}$ is the first row of $P^{(r+1)}$. Step 10 sets $P^{(r+1)} = P^{(\mathrm{I},r+1)} P^{(\mathrm{III},r)} P^{(\mathrm{II},r)} P^{(r)}$. Lemma 11 implies $P^{(\mathrm{I},r+1)}_{1,*} = e_1$; in addition, by Lemma 14, $P^{(\mathrm{II},r)}$ has a block decomposition as in the theorem statement. So, $u^{(r+1)} = P^{(\mathrm{III},r)}_{1,*} \left(\begin{smallmatrix} I & 0 \\ A^{(r)} & I \end{smallmatrix}\right) \left(\begin{smallmatrix} \Delta^{k_r}(u^{(r)}) \\ Q^{(r)} \end{smallmatrix}\right)$. The proof is now reduced to the existence of $h > k_r$ such that $P^{(\mathrm{III},r)}_{1,*} = e_h$.

In Algorithm $\mathsf{DBZ}^{\mathrm{III}}$, the matrices involved in the gauge transformations at Step 6 commute with one another. Let S denote their product. Set h to the integer defined at Step 2, then $h > k_r$ and $P^{(\mathrm{III},r)} = \mathsf{Rot}\, S\, \mathsf{Inv}^{(h,n)}$. By construction, $\mathsf{Rot}_{1,*} = e_n$, $S_{n,*} = we_n$ for a certain rational function w defined at Step 4, and $\mathsf{Inv}^{(h,n)}_{n,*} = e_h$. This ends the proof. ☐

We now express the growth of $\deg(u^{(r)})$ with respect to r. Let $d_{\mathrm{I},r}$ denote the degrees of the numerators and denominators of $P^{(r)}[M]$, so in particular a bound for $u^{(r)}$. Now, Proposition 15 implies that the degree of the numerators and denominators of $A^{(r)}$ are $\mathcal{O}(k_r d_{\mathrm{I},r}) = \mathcal{O}(k_r^3 d + k_r^2 \deg(u^{(r)}))$.

The rational function w of the theorem is the inverse of an element of $M^{(\mathrm{II},r)}$, so the degree of its numerator and denominator are $\mathcal{O}(k_r^3 d + k_r^2 \deg(u^{(r)}))$. Combined with Proposition 16, this implies $\deg(u^{(r+1)}) = \mathcal{O}(k_r^3 d + k_r^2 \deg(u^{(r)}))$.

We could not deduce from Proposition 16 any polynomial bound on the degree of the numerator of $u^{(r)}$, but we get $\deg(u^{(r)}) = \mathcal{O}(rdn^{2r+3} \deg(u^{(0)}))$. The worst case of this bound is obtained when $r = n - 1$.

3.5 Link with the Abramov-Zima algorithm

In [1], Abramov and Zima presented an algorithm, denoted by AZ in the following, that computes the solutions of inhomogeneous linear systems $Y' = MY + R$ in a general Ore polynomial ring setting. It starts by a partial uncoupling to obtain a differential equation that cancels Y_1, solves it and injects the solutions in the initial system. We reinterpret here its computations of a partial uncoupling, focusing on the case of systems $Y' = MY$ where M is a polynomial matrix, and we analyse the complexity in the *generic case*.

Step 1. Introduce a new vector Z of dimension $\ell \leq n$ (generically with equality), such that $Z_1 = Y_1$ and, for $i > 1$, Z_i is a linear combination of $Y_i, \ldots, Y_n$, such that where β is a lower-triangular matrix augmented by 1s on its upper-diagonal: $\beta_{i,i+1} = 1$ for all $1 \leq i \leq \ell - 1$.

Step 2. Eliminate the variables $Z_2, \ldots, Z_\ell$ by linear combinations on the system obtained in Step 1 to get a differential equation of order ℓ that cancels Y_1.

Theorem 17 *Let M be a generic matrix of dimension n with polynomial coefficients of degree d. Then, the complexity of AZ to uncouple the system $Y' = MY$ is $\tilde{\mathcal{O}}(n^5 d)$.*

Proof. When M is generic, the minimal monic differential equation that cancels Y_1 has order n and its coefficients of orders 0 to $n-1$ are the coefficients of the vector $\delta^n(e_1)P^{-1}$, where we have set $P = \Delta^n(e_1)$. Thus, for generic M, the integer ℓ defined in Step 1 is equal to n.

Step 1 implies that there is an upper-triangular matrix U such that $U_1 = e_1$, $Z = UY$ and $U[M] = \beta$. At Step 2, the eliminations of the variables $(Z_i)_{2 \leq i \leq n}$ are carried out by pivot operations. They transform the system $Z' = \beta Z$ into a new system $W' = CW$ with $W_1 = Z_1 = Y_1$ and C is a companion matrix, which is equal to $P[M]$. Because of the particular shape, the matrices matching those pivots operations are lower-triangular with 1s on their diagonal. Their product is a matrix L such that $W = LZ$; it is also lower-triangular with 1s on its diagonal. Since $P[M] = C = L[\beta]$ and $\beta = U[M]$, $P = LU$.

By construction, the degree of P is $\mathcal{O}(nd)$. The matrices L, U, and L^{-1} of its LU decomposition have degrees $\mathcal{O}(n^2 d)$ [4]. Thus, the degree of $\beta = U[M] = L^{-1}[P[M]]$ is $\mathcal{O}(n^2 d)$. Steps 1 and 2 of AZ compute $\mathcal{O}(n^2)$ pivot operations, each one involving $\mathcal{O}(n)$ manipulations (additions and products) of polynomial coefficients of degree $\mathcal{O}(n^2 d)$. This leads to the announced complexity for AZ. ☐

It can be proved that the product $L \cdot U$ in this proof is the LU-decomposition of P; for a non-generic M, [1] implicitly obtains an LUP-decomposition.

The degree bounds in the previous proof are reached in our experiments: the maximal degrees of the numerator of the matrices β computed for random matrices M of dimension n from 1 to 6 with polynomial coefficients of degree 1 are, respectively, 1, 2, 5, 10, 17, and 26.

4. IMPLEMENTATION

We have implemented the DBZ algorithm and several variants of the CVM algorithm to evaluate the pertinence of our theoretical complexity analyses and the practical efficiency of our algorithmic improvements. Because of its fast implementations of polynomial and matrix multiplications, we

Algorithm	c	e	p		$n = 100$, $d = 1$	$n = 5$, $d = 100$	$n = 30$, $d = 30$
CVM	$6.8\ 10^{-7}$	1.81	$\theta + 1$	3.88	103.11	3.53	155.41
DBZ	$7.5\ 10^{-8}$	1.61	5	6.01	∞	2.3	14409
BalConstr	$2.4\ 10^{-6}$	1.01	$\theta + 1$	3.00	12.55	0.5	2.7
NaiveConstr	$3.3\ 10^{-9}$	1.90	4	4.00	1.24	0.2	1.64
StorjohannSolve	$8.2\ 10^{-7}$	1.75	$\theta + 1$	3.87	83.60	3.48	153.16
NaiveSolve	$4.8\ 10^{-8}$	1.52	5	6.22	106352	0.85	13806
Output size					1000300	13010	810960

Table 1: Experimental complexity of DBZ, CVM, and their sub-algorithms; common output size match $n^3 d$

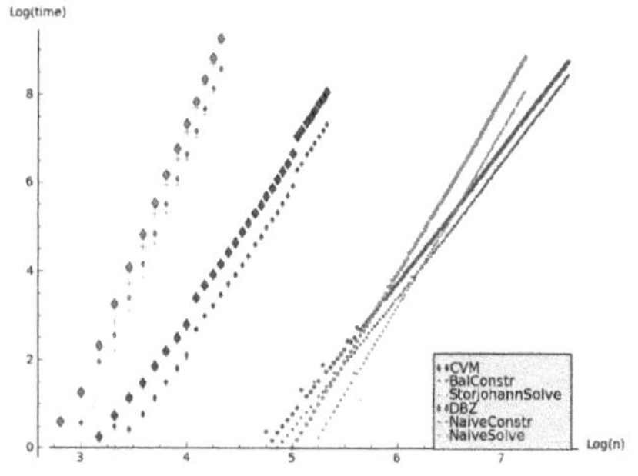

Figure 1: Timings for DBZ and CVM on input matrices M of dimension n and coefficients with fixed degree $d = 15$ (smaller marks) or $d = 20$ (larger marks)

chose the system Magma, using its release V2.16-7 on Intel Xeon 5160 processors (3 GHz) and 8 GB of RAM.

Our results are summarised in Table 1 and Figure 1. We fed our algorithms with matrices of size n and coefficients of degree d over $\mathbb{Z}/1048583\mathbb{Z}$. Linear regression on the logarithmic rescaling of the data was used to obtain parameters c, e, and p that express the practical complexity of the algorithms in the form $cd^e n^p$. For the exponents p, both theoretical and experimental values are shown for comparison. Sample timings for particular (n, d) are also given. BalConstr and NaiveConstr (resp. StorjohannSolve and NaiveSolve) compute the matrix P (resp. C) of Algorithm 1 with or without the algorithmic improvements introduced in Theorem 8.

In Figure 1, each algorithm shows two parallel straight lines, for $d = 15$ and 20, as was expected on a logarithmic scale. The improved algorithms are more efficient than their simpler counterparts when n and d are large enough.

The theory predicts $e = 1$ for all algorithms. Observing different values suggests that too low values of d have been used to reach the asymptotic regime. We also remark that, with respect to d, DBZ, BalConstr, and StorjohannSolve have slightly better practical complexity than their respective couterparts CVM, NaiveConstr, and NaiveSolve.

The practical exponent $p = 3.00$ of BalConstr is smaller than $\theta + 1$. The algorithm consists of n executions of a loop that contains a constant number of scans of matrices and matrix multiplications. By analysing their contributions to the complexity separately, we obtain $2.5\ 10^{-6}d^{0.97}n^{3.00}$ for the former, and $5.2\ 10^{-8}d^{1.34}n^{3.19}$ for the latter. In the range of n we are analysing, the first contribution dominates because of its constant, and its exponent 3.00 is the only one visible on the experimental complexity of BalConstr.

It is also visible that StorjohannSolve is the dominating sub-algorithm of CVM. Besides, our implementation of StorjohannSolve is limited by memory and cannot handle matrices of dimension n over 130. This bounds the size of the inputs manageable by our CVM implementation. A native Magma implementation of Storjohann's algorithm should improve the situation. However, our implementation already beats the naive matrix inversion, so that the experimental exponent 3.88 of StorjohannSolve is close to $\theta + 1$.

The experimental exponent p of DBZ is 6.01 instead of 5. This may be explained by the fact that the matrix coefficients that DBZ handles are fractions. Instead, in BalConstr, the coefficients are polynomial: denominators are extracted at the start of the algorithm and reintroduced at the end.

5. CONCLUSION

It would be interesting to study the relevance of uncoupling applied to system solving, and to compare this approach to direct methods. It would also be interesting to combine CVM and DBZ into a hybrid algorithm, merging speed of CVM and generality of DBZ.

6. REFERENCES

[1] S. Abramov and E. Zima. A universal program to uncouple linear systems. In *Proceedings of CMCP'96*, pages 16–26, 1997.

[2] S. A. Abramov. EG-eliminations. *J. Differ. Equations Appl.*, 5(4-5):393–433, 1999.

[3] K. Adjamagbo. Sur l'effectivité du lemme du vecteur cyclique. *C. R. Acad. Sci. Paris Sér. I Math.*, 306(13):543–546, 1988.

[4] E. H. Bareiss. Sylvester's identity and multistep integer-preserving Gaussian elimination. *Math. Comp.*, 22:565–578, 1968.

[5] M. A. Barkatou. An algorithm for computing a companion block diagonal form for a system of linear differential equations. *Appl. Algebra Engrg. Comm. Comput.*, 4(3):185–195, 1993.

[6] M. Bronstein and M. Petkovšek. An introduction to pseudo-linear algebra. *Theor. Comput. Sci.*, 157(1):3–33, 1996.

[7] D. G. Cantor and E. Kaltofen. On fast multiplication of polynomials over arbitrary algebras. *Acta Inform.*, 28(7):693–701, 1991.

[8] R. C. Churchill and J. Kovacic. Cyclic vectors. In *Differential algebra and related topics*, pages 191–218. 2002.

[9] T. Cluzeau. Factorization of differential systems in characteristic p. In *Proc. ISSAC'03*, pages 58–65. ACM, 2003.

[10] F. T. Cope. Formal Solutions of Irregular Linear Differential Equations. Part II. *Amer. J. Math.*, 58(1):130–140, 1936.

[11] A. Dabèche. Formes canoniques rationnelles d'un système différentiel à point singulier irrégulier. In *Équations différentielles et systèmes de Pfaff dans le champ complexe*, volume 712 of *Lecture Notes in Math.*, pages 20–32. 1979.

[12] A. M. Danilevski. The numerical solution of the secular equation. *Matem. sbornik*, 44(2):169–171, 1937. (in Russian).

[13] P. Deligne. *Équations différentielles à points singuliers réguliers*. Lecture Notes in Math., Vol. 163. Springer, 1970.

[14] L. E. Dickson. *Algebras and their arithmetics*. Chicago, 1923.

[15] B. Dwork and P. Robba. Effective p-adic bounds for solutions of homogeneous linear differential equations. *Trans. Amer. Math. Soc.*, 259(2):559–577, 1980.

[16] S. Gerhold. Uncoupling systems of linear Ore operator equations. Master's thesis, RISC, J. Kepler Univ. Linz, 2002.

[17] M. Giesbrecht and A. Heinle. A polynomial-time algorithm for the Jacobson form of a matrix of Ore polynomials. In *CASC'12*, volume 7442 of *LNCS*, pages 117–128. Springer, 2012.

[18] A. Hilali. Characterization of a linear differential system with a regular singularity. In *Computer algebra*, volume 162 of *Lecture Notes in Comput. Sci.*, pages 68–77. Springer, 1983.

[19] N. Jacobson. Pseudo-linear transformations. *Ann. of Math. (2)*, 38(2):484–507, 1937.

[20] A. Loewy. Über lineare homogene Differentialsysteme und ihre Sequenten. *Sitzungsb. d. Heidelb. Akad. d. Wiss., Math.-naturw. Kl.*, 17:1–20, 1913.

[21] A. Loewy. Über einen Fundamentalsatz für Matrizen oder lineare homogene Differentialsysteme. *Sitzungsb. d. Heidelb. Akad. d. Wiss., Math.-naturw. Kl.*, 5:1–20, 1918.

[22] L. Pech. Algorithmes pour la sommation et l'intégration symboliques. Master's thesis, 2009.

[23] E. G. C. Poole. *Introduction to the theory of linear differential equations*. Oxford Univ. Press, London, 1936.

[24] L. Schlesinger. *Vorlesungen über lineare Differentialgleichungen*. B. G. Teubner, Leipzig, 1908.

[25] A. Schönhage and V. Strassen. Schnelle Multiplikation großer Zahlen. *Computing*, 7:281–292, 1971.

[26] A. Storjohann. High-order lifting and integrality certification. *J. Symbolic Comput.*, 36(3-4):613–648, 2003.

[27] V. Strassen. Gaussian elimination is not optimal. *Numerische Mathematik*, 13:354–356, 1969.

[28] V. Vassilevska Williams. Multiplying matrices faster than Coppersmith-Winograd. In *STOC'12*, pages 887–898, 2012.

[29] J. H. M. Wedderburn. Non-commutative domains of integrity. *J. Reine Angew. Math.*, 167:129–141, 1932.

[30] B. Zürcher. Rationale Normalformen von pseudo-linearen Abbildungen. Master's thesis, ETH Zürich, 1994.

Creative Telescoping for Rational Functions Using the Griffiths–Dwork Method*

Alin Bostan
Inria (France)
alin.bostan@inria.fr

Pierre Lairez
Inria (France)
pierre.lairez@inria.fr

Bruno Salvy
Inria (France)
bruno.salvy@inria.fr

ABSTRACT

Creative telescoping algorithms compute linear differential equations satisfied by multiple integrals with parameters. We describe a precise and elementary algorithmic version of the Griffiths–Dwork method for the creative telescoping of rational functions. This leads to bounds on the order and degree of the coefficients of the differential equation, and to the first complexity result which is single exponential in the number of variables. One of the important features of the algorithm is that it does not need to compute certificates. The approach is vindicated by a prototype implementation.

Categories and Subject Descriptors:
I.1.2 [**Computing Methodologies**]: Symbolic and Algebraic Manipulations — *Algebraic Algorithms*

General Terms: Algorithms, Theory.

Keywords: Integration, creative telescoping, algorithms, complexity, Picard-Fuchs equation, Griffiths–Dwork method

1. INTRODUCTION

In computer algebra, *creative telescoping* is an approach introduced by Zeilberger to address definite summation and integration of a large class of functions and sequences [28, 29, 27]. Its vast scope includes the computation of differential equations for multiple integrals of rational or algebraic functions with parameters. Within this class, creative telescoping is similar to well-studied older approaches whose key notion is the Picard–Fuchs differential equation, see *e.g.* [23].

We study the multivariate rational case: Given a rational function $F(t, x_1, \dots, x_n)$, we aim at finding n rational functions $A_i(t, x_1, \dots, x_n)$ and a differential operator T with polynomial coefficients, say $\sum_{j=0}^{r} c_j(t)\partial_t^j$, such that

$$T(F) \stackrel{\text{def}}{=} \sum_{j=0}^{r} c_j(t)\partial_t^j F = \sum_{i=1}^{n} \partial_i A_i, \qquad (1)$$

*This work has been supported in part by the Microsoft Research – Inria Joint Centre.

ISSAC'13, June 26–29, 2013, Boston, Massachusetts, USA.

where ∂_t^j denotes $\frac{\partial^j}{\partial t^j}$ and ∂_i denotes $\frac{\partial}{\partial x_i}$. The operator T is a *telescoper* and the tuple $(A_1, \dots, A_n)$ is a *certificate* for T. The integer r is the *order* of T and $\max_j \deg c_j$ is its *degree*.

Throughout the article, the constant field k of F is assumed to be of characteristic zero. Under suitable additional hypotheses, $T(I) = 0$ is a differential equation satisfied by an integral $I(t) = \int F \mathrm{d}x$ over a domain γ, without boundaries, where F has no pole. A misbehavior may occur when the certificate has poles outside those of F: it may not be possible to integrate term by term the right-hand side of Equation (1), see §4.1. The certificate is called *regular* when it does not contain poles other than those of F. For integration, there is no need to compute the certificate provided that it is regular.

Several methods are known that can find a telescoper and the corresponding certificate [17, 26, 7, 15]. However, the practical cost of using these methods in multivariate problems remains high and a better understanding of the size or complexity of the objects of creative telescoping is clearly needed. The present work is part of the on-going effort in this direction [2, 3, 4]. The study of the rational case is motivated both by its fundamental nature and by its applications to the computation of diagonals in combinatorics, number theory and physics [17, 6, 20]. The rational case with n variables also includes the algebraic case with $n-1$ variables [4].

Previous works. An obviously related problem is, given a rational function $F(x_1, \dots, x_n)$, to decide whether there exist rational functions $A_1, \dots, A_n$ such that F equals $\sum_{i=1}^{n} \partial_i A_i$.

When $n = 1$, this question is easily solved by Hermite reduction. This is the basis of an algorithm for creative telescoping [3] that we outline in §2.1. Picard [25, chap. 7] gave methods when $n = 2$ from which he deduced that a telescoping equation exists in that case [24]. This too has led to an algorithm [4]. The Griffiths–Dwork method [8, §3; 9, §8; 12] solves the problem for a general n, in the setting of de Rham cohomology and under a regularity assumption. The method can be viewed as a generalization of Hermite reduction. Independently, Christol used a similar method to prove that diagonals of rational functions, under a regularity hypothesis, are differentially finite [5]; then he applied a deformation technique, for which he credits Dwork, to handle singular cases [6]. The Griffiths–Dwork method is also used in point counting [1, 11] and the study of mirror maps [20].

In terms of complexity, in more than two variables, not much is known. If a rational function $F(t, x_1, \dots, x_n)$ has degree d, a study of Lipshitz's argument [17] shows that there exists a telescoper of order and degree $d^{O(n)}$ with a regular certificate of size $d^{O(n^2)}$. Most algorithms [17, 28, 26, 7, 2, 15] cannot avoid the computation of the certificate, which impacts their complexity. The complexity of Lipshitz's algorithm is $d^{O(n^2)}$ operations in k; the complexity of no other algorithm is known. Pancratz [22] developed an approach similar to ours, under a restrictive hypothesis, much stronger than Griffiths' regularity assumption. He proceeds to a complexity analysis of his algorithm but in terms of operations in $k(t)$ rather than in the base field k. Algorithms based on non-commutative Gröbner bases and elimination [28, 26] or based on the search of rational solutions to differential equations [7] resist complexity analysis. The method of Apagodu and Zeilberger [2] requires a generic exponent and specialization seems problematic.

For the restricted class of diagonals of rational functions, there is a heuristic based on series expansion and differential approximation [14]; it does not need to compute a certificate. However, even using the bounds in $d^{O(n)}$, its direct implementation has a complexity of $d^{O(n^2)}$ operations in k.

Contributions. Our main result, obtained with the Griffiths–Dwork method and a deformation technique, is the existence of a telescoper with regular certificate of order at most d^n and degree $d^{O(n)}$ that can be computed in $d^{O(n)}$ arithmetic operations in k. For generic homogeneous rational functions, the telescoper computed is the minimal order telescoper with regular certificate. Theorems 6, 10 and 12 state precise complexity and size estimates. To the best of our knowledge, the bounds on the order and degree are better than what was known and it is the first time that a complexity single exponential in n is reached. For a generic rational function, every pair (telescoper, regular certificate) has a size larger than $d^{O(n)}$, see Remark 11, but our algorithm does not need to compute the certificate. A prototype implementation shows that this algorithm can lead to a spectacular improvement over previous methods, though the domain of improvement is not satisfactory yet.

Acknowledgement. We are grateful to G. Christol for many rewarding discussions, and we thank G. Villard and W. Zhou for communicating their complexity results in linear algebra.

2. OVERVIEW OF THE METHOD

In this section we introduce the basics of the Griffiths–Dwork method. In dimension 1, this method coincides with classical Hermite reduction, which we first recall.

2.1 Dimension one: Hermite reduction

Let F be a rational function in x, over a field L, written as a/f^ℓ, with a and f two polynomials not necessarily coprime, the latter being square-free, *i.e.* the polynomials $\partial_x f$ and f are coprime. In particular a equals $uf + v\partial_x f$ for some polynomials u and v. Then, if $\ell > 1$, the function F rewrites

$$F = \frac{u + \frac{1}{\ell-1}\partial_x v}{f^{\ell-1}} + \partial_x\left(\frac{-v}{(\ell-1)f^{\ell-1}}\right).$$

Iterating this reduction step ℓ times gives F as $\frac{U}{f} + \partial_x \frac{V}{f^{\ell-1}}$ for some polynomials U and V. Next, Euclidean division allows to write U as $r + sf$, with r of degree less than the degree of f, yielding the additive decomposition

$$F = \frac{r}{f} + \partial_x\left(\frac{V}{f^{\ell-1}} + \int s\right).$$

The rational function r/f is the *reduced form* of F and is denoted by $[F]$. This form features important properties:

(Linearity) f being fixed, $[F]$ depends linearly on F;

(Soundness) if $[F]$ is zero, then F is a derivative w.r.t. x;

(Confinement) $[F]$ lies in a finite-dimensional vector space over L depending only on f (with dimension $\deg_x f$);

(Normalization) if F is a derivative w.r.t. x, then $[F]$ is zero.

These properties are enough to compute a telescoper: Assume now that L is $k(t)$ for a field k. If for some elements of L, say $a_0, \dots, a_p$, the reduced form $\left[\sum_i a_i \partial_t^i F\right]$ vanishes, then the operator $\sum_i a_i \partial_t^i$ is a telescoper, thanks to the soundness property. Thanks to the linearity property, this is equivalent to the vanishing of $\sum_i a_i \left[\partial_t^i F\right]$. Thanks to the confinement property, it is always possible to find such a relation. Thanks to the normalization property, every telescoper arises in this way. In particular, so does the telescoper of minimal order.

2.2 Griffiths–Dwork reduction

Let F be a rational function in n variables $x_1, \dots, x_n$, written as a/f^ℓ, with f a square-free polynomial. If $\ell > 1$ and if a lies in the ideal of $L[x_1, \dots, x_n]$ generated by f and its derivatives $\partial_i f$, then we can write a as $uf + \sum_i v_i \partial_i f$, for some polynomials $u, v_1, \dots, v_n$, and F rewrites

$$F = \frac{u + \frac{1}{\ell-1}\sum_{i=1}^n \partial_i v_i}{f^{\ell-1}} + \sum_{i=1}^n \partial_i\left(\frac{-v_i}{(\ell-1)f^{\ell-1}}\right).$$

Provided that this ideal contains 1, any F can be reduced to a function with simple poles by iteration of this identity. The soundness and linearity properties are naturally satisfied, but extending further the reduction to obtain at least the confinement property is not straightforward and requires stronger assumptions [21, §4]. A difficulty with this approach is that the degrees of the cofactors v_i at each reduction step are poorly controlled: we lack the Euclidean division step and we reduce poles at finite distance at the cost of making worse the pole at infinity. This difficulty is overcome by working in the projective space. The translation between affine and projective is discussed more precisely in Section 7.

Now, assume that a and f are *homogeneous* polynomials in $L[\mathbf{x}] = L[x_0, \dots, x_n]$, with f of degree d. A central role is played by the *Jacobian ideal* $\operatorname{Jac} f$ *of* f, the ideal generated by the partial derivatives $\partial_0 f, \dots, \partial_n f$. Note that since f is homogeneous, Euler's relation, which asserts that f equals $\frac{1}{d}\sum_{i=0}^n x_i \partial_i f$ implies that $f \in \operatorname{Jac} f$.

We now decompose a as $r + \sum_i v_i \partial_i f$. In contrast with the affine case, each nonzero v_i can be chosen homogeneous of degree precisely $\deg a - \deg \partial_i f$. If $\ell > 1$, we obtain

$$F = \frac{r}{f^\ell} + \underbrace{\frac{\frac{1}{\ell-1}\sum_{i=0}^n \partial_i v_i}{f^{\ell-1}}}_{F_1} + \sum_{i=0}^n \partial_i\left(\frac{-v_i}{(\ell-1)f^{\ell-1}}\right). \quad (2)$$

If r is not zero, the order of the pole need not decrease, contrary to the affine case, but r is reduced to a normal form modulo $\operatorname{Jac} f$; this will help us obtain the confinement property, see Proposition 2. The reduction process proceeds recursively on F_1, which has pole order $\ell - 1$, and stops when $\ell = 1$. This procedure is summarized in Algorithm 1.

3. PROPERTIES OF THE GRIFFITHS–DWORK REDUCTION

Let f in $L[\mathbf{x}]$ be a homogeneous polynomial of degree d, where L is a field of characteristic zero. It is clear that the

---

Input $F = a/f^\ell$ a rational function in $x_0, \dots, x_n$
Output $[F]$ such that there exist rational functions $A_0, \dots, A_n$ such that $F = [F] + \sum_i \partial_i A_i$

Precompute a Gröbner basis G for $(\partial_0 f, \dots, \partial_n f)$
procedure REDUCE(a/f^ℓ)
  if $\ell = 1$ **then return** a/f^ℓ
  Decompose a as $r + \sum_i v_i \partial_i f$ using G
  $F_1 \leftarrow \frac{1}{\ell-1} \sum_i \frac{\partial_i v_i}{f^{\ell-1}}$
  return $\frac{r}{f^\ell}$ + REDUCE(F_1)

Algorithm 1. Griffiths–Dwork reduction

reduction procedure satisfies the soundness and the linearity properties. Analogues of confinement and normalization hold under the following regularity hypothesis:

$$L[\mathbf{x}]/\operatorname{Jac} f \text{ is finite-dimensional over } L. \qquad \text{(H)}$$

Geometrically, this hypothesis means that the hypersurface defined by f in $\mathbb{P}^n$ is smooth. In particular f is irreducible.

The ring of rational functions in $L(\mathbf{x})$ whose denominator is a power of f is denoted by $L[\mathbf{x}, \frac{1}{f}]$. Let $L[\mathbf{x}, \frac{1}{f}]_p$ denote the subspace of homogeneous functions of degree p, *i.e.* the set of F in $L[\mathbf{x}, \frac{1}{f}]$ such that $F(\lambda \mathbf{x})$ equals $\lambda^p F(\mathbf{x})$. Note that each derivation ∂_i induces a map from $L[\mathbf{x}, \frac{1}{f}]_p$ to $L[\mathbf{x}, \frac{1}{f}]_{p-1}$. Let D_f denote the subspace of $L[\mathbf{x}, \frac{1}{f}]$ consisting of rational functions $\sum_i \partial_i A_i$ for some A_i in $L[\mathbf{x}, \frac{1}{f}]_{-n}$. A major character of this study is the quotient space $L[\mathbf{x}, \frac{1}{f}]_{-n-1}/D_f$, denoted by H_f^{pr}.

The reduced form of F in $L[\mathbf{x}, \frac{1}{f}]_{-n-1}$ is denoted by $[F]$. It is by definition the output of the algorithm REDUCE. It depends on a choice of a Gröbner basis of $\operatorname{Jac} f$, but its vanishing does not, see Theorem 1 below.

The choice of the space $L[\mathbf{x}, \frac{1}{f}]_{-n-1}$ and the degree $-n-1$ may seem arbitrary. It is motivated by it being isomorphic to the space of regular differential n-forms on $\mathbb{P}^n \setminus V(f)$. The evaluation of x_0 to 1 is the restriction map to $\mathbb{A}^n \setminus V(f)$. The space H_f^{pr} is the nth de Rham cohomology space of the algebraic variety $\mathbb{P}^n \setminus V(f)$ over L.

THEOREM 1 (GRIFFITHS [12, §4]). *If f satisfies Hypothesis* (H), *then for all F in $L[\mathbf{x}, \frac{1}{f}]_{-n-1}$, the reduced form $[F]$ vanishes if and only if F is in D_f.*

Theorem 1 gives access to the dimension of H_f^{pr}. Let A be the finite dimensional vector space $L[\mathbf{x}]/\operatorname{Jac} f$. For a positive integer ℓ, let A_ℓ denote the linear subspace of A generated by homogeneous polynomials of degree $\ell d - (n+1)$. Let B denote $\oplus_\ell A_\ell$. Finally, for $\ell > 0$ let $(g_{\ell,i})_{1 \leqslant i \leqslant n_\ell}$ be a basis of A_ℓ, with $n_\ell = \dim_L A_\ell$.

PROPOSITION 2. *Under Hypothesis* (H), *the family of rational functions $\left(g_{\ell,i}/f^\ell\right)_{0<\ell, i \leqslant n_\ell}$ induces a basis of H_f^{pr}.*

PROOF. Suppose there exists a linear relation between the $g_{\ell,i}/f^\ell$ modulo D_f, that is $\sum_{\ell,i} u_{\ell,i} g_{\ell,i}/f^\ell$, denoted by F, lies in D_f for some elements $u_{\ell,i}$ of L, not all zero. Let ℓ_0 be the maximum ℓ such that $u_{\ell,i}$ is not zero for at least one i. By Theorem 1, $[F] = 0$ so that $\sum_{\ell,i} u_{\ell,i} g_{\ell,i} f^{\ell_0 - \ell}$, the numerator of F, lies in $\operatorname{Jac} f$. Since f itself is in $\operatorname{Jac} f$, so is the sum $\sum_i u_{\ell_0,i} g_{\ell_0,i}$, contradicting the fact that the $g_{\ell_0,i}$ are a basis of A_{ℓ_0}. Thus the $g_{\ell,i}/f^\ell$ form a free family.

To prove that this family generates H_f^{pr}, we first notice that the family of all the fractions $[F]$, for F in $L[\mathbf{x}, \frac{1}{f}]_{-n-1}$, generates H_f^{pr} since $[F]$ equals F modulo D_f. Now we assume for a moment that each $g_{\ell,i}$ is reduced with respect to a Gröbner basis G of $\operatorname{Jac} f$. Then each polynomial of $L[\mathbf{x}]$ of degree $\ell d - n - 1$ which is reduced with respect to G is a linear combination of the $g_{\ell,i}$. Thus for all $F = a/f^\ell$ in $L[\mathbf{x}, \frac{1}{f}]_{-n-1}$, the reduction $[F]$ is in the span of all the $g_{\ell,i}/f^\ell$. This makes the $g_{\ell,i}/f^\ell$ a system of generators of H_f^{pr} and by the previous paragraph a basis of it. Thus H_f^{pr} has the same dimension as B and any free family of H_f^{pr} of cardinal $\dim_L B$ is a basis of H_f^{pr}. In particular, the $g_{\ell,i}/f^\ell$ form a basis even if the $g_{\ell,i}$ are not reduced with respect to G. □

COROLLARY 3. *Under Hypothesis* (H), *H_f^{pr} has dimension*

$$\frac{1}{d}\left((d-1)^{n+1} + (-1)^{n+1}(d-1)\right) \quad (\leqslant d^n).$$

PROOF. It has the dimension of B, see [19, thm. 8.3] for its computation. The inequality is clear. □

4. CREATIVE TELESCOPING

We now introduce an algorithm, based on the Griffiths–Dwork reduction, that computes a telescoper of a rational function under Hypothesis (H).

In Equation (1), the telescoper T is said to have a *regular certificate* if the irreducible factors of the denominators of the A_i's, as rational functions over $k(t)$, divide the denominator of F; in other words, the A_i's have no pole outside those of F, over $k(t)$. Algorithm 2, described in § 4.2, returns the telescoper of minimal order having regular certificate. For the application of creative telescoping to integration, this class of telescopers is more interesting than the general one; that is the object of § 4.1.

4.1 Telescopers with regular certificate

Back to the affine case, let $F(t, x_1, \dots, x_n)$ be a rational function over $\mathbb{C}$ and γ be a n-cycle in $\mathbb{C}^n$ over which F has no pole for a generic t in $\mathbb{C}$. A common use of creative telescoping is the computation of a differential equation satisfied by the one-parameter integral $I(t) = \int_\gamma F \mathrm{d}\mathbf{x}$. As mentioned in the introduction, it is not always possible to deduce from the telescoping equation (1) that $T(I)$ vanishes. It may happen that the polar locus of the certificate meets γ for all $t \in \mathbb{C}$, and so some $\int_\gamma \partial_i A_i \mathrm{d}\mathbf{x}$ need not be zero. An example of this phenomenon is given by Picard [23] for a bivariate algebraic function and translated here into a rational example, using the method in [4, Lemma 4]:

$$\frac{x-y}{z^2 - P_t(x)P_t(y)} = \partial_x \frac{2P_t(x)}{(x-y)\left(z^2 - P_t(x)P_t(y)\right)} + \partial_y \frac{2P_t(y)}{(x-y)\left(z^2 - P_t(x)P_t(y)\right)} + \partial_z \frac{3\left(x^2+y^2\right)z}{(x-y)\left(z^2 - P_t(x)P_t(y)\right)}, \qquad (3)$$

where $P_t(u) = u^3 + t$. Note the factor $x-y$ in the denominator of the certificate. The operator 1 is a telescoper of the left-hand side F, however there exists a 3-cycle γ on which F has no pole and such that $\int_\gamma F \mathrm{d}\mathbf{x}$ is not zero. It is thus impossible to find a regular certificate for the telescoper 1.

Nevertheless, a differential equation for $I(t)$ can be obtained in two ways. First, one can carefully study the integral $\sum_i \int_\gamma \partial_i A_i \mathrm{d}\mathbf{x}$ and compute a differential equation for it. Usually this includes the analysis of the poles of the A_i's, and

Input $F = a/f^\ell$ a rational function in $L[\mathbf{x}, \frac{1}{f}]_{-n-1}$, with f satisfying (H)
Output $T(t, \partial_t)$ an operator such that $T(F) = \sum_i \partial_i A_i$ for some rational functions A_i

```
procedure TELESC(F)
    G_0 ← REDUCE(F)
    i ← 0
    loop
        if rank_L(G_0, ..., G_i) < i + 1 then
            solve Σ_{k=0}^{i-1} a_k G_k = G_i w.r.t. a_0, ..., a_{i-1} in L
            return ∂_t^i − Σ_k a_k ∂_t^k
        else
            G_{i+1} ← REDUCE(∂_t G_i)
            i ← i + 1
```

Algorithm 2. Creative telescoping, regular case

the search of a telescoper for some rational function with one variable less. The second way is to find a telescoper for F such that the certificate does not contain new poles, a *telescoper with regular certificate.* Contrary to the telescoper (3), the operator ∂_t is a telescoper with regular certificate:

$$\partial_t F = \partial_x\left(-\tfrac{x}{3t}F\right) + \partial_y\left(-\tfrac{y}{3t}F\right) + \partial_z\left(-\tfrac{z}{t}F\right).$$

This proves that $\partial_t I = 0$. More generally we have:

PROPOSITION 4. *If $T \in \mathbb{C}(t)\langle\partial_t\rangle$ is a telescoper of F with regular certificate, then $T(I)$ is zero.*

In this case, the certificate itself is not needed to prove the conclusion, its existence and regularity are sufficient. The Griffiths–Dwork method always produces a telescoper with regular certificate, see Equation (2).

4.2 Algorithm

In this section L is $k(t)$ for some field k and f is a homogeneous polynomial over L of degree d satisfying Hypothesis (H). For F a rational function in $L[\mathbf{x}, \frac{1}{f}]_{-n-1}$ we want to find a nonzero operator T in $L\langle\partial_t\rangle$ such that $T(F)$ lies in D_f. Algorithm 2 describes the procedure TELESC that outputs such a telescoper. Note that $L[\mathbf{x}, \frac{1}{f}]_{-n-1}$ is stable with respect to the derivation ∂_t.

PROPOSITION 5. *Algorithm 2 terminates and outputs the minimal telescoper of F that has regular certificate.*

PROOF. The sequence (G_k) is defined by $G_0 = [F]$ and the recurrence relation $G_{k+1} = [\partial_t G_k]$. We show by induction that for all k the fraction G_k equals $[\partial_t^k F]$. This is clear for $k = 0$. Assume that G_k equals $[\partial_t^k F]$. By the soundness of the reduction the operator $G_k - \partial_t^k F$ lies in D_f. And then so does $\partial_t G_k - \partial_t^{k+1} F$ since ∂_t commutes with the ∂_i's. By Theorem 1 and linearity, this implies that $[\partial_t G_k]$ equals $[\partial_t^{k+1} F]$.

At the ith step of the loop the algorithm is looking for a linear relation between $[F], \dots, [\partial_t^i F]$. By Theorem 1, there is one if and only if there is a telescoper with regular certificate of order i. If there is such a relation, the algorithm computes it and returns the corresponding telescoper. By Proposition 2, the algorithm terminates. The telescoper admits a regular certificate by design, see Equation (2). □

5. EFFECTIVE BOUNDS FOR CREATIVE TELESCOPING

We now review the steps of the algorithm with the aim of bounding the degrees and orders of all polynomials and operators that are constructed. This is then used in the next section to assess the complexity of this approach.

For the needs of Section 7, we track the degrees not only with respect to the parameter t but also to another free variable ε of the base field. In other words, we assume that L is $k(t, \varepsilon)$. For p a polynomial in $k[t, \varepsilon]$, the bidegree $(\deg_t p, \deg_\varepsilon p)$ of p is denoted by $\delta(p)$. If $p = \sum_I p_I \mathbf{x}^I$ is a polynomial in t, ε and $\mathbf{x}$, then $\delta(p)$ denotes the supremum of the $\delta(p_I)$'s, component by component.

THEOREM 6. *Let $f \in L[\mathbf{x}]$ be homogeneous of degree d satisfying (H). Let a/f^ℓ in $L[\mathbf{x}, \frac{1}{f}]_{-n-1}$ be a rational function, with a a polynomial in t and ε. The minimal telescoper of a/f^ℓ with regular certificate has degree*

$$\mathcal{O}\left(d^n\delta(a) + \left(\ell d^{2n} + d^{3n}\right)\mathrm{e}^n\delta(f)\right),$$

uniformly in all the parameters. It has order at most d^n.

The last part of the theorem is a direct consequence of the confinement property of Corollary 3. We now study more precisely the decomposition used in Algorithm 1 in order to control the degree of the telescoper and complete the proof.

The notation $a(\mathbf{n}) = \mathcal{O}(b(\mathbf{n}))$, for a tuple $\mathbf{n}$, means that there exists $C > 0$ such that for all $\mathbf{n} \geqslant 1$, with at most a finite number of exceptions, we have $a(\mathbf{n}) \leqslant Cb(\mathbf{n})$. The notation $a(\mathbf{n}) = \tilde{\mathcal{O}}(b(\mathbf{n}))$ means that $a(\mathbf{n}) = \mathcal{O}(b(\mathbf{n}) \log^k b(\mathbf{n}))$ for some integer k. We emphasize that when there are several parameters in a $\mathcal{O}$, the constant is uniform in all the parameters and there is at most a finite number of exceptions.

5.1 Reduction modulo the Jacobian ideal

An important ingredient of the Griffiths–Dwork reduction is the computation of a decomposition $r + \sum_i u_i \partial_i f$ of a homogeneous polynomial a. This can be done by means of a Gröbner basis of $\operatorname{Jac} f$, but instead of following the steps of a Gröbner basis algorithm, we cast the computation into a linear algebra framework using Macaulay's matrices, for which Cramer's rule and Hadamard's bound can then be used. While not strictly equivalent, both methods ensure that r depends linearly on a and vanishes when a is in $\operatorname{Jac} f$.

For a positive integer q, let φ_q denote the linear map

$$\varphi_q : (u_i) \in L[\mathbf{x}]_{q-d-n}^{n+1} \longrightarrow \sum_{i=0}^{n} u_i \partial_i f \in L[\mathbf{x}]_{q-n-1}.$$

Let $\operatorname{Mat}\varphi_q$ be the matrix of φ_q in a monomial basis. It has dimension $R_q \times C_q$, where R_q denotes $\binom{q-1}{n}$ and C_q denotes $(n+1)\binom{q-d}{n}$, and we note for future use that $C_q \leqslant R_q$ for all positive integers n and $d > 2$. Up to a change of ordering of the bases of the domain and codomain, $\operatorname{Mat}\varphi_q$ has the form $\left(\begin{smallmatrix} A & B \\ C & D \end{smallmatrix}\right)$, where A is a square submatrix of maximal rank. Note that D is necessarily $CA^{-1}B$. Then, the endomorphism ψ_q defined by the matrix $\left(\begin{smallmatrix} A^{-1} & 0 \\ 0 & 0 \end{smallmatrix}\right)$ satisfies $\varphi_q\psi_q\varphi_q = \varphi_q$; it is called a *split* of φ_q. It depends on the choice of the maximal rank minor. The map $\mathrm{id} - \varphi_q\psi_q$, denoted by π_q, performs the reduction in degree $q - n - 1$: it is idempotent; if a of degree $q - n - 1$ is in $\operatorname{Jac} f$ then it equals $\varphi_q(b)$ for some b and thus $\pi_q(a)$ vanishes; and for all a

in $L[\mathbf{x}]_{q-n-1}$ it gives a decomposition

$$a = \pi_q(a) + \sum_i \psi_q(a)_i \partial_i f.$$

Under Hypothesis (H), the map φ_q is surjective when q is at least $(n+1)d-n$. Let D denote this bound, known as *Macaulay's bound* [18, chap. 1; 16, corollaire, p. 169].

For q larger than D, a split of φ_q can be obtained from a split ψ_D of φ_D in the following way. Let S be the set of monomials in $\mathbf{x}$ of total degree $q-D$. Choose a linear map μ from $L[\mathbf{x}]_{q-n-1}$ to $L[\mathbf{x}]^S_{D-n-1}$ such that each a in $L[\mathbf{x}]_{q-n-1}$ equals $\sum_{m\in S} m\mu_m(a)$. Then a split of φ_d is defined by

$$\psi_d(a) = \sum_{m\in S} m\psi_D(\mu_m(a)).$$

Let q be a positive integer and let E_q be the least common multiple of the denominators of the entries of $\operatorname{Mat}\psi_q$. The entries of $\operatorname{Mat}\psi_q$ and $\operatorname{Mat}\pi_q$ are rational functions of the form p/E_q, with p polynomial. Let δ_E denote the supremum of all $\delta(p)$ and all $\delta(E_q)$, for $q \in \mathbb{N}\setminus\{0\}$.

PROPOSITION 7. *The supremum δ_E is finite and bounded above by $e^n d^n \delta(f)$. Moreover, if $q > D$ then E_q equals E_D.*

PROOF. Assume first that $q > D$. In this case, the entries of $\operatorname{Mat}\psi_q$ are entries of $\operatorname{Mat}\psi_D$ and π_q is zero. Thus the inequalities will follow from the case where $q \leqslant D$. Let $\operatorname{Mat}\psi_q$ and $\operatorname{Mat}\pi_q$ be written respectively as N/E_q and P/E_q with N and P polynomial matrices. Let r be the rank of ψ_q. The maximal rank minor A in the construction of ψ_q has dimension r. Cramer's rule and Hadamard's bound ensure that $\delta(N)$ is at most $(r-1)\delta(f)$ and that $\delta(E_q)$ is at most $r\delta(f)$. Since P equals $E_q \operatorname{id} - (\operatorname{Mat}\varphi_q)N$ and $\delta(\operatorname{Mat}\varphi_d)$ equals $\delta(f)$, the degree $\delta(P)$ is also at most $r\delta(f)$.

Next, r is bounded by R_q, the row dimension of $\operatorname{Mat}\phi_d$. Since $q \leqslant D$, we have $R_q \leqslant R_D$ and we conclude using inequality $\binom{p}{n} \leqslant \left(\frac{p\,e}{n+1}\right)^n$, with $p \geqslant n$ an integer. □

Algorithm 3 is a slightly modified version of Algorithm 1 which uses the construction above. Its output is in general not equal to the output of the former version, for any monomial order, but of course it satisfies Theorem 1. In particular the output of the algorithm TELESC does not depend on the reduction method in REDUCE. From now on the brackets $[\cdot]$ denote the output of Algorithm 3.

5.2 Degree bounds for the reduction

PROPOSITION 8. *Let $a/f^\ell \in L[\mathbf{x}, \frac{1}{f}]_{-n-1}$, with a a polynomial in t and ε. Then*

$$\left[\frac{a}{f^\ell}\right] = \frac{1}{P_\ell}\sum_{k=1}^{n}\frac{b_k}{f^k},$$

where $P_\ell = \prod_{i=1}^{\ell} E_{id}$ and b_k in $L[\mathbf{x}]_{kd-n-1}$ is a polynomial in t and ε such that $\delta(b_k) \leqslant \delta(a) + \ell\delta_E$, for $1 \leqslant k \leqslant n$.

PROOF. Using Algorithm 3, we obtain

$$\left[\frac{a}{f^\ell}\right] = \frac{p}{E_{\ell d} f^\ell} + \frac{1}{E_{\ell d}}\left[\frac{g}{f^{\ell-1}}\right],$$

where g and p are polynomials in $\mathbf{x}$, t and ε, with $\delta(p)$ and $\delta(g)$ at most $\delta(a) + \delta_E$. Induction over ℓ yields

$$\left[\frac{a}{f^\ell}\right] = \sum_{k=1}^{\ell}\frac{p_k}{f^k\prod_{j=k}^{\ell} E_{jd}}$$

Input $F = a/f^\ell$ a rational function in $L[\mathbf{x}, \frac{1}{f}]_{-n-1}$, with f of degree d

Output $[F]$ such that there exist rational functions $A_0, \dots, A_n$ such that $F = [F] + \sum_i \partial_i A_i$

For all $1 \leqslant i \leqslant \ell$, precompute a split ψ_{id} of φ_{id} (§5.1)
procedure REDUCE(a/f^ℓ)
  if $\ell = 1$ **then return** a/f^ℓ
  $F_1 \leftarrow \dfrac{1}{\ell-1}\displaystyle\sum_i \frac{\partial_i\psi_{\ell d}(a)_i}{f^{\ell-1}}$
  return $\frac{\pi_{\ell d}(a)}{f^\ell}$ + REDUCE(F_1)

Algorithm 3. Griffiths–Dwork reduction, linear algebra variant

with p_k polynomials such that $\delta(p_k) \leqslant \delta(a) + (\ell - k + 1)\delta_E$. For $k > n$, and hence $kd > D$, the map π_{kd} is 0 and thus so is p_k. Thus

$$\left[\frac{a}{f^\ell}\right] = \frac{1}{\prod_{j=1}^{\ell} E_{jd}}\sum_{k=1}^{\min(\ell,n)}\frac{p_k\prod_{j=1}^{k-1}E_{jd}}{f^k}. \quad \square$$

This proposition applied to $\partial_t^i(a/f^\ell)$ asserts that

$$\left[\partial_t^i\frac{a}{f^\ell}\right] = \frac{1}{P_{\ell+i}}\sum_{k=1}^{n}\frac{b_{i,k}}{f^k} = \frac{1}{P_{\ell+i}}\frac{b_i'}{f^n} \tag{4}$$

for some polynomials $b_{i,k}$ and b_i' such that

$$\delta(b_{i,k}) \leqslant \delta(a) + i\delta(f) + (i+\ell)\delta_E, \tag{5}$$

$$\text{and}\quad \delta(b_i') \leqslant \delta(a) + (i+n)\delta(f) + (i+\ell)\delta_E. \tag{6}$$

5.3 Degree bounds for the telescoper

PROPOSITION 9. *Let $T = \sum_{i=0}^{r} c_i\partial_t^i$, with coefficients c_i in $k[t,\varepsilon]$, be the minimal telescoper with regular certificate of a/f^ℓ. Then*

$$\delta(c_i) \leqslant r\delta(a) + \left(r^2 + r\ell\right)e^n d^n \delta(f).$$

PROOF. The operator T is the output of TELESC(a/f^ℓ). The rational functions c_i/c_r form the unique solution to the following system of inhomogeneous linear equations over L, with the Y_i's as unknown variables:

$$\sum_{i=0}^{r-1}\left[\partial_t^i\frac{a}{f^\ell}\right]Y_i = -\left[\partial_t^r\frac{a}{f^\ell}\right].$$

We write each $b_{i,k}$ in (4) as $\sum_{m\in S} b_{i,k,m}m$, where S is the set of all monomials in the variables $\mathbf{x}$ of degree at most $nd-n-1$. The previous system rewrites as

$$\forall m \in S, \forall k \in \{1,\dots,n\},\quad \sum_{i=0}^{r-1} Y_i\frac{b_{i,k,m}}{P_{\ell+i}} = -\frac{b_{r,k,m}}{P_{\ell+r}}$$

There is a set I of r indices $\{(k_0,m_0),\dots\}$ such that the square system formed by the corresponding equations admits a unique solution. We apply Cramer's rule to this system. Let B be the square matrix $(b_{i,k_j,m_j})_{i,j}$, for $0 \leqslant i,j < r$. Let B_i be the matrix obtained by replacing the row number i of B by the vector $(b_{r,k_j,m_j})_j$. We get, after simplification of the factors $P_{\ell+*}$ by multilinearity of the determinant,

$$\frac{c_i}{c_r} = \frac{\frac{P_{\ell+i}}{P_\ell}\det B_i}{\frac{P_{\ell+r}}{P_\ell}\det B}. \tag{7}$$

So, for all i, the polynomial c_i divides $\frac{P_{\ell+i}}{P_\ell}\det B_i$ and thus

$$\delta(c_i) \leqslant i\delta_E + \sum_{j=0, j\neq i}^{r} \delta(b_j).$$

With the previous bound (5) on $\delta(b_i)$ we get

$$\delta(c_i) \leqslant r\delta(a) + \frac{r(r+1)}{2}(\delta(f)+\delta_E) + r\ell\delta_E,$$

which gives the result with Proposition 7. □

6. COMPLEXITY

We assume that L is the field $k(t)$ and we evaluate the algebraic complexity of the steps of REDUCE and TELESC in terms of number of arithmetic operations in k. All the algorithms are deterministic. For univariate polynomial computations, we use the quasi-optimal algorithms in [10]. For simplicity, we assume that $d > 2$ so that several simplifications occur in the inequalities since $C_q \leqslant R_q$ and $d > \mathrm{e} \approx 2.72$.

THEOREM 10. *Under Hypothesis* (H) *and assuming that* $d > 2$, *Algorithm* TELESC *run with input* a/f^ℓ *takes*

$$\tilde{O}\left(\left(d^{5n} + d^{4n}\ell + d^{3n}\ell^2\left(\tfrac{\ell}{n}\right)^{2n}\right)\mathrm{e}^{3n}\delta\right)$$

arithmetic operations in k, *where* δ *is the larger of* $\delta(a)$ *and* $\delta(f)$, *uniformly in all the parameters. Asymptotically with* ℓ *and* n *fixed, this is* $\tilde{O}\left(d^{5n}\delta\right)$.

Note that while this may seem a huge complexity, it is not so bad when compared to the size of the output, which seems to be, empirically, comparable to $d^{3n}\delta$, with n fixed and $\ell = 1$. Note also that for $n = 1$, the complexity improves over that of the algorithm based on Hermite's reduction studied in [3], thanks to our avoiding too many rank computations.

REMARK 11. Let a/f be a generic fraction with a telescoper T and a regular certificate A. We claim that the size of A is asymptotically bounded below by $d^{(1-o(1))n^2}\delta$, making it crucial to avoid the computation of certificates. Indeed, the fraction $T(a/f)$ writes b/f^{r+1}, where r is the order of T. The number of monomials of b in $\mathbf{x}$ is $\binom{(r+1)d-1}{n} \approx (rd)^n/n!$. If a is generic then r is at least $\dim H_f^{\mathrm{pr}}$, by the Cyclic Vector Theorem; and if f is generic, it satisfies (H) and $\dim H_f^{\mathrm{pr}}$ is about d^n, by Corollary 3. Since $T(a/f)$ equals $\sum_i \partial_i(A_i)$, the size of the A_i has at least the same order of magnitude than that of $T(a/f)$; hence the claim.

6.1 Primitives of linear algebra

The complexity of Algorithm 2 lies in operations on matrices with polynomial coefficients. Let $A \in k[t]^{n\times m}$ have rank r and coefficients of degree at most d. One can compute r, a basis of $\ker A$ and a maximal rank minor in $\tilde{O}(nmr^{\omega-2}d)$ operations in k [30]. A maximal rank minor can be inverted in complexity $\tilde{O}(r^3 d)$ [13]. In particular, a matrix B such that $ABA = A$ can be computed in $\tilde{O}(nmr^{\omega-2}d + r^3 d)$ operations in k, or $\tilde{O}(n^2 md)$, using $r \leqslant n, m$ and $\omega \leqslant 3$.

A matrix A with rational entries is represented with the l.c.m. g of the entries and the polynomial matrix gA.

6.2 Precomputation

Algorithm 2 needs the splits ψ_{id} for i from 1 to the larger of $n+1$ and ℓ. Following §5.1, it is enough to compute ψ_{id} for i between 1 and $n+1$, each for a cost of $\tilde{O}(R_{id}C_{id}^2\delta(f))$ operations in k, and then ψ_{id} can be obtained with no further arithmetic operation for $i > n+1$. Thus the precomputation needs $\tilde{O}\left(\mathrm{e}^{3n}d^{3n}\delta(f)\right)$ operations in k.

6.3 Reduction

Let $\rho(\ell, \delta(a))$ be the complexity of the variant of the algorithm REDUCE based on linear algebra with input a rational function a/f^ℓ. The procedure first computes $\psi_{\ell d}(a)$. Since $\psi_{\ell d}$ is precomputed, it is only the product of a matrix of dimensions $C_{\ell d}$ by $R_{\ell d}$ with the vector of coefficients of a in a monomial basis. The elements of the matrix have degree at most δ_E and the elements of the vector have degree at most $\delta(a)$. Thus the product has complexity $\tilde{O}(R_{\ell d}C_{\ell d}(\delta(a)+\delta_E))$. Secondly, the procedure computes r as $\pi_{\ell d}(a)$ knowing $\psi_{\ell d}(a)$; this has the same complexity. Thirdly, it computes F_1, computation whose complexity is dominated by that of the first step. And lastly it computes REDUCE(F_1), which has complexity bounded by $\rho(\ell-1, \delta(a)+\delta_E)$. Unrolling the recurrence leads to

$$\rho(\ell, \delta(a)) = \tilde{O}\left(\ell\left(\frac{\mathrm{e}d\ell}{n+1}\right)^{2n}(\delta(a)+\ell\delta_E)\right).$$

6.4 Main loop

The computation of G_0 has complexity $\rho(\ell, \delta(a))$. Next, G_i has shape given by (4), and is differentiated before being reduced, so that the cost of the computation of G_{i+1} is at most $\rho(n+1, \delta(a)+(i+2n)\delta(f)+(i+\ell)\delta_E)$. Summing up, the computation of $G_0, \dots, G_r$ has a complexity

$$\rho(\ell, \delta(a)) + \tilde{O}\left((\mathrm{e}d)^{2n} r\left(\delta(a)+r\delta(f)+(r+\ell)\delta_E\right)\right). \quad (8)$$

During the ith step, the procedure computes the rank of $i+1$ vectors with $\mathcal{O}(\mathrm{e}^n d^n)$ coefficients of degree $\delta(b_i')$ and computes a linear dependence relation if there is one. This is done in complexity $\tilde{O}\left(i^{\omega-1}\mathrm{e}^n d^n \delta(b_i')\right)$. This step is quite expensive and doing it for all i up to r would ruin the complexity. It is sufficient to perform this computation only when i is a power of 2 so that the maximal i which is used is smaller than $2r$. When the rank of the family is not full, we deduce from it the exact order r and perform the computation in that order. Indeed, the rank over L of $G_0, \dots, G_i$ is the least of r and i. This way, finding the rank and solving has cost $\tilde{O}(r^{\omega-1}\mathrm{e}^n d^n(\delta(b_r') + \delta(b_{2r}')))$. In view of (6) and since $r \leqslant d^n$ and $\omega \leqslant 3$, the complexity of that step is bounded by (8). Adding the cost of the precomputation and using the bounds of the previous section leads to Theorem 10.

7. AFFINE SINGULAR CASE

Let L denote the field $k(t)$. Let F_{aff} be a rational function in $L(x_1, \dots, x_n)$, written as a/f_{aff}. We do not assume that F_{aff} is homogeneous, nor that f_{aff} satisfies a regularity property. Let d_{aff} be the total degree of f_{aff} w.r.t. $\mathbf{x}$.

In this section we show a deformation technique that regularizes singular cases. In particular, it allows to transfer the previous results to the general case and obtain the following bounds. The algorithm is again based on linear algebra.

THEOREM 12. *The function* F_{aff} *admits a telescoper, with regular certificate, of order at most* d^n *and degree*

$$\mathcal{O}\left(d_{\mathrm{pr}}^n\delta(a) + d_{\mathrm{pr}}^{3n}\mathrm{e}^n\delta(f_{\mathrm{aff}})\right),$$

where d_{pr} *is* $\max(d_{\mathrm{aff}}, \deg_{\mathbf{x}} a + n + 1)$. *This telescoper can be computed in complexity* $\tilde{O}\left(\mathrm{e}^{3n}d_{\mathrm{pr}}^{8n}\delta\right)$, *with* δ *the larger of* $\delta(a)$ *and* $\delta(f_{\mathrm{aff}})$.

degree of f	3	4	5	6
order of telesc.	2	6	12	20
degree of telesc. $\delta = 1$	32 *(68)* 0.4s	153 *(891)* 46s	480 *(5598)* 2h	1175 *(23180)* 150h
— , $\delta = 2$	66 *(136)* 0.6s	336 *(1782)* 140s	1092 *(11196)* 7h	? *(46360)* ∅
— , $\delta = 3$	100 *(204)* 0.9s	519 *(2673)* 270s	1704 *(16794)* 13h	? *(69540)* ∅

Table 1. Empirical order and degree of the minimal telescoper with regular certificate of a random rational function a/f^2 in $\mathbb{Q}(t, x_0, x_1, x_2)$, with f and a homogeneous in $\mathbf{x}$ satisfying $\deg_{\mathbf{x}} a + 3 = 2 \deg_{\mathbf{x}} f$ and $\delta(a)$ and $\delta(f)$ equal to δ; together with a proved upper bound (with a version of Theorem 9 without simplification) and mean computation time (CPU time).

It is easy to see that the *bit* complexity is also polynomial in d_{pr}^n. The dependence in n of the complexity, with $\deg_{\mathbf{x}} a$ and d_{aff} fixed, can be improved to $e^{O(n)}$ rather than $n^{O(n)}$ with a more careful analysis.

7.1 Homogenization and deformation

The regularization proceeds in two steps. First, let F_{pr} be the homogenization of F_{aff} in degree $-n-1$, that is

$$F_{\text{pr}} = x_0^{-n-1} F_{\text{aff}}\left(\tfrac{x_1}{x_0}, \dots, \tfrac{x_n}{x_0}\right),$$

which we write b/f_{pr} for some homogeneous polynomials b and f_{pr}. Let d_{pr} denote the degree of f_{pr}; it is given by Theorem 12. The degrees of b and f_{pr} satisfy the hypothesis of Theorem 6, by construction, but in general f_{pr} does not satisfy Hypothesis (H). (Although it does generically, as long as d_{pr} equals d_{aff}.) We consider a new indeterminate ε, the polynomial f_{reg} defined by

$$f_{\text{reg}} = f_{\text{pr}} + \varepsilon \sum_{i=0}^{n} x_i^{d_{\text{pr}}},$$

and the rational function F_{reg} defined by b/f_{reg}. We could also have perturbed the square-free part of f_{pr} rather than f_{pr}, leading to an improvement of the complexity in Theorem 12 at the cost of more technical details.

LEMMA 13. *The polynomial f_{reg} satisfies Hypothesis (H) over $L(\varepsilon)$, that is $L(\varepsilon)[\mathbf{x}]/\operatorname{Jac} f_{\text{reg}}$ has finite dimension.*

PROOF. This is true for $\varepsilon = \infty$, so it is generically true. ☐

Now, Theorem 6 gives bounds on the order and degree of a telescoper of F_{reg}, which is in $L(\varepsilon)[\mathbf{x}, \frac{1}{f_{\text{reg}}}]_{-n-1}$. The proof of Theorem 12 is concluded by the following.

PROPOSITION 14. *If T in $L[\varepsilon]\langle\partial_t\rangle$ is a telescoper of F_{reg} with regular certificate, then so is $T|_{\varepsilon=0}$ for F_{aff}.*

PROOF. By assumption, $T(F_{\text{reg}})$ equals $\sum_{i=0}^{n} \partial_i g_i / f_{\text{reg}}^p$ for some integer p and polynomials g_i in $L(\varepsilon)[\mathbf{x}]$. Each g_i/f_{reg}^p can be expanded in Laurent series in ε as $\sum_{j \geqslant N} h_{ij}\varepsilon^j$ for some possibly negative integer N and rational functions h_{ij} in $L[\mathbf{x}, \frac{1}{f_{\text{pr}}}]_{-n}$. Similarly, we can write the operator $T(F_{\text{reg}})$ as $T|_{\varepsilon=0}(F_{\text{pr}}) + \varepsilon \sum_{j \geqslant 0} b_j \varepsilon^j$ for some rational functions b_j in $L[\mathbf{x}, \frac{1}{f_{\text{pr}}}]$. Since the derivations ∂_i commute with ε, it is clear that $T|_{\varepsilon=0}(F_{\text{pr}})$ equals $\sum_{i=0}^{n} \partial_i h_{i0}$. Next, in this equality, x_0 can be evaluated to 1 to give

$$T|_{\varepsilon=0}(F_{\text{aff}}) = (\partial_0 h_{00})|_{x_0=1} + \sum_{i=1}^{n} \partial_i (h_{i0}|_{x_0=1}).$$

Euler's relation for h_{00} gives (with the index 00 dropped)

$$(\partial_0 h)|_{x_0=1} = -\sum_{i=1}^{n} \partial_i (x_i h|_{x_0=1}),$$

proving that $(\partial_0 h)|_{x_0=1}$ is in $D_{f_{\text{aff}}}$. Thus, so is $T|_{\varepsilon=0}(F_{\text{aff}})$ and the proof is complete. ☐

Nevertheless, a telescoper obtained in this way does not need to be minimal, even starting from a minimal one for the perturbed function F_{reg}. This is unfortunate because in presence of singularities the dimension of H_f^{pr} can collapse when compared to the generic order given by Corollary 3.

7.2 Algorithm and complexity

The algorithm is based on Proposition 14. We use an evaluation-interpolation scheme to control the complexity. Let the operator T in $k(t,\varepsilon)\langle\partial_t\rangle$ be the minimal telescoper of F_{reg}, written as $\partial_t^r + \sum_{k=0}^{r-1} \frac{c_k}{c_r} \partial_t^k$. It is the output of TELESC applied to F_{reg}. We aim at computing $(\varepsilon^\alpha T)|_{\varepsilon=0}$, where α is such that this evaluation is finite and not zero.

Proposition 14, slightly adapted, shows that $T|_{\varepsilon=u}$ is a telescoper with regular certificate of $F_{\text{reg}}|_{\varepsilon=u}$ whenever $c_r(t,u)$ is not zero, even if $f_{\text{reg}}|_{\varepsilon=u}$ does not satisfy (H). When it does, the specialization gives the minimal one:

LEMMA 15. *If $f_{\text{reg}}|_{\varepsilon=u}$ satisfies hypothesis (H) and if u does not cancel c_r, then $T|_{\varepsilon=u}$ is the minimal telescoper with regular certificate of $F_{\text{reg}}|_{\varepsilon=u}$.*

PROOF. We use the notation of Section 5, replacing f by f_{reg} and L by $L(\varepsilon)$. The operator T is the output of Algorithm 2 applied to F_{reg}. Since $f_{\text{reg}}|_{\varepsilon=u}$ satisfies (H), for all d the matrix $\operatorname{Mat}\varphi_d$, with coefficients in $L[\varepsilon]$, has the same rank as its specialization with $\varepsilon = u$ [18, §58]. Thus, to compute the splits ψ_d we can choose maximal rank minors of $\operatorname{Mat}\varphi_d$ that are also maximal rank minors of the specialization. When doing so, the reduction $[\cdot]$ commutes with the evaluation $\cdot|_{\varepsilon=u}$. In particular, the polynomials E_q do not vanish for $\varepsilon = u$.

In the proof of Prop. 9, Eq. (7) shows that c_r, the leading coefficient of T, divides $P_{\ell+r} \det B$. The polynomial $P_{\ell+r}$ is a product of several E_d's, in particular $P_{\ell+r}|_{\varepsilon=u}$ is not zero. Since $c_r|_{\varepsilon=u} \neq 0$, the determinant of $B|_{\varepsilon=u}$ is not zero either. Looking at the definition of B in the proof of Prop. 9, this implies that the $[\partial_t^i F_{\text{reg}}]|_{\varepsilon=u}$, for i between 0 and $r-1$ are free over $L(\varepsilon)$. In particular, a telescoper with regular certificate of $F_{\text{reg}}|_{\varepsilon=u}$ has order at least r. Since $T|_{\varepsilon=u}$ is a telescoper of order is r, it is the minimal one. ☐

We now present the algorithm. Let N be $e^n(d^{3n}+d^{2n}+d^n)$. By Proposition 9, the polynomials c_k have degree at most N in ε, and at most $N\delta$ in t. Choose a set U of $4N+1$ elements of k. Determine the set U' of elements u of U such that $f_{\text{reg}}|_{\varepsilon=u}$ satisfies (H). This step has complexity $\tilde{O}((ed)^{n\omega}\delta|U|)$: The polynomial $f_{\text{reg}}|_{\varepsilon=u}$ satisfies (H) if and only if $(\operatorname{Mat}\varphi_D)|_{\varepsilon=u}$ is full rank. In particular, if $f_{\text{reg}}|_{\varepsilon=u}$ does not satisfy (H), then $E_D|_{\varepsilon=u}$ vanishes. The polynomial E_D has degree at most $e^n d^n$ in ε, by Proposition 7, so $U \setminus U'$ has at most $e^n d^n$ elements. For each u in U', compute TELESC$(f_{\mathrm{reg}}|_{\varepsilon=u})$ with leading coefficient normalized to 1, denoted by T_u. This step has complexity $\tilde{\mathcal{O}}(d^{5n}e^{3n}\delta|U'|)$, by Theorem 10. Determine the subset U'' of U' where the order of T_u is maximal. By Lemma 15, the complement $U' \setminus U''$ is formed by u such that $c_r(t,u) = 0$. It has at most N elements since c_r has degree at most N in ε. For all u in U'' the operators T_u and $T|_{\varepsilon=u}$ coincide. Thus U'' has at most $2N+1$ elements.

The r rational functions $\frac{c_k(t,0)}{c_r(t,0)}$ can be computed using Lemma 16 in total complexity $\tilde{\mathcal{O}}(N^2 r\delta)$. If $c_r(t,0)$ is zero, we look for the positive integer α such that the functions $\varepsilon^\alpha \frac{c_k(t,\varepsilon)}{c_r(t,\varepsilon)}$ are finite for $\varepsilon = 0$ but not zero for at least one k. The integer α is at most N and thus can be found with a binary search, using at most $\log_2 N + 1$ times Lemma 16.

LEMMA 16. *Let R in $k(x,y)$ be written P/Q, with P and Q polynomials of degree less than d_x in x and d_y in y. Given evaluations $R(x,v)$, for $2d_y+1$ elements v of k, the function $R(x,0)$ (or ∞ if $Q(x,0)$ vanishes) can be computed using $\tilde{\mathcal{O}}(d_x d_y)$ arithmetic operations in k.*

PROOF. Let V the set of evaluation points. Choose a set U of $2d_x+1$ points of k. Compute $R(u,v)$ for $u \in U$ and $v \in V$ in $\tilde{\mathcal{O}}(d_x d_y)$ operations. Note that there is no need to check that the elements of U are not poles of the $R(x,v)$: univariate rational reconstruction can handle that. Use univariate rational reconstruction to compute $R(u,y)$, for u in U, in complexity $\tilde{\mathcal{O}}(d_y|U|)$ operations. Reconstruct $R(x,0)$ in complexity $\tilde{\mathcal{O}}(d_x)$ from the evaluations $R(u,0)$. □

8. EXPERIMENTS

A basic implementation of the algorithm TELESC has been written in Maple 16. As it uses only Maple primitives to compute with polynomial matrices, it is certainly too basic to reflect the complexity given in Theorem 10.

Table 1 presents empirical results for some generic rational functions, with $n = 2$. The bound on the order are generically exact as expected; however the bound on the degree is not very sharp. For $n = 1$ and $\delta(a)$ fixed, a careful study [3] proves that the degree of the minimal telescoper is $\mathcal{O}(d^2\delta)$, which is tighter than the $\mathcal{O}(d^3\delta)$ given by Theorem 6. Analogy, as well as numerical evidence and theoretical clues, lead us to think that for general n, the asymptotic behavior can be improved from $\mathcal{O}(d^{3n}\delta)$ to $\mathcal{O}(d^{2n}\delta)$.

The relative cost of each step of Algorithm 2 in the computation of telescopers of Table 1, on the example of the telescoper of degree 12 and degree 1092 of a generic function a/f^2 as described in Table 1, that is computed in about 7 hours breaks down as follows: The computation of splits of Macaulay matrices takes about 1% of the time, the reduction steps about 40%, and the final solving about 60% of the time. More efficient matrix multiplication and system resolution over univariate polynomials could improve speed dramatically. We have not been able to compute more than the first column of Table 1 with methods and programs in [15, 4].

On the other hand, the regularity hypothesis (H) is restrictive in applications: Even though generic polynomials satisfy this hypothesis, examples with physical or combinatorial meaning usually do not. The method shown in Section 7 is only of theoretical interest. By contrast, the algorithm for the regular case is very efficient in practice.

9. REFERENCES

[1] T. G. Abbott, K. S. Kedlaya, and D. Roe. Bounding Picard numbers of surfaces using p-adic cohomology. In *AGCT 2005*, volume 21 of *Sémin. Congr.*, pages 125–159. SMF, Paris, 2010.

[2] M. Apagodu and D. Zeilberger. Multi-variable Zeilberger and Almkvist-Zeilberger algorithms and the sharpening of Wilf-Zeilberger theory. *Adv. in Appl. Math.*, 37(2):139–152, 2006.

[3] A. Bostan, S. Chen, F. Chyzak, and Z. Li. Complexity of creative telescoping for bivariate rational functions. In *ISSAC'10*, pages 203–210. ACM, 2010.

[4] S. Chen, M. Kauers, and M. F. Singer. Telescopers for rational and algebraic functions via residues. In *ISSAC'12*, pages 130–137. ACM, 2012.

[5] G. Christol. Diagonales de fractions rationnelles et equations différentielles. In *Groupe de travail d'analyse ultramétrique, 1982/83*, volume 12, issue 2, exp. 18, pages 1–10. Paris, 1984.

[6] G. Christol. Diagonales de fractions rationnelles et équations de Picard-Fuchs. In *Groupe de travail d'analyse ultramétrique, 1984/85*, volume 12, issue 1, exp. 13, pages 1–12. Paris, 1985.

[7] F. Chyzak. An extension of Zeilberger's fast algorithm to general holonomic functions. *Disc. Math.*, 217(1-3):115–134, 2000.

[8] B. Dwork. On the zeta function of a hypersurface. *IHES Publ. Math.*, (12):5–68, 1962.

[9] B. Dwork. On the zeta function of a hypersurface. II. *Ann. of Math. (2)*, 80:227–299, 1964.

[10] J. von zur Gathen and J. Gerhard. *Modern Computer Algebra*. Cambridge University Press, Cambridge, second edition, 2003.

[11] R. Gerkmann. Relative rigid cohomology and deformation of hypersurfaces. *Int. Math. Res. Pap.*, (1):Art. 3, 1–67, 2007.

[12] P. A. Griffiths. On the periods of certain rational integrals. I, II. *Ann. of Math*, 90(3):460–495, 496–541, 1969.

[13] C.-P. Jeannerod and G. Villard. Essentially optimal computation of the inverse of generic polynomial matrices. *J. Complexity*, 21(1):72–86, 2005.

[14] M. Kauers and D. Zeilberger. The computational challenge of enumerating high-dimensional rook walks. *Adv. in Appl. Math.*, 47(4):813–819, 2011.

[15] C. Koutschan. A fast approach to creative telescoping. *Math. Comput. Sci.*, 4(2-3):259–266, 2010.

[16] D. Lazard. Algèbre linéaire sur $K[X_1, \cdots, X_n]$, et élimination. *Bull. Soc. Math. France*, 105(2):165–190, 1977.

[17] L. Lipshitz. The diagonal of a D-finite power series is D-finite. *J. Algebra*, 113(2):373–378, 1988.

[18] F. S. Macaulay. *The algebraic theory of modular systems*, volume XXXI of *Cambridge Mathematical Library*. Cambridge University Press, 1994. Revised reprint of the 1916 original.

[19] P. Monsky. *p-adic analysis and zeta functions*, volume 4 of *Lectures in Mathematics, Department of Mathematics, Kyoto University*. Kinokuniya Book-Store Co. Ltd., Tokyo, 1970.

[20] D. R. Morrison. Picard-Fuchs equations and mirror maps for hypersurfaces. In *Essays on mirror manifolds*, pages 241–264. Int. Press, Hong Kong, 1992.

[21] H. Movasati. *Multiple integrals and modular differential equations*. IMPA Mathematical Publications. Instituto Nacional de Matemática Pura e Aplicada (IMPA), Rio de Janeiro, 2011.

[22] S. Pancratz. Computing Gauss-Manin connections for families of projectives hypersurfaces. A thesis submitted for the Transfer of Status, Michaelmas, 2009.

[23] É. Picard. Quelques remarques sur les intégrales doubles de seconde espèce dans la théorie des surfaces algébriques. *C. R. Acad. Sci. Paris*, 129:539–540, 1899.

[24] É. Picard. Sur les périodes des intégrales doubles et sur une classe d'équations différentielles linéaires. *C. R. Acad. Sci. Paris*, 134:69–71, 1902.

[25] É. Picard and G. Simart. *Théorie des fonctions algébriques de deux variables indépendantes*. Gauthier-Villars, 1906. Tome II.

[26] N. Takayama. An algorithm of constructing the integral of a module — an infinite dimensional analog of Gröbner basis. In *ISSAC'90*, pages 206–211. ACM, 1990.

[27] H. S. Wilf and D. Zeilberger. An algorithmic proof theory for hypergeometric (ordinary and "q") multisum/integral identities. *Invent. Math.*, 108(3):575–633, 1992.

[28] D. Zeilberger. A holonomic systems approach to special functions identities. *J. Comput. Appl. Math.*, 32(3):321–368, 1990.

[29] D. Zeilberger. The method of creative telescoping. *J. Symb. Comp.*, 11(3):195–204, 1991.

[30] W. Zhou. *Fast Order Basis and Kernel Basis Computation and Related Problems*. PhD thesis, Univ. of Waterloo, 2013.

On the Integration of Differential Fractions

François Boulier
Université Lille 1
Villeneuve d'Ascq, France
Francois.Boulier@univ-lille1.fr

François Lemaire*
Université Lille 1
Villeneuve d'Ascq, France
Francois.Lemaire@lifl.fr

Georg Regensburger
RICAM
Linz, Austria
Georg.Regensburger@oeaw.ac.at

Markus Rosenkranz†
University of Kent
Canterbury, United Kingdom
M.Rosenkranz@kent.ac.uk

ABSTRACT

In this paper, we provide a differential algebra algorithm for integrating fractions of differential polynomials. It is not restricted to differential fractions that are the derivatives of other differential fractions. The algorithm leads to new techniques for representing differential fractions, which may help converting differential equations to integral equations (as for example used in parameter estimation).

Categories and Subject Descriptors

I.1.1 [**Symbolic and Algebraic Manipulation**]: Expressions and Their Representation—*simplification of expressions*; I.1.2 [**Symbolic and Algebraic Manipulation**]: Algorithms—*algebraic algorithms*

Keywords

Differential algebra, differential fractions, integration

1. INTRODUCTION

In this paper we present an algorithm that solves the following problem in differential algebra: Given a differential fraction F, i.e., a fraction P/Q where P and Q are differential polynomials, and a derivation δ, compute a finite sequence $F_1, F_2, F_3, \ldots, F_t$ of differential fractions such that

$$F \quad = \quad F_1 + \delta\, F_2 + \delta^2\, F_3 + \cdots + \delta^t\, F_t\,, \tag{1}$$

the differential fractions $\delta^\ell\, F_\ell$ have rank less than or equal to F (the result is thus ranking dependent) and the δ^ℓ differential operators are as much "factored out" as possible. In particular, if there exists some differential fraction G such that $F = \delta\, G$, then $F_1 = 0$: the algorithm recognizes first integrals.

This work originated from our attempts to generalize the construction of integro-differential polynomials [14] to the case of several independent and dependent variables. The results presented here constitute the most important step in this construction: the decomposition of an arbitrary differential polynomial—or differential fraction—into a total derivative and a remainder (see Example 1 with `iterated = false`). For ordinary differential polynomials, such a decomposition is described in [1]. See also [8, 2] and the recent dissertation [12] for further references.[1]

If the remainder is zero, the given differential fraction F is recognized to be the total derivative of another differential fraction G, so the latter appears as a *first integral* of F. Algorithms for determining such first integrals are known (see [15, 16]), and one such algorithm is implemented in Maple via the function `DEtools[firint]`.

Handling differential fractions rather than polynomials may be very important for the range of application of Algorithms 3 and 4 since differential equations may be much easier to integrate when multiplied, or divided, by an integrating factor. Beyond its theoretical interest, the algorithm presented in this paper may also be useful in practice:

1. Sometimes (though not always), a differential fraction is shorter in the representation (1) than in expanded form. Thus our algorithm provides new facilities for representing differential equations.
2. The representation (1) may also be more convenient than the expanded form if one wants to convert differential equations to integral equations. This may be a very important feature for the problem of estimating parameters of dynamical systems from their input-output behaviour (see Example 5).

The paper is organized as follows: In Section 2, we review some standard definitions for differential polynomials and generalize them to differential fractions. In Section 3, the main result of this paper is stated and proved (Algorithm 3 and Proposition 6). In Section 4, we describe an implementation along with a few worked-out examples.

*The first two authors acknowledge partial support by the French ANR-10-BLAN-0109 LEDA project.

†The fourth author acknowledges partial support by the EPSRC First Grant EP/I037474/1.

[1]The authors would like to thank the reviewers for pointing out these references.

ISSAC'13, June 26–29, 2013, Boston, Massachusetts, USA.

2. BASICS OF DIFFERENTIAL ALGEBRA

This paper is concerned with *differential fractions*, i.e., fractions of *differential polynomials*. A key problem with such fractions is to reduce them, which requires computing the gcd of multivariate polynomials, which is possible whenever the base field is computable.

The reference books are [13] and [11]. A *differential ring* $\mathscr{R}$ is a ring endowed with finitely many, say m, *derivations* $\delta_1, \ldots, \delta_m$, i.e., unary operations satisfying the following axioms, for all $a, b \in \mathscr{R}$:

$$\delta(a + b) = \delta(a) + \delta(b)\,, \quad \delta(a\,b) = \delta(a)\,b + a\,\delta(b)\,,$$

and which commute pairwise. To each derivation δ_i an *independent variable* x_i is associated such that $\delta_i\,x_j = 1$ if $i = j$ and 0 otherwise. The set of independent variables is denoted by $\mathscr{X} = \{x_1, \ldots, x_m\}$. The derivations generate a commutative monoid w.r.t. composition denoted by

$$\Theta = \{\delta_1^{a_1} \cdots \delta_m^{a_m} \mid a_1, \ldots, a_m \in \mathbb{N}\}\,,$$

where $\mathbb{N}$ stands for the nonnegative integers. The elements of Θ are called *derivation operators*. If $\theta = \delta_1^{a_1} \cdots \delta_m^{a_m}$ is a derivation operator then $\mathrm{ord}(\theta) = a_1 + \cdots + a_m$ denotes its *order*, with a_i being the order of θ w.r.t. derivation δ_i (or x_i).

In order to form differential polynomials, one introduces a set $\mathscr{U} = \{u_1, \ldots, u_n\}$ of n *differential indeterminates*. The monoid Θ acts on $\mathscr{U}$, giving the infinite set $\Theta\mathscr{U}$ of *derivatives*. For readability, we often index derivations by letters like δ_x and δ_y, denoting also the corresponding derivatives by these subscripts, so u_{xy} denotes $\delta_x\,\delta_y\,u$.

For applications, it is crucial that one can also handle *parametric* differential equations. Parameters are nothing but symbolic *constants*, i.e., symbols whose derivatives are zero. Let $\mathscr{C}$ denote the set of constants.

The differential fractions considered in this paper are ratios of differential polynomials taken from the differential ring $\mathscr{R} = \mathbb{Z}[\mathscr{X} \cup \mathscr{C}]\{\mathscr{U}\} = \mathbb{Z}[\mathscr{X} \cup \mathscr{C} \cup \Theta\mathscr{U}]$. A differential fraction is said to be *reduced* if its numerator and denominator do not have any common factor. A differential polynomial (respectively a differential fraction) is said to be *numeric* if it is an element of $\mathbb{Z}$ (resp. of $\mathbb{Q}$). It is said to be a *coefficient* if it is an element of $\mathbb{Z}[\mathscr{X} \cup \mathscr{C}]$ (resp. of $\mathbb{Q}(\mathscr{X} \cup \mathscr{C})$). The elements of $\mathscr{X} \cup \mathscr{C} \cup \Theta\mathscr{U}$ are called *variables*.

A *ranking* is a total ordering on $\Theta\mathscr{U}$ that satisfies the two following axioms:

1. $v \leq \theta v$ for every $v \in \Theta\mathscr{U}$ and $\theta \in \Theta$,
2. $v < w \Rightarrow \theta v < \theta w$ for every $v,\, w \in \Theta\mathscr{U}$ and $\theta \in \Theta$.

Rankings are well-orderings, i.e., every strictly decreasing sequence of elements of $\Theta\mathscr{U}$ is finite [11, §I.8]. Rankings such that $\mathrm{ord}(\theta) < \mathrm{ord}(\phi) \Rightarrow \theta u < \phi v$ for every $\theta,\, \phi \in \Theta$ and $u,\, v \in \mathscr{U}$ are called *orderly*. In this paper, it is convenient to extend rankings to the sets $\mathscr{X}$ and $\mathscr{C}$. For all rankings, we will assume that any element of $\mathscr{X} \cup \mathscr{C}$ is less than any element of $\Theta\mathscr{U}$; see Remark 5 for a brief discussion why we make this assumption.

Fix a ranking and consider some non-numeric differential polynomial P. The highest variable v w.r.t. the ranking such that $\deg(P, v) > 0$ is called the *leading variable* or *leading derivative* (though it may not be a derivative) of P. It is denoted by $\mathrm{ld}(P)$. The monomial $v^{\deg(P,v)}$ is called the *rank* of P. The leading coefficient of P w.r.t. v is called the *initial* of P. The differential polynomial $\partial P/\partial v$ is called the *separant* of P. More generally, if w is any variable, the differential polynomial $\partial P/\partial w$ is called the separant of P w.r.t. w and is denoted by $\mathrm{separant}(P, w)$.

2.1 Extension to differential fractions

In this section, some of the definitions introduced above are reformulated, to cover the case of differential fractions. For differential polynomials, these new definitions agree with the ones given before.

Remark 1. This section deals with differential fractions $F = P/Q$. However, the definitions stated below do not require F to be reduced, i.e., P and Q to be relatively prime.

Definition 1. Let F be a non-numeric differential fraction. The *leading variable* or *leading derivative* of F is defined as the highest variable v such that $\partial F/\partial v \neq 0$. It is denoted by $\mathrm{ld}(F)$.

PROPOSITION 1. *Let $F = P/Q$ be a non-numeric differential fraction. If P and Q have distinct leading derivatives or, if P or Q is numeric, then* $\mathrm{ld}(F)$ *is the highest variable v such that* $\deg(P, v) > 0$ *or* $\deg(Q, v) > 0$.

PROOF. First observe $\partial F/\partial w = 0$ for any $w > v$. It is thus sufficient to show that $\partial F/\partial v \neq 0$. Indeed, since v does not occur in both the numerator and the denominator of F, we have $\partial F/\partial v = (\partial P/\partial v)/Q$ or $\partial F/\partial v = -P\,(\partial Q/\partial v)/Q^2$. In each case, the corresponding fraction is nonzero. □

Definition 2. The *separant* of a non-numeric differential fraction F is defined as $\partial F/\partial v$, where $v = \mathrm{ld}(F)$.

Definition 3. Let $F = P/Q$ be a non-numeric differential fraction and write $v = \mathrm{ld}(F)$. The *degree* of F is defined as $\deg(F) = \deg(P, v) - \deg(Q, v)$. The *rank* of F is defined as the pair $(v, \deg(F))$.

Definition 4. A rank (v, d) is said to be lower than a rank (w, e) if $v < w$ or if $v = w$ and $d < e$.

The above definitions are a bit more complicated than in the polynomial case, because of differential fractions of degree 0. In particular, we wish to distinguish ranks $(v, 0)$ and $(w, 0)$, which allows us to state the following proposition.

PROPOSITION 2. *If F is a non-numeric differential fraction, then the separant of F is either numeric or has lower rank than F.*

PROOF. Assume the separant non-numeric. If its leading derivative is different from the leading derivative v of F, then it is lower than v and the Proposition is clear. Otherwise, the degree in v of the separant is less than or equal to $\deg(P, v) - \deg(Q, v) - 1$ and the Proposition is proved. □

Recall that if a polynomial P does not depend on a variable v, then $\mathrm{lcoeff}(P, v) = P$.

Definition 5. The *initial* of a non-numeric differential fraction $F = P/Q$ is defined as $\mathrm{lcoeff}(P, v)/\mathrm{lcoeff}(Q, v)$, where $v = \mathrm{ld}(F)$.

PROPOSITION 3. *Let F be a differential fraction which is not a coefficient, $v = \mathrm{ld}(F)$ and δ be a derivation. Then the leading derivative of $\delta\,F$ is $\delta\,v$, this derivative occurs in the numerator of $\delta\,F$ only, with degree 1, and the initial of $\delta\,F$ is the separant of F.*

PROOF. The first claim comes from the axioms of rankings. The two other ones are clear. □

3. MAIN RESULT

In this section, we write numer(F) and denom(F) for the numerator and denominator of a differential fraction F, both viewed as differential polynomials of the ring $\mathscr{R}$. Our result is Algorithm 3. It relies on two sub-algorithms (Algorithms 1 and 2), which are purely algebraic (i.e., they do not make use of derivations, in the sense of the differential algebra). These two sub-algorithms are related to the integration problem of rational fractions. They are either known or very close to known methods, such as those described in [7]. We state them in this paper, because our current implementation is actually based on them, the paper becomes self-contained, and the tools required by Algorithm 3 appear clearly.

Remark 2. In the three presented algorithms, all differential fractions are supposed to be reduced.

Algorithm 1 The prepareForIntegration algorithm

Require: F is a reduced differential fraction, v is a variable
Ensure: Three polynomials cont_F, N, B satisfying Prop. 4

1: **if** denom(F) is numeric **then**
2: $\quad \mathrm{cont}_F,\, N,\, B := \mathrm{denom}(F),\, \mathrm{numer}(F),\, 1$
3: **else**
4: $\quad \mathrm{cont}_F :=$ the gcd of all coeffs of denom(F) w.r.t. v
$\qquad\qquad$ { all gcd are of multivariate polynomials }
5: $\quad P_0 := \mathrm{numer}(F)$
6: $\quad Q_0 := \mathrm{denom}(F)/\mathrm{cont}_F \quad$ { all divisions are exact }
7: $\quad A_0 := \gcd(Q_0, \mathrm{separant}(Q_0, v))$
8: $\quad B_0 := Q_0/A_0$
9: $\quad C_0 := \gcd(A_0, B_0)$
10: $\quad D_0 := B_0/C_0$
11: $\quad A_1 := \gcd(A_0, \mathrm{separant}(A_0, v))$
12: $\quad N,\, B := P_0 \cdot D_0 \cdot A_1,\, D_0 \cdot A_0$
13: **end if**
14: **return** cont_F, N, B

Proposition 4 (Specification of Algorithm 1). *Let $F = P_0/(\mathrm{cont}_F\, Q_0)$ be a differential fraction (P_0 and Q_0 being relatively prime, Q_0 primitive w.r.t. v, and cont_F denoting the content of* denom(F) *w.r.t. v). Then the polynomials returned by Algorithm 1 satisfy*

$$F = \frac{N}{\mathrm{cont}_F\, B^2}\,, \qquad B = F_1\, F_2\, F_3^2 \cdots F_n^{n-1}\,,$$

where $Q_0 = F_1\, F_2^2\, F_3^3 \cdots F_n^n$ is the squarefree factorization of Q_0 w.r.t. v.

Proof. The case of F being a polynomial is clear. Assume F has a non-numeric denominator. We have $A_0 = F_2\, F_3^2 \cdots F_n^{n-1}$ at Line 8, $B_0 = F_1\, F_2\, F_3 \cdots F_n$ at Line 9, $C_0 = F_2\, F_3 \cdots F_n$ at Line 10, $D_0 = F_1$ at Line 11, $A_1 = F_3\, F_4^2 \cdots F_n^{n-2}$ at Line 12, $N = P_0\, F_1\, F_3\, F_4^2 \cdots F_n^{n-2}$ and $B = F_1\, F_2\, F_3^2 \cdots F_n^{n-1}$. □

The following Lemma, which is easy to see, establishes the relationship between a well-known fact on the integration of rational fractions and Algorithm 1.

Lemma 1. *With the same notations as in Proposition 4, if there exists a reduced differential fraction R such that $F =$* separant(R, v)*, then $F_1 = 1$ and the denominator of R is $\mathrm{cont}_R\, B$, where cont_R has degree 0 in v.*

Algorithm 2 The integrateWithRemainder algorithm

Require: F_0 is a differential fraction, v is a variable
Ensure: Two differential fractions R and W such that

1. $F_0 = \mathrm{separant}(R, v) + W$
2. W is zero iff there exists R s.t. $F_0 = \mathrm{separant}(R, v)$

1: $R,\, W := 0,\, 0$
2: $F := F_0$
3: **while** $F \neq 0$ **do**
4: $\qquad$ { invariant: $F_0 = F + \mathrm{separant}(R, v) + W$ }
5: $\quad \mathrm{cont}_F,\, N,\, B :=$ prepareForIntegration(F, v)
6: $\quad$ **if** $\deg(B, v) = 0$ **then**
7: $\qquad P :=$ the primitive of F w.r.t. v, with a 0 int. cst.
$\qquad\qquad$ { this amounts to integrate a polynomial }
8: $\qquad R := R + P$
9: $\qquad F := 0$
10: $\quad$ **else**
11: $\qquad$ { look for A such that
$\qquad\qquad \mathrm{separant}(A/(\mathrm{cont}_R\, B), v) = N/(\mathrm{cont}_F\, B^2)$ }
12: $\qquad c_B,\, c_N := \mathrm{lcoeff}(B, v),\, \mathrm{lcoeff}(N, v)$
13: $\qquad d_B,\, d_N,\, \bar{d}_A := \deg(B, v),\, \deg(N, v),\, d_N - d_B + 1$
14: $\qquad$ **if** $d_N = 2\, d_B - 1$ or $\bar{d}_A < 0$ **then**
15: $\qquad\quad H := c_N\, v^{d_N}$
16: $\qquad\quad W := W + H/(\mathrm{cont}_F\, B^2)$
17: $\qquad\quad F := F - H/(\mathrm{cont}_F\, B^2)$
18: $\qquad$ **else**
19: $\qquad\quad R_2 := c_N\, v^{\bar{d}_A}/((\bar{d}_A - d_B)\, \mathrm{cont}_F\, c_B\, B)$
20: $\qquad\quad R := R + R_2$
21: $\qquad\quad F := F - \mathrm{separant}(R_2, v)$
22: $\qquad$ **end if**
23: $\quad$ **end if**
24: **end while**
25: **return** R, W

Remark 3. Algorithm 2 relies on Algorithm 1 for computing cont_F, N and B. However, it does not need N. It only needs its leading coefficient c_N and its degree d_N. Our formulation improves the readability of our algorithm.

Proposition 5. *Algorithm 2 is correct.*

Proof. First suppose Algorithm 2 terminates.

The loop invariant stated in the algorithm is clear, since it is satisfied at the beginning of the first loop and maintained in each of the three cases considered by the algorithm. Combined with the loop condition, this implies that the first ensured condition is satisfied at the end of the algorithm.

Let us now show that the second ensured condition is satisfied. We assume that

$$\exists\, R \text{ s.t. } \quad F \quad = \quad \mathrm{separant}(R, v)\,. \tag{2}$$

We prove that Lines 15–17 are not performed. This is sufficient to prove the second ensured condition since, if Lines 15–17 are not performed and (2) holds then, after one iteration, F is modified either at Line 9 or 21. In both cases, the new value of F satisfies (2) again.

Assume (2) holds. By Lemma 1, the denominator of R is $\mathrm{cont}_R\, B$. Let $A = c_A\, v^{d_A} + q_A$ be its numerator and $B = c_B\, v^{d_B} + q_B$ (we only need to consider the case $d_B > 0$). One can assume that $d_A \neq d_B$. Indeed, if $d_A = d_B$, one can take $\bar{R} = (A - c_A/c_B\, B)/B$ since $\mathrm{separant}(\bar{R}, v) =$

separant$(R, v) = F$ and $\deg(A - c_A/c_B B) < d_B$. Differentiating the fraction $A/(\mathrm{cont}_R B)$ w.r.t. v and identifying with $F = N/(\mathrm{cont}_F B^2)$ (Proposition 4), we see that $N = (d_A - d_B) c_A c_B \mathrm{cont}_F/\mathrm{cont}_R v^{d_A+d_B-1} + \cdots$ where the dots hide terms of degree, in v, less than $d_A + d_B - 1$. Since $d_A \neq d_B$, the degree of N satisfies $d_N = d_A + d_B - 1$, thus $d_N \neq 2 d_B - 1$. Moreover, the variable $\bar{d}_A$ is equal to d_A since $d_N - d_B + 1 = d_A$. Summarizing, $d_N \neq 2 d_B - 1$ and $\bar{d}_A \geq 0$, so Lines 15–17 are not performed.

Termination of Algorithm 2 follows from the fact that the degrees of N and B, in v, decrease (strictly, in the case of N). Indeed, at each iteration, one of the Lines 9, 17 or 21 is performed. If Line 9 is performed, then the algorithm stops immediately. The key observation is that, at Line 21, the fraction R_2 is chosen such that

$$\frac{\partial R_2}{\partial v} = \frac{c_N v^{d_N} + \cdots}{\mathrm{cont}_F B^2},$$

where the dots hide terms of degree, in v, less than d_N. Thus, if Line 17 or 21 is performed, the algorithm subtracts from F a fraction which admits $\mathrm{cont}_F B^2$ as a denominator and the degree of N decreases strictly. If the numerator of the new fraction has a common factor with B, this can only decrease the degrees of both the numerator and the denominator. $\square$

PROPOSITION 6. *Algorithm 3 is correct.*

PROOF. Termination is guaranteed, essentially, by the fact that rankings are well-orderings. Here are a few more details. The algorithm considers eight cases. In Cases 1, 2, 3 and 7, F is assigned 0. In Cases 4 and 8, the new value of F has lower leading derivative than the old one. In Case 5, the new value of F has lower rank than the old one. In Case 6, the rank of F does not necessarily change (if it does, it decreases) but, at the next iteration, another case than 6 will be entered.

The loop invariant stated in the algorithm is clear, since it is satisfied at the beginning of the loop and maintained in each of the eight cases. Combined with the loop condition, it implies that, at the end of the algorithm, the first ensured condition is satisfied.

Let us now address the second ensured condition. The implication from left to right is clear, so we must show that W is zero if F_0 is a total derivative. Since derivations commute with sums, it suffices to prove that W remains zero in each pass through the main loop as long as F is a total derivative. Hence assume $F = \delta_x G$ for a differential fraction G. If G is a coefficient, so is F, which is handled by Case 1; the ensured condition is then guaranteed by the properties of Algorithm 2 (Proposition 5). Now assume G is not a coefficient and let $v \in \Theta \mathcal{U}$ be its leading derivative. By the axioms of rankings, the variable $v_x = \delta_x v$ is the leading derivative of F, it has positive order w.r.t. x, degree 1, and we have

$$F = \frac{\partial G}{\partial v} v_x + \cdots,$$

where the dots hide terms that do not depend on v_x. The initial of F is thus the separant of G, and it depends on variables less than or equal to v. Writing $F = P/Q$, it is clear that v_x occurs only in P and so must coincide with v_N. Hence Cases 2–5 are excluded, and we have $F_2 = \partial G/\partial v$. Since the latter cannot involve variables greater than v, also

---

Algorithm 3 The integrate algorithm

Require: F_0 is a differential fraction, x is an independent variable

Ensure: Two differential fractions R and W such that

1. $F_0 = \delta_x R + W$, where δ_x is the derivation w.r.t. x

2. W is zero iff there exists R such that $F_0 = \delta_x R$

3. Unless F_0 is a coefficient, $\delta_x R$ and W have ranks lower than or equal to F_0

```
 1: R, W := 0, 0
 2: F := F_0
 3: while F ≠ 0 do
 4:              { Invariant: F_0 = F + δ_x R + W }
 5:    if F is a coefficient then
 6:       R_2, W_2 := integrateWithRemainder(F, x)
 7:       R := R + R_2                            { Case 1 }
 8:       W := W + W_2
 9:       F := 0
10:    else if numer(F) is a coefficient then
11:       W := W + F                             { Case 2 }
12:       F := 0
13:    else
14:       denote v_N the leading derivative of numer(F)
15:       denote v_B the one of denom(F) (if not a coefficient)
16:       if denom(F) is not a coefficient and v_N ≤ v_B then
17:          W := W + F                          { Case 3 }
18:          F := 0
19:       else if v_N has order zero w.r.t. x then
20:          view numer(F) as a sum of monomials m_i
21:          denote H the sum of the m_i s.t. deg(m_i, v_N) > 0
22:          W := W + H/denom(F)                 { Case 4 }
23:          F := F − H/denom(F)
24:       else if deg(numer(F), v_N) ≥ 2 then
25:          view numer(F) as a sum of monomials m_i
26:          denote H the sum of the m_i s.t. deg(m_i, v_N) ≥ 2
27:          W := W + H/denom(F)                 { Case 5 }
28:          F := F − H/denom(F)
29:       else
30:              { we have: deg(numer(F), v_N) = 1 }
31:          let v be such that δ_x v = v_N
32:          F_2 := lcoeff(numer(F), v_N)/denom(F)
                  {recall F_2 is supposed to be reduced}
33:          if ∃ w > v such that deg(numer(F_2), w) > 0
              then
34:             view numer(F_2) as a sum of monomials m_i
35:             denote H the sum of the m_i such that
                     for some w > v, we have deg(m_i, w) > 0
36:             W := W + (H/denom(F_2)) · v_N    { Case 6 }
37:             F := F − (H/denom(F_2)) · v_N
38:          else if ∃ w > v s.t. deg(denom(F_2), w) > 0 then
39:             W := W + F                       { Case 7 }
40:             F := 0
41:          else
42:             R_2, W_2 := integrateWithRemainder(F_2, v)
43:             W := W + W_2 · v_N               { Case 8 }
44:             R := R + R_2
45:             F := F − δ_x R_2 − W_2 · v_N
46:          end if
47:       end if
48:    end if
49: end while
50: return W, R
```

Cases 6 and 7 are excluded. The remaining Case 8 is again handled by the properties of Algorithm 2 (Proposition 5).

Let us address the third ensured condition. Assume this condition is satisfied by R and W at the beginning of some loop. In Case 1, unless F_0 is a coefficient, R_2 and W_2 are assigned coefficients and have thus lower rank than F_0. The third condition is thus satisfied again. Cases 2–7 are clear. In Case 8, we show that the contribution $\delta_x R_2$ to the new value of F does not increase the rank of F. Recall that v_N, which is the leading derivative of numer(F), is also the leading derivative of F (by virtue of Case 3). Moreover, there exists a derivative v such that $\delta_x v = v_N$. An increase of the rank of F could only happen if F_2 involved derivatives w such that $\delta_x w > v_N$ and hence $w > v$ by the axioms of rankings. This situation is however impossible, due to Cases 6 and 7. Thus the third ensured condition is satisfied, and the Proposition is proved. □

Remark 4. The second ensured condition of Algorithm 3 could be made stronger. Indeed, the algorithm does not only ensure that W is zero whenever it is possible, it also makes W as small as possible, storing in this variable at each iteration, a "small part" of F that cannot be integrated (Cases 5 and 6).

In the next section, we use an "iterated" version of Algorithm 3, stated in Algorithm 4.

Algorithm 4 The integrate algorithm (iterated version)

Require: F_0 is a differential fraction which is not a coefficient, x is an independent variable

Ensure: A possibly empty list $[W_0, W_1, \ldots, W_t]$ of differential fractions such that

1. W_t is nonzero
2. $F_0 = W_0 + \delta_x W_1 + \cdots + \delta_x^t W_t$
3. $W_0, W_1, \ldots, W_i$ are zero if, and only if there exists R such that $F_0 = \delta_x^{i+1} R$
4. The differential fractions $W_0, \delta_x W_1, \ldots, \delta_x^t W_t$ have ranks lower than or equal to F_0

$L :=$ the empty list
$R := F_0$
while R is not a coefficient **do**
  $W, R :=$ integrate(R, x)
  append W at the end of L
end while
if $R \neq 0$ **then**
  append R at the end of L
end if
return L

PROPOSITION 7. *Algorithm 4 is correct.*

PROOF. The first ensured condition is clear from the code. The other ones follow the specifications of Algorithm 3. The number of iterations, t, is bounded by the total order of F_0. □

4. IMPLEMENTATION AND EXAMPLES

A first version of Algorithm 3 was implemented in Maple. More recently, this algorithm was implemented in C, within the BLAD libraries, version 3.10. See [3, `bap` library, file `bap_rat_bilge_mpz.c`]. The following computations were performed by the C version, through a testing version[2] of the Maple `DifferentialAlgebra` package [5].

Example 1. The first example shows that Algorithm 3 permits to decide if a differential fraction F is the derivative of some other differential fraction G. Observe that this test was already implemented in Maple, by the `DEtools[firint]` function.

The variable *Ring* receives a description of the differential ranking. The variable G receives a differential fraction, F receives its derivative w.r.t. x.

```
> with (DifferentialAlgebra):
> with (Tools):
> integrate := DifferentialAlgebra0:-Integrate:
> Ring := DifferentialRing
           (derivations = [x,y],
            blocks = [[u,v],w]);
             Ring := differential_ring

> G := u[x]^2 + w[y]/w^2 + w[x,x,y];

                      2   w[y]
           G := u[x]  + ---- + w[x, x, y]
                          2
                         w
> F := Differentiate (G, x, Ring);
                         4                     4
F := (2 u[x, x] u[x] w  + w[x, x, x, y] w

                 2                         / 4
    + w[x, y] w   - 2 w[x] w[y] w)  /  w
                                         /
```

Algorithm 3, implemented here using the name `integrate`, is applied to F and x. It returns the list $[W, R]$. We get $W = 0$, indicating that $F = \delta_x R$.

```
> L := integrate (F, x, Ring, iterated=false);
                    2  2                      2
                u[x]   w   + w[y] + w[x, x, y] w
      L := [0, -------------------------------]
                              2
                             w

> normal (L[2] - G);
                             0
```

If the optional parameter `iterated` is left to its default value (true), Algorithm 4 is called.

```
> L := integrate (F, x, Ring);
                     2  2
                 u[x]   w   + w[y]
         L := [0, ---------------, 0, w[y]]
                         2
                        w
```

Example 2. This variant of the previous example shows the differences between `DEtools[firint]` and Algorithm 3. Since `DEtools[firint]` does not handle PDE, we switch to ODE and to the standard Maple `diff` notation.

[2]This testing version, called `DifferentialAlgebra0`, is available at [4].

```
> Ring := DifferentialRing
          (derivations = [x],
           blocks = [[u,v],w]);
            Ring := differential_ring

> G := diff (u(x),x)^2 + v(x)/w(x)^2 + diff(v(x),x,x);
                                     / 2      \
              /d       \2   v(x)     |d       |
         G := |-- u(x)|  + ----- + |--- v(x)|
              \dx      /       2     |  2     |
                            w(x)     \dx      /

>  F := diff(G,x):

> DEtools[firint](F, u(x));
                              / 2      \
     /d       \2   v(x)       |d       |
     |-- u(x)|  + ----- + |--- v(x)| + _C1 = 0
     \dx      /       2     |  2     |
                   w(x)     \dx      /
```

Both methods recognize that F is a total derivative.

```
> integrate (F, x, Ring, notation=diff,
                           iterated=false);
        / 2      \
        |d       |      2   /d       \2     2
        |--- v(x)| w(x)  + |-- u(x)|  w(x)  + v(x)
        |  2     |          \dx      /
        \dx      /
    [0, -----------------------------------------]
                                 2
                              w(x)
```

However, if one adds a term $u(x)$ to F, one gets a differential fraction which is not a total derivative (which is not *exact*, in **DEtools[firint]** terminology). Our algorithm produces a decomposition while **DEtools[firint]** gives up (which is its expected behaviour).

```
> DEtools[firint](F+u(x), u(x));
Error, (in ODEtools/firint) the given ODE is not exact
> integrate (F+u(x), x, Ring, notation=diff,
                                iterated=false);
          / 2      \
          |d       |      2   /d       \2     2
          |--- v(x)| w(x)  + |-- u(x)|  w(x)  + v(x)
          |  2     |          \dx      /
          \dx      /
  [u(x), -----------------------------------------]
                                   2
                                w(x)
```

Example 3. The following example illustrates Algorithm 4 on differential polynomials. This very simple example shows that the result of Algorithm 3 is ranking dependent. Any derivative of u is greater than any derivative of w. Thus we get

$$u_x\, w_x = \delta_x(u\, w_x) - u\, w_{xx}\,.$$

```
> integrate (u[x]*w[x], x, Ring);

                [-u w[x, x], u w[x]]
```

However, the ranking between u and v is orderly, so that $v_{xx} > u_x$. For this reason, Algorithm 3 behaves differently.

```
> integrate (u[x]*v[x], x, Ring);

                      [u[x] v[x]]
```

Example 4. The following example shows that Algorithm 3 does not commute with sums. The issue is related to the existence of simple factors in the squarefree decompositions of denominators. For readability, results are displayed in factored form.

```
> F1 := u[x]/(u+1) - u[x]/(u+2)^2;
                       u[x]      u[x]
              F1 := ----- - --------
                      u + 1          2
                              (u + 2)

> F2 := -u[x]/(u+1) + u[x]/(u+3)^2;
                        u[x]      u[x]
             F2 := - ----- + --------
                       u + 1          2
                               (u + 3)

> F3 := F1+F2;

                      u[x]        u[x]
           F3 := - -------- + --------
                           2          2
                    (u + 2)    (u + 3)

> L1 := factor (
        integrate (F1, x, Ring, iterated=false));
                      3
             u[x] (u  + 2)            4 u + 3
     L1 := [-----------------, - ---------------]
                      2        2  (u + 2) (u + 1)
             (u + 2)  (u + 1)

> L2 := factor (
        integrate (F2, x, Ring, iterated=false));
                        3
                u[x] u             12 u + 13
   L2 := [- -----------------, -----------------]
                      2        2  2 (u + 3) (u + 1)
             (u + 3)  (u + 1)

> L3 := factor (
        integrate (F3, x, Ring, iterated=false));
                                   1
              L3 := [0, ---------------]
                         (u + 3) (u + 2)

> L12 := factor (L1+L2);
                 4      3      2
          u[x] (2 u  + 5 u  + 2 u  + 12 u + 18)
L12 := [-------------------------------------,
                    2          2          2
             (u + 3)  (u + 1)  (u + 2)

            2
         4 u  + 7 u + 8
    -------------------------]
     2 (u + 3) (u + 1) (u + 2)
```

Though $L_{12} \neq L_3$, it is possible to recover L_3 from L_{12} by applying Algorithm 3 once more.

```
> L := factor (
    integrate (L12[1], x, Ring, iterated=false));
                             2
                          4 u  + 5 u + 6
          L := [0, - -------------------------]
                      2 (u + 3) (u + 1) (u + 2)

> factor (L3 - (L + [0, L12[2]]));

                          [0, 0]
```

Example 5. This example, inspired from [9], illustrates the usefulness of Algorithm 4 for simplifying equations produced by differential elimination methods. It features a compartmental model with two compartments, 1 and 2, with a same unit volume. There are two state variables x_1 and x_2 (one per compartment). State variable x_i represents the concentration of some drug in compartment i. The exchanges between the two compartments are supposed to follow linear laws (depending on parameters k_1 and k_2). The drug is supposed to exit from the model, from compartment 1, following a Michaelis-Menten type law (depending on parameters V_e, k_e). The output of the system, denoted y, is the state variable x_1.

```
> Ring := DifferentialRing (
     derivations = [t],
     blocks = [y,[x1,x2],[k1(),k2(),ke(),Ve()]]);

               Ring := differential_ring

> S := [ x1[t] = -k1*x1 + k2*x2 - (Ve*x1)/(ke+x1),
         x2[t] = k1*x1 - k2*x2,
            y = x1];

                                       Ve x1
S := [x1[t] = -k1 x1 + k2 x2 - -------,
                                      ke + x1

    x2[t] = k1 x1 - k2 x2, y = x1]
```

By means of differential elimination [6], one computes the so-called input-output equation of the system (though there is no input).

```
> ideal := RosenfeldGroebner
    (S, Ring,
     basefield =
        field (generators = [k1,k2,ke,Ve]));

      ideal := [regular_differential_chain]

> io_ideal := RosenfeldGroebner
    (ideal[1],
     blocks = [[x1,x2],y,[k1(),k2(),ke(),Ve()]]):
> io_eq := Equations
    (io_ideal, leader = derivative(y))[1];

                     2                            2
io_eq := y[t, t] y  + 2 y[t, t] y ke + y[t, t] ke

             2          2
    + y[t] y  k1 + y[t] y  k2 + 2 y[t] y k1 ke

                                        2          2
    + 2 y[t] y k2 ke + y[t] k1 ke  + y[t] k2 ke

                        2
    + y[t] ke Ve + y  k2 Ve + y k2 ke Ve
```

Using Algorithm 3, we obtain

```
> integrate (io_eq / Initial (io_eq, Ring), t, Ring);

          2       2           2         2
 y k2 Ve  y  k1 + y  k2 - k1 ke  - k2 ke  - ke Ve
[-------, --------------------------------------, y]
 y + ke                  y + ke
```

This list can be translated into the following equation, whose structure is now much clearer:

$$\frac{y(t)\,k_2\,V_e}{y(t)+k_e} + \frac{\mathrm{d}}{\mathrm{d}t}\,\frac{(y(t)^2 - k_e^2)\,(k_1+k_2) - k_e\,V_e}{y(t)+k_e} + \frac{\mathrm{d}^2}{\mathrm{d}t^2}\,y(t)\,.$$

This example also shows that integrating differential fractions (as we did above), may yield better formulas than integrating differential polynomials. Indeed:

```
> integrate (io_eq, t, Ring);

          2             2         2
[-2 y[t]  y - 2 y[t]  ke + y  k2 Ve + y k2 ke Ve,

   3       3
  y  k1   y  k2    2             2               2
  ----- + ----- + y  k1 ke + y  k2 ke + y k1 ke
    3       3

             2                  3    2         2
   + y k2 ke  + y ke Ve, 1/3 y  + y  ke + y ke ]
```

While the fractional equation is simpler than that from an intuitive point of view, it would be interesting to investigate also their numerical properties from a more systematic viewpoint. In the case of differential-algebraic equations, the usual elimination methods are known to produce typically very lengthy polynomial differential equations whose numerical treatment may be more costly than that of the corresponding fractional differential equation.

Remark 5. The four parameters were placed at the bottom of the ranking, following the assumptions stated in Section 2. However, the **DifferentialAlgebra** package, which handles parameters as differential indeterminates constrained by implicit equations (their derivatives are zero), does not require parameters to lie at the bottom of rankings. With such rankings, the second ensured condition of Algorithm 3 would not hold anymore.

5. CONCLUSION

The algorithm presented in this paper may be extended in many different ways: to handle more complicated differential operators such as $\delta_x + \delta_y$, possibly with differential polynomials for coefficients, for instance. Some prototypes are currently under study.

In this paper, we have not addressed the interest of the third ensured condition of Algorithm 3. This topic will be covered in a further paper. It is related to the issue of characterizing the differential fractions which are returned "as is" by our method.

As mentioned in the Introduction, the decomposition of a differential polynomial or a differential fraction into a total derivative and a remainder is crucial for constructing the ring of integro-differential polynomials. Formalizing this idea in the general category of commutative integro-differential algebras over rings leads to the notion of quasi-antiderivative [10]. For ordinary differential polynomials in one indeterminate and for univariate rational functions, a quasi-antiderivative is exhibited in [10] but the case of partial differential polynomials or differential fractions remains open. While Algorithm 3 comes close to providing such a quasi-antiderivative, it falls short of being linear (Example 4). This question will be addressed in a future paper. It would allow us to make progress towards our original goal: defining and computing Gröbner bases for ideals of integro-differential polynomials of various kinds.

6. ACKNOWLEDGMENTS

The authors thank the referees for their detailed comments, suggestions, and references.

7. REFERENCES

[1] A. H. Bilge. A REDUCE program for the integration of differential polynomials. *Computer Physics Communications*, 71:263–268, 1992.

[2] G. W. Bluman and S. C. Anco. *Symmetry and Integration Methods for Differential Equations.* Springer Verlag, New York, 2002.

[3] F. Boulier. The BLAD libraries. `http://www.lifl.fr/~boulier/BLAD`, 2004.

[4] F. Boulier. The BMI library. `http://www.lifl.fr/~boulier/BMI`, 2010.

[5] F. Boulier and E. Cheb-Terrab. Help Pages of the DifferentialAlgebra Package. In MAPLE 14, 2010.

[6] F. Boulier, F. Lemaire, and M. Moreno Maza. Computing differential characteristic sets by change of ordering. *Journal of Symbolic Computation*, 45(1):124–149, 2010.

[7] M. Bronstein. *Symbolic Integration I.* Springer Verlag, Berlin, Heidelberg, New York, 1997.

[8] G. H. Campbell. Symbolic integration of expressions involving unspecified functions. *SIGSAM Bulletin*, 22(1):25–27, 1988.

[9] L. Denis-Vidal, G. Joly-Blanchard, and C. Noiret. System identifiability (symbolic computation) and parameter estimation (numerical computation). In *Numerical Algorithms*, volume 34, pages 282–292, 2003.

[10] L. Guo, G. Regensburger, and M. Rosenkranz. On integro-differential algebras. Preprint available at `http://arxiv.org/abs/1212.0266.`, 2012.

[11] E. R. Kolchin. *Differential Algebra and Algebraic Groups.* Academic Press, New York, 1973.

[12] C. G. Raab. *Definite Integration in Differential Fields.* PhD thesis, University of Linz, 2012.

[13] J. F. Ritt. *Differential Algebra*, volume 33 of *American Mathematical Society Colloquium Publications.* American Mathematical Society, New York, 1950.

[14] M. Rosenkranz and G. Regensburger. Integro-differential polynomials and operators. In *ISSAC'08: Proceedings of the twenty-first international symposium on Symbolic and algebraic computation*, pages 261–268, New York, NY, USA, 2008. ACM.

[15] F. Schwarz. An algorithm for determining polynomial first integrals of autonomous systems of ordinary differential equations. *Journal of Symbolic Computation*, 1(2):229–233, 1985.

[16] W. Y. Sit. On Goldman's algorithm for solving first-order multinomial autonomous systems. In *Proceedings of the 6th International Conference, on Applied Algebra, Algebraic Algorithms and Error-Correcting Codes*, AAECC-6, pages 386–395, London, UK, 1989. Springer-Verlag.

Rational Univariate Representations of Bivariate Systems and Applications

Yacine Bouzidi
INRIA Nancy Grand Est
LORIA, Nancy, France
Yacine.Bouzidi@inria.fr

Sylvain Lazard
INRIA Nancy Grand Est
LORIA, Nancy, France
Sylvain.Lazard@inria.fr

Marc Pouget
INRIA Nancy Grand Est
LORIA, Nancy, France
Marc.Pouget@inria.fr

Fabrice Rouillier
INRIA Paris-Rocquencourt
IMJ, Paris, France
Fabrice.Rouillier@inria.fr

ABSTRACT

We address the problem of solving systems of two bivariate polynomials of total degree at most d with integer coefficients of maximum bitsize τ. We suppose known a linear separating form (that is a linear combination of the variables that takes different values at distinct solutions of the system) and focus on the computation of a Rational Univariate Representation (RUR).

We present an algorithm for computing a RUR with worst-case bit complexity in $\widetilde{O}_B(d^7 + d^6\tau)$ and bound the bitsize of its coefficients by $\widetilde{O}(d^2 + d\tau)$ (where O_B refers to bit complexities and $\widetilde{O}$ to complexities where polylogarithmic factors are omitted). We show in addition that isolating boxes of the solutions of the system can be computed from the RUR with $\widetilde{O}_B(d^8+d^7\tau)$ bit operations. Finally, we show how a RUR can be used to evaluate the sign of a bivariate polynomial (of degree at most d and bitsize at most τ) at one real solution of the system in $\widetilde{O}_B(d^8 + d^7\tau)$ bit operations and at all the $\Theta(d^2)$ solutions in only $O(d)$ times that for one solution.

Categories and Subject Descriptors

F.2 [**Analysis of Algorithms and Problem Complexity**]: Nonnumerical Algorithms and Problems

Keywords

Bivariate system; Rational univariate representation

1. INTRODUCTION

There exists many algorithms, in the literature, for "solving" algebraic systems of equations. Some focus on computing "*formal solutions*" such as rational parameterizations, Gröbner bases, and triangular sets, others focus on isolating the solutions. By isolating the solution, we mean computing isolating axis-parallel boxes sets such that every real solution lies in a unique box and conversely. In this paper, we focus on the worst-case bit complexity of these methods (in the RAM model) for systems of **two bivariate polynomials of total degree d with integer coefficients of bitsize τ.**

For isolating the real solutions of systems of two bivariate polynomials, the algorithm with best known bit complexity was recently analyzed by Emeliyanenko and Sagraloff [9]. They solve the problem in $\widetilde{O}_B(d^8+d^7\tau)$ bit operations (where $\widetilde{O}$ refers to complexities where polylogarithmic factors are omitted and O_B refers to bit complexities). Furthermore, the isolating boxes can easily be refined because the algorithm computes the univariate polynomials that correspond to the projections of the solutions on each axis (that is, the resultants of the two input polynomials with respect to each of the variables).

Other widespread approaches that solve systems are those that compute rational parameterizations of the (complex) solutions. Recall that such a rational parameterization is a set of univariate polynomials and associated rational one-to-one mappings that send the roots of the univariate polynomials to the solutions of the system. The algorithm with the best known complexity for solving such systems via rational parameterizations was, in essence, first introduced by Gonzalez-Vega and El Kahoui [11]. The algorithm first applies a generic linear change of variables to the input polynomials, computes a rational parameterization using the subresultant sequence of the sheared polynomials and finally computes the isolating boxes of the solutions. Its initial bit complexity of $\widetilde{O}_B(d^{16} + d^{14}\tau^2)$ was improved by Diochnos et al. [8, Theorem 19] to (i) $\widetilde{O}_B(d^{10} + d^9\tau)$ for computing a generic shear (i.e., a separating linear form), to (ii) $\widetilde{O}_B(d^7+d^6\tau)$ for computing a rational parameterization and to (iii) $\widetilde{O}_B(d^{10} + d^9\tau)$ for the isolation phase with a modification of the initial algorithm.[1]

Main results. We addressed in [4] the first phase of the above algorithm and proved that a separating linear form

[1]The complexity of the isolation phase in [8, Theorem 19] is stated as $\widetilde{O}_B(d^{12}+d^{10}\tau^2)$ but it trivially decreases to $\widetilde{O}_B(d^{10}+d^9\tau)$ with the recent result of Sagraloff [17] which improves the complexity of isolating the real roots of a univariate polynomial.

ISSAC'13, June 26–29, 2013, Boston, Massachusetts, USA.

can be computed in $\widetilde{O}_B(d^8+d^7\tau+d^5\tau^2)$ bit operations.[2] We address in this paper the second and third phase of the above algorithm, that is the computation of a rational parameterization and the isolation of the solutions of the system. We also consider an important related problem, namely, the evaluation of the sign of a polynomial at a real solution of a system, referred to as the *sign_at* operation.

We first show that the Rational Univariate Representation (RUR for short) of Rouillier [15] (i) can be expressed with simple polynomial formulas, that (ii) it has a total bitsize which is asymptotically smaller than that of Gonzalez-Vega and El Kahoui by a factor d, and that (iii) it can be computed with the same complexity, that is $\widetilde{O}_B(d^7+d^6\tau)$ (Theorem 13). Namely, we prove that the RUR consists of four polynomials of degree $O(d^2)$ and bitsize $\widetilde{O}(d^2+d\tau)$ (instead of $O(d)$ polynomials with the same asymptotic degree and bitsize for Gonzalez-Vega and El Kahoui parameterization).

For the next two applications, we focus for simplicity on a parameterization given by the RUR as defined in [15], but the complexity results also hold for the one defined in [11].

We show that, given a RUR, isolating boxes of the solutions of the system can be computed with $\widetilde{O}_B(d^8+d^7\tau)$ bit operations (Proposition 16). This decreases by a factor d^2 the best known complexity for the isolation phase of the algorithm (see the discussion above). Globally, this bounds the overall bit complexity of all three phases of the algorithm by $\widetilde{O}_B(d^8+d^7\tau)$, if $\tau \in \widetilde{O}(d^2)$.

Finally, we show how a RUR can be used to perform efficiently the *sign_at* operation. Given a polynomial F of total degree at most d with integer coefficients of bitsize at most τ, we show that the sign of F at one real solution of the system can be computed in $\widetilde{O}_B(d^8+d^7\tau)$ bit operations, while the complexity of computing its sign at all the $\Theta(d^2)$ solutions of the system is only $O(d)$ times that for one real solution (Theorem 19). This improves the best known complexities of $\widetilde{O}_B(d^{10}+d^9\tau)$ and $\widetilde{O}_B(d^{12}+d^{11}\tau)$ for these respective problems (see [8, Th. 14 & Cor. 24] with the improvement of [17] for the root isolation).

2. NOTATION AND PRELIMINARIES

The bitsize of an integer p is the number of bits needed to represent it, that is $\lfloor \log p \rfloor + 1$ (log refer to the logarithm in base 2). For rational numbers, we refer to the bitsize as to the maximum bitsize of its numerator and denominator. The bitsize of a polynomial with integer or rational coefficients is the *maximum* bitsize of its coefficients.

We denote by $\mathbb{D}$ a unique factorization domain, typically $\mathbb{Z}[X,Y]$, $\mathbb{Z}[X]$ or $\mathbb{Z}$. We also denote by $\mathbb{F}$ a field, typically $\mathbb{Q}$, $\mathbb{C}$. For any polynomial $P \in \mathbb{D}[X]$, let $Lc_X(P)$ denote its leading coefficient with respect to the variable X (or simply $Lc(P)$ in the univariate case), $d_X(P)$ its degree with respect to X, and $\overline{P}$ its squarefree part. The ideal generated by two polynomials P and Q is denoted $\langle P, Q\rangle$, and the affine variety of an ideal I is denoted by $V(I)$. The solutions are always considered in the algebraic closure of the fraction field of $\mathbb{D}$, unless specified otherwise. For a point $\sigma \in V(I)$, $\mu_I(\sigma)$ denotes the multiplicity of σ in I. For simplicity, we refer indifferently to the ideal $\langle P, Q\rangle$ and to the corresponding system of polynomials.

[2]This improves the previous complexity $\widetilde{O}_B(d^{10}+d^9\tau)$ by a factor $\min(d^2, \frac{d^4}{\tau})$ if $\tau \in \widetilde{O}(d^4)$.

We finally introduce the following notation which are extensively used throughout the paper. Given the two input polynomials P and Q, we consider the "generic" change of variables $X = T - SY$, and we define the "sheared" polynomials $P(T-SY,Y)$, $Q(T-SY,Y)$, and their resultant with respect to Y,

$$R(T,S) = Res_Y(P(T-SY,Y), Q(T-SY,Y)). \quad (1)$$

We introduce

$$L_P(S) = Lc_Y(P(T-SY,Y))$$
$$L_Q(S) = Lc_Y(Q(T-SY,Y)), \quad L_R(S) = Lc_T(R(T,S)) \quad (2)$$

and remark that these polynomials do not depend on T.

Complexity. In the sequel, we often consider the gcd of two univariate polynomials P and Q and the gcd-free part of P with respect to Q, that is, the divisor D of P such that $P = \gcd(P,Q)D$. Note that when $Q = P'$, D is the squarefree part $\overline{P}$ of P.

LEMMA 1 ([2, REMARK 10.19]). *Two polynomials P, Q in $\mathbb{Z}[X]$ with maximum degree d and bitsize at most τ have a gcd (in $\mathbb{Q}[X]$ or in $\mathbb{Z}[X]$) with coefficients of bitsize in $\widetilde{O}(d+\tau)$ which can be computed with $\widetilde{O}_B(d^2\tau)$ bit operations. The same bounds hold for the bitsize and the computation of the gcd-free part of P with respect to Q. If P and Q are in $\mathbb{Q}[X]$ and there exists $c \in \mathbb{Z}$ of bitsize in $O(\tau)$ such that cP and cQ are in $\mathbb{Z}[X]$ with coefficients of bitsize in $O(\tau)$, then the same results also hold.*

We now state a bound on the complexity of evaluating a univariate polynomial which is straightforward and ought to be known, even though we were not able to find a proper reference for it (see [3] for details).

LEMMA 2. *Let a be a rational of bitsize τ_a, the evaluation at a of a univariate polynomial f of degree d and rational coefficients of bitsize τ can be done in $\widetilde{O}_B(d(\tau+\tau_a))$ bit operations, while the value $f(a)$ has bitsize in $O(\tau+d\tau_a)$.*

As we often use the "sheared" polynomials $P(T-SY,Y)$ and $Q(T-SY,Y)$ we also recall some related complexities.

LEMMA 3 ([4, LEMMA 5]). *Let $P, Q \in \mathbb{Z}[X,Y]$ of total degree d and maximum bitsize τ. The sheared polynomials $P(T-SY,Y)$ and $Q(T-SY,Y)$ can be expanded in $\widetilde{O}_B(d^4+d^3\tau)$ and their bitsizes are in $\widetilde{O}(d+\tau)$. The resultant $R(T,S)$ can be computed in $\widetilde{O}_B(d^7+d^6\tau)$ bit operations; its degree is at most $2d^2$ in each variable and its bitsize is in $\widetilde{O}(d^2+d\tau)$.*

3. RATIONAL UNIV. REPRESENTATIONS

The idea of this section is to express the polynomials of a RUR of two polynomials in terms of a resultant defined from these polynomials. Given a separating form, this yields a new algorithm to compute a RUR (Section 3.1) and it also enables us to derive the bitsize of the polynomials of a RUR (Section 3.2). Throughout this section we assume that the two input polynomials P and Q are coprime in $\mathbb{Z}[X,Y]$, that their maximum total degree d is at least 2 and that their coefficients have maximum bitsize τ. We first recall the definition and main properties of Rational Univariate Representations. In the following, for any polynomial $v \in \mathbb{Q}[X,Y]$ and $\sigma = (\alpha,\beta) \in \mathbb{C}^2$, we denote by $v(\sigma)$ the image of σ by the polynomial function v (e.g. $X(\alpha,\beta) = \alpha$).

Definition 4 ([15]). *Let $I \subset \mathbb{Q}[X,Y]$ be a zero-dimensional ideal, $V(I) = \{\sigma \in \mathbb{C}^2, v(\sigma) = 0, \forall v \in I\}$ its associated variety, and a linear form $T = X + aY$ with $a \in \mathbb{Q}$. The RUR-candidate of I associated to $X + aY$ (or simply, to a), denoted $RUR_{I,a}$, is the following set of four univariate polynomials in $\mathbb{Q}[T]$*

$$f_{I,a}(T) = \prod_{\sigma \in V(I)} (T - X(\sigma) - aY(\sigma))^{\mu_I(\sigma)} \qquad (3)$$

$$f_{I,a,v}(T) = \sum_{\sigma \in V(I)} \mu_I(\sigma) v(\sigma) \prod_{\varsigma \in V(I), \varsigma \neq \sigma} (T - X(\varsigma) - aY(\varsigma))$$

for $v \in \{1, X, Y\}$.

If $(X,Y) \mapsto X + aY$ is injective on $V(I)$, we say that the linear form $X + aY$ separates $V(I)$ (or is separating for I), $RUR_{I,a}$ is called a RUR (the RUR of I associated to a) and it defines a bijection between $V(I)$ and $V(f_{I,a}) = \{\gamma \in \mathbb{C}, f_{I,a}(\gamma) = 0\}$:

$$\begin{array}{ccc} V(I) & \to & V(f_{I,a}) \\ (\alpha, \beta) & \mapsto & \alpha + a\beta \\ \left(\frac{f_{I,a,X}}{f_{I,a,1}}(\gamma), \frac{f_{I,a,Y}}{f_{I,a,1}}(\gamma)\right) & \leftarrow & \gamma \end{array}$$

Moreover, this bijection preserves the real roots and the multiplicities.

3.1 RUR computation

We show here that the polynomials of a RUR can be expressed as combinations of specializations of the resultant R and its partial derivatives. The seminal idea has already been used by several authors in various contexts (see e.g. [6, 1, 18]) for computing rational parameterizations of the radical of a given zero-dimensional ideal and mainly for bounding the size of a Chow form. Based on the same idea but keeping track of multiplicities, we present a simple new formulation for the polynomials of a RUR, given a separating form.

Proposition 5. *For any $a \in \mathbb{Q}$ such that $L_P(a)L_Q(a) \neq 0$ and such that $X + aY$ is a separating form of $\langle P, Q \rangle$, the RUR of $\langle P, Q \rangle$ associated to a is as follows:*

$$f_{I,a}(T) = \frac{R(T,a)}{L_R(a)}$$

$$f_{I,a,1}(T) = \frac{f'_{I,a}(T)}{\gcd(f_{I,a}(T), f'_{I,a}(T))}$$

$$f_{I,a,Y}(T) = \frac{\frac{\partial R}{\partial S}(T,a) - f_{I,a}(T)\frac{\partial L_R}{\partial S}(a)}{L_R(a)\gcd(f_{I,a}(T), f'_{I,a}(T))}$$

$$f_{I,a,X}(T) = T f_{I,a,1}(T) - d_T(f_{I,a}) f_{I,a}(T) - a f_{I,a,Y}(T).$$

We postpone the proof of Proposition 5 and first analyze the complexity of the computation of the expressions therein. Note that a separating form $X + aY$ as in Proposition 5, with $0 \leqslant a < 2d^4$, can be computed in $\widetilde{O}_B(d^8 + d^7\tau + d^5\tau^2)$ [4] or $\widetilde{O}_B(d^{10} + d^9\tau)$ [8] bit operations.

Proposition 6. *Computing the polynomials in Proposition 5 can be done with $\widetilde{O}_B(d^7 + d^6(\tau + \tau_a))$ bit operations, where τ_a is the bitsize of a.*

Proof. According to Lemma 3, the resultant $R(T,S)$ of $P(T - SY, Y)$ and $Q(T - SY, Y)$ with respect to Y has degree $O(d^2)$ in T and S, has bitsize in $\widetilde{O}(d(d + \tau))$, and can be computed in $\widetilde{O}_B(d^6(d + \tau))$ bit operations. Specializing $R(T,S)$ at $S = a$ can be done by evaluating $O(d^2)$ polynomials in S, each of degree in $O(d^2)$ and bitsize in $\widetilde{O}(d^2 + d\tau)$. By Lemma 2, each of the $O(d^2)$ evaluations can be done in $\widetilde{O}_B(d^2(d^2 + d\tau + \tau_a))$ bit operations and each result has bitsize in $\widetilde{O}(d^2 + d\tau + d^2\tau_a)$. Hence, $R(T,a)$ and $f_{I,a}(T)$ have degree in $O(d^2)$, bitsize in $\widetilde{O}(d^2 + d\tau + d^2\tau_a)$, and they can be computed with $\widetilde{O}_B(d^4(d^2 + d\tau + \tau_a))$ bit operations.

The complexity of computing the numerators of $f_{I,a,1}(T)$ and $f_{I,a,Y}(T)$ is clearly dominated by the computation of $\frac{\partial R}{\partial S}(T,a)$. Indeed, computing the derivative $\frac{\partial R}{\partial S}(T,S)$ can trivially be done in $O(d^4)$ arithmetic operations of complexity $\widetilde{O}_B(d^2 + d\tau)$, that is in $\widetilde{O}_B(d^6 + d^5\tau)$. Then, as for $R(T,a)$, $\frac{\partial R}{\partial S}(T,a)$ has degree in $O(d^2)$, bitsize in $\widetilde{O}(d^2 + d\tau + d^2\tau_a)$, and it can be computed within the same complexity as the computation of $R(T,a)$.

On the other hand, since $f_{I,a}(T)$ and $f'_{I,a}(T)$ have degree in $O(d^2)$ and bitsize in $\widetilde{O}(d^2 + d\tau + d^2\tau_a)$, and $f_{I,a}(T) = \frac{R(T,a)}{L_R(a)}$, one can multiply these two polynomials by the product of $L_R(a)$ and the denominator of the rational a to the power of $d_S(R(T,S))$ which is an integer of bitsize in $O(d^2\tau_a)$, to obtain polynomials with coefficients in $\mathbb{Z}$. Hence, according to Lemma 1, their gcd can be computed with $\widetilde{O}_B(d^4(d^2 + d\tau + d^2\tau_a))$ bit operations, and has bitsize in the same class of complexity.

$f_{I,a,1}(T)$ and $f_{I,a,Y}(T)$ are then obtained by dividing the numerators by the above gcd which can be done with $\widetilde{O}_B(d^4 (d^2 + d\tau + d^2\tau_a))$ bit operations, according to [20, Theorem 9.6 and subsequent discussion]. Finally, computing $f_{I,a,X}(T)$ can be done within the same complexity as for $f_{I,a,1}(T)$ and $f_{I,a,Y}(T)$ since it is dominated by the computation of the squarefree part of $f_{I,a}(T)$.

The overall complexity is thus that of computing the resultant which is in $\widetilde{O}_B(d^6(d + \tau))$ plus that of computing the above gcd and Euclidean division which is in $\widetilde{O}_B(d^4(d^2 + d\tau + d^2\tau_a))$. This gives a total of $\widetilde{O}_B(d^7 + d^6(\tau + \tau_a))$. □

Proof of Proposition 5. Proposition 5 expresses the polynomials $f_{I,a}$ and $f_{I,a,v}$ of a RUR in terms of specializations (by $S = a$) of the resultant $R(T,S)$ and its partial derivatives. Since the specializations are done after considering the derivatives of R, we study the relations between these entities before specializing S by a.

For that purpose, we introduce the following polynomials which are exactly the polynomials $f_{I,a}$ and $f_{I,a,v}$ of (3) where the parameter a is replaced by the variable S. These polynomials can be seen as the RUR polynomials of the ideal I with respect to a "generic" linear form $X + SY$.

$$f_I(T,S) = \prod_{\sigma \in V(I)} (T - X(\sigma) - SY(\sigma))^{\mu_I(\sigma)} \qquad (4)$$

$$f_{I,v}(T,S) = \sum_{\sigma \in V(I)} \mu_I(\sigma) v(\sigma) \prod_{\varsigma \in V(I), \varsigma \neq \sigma} (T - X(\varsigma) - SY(\varsigma))$$

for $v \in \{1, X, Y\}$.

These polynomials are obviously in $\mathbb{C}[T,S]$, but they are actually in $\mathbb{Q}[T,S]$ because, when S is specialized at any rational value a, the specialized polynomials are those of $RUR_{I,a}$ which are known to be in $\mathbb{Q}[T]$ (see e.g. [15]). We express the derivatives of $f_I(T,S)$ in terms of $f_{I,v}(T,S)$, in Lemma 7, and show that $f_I(T,S)$ is the monic form of the resultant $R(T,S)$, seen as a polynomial in T, in Lemma 9. Let

$$g_I(T,S) = \prod_{\sigma\in V(I)} (T - X(\sigma) - SY(\sigma))^{\mu_I(\sigma)-1}.$$

LEMMA 7. *We have*

$$\frac{\partial f_I}{\partial T}(T,S) = g_I(T,S) f_{I,1}(T,S), \quad (5)$$

$$\frac{\partial f_I}{\partial S}(T,S) = g_I(T,S) f_{I,Y}(T,S). \quad (6)$$

PROOF. It is straightforward that the derivative of f_I with respect to T is $\sum_{\sigma\in V(I)} \mu_I(\sigma)(T-X(\sigma)-SY(\sigma))^{\mu_I(\sigma)-1}$ $\prod_{\varsigma\in V(I),\varsigma\neq\sigma}(T - X(\varsigma) - SY(\sigma))^{\mu_I(\varsigma)}$, which can be rewritten as the product of $\prod_{\sigma\in V(I)}(T - X(\sigma) - SY(\sigma))^{\mu_I(\sigma)-1}$ and $\sum_{\sigma\in V(I)} \mu_I(\sigma) \prod_{\varsigma\in V(I),\varsigma\neq\sigma}(T - X(\varsigma) - SY(\varsigma))$ which is exactly the product of $g_I(T,S)$ and $f_{I,1}(T,S)$.

The expression of the derivative of f_I with respect to S is similar to that with respect to T except that the derivative of $T - X(\sigma) - SY(\sigma)$ is now $Y(\sigma)$ instead of 1. It follows that $\frac{\partial f_I}{\partial S}$ is the product of $\prod_{\sigma\in V(I)}(T-X(\sigma)-SY(\sigma))^{\mu_I(\sigma)-1}$ and $\sum_{\sigma\in V(I)} \mu_I(\sigma)Y(\sigma) \prod_{\varsigma\in V(I),\varsigma\neq\sigma}(T - X(\varsigma) - SY(\varsigma))$ which is the product of $g_I(T,S)$ and $f_{I,Y}(T,S)$. □

For the proof of Lemma 9, we will use the following lemma which states that when two polynomials have no common solution at infinity in some direction, the roots of their resultant with respect to this direction are the projections of the solutions of the system with cumulated multiplicities.

LEMMA 8 ([5, PROP. 2 AND 5]). *Let $P, Q \in \mathbb{F}[X,Y]$ defining a zero-dimensional ideal $I = \langle P, Q\rangle$, such that their leading terms $Lc_Y(P)$ and $Lc_Y(Q)$ do not have common roots. Then $Res_Y(P,Q) = c\prod_{\sigma\in V(I)}(X - X(\sigma))^{\mu_I(\sigma)}$ where c is nonzero in $\mathbb{F}$.*

LEMMA 9. *$R(T,S) = L_R(S) f_I(T,S)$ and, for any $a \in \mathbb{Q}$, $L_P(a)L_Q(a) \neq 0$ implies that $L_R(a) \neq 0$.*

PROOF. The proof is organized as follows. We first prove that for any rational a such that $L_P(a)L_Q(a)$ does not vanish, $R(T,a) = c(a) f_I(T,a)$ where $c(a) \in \mathbb{Q}$ is a nonzero constant depending on a. This is true for infinitely many values of a and, since $R(T,S)$ and $f_I(T,S)$ are polynomials, we can deduce that $R(T,S) = L_R(S)f_I(T,S)$. This will also implies the second statement of the lemma since, if $L_P(a)L_Q(a) \neq 0$, then $R(T,a) = c(a)f_I(T,a) = L_R(a)f_I(T,a)$ with $c(a) \neq 0$, thus $L_R(a) \neq 0$ (since $f_I(T,a)$ is monic).

Since a is such that $L_P(a)L_Q(a) \neq 0$, the resultant $R(T,S)$ can be specialized at $S = a$: $R(T,a)$ is equal to the resultant of $P(T - aY, Y)$ and $Q(T - aY, Y)$ with respect to Y[2, Proposition 4.20]. The polynomials $P(T - aY, Y)$ and $Q(T - aY, Y)$ satisfy the hypotheses of Lemma 8: first, their leading coefficients (in Y) do not depend on T, hence they have no common root in $\mathbb{Q}[T]$; second, the polynomials $P(T - aY, Y)$ and $Q(T - aY, Y)$ are coprime because $P(X,Y)$ and $Q(X,Y)$ are coprime by assumption and the change of variables $(X,Y) \mapsto (T = X + aY, Y)$ is a one-to-one mapping. Hence Lemma 8 yields that $R(T,a) = c(a)\prod_{\sigma\in V(I_a)}(T - T(\sigma))^{\mu_{I_a}(\sigma)}$, where $c(a) \in \mathbb{Q}$ is a nonzero constant depending on a, and I_a is the ideal generated by $P(T - aY, Y)$ and $Q(T - aY, Y)$.

We now observe that $\prod_{\sigma\in V(I_a)}(T - T(\sigma))^{\mu_{I_a}(\sigma)}$ is equal to $f_I(T,a) = \prod_{\sigma\in V(I)}(T - X(\sigma) - aY(\sigma))^{\mu_I(\sigma)}$ since any solution (α,β) of $P(X,Y)$ is in one-to-one correspondence with the solution $(\alpha + a\beta, \beta)$ of $P(T - aY, Y)$ (and similarly for Q) and the multiplicities of the solutions also match, i.e. $\mu_I(\sigma) = \mu_{I_a}(\sigma_a)$ when σ and σ_a are in correspondence through the mapping [10, §3.3 Prop. 3 and Thm. 3]. Hence,

$$L_P(a)L_Q(a) \neq 0 \Rightarrow R(T,a) = c(a)f_I(T,a) \text{ with } c(a) \neq 0. \quad (7)$$

Since there is finitely many values of rational a such that $L_P(a)L_Q(a)L_R(a) = 0$ and since $f_I(T,S)$ is monic with respect to T, (7) implies that $R(T,S)$ and $f_I(T,S)$ have the same degree in T, say D. We write these polynomials as

$$R(T,S) = L_R(S)T^D + \sum_{i=0}^{D-1} r_i(S)T^i,$$
$$f_I(T,S) = T^D + \sum_{i=0}^{D-1} f_i(S)T^i. \quad (8)$$

If a is such that $L_P(a)L_Q(a)L_R(a) \neq 0$, (7) and (8) imply that $L_R(a) = c(a)$ and $r_i(a) = L_R(a)f_i(a)$, for all i. These equalities hold for infinitely many values of a, and $r_i(S)$, $L_R(S)$ and $f_i(S)$ are polynomials in S, thus $r_i(S) = L_R(S)f_i(S)$ and, by (8), $R(T,S) = L_R(S)f_I(T,S)$. □

PROOF OF PROPOSITION 5. Lemma 9 immediately gives the first formula. Equation 5 states that $f_{I,1}(T,S)g_I(T,S) = \frac{\partial f_I(T,S)}{\partial T}$, with $g_I(T,S) = \prod_{\sigma\in V(I)}(T-X(\sigma)-SY(\sigma))^{\mu_I(\sigma)-1}$. In addition, g_I being monic in T, it never identically vanishes when S is specialized, thus the preceding formula yields after specialization: $f_{I,a,1}(T) = \frac{f'_{I,a}(T)}{g_I(T,a)}$. Furthermore, $g_I(T,a) = \gcd(f_{I,a}(T), f'_{I,a}(T))$. Indeed, $f_{I,a}(T) = \prod_{\sigma\in V(I)}(T-X(\sigma)-aY(\sigma))^{\mu_I(\sigma)}$ and all values $X(\sigma) + aY(\sigma)$, for $\sigma \in V(I)$, are pairwise distinct since $X + aY$ is a separating form, thus the gcd of $f_{I,a}(T)$ and its derivative is $g_I(T,a) = \prod_{\sigma\in V(I)}(T - X(\sigma) - aY(\sigma))^{\mu_I(\sigma)-1}$. This proves the formula for $f_{I,a,1}$.

Concerning the third equation, Lemma 9 together with Equation 6 implies:

$$f_{I,Y}(T,S) = \frac{\frac{\partial f_I(T,S)}{\partial S}}{g_I(T,S)} = \frac{\frac{\partial (R(T,S)/L_R(S))}{\partial S}}{g_I(T,S)} = \frac{\frac{\partial R(T,S)}{\partial S} - f_I(T,S)\frac{\partial L_R(S)}{\partial S}}{L_R(S)g_I(T,S)}.$$

As argued above, when specialized, $g_I(T,a)$ is equal to the gcd of $f_{I,a}(T)$ and $f'_{I,a}(T)$), and it does not identically vanish. By Lemma 9, $L_R(a)$ does not vanish either, and the formula for $f_{I,a,Y}$ follows.

It remains to compute $f_{I,a,X}$.

Definition 4 implies that, for any root γ of $f_{I,a}$: $\gamma = \frac{f_{I,a,X}}{f_{I,a,1}}(\gamma) + a\frac{f_{I,a,Y}}{f_{I,a,1}}(\gamma)$, and thus $f_{I,a,X}(\gamma) + af_{I,a,Y}(\gamma) - \gamma f_{I,a,1}(\gamma) = 0$. Replacing γ by T, we have that the polynomial $f_{I,a,X}(T) + af_{I,a,Y}(T) - Tf_{I,a,1}(T)$ vanishes at every root of $f_{I,a}$, thus the squarefree part of $f_{I,a}$ divides that polynomial. In other words, $f_{I,a,X}(T) = Tf_{I,a,1}(T) - af_{I,a,Y}(T) \bmod \overline{f_{I,a}(T)}$. We now compute $Tf_{I,a,1}(T)$ and $af_{I,a,Y}(T)$ modulo $\overline{f_{I,a}(T)}$.

Equation (3) implies that $f_{I,a,v}(T)$ is equal to $T^{\#V(I)-1}$ $\sum_{\sigma\in V(I)} \mu_I(\sigma)v(\sigma)$ plus some terms of lower degree in T, and that the degree of $\overline{f_{I,a}(T)}$ is $\#V(I)$ (since $X + aY$ is a separating form). First, for $v = Y$, this implies that $d_T(f_{I,a,Y}) < d_T(\overline{f_{I,a}})$, and thus that $af_{I,a,Y}(T)$ is already reduced modulo $\overline{f_{I,a}(T)}$. Second, for $v = 1$, $\sum_{\sigma\in V(I)} \mu_I(\sigma)$

is nonzero and equal to $d_T(f_{I,a})$. Thus, $Tf_{I,a,1}(T)$ and $\overline{f_{I,a}(T)}$ are both of degree $\#V(I)$, and their leading coefficients are $d_T(f_{I,a})$ and 1, respectively. Hence $Tf_{I,a,1}(T) \bmod \overline{f_{I,a}(T)} = Tf_{I,a,1}(T) - d_T(f_{I,a})\overline{f_{I,a}(T)}$. We thus obtain the last equation of Proposition 5. □

3.2 RUR bitsize

We prove here a new bound on the bitsize of the coefficients of the polynomials of a RUR. This bound is interesting in its own right and is instrumental for our analysis of the complexity of computing isolating boxes of the solutions of the input system, as well as for performing *sign_at* evaluations. Note that we state our bound for RUR-*candidates*, that is even when the linear form $X + aY$ is not separating. In this paper, we only use this result when the form is separating but the general result is interesting in a probabilistic context when a RUR-candidate is computed with a random linear form.

PROPOSITION 10. *Let $P, Q \in \mathbb{Z}[X, Y]$ be two coprime polynomials of total degree at most d and maximum bitsize τ, and let a be a rational of bitsize τ_a. The polynomials of the RUR-candidate of $\langle P, Q\rangle$ associated to a have bitsize in $\widetilde{O}(d^2\tau_a + d\tau)$. Moreover, there exists an integer of bitsize in $\widetilde{O}(d^2\tau_a + d\tau)$ such that the product of this integer with the polynomials of the RUR-candidate yields polynomials with integer coefficients.*

Before proving Proposition 10, we state a corollary of Mignotte's lemma [13, Theorem 4bis]. We also introduce a notion of primitive part for polynomials in $\mathbb{Q}[X, Y]$ and some of its properties.

LEMMA 11. *Let $P \in \mathbb{Z}[X, Y]$ be of degree at most d in each variable with coefficients bitsize at most τ. If $P = Q_1Q_2$ with Q_1, Q_2 in $\mathbb{Z}[X, Y]$, then the bitsize of Q_i, $i = 1, 2$, is in $\widetilde{O}(d + \tau)$.*

Primitive part. Consider a polynomial P in $\mathbb{Q}[X, Y]$ of degree at most d in each variable. It can be written $P = \sum_{i,j=0}^{d} \frac{a_{ij}}{b_{ij}} X^iY^j$ with a_{ij} and b_{ij} coprime in $\mathbb{Z}$ for all i, j. We define the *primitive part* of P, denoted $pp(P)$, as P divided by the gcd of the a_{ij} and multiplied by the least common multiple (lcm) of the b_{ij}. (Note that this definition is not entirely standard since we do not consider contents that are polynomials in X or in Y.) We also denote by τ_P the bitsize of P (that is, the maximum bitsize of all the a_{ij} and b_{ij}). We prove three properties of the primitive part which will be useful in the proof.

LEMMA 12. *For any two polynomials P and Q in $\mathbb{Q}[X, Y]$, we have the following properties: (i) $pp(PQ) = pp(P)\,pp(Q)$. (ii) If P is monic then $\tau_P \leqslant \tau_{pp(P)}$ and, more generally, if P has one coefficient, ξ, of bitsize τ_ξ, then $\tau_P \leqslant \tau_\xi + \tau_{pp(P)}$. (iii) If P has coefficients in $\mathbb{Z}$, then $\tau_{pp(P)} \leqslant \tau_P$.*

PROOF. Gauss Lemma states that if two univariate polynomials with integer coefficients are primitive, so is their product. This lemma can straightforwardly be extended to be used in our context by applying a change of variables of the form $X^iY^j \to Z^{ik+j}$ with $k > 2\max(d_Y(P), d_Y(Q))$. Thus, if P and Q in $\mathbb{Q}[X, Y]$ are primitive (i.e., each of them has integer coefficients whose common gcd is 1), their product is primitive. It follows that $pp(PQ) = pp(P)\,pp(Q)$ because, writing $P = \alpha\, pp(P)$ and $Q = \beta\, pp(Q)$, we have $pp(PQ) = pp(\alpha\, pp(P)\, \beta\, pp(Q)) = pp(pp(P)\, pp(Q))$ which is equal to $pp(P)\, pp(Q)$ since the product of two primitive polynomials is primitive.

Second, if $P \in \mathbb{Q}[X, Y]$ has one coefficient, ξ, of bitsize τ_ξ, then $\tau_P \leqslant \tau_\xi + \tau_{pp(P)}$. Indeed, We have $P = \xi\frac{P}{\xi}$ thus $\tau_P \leqslant \tau_\xi + \tau_{\frac{P}{\xi}}$. Since $\frac{P}{\xi}$ has one of its coefficients equal to 1, its primitive part is $\frac{P}{\xi}$ multiplied by an integer (the lcm of the denominators), thus $\tau_{\frac{P}{\xi}} \leqslant \tau_{pp(\frac{P}{\xi})}$ and $pp(\frac{P}{\xi}) = pp(P)$ by definition, which implies the claim.

Third, if P has coefficients in $\mathbb{Z}$, then $\tau_{pp(P)} \leqslant \tau_P$ since $pp(P)$ is equal to P divided by an integer (the gcd of the integer coefficients). □

PROOF OF PROPOSITION 10. The idea of the proof is to use the equations of Lemmas 9 and 7 which say, roughly speaking, that the polynomials of the RUR-candidate before specialization at $S = a$ are factors of the resultant $R(T, S)$ and some of its derivatives. The bounds are then derived using Lemma 11. More formally, we prove that the polynomials $f_I, f_{I,v} \in \mathbb{Q}[T, S]$, $v \in \{1, X, Y\}$ (see Equation (4)) have bitsize in $\widetilde{O}(d^2 + d\tau)$. We then specialize these polynomials at $S = a$ which yields the result.

Bitsize of f_I. We apply Lemma 11 to the primitive part of both sides of the equation $R(T, S) = L_R(S)f_I(T, S)$ of Lemma 9, where $R, L_R \in \mathbb{Z}[T, S]$, $f_I \in \mathbb{Q}[T, S]$ is monic with respect to T (see Equation (4)). By Lemma 3, R has bitsize in $\widetilde{O}(d(d+\tau))$ and degree at most $2d^2$ in each variable. Note that this directly implies that, when $L_R(a) \neq 0$, the bitsize of $f_{I,a}(T) = f_I(T, a) = R(T, a)/L_R(a)$ is in $\widetilde{O}(d^2\tau_a + d\tau)$. For any value of $L_R(a)$, we show that $f_I(T, S)$ has bitsize in $\widetilde{O}(d^2 + d\tau)$ and we specialize S by a afterward. Indeed, Lemma 12 implies that $pp(R)$ also has bitsize $\widetilde{O}(d(d + \tau))$. Since f_I is monic, $\tau_{f_I} \leqslant \tau_{pp(f_I)}$ which is, by Lemma 11, in $\widetilde{O}(d^2 + d\tau)$. Hence, f_I has bitsize in $\widetilde{O}(d^2 + d\tau)$ and its degree in each variable is at most that of R, that is $2d^2$.

Moreover, since f_I is monic (in T), the corresponding coefficient of $pp(f_I)$ is equal to the lcm of the denominators of the coefficients of f_I, which we denote by Lcm_{f_I}. It follows that $\tau_{Lcm_{f_I}} \leqslant \tau_{pp(f_I)}$ which we proved is in $\widetilde{O}(d^2 + d\tau)$.

Bitsize of $f_{I,v}$, $v \in \{1, Y\}$. We consider the equations of Lemma 7. These equations can be written as $\frac{\partial f_I}{\partial u}(T, S) = g_I(T, S)f_{I,v}(T, S)$ where u is T or S, and v is 1 or Y, respectively. We first bound the bitsize of one coefficient, ξ, of $f_{I,v}$ so that we can apply Lemma 12 which states that $\tau_{f_{I,v}} \leqslant \tau_\xi + \tau_{pp(f_{I,v})}$. We consider the leading coefficient ξ of $f_{I,v}$ with respect to the lexicographic order (T, S). Since g_I is monic in T (see Lemma 7), the leading coefficient (with respect to the same ordering) of the product $g_If_{I,v} = \frac{\partial f_I}{\partial u}$ is ξ which thus has bitsize in $\widetilde{O}(\tau_{f_I})$ (since it is bounded by τ_{f_I} plus the log of the degree of f_I). It follows that $\tau_{f_{I,v}}$ is in $\widetilde{O}(d^2 + d\tau + \tau_{pp(f_{I,v})})$.

We now take the primitive part of the above equation (of Lemma 7), which gives $pp(\frac{\partial f_I}{\partial u}(T, S)) = pp(g_I(T, S))\, pp(f_{I,v}(T, S))$. By Lemma 11, $\tau_{pp(f_{I,v})}$ is in $\widetilde{O}(d^2 + \tau_{pp(\frac{\partial f_I}{\partial u})})$. In order to bound the bitsize of $pp(\frac{\partial f_I}{\partial u})$ we multiply $\frac{\partial f_I}{\partial u}$ by Lcm_{f_I} so that it has integer coefficients (multiplying by a constant does not change the primitive part). The bitsize of $pp(\frac{\partial f_I}{\partial u}) = pp(Lcm_{f_I}\frac{\partial f_I}{\partial u})$ is thus at most that of $Lcm_{f_I}\frac{\partial f_I}{\partial u}$ which is bounded by the sum of the bitsizes of Lcm_{f_I} and

$\frac{\partial f_I}{\partial u}$. We proved that $\tau_{Lcm_{f_I}}$ and τ_{f_I} are in $\widetilde{O}(d^2+d\tau)$, thus the bitsize of $pp(\frac{\partial f_I}{\partial u})$ is in $\widetilde{O}(d^2+d\tau)$. It follows that $\tau_{pp(f_{I,v})}$ and $\tau_{f_{I,v}}$ are also in $\widetilde{O}(d^2+d\tau)$ for $v \in \{1, Y\}$.

Bitsize of $f_{I,X}$. We obtain the bound for $f_{I,X}$ by symmetry. Similarly as we proved that $f_{I,Y}$ has bitsize in $\widetilde{O}(d^2+d\tau)$, we get, by exchanging the role of X and Y in Equation (4) and Lemma 7, that $\sum_{\sigma\in V(I)} \mu_I(\sigma)X(\sigma)\prod_{\varsigma\in V(I),\varsigma\neq\sigma}(T - Y(\varsigma) - SX(\varsigma))$ has bitsize in $\widetilde{O}(d^2+d\tau)$. This polynomial is of degree $O(d^2)$ in T and S, thus after replacing S by $\frac{1}{S}$ and then T by $\frac{T}{S}$, the polynomial is of degree $O(d^2)$ in T and $\frac{1}{S}$. We multiply it by S to the power of $\frac{1}{S}$ and obtain $f_{I,X}$ which is thus of bitsize $\widetilde{O}(d^2+d\tau)$.

Specialization at $S = a$. To bound the bitsize of the polynomials of $RUR_{I,a}$ (Definition 4), it remains to evaluate the polynomials f_I and $f_{I,v}$, $v \in \{1, X, Y\}$, at the rational value $S = a$ of bitsize τ_a. Since these polynomials have degree in S in $O(d^2)$ and bitsize in $\widetilde{O}(d^2+d\tau)$, it is straightforward that their specializations at $S = a$ have bitsize in $\widetilde{O}(d^2+d\tau+d^2\tau_a) = \widetilde{O}(d^2\tau_a + d\tau)$.

RUR-candidate with integer coefficients. With the above notation, we set $l = Lcm(Lcm_{f_I}, Lcm_{f_{I,v}}, v \in 1, X, Y)$. Similarly as above, it is straightforward to prove that l has bitsize in $\widetilde{O}(d^2+d\tau)$ and that the product of l with f_I and $f_{I,v}$, $v \in \{1, X, Y\}$ yields polynomials with integer coefficients of bitsize in $\widetilde{O}(d^2+d\tau)$. Moreover, by Equation 4, these polynomials have degree in S bounded by d^2, therefore, multiplying their specialization at a by $l \times denom(a)^{d^2}$ where $denom(a)$ is the denominator of a, yields polynomials with integer coefficients. This concludes the proof, since $l \times denom(a)^{d^2}$ has bitsize in $\widetilde{O}(d^2\tau_a + d\tau)$. □

It is known that there exists a separating form $X + aY$ with a an integer in $O(d^4)$. Moreover, such a separating form, with $a < 2d^4$, can be computed in $\widetilde{O}_B(d^8+d^7\tau+d^5\tau^2)$ bit operations [4]. As a direct consequence of Propositions 6 and 10, we get the following result.

THEOREM 13. *Let $P, Q \in \mathbb{Z}[X, Y]$ be two coprime bivariate polynomials of total degree at most d and maximum bitsize τ. Given a separating form $X + aY$ with integer a of bitsize[3] $\widetilde{O}(1)$, the RUR of $\langle P, Q\rangle$ associated to a can be computed using Proposition 5 with $\widetilde{O}_B(d^7+d^6\tau)$ bit operations. Furthermore, the polynomials of this RUR have degree at most d^2 and bitsize in $\widetilde{O}(d^2+d\tau)$.*

4. APPLICATIONS

In this section, we present two important applications of the RUR, that is, computing boxes with rational coordinates that isolate the real solutions of the system and evaluating the sign of a bivariate polynomial at these solutions. For simplicity we focus on a parameterization given by a RUR, but the complexity results also hold for the classical one via subresultants.

We start by recalling the complexity of isolating the real roots of a univariate polynomial. Here, f denotes a univariate polynomial of degree d with integer coefficients of bitsize at most τ.

[3]By abuse of notation, $\widetilde{O}(1)$ refers to O of any polylogarithmic function in d in τ.

LEMMA 14 ([17, THEOREM 10]). *Let f be squarefree. Isolating intervals of all the real roots of f can be computed and refined up to a width less than 2^{-L} with $\widetilde{O}_B(d^3\tau + d^2L)$ bit operations.*

LEMMA 15 ([16, THEOREM 4]). *Let the minimum root separation bound of f (or simply the separation bound of f) be the minimum distance between two different complex roots of f: $sep(f) = \min_{\{\gamma,\,\delta \text{ roots of } f,\ \gamma\neq\delta\}} |\gamma - \delta|$. One has $sep(f) > 1/(2d^{d/2+2}(d2^\tau+1)^d)$, hence $sep(f) > 2^{-\widetilde{O}(d\tau)}$.*

4.1 Computation of isolating boxes

Given a RUR of the ideal I, $\{f_{I,a}, f_{I,a,1}, f_{I,a,X}, f_{I,a,Y}\}$, isolating boxes for the real solutions can be computed by first computing isolating intervals for the real roots of the univariate polynomial $f_{I,a}$ and then, evaluating the rational fractions $\frac{f_{I,a,X}}{f_{I,a,1}}$ and $\frac{f_{I,a,Y}}{f_{I,a,1}}$ by interval arithmetic. However, for the simplicity of the proof, instead of evaluating by interval each of these fractions of polynomials, we compute the product of its numerator with the inverted denominator modulo $f_{I,a}$, and then evaluate this resulting polynomial on the isolating intervals of the real roots of $f_{I,a}$ (note that we obtain the same complexity bound if we directly evaluate the fractions, but the proof is rather technical, although not difficult). When these isolating intervals are sufficiently refined, the computed boxes are necessarily disjoint and thus isolating. The following proposition analyzes the bit complexity of this algorithm.

PROPOSITION 16. *Given a RUR of $\langle P, Q\rangle$, isolating boxes for the solutions of $\langle P, Q\rangle$ can be computed in $\widetilde{O}_B(d^8+d^7\tau)$ bit operations, where d bounds the total degree of P and Q, and τ bounds the bitsize of their coefficients.*

PROOF. For every real solution α of $I = \langle P, Q\rangle$, let $J_{X,\alpha} \times J_{Y,\alpha}$ be a box containing it. A sufficient condition for these boxes to be isolating is that the width of every interval $J_{X,\alpha}$ and $J_{Y,\alpha}$ is less than half the separation bound of the resultant of P and Q with respect to X and Y, respectively. Such a resultant has degree at most $2d^2$ and bitsize in $\widetilde{O}(d\tau)$, and we furthermore have an explicit upper bound on this bitsize which is $2d(\tau + \log 2d + 1) + \log(2d^2+1) + 1$ [2, Proposition 8.46]. Lemma 15 thus yields an explicit lower bound of $2^{-\varepsilon}$ with ε in $\widetilde{O}(d^3\tau)$ on the separating bound of such a resultant. It is thus sufficient to analyze the complexity of computing, for every α, a box $J_{X,\alpha} \times J_{Y,\alpha}$ that contains α and such that the widths of these intervals are smaller than half of $2^{-\varepsilon}$.[4] For technical reasons, we require that the interval widths are smaller than $2^{-\varepsilon'}$ with $\varepsilon' = \varepsilon + 2$.

Given a RUR $\{f_{I,a}, f_{I,a,1}, f_{I,a,X}, f_{I,a,Y}\}$ of I, we first show how to modify the rational mapping induced by this RUR into a polynomial one. Second, we bound, in terms of the width of J_γ, the side length of the box obtained by interval arithmetic as the image of J_γ through the mapping. We will then deduce an upper bound on the width of J_γ that ensures that the side length of its box image is less than $2^{-\varepsilon'}$, and the result will follow.

Polynomial mapping. Since $f_{I,a}$ and $f_{I,a,1}$ are coprime (see Proposition 5), the rational mapping can be transformed

[4]For clarity, we use an explicit bound for ε but the refinement can be stopped when all the boxes are pairwise disjoint, independently of their side length, and without changing the overall complexity.

into a polynomial one by replacing $\frac{1}{f_{I,a,1}}$ by $\frac{1}{f_{I,a,1}} \mod f_{I,a}$. This polynomial mapping still maps the real roots of $f_{I,a}$ to those of I, since $\frac{1}{f_{I,a,1}}$ and $\frac{1}{f_{I,a,1}} \mod f_{I,a}$ coincide when $f_{I,a}$ vanishes (by Bézout's identity). This mapping can be computed in $\widetilde{O}_B(d^6 + d^5\tau)$ bit operations and leads polynomials with degree less than $4d^2$ and bitsize in $\widetilde{O}(d^4 + d^3\tau)$. Indeed, $f_{I,a}$ and $\frac{1}{f_{I,a,1}}$ have degree at most d^2 and bitsize $\widetilde{O}(d^2+d\tau)$ (Theorem 13), thus $\frac{1}{f_{I,a,1}} \mod f_{I,a}$ has bitsize in $\widetilde{O}(d^2(d^2+d\tau))$ and it can be computed with $\widetilde{O}_B((d^2)^2(d^2 + d\tau))$ bit operations [20, Corollaries 11.11(ii) & 6.52]. Thus, its product with $f_{I,a,X}$ or $f_{I,a,Y}$ can be computed in complexity $\widetilde{O}_B(d^2(d^4 + d^3\tau))$ [20, Corollary 8.27]. The result follows since the degree of the inverse modulo $f_{I,a}$ is less than that of $f_{I,a}$ and all the polynomials of the RUR have degrees at most d^2 by Theorem 13.

Width expansion through interval arithmetic evaluation. We consider here exact interval arithmetic, that is, interval arithmetic where operations on the interval boundaries are done exactly (with arbitrary precision). Let $J = [a, b]$ be an interval with rational endpoints such that $\max(|a|, |b|) \leqslant 2^\sigma$ and let $f \in \mathbb{Z}[T]$ be a polynomial of degree d_f with coefficients of bitsize τ_f. Denoting the width of J by $w(J) = |b-a|$, $f(J)$ can be evaluated by interval arithmetic into an interval $f_\square(J)$ whose width is at most $2^{\tau_f + d_f\sigma} d_f^2 w(J)$ (see [7, Lemma 8]). In other words, if $w(J) \leqslant 2^{-\varepsilon' - \tau_f - d_f\sigma - 2\log d_f}$, then $w(f_\square(J)) \leqslant 2^{-\varepsilon'}$.

Computing isolating boxes. We now apply the previous property on the polynomials of the mapping evaluated on isolating intervals of $f_{I,a}$. We denote by d_f and τ_f the maximum degree and bitsize of the polynomials of the mapping; as shown above $d_f < 4d^2$ and $\tau_f \in \widetilde{O}(d^4 + d^3\tau)$.

The polynomial $f_{I,a}$ has bitsize in $\widetilde{O}(d^2 + d\tau)$ (Theorem 13), thus, by Cauchy's bound (see e.g. [21, §6.2]) and considering intervals of isolation for $f_{I,a}$ of widths upper bounded by some constant, we have that the maximum absolute value of the boundaries of the isolating intervals are smaller than 2^σ with $\sigma = \widetilde{O}(d^2 + d\tau)$. Now, if all isolating intervals of $f_{I,a}$ are of width less than $2^{-\varepsilon' - \tau_f - d_f\sigma - 2\log d_f}$, the above property implies that the boxes evaluated by the polynomial mapping have side width less than $2^{-\varepsilon'}$ and are hence isolating. By Lemma 1, the squarefree part of $f_{I,a}$ has degree $O(d^2)$ and bitsize $\widetilde{O}(d^2 + d\tau)$. Lemma 14 thus implies that, for all the real roots of $\overline{f_{I,a}}$, isolating intervals of width less than $2^{-\varepsilon' - \tau_f - d_f\sigma - 2\log d_f} = 2^{-\widetilde{O}(d^4 + d^3\tau)}$ can be computed with $\widetilde{O}_B(d^8 + d^7\tau)$ bit operations.

It remains to evaluate by interval arithmetic the polynomials of the mapping which have degree $O(d^2)$ and bitsize $\widetilde{O}(d^4 + d^3\tau)$ on each of these $O(d^2)$ isolating intervals whose endpoints have bitsize at most in $\widetilde{O}(d^4 + d^3\tau)$. By Lemma 2, this can be done with $\widetilde{O}_B(d^2 d^2(d^4 + d^3\tau))$ bit operations. Therefore, we can compute isolating boxes for the solutions of $\langle P, Q\rangle$ in $\widetilde{O}_B(d^8 + d^7\tau)$ bit operations. □

4.2 Sign evaluation

This section addresses the problem of computing the sign ($+$, $-$ or 0) of a given polynomial F at the solutions of a bivariate system defined by two polynomials P and Q. We consider in the following that all input polynomials P, Q, and F are in $\mathbb{Z}[X, Y]$ and have degree at most d and coefficients of bitsize at most τ. We assume without loss of generality that the bound d is *even*.

Once the RUR $\{f_{I,a}, f_{I,a,1}, f_{I,a,X}, f_{I,a,Y}\}$ of $I = \langle P, Q\rangle$ is computed, we can use it to translate a bivariate sign computation into a univariate sign computation. Indeed, let $F(X, Y)$ be the polynomial to be evaluated at the solution (α, β) of I that is the image of the root γ of $f_{I,a}$ by the RUR mapping. We define the polynomial $f_F(T)$ roughly as the numerator of the rational fraction obtained by substituting $X = \frac{f_{I,a,X}(T)}{f_{I,a,1}(T)}$ and $Y = \frac{f_{I,a,Y}(T)}{f_{I,a,1}(T)}$ in the polynomial $F(X, Y)$, so that the sign of $F(\alpha, \beta)$ is the same as that of $f_F(\gamma)$.

LEMMA 17. *The primitive part of $f_F(T) = f_{I,a,1}^d(T)F(T - aY, Y)$, with $Y = \frac{f_{I,a,Y}(T)}{f_{I,a,1}(T)}$, has degree $O(d^3)$, bitsize in $\widetilde{O}(d^3 + d^2\tau)$, and it can be computed with $\widetilde{O}_B(d^7 + d^6\tau)$ bit operations. The sign of F at a real solution of $I = \langle P, Q\rangle$ is equal to the sign of $pp(f_F)$ at the corresponding root of $f_{I,a}$.*

PROOF. We first compute the polynomial $F(T - aY, Y)$ in the form $\sum_{i=0}^d a_i(T)Y^i$. Then, $f_F(T)$ is equal to $\sum_{i=0}^d a_i(T) f_{I,a,Y}(T)^i f_{I,a,1}(T)^{d-i}$. Thus, computing an expanded form of $f_F(T)$ can be done by computing the $a_i(T)$, the powers $f_{I,a,Y}(T)^i$ and $f_{I,a,1}(T)^i$, and their appropriate products and sum.

Computing $a_i(T)$. According to Lemma 3, $F(T - SY, Y)$ can be expanded with $\widetilde{O}_B(d^4 + d^3\tau)$ bit operations and its bitsize is in $\widetilde{O}(d+\tau)$. Then, $F(T - aY, Y)$ is obtained by substituting S by a. Writing $F(T - SY, Y) = \sum_{i=0}^d f_i(T, Y)S^i$, this substitution amounts to compute for each i, a^i and $f_i(T, Y)a^i$ and then, summing all the $f_i(T, Y)a^i$. The dominating bit complexity is that of computing all the $f_i(T, Y)a^i$ and it is in $\widetilde{O}_B(d^4 + d^3\tau)$.

Computing $f_{I,a,Y}(T)^i$ and $f_{I,a,1}(T)^i$. $f_{I,a,Y}(T)$ has degree $O(d^2)$ and bitsize $\widetilde{O}(d^2 + d\tau)$ (by Theorem 13), thus $f_{I,a,Y}(T)^i$ has degree in $O(d^3)$ and bitsize in $\widetilde{O}(d^3 + d^2\tau)$. Computing all the $f_{I,a,Y}(T)^i$ can be done with $O(d)$ multiplications between these polynomials. Every multiplication can be done with $\widetilde{O}_B(d^3(d^3 + d^2\tau))$ bit operations [20, Corollary 8.27], thus all the multiplications can be done with $\widetilde{O}_B(d^4(d^3 + d^2\tau))$ bit operations in total. It follows that all the $f_{I,a,Y}(T)^i$, and similarly all the $f_{I,a,1}(T)^i$, can be computed using $\widetilde{O}_B(d^7 + d^6\tau)$ bit operations and their bitsize is in $\widetilde{O}(d^3 + d^2\tau)$.

Computing $f_F(T)$. Computing, for $i = 0, \ldots, d$, all $a_i(T) f_{I,a,Y}(T)^i f_{I,a,1}(T)^{d-i}$ amounts to multiplying $O(d)$ times, univariate polynomials of degree $O(d^3)$ and bitsize $\widetilde{O}(d^3 + d^2\tau)$, which can be done, similarly as above, with $\widetilde{O}(d^7 + d^6\tau)$ bit operations. Finally, their sum is the sum of d univariate polynomials of degree $O(d^3)$ and bitsize $\widetilde{O}(d^3 + d^2\tau)$, which can also be computed within the same bit complexity. Hence, $f_F(T)$ can be computed with $\widetilde{O}(d^7 + d^6\tau)$ bit operations and its coefficients have bitsize in $\widetilde{O}(d^3 + d^2\tau)$.

Primitive part of $f_F(T)$. By Proposition 10, there exists an integer r of bitsize in $\widetilde{O}(d^2+d\tau)$ such that its product with the RUR polynomials gives polynomials in $\mathbb{Z}[T]$ of bitsize in $\widetilde{O}(d^2 + d\tau)$. We consider the polynomial $r^d f_F(T) = (r f_{I,a,1}(T))^d F(T - aY, Y)$ with $Y = \frac{r f_{I,a,Y}(T)}{r f_{I,a,1}(T)}$. This polynomial has its coefficients in $\mathbb{Z}$ since $r f_{I,a,Y}(T)$ and $r f_{I,a,1}(T)$

are in $\mathbb{Z}[T]$. Moreover, since $rf_{I,a,Y}(T)$ and $rf_{I,a,1}(T)$ have bitsize in $\widetilde{O}(d^2 + d\tau)$, $r^d f_F(T)$ can be computed, similarly as above, in $\widetilde{O}_B(d^7 + d^6\tau)$ and it has bitsize in $\widetilde{O}(d^3 + d^2\tau)$. The primitive part of $f_F(T)$ has also bitsize in $\widetilde{O}(d^3 + d^2\tau)$ (since it is smaller than or equal to that of $r^d f_F(T)$) and it can be computed from $r^d f_F(T)$ with $\widetilde{O}_B(d^3(d^3 + d^2\tau))$ bit operations by computing $O(d^3)$ gcd of coefficients of bitsize $\widetilde{O}(d^3 + d^2\tau)$ [21, §2.A.6].

Signs of F and f_F. By Definition 4, there is a one-to-one mapping between the roots of $f_{I,a}$ and those of $I = \langle P, Q\rangle$ that maps a root γ of $f_{I,a}$ to a solution $(\alpha, \beta) = (\frac{f_{I,a,X}(\gamma)}{f_{I,a,1}(\gamma)}, \frac{f_{I,a,Y}(\gamma)}{f_{I,a,1}(\gamma)})$ of I such that $\gamma = \alpha + a\beta$ and $f_{I,a,1}(\gamma) \neq 0$. For any such pair of γ and (α,β), $f_F(\gamma) = f_{I,a,1}^d(\gamma) F(\gamma - a\frac{f_{I,a,Y}(\gamma)}{f_{I,a,1}(\gamma)}, \frac{f_{I,a,Y}(\gamma)}{f_{I,a,1}(\gamma)})$ by definition of $f_F(T)$, and thus $f_F(\gamma) = f_{I,a,1}^d(\gamma)F(\alpha,\beta)$. It follows that $f_F(\gamma)$ and $F(\alpha,\beta)$ have the same sign since $f_{I,a,1}(\gamma) \neq 0$ and d is even by hypothesis. □

An algorithm for evaluating the sign of a univariate polynomial (here f_F) at the roots of a squarefree univariate polynomial (here $\overline{f_{I,a}}$) is analysed in [8, Corollary 5, Lemma 7]. The idea of this algorithm comes originally from [12], where the Cauchy index of two polynomials is computed by means of sign variations of a particular remainder sequence called the Sylvester-Habicht sequence. In [8], this approach is slightly adapted to deduce the sign from the Cauchy index ([21, Thm. 7.3]) and the bit complexity is given in terms of the two initial degrees and bitsizes. Unfortunately, the corresponding proof is problematic because the authors refer to two complexity results for computing parts of the Sylvester-Habicht sequences and none of them actually applies. Their approach is however correct in spirit and we state below a corrected (weaker) version of their complexity result, see [3] for details.

LEMMA 18. *Let $f \in \mathbb{Z}[X]$ be a squarefree polynomial of degree d_f and bitsize τ_f, and (a,b) be an isolating interval of one of its real roots γ with a and b distinct rationals of bitsize in $\widetilde{O}(d_f\tau_f)$ and $f(a)f(b) \neq 0$. Let $g \in \mathbb{Z}[X]$ be of degree d_g and bitsize τ_g. The sign of $g(\gamma)$ can be computed in $\widetilde{O}_B((d_f^3 + d_g^2)\tau_f + (d_f^2 + d_f d_g)\tau_g)$ bit operations. The sign of g at all the real roots of f can be computed with $\widetilde{O}_B((d_f^3 + d_f^2 d_g + d_g^2)\tau_f + (d_f^3 + d_f d_g)\tau_g)$ bit operations.*

THEOREM 19. *Given a RUR $\{f_{I,a}, f_{I,a,1}, f_{I,a,X}, f_{I,a,Y}\}$ of $I = \langle P, Q\rangle$ (satisfying the bounds of Theorem 13), the sign of F at a real solution of I can be computed with $\widetilde{O}_B(d^8 + d^7\tau)$ bit operations. The sign of F at all the solutions of I can be computed with $\widetilde{O}_B(d^9 + d^8\tau)$ bit operations.*

PROOF. By Lemma 17, the sign of F at the real solutions of I, is equal to the sign of $pp(f_F)$ at the corresponding roots of $pp(\overline{f_{I,a}})$. By Theorem 13 and Proposition 10, $f_{I,a}$ has degree at most d^2 and its primitive part has bitsize in $\widetilde{O}(d^2 + d\tau)$. Its primitive squarefree part $pp(\overline{f_{I,a}})$ can thus be computed in $\widetilde{O}_B(d^4(d^2 + d\tau))$ bit operations and has bitsize in $\widetilde{O}(d^2 + d\tau)$, by Lemma 1. By Lemmas 14 and 15, the isolating intervals (if not given) of $pp(\overline{f_{I,a}})$ can be computed in $\widetilde{O}_B((d^2)^3(d^2 + d\tau))$ bit operations with intervals boundaries of bitsize satisfying the hypotheses of Lemma 18. Applying this lemma then concludes the proof. □

5. REFERENCES

[1] M.-E. Alonso, E. Becker, M.-F. Roy, and T. Wörmann. Multiplicities and idempotents for zerodimensional systems. In *Algorithms in Algebraic Geometry and Applications*, vol. 143 of *Progress in Mathematics*, pp. 1–20, 1996.

[2] S. Basu, R. Pollack, and M.-F. Roy. *Algorithms in Real Algebraic Geometry*, volume 10 of *Algorithms and Computation in Mathematics*. Springer-Verlag, 2006.

[3] Y. Bouzidi, S. Lazard, M. Pouget, and F. Rouillier. Rational Univariate Representations of Bivariate Systems and Applications. Research Report RR-8262, INRIA, 2013.

[4] Y. Bouzidi, S. Lazard, M. Pouget, and F. Rouillier. Separating linear forms for bivariate systems. In Proc. *ISSAC*, 2013.

[5] L. Busé, H. Khalil, and B. Mourrain. Resultant-based methods for plane curves intersection problems. In *Computer Algebra in Scientific Computing (CASC)*, vol. 3718 of *LNCS*, pp. 75–92, 2005.

[6] J. Canny. A new algebraic method for robot motion planning and real geometry. In Proc. *FoCS*, pp. 39–48, 1987.

[7] J. Cheng, S. Lazard, L. Peñaranda, M. Pouget, F. Rouillier, and E. Tsigaridas. On the topology of real algebraic plane curves. *Mathematics in Computer Science*, 4:113–137, 2010.

[8] D. I. Diochnos, I. Z. Emiris, and E. P. Tsigaridas. On the asymptotic and practical complexity of solving bivariate systems over the reals. *J. Symb. Comput.*, 44(7):818–835, 2009.

[9] P. Emeliyanenko and M. Sagraloff. On the complexity of solving a bivariate polynomial system. In Proc. *ISSAC*, pp. 154–161, 2012.

[10] W. Fulton. *Algebraic curves: an introduction to algebraic geometry*. 2008. `http://www.math.lsa.umich.edu/~wfulton/CurveBook.pdf`

[11] L. González-Vega and M. El Kahoui. An improved upper complexity bound for the topology computation of a real algebraic plane curve. *J. of Complexity*, 12(4):527–544, 1996.

[12] T. Lickteig and M.-F. Roy. Sylvester-Habicht Sequences and Fast Cauchy Index Computation. *J. Symb. Comput.*, 31(3):315–341, 2001.

[13] M. Mignotte. *Mathématiques pour le calcul formel*. Presses Universitaires de France, 1989.

[14] V. Y. Pan. Solving a polynomial equation: some history and recent progress. *SIAM Review*, 39(2):187–220, 1997.

[15] F. Rouillier. Solving zero-dimensional systems through the rational univariate representation. *J. of Applicable Algebra in Engineering, Communication and Computing*, 9(5):433–461, 1999.

[16] S. M. Rump. Polynomial minimum root separation. *Mathematics of Computation*, 33(145):327–336, 1979.

[17] M. Sagraloff. When Newton meets Descartes: A Simple and Fast Algorithm to Isolate the Real Roots of a Polynomial. In Proc. *ISSAC*, pp.297–304 2012.

[18] E. Schost. *Sur la Résolution des Systèmes Polynomiaux à Paramètres*. PhD thesis, Ecole Polytechnique, France, 2001.

[19] A. Schönhage. The fundamental theorem of algebra in terms of computational complexity. Manuscript, 1982.

[20] J. von zur Gathen and J. Gerhard. *Modern Computer Algebra*. Cambridge Univ. Press, Cambridge, U.K., 2003.

[21] C. Yap. *Fundamental Problems of Algorithmic Algebra*. Oxford University Press, Oxford-New York, 2000.

Separating Linear Forms for Bivariate Systems

Yacine Bouzidi
INRIA Nancy Grand Est
LORIA, Nancy, France
Yacine.Bouzidi@inria.fr

Sylvain Lazard
INRIA Nancy Grand Est
LORIA, Nancy, France
Sylvain.Lazard@inria.fr

Marc Pouget
INRIA Nancy Grand Est
LORIA, Nancy, France
Marc.Pouget@inria.fr

Fabrice Rouillier
INRIA Paris-Rocquencourt
IMJ, Paris, France
Fabrice.Rouillier@inria.fr

ABSTRACT

We present an algorithm for computing a separating linear form of a system of bivariate polynomials with integer coefficients, that is a linear combination of the variables that takes different values when evaluated at distinct (complex) solutions of the system. In other words, a separating linear form defines a shear of the coordinate system that sends the algebraic system in generic position, in the sense that no two distinct solutions are vertically aligned. The computation of such linear forms is at the core of most algorithms that solve algebraic systems by computing rational parameterizations of the solutions and, moreover, the computation of a separating linear form is the bottleneck of these algorithms, in terms of worst-case bit complexity.

Given two bivariate polynomials of total degree at most d with integer coefficients of bitsize at most τ, our algorithm computes a separating linear form in $\widetilde{O}_B(d^8 + d^7\tau + d^5\tau^2)$ bit operations in the worst case, where the previously known best bit complexity for this problem was $\widetilde{O}_B(d^{10} + d^9\tau)$ (where $\widetilde{O}$ refers to the complexity where polylogarithmic factors are omitted and O_B refers to the bit complexity).

Categories and Subject Descriptors

F.2 [**Analysis of Algorithms and Problem Complexity**]: Nonnumerical Algorithms and Problems

Keywords

Bivariate system; Separating Linear Form

1. INTRODUCTION

One approach, that can be traced back to Kronecker, to solve a system of polynomials with a finite number of solutions is to compute a rational parameterization of its solutions. Such a representation of the (complex) solutions of a system is given by a set of univariate polynomials and associated rational one-to-one mappings that send the roots of the univariate polynomials to the solutions of the system. Such parameterizations enable to reduce computations on the system to computations with univariate polynomials and thus ease, for instance, the isolation of the solutions or the evaluation of other polynomials at the solutions.

The computation of such parameterizations has been a focus of interest for a long time; see for example [1, 9, 13, 8, 3, 6] and references therein. Most algorithms first shear the coordinate system, with a linear change of variables, so that the input algebraic system is in generic position, that is such that no two solutions are vertically aligned. These algorithms thus need a *linear separating form*, that is a linear combination of the coordinates that takes different values when evaluated at different solutions of the system. Since a random linear form is separating with probability one, probabilist Monte-Carlo algorithms can overlook this issue. However, for deterministic algorithms, computing a linear separating form is critical, especially because this is, surprisingly, the current bottleneck for bivariate systems, as discussed below.

We restrict our attention to systems of two bivariate polynomials of total degree bounded by d with integer coefficients of bitsize bounded by τ. For such systems, the approach with best known worst-case bit complexity for computing a rational parameterization was first introduced by Gonzalez-Vega and El Kahoui [9]: their initial analysis of $\widetilde{O}_B(d^{16}+d^{14}\tau^2)$ was improved by Diochnos et al. [6, Lemma 16 & Thm. 19][1] to (i) $\widetilde{O}_B(d^{10} + d^9\tau)$ for computing a separating linear form and then (ii) $\widetilde{O}_B(d^7 + d^6\tau)$ for computing a parameterization. Computing a separating linear form is thus the bottleneck of the computation of the rational parameterization. This is still true even when considering the additional phase of computing isolating boxes of the solutions (from the rational parameterization), which state-of-the-art complexity is in $\widetilde{O}_B(d^8 + d^7\tau)$ [4].

Main results. Our main contribution is a new deterministic algorithm of worst-case bit complexity $\widetilde{O}_B(d^8 + d^7\tau + d^5\tau^2)$ for computing a separating linear form of a system of two bivariate polynomials of total degree at most d and integer coefficients of bitsize at most τ (Thm. 18). The system should be zero dimensional but this is tested in our

[1]The overall bit complexity stated in [6, Thm. 19] is $\widetilde{O}_B(d^{12} + d^{10}\tau^2)$ because it includes the isolation of the solutions of the system.

algorithm. When $\tau \in \widetilde{O}(d^2)$, this gives a complexity in $\widetilde{O}_B(d^8+d^7\tau)$ which decreases by a factor d^2 the best known complexity for this problem (see the discussion above). Note furthermore that, while τ is asymptotically negligible compared to d^4 (modulo polylogarithmic factors), i.e. $\tau \in \widetilde{o}(d^4)$, the complexity of our algorithm is asymptotically better than the best known complexity for this problem, i.e. $\widetilde{O}(d^8+d^7\tau+d^5\tau^2)$ is in $\widetilde{o}(d^{10}+d^9\tau)$.

As a direct consequence, using our algorithm for computing a separating linear form directly yields a rational parameterization within the same overall complexity as our algorithm, both in the approach of Gonzalez-Vega et al. [9, 6] and in that of Bouzidi et al. [4] for computing the alternative rational parameterization as defined in [13]. Moreover, this contribution is likely to impact the complexity of algorithms studying plane algebraic curves that require finding a shear that ensures the curves to be in "generic" position (such as [9, 10]). In particular, it is hopeful that this result will improve the complexity of computing the topology of an algebraic plane curve.

As a byproduct, we obtain an algorithm for computing the number of distinct solutions of such systems within the same complexity, i.e. $\widetilde{O}_B(d^8+d^7\tau+d^5\tau^2)$.

2. OVERVIEW AND ORGANIZATION

Let P and Q be two bivariate polynomials of total degree bounded by d and integer coefficients of maximum bitsize τ. Let $I = \langle P, Q\rangle$ be the ideal they define and suppose that I is zero-dimensional. The goal is to find a linear form $T = X + aY$, with $a \in \mathbb{Z}$, that separates the solutions of I.

We first outline a classical algorithm which is essentially the same as those proposed, for instance, in [6, Lemma 16] and [10, Thm. 24][2] and whose complexity, in $\widetilde{O}_B(d^{10}+d^9\tau)$, is the best known so far for this problem. This algorithm serves two purposes: it gives some insight on the more involved $\widetilde{O}_B(d^8+d^7\tau+d^5\tau^2)$-time algorithm that follows and it will be used in that algorithm but over $\mathbb{Z}/\mu\mathbb{Z}$ instead of $\mathbb{Z}$.

Known $\widetilde{\mathbf{O}}_{\mathbf{B}}(\mathbf{d^{10}}+\mathbf{d^9}\tau)$-time algorithm for computing a separating linear form. The idea is to work with a "generic" linear form $T = X + SY$, where S is an indeterminate, and find conditions such that the specialization of S by an integer a gives a separating form. We thus consider $P(T-SY,Y)$ and $Q(T-SY,Y)$, the "generic" sheared polynomials associated to P and Q, and $R(T,S)$ their resultant with respect to Y. This polynomial has been extensively used and defined in several context; see for instance the related u-resultant [14].

It is known that, in a set $\mathcal{S}$ of d^4 integers, there exists at least one integer a such that $X + aY$ is a separating form for I since I has at most d^2 solutions which define at most $\binom{d^2}{2}$ directions in which two solutions are aligned. Hence, a separating form can be found by computing, for every a in $\mathcal{S}$, the degree of the squarefree part of $R(T,a)$ and by choosing one a for which this degree is maximum. Indeed, for any (possibly non-separating) linear form $X + aY$, the number of distinct roots of $R(T,a)$, which is the degree of its squarefree part, is always smaller than or equal to the number of distinct solutions of I, and equality is attained when the linear form $X+aY$ is separating (Lemma 8). The complexity of this algorithm is in $\widetilde{O}_B(d^{10}+d^9\tau)$ because, for d^4 values of a, the polynomial $R(T,a)$ can be shown to be of degree $O(d^2)$ and bitsize $\widetilde{O}(d^2+d\tau)$, and its squarefree part can be computed in $\widetilde{O}_B(d^6+d^5\tau)$ time.

$\widetilde{\mathbf{O}}_{\mathbf{B}}(\mathbf{d^8}+\mathbf{d^7}\tau+\mathbf{d^5}\tau^2)$-time algorithm for computing a separating linear form. To reduce the complexity of the search for a separating form, one can first consider to perform naively the above algorithm on the system $I_\mu = \langle P \bmod \mu, Q \bmod \mu\rangle$ in $\mathbb{Z}_\mu = \mathbb{Z}/\mu\mathbb{Z}$, where μ is a prime number upper bounded by some polynomial in d and τ (so that the bit complexity of arithmetic operations in $\mathbb{Z}_\mu$ is polylogarithmic in d and τ). The resultant $R_\mu(T,S)$ of $P(X-SY,Y) \bmod \mu$ and $Q(X-SY,Y) \bmod \mu$ with respect to Y can be computed in $\widetilde{O}_B(d^6+d^5\tau)$ bit operations and, since its degree is at most $2d^2$ in each variable, evaluating it at $S=a$ in $\mathbb{Z}_\mu$ can be easily done in $\widetilde{O}_B(d^4)$ bit operations. Then, the computation of its squarefree part does not suffer anymore from the coefficient growth, and it becomes softly linear in its degree, that is $\widetilde{O}_B(d^2)$. Considering d^4 choices of a, we get an algorithm that computes a separating form for I_μ in $\widetilde{O}_B(d^8)$ time in $\mathbb{Z}_\mu$. However, a serious problem remains, that is to ensure that a separating form for I_μ is also a separating form for I. This issue requires to develop a more subtle algorithm.

We first show, in Section 4.1, a critical property (Prop. 7) which states that a separating linear form over $\mathbb{Z}_\mu$ is also separating over $\mathbb{Z}$ when μ is a *lucky* prime number, which is, essentially, a prime such that the number of solutions of $\langle P, Q\rangle$ is the same over $\mathbb{Z}$ and over $\mathbb{Z}_\mu$. We then show in Sections 4.2 to 4.4 how to compute such a lucky prime number. We do that by first proving in Section 4.2 that, under mild conditions on μ, the number of solutions of I_μ is always less than or equal to the number of solutions of I (Prop. 11) and then by computing a bound on the number of unlucky primes (Prop. 12). Computing a lucky prime can then be done by choosing a μ that maximizes the number of solutions of I_μ among a set of primes of cardinality $\widetilde{O}(d^4+d^3\tau)$. For that purpose, we present in Section 4.3 a new algorithm, of independent interest, for computing in $\widetilde{O}(d^4)$ arithmetic operations in $\mathbb{Z}_\mu$ the number of distinct solutions of the system I_μ; this algorithm is based on a classical triangular decomposition. This yields, in Section 4.4, a $\widetilde{O}_B(d^8+d^7\tau+d^5\tau^2)$-time algorithm for computing a lucky prime μ in $\widetilde{O}(d^4+d^3\tau)$ (the $d^5\tau^2$ term results from the fact that we need to check that some coefficients do not vanish modulo μ). Now, μ is fixed, and we can apply the algorithm outlined above for computing a separating form for I_μ in $\mathbb{Z}_\mu$ in $\widetilde{O}_B(d^8)$ time (Section 4.5). This form, which is also separating for I, is thus obtained with a total bit complexity of $\widetilde{O}_B(d^8+d^7\tau+d^5\tau^2)$ (Thm. 18).

3. NOTATION AND PRELIMINARIES

We introduce notation and recall some classical material.

The bitsize of an integer p is the number of bits needed to represent it, that is $\lfloor \log p \rfloor + 1$ (log refers to the logarithm in base 2). For rational numbers, we refer to the bitsize as to the maximum bitsize of its numerator and denominator. The bitsize of a polynomial with integer or rational coefficients is the *maximum* bitsize of its coefficients. As mentioned earlier, O_B refers to the bit complexity and $\widetilde{O}$

[2]The stated complexity of [10, Thm. 24] is $\widetilde{O}_B(d^9\tau)$, but it seems the fact that the sheared polynomials have bitsize in $\widetilde{O}(d+\tau)$ (see Lemma 5) instead of $\widetilde{O}(\tau)$ has been overlooked in their proof.

and $\widetilde{O}_B$ refer to complexities where polylogarithmic factors are omitted, see [15, Definition 25.8] for details.

In the following, μ is a prime number and we denote by $\mathbb{Z}_\mu$ the quotient $\mathbb{Z}/\mu\mathbb{Z}$. We denote by $\phi_\mu\colon \mathbb{Z} \to \mathbb{Z}_\mu$ the reduction modulo μ, and extend this definition to the reduction of polynomials with integer coefficients. We denote by $\mathbb{D}$ a unique factorization domain, typically $\mathbb{Z}[X,Y]$, $\mathbb{Z}[X]$, $\mathbb{Z}_\mu[X]$, $\mathbb{Z}$ or $\mathbb{Z}_\mu$. We also denote by $\mathbb{F}$ a field, typically $\mathbb{Q}$, $\mathbb{C}$, or $\mathbb{Z}_\mu$.

For any polynomial $P \in \mathbb{D}[X]$, let $Lc_X(P)$ denote its leading coefficient with respect to the variable X, $d_X(P)$ its degree with respect to X, and $\overline{P}$ its squarefree part. The ideal generated by two polynomials P and Q is denoted $\langle P,Q\rangle$, and the affine variety of an ideal I is denoted by $V(I)$; in other words, $V(I)$ is the set of distinct solutions of the system $\{P,Q\}$. The solutions are always considered in the algebraic closure of $\mathbb{D}$ and the number of distinct solutions is denoted by $\#V(I)$. For a point $\sigma \in V(I)$, $\mu_I(\sigma)$ denotes the multiplicity of σ in I. For simplicity, we refer indifferently to the ideal $\langle P,Q\rangle$ and to the system $\{P,Q\}$.

We finally introduce the following notation which are extensively used throughout the paper. Given the two input polynomials P and Q, we consider the "generic" change of variables $X = T - SY$, and define the "sheared" polynomials $P(T-SY,Y)$, $Q(T-SY,Y)$, and their resultant with respect to Y,

$$R(T,S) = Res_Y(P(T-SY,Y), Q(T-SY,Y)). \quad (1)$$

The complexity bounds on the degree, bitsize and computation of these polynomials are analyzed at the end of this section in Lemma 5. We introduce

$$\begin{array}{c} L_P(S) = Lc_Y(P(T-SY,Y)) \\ L_Q(S) = Lc_Y(Q(T-SY,Y)), \quad L_R(S) = Lc_T(R(T,S)) \end{array} \quad (2)$$

and remark that these polynomials do not depend on T.

Subresultant sequences. We first recall the concept of *polynomial determinant* of a matrix which is used in the definition of subresultants. Let M be an $m\times n$ matrix with $m \leqslant n$ and M_i be the square submatrix of M consisting of the first $m-1$ columns and the i-th column of M, for $i = m,\dots,n$. The *polynomial determinant* of M is the polynomial defined as $\det(M_m)Y^{n-m} + \det(M_{m+1})Y^{n-(m+1)} + \ldots + \det(M_n)$.

Let $P = \sum_{i=0}^{p} a_i Y^i$ and $Q = \sum_{i=0}^{q} b_i Y^i$ be two polynomials in $\mathbb{D}[Y]$ and assume without loss of generality that $p \geqslant q$. The Sylvester matrix of P and Q, $Sylv(P,Q)$ is the $(p+q)$-square matrix whose rows are $Y^{q-1}P,\dots,P,Y^{p-1}Q,\dots,Q$ considered as vectors in the basis $Y^{p+q-1},\dots,Y,1$.

Definition 1. *([7, §3]). For $i = 0,\dots,\min(q,p-1)$, let $Sylv_i(P,Q)$ be the $(p+q-2i)\times(p+q-i)$ matrix obtained from $Sylv(P,Q)$ by deleting the i last rows of the coefficients of P, the i last rows of the coefficients of Q, and the i last columns.*

For $i = 0,\dots,\min(q,p-1)$, the i-th polynomial subresultant of P and Q, denoted by $Sres_{Y,i}(P,Q)$ is the polynomial determinant of $Sylv_i(P,Q)$. When $q = p$, the q-th polynomial subresultant of P and Q is $b_q^{-1}Q$.

$Sres_{Y,i}(P,Q)$ has degree at most i in Y, and the coefficient of its monomial of degree i in Y, denoted by $sres_{Y,i}(P,Q)$, is called the i-th *principal subresultant coefficient*. Note that $Sres_{Y,0}(P,Q) = sres_{Y,0}(P,Q)$ is the *resultant* of P and Q with respect to Y, which we also denote by $Res_Y(P,Q)$.

We state below a fundamental property of subresultants which is instrumental in the triangular decomposition algorithm used in Section 4.3. For clarity, we state this property for bivariate polynomials $P = \sum_{i=0}^{p} a_i Y^i$ and $Q = \sum_{i=0}^{q} b_i Y^i$ in $\mathbb{D}[X,Y]$, with $p \geqslant q$. Note that this property is often stated with a stronger assumption that is that *none* of the leading terms $a_p(\alpha)$ and $b_q(\alpha)$ vanishes. This property is a direct consequence of the specialization property of subresultants and of the gap structure theorem; see for instance [7, Lemmas 2.3, 3.1 and Cor. 5.1].

Lemma 2. *For any α such that $a_p(\alpha)$ and $b_q(\alpha)$ do not both vanish, the first $Sres_{Y,k}(P,Q)(\alpha,Y)$ (for k increasing) that does not identically vanish is of degree k and it is the gcd of $P(\alpha,Y)$ and $Q(\alpha,Y)$ (up to a nonzero constant in the fraction field of $\mathbb{D}(\alpha)$).*

Complexity. We recall complexity results, using fast algorithms, on subresultants and gcd computations. We also state complexities related to the computation of the "sheared" polynomials and their resultant.

Lemma 3 ([2, Prop. 8.46] [12, §8]). *Let P and Q be in $\mathbb{Z}[X_1,\dots,X_n][Y]$ (n fixed) with coefficients of bitsize at most τ such that their degrees in Y are bounded by d_Y and their degrees in the other variables are bounded by d.*

- *The coefficients of $Sres_{Y,i}(P,Q)$ have bitsize in $\widetilde{O}(d_Y\tau)$.*
- *The degree in X_j of $Sres_{Y,i}(P,Q)$ is at most $2d(d_Y-i)$.*
- *Any subresultant $Sres_{Y,i}(P,Q)$ can be computed in $\widetilde{O}(d^n d_Y^{n+1})$ arithmetic operations, and $\widetilde{O}_B(d^n d_Y^{n+2}\tau)$ bit operations.*

In the sequel, we often consider the gcd of two univariate polynomials P and Q and the gcd-free part of P with respect to Q, that is, the divisor D of P such that $P = \gcd(P,Q)D$. Note that when $Q = P'$, the latter is the squarefree part $\overline{P}$.

Lemma 4 ([2, Rem. 10.19]). *Let P and Q in $\mathbb{F}[X]$ of degree at most d. $\gcd(P,Q)$ or the gcd-free part of P with respect to Q can be computed with $\widetilde{O}(d)$ operations in $\mathbb{F}$.*

Lemma 5. *Let P and Q in $\mathbb{Z}[X,Y]$ of total degree d and maximum bitsize τ. The sheared polynomials $P(T-SY,Y)$ and $Q(T-SY,Y)$ can be expanded in $\widetilde{O}_B(d^4+d^3\tau)$ and their bitsizes are in $\widetilde{O}(d+\tau)$. The resultant $R(T,S)$ can be computed in $\widetilde{O}_B(d^7+d^6\tau)$ bit operations and $\widetilde{O}(d^5)$ arithmetic operations in $\mathbb{Z}$; its degree is at most $2d^2$ in each variable and its bitsize is in $\widetilde{O}(d^2+d\tau)$.*

Proof. Writing $P(T-SY,Y)$ as $\sum_{i=0}^{d} p_i(Y)(T-SY)^i$ and considering the bitsize of the binomial coefficients, we easily get the first statement of the lemma. The second statement is a direct application of Lemma 3 on trivariate polynomials of partial degree at most d in each variable. □

4. SEPARATING LINEAR FORM

Throughout this section, we assume that the two input polynomials P and Q are coprime in $\mathbb{Z}[X,Y]$, that they define the ideal I, that their maximum total degree d is at least 2 and that their coefficients have maximum bitsize τ. Note that the coprimality of P and Q is implicitly tested during Algorithm 4 because they are coprime if and only if $R(T,S)$ does not identically vanish. By abuse of notation, some complexity $\widetilde{O}_B(d^k)$ may refer to a complexity in which polylogarithmic factors in d *and in* τ are omitted. $I_\mu = \langle P_\mu, Q_\mu\rangle$ denotes the ideal generated by $P_\mu = \phi_\mu(P)$ and $Q_\mu = \phi_\mu(Q)$.

Similarly as in Equation (1), we define $R_\mu(T,S)$ as the resultant of $P_\mu(T-SY,Y)$ and $Q_\mu(T-SY,Y)$ with respect to Y, and we define $L_{P_\mu}(S)$, $L_{Q_\mu}(S)$, and $L_{R_\mu}(S)$, similarly as in (2). We refer to the overview in Section 2 for the organization of this section.

4.1 Separating linear form over $\mathbb{Z}_\mu$ versus $\mathbb{Z}$

We first introduce the notion of lucky prime numbers μ which are, roughly speaking, primes μ for which the number of distinct solutions of $\langle P,Q\rangle$ does not change when considering the polynomials modulo μ. We then show the critical property that, if a linear form is separating modulo such a μ, then it is also separating over $\mathbb{Z}$.

Definition 6. *A prime number μ is said to be* **lucky** *for an ideal $I=\langle P,Q\rangle$ if it is larger than $2d^4$ and satisfies*

$$\phi_\mu(L_P(S))\,\phi_\mu(L_Q(S))\,\phi_\mu(L_R(S))\not\equiv 0 \text{ and } \#V(I)=\#V(I_\mu).$$

Proposition 7. *Let μ be a lucky prime for the ideal $I=\langle P,Q\rangle$ and let $a<\mu$ be an integer such that*

$$\phi_\mu(L_P(a))\,\phi_\mu(L_Q(a))\,\phi_\mu(L_R(a))\neq 0.$$

If $X+aY$ separates $V(I_\mu)$, it also separates $V(I)$.

The key idea of the proof of Prop. 7, as well as Prop. 11 and 12, is to prove the following inequalities (under the hypothesis that various leading terms do not vanish)

$$\#V(I_\mu)\geqslant d_T(\overline{R_\mu(T,a)})\leqslant d_T(\overline{R(T,a)})\leqslant \#V(I) \qquad (3)$$

and argue that the first (resp. last) one is an equality if $X+aY$ separates $V(I_\mu)$ (resp. $V(I)$), and that the middle one is an equality except for finitely many μ. We establish these claims in Lemmas 8 and 10. As mentioned in Section 2, Lemma 8 is the key property in the classical algorithm for computing a separating form for I, which algorithm we will use over $\mathbb{Z}_\mu$ to compute a separating form for I_μ in Section 4.5. We refer to [6, Lemma 16] or [2, Prop. 11.23] for a proof. Recall that P and Q are assumed to be coprime but not P_μ and Q_μ; we address this issue in Lemma 9.

Lemma 8. *If $a\in\mathbb{Z}$ is such that $L_P(a)\,L_Q(a)\neq 0$ then $d_T(\overline{R(T,a)})\leqslant\#V(I)$ and they are equal if and only if $X+aY$ separates $V(I)$. The same holds over $\mathbb{Z}_\mu$, that is for P_μ, Q_μ, R_μ and I_μ, provided that P_μ and Q_μ are coprime.*

Lemma 9. *If $\phi_\mu(L_P(S))\,\phi_\mu(L_Q(S))\,\phi_\mu(L_R(S))\not\equiv 0$ and $\mu>4d^2$ then P_μ and Q_μ are coprime in $\mathbb{Z}_\mu[X,Y]$.*

Proof. Since $\phi_\mu(L_P(S))\,\phi_\mu(L_Q(S))\not\equiv 0$, the property of specialization of resultants [2, Prop. 4.20] yields that $\phi_\mu(R(T,S))=R_\mu(T,S)$ and $\phi_\mu(L_R(S))\not\equiv 0$ implies that $R_\mu(T,S)\not\equiv 0$. We can thus choose a value $S=a\in\mathbb{Z}_\mu$ so that $R_\mu(T,a)\not\equiv 0$ and $L_{P_\mu}(a)\,L_{Q_\mu}(a)\neq 0$; indeed, $\mu>4d^2$ and $\phi_\mu(L_R(S))$, $L_{P_\mu}(S)$ and $L_{Q_\mu}(S)$ have degree at most $2d^2$, d and d respectively (Lemma 3). For such a value, the resultant of $P_\mu(T-aY,Y)$ and $Q_\mu(T-aY,Y)$ is $R_\mu(T,a)$. This resultant is not identically zero, the leading coefficients (in Y) $L_{P_\mu}(a)$ and $L_{Q_\mu}(a)$ do not depend on T (see Eq. (2)) and are not zero, thus $P_\mu(T-aY,Y)$ and $Q_\mu(T-aY,Y)$ are coprime. The result follows. □

The following lemma is a direct consequence of the property of specialization of resultants [2, Prop. 4.20] and of the fact that the degree of the gcd cannot decrease when the polynomials are reduced modulo μ [16, Lemma 4.8].

Lemma 10. *Let μ be a prime and a be an integer such that $\phi_\mu(L_P(a))\,\phi_\mu(L_Q(a))\,\phi_\mu(L_R(a))\neq 0$, then $d_T(\overline{R_\mu(T,a)})\leqslant d_T(\overline{R(T,a)})$.*

Proof of Prop. 7. By Lemmas 8, 9 and 10, if μ is a prime and a is an integer such that $X+aY$ separates $V(I_\mu)$ and $\phi_\mu(L_P(a))\,\phi_\mu(L_Q(a))\,\phi_\mu(L_R(a))\neq 0$, then

$$\#V(I_\mu)=d_T(\overline{R_\mu(T,a)})\leqslant d_T(\overline{R(T,a)})\leqslant\#V(I).$$

Since μ is lucky, $\#V(I_\mu)=\#V(I)$ thus $d_T(\overline{R(T,a)})=\#V(I)$ and by Lemma 8, $X+aY$ separates $V(I)$. □

4.2 Number of solutions of I_μ versus I

As shown in Prop. 7, the knowledge of a lucky prime permits to search for separating linear forms over $\mathbb{Z}_\mu$ rather than over $\mathbb{Z}$. We prove here two propositions that are critical for computing a lucky prime, which state that the number of solutions of $I_\mu=\langle P_\mu,Q_\mu\rangle$ is always at most that of $I=\langle P,Q\rangle$ and give a bound on the number of unlucky primes.

Proposition 11. *Let $I=\langle P,Q\rangle$ be a zero-dimensional ideal in $\mathbb{Z}[X,Y]$. If a prime μ is larger than $2d^4$ and*

$$\phi_\mu(L_P(S))\,\phi_\mu(L_Q(S))\,\phi_\mu(L_R(S))\not\equiv 0$$

then $\#V(I_\mu)\leqslant\#V(I)$.

Proof. Let μ be a prime that satisfies the hypotheses of the proposition. We also consider an integer $a<\mu$ such that $\phi_\mu(L_P(a))\,\phi_\mu(L_Q(a))\,\phi_\mu(L_R(a))\neq 0$ and such that the linear form $X+aY$ is separating for I_μ. Such an integer exists because (i) $\phi_\mu(L_P(S))$, $\phi_\mu(L_Q(S))$, and $\phi_\mu(L_R(S))$ are not identically zero by hypothesis and they have degree at most d or $2d^2$ (Lemma 3) and, as mentioned earlier, (ii) I_μ is zero dimensional (Lemma 9) and it has at most d^2 solutions which define at most $\binom{d^2}{2}$ directions in which two solutions are aligned. Since $2d+2d^2+\binom{d^2}{2}<2d^4$ (for $d\geqslant 2$), there exists such an integer $a\leqslant 2d^4<\mu$. With such an a, we can apply Lemmas 8 and 10 which imply that $\#V(I_\mu)=d_T(\overline{R_\mu(T,a)})\leqslant d_T(\overline{R(T,a)})\leqslant\#V(I)$. □

Proposition 12. *An upper bound on the number of unlucky primes for the ideal $\langle P,Q\rangle$ can be explicitly computed in terms of d and τ, and this bound is in $\widetilde{O}(d^4+d^3\tau)$.*

Proof. According to Def. 6, a prime μ is unlucky if it is smaller than $2d^4$, if $\phi_\mu(L_P(S)L_Q(S)L_R(S))=0$, or if $\#V(I)\neq\#V(I_\mu)$. In the following, we consider $\mu>2d^4$. We first determine some conditions on μ that ensure that $\#V(I)=\#V(I_\mu)$, and we then bound the number of μ that do not satisfy these conditions. As we will see, under these conditions, $L_P(S)$, $L_Q(S)$, and $L_R(S)$ do not vanish modulo μ and thus this constraint is redundant.

The first part of the proof is similar in spirit to that of Prop. 11 in which we first fixed a prime μ and then specialized the polynomials at $S=a$ such that the form $X+aY$ was separating for I_μ. Here, we first choose a such that $X+aY$ is separating for I. With some conditions on μ, Lemmas 8 and 10 imply Equation (4) and we determine some more conditions on μ such that the middle inequality of (4) is an equality. We thus get $\#V(I_\mu)\geqslant\#V(I)$ which is the converse of that of Prop. 11 and thus $\#V(I_\mu)=\#V(I)$. In the second part of the proof, we bound the number of μ that violate the conditions we considered.

Prime numbers such that $\#V(I) \neq \#V(I_\mu)$. Let a be such that the form $X + aY$ separates $V(I)$ and $L_P(a)\, L_Q(a)\, L_R(a) \neq 0$. Similarly as in the proof of Prop. 11, we can choose $a \leqslant 2d^4$.

We consider any prime μ *such that* $\phi_\mu(L_P(a))\, \phi_\mu(L_Q(a))\, \phi_\mu(L_R(a)) \neq 0$, *so that we can apply* Lemmas 8 and 10. Since $X + aY$ separates $V(I)$, these lemmas yield that

$$\#V(I_\mu) \geqslant d_T(\overline{R_\mu(T,a)}) \leqslant d_T(\overline{R(T,a)}) = \#V(I). \qquad (4)$$

Now, $d_T(\overline{R(T,a)}) = d_T(R(T,a)) - d_T(\gcd(R(T,a), R'(T,a)))$, and similarly for $R_\mu(T,a)$. The leading coefficient of $R(T,S)$ with respect to T is $L_R(S)$, and since it does not vanish at $S = a$, $L_R(a)$ is the leading coefficient of $R(T,a)$. In addition, we have $R_\mu(T,a) = \phi_\mu(R(T,a))$, hence the hypothesis $\phi_\mu(L_R(a)) \neq 0$ implies that $R_\mu(T,a)$ and $R(T,a)$ have the same degree. It follows that, *if* μ *is such that the degree of* $\gcd(R(T,a), R'(T,a))$ *does not change when* $R(T,a)$ *and* $R'(T,a)$ *are reduced modulo* μ, we have

$$\#V(I_\mu) \geqslant d_T(\overline{R_\mu(T,a)}) = d_T(\overline{R(T,a)}) = \#V(I).$$

Since $\phi_\mu(L_P(a))\, \phi_\mu(L_Q(a))\, \phi_\mu(L_R(a)) \neq 0$, we can apply Prop. 11 which yields that $\#V(I_\mu) \leqslant \#V(I)$ and thus $\#V(I_\mu) = \#V(I)$.

Therefore, the primes μ such that $\#V(I_\mu) \neq \#V(I)$ are among those such that $L_P(a)$, $L_Q(a)$ or $L_R(a)$ vanishes modulo μ or such that the degree of $\gcd(R(T,a), R'(T,a))$ changes when $R(T,a)$ and $R'(T,a)$ are reduced modulo μ. Note that if $L_P(a)$, $L_Q(a)$, and $L_R(a)$ do not vanish modulo μ, then $L_P(S)$, $L_Q(S)$, and $L_R(S)$ do not identically vanish modulo μ. It is straightforward to prove that we can compute an explicit bound, in $\widetilde{O}(d^2 + d\tau)$, on the number of prime divisors of $L_P(a)$, $L_Q(a)$, or $L_R(a)$.

Bounding the number of prime μ *such that the degree of* $\gcd(R(T,a), R'(T,a))$ *changes when* $R(T,a)$ *and* $R'(T,a)$ *are reduced modulo* μ. By [16, Lemma 4.12], given two univariate polynomials in $\mathbb{Z}[X]$ of degree at most d' and bitsize at most τ', the product of all μ, such that the degree of the gcd of the two polynomials changes when the polynomials are considered modulo μ, is bounded by $(2^{\tau'}\sqrt{d'+1})^{2d'+2}$. The number of such primes μ is bounded by the bitsize of this bound, and thus is bounded by $(d'+1)\,(2\tau' + \log(d'+1)) + 1$. Here $d' \leqslant 2d^2$ and τ' is in $\widetilde{O}(d^2 + d\tau)$ since our explicit bound on the bitsize of $L_R(a)$ holds as well for the bitsize of $R(T,a)$, and, since $R(T,a)$ is of degree at most $2d^2$, the bitsize of $R'(T,a)$ is bounded by that of $R(T,a)$ plus $1 + \log 2d^2$. We thus obtain an explicit bound in $\widetilde{O}(d^4 + d^3\tau)$ on the number of primes μ such that the degree of $\gcd(R(T,a), R'(T,a))$ changes when $R(T,a)$ and $R'(T,a)$ are reduced modulo μ.

The result follows by summing this bound with the bounds we obtained on the number of prime divisors of $L_P(a)$, $L_Q(a)$, or $L_R(a)$, and a bound (e.g. $2d^4$) on the number of primes smaller than $2d^4$. □

4.3 Counting the number of solutions of I_μ

For counting the number of (distinct) solutions of $I_\mu = \langle P_\mu, Q_\mu \rangle$, we use a classical algorithm for computing a triangular decomposition of an ideal defined by two bivariate polynomials. We first recall this algorithm, slightly adapted to our needs, and analyze its arithmetic complexity.

Triangular decomposition. Let P and Q be two polynomials in $\mathbb{F}[X,Y]$. A decomposition of the solutions of the system $\{P, Q\}$ using the subresultant sequence appears in the theory of triangular sets [11] and for the computation of topology of curves [9].

The idea is to use Lemma 2 which states that, after specialization at $X = \alpha$, the first (with respect to increasing i) nonzero subresultant $Sres_{Y,i}(P,Q)(\alpha, Y)$ is of degree i and is equal to the gcd of $P(\alpha, Y)$ and $Q(\alpha, Y)$. This induces a decomposition into triangular subsystems $(\{A_i(X), Sres_{Y,i}(P,Q)(X,Y)\})$ where a solution α of $A_i(X) = 0$ is such that the system $\{P(\alpha, Y), Q(\alpha, Y)\}$ admits exactly i roots (counted with multiplicity), which are exactly those of $Sres_{Y,i}(P,Q)(\alpha, Y)$. Furthermore, these triangular subsystems are regular chains, i.e., the leading coefficient of the bivariate polynomial (seen in Y) is coprime with the univariate polynomial. For clarity and self-containedness, we recall this decomposition in Algorithm 1, where, in addition, we restrict the solutions of the system $\{P, Q\}$ to those where some univariate polynomials $A(X)$ vanishes (A could be identically zero).

Algorithm 1 Triangular decomposition

Input: P, Q in $\mathbb{F}[X,Y]$ coprime such that $Lc_Y(P)$ and $Lc_Y(Q)$ are coprime, $d_Y(Q) \leqslant d_Y(P)$, and $A \in \mathbb{F}[X]$ squarefree.
Output: Triangular decomp. $\{(A_i(X), B_i(X,Y))\}_{i\in\mathcal{I}}$ such that $V(\langle P, Q, A\rangle)$ is the disjoint union of the sets $V(\langle A_i(X), B_i(X,Y)\rangle)_{i\in\mathcal{I}}$

1: Compute the subresultant sequence of P and Q with respect to Y: $B_i = Sres_{Y,i}(P,Q)$
2: $G_0 = \gcd(\overline{Res_Y(P,Q)}, A)$ and $\mathcal{T} = \emptyset$
3: **for** $i = 1$ **to** $d_Y(Q)$ **do**
4: $\quad G_i = \gcd(G_{i-1}, sres_{Y,i}(P,Q))$
5: $\quad A_i = G_{i-1}/G_i$
6: $\quad$ if $d_X(A_i) > 0$, add (A_i, B_i) to $\mathcal{T}$
7: **return** $\mathcal{T} = \{(A_i(X), B_i(X,Y))\}_{i\in\mathcal{I}}$

The following lemma states the correctness of Algorithm 1 which follows from Lemma 2 and from the fact that the solutions of P and Q project on the roots of their resultant.

LEMMA 13 ([9, 11]). *Algorithm 1 computes a triangular decomposition* $\{(A_i(X), B_i(X,Y))\}_{i\in\mathcal{I}}$ *such that*
(i) the set $V(\langle P, Q, A\rangle)$ *is the disjoint union of the sets* $V(\langle A_i(X), B_i(X,Y)\rangle)_{i\in\mathcal{I}}$,
(ii) $\prod_{i\in\mathcal{I}} A_i$ *is squarefree,*
(iii) $\forall \alpha \in V(A_i)$, $B_i(\alpha, Y)$ *is of degree* i *and is equal to* $\gcd(P(\alpha,Y), Q(\alpha,Y))$, *and*
(iv) $A_i(X)$ *and* $Lc_Y(B_i(X,Y))$ *are coprime.*

In the following lemma, we analyze the complexity of Algorithm 1 for P and Q of degree at most d_X in X and d_Y in Y and A of degree at most d^2, where d denotes a bound on the total degree of P and Q. We will use Algorithm 1 with polynomials with coefficients in $\mathbb{F} = \mathbb{Z}_\mu$ and we thus only consider its arithmetic complexity in $\mathbb{F}$. The bit complexity of this algorithm over $\mathbb{Z}$ is analyzed in [6, Thm. 19] and its arithmetic complexity is thus implicitly analyzed as well; see also [5].

LEMMA 14. *Algorithm 1 performs* $\widetilde{O}(d_X d_Y^3) = \widetilde{O}(d^4)$ *arithmetic operations in* $\mathbb{F}$.

Counting the number of solutions of I_μ. Algorithm 2 computes the number of distinct solutions of an ideal $I_\mu =$

Algorithm 2 Number of distinct solutions of $\langle P_\mu, Q_\mu\rangle$

Input: P_μ, Q_μ in $\mathbb{Z}_\mu[X,Y]$ coprime, μ larger than their total degree
Output: Number of distinct solutions of $\langle P_\mu, Q_\mu\rangle$

1: Shear P_μ and Q_μ by replacing X by $X-bY$ with $b\in\mathbb{Z}_\mu$ so that $Lc_Y(P_\mu(X-bY,Y))\in\mathbb{Z}_\mu$
2: Triangular decomposition: $\{(A_i(X), B_i(X,Y))\}_{i\in\mathcal{I}} =$ Algorithm 1 $(P_\mu, Q_\mu, 0)$
3: **for all** $i\in\mathcal{I}$ **do**
4: $\quad C_i(X) = Lc_Y(B_i(X,Y))^{-1} \bmod A_i(X)$
5: $\quad \tilde{B}_i(X,Y) = C_i(X)B_i(X,Y) \bmod A_i(X)$
6: $\quad$ Triangular decomp.: $\{(A_{ij}(X), B_{ij}(X,Y))\}_{j\in\mathcal{J}_i} =$ Algorithm 1 $\left(\tilde{B}_i(X,Y), \frac{\partial \tilde{B}_i(X,Y)}{\partial Y}, A_i(X)\right)$
7: **return** $\sum_{i\in\mathcal{I}}\left(i\, d_X(A_i) - \sum_{j\in\mathcal{J}_i} j\, d_X(A_{ij})\right)$

$\langle P_\mu, Q_\mu\rangle$ of $\mathbb{Z}_\mu[X,Y]$. Roughly speaking, this algorithm first performs one triangular decomposition with the input polynomials P_μ and Q_μ, and then performs a sequence of triangular decompositions with polynomials resulting from this decomposition. The result is close to a radical triangular decomposition and the number of solutions of I_μ can be read, with a simple formula, from the degrees of the polynomials in the decomposition.

LEMMA 15. *Algorithm 2 computes the number of distinct solutions of* $\langle P_\mu, Q_\mu\rangle$.

PROOF. The shear of Line 1 allows to fulfill the requirement of the triangular decomposition algorithm, called in Line 2, that the input polynomials have coprime leading coefficients. Once the generically sheared polynomial $P_\mu(X-SY,Y)$ is computed (in $\mathbb{Z}_\mu[S,X,Y]$), a specific shear value $b\in\mathbb{Z}_\mu$ can be selected by evaluating the univariate polynomial $L_{P_\mu}(S) = Lc_Y(P_\mu(X-SY,Y))$ at $d+1$ elements of $\mathbb{Z}_\mu$. The polynomial does not vanish at one of these values since it is of degree at most d and $d<\mu$. Note that such a shear clearly does not change the number of solutions.

According to Lemma 13, the triangular decomposition $\{(A_i(X), B_i(X,Y))\}_{i\in\mathcal{I}}$ computed in Line 2 is such that the solutions of $\langle P_\mu, Q_\mu\rangle$ is the disjoint union of the solutions of the $\langle A_i(X), B_i(X,Y)\rangle$, for $i\in\mathcal{I}$. It follows that the number of (distinct) solutions of $I_\mu = \langle P_\mu, Q_\mu\rangle$ is

$$\#V(I_\mu) = \sum_{i\in\mathcal{I}}\sum_{\alpha\in V(A_i)} d_Y(\overline{B_i(\alpha,Y)}).$$

Since $B_i(\alpha,Y)$ is a univariate polynomial in Y, $d_Y(\overline{B_i(\alpha,Y)})$ is equal to $d_Y(B_i(\alpha,Y)) - d_Y(\gcd(B_i(\alpha,Y), B_i'(\alpha,Y)))$, where $B_i'(\alpha,Y)$ is the derivative of $B_i(\alpha,Y)$, which is also equal to $\frac{\partial B_i}{\partial Y}(\alpha,Y)$. By Lemma 13, $d_Y(B_i(\alpha,Y)) = i$, and since the degree of the gcd is zero when $B_i(\alpha,Y)$ is squarefree, we have

$$\#V(I_\mu) = \sum_{i\in\mathcal{I}}\sum_{\alpha\in V(A_i)} i \qquad (5)$$
$$-\sum_{i\in\mathcal{I}}\sum_{\substack{\alpha\in V(A_i)\\ B_i(\alpha,Y)\text{ not sqfr.}}} d_Y(\gcd(B_i(\alpha,Y), \tfrac{\partial B_i}{\partial Y}(\alpha,Y))).$$

The polynomials $A_i(X)$ are squarefree by Lemma 13, so $\sum_{\alpha\in V(A_i)} i$ is equal to $i\, d_X(A_i)$.

We now consider the sum of the degrees of the gcds. The rough idea is to apply Algorithm 1 to $B_i(X,Y)$ and $\frac{\partial B_i}{\partial Y}(X,Y)$, for every $i\in\mathcal{I}$, which computes a triangular decomposition $\{(A_{ij}(X), B_{ij}(X,Y))\}_{j\in\mathcal{J}_i}$ such that, for $\alpha\in V(A_{ij})$, $d_Y(\gcd(B_i(\alpha,Y), \frac{\partial B_i}{\partial Y}(\alpha,Y))) = j$ (by Lemma 13), which simplifies Equation (5) into $\#V(I_\mu) = \sum_{i\in\mathcal{I}}(i\, d_X(A_i) - \sum_{j\in\mathcal{J}_i}\sum_{\alpha\in V(A_{ij})} j)$. However, we cannot directly apply Algorithm 1 to $B_i(X,Y)$ and $\frac{\partial B_i}{\partial Y}(X,Y)$ because their leading coefficients in Y have no reason to be coprime.

By Lemma 13, $A_i(X)$ and $Lc_Y(B_i(X,Y))$ are coprime, thus $Lc_Y(B_i(X,Y))$ is invertible modulo $A_i(X)$ (by Bézout's identity); let $C_i(X)$ be this inverse and define $\tilde{B}_i(X,Y) = C_i(X)B_i(X,Y) \bmod A_i(X)$ (such that every coefficient of $C_i(X)B_i(X,Y)$ with respect to Y is reduced modulo $A_i(X)$). The leading coefficient in Y of $\tilde{B}_i(X,Y)$ is equal to 1, so we can apply Algorithm 1 to $\tilde{B}_i(X,Y)$ and $\frac{\partial \tilde{B}_i}{\partial Y}(X,Y)$. Furthermore, if $A_i(\alpha)=0$, then $\tilde{B}_i(\alpha,Y) = C_i(\alpha)B_i(\alpha,Y)$ where $C_i(\alpha)\neq 0$ since $C_i(\alpha)Lc_Y(B_i(\alpha,Y)) = 1$. Equation (5) can thus be rewritten by replacing B_i by $\tilde{B}_i$.

By Lemma 13, for every $i\in\mathcal{I}$, Algorithm 1 computes a triangular decomposition $\{(A_{ij}(X), B_{ij}(X,Y))\}_{j\in\mathcal{J}_i}$ such that $V(\langle\tilde{B}_i, \frac{\partial\tilde{B}_i}{\partial Y}, A_i\rangle)$ is the disjoint union of the sets $V(\langle A_{ij}(X), B_{ij}(X,Y)\rangle)$, $j\in\mathcal{J}_i$, and for all $\alpha\in V(A_{ij})$, $d_Y(\gcd(\tilde{B}_i(\alpha,Y), \frac{\partial\tilde{B}_i}{\partial Y}(\alpha,Y))) = j$. Since the set of $\alpha\in V(A_i)$ such that $\tilde{B}_i(\alpha,Y)$ is not squarefree is the projection of the set of solutions $(\alpha,\beta)\in V(\langle\tilde{B}_i, \frac{\partial\tilde{B}_i}{\partial Y}, A_i\rangle)$ we get

$$\#V(I_\mu) = \sum_{i\in\mathcal{I}}\left(i\, d_X(A_i) - \sum_{j\in\mathcal{J}_i}\sum_{\alpha\in V(A_{ij})} j\right).$$

$A_{ij}(X)$ is squarefree (Lemma 13) so $\sum_{\alpha\in V(A_{ij})} j = j\, d_X(A_{ij})$, which concludes the proof. □

The next lemma gives the arithmetic complexity of the above algorithm.

LEMMA 16. *Given* P_μ, Q_μ *in* $\mathbb{Z}_\mu[X,Y]$ *of total degree at most* d, *Algorithm 2 performs* $\widetilde{O}(d^4)$ *operations in* $\mathbb{Z}_\mu$.

PROOF. According to Lemma 5, the sheared polynomials $P(T-SY,Y)$ and $Q(T-SY,Y)$ can be expanded in $\widetilde{O}_B(d^4+d^3\tau)$ bit operations in $\mathbb{Z}$. Thus the sheared polynomials $P_\mu(X-SY,Y)$ and $Q_\mu(X-SY,Y)$ can obviously be computed in $\widetilde{O}(d^4)$ arithmetic operations in $\mathbb{Z}_\mu$. The leading term $Lc_Y(P_\mu(X-SY,Y))\in\mathbb{Z}_\mu[S]$ is a polynomial of degree at most d and a value $b\in\mathbb{Z}_\mu$ that does not vanish it can be found by at most $d+1$ evaluations. Each evaluation can be done with $O(d)$ arithmetic operations, thus the shear value b can be computed in $\widetilde{O}(d^2)$ operations. It remains to evaluate the generically sheared polynomials at this value $S=b$. These polynomials have $O(d^2)$ monomials in X and Y, each with a coefficient in $\mathbb{Z}_\mu[S]$ of degree at most d; since the evaluation of each coefficient is softly linear in d, this gives a total complexity in $\widetilde{O}(d^4)$ for Line 1.

According to Lemma 14, the triangular decomposition in Line 2 can be done in $\widetilde{O}(d^4)$ arithmetic operations. In Lines 4 and 5, $C_i(X)$ and $\tilde{B}_i(X,Y)$ can be computed by first reducing modulo $A_i(X)$ every coefficient of $B_i(X,Y)$ (with respect to Y). There are at most i coefficients (by definition of subresultants) and the arithmetic complexity of every reduction is softly linear in the degree of the operands [15, Cor. 11.6], which is $\widetilde{O}(d^2)$ by Lemma 3. The reduction of

Algorithm 3 Number of distinct solutions and lucky prime for $\langle P, Q\rangle$

Input: P, Q in $\mathbb{Z}[X, Y]$ coprime of total degree at most d and bitsize at most τ
Output: Number of solutions and lucky prime μ for $\langle P, Q\rangle$

1: Compute $P(T-SY, Y)$, $Q(T-SY, Y)$, $R(T, S) = Res_Y(P(T-SY, Y), Q(T-SY, Y))$
2: Compute a set B of primes larger than $2d^4$ and of cardinality $\widetilde{O}(d^4+d^3\tau)$ that contains a lucky prime for $\langle P, Q\rangle$ (see Prop. 12)
3: **for all** μ in B **do**
4: **if** $\phi_\mu(L_P(S))\, \phi_\mu(L_Q(S))\, \phi_\mu(L_R(S)) \not\equiv 0$ **then**
5: Compute $N_\mu =$ Algorithm $2(\phi_\mu(P), \phi_\mu(Q))$
6: **return** (μ, N_μ) such that N_μ is maximum

$B_i(X, Y)$ modulo $A_i(X)$ can thus be done with $\widetilde{O}(d^3)$ arithmetic operations in $\mathbb{Z}_\mu$. Now, in Line 4, the arithmetic complexity of computing the inverse of one of these coefficients modulo $A_i(X)$ is softly linear in its degree [15, Cor. 11.8], that is $\widetilde{O}(d_i)$ where d_i denotes the degree of $A_i(X)$. Furthermore, computing the product modulo $A_i(X)$ of two polynomials which are already reduced modulo $A_i(X)$ can be done in $\widetilde{O}(d_i)$ arithmetic operations [15, Cor. 11.8]. Thus, in Line 5, the computation of $\widetilde{B}_i(X, Y)$ can be done with i such multiplications, and thus with $\widetilde{O}(id_i)$ arithmetic operations. Finally, in Line 6, the triangular decomposition can be done with $\widetilde{O}(i^3 d_i)$ arithmetic operations by Lemma 14. The complexity of Lines 4-6 is thus in $\widetilde{O}(d^3 + i^3 d_i)$ which is in $\widetilde{O}(d^3 + d^2 id_i)$. The total complexity of the loop in Line 3 is thus $\widetilde{O}(d^4 + d^2 \sum_i id_i)$ which is in $\widetilde{O}(d^4)$ because the number of solutions of the triangular system $(A_i(X), B_i(X, Y))$ is at most the degree of A_i times the degree of B_i in Y, that is id_i, and the total number of these solutions for $i \in \mathcal{I}$ is that of (P, Q), by Lemma 13, which is at most d^2 by Bézout's bound. This concludes the proof because the sum in Line 7 can obviously be done in linear time in the size of the triangular decompositions that are computed during the algorithm. □

4.4 Lucky prime and number of solutions of I

We now show how to compute the number of solutions of $I = \langle P, Q\rangle$ and a lucky prime for that ideal.

LEMMA 17. *Algorithm 3 computes the number of distinct solutions and a lucky prime for $\langle P, Q\rangle$ in $\widetilde{O}_B(d^8 + d^7\tau + d^5\tau^2)$ bit operations. Moreover, this lucky prime is upper bounded by $\widetilde{O}(d^4 + d^3\tau)$.*

PROOF. We first prove the correctness of the algorithm. Note first that for all $\mu \in B$ satisfying the constraint of Line 4, Lemma 9 implies that $\phi_\mu(P)$ and $\phi_\mu(Q)$ are coprime. It follows that Algorithm 2 computes the number of distinct solutions $N_\mu = \#V(I_\mu)$ of I_μ. By Prop. 11 and Def. 6, $N_\mu \leqslant \#V(I)$ and the equality holds if μ is lucky for I. Since the set B of considered primes contains a lucky one by construction, the maximum of the computed value of N_μ is equal to $\#V(I)$. Finally, the μ associated to any such maximum value of N_μ is necessarily lucky by the constraint of Line 4 and since μ is larger than $2d^4$.

We now prove the complexity of the algorithm. The polynomials $P(T-SY, Y), Q(T-SY, Y)$ and their resultant

Algorithm 4 Separating form for $\langle P, Q\rangle$

Input: P, Q in $\mathbb{Z}[X, Y]$ of total degree at most d and defining a zero-dimensional ideal I
Output: A linear form $X + aY$ that separates $V(I)$, with $a < 2d^4$ and $L_P(a)\, L_Q(a) \neq 0$

1: Apply Algorithm 3 to compute the number of solutions $\#V(I)$ and a lucky prime μ for I
2: Compute $P(T-SY, Y)$, $Q(T-SY, Y)$ and $R(T, S) = Res_Y(P(T-SY, Y), Q(T-SY, Y))$
3: Compute $R_\mu(T, S) = \phi_\mu(R(T, S))$
4: Compute $\Upsilon_\mu(S) = \phi_\mu(L_P(S))\, \phi_\mu(L_Q(S))\, \phi_\mu(L_R(S))$
5: $a := 0$
6: **repeat**
7: Compute the degree N_a of the squarefree part of $R_\mu(T, a)$
8: $a := a + 1$
9: **until** $\Upsilon_\mu(a) \neq 0$[3]and $N_a = \#V(I)$
10: **return** The linear form $X + aY$

$R(T, S)$ can be computed in $\widetilde{O}_B(d^7 + d^6\tau)$ bit operations, by Lemma 5.

Prop. 12 states that we can compute an explicit bound $\Xi(d, \tau)$ in $\widetilde{O}(d^4 + d^3\tau)$ on the number of unlucky primes for $\langle P, Q\rangle$. We want to compute in Line 2 a set B of at least $\Xi(d, \tau)$ primes (plus one) that are larger than $2d^4$. For computing B, we can thus compute the first $\Xi(d, \tau) + 2d^4 + 1$ prime numbers and reject those that are smaller than $2d^4$. The bit complexity of computing the r first prime numbers is in $\widetilde{O}(r)$ and their maximum is in $\widetilde{O}(r)$ [15, Thm. 18.10]. We can thus compute the set of primes B with $\widetilde{O}_B(d^4 + d^3\tau)$ bit operations and these primes are in $\widetilde{O}(d^4 + d^3\tau)$.

In Line 4, we test to zero the reduction modulo μ of three polynomials in $\mathbb{Z}[S]$ which have been computed in Line 1 and which are of degree $O(d^2)$ and bitsize $O(d^2 + d\tau)$ in the worst case (by Lemma 5). For each of these polynomials, the test to zero can be done by first computing (once for all) the gcd of its $O(d^2)$ integer coefficients of bitsize $O(d^2 + d\tau)$. Each gcd can be computed with a bit complexity that is softly linear in the bitsize of the integers [16, §2.A.6] (and the bitsize clearly does not increase), hence all the gcds can be done with a bit complexity of $\widetilde{O}_B(d^2(d^2 + d\tau))$. Then the reduction of each of the three gcds modulo μ is performed, for each of the $\widetilde{O}(d^4 + d^3\tau)$ choices of μ, in a bit complexity that is softly linear in the maximum bitsize, that is in $\widetilde{O}_B(d^2 + d\tau)$ [15, Thm. 9.8] since μ has bitsize in $O(\log(d^4 + d^3\tau))$. Hence, the tests in Line 4 can be done with a total bit complexity in $\widetilde{O}_B((d^4 + d^3\tau)(d^2 + d\tau)) = \widetilde{O}_B(d^6 + d^4\tau^2)$.

In Line 5, we compute, for $\widetilde{O}(d^4 + d^3\tau)$ prime numbers μ, $\phi_\mu(P)$ and $\phi_\mu(Q)$ and call Algorithm 2 to compute the number of their common solutions. For every μ, the computation of $\phi_\mu(P)$ and $\phi_\mu(Q)$ can be done with $\widetilde{O}_B(d^2\tau)$ bit operations, since the reduction modulo μ of each of the $O(d^2)$ coefficients is softly linear in its bitsize. By Lemma 16, the bit complexity of Algorithm 2 is in $\widetilde{O}_B(d^4)$. Hence, the total bit complexity of Line 5 is in $\widetilde{O}_B(d^8 + d^7\tau + d^5\tau^2)$, and so is the overall bit complexity of Algorithm 3. □

4.5 Computing a separating linear form

Using Algorithm 3, we now present our algorithm for computing a linear form that separates the solutions of $\langle P, Q\rangle$.

THEOREM 18. *Algorithm 4 computes a separating linear form $X + aY$ for $\langle P, Q\rangle$ with $a < 2d^4$. The bit complexity of the algorithm is in $\widetilde{O}_B(d^8 + d^7\tau + d^5\tau^2)$.*

PROOF. We first prove the correctness of the algorithm. We start by proving that the value a returned by the algorithm is the smallest nonnegative integer such that $X + aY$ separates $V(I_\mu)$ with $\Upsilon_\mu(a) \neq 0$. Note first that, in Line 3, $\phi_\mu(R(T, S))$ is indeed equal to $R_\mu(T, S)$ which is defined as $Res_Y(P_\mu(T - SY, Y), Q_\mu(T - SY, Y))$ since the leading coefficients $L_P(S)$ and $L_Q(S)$ of $P(T - SY, Y)$ and $Q(T - SY, Y)$ do not identically vanish modulo μ (since μ is lucky), and thus $L_{P_\mu}(S) = \phi_\mu(L_P(S))$, similarly for Q, and the resultant can be specialized modulo μ [2, Prop. 4.20]. Now, Line 9 ensures that the value a returned by the algorithm satisfies $\Upsilon_\mu(a) \neq 0$, and we restrict our attention to nonnegative such values of a. Note that $\Upsilon_\mu(a) \neq 0$ implies that $\phi_\mu(L_P(a))\ \phi_\mu(L_Q(a))\ \phi_\mu(L_R(a)) \neq 0$ because the specialization at $S = a$ and the reduction modulo μ commute (in $\mathbb{Z}_\mu$). For the same reason, $L_{P_\mu}(S) = \phi_\mu(L_P(S))$ implies $L_{P_\mu}(a) = \phi_\mu(L_P(a))$ and thus $L_{P_\mu}(a) \neq 0$ and, similarly, $L_{Q_\mu}(a) \neq 0$. On the other hand, Line 9 implies that the value a is the smallest that satisfies $d_T(\overline{R_\mu(T, a)}) = \#V(I)$, which is also equal to $\#V(I_\mu)$ since μ is lucky. Lemma 8 thus yields that the returned value a is the smallest nonnegative integer such that $X + aY$ separates $V(I_\mu)$ and $\Upsilon_\mu(a) \neq 0$, which is our claim.

This property first implies that $a < 2d^4$ because the degree of Υ_μ is bounded by $2(d^2 + d)$, the number of non-separating linear forms is bounded by $\binom{d^2}{2}$ (the maximum number of directions defined by any two of d^2 solutions), and their sum is less than $2d^4$ for $d \geqslant 2$. Note that, since μ is lucky, $2d^4 < \mu$ and thus $a < \mu$. The above property thus also implies, by Prop. 7, that $X + aY$ separates $V(I)$. This concludes the proof of correctness of the algorithm since $a < 2d^4$ and $L_P(a)\, L_Q(a) \neq 0$ (since $\Upsilon_\mu(a) \neq 0$).

We now focus on the complexity of the algorithm. By Lemma 17, the bit complexity of Line 1 is in $\widetilde{O}_B(d^8 + d^7\tau + d^5\tau^2)$. The bit complexity of Lines 2 to 5 is in $\widetilde{O}_B(d^7 + d^6\tau)$. Indeed, by Lemma 5, $R(T, S)$ has degree $O(d^2)$ in T and in S, bitsize $\widetilde{O}(d^2 + d\tau)$, and it can be computed in $\widetilde{O}_B(d^7 + d^6\tau)$ time. Computing $R_\mu(T, S) = \phi_\mu(R(T, S))$ can thus be done in reducing $O(d^4)$ integers of bitsize $\widetilde{O}(d^2 + d\tau)$ modulo μ. Each reduction is softly linear in the maximum of the bitsizes [15, Thm. 9.8] thus the reduction of $R(T, S)$ can be computed in $\widetilde{O}_B(d^4(d^2 + d\tau))$ time. The computation of Υ_μ can clearly be done with the same complexity since each reduction is easier than the one in Line 3, and the product of the polynomials can be done with a bit complexity that is softly linear in the product of the maximum degrees and maximum bitsizes [15, Cor. 8.27].

We proved that the value a returned by the algorithm is less than $2d^4$, thus the loop in Line 6 is performed at most $2d^4$ times. Each iteration consists of computing the squarefree part of $R_\mu(T, a)$ which requires $\widetilde{O}_B(d^4)$ bit operations. Indeed, computing $R_\mu(T, S)$ at $S = a$ amounts to evaluating, in $\mathbb{Z}_\mu$, $O(d^2)$ polynomials in S, each of degree $O(d^2)$ (by Lemma 5). Note that a does not need to be reduced modulo μ because $a < 2d^4$ and $2d^4 < \mu$ since μ is lucky. Thus, the bit complexity of evaluating in $\mathbb{Z}_\mu$ each of the $O(d^2)$ polynomials in S is the number of arithmetic operations in $\mathbb{Z}_\mu$, which is linear the degree that is $O(d^2)$, times the (maximum) bit complexity of the operations in $\mathbb{Z}_\mu$, which is in $O_B(\log d\tau)$ since μ is in $\widetilde{O}(d^4 + d^3\tau)$ by Lemma 17. Hence, computing $R_\mu(T, a)$ can be done in $\widetilde{O}_B(d^4)$ bit operations. Once $R_\mu(T, a)$ is computed, the arithmetic complexity of computing its squarefree part in $\mathbb{Z}_\mu$ is softly linear in its degree (Lemma 4), that is $\widetilde{O}(d^2)$, which yields a bit complexity in $\widetilde{O}_B(d^2)$ since, again, μ is in $\widetilde{O}(d^4 + d^3\tau)$. This leads to a total bit complexity of $\widetilde{O}_B(d^8)$ for the loop in Lines 6 to 9, and thus to a total bit complexity for the algorithm in $\widetilde{O}_B(d^8 + d^7\tau + d^5\tau^2)$. □

[3] $\Upsilon_\mu(S) \in \mathbb{Z}_\mu[S]$ and we consider $\Upsilon_\mu(a)$ in $\mathbb{Z}_\mu$.

5. REFERENCES

[1] M.-E. Alonso, E. Becker, M.-F. Roy, and T. Wörmann. Multiplicities and idempotents for zerodimensional systems. In *Algorithms in Algebraic Geometry and Applications*, volume 143 of *Progress in Mathematics*, pages 1–20. Birkhäuser, 1996.

[2] S. Basu, R. Pollack, and M.-F. Roy. *Algorithms in Real Algebraic Geometry*, volume 10 of *Algorithms and Computation in Mathematics*. Springer-Verlag, 2nd edition, 2006.

[3] A. Bostan, B. Salvy, and É. Schost. Fast algorithms for zero-dimensional polynomial systems using duality. *Applicable Algebra in Engineering, Communication and Computing*, 14(4):239–272, 2003.

[4] Y. Bouzidi, S. Lazard, M. Pouget, and F. Rouillier. Rational Univariate Representations of Bivariate Systems and Applications. In *ISSAC*, 2013.

[5] Y. Bouzidi, S. Lazard, M. Pouget, and F. Rouillier. Separating linear forms for bivariate systems. Research Report RR-8261, INRIA, Mar. 2013.

[6] D. I. Diochnos, I. Z. Emiris, and E. P. Tsigaridas. On the asymptotic and practical complexity of solving bivariate systems over the reals. *J. Symb. Comput.*, 44(7):818–835, 2009.

[7] M. El Kahoui. An elementary approach to subresultants theory. *J. Symb. Comput.*, 35(3):281–292, 2003.

[8] M. Giusti, G. Lecerf, and B. Salvy. A Gröbner free alternative for solving polynomial systems. *J. of Complexity*, 17(1):154–211, 2001.

[9] L. González-Vega and M. El Kahoui. An improved upper complexity bound for the topology computation of a real algebraic plane curve. *J. of Complexity*, 12(4):527–544, 1996.

[10] M. Kerber and M. Sagraloff. A worst-case bound for topology computation of algebraic curves. *J. Symb. Comput.*, 47(3):239–258, 2012.

[11] X. Li, M. Moreno Maza, R. Rasheed, and É. Schost. The modpn library: Bringing fast polynomial arithmetic into maple. *J. Symb. Comput.*, 46(7):841–858, 2011.

[12] D. Reischert. Asymptotically fast computation of subresultants. In ISSAC, pp. 233–240, 1997.

[13] F. Rouillier. Solving zero-dimensional systems through the rational univariate representation. *J. of Applicable Algebra in Engineering, Communication and Computing*, 9(5):433–461, 1999.

[14] B.-L. Van der Waerden. *Moderne Algebra I*. Springer, Berlin, 1930.

[15] J. von zur Gathen and J. Gerhard. *Modern Computer Algebra*. Cambridge Univ. Press, Cambridge, U.K., 2nd edition, 2003.

[16] C. Yap. *Fundamental Problems of Algorithmic Algebra*. Oxford University Press, Oxford-New York, 2000.

Cylindrical Algebraic Decompositions for Boolean Combinations

Russell Bradford
University of Bath
R.J.Bradford@bath.ac.uk

James H. Davenport
University of Bath
J.H.Davenport@bath.ac.uk

Matthew England
University of Bath
M.England@bath.ac.uk

Scott McCallum
Macquarie University
Scott.McCallum@mq.edu.au

David Wilson
University of Bath
D.J.Wilson@bath.ac.uk

ABSTRACT

This article makes the key observation that when using cylindrical algebraic decomposition (CAD) to solve a problem with respect to a set of polynomials, it is not always the signs of those polynomials that are of paramount importance but rather the truth values of certain quantifier free formulae involving them. This motivates our definition of a Truth Table Invariant CAD (TTICAD). We generalise the theory of equational constraints to design an algorithm which will efficiently construct a TTICAD for a wide class of problems, producing stronger results than when using equational constraints alone. The algorithm is implemented fully in MAPLE and we present promising results from experimentation.

Categories and Subject Descriptors

I.1.2 [**Symbolic and Algebraic Manipulation**]: Algorithms—*Algebraic algorithms, Analysis of algorithms*

General Terms

Algorithms, Experimentation, Theory

Keywords

cylindrical algebraic decomposition; equational constraint

1. INTRODUCTION

Cylindrical algebraic decompositions (CADs) are a key tool in real algebraic geometry, both for their original motivation, solving quantifier elimination problems, but also for use in many other applications ranging from robot motion planning [22, etc.] to programming with complex functions [12, etc.]. Traditionally CADs are produced sign-invariant to a given set of polynomials, (the signs of the polynomials do not vary on the cells of the decomposition). However, this gives far more information than required for most problems. The idea of a truth invariant CAD (the truth of a formula does not vary on each cell) was defined in [2] for use in simplifying CADs. The key contribution of this paper is an approach to construct CADs which are truth invariant without having to first build a sign-invariant CAD. Actually, we directly build CADs which are truth table invariant, (the truth values of various quantifier free formulae do not vary).

We present an algorithm to efficiently produce TTICADs for a wide class of problems, utilising the theory of equational constraints [19]. The algorithm goes further than equational constraints by allowing the creation of smaller CADs in a wider variety of cases; for example disjunctive normal form where each individual conjunction has an equational constraint but no single explicit equational constraint is present for the formula. The problem of decomposing complex space according to a set of branch cuts for the purpose of algebraic simplification ([21, etc.]) is of this case.

1.1 Background on CAD

We briefly remind the reader about the theory of CAD, first proposed by Collins in [9].

DEFINITION 1. *A Tarski formula $F(x_1, \ldots, x_n)$ is a Boolean combination $(\wedge, \vee, \neg)$ of statements about the signs, $(= 0, > 0, < 0$, but therefore $\neq 0, \geq 0, \leq 0$ as well), of certain integral polynomials $f_i(x_1, \ldots, x_n)$. We use* QFF *to denote a quantifier free Tarski formula.*

CAD was developed as a tool for the problem of quantifier elimination over the reals: given a quantified Tarski formula

$$Q_{k+1}x_{k+1} \ldots Q_n x_n F(x_1, \ldots, x_n) \tag{1}$$

(where $Q_i \in \{\forall, \exists\}$ and F is a QFF), produce an equivalent QFF $\psi(x_1, \ldots, x_k)$. Collins proposed to decompose $\mathbb{R}^n$ cylindrically such that each cell was sign-invariant for all f_i occurring in F. Then ψ would be the disjunction of the defining formulae of those cells c_i in $\mathbb{R}^k$ such that (1) was true over the whole of c_i, which is the same as saying that (1) is true at any one "sample point" of c_i.

Collins' algorithm has two phases. The first, *projection*, applies a projection operator repeatedly to a set of polynomials, each time producing another set in one fewer variables. Together these sets contain the *projection polynomials*. These are then used in the second phase, *lifting*, to build the CAD incrementally. First $\mathbb{R}$ is decomposed into cells which are points and intervals corresponding to the real roots of the univariate polynomials. Then $\mathbb{R}^2$ is decomposed by repeating the process over each cell using the bivariate

ISSAC'13, June 26–29, 2013, Boston, Massachusetts, USA.

polynomials at a sample point. The output for each cell consists of *sections* (where a polynomial vanishes) and *sectors* (the regions between). Together these form a *stack* over the cell, and taking the union of these stacks gives the CAD of $\mathbb{R}^2$. This is repeated until a CAD of $\mathbb{R}^n$ is produced.

To conclude that a CAD produced in this way is sign-invariant we need delineability. A polynomial is *delineable* in a cell if the portion of its zero set in the cell consists of disjoint sections. A set of polynomials are *delineable* in a cell if each is delineable and the sections of different polynomials in the cell are either identical or disjoint. The projection operator used must be defined so that over each cell of a sign-invariant CAD for projection polynomials in r variables, the polynomials in $r+1$ variables are delineable.

The output of a CAD algorithm depends on the variable ordering. We usually work with polynomials in $\mathbb{Z}[x_1, \ldots, x_n]$ with the variables, $\mathbf{x}$, in ascending order (so we first project with respect to x_n and continue to reach univariate polynomials in x_1). The *main variable* of a polynomial (mvar) is the greatest variable present with respect to the ordering.

Major directions of work since 1975 includes the following:

1. Improvements in Collins' main algorithms by [17, and many others]. These have focussed on reducing the projection sets required as discussed further later.
2. Complexity theory of CAD [5, 13].
3. Partial CAD, introduced in [11], where the structure of F is used to lift only when required to deduce ψ.
4. The theory of equational constraints, [19, 20, 6] discussed in Section 2.1. This is related to the previous direction but differs by using more efficient projections.
5. CAD via Triangular Decomposition [8]: a radically different approach for computing a sign-invariant CAD which is used for MAPLE's inbuilt CAD command.

1.2 TTICAD

We define a new type of CAD, the topic of this paper.

DEFINITION 2. *Let $\Phi = \{\phi_i\}_{i=1}^t$ be a list of QFFs. We say a cylindrical algebraic decomposition $\mathcal{D}$ is a* Truth Table Invariant *CAD for Φ (TTICAD) if the Boolean value of each ϕ_i is constant (either true or false) on each cell of $\mathcal{D}$.*

A full sign-invariant CAD for the set of polynomials occurring in the formulae of Φ would clearly be a TTICAD. However, we aim to produce an algorithm that will construct smaller TTICADs for certain Φ. We will achieve this using the theory of equational constraints (first suggested in [10] with the key theory developed in [19]).

DEFINITION 3. *Suppose some quantified formula is given:*

$$\phi^* = (Q_{k+1}x_{k+1}) \cdots (Q_n x_n)\phi(\mathbf{x}).$$

where the Q_i are quantifiers and ϕ is quantifier free. An equation $f = 0$ is called an **equational constraint** *of ϕ^* if $f = 0$ is logically implied by ϕ (the quantifier-free part of ϕ^*). Such a constraint may be either explicit or implicit.*

We suppose that we are given a formula list Φ in which every QFF ϕ_i has a designated explicit equational constraint $f_i = 0$. We will construct TTICADs by generalising McCallum's reduced projection operator for equational constraints (as in [19]) so that we may make use of the equational constraints.

1.3 Worked Example

We will provide details for the following worked example.

Figure 1: The polynomials from Section 1.3.

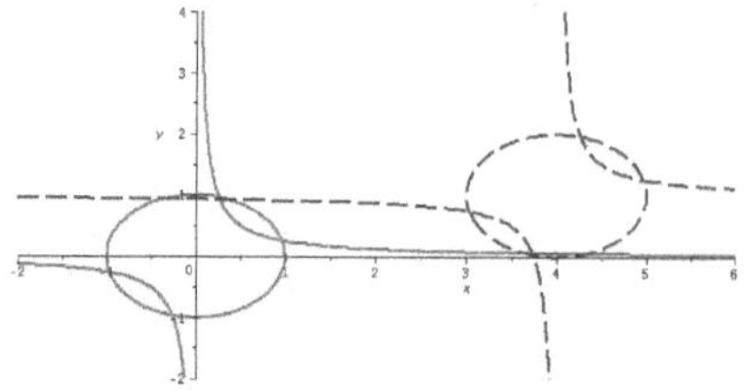

Consider the polynomials:

$$f_1 := x^2 + y^2 - 1 \qquad g_1 := xy - \tfrac{1}{4}$$
$$f_2 := (x-4)^2 + (y-1)^2 - 1 \quad g_2 := (x-4)(y-1) - \tfrac{1}{4}$$

which are plotted in Figure 1. We wish to solve the following problem: find the regions of $\mathbb{R}^2$ where the formula

$$\Phi := (f_1 = 0 \land g_1 < 0) \lor (f_2 = 0 \land g_2 < 0)$$

is true. Assume that we are using the variable ordering $y \succ x$ (so the 1-dimensional CAD is with respect to x).

Both QEPCAD [3] and MAPLE 16 [8] produce a full sign-invariant CAD for the polynomials with 317 cells. At first glance it seems that the theory of equational constraints [19, 20, 6] is not applicable here as neither $f_1 = 0$ nor $f_2 = 0$ is logically implied by Φ. However, while there is no explicit equational constraint we can observe that $f_1 f_2 = 0$ is an *implicit* constraint of Φ. Using QEPCAD with this declared gives a CAD with 249 cells. Later, in Section 2.3 we demonstrate how a TTICAD with 105 cells can be produced.

2. PROJECTION OPERATORS

2.1 Equational Constraints

We use two key theorems from McCallum's work on projection and equational constraints. Both theorems use CADs which are not just sign-invariant but have the stronger property of order-invariance. A CAD is *order-invariant* with respect to a set of polynomials if each polynomial has constant order of vanishing within each cell.

Let P be the McCallum projection operator [17], which produces coefficients, discriminant and cross resultants from a set of polynomials. We assume the usual trivial simplifications such as removal of constants, exclusion of entries identical to a previous entry (up to constant multiple), and using only the necessary coefficients. Recall that a set $A \subset \mathbb{Z}[\mathbf{x}]$ is an *irreducible basis* if the elements of A are of positive degree in the main variable, irreducible and pairwise relatively prime. The main theorem underlying P follows.

THEOREM 1 ([18]). *Let A be an irreducible basis in $\mathbb{Z}[\mathbf{x}]$ and let S be a connected submanifold of $\mathbb{R}^{n-1}$. Suppose each element of $P(A)$ is order-invariant in S. Then each element of A either vanishes identically on S or is analytic delineable on S, (a slight variant on traditional delineability, see [18]). The sections of A not identically vanishing are pairwise disjoint, and each element of A not identically vanishing is order-invariant in such sections.*

The main mathematical result underlying the reduction of P in the presence of an equational constraint f is as follows.

THEOREM 2 ([19]). *Let $f(\mathbf{x}), g(\mathbf{x})$ be integral polynomials with positive degree in x_n, let $r(x_1, \ldots, x_{n-1})$ be their*

Figure 2: Graphical representation of Theorem 2

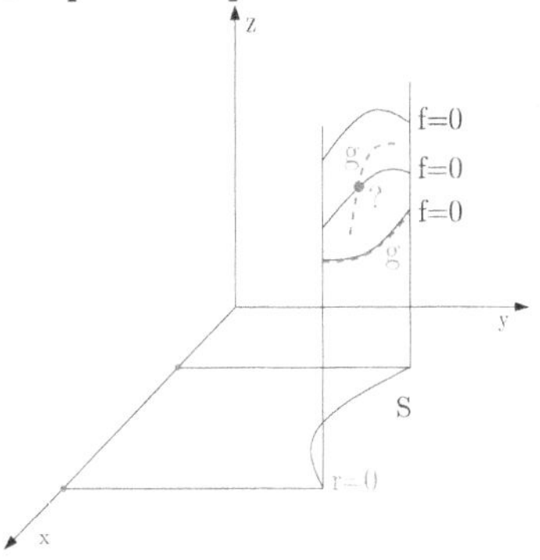

resultant, and suppose $r \neq 0$. *Let* S *be a connected subset of* $\mathbb{R}^{n-1}$ *such that* f *is delineable on* S *and* r *is order-invariant in* S. *Then* g *is* sign-invariant *in every section of* f *over* S.

Figure 2 gives a graphical representation of the question answered by Theorem 2. Here we consider polynomials $f(x, y, z)$ and $g(x, y, z)$ of positive degree in z whose resultant r is non-zero, and a connected subset $S \subset \mathbb{R}^2$ in which r is order-invariant. We further suppose that f is delineable on S (noting that Theorem 1 with $n = 3$ and $A = \{f\}$ provides sufficient conditions for this). We ask whether g is sign-invariant in the sections of f over S. Theorem 2 answers this question affirmatively: the real variety of g either aligns with a given section of f exactly (as for the bottom section of f in Figure 2), or has no intersection with such a section (as for the top). The situation at the middle section of f cannot happen. Theorem 2 thus suggests a reduction of the projection operator P relative to an equational constraint $f = 0$ for the first projection step, as in [19].

2.2 A Projection Operator for TTICAD

In [19] the central concept is that of the reduced projection of a set A of integral polynomials relative to a nonempty subset E of A and it is an extension of this which is central here. For simplicity in [19], the concept is first defined for the case when A is an irreducible basis and by analogy we start with a similar special case. Let $\mathcal{A} = \{A_i\}_{i=1}^{t}$ be a list of irreducible bases A_i and let $\mathcal{E} = \{E_i\}_{i=1}^{t}$ be a list of nonempty subsets $E_i \subseteq A_i$. Put $A = \bigcup_{i=1}^{t} A_i$ and $E = \bigcup_{i=1}^{t} E_i$ (we will use the convention of uppercase Roman letters for sets and calligraphic letters for sequences).

DEFINITION 4. *We define the* reduced projection of $\mathcal{A}$ with respect to $\mathcal{E}$, *denoted by* $P_{\mathcal{E}}(\mathcal{A})$, *as follows:*

$$P_{\mathcal{E}}(\mathcal{A}) := \bigcup_{i=1}^{t} P_{E_i}(A_i) \cup \mathrm{Res}^{\times}(\mathcal{E}) \tag{2}$$

where

$$P_{E_i}(A_i) = P(E_i) \cup \{\mathrm{res}_{x_n}(f, g) \mid f \in E_i, g \in A_i, g \notin E_i\}$$

$$\mathrm{Res}^{\times}(\mathcal{E}) = \{\mathrm{res}_{x_n}(f, \hat{f}) \mid \exists i, j : f \in E_i, \hat{f} \in E_j, i < j, f \neq \hat{f}\}$$

In Section 3.1 we build Algorithm 1 to apply the reduced projection operator for less special input sets by considering contents and irreducible factors of positive degree.

DEFINITION 5. *The* excluded projection polynomials *of* (A_i, E_i) *are those in* $P(A)$ *but excluded from* $P_{\mathcal{E}}(\mathcal{A})$:

$$\mathrm{ExclP}_{E_i}(A_i) := P(A_i) \setminus P_{E_i}(A_i) \tag{3}$$

$$= \{\mathrm{coeffs}(g), \mathrm{disc}_{x_n}(g), \mathrm{res}_{x_n}(g, \hat{g}) \mid g, \hat{g} \in A_i \setminus E_i, g \neq \hat{g}\}.$$

The total set of excluded polynomials, denoted $\mathrm{ExclP}_{\mathcal{E}}(\mathcal{A})$, *consists of all the* $\mathrm{ExclP}_{E_i}(A_i)$, *along with the cross resultants of* g_i *with all of* A_j *for* $i \neq j$.

The following theorem is an analogue of Theorem 2.3 of [19], and provides the foundation for our algorithm in Section 3.1.

THEOREM 3. *Let* S *be a connected submanifold of* $\mathbb{R}^{n-1}$. *Suppose each element of* $P_{\mathcal{E}}(\mathcal{A})$ *is order invariant in* S. *Then each* $f \in E$ *either vanishes identically on* S *or is analytically delineable on* S, *the sections over* S *of the* $f \in E$ *which do not vanish identically are pairwise disjoint, and each element* $f \in E$ *which does not vanish identically is order-invariant in such sections.*

Moreover, *for each* i, *with* $1 \leq i \leq t$, *every* $g \in A_i \setminus E_i$ *is sign-invariant in each section over* S *of every* $f \in E_i$ *which does not vanish identically.*

PROOF. The crucial observation is that $P(E) \subseteq P_{\mathcal{E}}(\mathcal{A})$. To see this, recall equation (2) and note that we can write

$$P(E) = \textstyle\bigcup_i P(E_i) \cup \mathrm{Res}^{\times}(\mathcal{E}).$$

We can therefore apply Theorem 1 to the set E and obtain the first three conclusions immediately.

There remains the final conclusion to prove. Let i be in the range $1 \leq i \leq t$, let $g \in A_i \backslash E_i$ and let $f \in E_i$; suppose f does not vanish identically on S. Now $\mathrm{res}_{x_n}(f, g) \in P_{\mathcal{E}}(\mathcal{A})$, and so is order-invariant in S by hypothesis. Further, we already concluded that f is delineable. Therefore by Theorem 2, g is sign-invariant in each section of f over S. □

In the following section we can use Theorem 3 as the key tool for our implementation of TTICAD, so long as the equational constraint f does not vanish identically on the lower dimensional manifold, S. When working with a polynomial f considered in r variables that vanishes identically at a point $\alpha \in \mathbb{R}^{r-1}$ we say that f is *nullified* at α.

REMARK 4. *It is clear that the reduced projection* $P_{\mathcal{E}}(\mathcal{A})$ *will lead to fewer (or the same) projection polynomials than the full projection* P. *One may consider instead using the reduced projection* $P_E(A)$ *of [19], (with* $E = \cup_i E_i$ *and* $A = \cup_i A_i$ *as above). In the context of Section 1.2 this corresponds to using* $\prod_i f_i$ *as an implicit equational constraint for a single formula. Note that* $P_{\mathcal{E}}(\mathcal{A})$ *also contains fewer polynomials than* $P_E(A)$ *in general since* $P_E(A)$ *contains all resultants* res(f, g) *where* $f \in E_i, g \in A_j$ *(and* $g \notin E$*), while* $P_{\mathcal{E}}(\mathcal{A})$ *contains only those with* $i = j$ *(and* $g \notin E_i$*).*

2.3 Worked Example

In Section 3 we will discuss how to use these results to define an algorithm for TTICAD. First we illustrate the potential savings with our worked example from Section 1.3.

In the notation introduced above we have:

$$A_1 := \{f_1, g_1\},\ E_1 := \{f_1\};\ A_2 := \{f_2, g_2\},\ E_2 := \{f_2\}.$$

We construct the reduced projection sets for each ϕ_i,

$$P_{E_1}(A_1) = \left\{x^2 - 1, x^4 - x^2 + \tfrac{1}{16}\right\},$$

$$P_{E_2}(A_2) = \left\{x^2 - 8x + 15, x^4 - 16x^3 + 95x^2 - 248x + \tfrac{3841}{16}\right\}$$

and the cross-resultant set

$$\mathrm{Res}^{\times}(\mathcal{E}) = \{\mathrm{res}_y(f_1, f_2)\} = \{68x^2 - 272x + 285\}.$$

Figure 3: The polynomials from the worked example along with the solutions to the projection sets.

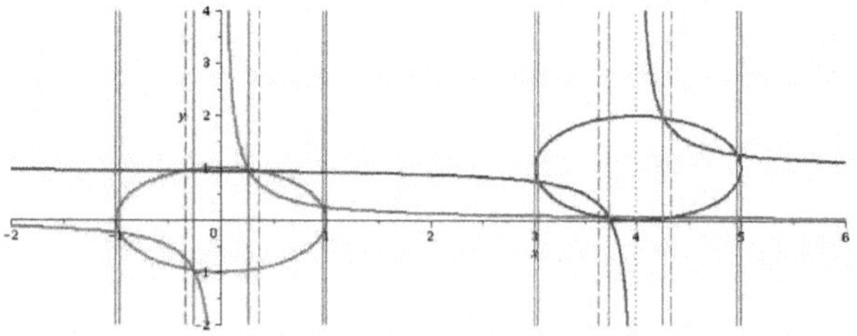

Figure 4: Magnified region of Figure 3

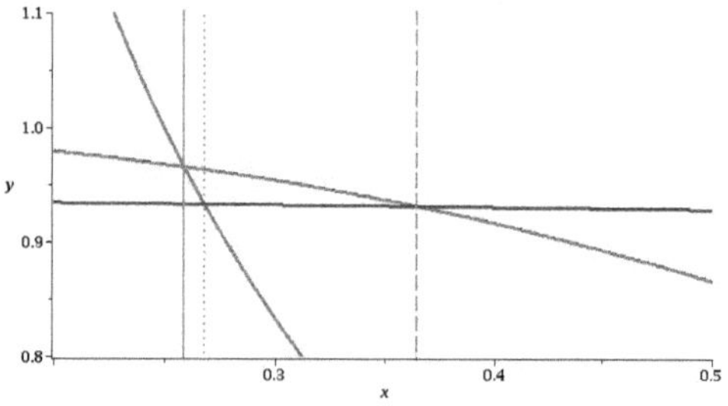

$P_{\mathcal{E}}(\mathcal{A})$ is then the union of these three sets. In Figure 3 we plot the polynomials (solid curves) and identify the 12 real solutions of $P_{\mathcal{E}}(\mathcal{A})$ (solid vertical lines). We can see the solutions align with the asymptotes of the fs and the important intersections (those of f_1 with g_1 and f_2 with g_2).

If we were to instead use a projection operator based on an implicit equational constraint $f_1 f_2 = 0$ then in the notation above we would construct $P_E(A)$ from $A = \{f_1, f_2, g_1, g_2\}$ and $E = \{f_1, f_2\}$. This set provides an extra 4 solutions (the dashed vertical lines) which align with the intersections of f_1 with g_2 and f_2 with g_1. Finally, if we were to consider $P(A)$ then we gain another 4 solutions (the dotted vertical lines) which align with the intersections of g_1 and g_2 and the asymptotes of the gs. In Figure 4 we magnify a region to show explicitly that the point of intersection between f_1 and g_1 is identified in $P_{\mathcal{E}}(\mathcal{A})$, whereas the intersection points of g_2 with both f_1 and g_1 are ignored.

Hence the 1-dimensional CAD produced using $P_{\mathcal{E}}(\mathcal{A})$ has 25 cells compared to 33 when using $P_E(A)$ and 41 when using $P(A)$. However, it is important to note that this reduction is amplified after lifting (using Theorem 3 and and Algorithm 1). The full dimensional TTICAD has 105 cells, the CAD invariant with respect to the implicit equational constraint has 249 cells and the full sign-invariant CAD has 317.

3. IMPLEMENTATION

3.1 Algorithm Description and Proof

We describe carefully Algorithm 1. This will create a TTICAD of $\mathbb{R}^n$ for a list of QFFs, $\Phi = \{\phi_i\}_{i=1}^t$, in variables $\mathbf{x} = x_1 \prec x_2 \prec \cdots \prec x_n$ where each ϕ_i has a designated equational constraint $f_i = 0$ of positive degree. We use a subalgorithm CADW, fully specified and validated in [18]. The input of CADW is: r, a positive integer and A, a set of r-variate integral polynomials. The output is a Boolean w which if true is accompanied by an order-invariant CAD for A (a list of indices I and sample points S).

Let A_i be the set of all polynomials occurring in ϕ_i, put $E_i = \{f_i\}$, and let $\mathcal{A}$ and $\mathcal{E}$ be the lists of the A_i and E_i, respectively. Our algorithm effectively defines the reduced projection of $\mathcal{A}$ with respect to $\mathcal{E}$ using the special case of this definition from the previous section. The definition amounts to using $\mathfrak{P} := C \cup P_{\mathcal{F}}(\mathcal{B})$ for $P_{\mathcal{E}}(\mathcal{A})$, where C is the set of contents of all the elements of all the A_i, $\mathcal{B}$ is the list $\{B_i\}_{i=1}^t$, such that B_i is the finest squarefree basis for the set $\mathrm{prim}(A_i)$ of primitive parts of elements of A_i which have positive degree, and $\mathcal{F}$ is the list $\{F_i\}_{i=1}^t$, such that F_i is the finest squarefree basis for $\mathrm{prim}(E_i)$. (The reader will notice that this notation and the definition of $P_{\mathcal{E}}(\mathcal{A})$ is analogous to the work in Section 5 of [19].)

Algorithm 1: TTICAD Algorithm

Input : A list of quantifier-free formulae $\Phi = \{\phi_i\}_{i=1}^t$ in variables $x_1, \ldots, x_n$. Each ϕ_i has a designated equational constraint $f_i = 0$.
Output: Either • $\mathcal{D}$: A TTICAD of $\mathbb{R}^n$ for Φ (described by lists I and S of cell indices and sample points, respectively); or
• **FAIL**: If Φ is not well oriented (Def. 6).

for $i = 1 \ldots t$ **do**
  Set $E_i \leftarrow \{f_i\}$. Compute the finest squarefree basis F_i for $\mathrm{prim}(E_i)$;
Set $F \leftarrow \cup_{i=1}^t F_i$;
if $n = 1$ **then**
  Isolate in (I, S) the real roots of the product of the polynomials in F;
  return I and S for $\mathcal{D}$;
else
  for $i = 1 \ldots t$ **do**
    Extract the set A_i of polynomials in ϕ_i ;
    Compute the set C_i of contents of the elements of A_i; Compute the set B_i, the finest squarefree basis for $\mathrm{prim}(A_i)$;
  Set $C \leftarrow \cup_{i=1}^t C_i$, $\mathcal{B} \leftarrow (B_i)_{i=1}^t$ and $\mathcal{F} \leftarrow (F_i)_{i=1}^t$;
  Construct the projection set: $\mathfrak{P} \leftarrow C \cup P_{\mathcal{F}}(\mathcal{B})$;
  Attempt to construct a lower-dimensional CAD: $w', I', S' \leftarrow$ CADW$(n-1, \mathfrak{P})$;
  if $w' = \mathit{false}$ **then**
    return FAIL ($\mathfrak{P}$ not well oriented);
  $I \leftarrow \emptyset$; $S \leftarrow \emptyset$;
  for *each cell* $c \in \mathcal{D}'$ **do**
    $L_c \leftarrow \{\}$;
    for $i = 1, \ldots t$ **do**
      if f_i *is nullified on* c **then**
        if $\dim(c) > 0$ **then**
          return FAIL (Φ not well oriented);
        else
          $L_c \leftarrow L_c \cup B_i$;
      else
        $L_c \leftarrow L_c \cup F_i$;
    Lift over c using L_c: construct cell indices and sample points for the stack over c of the polynomials in L_c, adding them to I and S;
  return I and S for $\mathcal{D}$;

We shall prove that, provided $\mathcal{A}$ and $\mathcal{E}$ satisfy the condition of well-orientedness given in Definition 6, the output of Algorithm 1 is indeed a TTICAD for Φ. Note that this condition is specialised and new, introduced for this paper. Its requirement is due to both the use of `CADW` from [18] and the introduction of our new reduced projection operator.

We first recall the more general notion of well-orientedness from [18]. A set A of n-variate polynomials is said to be *well oriented* if whenever $n > 1$, every $f \in \text{prim}(A)$ is nullified by at most a finite number of points in $\mathbb{R}^{n-1}$, and (recursively) $P(A)$ is well-oriented. The Boolean output of `CADW` is false if the input set was not well-oriented in this sense. Now we define our new notion of well-orientedness for the set lists $\mathcal{A}$ and $\mathcal{E}$ defined above, and hence Φ.

DEFINITION 6. *We say $\mathcal{A}$ is* well oriented with respect to *$\mathcal{E}$ (and that Φ is* well oriented*) if whenever $n > 1$, every constraint polynomial f_i is nullified by at most a finite number of points in $\mathbb{R}^{n-1}$, and $P_{\mathcal{E}}(\mathcal{A})$ (hence $\mathfrak{P}$ in the algorithm) is well-oriented in the sense of [18].*

THEOREM 5. *The output of Algorithm 1 is as specified.*

PROOF. We must show that when Φ is well-oriented the output is a Truth Table Invariant CAD, (each ϕ_i has constant truth value in each cell of $\mathcal{D}$), and **FAIL** otherwise.

If the input was univariate then it is trivially well-oriented. The algorithm will construct a CAD $\mathcal{D}$ of $\mathbb{R}^1$ using the roots of the irreducible factors of the constraint polynomials (steps 5 to 6). At each 0-cell all the polynomials in each ϕ_i trivially have constant signs, and hence every ϕ_i has constant truth value. In each 1-cell no constraint polynomial has a root, so every ϕ_i has constant truth value *false*.

Now suppose $n > 1$. If $\mathfrak{P}$ is not well-oriented in the sense of [18] then `CADW` returns w' as false. In this case the input Φ is not well oriented in the sense of Definition 6 and Algorithm 1 correctly returns **FAIL**. Otherwise, $\mathfrak{P}$ is well-oriented and at step 13 we have $w' = true$. Further, I' and S' specify a CAD, $\mathcal{D}'$, order-invariant with respect to $\mathfrak{P}$. Let c, a submanifold of $\mathbb{R}^{n-1}$, be a cell of $\mathcal{D}'$.

Suppose first that the dimension of c is positive. If any constraint polynomial f_i vanishes identically on c then Φ is not well oriented in the sense of Definition 6 and the algorithm correctly returns **FAIL** at step 22. Otherwise, we know that Φ is certainly well-oriented. Since no constraint polynomial f_i vanishes then no element of the basis F vanishes identically on c either. Hence, by Theorem 3, applied with $\mathcal{A} = \mathcal{B}$ and $\mathcal{E} = \mathcal{F}$, each element of F is delineable on c, and the sections over c of the elements of F are pairwise disjoint. Thus the sections and sectors over c of the elements of F comprise a stack Σ over c. Furthermore, Theorem 3 assures us that, for each i, every element of $B_i \setminus F_i$ is sign-invariant in each section over c of every element of F_i.

Let $1 \le i \le t$. Consider first a section σ of the stack Σ. We shall show that ϕ_i has constant truth value in σ. Now the constraint polynomial f_i is a product of its content $\text{cont}(f_i)$ and some elements of the basis F_i. But $\text{cont}(f_i)$, an element of $\mathfrak{P}$, is sign-invariant in the whole cylinder $c \times \mathbb{R}$ which includes σ. Moreover all of the elements of F_i are sign-invariant in σ, as noted previously. Therefore f_i is sign-invariant in σ. If f_i is positive or negative in σ then ϕ_i has constant truth value *false* in σ.

Suppose that $f_i = 0$ throughout σ. It follows that σ must be a section of some element of the basis F_i. Let $g \in A_i \setminus E_i$ be a non-constraint polynomial in A_i. Now, by the definition of B_i, we see g can be written as $g = \text{cont}(g) h_1^{p_1} \cdots h_k^{p_k}$ where $h_j \in B_i, p_j \in \mathbb{N}$. But $\text{cont}(g)$, in $\mathfrak{P}$, is sign-invariant in the whole cylinder $c \times \mathbb{R}$ including σ. Moreover each h_j is sign-invariant in σ, as noted previously. Hence g is sign-invariant in σ. (Note that in the case where g does not have main variable x_n then $g = \text{cont}(g)$ and the conclusion still holds). Since g was an arbitrary element of $A_i \setminus E_i$, it follows that all polynomials in A_i are sign-invariant in σ, and hence that ϕ_i has constant truth value in σ.

Next consider a sector σ of the stack Σ, and notice that at least one such sector exists. As observed above, $\text{cont}(f_i)$ is sign-invariant in c, and f_i does not vanish identically on c. Hence $\text{cont}(f_i)$ is non-zero throughout c. Moreover each element of the basis F_i is delineable on c. Hence the constraint polynomial f_i is nullified by no point of c. It follows from this that the algorithm does not return **FAIL** during the lifting phase. It follows also that $f_i \neq 0$ throughout σ. Therefore ϕ_i has constant truth value *false* in σ.

It remains to consider the case in which the dimension of c is 0. In this case the roots of the polynomials in the lifting set L_c constructed by the algorithm determine a stack Σ over c. Each ϕ_i trivially has constant truth value in each section (0-cell) of this stack, and the same can routinely be shown for each sector (1-cell) of this stack. □

REMARK 6. *When the input to Algorithm 1 is a single QFF then it produces a CAD which is invariant with respect to the sole equational constraint. This may be shown using the results of [19] alone. However, we note that Algorithm 1 is actually more efficient in the lifting stage than the modified QEPCAD algorithm discussed in [19] since the lifting set excludes some non-equational constraint input polynomials.*

Algorithm 1 and Definition 6 have been kept conceptually simple to aid readability. However in practice the algorithm may sometimes be unnecessarily cautious. In [4], several cases where non-well oriented input can still lead to an order-invariant CAD are discussed. Similarly here, we can sometimes allow the nullification of an equational constraint on a positive dimensional cell.

LEMMA 7. *Let f_i be an equational constraint which vanishes identically on a cell $c \in \mathcal{D}'$ constructed during Algorithm 1. If all polynomials in $\text{ExclP}_{E_i}(A_i)$ are constant on c then any $g \in A_i \setminus E_i$ will be delineable over c.*

PROOF. Suppose first that A_i and E_i satisfy the simplifying conditions from Section 2.2. Rearranging (3) we see

$$P(A_i) = P_{E_i}(A_i) \cup \text{ExclP}_{E_i}(A_i).$$

However, given the conditions of the lemma, this is equivalent (after the removal of constants which do not affect CAD construction) to $P_{E_i}(A_i)$ on c. So here $P(A_i)$ is a subset of $P_{\mathcal{E}}(\mathcal{A})$ and we can conclude by Theorem 1 that all elements of A_i vanish identically on c or are delineable over c.

In the more general case we can still draw the same conclusion because $P(A_i) = C_i \cup P_{F_i}(B_i) \cup \text{ExclP}_{F_i}(B_i) \subseteq \mathfrak{P}$. □

Hence we can use Lemma 7 to safely extend step 24 to also apply in such cases. In particular, we can allow equational constraints f_i which do not have main variable x_n in such cases. We have included this in our implementation discussed in Section 3.3. In theory, we may be able to go further and allow step 24 to apply in cases where the polynomials in $\text{ExclP}_{E_i}(A_i)$ are not necessarily all constant, but have no real roots within the cell c. However, identifying such cases would require answering a separate quantifier elimination question, which may not be trivial.

3.2 TTICAD via the ResCAD Set

In Algorithm 1 the lifting stage (steps 16 to 27) varies according to whether an equational constraint is nullified. When this does not occur there is an alternative implementation of TTICAD which would be simpler to introduce into existing CAD algorithms. Define the *ResCAD Set* of Φ as

$$\mathcal{R}(\Phi) = E \cup \bigcup_{i=1}^{t} \{\mathrm{res}_{x_n}(f,g) \mid f \in E_i, g \in A_i, g \notin E_i\}.$$

THEOREM 8. *Let $\mathcal{A} = (A_i)_{i=1}^t$ be a list of irreducible bases A_i and let $\mathcal{E} = (E_i)_{i=1}^t$ be a list of non-empty subsets $E_i \subseteq A_i$. For the McCallum projection operator P, [17] we have:*

$$P(\mathcal{R}(\Phi)) = P_{\mathcal{E}}(\mathcal{A}).$$

The proof is straightforward and so omitted here.

COROLLARY 9. *If no f_i is nullified by a point in $\mathbb{R}^{n-1}$ then inputting $\mathcal{R}(\Phi)$ into any algorithm which produces a sign-invariant CAD using McCallum's projection operator, will result in the TTICAD for Φ produced by Algorithm 1.*

Hence Corollary 9 gives us a simple way to compute TTICADs using existing CAD implementations, such as QEPCAD, but this cannot be applied as widely as Algorithm 1.

3.3 Implementation in Maple

There are various implementations of CAD available but none guarantee order-invariance, required for proving the validity of our TTICAD algorithm. Hence we needed to construct our own implementation to obtain experimental results. We built an implementation of McCallum projection, so that we could reproduce `CADW` and modified the existing stack generation commands in MAPLE from [8] so they could be used more widely. Together these allowed us to fully implement Algorithm 1. The CAD implementation grew to a MAPLE package `ProjectionCAD` which gathers together algorithms for producing CADs via projection and lifting to complement the existing CAD commands in MAPLE which use triangular decomposition, giving the same representation of sample points using regular chains. For further details (along with free access to the code) see [15].

3.4 Formulating a Problem for TTICAD

When formulating a problem for TTICAD there may be choices for the input, such as choosing which equational constraint to designate in a QFF when more than one is present. Other possibilities include choosing whether conjunctions of formulae should be split into separate QFFs. Usually it will be preferable to minimise the number of QFFs, but if for example a designated equational constraint has many intersections with another polynomial which could be ignored by using separate QFFs, then the cost of the extra polynomials in the projection set may be outweighed by the complexity of those removed. Hence it is worth taking care in how we formulate the TTICAD. A simple problem of the form

$$f_1 = 0 \wedge f_2 = 0 \wedge g_1 < 0 \wedge g_2 < 0$$

has six acceptable choices for the composition of Φ.

We have started exploring heuristics for choosing the best composition. The metric `sotd` (sum of total degrees) as defined in [14] may be used to approximate the complexity of polynomials. We first considered using $\mathtt{sotd}(\mathfrak{P})$ and found that while it was fairly well correlated with the number of cells produced by Algorithm 1 it was not always fine enough to separate compositions leading to TTICADs with significantly different numbers of cells. Hence we prefer a stronger heuristic, $\mathtt{sotd}(\mathfrak{P} \cup \overline{P}(\mathfrak{P}))$ where $\overline{P}$ is the complete set of projection polynomials obtained by repeatedly applying P.

For the problems in Section 4 we used the QFFs imposed by the disjunctions of formulae using this heuristic to choose which equational constraints are designated when there was a choice. For these problems the heuristic computation time was negligible compared to the overall time, but for larger problems this would not be the case. Work on heuristics is ongoing with a more detailed report available in [1].

4. EXPERIMENTAL RESULTS

4.1 Description of experiments

Our timings were obtained on a Linux desktop (3.1GHz Intel processor, 8.0Gb total memory) with MAPLE 16 (command line interface), MATHEMATICA 9 (graphical interface) and QEPCAD-B 1.69. For each experiment we produce a CAD and give the time taken and number of cells (cell count). The first is an obvious metric while the second is crucial for applications performing operations on each cell.

For QEPCAD the options `+N500000000` and `+L200000` were provided, the initialization included in the timings and explicit equational constraints declared when present with the product of those from the individual QFFs declared otherwise. In MATHEMATICA the output is not a CAD but a formula constructed from one [24], with the actual CAD not available to the user. Cell counts for the algorithms were provided by the author of the MATHEMATICA code.

TTICADs are calculated using our implementation described in Section 3.3, which is simple and not optimized. The results in this section are not presented to claim that our implementation is state of the art, but to demonstrate the power of the TTICAD theory over the the conventional theory, and how it can allow even a simple implementation to compete. Hence the cell counts are of most interest.

The time is measured to the nearest tenth of a second, with a time out (T/O) set at 5000 seconds. When **F** occurs it indicates failure due to a theoretical reason such as not well-oriented (in either sense). The occurrence of Err indicates an error in an internal subroutine of MAPLE's **RegularChains** package, used by **ProjectionCAD**. This error is not theoretical but a bug, beyond our control.

We considered examples originating from [7]. However these problems (and most others in the literature) involve conjunctions of conditions, chosen as such to make them amenable to existing technologies. These problems can be tackled using TTICAD, but they do not demonstrate its full strength. Hence we introduced new examples, denoted with a †, which are adapted from [7] to have disjuncted QFFs.

Two examples came from the application of branch cut analysis for simplification. These problems require a decomposition according to branch cuts of the form $f = 0 \wedge g < 0$, and then go on to test the validity of a simplification on each cell, [21, etc.]. We need to consider the disjunction of the branch cuts making such problems suitable for Algorithm 1. We included a key example from Kahan [16], along with the problem induced by considering the validity of the double angle formulae for arcsin. Finally we considered the worked example from Section 1.3 and its generalisation to three dimensions. Note that A and B following the problem name indicate different variable orderings. Full details for all examples can all be found in the CAD repository [25] available freely online at `http://opus.bath.ac.uk/29503`.

4.2 Results

We present our results in Table 1. For each problem we give the name used in the repository, n the number of variables, d the maximum degree of polynomials involved and t the number of QFFs used for TTICAD. We then give the time taken and number of cells produced by each algorithm.

We first compare our TTICAD implementation with the sign-invariant CAD generated using `ProjectionCAD` with McCallum's projection operator [15]. Since these use the same architecture the comparison makes clear the benefits of the TTICAD theory. The experiments confirm the fact that the cell count for TTICAD will always be less than or equal to that of a sign-invariant CAD produced using the same implementation. Ellipse† A is not well-oriented in the sense of [18], and so both methods return **FAIL**. Solotareff† A and B are well-oriented in this sense but not in the stronger sense of Definition 6 and hence TTICAD fails while the full sign-invariant CADs can be produced. The only example with equal cell counts is Collision† A in which the non-equational constraints were so simple that the projection polynomials were unchanged. Examining the results for the worked example and its generalisation we start to see the true power of TTICAD. In 3D Example A we see a 759-fold reduction in time and a 50-fold reduction in cell count.

We next compare our implementation of TTICAD with the state of the art in CAD: QEPCAD [3], MAPLE [8] and MATHEMATICA [23, 24]. MATHEMATICA is the quickest, however TTICAD often produces fewer cells. We note that MATHEMATICA's algorithm uses powerful heuristics and so actually used Gröbner bases on the first two problems, causing the cell counts to be so low. When all implementations succeed TTICAD usually produces far fewer cells than QEPCAD or MAPLE, especially impressive given QEPCAD is producing partial CADs for the quantified problems, while TTICAD is only working with the polynomials involved. For Collision† A the TTICAD theory offers no benefit allowing the better optimized alternatives to have a lower cell count.

Reasons for the TTICAD implementation struggling to compete on speed in general are that the MATHEMATICA and QEPCAD algorithms are largely implemented directly in `C`, have had far more optimization, and in the case of MATHEMATICA use validated numerics for lifting [23]. However, the strong performance in cell counts is very encouraging, both due its importance for applications where CAD is part of a wider algorithm (such as branch cut analysis) and for the potential if TTICAD theory were implemented elsewhere.

5. CONCLUSIONS

We have defined Truth Table Invariant CADs, which can be more closely aligned to the needs of problems than traditional sign-invariant CADs. Theorem 3 extended the theory of equational constraints allowing us to develop Algorithm 1 to construct TTICADs efficiently for a large range of problems. The algorithm has been implemented in MAPLE giving promising experimental results. TTICADs in general have less cells than full sign-invariant CADs using the same implementation and we showed that this allows even a simple implementation of TTICAD to compete with the state of the art CAD implementations. It is anticipated that future implementations of TTICAD could be far better optimized leading to lower times for the same cell counts. We also note that the benefits of TTICAD increase with the number of QFFs in a problem and so larger problems may be susceptible to TTICAD when other approaches fail.

We hope that these results inspire other implementations of TTICAD, with Corollary 9 showing a particularly easy way to adapt existing CAD implementations.

5.1 Future Work

There is scope for optimizing the algorithm and extending it to allow less restrictive input. Lemma 7 gives one extension that is included in our implementation while other possibilities include removing some of the caution implied by well-orientedness, analogous to [4]. Also, work developing heuristics for composing the input is underway in [1].

Of course, the implementation of TTICAD used here could be improved in many ways, but perhaps more desirable would be for TTICAD to be incorporated into existing state of the art CAD implementations. In particular, we would like to use the existing MAPLE CAD commands [8] but this requires first understanding when they give order-invariance, a key question currently under consideration. We see several possibilities for the theoretical development of TTICAD:

- Can we apply the theory recursively instead of only at the top level? For example by widening the projection operator to conclude order-invariance, as in [20].
- Can we apply TTICAD to forms of QFF other than "one equality and other items"? For example, can we generalise the theory of bi-equational constraints?
- Can we make use of the ideas behind partial CAD to avoid unnecessary lifting once the truth value of a QFF on a cell is determined?
- Can anything be done when Φ is not well oriented?
- Can we implement the lifting algorithm in parallel?

Acknowledgements

We are grateful to A. Strzeboński for assistance in performing the Mathematica tests and to the anonymous referees for useful comments. We also thank the rest of the Triangular Sets seminar at Bath (A. Locatelli, G. Sankaran and N. Vorobjov) for their input, and the team at Western University (C. Chen, M. Moreno Maza, R. Xiao and Y. Xie) for access to their MAPLE code and helpful discussions. This work was supported by the EPSRC grant: EP/J003247/1.

6. REFERENCES

[1] R. Bradford, J.H. Davenport, M. England, and D. Wilson. Optimising Problem Formulation for Cylindrical Algebraic Decomposition. In Press: *Proc. CICM '13*. Preprint: http://opus.bath.ac.uk/34373.

[2] C.W. Brown. Simplification of truth-invariant cylindrical algebraic decompositions. *Proc. ISSAC '98*, pages 295–301, 1998.

[3] C.W. Brown. QEPCAD B: A program for computing with semi-algebraic sets using CADs. *ACM SIGSAM Bulletin*, 37:4, pages 97–108, 2003.

[4] C.W. Brown. The McCallum projection, lifting, and order-invariance. Technical report, U.S. Naval Academy, Computer Science Department, 2005.

[5] C.W. Brown and J.H. Davenport. The Complexity of Quantifier Elimination and Cylindrical Algebraic Decomposition. *Proc. ISSAC '07*, pages 54–60, 2007.

Table 1: Comparing TTICAD to the full CAD built with the same architecture and other CAD algorithms.

Problem				Full-CAD		TTICAD		QEPCAD		MAPLE		MATHEMATICA	
Name	n	d	t	Time	Cells	Time	Cells	Time	Cells	Time	Cells	Time	Cells
Intersection A	3	2	1	360.1	3707	1.7	269	4.5	825	—	Err	0.0	3
Intersection B	3	2	1	332.2	2985	1.5	303	4.5	803	50.2	2795	0.0	3
Random A	3	3	1	268.5	2093	4.5	435	4.6	1667	23.0	1267	0.1	657
Random B	3	3	1	442.7	4097	8.1	711	5.4	2857	48.1	1517	0.0	191
Intersection† A	3	2	2	360.1	3707	68.7	575	4.8	3723	—	Err	0.1	601
Intersection† B	3	2	2	332.2	2985	70.0	601	4.7	3001	50.2	2795	0.1	549
Random† A	3	3	2	268.5	2093	223.4	663	4.6	2101	23.0	1267	0.2	808
Random† B	3	3	2	442.7	4097	268.4	1075	142.4	4105	48.1	1517	0.2	1156
Ellipse† A	5	4	2	—	**F**	—	**F**	291.6	500609	1940.1	81193	11.2	80111
Ellipse† B	5	4	2	T/O	—	T/O	—	T/O	—	T/O	—	2911.2	16603131
Solotareff† A	4	3	2	677.6	54037	46.1	**F**	4.9	20307	1014.2	54037	0.1	260
Solotareff† B	4	3	2	2009.2	154527	123.8	**F**	6.3	87469	2951.6	154527	0.1	762
Collision† A	4	4	2	264.6	8387	267.7	8387	5.0	7813	376.4	7895	3.6	7171
Collision† B	4	4	2	—	Err	—	Err	T/O	—	T/O	—	591.5	1234601
Kahan A	2	4	7	10.7	409	0.3	55	4.8	261	15.2	409	0.0	72
Kahan B	2	4	7	87.9	1143	0.3	39	4.8	1143	154.9	1143	0.1	278
Arcsin A	2	4	4	2.5	225	0.3	57	4.6	225	3.3	225	0.0	175
Arcsin B	2	4	4	6.5	393	0.2	25	4.5	393	7.8	393	0.0	79
2D Example A	2	2	2	5.7	317	1.2	105	4.7	249	6.3	317	0.0	24
2D Example B	2	2	2	6.1	377	1.5	153	4.5	329	7.2	377	0.0	175
3D Example A	3	3	2	3795.8	5453	5.0	109	5.3	739	—	Err	0.1	44
3D Example B	3	3	2	3404.7	6413	5.8	153	5.7	1009	—	Err	0.1	135

[6] C.W. Brown and S. McCallum. On using bi-equational constraints in CAD construction. *Proc. ISSAC '05*, pages 76–83, 2005.

[7] B. Buchberger and H. Hong. Speeding-up Quantifier Elimination by Gröbner Bases. RISC Technical Report 91-06, 1991.

[8] C. Chen, M. Moreno Maza, B. Xia, and L. Yang. Computing Cylindrical Algebraic Decomposition via Triangular Decomposition. *Proc. ISSAC '09*, pages 95–102, 2009.

[9] G.E. Collins. Quantifier Elimination for Real Closed Fields by Cylindrical Algebraic Decomposition. *Proc. 2nd. GI Conference Automata Theory & Formal Languages*, pages 134–183, 1975.

[10] G.E. Collins. Quantifier elimination by cylindrical algebraic decomposition — twenty years of progess. *Quantifier Elimination and Cylindrical Algebraic Decomposition*, pages 8–23, 1998.

[11] G.E. Collins and H. Hong. Partial Cylindrical Algebraic Decomposition for Quantifier Elimination. *J. Symbolic Comp.*, 12:3, pages 299–328, 1991.

[12] J.H. Davenport, R. Bradford, M. England, and D. Wilson. Program verification in the presence of complex numbers, functions with branch cuts etc. *Proc. SYNASC '12*, pages 83–88, 2012.

[13] J.H. Davenport and J. Heintz. Real Quantifier Elimination is Doubly Exponential. *J. Symbolic Comp.*, 5:1-2, pages 29–35, 1988.

[14] A. Dolzmann, A. Seidl, and Th. Sturm. Efficient Projection Orders for CAD. *Proc. ISSAC '04*, pages 111–118, 2004.

[15] M. England. An implementation of CAD utilising McCallum projection in Maple. *University of Bath, Dept. Computer Science Technical Report Series*, 2013:2. http://opus.bath.ac.uk/33180, 2013.

[16] W. Kahan. Branch Cuts for Complex Elementary Functions. A. Iserles and M.J.D. Powell, editorss, *Proc. The State of the Art in Numerical Analysis*, pages 165–211, 1987.

[17] S. McCallum. An Improved Projection Operation for Cylindrical Algebraic Decomposition of Three-dimensional Space. *J. Symbolic Comp.*, 5:1-2, pages 141–161, 1988.

[18] S. McCallum. An Improved Projection Operation for Cylindrical Algebraic Decomposition. *Quantifier Elimination and Cylindrical Algebraic Decomposition*, pages 242–268, 1998.

[19] S. McCallum. On Projection in CAD-Based Quantifier Elimination with Equational Constraints. *Proc. ISSAC '99*, pages 145–149, 1999.

[20] S. McCallum. On Propagation of Equational Constraints in CAD-Based Quantifier Elimination. *Proc. ISSAC '01*, pages 223–230, 2001.

[21] N. Phisanbut, R.J. Bradford, and J.H. Davenport. Geometry of Branch Cuts. *Communications in Computer Algebra*, 44:132–135, 2010.

[22] J.T. Schwartz and M. Sharir. On the "Piano-Movers" Problem: II. General Techniques for Computing Topological Properties of Real Algebraic Manifolds. *Adv. Appl. Math.*, 4:298–351, 1983.

[23] A. Strzeboński. Cylindrical algebraic decomposition using validated numerics. *Journal of Symbolic Computation*, 41:9, pages 1021–1038, 2006.

[24] A. Strzeboński. Computation with semialgebraic sets represented by cylindrical algebraic formulas. *Proc. ISSAC '10*, pages 61–68. ACM, 2010.

[25] D.J. Wilson, R.J. Bradford, and J.H. Davenport. A Repository for CAD Examples. *ACM Communications in Computer Algebra*, 46:3 pages 67–69, 2012.

Constructing a Single Open Cell in a Cylindrical Algebraic Decomposition

Christopher W. Brown
Computer Science Department, Stop 9F
United States Naval Academy
572M Holloway Road
Annapolis, MD 21402
wcbrown@usna.edu

ABSTRACT

This paper presents an algorithm that, roughly speaking, constructs a single open cell from a cylindrical algebraic decomposition (CAD). The algorithm takes as input a point and a set of polynomials, and computes a description of an open cylindrical cell containing the point in which the input polynomials have constant non-zero sign, provided the point is sufficiently generic. The paper reports on a few example computations carried out by a test implementation of the algorithm, which demonstrate the functioning of the algorithm and illustrate the sense in which it is more efficient than following the usual "open CAD" approach. Interest in the problem of computing a single cell from a CAD is motivated by a 2012 paper of Jovanovic and de Moura that require solving this problem repeatedly as a key step in NLSAT system. However, the example computations raise the possibility that repeated application of the new method may in fact be more efficient than the usual open CAD approach, both in time and space, for a broad range of problems.

Categories and Subject Descriptors

G.4 [**Mathematics of Computation**]: Mathematical software—*Algorithm design and analysis*

Keywords

cylindrical algebraic decomposition; polynomial inequalities

1. INTRODUCTION

In a 2012 paper [3], Jovanovic and de Moura present NLSAT, a novel algorithm that uses Conflict-Driven Clause Learning (CDCL)-style search to decide the satisfiability of systems of polynomial equalities and inequalities over the real numbers. An essential step in this algorithm is to take an assignment of real values that does not satisfy the original system, and generalize it to a larger region in which all points fail to satisfy the original system. They do this by constructing a single cell in a Cylindrical Algebraic Decomposition (CAD) defined from a set of polynomials — namely the cell that contains the unsatisfying assignment. This problem of constructing a single cell in a CAD efficiently has not, to the best of our knowledge, been addressed in the literature prior to Jovanovic and de Moura's work. The problem is important in light of the success NLSAT has already demonstrated, but also, as will be described later in the paper, it is a problem that has the potential to be important in other contexts as well. This paper takes the first steps towards a dedicated study of this problem, providing an efficient algorithm for constructing and computing with individual *open* cells in CADs, and reporting on the performance of a test implementation of this algorithm. Adapting and extending the algorithm to lower-dimensional cells is an interesting subject for future work.

Roughly speaking, the problem we consider is this: given a set P of polynomials in $\mathbb{R}[x_1, \ldots, x_n]$ and a point $\alpha \in \mathbb{R}^n$, compute the full dimensional cell from the CAD defined by P that contains point α, assuming α is not a zero of any of the polynomials in P. Jovanovic and de Moura gain an advantage over a straight-forward application of CAD to determine the satisfiability of a formula in several ways, but the two that are relevant here are that: 1) the set of polynomials they start with when they build a single CAD cell to generalize conflicting assignment is often smaller than the full set of polynomials in the input formula[1], and 2) in lifting they only lift above the one cell at each level containing the given point α. The fundamental idea behind this paper is that the projection operator can be substantially improved when we restrict ourselves to constructing a single cell containing the point α. In particular, the present paper describes how this can be done when the goal is to construct an open CAD cell containing α. Restricting the problem to open cells simplifies the algorithm, correctness proof, and implementation. However, the approach should be generalizable to constructing cells of lower dimension, and such a generalization is planned as future work.

A precise formulation of the problem appears below. One of the issues that must be addressed in a precise formulation is that smaller projection factor sets generally result in larger CAD cells. Thus, the cell that we construct is often a superset of the cell containing α in the full CAD

[1] A more strictly correct statement of this requires comparing the projection closure of the set of polynomials appearing in the input formula to the set of polynomials they start with in building the single CAD cell.

ISSAC'13, June 26–29, 2013, Boston, Massachusetts, USA.
ACM 978-1-4503-2059-7/13/06.

constructed in the usual way from the same initial set of polynomials. Although this complicates the description of the problem, it is a good thing! In the context of Jovanovic and de Moura's NLSAT procedure, for instance, a larger cell means a stronger generalization of the conflicting assignment.

1.1 The precise problem

This paper assumes some familiarity with CAD (Cylindrical Algebraic Decomposition) on the part of the reader. In particular, the paper requires the concepts behind and algorithms for constructing an "Open CAD" for a set of polynomials (see for example [4] and [6]), which are actually much simpler than their counterparts for standard CADs. Those unfamiliar with the subject of CAD, may wish to look at the introduction given in [1] which, though quite old, provides a clearly-written, well-explained starting point for the subject. For those familiar with CAD, but not Open CAD, we give a very brief overview.

Generally, CAD construction starts with a set A of input polynomials over some ordered set of variables, $x_1 \prec x_2 \prec \cdots \prec x_n$. The goal is to construct a decomposition of $\mathbb{R}^n$ into cells in which the elements of A have constant sign, and to do it in such a way that the regions in the decomposition have a special "cylindrical" arrangement with respect to the variable ordering. In the case of an *Open CAD* all cells are open sets, and we do not require $\mathbb{R}^n$ to be decomposed in the strict sense, rather we allow some points to be "missed", as long as the set of points in $\mathbb{R}^n$ that are not contained in any cell has measure zero. Usually, CAD construction proceeds in two stages: *projection*, and *lifting*. Projection produces a set $P \subseteq A$, called the projection factor set, that implicitly defines a CAD. Lifting produces an explicit data structure representing the CAD implicitly defined by P. This data structure can be queried to provide a sample point from each cell in the CAD, and a semi-algebraic description of each cell in the CAD.

The projection process is usually defined in terms of a *projection operator*, which eliminates a variable from a set of projection factors, thereby creating new projection factors. The projection operator is applied recursively until all but one variable is eliminated, and the set of input polynomials along with all the polynomials created by applications of the projection operator is the projection factor set. In the case of Open CAD we would use the "Open-McCallum projection" operator, which is defined as follows:[2] If P_k is a set of irreducible polynomials in $\mathbb{R}[x_1, \ldots, x_k]$ of positive degree in x_k, then $\mathrm{ProjMOpen}_k(P_k)$ is the set of irreducible factors of

$$\bigcup_{p \in P_k} \mathrm{ldcf}_{x_k}(p) \cup \bigcup_{p \in P_k} \mathrm{disc}_{x_k}(p) \cup \bigcup_{p,q \in P_k, p \neq q} \mathrm{res}_{x_k}(p, q).$$

The closure of a set of polynomials $P \subset \mathbb{R}[x_1, \ldots, x_n]$ under the Open-McCallum projection is the set P^* computed as follows:

$P^* := P$
for k from n down to 2 do
  $P^* := P^* \cup \mathrm{ProjMOpen}_k(\{p \in P^* | \text{the level of } p \text{ is } k\})$

Given a finite set of input polynomials $P \subset \mathbb{R}[x_1, \ldots, x_n]$, an Open CAD for P is constructed by computing P^*, the closure of P under the Open-McCallum projection, and following the usual CAD lifting procedure with projection factor set P^* with the restriction that lifting is only done over full-dimensional cells, i.e. cells whose level and dimension are equal. The full-dimensional cells of level n comprise the cells of the Open CAD.

[2]This description uses the term *level* (see Definition 1).

We are now ready to give a precise statement of the problem we want to solve:

The Open-Cell Problem Given a set P of polynomials in $\mathbb{R}[x_1, \ldots, x_n]$ and a point $\alpha \in \mathbb{R}^n$, compute a description of an open cylindrical algebraic cell $C \subseteq \mathbb{R}^n$ containing point α and in which the elements of P have constant, non-zero sign such that, if F is the projection closure of P under the Open-McCallum projection, C contains at least one open maximal connected region in which the elements of F have constant non-zero sign. If any element of F is zero at α, FAIL is an acceptable result.

1.2 This paper's contribution

The problem of constructing a single cell of a CAD does not seem to have been considered in the literature prior Jovanovic and de Moura's paper. As described earlier, the approach we take goes beyond what they describe in their paper because it actually reduces the size of the set of projection polynomials, which both speeds up the process of constructing the cell and produces a larger cell — both of which are improvements. The specific contributions of this paper are:

1. OC-CONSTRUCT a simple, efficient algorithm for the Open-Cell problem,
2. a formal proof of correctness for OC-CONSTRUCT, and
3. empirical examples of the operation of OC-CONSTRUCT based on a test implementation.

The remainder of this paper is organized as follows: Section 2 discusses the basic ideas that lead to the algorithms presented later in the paper. An intuitive feeling for these ideas will aid in reading the remainder of the paper. Section 3 describes the OpenCell data structure — which is what the Open-Cell Problem is asking us to produce. Some important properties of this data structure are proved. Section 4 gives the Algorithm OC-CONSTRUCT and proves its correctness. Finally, Section 5 provides some example computations with our test implementation that illustrate how OC-CONSTRUCT works, and why it has the real potential to be the basis for more efficient algorithms to solve some problems than the usual CAD approach can provide.

2. THE BASIC IDEA

The fundamental idea behind the algorithms proposed in this paper is best understood from a simple example. Suppose that we want to construct a single open cell from the point $\alpha = (-1/2, 1/2)$ and input polynomials $P = \{y^2 - 1 - x^2, y - x - 4, y + x - 5\}$. The CAD that would normally be constructed from this set, is shown on the left in Figure 1, with the cell containing $(-1/2, 1/2)$ highlighted.

The three input polynomials appear as the line with positive slope, the line with negative slope and the "hour glass" curve. The vertical lines are defined by the zeros of the polynomials resulting from the projection of the input polynomials.

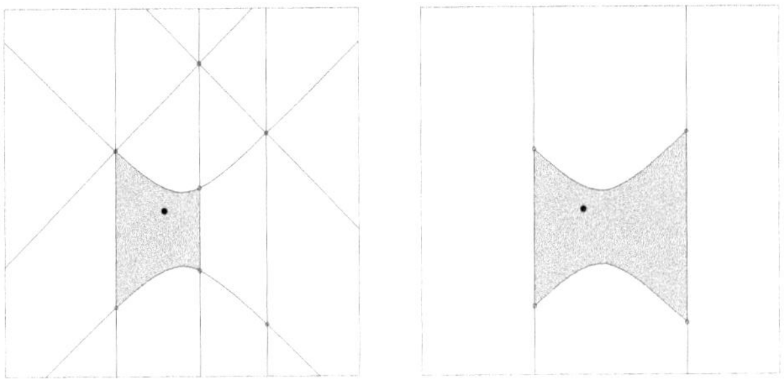

Figure 1: On the left is the full CAD. On the right is the single open cell constructed by OC-Construct. In both plots, the point α is shown by a black dot.

The important observation is that the right boundary of the cell containing α is defined by the intersection of the two non-vertical lines, but that intersection point lies outside of any region containing α in which the signs of the input polynomials stay constant. The algorithm we present recognizes that neither $V(y-x-4)$ nor $V(y+x-5)$ are part of the boundary of the cell containing α we will ultimately construct, and therefore their intersections with one another are irrelevant. Thus the algorithm deduces that the resultant of $y-x-4$ and $y+x-5$ can be left out of the projection. The right side of Figure 1 shows the cell OC-CONSTRUCT constructs from this input. The cell is larger, and fewer distinct polynomials are required to define its boundaries.

Making this optimization requires us to be able to recognize which polynomials form the boundaries of the cell we are constructing. This is easy to do because we start with the point α that identifies which particular cell we want to construct. Continuing with our example problem, we substitute all but the last coordinate of α into our input polynomials to get univariate polynomials in the main variable – y in this case. We isolate the roots of the three resulting univariate polynomials, and order them and the last coordinate of α. The result is shown in Figure 2.

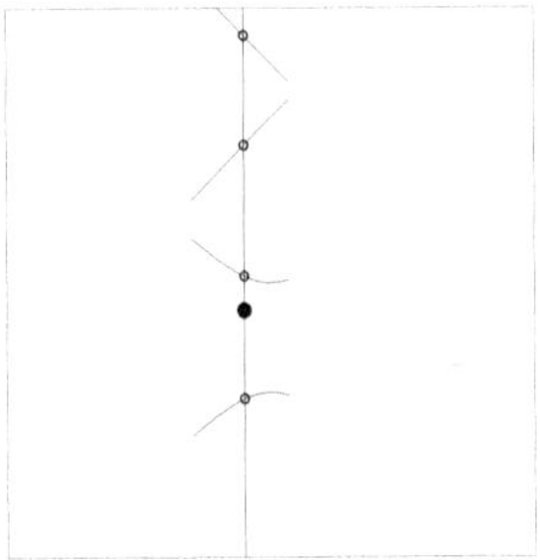

Figure 2: Ordered roots of the input polynomials (circles) and α (filled dot).

The roots immediately above and below the final coordinate of α identify which of the original input polynomials will form the upper and lower boundary of the cell we ultimately construct. (In higher dimensions, we can do the exact same thing — still only ever requiring univariate root isolation.) Generalizing the example, intersections of the varieties of polynomials that do not define the upper and/or lower boundaries of the cell we are trying to construct are irrelevant, and therefore pairwise resultants of those polynomials need not be included in the projection.

3. THE OPENCELL DATA STRUCTURE

Our goal is to construct a representation of a single open cell containing the point α. Before we can give an algorithm for computing this, we must define the representation — we must define the data structure we will use to represent the cell. We call this the *OpenCell data structure*, and in this section we define the structure and prove several important properties about it.

CADs are defined with respect to some variable ordering. In this paper, without loss of generality, we assume the order $x_1 \prec x_2 \prec \cdots \prec x_n$.

DEFINITION 1. *The* level *of a non-constant polynomial* $p \in \mathbb{R}[x_1, \ldots, x_n]$ *is the largest* k *such that* $deg_{x_k}(p) > 0$.

DEFINITION 2. *For* $p \in \mathbb{R}[x_1, \ldots, x_n]$ *and non-negative integer* i, *we define the* indexed root expression *root*(p, i, x_k) *at the point* $\gamma = (\gamma_1, \ldots, \gamma_n)$ *as the ith distinct real root of* $p(\gamma_1, \ldots, \gamma_{k-1}, x_k)$ *(ordered smallest to largest). If the level of* p *is not* k, *if polynomial* $p(\gamma_1, \ldots, \gamma_{k-1}, x_k)$ *is the zero polynomial or if it has fewer than* i *distinct real roots, the expression has value* UNDEF. *Note that this is a more restricted notation than the notation introduced in [2]. We also allow the special indexed root expressions* root$(+\infty, 1, x_k)$ *and* root$(-\infty, 1, x_k)$ *to represent positive and negative infinity, respectively. We won't do any arithmetic with these expressions, so those semantics won't be addressed.*

We formally extend the usual relational operators $\{<, >, \leq, \geq, =, \neq\}$ to be false if either the left or right hand sides is *undef*, and we allow the expressions root$(+\infty, 1, x_k)$ and root$(-\infty, 1, x_k)$ on the left and right side of the relational operators with the obvious semantics.

DEFINITION 3. *A* RealAlgNum *is a 4-tuple* (p, I, t, j) *where* p *is a squarefree univariate polynomial,* I *is an isolating interval for the* j*th distinct real root of* p *ordered from smallest to largest, and* t *is the trend of* p *at the root in* I, *i.e. the sign of* p' *at the root. If* A *is a RealAlgNum we adopt the notation* $A.p$, $A.I$, $A.t$ *and* $A.j$ *to refer to the four components of the tuple. We refer to the real number that is the* $A.j$*th distinct real root of* $A.p$ *as val*(A). *To simplify the presentation below, we allow* $A.p$ *to be* $+\infty$ *or* $-\infty$, *in which case val*(A) *is* $+\infty$ *or* $-\infty$, *respectively.*

DEFINITION 4. *An* OpenCell Data Structure D *containing rational point* $\alpha = (\alpha_1, \ldots, \alpha_k) \in \mathbb{R}^k$ *is a list* $D[1], \ldots, D[k]$ *of 4-tuples* (l, L, u, U), *where* $l \in \mathbb{R}[x_1, \ldots, x_k]$ *and* L *is a RealAlgNum or* $l = -\infty$, *and* $u \in \mathbb{R}[x_1, \ldots, x_k]$ *and* U *is a RealAlgNum or* $u = +\infty$, *and the following additional requirements hold: Note:* $S(D)$ *and* $F(D)$ *are defined below, but used in this definition.*

1. $D[1], \ldots, D[k-1]$ *is an Open Cell Data Structure containing the point* $(\alpha_1, \ldots, \alpha_{k-1})$.
2. $D[k].l = -\infty$ *and* $D[k].L = (-\infty, [0, 0], 0, 1)$ *or*
   (a) $D[k].l$ *is irreducible and of positive degree in* x_k *and* $D[k].l(\alpha_1, \ldots, \alpha_{k-1}, x_k) = D[k].L.p(x_k)$.
   (b) $disc_{x_k}(D[k].l)$ *and* $ldcf_{x_k}(D[k].l)$ *have constant non-zero sign on* $S(D[1], \ldots, D[k-1])$.

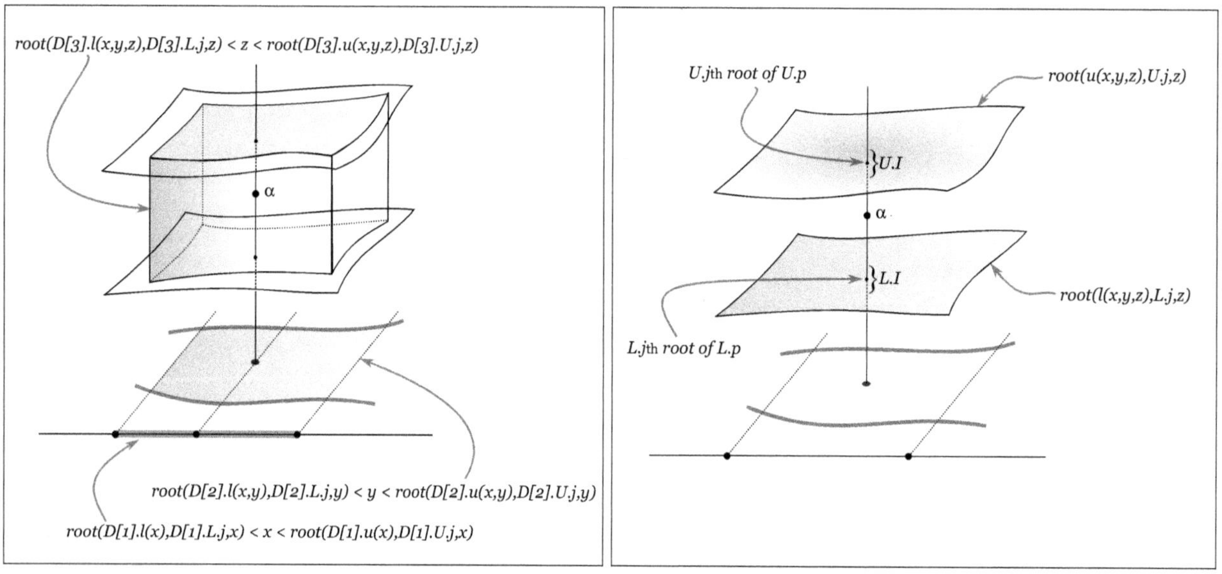

Figure 3: The diagram on the left shows how the elements of an OpenCell data structure D taken together define an open cell. The shaded regions are $S(D[1])$, $S(D[1], D[2])$ and $S(D[1], D[2], D[3])$. The diagram on the right shows how univariate root isolation and the components of the OpenCell data structure identify which sections of which polynomials define the upper and lower boundaries of the open cell defined by the data structure.

   (c) *$D[k].L.p$ has no root in the interval $(val(D[k].L), val(D[k].U))$.*

3. *$D[k].u = +\infty$ and $D[k].U = (+\infty, [0, 0], 0, 1)$ or*

   (a) *$D[k].u$ is irreducible and of positive degree in x_k and $D[k].u(\alpha_1, \ldots, \alpha_{k-1}, x_k) = D[k].U.p(x_k)$.*

   (b) *$disc_{x_k}(D[k].u)$ and $ldcf_{x_k}(D[k].u)$ have constant non-zero sign on $S(D[1], \ldots, D[k-1])$.*

   (c) *$D[k].U.p$ has no root in the interval $(val(D[k].L), val(D[k].U))$.*

4. *If $D[k].l \neq -\infty$ and $D[k].u \neq +\infty$, $res_{x_k}(D[k].l, D[k].u)$ has constant non-zero sign on $S(D[1], \ldots, D[k-1])$.*

5. *The following formula is satisfied at point α:*

$$root(D[k].l, D[k].L.j, x_k) < \alpha_k < root(D[k].u, D[k].U.j, x_k)$$

Figure 3 provides two diagrams that illustrate how the OpenCell data structure defines an actual open cell containing point α.

DEFINITION 5. *Let D be an OpenCell data structure containing point $\alpha \in \mathbb{R}^k$. We define $F(D)$ to be true if D is the empty list, and otherwise*

$$F(D[1], \ldots, D[k-1]) \wedge$$
$$root(D[k].l, D[k].L.j, x_k) < x_k < root(D[k].u, D[k].U.j, x_k).$$

We define $S(D)$ as $\mathbb{R}^0$ if D is the empty list, and otherwise

$$S(D) = \{\gamma \in \mathbb{R}^n \mid F(D) \text{ is satisfied at } \gamma\}.$$

THEOREM 1. *Let D be an OpenCell data structure containing point $\alpha = (\alpha_1, \ldots, \alpha_k) \in \mathbb{R}^k$. The following properties hold:*

1. *$D[k].l$ and $D[k].u$ are delineable over $S(D[1], \ldots, D[k-1])$.*

2. *$S(D)$ is an open cylindrical subset of $\mathbb{R}^k$ containing α.*

3. *$S(D)$ is the maximal connected region containing α in which $D[1].l, D[1].u, \ldots, D[k].l, D[k].u$ have constant, non-zero sign.*

PROOF. We proceed inductively on k. Consider the case $k = 1$. Univariate polynomials are by definition delineable, which takes care of Property 1. When $k = 1$, $S(D)$ is an open interval, which is an open cylindrical subset of $\mathbb{R}^1$. This takes care of Property 2. Finally, when $k = 1$, $S(D)$ is precisely the interval $(\text{val}(D[k].L), \text{val}(D[k].U))$ in which, by definition, neither $D[k].L.p = D[k].l$ nor $D[k].U.p = D[k].u$ are zero. Since the zeros of $D[k].l$ and $D[k].u$ are the endpoints of this interval, it is a maximal connected region in which $D[k].l$ and $D[k].u$ have constant non-zero sign. This takes care of Property 3.

Suppose $k > 1$. We will only consider the case in which $D[k].l \neq -\infty$ and $D[k].u \neq +\infty$, since the other three cases are similar but easier. By induction, all three points hold for the OpenCell data structure $D[1], \ldots, D[k-1]$ containing point $(\alpha_1, \ldots, \alpha_{k-1})$. By definition, the leading coefficients and discriminants of $D[k].l$ and $D[k].u$, as well as their resultant, all have constant non-zero sign in $S(D[1], \ldots, D[k])$, which is an open cylindrical cell and, since by Property 3 its boundaries are semi-algebraic, it is an open cylindrical algebraic cell — in fact, an analytic submanifold. By Theorem 2 of McCallum's paper defining his projection operator ([5]), $D[k].l$ and $D[k].u$ are delineable over $S(D[1], \ldots, D[k])$. This takes care of Property 1. McCallum's Theorem also assures us that the sections $\text{root}(D[k].l, D[k].L.j, x_k)$ and $\text{root}(D[k].u, D[k].U.j, x_k)$ are the graphs of non-intersecting real-valued algebraic functions over $S(D[1], \ldots, D[k])$, so the fact that (by the definition of the OpenCell data-structure)

$$\mathrm{root}(D[k].l, D[k].L.j, x_k) < \mathrm{root}(D[k].u, D[k].U.j, x_k)$$

at $(\alpha_1, \ldots, \alpha_{k-1})$ implies that the same inequality holds over the entire region $S(D[1], \ldots, D[k-1])$. This proves that $S(D)$ is an open cylindrical subset of $\mathbb{R}^k$ which, since it's obvious that $\alpha \in S(D)$, proves Property 2.

Finally, we prove Property 3. Since we have shown that $S(D)$ is an open cylindrical subset of $\mathbb{R}^k$ containing α, it suffices to show that if a sequence of points in $S(D)$ converges to a point $\gamma \notin S(D)$, then some polynomial in the list $D[1].l, D[1].u, D[2].l, D[2].u, \ldots, D[k].l, D[k].u$ vanishes at γ.

Suppose there is an infinite sequence $\gamma^1, \gamma^2, \ldots$ of points in $S(D)$ that converges to point γ not in $S(D)$. Let π_{k-1} denote the projection of a point in real n-dimensional space, where $n \geq k$, onto $(k-1)$-dimensional space. We have two cases. First, suppose $\pi_{k-1}(\gamma) \notin S(D[1], \ldots, D[k-1])$. Then since $\pi_{k-1}(\gamma^1), \pi_{k-1}(\gamma^2), \ldots$ converges to $\pi_{k-1}(\gamma)$, we are guaranteed by induction that one of $D[1].l$, $D[1].u$, $D[2].l$, $D[2].u$, ..., $D[k-1].l$, $D[k-1].u$ is zero at $\pi_{k-1}(\gamma)$. So the second case is $\pi_{k-1}(\gamma) \in S(D[1], \ldots, D[k-1])$. In this case, $\gamma \notin S(D)$ implies that

$$\mathrm{root}(D[k].l, D[k].L.j, x_k) < x_k < \mathrm{root}(D[k].u, D[k].U.j, x_k) \quad (1)$$

fails to hold at γ. Since (1) holds for all points in the sequence $\gamma^1, \gamma^2, \ldots$, this implies that either $D[k].l$ or $D[k].u$ is zero at γ, which completes the proof of Property 3. □

4. THE OC-CONSTRUCT ALGORITHM

In this section we describe *OC − Construct*, a simple, efficient algorithm for the Open-Cell problem. The majority of work is done by a sub-algorithm O-P-MERGE which we define first.

THEOREM 2. *Algorithm* O-P-MERGE *meets its specification.*

PROOF. We fix n and proceed inductively on k. Clearly, in Step 1, FAIL is returned if $P(\alpha) = 0$ as required by the specification. So we continue under the assumption that $P(\alpha) \neq 0$. As in the proof of Theorem 1, we will assume that $D[k].l \neq -\infty$ and $D[k].u \neq +\infty$, since the proof for cases in which this does not hold is similar but simpler.

Case $k = 0$ and $k = 1$. If $k = 0$, P is a non-zero constant, so returning $D' = D$ meets the specification. When $k = 1$ we skip Steps 4–7. Those steps basically refine the cell defined by D' to ensure the delineability of P, $D[k].l$, and $D[k].u$. Since 1-level polynomials are always delineable, those steps aren't needed. When $k = 1$ the cell represented by D is an open interval. Steps 8–10 orders the roots of P, $D[1].l$ and $D[1].u$ and chooses the nearest root below α_1 as the (possibly new) lower bound, and the nearest root above α_1 as the (possibly new) upper bound. Thus Point 1 of the output requirements is satisfied. When $k = 1$, $F = \{P, D[1].l, D[1].u\}$ and, as discussed above $D'[1].l$ and $D'[1].u$ are taken from this set, so Point 3 of the output requirements is satisfied. It is also clear from the way $D'[1].l$ and $D'[1].u$ are chosen that Point 2 also holds. Finally, we must be sure that D' has the properties required of an OpenCell data structure. All but parts 2c, 3c and 5 of Definition 4 hold trivially, because they deal with properties of $D[1], \ldots, D[k-1]$, which is the empty list when $k = 1$. We

Algorithm 1 OpenCell–Polynomial Merge (O-P-MERGE)

Input:

α : a rational point in $\mathbb{R}^n$, $\alpha = (\alpha_1, \ldots, \alpha_n)$
D : an OpenCell data structure containing α
P : an integral polynomial of level k, where $k < n$

Output:

D' : an OpenCell data structure containing α such that (denote by F is the closure under the Open-McCallum projection of $\{P, D[1].l, D[1].u, \ldots, D[k].l, D[k].u\}$)

1. $S(D') \subseteq S(D)$
2. $S(D')$ contains at least one maximal connected region in which the elements of F have constant, non-zero sign, and
3. $\{D'[1].l, D'[1].u, \ldots, D'[k].l, D'[k].u\} \subseteq F$

-or-

FAIL whenever $P(\alpha) = 0$. Outputting FAIL is also allowed, but not required, whenever any element of F is zero at α.

1: if $P(\alpha) = 0$ return FAIL
2: set D' to a copy of D
3: if $k = 0$ return, if $k = 1$ goto Step 8
4: set $F = \{\mathrm{ldcf}_{x_k}(P), \mathrm{disc}_{x_k}(P)\}$
5: if $D[k].l \neq -\infty$ set $F = F \cup \{\mathrm{res}_{x_k}(P, D[k].l)\}$
6: if $D[k].u \neq +\infty$ set $F = F \cup \{\mathrm{res}_{x_k}(P, D[k].u)\}$
7: for each f from the set of irreducible factors of the elements of F do
    set $R = $ O-P-MERGE(α, D', f)
    if $R = $ FAIL return FAIL, else set $D' = R$
8: isolate the real roots of $P(\alpha_1, \ldots, \alpha_{k-1}, x_k)$ obtaining the ordered list of RealAlgNum's $B_1, \ldots, B_s$, and merge $D[k].L$, $D[k].U$ and α_k into that list.
9: if, for some i, B_i, α_k appear consecutively in the list, set $D'[k].l = P$ and $D'[k].L = B_i$
10: if, for some j, α_k, B_j appear consecutively in the list, set $D'[k].u = P$ and $D'[k].U = B_j$
11: return

choose $D'[1].l, D'[1].L$ to be the nearest root below α_1 and $D'[1].u, D'[1].U$ to be the nearest root above α_1. So $S(D')$ is the interval $(val(D'[1].L), val(D'[1].U))$ and α_1 is in that interval. This shows that part 5 of Definition 4 holds, and since we chose the nearest roots as upper and lower bounds for this interval, parts 2c and 3c hold as well.

Case $k > 0$. Denote by $f_1, \ldots, f_r$ the irreducible factors of the elements of F, in the order in which they are added in Step 7.[3] Each of the f_i's is of level less than k, since they are resultants, discriminants and coefficients, so by induction we may assume that the calls to O-P-MERGE return results meeting its specification.

Suppose that $i - 1$ iterations of the loop in Step 7 have completed without FAIL having been returned, and consider the call O-P-MERGE(α, D', f_i) in the ith iteration. Denote by F' the closure under the Open-McCallum projection of

$$\{f_i, D'[1].l, D'[1].u, \ldots, D'[k-1].l, D'[k-1].u\}.$$

[3]Notice that the order in which these factors are added is not specified by the algorithm. Different choices of order can result in different cells — some larger, some smaller, some more quickly computed, some more slowly. Investigating good orders is a possible avenue for future work.

Polynomial f_i is a factor of $\text{res}_{x_k}(P, D[k].l)$, $\text{res}_{x_k}(P, D[k].u)$, $\text{disc}_{x_k}(P)$, or $\text{ldcf}_{x_k}(P)$, which are all elements of the Open-McCallum projection of

$$\{P, D[1].l, D[1].u, D[2].l, D[2].u, \ldots, D[k].l, D[k].u\}$$

from which it follows that $F' \subseteq F$.

Suppose FAIL is returned by the recursive call. This means that some element F' vanishes at α, which in turn means that some element of F vanishes at α, in which case the specification allows the result of O-P-Merge(α, D, P) to be FAIL.

Next, suppose that the recursive call does not return FAIL, but instead returns the OpenCell data structure R. By induction, $S(R[1], \ldots, R[k-1]) \subseteq S(D'[1], \ldots, D'[k-1])$ which means that any polynomial that has constant non-zero sign in $S(D'[1], \ldots, D'[k-1])$ also has constant non-zero sign in $S(R[1], \ldots, R[k-1])$.

Assuming that all iterations of the loop at Step 7 are completed without FAIL being returned, we are left after Step 7 with the OpenCell data structure D' such that $\text{disc}_{x_k}(P)$, $\text{ldcf}_{x_k}(P)$, $\text{res}_{x_k}(P, D[k].l)$ and $\text{res}_{x_k}(P, D[k].u)$ all have constant non-zero sign in $S(D'[1], \ldots, D'[k-1])$. Thus, by Theorem 2 of McCallum's paper on projection [5], the zeros of $D[k].l$, $D[k].u$ and P over $S(D'[1], \ldots, D'[k-1])$ are delineable, i.e. their zeros define a finite set of non-intersecting graphs of real-valued algebraic functions over $S(D'[1], \ldots, D'[k-1])$. Steps 8,9,10 simply use this fact to deduce which section of $D[k].l$, $D[k].u$ and P is immediately below α and which section is immediately above α based on which root of $D[k].l(\alpha_1, \ldots, \alpha_{k-1}, x_k)$, $D[k].u(\alpha_1, \ldots, \alpha_{k-1}, x_k)$ and $P(\alpha_1, \ldots, \alpha_{k-1}, x_k)$ is immediately below α_k and which is immediately above. This is, in fact, the usual concept of a "sample point" in CAD. Step 9 sets $D'[k].l$ and $D'[k].L$ to the nearest lower-bound, which guarantees us that

$$\text{root}(D[k].l, D[k].L.j, x_k) \leq \text{root}(D'[k].l, D'[k].L.j, x_k) < \alpha_k.$$

Step 10 sets $D'[k].u$ and $D'[k].U$ to the nearest upper-bound, which guarantees us that

$$\alpha_k < \text{root}(D'[k].u, D'[k].U.j, x_k) \leq \text{root}(D[k].u, D[k].U.j, x_k).$$

Combining these facts with

$$S(D'[1], \ldots, D'[k-1]) \subseteq S(D[1], \ldots, D[k-1])$$

which we know by induction, shows that $S(D') \subseteq S(D)$. Thus Point 1 of the output requirements has been shown to hold. To address Point 3 of the output requirements, we note that we have already shown that

$$\{D'[1].l, D'[1].u, \ldots, D'[k-1].l, D'[k-1].u\} \subseteq F$$

and that $D'[k].l$ and $D'[k].u$ are both elements of $\{D[k].l, D[k].u, P\} \subseteq F$.

At this point we note that the preceding observations also show that the structure D' returned in this case satisfies the five properties enumerated in Definition 4 as requirements of an OpenCell data structure — which, of course, is also a requirement of the algorithm.

Finally, we consider Point 2 of the output requirements. Since $S(D')$ is an open set, it must contain a point at which all elements of F are non-zero in which all elements of F are non-zero. Let T be a maximal connected region containing such a point. The polynomials defining the boundaries of the cell $S(D')$ are all elements of F, from which it follows that $T \subseteq S(D')$. □

Algorithm 2 OpenCell Construct (OC-CONSTRUCT)

Input:

α : a rational point in $\mathbb{R}^n$, $\alpha = (\alpha_1, \ldots, \alpha_n)$

P : a set $\{p_1, \ldots, p_r\}$ of irreducible elements of $\mathbb{R}[x_1, \ldots, x_n]$

Output:

D : an OpenCell data structure containing α such that (denoting by F the closure of P under the Open-McCallum projection operator)

1. $\{D[1].l, D[1].u, \ldots, D[n].l, D[n].u\} \subseteq F$
2. the elements of P have constant, non-zero sign in $S(D)$

-or-

FAIL whenever any element of P is zero at α. Outputting FAIL is also allowed, but not required, whenever any element of F is zero at α.

1: if $P = \emptyset$ and $n = 0$ return the empty list
2: if $P = \emptyset$ and $n > 0$
    set $D' = $ OC-CONSTRUCT $((\alpha_1, \ldots, \alpha_{n-1}), \emptyset)$
    set $B = (-\infty, (-\infty, (0,0), 0, 1), +\infty, (+\infty, (0,0), 0, 1))$
    set $D = D'[1], \ldots, D'[n-1], B$
3: if $P = \{p_r\} \cup \{p_1, \ldots, p_{r-1}\}$, where $r > 0$
    set $D' = $ OC-Construct$(\alpha, \{p_1, \ldots, p_{r-1}\})$
    if $D' = $ FAIL return FAIL
    set $D = $ O-P-MERGE(α, D', p_r)
4: return D

THEOREM 3. *Algorithm* OC-CONSTRUCT *meets its specification.*

This we state without proof, since the correctness of Algorithm OC-CONSTRUCT follows quite directly from the correctness of Algorithm O-P-MERGE.

The Algorithm OC-CONSTRUCT solves the Open-Cell problem. In particular, if it returns an OpenCell data structure D, the formula $F(D)$ describes a cell with the required properties.

5. EXAMPLE COMPUTATIONS

In this section we present a few example computations. These are for the purpose of explaining the algorithm and illustrating where and how it avoids some computations and produces larger — i.e. more general — cells than would be produced computing a complete CAD in the usual way.

This paper does not attempt to provide an exhaustive comparison of the efficiency of OC-CONSTRUCT versus Open-CAD construction. For starters, an investigation of the order in which polynomials are added should be made first. Right now, OC-CONSTRUCT and O-P-MERGE add polynomials in an arbitrary order. Secondly, it is not clear that comparing the two algorithms outside of the context of a particular application is possible. Jovanovic and de Moura's NLSAT algorithm is an example of one such context.

5.1 A 2D problem

Consider the Open-Cell Problem:

$$\left(-\frac{1}{3}, \frac{1}{3}\right), \left\{x^2 + y^2 - 1, 2y^2 - x^2(2x+3), y + \frac{1}{2}x - \frac{1}{2}\right\}.$$

The OC-CONSTRUCT algorithm constructs a OpenCell data structure for α and x^2+y^2-1, then merges $2y^2-x^2(2x+3)$ with that OpenCell, and then merges the resulting OpenCell with the polynomial $y+1/2x-1/2$. This is illustrated in the sequence of diagrams in Figure 4. The important

Figure 4: The sequence of OpenCell's constructed by OC-CONSTRUCT for this example. Each is superimposed on the full CAD that would have been constructed from the polynomials involved at that point. Especially important to note is that the transition from the middle to the right-most OpenCell (i.e. merging $y+1/2x-1/2$) does not change the OpenCell data-structure.

thing to see in this example is that merging $y+1/2x-1/2$ with the current OpenCell in the final step (going from the middle to the right-most OpenCell in Figure 4) does not actually change the OpenCell. In this step it is deduced that $y+1/2x-1/2$ has constant non-zero sign in the current OpenCell, so something is gained, but the OpenCell itself is not refined. Moreover, we did not even have to compute the resultant of x^2+y^2-1 and $y+1/2x-1/2$, which is something the regular CAD algorithm would have constructed given the same polynomials.

Another important observation to be made about this example is that adding the polynomials in a different order makes a difference in which polynomials get computed and which computations can be avoided. Although it is not the case here, examples can be constructed in which the actual cell that gets computed is affected by the order in which the OC-CONSTRUCT and O-P-MERGE add polynomials. Investigating the impact of different criteria for ordering polynomials to add is a direction for future work.

5.2 A 4D example

Next we consider the 4-variable polynomial p

$$b^4t_1^4t_2^4+2a^2b^2t_1^4t_2^4-b^2t_1^4t_2^4+a^4t_1^4t_2^4-a^2t_1^4t_2^4+2b^4t_1^2t_2^4+ 2a^2b^2t_1^2t_2^4-2b^2t_1^2t_2^4+b^4t_2^4-b^2t_2^4+4ab^3t_1^3t_2^3+4a^3bt_1^3t_2^3+ 4ab^3t_1t_2^3+2a^2b^2t_1^4t_2^2+2a^4t_1^4t_2^2-2a^2t_1^4t_2^2+6a^2b^2t_1^2t_2^2+ 4a^3bt_1^3t_2+a^4t_1^4-a^2t_1^4$$

and variable order $b \prec a \prec t_2 \prec t_1$. This example is for purpose of illustration, so there's no claim that we can extrapolate much from this problem, just that it helps understand what the algorithm does. This polynomial/order was chosen because a) it comes from an application, b) it is a 4-variable problem, which allows us to observe behavior that only occurs when there are repeated projections, and c) this polynomial is small enough that that a full open CAD can be relatively easily computed for it. With a single input polynomial the first projection step performed by OC-CONSTRUCT is identical to first projection step performed in constructing a complete open CAD. That means there is no obvious reason why fewer projection factors would be computed by OC-CONSTRUCT — unlike the 2D example from the previous section.

The program QEPCADB computes a complete Open-CAD for this problem in 88s — almost all of this time is spent computing and factoring resultants and discriminants. That CAD consists of 824 full-dimensional cells. The projection factor set consists of 23 one-level polynomials, 9 two-level polynomials, 3 three-level polynomials and 1 four-level polynomial.

Consider the Open-Cell Problem instance $\left(\frac{2}{7}, \frac{-3}{11}, \frac{5}{9}, \frac{12}{23}\right), p$. (This point was chosen arbitrarily, and has no special significance.) Our test implementation constructs an Open-Cell data structure for this in .57s, with description

$$\begin{array}{rcccl} -\infty & < & t_1 & < & +\infty \\ 0 & < & t_2 & < & +\infty \\ root(q,1,a) & < & a & < & 0 \\ 0 & < & b & < & 1 \end{array}$$

where $q = a^6+3b^2a^4-3a^4+3b^4a^2+21b^2a^2+3a^2+b^6-3b^4+3b^2-1$. Comparing times is of limited meaning at this point, because of the many implementation differences that may obscure algorithmic differences. For example, the test implementation uses Maple to factor and compute resultants and discriminants, while QEPCADB uses Singular. Moreover, OC-CONSTRUCT computes a lot less than QEPCADB, because QEPCADB computes all open cells, there being no way to instruct it to only compute what is needed to construct the cell with sample point α. However, one meaningful comparison is the number of projection factors that get computed. There are only five projection factors in the OpenCell data structure that the test implementation produces. However, projection factors are constructed and then discarded. Accounting for that, the test implementation computes 10 one-level, 8 two-level, 3 three-level and 1 four-level projection factors. This is substantially less than is required to construct a complete open CAD. It should be noted that the number of four and three-level projection factors is necessarily the same for this example, because there is a single input polynomial.

As previously noted, comparing QEPCADB and OC-CONSTRUCT as in the above example is not terribly meaningful, because QEPCADB computes so much less. To better understand how they really compare, we performed the following test. We called the test implementation on polynomial p for the sample points of each of the 824 open cells constructed by QEPCADB. This took a combined total of 15s.[4] The results show that the set of all projection factors constructed in the 824 calls to the test implementation consists of 15 one-level polynomials, 9 two-level polynomials, 3 three-level polynomials and 1 four-level polynomial. In other words, eight of the 23 one-level projection factors constructed by the usual Open-McCallum projection were never produced at any point by the 824 calls to the test implementation. Hand-in-hand with this, close inspection of the cells constructed by the test implementation shows that most of them contain multiple open cells from the complete open CAD. In fact, there are only 352 distinct cells constructed, and they combine to cover the same space as the 824 open cells in the complete open CAD. All of this raises

[4]The test implementation memoizes discriminant, resultant and factorization computations, so that time isn't wasted in such a scenario recomputing the same factors.

the possibility that covering $\mathbb{R}^4$ — up to some lower dimensional set — with (possibly overlapping) open cells might be accomplished faster and with fewer cells using an approach based on OC-CONSTRUCT than by computing a complete open CAD in the usual way.

6. CONCLUSION & FUTURE WORK

This paper has considered the *Open-Cell Problem* which, roughly speaking, is the problem of starting with a point α and set P of irreducible polynomials, and computing from them a description of an open cylindrical cell containing α in which the elements of P have constant non-zero sign. Even more roughly speaking, it could be described as constructing a single open cell from a CAD. The algorithm OC-CONSTRUCT, which solves this problem, was introduced and proved correct. Finally, some example computations using a test implementation were described. They demonstrated the algorithm, and illustrated how OC-CONSTRUCT constructs its one cell more efficiently than does applying open CAD techniques naively, and how the cell constructed by OC-CONSTRUCT is often larger ("more general") than the cell from the complete open CAD that contains the point α. The last example computation illustrated why it may be more efficient to apply OC-CONSTRUCT over and over to cover (up to a measure zero set) the input space with open cells than to apply the usual open CAD, which offers some exciting possibilities.

This paper has introduced a new approach to constructing CAD cells — basically allowing us to compute CAD projection "locally", meaning around the input point α. The original motivation to even consider the problem was the paper by Jovanovic and de Moura [3] that introduced NLSAT. A natural starting point for future work building on the present paper would be to try to formulate an "open-NLSAT" and base it around OC-CONSTRUCT. Other applications could also be considered. The approach itself could most likely be improved in a number of ways, the one highlighted here being the order in which polynomials are "merged" with the OpenCell data structure. Indeed, the approach could be formulated in terms of merging OpenCell data structures rather than merging a polynomial and an OpenCell data structure, which would provide even more choices. Finally, this paper has demonstrated that the approach followed by OC-CONSTRUCT works, meaning that we gain both in efficiency and quality of solution. The approach needs to be extended to lower-dimensional cells so that the same gains can be made in those cases. These last two points will be explained in more detail below.

Choices and Merging Step 7 of Algorithm 1 refines the OpenCell data structure D' by iteratively merging the elements of F. The order in which the elements of F are chosen affects the OpenCell resulting from the process. Therefore, one should investigate ordering criteria. In fact, many variations on Algorithm 1 might be considered. An interesting variation is to recast the process as merging two OneCell data structures, rather than a polynomial and a OneCell data structure. In this case, we arrive at a natural divide & conquer algorithm that does a projection step, produces OneCell data structures recursively for each of the resulting projection factors (using the same point for all of them), and then merges all the OneCell data structures together. Preliminary work on this approach shows promise.

Generalizing to Lower-Dimensional Cells Algorithms 1 & 2 return FAIL if point α is a zero of a projection factor. A generalization of this work would have them return a lower-dimensional cell in which projection factors have constant sign, but not necessarily non-zero sign. This necessitates considering a more complicated projection operator, and performing root isolation with algebraic numbers. However, it also offers far more opportunities for improvements in efficiency compared to the standard CAD approach.

Replacing the Usual CAD Approach As already noted, constructing a OneCell data structure for each of the sample points in a CAD can be substantially faster and more memory efficient than constructing a regular CAD of $\mathbb{R}^n$. In many applications it would be possible to replace the usual CAD construction with the construction of many, possibly overlapping, OneCell data structures — as long as it could be demonstrated that $\mathbb{R}^n$ was covered by the union of these cells (at least up to some measure zero set). To make this work, some kind of control must be provided that chooses points and applies OC-CONSTRUCT in such a way that it terminates after a finite number of steps having provably covered $\mathbb{R}^n$, up to a measure zero set. Jovanovic and de Moura approach provides such a control, though it would be interesting to consider other, more geometric rather that logic-oriented ways of doing it.

7. ACKNOWLEDGEMENTS

This work arose out of discussions at the Schloss Dagstuhl Seminar 12462, *Symbolic Methods for Chemical Reaction Networks*, held in November 2012. I would like to thank the organizers of the event and Leibniz Zentrum für Mathematik for a wonderful scientific program. Particularly, I would like to thank Thomas Sturm and Marek Kosta for introducing me to Jovanovic and de Moura's paper, and for helping me understand it during our several discussions at Schloss Dagstuhl along with Carsten Conradi and Andreas Eggers.

8. REFERENCES

[1] ARNON, D. S., COLLINS, G. E., AND MCCALLUM, S. Cylindrical algebraic decomposition I: The basic algorithm. *SIAM Journal on Computing 13*, 4 (1984), 865–877.

[2] BROWN, C. W. *Solution Formula Construction for Truth Invariant CAD's*. PhD thesis, University of Delaware, 1999.

[3] JOVANOVIC, D., AND DE MOURA, L. M. Solving non-linear arithmetic. In *IJCAR* (2012), pp. 339–354.

[4] MCCALLUM, S. Solving polynomial strict inequalities using cylindrical algebraic decomposition. *The Computer Journal 36*, 5 (1993), 432–438.

[5] MCCALLUM, S. An improved projection operator for cylindrical algebraic decomposition. In *Quantifier Elimination and Cylindrical Algebraic Decomposition* (1998), B. Caviness and J. Johnson, Eds., Texts and Monographs in Symbolic Computation, Springer-Verlag, Vienna.

[6] STRZEBONSKI, A. Solving systems of strict polynomial inequalities. *Journal of Symbolic Computation 29* (2000), 471–480.

Factoring Bivariate Lacunary Polynomials without Heights*

Arkadev Chattopadhyay
School of Technology and Computer Science
Tata Institute for Fundamental Research, Mumbai, India
arkadev.c@tifr.res.in

Bruno Grenet
LIP – UMR 5668 – École Normale Supérieure de Lyon
Université de Lyon, France
ENSL-CNRS-UCBL-Inria
bruno.grenet@ens-lyon.fr

Pascal Koiran
LIP – UMR 5668 – École Normale Supérieure de Lyon
Université de Lyon, France
ENSL-CNRS-UCBL-Inria
pascal.koiran@ens-lyon.fr

Natacha Portier
LIP – UMR 5668 – École Normale Supérieure de Lyon
Université de Lyon, France
ENSL-CNRS-UCBL-Inria
natacha.portier@ens-lyon.fr

Yann Strozecki
PRiSM
Université de Versailles Saint-Quentin, France
yann.strozecki@prism.uvsq.fr

ABSTRACT

We present an algorithm which computes the multilinear factors of bivariate lacunary polynomials. It is based on a new Gap theorem which allows to test whether $P(X) = \sum_{j=1}^{k} a_j X^{\alpha_j}(1+X)^{\beta_j}$ is identically zero in polynomial time. The algorithm we obtain is more elementary than the one by Kaltofen and Koiran (ISSAC'05) since it relies on the valuation of polynomials of the previous form instead of the height of the coefficients. As a result, it can be used to find some linear factors of bivariate lacunary polynomials over a field of large finite characteristic in probabilistic polynomial time.

Categories and Subject Descriptors

I.1.2 [**Computing Methodologies**]: Symbolic and Algebraic Manipulation—*Algorithms*; F.2.2 [**Theory of Computation**]: Analysis of Algorithms and Problem Complexity—*Nonnumerical Algorithms and Problems*

General Terms

algorithms, theory

Keywords

lacunary polynomials, bivariate polynomials, polynomial factorization, polynomial-time complexity, finite fields, Wronskian determinant

*Part of this work was done while the authors were visiting the University of Toronto.

1. INTRODUCTION

The *lacunary*, or *supersparse*, representation of a polynomial

$$P(X_1, \dots, X_n) = \sum_{j=1}^{k} a_j X_1^{\alpha_{1,j}} \cdots X_n^{\alpha_{n,j}}$$

is the list of the tuples $(a_j, \alpha_{1,j}, \dots, \alpha_{n,j})$ for $1 \le j \le k$. This representation allows very high degree polynomials to be represented in a concise manner. The factorization of lacunary polynomials has been investigated in a series of papers. Cucker, Koiran and Smale first proved that integer roots of univariate integer lacunary polynomials can be found in polynomial time [4]. This result was generalized by Lenstra who proved that low-degree factors of univariate lacunary polynomials over algebraic number fields can also be found in polynomial time [22]. More recently, Kaltofen and Koiran generalized Lenstra's results to bivariate and then multivariate lacunary polynomials [11, 12]. A common point to these algorithms is that they all rely on a so-called *Gap Theorem*: If F is a factor of $P(\bar{X}) = \sum_{j=1}^{t} a_j \bar{X}^{\bar{\alpha}_j}$, then there exists k_0 such that F is a factor of both $\sum_{j=1}^{k_0} a_j \bar{X}^{\bar{\alpha}_j}$ and $\sum_{j=k_0+1}^{k} a_j \bar{X}^{\bar{\alpha}_j}$. Moreover, the different Gap Theorems in these papers are all based on the notion of height of an algebraic number, and some of them use quite sophisticated results of number theory.

In this paper, we are interested in more elementary proofs for some of these results. We focus on Kaltofen and Koiran's first paper [11] dealing with linear factors of bivariate lacunary polynomials. We show how a Gap Theorem that does not depend on the height of an algebraic number can be proved. In particular, our Gap Theorem is valid for any field of characteristic zero. As a result, we get a new, more elementary algorithm for finding linear factors of bivariate lacunary polynomials over an algebraic number field. In particular, this new algorithm is easier to implement since there is no need to explicitly compute some constants from number theory, and the use of the Gap Theorem does not require to evaluate the heights of the coefficients of the polynomial. Moreover we use the same methods to prove a Gap Theorem for polynomials over some fields of positive characteristic, yielding an algorithm to find linear factors of bivariate lacunary polynomials of the form $(uX + vY + w)$ with $uvw \neq 0$. Finding linear factors with $u = 0$ is NP-hard, and the same is true for linear factors with $v = 0$ or $w = 0$. This follows from the fact that finding univariate linear factors over finite fields is NP-hard [18, 1, 13]. In algebraic number fields we can find *all* linear factors in polynomial time, even those with $uvw = 0$. For this we rely as Kaltofen and Koiran on Lenstra's univariate algorithm [22].

Our Gap Theorem is based on the valuation of a univariate polynomial, defined as the maximum integer v such that X^v divides the polynomial. We give an upper bound on the valuation of a nonzero polynomial

$$P(X) = \sum_{j=1}^{k} a_j X^{\alpha_j}(1+X)^{\beta_j}.$$

This bound can be viewed as an extension of a result due to Hajós [9, 24]. We also note that Kayal and Saha recently used the valuation of square roots of polynomials to make some progress on the "Sum of Square Roots" problem [17].

Lacunary polynomials have also been studied with respect to other computational tasks. For instance, Plaisted showed the NP-hardness of computing the greatest common divisor (GCD) of two univariate integer lacunary polynomials [26], and his results were extended to finite fields [28, 16, 11]. On the other hand, some important special cases were identified for which the GCD of two lacunary polynomials can be computed in polynomial time [5]. Other efficient algorithms for lacunary polynomials have been recently given, for instance for the detection of perfect powers [7, 8] or interpolation [14].

2. BOUND ON THE VALUATION

In this section, we consider a field $\mathbb{K}$ of characteristic zero and polynomials over $\mathbb{K}$.

THEOREM 1. *Let $P = \sum_{j=1}^{k} a_j X^{\alpha_j}(1+X)^{\beta_j}$ with $\alpha_1 \leq \cdots \leq \alpha_k$. If P is not identically zero then its valuation is at most $\max_j(\alpha_j + \binom{k+1-j}{2})$.*

A lower bound for the valuation of P is clearly α_1 (and it is attained when $\alpha_2 > \alpha_1$ for instance). If the family $(X^{\alpha_j}(1+X)^{\beta_j})_{1\leq j\leq k}$ is linearly independent over $\mathbb{K}$, the upper bound we get is actually $\alpha_1 + \binom{k}{2}$: At most the first $\binom{k}{2}$ lowest-degree monomials can be cancelled. If $\alpha_j = \alpha_1$ for all j, Hajós' Lemma [9, 24] gives the better bound $\alpha_1 + (k-1)$. (This bound can be shown to be tight by expanding $X^{k-1} = (-1 + (X+1))^{k-1}$ with the binomial formula.) This is not true anymore when the α_j's are not all equal. One can show that the valuation can be as large as $\alpha_1 + (2k-3)$ (see Proposition 5). The exact bound remains unknown, and whether this bound is still linear as in Hajós' Lemma or quadratic is open.

Our proof of Theorem 1 is based on the so-called *Wronskian* of a family of polynomials. This is a classical tool for the study of differential equations but it has recently been used to bound the valuation of a sum of square roots of polynomials [17] and also to bound the number of real roots of some sparse-like polynomials [19].

Definition 1. Let $f_1, \ldots, f_k \in \mathbb{K}[X]$. Their *Wronskian* is the determinant of the *Wronskian matrix*

$$\mathsf{W}(f_1,\ldots,f_k) = \det\begin{bmatrix} f_1 & f_2 & \cdots & f_k \\ f_1' & f_2' & \cdots & f_k' \\ \vdots & \vdots & & \vdots \\ f_1^{(k-1)} & f_2^{(k-1)} & \cdots & f_k^{(k-1)} \end{bmatrix}.$$

The main property of the Wronskian is its relation to linear independence. The following result is classical (see [2] for a simple proof of this fact).

PROPOSITION 2. *The Wronskian of $f_1, \ldots, f_k$ is nonzero if and only if the f_j's are linearly independent over $\mathbb{K}$.*

The next two lemmas are our main ingredients to give a bound on the valuation for P, using a bound on the valuation of some Wronskian.

LEMMA 3. *Let $f_1, \ldots, f_k \in \mathbb{K}[X]$. Then*

$$\mathrm{Val}(\mathsf{W}(f_1,\ldots,f_k)) \geq \sum_{j=1}^{k} \mathrm{Val}(f_j) - \binom{k}{2}.$$

PROOF. Each term of the determinant is a product of k terms, one from each column and one from each row. The valuation of such a term is at least $\sum_j \mathrm{Val}(f_j) - \sum_{i=1}^{k-1} i$ since for all i, j, $\mathrm{Val}(f_j^{(i)}) \geq \mathrm{Val}(f_j) - i$. The result follows. □

We can slightly refine the bound in this lemma. The term of valuation $\sum_j \mathrm{Val}(f_j) - \binom{k}{2}$ in the Wronskian is indeed the determinant of the matrix made of the smallest degree monomials of each $f_j^{(i)}$. This determinant can vanish. In fact, one can easily see that this is the case if two f_j's have the same valuation since this yields two proportional columns in the matrix. To use this idea more generally, consider that the f_j's are ordered by increasing valuation. We define a *plateau* to be a set $\{f_{j_0}, \ldots, f_{j_0+s}\}$ such that for $0 < t \leq s$, $\mathrm{Val}(f_{j_0+t}) \leq \mathrm{Val}(f_{j_0}) + t - 1$. The f_j's are naturally partitioned into plateaux. Suppose that there are $(m+1)$ plateaux, of length $p_0, \ldots, p_m$ respectively, and let $f_{j_0}, \ldots, f_{j_m}$ their respective first elements. Generalizing the previous remark to plateaux, it can be shown that

$$\mathrm{Val}(\mathsf{W}(f_1,\ldots,f_k)) \geq \sum_{i=0}^{m} \left(p_i \mathrm{Val}(f_{j_i}) + \binom{p_i}{2}\right) - \binom{k}{2}. \tag{1}$$

This bound is at least as large as in the lemma. If all the f_j's have a different valuation, then the bound is equal to the bound stated in the lemma since there are in this case k plateaux, each of length 1. On the other side, if they all have the same valuation α, there is one plateau of length k and the bound is $\mathrm{Val}(\mathsf{W}(f_1,\ldots,f_k)) \geq k\alpha$. We investigate the implications of this refinement after the proof of Theorem 1.

LEMMA 4. *Let $f_j = X^{\alpha_j}(1+X)^{\beta_j}$, $1 \leq j \leq k$, such that $\alpha_j, \beta_j \geq k$ for all j. If the f_j's are linearly independent, then*

$$\mathrm{Val}(\mathsf{W}(f_1,\ldots,f_k)) \leq \sum_{j=1}^{k} \alpha_j.$$

PROOF. By Leibniz rule, for all i, j

$$f_j^{(i)}(X) = \sum_{t=0}^{i} \binom{i}{t} (\alpha_j)_t (\beta_j)_{i-t} X^{\alpha_j - t}(1+X)^{\beta_j - i + t} \tag{2}$$

where $(m)_n = m(m-1)\cdots(m-n+1)$ is the falling factorial. Since $\alpha_j - t \geq \alpha_j - i$ and $\beta_j - i + t \geq \beta_j - i$ for all t,

$$f_j^{(i)}(X) = X^{\alpha_j - i}(1+X)^{\beta_j - i} \times \sum_{t=0}^{i} \binom{i}{t} (\alpha_j)_t (\beta_j)_{i-t} X^{i-1}(1+X)^t.$$

Furthermore, since $\alpha_j \geq k \geq i$, we can write $X^{\alpha_j - i} = X^{\alpha_j - k} X^{k-i}$ and since $\beta_j \geq k \geq i$, $(1+X)^{\beta_j - i} = (1+X)^{\beta_j - k}(1+X)^{k-i}$. Thus, $X^{\alpha_j - k}(1+X)^{\beta_j - k}$ is a common factor of the entries of the j-th column of the Wronskian matrix, and $X^{k-i}(1+X)^{k-i}$ is a common factor of the entries of the i-th row. Together, we get

$$\mathsf{W}(f_1, \ldots, f_k) = X^{\sum_j \alpha_j - \binom{k}{2}}(1+X)^{\sum_j \beta_j - \binom{k}{2}} \det(M)$$

where the matrix M is defined by

$$M_{i,j} = \sum_{t=0}^{i} \binom{i}{t} (\alpha_j)_t (\beta_j)_{i-t} X^{i-t}(1+X)^t.$$

The polynomial $\det(M)$ is nonzero since the f_j's are supposed linearly independent and its degree is at most $\binom{k}{2}$. Therefore its valuation cannot be larger than its degree and is bounded by $\binom{k}{2}$.

Altogether, the valuation of the Wronskian is bounded by $\sum_j \alpha_j - \binom{k}{2} + \binom{k}{2} = \sum_j \alpha_j$. □

Proof of Theorem 1. Let $P = \sum_j a_j X^{\alpha_j}(1+X)^{\beta_j}$, and let $f_j = X^{\alpha_j}(1+X)^{\beta_j}$. We assume first that $\alpha_j, \beta_j \geq k$ for all j, and that the f_j's are linearly independent. Note that $\mathrm{Val}(f_j) = \alpha_j$ for all j.

Let W denote the Wronskian of the f_j's. We can replace f_1 by P in the first column of the Wronskian matrix using column operations which multiply the determinant by a_1 (its valuation does not change). The matrix we obtain is the Wronskian matrix of $P, f_2, \ldots, f_k$. Now using Lemma 3, we get

$$\mathrm{Val}(\mathsf{W}) \geq \mathrm{Val}(P) + \sum_{j \geq 2} \alpha_j - \binom{k}{2}.$$

This inequality combined with Lemma 4 shows that

$$\mathrm{Val}(P) \leq \alpha_1 + \binom{k}{2}. \tag{3}$$

We now aim to remove our two previous assumptions. If the f_j's are not linearly independent, we can extract from this family a basis $f_{j_1}, \ldots, f_{j_d}$. Then P can be expressed in this basis as $P = \sum_{l=1}^{d} \tilde{a}_l f_{j_l}$. We can apply Equation (3) to $f_{j_1}, \ldots, f_{j_d}$ and obtain $\mathrm{Val}(P) \leq \alpha_{j_1} + \binom{d}{2}$. Since $j_d \leq k$, we have $j_1 + d - 1 \leq k$ and $\mathrm{Val}(P) \leq \alpha_{j_1} + \binom{k+1-j_1}{2}$. The value of j_1 being unknown, we conclude that

$$\mathrm{Val}(P) \leq \max_{1 \leq j \leq k} \left(\alpha_j + \binom{k+1-j}{2}\right). \tag{4}$$

The second assumption is that $\alpha_j, \beta_j \geq k$. Given P, consider $\tilde{P} = X^k(1+X)^k P = \sum_j a_j X^{\tilde{\alpha_j}}(1+X)^{\tilde{\beta_j}}$. Then $\tilde{P}$ satisfies $\tilde{\alpha_j}, \tilde{\beta_j} \geq k$, whence by Equation (4), $\mathrm{Val}(\tilde{P}) \leq \max_j(\tilde{\alpha_j} + \binom{k+1-j}{2})$. Since $\mathrm{Val}(\tilde{P}) = \mathrm{Val}(P) + k$ and $\tilde{\alpha_j} = \alpha_j + k$, the result follows. □

Remark 1. In Theorem 1, we can replace $(1+X)$ by $(uX+v)$ for any $u, v \neq 0$. Indeed, we can write $uX+v = v(\frac{u}{v}X+1)$ and then use the change of variables $Y = \frac{u}{v}X$. This gives us a polynomial of the same form as in the theorem, with the same valuation as the original one.

Remark 2. Theorem 1 does not hold in positive characteristic as shown by the equality $(1+X)^{2^n} + (1+X)^{2^{n+1}} = X^{2^n}(1+X) \mod 2$. Section 5 investigates the case of positive characteristic in more details.

We argued after Lemma 3 that it can be refined. In the previous proof, it is used with P, f_2, …, f_k. If all the f_j's have the same valuation α, Equation (1) gives the bound $\mathrm{Val}(\mathsf{W}) \geq \mathrm{Val}(P) + ((k-1)\alpha + \binom{k-1}{2}) - \binom{k}{2}$, whence $\mathrm{Val}(P) \leq \alpha + (k-1)$. In this case, replacing Lemma 3 by Equation (1) gives us a new proof of Hajós' Lemma, with the correct bound.

On the other hand, if the f_j's have pairwise distinct valuations, Equation 1 gives the same bound as Lemma 3. Yet in this case Lemma 4 can be refined to obtain the bound $\mathrm{Val}(\mathsf{W}) \leq \sum_j \alpha_j - \binom{k}{2}$. Again, we find the optimal bound for the valuation, that is $\mathrm{Val}(P) = \alpha_1$ here.

The refinement of Lemma 3 alone is not sufficient to improve Theorem 1 in the general case. To this end, one needs to improve Lemma 4 as well. As already mentioned, it is an open problem to determine the best achievable bound for Theorem 1. The next proposition shows that it cannot be as low as in Hajós' Lemma.

Proposition 5. *For $k \geq 3$, there exists a linearly independent family of polynomials $(X^{\alpha_j}(1+X)^{\beta_j})_{1 \leq j \leq k}$, $\alpha_1 \leq \cdots \leq \alpha_k$ and a family of rational coefficients $(a_j)_{1 \leq j \leq k}$ such that the polynomial*

$$P(X) = \sum_{j=1}^{k} a_j X^{\alpha_j}(1+X)^{\beta_j}$$

is nonzero and has valuation $\alpha_1 + (2k-3)$.

Proof. A polynomial that achieves this bound is

$$P_k(X) = -1 + (1+X)^{2k+3} - \sum_{j=0}^{k} a_j X^{2j+1}(1+X)^{k+1-j},$$

where

$$a_j = \frac{2k+3}{2j+1}\binom{k+1+j}{k+1-j}.$$

We aim to prove that $P_k(X) = X^{2k+3}$. Since it has $(k+3)$ terms and $\alpha_1 = 0$, this proves the proposition. To prove the result for an arbitrary value of α_1, it is sufficient to multiply P_k by some power of X.

It is clear that P_k has degree $(2k+3)$ and is monic. Let $[X^m]P_k$ be the coefficient of the monomial X^m in P_k. Then for $m > 0$

$$[X^m]P_k = \binom{2k+3}{m} - \sum_{j=0}^{k} a_j \binom{k+1-j}{m-2j-1}.$$

We aim to prove that $[X^m]P_k = 0$ as soon as $m < 2k+3$. Using the definition of the a_j's, this is equivalent to proving

$$\sum_{j=0}^{k} \frac{2k+3}{2j+1}\binom{k+1+j}{k+1-j}\binom{k+1-j}{m-2j-1} = \binom{2k+3}{m}. \tag{5}$$

To prove this equality, we rely on Wilf and Zeilberger's algorithm [25], and its implementation in the Maple package EKHAD of Doron Zeilberger (see [25] for more on this package). The program asserts the correctness of the equality and provides a recurrence relation satisfied by the summand that we can verify by hand.

Let $F(m,j)$ be the summand in equation (5) divided by $\binom{2k+3}{m}$. We thus want to prove that $\sum_{j=0}^{k} F(m,j) = 1$. The EKHAD package provides

$$R(m,j) = \frac{2j(2j+1)(k+j+2-m)}{(2k+3-m)(2j-m)}$$

and claims that

$$mF(m+1,j) - mF(m,j) \\ = F(m,j+1)R(m,j+1) - F(m,j)R(m,j). \quad (6)$$

In the rest of the proof, we show why this claim implies Equation (5), and then that the claim holds.

Suppose first that Equation (6) holds and let us prove Equation (5). If we sum Equation (6) for $j = 0$ to k, we obtain

$$m(\sum_{j=0}^{k} F(m+1,j) - F(m,j)) \\ = F(m,k+1)R(m,k+1) - F(m,0)R(m,0).$$

Since $R(m,0) = 0$ and $F(m,k+1) = 0$, $\sum_j F(m,j)$ is constant with respect to m. One can easily check that the sum is 1 when $m = 2k+2$. (Actually the only nonzero term in this case is for $j = k$.) Therefore, we deduce that for all $m < 2k+3$,[1] $\sum_j F(m,j) = 1$, that is Equation (5) is true.

To prove Equation (6), note that

$$\frac{F(m+1,j)}{F(m,j)} = \frac{(j+k+2-m)(m+1)}{(m-2j)(2k+3-m)}$$

and

$$\frac{F(m,j+1)}{F(m,j)} = \frac{(k+2-j)(m-2j-1)(m-2j-2)}{(2j+2)(2j+3)(j+k+3-m)}.$$

Therefore, to prove the equality, it is sufficient to check that

$$0 = m\frac{j+k+2-m}{m-2j}\frac{m+1}{2k+3-m} - m + R(m,j) \\ - \frac{(k+2-j)(m-2j-1)(m-2j-2)}{(2j+2)(2j+3)(j+k+3-m)}R(m,j+1).$$

This is done by a mere computation. □

From Theorem 1, we can deduce the following Gap Theorem.

THEOREM 6 (GAP THEOREM). *Let*

$$P = \sum_{j=1}^{k} a_j X^{\alpha_j}(uX+v)^{\beta_j}$$

with $u, v \neq 0$ and $\alpha_{j+1} \geq \alpha_j$, $0 \leq j < k$. Assume that there exists ℓ such that

$$\alpha_{\ell+1} > \max_{1\leq j\leq \ell}\left(\alpha_j + \binom{\ell+1-j}{2}\right). \quad (7)$$

[1]The bound on m is given by the fact that $R(m,j)$ is undefined for $m = 2k+3$.

Then P is identically zero if and only if the polynomials $\sum_{j=1}^{\ell} a_j X^{\alpha_j}(uX+v)^{\beta_j}$ and $\sum_{j=\ell+1}^{k} a_j X^{\alpha_j}(uX+v)^{\beta_j}$ are both identically zero.

In particular, the smallest ℓ satisfying (7) is the smallest ℓ satisfying

$$\alpha_{\ell+1} > \alpha_1 + \binom{\ell}{2}.$$

PROOF. Let $Q = \sum_{j=1}^{\ell} a_j X^{\alpha_j}(uX+v)^{\beta_j}$ and $R = P - Q$. Suppose that Q is not identically zero. By Theorem 1, its valuation is at most $\max_j(\alpha_j + \binom{\ell+1-j}{2})$. Since $\alpha_j \geq \alpha_{\ell+1}$ for $j > \ell$, the valuation of R is at least $\alpha_{\ell+1} > \max_j(\alpha_j + \binom{\ell+1-j}{2})$. Therefore, if Q is not identically zero, its monomial of lowest degree cannot be canceled by a monomial of R. In other words, $P = Q + R$ is not identically zero.

For the second part of the theorem, consider the smallest ℓ satisfying Equation (7). It is clear that $\alpha_{\ell+1} > \alpha_1 + \binom{\ell}{2}$. Moreover for all $j \leq \ell$, $\alpha_{j+1} \leq \max_{i\leq j}(\alpha_i + \binom{j+1-i}{2})$. We now prove by induction on j that $\alpha_j \leq \alpha_1 + \binom{j-1}{2}$ for all $j \leq \ell$. This is obviously true for $j = 1$. Let $j < \ell$ and suppose that for all $i \leq j$, $\alpha_i \leq \alpha_1 + \binom{i-1}{2}$. Then

$$\alpha_{j+1} \leq \max_{i<j}\left(\alpha_i + \binom{j+1-i}{2}\right) \\ \leq \alpha_1 + \max_{i<j}\left(\binom{i-1}{2} + \binom{j-(i-1)}{2}\right).$$

To conclude, we remark that $\binom{i-1}{2} + \binom{j-(i-1)}{2} \leq \binom{j}{2}$ for all $i < j$. □

It is straightforward to extend this theorem to more *gaps*. The theorem can be recursively applied to Q and R (as defined in the proof). Then, if $P = P_1 + \ldots + P_s$ where there is a *gap* between P_t and P_{t+1} for $1 \leq t < s$, then P is identically zero if and only if each P_t is zero.

3. ALGORITHMS

In this section, we prove that there exists a deterministic polynomial-time algorithm to test if a polynomial of the form

$$P = \sum_{j=1}^{k} a_j X^{\alpha_j}(uX+v)^{\beta_j}, \quad (8)$$

is identically zero and give a deterministic polynomial-time algorithm to compute the linear factors of a lacunary bivariate polynomial. The size of P is defined by

$$\text{size}(P) = \text{size}(u) + \text{size}(v) + \sum_{j=1}^{k}(\text{size}(a_j) + \log(\alpha_j\beta_j)). \quad (9)$$

The algorithms use Lenstra's algorithm [22] or a variant of it for treating some special cases. This use of Lenstra's algorithm implies some restrictions on the field $\mathbb{K}$ in which the coefficients of the polynomials lie. In this section, $\mathbb{K}$ is an algebraic number field, and it is represented as $\mathbb{K} = \mathbb{Q}[\xi]/\langle\varphi\rangle$ where $\varphi \in \mathbb{Z}[\xi]$ is a monic irreducible polynomial. Elements of $\mathbb{K}$ are given as vectors in the basis $(1, \xi, \ldots, \xi^{\deg\varphi-1})$. That is for $e \in \mathbb{K}$, $e = (e_0, \ldots, e_{\deg\varphi-1})$ with $e_t = n_t/d_t$ for each t where $n_t, d_t \in \mathbb{Z}$. Then

$$\text{size}(e) = \log(n_1 d_1) + \cdots + \log(n_{\deg\varphi-1} d_{\deg\varphi-1}).$$

The size of a polynomial defined as above is then approximately the number of bits needed to write down its binary representation.

Theorems 7 and 8 were already proven in [11]. We give here new proofs based on our Gap Theorem. The structures of the algorithms we propose are the same as in [11]. The only differences are the ones induced by the use of a different Gap Theorem. This implies some differences in terms of the complexity that are discussed at the end of this section.

THEOREM 7. *There exists a deterministic polynomial-time algorithm to decide if a polynomial of the form (8) is identically zero.*

PROOF. We assume without loss of generality that $\alpha_{j+1} \geq \alpha_j$ for all j and $\alpha_1 = 0$. If α_1 is nonzero, X^{α_1} is a factor of P and we consider P/X^{α_1}.

Suppose first that $u = 0$. Then P is given as a sum of monomials, and we only have to test each coefficient for zero. Note that the α_j's are not distinct. Thus the coefficients are of the form $\sum_j a_j v^{\beta_j}$. Lenstra [22] gives an algorithm to find low-degree factors of univariate lacunary polynomials. It is easy to deduce from his algorithm an algorithm to test such sums for zero. A strategy could be to simply apply Lenstra's algorithm to $\sum_j a_j X^{\beta_j}$ and then check whether $(X - v)$ is a factor, but one can actually improve the complexity by extracting from his algorithm the relevant part. The case $v = 0$ is similar.

We assume now that $u, v \neq 0$. We split P into small parts $P = P_1 + \cdots + P_s$, such that according to the Gap Theorem, P is identically zero if and only if each part P_t is identically zero. Formally, let $I_1, \ldots, I_s$ be the (unique) partition of $\{1, \ldots, k\}$ into intervals defined recursively as follows. Let $1 \in I_1$. For $1 \leq j < k$, suppose that $\{1, \ldots, j\}$ has been partitioned into $I_1, \ldots, I_t$, and let i_t be the smallest element of I_t. Then $(j+1) \in I_t$ if $\alpha_{j+1} \leq \alpha_{i_t} + \binom{j-i_t+1}{2}$, and $(j+1) \in I_{t+1}$ otherwise. The polynomials $P_t = \sum_{j \in I_t} a_j X^{\alpha_j}(1 + X)^{\beta_j}$ satisfy the conditions of Theorem 6. Therefore, we are left with testing if the P_t's are identically zero. Moreover, $X^{\alpha_{i_t}}$ divides P_t for each t and it is thus equivalent to be able to test if each $P_t/X^{\alpha_{i_t}}$ is identically zero.

To this end, let Q be a polynomial of the form (8) satisfying $\alpha_1 = 0$ and $\alpha_{j+1} \leq \binom{j}{2}$ for all j. In particular, $\alpha_k \leq \binom{k-1}{2}$. Consider the change of variables $Y = uX + v$. Then

$$Q(Y) = \sum_{j=1}^{k} a_j u^{-\alpha_j} (Y - v)^{\alpha_j} Y^{\beta_j}$$

is identically zero if and only if $Q(X)$ is. We can express $Q(Y)$ as a sum of powers of Y:

$$Q(Y) = \sum_{j=1}^{k} \sum_{\ell=0}^{\alpha_j} a_j u^{-\alpha_j} \binom{\alpha_j}{\ell} (-v)^{\ell} Y^{\alpha_j + \beta_j - l}.$$

There are at most $k\binom{k-1}{2} = \mathcal{O}(k^3)$ monomials. Then, testing if $Q(Y)$ is identically zero consists in testing each coefficient for zero. Moreover, each coefficient has the form $\sum_j \binom{\alpha_j}{\ell_j} a_j u^{-\alpha_j} (-v)^{\ell_j}$ where the sum ranges over at most k indices. Since $\ell_j, \alpha_j \leq \binom{k-1}{2}$ for all j, the terms in these sums have polynomial bit-lengths. Therefore, the coefficients can be tested for zero in polynomial time.

Altogether, this gives a polynomial-time algorithm to test if P is identically zero. □

THEOREM 8. *Let*

$$P(X, Y) = \sum_{j=1}^{k} a_j X^{\alpha_j} Y^{\beta_j} \in \mathbb{K}[X, Y].$$

There exists a deterministic polynomial-time algorithm that finds all the linear factors of P, together with their multiplicities.

PROOF. A linear factor of P is either of the form $(Y - uX - v)$ or $(X - a)$. To search factors of the form $(X - a)$, we see P as a univariate polynomial in Y whose coefficients are univariate polynomials in X. Then, $(X - a)$ is a factor of P if and only if it is a factor of all the coefficients of P viewed as a polynomial in Y. Lenstra gives an algorithm to compute linear factors of univariate lacunary polynomials [22]. Thus, we can find all the factors of the form $(X - a)$ and their multiplicities using his algorithm.

Now $(Y - uX - v)$ is a factor of P if and only if $P(X, uX + v)$ vanishes identically. We can assume that $u \neq 0$. If $v = 0$, $P(X, uX) = \sum_j a_j u^{\beta_j} X^{\alpha_j + \beta_j}$. Therefore, it vanishes if and only if each coefficient vanishes. But a coefficient of this polynomial is of the form $\sum_j a_j u^{\beta_j}$. Testing such a coefficient for zero is done in polynomial time using Lenstra's algorithm as in the proof of Theorem 7, and there are at most k of them to test.

Suppose now that $u, v \neq 0$. Since $P(X, uX + v)$ is of the form (8), we can use our Gap Theorem (Theorem 6) as in the proof of Theorem 7: Let $P = \sum_{i=1}^{s} X^{\alpha(i)} P_i$ where each P_i is of the form (8) and satisfies $\alpha_1 = 0$ and $\alpha_k \leq \binom{k-1}{2}$. Then by Theorem 6, $P(X, uX + v)$ vanishes if and only if $P_i(X, uX + v)$ vanishes for every i. Now apply the same transformation to each P_i, inverting the roles of X and Y. Then each P_i can be written as the sum $\sum_{\ell=1}^{s_i} Y^{\beta(\ell)} P_{i\ell}$ where each $P_{i\ell}$ is of the form (8) and satisfies $\alpha_1 = \beta_1 = 0$ and $\alpha_k, \beta_k \leq \binom{k-1}{2}$. Furthermore, $P(X, uX + v)$ vanishes if and only if all the $P_{i\ell}(X, uX + v)$ vanish.

Since the $P_{i\ell}$'s are low-degree polynomials, and there are at most k of them, one can find all their linear factors. This relies on one of the numerous deterministic polynomial-time algorithms to factor dense multivariate polynomials that appear in the literature, from [10, 21] to [6, 20]. By the above discussion, the linear factors of P are exactly the linear factors that all the $P_{i\ell}$'s have in common. Several strategies can be used to find these linear factors: Either we search the linear factors of all the $P_{i\ell}$'s and keep only the ones they have in common, or we search the linear factors of one particular $P_{i\ell}$ (for instance the one of smallest degree) and test if they are factors of the other $P_{i\ell}$'s using our PIT algorithm, or we compute the gcd of all the $P_{i\ell}$'s and then search its linear factors. In particular, this last solution directly gives the multiplicities of the factors of P, since it is the same as their multiplicities in the gcd. □

As Kaltofen and Koiran's algorithm [11], our algorithm uses Lenstra's algorithm for univariate lacunary polynomials [22] to find univariate factors of the input polynomial. To compare both algorithms, let us thus focus on the task on finding truly bivariate linear factors, that is of the form $(Y - uX - v)$ with $uv \neq 0$.

A first remark concerns the simplicity of the algorithm. The computation of the gap function is much simpler in our case since we do not have to compute the height of the coefficients. This means that the task of finding the gaps in the input polynomial is reduced to completely combinatorial considerations.

Both our and Kaltofen and Koiran's algorithms use a dense factorization algorithm as a subroutine. This is in both cases the main computational task since the rest of the algorithm is devoted to the computation of the gaps in the input polynomial. Thus, a relevant measure to estimate the complexity of these algorithms is the maximum degree of the polynomials given as input to the dense factorization algorithm. This maximum degree is given by the values of the gaps in the two Gap Theorems. In our algorithm, the maximum degree is $\binom{k}{2}$. In Kaltofen and Koiran's, it is $\mathcal{O}(k \log k + k \log h_P)$ where h_P is the *height* of the polynomial P and the value $\log(h_P)$ is a bound on the size of the coefficients of P. For instance, if the coefficients of P are integers, then h_P is the maximum of their absolute values. Therefore, our algorithm has a better asymptotic complexity as soon as the size of the coefficients exceeds the number k of terms. Furthermore, the hidden constant in the bound for Kaltofen and Koiran's algorithm is only known to be bounded by approximately 15 while the corresponding constant in our case is 1/2.

Note that an improvement of Theorem 1 to a linear bound instead of a quadratic one would give us a better complexity than Kaltofen and Koiran's algorithm for all polynomials. Finally, it is naturally possible to combine both Gap Theorems in order to obtain the best complexity in all cases.

4. GENERALIZATIONS

In this section, we aim to prove some generalizations of the results obtained in Sections 2 and 3. The field $\mathbb{K}$ is still supposed to be an algebraic number field as in Section 3, unless otherwise stated.

Our first generalization shows that the identity test algorithm of Theorem 7 can be extended to a slightly more general family of polynomials. Namely, the linear polynomial $(uX + v)$ can be replaced by any 2-sparse polynomial.

THEOREM 9. *Let $P = \sum_{j=1}^{k} a_j X^{\alpha_j}(uX^d + v)^{\beta_j}$. There exists a deterministic polynomial-time algorithm to decide if the polynomial P is identically zero.*

In the theorem, $(uX^d + v)$ could be replaced by the seemingly more general expression $(uX^d + vX^{d'})$ with $d > d' > 0$. Yet, in this case we can factor out $X^{d'}$. A term $X^{\alpha_j}(uX^d + vX^{d'})^{\beta_j}$ can thus be written $X^{\alpha_j + d'\beta_j}(uX^{d-d'} + v)^{\beta_j}$. This has the same form as in the theorem, replacing α_j by $(\alpha_j + d'\beta_j)$ and d by $(d - d')$.

The size of the polynomial in the statement of the theorem is defined as in Equation (9) of Section 3 with the additional term $\log d$ in the sum. This means that the complexity of the algorithm is still polylogarithmic in the degree.

PROOF. For all j we consider the Euclidean division of α_j by d: $\alpha_j = q_j d + r_j$ with $r_j < d$. We rewrite P as

$$P = \sum_{j=1}^{k} a_j X^{r_j} (X^d)^{q_j} (uX^d + v)^{\beta_j}.$$

Let us group in the sum all the terms with a common r_j. That is, let

$$P_i(Y) = \sum_{\substack{1 \le j \le k \\ r_j = i}} a_j Y^{q_j} (uY + v)^{\beta_j}$$

for $0 \le i < d$. We remark that regardless of the value of d, the number of nonzero P_i's is bounded by k. We have $P(X) = \sum_{i=0}^{d-1} X^i P_i(X^d)$. Each monomial X^α of $X^i P_i(X^d)$ satisfies $\alpha \equiv i \mod d$. Therefore, P is identically zero if and only if all the P_i's are identically zero.

Since each P_i is of the form (8), and there are at most k of them, we can apply the algorithm of Theorem 7 to each of them. $\square$

We now state a generalization of Theorem 1. A special case of this generalization is used in the following to extend our factorization algorithm of Theorem 8. It is not known whether the most general version of the theorem can be used to further extend our algorithms to be able to find small-degree factors of lacunary polynomials.

Note that this result holds whatever field $\mathbb{K}$ of characteristic zero is considered.

THEOREM 10. *Let $(\alpha_{ij}) \in \mathbb{Z}_+^{m \times k}$ and*

$$P = \sum_{j=1}^{k} a_j \prod_{i=1}^{m} f_i^{\alpha_{ij}} \in \mathbb{K}[X],$$

where the degree of $f_i \in \mathbb{K}[X]$ is d_i for all i. Let $\xi \in \mathbb{K}$ and denote by μ_i the multiplicity of ξ as a root of f_i. Then the multiplicity $\mu_P(\xi)$ of ξ as a root of P satisfies

$$\mu_P(\xi) \le \max_{1 \le j \le k} \sum_{i=1}^{m} \left(\mu_i \alpha_{ij} + (d_i - \mu_i) \binom{k+1-j}{2} \right).$$

We refer the reader to the long version of this article for a proof of this theorem [3]. Note that it can be stated in the more general settings of rational exponents α_{ij}. It can then be seen as a generalization of a result of Kayal and Saha [17, Theorem 2.1].

The following corollary, used to find multilinear factors of bivariate lacunary polynomials, is a direct consequence of the theorem.

COROLLARY 11. *Let $P = \sum_{j=1}^{k} a_j X^{\alpha_j}(uX + v)^{\beta_j}(wX + t)^{\gamma_j}$, $uvwt \neq 0$. If P is nonzero, its valuation is at most*

$$\max_{1 \le j \le k} \left(\alpha_j + 2\binom{k+1-j}{2} \right).$$

We now describe how to use this corollary to get a new factorization algorithm. Compared to Theorem 8, we are now able to find the multilinear factors instead of the linear ones.

THEOREM 12. *Let $P = \sum_{j=1}^{k} a_j X^{\alpha_j} Y^{\beta_j}$. There exists a deterministic polynomial time algorithm to compute all the* multilinear *factors of P, with multiplicity.*

PROOF SKETCH. The proof goes along the same lines as the proof of Theorem 8. Suppose that $XY - (aX - bY + c)$ is a factor of P. Then the rational function $P(X, \frac{aX+c}{X+b})$ vanishes identically. Let us assume for simplicity that $a, b, c \neq 0$. (The other cases can be handled separately, as in the proof of Theorem 8.) Let

$$\begin{aligned} Q(X) &= (X+b)^{\max_i \beta_i} P(X, \frac{aX+c}{X+b}) \\ &= \sum_{j=1}^{k} a_j X^{\alpha_j} (aX+c)^{\beta_j} (X+b)^{\gamma_j} \end{aligned}$$

where $\gamma_j = \max_i(\beta_i) - \beta_j$. Then Q is a polynomial and it vanishes if and only if the rational function $P(X, \frac{aX+c}{X+b})$ does. By Corollary 11, if Q is nonzero its valuation is at most $\max_j(\alpha_j + 2\binom{k+1-j}{2})$. We can deduce a Gap Theorem: For $1 \leq k_0 \leq k$, let

$$Q_0(X) = \sum_{j=1}^{k_0} a_j X^{\alpha_j}(aX+c)^{\beta_j}(X+b)^{\gamma_j}$$

and $Q_1 = Q - Q_0$. Suppose that $\alpha_{k_0+1} > \max_{1\leq j\leq k_0}(\alpha_j + 2\binom{k_0+1-j}{2})$. Then Q vanishes identically if and only if Q_0 and Q_1 both vanish identically. Hence, $XY - (aX - bY + c)$ is a factor of P if and only if it is a factor of both P_0 and P_1, defined by analogy with Q_0 and Q_1: P_0 is the sum of the k_0 first terms of P and P_1 the sum of the $(k - k_0)$ last terms.

This proves that P can be written as a sum of $P_{i\ell}$'s as in the proof of Theorem 8 such that the multilinear factors of P are the common multilinear factors of the $P_{i\ell}$'s, and such that each $P_{i\ell}$ is of the same form as P and satisfies $\alpha_k, \beta_k \leq 2\binom{k-1}{2}$. It thus remains to find the common multilinear factors of some low-degree polynomials. Since there are at most k of them, this can be done in polynomial time. □

5. POSITIVE CHARACTERISTIC

As mentioned earlier, Theorem 1 does not hold in positive characteristic. We considered the polynomial $(1+X)^{2^n} + (1+X)^{2^{n+1}} = X^{2^n}(X+1)$ in characteristic 2. It only has two terms, but its valuation equals 2^n. Therefore, its valuation cannot be bounded by a function of the number of terms. Note that this can be generalized to any positive characteristic. In characteristic p, one can consider the polynomial $\sum_{i=1}^{p}(1+X)^{p^{n+i}}$.

Nevertheless, the exponents used in all the examples depend on the characteristic. In particular, the characteristic is always smaller than the largest exponent that appears. We shall show that in large characteristic, Theorem 1 still holds. This contrasts with the previous result [11] that uses the notion of height of an algebraic number, and which is thus not valid in any positive characteristic.

In fact, Theorem 1 holds as soon as $\mathsf{W}(f_1, \ldots, f_k)$ does not vanish. The difficulty in positive characteristic is that Proposition 2 does not hold anymore. Yet, the Wronskian is still related to linear independence by the following result (see [15]):

Proposition 13. *Let $\mathbb{K}$ be a field of characteristic p and $f_1, \ldots, f_k \in \mathbb{K}[X]$. Then $f_1, \ldots, f_k$ are linearly independent over $\mathbb{K}[X^p]$ if and only if their Wronskian does not vanish.*

This allows us to give an equivalent of Theorem 1 in large positive characteristic.

Theorem 14. *Let $P = \sum_{j=1}^{k} a_j X^{\alpha_j}(1+X)^{\beta_j} \in \mathbb{K}[X]$ with $\alpha_1 \leq \cdots \leq \alpha_k$. If the characteristic p of $\mathbb{K}$ satisfies $p > \max_j(\alpha_j + \beta_j)$, then the valuation of P is at most $\max_j(\alpha_j + \binom{k+1-j}{2})$, provided P does not vanish identically.*

Proof. Let $f_j = X^{\alpha_j}(1+X)^{\beta_j}$ for $1 \leq j \leq k$. The proof of Theorem 1 has two steps: We prove that we can assume that the Wronskian of the f_j's does not vanish, and then under this assumption we get a bound of the valuation of the polynomial. The second part only uses the non-vanishing of the Wronskian and can be used here too. We are left with proving that the Wronskian of the f_j's can be assumed to not vanish when the characteristic is large enough.

Assume that the Wronskian of the f_j's is zero: By Proposition 2, there is a vanishing linear combination of the f_j's with coefficients b_j in $\mathbb{K}[X^p]$. Let us write $b_j = \sum b_{i,j} X^{ip}$. Then $\sum_i X^{ip} \sum_j b_{i,j} f_j = 0$. Since $\deg f_j = \alpha_j + \beta_j < p$, $\sum_j b_{i,j} f_j = 0$ for all i. We have thus proved that there is a linear combination of the f_j's equal to zero with coefficients in $\mathbb{K}$. Therefore, we can assume we have a basis of the f_j's whose Wronskian is nonzero and use the same argument as for the characteristic zero. □

Based on this result, the algorithms we develop in characteristic zero for PIT and factorization can be used for large enough characteristics. Computing with lacunary polynomials in positive characteristic has been shown to be hard in many cases [28, 16, 18, 11, 1, 13]. In particular, it is shown in a very recent paper that it is NP-hard to find roots in $\mathbb{F}_p$ for polynomials over $\mathbb{F}_p$ [1].

Let $\mathbb{F}_{p^s}$ be the field with p^s elements for p a prime number and $s > 0$. In the algorithms, it is given as $\mathbb{F}_p[\xi]/\langle\varphi\rangle$ where φ is a monic irreducible polynomial of degree s with coefficients in $\mathbb{F}_p$.

Theorem 15. *Let $P = \sum_{j=1}^{k} a_j X^{\alpha_j}(uX+v)^{\beta_j} \in \mathbb{F}_{p^s}[X]$, where $p > \max_j(\alpha_j + \beta_j)$. There exists a polynomial-time deterministic algorithm to test if P vanishes identically.*

The proof of this theorem is very similar to the proof of Theorem 7, using Theorem 14 instead of Theorem 1. The main difference occurs when $u = 0$ or $v = 0$. In these cases, we rely in characteristic zero on an external algorithm to test sums of the form $\sum_j a_j v^{\beta_j}$ for zero. This external algorithm does not work in positive characteristic, but these tests are actually much simpler. These sums can be evaluated using repeated squaring in time polynomial in $\log \beta_j$, that is polynomial in the input length.

Note that the condition $p > \max_j(\alpha_j + \beta_j)$ means that p has to be greater than the degree of P. This condition is a fairly natural condition for many algorithms dealing with polynomials over finite fields, especially prime fields, for instance for root finding algorithms [1].

The basic operations in the algorithm are operations in the ground field $\mathbb{F}_p$. Therefore, the result also holds if bit operations are considered. The only place where computations in $\mathbb{F}_{p^s}$ have to be performed in the algorithm is the tests for zero of coefficients of the form $\sum_j \binom{\alpha_j}{\ell_j} a_j u^{-\alpha_j}(-v)^{\ell_j}$ where the α_j's and ℓ_j's are integers and $a_j \in \mathbb{F}_{p^s}$, and the sum has at most k terms. The binomial coefficient is to be computed *modulo* p using for instance Lucas' Theorem [23].

We now turn to the problem of finding linear factors of lacunary bivariate polynomials.

Theorem 16. *Let $P = \sum_j a_j X^{\alpha_j} Y^{\beta_j} \in \mathbb{F}_{p^s}[X, Y]$, where $p > \max_j(\alpha_j + \beta_j)$. There exists a probabilistic polynomial-time algorithm to find all the linear factors of P of the form $(uX + vY + w)$ with $uvw \neq 0$.*

Furthermore, deciding the existence of factors of the form $(X - w)$, $(Y - w)$ or $(X - wY)$ with $w \neq 0$ is NP-hard under randomized reductions.

Proof. The second part of the theorem is the consequence of the NP-hardness (under randomized reductions) of finding roots in $\mathbb{F}_{p^s}$ of lacunary univariate polynomials with coefficients in $\mathbb{F}_{p^s}$ [18, 1, 13]: Let Q be a lacunary univariate polynomial over $\mathbb{F}_{p^s}$, and define $P(X,Y) = Q(X)$. Then P has the same form as in the theorem with $\beta_j = 0$ for all j, and factors of the form $(X - w)$ of P are in one-to-one correspondence with roots w of Q. Thus, detecting such factors is NP-hard under randomized reductions. The same applies to factors of the form $(Y - w)$. Finally, let us now define P as the homogeneization of Q, that is $P(X,Y) = Y^{\deg(Q)}Q(X/Y)$. Then, $P(wY,Y) = Y^{\deg(Q)}P(w,1) = Y^{\deg(Q)}Q(w)$. In other words, factors of P of the form $(X - wY)$ correspond to roots w of Q. Thus detecting such factors is also NP-hard under randomized reduction.

For the first part, the algorithm we propose is actually the same as in characteristic zero (Theorem 8). This means that it relies on known results for factorization of dense polynomials. Yet, the only polynomial-time algorithms known for factorization in positive characteristic are probabilistic [27]. Therefore our algorithm is probabilistic and not deterministic as in characteristic zero. □

6. ACKNOWLEDGMENTS

We wish to thank Sébastien Tavenas for his help on Proposition 5, and Erich L. Kaltofen for pointing us out a mistake in Theorem 16 in a previous version of this paper.

7. REFERENCES

[1] J. Bi, Q. Cheng, and J. M. Rojas. Sub-Linear Root Detection, and New Hardness Results, for Sparse Polynomials Over Finite Fields. In *Proc. ISSAC*, 2013. `arXiv:1204.1113`.

[2] A. Bostan and P. Dumas. Wronskians and linear independence. *Am. Math. Mon.*, 117(8):722–727, 2010.

[3] A. Chattopadhyay, B. Grenet, P. Koiran, N. Portier, and Y. Strozecki. Factoring bivariate lacunary polynomials without heights. `arXiv:1206.4224`, 2012.

[4] F. Cucker, P. Koiran, and S. Smale. A polynomial time algorithm for Diophantine equations in one variable. *J. Symb. Comput.*, 27(1):21–30, 1999.

[5] M. Filaseta, A. Granville, and A. Schinzel. Irreducibility and Greatest Common Divisor Algorithms for Sparse Polynomials. In *Number Theory and Polynomials*, volume 352 of *P. Lond. Math. Soc.*, pages 155–176. Camb. U. Press, 2008.

[6] S. Gao. Factoring multivariate polynomials via partial differential equations. *Math. Comput.*, 72(242):801–822, 2003.

[7] M. Giesbrecht and D. S. Roche. On lacunary polynomial perfect powers. In *Proc. ISSAC'08*, pages 103–110. ACM, 2008.

[8] M. Giesbrecht and D. S. Roche. Detecting lacunary perfect powers and computing their roots. *J. Symb. Comput.*, 46(11):1242 – 1259, 2011.

[9] G. Hajós. [solution to problem 41] (in hungarian). *Mat. Lapok*, 4:40–41, 1953.

[10] E. Kaltofen. Polynomial-Time Reductions from Multivariate to Bi- and Univariate Integral Polynomial Factorization. *SIAM J. Comput.*, 14(2):469–489, 1985.

[11] E. Kaltofen and P. Koiran. On the complexity of factoring bivariate supersparse (lacunary) polynomials. In *Proc. ISSAC'05*, pages 208–215. ACM, 2005.

[12] E. Kaltofen and P. Koiran. Finding small degree factors of multivariate supersparse (lacunary) polynomials over algebraic number fields. In *Proc. ISSAC'06*, pages 162–168. ACM, 2006.

[13] E. L. Kaltofen and G. Lecerf. Factorization of Multivariate Polynomials. In *Handbook of Finite Fields*, Disc. Math. Appl. CRC Press, 2013. To appear.

[14] E. L. Kaltofen and M. Nehring. Supersparse black box rational function interpolation. In *Proc. ISSAC'11*, pages 177–186. ACM, 2011.

[15] I. Kaplansky. *An introduction to differential algebra.* Actualités scientifiques et industrielles. Hermann, 1976.

[16] M. Karpinski and I. Shparlinski. On the computational hardness of testing square-freeness of sparse polynomials. In *Applied Algebra, Algebraic Algorithms and Error-Correcting Codes*, volume 1719 of *LNCS*, pages 731–731. Springer, 1999.

[17] N. Kayal and C. Saha. On the Sum of Square Roots of Polynomials and Related Problems. In *Proc. CCC'11*, pages 292–299. IEEE, 2011.

[18] A. Kipnis and A. Shamir. Cryptanalysis of the HFE public key cryptosystem by relinearization. In *Proc. CRYPTO*, pages 19–30. Springer, 1999.

[19] P. Koiran, N. Portier, and S. Tavenas. A Wronskian approach to the real τ-conjecture. `arXiv:1205.1015`, 2012. Accepted for oral presentation at MEGA 2013.

[20] G. Lecerf. Improved dense multivariate polynomial factorization algorithms. *J. Symb. Comput.*, 42(4):477–494, 2007.

[21] A. K. Lenstra. Factoring Multivariate Polynomials over Algebraic Number Fields. *SIAM J. Comput.*, 16(3):591–598, 1987.

[22] H. Lenstra Jr. Finding small degree factors of lacunary polynomials. In *Number theory in progress*, pages 267–276. De Gruyter, 1999.

[23] É. Lucas. Théorie des fonctions numériques simplement périodiques. *Amer. J. Math.*, 1(2–4):184–240,289–321, 1878.

[24] H. Montgomery and A. Schinzel. Some arithmetic properties of polynomials in several variables. In *Transcendence Theory: Advances and Applications*, chapter 13, pages 195–203. Academic Press, 1977.

[25] M. Petkovšek, H. S. Wilf, and D. Zeilberger. *A=B*. AK Peters, 1996.

[26] D. Plaisted. Sparse complex polynomials and polynomial reducibility. *J. Comput. Syst. Sci.*, 14(2):210–221, 1977.

[27] J. von zur Gathen and J. Gerhard. *Modern Computer Algebra*. Camb. U. Press, 2nd edition, 2003.

[28] J. von zur Gathen, M. Karpinski, and I. Shparlinski. Counting curves and their projections. *Comput. Complex.*, 6(1):64–99, 1996.

A New View on HJLS and PSLQ: Sums and Projections of Lattices

Jingwei Chen[†,‡]
[†]Chengdu Institute of Computer Application, CAS
[‡]CNRS, ENS de Lyon, UCBL, Université de Lyon
Laboratoire LIP
velen.chan@163.com

Damien Stehlé
CNRS, ENS de Lyon, Inria
UCBL, Université de Lyon
Laboratoire LIP
damien.stehle@gmail.com

Gilles Villard
CNRS, ENS de Lyon, Inria,
UCBL, Université de Lyon
Laboratoire LIP
gilles.villard@ens-lyon.fr

ABSTRACT

The HJLS and PSLQ algorithms are the de facto standards for discovering non-trivial integer relations between a given tuple of real numbers. In this work, we provide a new interpretation of these algorithms, in a more general and powerful algebraic setup: we view them as special cases of algorithms that compute the intersection between a lattice and a vector subspace. Further, we extract from them the first algorithm for manipulating finitely generated additive subgroups of a euclidean space, including projections of lattices and finite sums of lattices. We adapt the analyses of HJLS and PSLQ to derive correctness and convergence guarantees.

Categories and Subject Descriptors

I.1.2 [**Symbolic and Algebraic Manipulation**]: Algorithms—*Algebraic algorithms*

Keywords

integer relation, lattice, HJLS, PSLQ, LLL

1. INTRODUCTION

A vector $\boldsymbol{m} \in \mathbb{Z}^n \setminus \{\mathbf{0}\}$ is called an *integer relation* for $\boldsymbol{x} \in \mathbb{R}^n$ if $\boldsymbol{x} \cdot \boldsymbol{m}^T = 0$. The HJLS algorithm [7, Sec. 3], proposed by Håstad, Just, Lagarias and Schnorr in 1986, was the first algorithm for discovering such a relation (or proving that no small relation exists) that consumed a number of real arithmetic operations polynomial in n and the bit-size of the relation bound. In 1992, Ferguson and Bailey published the other de facto standard algorithm for this task, the PSLQ algorithm [5] (see also [6] for a simplified analysis). We refer to the introduction of [7], and to [6, Sec. 9] for a historical perspective on integer relation finding. Our computational model will assume exact operations on real numbers. In this model, Meichsner has shown in [10, Sec. 2.3.1] that PSLQ is essentially equivalent to HJLS (see also [2, App. B, Th. 7] and the comments in Section 2).

Given as input $\boldsymbol{x} \in \mathbb{R}^n$, HJLS aims at finding a nonzero element in the intersection between the integer lattice $\Lambda = \mathbb{Z}^n$ and the $(n-1)$-dimensional vector subspace $E = \text{Span}(\boldsymbol{x})^\perp \subseteq \mathbb{R}^n$. It proceeds as follows. (1) It first projects the rows of the identity matrix (which forms a basis of Λ) onto E. This leads to n vectors belonging to a vector space of dimension $n-1$. The set of all integer linear combinations of these n vectors may not be a lattice: in full generality, it is only guaranteed to be a *finitely generated additive subgroup*, or *fgas* for short, of $\mathbb{R}^n$ (fgas's are studied in detail in Section 3). (2) It performs unimodular operations (swaps and integral translations) on these n vectors, in a fashion akin to (though different from) the LLL algorithm [8]. This aims at removing the linear dependencies between the fgas generators. (3) It stops computing with the fgas if it finds $n-1$ vectors belonging to the same $(n-2)$-dimensional vector subspace and an n-th vector that is linearly independent with those first $n-1$ vectors. This n-th vector contains a component that cannot be shortened any further using any linear combination of the previous vectors. At this stage, the inverse of the unimodular transformation matrix contains a non-trivial integer relation for $\boldsymbol{x}$. The computationally expensive step of HJLS is the second one, i.e., the manipulation of the fgas representation.

Our results. Our first contribution is to propose a new view on HJLS, and hence PSLQ, in a more general algebraic setup. It (partially) solves a special case of the following lattice and vector space intersection problem `Intersect`: given as inputs a basis of a lattice $\Lambda \subseteq \mathbb{R}^m$ and a basis of the vector subspace $E \subseteq \mathbb{R}^m$, the goal is to find a basis of the lattice $\Lambda \cap E$ (i.e., in the case of HJLS, the lattice of all integer relations). The main step of HJLS for (partially) solving (a particular case of) this problem, i.e., Step (2), is itself closely related to the following structural problem on fgas's. The topological closure $\overline{\mathcal{S}}$ of any fgas $\mathcal{S} \subseteq \mathbb{R}^m$ is the orthogonal sum of a unique lattice component Λ and a unique vector subspace component E, i.e., $\overline{\mathcal{S}} = \Lambda \oplus E$. The `Decomp` problem takes as input an fgas $\mathcal{S}$ described by a generating set and returns bases of Λ and E. We exhibit a duality relationship between the `Intersect` and `Decomp` problems that was somewhat implicit in HJLS.

Apart from putting HJLS in a broader context, this new view leads to the first algorithm, which we call `Decomp_HJLS`, for decomposing fgas's. Prior to this work, only special cases

ISSAC'13, June 26–29, 2013, Boston, Massachusetts, USA.

were handled: Pohst's MLLL algorithm [12] (see also [7, Sec. 2]) enables the computation of a basis of a lattice given by linearly dependent lattice vectors; and special cases of fgas's, corresponding to integer relations detection instances, were handled by HJLS and PSLQ. We describe the `Decomp_HJLS` algorithm in details, provide a correctness proof and analyze its convergence by adapting similar analyzes from [6] (which are essentially the same as in [7]). We show that it consumes a number of iterations (akin to LLL swaps) that is $\mathcal{O}(r^3 + r^2 \log \frac{X}{\lambda_1(\Lambda)})$, where r is the rank of the input fgas, X is an upper bound on the euclidean norms of the input generators and $\lambda_1(\Lambda)$ is the minimum of the lattice component Λ. For an fgas $\mathcal{S} \subseteq \mathbb{R}^m$ with n generators, an iteration consumes $\mathcal{O}(nm^2)$ arithmetic operations. Additionally, we prove that the returned lattice basis is reduced, for a notion of reduction that is similar to the LLL reduction.

Finally, we investigate a folklore strategy for solving problems similar to `Decomp`. This approach can be traced back to the original LLL article [8, p. 525]. It consists in embedding the input fgas into a higher-dimensional lattice, and calls the LLL algorithm. In order to ensure that the lattice component of the fgas can be read from the LLL output, we modify the underlying inner product by multiplying a subpart of the LLL input basis by a very small weight. More specifically, if we aim at decomposing the fgas spanned by the rows of a matrix $A \in \mathbb{R}^{n\times m}$, the `Decomp_LLL` algorithm will call LLL on the lattice basis $(c^{-1} \cdot I_n|A)$, where I_n denotes the n-dimensional identity matrix and $c > 0$. For a sufficiently large c, it is (heuristically) expected the lattice component of the fgas will appear in the bottom right corner of the LLL output.

Notation. All our vectors are row vectors and are denoted in bold. If $\boldsymbol{b}$ is a vector, then $\|\boldsymbol{b}\|$ denotes its euclidean norm. We let $\langle \boldsymbol{b}, \boldsymbol{c}\rangle$ denote the usual inner product between two real vectors $\boldsymbol{b}$ and $\boldsymbol{c}$ sharing the same dimension. If $\boldsymbol{b} \in \mathbb{R}^n$ is a vector and $E \subseteq \mathbb{R}^n$ is a vector space, we let $\pi(\boldsymbol{b}, E)$ denote the orthogonal projection of $\boldsymbol{b}$ onto E. Throughout this paper, we assume exact computations on real numbers. The unit operations are addition, substraction, multiplication, division, comparison of two real numbers, and the floor and square root functions.

2. REMINDERS

We give some brief reminders on lattices, and on the HJLS and PSLQ algorithms. For a comprehensive introduction to lattices, we refer the reader to [13].

LQ decomposition. Let $A \in \mathbb{R}^{n\times m}$ be a matrix of rank r. It has a unique LQ decomposition $A = L \cdot Q$, where the Q-factor $Q \in \mathbb{R}^{r\times m}$ has orthonormal rows (i.e., $QQ^T = I_r$), and the L-factor $L \in \mathbb{R}^{n\times r}$ satisfies the following property: there exist *diagonal indices* $1 \le k_1 < \ldots < k_r \le n$, such that $l_{i,j} = 0$ for all $i < k_j$, and $l_{k_j,j} > 0$ for all $j \le r$ (when $n = r$, the L-factor is lower-triangular with positive diagonal coefficients). The LQ decomposition of A is equivalent to the more classical QR decomposition of A^T.

DEFINITION 2.1. *Let $L = (l_{i,j}) \in \mathbb{R}^{n\times r}$ be a lower trapezoidal matrix with rank r and diagonal indices $k_1 < \ldots < k_r$. We say L is* size-reduced *if $|l_{i,j}| \le \frac{1}{2}\left|l_{k_j,j}\right|$ holds for $i > k_j$.*

Given L, it is possible to find a unimodular matrix $U \in \mathrm{GL}_n(\mathbb{Z})$ such that $U \cdot L$ is size-reduced. Computing U and updating $U \cdot L$ can be achieved within $\mathcal{O}(n^3)$ real arithmetic operations.

Lattices. A euclidean *lattice* $\Lambda \subseteq \mathbb{R}^m$ is a discrete (additive) subgroup of $\mathbb{R}^m$. A *basis* of Λ consists of n linearly independent vectors $\boldsymbol{b}_1, \cdots, \boldsymbol{b}_n \in \mathbb{R}^m$ such that $\Lambda = \sum_i \mathbb{Z}\boldsymbol{b}_i$. We say that $B = (\boldsymbol{b}_1^T, \ldots, \boldsymbol{b}_n^T)^T \in \mathbb{R}^{n\times m}$ is a basis matrix of Λ. The integer n is called the *dimension* of Λ. If $n \ge 2$, then Λ has infinitely many bases, that exactly consist in the rows of $U \cdot B$ where B is an arbitrary basis matrix of Λ and U ranges over $\mathrm{GL}_n(\mathbb{Z})$. The *i-th successive minimum* $\lambda_i(\Lambda)$ (for $i \le n$) is defined as the radius of the smallest ball that contains i linearly independent vectors of Λ. The *dual lattice* $\widehat{\Lambda}$ of Λ is defined as $\widehat{\Lambda} = \{\boldsymbol{x} \in \mathrm{Span}(\Lambda) : \forall \boldsymbol{b} \in \Lambda, \langle \boldsymbol{b}, \boldsymbol{x}\rangle \in \mathbb{Z}\}$. If B is a basis matrix of Λ, then $(BB^T)^{-1}B$ is a basis of $\widehat{\Lambda}$, called the *dual basis* of B.

Weakly-reduced bases. Weak reduction is a weakening of the classical notion of LLL reduction. It is very similar to the semi-reduction of [14]. Let $B = (\boldsymbol{b}_1^T, \cdots, \boldsymbol{b}_n^T)^T \in \mathbb{R}^{n\times m}$ be the basis matrix of a lattice Λ, and $L = (l_{i,j})$ be its L-factor. We say the basis $\boldsymbol{b}_1, \cdots, \boldsymbol{b}_n$ is *weakly-reduced* with parameters $\gamma > 2/\sqrt{3}$ and $C \ge 1$ if L is size-reduced and satisfies the (generalized) Schönhage condition $l_{j,j} \le C \cdot \gamma^i \cdot l_{i,i}$ for $1 \le j \le i \le n$. Note that a LLL-reduced basis is always weakly-reduced, with $C = 1$. If a lattice basis is weakly-reduced, then

$$\begin{gathered} \|\boldsymbol{b}_i\| \le \sqrt{n} C\gamma^i \cdot l_{i,i}, \\ (\sqrt{n}C^2\gamma^{2i})^{-1} \cdot \lambda_i(\Lambda) \le \|\boldsymbol{b}_i\| \le \sqrt{n}C^2\gamma^{2n} \cdot \lambda_i(\Lambda). \end{gathered} \tag{2.1}$$

HJLS-PSLQ. We recall HJLS [7, Sec. 3] using the PSLQ setting [6]. We call the resulting algorithm HJLS-PSLQ (Algorithm 1). Given $\boldsymbol{x} = (x_1, \cdots, x_n) \in \mathbb{R}^n$, HJLS-PSLQ either returns an integer relation for $\boldsymbol{x}$, or gives a lower bound on $\lambda_1(\Lambda_x)$, where Λ_x is the lattice of all integer relations for $\boldsymbol{x}$, and $\lambda_1(\Lambda_x)$ is the norm of any shortest nonzero vector in Λ_x. The updates of U and Q at Steps 1b, 2a, 2b and 2c are implemented so as to maintain the relationship $UL_x = LQ$ at any stage of the execution. Note that storing and updating Q is not necessary for the execution of the algorithm (and does not appear in [6]). It has been added for easing explanations in Section 4.

Algorithm 1 (HJLS-PSLQ).

Input: $\boldsymbol{x} = (x_1, \cdots, x_n) \in \mathbb{R}^n$ with $x_i \neq 0$ for $i \le n$, $M > 0$ and $\gamma > 2/\sqrt{3}$.

Output: Either return an integer relation for $\boldsymbol{x}$, or claim that $\lambda_1(\Lambda_x) > M$.

1. (a) Normalize $\boldsymbol{x}$, i.e., set $\boldsymbol{x} := \boldsymbol{x}/\|\boldsymbol{x}\|$; set $U := I_n$ and $Q = I_{n-1}$.
   (b) Compute the Q-factor $(\boldsymbol{x}^T|L_x)^T$ of $(\boldsymbol{x}|I_n)^T$; set $L := L_x$; size-reduce L and update U.
2. While $l_{n-1,n-1} \neq 0$ and $\max_i l_{i,i} \ge 1/M$ do
   (a) Choose k such that $\gamma^k \cdot l_{k,k} = \max_{j<n} \gamma^j \cdot l_{j,j}$; swap the k-th and $(k+1)$-th rows of L and update U;
   (b) Compute the LQ decomposition of L; replace L by its L-factor and update Q.
   (c) Size-reduce L and update U.
3. If $l_{n-1,n-1} \neq 0$, return "$\lambda_1(\Lambda_x) > M$". Else return the last column of U^{-1}.

For the proof of termination, it suffices to enforce the partial size-reduction condition $|l_{k+1,k}| \le \frac{1}{2} l_{k,k}$ before swapping

(HJLS), instead of full size-reduction (PSLQ). Along with a stronger size-reduction, PSLQ may have a slightly faster termination for specific cases, due to a refined while loop test. PSLQ has been proposed with the additional nullity test of $(\boldsymbol{x}\cdot U^{-1})_j$ for some $j \leq n$, possibly leading to the early output of the j-th column of U^{-1}. Apart from the latter test, if HJLS is implemented with full size-reduction then PSLQ is equivalent to HJLS [10, Sec. 2.3.1] (see also [2, App. B, Th. 7]). Full size-reduction is essentially irrelevant in the exact real number model. Indeed, HJLS works correctly and consumes $\mathcal{O}(n^3 + n^2\lambda_1(\Lambda_x))$ iterations. The same bound has been established for PSLQ. We note that without loss of generality HJLS has been initially stated with $\gamma = \sqrt{2}$. Both the roles of the size-reduction and the parameter γ may be important in a bit complexity model for keeping integer bit sizes small [8, 7], or in a model based on approximate real number operations for mastering the required number precision [6]. This is outside the scope of the present paper.

3. DECOMP AND INTERSECT

In this section, we provide efficient reductions in both directions between the problem of computing the decomposition of an fgas (Decomp) and the problem of computing the intersection of a lattice and a vector subspace (Intersect), assuming exact computations over the reals.

3.1 FGAS of a euclidean space

Given $\boldsymbol{a}_1, \cdots, \boldsymbol{a}_n \in \mathbb{R}^m$, the *finitely generated additive subgroup* (*fgas* for short) spanned by the $\boldsymbol{a}_i$'s is the set of all integral linear combinations of the $\boldsymbol{a}_i$'s:

$$\mathcal{S} = \sum_{i=1}^{n} \mathbb{Z}\boldsymbol{a}_i = \left\{ \sum_{i=1}^{n} z_i\boldsymbol{a}_i : z_i \in \mathbb{Z} \right\} \subseteq \mathbb{R}^m. \quad (3.1)$$

Given an fgas $\mathcal{S}$ as in (3.1), the matrix $A \in \mathbb{R}^{n\times m}$ whose i-th row is $\boldsymbol{a}_i$ is called a *generating matrix* of $\mathcal{S}$. The rank of A is called the *rank* of the fgas. If a matrix $U \in \mathrm{GL}_n(\mathbb{Z})$, then $U \cdot A$ is also a generating matrix of $\mathcal{S}$.

When the vectors $\boldsymbol{a}_i$ are linearly independent, then the set $\mathcal{S}$ is a lattice and the $\boldsymbol{a}_i$'s form a basis of the lattice. If the $\boldsymbol{a}_i$'s are linearly dependent, but $\mathcal{S}$ can be written as $\mathcal{S} = \sum_{i=1}^{d} \mathbb{Z}\boldsymbol{b}_i$ for some linearly independent $\boldsymbol{b}_i$'s, then $\mathcal{S}$ is also a lattice. In this case, the $\boldsymbol{a}_i$'s are not a basis of $\mathcal{S}$ and $\dim(\mathcal{S}) < n$.

The situation that we are mostly interested in the present work is when $\mathcal{S}$ is not a lattice. The simplest example may be the fgas $\mathbb{Z}+\alpha\mathbb{Z}$ with $\alpha \notin \mathbb{Q}$: it contains non-zero elements that are arbitrarily close to 0, and thus cannot be a lattice. More generally, an fgas can always be viewed as a *finite sum of lattices*.

Fgas's can also be viewed as *orthogonal projections of lattices onto vector subspaces*. Let $\Lambda = \sum_i \mathbb{Z}\boldsymbol{b}_i \subseteq \mathbb{R}^m$ be a lattice and $E \subseteq \mathbb{R}^m$ be a vector subspace. The *orthogonal projection* of Λ onto E, i.e., the set $\pi(\Lambda, E) = \{\boldsymbol{v}_1 \in E : \exists \boldsymbol{v}_2 \in E^\perp, \boldsymbol{v}_1 + \boldsymbol{v}_2 \in \Lambda\}$, is an fgas of $\mathbb{R}^m$: it is spanned by the projections of the $\boldsymbol{b}_i$'s. Conversely, given an fgas $\mathcal{S}$ with generating matrix $A \in \mathbb{R}^{n\times m}$, let $\Lambda \subseteq \mathbb{R}^{n+m}$ be the lattice generated by the rows of $(I_n|A)$ and $E = \mathrm{Span}(I_n|0)^\perp \subseteq \mathbb{R}^{n+m}$. Then $\mathcal{S} = \pi(\Lambda, E)$.

3.2 The Decomp and Intersect problems

Consider the topological closure $\overline{\mathcal{S}}$ of an fgas $\mathcal{S} \subseteq \mathbb{R}^m$, i.e., the set of all limits of converging sequences of $\mathcal{S}$ (which is hence a closed additive subgroup in $\mathbb{R}^m$). By [9, Th. 1.1.2] (see also [3, Chap. VII, Th. 2]), there exists a unique lattice $\Lambda \subseteq \mathbb{R}^m$ and a unique vector subspace $E \subseteq \mathbb{R}^m$ such that their direct sum is $\overline{\mathcal{S}}$, and the vector space $\mathrm{Span}(\Lambda)$ spanned by Λ is orthogonal to E. We denote the latter decomposition by $\overline{\mathcal{S}} = \Lambda \bigoplus E$. More explicitly, if $\mathrm{rank}\,\mathcal{S} = \dim(\mathrm{Span}(\mathcal{S})) = r \leq m$, then there exist $0 \leq d \leq r$, $(\boldsymbol{b}_i)_{i\leq d}$ and $(\boldsymbol{e}_i)_{i\leq r-d}$ in $\mathbb{R}^m$ such that:

- $\overline{\mathcal{S}} = \sum_{i\leq d} \mathbb{Z}\boldsymbol{b}_i + \sum_{i\leq r-d} \mathbb{R}\boldsymbol{e}_i$;
- the r vectors $\boldsymbol{b}_i$ ($i \leq d$) and $\boldsymbol{e}_i$ ($i \leq r-d$) are linearly independent;
- for any $i \leq d$ and $j \leq r-d$, we have $\langle \boldsymbol{b}_i, \boldsymbol{e}_j\rangle = 0$.

Then $(\boldsymbol{b}_i)_{i\leq d}$ and $(\boldsymbol{e}_i)_{i\leq r-d}$ are bases of Λ and E, respectively. We call Λ and E the *lattice and vector space components* of $\mathcal{S}$, respectively, and define the *ΛE decomposition* of $\mathcal{S}$ as (Λ, E). The Decomp problem is the associated computational task.

Definition 3.1. *The* Decomp *problem is as follows: Given as input a finite generating set of an fgas $\mathcal{S}$, the goal is to compute its ΛE decomposition, i.e., find bases for the lattice and vector space components Λ and E.*

The following result, at the core of the correctness analysis of our decomposition algorithm of Section 5, reduces Decomp to the task of obtaining an fgas generating set that contains sufficiently many linear independencies.

Lemma 3.2. *Let $(\boldsymbol{a}_i)_{i\leq n}$ be a generating set of an fgas $\mathcal{S}$ with ΛE decomposition $\overline{\mathcal{S}} = \Lambda \bigoplus E$. Define $\boldsymbol{a}'_j$ as the projection of $\boldsymbol{a}_j$ orthogonally to $\mathrm{Span}(\boldsymbol{a}_i)_{i\leq n-k}$, for $n-k+1 \leq j \leq n$ and some $k < n$, and assume the $\boldsymbol{a}'_j$'s are linearly independent. Then $\boldsymbol{a}'_{n-k+1}, \ldots, \boldsymbol{a}'_n$ form a basis of a projection of Λ and $E \subseteq \mathrm{Span}(\boldsymbol{a}_i)_{i\leq n-k}$. Further, if $k = \dim \Lambda$, then $\Lambda = \sum_{n-k+1\leq i\leq n} \mathbb{Z}\cdot\boldsymbol{a}'_i$ and $E = \mathrm{Span}(\boldsymbol{a}_i)_{i\leq n-k}$.*

The proof derives from the definition of the ΛE decomposition. The vector space component E is the largest vector subspace of $\mathrm{Span}(\mathcal{S})$ that is contained in $\overline{\mathcal{S}}$. This characterisation of E implies that it is contained in $\mathrm{Span}(\boldsymbol{a}_i)_{i\leq n-k}$. Indeed, the projections $\boldsymbol{a}'_j$ are linearly independent and lead to a discrete subgroup that must be orthogonal to E. By unicity of the ΛE decomposition, the vectors $\boldsymbol{a}'_{n-k+1}, \ldots, \boldsymbol{a}'_n$ form a basis of a projection of the lattice component Λ.

We now introduce another problem, Intersect, which generalizes the integer relation finding problem.

Definition 3.3. *The* Intersect *problem is as follows: Given as inputs a basis of a lattice Λ and a basis of a vector subspace E, the goal is to find a basis of the lattice $\Lambda \cap E$.*

Finding a non-zero integer relation corresponds to taking $\Lambda = \mathbb{Z}^n$ and $E = \mathrm{Span}(\boldsymbol{x})^\perp$, and asking for one vector in $\Lambda \cap E$. In that case, Intersect aims at finding a description of all integer relations for $\boldsymbol{x}$. When E is arbitrary but Λ remains $\mathbb{Z}^n$, Intersect corresponds to the task of finding all *simultaneous integer relations*. These special cases are considered in [7].

3.3 Relationship between the problems

The Decomp and Intersect problems turn out to be closely related. To explain this relationship, we need the concept of dual lattice of an fgas. The facts of this subsection are adapted from basic techniques on lattices (see, e.g.,[4]).

Definition 3.4. *The* dual lattice $\widehat{\mathcal{S}}$ *of an fgas* $\mathcal{S}$ *is defined as* $\widehat{\mathcal{S}} = \{\boldsymbol{x} \in \mathrm{Span}(\mathcal{S}) : \forall \boldsymbol{b} \in \mathcal{S}, \langle \boldsymbol{x}, \boldsymbol{b} \rangle \in \mathbb{Z}\}$.

We could equivalently define $\widehat{\mathcal{S}}$ as $\{\boldsymbol{x} \in \mathrm{Span}(\mathcal{S}) : \forall \boldsymbol{b} \in \overline{\mathcal{S}}, \langle \boldsymbol{x}, \boldsymbol{b} \rangle \in \mathbb{Z}\}$. Indeed, for all $\boldsymbol{b} \in \overline{\mathcal{S}}$, there exists a converging sequence $(\boldsymbol{b}_i)_i$ in $\mathcal{S}$ such that $\boldsymbol{b}_i \to \boldsymbol{b}$ as $i \to \infty$. Thus, for all $\boldsymbol{x} \in \widehat{\mathcal{S}}$, we have $\langle \boldsymbol{x}, \boldsymbol{b} \rangle = \langle \boldsymbol{x}, \lim_i \boldsymbol{b}_i \rangle = \lim_i \langle \boldsymbol{x}, \boldsymbol{b}_i \rangle \in \mathbb{Z}$. We will freely use both definitions.

Note further that if $\mathcal{S}$ is a lattice, then $\widehat{\mathcal{S}}$ is exactly the dual lattice of $\mathcal{S}$. Interestingly, $\widehat{\mathcal{S}}$ is always a lattice, even if $\mathcal{S}$ is not a lattice.

Lemma 3.5. *Let* $\mathcal{S}$ *be an fgas and* Λ *its lattice component. Then* $\widehat{\Lambda} = \widehat{\mathcal{S}}$.

Proof. Let $\overline{\mathcal{S}} = \Lambda \bigoplus E$ be the ΛE decomposition of $\mathcal{S}$. Recall that for any $\boldsymbol{x} \in \overline{\mathcal{S}}$, there exist unique $\boldsymbol{x}_\Lambda \in \Lambda$ and $\boldsymbol{x}_E \in E$ such that $\boldsymbol{x} = \boldsymbol{x}_\Lambda + \boldsymbol{x}_E$ and $\langle \boldsymbol{x}_\Lambda, \boldsymbol{x}_E \rangle = 0$.

We first prove that $\widehat{\Lambda} \subseteq \widehat{\mathcal{S}}$. For all $\widehat{\boldsymbol{x}} \in \widehat{\Lambda}$ and all $\boldsymbol{x} \in \overline{\mathcal{S}}$, we have $\langle \widehat{\boldsymbol{x}}, \boldsymbol{x} \rangle = \langle \widehat{\boldsymbol{x}}, \boldsymbol{x}_\Lambda \rangle + \langle \widehat{\boldsymbol{x}}, \boldsymbol{x}_E \rangle = \langle \widehat{\boldsymbol{x}}, \boldsymbol{x}_\Lambda \rangle \in \mathbb{Z}$, where the second equality follows from the orthogonality between the vector subspaces E and $\mathrm{Span}(\Lambda)$, and $\langle \widehat{\boldsymbol{x}}, \boldsymbol{x}_\Lambda \rangle \in \mathbb{Z}$ derives from the definition of $\widehat{\Lambda}$.

Further, for all $\widehat{\boldsymbol{x}} \in \widehat{\mathcal{S}}$ and all $\boldsymbol{x} \in \Lambda \subseteq \overline{\mathcal{S}}$, it follows from the second definition of $\widehat{\mathcal{S}}$ that $\langle \widehat{\boldsymbol{x}}, \boldsymbol{x} \rangle \in \mathbb{Z}$, i.e., we have $\widehat{\boldsymbol{x}} \in \widehat{\Lambda}$. This completes the proof. □

From Lemma 3.5, we derive the following alternative definition of the lattice component of an fgas.

Lemma 3.6. *Let* $\mathcal{S}$ *be an fgas and* Λ *its lattice component. Then* $\Lambda = \widehat{\widehat{\mathcal{S}}}$.

Let $\Lambda \subseteq \mathbb{R}^m$ be a lattice and $E \subseteq \mathbb{R}^m$ a vector subspace. If $\pi(\widehat{\Lambda}, E)$ happens to be a lattice, then it is exactly $\widehat{\Lambda \cap E}$ (see, e.g., [9, Prop. 1.3.4]). However, in general, the fgas $\pi(\widehat{\Lambda}, E)$ may not be a lattice. Using Definition 3.4, we can prove the following result, which plays a key role in the relationship between `Intersect` and `Decomp`.

Lemma 3.7. *For any lattice* $\Lambda \subseteq \mathbb{R}^m$ *and a vector subspace* $E \subseteq \mathbb{R}^m$, *we have* $\Lambda \cap E = \widehat{\pi(\widehat{\Lambda}, E)}$.

Proof. Let $\boldsymbol{b} \in \Lambda \cap E$ and $\boldsymbol{y} \in \pi(\widehat{\Lambda}, E)$. There exist $\widehat{\boldsymbol{b}} \in \widehat{\Lambda}$ and $\boldsymbol{y}' \in E^\perp$ such that $\widehat{\boldsymbol{b}} = \boldsymbol{y} + \boldsymbol{y}'$. Then

$$\langle \boldsymbol{b}, \boldsymbol{y} \rangle = \langle \boldsymbol{b}, \widehat{\boldsymbol{b}} \rangle - \langle \boldsymbol{b}, \boldsymbol{y}' \rangle = \langle \boldsymbol{b}, \widehat{\boldsymbol{b}} \rangle \in \mathbb{Z}.$$

Hence $\Lambda \cap E \subseteq \widehat{\pi(\widehat{\Lambda}, E)}$.

Now, let $\boldsymbol{b} \in \widehat{\pi(\widehat{\Lambda}, E)}$. By definition, we have

$$\boldsymbol{b} \in \mathrm{Span}(\pi(\widehat{\Lambda}, E)) \subseteq E.$$

Moreover, for all $\widehat{\boldsymbol{b}} \in \widehat{\Lambda}$, using $\widehat{\boldsymbol{b}} = \pi(\widehat{\boldsymbol{b}}, E) + \pi(\widehat{\boldsymbol{b}}, E^\perp)$:

$$\langle \boldsymbol{b}, \widehat{\boldsymbol{b}} \rangle = \langle \boldsymbol{b}, \pi(\widehat{\boldsymbol{b}}, E) \rangle + \langle \boldsymbol{b}, \pi(\widehat{\boldsymbol{b}}, E^\perp) \rangle = \langle \boldsymbol{b}, \pi(\widehat{\boldsymbol{b}}, E) \rangle \in \mathbb{Z}.$$

Hence $\boldsymbol{b} \in \widehat{\widehat{\Lambda}} = \Lambda$. We obtain that $\boldsymbol{b} \in \Lambda \cap E$, which completes the proof. □

Reducing `Decomp` to `Intersect`. Suppose we are given a generating set $\boldsymbol{a}_1, \cdots, \boldsymbol{a}_n \in \mathbb{R}^m$ of an fgas $\mathcal{S}$. Our goal is to find the ΛE decomposition of $\mathcal{S}$, using an oracle that solves `Intersect`. It suffices to find a basis of the lattice component Λ, which, by Lemma 3.6, satisfies $\Lambda = \widehat{\widehat{\mathcal{S}}}$.

Recall that we can construct a lattice Λ' and a vector space E such that (see the end of Section 3.1) $\mathcal{S} = \pi(\Lambda', E)$. From Lemma 3.7, the lattice component Λ of $\mathcal{S}$ is the dual lattice of $\widehat{\Lambda'} \cap E$. This means we can get a basis of $\widehat{\Lambda}$ by calling the `Intersect` oracle on $\widehat{\Lambda'}$ and E, and then computing the dual basis of the returned basis.

Reducing `Intersect` to `Decomp`. Assume we are given a basis $(\boldsymbol{b}_i)_i$ of a lattice $\Lambda \subseteq \mathbb{R}^m$ and a basis $(\boldsymbol{e}_i)_i$ of a vector subspace $E \subseteq \mathbb{R}^m$. We aim at computing a basis of the lattice $\Lambda \cap E$, using an oracle that solves `Decomp`.

We first compute the dual basis $(\widehat{\boldsymbol{b}}_i)_i$ of $(\boldsymbol{b}_i)_i$. Then we compute the projections $\widehat{\boldsymbol{b}}'_i = \pi(\widehat{\boldsymbol{b}}_i, E)$, for all i. Let $\mathcal{S}$ denote the fgas spanned by $(\widehat{\boldsymbol{b}}'_i)_i$. We now use the `Decomp` oracle on $\mathcal{S}$ to obtain a basis of the lattice component Λ' of $\mathcal{S}$. Then, by Lemma 3.7, the dual basis of the oracle output is a basis of $\Lambda \cap E$.

4. A NEW VIEW ON HJLS-PSLQ

We explain the principle of HJLS-PSLQ described in Section 2, by using the results of Section 3. At a high level, HJLS-PSLQ proceeds as in the `Intersect` to `Decomp` reduction from Section 3. The algorithm in Section 2 halts as soon as a relation is found. Hence it only partially solves `Decomp`, on the specific input under scope, and, as a result, only partially solves `Intersect`. The full decomposition will be studied in Section 5.

Step 1 revisited: Projection of $\mathbb{Z}^n$ on $\mathrm{Span}(\boldsymbol{x})^\perp$. The reduction from `Intersect` to `Decomp` starts by projecting $\widehat{\Lambda}$ onto E. In our case, we have $\Lambda = \widehat{\Lambda} = \mathbb{Z}^n$ (the lattice under scope is self-dual) and $E = \mathrm{Span}(\boldsymbol{x})^\perp$. The start of the reduction matches with the main component of Step 1, which is the computation of the Q-factor $Q_x := (\boldsymbol{x}^T | L_x)^T$ of $(\boldsymbol{x}^T | I_n)^T$, considering $\boldsymbol{x}$ after normalization. Using that $Q_x^T \cdot Q_x = I_n$ we now observe that L_x satisfies the following equation

$$\begin{pmatrix} \boldsymbol{x} \\ I_n \end{pmatrix} = \begin{pmatrix} 1 & \\ \boldsymbol{x}^T & L_x \end{pmatrix} \cdot \begin{pmatrix} \boldsymbol{x} \\ L_x^T \end{pmatrix}.$$

By construction, the matrix L_x is lower trapezoidal. Indeed, since the i-th row of Q_x is orthogonal to the linear span of the first $i-1$ rows, and as this linear span contains the first $i-2$ unit vectors, the first $i-2$ coordinates of this i-th row of Q_x are zero. Hence the equation above provides the LQ decomposition of $(\boldsymbol{x}^T | I_n)^T$. It is worth noticing the unusual fact that L_x is involved in both the L-factor and the Q-factor. Also, as a consequence of the equation above, we have that the matrix $\pi_x = I_n - \boldsymbol{x}^T \boldsymbol{x} = L_x L_x^T$ corresponds to the orthogonal projection that maps $\mathbb{R}^n$ to $\mathrm{Span}(\boldsymbol{x})^\perp$. Therefore the rows of L_x are the coordinate vectors of the rows of π_x with respect to the normalized orthogonal basis of $\mathrm{Span}(\boldsymbol{x})^\perp$ given by the $n-1$ rows of L_x^T. Overall, we obtain that $(\boldsymbol{0} | L_x) \cdot Q_x$ is a generating matrix of the fgas $\mathcal{S}_x = \pi(\mathbb{Z}^n, \mathrm{Span}(\boldsymbol{x})^\perp)$.

Step 2 revisited: A partial solution to `Decomp`. Since $(\boldsymbol{0} | L_x) \cdot Q_x$ is a generating matrix of the fgas $\mathcal{S}_x$, the while loop of HJLS-PSLQ only considers this fgas (in fact, HJLS-PSLQ only works on L_x since its only requires U). In Section 5, we will show that a generalization of the while loop may be used to solve the `Decomp` problem. By Lemma 3.7, finding a basis of the lattice component of $\mathcal{S}_x$ suffices to find all integer relations of $\boldsymbol{x}$: indeed, the dual basis is a basis of the integer relation lattice. However, when HJLS-PSLQ terminates, we may not have the full lattice component Λ' of $\mathcal{S}_x$. If the loop stops because $l_{n-1,n-1} = 0$, then we have found a projection to a 1-dimensional subspace of a vector belonging to the lattice component. In this sense, Step 2 of HJLS-PSLQ partially solves `Decomp` on input $\mathcal{S}_x$. It gets the full solution only when $\dim(\mathbb{Z}^n \cap \mathrm{Span}(\boldsymbol{x})^\perp) = \dim(\Lambda') = 1$.

Step 3 revisited: Getting back to `Intersect`. Suppose HJLS-PSLQ exits the while loop because $l_{n-1,n-1} = 0$. Because of the shape of L (see Lemma 3.2), it has found a 1-dimensional projection of a non-zero basis vector of Λ', orthogonally to the first vectors of that basis of Λ'. This vector is:

$$\boldsymbol{b} := (\mathbf{0}|l_{n-1,n-1}) \cdot \mathrm{diag}(1, Q) \cdot Q_x.$$

Its dual, when considered as a basis, is

$$\widehat{\boldsymbol{b}} = \boldsymbol{b}/\|\boldsymbol{b}\|^2 = (\mathbf{0}|l_{n-1,n-1}^{-1}) \cdot \mathrm{diag}(1, Q) \cdot Q_x.$$

As $\widehat{\boldsymbol{b}}$ is a projection of a non-zero basis vector of Λ', orthogonally to the first vectors of that basis, we have that $\widehat{\boldsymbol{b}}$ belongs to $\widehat{\Lambda}' = \mathbb{Z}^n \cap \mathrm{Span}(\boldsymbol{x})^\perp$. Because of the specific shape of Q_x, we obtain

$$\begin{aligned}\widehat{\boldsymbol{b}} &= (\mathbf{0}|l_{n-1,n-1}^{-1}) \cdot \left(\begin{array}{c|c} 1 & \\ \hline & Q \end{array}\right) \cdot \left(\begin{array}{c} \boldsymbol{x} \\ \hline L_x^T \end{array}\right) \\ &= (\mathbf{0}|l_{n-1,n-1}^{-1}) \cdot \left(\begin{array}{c} \boldsymbol{x} \\ \hline QL_x^T \end{array}\right).\end{aligned}$$

Now, as $UL_x = LQ$, we obtain that $\widehat{\boldsymbol{b}} = (\mathbf{0}|l_{n-1,n-1}^{-1}) \cdot (\boldsymbol{x}^T|U^{-1}L)^T = (\mathbf{0}|1)U^{-T}$. This explains why the relation is embedded in the inverse of the transformation matrix. Note that this is somewhat unexpected, and derives from the uncommon similarity between L_x and Q_x.

A numerical example. Consider the input $(1, \sqrt{2}, 2)$. After normalization, it becomes $\boldsymbol{x} = (\frac{1}{\sqrt{7}}, \frac{\sqrt{2}}{\sqrt{7}}, \frac{2}{\sqrt{7}})$. At the beginning, we have

$$L_x = \begin{pmatrix} \frac{6}{\sqrt{42}} & 0 \\ -\frac{\sqrt{2}}{\sqrt{42}} & \frac{\sqrt{2}}{\sqrt{3}} \\ -\frac{2}{\sqrt{42}} & -\frac{1}{\sqrt{3}} \end{pmatrix} \text{ and } Q_x = \begin{pmatrix} \frac{1}{\sqrt{7}} & \frac{\sqrt{2}}{\sqrt{7}} & \frac{2}{\sqrt{7}} \\ \frac{\sqrt{6}}{\sqrt{42}} & -\frac{\sqrt{2}}{\sqrt{42}} & -\frac{2}{\sqrt{42}} \\ 0 & \frac{\sqrt{2}}{\sqrt{3}} & -\frac{1}{\sqrt{3}} \end{pmatrix}.$$

The matrix $(\mathbf{0}|L_x) \cdot Q_x$ is a generating matrix of the fgas $\mathcal{S}_x = \pi(\mathbb{Z}^n, \mathrm{Span}(\boldsymbol{x})^\perp)$. After 5 loop iterations, HJLS-PSLQ terminates. At that stage, we obtain

$$L = \begin{pmatrix} \frac{15-10\sqrt{2}}{\sqrt{35}} & 0 \\ -\frac{5(-41+29\sqrt{2})}{\sqrt{35}(-3+2\sqrt{2})} & 0 \\ \frac{41\sqrt{2}-58}{\sqrt{35}(-3+2\sqrt{2})} & \frac{1}{\sqrt{5}} \end{pmatrix},$$

$$U = \begin{pmatrix} -2 & -3 & -4 \\ 5 & 7 & 10 \\ -1 & -2 & -3 \end{pmatrix}, \; Q = \begin{pmatrix} \frac{-4+3\sqrt{2}}{\sqrt{30}(3-2\sqrt{2})} & -\frac{\sqrt{14}}{\sqrt{15}} \\ \frac{\sqrt{14}}{\sqrt{15}} & \frac{-24+17\sqrt{2}}{\sqrt{30}(17-12\sqrt{2})} \end{pmatrix}.$$

Thanks to the shape of L, the ΛE decomposition $\overline{\mathcal{S}}_x = \Lambda \bigoplus E$ can be derived from $(\mathbf{0}|L)$. In this precise case, HJLS-PSLQ discloses the full lattice component. Thanks to Lemma 3.7, we have $\Lambda = \widehat{\Lambda}_x$, and hence $\dim(\Lambda) = \dim(\widehat{\Lambda}_x) = \dim(\Lambda_x) = 1$ (as $\boldsymbol{x}$ contains two rational entries and one irrational entry). Using the matrix factorisation above, we obtain

$$\Lambda = \mathbb{Z} \cdot (0, 0, 1/\sqrt{5}) \cdot \mathrm{diag}\,(1, Q) \cdot Q_x = \mathbb{Z} \cdot (2/5, 0, -1/5)$$

and $E = (0, 1, 0) \cdot \mathrm{diag}(1, Q) \cdot Q_x$. By Lemma 3.7, we obtain $\mathbb{Z}^3 \cap \mathrm{Span}(\boldsymbol{x})^\perp = \widehat{\Lambda} = \mathbb{Z} \cdot (2, 0, -1)$. Note that we recovered the last column vector of U^{-1}.

5. SOLVING DECOMP *À LA* HJLS

Let $A \in \mathbb{R}^{n \times m}$ be a generating matrix of an fgas $\mathcal{S}$ and $\overline{\mathcal{S}} = \Lambda \bigoplus E$ be the ΛE decomposition of $\mathcal{S}$ with $\dim(\Lambda) = d$. In this section, we present and analyze an algorithm, named `Decomp_HJLS`, for solving the `Decomp` problem.

Note that `Decomp_HJLS` requires as input the dimension d of the lattice component. One might ask whether there exists an algorithm, based on the unit cost model over the reals, solving the problem without knowing d before. This is actually not the case: In [1], Babai, Just and Meyer auf der Heide showed that, in this model, it is not possible to decide whether there exists a relation for given input $\boldsymbol{x} \in \mathbb{R}^n$. Computing the dimension of the lattice component of an fgas would allow us to solve that decision problem.

5.1 The `Decomp_HJLS` algorithm

`Decomp_HJLS`, given as Algorithm 2, is a full fgas decomposition. It is derived, thanks to the new algebraic view, from the Simultaneous Relations Algorithm in [7, Sec. 5]. The latter is a generalization of the Small Integer Relation Algorithm of Section 2 which contains, as we have seen, a partial decomposition algorithm. We keep using the PSLQ setting and follow the lines of [11, Sec. 2.5]. In particular we adopt a slight change, with respect to [7, Sec. 5], in the swapping strategy. (The index κ' we select, hereafter at Step 2c of Algorithm 2, may differ from $\kappa + 1$.) However, as for differences between HJLS and PSLQ we have seen in Section 2, there is no impact on the asymptotic number of iterations.

We introduce the next definition to describe different stages in the execution of the algorithm, using the shape of the current L-factor L.

Definition 5.1. *Let $0 \leq \ell \leq r$. If a lower trapezoidal matrix $L \in \mathbb{R}^{n \times r}$ can be written as*

$$L = \begin{pmatrix} M & \\ F & \\ G & N \end{pmatrix},$$

with $F \in \mathbb{R}^{(n-r)\times(r-\ell)}$, $G \in \mathbb{R}^{\ell \times (r-\ell)}$, and both $N \in \mathbb{R}^{\ell \times \ell}$ and $M \in \mathbb{R}^{(r-\ell)\times(r-\ell)}$ are lower triangular with positive diagonal coefficients, then we say that L has shape $\mathrm{Trap}(\ell)$.

`Decomp_HJLS` takes as input an fgas generating matrix. It also requires the dimension of the lattice component (see the end of Section 3.2). Without loss of generality, we may assume that the initial L-factor $L^{(0)}$ has shape $\mathrm{Trap}(0)$ (this is provided by Step 1a). The objective of `Decomp_HJLS` is to apply unimodular transformations (namely, size-reductions and swaps) to a current generating matrix $L \cdot Q$ of the input fgas, in order to eventually obtain an L-factor that has shape $\mathrm{Trap}(d)$, where d is the dimension of the lattice component. These unimodular transformations are applied through successive loop iterations (Step 2), that progressively modify the shape of the current L-factor from $\mathrm{Trap}(0)$

to Trap(1), ..., and eventually to Trap(d). When the latter event occurs, the algorithm exits the while loop and moves on to Step 3: the lattice component can now be extracted by taking the last d rows of L and cancelling their first $r-d$ columns, where r is the rank of L.

Algorithm 2 (Decomp_HJLS).

Input: A generating matrix $A = (\boldsymbol{a}_1^T, \cdots, \boldsymbol{a}_n^T)^T \in \mathbb{R}^{n\times m}$ of an fgas $\mathcal{S}$ with $\max_{i\leq m} \|\boldsymbol{a}_i\|_2 \leq X$; a positive integer d as the dimension of the lattice component Λ of $\mathcal{S}$; a parameter $\gamma > 2/\sqrt{3}$.

Output: A basis matrix of Λ.

1. (a) Compute $r = \text{rank}(A)$. If $d = r$, then return $\boldsymbol{a}_1, \cdots, \boldsymbol{a}_r$. Else, using row pivoting, ensure that the first r rows of A are linearly independent.
   (b) Compute the LQ decomposition $A = L_0 \cdot Q_0$.
   (c) Set $L := L_0$ and size-reduce it; set $Q := Q_0$ and $\ell := 0$.
2. While $l_{r-d+1,r-d+1} \neq 0$ do
   (a) Choose κ such that $\gamma^\kappa \cdot l_{\kappa,\kappa} = \max_{k\leq r-\ell} \gamma^k \cdot l_{k,k}$.
   (b) If $\kappa < r - \ell$, then swap the κ-th and the $(\kappa+1)$-th rows of L; compute the LQ decomposition of L; replace L by its L-factor and update Q.
   (c) Else swap the κ-th and κ'-th rows of L, where $\kappa' \geq \kappa + 1$ is the largest index such that $|l_{\kappa',\kappa}| = \max_{\kappa+1\leq k\leq n-\ell} |l_{k,\kappa}|$. If $l_{\kappa,\kappa} = 0$, set $\ell := \ell + 1$.
   (d) Size-reduce L.
3. Return $\left(\mathbf{0}_{d\times(r-d)} | (l_{i,j})_{i\in[n-d+1,n], j\in[r-d+1,r]}\right) \cdot Q$.

In the remainder of this section, we let $L^{(t)} = (l_{i,j}^{(t)})$ denote the matrix L at the beginning of the t-th loop iteration of `Decomp_HJLS`. We also let $\ell(t)$ and $\kappa(t)$ respectively denote the values of ℓ and κ at the end of Step 2a of the t-th loop iteration. We let τ denote the total number of loop iterations and $L^{(\tau+1)}$ and $Q^{(\tau+1)}$ respectively denote the values of L and Q at Step 3.

5.2 The correctness of `Decomp_HJLS`

Note that if $\kappa(t) = r - \ell(t)$ and $l_{\kappa'(t),r-\ell(t)}^{(t)} = 0$, then $l_{r-\ell,r-\ell}^{(t+1)} = 0$. This is the only situation that transforms L from shapes Trap(ℓ) to Trap($\ell+1$), i.e., that decrements (resp. increments) the dimension of the triangular matrix M (resp. N) from Definition 5.1. Indeed, LQ decompositions, size-reductions and swaps of consecutive vectors of indices $\kappa < r - \ell$ preserve the trapezoidal shape of L.

The two lemmas below give insight on the execution of the algorithm. They will be useful especially for proving that `Decomp_HJLS` terminates, and bounding the number of iterations. On the one hand, the maximum of the diagonal coefficients of the M-part of the current L-factor does not increase during the successive loop iterations (Lemma 5.2). On the other hand, because of the existence of the lattice component, which is linearly independent from the vector space component, these diagonal coefficients cannot decrease arbitrarily while maintaining the dimension of M. As long as the lattice component has not been fully discovered, this maximum must remain larger than the first minimum of that lattice (Lemma 5.3).

LEMMA 5.2. *For any $t \in [1, \tau]$, we have $\max_i l_{i,i}^{(t+1)} \leq \max_i l_{i,i}^{(t)}$, where i ranges over $[1, r-\ell(t+1)]$ and $[1, r-\ell(t)]$ respectively.*

The proof is standard. The only $l_{i,i}$'s that may change are those that correspond to the swapped vectors, and the non-increase of the maximum of this or these $l_{i,i}$'s originates from the choice of the swapping index.

LEMMA 5.3. *Let Λ be the lattice component of the input fgas, and $d = \dim(\Lambda) \geq 1$. Then, for any $t \in [1, \tau]$, we have*

$$\lambda_1(\Lambda) \leq \max_{i\leq r-\ell(t)} l_{i,i}^{(t)}.$$

PROOF. The matrix $L^{(\tau+1)}$ has shape Trap(d), and

$$\left(\mathbf{0}^{r-d}, l_{n-d+1,r-d+1}^{(\tau+1)}, \mathbf{0}^{d-1}\right) \cdot Q^{(\tau+1)}$$

belongs to Λ (by Lemma 3.2). As the matrix $Q^{(\tau+1)}$ is orthogonal, it has norm $l_{n-d+1,r-d+1}^{(\tau+1)}$. We thus have $\lambda_1(\Lambda) \leq l_{n-d+1,r-d+1}^{(\tau+1)}$. Now, as τ is the last loop iteration, Step 2c must have been considered at that loop iteration, with a swap between rows $\kappa(\tau) = r - d + 1$ and $n - d + 1$ of $L^{(\tau)}$. We thus obtain:

$$\begin{aligned}\lambda_1(\Lambda) \leq l_{n-d+1,r-d+1}^{(\tau+1)} &= l_{r-d+1,r-d+1}^{(\tau)} \\ \leq \max_{i\leq r-d+1} l_{i,i}^{(\tau)} &\leq \max_{i\leq r-\ell(t)} l_{i,i}^{(t)}.\end{aligned}$$

The last inequality follows from Lemma 5.2. □

We now prove the correctness of the `Decomp_HJLS` algorithm, i.e., that it returns a basis of the lattice component of the input fgas. We also prove that the returned lattice basis is weakly-reduced (see Section 2), and hence that the successive basis vectors are relatively short compared to the successive lattice minima (by Equation (2.1)).

THEOREM 5.4. *If the `Decomp_HJLS` algorithm terminates (which will follow from Theorem 5.6), then it is correct: given a generating matrix of a rank r fgas $\mathcal{S}$ as input and the dimension d of its lattice component, it returns a weakly-reduced basis, with parameters γ and $C = \gamma^{r-d}$, of the lattice component of $\mathcal{S}$.*

PROOF. At the end of the while loop in `Decomp_HJLS`, the L-factor $L^{(\tau+1)}$ has shape Trap(d), where $d = \dim(\Lambda)$. As we only apply unimodular operations to the row vectors, the fgas $\mathbb{Z}^n \cdot L^{(\tau+1)} \cdot Q^{(\tau+1)}$ matches the input fgas $\mathbb{Z}^n \cdot A$. Let $\Lambda' = \mathbb{Z}^d \cdot \left(\mathbf{0}_{d\times(r-d)} | (l_{i,j}^{(\tau+1)})_{i\in[n-d+1,n], j\in[r-d+1,r]}\right) \cdot Q^{(\tau+1)}$ denote the output of `Decomp_HJLS`. By Lemma 3.2, the lattice Λ' is exactly the lattice component Λ.

Let $L' \in \mathbb{R}^{d\times d}$ be the matrix corresponding to the bottom right d rows and d columns of L. We now check that L' is size-reduced and satisfies the Schönhage conditions. Thanks to the size-reductions of Steps 1c and 2d, the whole matrix $L^{(\tau+1)}$ is size-reduced. It remains to show that $l'_{j,j} \leq \gamma^{r-d+i} \cdot l'_{i,i}$ for all $1 \leq j < i \leq d$. For this purpose, we consider two moments $t_i < t_j$ during the execution of the algorithm: the t_i-th (resp. t_j-th) loop iteration is the first one such that $L^{(t)}$ has shape Trap($d-i+1$) (resp. Trap($d-j+1$)). By construction of t_i and t_j, we have:

$$l'_{i,i} = l_{n-d+i,r-d+i}^{(\tau+1)} = l_{n-d+i,r-d+i}^{(t_i)},$$

$$l'_{j,j} = l_{n-d+j,r-d+j}^{(\tau+1)} = l_{n-d+j,r-d+j}^{(t_j)}.$$

As t_i and t_j are chosen minimal, Step 2c was considered at iterations $t_i - 1$ and $t_j - 1$. We thus have $\kappa(t_i - 1) = r - d + i$

and $\kappa(t_j-1) = r-d+j$. Thanks to the choice of κ at Step 2a, we have (recall that $\ell(t_i-1) = d-i$ and $\ell(t_j-1) = d-j$):

$$l'_{i,i} = l^{(t_i)}_{n-d+i,r-d+i} = \gamma^{-(r-d+i)} \cdot \max_{k \leq r-d+i} \gamma^k \cdot l^{(t_i-1)}_{k,k},$$

$$l'_{j,j} = l^{(t_j)}_{n-d+i,r-d+i} = \gamma^{-(r-d+j)} \cdot \max_{k \leq r-d+j} \gamma^k \cdot l^{(t_j-1)}_{k,k}.$$

Using Lemma 5.2 and the fact that $t_i < t_j$, we conclude that:

$$l'_{j,j} \leq \max_{k \leq r-d+j} l^{(t_j-1)}_{k,k} \leq \max_{k \leq r-d+i} l^{(t_i-1)}_{k,k}$$
$$\leq \max_{k \leq r-d+i} \gamma^k \cdot l^{(t_i-1)}_{k,k} = \gamma^{r-d+i} \cdot l'_{i,i},$$

which completes the proof. □

Integer relation algorithms may not have any a priori information on the set of solutions. As mentioned previously, under the exact real arithmetic model, it is impossible to decide whether there exists an integer relation for a given $\boldsymbol{x} \in \mathbb{R}^n$. Hence, as we have seen with HJLS-PSLQ, they have been designed for only ruling out the existence of small relations. Similarly, in our more general context, we can rule out the existence of some large invariants in the lattice component. If the target dimension is not known in advance, then `Decomp_HJLS` may not return a basis of the lattice component. However, if the input integer d is smaller than the dimension d' of the lattice component Λ, then `Decomp_HJLS` returns a d-dimensional lattice that is a projection of Λ orthogonally to a sublattice of Λ and one can prove that:

$$\lambda_{d'-d}(\Lambda) \leq \sqrt{2r}\gamma^{2r} \cdot \max_{k \leq r-d} l^{(\tau+1)}_{k,k}, \qquad (5.1)$$

where τ is the total number of iterations of `Decomp_HJLS` with integer input d.

5.3 Speed of convergence of `Decomp_HJLS`

We adapt the convergence analyses from [7] to `Decomp_HJLS`. We will use the following notations. For each iteration $t \geq 1$, we define

$$\pi^{(t)}_j := \begin{cases} l^{(t)}_{j,j} & \text{if } l^{(t)}_{j,j} \neq 0, \\ l^{(t)}_{n-r+j,j} & \text{if } l^{(t)}_{j,j} = 0 \end{cases}$$

and

$$\Pi(t) := \prod_{i=1}^{r-1} \prod_{j=1}^{i} \max\left(\pi^{(t)}_j, \gamma^{-r-1} \cdot \lambda_1(\Lambda)\right).$$

The following result allows us to quantify progress during the execution of the algorithm: at every loop iteration, the potential function $\Pi(t)$ decreases significantly.

LEMMA 5.5. *Let $\beta = 1/\sqrt{1/\gamma^2 + 1/4} > 1$. Then for any loop iteration $t \in [1, \tau]$, we have $\Pi(t) \geq \beta \cdot \Pi(t+1)$. Further, we also have $\Pi(1) \leq X^{\frac{r(r-1)}{2}}$ with $X = \max_{i \leq n} \|\boldsymbol{a}_i\|$ and $\Pi(\tau+1) \geq (\gamma^{r+1})^{-\frac{r(r-1)}{2}} \cdot \lambda_1(\Lambda)^{\frac{r(r-1)}{2}}$.*

PROOF. The proof of the first claim is similar to the proofs of [7, Th. 3.2] and [6, Lem. 9]. We omit it here. The upper bound on $\Pi(1)$ follows from

$$\gamma^{-r-1} \cdot \lambda_1(\Lambda) \leq \lambda_1(\Lambda) \leq \max_{i \leq r} l^{(1)}_{i,i} \leq \max_i \|\boldsymbol{a}_i\| = X,$$

where the second inequality follows from Lemma 5.3 with $t = 1$. The last item follows from $\max(\pi^{(\tau+1)}_j, \gamma^{-r-1} \cdot \lambda_1(\Lambda)) \geq \gamma^{-r-1} \cdot \lambda_1(\Lambda)$. □

The following result directly follows from Lemma 5.5.

THEOREM 5.6. *The number of loop iterations consumed by `Decomp_HJLS` is $\mathcal{O}(r^3 + r^2 \log \frac{X}{\lambda_1(\Lambda)})$. The number of arithmetic operations consumed at each loop iteration is $\mathcal{O}(nm^2)$.*

6. USING LLL TO SOLVE `DECOMP`

In this section, we provide elements of analysis for a folklore method to solve problems akin to `Decomp` using lattice reduction, such as LLL.

Directly calling LLL on the input fgas generating matrix does not work: it may launch an infinite loop and fail to disclose the lattice component. For instance, consider

$$A = \begin{pmatrix} 1 & 0 \\ \sqrt{2} & 0 \\ x & 2 \end{pmatrix},$$

where $x \in \mathbb{R}$ is arbitrary. LLL keeps swapping (and size-reducing) the first two rows forever, and fails to disclose the lattice component $\mathbb{Z} \cdot (0, 2)$. A crucial point here (see the discussion in [7, Sec. 1-2]) is the fact that the swap strategy is not global enough.

6.1 The `Decomp_LLL` algorithm

In `Decomp_LLL`, we lift the input fgas generating matrix $A \in \mathbb{R}^{n \times m}$ to a lattice basis $A_c := (c^{-1} I_n | A) \in \mathbb{R}^{n \times (m+n)}$, where $c > 0$ is a parameter. LLL will be called on A_c. This creates a unimodular matrix U such that $U \cdot A_c$ is LLL-reduced. The output matrix $U \cdot A_c$ is of the shape $(c^{-1} U | U \cdot A)$, and hence the right hand side $U \cdot A$ of the LLL output is a generating matrix for the input fgas. The goal is to set c sufficiently large so that in A_c there exists a gap between those vectors corresponding the lattice component and those vectors corresponding the vector space component. Ideally, the first vectors of $U \cdot A$ should be very small, because of the LLL-reduction of $U \cdot A_c$: for a large c, the matrix U can get quite large to decrease the right hand side of A_c. Oppositely, by linear independence, the vectors belonging to the lattice component of the input fgas cannot be shortened arbitrarily. They will always lead to large vectors in the lattice spanned by A_c, even for very large values of c. Overall, the key point in `Decomp_LLL` is the choice of the parameter c.

Algorithm 3 (Decomp_LLL).

Input: A generating matrix $A = (\boldsymbol{a}_1^T, \cdots, \boldsymbol{a}_n^T)^T \in \mathbb{R}^{n \times m}$ of an fgas $\mathcal{S}$; the dimension d of the lattice component Λ of $\mathcal{S}$; a parameter $c > 0$.

Output: Hopefully, a basis of Λ.

1. Define $A_c := (c^{-1} \cdot I_n | A)$.
2. Call LLL on input A_c; let A'_c denote the output basis.
3. Let $\pi_m(A'_c)$ denote the submatrix of A'_c consisting in the last m columns. Compute the LQ decomposition of $\pi_m(A'_c) = L \cdot Q$; define
$$L' := \left(\mathbf{0}_{d \times (r-d)} | (l_{i,j})_{i \in [n-d+1,n], j \in [r-d+1,r]}\right)$$
with $r = \operatorname{rank}(A)$; return $L' \cdot Q$.

THEOREM 6.1. *Let $c > 0$ and Λ_c denote the lattice spanned by the A_c matrix of Step 1. If $2^{\frac{n-1}{2}} \cdot \lambda_{n-d}(\Lambda_c) < \lambda_1(\Lambda)$, then Algorithm 3 works correctly: it outputs a basis of the lattice component of the input fgas. Further, for any input A,* *there exists a threshold* $c_0 > 0$ *such that* $2^{\frac{n-1}{2}} \cdot \lambda_{n-d}(\Lambda_c) < \lambda_1(\Lambda)$ *holds for any* $c > c_0$. *Finally, Algorithm 3 consumes* $\mathcal{O}(n^2 \log(cX))$ LLL *swaps, where* $X = \max_i \|\boldsymbol{a}_i\|$.

PROOF. Since $\dim(\Lambda) = d$, there is a unimodular matrix U such that the L-factor of UA has shape Trap(d) (see Definition 5.1) and the first $n - d$ vectors of UA have norms $\leq 2^{-n}\lambda_1(\Lambda)$. For example, we can use the while loop in `Decomp_HJLS` to generate such a U that makes the M-part small enough (using the notation from Definition 5.1). Then, choosing $c > \max_{i \leq n-d}(2^n\|\boldsymbol{u}_i\|/\lambda_1(\Lambda))$ implies that $\lambda_{n-d}(\Lambda_c) < 2^{\frac{1-n}{2}} \cdot \lambda_1(\Lambda)$, where $\boldsymbol{u}_i$ is the i-th row of U.

Write $A'_c = (\boldsymbol{a}_1'^T, \cdots, \boldsymbol{a}_n'^T)^T$. Since the basis $\boldsymbol{a}'_1, \cdots, \boldsymbol{a}'_n$ is LLL-reduced, it follows that

$$\forall i \leq n-d: \quad \|\boldsymbol{a}'_i\| \leq 2^{(n-1)/2} \cdot \lambda_i(\Lambda_c) \leq 2^{(n-1)/2} \cdot \lambda_{n-d}(\Lambda_c).$$

Hence the condition on c implies that $\|\boldsymbol{a}'_i\| < \lambda_1(\Lambda)$ for $1 \leq i \leq n-d$. Let $\pi_m(\boldsymbol{a}'_i)$ denote the vector in $\mathbb{R}^m$ consisting in keeping only the last m components of $\boldsymbol{a}'_i$. Then for $i \leq n-d$, it follows that $\pi_m(\boldsymbol{a}'_i) \in \mathcal{S}$ and $\|\pi_m(\boldsymbol{a}'_i)\| < \lambda_1(\Lambda)$. Thus $\pi_m(\boldsymbol{a}'_i) \in E$, where E is the vector space component of $\overline{\mathcal{S}} = \Lambda \bigoplus E$. Since $\dim(\Lambda) = d$, it follows from Lemma 3.2 that $E = \mathrm{Span}_{i \leq n-d}(\pi_m(\boldsymbol{a}'_i))$, and that the output is exactly a basis of the lattice component Λ. Recall that in the classical LLL analysis for integral inputs, the number of iterations is at most $\mathcal{O}(n^2 \log K)$, where K is the maximum of the norms of the input vectors. For Algorithm 3, we can map the matrix A_c to $c \cdot A_c$, and then the new vectors have norms less than cX. □

In practice, the parameter c may need to be arbitrary large. Consider the fgas generated by the rows of

$$A = \begin{pmatrix} 0 & 1 \\ 1/c_0 & 1 \\ 3 & 0 \end{pmatrix}$$

with c_0 a large irrational number. Its lattice component is $\mathbb{Z} \cdot (0, 1)$. If we choose $2 \leq c \leq c_0$ in Algorithm 3, then after LLL reduction, the first two rows of the submatrix UA of $(c^{-1}U|UA)$ will be $(1/c_0, 0)$ and $(0, 1)$. In this case, `Decomp_LLL` fails to disclose the lattice component, which means that we should choose $c > c_0$. Thus, when c_0 tends to infinity, the required parameter c will be arbitrary large, even for bounded input norms: `Decomp_LLL` may hide singularities when appending the scaled identity matrix.

7. OPEN PROBLEMS

We restricted ourselves to describing and analyzing algorithms with exact real arithmetic operations, and we did not focus on lowering the cost bounds. A natural research direction is to analyze the numerical behavior of these algorithms when using floating-point arithmetic and to bound their bit-complexities. It has been experimentally observed (see, e.g., [5]) that the underlying QR-factorisation algorithm and the choice of full size-reduction impact the numerical behavior. However, to the best of our knowledge, there is no theoretical study of those experimental observations, nor bit-complexity analysis.

An intriguing aspect of HJLS-PSLQ is that it solves (a variant of) `Intersect` via a reduction to `Decomp` and (partially) solving `Decomp`. Designing a more direct approach for `Intersect` is an interesting open problem.

Acknowledgments. We thank D. Dadush, G. Hanrot and G. Lecerf for helpful discussions. We also thank the reviewers for helpful comments, and for pointing out the result of Babai et al. [1] on the impossibility of deciding whether there exists an integer relation among real numbers. This work was partly supported by the ANR HPAC project, the CAS-CNRS Joint Doctoral Promotion Programme, NSFC (11001040, 11171053) and NKBRPC (2011CB302400). Part of this research was undertaken while the first author was visiting École Normale Supérieure de Lyon, whose hospitality is gratefully acknowledged.

8. REFERENCES

[1] L. Babai, B. Just, and F. Meyer auf der Heide. On the limits of computations with the floor function. *Inf. Comput.*, 78(2):99–107, 1988.

[2] P. Borwein. *Computational Excursions in Analysis and Number Theory.* Springer, New York, 2002.

[3] N. Bourbaki. *Elements of Mathematics: General Topology, Part II.* Addison-Wesley, Massachusetts, 1967. A translation of *Éléments de Mathématique : Topologie Générale*, Hermann, Paris, 1966.

[4] D. Dadush and O. Regev. Lattices, convexity and algorithms: Dual lattices and lattice membership problems, 2013. Notes of the second lecture of a course taught at New York University. Available from `http://cs.nyu.edu/~dadush/`.

[5] H. Ferguson and D. Bailey. A polynomial time, numerically stable integer relation algorithm. Technical Report RNR-91-032, SRC-TR-92-066, NAS Applied Research Branch, NASA Ames Research Center, July 1992.

[6] H. Ferguson, D. Bailey, and S. Arno. Analysis of PSLQ, an integer relation finding algorithm. *Math. Comput.*, 68(225):351–369, 1999.

[7] J. Håstad, B. Just, J. Lagarias, and C. Schnorr. Polynomial time algorithms for finding integer relations among real numbers. *SIAM J. Comput.*, 18(5):859–881, 1989. Preliminary version: Proceedings of STACS'86, pp. 105–118, 1986.

[8] A. Lenstra, H. Lenstra, and L. Lovász. Factoring polynomials with rational coefficients. *Math. Ann.*, 261(4):515–534, 1982.

[9] J. Martinet. *Perfect Lattices in Euclidean Spaces.* Springer, Berlin, 2003.

[10] A. Meichsner. Integer Relation Algorithms and the Recognition of Numerical Constants. Master's thesis, Simon Fraser University, 2001.

[11] A. Meichsner. *The Integer Chebyshev Problem: Computational Explorations.* PhD thesis, Simon Fraser University, 2009.

[12] M. Pohst. A modification of the LLL reduction algorithm. *J. Symb. Comput.*, 4(1):123–127, 1987.

[13] O. Regev. Lattices in Computer Science, 2004. Lecture notes of a course taught at Tel Aviv University. Available from `http://www.cims.nyu.edu/~regev/`.

[14] A. Schönhage. Factorization of univariate integer polynomials by Diophantine approximation and an improved basis reduction algorithm. In *Proceedings of ICALP*, volume 172 of *LNCS*, pages 436–447. Springer, 1984.

Desingularization Explains Order-Degree Curves for Ore Operators

Shaoshi Chen*
Dept. of Mathematics / NCSU
Raleigh, NC 27695, USA
schen21@ncsu.edu

Maximilian Jaroschek†
RISC / Joh. Kepler University
4040 Linz, Austria
mjarosch@risc.jku.at

Manuel Kauers†
RISC / Joh. Kepler University
4040 Linz, Austria
mkauers@risc.jku.at

Michael F. Singer*
Dept. of Mathematics / NCSU
Raleigh, NC 27695, USA
singer@ncsu.edu

ABSTRACT

Desingularization is the problem of finding a left multiple of a given Ore operator in which some factor of the leading coefficient of the original operator is removed. An order-degree curve for a given Ore operator is a curve in the (r, d)-plane such that for all points (r, d) above this curve, there exists a left multiple of order r and degree d of the given operator. We give a new proof of a desingularization result by Abramov and van Hoeij for the shift case, and show how desingularization implies order-degree curves which are extremely accurate in examples.

Categories and Subject Descriptors

I.1.2 [**Computing Methodologies**]: Symbolic and Algebraic Manipulation—*Algorithms*

General Terms

Algorithms

Keywords

Ore Operators, Singular Points

1. INTRODUCTION

We consider linear operators of the form

$$L = \ell_0 + \ell_1\partial + \cdots + \ell_r\partial^r,$$

*Supported by the National Science Foundation (NSF) grant CCF-1017217.

†Supported by the Austrian Science Fund (FWF) grant Y464-N18.

ISSAC'13, June 26–29, 2013, Boston, Massachusetts, USA.

where $\ell_0, \dots, \ell_r$ are polynomials or rational functions in x, and ∂ denotes, for instance, the derivation $\frac{d}{dx}$ or the shift operator $x \mapsto x + 1$. (Formal definitions are given later.) Operators act in a natural way on functions. They are used in computer algebra to represent the functions f which they annihilate, i.e., $L \cdot f = 0$.

Multiplication of operators is defined in such a way that the product of two operators acts on a function like the two operators one after the other: $(PL)\cdot f = P\cdot(L\cdot f)$. Therefore, if L is an annihilating operator for some function f, and if P is any other operator, then PL is also an annihilating operator for f.

We are interested in turning a given operator L into a "nicer" one by multiplying it from the left by a suitable P, for two different flavors of "nice". First, we consider the problem of removing factors from the leading coefficient ℓ_r of L. This is known as desingularization and it is needed for computing the values of f at the roots of ℓ_r (provided it is defined there). Desingularization of differential operators is classical [9], and for difference operators, Abramov and van Hoeij [2, 1] give an algorithm for doing it. We give below a new proof of (a slightly generalized version of) their results.

Secondly, we consider the problem of producing left multiples with polynomial coefficients of low degree. Unlike the situation for commutative polynomials, a left multiple PL of L may have polynomial coefficients even if P has rational function coefficients with nontrivial denominators and the polynomial coefficients of L have no common factors. In such situations, it may happen that the degrees of the polynomial coefficients in PL are strictly less than those in L. This phenomenon can be exploited in the design of fast algorithms because a small increase of the order can allow for a large decrease in degree and therefore yield a smaller total size of the operator ("trading order for degree"). Degree estimates supporting this technique have been recently given for a number of different computational problems [4, 7, 6, 3]. Although limited to special situations, these estimates can overshoot by quite a lot. Below we derive a general estimate for the relation between orders and degrees of left multiples of a given operator L from the results about desingularization. This estimate is independent of the context from which the operator L arose, and it is fairly accurate in examples.

2. OVERVIEW

Before discussing the general case, let us illustrate the concepts of desingularization and trading order for degree on a concrete example. Consider the differential operator

$$\begin{aligned}L = &-(45 + 25x - 35x^2 - x^3 + 2x^4)\\ &+ 2(33 - 9x - 3x^2 - x^3)\partial\\ &(1 + x)(23 - 20x - x^2 + 2x^3)\partial^2 \in \mathbb{Q}[x][\partial],\end{aligned}$$

where $\partial = \frac{d}{dx}$. That L is desingularizable at (a root of) $p := 23 - 20x - x^2 + 2x^3$ means that there is some other operator $P \in \mathbb{Q}(x)[\partial]$ such that PL has coefficients in $\mathbb{Q}[x]$ and its leading coefficient no longer contains p as factor. Such a P is called a desingulariz*ing* operator for L at p and PL the corresponding desingulariz*ed* operator. In our example,

$$P = \frac{299}{p}\partial + \frac{1035 - 104x - 136x^2}{p} \in \mathbb{Q}(x)[\partial]$$

is a desingularizing operator for L at p, the desingularized operator is

$$\begin{aligned}PL = &(-2350 - 2055x + 104x^2 + 136x^3)\\ &+ (2151 + 281x + 136x^2)\partial\\ &+ (1932 + 931x - 240x^2 - 136x^3)\partial^2 + 299(1 + x)\partial^3.\end{aligned}$$

A desingularizing operator need not exist. For example, it is impossible to remove the factor $x + 1$ from the leading coefficient of L by means of desingularization. In Section 3 we explain how to check for a given operator L and a factor p of its leading coefficient whether a desingularizing operator exists, and if so, how to compute it.

Desingularization causes a degree drop in the leading coefficient but may affect the other coefficients of the operator in an arbitrary fashion. However, a desingularizing operator can be turned into an operator which lowers the degrees of all the coefficients. To this end, multiply P from the left by some polynomial $q \in \mathbb{Q}[x]$ for which the coefficients of pqP have low degree modulo p, i.e., for which $qP = \frac{1}{p}P_1 + P_2$ where $P_1, P_2 \in \mathbb{Q}[x][\partial]$ and P_1 has low degree coefficients. In our example, a good choice is $q = (-43 + 34x)/299$, i.e.

$$P_1 = (-22x + 29) + (-43 + 34x)\partial, \quad P_2 = -\tfrac{2312}{299}.$$

Since PL has polynomial coefficients, so does

$$\begin{aligned}\frac{1}{p}P_1 L &= qPL - P_2 L\\ &= (-10 - 165x + 22x^2) + (201 + 65x - 34x^2)\partial\\ &\quad + (-100 + 109x - 22x^2)\partial^2 - (1 + x)(43 - 34x)\partial^3.\end{aligned}$$

This operator has degree $\deg_x(L) + \deg_x(P_1) - \deg_x(p) = 2$, compared to $\deg_x(L) + \deg_x(P) = 3$ achieved with the original desingularizing operator. There is no left multiple of L of order 3 and degree less than 2. There is also none of order 4 and degree less than 2, but there does exist a left multiple of degree 1 and order 5. It can be obtained from P by multiplying from the left by an operator $q_0 + q_1\partial + q_2\partial^2 \in \mathbb{Q}[x][\partial]$ of order 2 for which the coefficients of $p^3(q_0 + q_1\partial + q_2\partial^2)P$ have low degrees modulo p^3: Taking

$$q_0 = \tfrac{633+64x-88x^2}{89401}, \quad q_1 = \tfrac{8(17x^2+13x-92)}{89401}, \quad q_2 = \tfrac{1}{299}$$

we have $(q_0 + q_1\partial + q_2\partial^2)P = \frac{1}{p^3}Q_1 + Q_2$, where

$$\begin{aligned}Q_1 = &(841 + 580x - 436x^2 - 148x^3 + 59x^4 + 12x^5 - 4x^6)\\ &+ (1697 - 528x - 752x^2 + 120x^3 + 127x^4 - 12x^5\\ &- 4x^6)\partial + (x - 7)(-9 + x + 2x^2)p\,\partial^2 + p^2\partial^3,\\ Q_2 = &\tfrac{16(779+374x)}{89401} - \tfrac{272(69+34x)}{89401}\partial.\end{aligned}$$

Set $Q := \frac{1}{p^3}Q_1$. Then, since PL has polynomial coefficients, so does

$$\begin{aligned}QL &= (q_0 + q_1\partial + q_2\partial^2)PL - Q_2 L\\ &= (2 + x) + (-3 + x)\partial - (8 + 2x)\partial^2\\ &\quad + (2 - 2x)\partial^3 + (6 + x)\partial^4 + (1 + x)\partial^5.\end{aligned}$$

Its degree is $\deg_x(L) + \deg_x(Q_1) - 3\deg_x(p) = 1$.

As the factor $x + 1$ cannot be removed from L, we cannot hope to reduce the degree even further. We have thus found that the region of all points $(r, d) \in \mathbb{N}^2$ such that there is a left $\mathbb{Q}(x)[\partial]$-multiple of L of order r and with polynomial coefficients of degree at most d is given by $((2, 4) + \mathbb{N}^2) \cup ((3, 2) + \mathbb{N}^2) \cup ((5, 1) + \mathbb{N}^2)$.

In Section 4 we explain the construction of the operators Q that turn a desingularizing operator into one that lowers all the degrees as far as possible, and we give a formula that describes the points (r, d) for which such a Q exists.

3. PARTIAL DESINGULARIZATION

In this section we discuss under which circumstances an operator L admits a left multiple PL in which a factor of the leading coefficient of L is removed. This is of interest in its own right, and will also serve as the starting point for the construction described in the following section. In view of this latter application, we cover here a slightly generalized variant of desingularization, which not only applies to the case where a factor can be completely removed, but also cases where only the multiplicity of the factor can be lowered.

Example 1. *In the shift case (i.e., $\partial x = (x+1)\partial$), consider the operator*

$$\begin{aligned}L = &(3 + x)(9 + 7x + x^2) - (33 + 70x + 47x^2 + 12x^3 + x^4)\partial\\ &+ (2 + x)^2(3 + 5x + x^2)\partial^2.\end{aligned}$$

The factor $(x + 2)^2$ in the leading coefficient cannot be removed completely. Yet we can find a multiple in which $x + 2$ appears in the leading coefficient (in shifted form) with multiplicity one only. One such left multiple of L is

$$\begin{aligned}&(402 + 208x + 25x^2) - (514 + 743x + 258x^2 + 25x^3)\partial\\ &+ (233 + 378x + 183x^2 + 25x^3)\partial^2 - 9(3 + x)\partial^3.\end{aligned}$$

We speak in this case of a partial desingularization. The general definition is as follows. We formulate it for operators in an arbitrary Ore algebra $\mathbb{O} := A[\partial] := A[\partial; \sigma, \delta]$ where A is a $\mathbb{K}$-algebra (in our case typically $A = \mathbb{K}[x]$ or $A = \mathbb{K}(x)$), $\mathbb{K}$ is a field, $\sigma\colon A \to A$ is an automorphism and $\delta\colon A \to A$ a σ-derivation, i.e., a $\mathbb{K}$-linear map satisfying the skew Leibniz rule $\delta(pq) = \delta(p)q + \sigma(p)\delta(q)$ for $p, q \in A$. For any $f \in A$, the multiplication rule in $A[\partial; \sigma, \delta]$ is $\partial f = \sigma(f)\partial + \delta(f)$. We write $\deg_\partial(L)$ for the *order* of $L \in A[\partial]$, and if $A = \mathbb{K}[x]$, we write $\deg_x(L)$ for the maximum degree among the polynomial coefficients of L. For general information about Ore algebras, see [5].

Definition 2. *Let $L \in \mathbb{K}[x][\partial;\sigma,\delta]$ and let $p \in \mathbb{K}[x]$ be such that $p \mid \operatorname{lc}_\partial(L) \in \mathbb{K}[x]$. We say that p is* removable *from L at order n if there exists some $P \in \mathbb{K}(x)[\partial]$ with $\deg_\partial(P) = n$ and some $w, v \in \mathbb{K}[x]$ with $\gcd(p,w) = 1$ such that $PL \in \mathbb{K}[x][\partial]$ and $\sigma^{-n}(\operatorname{lc}_\partial(PL)) = \frac{w}{vp}\operatorname{lc}_\partial(L)$. We then call P a p-removing operator for L, and PL the corresponding p-removed operator. p is simply called* removable *from L if it is removable at order n for some $n \in \mathbb{N}$.*

If $\gcd(p, \operatorname{lc}_\partial(L)/(vp)) = 1$, we say desingulariz[able|ing|ed] *instead of remov[able|ing|ed], respectively.*

The backwards shift σ^{-n} in the definition above is introduced in order to compensate the effect of the term ∂^n in P on the leading coefficient on L (i.e., $\operatorname{lc}_\partial(\partial^n L) = \sigma^n(\operatorname{lc}_\partial(L))$.) Moreover, observe that in this definition, removing a polynomial p does not necessarily mean that the p-removed operator has no roots of (some shift of) p in its leading coefficient. If L contains some factors of higher multiplicity, as in the example above, then removal of a polynomial is defined so as to respect multiplicities. Also observe that in the definition we allow that some new factors w are introduced when p is removed. This is only a matter of convenience. We will see below that we may always assume $v = w = 1$, i.e., if something can be removed at the cost of introducing new factors into the leading coefficient, then it can also be removed without introducing new factors. The justification rests on the following lemma.

Lemma 3. *Let $L \in \mathbb{K}[x][\partial;\sigma,\delta]$, let $p \in \mathbb{K}[x]$ with $p \mid \operatorname{lc}_\partial(L)$ be removable from L, and let $P \in \mathbb{K}(x)[\partial;\sigma,\delta]$ be a p-removing operator for L with $\deg_\partial(P) = n$.*

1. *If $U \in \mathbb{K}[x][\partial]$ with $\gcd(\operatorname{lc}_\partial(U), \sigma^{n+\deg_\partial(U)}(p)) = 1$, then UP is also a p-removing operator for L.*
2. *If $P = P_1 + P_2$ for some $P_1 \in \mathbb{K}(x)[\partial]$ with $\deg_\partial(P_1) = n$ and $P_2 \in \mathbb{K}[x][\partial]$, then P_1 is also a p-removing operator for L.*
3. *There exists a p-removing operator P' with $\deg_\partial(P') = n$ and with $p\sigma^{-n}(\operatorname{lc}_\partial(P'L)) = \operatorname{lc}_\partial(L)$.*

Proof. Let $v, w \in \mathbb{K}[x]$ be as in Definition 2, i.e., $\gcd(p, w) = 1$ and $vp\sigma^{-n}(\operatorname{lc}_\partial(PL)) = w\operatorname{lc}_\partial(L)$.

1. Since PL is an operator with polynomial coefficients, so is UPL. Furthermore, with $u = \operatorname{lc}_\partial(U)$ and $m = \deg_\partial(U)$ we have

$$vp\sigma^{-n-m}(\operatorname{lc}_\partial(UPL)) = \sigma^{-n-m}(u)w\operatorname{lc}_\partial(L).$$

Since $\gcd(u, \sigma^{n+m}(p)) = 1$, we have $\gcd(\sigma^{-n-m}(u)w, p) = 1$, as required.

2. Clearly, $P_2 \in \mathbb{K}[x][\partial]$ implies $P_2L \in \mathbb{K}[x][\partial]$. Since also $PL \in \mathbb{K}[x][\partial]$, it follows that

$$P_1L = (P - P_2)L = PL - P_2L \in \mathbb{K}[x][\partial].$$

If $\deg_\partial(P_2) < n$, then we have $\operatorname{lc}_\partial(PL) = \operatorname{lc}_\partial(P_1L)$, so there is nothing else to show. If $\deg_\partial(P_2) = n$, then $\operatorname{lc}_\partial(P_1L) = \operatorname{lc}_\partial(PL) - \operatorname{lc}_\partial(P_2L)$ and therefore

$$\begin{aligned} vp\sigma^{-n}(\operatorname{lc}_\partial(P_1L)) &= vp\sigma^{-n}(\operatorname{lc}_\partial(PL) - \operatorname{lc}_\partial(P_2L)) \\ &= (w - vp\sigma^{-n}(\operatorname{lc}_\partial(P_2)))\operatorname{lc}_\partial(L). \end{aligned}$$

Since $\gcd(p, w - vp\sigma^{-n}(\operatorname{lc}_\partial(P_2))) = \gcd(p, w) = 1$, the claim follows.

3. By the extended Euclidean algorithm we can find $s, t \in \mathbb{K}[x]$ with $1 = sw + tpv$. Then $\sigma^n(s)P$ is p-removing of order n by part 1 ($\sigma^n(s)$ is obviously coprime to $\sigma^n(p)$), and its leading coefficient is

$$\sigma^n\left(\frac{sw}{pv}\right) = \frac{1}{\sigma^n(pv)} - \sigma^n(t).$$

By part 2 we may discard the polynomial part $\sigma^n(t)$, obtaining a p-removing operator P' with $\deg_\partial(P') = n$ and $p\sigma^{-n}(\operatorname{lc}_\partial(P'L)) = v\operatorname{lc}_\partial(L)$. Using part 1, we can obtain from here an operator with the desired property. ∎

The lemma implies that if there is a p-removing operator at all, then there is also one in which all the denominators are powers of $\sigma^n(p)$ (because any factors coprime with p can be cleared according to part 1), and where all numerators have smaller degree than the corresponding denominators (because polynomial parts can be removed according to part 2).

Similarly as in the proof of part 3, we can also reduce the problem of removing a composite polynomial to the problem of removing powers of irreducible polynomials. For example, if $p = p_1p_2$ is removable from L, where $p_1, p_2 \in \mathbb{K}[x]$ are coprime, then obviously both p_1 and p_2 are removable. Conversely, if p_1 and p_2 are removable, and if P_1, P_2 are removing operators of orders n_1, n_2 with $\operatorname{lc}_\partial(P_1) = 1/\sigma^{n_1}(p_1)$ and $\operatorname{lc}_\partial(P_2) = 1/\sigma^{n_1}(p_2)$, then for $n = \max\{n_1, n_2\}$ and $u_1, u_2 \in \mathbb{K}[x]$ with

$$u_1\sigma^n(p_2) + u_2\sigma^n(p_1) = \gcd(\sigma^n(p_1), \sigma^n(p_2)) = 1$$

the operator $P := u_1\partial^{n-n_1}P_1 + u_2\partial^{n-n_2}P_2 \in \mathbb{K}(x)[\partial]$ is such that $PL \in \mathbb{K}[x][\partial]$ and $\operatorname{lc}_\partial(PL) = \operatorname{lc}_\partial(L)/\sigma^n(p)$.

In summary, in order to determine whether a polynomial $p = p_1^{k_1}p_2^{k_2}\cdots p_m^{k_m}$ is removable from an operator L, it suffices to be able to check for an irreducible polynomial p_i and a given $k_i \geq 1$ whether $p_i^{k_i}$ is removable. Let now p be an irreducible polynomial and $k \geq 1$. If there exists a p^k-removing operator, then it can be assumed to be of the form

$$P = \frac{p_0}{\sigma^n(p)^{e_0}} + \frac{p_1}{\sigma^n(p)^{e_1}}\partial + \cdots + \frac{p_{n-1}}{\sigma^n(p)^{e_{n-1}}}\partial^{n-1} + \frac{1}{\sigma^n(p)^k}\partial^n,$$

for some $e_0, \ldots, e_{n-1} \in \mathbb{N}$, and $p_0, \ldots, p_{n-1} \in \mathbb{K}[x]$ with $\deg_x(p_i) < e_i\deg_x(p)$. In order to decide whether such an operator exists, it is now enough to know a bound on n as well as a bound e on the exponents e_i, for if n and e are known, we can make an ansatz $p_i = \sum_{j=0}^{e-1} p_{i,j}x^j$ with undetermined coefficients $p_{i,j}$, then calculate PL and rewrite all its coefficients in the form $a/\sigma^n(p)^e + b$ for some polynomials a, b depending linearly on the undetermined $p_{i,j}$, then compare the coefficients of the various a's with respect to x to zero and solve the resulting linearly system for the $p_{i,j}$.

How the bounds on n and e are derived depends on the particular Ore algebra at hand. In this paper, we give a complete treatment of the shift case ($(\sigma p)(x) = p(x+1)$, $\delta = 0$) and make some remarks about the differential case ($\sigma = \mathrm{id}$, $\delta = \frac{d}{dx}$). For other cases, see the preprint [8].

3.1 Shift Case

In this section, let $\mathbb{K}[x][\partial]$ denote the Ore algebra of recurrence operators, i.e., σ is the automorphism mapping x

to $x+1$ and δ is the zero map. This case was studied by Abramov and van Hoeij [2, 1]. We give below a new proof of their result, and extend it to the case of partial desingularization. For consistency with the differential case, we formulate the result for the leading coefficients, while Abramov and van Hoeij consider the analogous for the trailing coefficients. Of course, this difference is immaterial.

We proceed in two steps. First we give a bound on the order of a removing operator (Lemma 4), and then, in a second step, we provide a bound on the exponents in the denominators (Theorem 5). As explained above, it is sufficient to consider the case of removing powers of irreducible polynomials, and we restrict to this case.

Lemma 4. *In the operator algebra $\mathbb{K}[x][\partial]$ with the commutation rule $\partial x = (x+1)\partial$, let $L = \ell_0 + \ell_1\partial + \cdots + \ell_r\partial^r \in \mathbb{K}[x][\partial]$ with $\ell_0, \ell_r \neq 0$, and let p be an irreducible factor of $\operatorname{lc}_\partial(L)$ such that p^k is removable from L for some $k \geq 1$. Let $n \in \mathbb{N}$ be s.t. $\gcd(\sigma^n(p), \ell_0) \neq 1$ and $\gcd(\sigma^m(p), \ell_0) = 1$ for all $m > n$. Then p^k is removable at order n from L.*

Proof. By assumption on L, there exists a p^k-removing operator P, say of order m, and by the observations following Lemma 3 we may assume that

$$P = \frac{p_0}{\sigma^m(p)^{e_0}} + \frac{p_1}{\sigma^m(p)^{e_1}}\partial + \cdots + \frac{p_m}{\sigma^m(p)^{e_m}}\partial^m,$$

for $e_i \in \mathbb{N}$ and $p_i \in \mathbb{K}[x]$ with $\deg_x(p_i) < e_i \deg_x(p)$ ($i = 0, \ldots, m$). We may further assume $\gcd(\sigma^m(p), p_i) = 1$ for $i = 0, \ldots, m$ (viz. that the e_i are chosen minimally).

Suppose that $m > n$. We show by induction that then $e_0 = e_1 = \cdots = e_{m-n-1} = 0$, so that $p_i = 0$ for $i = 0, \ldots, m-n-1$, i.e., the operator P has in fact the form

$$P = \frac{p_{m-n}}{\sigma^m(p)^{e_{m-n}}}\partial^{m-n} + \cdots + \frac{p_m}{\sigma^m(p)^{e_m}}\partial^m.$$

Thus $\partial^{n-m}P \in \mathbb{K}(x)[\partial]$ is a p^k-removing operator of order n.

Consider the operator $T := \sum_{i=0}^{r+m} t_i\partial^i := PL \in \mathbb{K}[x][\partial]$. From $\frac{p_0}{\sigma^m(p)^{e_0}}\ell_0 = t_0 \in \mathbb{K}[x]$ it follows that $e_0 = 0$, because

$$\gcd(\sigma^m(p), p_0) = \gcd(\sigma^m(p), \ell_0) = 1$$

by the choice of p_0 and the assumption in the lemma, respectively, and this leaves no possibility for cancellation.

Assume now, as induction hypothesis, that $e_0 = e_1 = \cdots = e_{i-1} = 0$ for some $i < m-n$. Then from

$$\begin{aligned} t_i &= \frac{p_i}{\sigma^m(p)^{e_i}}\sigma^i(\ell_0) + \frac{p_{i-1}}{\sigma^m(p)^{e_{i-1}}}\sigma^{i-1}(\ell_1) + \cdots + \frac{p_0}{\sigma^m(p)^{e_0}}\ell_i \\ &= \frac{p_i}{\sigma^m(p)^{e_i}}\sigma^i(\ell_0) \end{aligned}$$

it follows that $\sigma^m(p)^{e_i} \mid p_i\sigma^i(\ell_0)$. By the choice of p_i we have $\gcd(\sigma^m(p), p_i) = 1$ and by the assumption in the lemma we have $\gcd(\sigma^{m-i}(p), \ell_0) = 1$ (because $m-i > n$), so it follows that $e_i = 0$. Inductively, we obtain $e_0 = e_1 = \cdots = e_{m-n-1} = 0$, which completes the proof. ∎

It can be shown that p cannot be removed from L if $\sigma^n(p)$ is coprime with the trailing coefficient of L for all $n \in \mathbb{N}$ by a variant of [2, Lemma 3.], so the above lemma covers all situations where removing of a factor is possible.

In order to formulate the result about the possible exponents in the denominator, it is convenient to first introduce some notation. Let us call two irreducible polynomials $p, q \in \mathbb{K}[x] \setminus \{0\}$ *equivalent* if there exists $n \in \mathbb{Z}$ such that $\sigma^n(p)/q \in \mathbb{K}$. We write $[q]$ for the equivalence class of $q \in \mathbb{K}[x] \setminus \{0\}$. If p, q are equivalent in this sense, we write $p \leq q$ if $\sigma^n(p)/q \in \mathbb{K}$ for some $n \geq 0$, and $p > q$ otherwise.

The irreducible factors of a polynomial $u \in \mathbb{K}[x]$ can be grouped into equivalence classes, for example

$$u = \underbrace{(x-4)(x-1)^3x(x+1)^2}\,\underbrace{(2x-5)(2x+3)^2(2x+9)} \times \underbrace{(x^2+5x+1)(x^2+11x+25)^3}.$$

For any monic irreducible factor p of $u \in \mathbb{K}[x]$, let $v_p(u)$ denote the multiplicity of p in u, and define

$$v_{<p}(u) := \max\{\, v_q(u) \mid q \in [p] : p > q \,\}.$$

For example, for the particular u above we have $v_{x-4}(u) = 1$, $v_{<x-4}(u) = 0$, $v_{<x+1}(u) = 3$, and so on.

Besides being applicable not only to desingularization but also removal of any factors, the following theorem also refines the corresponding result of Abramov and van Hoeij in so far as their version only covers the case of desingularizing L at some p with $v_{<p}(\operatorname{lc}_\partial(L)) = 0$ whereas we do not need this assumption.

Theorem 5. *In the operator algebra $\mathbb{K}[x][\partial]$ with the commutation rule $\partial x = (x+1)\partial$, let $L = \ell_0 + \ell_1\partial + \cdots + \ell_r\partial^r \in \mathbb{K}[x][\partial]$ with $\ell_0, \ell_r \neq 0$, and let p be an irreducible factor of ℓ_r such that p^k is removable from L for some $k \geq 1$. Let $n \in \mathbb{N}$ be such that $\gcd(\sigma^n(p), \ell_0) \neq 1$ and $\gcd(\sigma^m(p), \ell_0) = 1$ for all $m > n$. Then there exists a p^k-removing operator P for L and p of the form*

$$P = \frac{p_0}{\sigma^n(p)^{e_0}} + \frac{p_1}{\sigma^n(p)^{e_1}}\partial + \cdots + \frac{p_n}{\sigma^n(p)^{e_n}}\partial^n,$$

for some $e_i \in \mathbb{N}$ and $p_i \in \mathbb{K}[x]$ with

1. *$\deg_x(p_i) < e_i \deg_x(p)$ and $\gcd(\sigma^n(p), p_i) = 1$, and*
2. *$e_i \leq k + n\, v_{<p}(\operatorname{lc}_\partial(L))$*

for $i = 0, \ldots, n-1$, and $p_n = 1$, $e_n = k$.

Proof. Lemmas 3 and 4 imply the existence of an operator P with all the required properties except possibly the exponent estimate in item 2. Let P be such an operator, and consider the operator $T := \sum_{i=0}^{r+n} t_i\partial^i := PL \in \mathbb{K}[x][\partial]$.

Let $e = \max\{e_1, \ldots, e_n\}$ and $\bar{P} := \sum_{i=0}^n \bar{p}_i\partial^i := \sigma^n(p)^e P$. Then $\bar{p}_i = \sigma^n(p)^{e-e_i}p_i$ ($i = 0, \ldots, n$) and $\sigma^n(p)^e T = \bar{P}L$ and $\gcd(\bar{p}_0, \ldots, \bar{p}_n, \sigma^n(p)) = 1$.

Abbreviating $v := v_{<p}(\operatorname{lc}_\partial(L))$, assume that $e > k + nv$. We will show by induction that then $\bar{p}_i$ contains $\sigma^n(p)$ with multiplicity more than iv for $i = n, n-1, \ldots, 0$, which is inconsistent with $\gcd(\bar{p}_0, \ldots, \bar{p}_n, \sigma^n(p)) = 1$.

First it is clear that $\bar{p}_n = \sigma^n(p)^e p_n \sigma^n(\ell_r)$ contains $\sigma^n(p)$ with multiplicity $\geq e - k > nv$, because P is p^k-removing. Suppose now as induction hypothesis that there is an $i \geq 0$ such that $\sigma^n(p)^{j\,v+1} \mid \bar{p}_j$ for $j = n, n-1, \ldots, i+1$. Consider the equality

$$\sigma^n(p)^e t_{i+r} = \bar{p}_i\sigma^i(\ell_r) + \bar{p}_{i+1}\sigma^{i+1}(\ell_{r-1}) + \cdots + \bar{p}_n\ell_{r-n},$$

where we use the convention $\ell_j := 0$ for $j < 0$. The induction hypothesis implies that $\sigma^n(p)^{(i+1)v+1} \mid \bar{p}_j$ for $j = n, n-1, \ldots, i+1$. Furthermore, since $(i+1)v \leq nv < e$, we have $\sigma^n(p)^{(i+1)v+1} \mid \sigma^n(p)^e t_{i+r}$. Both facts together imply $\sigma^n(p)^{(i+1)v+1} \mid \bar{p}_i\sigma^i(\ell_r)$. The definition of v ensures that

$\sigma^n(p)$ is contained in $\sigma^i(\ell_r)$ with multiplicity at most v, so it must be contained in $\bar{p}_i$ with multiplicity more than $(i+1)v - v = i\,v$, as claimed. ■

A referee comments that one can improve the denominator bounds of terms after the leading term of Theorem 5 as in the code given in the paper [2]. However, this refined bound would not affect our main result, Theorem 9 below.

3.2 Differential Case

In this section $\mathbb{K}[x][\partial]$ refers to the Ore algebra of differential operators, i.e., $\sigma = \mathrm{id}$ and $\delta = \frac{d}{dx}$. Let $L \in \mathbb{K}[x][\partial]$ and suppose for simplicity that $p = x$ is a factor of $\mathrm{lc}_\partial(L)$. In [1], the authors show that L can be desingularized at x if and only if $x = 0$ is an *apparent singularity,* that is, if and only if $L(y) = 0$ admits $\deg_\partial(L)$ linearly independent formal power series solutions. The authors furthermore give an algorithm to find an operator P such that if ξ is either an ordinary point of L or an apparent singularity of L, then ξ is an ordinary point of PL. Therefore this algorithm desingularizes all the points that can be desingularized. The authors also give a sharp bound for $\deg_\partial(P)$. The authors furthermore give some indications concerning partial desingularizations. Additional information concerning desingularization and the degrees of initial terms can also be found in [10].

4. ORDER-DEGREE CURVES

We now turn to the construction of left multiples of L with polynomial coefficients of small degree, and to the question of how small these degrees can be made. As already indicated in Section 2, we start from an operator P which removes some factor from the leading coefficient of L, say it removes a polynomial p of degree k. According to Lemma 3, we may assume that $\mathrm{lc}_\partial(P) = 1/\sigma^{\deg_\partial(P)}(p)$ and that all other coefficients of P are rational functions whose numerators have lower degree than the corresponding denominators. Thus we already have $\deg_x(PL) \leq \deg_x(L) - 1$. Furthermore, if q is any polynomial with $\deg_x(q) < \deg_x(p) = k$, then multiplying P by q (from left) and removing polynomial parts by Lemma 3 (part 2) gives another operator Q with $\deg_x(QL) \leq \deg_x(L) - 1$. All the operators Q obtained in this way form a $\mathbb{K}$-vector space of dimension k. Within this vector space we search for elements where $\deg_x(QL)$ is as small as possible. Forcing the coefficients of the highest degrees to zero gives a certain number of linear constraints which can be balanced with the number of degrees of freedom offered by the coefficients of q, as illustrated in the figure below. As long as we force fewer than k terms to zero, we will find a nontrivial solution.

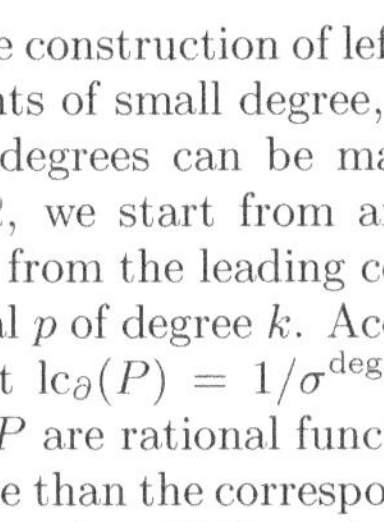

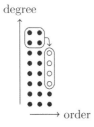

If we want to eliminate k terms or more in order to reduce the degree even further, we need more variables. We can create k more variables if instead of an ansatz qP we make an ansatz $(q_0 + q_1\partial)P$ for some $q_0, q_1 \in \mathbb{K}[x]$ with $\deg_x(q_0), \deg_x(q_1) < k$. Again removing all polynomial parts from the rational function coefficients we obtain a vector space of operators Q with $\deg_x(QL) \leq \deg_x(L) - 1$ whose dimension is $2k$. The additional degrees of freedom can be used to eliminate more high degree terms, the result being an operator of lower degree but higher order. If we let the order increase further and for each fixed order use all the available degrees of freedom to reduce the degrees to minimize the degrees of the polynomial coefficients, a hyperbolic relationship between the order and the degree of QL emerges. In Theorem 9 below, we make this relationship precise, taking into account that for a given operator L the leading coefficient may contain several factors p that are removable at different orders n. The resulting region of all points $(r, d) \in \mathbb{N}^2$ for which there exists a left multiple of L of order r with polynomial coefficients of degree at most d is then given by an overlay of a finite number of hyperbolas.

Before turning to the proof of this theorem, let us illustrate its basic idea with the example operators from Section 2.

Example 6. *Let $L \in \mathbb{Q}[x][\partial]$, $p \in \mathbb{Q}[x]$, and $P \in \mathbb{Q}(x)[\partial]$ be as in Section 2. Recall that p is an irreducible cubic factor of $\mathrm{lc}_\partial(L)$ and that P is a p-removing operator for L. We have $P = \frac{p_1}{p}\partial + \frac{p_0}{p}$ for some $p_1, p_0 \in \mathbb{Q}[x]$ with $\deg_x(p_1) = 0$ and $\deg_x(p_0) = 2$. We have seen in Section 2 that there is an operator $Q \in \mathbb{Q}(x)[\partial]$ of order 3 such that $QL \in \mathbb{Q}[x][\partial]$ and $\deg_x(QL) = 1$. Our goal here is to explain why this operator exists.*

Make an ansatz $Q_1 = (q_0 + q_1\partial + q_2\partial^2)P$ with undetermined polynomials q_0, q_1, q_2. After expanding the product and applying commutation rules, Q_1 has the form

$$\frac{p_1 q_2}{p}\partial^3 + \frac{(\ldots)q_2 + (\ldots)q_1}{p^2}\partial^2 + \frac{(\ldots)q_2 + (\ldots)q_1 + (\ldots)q_0}{p^3}\partial + \frac{(\ldots)q_2 + (\ldots)q_1 + (\ldots)q_0}{p^3},$$

where the $(\ldots)$ are certain polynomials whose precise form is irrelevant for our purpose.

Note that by Lemma 3 (part 1), $Q_1 L \in \mathbb{Q}[x][\partial]$ regardless of the choice of q_0, q_1, q_2, and that by Lemma 3 (part 2), this property is not lost if we add to Q_1 some operator in $\mathbb{Q}[x][\partial]$ of our choice. Therefore, if $Q_2 \in \mathbb{Q}[x][\partial]$ is the operator obtained from $p^3 Q_1 \in \mathbb{Q}[x][\partial]$ by reducing all the coefficients modulo p^3, then $p^{-3}Q_2 L \in \mathbb{Q}[x][\partial]$, still regardless of the choice of q_0, q_1, q_2.

The coefficients of Q_2 depend linearly on the undetermined polynomials q_0, q_1, q_2. If we choose their degree to be $\deg_x(p) - 1 = 2$, then we have $3(2+1) = 9$ variables for the coefficients of q_0, q_1, q_2. Choosing a higher degree would give more variables but also introduce undesired solutions such as $q_0 = q_1 = q_2 = p$, for which the reduction modulo p^3 leads to the useless result $Q_2 = 0$. This cannot happen if we enforce $\deg_x(q_i) < \deg_x(p)$.

The operator $p^{-3}Q_2 L$ has degree

$$\deg_x(Q_2) + \deg_x(L) - 3\deg_x(p) = \deg_x(Q_2) - 5,$$

which is equal to 1 if $\deg_x(Q_2) = 6$. A priori, the degree of Q_2 in x may be up to $\deg_x(p^3) - 1 = 8$. In order to bring it down to 6, we equate the coefficients of $x^i\partial^j$ for $i = 7, 8$ and $j = 0, \ldots, 3$ to zero. This gives 8 equations. As there *are more variables than equations, there must be a nontrivial solution.*

For formulating the proof of the general statement, it is convenient to work with an alternative formulation of removability, which is provided in the following lemma. Throughout the section, $\mathbb{K}[x][\partial] = \mathbb{K}[x][\partial;\sigma,\delta]$ is an arbitrary Ore algebra.

Lemma 7. *$p \in \mathbb{K}[x]$ is removable from $L \in \mathbb{K}[x][\partial]$ at order n if and only if there exists $P \in \mathbb{K}[x][\partial]$ with $\deg_\partial(P) = n$ and $PL \in \sigma^n(p)\operatorname{lc}_\partial(P)\mathbb{K}[x][\partial]$.*

Proof. "$\Leftarrow$": $P_0 = \frac{1}{\sigma^n(p)\operatorname{lc}_\partial(P)}P$ is a p-removing operator.

"$\Rightarrow$": Start from a p-removing operator of the form

$$P_0 = \sum_{i=0}^{n-1} \frac{p_i}{\sigma^n(p)^{e_i}}\partial^i + \frac{1}{\sigma^n(p)}\partial^n,$$

and set $P = \sigma^n(p)^e P_0$ where $e = \max\{e_0,\dots,e_{n-1},1\} \geq 1$. Because of $P_0L \in \mathbb{K}[x][\partial]$ it follows that

$$PL \in \sigma^n(p)^e\mathbb{K}[x][\partial] = \sigma^n(p)\operatorname{lc}_\partial(P)\mathbb{K}[x][\partial]. \qquad \blacksquare$$

The next lemma is a generalization of Bezout's relation to more than two coprime polynomials, which we will also need in the proof.

Lemma 8. *Let $u_1,\dots,u_m \in \mathbb{K}[x]$ be pairwise coprime and $u = u_1u_2\cdots u_m$, and let $v_1,\dots,v_m \in \mathbb{K}[x]$ be such that $\deg_x(v_i) < \deg_x(u_i)$ $(i = 1,\dots,m)$. If*

$$\sum_{i=1}^m v_i\frac{u}{u_i} = 0$$

then $v_1 = v_2 = \dots = v_m = 0$.

Proof. Since the u_i are pairwise coprime, $u_i \nmid u/u_i$ for all i. However, $u_i \mid u/u_j$ for all $j \neq i$. Both facts together with $\sum_{i=1}^m v_iu/u_i = 0$ imply that $u_i \mid v_i$ for all i. Since $\deg_x(v_i) < \deg_x(u_i)$, the claim follows. $\blacksquare$

Theorem 9. *Let $L \in \mathbb{K}[x][\partial]$, and let $p_1,\dots,p_m \in \mathbb{K}[x]$ be factors of $\operatorname{lc}_\partial(L)$ which are removable at orders $n_1,\dots,n_m$, respectively, so that the $\sigma^{n_i}(p_i)$ are pairwise coprime. Let $r \geq \deg_\partial(L)$ and*

$$d \geq \deg_x(L) - \left\lceil \sum_{i=1}^m \Bigl(1 - \frac{n_i}{r - \deg_\partial(L) + 1}\Bigr)^+ \deg_x(p_i) \right\rceil,$$

where we use the notation $(x)^+ := \max\{x,0\}$. Then there exists an operator $Q \in \mathbb{K}(x)[\partial]\setminus\{0\}$ such that $QL \in \mathbb{K}[x][\partial]$ and $\deg_\partial(QL) = r$ and $\deg_x(QL) = d$.

Proof. Let $r \geq \deg_\partial(L)$, and set $s := r - \deg_\partial(L)$ so that $s = \deg_\partial(Q)$. We may assume without loss of generality that s is such that $1 - \frac{n_i}{r-\deg_\partial(L)+1} = 1 - \frac{n_i}{s+1} > 0$ for all i by simply removing all the p_i for which $1 - \frac{n_i}{s+1} \leq 0$ from consideration. We thus have $s \geq n_i$ for all i.

Lemma 7 yields operators $P_i \in \mathbb{K}[x][\partial]$ of order n_i with $P_iL \in \sigma^{n_i}(p_i)\operatorname{lc}(P_i)\mathbb{K}[x][\partial]$. Set

$$q = \prod_{i=1}^m \prod_{j=0}^{s-n_i} \sigma^{j+n_i}(p_i)\sigma^j(l_i),$$

where $l_i = \operatorname{lc}_\partial(P_i)$. Consider the ansatz

$$Q_1 = \sum_{i=1}^m \sum_{j=0}^{s-n_i} q_{i,j}\frac{q}{\sigma^{j+n_i}(p_i)\sigma^j(l_i)}\partial^j P_i$$

for undetermined polynomial coefficients $q_{i,j}$ $(i = 1,\dots,m;$ $j = 0,\dots,s-n_i)$ of degree less than $\deg_x(p_i)$. Regardless of the choice of these coefficients, we will always have $Q_1 \in \mathbb{K}[x][\partial]$ and $Q_1L \in q\mathbb{K}[x][\partial]$. Also, for arbitrary $R \in \mathbb{K}[x][\partial]$ and $Q_2 = Q_1 - qR$ we have $Q_2 \in \mathbb{K}[x][\partial]$ and $Q_2L \in q\mathbb{K}[x][\partial]$. This means that we can replace the coefficients in Q_1 by their remainders upon division by q without violating any of the mentioned properties of Q_1.

Also observe that any operator Q_2 obtained in this way is nonzero unless all the $q_{i,j}$ are zero, because if k is maximal such that at least one of the $q_{i,k}$ is nonzero, then

$$\operatorname{lc}_\partial(Q_1) = \sum_{i=1}^m q_{i,k}\frac{q}{\sigma^{k+n_i}(p_i)\sigma^k(l_i)}\sigma^k(l_i) = \sum_{i=1}^m q_{i,k}\frac{q}{\sigma^{k+n_i}(p_i)}$$

is nonzero by Lemma 8. Furthermore, $\operatorname{lc}_\partial(Q_1) \not\equiv 0 \bmod q$ because $\deg_x(q_{i,k}) < \deg_x(p_i)$ implies $\deg_x(\operatorname{lc}_\partial(Q_1)) < \deg_x(q)$.

The ansatz for the $q_{i,j}$ gives $\sum_{i=1}^m (s-n_i+1)\deg_x(p_i)$ variables. Plug this ansatz into Q_1 and reduce all the polynomial coefficients modulo q to obtain an operator Q_2 of degree less than $\deg_x(q) = \sum_{i=1}^m (s-n_i+1)(\deg_x(p_i)+\deg_x(l_i))$. Then for each of the $s+1$ polynomial coefficients in Q_2 equate the coefficients of the terms x^j for

$$j > \sum_{i=1}^m \Bigl((s-n_i)\deg_x(p_i) + (s-n_i+1)\deg_x(l_i)\Bigr) + \left\lfloor \frac{\sum_{i=1}^m n_i\deg_x(p_i)}{s+1} \right\rfloor$$

to zero. This gives altogether

$$\begin{aligned}
&(s+1)\Bigl(\sum_{i=1}^m (s-n_i+1)\bigl(\deg_x(p_i)+\deg_x(l_i)\bigr) \\
&\quad - \sum_{i=1}^m \Bigl((s-n_i)\deg_x(p_i) + (s-n_i+1)\deg_x(l_i)\Bigr) \\
&\quad - \left\lfloor \frac{\sum_{i=1}^m n_i\deg_x(p_i)}{s+1} \right\rfloor - 1\Bigr) \\
&= (s+1)\Bigl(\sum_{i=1}^m \deg_x(p_i) - \left\lfloor \frac{\sum_{i=1}^m n_i\deg_x(p_i)}{s+1} \right\rfloor - 1\Bigr)
\end{aligned}$$

equations. The resulting linear system has a nontrivial solution because

$$\begin{aligned}
&\#\text{vars} - \#\text{eqns} \\
&= \sum_{i=1}^m (s-n_i+1)\deg_x(p_i) \\
&\quad - (s+1)\Bigl(\sum_{i=1}^m \deg_x(p_i) - \left\lfloor \frac{\sum_{i=1}^m n_i\deg_x(p_i)}{s+1} \right\rfloor - 1\Bigr) \\
&= -\sum_{i=1}^m n_i\deg_x(p_i) - (s+1)\Bigl(-\left\lfloor \frac{\sum_{i=1}^m n_i\deg_x(p_i)}{s+1} \right\rfloor - 1\Bigr) \\
&> -\sum_{i=1}^m n_i\deg_x(p_i) + \frac{s+1}{s+1}\sum_{i=1}^m n_i\deg_x(p_i) = 0.
\end{aligned}$$

By construction, the solution gives rise to an operator $Q_2 \in \mathbb{K}[x][\partial]$ of order at most n with polynomial coefficients of degree at most

$$\sum_{i=1}^{m}\big((s-n_i)\deg_x(p_i)+(s-n_i+1)\deg_x(l_i)\big) + \left\lfloor\frac{\sum_{i=1}^m n_i\deg_x(p_i)}{s+1}\right\rfloor,$$

for which $Q_2 L \in q\mathbb{K}[x][\partial]$. Thus if we set $Q = \frac{1}{q}Q_2 \in \mathbb{K}(x)[\partial]$, we have $\deg_\partial(QL) = \deg_\partial(L)+s = r$ and $\deg_x(QL)$ is at most

$$\begin{aligned}
&\deg_x(L)+\deg_x(Q_2)-\deg_x(q)\\
&\le \deg_x(L)+\sum_{i=1}^{m}\big((s-n_i)\deg_x(p_i)+(s-n_i+1)\deg_x(l_i)\big)\\
&\quad+\left\lfloor\frac{\sum_{i=1}^m n_i\deg_x(p_i)}{s+1}\right\rfloor\\
&\quad-\sum_{i=1}^{m}(s-n_i+1)(\deg_x(p_i)+\deg_x(l_i))\\
&=\deg_x(L)-\sum_{i=1}^{m}\deg_x(p_i)+\left\lfloor\frac{\sum_{i=1}^m n_i\deg_x(p_i)}{s+1}\right\rfloor\\
&=\deg_x(L)-\left\lceil\sum_{i=1}^{m}\Big(1-\frac{n_i}{s+1}\Big)\deg_x(p_i)\right\rceil,
\end{aligned}$$

as required. (The final step uses the facts $\lfloor -x\rfloor = -\lceil x\rceil$ and $\lceil x+n\rceil = \lceil x\rceil + n$ for $x\in\mathbb{R}$ and $n\in\mathbb{Z}$.) ∎

Example 10. *1. Consider again the example from Section 2. There we started from an operator $L \in \mathbb{K}[x][\partial]$ of order 2 and degree 4 for which there exists a desingularizing operator P of order 1 which removes a polynomial p of degree 3. According to the theorem, for every $r \ge 2$ exists an operator $Q \in \mathbb{K}[x][\partial]$ with $QL \in \mathbb{K}[x][\partial]$, $\deg_\partial(QL) \le r$ and*

$$d := \deg_x(QL) \le 4 - \Big(1-\frac{1}{r-2+1}\Big)^{+} 3 = \frac{r+2}{r-1}.$$

This hyperbola precisely predicts the order-degree pairs we found in Section 2:

r	2	3	4	5	6	7	8	9
d	4	2	2	1	1	1	1	1

2. Consider the minimal order telescoper L for the hypergeometric term in Example 6.1 of [6],

$$\begin{aligned}L &= 9(x+1)(3x+1)(3x+2)^2(3x+4)p(x+1)\\ &\quad+(\ldots\text{degree }16\ldots)\partial+(\ldots\text{degree }15\ldots)\partial^2\\ &\quad-10x(8+5x)(9+5x)(11+5x)(12+5x)p(x)\partial^3,\end{aligned}$$

where ∂ represents the shift operator and p is a certain irreducible polynomial of degree 10. This polynomial is removable of order 1. Therefore, by the theorem, we expect left multiples of L of order r and degree bounded by

$$16 - \Big(1-\frac{1}{r-3+1}\Big)^{+} 10 = \frac{6r-2}{r-2}.$$

In the figure below, the curve $d = \frac{6r-2}{r-2}$ (solid) is contrasted with the estimate $d = \frac{8r-1}{r-2}$ (dashed) derived last year for this example as well as the region of all points (r,d) for which a left multiple of L of order r and degree d exists (gray). The new curve matches precisely the boundary of the gray region, even including the very last degree drop (which is not clearly visible on the figure): for $r = 12$ we have $\frac{6r-2}{r-2} = 7$ and for $r = 13$ we have $\frac{6r-2}{r-2} \approx 6.9 < 7$.

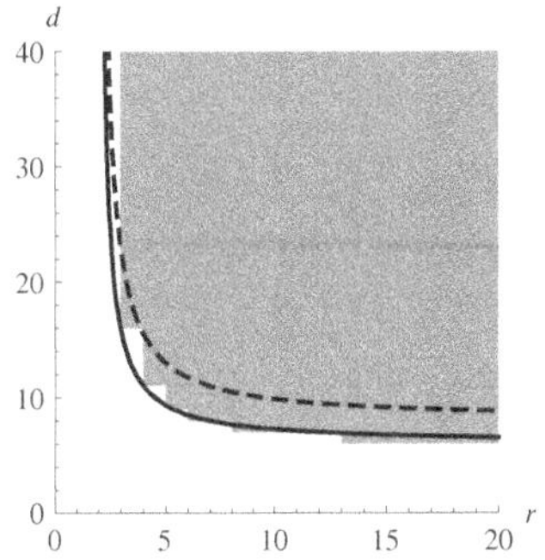

3. Consider the minimal order telescoper L for the hyperexponential term in Example 15.2 in [7]. It has order 3 and degree 40. The leading coefficient contains an irreducible polynomial p of degree 23 at order 1 and otherwise only non-removable factors. Theorem 9 therefore predicts left multiples of L of order r and degree

$$40 - \Big(1-\frac{1}{r-3+1}\Big)^{+} 23 = \frac{17r-11}{r-2}$$

for all $r \in \mathbb{N}$. Again, this estimate is accurate, while the estimate $\frac{24r-9}{r-2}$ derived in [7] overshoots.

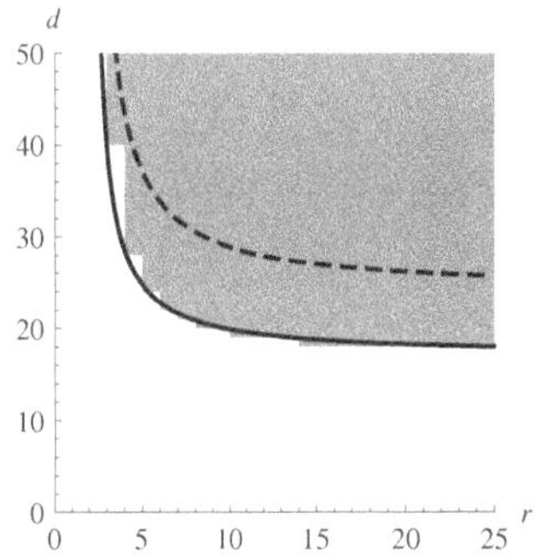

4. Operators coming from applications tend to have leading coefficients that contain a single irreducible polynomial of large degree which can be removed at order 1, besides factors that are not removable. But Theorem 9 also covers the more general situation of factors that are only removable of higher order, and even the case of several polynomials that are removable at several orders. As an example for this general situation, consider the operator

$$\begin{aligned}L &= 8(1+x)(1+2x)^3(37+3z)^7(14+32x+26x^2+7x^3)^7\\ &\quad-9(1+3x)^9(2+3x)^2(1+x+5x^2+7x^3)^7\partial,\end{aligned}$$

where ∂ represents the shift operator. From its leading coefficient, the polynomial $(1+x+5x^2+7x^3)^7$ is removable at order 1, and in addition, $(1+3x)^7$ is removable at order 12. The remaining factors are not removable. According to Theorem 9 we expect that L

admits left multiples of order r and degree

$$32 - 21\Big(1 - \frac{1}{r}\Big)^+ - 7\Big(1 - \frac{3}{1}\Big)^+,$$

for all $r \in \mathbb{N}$. It turns out that this prediction is again accurate for every r. Observe that in this example the curve is a superposition of two hyperbolas.

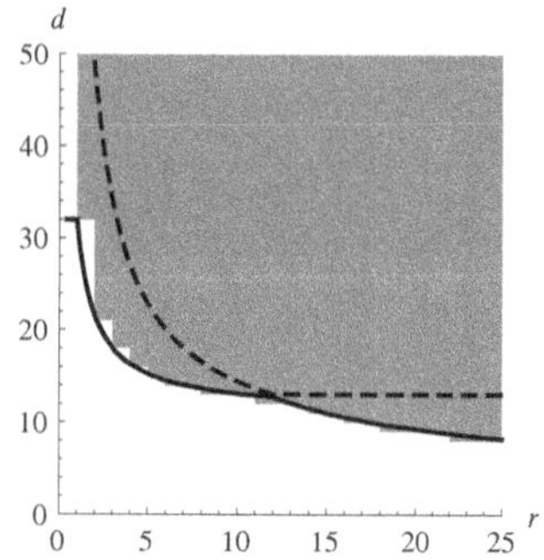

5. *Although the bound of Theorem 9 appears to be tight in many cases, it is not always tight. The operator*

$$(x-1)(x+1)^2(x+3)^2(x+5)^2(2x-1)$$
$$+ x^2(x+2)^2(x+4)^2(x+6)(2x-3)\partial,$$

in which ∂ denotes the shift, is an example: It has a left multiple of order 2 and degree 3 although according to Theorem 9 we would expect a multiple of order 2 to have degree $8 - (1 - \frac{1}{2-1})^+ 4 = 4$. (Thanks to Mark van Hoeij for sharing this example.)

5. CONCLUSION

We believe that removable factors provide a universal explanation for all the order-degree curves that have been observed in recent years for various different contexts. We have derived a formula for the boundary of the gray region associated to a fixed operator L, which, although formally only a bound, seems to be exact in examples coming from applications. This does not immediately imply better complexity estimates or faster variants of algorithms exploiting the phenomenon of order-degree curves, because usually L is not known in advance but rather the desired output of a calculation, and therefore we usually have no information about the removable factors of $\mathrm{lc}_\partial(L)$. However, we now know what we have to look at: in order to improve algorithms based on trading order for degree, we need to develop a theory which provides a priori information about the removable factors of $\mathrm{lc}_\partial(L)$. In other words, our result reduces the task of better understanding order-degree curves to the task of better understanding what causes the appearance of removable factors in operators coming from applications.

6. ACKNOWLEDGEMENT

We thank the referees for their valuable remarks.

7. REFERENCES

[1] Sergei A. Abramov, Moulay A. Barkatou, and Mark van Hoeij. Apparent singularities of linear difference equations with polynomial coefficients. *AAECC*, 17:117–133, 2006.

[2] Sergei A. Abramov and Mark van Hoeij. Desingularization of linear difference operators with polynomial coefficients. In *Proceedings of ISSAC'99*, pages 269–275, 1999.

[3] Alin Bostan, Frederic Chyzak, Ziming Li, and Bruno Salvy. Fast computation of common left multiples of linear ordinary differential operators. In *Proceedings of ISSAC'12*, pages 99–106, 2012.

[4] Alin Bostan, Frédéric Chyzak, Bruno Salvy, Grégoire Lecerf, and Éric Schost. Differential equations for algebraic functions. In *Proceedings of ISSAC'07*, pages 25–32, 2007.

[5] Manuel Bronstein and Marko Petkovšek. An introduction to pseudo-linear algebra. *Theoretical Computer Science*, 157(1):3–33, 1996.

[6] Shaoshi Chen and Manuel Kauers. Order-degree curves for hypergeometric creative telescoping. In *Proceedings of ISSAC'12*, pages 122–129, 2012.

[7] Shaoshi Chen and Manuel Kauers. Trading order for degree in creative telescoping. *Journal of Symbolic Computation*, 47(8):968–995, 2012.

[8] Frederic Chyzak, Philippe Dumas, Ha Le, Jose Martin, Marni Mishna, and Bruno Salvy. Taming apparent singularities via Ore closure. in preparation.

[9] E. L. Ince. *Ordinary Differential Equations*. Dover, 1926.

[10] Harrison Tsai. Weyl closure of a linear differential operator. *Journal of Symbolic Computation*, 29(4–5):747–775, 2000.

Fast Algorithms for ℓ-adic Towers over Finite Fields

Luca De Feo
Laboratoire PRiSM
Université de Versailles
luca.de-feo@uvsq.fr

Javad Doliskani
Computer Science Department
Western University
jdoliska@uwo.ca

Éric Schost
Computer Science Department
Western University
eschost@uwo.ca

ABSTRACT

Inspired by previous work of Shoup, Lenstra-De Smit and Couveignes-Lercier, we give fast algorithms to compute in (the first levels of) the ℓ-adic closure of a finite field. In many cases, our algorithms have quasi-linear complexity.

Categories and Subject Descriptors

F.2.1 [**Theory of computation**]: Analysis of algorithms and problem complexity—*Computations in finite fields*; G.4 [**Mathematics of computing**]: Mathematical software

General Terms

Algorithms,Theory

Keywords

Finite fields, irreducible polynomials, extension towers, algebraic tori, Pell's equation, elliptic curves.

1. INTRODUCTION

Building arbitrary finite extensions of finite fields is a fundamental task in any computer algebra system. For this, an especially powerful system is the "compatibly embedded finite fields" implemented in Magma [2, 3], capable of building extensions of any finite field and keeping track of the embeddings between the fields.

The system described in [3] uses linear algebra to describe the embeddings of finite fields. From a complexity point of view, this is far from optimal: one may hope to compute and apply the morphisms in quasi-linear time in the degree of the extension, but this is usually out of reach of linear algebra techniques. Even worse, the quadratic memory requirements make the system unsuitable for embeddings of large degree extensions. Although the Magma core has evolved since the publication of the paper, experiments in Section 5 show that embeddings of large extension fields are still out of reach.

In this paper, we discuss an approach based on polynomial arithmetic, rather than linear algebra, with much better performance. We consider here one aspect of the question, ℓ-adic towers; we expect that this will play an important role towards a complete solution.

Let q be a power of a prime p, let $\mathbb{F}_q$ be the finite field with q elements and let ℓ be a prime. Our main interest in this paper is on the algorithmic aspects of the *ℓ-adic closure* of $\mathbb{F}_q$, which is defined as follows. Fix arbitrary embeddings

$$\mathbb{F}_q \subset \mathbb{F}_{q^\ell} \subset \mathbb{F}_{q^{\ell^2}} \subset \cdots;$$

then, the ℓ-adic closure of $\mathbb{F}_q$ is the infinite field defined as

$$\mathbb{F}_q^{(\ell)} = \bigcup_{i\geq 0} \mathbb{F}_{q^{\ell^i}}.$$

We also call an *ℓ-adic tower* the sequence of extensions $\mathbb{F}_q, \mathbb{F}_{q^\ell}, \ldots$ In particular, they allow us to build the algebraic closure $\bar{\mathbb{F}}_q$ of $\mathbb{F}_q$, as there is an isomorphism

$$\bar{\mathbb{F}}_q \cong \bigotimes_{\ell \text{ prime}} \mathbb{F}_q^{(\ell)}, \tag{1}$$

where the tensor products are over $\mathbb{F}_q$; we will briefly mention below the algorithmic counterpart of this equality.

We present here algorithms that allow us to "compute" in the first levels of ℓ-adic towers (in a sense defined hereafter); at level i, our goal is to be able to perform all basic operations in quasi-linear time in the extension degree ℓ^i. We do not discuss the representation of the base field $\mathbb{F}_q$, and we count operations $\{+,-,\times,\div\}$ in $\mathbb{F}_q$ at unit cost.

Our techniques are inspired by those in [4, 5, 8], which dealt with the Artin-Schreier case $\ell = p$ (see also [9], which reused these ideas in the case $\ell = 2$): we construct families of irreducible polynomials with special properties, then give algorithms that exploit the special form of those polynomials to apply the embeddings. Because they are treated in the references [8, 9], *we exclude the cases $\ell = p$ and $\ell = 2$.*

The field $\mathbb{F}_{q^{\ell^i}}$ will be represented as $\mathbb{F}_q[X_i]/\langle Q_i\rangle$, for some irreducible polynomial $Q_i \in \mathbb{F}_q[X_i]$. Letting x_i be the residue class of X_i modulo Q_i endows $\mathbb{F}_{q^{\ell^i}}$ with the monomial basis

$$\mathbf{U}_i = (1, x_i, x_i^2, \ldots, x_i^{\ell^i-1}). \tag{2}$$

Let $\mathsf{M} : \mathbb{N} \to \mathbb{N}$ be such that polynomials in $\mathbb{F}_q[X]$ of degree less than n can be multiplied in $\mathsf{M}(n)$ operations in $\mathbb{F}_q$, under the assumptions of [33, Ch. 8.3]; using FFT multiplication, one can take $\mathsf{M}(n) \in O(n\log(n)\log\log(n))$. Then, multiplications and inversions in $\mathbb{F}_q[X_i]/\langle Q_i\rangle$ can be done in respectively $O(\mathsf{M}(\ell^i))$ and $O(\mathsf{M}(\ell^i)\log(\ell^i))$ operations in $\mathbb{F}_q$ [33, Ch. 9-11]. This is almost optimal, as both results are quasi-linear in $[\mathbb{F}_{q^{\ell^i}} : \mathbb{F}_q] = \ell^i$.

ISSAC'13, June 26–29, 2013, Boston, Massachusetts, USA.

Condition	Initialization	Q_i, T_i	Lift, push
$q = 1 \bmod \ell$	$O_e(\log(q))$	$O(\ell^i)$	$O(\ell^i)$
$q = -1 \bmod \ell$	$O_e(\log(q))$	$O(\ell^i)$	$O(\mathsf{M}(\ell^i)\log(\ell^i))$
–	$O_e(\ell^2 + \mathsf{M}(\ell)\log(q))$	$O(\mathsf{M}(\ell^{i+1})\mathsf{M}(\ell)\log(\ell^i)^2)$	$O(\mathsf{M}(\ell^{i+1})\mathsf{M}(\ell)\log(\ell^i))$
$4\ell \le q^{1/4}$	$O_e^{\sim}(\ell\log^5(q) + \ell^3)$ (bit)	$O_e(\ell^2 + \mathsf{M}(\ell)\log(\ell q) + \mathsf{M}(\ell^i)\log(\ell^i))$	$O(\mathsf{M}(\ell^i)\log(\ell^i))$
$4\ell \le q^{1/4}$	$O_e^{\sim}(\ell\log^5(q))$ (bit) $+ O_e(\mathsf{M}(\ell)\sqrt{q}\log(q))$	$O_e(\log(q) + \mathsf{M}(\ell^i)\log(\ell^i))$	$O(\mathsf{M}(\ell^i)\log(\ell^i))$

Table 1: Summary of results

Computing embeddings requires more work. For this problem, it is enough consider a pair of consecutive levels in the tower, as any other embedding can be done by applying repeatedly this elementary operation. Following again [8], we introduce two slightly more general operations, *lift* and *push*.

To motivate them, remark that for $i \ge 2$, $\mathbb{F}_{q^{\ell^i}}$ has two natural bases as a vector space over $\mathbb{F}_q$. The first one is via the monomial basis $\mathbf{U}_i$ seen above, corresponding to the univariate model $\mathbb{F}_q[X_i]/\langle Q_i\rangle$. The second one amounts to seeing $\mathbb{F}_{q^{\ell^i}}$ as a degree ℓ extension of $\mathbb{F}_{q^{\ell^{i-1}}}$, that is, as

$$\mathbb{F}_q[X_{i-1}, X_i]/\langle Q_{i-1}(X_{i-1}), T_i(X_{i-1}, X_i)\rangle, \qquad (3)$$

for some polynomial T_i monic of degree ℓ in X_i, and of degree less than ℓ^{i-1} in X_{i-1}. The corresponding basis is bivariate and involves x_{i-1} and x_i:

$$\mathbf{B}_i = (1, \dots, x_{i-1}^{\ell^{i-1}-1}, \dots, x_i^{\ell-1}, \dots, x_{i-1}^{\ell^{i-1}-1}x_i^{\ell-1}). \qquad (4)$$

Lifting corresponds to the change of basis from $\mathbf{B}_i$ to $\mathbf{U}_i$; *pushing* is the inverse transformation.

Lift and push allow us to perform embeddings as a particular case, but they are also the key to many further operations. We do not give details here, but we refer the reader to [8, 9, 18] for examples such as the computation of relative traces, norms or characteristic polynomials, and applications to solving Artin-Schreier or quadratic equations, given in [8] and [9] for respectively $\ell = p$ and $\ell = 2$.

Table 1 summarizes our main results. Under various assumptions, it gives costs (counted in terms of operations in $\mathbb{F}_q$) for initializing the construction, building the polynomials Q_i and T_i from Eq.(3), and performing lift and push. $O_e(\ )$ indicates probabilistic algorithms with expected running time, and $O_e^{\sim}(\ )$ indicates the additional omission of logarithmic factors. Two entries mention bit complexity, as they use an elliptic curve point counting algorithm.

In all cases, our results are close to being linear-time in ℓ^i, up to sometimes the loss of a factor polynomial in ℓ. Except for the (very simple) case where $q = 1 \bmod \ell$, these results are new, to the best of our knowledge. To otbain them, we use two constructions: the first one (Section 2) uses cyclotomy and descent algorithms; the second one (Section 3) relies on the construction of a sequence of fibers of isogenies between algebraic groups.

These constructions are inspired by previous work due to respectively Shoup [27, 28] and Lenstra / De Smit [21], and Couveignes / Lercier [6]. We briefly discuss them here and give more details in the further sections.

Lenstra and De Smit [21] address a question similar to ours, the construction of the ℓ-adic closure of $\mathbb{F}_q$ (and of its algebraic closure), with the purpose of standardizing it. The resulting algorithms run in polynomial time, but (implicitly) rely on linear algebra and multiplication tables, so quasi-linear time is not directly reachable. References [27, 28, 6] discuss a related problem, the construction of irreducible polynomials over $\mathbb{F}_q$; the question of computing embeddings is not considered. The results in [6] are *quasi-linear*, but they rely on an algorithm by Kedlaya and Umans [14] that works only in a boolean model.

To conclude the introduction, let us mention a few applications of our results. A variety of computations in number theory and algebraic geometry require constructing new extension fields and moving elements from one to the other. As it turns out, in many cases, the ℓ-adic constructions considered here are sufficient: two examples are [7, 11], both in relation to torsion subgroups of Jacobians of curves.

The main question remains of course the cost of computing in arbitrary extensions. As showed by Eq. (1), this boils down to the study of ℓ-adic towers, as done in this paper, together with algorithms for computing in *composita*. References [27, 28, 6] deal with related questions for the problem of computing irreducible polynomials; a natural follow-up to the present work is to study the cost of embeddings and similar changes of bases in this more general context.

2. QUASI-CYCLOTOMIC TOWERS

In this section, we discuss a construction of the ℓ-adic tower over $\mathbb{F}_q$ inspired by previous work of Shoup [27, 28], Lenstra-De Smit [21] and Couveignes-Lercier [6]. The results of this section establish rows 1 and 3 of Table 1.

The construction starts by building an extension $\mathbb{K}_0 = \mathbb{F}_q[Y_0]/\langle P_0\rangle$ obtained by adjoining an ℓth root of unity to $\mathbb{F}_q$, such that the residue class y_0 of Y_0 is a non ℓ-adic residue in $\mathbb{K}_0$ (we discuss this in more detail in the first subsection); we let r be the degree of P_0. By [17, Th. VI.9.1], for $i \ge 1$, the polynomial $Y_i^{\ell^i} - y_0 \in \mathbb{K}_0[Y_i]$ is irreducible, so $\mathbb{K}_i = \mathbb{K}_0[Y_i]/\langle Y_i^{\ell^i} - y_0\rangle$ is a field with $q^{r\ell^i}$ elements. Letting y_i be the residue class of Y_i in $\mathbb{K}_i$, these fields are naturally embedded in one another by the isomorphism $\mathbb{K}_{i+1} \simeq \mathbb{K}_i[Y_{i+1}]/\langle Y_{i+1}^{\ell} - y_i\rangle$; in particular, we have $y_{i+1}^{\ell} = y_i$.

In order to build $\mathbb{F}_{q^{\ell^i}}$, we apply a descent process, for which we follow an idea of Shoup's. For $i \ge 0$, let x_i be the trace of y_i over a subfield of index r:

$$x_i \;=\; \textstyle\sum_{0\le j\le r-1} y_i^{q^{\ell^i j}}.$$

Then, [27, Th. 2.1] proves that $\mathbb{F}_q(x_i) = \mathbb{F}_{q^{\ell^i}}$ (see Figure 1). In particular, the minimal polynomials of $x_1, x_2, \dots$ over $\mathbb{F}_q$ are the irreducible polynomials Q_i we are interested in.

We show here how to compute these polynomials, the polynomials T_i of Eq. (3) and how to perform lift and push. To this effect, we will define more general minimal polynomials: for $0 \le j \le i$, we will let $Q_{i,j} \in \mathbb{F}_q(x_j)[X_i]$ be the minimal polynomial of x_i over $\mathbb{F}_q(x_j)$, so that $Q_{i,j}$ has degree ℓ^{i-j}, with in particular $Q_{i,0} = Q_i$ and $Q_{i,i-1} = T_i(x_{i-1}, X_i)$.

In Subsections 2.2 and 2.3, we discuss favorable cases,

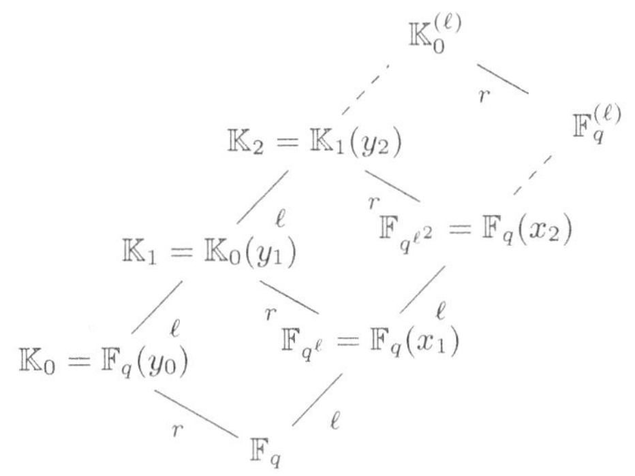

Figure 1: The ℓ-adic towers over $\mathbb{F}_q$ and $\mathbb{K}_0$.

where ℓ divides respectively $q-1$ and $q+1$. The first case is folklore; it yields the fastest and simplest algorithms. Our results for the second case are related to known facts about Chebyshev polynomials [30, § 6.2], but, to the best of our knowledge, are new. We will revisit these cases in Section 3 and account for their naming convention. Our results in the general case (Subsection 2.4) are slower, but still quasi-linear in ℓ^i, up to a factor polynomial in ℓ.

Shoup used this setup to compute Q_i in time quadratic in ℓ^i [28, Th. 11]. It is noted there that using *modular composition* techniques [33, Ch. 12], this could be improved to get a subquadratic exponent in ℓ^i, up to an extra cost polynomial in ℓ. For $\ell = 3$ (where we are in one the first two cases), Couveignes and Lercier make a similar remark in [6, § 2.4]; using a result by Kedlaya and Umans [14] for modular composition, they derive for any $\varepsilon > 0$ a cost of $3^{i(1+\varepsilon)}O(\log(q))$ *bit* operations, up to polynomial terms in $\log\log(q)$.

In this section, and in the rest of this paper, if L/K is a field extension, we write $\mathrm{Tr}_{L/K}$, $\mathrm{N}_{L/K}$ and $\mathrm{Gal}_{L/K}$ for the trace, norm and Galois group of the extension.

2.1 Finding P_0

To determine P_0, we compute the ℓ-th cyclotomic polynomial $\Phi_\ell \in \mathbb{Z}[X_0]$ and factor it over $\mathbb{F}_q[X_0]$: by [28, Th. 9], this takes $O_e(\mathsf{M}(\ell)\log(\ell q))$ operations in $\mathbb{F}_q$.

Over $\mathbb{F}_q[X_0]$, Φ_ℓ splits into irreducible factors of the same degree r, where r is the order of q in $\mathbb{Z}/\ell\mathbb{Z}$ (so r divides $\ell-1$); let F_0 be one of these factors. By construction, there exist non ℓ-adic residues in $\mathbb{F}_q[X_0]/\langle F_0\rangle$. Once such a non-residue y_0 is found, we simply let P_0 be its minimal polynomial over $\mathbb{F}_q$ (which still has degree r); given y_0, computing P_0 takes $O(r^2)$ operations in $\mathbb{F}_q$ [28, Th. 4].

Following [27, 28, 6], we pick y_0 at random: we expect to find a non-residue after $O(1)$ trials; by [28, Lemma 15], each takes $O_e(\mathsf{M}(\ell)\log(r) + \mathsf{M}(r)\log(\ell)\log(r) + \mathsf{M}(r)\log(q))$ operations in $\mathbb{F}_q$. An alternative due to Lenstra and De Smit is to take iterated ℓ-th roots of $X_0 \bmod F_0$ until we find a non-residue: this idea is helpful in making the construction canonical, but more costly, so we do not consider it.

2.2 $\mathbb{G}_m$-type extensions

We consider here the simplest case, where ℓ divides $q-1$; the (classical) facts below give the first row of Table 1. In this case, Φ_ℓ splits into linear factors over $\mathbb{F}_q$ (so $r=1$). The polynomial P_0 is of the form $Y_0 - y_0$, where y_0 is a non ℓ-adic residue in $\mathbb{F}_q$; since we can bypass the factorization of Φ_ℓ, the cost of initialization is $O_e(\log(q))$ operations in $\mathbb{F}_q$. Besides, no descent is required: for $i \ge 0$, we have $\mathbb{K}_i = \mathbb{F}_{q^{\ell^i}}$ and $x_i = y_i$; the families of polynomials we obtain are

$$Q_i = X_i^{\ell^i} - y_0 \quad \text{and} \quad T_i = X_i^{\ell} - X_{i-1}. \qquad (5)$$

Lift and push use no operation in $\mathbb{F}_q$, only exponent arithmetic. Lift takes $F = \sum_{0\le j<\ell^{i+1}} f_j x_{i+1}^j$ and rewrites it as a bivariate polynomial in x_i, x_{i+1} and push does the converse operation, using the rules

$$x_{i+1}^j = x_i^{j \text{ div } \ell} x_{i+1}^{j \bmod \ell} \quad \text{and} \quad x_i^e x_{i+1}^f = x_{i+1}^{e\ell+f}.$$

2.3 Chebyshev-type extensions

Consider now the case where ℓ divides $q+1$: then, Φ_ℓ splits into quadratic factors over $\mathbb{F}_q$ and $r=2$. We also require that y_0 has norm 1 over $\mathbb{F}_q$ (see below for a discussion); we deduce an expression for the polynomials $Q_{i,j} \in \mathbb{F}_q(x_j)[X_i]$.

PROPOSITION 1. *For $1 \le j < i$, $Q_{i,j}$ satisfies*

$$Q_{i,j}(X_i) = Y^{\ell^{i-j}} + Y^{-\ell^{i-j}} - x_j \mod Y^2 - X_iY + 1. \quad (6)$$

PROOF. Since $\mathrm{N}_{\mathbb{K}_0/\mathbb{F}_q}(y_0) = 1$, $\mathrm{N}_{\mathbb{K}_i/\mathbb{F}_q(x_i)}(y_i)$ is an ℓ^i-th root of unity. But ℓ does not divide $q-1$, so 1 is the only such root in $\mathbb{F}_q$, and by induction on i it also is the only root in $\mathbb{F}_q(x_i)$; hence, the minimal polynomial of y_i over $\mathbb{F}_q(x_i)$ is $Y_i^2 - x_iY_i + 1$. By composition, it follows that the minimal polynomial of y_i over $\mathbb{F}_q(x_j)$ is $Y_i^{2\ell^{i-j}} - x_jY_i^{\ell^{i-j}} + 1$. Taking a resultant to eliminate Y_i between these two polynomials gives the following relation between x_j and x_i:

$$Q_{i,j}(X_i)^2 = \mathrm{Res}_{Y_i}(Y_i^{2\ell^{i-j}} - x_jY_i^{\ell^{i-j}} + 1,\ Y_i^2 - X_iY_i + 1).$$

By direct calculation, this is equivalent to Eq. (6). □

As a result, we can compute $Q_{i,j}$ in time $O(\mathsf{M}(\ell^{i-j}))$ by repeated squaring, but we give a better algorithm in Section 3.1 (and show how to find a y_0 satisfying the hypotheses); we leave the algorithms for lift and push to Section 4.

2.4 The general case

Finally, we discuss the general case, with no assumption on the behavior of Φ_ℓ in $\mathbb{F}_q[X]$. This completes the third row of Table 1, using the bound $r \in O(\ell)$. Because $r = [\mathbb{K}_0 : \mathbb{F}_q]$ divides $\ell - 1$, it is coprime with ℓ. Thus, Q_i remains the minimal polynomial of x_i over $\mathbb{K}_0$, and more generally $Q_{i,j}$ remains the minimal polynomial of x_i over $\mathbb{K}_j$; this will allow us to replace $\mathbb{F}_q$ by $\mathbb{K}_0$ as our base field. We will measure all costs by counting operations in $\mathbb{K}_0$, and we will deduce the cost over $\mathbb{F}_q$ by adding a factor $O(\mathsf{M}(r)\log(r))$ to account for the cost of arithmetic in $\mathbb{K}_0$.

For $i \ge 0$, since $\mathbb{K}_i = \mathbb{K}_0[Y_i]/\langle Y_i^{\ell^i} - y_0\rangle$, we represent its elements on the basis $\{y_i^e \mid 0 \le e < \ell^i\}$; e.g., x_i is written as

$$x_i \;=\; \sum_{0\le j\le r-1} y_i^{q^{\ell^i j} \bmod \ell^i} y_0^{q^{\ell^i j} \text{ div } \ell^i}.$$

Our strategy is to convert between two univariate bases of $\mathbb{K}_i$, $\{y_i^e \mid 0 \le e < \ell^i\}$ and $\{x_i^e \mid 0 \le e < \ell^i\}$. In other words, we show how to apply the isomorphism

$$\Psi_i : \mathbb{K}_i = \mathbb{K}_0[Y_i]/\langle Y_i^{\ell^i} - y_0\rangle \to \mathbb{K}_0[X_i]/\langle Q_{i,0}\rangle$$

and its inverse; we will compute the required polynomials $Q_{i,0}$ and $Q_{i,i-1}$ as a byproduct. In a second time, we will use Ψ_i to perform push and lift between the monomial basis in x_i and the bivariate basis in (x_{i-1}, x_i).

We will factor Ψ_i into elementary isomorphisms

$$\Psi_{i,j} : \mathbb{K}_j[X_i]/\langle Q_{i,j}\rangle \to \mathbb{K}_{j-1}[X_i]/\langle Q_{i,j-1}\rangle, \quad j = i, \ldots, 1.$$

To start the process, with $j = i$, we let $Q_{i,i} = X_i - x_i \in \mathbb{K}_i[X_i]$, so that $\mathbb{K}_i = \mathbb{K}_i[X_i]/\langle Q_{i,i}\rangle$. Take now $j \le i$ and suppose that $Q_{i,j}$ is known. We are going to factor $\Psi_{i,j}$ further as $\Phi''_{i,j} \circ \Phi'_{i,j} \circ \Phi_{i,j}$, by introducing first the isomorphism

$$\varphi_j : \mathbb{K}_j \to \mathbb{K}_{j-1}[Y_j]/\langle Y_j^\ell - y_{j-1}\rangle.$$

The forward direction is a push from the monomial basis in y_j to the bivariate basis in (y_{j-1}, y_j) and the inverse is a lift; none of them involves any arithmetic operation (see Subsection 2.2). Then, we deduce the isomorphism

$$\Phi_{i,j} : \mathbb{K}_j[X_i]/\langle Q_{i,j}\rangle \to \mathbb{K}_{j-1}[Y_j, X_i]/\langle Y_j^\ell - y_{j-1}, Q^\star_{i,j}\rangle,$$

where $Q^\star_{i,j}$ is obtained by applying φ_j to all coefficients of $Q_{i,j}$. Since $\Phi_{i,j}$ consists in a coefficient-wise application of φ_j, applying it or its inverse costs no arithmetic operations.

Next, changing the order of Y_j and X_i, we deduce that there exists $S_{i,j}$ in $\mathbb{K}_{j-1}[X_j]$ and an isomorphism

$$\Phi'_{i,j} : \mathbb{K}_{j-1}[Y_j, X_i]/\langle Y_j^\ell - y_{j-1}, Q^\star_{i,j}\rangle \to \mathbb{K}_{j-1}[X_i, Y_j]/\langle Q_{i,j-1}, Y_j - S_{i,j}\rangle,$$

where $\deg(Q^\star_{i,j}, X_i) = \ell^{i-j}$ and $\deg(Q_{i,j-1}, X_i) = \ell^{i-j+1}$.

LEMMA 2. *From $Q^\star_{i,j}$, we can compute $Q_{i,j-1}$ and $S_{i,j}$ in $O(\mathsf{M}(\ell^{i+1})\log(\ell^i))$ operations in $\mathbb{K}_0$. Once this is done, we can apply $\Phi'_{i,j}$ or its inverse in $O(\mathsf{M}(\ell^{i+1}))$ operations in $\mathbb{K}_0$.*

PROOF. We obtain $Q_{i,j-1}$ and $S_{i,j}$ from the resultant and degree-1 subresultant of $Y_j^\ell - y_{j-1}$ and $Q^\star_{i,j}$ with respect to Y_j, computed over the polynomial ring $\mathbb{K}_{j-1}[X_i]$. This is done by the algorithms of [24, 22], using $O(\mathsf{M}(\ell^{i+1})\log(\ell))$ operations in $\mathbb{K}_0$ (for this analysis, and all others in this proof, we assume that we use Kronecker's substitution for multiplications). To obtain $S_{i,j}$, we invert the leading coefficient of the degree-1 subresultant modulo the resultant $Q_{i,j-1}$; this takes $O(\mathsf{M}(\ell^i)\log(\ell^i))$ operations in $\mathbb{K}_0$.

Applying $\Phi'_{i,j}$ amounts to taking a polynomial $A(Y_j, X_i)$ reduced modulo $\langle Y_j^\ell - y_{j-1}, Q^\star_{i,j}\rangle$ and reducing it modulo $\langle Q_{i,j-1}, Y_j - S_{i,j}\rangle$. This is done by computing $A(S_{i,j}, X_i)$, doing all operations modulo $Q_{i,j-1}$. Using Horner's scheme, this takes $O(\ell)$ operations $(+, \times)$ in $\mathbb{K}_{j-1}[X_i]/\langle Q_{i,j-1}\rangle$, so the complexity claim follows.

Conversely, we start from $A(X_i)$ reduced modulo $Q_{i,j-1}$; we have to reduce it modulo $\langle Y_j^\ell - y_{j-1}, Q^\star_{i,j}\rangle$. This is done using the fast Euclidean division algorithm with coefficients in $\mathbb{K}_{j-1}[Y_j]/\langle Y_j^\ell - y_{j-1}\rangle$ for $O(\mathsf{M}(\ell^{i+1}))$ operations in $\mathbb{K}_0$. □

The last isomorphism $\Phi''_{i,j}$ is trivial:

$$\Phi''_{i,j} : \mathbb{K}_{j-1}[X_i, Y_j]/\langle Q_{i,j-1}, Y_j - S_{i,j}\rangle \to \mathbb{K}_{j-1}[X_i]/\langle Q_{i,j-1}\rangle$$

forgets the variable Y_j; it requires no arithmetic operation.

Taking $j = i, \dots, 1$ allows us to compute $Q_{i,i-1}$ and $Q_{i,0}$ for $O(i^2\mathsf{M}(\ell^{i+1})\log(\ell))$ operations in $\mathbb{K}_0$. Composing the maps $\Psi_{i,j}$, we deduce further that we can apply Ψ_i or its inverse for $O(i\mathsf{M}(\ell^{i+1}))$ operations in $\mathbb{K}_0$.

We claim that we can then perform push and lift between the monomial basis in x_i and the bivariate basis in (x_{i-1}, x_i) for the same cost. Let us for instance explain how to lift.

We start from A written on the bivariate basis in (x_{i-1}, x_i); that is, A is in $\mathbb{K}_0[X_{i-1}, X_i]/\langle Q_{i-1}, T_i\rangle$. Apply Ψ_{i-1} to its coefficients in $x_i^0, \dots, x_i^{\ell-1}$, to rewrite A as an element of

$$\mathbb{K}_0[Y_{i-1}, X_i]/\langle Y_{i-1}^{\ell^{i-1}} - y_{i-2}, T_i\rangle = \mathbb{K}_{i-1}[X_i]/\langle Q_{i,i-1}\rangle.$$

Applying $\Psi_{i,i}^{-1}$ gives us the image of A in $\mathbb{K}_i$, and applying Ψ_i finally brings it to $\mathbb{K}_0[X_i]/\langle Q_i\rangle$.

3. TOWERS FROM IRREDUCIBLE FIBERS

In this section we discuss another construction of the ℓ-adic tower based on work of Couveignes and Lercier [6]. The results of this section are summarized in rows 2, 4 and 5 of Table 1. This construction is not unrelated to the ones of the previous section, and indeed we will start by showing how those of Sections 2.2 and 2.3 reduce to it.

Here is the bottom line of Couveignes' and Lercier's idea. Let G, G' be integral algebraic $\mathbb{F}_q$-groups of the same dimension and let $\phi : G' \to G$ be a surjective, separable algebraic group morphism. Let ℓ be the degree of ϕ; then, the set of points $x \in G$ with fiber G'_x of cardinality ℓ is a nonempty open subset $U \subset G$. If the induced homomorphism $G'(\mathbb{F}_q) \to G(\mathbb{F}_q)$ of groups is not surjective then there are points of $G(\mathbb{F}_q)$ with fibers lying in algebraic extensions of $\mathbb{F}_q$. Assume that we are able to choose ϕ so that we can find one of these points contained in U, with an irreducible fiber, and apply a linear projection to this fiber (e.g., onto an axis). The resulting polynomial is irreducible of degree dividing ℓ (and expectedly equal to ℓ). If we can repeat the construction with a new map $\phi' : G'' \to G'$, and so on, the sequence of extensions makes an ℓ-adic tower over $\mathbb{F}_q$.

3.1 Towers from algebraic tori

In [6], Couveignes and Lercier explain how their idea yields the tower of Section 2.2. Consider the multiplicative group $\mathbb{G}_m$: this is an algebraic group of dimension one, and $\mathbb{G}_m(\mathbb{F}_q)$ has cardinality $q - 1$. The ℓ-th power map defined by $\phi : X \mapsto X^\ell$ is a degree ℓ algebraic endomorphism of $\mathbb{G}_m$, surjective over the algebraic closure.

Suppose that ℓ divides $q - 1$, and let η be a non ℓ-adic residue in $\mathbb{F}_q$ (η plays here the same role as y_0 in Section 2). For any $i > 0$, the fiber $\phi^{-i}(\eta)$ is defined by $X^{\ell^i} - \eta$: we recover the construction of Subsection 2.2.

More generally, following [26, 34], we let $k = \mathbb{F}_q$, $L = \mathbb{F}_{q^n}$ and $k \subset F \subsetneq L$. The Weil restriction $\mathrm{Res}_{L/k}\,\mathbb{G}_m$ is an *algebraic torus*, and the norm $\mathrm{N}_{L/F}$ induces a map $\mathrm{Res}_{L/k}\,\mathbb{G}_m \to \mathrm{Res}_{F/k}\,\mathbb{G}_m$. Define the *maximal torus* $\mathbb{T}_n$ as the intersection of the kernels of the maps $\mathrm{N}_{L/F}$ for all subfields F. Then $\mathbb{T}_n$ has dimension $\varphi(n)$, is isomorphic to $\mathbb{G}_m^{\varphi(n)}$ over the algebraic closure, and its k-rational points form a group of cardinality $\Phi_n(q)$:

$$\mathbb{T}_n(k) \cong \{\alpha \in L^* \mid \mathrm{N}_{L/F}(\alpha) = 1 \text{ for all } k \subset F \subsetneq L\}. \quad (7)$$

We now detail how the construction of Section 2.3 can be obtained by considering the torus $\mathbb{T}_2$; this will allow us to start completing the second row in Table 1.

LEMMA 3. *Let $\Delta \in \mathbb{F}_q$ be a quadratic non-residue if $p \neq 2$, or such that $\mathrm{Tr}_{\mathbb{F}_q/\mathbb{F}_2}(\Delta) = 1$ otherwise. Let $\delta = \sqrt{\Delta}$ or $\delta^2 + \delta = \Delta$ accordingly. The maximal torus $\mathbb{T}_2$ is isomorphic to the* Pell conic

$$C : \begin{cases} x^2 - \Delta y^2 = 4 & \text{if } p \neq 2, \\ x^2\Delta + xy + y^2 = 1 & \text{if } p = 2. \end{cases} \quad (8)$$

Multiplication in $\mathbb{T}_2$ induces a group law on C. The neutral element is $(2, 0)$ if $p \neq 2$, or $(0, 1)$ if $p = 2$. The sum of two points $P = (x_1, y_1)$ and $Q = (x_2, y_2)$ is defined by

$$P \oplus Q = \begin{cases} \left(\dfrac{x_1x_2 + \Delta y_1y_2}{2}, \dfrac{x_1y_2 + x_2y_1}{2}\right) & \text{if } p \neq 2, \\ (x_1x_2 + x_1y_2 + x_2y_1,\ x_1x_2\Delta + y_1y_2) & \text{if } p = 2. \end{cases}$$

PROOF. The isomorphism follows by Weil restriction to $\mathbb{F}_q(\delta)/\mathbb{F}_q$ with respect to the basis $(1/2, \delta/2)$ if $p \neq 2$, or $(\delta, 1)$ if $p = 2$. Indeed, by virtue of Eq. (7), an element (x, y) of $\mathbb{F}_q(\delta)$ belongs to $\mathbb{T}_2$ if and only if its norm over $\mathbb{F}_q$ is 1. Let σ be the generator of $\mathrm{Gal}_{\mathbb{F}_q(\delta)/\mathbb{F}_q}$. For $p \neq 2$, clearly $\delta^\sigma = -\delta$. For $p = 2$, by Artin-Schreier theory, $\mathrm{Tr}_{\mathbb{F}_q(\delta)/\mathbb{F}_q}(\delta) = \mathrm{Tr}_{\mathbb{F}_q/\mathbb{F}_2}(\Delta) = 1$, hence $\delta^\sigma = 1 + \delta$. In both cases, Eq. (8) follows. The group law is obtained by direct calculation. □

Pell conics are a classic topic in number theory[20] and computer science, with applications to primality proving, factorization [19, 12] and cryptography [25].

As customary, we denote by $[n](x, y)$ the n-th scalar multiple of a point (x, y). $[n]$ is an endomorphism of C of degree n, separable if and only if $(n, p) = 1$.

LEMMA 4. *Let $P = (\alpha, \beta)$ be a point of C. The abscissa of $[n]P$ is given by $C_n(\alpha)$, where $C_n \in \mathbb{Z}[X]$ is the n-th Chebyshev polynomial, defined by $C_0 = 2$, $C_1 = X$, and*

$$C_{n+1} = XC_n - C_{n-1}. \tag{9}$$

PROOF. Induction on n. A detailed proof can be found in [30, Prop. 6.6]. □

THEOREM 5. *Let $\eta \in \mathbb{F}_q(\delta)$ be a non ℓ-adic residue in $\mathbb{T}_2$, and let $P = (\alpha, \beta)$ be its image in $C/\mathbb{F}_q$. For any $i > 0$, the polynomials $C_{\ell^i} - \alpha$ are irreducible. Their roots are the abscissas of the images in $C/\mathbb{F}_{q^{\ell^i}}$ of the ℓ^i-th roots of η.*

PROOF. By [17, Th. VI.9.1], the polynomial $X^{\ell^i} - \eta$ is irreducible. Its roots correspond to the fiber $[\ell^i]^{-1}(P)$, and the Galois group of $\mathbb{F}_{q^{\ell^i}}/\mathbb{F}_q$ acts transitively on them.

Two points of C have the same abscissa if and only if they are opposite. But η is a non ℓ-adic residue, hence $\eta \neq \eta^{-1}$, and all the points in $[\ell^i]^{-1}(P)$ have distinct abscissa. By Lemma 4, $C_{\ell^i} - \alpha$ vanishes precisely on those abscissas and is thus irreducible. □

We can now apply our results to the computation of the polynomials Q_i and T_i of Section 2.3.

COROLLARY 6. *The polynomials $Q_{i,j}$ of Prop. 1 satisfy*

$$Q_{i,j}(X_i) = C_{\ell^{i-j}}(X_i) - x_j.$$

PROOF. We have already shown that $\mathrm{N}_{\mathbb{K}_j/\mathbb{F}_q(x_j)}(y_j) = 1$ for any j, thus y_j is a non ℓ-adic residue in $\mathbb{T}_2/\mathbb{F}_q(x_j)$. Independently of the characteristic and of the element $\Delta \in \mathbb{F}_q(x_j)$ chosen, the abscissa of the image of y_j in $C/\mathbb{F}_q(x_j)$ is $\mathrm{Tr}_{\mathbb{K}_j/\mathbb{F}_q(x_j)}\, y_j = x_j$. The statement follows from the previous theorem. □

There is a folklore algorithm computing the n-th Chebyshev polynomial using $O(n)$ operations in $\mathbb{Z}$ [15]. We shall need a slightly better algorithm working modulo p.

COROLLARY 7. *The polynomials $Q_{i,j}$ can be computed using $O(\ell^{i-j})$ operations in $\mathbb{F}_p$.*

PROOF. Let $C_n = \sum_i c_{n,i} X^{n-i}$. It is well known that $|c_{n+k,2k}|$ are the coefficients of the $(1, 2)$-Pascal triangle, also called Lucas' triangle (see [30, Prop. 6.6] and [1]). It follows that

$$\frac{c_{n,2k+2}}{c_{n,2k}} = -\frac{(n-2k)(n-2k-1)}{(n-k-1)(k+1)},$$

which immediately gives the algorithm. Indeed, since we know the $c_{n,2k}$'s are the image mod p of integers, we compute them using multiplications and divisions in $\mathbb{Q}_p$ with relative precision 1. □

We are left with the problem of finding the non ℓ-adic residue η to initialize the tower. As before, this will be done by random sampling and testing.

LEMMA 8. *Let $P = (\alpha, \beta)$ be a point on C. For any n, there is a formula to compute the abscissa of $[\pm n]P$, using $O(\log n)$ operations in $\mathbb{F}_q$, and not involving β.*

PROOF. Observe that if $n = 2$, the abscissa of $[\pm 2]P$ is $\alpha^2 - 2$ (for any p). Let $P' = (\alpha', \beta')$, and let γ be the abscissa of $P \ominus P'$. By direct computation we find that the abscissa of $P \oplus P'$ is $\alpha\alpha' - \gamma$ (for any p); this formula is called a *differential addition*. Thus, $O(1)$ operations are needed for a doubling or a differential addition. To compute the abscissa of $[\pm n]P$, we use the ladder algorithm of [23], requiring $O(\log n)$ doublings and differential additions. □

PROPOSITION 9. *The abscissa of a point $P \in C/\mathbb{F}_q$ satisfying the conditions of Theorem 5 can be found using $O_e(\log q)$ operations in $\mathbb{F}_q$.*

PROOF. We randomly select $\alpha \in \mathbb{F}_q$ and test that it belongs to C. If $p \neq 2$, this amounts to testing that $\alpha^2 - 4$ is a quadratic non-residue in $\mathbb{F}_q$, a task that can be accomplished with $O(\log q)$ operations. If $p = 2$, by Artin-Schreier theory this is equivalent to $\mathrm{Tr}_{\mathbb{F}_q/\mathbb{F}_2}(1/\alpha^2) = 1$, which can be tested in $O(\log q)$ operations in $\mathbb{F}_q$.

Then we check that P is a non ℓ-adic residue by verifying that $[(q+1)/\ell]P$ is not the group identity. By Lemma 8, this computation requires $O(\log q)$ operations. About half of the points of $\mathbb{F}_q$ are quadratic non-residues, and about $1-1/\ell$ of them are the abscissas of points with the required order, thus we expect to find the required element after $O_e(1)$ trials. □

It is natural to ask whether a similar construction could be applied to any ℓ. If r is the order of q modulo ℓ, the natural object to look at is $\mathbb{T}_r$, but here we are faced with two problems. First, multiplication by ℓ is now a degree $\ell^{\varphi(r)}$ map, thus its fibers have too many points; instead, isogenies of degree ℓ should be considered. Second, it is an open question whether $\mathbb{T}_r$ can be parameterized using $\varphi(r)$ coordinates; but even assuming it can be, we are still faced with the computation of a univariate annihilating polynomial for a set embedded in a $\varphi(r)$-dimensional space, a problem not known to be feasible in quasi-linear time. Studying this generalization is another natural follow-up to the present work.

3.2 Towers from elliptic curves

Since it seems hard to deal with higher dimensional algebraic tori, it is interesting to look at other algebraic groups. Being one-dimensional, elliptic curves are good candidates. In this section, we quickly review Couveignes' and Lercier's construction, referring to [6] for details, and point out the modifications needed in order to build towers (as opposed to constructing irreducible polynomials).

Let ℓ be a prime different from p and not dividing $q - 1$. Let E_0 be an elliptic curve whose cardinality over $\mathbb{F}_q$ is a multiple of ℓ. By Hasse's bound, this is only possible if $\ell \leq q + 2\sqrt{q} + 1$. An *isogeny* is an algebraic group morphism between two elliptic curves that is surjective in the algebraic

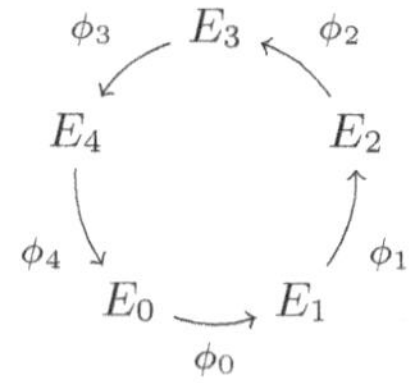

Figure 2: The isogeny cycle of E_0.

closure. It is said to be rational over $\mathbb{F}_q$ if it is invariant under the q-th power map; such an isogeny exists if and only if the curves have the same number of points over $\mathbb{F}_q$. An isogeny of degree n is separable if and only if n is prime to p, in which case its kernel contains exactly n points. Because of the assumptions on ℓ, there exists an $e \geq 1$ such that, for any curve E isogenous to E_0, the $\mathbb{F}_q$-rational part of $E[\ell]$ is cyclic of order ℓ^e.

Suppose for simplicity, that $p \neq 2, 3$ and let E_0 be expressed as the locus

$$E_0 \,:\, y^2 = x^3 + ax + b, \quad \text{with } a, b \in \mathbb{F}_q, \tag{10}$$

plus one point at infinity. We denote by H_0 the unique subgroup of $E_0/\mathbb{F}_q$ of order ℓ, and by ϕ_0 the unique isogeny whose kernel is H_0; we then label E_1 the image curve of ϕ_0. We go on denoting by H_i the unique subgroup of $E_i/\mathbb{F}_q$ of order ℓ, and by $\phi_i : E_i \to E_{i+1}$ the unique isogeny with kernel H_i. The construction is depicted in Figure 2.

LEMMA 10. *Let $E_0, E_1, \dots$ be defined as above, there exists $n \in O(\sqrt{q}\log(q))$ such that E_n is isomorphic to E_0.*

PROOF. It is shown in [6, § 4] that the isogenies ϕ_i are *horizontal* in the sense of [16], hence they necessarily form a cycle. Let t be the trace of E_0, the length of the cycle is bounded by the class number of $\mathbb{Q}[X]/(X^2 - tX - q)$, thus by Minkowski's bound it is in $O(\sqrt{q}\log(q))$. □

In what follows, the index i is to be understood modulo the length of the cycle. This is a slight abuse, because E_n is isomorphic but not equal to E_0, but it does not hide any theoretical or computational difficulty.

Under the former assumptions, it is proved in [6, § 4] that if P is a point of E_i of order divisible by ℓ^e, if $\psi = \phi_{i-1} \circ \phi_{i-2} \circ \cdots \circ \phi_j$, then the fiber $\psi^{-1}(P)$ is irreducible and has cardinality ℓ^{i-j}. Knowing E_i, Vélu's formulas [32] allow us to express the isogenies ϕ_i as rational fractions

$$\begin{aligned} \phi_i : E_i &\to E_{i+1}, \\ (x, y) &\mapsto \left(\frac{f_i(x)}{g_i(x)}, y \left(\frac{f_i(x)}{g_i(x)} \right)' \right), \end{aligned} \tag{11}$$

where g_i is the square polynomial of degree $\ell - 1$ vanishing on the abscissas of the affine points of H_i, and f_i is a polynomial of degree ℓ.

There is a subtle difference between our setting and Couveignes' and Lercier's. The goal of [6] is to compute an extension of degree ℓ^i of $\mathbb{F}_q$ for a fixed i: this can be done by going forward i times, then taking the fiber of a point of E_i by the isogenies $\phi_{i-1}, \dots, \phi_0$. In our case, we are interested in building extensions of degree ℓ^i *incrementally*, i.e. without any *a priori* bound on i. Thus, we have to walk *backwards* in the isogeny cycle: if $\eta \in \mathbb{F}_q$ is the abscissa of a point of E_0 of order $\ell^e \neq 2$, we will use the following polynomials to define the ℓ-adic tower:

$$\begin{aligned} T_1 &= f_{-1}(X_1) - \eta g_{-1}(X_1), \\ T_i &= f_{-i}(X_i) - X_{i-1} g_{-i}(X_i). \end{aligned}$$

The following theorem gives the time for building the tower; lift and push are detailed in the next section.

THEOREM 11. *Suppose $4\ell \leq q^{1/4}$, and under the above assumption. Initializing the ℓ-adic tower requires $O_{\tilde{e}}(\ell \log^5(q) + \ell^3)$ bit operations; and building the i-th level requires $O_e(\ell^2 + \mathsf{M}(\ell)\log(\ell q) + \mathsf{M}(\ell^i)\log(\ell^i))$ operations in $\mathbb{F}_q$.*

PROOF. For the initialization, [6, § 4.3] shows that if $4\ell \leq q^{1/4}$, a curve E_0 with the required number of points can be found in $O_{\tilde{e}}(\ell \log^5(q))$ bit operations. We also need to compute the ℓth modular polynomial Φ_ℓ mod p; for this, we compute it over $\mathbb{Z}$ with $\tilde{O}(\ell^3)$ bit operations [10], then reduce it modulo p.

To build the i-th level, we first need to find the equation of E_{-i}. For this, we evaluate Φ_ℓ at $j(E_{-i+1})$, using $O(\ell^2)$ operations. Lemma 10 implies that this polynomial has only two roots in $\mathbb{F}_q$, namely $j(E_{-i})$ and $j(E_{-i+2})$. We factor it using $O_e(\mathsf{M}(\ell)\log(\ell q))$ operations [33, Ch 14], and we take an arbitrary curve with j-invariant $j(E_{-i})$. Then we find an ℓ-torsion point using $O_e(\log q)$ operations, and apply Vélu's formulas to compute ϕ_{-i}. We deduce the polynomial T_i, and Q_i is obtained using $O(\mathsf{M}(\ell^i)\log(\ell^i))$ operations using Algorithm 1 given in the next section. □

Remark 1. Instead of computing the cycle step by step, we could compute it entirely during the initialization phase, by using Vélu's formulas alone to compute $E_1, E_2, \dots$ until we hit E_0 again. By doing so, we avoid using the modular polynomial Φ_ℓ at each new level. By Lemma 10, this requires $O_e(\ell\sqrt{q}\log(q))$ operations. This is not asymptotically good in q, but for practical values of q and ℓ the cycle is often small and this approach works well. This is accounted for in the last row of Table 1.

4. LIFTING AND PUSHING

The previous constructions of ℓ-adic towers based on irreducible fibers share a common structure that allows us to treat lifting and pushing in a unified way. Renaming the variables (X_{i-1}, X_i) as (X, Y), the polynomials (Q_{i-1}, Q_i, T_i) as (R, S, T), the extension at level i is described as

$$\mathbb{F}_q[Y]/\langle S(Y)\rangle \quad \text{and} \quad \mathbb{F}_q[X, Y]/\langle R(X), T(X, Y)\rangle,$$

with R of degree ℓ^{i-1}, S of degree ℓ^i, and where $T(X, Y)$ has the form $f(Y) - Xg(Y)$, with $\deg(f) = \ell$, $\deg(g) < \ell$ and $\gcd(f, g) = 1$; possibly, $g = 1$. In all this section, f, g and their degree ℓ are fixed.

Lift is the conversion from the bivariate basis associated to the right-hand side to the univariate basis associated to the left-hand side; push is the inverse. Using the special shape of the polynomial T, they reduce to composition and decomposition of rational functions, as we show next. These results fill in all missing entries in the lift / push column of Table 1.

Algorithm 1 Compose

Input: $P \in \mathbb{F}_q[X,Y]$, $f, g \in \mathbb{F}_q[Y]$, $n \in \mathbb{N}$

1: **if** $n = 1$ **then**
2:    **return** P
3: **else**
4:    $m \leftarrow \lceil n/2 \rceil$
5:    Let P_0, P_1 be such that $P = P_0 + X^m P_1$
6:    $Q_0 \leftarrow \text{Compose}(P_0, f, g, m)$
7:    $Q_1 \leftarrow \text{Compose}(P_1, f, g, n-m)$
8:    $Q \leftarrow Q_0 g^{n-m} + Q_1 f^m$
9:    **return** Q
10: **end if**

Algorithm 2 Decompose

Input: $Q, f, g, h \in \mathbb{F}_q[Y]$, $n \in \mathbb{N}$

1: **if** $n = 1$ **then**
2:    **return** Q
3: **else**
4:    $m \leftarrow \lceil n/2 \rceil$
5:    $u \leftarrow 1/g^{n-m} \bmod f^m$
6:    $Q_0 \leftarrow Qu \bmod f^m$
7:    $Q_1 \leftarrow (Q - Q_0 g^{n-m}) \text{ div } f^m$
8:    $P_0 \leftarrow \text{Decompose}(Q_0, f, g, h, m)$
9:    $P_1 \leftarrow \text{Decompose}(Q_1, f, g, h, n-m)$
10:    **return** $P_0 + X^m P_1$
11: **end if**

4.1 Lifting

Let P be in $\mathbb{F}_q[X,Y]$ and n be in $\mathbb{N}$, with $\deg(P,X) < n$. We define $P[f,g,n]$ as

$$P[f,g,n] = g^{n-1} P\left(\frac{f}{g}, Y\right) \in \mathbb{F}_q[X,Y].$$

If $P = \sum_{i=0}^{n-1} p_i(Y) X^i$, then $P[f,g,n] = \sum_{i=0}^{n-1} p_i f^i g^{n-1-i}$. We first give an algorithm to compute this expression, then show how to relate it to lifting; when $g = 1$, Algorithm 1 reduces to a well known algorithm for polynomial composition [33, Ex. 9.20].

THEOREM 12. *On input P, f, g, n, with $\deg(P,X) < n$ and $\deg(P,Y) < \ell$, Algorithm 1 computes $Q = P[f,g,n]$ using $O(\mathsf{M}(\ell n)\log(n))$ operations in $\mathbb{F}_q$.*

PROOF. If $n = 1$, the theorem is obvious. Suppose $n > 1$, then P_0 and P_1 have degrees less than m and $n - m$ respectively. By induction hypothesis,

$$\begin{aligned} Q_0 = P_0[f,g,m] &= \textstyle\sum_{i=0}^{m-1} p_i f^i g^{m-1-i}, \\ Q_1 = P_1[f,g,n-m] &= \textstyle\sum_{i=0}^{n-m-1} p_{i+m} f^i g^{n-m-1-i}. \end{aligned}$$

Hence,

$$\begin{aligned} Q &= \textstyle\sum_{i=0}^{m-1} p_i f^i g^{n-1-i} + \sum_{i=0}^{n-m-1} p_{i+m} f^{i+m} g^{n-m-1-i} \\ &= P[f,g,n]. \end{aligned}$$

The only step that requires a computation is Step 8, costing $O(\mathsf{M}(\ell n))$ operations in $\mathbb{F}_q$. The recursion has depth $\log(n)$, hence the overall complexity is $O(\mathsf{M}(\ell n)\log(n))$. $\square$

COROLLARY 13. *At level i, one can perform the lift operation using $O(\mathsf{M}(\ell^i)\log(\ell^i))$ operations in $\mathbb{F}_q$.*

PROOF. We start from an element α written on the bivariate basis, that is, represented as $A(X,Y)$ with $\deg(A,X) < n = \ell^{i-1}$ and $\deg(A,Y) < \ell$ (note that $\ell n = \ell^i$). We compute the univariate polynomials $A^\star = A[f,g,n]$ and $\gamma = g^{n-1}$ using $O(\mathsf{M}(\ell^i)\log(\ell^i))$ operations in $\mathbb{F}_q$; then the lift of α is $A^\star/\gamma$ modulo S. The inverse of γ is computed using $O(\mathsf{M}(\ell n)\log(\ell n))$ operations, and the multiplication adds an extra $O(\mathsf{M}(\ell n))$. $\square$

4.2 Pushing

We first deal with the inverse of the question dealt with in Theorem 12: starting from $Q \in \mathbb{F}_q[Y]$, reconstruct $P \in \mathbb{F}_q[X,Y]$ such that $Q = P[f,g,n]$. When $g = 1$, Algorithm 2 reduces to Algorithm 9.14 of [33].

THEOREM 14. *On input Q, f, g, h, n, with $\deg(Q) < \ell n$ and $h = 1/g \bmod f$, Algorithm 2 computes a polynomial $P \in \mathbb{F}_q[X,Y]$ such that $\deg(P,X) < n$, $\deg(P,Y) < \ell$ and $Q = P[f,g,n]$ using $O(\mathsf{M}(\ell n)\log(n))$ operations in $\mathbb{F}_q$.*

PROOF. We prove the theorem by induction. If $n = 1$, the statement is obvious, so let $n > 1$. The polynomials Q_0 and Q_1 verify $Q = Q_0 g^{n-m} + Q_1 f^m$. By construction, Q_0 has degree less than ℓm. Since $\deg(g) < \ell$, this implies that $Q_0 g^{n-m}$ has degree less than ℓn; thus, Q_1 has degree less than $\ell(n-m)$. By induction, P_0 and P_1 have degree less than m, resp. $n - m$, in X, and less than ℓ in Y, and

$$\begin{aligned} Q_0 = P_0[f,g,m] &= \textstyle\sum_{i=0}^{m-1} p_{0,i} f^i g^{m-1-i}, \\ Q_1 = P_1[f,g,n-m] &= \textstyle\sum_{i=0}^{n-m-1} p_{1,i} f^i g^{n-m-1-i}. \end{aligned}$$

Hence, $P = P_0 + X^m P_1$ has degree less than n in X and less than ℓ in Y, and the following proves correctness:

$$\begin{aligned} P[f,g,n] &= \textstyle\sum_{i=0}^{m-1} p_{0,i} f^i g^{n-1-i} + \sum_{i=m}^{n-1} p_{1,i-m} f^i g^{n-1-i} \\ &= P_0[f,g,m] g^{n-m} + P_1[f,g,n-m] f^m = Q. \end{aligned}$$

At Step 5, we do as follows: starting from $h = 1/g \bmod f$, we deduce $1/g^{n-m} \bmod f$ in time $O(\mathsf{M}(\ell)\log(n))$ by binary powering mod f. We also compute g^{n-m} in time $O(\mathsf{M}(\ell n))$ by binary powering, and we use Newton iteration (starting from $1/g^{n-m} \bmod f$) to deduce $1/g^{n-m} \bmod f^m$ in time $O(\mathsf{M}(\ell n))$. All other steps cost $O(\mathsf{M}(\ell n))$; the recursion has depth $\log(n)$, so the total cost is $O(\mathsf{M}(\ell n)\log(n))$. $\square$

COROLLARY 15. *At level i, one can perform the push operation using $O(\mathsf{M}(\ell^i)\log(\ell^i))$ operations in $\mathbb{F}_q$.*

PROOF. Given α represented by a univariate polynomial $A(Y)$ of degree less than ℓn, with $n = \ell^{i-1}$. We compute g^{n-1} and $A^\star = g^{n-1} A \bmod S$ using $O(\mathsf{M}(\ell^i))$ operations. Then, we compute $h = 1/g \bmod f$ in time $O(\mathsf{M}(\ell)\log(\ell))$ and apply Algorithm 2 to $A^\star$, f, g, h and n. The result is a bivariate polynomial B, representing α on the bivariate basis. The dominant phase is Algorithm 2, costing $O(\mathsf{M}(\ell^i)\log(\ell^i))$ operations in $\mathbb{F}_q$. $\square$

5. IMPLEMENTATION

To demonstrate the interest of our constructions, we made a very basic implementation of the towers of Sections 3.1 and 3.2 in Sage [31]. It relies on Sage's default implementation of quotient rings of $\mathbb{F}_p[X]$, which itself uses NTL [29] for $p = 2$ and FLINT [13] for other primes. Towers based on elliptic curves are constructed using the algorithm

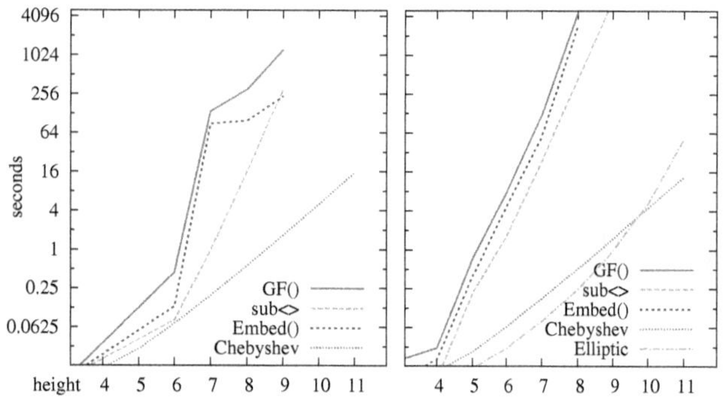

Figure 3: Times for building 3-adic towers on top of $\mathbb{F}_2$ (left) and $\mathbb{F}_5$ (right), in Magma (first three lines) and using our code.

described in Remark 1. The source code is available at `http://defeo.github.io/towers`

We compare our implementation to three ways of constructing ℓ-adic towers in Magma. First, one may construct the levels from bottom to top using the finite field constructor `GF()`. For the parameters we used, Magma uses tables of precomputed Conway polynomials and automatically computes embeddings on creation, see `http://magma.maths.usyd.edu.au/magma/releasenotes/2/14`. The second approach constructs the highest level of the tower first, then all the lower levels using the `sub<>` constructor. The last one constructs the levels from bottom to top using random dense polynomials and calls the `Embed()` function; we do not count the time for finding the irreducible polynomials.

We ran tests on an Intel Xeon E5620 clocked at 2.4 GHz, using Sage 5.5 and Magma 2.18.12. The time required for the creation of 3-adic towers of increasing height is summarized in Figure 3; the timings of our algorithms are labeled Chebyshev and Elliptic. Computations that took more than 4GB RAM were interrupted.

Despite its simplicity, our code consistently outperforms Magma on creation time. On the other hand, lift and push operations take essentially no time in Magma, while in all the tests of Figure 3 we measured a running time almost perfectly linear for one push followed by one lift, taking approximately $70\mu s$ per coefficient (this is in the order of a second around level 10). Nevertheless, the large gain in creation time makes the difference in lift and push tiny, and we are convinced that an optimized C implementation of the algorithms of Section 4 would match Magma's performances.

Acknowledgments. We acknowledge support from NSERC, the CRC program, and ANR through the ECLIPSES project under Contract ANR-09-VERS-018. De Feo would like to thank Antoine Joux and Jérôme Plût for fruitful discussions. We are grateful to the reviewers for their remarks.

6. REFERENCES

[1] A. T. Benjamin. The Lucas triangle recounted. In *Congressus Numerantium*, volume 200, pages 169–177, 2010.

[2] W. Bosma, J. Cannon, and C. Playoust. The MAGMA algebra system I: the user language. *J. Symbolic Comput.*, 24(3-4):235–265, 1997.

[3] W. Bosma, J. Cannon, and A. Steel. Lattices of compatibly embedded finite fields. *J. Symbolic Comput.*, 24(3-4):351–369, 1997.

[4] D. G. Cantor. On arithmetical algorithms over finite fields. *J. Combin. Theory Ser. A*, 50(2):285–300, 1989.

[5] J.-M. Couveignes. Isomorphisms between Artin-Schreier towers. *Math. Comp.*, 69(232):1625–1631, 2000.

[6] J.-M. Couveignes and R. Lercier. Fast construction of irreducible polynomials over finite fields. *To appear in the Israel Journal of Mathematics*, July 2011.

[7] L. De Feo. Fast algorithms for computing isogenies between ordinary elliptic curves in small characteristic. *Journal of Number Theory*, 131(5):873–893, May 2011.

[8] L. De Feo and É. Schost. Fast arithmetics in Artin-Schreier towers over finite fields. *J. Symbolic Comput.*, 47(7):771–792, 2012.

[9] J. Doliskani and É. Schost. A note on computations in degree 2^k-extensions of finite fields, 2012. Manuscript.

[10] A. Enge. Computing modular polynomials in quasi-linear time. *Math. Comp.*, 78(267):1809–1824, 2009.

[11] P. Gaudry and É. Schost. Point-counting in genus 2 over prime fields. *J. Symbolic Comput.*, 47(4):368–400, 2012.

[12] S. A. Hambleton. Generalized Lucas-Lehmer tests using Pell conics. *Proceedings of the American Mathematical Society*, 140:2653–2661, 2012.

[13] W. Hart. Fast library for number theory: an introduction. *Mathematical Software–ICMS 2010*, pages 88–91, 2010.

[14] K. S. Kedlaya and C. Umans. Fast polynomial factorization and modular composition. *SIAM J. Computing*, 40(6):1767–1802, 2011.

[15] W. Koepf. Efficient computation of chebyshev polynomials in computer algebra. *Computer Algebra Systems: A Practical Guide.*, pages 79–99, 1999.

[16] D. Kohel. *Endomorphism rings of elliptic curves over finite fields.* PhD thesis, University of California at Berkley, 1996.

[17] S. Lang. *Algebra.* Springer, 3rd edition, Jan. 2002.

[18] R. Lebreton and É. Schost. Algorithms for the universal decomposition algebra. In *ISSAC'12*, pages 234–241. ACM, 2012.

[19] F. Lemmermeyer. Conics - a Poor Man's Elliptic Curves, 2003.

[20] H. W. Lenstra. Solving the Pell equation. *Notices of the AMS*, 49(2):182–192, 2002.

[21] H. W. Lenstra and B. De Smit. Standard models for finite fields: the definition, 2008.

[22] T. Lickteig and M. Roy. Sylvester–habicht sequences and fast cauchy index computation. *J. Symbolic Comput.*, 31(3):315 – 341, 2001.

[23] P. L. Montgomery. Speeding the pollard and elliptic curve methods of factorization. *Math. Comp.*, 48(177), 1987.

[24] D. Reischert. Asymptotically fast computation of subresultants. In *ISSAC*, pages 233–240. ACM, 1997.

[25] K. Rubin and A. Silverberg. Torus-Based cryptography. In D. Boneh, editor, *Advances in Cryptology - CRYPTO 2003*, volume 2729 of *Lecture Notes in Computer Science*, pages 349–365, Berlin, Heidelberg, 2003. Springer Berlin / Heidelberg.

[26] K. Rubin and A. Silverberg. Algebraic tori in cryptography. In *In High Primes and Misdemeanours: Lectures in Honour of the 60th birthday of Hugh Cowie Williams*, volume 41 of *Fields Institute Communications*. AMS, 2004.

[27] V. Shoup. New algorithms for finding irreducible polynomials over finite fields. *Math. Comp.*, 54:435–447, 1990.

[28] V. Shoup. Fast construction of irreducible polynomials over finite fields. *J. Symbolic Comput.*, 17(5):371–391, 1994.

[29] V. Shoup. NTL: A library for doing number theory. `http://www.shoup.net/ntl`, 2003.

[30] J. H. Silverman. *The arithmetic of dynamical systems*, volume 241 of *Graduate Texts in Mathematics*. Springer, 2007.

[31] W. A. Stein and Others. *Sage Mathematics Software (Version 5.5).* The Sage Development Team, 2013.

[32] J. Vélu. Isogénies entre courbes elliptiques. *Comptes Rendus de l'Académie des Sciences de Paris*, 273:238–241, 1971.

[33] J. von zur Gathen and J. Gerhard. *Modern computer algebra.* Cambridge University Press, New York, NY, USA, 1999.

[34] V. E. Voskresenskiĭ. *Algebraic groups and their birational invariants*, volume 179. American Mathematical Society, 1998.

Combinatorics of 4-dimensional Resultant Polytopes

Alicia Dickenstein [*]
Dto. de Matemática, FCEN
Universidad de Buenos Aires,
& IMAS-CONICET, Argentina
alidick@dm.uba.ar

Ioannis Z. Emiris [†]
Department of Informatics
and Telecommunications
University of Athens, Greece
emiris@di.uoa.gr

Vissarion Fisikopoulos [**]
Department of Informatics
and Telecommunications
University of Athens, Greece
vissarion@di.uoa.gr

ABSTRACT

The Newton polytope of the resultant, or *resultant polytope*, characterizes the resultant polynomial more precisely than total degree. The combinatorics of resultant polytopes are known in the Sylvester case [5] and up to dimension 3 [9]. We extend this work by studying the combinatorial characterization of 4-dimensional resultant polytopes, which show a greater diversity and involve computational and combinatorial challenges. In particular, our experiments, based on software `respol` for computing resultant polytopes, establish lower bounds on the maximal number of faces. By studying mixed subdivisions, we obtain tight upper bounds on the maximal number of facets and ridges, thus arriving at the following maximal f-vector: $(22, 66, 66, 22)$, i.e. vector of face cardinalities. Certain general features emerge, such as the symmetry of the maximal f-vector, which are intriguing but still under investigation. We establish a result of independent interest, namely that the f-vector is maximized when the input supports are sufficiently generic, namely full dimensional and without parallel edges. Lastly, we offer a classification result of all possible 4-dimensional resultant polytopes.

Categories and Subject Descriptors

G.2.1 [**Discrete Mathematics**]: Combinatorics—*Counting problems*; I.1.2 [**Symbolic and Algebraic Manipulation**]: Algorithms—*Algebraic algorithms*

Keywords

Resultant, f-vector, Mixed subdivision, Secondary polytope

[*]Partially supported by UBACYT 20020100100242, CONICET PIP 112-200801-00483 and ANPCyT PICT 2008-0902, Argentina.

[†]Partially supported by "Computational Geometric Learning", financed within the 7th Framework Program for research of the European Commission, under FET-Open grant number 255827.

1. INTRODUCTION

Let $\mathcal{A} = (A_0, \ldots, A_n)$ be a family of subsets of $\mathbb{Z}^n$ and let $f_0, \ldots, f_n \in \mathbb{C}[x_1, \ldots, x_n]$ be polynomials with this family of sets as supports, and symbolic coefficients $c_{ij} \neq 0$, $i = 0, \ldots, n$, $j = 1, \ldots, |A_i|$, i.e. $f_i = \sum_{a \in A_i} c_{ij} x^a$. The family $\mathcal{A}$ is *essential* if these sets jointly affinely span $\mathbb{Z}^n$, and every subfamily of A_i's of cardinality j, $1 \leq j < n$, spans an affine space of dimension $\geq j$. In this paper we assume that $\mathcal{A}$ is essential. The *sparse (or toric) resultant* $\mathcal{R} = \mathcal{R}_{\mathcal{A}}$ of $f_0, \ldots, f_n$ is then a non-constant irreducible polynomial in $\mathbb{Z}[c_{ij} : i = 0, \ldots, n,\ j = 1, \ldots, |A_i|]$, defined up to sign, which vanishes if $f_0 = f_1 = \cdots = f_n = 0$ has a solution in $(\mathbb{C}^*)^n$, $\mathbb{C}^* = \mathbb{C} \setminus \{0\}$. The Newton polytope $N(\mathcal{R})$ of the resultant, that is, the convex hull of the exponents occurring in $\mathcal{R}$ with non-zero coefficient, is a lattice polytope called *resultant polytope*. A famous example is the Birkhoff polytope of a linear system, cf Example 1.

The resultant has $m = \sum_{i=0}^{n} |A_i|$ variables, hence $N(\mathcal{R})$ lies in $\mathbb{R}^m$. However, for essential families, $\mathcal{R}$ satisfies $n+(n+1)$ natural homogeneities [4], so its dimension is $\dim(N(\mathcal{R})) = m - 2n - 1$. If $\mathcal{A}$ is not essential, but contains a single essential subfamily, the resultant depends only on the coefficients of the polynomials in this subfamily. Otherwise, the resultant locus has codimension bigger than one, and then the sparse resultant is defined to be the constant 1.

Previous work. In [3] an algorithm is described for computing the vertex and facet representations of $N(\mathcal{R})$; the algorithm also produces a triangulation of the polytope's interior into simplices. The input to this algorithm is an oracle for computing the extreme resultant vertex given a direction. The software implementation is called `respol`[1] and is the one used in our experiments. This method is readily generalized to compute the discriminant and secondary polytopes, although for the latter there exists a faster method to enumerate the vertices, when only these are needed [8]. An alternative way for computing resultant polytopes exploits tropical geometry [6] and is implemented based on the software library `Gfan`.

The combinatorics of resultant polytopes is known only in small cases, namely for linear systems (Example 1), in the Sylvester case ($n = 1$), and when $\dim N(\mathcal{R}) = 3$. The univariate case is fully described in [5]: $N(\mathcal{R})$ is combinatorially isomorphic to a polytope denoted by $\mathcal{N}_{k_0,k_1}$, of dimension $k_0 + k_1 - 1$, where the A_i may be multisets with cardinality k_i. They may lead to polytopes in any dimension if one picks the A_i accordingly. In [4] they show that $\mathcal{N}_{k_0,k_1}$ has

[1]http://respol.sourceforge.net

$\binom{k_0+k_1}{k_0}$ vertices and, when both $k_i \geq 2$, it has k_0k_1+3 facets. Sturmfels [9] classifies all resultant polytopes up to dimension 3. In his notation, the 3-dimensional polytope $\mathcal{N}_{1111,111}$, denoted by $\mathcal{N}_{3,2}$ in [5], depicted in Fig. 1 with the resultant label, has maximal f-vector.

PROPOSITION 1. [9, Sect. 6] *Assume $\mathcal{A}$ is an essential family. Then, $N(\mathcal{R})$ is 1-dimensional iff $|A_i| = 2$, for all i. The only planar resultant polytope is the triangle. The only 3-dimensional $N(\mathcal{R})$ are, combinatorially: (a) the tetrahedron, (b) the square-based pyramid, and (c) the polytope $\mathcal{N}_{3,2}$.*

In [6] the authors raise explicitly the open question of describing 4-dimensional resultant polytopes, which we undertake in this paper.

Our contribution. We study the combinatorial characterization of 4-dimensional resultant polytopes. To bound the maximum number of faces, we prove that it suffices to focus on one case, which corresponds to 3 Newton polygons with support cardinalities $|A_i| = 3$, thus $m = 9$ and $n = 2$. We further show it is enough to consider sufficiently generic polygons, namely where they all have non-zero area, and no parallel edges exist among them. Our experiments, based on `respol` [3], establish lower bounds on the maximal number of faces (Table 1). By studying mixed subdivisions, we obtain tight upper bounds on the maximal number of facets and ridges, thus raising new conjectures, the most important of which is that the maximal f-vector is $(22, 66, 66, 22)$ for 4-dimensional $N(\mathcal{R})$. These results are summarized in Thm 16. Our (loose) upper bound on the number of vertices, namely 28, significantly improves the known bound of 6608 [9, Cor.6.2]. Certain general features emerge, such as the symmetry of the maximal f-vector, which are intriguing but still under investigation. However, the Newton polytopes are not *self-dual*. Our main result is Thm 17, where we offer a characterization of all possible 4-dimensional resultant polytopes.

The rest of the paper is organized as follows. The next section introduces 4-dimensional resultant polytopes. Sect. 3 focuses on three 2d-triangles with non-parallel edges, which maximizes the number of faces, and upper bounds the number of facets and ridges in $N(\mathcal{R})$ by combinatorial arguments. Sect. 4 classifies all 4-dimensional resultant polytopes, and proves we can ignore parallel edges when maximizing the number of faces. We conclude with open questions and generalizations.

2. RESULTANT POLYTOPES

We start with previous results from [9, 4]. The main tool for computing sparse resultants are the *regular mixed subdivisions* of the convex hull of Minkowski sum $P = \sum_i A_i$. By abuse of notation, we may also refer to this sum as $\sum_i P_i$, where P_i denotes the convex hull of A_i and it is understood that the information from the A_i's is preserved. A *subdivision* of P is a collection of subsets of P, the cells of the subdivision, such that the union of the cells' convex hulls equals the convex hull of P and every pair of convex hulls of cells intersect at a common face. *Maximal* cells are those with dimension equal to the dimension of the subdivision. *Fine (or tight)* are those whose dimension equals the sum of its summands' dimensions. A subdivision is *regular* if it can be obtained as the projection of the lower hull of the Minkowski sum $\sum_i \widehat{A}_i$ of the lifted point sets $\widehat{A}_i$, for some lifting to $\mathbb{R}^{n+1}$. A subdivision is *mixed* when its cells have (unique) expressions as Minkowski sums of convex hulls of point subsets in the A_i's. It is *fine (or tight)* if all its cells are fine. Maximal cells are *mixed* if the dimension of every summand, except possibly one, equals one. In the sequel, we focus only on *regular* subdivisions, thus we will omit the word regular in general.

Given a family $\mathcal{A}$, the associated Cayley configuration $\mathcal{C}$ is the lattice configuration in $\mathbb{Z}^{n+1} \times \mathbb{Z}^n = \mathbb{Z}^{2n+1}$ defined by

$$\{e_0\} \times A_0 \cup \cdots \cup \{e_n\} \times A_n,$$

where $e_0, \ldots, e_n$ denotes the canonical basis in $\mathbb{Z}^{n+1}$. We denote by Q its convex hull. Regular fine mixed subdivisions of P are in bijection with regular triangulations of Q. Indeed, there is a bijection of maximal cells given as follows: any maximal cell (simplex) σ_T in a given regular triangulation $T = T_w$ of Q (with vertices in $\mathcal{C}$) has $2n$ vertices; the corresponding maximal cell in the associated regular fine subdivision $S = S_w$ of P has vertices of the form $\alpha_0 + \cdots + \alpha_n$, with (e_i, α_i) a vertex of σ_T. Note that for σ_T to be of maximal dimension, at least one of its vertices lies in $e_i \times A_i$, for all i. For more details about the translation between regular subdivisions of Q and regular mixed subdivisions of P, see [2, Sect. 9.2].

Let C be the $(2n+1) \times m$ associated Cayley matrix, i.e., the matrix whose columns are the points in the Cayley configuration $\mathcal{C}$. The inner product of any point in $N(\mathcal{R})$ with any vector in the rowspan of the Cayley matrix C is constant, and so $N(\mathcal{R})$ lies in a parallel translate to the nullspace of C. This explains why $\dim(N(\mathcal{R})) = m - 2n - 1$.

The faces (resp. vertices) of $N(\mathcal{R})$ can be obtained, by a many-to-one mapping, from the set of all regular (resp. fine) mixed subdivisions of P [9]. Given a mixed subdivision of P, every cell σ defines a subsystem of the $f_i|_\sigma$, where each polynomial is a restriction of f_i on the face of A_i appearing as a summand in σ. If the subdivision is the projection of the lower hull under a lifting w, then the face of $N(\mathcal{R})$ whose outer normal is w is

$$\prod_\sigma \mathcal{R}(f_0|_\sigma, \ldots, f_n|_\sigma)^{d_\sigma}, \qquad (1)$$

where $d_\sigma \in \mathbb{N}$ is specified in [9, Thm 4.1]. We shall call σ *essential* if the corresponding $f_i|_\sigma$ define an essential subsystem. Hence, all faces of $N(\mathcal{R})$ are Minkowski sums of lower-dimensional resultant polytopes, corresponding to essential subsystems. These lower-dimensional resultant polytopes correspond to subsets of the cells of the subdivision defining the face of $N(\mathcal{R})$. In particular, resultant vertices are obtained when all resultants in (1) are monomials, hence all σ are mixed; then d_σ is the normalized volume of σ.

We call *flip* the transformation of a fine mixed subdivision of P to another fine mixed subdivision of P. Following [9], if these subdivisions correspond to different vertices of $N(\mathcal{R})$ we call this flip *cubical*. In other words, a cubical flip corresponds to a resultant edge. The above discussion yields the following which, will be our basic tool for counting the faces of $N(\mathcal{R})$.

PROPOSITION 2. *[9, Thm.4.1] A mixed subdivision S of P corresponds to a face of $N(\mathcal{R})$, which is the Minkowski sum of the resultant polytopes of the cells of S.*

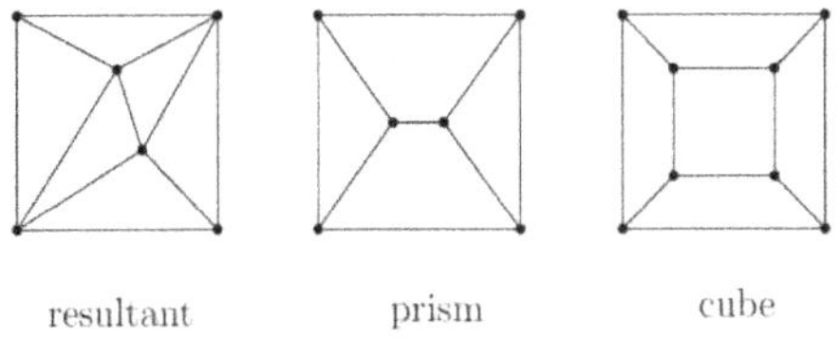

Figure 1: Graph skeleta of 3d facets of 4d resultant polytopes of 3 triangles with non-parallel edges.

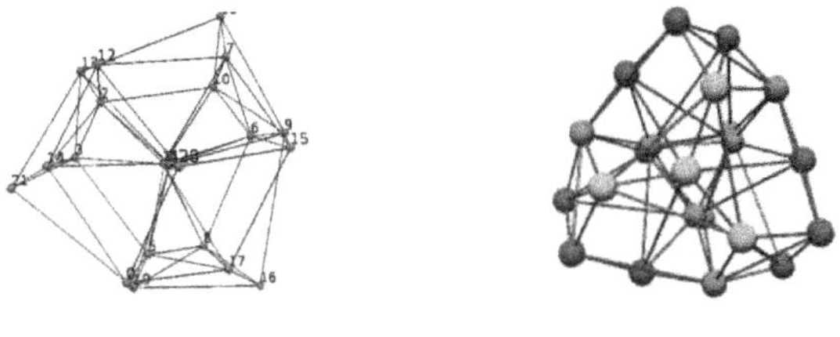

Figure 2: Vertex graph and facet graph (courtesy of M. Joswig) of the resultant polytope in Example 2.

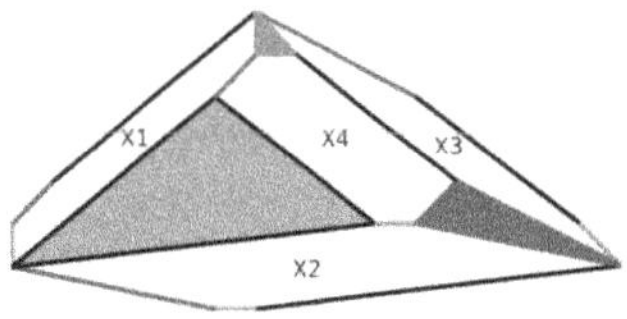

Figure 3: A non-regular subdivision of P of Example 2. There are 4 hexagons X1, X2, X3, X4.

2.1 4-dimensional resultant polytopes.

In this case, $m = 2n + 5$, where $m = \sum_{i=0}^{n} |A_i|$, and $|A_i| \geq 2$. So, there are only 3 cases, up to reordering:

(i) All $|A_i| = 2$, except for one with cardinality 5.

(ii) All $|A_i| = 2$, except for two with cardinalities 3 and 4.

(iii) All $|A_i| = 2$, except for three with cardinality 3.

Cases (i) and (ii) are similar to the study of 3d-resultant polytopes in [9], cf Thm 17. So, we concentrate on the new case (iii) and, more precisely, on the main case $n = 2$ and each $|A_i| = 3$, which we term the case $(3, 3, 3)$. This is done without loss of generality, by the following:

THEOREM 3. *[9, Thm.6.2] Every resultant polytope of an essential family is affinely isomorphic to a resultant polytope of an essential family $(A_0, \ldots, A_n)$ with $|A_i| \geq 3$, for all $i = 0, \ldots, n$.*

The proof is an algorithm to produce this reduction: up to an affine change of variables and reordering, we can assume that $A_i = \{0, \nu_i e_{i+1}\}, i = 0, \ldots, n-3$, so we can solve (with rational powers) the first $n-2$ variables and replace them in the last 3 polynomials. Then, $N(\mathcal{R})$ has the same combinatorial type as an essential $(3, 3, 3)$ configuration, where we could have repeated points or parallel edges, even if they were not present in A_{n-2}, A_{n-1} and A_n (and some coefficients could be equal to the sum of Laurent monomials in the original coefficients).

Let us focus on $N(\mathcal{R})$ for an essential $(3, 3, 3)$ configuration $\mathcal{A} = (A_0, A_1, A_2)$, and consider $A_2' \subset A_2$.

LEMMA 4. *[9] When there are no parallel edges in the subfamily (A_0, A_1, A_2'), $|A_2'| = 2$, their corresponding resultant polytope gives a facet of $N(\mathcal{R})$, which is combinatorially equivalent to the polytope of Fig 1 (resultant).*

3. THE CASE $(3, 3, 3)$

In this section, we assume that we have an essential family with $n = 3, m = 9$ and each A_i has cardinality 3. Our study shows the richness of possible polytopes, in contrast to the case of resultant polytopes with dimension ≤ 3.

We write the f-vectors as (f_0, f_1, f_2, f_3), omitting the 1 corresponding to the unique 4-face. We define the *minimum* and *maximum* f-vector to be the one with minimum and maximum *number of facets*, i.e. with minimum or maximum value of f_3. The minimum f-vector of the $(3, 3, 3)$ case when all P_i have dimension 2 but we admit parallel edges, is $(6, 15, 18, 9)$ and is attained by the following example.

EXAMPLE 1 (BIRKHOFF POLYTOPE). *Let $A_0 = A_1 = A_2 = \{(0, 0), (1, 0), (0, 1)\}$. Then $N(\mathcal{R})$ is the 4-dimensional Birkhoff polytope [10] which has f-vector (6,15,18,9).*

A complete list of f-vectors when $A_0 = \{(0, 0), (0, 1), (1, 0)\}$ and A_1, A_2 take all the possible values from the set $\{(0, 0), a_1, a_2\}$ where $a_1, a_2 \in \{(j, k) \mid j, k \in \mathbb{N} \wedge j, k \leq 5\}$ is presented in Table 1. There is a unique f-vector, $(22, 66,66, 22)$, which is maximal, and corresponds to more than one input family of supports. Highlighted f-vectors correspond to triangles that share no parallel edges between any of them. The computations have been performed using `respol` and last several days.

(6, 15, 18, 9)	(13, 37, 37, 13)	(16, 43, 40, 13)	(18, 52, 50, 16)
(8, 20, 21, 9)	(14, 35, 32, 11)	(16, 43, 41, 14)	(18, 52, 51, 17)
(9, 22, 21, 8)	(14, 36, 33, 11)	(16, 44, 41, 13)	(18, 53, 51, 16)
(9, 24, 25, 10)	(14, 36, 34, 12)	(16, 44, 42, 14)	(18, 53, 53, 18)
(10, 24, 23, 9)	(14, 37, 34, 11)	(16, 45, 43, 14)	(18, 54, 54, 18)
(10, 25, 24, 9)	(14, 37, 35, 12)	(16, 45, 44, 15)	(19, 54, 52, 17)
(10, 25, 25, 10)	(14, 37, 36, 13)	(16, 46, 45, 15)	(19, 55, 51, 15)
(10, 26, 25, 9)	(14, 38, 36, 12)	(16, 46, 46, 16)	(19, 55, 52, 16)
(11, 28, 27, 10)	(14, 38, 37, 13)	(17, 46, 43, 14)	(19, 55, 54, 18)
(11, 29, 28, 10)	(14, 38, 38, 14)	(17, 47, 43, 13)	(19, 56, 54, 17)
(11, 29, 29, 11)	(14, 40, 40, 14)	(17, 47, 44, 14)	(19, 56, 56, 19)
(12, 29, 26, 9)	(15, 39, 36, 12)	(17, 47, 45, 15)	(19, 57, 57, 19)
(12, 30, 27, 9)	(15, 40, 36, 11)	(17, 48, 45, 14)	(20, 58, 54, 16)
(12, 30, 28, 10)	(15, 40, 37, 12)	(17, 48, 46, 15)	(20, 59, 57, 18)
(12, 32, 31, 11)	(15, 40, 38, 13)	(17, 48, 47, 16)	(20, 60, 60, 20)
(12, 33, 33, 12)	(15, 41, 39, 13)	(17, 49, 47, 15)	(21, 62, 60, 19)
(13, 32, 29, 10)	(15, 41, 40, 14)	(17, 49, 48, 16)	(21, 63, 63, 21)
(13, 33, 30, 10)	(15, 42, 41, 14)	(17, 49, 49, 17)	(22, 66, 66, 22)
(13, 33, 31, 11)	(15, 42, 42, 15)	(17, 50, 50, 17)	
(13, 34, 32, 11)	(16, 42, 39, 13)	(18, 51, 48, 15)	
(13, 34, 33, 12)	(16, 43, 39, 12)	(18, 51, 49, 16)	

Table 1: The largest f-vectors of 4d $N(\mathcal{R})$ computed: 9 highlighted f-vectors correspond to triangles without parallel edges.

A particular extremal case follows:

EXAMPLE 2. *Let $A_0 = \{(0, 0), (1, 0), (0, 1)\}$, $A_1 = \{(0, 0), (5, 4), (9, 1)\}$, $A_2 = \{(5, 0), (0, 1), (1, 2)\}$. Then, $N(\mathcal{R})$ has f-vector (22,66,66,22); the vertex and facet graphs are in Fig. 2. A non-regular subdivision of $P = \sum_i P_i$ is depicted in Fig. 3 (see also Rem. 1).*

On the other hand, the following example concerns the case of three 1-dimensional configurations, excluded from the above list.

EXAMPLE 3. *Let $A_0 = \{(0, 0), (0, 1), (0, 2)\}$, $A_1 = \{(0, 0), (1, 0), (2, 0)\}$, $A_2 = \{(0, 0), (1, 1), (2, 2)\}$. Then, $N(\mathcal{R})$ has f-vector (20,57,51,14).*

If we replace the configuration (A_0, A_1, A_2) in Example 3 by the configuration $(A_0, A_1, \{(0,0),(1,1),(2,3)\})$, where two parallelisms are broken without introducing any new one, the f-vector becomes $(20, 58, 54, 16)$. Note that we get higher values for the different f_i.

3.1 Non-parallel edges

All along this section, we consider essential $(3,3,3)$ families with $\dim(P_i) = 2$, for all i, where moreover no edges coming from two different P_i are parallel. Even in this case, there are several different relevant polytopes (cf highlighted f-vectors in Table 1).

We describe the subsets $A_i' \subset A_i$, which form *subsystems*, that define cells of a subdivision S of P, and their connection to the faces of $N(\mathcal{R})$. The simpler non-trivial subsystem is 2-element subsets A_i' in each A_i (namely edges). Such a subsystem is essential when no two of the convex hulls of the A_i' are parallel. In this case, the cell is a Minkowski sum of 3 edges from the different A_i's which we call a *hexagon*. Every hexagon can be refined in two possible ways to a regular mixed decomposition and corresponds to an edge of $N(\mathcal{R})$ (see the cubical flips discussed above). A *heptagon* cell is a Minkowski sum of an A_i, w.l.o.g. A_0, and 2 edges A_1', A_2' from A_1, A_2 resp., which form an essential subfamily $\mathcal{A}' = (A_0, A_1', A_2')$. Every heptagon has up to 3 mixed refinements, each of which has a hexagon cell which is the sum of A_1', A_2' and one edge of A_0, not parallel to any of them. In this general case, a heptagon corresponds to a triangular 2-face of $N(\mathcal{R})$. Two hexagons in S might give rise to the Minkowski sum of two $N(\mathcal{R})$ edges which forms a parallelogram 2-face of $N(\mathcal{R})$, again in case of non-parallelism. An *octagon* cell is a Minkowski sum of two A_i, w.l.o.g. A_0, A_1, and an edge A_2' of A_2. An octagon corresponds to a 3d resultant polytope, thus to a facet of $N(\mathcal{R})$. We call these facets *resultant* facets (Fig. 1). A heptagon and a hexagon in S correspond to a facet which equals the Minkowski sum of a segment and a triangle, i.e., the sum of a 1d and 2d resultant polytopes. In generic position, this is a *prism* facet (Fig. 1). Finally, 3 hexagons in S correspond to a facet which is the Minkowski sum of 3 segments. In generic position this is a *cube* facet (Fig. 1).

The number of facets. We describe resultant polytopes corresponding to maximum facet cardinality. We start with several lemmata that serve as tools later. Assume we have a hexagon $X = s_0 + s_1 + s_2$ where $s_i \subset A_i$ of cardinality 2, and a heptagon $H = A_0 + s_1' + s_2'$ where $s_i' \subset A_i$ is of cardinality 2 for $i \in \{1, 2\}$, with the corresponding support sets being essential.

LEMMA 5. *A heptagon and a hexagon in a subdivision have exactly one common edge from A_i for some $i \in \{0, 1, 2\}$.*

PROOF. Observe that X, H always share the edge $s_0 \subset A_0$. Then if they have one more common edge, this should be from A_1 or A_2. W.l.o.g. it is an edge from A_1 and we call it s_1. Then we can construct a subdivision of $A_0 + s_1 + A_2$ that contains X, H. This yields a prism as 3d resultant polytope, which cannot exist. □

LEMMA 6. *There are no two subdivisions which contain the same hexagon and heptagon.*

PROOF. We will show that by fixing the common edge $s_0 \subset A_0$, there is a unique way to construct a regular mixed subdivision with X and H. Assume there are two such subdivisions: one with X, H and one with X^*, H^*, which both share the segment s_0. We can always subdivide H in a way s.t. it has the hexagon $s_0 + s_1' + s_2'$ as a cell. Similarly, H^* has $s_0 + s_1^* + s_2^*$ as a cell, where $s_i^*, s_i^* \subset A_i$, for $i \in \{1, 2\}$. This implies that $s_0 + A_1 + A_2$ can have two different pairs of hexagons, namely $s_0 + s_1' + s_2'$ and $s_0 + s_1^* + s_2^*$. Thus, the 3d resultant polytope of $s_0 + A_1 + A_2$ has two different parallelogram facets, which is impossible. □

COROLLARY 7. *There is no subdivision with an octagon and a hexagon cell or with two heptagon cells.*

We now introduce a technical tool. Let G be the dual graph of a regular subdivision S (not necessarily tight); G is planar. The cells, edges and vertices of S correspond, resp., to nodes, edges and cells of G. The graph G also contains a special node, which can be placed at infinity, corresponding to the complement of S in the plane. The cells and edges incident to this node are called unbounded. There are 9 unbounded edges incident to this node. The unbounded edges may form multi-edges if at least 2 of them are incident to the same non-infinity node. An example is in Fig. 4, corresponding to the non-regular subdivision S of Fig. 3.

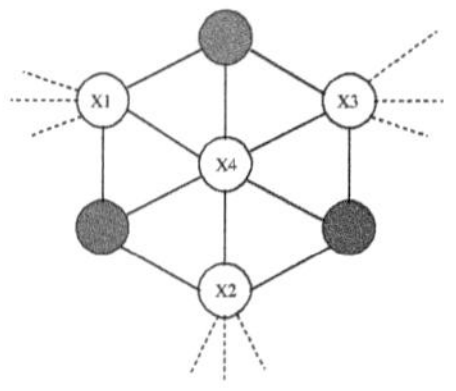

Figure 4: The graph G dual to S of Fig. 3, with labels on the 4 nodes dual to hexagons. Nodes dual to X1, X2, X3 and the triangles (colored) define the perimeter. Unbounded edges are dashed.

LEMMA 8. *The dual graph G has 9 unbounded cells, each defined by two unbounded edges and up to 2 bounded edges.*

PROOF. G has as many unbounded cells as unbounded edges, which are as many as the boundary edges of P. Each unbounded cell is defined by two unbounded edges and, possibly, some bounded edges. There cannot be ≥ 3 bounded edges in an unbounded cell because this would imply having ≥ 2 cells in S with no edge on the boundary ∂P. These would have to be triangles sharing an edge, which is impossible if S is obtained by a sufficiently generic lifting. □

Unbounded cells with no bounded edges are defined by an unbounded double-edge between the dual of a hexagon cell and the node at infinity. For unbounded cells with 2 bounded edges, the common node between the bounded edges is incident to no unbounded edge hence it is dual to a triangle that intersects ∂P at a vertex; see e.g., the 3 nodes in Fig. 4 dual to triangles in S.

Given 3 triangles, no subdivision can have more than 4 hexagons, but the following can occur.

LEMMA 9. *There exists an essential $(3,3,3)$ family of 2-dimensional triangles with no parallel edges, which has a unique subdivision with 4 hexagons. This coarse subdivision is not regular, but it gets refined to regular subdivisions defining different faces of $N(\mathcal{R})$.*

PROOF. Existence follows from Example 2 and the subdivision of Fig. 3, with dual graph in Fig. 4. The 4 segments corresponding to the Newton polytopes of the 4 associated resultants of 3 binomials in dimension 2 are linearly independent, and thus cannot define any boundary face of $N(\mathcal{R})$. We detail in Remark 1 its refinements giving rise to faces of $N(\mathcal{R})$. □

LEMMA 10. *For any* (A_0, A_1, A_2) *the number of triplets of hexagons* X_1, X_2, X_3 *over all subdivisions of* $\sum_{i=0}^{2} A_i$ *is:*

1. *at most 1, if any two* X_i*'s have empty intersection,*
2. *0, if more than two pairs of* X_i*'s have a non-empty intersection.*
3. *0, if exactly two* X_i*'s have a non-empty intersection,*
4. *at most 3, if exactly two pairs of* X_i*'s have a non-empty intersection,*

PROOF. The intersection of 2 X_i's is empty or a segment.

Case 1. The hexagons must intersect the boundary: observe that the hexagons use all the edges of P_i thus their convex hull should be P. Then, there is only one way to put 3 hexagons to intersect the boundary. There is one way to color the edges of the graph in Fig. 4 such that the edges of the hexagons alternate and the edges of the triangles have the same color.

Case 2. If all 3 pairs of X_1, X_2, X_3 have non-empty intersection, their convex hull is a 9-gon that contain the 3 hexagons and a triangle in the middle, i.e. it intersects all hexagons but not the boundary of the 9-gon. The dual graph of this subdivision is a 4-clique. Assume that A_i have colors, black, blue, red. Observe that we cannot color the edges of the 4-clique in a way s.t. the edges of the triangle are colored with the same color and no two bounded edges of each hexagons are colored with the same color. The claim follows from the fact that a hexagon is a sum of 3 edges of different colors. Case 3 holds with similar, more complicated arguments.

Case 4. Let $X_1 = \{s_0, s_1, s_2\}$, $X_2 = \{s'_0, s_1, s'_2\}$, $X_3 = \{s''_0, s'_1, s_2\}$ for segments $s_i, s'_i, s''_i \subset A_i$. The first property is that there is no other triplet of hexagons sharing s_1, s_2, otherwise $A_0 + s_1 + A_2$ would have another pair of hexagons different than X_1, X_2 which implies that a 3d $N(\mathcal{R})$ has 2 parallelogram facets, which is impossible. Thus, if there exists another triplet of hexagons that uses all segments of A_0, it shares $s'_1 \neq s_1$ and $s'_2 \neq s_2$. The second property is that the hexagons use all the segments of the triangle P_0, hence P_0 should be placed in the middle of them. Thus, each hexagon can form a heptagon with A_0, namely $A_0 + s_1 + s_2$, $A_0 + s_1 + s'_2$, $A_0 + s'_1 + s_2$. Each of these 3 heptagons corresponds to a prism facet. By the first property above we observe that all prism facets produced by different triplets of hexagons are all different. There are at most 9 prism facets by Lem.13, so there are at most 3 cube facets. □

Note that the subdivision of Fig. 3 has 4 triplets of hexagons and thus satisfies the statements of Lem. 10 at their extreme.

We are now ready to bound the maximum number of facets.

LEMMA 11. *If the 3 triangles* A_i *share no parallel edges, then the only possible facets are the ones depicted in Fig 1.*

PROOF. The uniqueness of the type of the resultant facet comes from Lem. 4. Combining this with Cor. 7 we observe that we cannot have a resultant facet as a Minkowski sum that has a 3-dimensional summand or two 2-dimensional summands. The other possibilities are the Minkowski sum of a segment and a triangle (i.e. a prism facet), or three affinely independent segments (i.e. a cube facet), or four affinely dependent segments. The latter cannot occur since the only subdivision that corresponds to a Minkowski sum of 4 segments gives raise to affinely independent segments in the non parallel case. □

We start by bounding the number of *resultant facets*, they are as the one in Fig. 1(resultant).

LEMMA 12. *There can be at most 9 resultant facets in the* $(3,3,3)$ *case, and this is tight.*

PROOF. The resultant facet is a 3d resultant polytope, corresponding to a subsystem with no parallel edges and support cardinalities $(3,3,2)$. This subsystem, comprised of two triangles and an edge, defines a Minkowski sum equal to an octagon. Consider a coarse mixed subdivision S of A_0, A_1, A_2, containing this octagon as a cell. All the other cells of S correspond to non-essential subsystems, hence their resultant is a monomial. There are 9 different subsystems with support cardinalities $3, 3, 2$, because there are 3 ways to choose the A_i contributing an edge, and 3 ways to specify this edge. This bound is tight because it is achieved in Example 2. □

The facet in Fig. 1 (resultant) contains 6 vertices, 11 edges, and 7 ridges: 6 triangular and one parallelogram.

A prism facet is the Minkowski sum of a triangular ridge T and an edge E of $N(\mathcal{R})$, see Fig. 1(prism), where T, E are resultant polytopes of subsystems with cardinalities $(3,2,2)$ and $(2,2,2)$ resp. This type of facet has 6 vertices. The ridges are two translates of T, and 3 Minkowski sums of E with every edge of T. Each prism facet has 9 edges: 3 translates of E, and two translates of each edge of T. The subdivision of P which corresponds to a prism facet should contain a hexagon X and a heptagon H, where X corresponds to E and H to T.

LEMMA 13. *There can be at most 9 prism facets in the* $(3,3,3)$ *case, and this is tight.*

PROOF. Consider a hexagon $X = s_0 + s_1 + s_2$, where $s_i \in A_i$, and a heptagon H, which is the Minkowski sum of A_0 and segments s'_1, s'_2, where $s'_i \subset A_i$, for $i \in \{1,2\}$. By Lem. 5, X and H should have a common edge from A_0. By Lem. 6, if we fix $s_0 \subset A_0$, there is a unique way to construct a regular mixed subdivision with X and H. Hence, there are at most 9 such subdivisions, which shows there are at most 9 prism facets. This is tight because it is achieved in Example 2. □

The cube facet is a Minkowski sum of 3 $N(\mathcal{R})$ edges, each corresponding to a hexagon, see Fig. 1(cube). The corresponding (regular) mixed subdivision of P contains 3 hexagons, yielding 3 resultant edges whose Minkowski sum is the cube facet. Each cube facet contains 8 vertices, 12 edges, and 6 parallelogram ridges.

LEMMA 14. *There can be at most 4 cube facets in the* $(3,3,3)$ *case, and this is tight.*

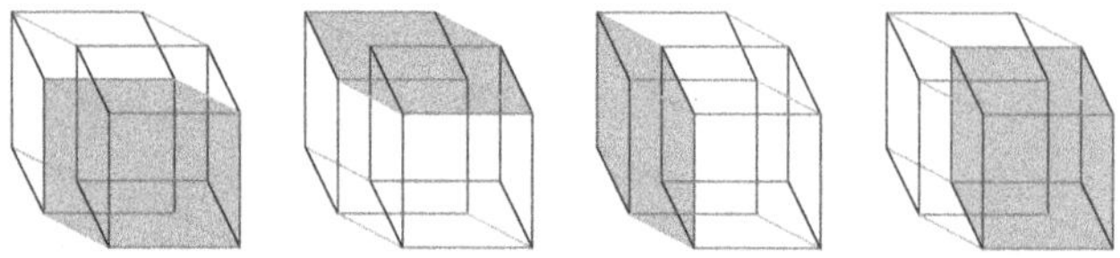

Figure 5: The complex of 4 cube facets: a Minkowski sum of 4 affinely independent segments, each associated to a hexagon in the subdivision; each subfigure highlights a cube.

PROOF. In order to have a cube facet we need 3 hexagons in a subdivision. By Lem. 10 the largest number of triplets of hexagons in all subdivisions of a fixed family (A_0, A_1, A_2) is 4. Thus, there are at most 4 cube facets. The maximum cardinality is achieved by our maximal instance in Ex. 2 where the 4 cube facets are constructed by selecting the 4 different triplets of hexagons from the subdivision of Fig. 3. □

REMARK 1. *The subdivision of Fig. 3, attained in our maximal instance, corresponds to a Minkowski sum of 4 affinely independent segments, each corresponds to a subdivision's hexagon. As we remarked, this Minkowski sum is 4-dimensional and thus the subdivision could not be regular because it neither corresponds to the whole 4-dimensional polytope nor to any of its faces. We indicate now the topology of the cube facets correspond to refinements of this subdivision. Let $(a, b, c, d) \in \{0, 1\}^4$ stand for the two possible flips in the 4 hexagons. There are 16 fine subdivisions of S: those which are regular correspond to resultant vertices. Let us denote, w.l.o.g., by*

$$(0bcd), (a0cd), (ab0d), (abc0) \subset \{0, 1\}^4,$$

the subsets of regular fine subdivisions defining cubical facets, each with cardinality 8. The flip graph corresponds to 4 cubical facets, each defined by all possible flips in 3 of the hexagons. Hence, each is a neighbor of the other 3, as shown in Fig. 5, with a parallelogram in common. The facet graph is a 4-clique. Overall, 15 fine subdivisions are involved, hence regular, while one fine subdivision, namely (1111), is non-regular and not contained in any of the 4 facets. Of course, each cube has another 3 parallelograms in common with non-cube facets.

3.2 The number of faces

We denote by $\tilde{f}_i$ the maximum number of faces of dimension i of any $(3, 3, 3)$ resultant polytope. It follows from Thm 19 that it is enough to bound the maximal number of faces in the generic case with no parallel edges, considered in Sect. 3.1.

We will make use of a powerful result extending Barnette's Lower bound to non-simplicial polytopes:

PROPOSITION 15. [7, thm.1.4] *For d-dimensional polytopes:*

$$f_1 + \sum_{i \geq 4} (i-3) f_2^i \geq d f_0 - \binom{d+1}{2},$$

where f_2^i is the number of 2-faces which are i-gons.

The following theorem summarizes our results on the maximum numbers $\tilde{f}_i$.

THEOREM 16. . *The maximal number of ridges of a $(3, 3, 3)$ resultant polytope is $\tilde{f}_2 = 66$ and the maximal number of facets is $\tilde{f}_3 = 22$. Moreover, $\tilde{f}_1 = \tilde{f}_0 + 44$, $22 \leq \tilde{f}_0 \leq 28$, and $66 \leq \tilde{f}_1 \leq 72$. The lower bounds are tight.*

PROOF. Assume that we have a non parallel $(3, 3, 3)$ configuration and let us relate f_2 and f_3. Let ϕ_1, ϕ_2, ϕ_3 be the number of resultant, prism and cube facets resp.; i.e. ϕ_i is the number of facets with i summands. By Lemma 11, the total number of facets is $f_3 = \phi_1 + \phi_2 + \phi_3$. We observe that there are only triangular and parallelogram ridges, whose cardinalities are at most 36 and 30, resp.:

$$\frac{1}{2}(6\phi_1 + 2\phi_2) = 3\phi_1 + \phi_2 \leq 36,$$

$$\frac{1}{2}(\phi_1 + 3\phi_2 + 6\phi_3) = \frac{1}{2}(\phi_1 + 3\phi_2) + 3\phi_3 \leq 30.$$

The total number of ridges is then

$$f_2 = \frac{1}{2}(7\phi_1 + 5\phi_2) + 3\phi_3 \leq 66. \tag{2}$$

Thus, $\tilde{f}_2 \leq 66$ and our maximal instance establishes the lower bound.

With respect to the number of facets, there are at most 9 resultant, 9 prism, and 4 cubical facets by Lemmata 12, 13, and 14. Thus $\tilde{f}_3 \leq 22$ and again, our maximal instance in Example 2 establishes the lower bound.

By Euler's equality, for any resultant polytope we have $f_0 + f_2 = f_1 + f_3 \leq \tilde{f}_1 + \tilde{f}_3$, therefore $\tilde{f}_0 + \tilde{f}_2 \leq \tilde{f}_1 + \tilde{f}_3$. By symmetry, we get $\tilde{f}_0 + \tilde{f}_2 = \tilde{f}_1 + \tilde{f}_3$. Then,

$$\tilde{f}_1 - \tilde{f}_0 = \tilde{f}_2 - \tilde{f}_3 = 44 \tag{3}$$

With respect to the two last inequalities in the statement, t he lower bounds are given by our maximal instance and by equality (3), it is enough to prove $\tilde{f}_0 \leq 28$. Again, assume we are in the non parallel case. In the resultant polytope with maximal number of facets, the 2-faces are either triangles or parallelograms and there are $\tilde{f}_2^4 = 30$ parallelograms. Prop. 15 becomes $\tilde{f}_1 + 30 \geq 4\tilde{f}_0 - 10$. Then,

$$\tilde{f}_1 + 40 = \tilde{f}_0 + 84 \geq 4\tilde{f}_0,$$

and the desired bound follows. □

4. CLASSIFICATION

Let us summarize the characterization of 4d resultant polytopes. We need to consider 3 special instances, corresponding to 3 possible cardinalities of supports in Sect. 2.1. As mentioned before, the cases $n = 0, 1$ are similar to those in [9], so we concentrate on $(3, 3, 3)$. We fix $n = 2$ and $m = 9 = 3+3+3$ and consider such families. The associated *mixed Grassmannian* $G(2; 3, 3, 3; \mathbb{Q})$, defined in [1], is the linear subvariety of the Grassmanian of 5-dimensional subspaces in $\mathbb{Q}^9$ which contain the vectors $e_1+e_2+e_3, e_4+e_5+e_6$, and $e_7 + e_8 + e_9$. Given a $(3, 3, 3)$ family $\mathcal{A}$, its associated Cayley matrix C represents (via its rowspan) an element in $G(2; 3, 3, 3; \mathbb{Q})$. All 5×9 matrices representing an element in $G(2; 3, 3, 3; \mathbb{Q})$ are affinely equivalent to an integer Cayley matrix of an integer $(3, 3, 3)$ family and have some structural vanishing minors. In the case of Cayley matrices of essential configurations, not too many minors can be 0, but there could be parallel vectors and repeated points. In Sturmfels' notation [9], the Newton polytope $\mathcal{N}_{12,111}$ corresponds to two univariate configurations of multisets of 3 points, but in the first, two of the points coincide: this is the square-based pyramid in Fig. 1(b). Thus, this is a degeneration of $\mathcal{N}_{111,111}$, which is the Newton polytope for two univariate configurations with 3 different points each, cf Fig. 1. Note that from the point of view of the Cayley matrix C, having a configuration with a repeated point is just an occurrence of the fact that some minors of C vanish, similarly to the existence of parallel edges in $\mathcal{A}$, which is the new feature that we have encountered in the study of 4-d resultant polytopes.

THEOREM 17. *Assume we have an essential family $\mathcal{A}$ of $n+1$ (finite) lattice point configurations in $\mathbb{R}^n$ with $N(\mathcal{R})$ of dimension 4. Then, up to reordering, we are in one of the situations (i), (ii) or (iii) in Sect. 2.1. These resultant polytopes are, resp., a degeneration of the following:*

1. *$n = 0$, $|A_0| = 5$, which is a 4-simplex with f-vector $(5, 10, 10, 5)$,*
2. *$n = 1$, $|A_0| = 3, |A_1| = 4$, which is a Sylvester case, with f-vector $(10, 26, 25, 9)$,*
3. *$n = 2$, $|A_0| = |A_1| = |A_2| = 3$, which are the polytopes described in Sect. 3.*

In particular, no resultant polytope of dimension 4 can have more than 22 facets and 66 ridges.

PROOF. By Thm 3, we restrict our attention to cases 1 to 3. We discuss case 3 because cases 1 and 2 are settled, resp., in [9] and [4, ch.12], cf also the 8th instance in Table 1.

We can perturb (with values in $\mathbb{Q}$), e.g. a point $p \in A_0$ to a nearby rational point $p*$. We get a perturbed matrix $C'_{\mathbb{Q}} \in G(2; 3, 3, 3; \mathbb{Q})$ of the Cayley matrix C. The resultant is an affine invariant of a configuration or its Cayley matrix, so we can left multiply $C'_{\mathbb{Q}}$ by an invertible matrix M in block form with a 3×3 identity matrix in the upper left corner and an integer 2×2 integer matrix M' with non-zero determinant in the lower right corner, to get an integer matrix $C' = MC'_{\mathbb{Q}}$, which corresponds to the same point in the mixed Grassmanian. Then, C' is the Cayley matrix of an essential integer family $\mathcal{A}'$. We can say that $\mathcal{A}$ is then a degeneration of this new integer family $\mathcal{A}'$, which is the image of the family $\mathcal{A}'_{\mathbb{Q}} = (A_0 - \{p\} \cup \{p*\}, A_1, A_2)$ by M'.

Given a regular mixed subdivision S of $\mathcal{A}$ associated to a generic lifting vector w (i.e., w is generic among the vectors that produce the same regular subdivision), we consider the regular subdivision $S'_{\mathbb{Q}}$ that w induces on the perturbed configuration $\mathcal{A}_{\mathbb{Q}}$. We then translate $S'_{\mathbb{Q}}$ via multiplication by M' to a combinatorially equivalent regular subdivision S' of $\mathcal{A}'$. It follows from Thm 19 that the number of facets of $N(\mathcal{R})$ cannot exceed the number of facets of $N(\mathcal{R}')$, and we conclude by Thm 16. □

EXAMPLE 4. *Let us consider degeneracy when $n = 1$, i.e. points are repeated: $A_0 = \{0, 1\}, A_1 = \{0, 1, 1, 2\}$. We perturb A_1 and get $A_1^* = \{0, 1, 101/100, 2\}$. We dilate by 100 (multiply a row of the Cayley matrix) and get $B_0 = \{0, 100\}, B_1 = \{0, 100, 101, 200\}$, which span $\mathbb{Z}$. The resultant polytopes for $\mathcal{A}$, $\mathcal{B}$ are combinatorially equivalent, although the former resultant has total degree $2 + 1 = 3$, and the latter $200 + 100 = 300$.*

4.1 Input genericity maximizes complexity

Given a polytope $Q \subset \mathbb{R}^3$ and a direction $u \in \mathbb{R}^3$, its lower hull along u, denoted $\mathrm{LH}_u Q$, is the union of all facets whose outer normal has negative or zero inner product with u. In the case of zero inner product, the facet is called degenerate and its projection is not a maximal cell. We assume that the triangles $A_i = \{p_{ij}, j = 0, 1, 2\}$, $i = 0, 1, 2$, have 2d convex hulls P_i. Let S be a regular subdivision of P, and $\widehat{A}_i, \widehat{P}$ be the lifted Newton polytopes and their Minkowski sum; $\mathrm{LH}_{e_3}\widehat{P}$, where e_3 is the unit vector on the x_3-axis, is in bijection with S. Consider edges E_0, E_1 with the same outer normal v:

$$E_0 = (p_{00}, p_{01}) \subset P_0,\; E_1 = (p_{10}, p_{11}) \subset P_1.$$

For some vertex $p_{2k} \in A_2$, $E_1 + E_2 + p_{2k}$ is an edge of P with outer normal v. Thus, their lifting $\widehat{E}_1 + \widehat{E}_2 + \widehat{p}_{2k}$ has outer normal $(v, 0)$ and yields one or two facets of $\widehat{P}$, yielding one or two degenerate facets on $\mathrm{LH}_{e_3}\widehat{P}$, i.e. segments, depending on whether the lifting leads, resp., to a coarse or fine subdivision. In the latter case, the two segments are collinear but their union has been subdivided into one of two possible mixed subdivisions, each with two cells. W.l.o.g., these are:

$$\{p_{00}+E_1+p_{2k}, E_0+p_{11}+p_{2k}\}, \{E_0+p_{10}+p_{2k}, p_{01}+E_1+p_{2k}\} \tag{4}$$

We consider a perturbation in the direction of v

$$p_{00}^* := p_{00} + \epsilon v, \tag{5}$$

with indeterminate $\epsilon \to 0^+$. Since we are considering a finite process that branches on signs of algebraic expressions, namely Cayley minors, ϵ can take sufficiently small positive rational values, as is the case in standard symbolic perturbation methods.

LEMMA 18. *With the above hypotheses and notation, let*

$$\mathcal{A}^* := (\{p_{00}^*, p_{01}, p_{02}\}, A_1, A_2),$$

and $P^* := A_0^* + A_1 + A_2$ be the family and Minkowski sum associated with a perturbation (5). Let S be a (regular fine) mixed subdivision of P associated with a generic weight vector w, and S^* the regular subdivision of P^* associated to the same vector w. Then, S^* is mixed and contains at most one more cell σ than does S. There is a bijection between all cells of S^* (except σ, if it exists) and the cells of S, which associates combinatorially equivalent cells.

We expect this lemma to extend to any dimension.

PROOF. Eq. (5) defines $p_{00}^* \in \mathbb{Q}^2$. As in the proof of Theorem 17, by an appropriate dilation we define a family of supports in $\mathbb{Z}^2$. By abuse of notation, we denote the latter by $\mathcal{A}^*$. To prove the lemma for any S, we consider two cases according to the subdivisions of $E_0 + E_1 + p_{2k}$ in (4). In the first case, $p_{00} + p_{11} + p_{2k}$ is not a vertex of P but $p_{00}^* + p_{11} + p_{2k}$ is a vertex of P^*: the perturbation has moved outward the middle point of $E_0 + E_1 + p_{2k}$. $\mathrm{LH}_{e_3}\widehat{P}$ is combinatorially equivalent to $\mathrm{LH}_{e_3+\epsilon v}\widehat{P}$, where the latter is defined by shifting our viewpoint by an infinitesimal amount: the two degenerate facets whose union is $\widehat{E}_0+\widehat{E}_1+\widehat{p}_{2k}$ appear in both lower hulls (the subdivision to two facets occurs because S is fine). The non-degenerate facets are clearly combinatorially equivalent in both lower hulls. Formally, non-degenerate facets on $\mathrm{LH}_{e_3}\widehat{P}$, i.e. with positive area, have outer normal $(w, -1)$ and we claim that

$$(w, -1) \cdot (e_3 + \epsilon(v, 0)) = -1 + \epsilon w \cdot v < 0,$$

for sufficiently small $\epsilon > 0$. Thus, these facets also lie on $\mathrm{LH}_{e_3+\epsilon v}\widehat{P}$. Non-degenerate facets of $\widehat{P}$ but not on $\mathrm{LH}_{e_3}\widehat{P}$ have outer normal $(w, 1)$ and we claim that

$$(w, 1) \cdot (e_3 + \epsilon(v, 0)) = 1 + \epsilon w \cdot v > 0,$$

for sufficiently small $\epsilon > 0$. So, these facets do not lie on $\mathrm{LH}_{e_3+\epsilon v}\widehat{P}$. We now show $\mathrm{LH}_{e_3}\widehat{P}^*$ is combinatorially equivalent to $\mathrm{LH}_{e_3+\epsilon v}\widehat{P}$. Any facet except the degenerate ones in $\mathrm{LH}_{e_3+\epsilon v}\widehat{P}$ clearly corresponds to a combinatorially equivalent facet in $\mathrm{LH}_{e_3}\widehat{P}^*$. The degenerate facets give rise to two edges in $\widehat{P}^*$, which proves that S^* is fine, hence a mixed subdivision; moreover, these edges are combinatorially equivalent to those on $\mathrm{LH}_{e_3+\epsilon v}\widehat{P}$. Thus the lemma is proved in the case no new cell is created.

In the second subdivision of $E_0 + E_1 + p_{2k}$, the middle point is $p_{01}+p_{10}+p_{2k}$; this point is perturbed to the relative interior of P^*. The perturbation creates an extra (mixed) cell $E_0^* + E_1 + p_{2k}$ which intersects ∂P^*. For all other cells in S^* the discussion for the above case holds. This settles the case a new cell is created. □

THEOREM 19. *For any family $\mathcal{A}$ whose triangles have one or more pairs of parallel edges, there exists a family of triangles $\mathcal{A}^*$ without any parallel edges as in section 3.1, whose resultant polytope $N(\mathcal{R}^*)$ has at least as many faces of any dimension as those in the polytope $N(\mathcal{R})$ of $\mathcal{A}$.*

PROOF. We first assume all P_i's have non-zero area. Given $\mathcal{A}$ with strongly parallel edges $E_0 \subset P_0$, $E_1 \subset P_1$, perturbation (5) defines $\mathcal{A}^*$, where the corresponding edges are not parallel. In the case of other strongly parallel edges, we apply the same procedure sufficiently many times. For every mixed subdivision S of $\mathcal{A}$ the same lifting defines a mixed subdivision S^*, as in Lem. 18. This shows that the vertices of $N(\mathcal{R})$ can be mapped in a 1-1 fashion to, possibly a subset of, vertices of $N(\mathcal{R}^*)$. Hence the number of vertices in $N(\mathcal{R}^*)$ is at least as large as that of $N(\mathcal{R})$.

To prove the statement for k-faces, $k \geq 1$, we extend Lem. 18 to an arbitrary (coarse) regular subdivision S and its perturbed counterpart S^*. The only difference is that S may contain a single 1d cell $E_0 + E_1 + p_{2k}$ and cells of the form $\sigma = E_0 + E_1 + F_2$, for a face $F_2 \subset P_2$. Each σ is subdivided to 3 or 2 cells in S^*, depending on whether a new cell is created or not. The subdivision follows one of the subdivisions of $E_0 + E_1 + p_{2k}$ discussed in the proof of Lem. 18. Now σ is not essential hence contributes a point summand to the $N(\mathcal{R})$ face corresponding to S. The $N(\mathcal{R}^*)$ face corresponding to S^* is an edge if σ^* is a hexagon, thus establishing the lemma for k-faces.

If parallel edges E_0, E_1 have antiparallel outer normals, no regular subdivision (even coarse) may contain a cell of the form $E_0 + E_1 + F_2$, though there may be adjacent cells $E_0 + p_{1j} + F_2, p_{0i} + E_1 + F_2$. Any infinitesimal perturbation, such as (5), yields S^* combinatorially equivalent to S.

When some P_i's have zero area, the result still holds in a similar way after a detailed study of each possible case (including repeated points), which we omit due to space restrictions. The key case is the following: $\mathcal{A}$ satisfies $|A_0| = |A_1| = |A_2| = 3$, $\dim P_0 = 1$, $\dim P_1 = \dim P_2 = 2$, then let $\mathcal{A}^* = (A_0^*, A_1, A_2)$ such that the middle point of A_0 is infinitesimally perturbed to yield $\dim P_0^* = 2$. Then there is an injection of regular subdivisions of $\mathcal{A}$ to those of $\mathcal{A}^*$, such that if S maps to S^* then S^* contains one more cell equal to $A_0^* + p_{1j} + p_{2k}$, for vertices $p_{1j} \in A_1, p_{2k} \in A_2$, and all other cells are combinatorially equivalent to the corresponding cells in S. □

5. OPEN PROBLEMS AND EXTENSIONS

OPEN PROBLEM 1. *Prove that either $f_0 \leq 22$ or $f_1 \leq 66$. That is, we conjecture that the maximum f-vector of a 4d-resultant polytope is $(22, 66, 66, 22)$.*

OPEN PROBLEM 2. *Is it true that, for maximal f-vectors, it holds $f_0 = f_3$? Is it always true that $f_1 \geq f_2$, if $f_0 \geq 10$?*

The proof of Thm 19 should extend to high dimensions. Lem. 20 generalizes Lem. 12 in any dimension and is proven analogously. It motivates us to raise Conj. 1.

LEMMA 20. *A d-dimensional resultant polytope has at most m resultant facets.*

CONJECTURE 1. *The number of vertices of a d-dimensional resultant polytope is bounded above by*

$$3 \cdot \sum_{\|S\|=d-1} \prod_{i \in S} \tilde{f}_0(i)$$

where S is any multiset with elements in $\{1, \ldots, d-1\}$, $\|S\| := \sum_{i \in S} i$, and $\tilde{f}_0(i)$ is the maximum number of vertices of a i-dimensional $N(\mathcal{R})$.

The only bound in terms of d is $(3d-3)^{2d^2}$ [9], yielding $\tilde{f}_0(5) \leq 12^{50}$ whereas our conjecture yields $\tilde{f}_0(5) \leq 231$.

Acknowledgments.

We thank J. de Loera, M. Joswig, F. Santos, and G. Ziegler for helpful discussions and anonymous reviewers for useful comments.

6. REFERENCES

[1] E. Cattani, M. A. Cueto, A. Dickenstein, S. Di Rocco, and B. Sturmfels. Mixed discriminants. *Math. Z.*, 2013. to appear; also in ArXiv 2011.

[2] J. de Loera, J.A. Rambau, and F. Santos. *Triangulations: Structures for Algorithms and Applications, vol. 25.* Springer-Verlag, 2010.

[3] I.Z. Emiris, V. Fisikopoulos, C. Konaxis, and L. Peñaranda. An output-sensitive algorithm for computing projections of resultant polytopes. In *Proc. ACM Symp. Comp. Geometry*, pages 179–188, 2012.

[4] I.M. Gelfand, M.M. Kapranov, and A.V. Zelevinsky. *Discriminants, Resultants and Multidimensional Determinants.* Birkhäuser, Boston, 1994.

[5] I.M. Gelfand, M.M. Kapranov, and A.V. Zelevinsky. Newton polytopes of the classical resultant and discriminant. *Advances in Math.*, 84:237–254, 1990.

[6] A. Jensen and J. Yu. Computing tropical resultants. *J. Algebra.*, 2013. to appear; also in ArXiv 2011.

[7] G. Kalai. Rigidity and the lower bound theorem 1. *Inventiones Mathematicae*, 88:125–151, 1987.

[8] J. Rambau. TOPCOM: Triangulations of point configurations and oriented matroids. In *Proc. Intern. Congress Math. Software*, pages 330–340, 2002.

[9] B. Sturmfels. On the Newton polytope of the resultant. *J. Algebr. Combin.*, 3:207–236, 1994.

[10] G.M. Ziegler. *Lectures on Polytopes.* Springer, 1995.

Simultaneous Computation of the Row and Column Rank Profiles*

Jean-Guillaume Dumas
Université de Grenoble
Laboratoire LJK
UMR CNRS, INRIA, UJF, UPMF, GINP
51, av. des Mathématiques,
F38041 Grenoble, France
Jean-Guillaume.Dumas@imag.fr

Clément Pernet
Université de Grenoble
INRIA, Laboratoire LIG
UMR CNRS, INRIA, UJF, UPMF, GINP
Inovallée, 655, av. de l'Europe,
F38334 St Ismier Cedex,
France
Clement.Pernet@imag.fr

Ziad Sultan
Université de Grenoble
Laboratoires LJK and LIG
UMR CNRS, INRIA, UJF, UPMF, GINP
Inovallée, 655, av. de l'Europe,
F38334 St Ismier Cedex,
France
Ziad.Sultan@imag.fr

ABSTRACT

Gaussian elimination with full pivoting generates a PLUQ matrix decomposition. Depending on the strategy used in the search for pivots, the permutation matrices can reveal some information about the row or the column rank profiles of the matrix. We propose a new pivoting strategy that makes it possible to recover at the same time both row and column rank profiles of the input matrix and of any of its leading sub-matrices. We propose a rank-sensitive and quad-recursive algorithm that computes the latter PLUQ triangular decomposition of an $m \times n$ matrix of rank r in $O\left(mnr^{\omega-2}\right)$ field operations, with ω the exponent of matrix multiplication. Compared to the LEU decomposition by Malashonock, sharing a similar recursive structure, its time complexity is rank sensitive and has a lower leading constant. Over a word size finite field, this algorithm also improves the practical efficiency of previously known implementations.

Categories and Subject Descriptors

G.4 [**Mathematics and Computing**]: Mathematical Software—*Algorithm Design and Analysis*; I.1.2 [**Computing Methodologies**]: Symbolic and Algebraic Manipulation

Keywords

FFLAS-FFPACK, Finite field, Gaussian elimination, Rank profile

1. INTRODUCTION

Triangular matrix decomposition is a fundamental building block in computational linear algebra. It is used to solve linear systems, compute the rank, the determinant, the nullspace or the row and column rank profiles of a matrix. The LU decomposition, defined for matrices whose leading principal minors are all nonsingular, can be generalized to arbitrary dimensions and ranks by introducing pivoting on sides, leading e.g. to the LQUP decomposition of [6] or the PLUQ decomposition [5, 8]. Many algorithmic variants exist allowing fraction free computations [8], in-place computations [2, 7] or sub-cubic rank-sensitive time complexity [11, 7]. More precisely, the pivoting strategy reflected by the permutation matrices P and Q is the key difference between these PLUQ decompositions. In numerical linear algebra [5], pivoting is used to ensure a good numerical stability, good data locality, and reduce the fill-in. In the context of exact linear algebra, the role of pivoting differs. Indeed, only certain pivoting strategies for these decompositions will reveal the rank profile of the matrix. The latter is crucial in many applications using exact Gaussian elimination, such as Gröbner basis computations [4] and computational number theory [10].

The *row rank profile* of an $m \times n$ matrix A with rank r is the lexicographically smallest sequence of r indices of linearly independent rows of A. Similarly the *column rank profile* is the lexicographically smallest sequence of r indices of linearly independent columns of A.

The common strategy to compute the row rank profile is to search for pivots in a row-major fashion: exploring the current row, then moving to the next row only if the current row is zero. Such a $\bar{P}L U \bar{Q}$ decomposition can be transformed into a CUP decomposition (where $P = \bar{Q}$ and $C = \bar{P}L$ is in column echelon form) and the first r values of the permutation associated to P are exactly the row rank profile [7]. A block recursive algorithm can be derived from this scheme by splitting the row dimension [6]. Similarly, the column rank profile can be obtained in a column major search: exploring the current column, and moving to the next column only if the current one is zero. The $\bar{P}LU\bar{Q}$ decomposition can be transformed into a PLE decomposition (where $P = \bar{P}$ and $E = U\bar{Q}$ is in row echelon form) and the first r values of Q are exactly the column rank profile [7]. The corresponding block recursive algorithm uses a splitting of the column dimension.

Recursive elimination algorithms splitting both row and column dimensions include the TURBO algorithm [3] and the LEU decomposition [9]. No connection is made to the computation of the rank profiles in any of these algorithms. The TURBO algorithm does not compute the lower triangular matrix L and performs five recursive calls. It therefore

*This work is partly funded by the HPAC project of the French Agence Nationale de la Recherche (ANR 11 BS02 013).

implies an arithmetic overhead compared to classic Gaussian elimination. The LEU avoids permutations but at the expense of many additional matrix products. As a consequence its time complexity is not rank-sensitive.

We propose here a pivoting strategy following a Z-curve structure and working on an incrementally growing leading sub-matrix. This strategy is first used in a recursive algorithm splitting both rows and columns which recovers simultaneously both row and column rank profiles. Moreover, the row and column rank profiles of *any leading sub-matrix* can be deduced from the P and Q permutations. We show that the arithmetic cost of this algorithm remains rank sensitive of the form $O(mnr^{\omega-2})$ where ω is the exponent of matrix multiplication. The best currently known upper bound for ω is 2.3727 [12]. To the best of our knowledge, this is the first reduction to matrix multiplication for the problem of computing the column and row rank profiles of all leading sub-matrices of an input matrix.

As for the CUP and PLE decompositions, this PLUQ decomposition can be computed in-place, at the same cost. Compared to the CUP and PLE decompositions, this new algorithm has the following new salient features:

- it computes *simultaneously* both rank profiles at the cost of one,
- it preserves the squareness of the matrix passed to the recursive calls, thus allowing more efficient use of the matrix multiplication building block,
- it uses less modular reductions in a finite field,
- a CUP and a PLE decompositions can be obtained from it, with row and column permutations only.

Compared to the LEU decomposition,

- it is in-place,
- its time complexity bound is rank sensitive and has a better leading constant,
- a LEU decomposition can be obtained from it, with row and column permutations.

In Section 2 we present the new block recursive algorithm. Section 3 shows the connection with the LEU decomposition and Section 4 states the main property about rank profiles. We then analyze the complexity of the new algorithm in terms of arithmetic operations: first we prove that it is rank sensitive in Section 5 and second we show in Section 6 that, over a finite field, it reduces the number of modular reductions when compared to state of the art techniques. We then propose an iterative variant in Section 7 to be used as a base-case to terminate the recursion before the dimensions get too small. Experiments comparing computation time and cache efficiency are presented in Section 8.

2. A RECURSIVE PLUQ ALGORITHM

We first recall the name of the main sub-routines being used: `MM` stands for matrix multiplication, `TRSM` for triangular system solving with matrix unknown (left and right variants are implicitly indicated by the parameter list), `PermC` for matrix column permutation, `PermR` for matrix row permutation, etc. For instance, we will use:

`MM`(C, A, B) to denote $C \leftarrow C - AB$,

`TRSM`(U, B) for $B \leftarrow U^{-1}B$ with U upper triangular,

`TRSM`(B, L) for $B \leftarrow BL^{-1}$ with L lower triangular.

We also denote by $T_{k,l}$ the transposition of indices k and l and by $L\backslash U$, the storage of the two triangular matrices L and U one above the other. Further details on these subroutines and notations can be found in [7]. In block decompositions, we allow for zero dimensions. By convention, the product of any $m \times 0$ matrix by an $0 \times n$ matrix is the $m \times n$ zero matrix. The notation $j = i : k$ being inclusive on the left only (i.e. $j = i : k$ means $j \in \mathbb{Z}$ and $i \leq j < k$).

We now present the block recursive Algorithm 1, computing a PLUQ decomposition.

It is based on a splitting of the matrix in four quadrants. A first recursive call is done on the upper left quadrant followed by a series of updates. Then two recursive calls can be made on the anti-diagonal quadrants if the first quadrant exposed some rank deficiency. After a last series of updates, a fourth recursive call is done on the bottom right quadrant. Figure 1 illustrates the position of the blocks computed in the course of Algorithm 1, before and after the final permutation with matrices S and T.

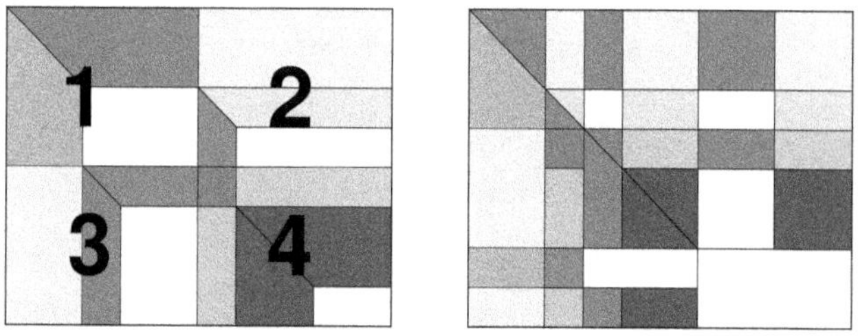

Figure 1: Block recursive Z-curve PLUQ decomposition and final block permutation.

This framework differs from the one in [3] by the order in which the quadrants are treated, leading to only four recursive calls in this case instead of five in [3]. We will show in Section 4 that this fact together with the special form of the block permutations S and T makes it possible to recover rank profile information. The correctness of Algorithm 1 is proven in Appendix A.

REMARK 1. *Algorithm 1 is in-place (as defined in [7, Definition 1]): all operations of the* `TRSM`*,* `MM`*,* `PermC`*,* `PermR` *subroutines work with $O(1)$ extra memory allocations except possibly in the course of fast matrix multiplications. The only constraint is for the computation of $J \leftarrow L_3^{-1}I$ which would overwrite the matrix I that should be kept for the final output. Hence a copy of I has to be stored for the computation of J. The matrix I has dimension $r_3 \times r_2$ and can be stored transposed in the zero block of the upper left quadrant (of dimension $(\frac{m}{2} - r_1) \times (\frac{n}{2} - r_1)$, as shown on Figure 1).*

Algorithm 1 PLUQ

Input: $A = (a_{ij})$ a $m \times n$ matrix over a field
Output: P, Q: $m \times m$ and $n \times n$ permutation matrices
Output: r: the rank of A
Output: $A \leftarrow \begin{bmatrix} L\backslash U & V \\ M & 0 \end{bmatrix}$ where L is $r \times r$ unit lower triangular, U is $r \times r$ upper triangular, and

$$A = P \begin{bmatrix} L \\ M \end{bmatrix} \begin{bmatrix} U & V \end{bmatrix} Q.$$

if m=1 **then**
  if $A = \begin{bmatrix} 0 & \dots & 0 \end{bmatrix}$ **then** $P \leftarrow \begin{bmatrix} 1 \end{bmatrix}, Q \leftarrow I_n, r \leftarrow 0$
  else
    $i \leftarrow$ the col. index of the first non zero elt. of A
    $P \leftarrow \begin{bmatrix} 1 \end{bmatrix}; Q \leftarrow T_{1,i}, r \leftarrow 1$
    Swap $a_{1,i}$ and $a_{1,1}$
  end if
  Return (P, Q, r, A)
end if
if n=1 **then**
  if $A = \begin{bmatrix} 0 & \dots & 0 \end{bmatrix}^T$ **then** $P \leftarrow I_m; Q \leftarrow \begin{bmatrix} 1 \end{bmatrix}, r \leftarrow 0$
  else
    $i \leftarrow$ the row index of the first non zero elt. of A
    $P \leftarrow \begin{bmatrix} 1 \end{bmatrix}, Q \leftarrow T_{1,i}, r \leftarrow 1$
    Swap $a_{i,1}$ and $a_{1,1}$
    for $j = i + 1 : m$ **do** $a_{j,1} \leftarrow a_{j,1} a_{1,1}^{-1}$
    end for
  end if
  Return (P, Q, r, A)
end if
  ▷ Splitting $A = \begin{bmatrix} A_1 & A_2 \\ A_3 & A_4 \end{bmatrix}$ where A_1 is $\lfloor \frac{m}{2} \rfloor \times \lfloor \frac{n}{2} \rfloor$.
Decompose $A_1 = P_1 \begin{bmatrix} L_1 \\ M_1 \end{bmatrix} \begin{bmatrix} U_1 & V_1 \end{bmatrix} Q_1$ ▷ $\texttt{PLUQ}(A_1)$
$\begin{bmatrix} B_1 \\ B_2 \end{bmatrix} \leftarrow P_1^T A_2$ ▷ $\texttt{PermR}(A_2, P_1^T)$
$\begin{bmatrix} C_1 & C_2 \end{bmatrix} \leftarrow A_3 Q_1^T$ ▷ $\texttt{PermC}(A_3, Q_1^T)$

Here $A = \left[\begin{array}{cc|c} L_1\backslash U_1 & V_1 & B_1 \\ M_1 & 0 & B_2 \\ \hline C_1 & C_2 & A_4 \end{array}\right].$

$D \leftarrow L_1^{-1} B_1$ ▷ $\texttt{TRSM}(L_1, B_1)$
$E \leftarrow C_1 U_1^{-1}$ ▷ $\texttt{TRSM}(C_1, U_1)$
$F \leftarrow B_2 - M_1 D$ ▷ $\texttt{MM}(B_2, M_1, D)$
$G \leftarrow C_2 - E V_1$ ▷ $\texttt{MM}(C_2, E, V_1)$
$H \leftarrow A_4 - E D$ ▷ $\texttt{MM}(A_4, E, D)$

Here $A = \left[\begin{array}{cc|c} L_1\backslash U_1 & V_1 & D \\ M_1 & 0 & F \\ \hline E & G & H \end{array}\right].$

Decompose $F = P_2 \begin{bmatrix} L_2 \\ M_2 \end{bmatrix} \begin{bmatrix} U_2 & V_2 \end{bmatrix} Q_2$ ▷ $\texttt{PLUQ}(F)$
Decompose $G = P_3 \begin{bmatrix} L_3 \\ M_3 \end{bmatrix} \begin{bmatrix} U_3 & V_3 \end{bmatrix} Q_3$ ▷ $\texttt{PLUQ}(G)$
$\begin{bmatrix} H_1 & H_2 \\ H_3 & H_4 \end{bmatrix} \leftarrow P_3^T H Q_2^T$ ▷ $\texttt{PermR}(H, P_3^T); \texttt{PermC}(H, Q_2^T)$
$\begin{bmatrix} E_1 \\ E_2 \end{bmatrix} \leftarrow P_3^T E$ ▷ $\texttt{PermR}(E, P_3^T)$
$\begin{bmatrix} M_{11} \\ M_{12} \end{bmatrix} \leftarrow P_2^T M_1$ ▷ $\texttt{PermR}(M_1, P_2^T)$
$\begin{bmatrix} D_1 & D_2 \end{bmatrix} \leftarrow D Q_2^T$ ▷ $\texttt{PermR}(D, Q_2^T)$
$\begin{bmatrix} V_{11} & V_{12} \end{bmatrix} \leftarrow V_1 Q_3^T$ ▷ $\texttt{PermR}(V_1, Q_3^T)$

Here $A = \left[\begin{array}{ccc|cc} L_1\backslash U_1 & V_{11} & V_{12} & D_1 & D_2 \\ M_{11} & 0 & 0 & L_2\backslash U_2 & V_2 \\ M_{12} & 0 & 0 & M_2 & 0 \\ \hline E_1 & L_3\backslash U_3 & V_3 & H_1 & H_2 \\ E_2 & M_3 & 0 & H_3 & H_4 \end{array}\right].$

$I \leftarrow H_1 U_2^{-1}$ ▷ $\texttt{TRSM}(H_1, U_2)$
$J \leftarrow L_3^{-1} I$ ▷ $\texttt{TRSM}(L_3, I)$
$K \leftarrow H_3 U_2^{-1}$ ▷ $\texttt{TRSM}(H_3, U_2)$
$N \leftarrow L_3^{-1} H_2$ ▷ $\texttt{TRSM}(L_3, H_2)$
$O \leftarrow N - J V_2$ ▷ $\texttt{MM}(N, J, V_2)$
$R \leftarrow H_4 - K V_2 - M_3 O$ ▷ $\texttt{MM}(H_4, K, V_2); \texttt{MM}(H_4, M_3, O)$
Decompose $R = P_4 \begin{bmatrix} L_4 \\ M_4 \end{bmatrix} \begin{bmatrix} U_4 & V_4 \end{bmatrix} Q_4$ ▷ $\texttt{PLUQ}(R)$
$\begin{bmatrix} E_{21} & M_{31} & 0 & K_1 \\ E_{22} & M_{32} & 0 & K_2 \end{bmatrix} \leftarrow P_4^T \begin{bmatrix} E_2 & M_3 & 0 & K \end{bmatrix}$ ▷ $\texttt{PermR}$
$\begin{bmatrix} D_{21} & D_{22} \\ V_{21} & V_{22} \\ 0 & 0 \\ O_1 & O_2 \end{bmatrix} \leftarrow \begin{bmatrix} D_2 \\ V_2 \\ 0 \\ O \end{bmatrix} Q_4^T$ ▷ $\texttt{PermC}$

Here $A = \left[\begin{array}{ccc|ccc} L_1\backslash U_1 & V_{11} & V_{12} & D_1 & D_{21} & D_{22} \\ M_{11} & 0 & 0 & L_2\backslash U_2 & V_{21} & V_{22} \\ M_{12} & 0 & 0 & M_2 & 0 & 0 \\ \hline E_1 & L_3\backslash U_3 & V_3 & I & O_1 & O_2 \\ E_{21} & M_{31} & 0 & K_1 & L_4\backslash U_4 & V_4 \\ E_{22} & M_{32} & 0 & K_2 & M_4 & 0 \end{array}\right].$

$$S \leftarrow \begin{bmatrix} I_{r_1+r_2} & & & \\ & & I_{k-r_1-r_2} & \\ & I_{r_3+r_4} & & \\ & & & I_{m-k-r_3-r_4} \end{bmatrix}$$

$$T \leftarrow \begin{bmatrix} I_{r_1} & & & & & \\ & & & I_{r_2} & & \\ & I_{r_3} & & & & \\ & & & & I_{r_4} & \\ & & I_{k-r_1-r_3} & & & \\ & & & & & I_{n-k-r_2-r_4} \end{bmatrix}$$

$P \leftarrow \mathrm{Diag}(P_1 \begin{bmatrix} I_{r_1} & \\ & P_2 \end{bmatrix}, P_3 \begin{bmatrix} I_{r_3} & \\ & P_4 \end{bmatrix}) S$
$Q \leftarrow T \mathrm{Diag}(\begin{bmatrix} I_{r_1} & \\ & Q_3 \end{bmatrix} Q_1, \begin{bmatrix} I_{r_2} & \\ & Q_4 \end{bmatrix} Q_2)$
$A \leftarrow S^T A T^T$ ▷ $\texttt{PermR}(A, S^T); \texttt{PermC}(A, T^T)$

Here $A = \begin{bmatrix} L_1\backslash U_1 & D_1 & V_{11} & D_{21} & V_{12} & D_{22} \\ M_{11} & L_2\backslash U_2 & 0 & V_{21} & 0 & V_{22} \\ E_1 & I & L_3\backslash U_3 & O_1 & V_3 & O_2 \\ E_{21} & K_1 & M_{31} & L_4\backslash U_4 & 0 & V_4 \\ M_{12} & M_2 & 0 & 0 & 0 & 0 \\ E_{22} & K_2 & M_{32} & M_4 & 0 & 0 \end{bmatrix}$

Return $(P, Q, r_1 + r_2 + r_3 + r_4, A)$

3. FROM PLUQ TO LEU

We now show how to compute the LEU decomposition of [9] from the PLUQ decomposition. The idea is to write

$$P \begin{bmatrix} L \\ M \end{bmatrix} \begin{bmatrix} U \, V \end{bmatrix} Q = \underbrace{P \begin{bmatrix} L & 0 \\ M & I_{m-r} \end{bmatrix} P^T}_{\overline{L}} \underbrace{P \begin{bmatrix} I_r & \\ & 0 \end{bmatrix} Q}_{E} \underbrace{Q^T \begin{bmatrix} U & V \\ & I_{n-r} \end{bmatrix} Q}_{\overline{U}}$$

and show that $\overline{L}$ and $\overline{U}$ are respectively lower and upper triangular. This is not true in general, but turns out to be satisfied by the P, L, M, U, V and Q obtained in Algorithm 1.

THEOREM 1. *Let $A = P \begin{bmatrix} L \\ M \end{bmatrix} [U\ V] Q$ be the* **PLUQ** *decomposition computed by Algorithm 1. Then for any unit lower triangular matrix Y and any upper triangular matrix Z, the matrix $P \begin{bmatrix} L & \\ M & Y \end{bmatrix} P^T$ is unit lower triangular and $Q^T \begin{bmatrix} U & V \\ & Z \end{bmatrix} Q$ is upper triangular.*

PROOF. Proceeding by induction, we assume that the theorem is true on all four recursive calls, and show that it is true for the matrices $P \left[\begin{smallmatrix} L & \\ M & Y \end{smallmatrix}\right] P^T$ and $Q^T \left[\begin{smallmatrix} U & V \\ & Z \end{smallmatrix}\right] Q$. Let $Y = \begin{bmatrix} Y_1 & \\ Y_2 & Y_3 \end{bmatrix}$ where Y_1 is unit lower triangular of dimension $k - r_1 - r_2$. From the correctness of Algorithm 1 (see e.g. Equation A), $S \begin{bmatrix} L & \\ M & Y \end{bmatrix} S^T = \left[\begin{array}{ccccc} L_1 & & & & \\ M_{11} & L_2 & & & \\ M_{12} & M_2 & Y_1 & & \\ \hline E_1 & I & & L_3 & \\ E_{21} & K_1 & & M_{31} & L_4 \\ E_{22} & K_2 & Y_2 & M_{32} & M_4 Y_3 \end{array}\right]$

Hence $P \begin{bmatrix} L & \\ M & Y \end{bmatrix} P^T$ equals

$$\begin{bmatrix} P_1 & \\ & P_3 \end{bmatrix} \begin{bmatrix} I_{r_1} & & & \\ & P_2 & & \\ & & I_{r_3} & \\ & & & P_4 \end{bmatrix} \left[\begin{array}{ccccc} L_1 & & & & \\ M_{11} & L_2 & & & \\ M_{12} & M_2 & Y_1 & & \\ \hline E_1 & I & & L_3 & \\ E_{21} & K_1 & & M_{31} & L_4 \\ E_{22} & K_2 & Y_2 & M_{32} & M_4 Y_3 \end{array}\right] \times$$
$$\begin{bmatrix} I_{r_1} & & & \\ & P_2^T & & \\ & & I_{r_3} & \\ & & & P_4^T \end{bmatrix} \begin{bmatrix} P_1^T & \\ & P_3^T \end{bmatrix}$$

By induction hypothesis, the matrices $\overline{L_2} = P_2 \begin{bmatrix} L_2 & \\ M_2 & Y_1 \end{bmatrix} P_2^T$, $\overline{L_4} = P_4 \begin{bmatrix} L_4 & \\ M_4 & Y_3 \end{bmatrix} P_4^T$, $P_1 \begin{bmatrix} L_1 & \\ M_1 & \overline{L_2} \end{bmatrix} P_1^T$ and $P_3 \begin{bmatrix} L_3 & \\ M_3 & \overline{L_4} \end{bmatrix} P_3^T$ are unit lower triangular. Therefore the matrix $P \left[\begin{smallmatrix} L & \\ M & Y \end{smallmatrix}\right] P^T$ is also unit lower triangular.

Similarly, let $Z = \begin{bmatrix} Z_1 & Z_2 \\ & Z_3 \end{bmatrix}$ where Z_1 is upper triangular of dimension $k - r_1 - r_2$. The matrix $T^T \begin{bmatrix} U & V \\ & Z \end{bmatrix} T$ equals

$$T^T \left[\begin{array}{ccc|ccc} U_1 & V_{11} & V_{12} & D_1 & D_{21} & D_{22} \\ & 0 & 0 & U_2 & V_{21} & V_{22} \\ & U_3 & V_3 & 0 & O_1 & O_2 \\ & & 0 & & U_4 & V_4 \\ & & Z_1 & & & Z_2 \\ & & & & & Z_3 \end{array}\right] = \left[\begin{array}{ccc|ccc} U_1 & V_{11} & V_{12} & D_1 & D_{21} & D_{22} \\ & U_3 & V_3 & & O_1 & O_2 \\ & & Z_1 & & & Z_2 \\ & 0 & 0 & U_2 & V_{21} & V_{22} \\ & & 0 & & U_4 & V_4 \\ & & & & & Z_3 \end{array}\right]$$

Hence $Q^T \begin{bmatrix} U & V \\ & Z \end{bmatrix} Q$ equals

$$\begin{bmatrix} Q_1^T & \\ & Q_2^T \end{bmatrix} \begin{bmatrix} I_{r_1} & & & \\ & Q_3^T & & \\ & & I_{r_2} & \\ & & & Q_4^T \end{bmatrix} \left[\begin{array}{ccc|ccc} U_1 & V_{11} & V_{12} & D_1 & D_{21} & D_{22} \\ & U_3 & V_3 & & O_1 & O_2 \\ & & Z_1 & & & Z_2 \\ & 0 & 0 & U_2 & V_{21} & V_{22} \\ & & 0 & & U_4 & V_4 \\ & & & & & Z_3 \end{array}\right] \times$$
$$\begin{bmatrix} I_{r_1} & & & \\ & Q_3 & & \\ & & I_{r_2} & \\ & & & Q_4 \end{bmatrix} \begin{bmatrix} Q_1 & \\ & Q_2 \end{bmatrix}.$$

By induction hypothesis, the matrices $\overline{U_3} = Q_3^T \begin{bmatrix} U_3 & V_3 \\ & Z_1 \end{bmatrix} Q_3$, $\overline{U_4} = Q_4^T \begin{bmatrix} U_4 & V_4 \\ & Z_3 \end{bmatrix} P_4^T$, $Q_1^T \begin{bmatrix} U_1 & V_1 \\ & \overline{U_3} \end{bmatrix} Q_1$ and $Q_2^T \begin{bmatrix} U_2 & V_2 \\ & \overline{U_4} \end{bmatrix} Q_2$ are upper triangular. Consequently the matrix $Q^T \left[\begin{smallmatrix} U & V \\ & Z \end{smallmatrix}\right] Q$ is upper triangular.

For the base case with $m = 1$. The matrix $\overline{L}$ has dimension 1×1 and is unit lower triangular. If $r = 0$, then $\overline{U} = I_n^T Z I_n$ is upper triangular. If $r = 1$, then $Q = T_{1,i}$ where i is the column index of the pivot and is therefore the column index of the leading coefficient of the row $[UV]\, Q$. Applying Q^T on the left only swaps rows 1 and i, hence row $[UV]\, Q$ is the ith row of $Q^T \begin{bmatrix} UV \\ Z \end{bmatrix} Q$. The latter is therefore upper triangular. The same reasoning can be applied to the case $n = 1$. □

CORROLARY 1. *Let $\overline{L} = P \begin{bmatrix} L & \\ M & I_{m-r} \end{bmatrix} P^T, E = P \begin{bmatrix} I_r & \\ & 0 \end{bmatrix} Q$ and $\overline{U} = Q^T \begin{bmatrix} UV \\ 0 \end{bmatrix} Q$. Then $A = \overline{L} E \overline{U}$ is a LEU decomposition of A.*

REMARK 2. *The converse is not always possible: given $A = L, E, U$, there are several ways to choose the last $m - r$ columns of P and the last $n - r$ rows of Q. The LEU algorithm does not keep track of these parts of the permutations.*

4. COMPUTING THE RANK PROFILES

We prove here the main feature of the PLUQ decomposition computed by Algorithm 1: it reveals the row and column rank profiles of all leading sub-matrices of the input matrix. We recall in Lemma 1 basic properties of rank profiles.

LEMMA 1. *For any matrix,*

1. *the row rank profile is preserved by right multiplication with an invertible matrix and by left multiplication with an invertible upper triangular matrix.*
2. *the column rank profile is preserved by left multiplication with an invertible matrix and by right multiplication with an invertible lower triangular matrix.*

LEMMA 2. *Let $A = PLUQ$ be the* **PLUQ** *decomposition computed by Algorithm 1. Then the row (resp. column) rank profile of any leading (k, t) submatrix of A is the row (resp. column) rank profile of the leading (k, t) submatrix of $P \begin{bmatrix} I_r & \\ & 0 \end{bmatrix} Q$.*

PROOF. With the notations of Corollary 1, we have:

$$A = P \begin{bmatrix} L & \\ M & I_{m-r} \end{bmatrix} \begin{bmatrix} I_r & \\ & 0 \end{bmatrix} \begin{bmatrix} U & V \\ & I_{n-r} \end{bmatrix} Q = \overline{L} P \begin{bmatrix} I_r & \\ & 0 \end{bmatrix} Q \overline{U}$$

Hence

$$[I_k 0]\, A \begin{bmatrix} I_t \\ 0 \end{bmatrix} = \overline{L_1}\, [I_k 0]\, P \begin{bmatrix} I_r & \\ & 0 \end{bmatrix} Q \overline{U_1},$$

where $\overline{L_1}$ is the $k \times k$ leading submatrix of $\overline{L}$ (hence it is an invertible lower triangular matrix) and $\overline{U_1}$ is the $t \times t$ leading submatrix of $\overline{U}$ (hence it is an invertible upper triangular matrix). Now, Lemma 1 implies that the rank profile of $[I_k 0]\, A \begin{bmatrix} I_t \\ 0 \end{bmatrix}$ is that of $[I_k 0]\, P \begin{bmatrix} I_r & \\ & 0 \end{bmatrix} Q \begin{bmatrix} I_t \\ 0 \end{bmatrix}$. □

From this lemma we deduce how to compute the row and column rank profiles of any (k,t) leading submatrix and more particularly of the matrix A itself.

CORROLARY 2. *Let $A = PLUQ$ be the* PLUQ *decomposition of a $m \times n$ matrix computed by Algorithm 1. The row (resp. column) rank profile of any (k,t)-leading submatrix of a A is the sorted sequence of the row (resp. column) indices of the non zero rows (resp. columns) in the matrix*

$$R = [I_k 0]\, P \begin{bmatrix} I_r \\ & 0 \end{bmatrix} Q \begin{bmatrix} I_t \\ 0 \end{bmatrix}$$

CORROLARY 3. *The row (resp. column) rank profile of A is the sorted sequence of row (resp. column) indices of the non zero rows (resp. columns) of the first r columns of P (resp. first r rows of Q).*

5. COMPLEXITY ANALYSIS

We study here the time complexity of Algorithm 1 by counting the number of field operations. For the sake of simplicity, we will assume here that the dimensions m and n are powers of two. The analysis can easily be extended to the general case for arbitrary m and n.

For $i = 1,2,3,4$ we denote by T_i the cost of the i-th recursive call to PLUQ, on a $\frac{m}{2} \times \frac{n}{2}$ matrix of rank r_i. We also denote by $T_{\texttt{TRSM}}(m,n)$ the cost of a call TRSM on a rectangular matrix of dimensions $m \times n$, and by $T_{\texttt{MM}}(m,k,n)$ the cost of multiplying an $m \times k$ by an $k \times n$ matrix.

THEOREM 2. *Algorithm 1, run on an $m \times n$ matrix of rank r, performs $O\left(mnr^{\omega-2}\right)$ field operations.*

PROOF. Let $T = T_{\texttt{PLUQ}}(m,n,r)$ be the cost of Algorithm 1 run on a $m \times n$ matrix of rank r. From the complexities of the subroutines given, e.g., in [2] and the recursive calls in Algorithm 1, we have:

$$\begin{aligned}
T =& T_1 + T_2 + T_3 + T_4 + T_{\texttt{TRSM}}(r_1, \tfrac{m}{2}) + T_{\texttt{TRSM}}(r_1, \tfrac{n}{2}) \\
&+ T_{\texttt{TRSM}}(r_2, \tfrac{m}{2}) + T_{\texttt{TRSM}}(r_3, \tfrac{n}{2}) + T_{\texttt{MM}}(\tfrac{m}{2} - r_1, r_1, \tfrac{n}{2}) \\
&+ T_{\texttt{MM}}(\tfrac{m}{2}, r_1, \tfrac{n}{2} - r_1) + T_{\texttt{MM}}(\tfrac{m}{2}, r_1, \tfrac{n}{2}) \\
&+ T_{\texttt{MM}}(r_3, r_2, \tfrac{n}{2} - r_2) + T_{\texttt{MM}}(\tfrac{m}{2} - r_3, r_2, \tfrac{n}{2} - r_2 - r_4) \\
&+ T_{\texttt{MM}}(\tfrac{m}{2} - r_3, r_3, \tfrac{n}{2} - r_2 - r_4) \\
\leq& T_1 + T_2 + T_3 + T_4 + K\Big(\frac{m}{2}(r_1^{\omega-1} + r_2^{\omega-1}) + \frac{n}{2}(r_1^{\omega-1} \\
&+ r_3^{\omega-1}) + \frac{m}{2}\frac{n}{2} r_1^{\omega-2} + \frac{m}{2}\frac{n}{2} r_2^{\omega-2} + \frac{m}{2}\frac{n}{2} r_3^{\omega-2}\Big) \\
\leq& T_1 + T_2 + T_3 + T_4 + K' mnr^{\omega-2}
\end{aligned}$$

for some constants K and K' (we recall that $a^{\omega-2} + b^{\omega-2} \leq 2^{3-\omega}(a+b)^{\omega-2}$ for $2 \leq \omega \leq 3$).

Let $C = max\{\frac{K'}{1-2^{4-2\omega}}; 1\}$.Then we can prove by a simultaneous induction on m and n that $T \leq Cmnr^{\omega-2}$.

Indeed, if $(r = 1, m = 1, n \geq m)$ or $(r = 1, n = 1, m \geq n)$ then $T \leq m - 1 \leq Cmnr^{\omega-2}$. Now if it is true for $m = 2^j, n = 2^i$, then for $m = 2^{j+1}, n = 2^{i+1}$, we have

$$\begin{aligned}
T \leq& \frac{C}{4} mn(r_1^{\omega-2} + r_2^{\omega-2} + r_3^{\omega-2} + r_4^{\omega-2}) + K' mnr^{\omega-2} \\
\leq& \frac{C(2^{3-\omega})^2}{4} mnr^{\omega-2} + K' mnr^{\omega-2} \\
\leq& K' \frac{2^{4-2\omega}}{1 - 2^{4-2\omega}} mnr^{\omega-2} + K' mnr^{\omega-2} \leq Cmnr^{\omega-2}.
\end{aligned}$$

□

In order to compare this algorithm with usual Gaussian elimination algorithms, we now refine the analysis to compare the leading constant of the time complexity in the special case where the matrix is square and has a generic rank profile: $r_1 = \frac{m}{2} = \frac{n}{2}, r_2 = 0, r_3 = 0$ and $r_4 = \frac{m}{2} = \frac{n}{2}$ at each recursive step.

Hence, with C_ω the constant of matrix multiplication, we have

$$\begin{aligned}
T_{\texttt{PLUQ}} =& 2T_{\texttt{PLUQ}}(\tfrac{n}{2}, \tfrac{n}{2}, \tfrac{n}{2}) + 2T_{\texttt{TRSM}}(\tfrac{n}{2}, \tfrac{n}{2}) + T_{\texttt{MM}}(\tfrac{n}{2}, \tfrac{n}{2}, \tfrac{n}{2}) \\
=& 2T_{\texttt{PLUQ}}(\tfrac{n}{2}, \tfrac{n}{2}, \tfrac{n}{2}) + 2\frac{C_\omega}{2^{\omega-1} - 2}\left(\frac{n}{2}\right)^\omega + C_\omega\left(\frac{n}{2}\right)^\omega
\end{aligned}$$

Writing $T_{\texttt{PLUQ}}(n,n,n) = \alpha n^\omega$, the constant α satisfies:

$$\alpha = C_\omega \frac{1}{(2^\omega - 2)}\left(\frac{1}{2^{\omega-2} - 1} + 1\right) = C_\omega \frac{2^{\omega-2}}{(2^\omega - 2)(2^{\omega-2} - 1)}.$$

which is equal to the constant of the CUP and LUP decompositions [7, Table 1]. In particular, it equals 2/3 when $\omega = 3, C_\omega = 2$, matching the constant of the classical Gaussian elimination.

6. COUNTING MODULAR REDUCTIONS OVER A PRIME FIELD

In the following we suppose that the operations are done with full delayed reduction for a single multiplication and any number of additions: operations of the form $\sum a_i b_i$ are reduced only once at the end of the addition, but $a \cdot b \cdot c$ requires two reductions. In practice, only a limited number of accumulations can be done on an actual mantissa without overflowing, but we neglect this in this section for the sake of simplicity. See e.g. [2] for more details. For instance, with this model, the number of reductions required by a classic multiplication of matrices of size $m \times k$ by $k \times n$ is simply: $m \cdot n$. We denote this by $R_{MM}(m,k,n) = mn$. This extends e.g. also for triangular solving:

THEOREM 3. *Over a prime field modulo p, the number of reductions modulo p required by $TRSM(m,n)$ with full delayed reduction is:*

$R_{UnitTRSM}(m,n) = mn$ *if the triangular matrix is unitary,*
$R_{TRSM}(m,n) = 2mn$ *in general.*

PROOF. If the matrix is unitary, then a fully delayed reduction is required only once after the update of each row of the result. In the generic case, we invert each diagonal element first and multiply each element of the right hand side by this inverse diagonal element, prior to the update of each row of the result. This gives mn extra reductions. □

Next we show that the new pivoting strategy is more efficient in terms of number of integer division.

THEOREM 4. *Over a prime field modulo p and on a full-rank square $m \times m$ matrix with generic rank profile, and m a power of two, the number of reductions modulo p required by the elimination algorithms with full delayed reduction is:*

$$R_{PLUQ}(m,m) = 2m^2 + o(m^2),$$
$$R_{PLE}(m,m) = R_{CUP}(m,m) = \left(1 + \tfrac{1}{4}\log_2(m)\right) m^2 + o(m^2)$$

PROOF. If the top left square block is full rank then PLUQ reduces to one recursive call, two square TRSM (one unitary, one generic) one square matrix multiplication and a final recursive call. In terms of modular reductions, this gives: $R_{\mathtt{PLUQ}}(m) = 2R_{\mathtt{PLUQ}}(\frac{m}{2}) + R_{\mathtt{UnitTRSM}}(\frac{m}{2},\frac{m}{2}) + R_{\mathtt{TRSM}}(\frac{m}{2},\frac{m}{2}) + R_{\mathtt{MM}}(\frac{m}{2},\frac{m}{2},\frac{m}{2})$. Therefore, using Theorem 3, the number of reductions within PLUQ satisfies $T(m) = 2T(\frac{m}{2}) + m^2$ so that it is $R_{PLUQ}(m,m) = 2m^2 - 2m$ if m is a power of two.

For row or column oriented elimination this situation is more complicated since the recursive calls will always be rectangular even if the intermediate matrices are full-rank. We in fact prove, by induction on m, the more generic:

$$R_{\mathtt{PLE}}(m,n) = \log_2(m)\left(\frac{mn}{2} - \frac{m^2}{4}\right) + m^2 + o(mn + m^2) \quad (1)$$

First $R_{\mathtt{PLE}}(1,n) = 0$ since $[1] \times [a_1,\ldots,a_n]$ is a triangular decomposition of the $1 \times n$ matrix $[a_1,\ldots,a_n]$. Now suppose that Equation (1) holds for $k = m$. Then we follow the row oriented algorithm of [2, Lemma 5.1] which makes two recursive calls, one TRSM and one MM to get $R_{\mathtt{PLE}}(2m,n) = R_{\mathtt{PLE}}(m,n) + R_{\mathtt{PLE}}(m,m) + R_{\mathtt{MM}}(m,m,n-m) + R_{\mathtt{PLE}}(m,n-m) = R_{\mathtt{PLE}}(m,n) + R_{\mathtt{PLE}}(m,n-m) + m(n+m)$. We then apply the induction hypothesis on the recursive calls to get

$$\begin{aligned} R_{\mathtt{PLE}}(2m,n) =& \frac{1}{2}\log_2(m)mn - \frac{1}{4}\log_2(m)m^2 + m^2 + \\ & \frac{1}{2}\log_2(m)m(n-m) - \frac{1}{4}\log_2(m)m^2 + m^2 + \\ & m(n+m) + o(mn+m^2) \\ =& \log_2(m)(mn - m^2) + 3m^2 + mn + o(mn + m^2). \end{aligned}$$

The latter is also obtained by substituting $m \leftrightarrow 2m$ in Equation (1) so that the induction is proven. □

This shows that the new algorithm requires fewer modular reductions, as soon as m is larger than 32. Over finite fields, since reductions can be much more expensive than multiplications or additions by elements of the field, this is a non negligible advantage. We show in Section 8 that this participates to the better practical performance of the PLUQ algorithm.

7. A BASE CASE ALGORITHM

We propose in Algorithm 2 an *iterative* algorithm computing the same PLUQ decomposition as Algorithm 1. The motivation is to offer an alternative to the recursive algorithm improving the computational efficiency on small matrix sizes. Indeed, as long as the matrix fits the cache memory, the number of page faults of the two variants are similar, but the iterative variant reduces the number of row and column permutations. The block recursive algorithm can then be modified so that it switches to the iterative algorithm whenever the matrix dimensions are below some threshold.

Unlike the common Gaussian elimination, where pivots are searched in the whole current row or column, the strategy is here to proceed with an incrementally growing leading sub-matrix. This implies a Z-curve type search scheme, as shown on Figure 2. This search strategy is meant to ensure the properties on the rank profile that have been presented in Section 4.

Algorithm 2 PLUQ iterative base case

Input: A a $m \times n$ matrix over a field
Output: P, Q: $m \times m$ and $n \times n$ permutation matrices
Output: r: the rank of A
Output: $A \leftarrow \begin{bmatrix} L\backslash UV \\ M \;\; 0 \end{bmatrix}$ where L is $r \times r$ unit lower triang., U is $r \times r$ upper triang. and such that $A = P \begin{bmatrix} L \\ M \end{bmatrix} [UV] Q$.

$r \leftarrow 0; i \leftarrow 0; j \leftarrow 0$
while $i < m$ or $j < n$ **do**
  ▷ Let $v = [A_{i,r} \ldots A_{i,j-1}]$ and $w = [A_{r,j} \ldots A_{i-1,r}]^T$
  if $j < n$ and $w \neq 0$ **then**
    $p \leftarrow$ row index of the first non zero entry in w
    $q \leftarrow j; j \leftarrow \max(j+1, n)$
  else if $i < m$ and $v \neq 0$ **then**
    $q \leftarrow$ column index of the first non zero entry in v
    $p \leftarrow i; i \leftarrow \max(i+1, m)$
  else if $i < m$ and $j < n$ and $A_{i,j} \neq 0$ **then**
    $(p,q) \leftarrow (i,j)$
    $i \leftarrow \max(i+1,m); j \leftarrow \max(j+1,n)$
  else
    $i \leftarrow \max(i+1,m); j \leftarrow \max(j+1,n)$
    continue
  end if ▷ At this stage, $A_{p,q}$ is a pivot
  for $k = p+1 : n$ **do**
    $A_{k,q} \leftarrow A_{k,p} A_{p,q}^{-1}$
    $A_{k,q+1:n} \leftarrow A_{k,q+1:n} - A_{k,q} A_{p,q+1:n}$
  end for
  ▷ Cyclic shifts of pivot column and row
  $A_{0:m,r:q} \leftarrow A_{0:m,r>>>_1 q}$
  $A_{r:p,0:n} \leftarrow A_{r>>>_1 p,0:n}$
  $P \leftarrow P_{r>>>_1 p,*}$;
  $Q \leftarrow Q_{*,r>>>_1 q}$
  $r \leftarrow r+1$
end while

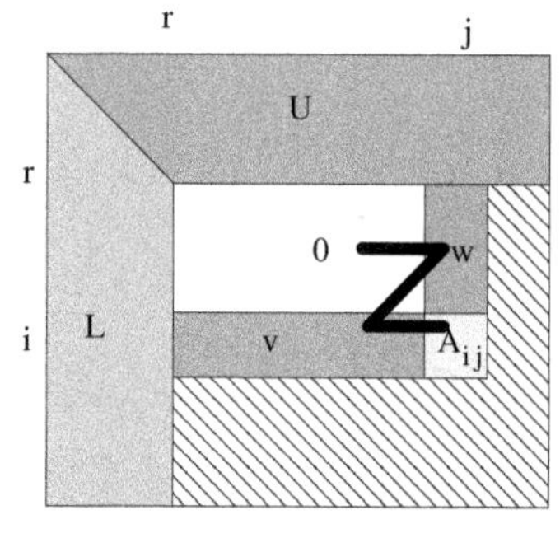

Figure 2: Iterative base case PLUQ decomposition

In order to perform the correct updates on the remaining parts, when a pivot is found its whole row and column have to be permuted to the current diagonal location, see Figure 2. But then, in order to preserve the row and column rank profiles, all the rows and column in between have to be shifted by 1 location. Therefore after the elimination step, the rows and columns of the matrix, as well as the rows of the left permutation matrix and the columns of the right permutation matrix have to be cyclically shifted accordingly. This is presented in the last steps of Algorithm 2, where the notation $A_{*,i>>>_1 j}$ means that in matrix A, columns i through j, both inclusive, have to be shifted by 1 location, cyclically to the right.

REMARK 3. *Applying the cyclic permutations in steps 22 to 25 may cost in worst case a cubic number of operations. Instead one can delay these permutations and leave the pivots at the position where they were found. These positions are then used to form the matrices P and Q, only after the end of the* `while` *loop. Then applying these permutations to the current matrix gives the final decomposition* $\begin{bmatrix} L\backslash U & V \\ M & 0 \end{bmatrix}$.

REMARK 4. *In order to further improve the data locality, this iterative algorithm can be transformed into a left-looking variant [1]. Over a finite field, this variant performs fewer modular operations: Step 19 of Algorithm 2 requires a modular reduction after each multiplication while a left-looking variant will delay these reductions within block operations.*

Updating Algorihtm 2 with Remarks 3 and 4 would be too technical to be presented here, but this is how we implemented the base case used for the experiments of Section 8.

8. EXPERIMENTS

Algorithm 1 combined with the base case Algorithm 2 has been implemented in the `FFLAS-FFPACK` library[1] and is available from revision svn@361. We present here experiments comparing its efficiency with the implementation of the CUP/PLE decomposition, called `LUdivine` in this same library. We ran our tests on a single core of an Intel Xeon E5-4620@2.20GHz using gcc-4.7.2.

In Figure 3, the matrices are dense, with full rank. The computation times are similar, the PLUQ algorithm with base case being slightly faster than `LUdivine`. In Figures 4

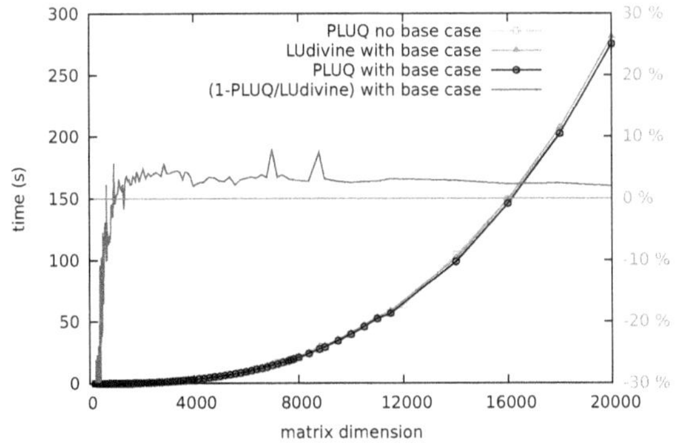

Figure 3: Dense full rank matrices modulo 1009

and 5, the matrices are square, dense with a rank equal to half the dimension. To ensure non trivial row and column rank profiles, they are generated from a LEU decomposition, where L and U are uniformly random non-singular lower and upper triangular matrices, and E is zero except on $r = n/2$ positions, chosen uniformly at random, set to one. The cutoff dimension for the switch to the base case has been set

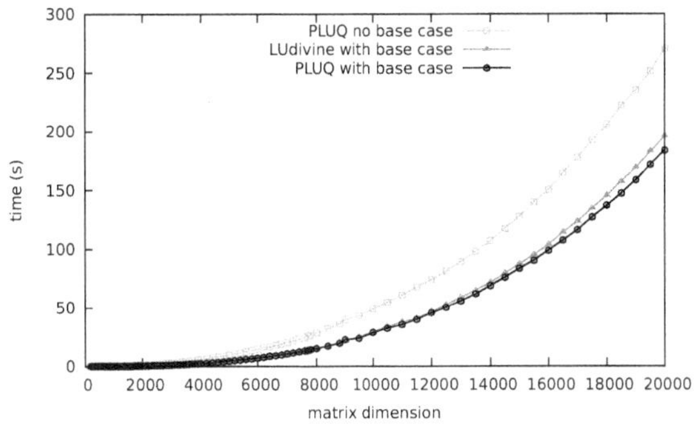

Figure 4: Computation time with dense rank deficient matrices (rank is half the dimension)

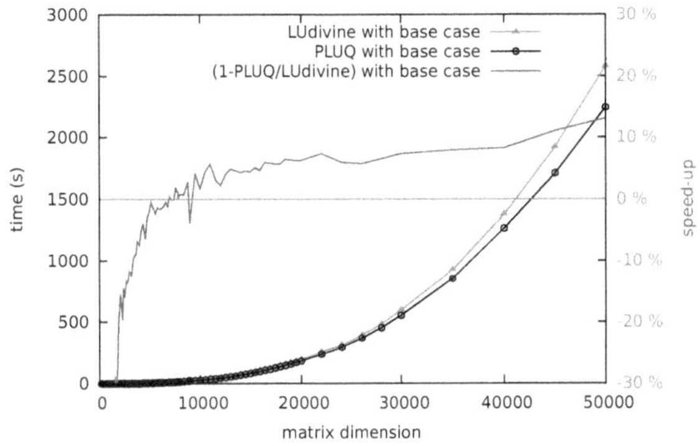

Figure 5: Computation time with dense rank deficient matrices of larger dimension (half rank)

to an optimal value of 288 by experiments. Figure 4 shows how the base case greatly improves the efficiency for PLUQ, presumably for it reduces the number of row and column permutations. More precisely, PLUQ becomes faster than `LUdivine` for dimensions above 7000. Figure 5 shows that, on larger matrices, PLUQ can be up to 13% faster.

Table 1 shows the cache misses reported by the callgrind tool (valgrind emulator version 3.8.1). We also report in the last column the corresponding computation time (without emulator). We used the same matrices as in Figure 4, with rank half the dimension. We first notice the impact of the base case on the PLUQ algorithm: although it does not change the number of cache misses, it strongly reduces the total number of memory accesses (fewer permutations), thus improving the computation time. Now as the dimension grows, the amount of memory accesses and of cache misses plays in favor of PLUQ which becomes faster than `LUdivine`.

9. CONCLUSION AND PERSPECTIVES

We showed the first reduction to matrix multiplication of the problem of computing both row and column rank profiles of all leading sub-matrices of an input matrix.

The decomposition that we propose can first be viewed as an improvement over the LEU decomposition, introducing a finer treatment of rank deficiency that reduces the number of arithmetic operations, makes the time complexity rank sensitive and allows to perform the computation in-place.

Second, viewed as a variant of the existing CUP/PLE decompositions, this new algorithm produces more information on the rank profile and is more efficient, as it deals with ma-

[1] `http://linalg.org/projects/fflas-ffpack`

Matrix	Algorithm	Accesses	L1 Misses	L3 Misses	L3/Accesses	Timing (s)
A4K	`LUdivine`	1.529E+10	1.246E+09	2.435E+07	.159	**2.31**
	PLUQ-no-base-case	1.319E+10	**7.411E+08**	1.523E+07	**.115**	5.82
	PLUQ-base-case	**8.105E+09**	7.467E+08	**1.517E+07**	.187	2.48
A8K	`LUdivine`	7.555E+10	9.693E+09	2.205E+08	.292	15.2
	PLUQ-no-base-case	6.150E+10	**5.679E+09**	1.305E+08	**.212**	28.4
	PLUQ-base-case	**4.067E+10**	5.686E+09	**1.303E+08**	.321	**15.1**
A12K	`LUdivine`	2.003E+11	3.141E+10	7.943E+08	.396	46.5
	PLUQ-no-base-case	1.575E+11	**1.911E+10**	4.691E+08	**.298**	73.9
	PLUQ-base-case	**1.111E+11**	1.913E+10	**4.687E+08**	.422	**45.5**
A16K	`LUdivine`	4.117E+11	7.391E+10	1.863E+09	.452	103
	PLUQ-no-base-case	3.142E+11	4.459E+10	1.092E+09	**.347**	150
	PLUQ-base-case	**2.299+11**	**4.458E+10**	**1.088E+09**	.473	**98.8**

Table 1: Cache misses for dense matrices with rank equal half of the dimension

trices of more evenly balanced dimensions. It also performs fewer modular reductions when computing over a finite field.

Overall the new algorithm is also faster in practice than previous implementations with large enough matrices.

Lastly, it also exhibits more parallelism than classical Gaussian elimination since the recursive calls in step 2 and 3 are independent. This is also the case for the TURBO algorithm of [3], but it has a higher arithmetic complexity. Further experiments and analysis of communication costs should be done in both shared and distributed memory settings.

10. ACKNOWLEDGEMENT

We are gratefull to Jean-Louis Roch and an anonymous referee for their helpful remarks and suggestions.

11. REFERENCES

[1] J. J. Dongarra, L. S. Duff, D. C. Sorensen, and H. A. V. Vorst. *Numerical Linear Algebra for High Performance Computers*. SIAM, 1998.

[2] J.-G. Dumas, P. Giorgi, and C. Pernet. Dense linear algebra over prime fields. *ACM TOMS*, 35(3):1–42, Nov. 2008. URL: `http://arxiv.org/abs/cs/0601133`.

[3] J.-G. Dumas and J.-L. Roch. On parallel block algorithms for exact triangularizations. *Parallel Computing*, 28(11):1531–1548, Nov. 2002.

[4] J.-C. Faugère. A new efficient algorithm for computing Gröbner bases (F4). *Journal of Pure and Applied Algebra*, 139(1–3):61–88, June 1999. URL: `http://www-salsa.lip6.fr/~jcf/Papers/F99a.pdf`.

[5] G. Golub and C. Van Loan. *Matrix Computations*. The Johns Hopkins University Press, third edition, 1996.

[6] O. H. Ibarra, S. Moran, and R. Hui. A generalization of the fast LUP matrix decomposition algorithm and applications. *J. of Algorithms*, 3(1):45–56, Mar. 1982.

[7] C.-P. Jeannerod, C. Pernet, and A. Storjohann. Rank profile revealing Gaussian elimination and the CUP matrix decomposition. *Journal of Symbolic Computations*, 2013. To appear. report, arXiV cs.SC/1112.5717.

[8] D. J. Jeffrey. LU factoring of non-invertible matrices. *ACM Comm. Comp. Algebra*, 44(1/2):1–8, July 2010. URL: `http://www.apmaths.uwo.ca/~djeffrey/Offprints/David-Jeffrey-LU.pdf`.

[9] G. I. Malaschonok. Fast generalized Bruhat decomposition. In *CASC'10*, volume 6244 of *LNCS*, pages 194–202. Springer-Verlag, Berlin, Heidelberg, 2010.

[10] W. Stein. *Modular forms, a computational approach*. Graduate studies in mathematics. AMS, 2007. URL: `http://wstein.org/books/modform/modform`.

[11] A. Storjohann. *Algorithms for Matrix Canonical Forms*. PhD thesis, ETH-Zentrum, Zürich, Switzerland, Nov. 2000. `doi:10.3929/ethz-a-004141007`.

[12] V. V. Williams. Multiplying matrices faster than Coppersmith-Winograd. In *STOC'12*, pages 887–898, New York, NY, USA, 2012. ACM. URL: `http://www.cs.berkeley.edu/~virgi/matrixmult.pdf`.

APPENDIX

A. CORRECTNESS OF ALGORITHM 1

First note that $S\begin{bmatrix} L \\ M \end{bmatrix} = \begin{bmatrix} L_1 & & & \\ M_{11} & L_2 & & \\ M_{12} & M_2 & 0 & \\ \hline E_1 & I & L_3 & \\ E_{21} & K_1 & M_{31} & L_4 \\ E_{22} & K_2 & M_{32} & M_4 0 0 \end{bmatrix}$

Hence $P\begin{bmatrix} L \\ M \end{bmatrix} = \begin{bmatrix} P_1 & \\ & P_3 \end{bmatrix} \begin{bmatrix} L_1 & & \\ M_1 & P_2\begin{bmatrix} L_2 \\ M_2 \end{bmatrix} & \\ \hline E_1 & I & L_3 \\ E_2 & K & M_3 P_4 \begin{bmatrix} L_4 \\ M_4 \end{bmatrix} \end{bmatrix}$

Similarly, $[UV]\,T = \left[\begin{array}{ccc|ccc} U_1 & V_{11} & V_{12} & D_1 & D_{21} & D_{22} \\ & 0 & 0 & U_2 & V_{21} & V_{22} \\ & U_3 & V_3 & 0 & O_1 & O_2 \\ & & & & U_4 & V_4 \\ & & & & & 0 \end{array}\right]$ and $[UV]\,Q = \left[\begin{array}{cc|cc} U_1 & V_1 & D_1 & D_2 \\ & 0 & U_2 & V_2 \\ & [U_3 V_3]\,Q_3 & 0 & O \\ & & & [U_4 V_4]\,Q_4 \end{array}\right] \begin{bmatrix} Q_1 & \\ & Q_2 \end{bmatrix}$.

Now as $H_1 = IU_2, H_2 = IV_2 + L_3 O, H_3 = KU_2$ and $H_4 = KV_2 + M_3 O + P_4 \begin{bmatrix} L_4 \\ M_4 \end{bmatrix} [U_4 V_4]\,Q_4$ we have

$$\begin{aligned} P\begin{bmatrix} L \\ M \end{bmatrix}[UV]\,Q = & \begin{bmatrix} P_1 & \\ & P_3 \end{bmatrix} \begin{bmatrix} L_1 & & \\ M_1 & P_2\begin{bmatrix} L_2 \\ M_2 \end{bmatrix} & \\ \hline E_1 & I & L_3 \\ E_2 & K & M_3 P_4 \begin{bmatrix} L_4 \\ M_4 \end{bmatrix} \end{bmatrix} \\ & \left[\begin{array}{cc|cc} U_1 & V_1 & D_1 & D_2 \\ & 0 & U_2 & V_2 \\ & [U_3 V_3]\,Q_3 & 0 & O \\ & & & [U_4 V_4]\,Q_4 \end{array}\right] \begin{bmatrix} Q_1 & \\ & Q_2 \end{bmatrix} \\ = & \begin{bmatrix} P_1 & \\ & P_3 \end{bmatrix} \begin{bmatrix} L_1 & & \\ M_1 & P_2\begin{bmatrix} L_2 \\ M_2 \end{bmatrix} & \\ \hline E_1 & I_{r_3} & \\ E_2 & & I_{m-k-r_3} \end{bmatrix} \\ & \left[\begin{array}{ccc|cc} U_1 & & V_1 & D_1 & D_2 \\ & & 0 & U_2 & V_2 \\ & L_3 & [U_3 V_3]\,Q_3 & H_1 & H_2 \\ & M_3 & [U_3 V_3]\,Q_3 & H_3 & H_4 \end{array}\right] \begin{bmatrix} Q_1 & \\ & Q_2 \end{bmatrix} \\ = & \begin{bmatrix} P_1 & \\ & I_{m-k} \end{bmatrix} \begin{bmatrix} L_1 & \\ M_1 & \\ \hline E & 0 I_{m-k} \end{bmatrix} \left[\begin{array}{cc|c} U_1 & V_1 & D \\ & 0 & F \\ & G & H \end{array}\right] \\ & \begin{bmatrix} Q_1 & \\ & I_{n-k} \end{bmatrix} \\ = & \begin{bmatrix} P_1 & \\ & I_{m-k} \end{bmatrix} \left[\begin{array}{cc|c} L_1 U_1 & L_1 V_1 & B_1 \\ M_1 U_1 & M_1 V_1 & B_2 \\ C_1 & C_2 & A_4 \end{array}\right] \begin{bmatrix} Q_1 & \\ & I_{n-k} \end{bmatrix} \\ = & A \end{aligned}$$

On the Complexity of Computing Gröbner Bases for Quasi-Homogeneous Systems

Jean-Charles Faugère[*] Jean-Charles.Faugere@inria.fr
Mohab Safey El Din[*‡] Mohab.Safey@lip6.fr
Thibaut Verron[†*] Thibaut.Verron@ens.fr

[*]INRIA, Paris-Rocquencourt Center, PolSys Project
UPMC, Univ. Paris 06, LIP6
CNRS, UMR 7606, LIP6
Case 169, 4, Place Jussieu, F-75252 Paris

[‡]Institut Universitaire de France

[†]École Normale Supérieure,
45, rue d'Ulm, F-75230, Paris

ABSTRACT

Let $\mathbb{K}$ be a field and $(f_1,\dots,f_n) \subset \mathbb{K}[X_1,\dots,X_n]$ be a sequence of quasi-homogeneous polynomials of respective weighted degrees $(d_1,\dots,d_n)$ w.r.t a system of weights $(w_1,\dots,w_n)$. Such systems are likely to arise from a lot of applications, including physics or cryptography.

We design strategies for computing Gröbner bases for quasi-homogeneous systems by adapting existing algorithms for homogeneous systems to the quasi-homogeneous case. Overall, under genericity assumptions, we show that for a generic zero-dimensional quasi-homogeneous system, the complexity of the full strategy is polynomial in the weighted Bézout bound $\prod_{i=1}^n d_i / \prod_{i=1}^n w_i$.

We provide some experimental results based on generic systems as well as systems arising from a cryptography problem. They show that taking advantage of the quasi-homogeneous structure of the systems allow us to solve systems that were out of reach otherwise.

Categories and Subject Descriptors

I.1.2 [**Symbolic and Algebraic Manipulation**]: Algorithms; F.2.2 [**Analysis of Algorithms and Problem Complexity**]: Nonnumerical Algorithms and Problems

Keywords

Gröbner bases; Polynomial system solving; Quasi-homogeneous polynomials

1. INTRODUCTION

Polynomial system solving is a very important problem in computer algebra, with a wide range of applications in theory (algorithmic geometry) or in real life (cryptography). For that purpose, Gröbner bases of polynomial ideals are a valuable tool, and practicable computation of the Gröbner bases of any given ideal is a major challenge of modern computer algebra. Since their introduction in 1965, many algorithms have been designed to compute Gröbner bases ([6, 9, 10, 11]), improving the efficiency of the computations.

ISSAC'13, June 26–29, 2013, Boston, Massachusetts, USA.

Systems arising from "real life" problems often have some structure. It has been observed that most of these structures can make the Gröbner basis easier to compute. For example, it is known that homogeneous systems, or systems with an important maximal homogeneous component, are better solved by using a degree-compatible order, and then applying a change of ordering. In this paper, we study a structure slightly more general than homogeneity, called *quasi-homogeneity*. More precisely, we will say that a polynomial $P(X_1,\dots,X_n)$ is quasi-homogeneous for the *system of weights* $W = (w_1,\dots,w_n)$, if the polynomial

$$Q(Y_1,\dots,Y_n) := P(Y_1^{w_1},\dots,Y_n^{w_n})$$

is homogeneous. Systems with such a structure are likely to arise for example from physics, where all measures are associated with a dimension which, to some extent, can be seen as a weight.

Let $F = (f_1,\dots,f_m)$ be a system of polynomials, in a polynomial algebra graded w.r.t the system of weights $W = (w_1,\dots,w_n)$. In the following, we will assume that F is quasi-homogeneous and generic, or more generally that its quasi-homogeneous components of maximal weighted degree are generic. It is possible to compute directly a Gröbner basis of the ideal generated by F . This strategy consists of running the classical algorithms F_5 ([10]) and FGLM ([11]) on F, while ignoring the quasi-homogeneous structure. However, to the best of our knowledge, there is no general way of evaluating the complexity of that strategy.

Another approach is to compute the *homogenized* system defined by $\widetilde{F} := (f_i(X_1^{w_1},\dots,X_n^{w_n}))$, and then compute a Gröbner basis of that system, using the usual strategies for the homogeneous structure. Experimentally, the first step of the computation is much faster than with the naive strategy. However, the number of solutions is increased by a factor of $\prod_{i=1}^n w_i$, slowing down the change of ordering, which thus becomes the main bottleneck of the computation.

Furthermore, to the best of our knowledge, the best complexity bounds for this computation are those we obtain for a homogeneous system of the same degree. However, experimentally, the first step of the computation proves faster for a homogenized quasi-homogeneous system with weighted degree $(d_1,\dots,d_n)$ than for a homogeneous system of total degree $(d_1,\dots,d_n)$.

Main results. We provide a complexity study of the above strategy, allowing us to quantify this speed-up, as well as to propose a workaround for the change of ordering. Overall, we prove that the known bounds for this strategy can be divided by $\prod_{i=1}^n w_i$ for a generic zero-dimensional W-homogeneous system with weights $W = (w_1,\dots,w_n)$.

More precisely, we assume the system $(f_1,\dots,f_m)$ to satisfy the two following generic assumptions:

H1. The sequence $f_1,\dots,f_m$ is regular;

H2. The sequence $f_1, \ldots, f_i$ is in Noether position w.r.t. $X_1, \ldots, X_i$, for any $1 \le i \le m$.

Under hypothesis **H1**, we adapt the classical results of the homogeneous case, using similar arguments based on Hilbert series, to estimate the degree of the ideal and the degree of regularity of the system:

$$\deg(I) = \prod_{i=1}^{n} \frac{d_i}{w_i} \ ; \ d_{\text{reg}}(F) \le \sum_{i=1}^{n} \left(d_i - w_i\right) + \max\{w_i\}.$$

We study the complexity of the F_5 algorithm through its matrix variant matrix-F_5. This is a usual approach, carried on for example in [14]. With minor changes, the matrix-F_5 algorithm for homogeneous systems can be adapted to quasi-homogeneous systems. A combinatorial result found in [1] shows that the number of columns of the matrices appearing in that variant of matrix-F_5 is approximately smaller by a factor of $\prod_{i=1}^{n} w_i$, when compared to the regular matrix-F_5 algorithm. Overall, we can obtain complexity bounds which are smaller by a factor of P^{ω} than the bounds we would obtain for a generic homogeneous system with same degrees, where $P = \prod_{i=1}^{n} w_i$ and ω is the exponent of the complexity of matrix multiplication. In the end, we show that for systems satisfying **H1**, our strategy, running F_5 on the homogenized system, dehomogenizing the result, and then running FGLM, performs in time polynomial in $\prod_{i=1}^{n} d_i / \prod_{i=1}^{n} w_i$, that is polynomial in the number of solutions.

Further assuming hypothesis **H2**, we also carry on the precise complexity analyses done in [2] for homogeneous systems, and adapt them to the quasi-homogeneous case to deduce a precise complexity bound for our quasi-homogeneous variant of Matrix-F_5. These new complexity bounds are also smaller by a factor of P^{ω} than similar bounds for a generic homogeneous system. Even though these bounds still do not match exactly the experimental complexity, they tend to confirm that overall, we are able to compute a LEX Gröbner basis for a generic quasi-homogeneous system in time reduced by a factor of P^{ω}, when compared with a generic homogeneous system with same degrees.

We have run benchmarks with the FGb library ([16]) and the Magma computer algebra software ([5]), on both generic systems and real-life systems arising in cryptography. Experimentally, in both cases, our strategy seems always faster than ignoring the quasi-homogeneous structure, and the speed-up increases with the considered weights.

Experiments have also shown that the order of the variables can have an impact on the performances of both strategies. Predicting this behavior seems to require more sophisticated tools and may be material for future research.

Prior works. Making use of the structure of polynomial systems to develop faster algorithms has been a general trend over the past few years: see for example [12], [7] or [15]. Polynomial algebras graded with respect to a system of weights have been studied by researchers in commutative algebra. Most notably, the Hilbert series of ideals defined by regular sequences, which we use several times in this paper, is well known, and could be found for example in [21]. The paper [20] defines many structures of polynomial algebras, including weighted gradings, in preparation for future algorithmic developments. Combinatorial objects arising when we try to estimate the number of monomials of a given W-degree are called *Sylvester denumerants*, and studied for example in [1].

When it comes to Gröbner bases, weighted gradings and related orderings have been described in early works such as [4]. However, as far as we know, the impact of a quasi-homogeneous structure on the complexity of Gröbner bases computations had never been studied.

Among the various computer algebra software able to compute Gröbner bases, it seems that only Magma has algorithms dedicated to quasi-homogeneous systems. Given a quasi-homogeneous system, it will detect the appropriate system of weights, and use the W-GREVLEX ordering to compute an intermediate basis before the change of ordering. However, this strategy is only available for quasi-homogeneous systems, while it can be useful in many other cases, for example systems of polynomials defined as the sum of a quasi-homogeneous component and a scalar.

Other computer algebra software (e.g. Singular) allow the user to compute $\widetilde{F}$ and to run the Gröbner basis algorithm on it. Since all these algorithms (most often Buchberger, F_4 or F_5) use S-pairs, they will show a similar speed-up. However, the user must notice that the computations may benefit from using a quasi-homogeneous structure of the system, and provide the system of weights.

We do not provide a way to know what is the "appropriate" system of weights for a given system, or even to detect systems which would benefit from taking into account the quasi-homogeneous structure. However, some systems obviously belong to that category (e.g quasi-homogeneous plus scalar), and the system of weights will then be easy to compute.

Structure of the paper. In section 2 we define more precisely quasi-homogeneous systems, and we compute their degree and degree of regularity assuming the above hypotheses. We also take this opportunity to show briefly that these hypotheses are generic. In section 3 we prove that the strategy consisting of modifying the system is correct, we explain how we can adapt matrix-F_5 and FGLM to quasi-homogeneous systems, and then we evaluate the complexity of these algorithms. In section 4, we briefly explain how these results for quasi-homogeneous systems still help in case the system was obtained from a quasi-homogeneous system by specializing one of the variables to 1. We also give an example of such a structure, as well as the associated algorithm. Finally, in section 5, we give some experimental results.

2. QUASI-HOMOGENEOUS SYSTEMS

2.1 Weighted degrees and polynomials

Let $\mathbb{K}$ be a field. We consider the algebra $A := \mathbb{K}[X_1, \ldots, X_n] = \mathbb{K}[\mathbf{X}]$. Even though one usually uses the total degree to grade the algebra A, there are other ways to define such a grading, as seen in [4], for example.

Definition 1. Let $W = (w_1, \ldots, w_n)$ be a vector of positive integers. Let $\alpha = (\alpha_1, \ldots, \alpha_n)$ be a tuple of nonnegative integers. Let the integer $\deg_W(\mathbf{X}^{\alpha}) = \sum_{i=1}^{n} w_i \alpha_i$ be the W-*degree*, or *weighted degree* of the monomial $\mathbf{X}^{\alpha} = X_1^{\alpha_1} \cdots X_n^{\alpha_n}$. Call the vector W a *system of weights*. We denote by $\mathbf{1}$ the system of weights defined by $(1, \ldots, 1)$, associated with the usual grading on A.

One can prove that any grading on $\mathbb{K}[\mathbf{X}]$ comes from such a system of weights ([4, sec. 10.2]). We denote by $(\mathbb{K}[\mathbf{X}], W)$ the W-graded structure on A, and in that case, to clear ambiguities, we use the adjective W-*homogeneous* for elements or ideals, or *quasi-homogeneous* or *weighted homogeneous* if W is clear in the context. The word *homogeneous* will be reserved for $\mathbf{1}$-homogeneous items.

PROPOSITION 1. *Let* $(\mathbb{K}[X_1, \ldots, X_n], W)$ *be a graded polynomial algebra. Then the application*

$$\begin{array}{cccc} \hom_W : & (\mathbb{K}[X_1, \ldots, X_n], W) & \to & (\mathbb{K}[t_1, \ldots, t_n], \mathbf{1}) \\ & f & \mapsto & f(t_1^{w_1}, \ldots, t_n^{w_n}) \end{array}$$

is an injective graded morphism, and in particular the image of a quasi-homogeneous polynomial is a homogeneous polynomial.

PROOF. It is an easy consequence of the definition of the grading w.r.t a system of weights. □

The above morphism also provides a quasi-homogeneous variant of the GREVLEX ordering (as found for example in [4]), which we call the W-GREVLEX *ordering*:

$$u <_{W\text{-grevlex}} v \iff \hom_W(v) <_{\text{grevlex}} \hom_W(v)$$

Given a W-homogeneous system F, one can build the homogeneous system $\hom_W(F)$, and then apply classical algorithms ([10, 11]) to that system to compute a GREVLEX (resp. LEX) Gröbner basis of the ideal generated by $\hom_W(F)$. We will prove in section 3 (prop. 7) that this basis is contained in the image of $\hom_W$, and that its pullback is a W-GREVLEX (resp. LEX) Gröbner basis of the ideal generated by F.

Let us end this paragraph with some notations and definitions. The *degree of regularity* of the system F is the highest degree $d_{\text{reg}}(F)$ reached in a run of F_5 to compute a GREVLEX Gröbner basis of $\hom_W(F)$. The *index of regularity* of an ideal I is the degree i_{reg} of the Hilbert series $\text{HS}_{A/I}$, defined as the difference of the degree of its numerator and the degree of its denominator.

Recall that given a homogeneous ideal I, we define its degree D as the degree of the projective variety $V(I)$, as introduced for example in [18]. This definition still holds for the quasi-homogeneous case. In case the projective variety is empty, that is if the affine variety is equal to $\{0\}$, we extend that definition by letting D be the multiplicity of the 0 point, that is the dimension of the $\mathbb{K}$-vector space A/I. Finally, from now on we will only consider *affine* varieties, even when the ideal is quasi-homogeneous. In particular, the dimension of $V(0)$ is n, and that a zero-dimensional variety will be defined by at least n polynomials.

2.2 Degree and degree of regularity

Zero-dimensional regular sequences. As in the homogeneous case, regular sequences are an important case to study, because it is a generic property which allows us to compute several key parameters and good complexity bounds. We first characterize the degree and bound the degree of regularity of a zero-dimensional ideal defined by a regular sequence.

THEOREM 2. *Let $W = (w_1,\dots,w_n)$ be a system of weights, and $F = (f_1,\dots,f_m)$ a regular sequence of W-homogeneous polynomials, of respective W-degrees $d_1,\dots,d_m$. Further assume that the set of solutions is zero-dimensional, that is $m = n$. We denote by I the quasi-homogeneous ideal generated by F. Then we have $\deg(I) = \prod_{i=1}^{n} \frac{d_i}{w_i}$ and $d_{\text{reg}}(F) \le \sum_{i=1}^{n}(d_i - w_i) + \max\{w_i\}$.*

PROOF. We will determine the degree and degree of regularity of the system from the Hilbert series (or Poincaré series) of the algebra A/I. A classical result which can be found for example in [21, cor. 3.3] states that, for regular sequences, this series is

$$\text{HS}_{A/I}(t) = \frac{(1-t^{d_1})\cdots(1-t^{d_m})}{(1-t^{w_1})\cdots(1-t^{w_n})}. \tag{1}$$

We assumed $n = m$, so the Hilbert series can be rewritten as

$$\text{HS}_{A/I}(t) = \frac{(1+\cdots+t^{d_1-1})\cdots(1+\cdots+t^{d_n-1})}{(1+\cdots+t^{w_1-1})\dots(1+\cdots+t^{w_n-1})}.$$

In the 0-dimensional case, recall that the Hilbert series is actually a polynomial, and has degree $i_{\text{reg}} = \sum_{i=1}^{n}(d_i - w_i)$. This means that all monomials of W-degree greater than i_{reg} are in the ideal, and as such, that the leading terms of the W-GREVLEX Gröbner basis of F need to divide all the monomials of W-degree greater than i_{reg}.This proves that all the polynomials in the Gröbner basis computed by F_5 have W-degree at most $i_{\text{reg}} + \max\{w_i\}$. And since the F_5 criterion ([10]) ensures that there is no reduction to zero in a run of F_5 on a regular sequence, the algorithm indeed stops in degree at most $i_{\text{reg}} + \max\{w_i\}$.

Furthermore, the degree of the ideal I is equal to the dimension of the vector space A/I, that is the value of the Hilbert series at $t = 1$, that is $\prod_{i=1}^{n} \frac{d_i}{w_i}$. □

Note that except for this inequality, not much is known about the degree of regularity of a quasi-homogeneous system. In particular, the above bound is nothing more than a bound, even in the generic case. Let us introduce some examples of the three cases one can observe with a quasi-homogeneous generic system:

1. $W = (3,2,1)$, generic system of W-degree $\mathbf{D} = (6,6,6)$: then $d_{\text{reg}} = i_{\text{reg}} + 1 = 13$;
2. $W = (1,2,3)$, generic system of W-degree $\mathbf{D} = (6,6,6)$: then $d_{\text{reg}} = 15 > i_{\text{reg}} + 1 = 13$;
3. $W = (2,3)$, generic system of W-degree $\mathbf{D} = (6,6)$: then $d_{\text{reg}} = 6 < i_{\text{reg}} = 7$.

Only the case 1 is observed with generic homogeneous systems. Furthermore, examples 1 and 2 show that the degree of regularity depends upon the order of the variables (chosen in the description of the system of weights). As the Hilbert series of a generic sequence doesn't depend on that order, it shows that we probably need to find a better tool in order to evaluate more precisely the degree of regularity in the quasi-homogeneous case. However, the above bound already leads to good improvements on the complexity bounds, as we will see in the following sections. Also note that these computations only hold when the system is 0-dimensional, we will discuss that restriction in section 2.3.

Genericity. We now prove that zero-dimensional W-homogeneous sequences of given W-degree are generically regular, under some assumptions on the W-degree. Let us start with the first part of this statement:

LEMMA 3. *Let n be a positive integer, and consider the algebra $A := \mathbb{K}[X_1,\dots,X_n]$, graded with respect to the system of weights $W = (w_1,\dots,w_n)$. Regular sequences of length n form a Zariski-open subset of all sequences of quasi-homogeneous polynomials of given W-degree in A.*

PROOF. Let $(d_1,\dots,d_m)$ be a family of W-degrees, we consider the set $V(\mathbb{K}[\mathbf{a}][\mathbf{X}])$ of all systems of quasi-homogeneous polynomials of W-degree $d_1,\dots,d_m$, where $\mathbf{a}$ is a set of variables representing the coefficients of the polynomials. We denote by $f_1,\dots,f_m$ the polynomials of the generic system, and by I the ideal they generate, in $\mathbb{K}[\mathbf{a}][\mathbf{X}]$.

Since the Hilbert series (1) characterizes regular sequences ([21, cor. 3.2]), the sequence (f_i) is regular if and only if the ideal I contains all monomials of W-degree between $i_{\text{reg}}(I) + 1$ and $i_{\text{reg}}(I) + \max\{w_i\}$, where $i_{\text{reg}}(I)$ is given by $\sum(d_i - w_i)$. This expresses that a given set of linear equations has solutions, and so it can be coded as some determinants being non-zero. □

There are some systems of W-degree for which there is no regular sequence. The reason is that because of the weights, for some systems of W-degrees, there exists no or very few monomials. For example, take $n = 2$, $W = (1,2)$ and $\mathbf{D} = (1,1)$. All quasi-homogeneous polynomials of W-degree 1 are in $\mathbb{K}X$, so there is no regular sequence of quasi-homogeneous polynomials with these W-degrees.

However, if we only consider "reasonable" systems of W-degrees, that is systems of W-degrees for which there exists a regular sequence, regular sequences form a Zariski-dense subset from the above.

Remark 1. A sufficient condition for example is to take weighted degrees such that d_1 is divisible by w_1, ..., d_n is divisible by w_n. Thus we can define the sequence $X_1^{d_1/w_1},\dots,X_n^{d_n/w_n}$, which is regular, and so for such systems of weight, the regularity condition is generic.

We only proved the genericity for quasi-homogeneous sequences of length n, the more general case of a sequence of length $m \le n$ will be proved in section 2.3 (remark 2).

2.3 Noether position

To compute the degree and degree of regularity of quasi-homogeneous systems of positive dimension, we will assume that the system $F = (f_1, \dots, f_m)$ we consider is in *Noether position* (as seen in [8, ch. 13, sec. 1] or [3, def. 2]), i.e. the ideal $I = \langle F \rangle$ satisfies the two following conditions:

- for $i \leq m$, the canonical image of X_i in $\mathbb{K}[\mathbf{X}]/I$ is an algebraic integer over $\mathbb{K}[X_{m+1}, \dots, X_n]$;
- $\mathbb{K}[X_{m+1}, \dots, X_n] \cap I = 0$.

LEMMA 4. *Let $F = f_1, \dots, f_m$ be a regular quasi-homogeneous sequence of polynomials in $\mathbb{K}[X_1, \dots, X_n]$. The sequence F is in Noether position if and only if $F_{\text{ext}} := f_1, \dots, f_m, X_{m+1}, \dots, X_n$ is a regular sequence.*

PROOF. Let I be the ideal generated by the f_i's. The geometric characterization of Noether position (see e.g. [19]) shows that the canonical projection onto the m first coordinates

$$\pi : V(I) \to V(\langle X_1, \dots, X_m \rangle)$$

is a surjective morphism with finite fibers. This implies that the variety $V(\langle F_{\text{ext}} \rangle)$, that is $\pi^{-1}(0)$, is zero-dimensional, and so the sequence is regular.

Conversely, assume F_{ext} is a regular sequence. Let $i \leq m$, we want to show that X_i is integral over the ring $\mathbb{K}[X_{m+1}, \dots, X_n]$. Since F_{ext} defines a zero-dimensional ideal, there exists $n_i \in \mathbb{N}$ such that $X_i^{n_i} = \text{LT}(f)$ with $f \in \langle F_{\text{ext}} \rangle$ for the GREVLEX ordering with $X_1 > \dots > X_n$. By definition of the GREVLEX ordering, we can assume that f simply belongs to I. This shows that every X_i is integral over $\mathbb{K}[X_{i+1}, \dots, X_n]/I$. We get the requested result by induction on i : first, this is clear if $i = m$. Now assume that we know that $\mathbb{K}[X_i, \dots, X_n]/I$ is an integral extension of $\mathbb{K}[X_{m+1}, \dots, X_n]$. From the above, we also know that X_{i-1} is integral over $\mathbb{K}[X_i, \dots, X_n]$, and so, since the composition of integral homomorphisms is integral, we get the requested result.

Finally, we want to check the second part of the definition of Noether position. Assume that there is a non-zero polynomial in $\mathbb{K}[X_{m+1}, \dots, X_n] \cap I$, since the ideal is quasi-homogeneous, we can assume this polynomial to be quasi-homogeneous. Either this polynomial is of degree 0, or it is a non-trivial syzygy between $X_{m+1}, \dots, X_n$. So in any case, it contradicts the regularity hypothesis. □

As we did for regular sequences, we first show how we can evaluate the degree and degree of regularity of a sequence in Noether position, and then we show that the Noether position property is generic under some assumptions on the W-degree of the polynomials.

THEOREM 5. *Let $W = (w_1, \dots, w_n)$ be a system of weights, and $f_1, \dots, f_m$ a regular sequence in Noether position, of quasi-homogeneous polynomials of W-degrees $(d_1, \dots, d_m)$. The same way we did above, we denote by I the ideal generated by the f_i's. Then we have $\deg(I) = \prod_{i=1}^{m} \frac{d_i}{w_i}$ and $d_{\text{reg}}(I) \leq \sum_{i=1}^{m} (d_i - w_i) + \max\{w_i\}$.*

PROOF. Let us denote by I' the ideal generated by F_{ext}. The degree of the ideal I' is the same as that of I, because the variety it defines is the intersection of $V(I)$ with some non-zero-divisor hyperplanes. Furthermore, all critical pairs appearing in a run of F_5 on F will also appear in a run of F_5 on F_{ext}, ensuring that $d_{\text{reg}}(F) \leq d_{\text{reg}}(F_{\text{ext}})$.

But since by Noether position, the family F_{ext} defines a zero-dimensional variety, we can use the previous computations to deduce its degree of regularity and the degree of I'. □

LEMMA 6. *Let n be a positive integer, and consider the algebra $A := \mathbb{K}[X_1, \dots, X_n]$, graded with respect to the system of weights $W = (w_1, \dots, w_n)$. Systems in Noether position form a Zariski-open subset of all systems of quasi-homogeneous polynomials of given W-degrees in A.*

PROOF. Let $F = (f_1, \dots, f_m)$ be m generic quasi-homogeneous polynomials, with coefficients in $\mathbb{K}[\mathbf{a}]$. We use the same characterization of a zero-dimensional regular sequence as we did in the proof of Lemma 3. It allows us to express the regularity condition for the sequence $(f_1, \dots, f_m, X_{m+1}, \dots, X_n)$ as some determinants being non-zero, which by definition, shows that the condition of being in Noether position is an open condition. □

Since a sequence in Noether position is in particular a regular sequence, we are confronted with the same problem as for the genericity of regular sequences, that is the possible emptiness of the condition. However, it is still true that for "reasonable" systems of W-degrees, i.e. systems of W-degrees for which there exists enough monomials, sequences in Noether position do exist, and thus form a Zariski-dense subset of all sequences. For example, since the sequence $X_1^{d_1/w_1}, \dots, X_m^{d_m/w_m}$ is in Noether position, the sufficient condition given in Remark 1 is also sufficient to ensure that sequences in Noether position are Zariski-dense.

Remark 2. Any sequence in Noether position is in particular a regular sequence, so Lemma 6 proves that, under the same assumption on the degree, regular sequences of length $m \leq n$ are generic among quasi-homogeneous sequences of given W-degree.

3. COMPUTING GRÖBNER BASES

3.1 Using the standard algorithms on the homogenized system

As we said before, in order to apply the F_5 algorithm to a quasi-homogeneous system, we may run it through hom_W. This is shown by the following proposition.

PROPOSITION 7. *Let $F = (f_1, \dots, f_m)$ be a family of polynomials in $\mathbb{K}[X_1, \dots, X_n]$, assumed to be quasi-homogeneous for a system of weights $W = (w_1, \dots, w_n)$. Let $<_1$ be a monomial order, G the reduced Gröbner basis of $\text{hom}_W(F)$ for this order, and $<_2$ the pullback of $<_1$ through hom_W. Then*

1. *all elements of G are in the image of hom_W;*
2. *the family $G' := \text{hom}_W^{-1}(G)$ is a reduced Gröbner basis of the system F for the order $<_2$.*

PROOF. The morphism hom_W preserves S-polynomials, in the sense that $\text{S-Pol}(\text{hom}_W(f), \text{hom}_W(g)) = \text{hom}_W(\text{S-Pol}(f, g))$. Recall that we can compute a reduced Gröbner basis by running the Buchberger algorithm, which involves only multiplications, additions, tests of divisibility and computation of S-polynomials. Since all these operations are compatible with hom_W, if we run the Buchberger algorithm on both F and $\text{hom}_W(F)$ simultaneously, they will follow exactly the same computations up to application of hom_W. The consequences on the final reduced Gröbner basis follow. □

In practice, if we want to compute a LEX Gröbner basis of F, we generate the system $\widetilde{F} = \text{hom}_W(F)$, we compute a GREVLEX basis $\widetilde{G_1}$ of $\widetilde{F}$ with F_5, and then we compute a LEX Gröbner basis $\widetilde{G_2}$ of $\widetilde{F}$ with FGLM. In the end, we get a LEX Gröbner basis of $\widetilde{F}$, which we turn into a LEX Gröbner basis of F via hom_W^{-1}.

3.2 Direct algorithms

We can now explain why algorithm FGLM becomes a bottleneck with the above strategy. Indeed, we have seen that going through hom_W increases the Bézout bound of the system by a factor $\prod_{i=1}^{n} w_i$, and recall that the complexity of the FGLM step is polynomial in that bound.

Here is a workaround. In the above process, we can apply hom_W^{-1} to the basis $\widetilde{G_1}$ and thus obtain a W-GREVLEX basis G_1 of F. We can then run FGLM on that basis to obtain a LEX basis of F. Thus, we can avoid the problem of a greater degree of the ideal on the complexity of the FGLM step.

Algorithm F_5 operates by computing S-pairs, and as such, the argument of the proof of proposition 7 can be adapted, showing that going through $\hom_W$ is equivalent to running a F_5 algorithm following weighted degree instead of total degree. However, to evaluate the complexity of the F_5 algorithm, we instead study a less-efficient variant called Matrix-F_5 (described for example in [14]), which needs to be adapted to the quasi-homogeneous case. All we need to do is change the algorithm a little, in order to consider directly the variables with their weight. The modified algorithm is algorithm 1 opposite. The function F5CRITERION$(\mu, i, \mathscr{M})$ implements the F_5-criterion described in [10]: it evaluates to false if and only if μ is the leading term of a line of the matrix $\mathscr{M}_{d-d_i,i-1}$. The function ECHELONFORM(M) reduces the matrix M to row-echelon form, not allowing any row swap.

Algorithm 1: Matrix-F_5 (W-homogeneous version)

Input: $f_1, \dots, f_m$ W-homogeneous polynomials with W-degrees $d_1, \dots, d_m$; $d_{\max} \in \mathbb{N}$

Output: G Gröbner basis of $\langle f_1, \dots, f_m \rangle$ up to W-degree $d_{\max}$

$G \leftarrow \{f_1, \dots, f_m\}$;
for $d = 1$ **to** $d_{\max}$ **do**
  $\mathscr{M}_{d,0} \leftarrow$ matrix with 0 lines;
  for $i = 1$ **to** m **do**
    if $d = d_i$ **then**
      $\mathscr{M}_{d,i} \leftarrow \widetilde{\mathscr{M}_{d,i-1}} \cup$ line f_i with label $(1, f_i)$;
    else if $d > d_i$ **then**
      $\mathscr{M}_{d,i} \leftarrow \widetilde{\mathscr{M}_{d,i-1}}$;
      for $j = 1$ **to** n **do**
        forall the lines f of $\widetilde{\mathscr{M}_{d-w_j,i}}$ with label (e, f_i) s.t. the biggest variable dividing e is x_j **do**
          for $k = n$ **downto** j **do**
            if F5CRITERION$(x_k e, i, \mathscr{M})$ **then**
              $\mathscr{M}_{d,i} \leftarrow \mathscr{M}_{d,i} \cup x_k f$ with label $(x_k e, f_i)$;
  $\widetilde{\mathscr{M}_{d,m}} \leftarrow$ ECHELONFORM$(\mathscr{M}_{d,m})$;
  For any line having been reduced to a non-zero polynomial, append it to G ;
return G

3.3 First complexity bounds

Let $F = (f_1, \dots, f_n)$ be a system of W-homogeneous polynomials in $\mathbb{K}[X_1, \dots, X_n]$, and let I be the ideal generated by F, D the degree of I, d_{reg} its degree of regularity and i_{reg} its index of regularity. The classical complexity bounds of Matrix-F_5 (for a regular system) and FGLM are

$$C_{F_5} = O\left(d_{\text{reg}} M_{d_{\text{reg}},W}(n)^{\omega}\right); \; C_{FGLM} = O\left(nD^3\right),$$

where $M_{d,W}(n)$ stands for the number of monomials of W-degree d in n variables (see for example [2] for F_5 and [11] for FGLM).

Assuming the system F is a regular sequence, we have already seen the following estimates:

$$d_{\text{reg}} \le i_{\text{reg}} + \max\{w_i\}\,; \qquad D = \frac{\prod_{i=1}^n d_i}{\prod_{i=1}^n w_i}.$$

If we compare these values with their equivalent with the system of weights $\mathbf{1}$, we notice a significant gain in theoretical complexity bounds for both the FGLM and F_5 algorithms.

But this gain in complexity for F_5 does not take into account the size of the computed matrices. That size is necessarily reduced, because the number of monomials of given W-degree is much smaller than the number of monomials of given $\mathbf{1}$-degree. The point of the following lemma is to evaluate this gain.

LEMMA 8. *Let $W = (w_1, \dots, w_n)$ be a system of weights, and for any i, let $W_i = (w_1, \dots, w_i)$. For any integer d, we denote by $M_{d,W}(n)$ the number of monomials of W-degree d, that is the size of the matrix of W-degree d. Let $\delta := \gcd(W)$, $P := \prod_{i=1}^n w_i$, S_i the integer defined recursively as following:*

$$S_1 = 0,\; S_i = S_{i-1} + w_i \cdot \frac{\gcd(W_{i-1})}{\gcd(W_i)} \; \textit{for } i \ge 2$$

and T_i the integer defined recursively as following:

$$T_1 = 0,\; T_i = T_{i-1} + w_i \cdot \left(\frac{\gcd(W_{i-1})}{\gcd(W_i)} - 1\right) - 1 \; \textit{for } i \ge 2.$$

Then the number of monomials of W-degree d is bounded above and below by:

$$\frac{\delta}{P} M_{d-T_n-n+1,\mathbf{1}}(n) \le M_{d,W}(n) \le \frac{\delta}{P} M_{d+S_n-n+1,\mathbf{1}}(n).$$

PROOF. This is a consequence of theorems 3.3 and 3.4 in [1], if we recall that $M_{d,\mathbf{1}}(n) = \binom{d+n-1}{d} = \binom{d+n-1}{n-1}$. □

Note that if $W = \mathbf{1}$, the bounds we get are trivial, which means the complexity bounds we will obtain with them will specialize without any difficulty to the known bounds for the homogeneous case.

Using the notation $S = S_n$, we get this new complexity bound for quasi-homogeneous Matrix-F_5:

$$\begin{aligned} C_{F_5} &= O\left(d_{\text{reg}} M_{d_{\text{reg}},W}(n)^{\omega}\right) \\ &= O\left(\left(i_{\text{reg}} + \max\{w_i\}\right) \cdot \left[\frac{\delta}{P}\binom{i_{\text{reg}} + \max\{w_i\} + S - 1}{n-1}\right]^{\omega}\right). \end{aligned} \tag{2}$$

On the other hand, the estimate on the degree of a quasi-homogeneous variety gives the following complexity bound for FGLM:

$$C_{FGLM} = O\left(n\left[\frac{\widetilde{D}}{P}\right]^3\right),$$

where $\widetilde{D} = \prod_{i=1}^n d_i$ is the degree of the ideal $\langle \hom_W(F) \rangle$. In the end, for the whole process, we can see that the complexity bound for our direct strategy is smaller by a factor of P^{ω}, when compared to the strategy of going through $\hom_W$.

3.4 Precise analysis of matrix-F_5

Let us now follow more closely the computations occurring in the Matrix-F_5 algorithm, and obtain more accurate complexity bounds. For this purpose, we take on the computations made in [2, ch. 3], without proving them whenever the proof is an exact transcription of the homogeneous case.

Let $W = (w_1, \dots, w_n)$ be a system of weights, and $f_1, \dots, f_m$ a system of quasi-homogeneous polynomials in $\mathbb{K}[X_1, \dots, X_n]$, which we assume satisfies the hypotheses **H1** and **H2**. We denote by $(d_1, \dots, d_m)$ the respective W-degrees of the polynomials $f_1, \dots, f_m$, and we will assume them to allow the existence of such systems.

We also denote by:

- $A_i = \mathbb{K}[X_1, \dots, X_i]$, and $A = A_n$;
- S_i the integer defined in Lemma 8, and $S = S_n$;
- $P_i = \prod_{j=1}^i w_j$, and $P = P_n$;
- $I_i = \langle f_1, \dots, f_i \rangle$, and $I = I_m$;
- $\widetilde{f_j} = \hom_W(f_j)$;
- $\widetilde{I_i} = \langle \widetilde{f_1}, \dots, \widetilde{f_i} \rangle$, and $\widetilde{I} = \widetilde{I_m}$;
- $D_i = \deg(I_i) = \prod_{j=1}^i (d_j / w_j)$;

- $\widetilde{D}_i = \deg(\widetilde{I}_i) = \prod_{j=1}^{i} d_j$;
- $d_{\mathrm{reg}}^{(i)}$ the degree of regularity of I_i (or of $\widetilde{I}_i$) ;
- G_i the W-GREVLEX Gröbner basis of I_i as given by Matrix-F_5.

With these notations, we are going to prove the following theorem:

THEOREM 9. *Let $W = (w_1, \ldots, w_n)$ be a system of weights, and $f_1, \ldots, f_m$ ($m \leq n$) a system of W-homogeneous polynomials satisfying **H1** and **H2**. Then the complexity of quasi-homogeneous Matrix-F_5 algorithm (algorithm 1) is:*

$$C_{\mathrm{F}_5} = O\left(\sum_{i=2}^{m} (D_{i-1} - D_{i-2}) M_{d_{\mathrm{reg}}^{(i)},W}(i) M_{d_{\mathrm{reg}}^{(i)},W}(n)\right)$$

We aim at computing precisely how many lines are reduced in a run of the Matrix-F_5 algorithm, that is, the number of polynomials in the returned Gröbner basis. This is done by the following proposition, which is a weak variant of [3, th. 10]:

PROPOSITION 10. *Let $(f_1, \ldots, f_m)$ be a W-homogeneous system (w.r.t a system of weights W) satisfying the hypotheses **H1** and **H2**. Let G_i be a reduced Gröbner basis of $(f_1, \ldots, f_i)$ for the W-GREVLEX monomial ordering, for $1 \leq i \leq m$. Then the number of polynomials of W-degree d in G_i whose leading term does not belong to $\mathrm{LT}(G_{i-1})$ is bounded by $b_{d,i}$, defined by the generating series*

$$B_i(z) = \sum_{d=0}^{\infty} b_{d,i} z^d = z^{d_i} \prod_{k=1}^{i-1} \frac{1 - z^{d_k}}{1 - z^{w_k}}.$$

PROOF. The proof of [3, th. 10] still holds in the quasi-homogeneous case, using formula (1) for the Hilbert series of a quasi-homogeneous regular sequence. □

So we can obtain a better bound for the number of elementary operations performed in a Matrix-F_5 run. Indeed, $B_i(1)$ represents the number of reduced polynomials in the computation of a Gröbner basis of $(f_1, \ldots, f_i, X_{i+1}, \ldots, X_n)$, that is as many as in the computation of a Gröbner basis of $(f_1, \ldots, f_i)$: since we only perform reductions under the pivot line, [3, prop. 9] shows that the lines coming from $X_{i+1}, \ldots, X_n$ will not add any reduction. Note that the above generating series is the same as the Hilbert series of $\langle f_1, \ldots, f_{i-1}, X_i, \ldots, X_n \rangle$, and so, that its value at $z = 1$ is the degree of that ideal, or D_{i-1}. Therefore, we know that the number of reduced polynomials with label (m, f_i) will be $D_{i-1} - D_{i-2}$ (with convention that $D_0 = 0$).

Now, let g be any polynomial of W-degree d being reduced in a run of the Matrix-F_5 algorithm on $(f_1, \ldots, f_i)$. From [3, prop. 9], we know that the leading term of g, after reduction, is in A_i. So overall, in W-degree d, we reduce by at most as many lines as there are monomials in A_i, that is $M_{d,W}(i)$. Furthermore, each reduction costs at most $O(M_{d,W}(n))$ elementary algebraic operations, since this is the length of the matrix lines. And we perform these reductions up to degree $d_{\mathrm{reg}}^{(i)}$. Note that, if $i = 1$, there clearly isn't any reduction in the computation, and we obtain the following formulas:

$$C_{\mathrm{F}_5} = O\left(\sum_{i=2}^{m} (D_{i-1} - D_{i-2}) M_{d_{\mathrm{reg}}^{(i)},W}(i) M_{d_{\mathrm{reg}}^{(i)},W}(n)\right) \quad (3)$$

$$= O\left(\sum_{i=2}^{m} \frac{1}{P_i P_n} \left(\frac{\widetilde{D_{i-1}}}{P_{i-1}} - \frac{\widetilde{D_{i-2}}}{P_{i-2}}\right) \cdot M_{d_{\mathrm{reg}}^{(i)}+S_i-i+1,\mathbf{1}}(i) \cdot M_{d_{\mathrm{reg}}^{(i)}+S_n-n+1,\mathbf{1}}(n)\right)$$

In comparison, the above reasoning for Matrix-F_5 applied to $\widetilde{F}$ would give

$$C_{\mathrm{F}_5} = O\left(\sum_{i=2}^{m} \left(\widetilde{D_{i-1}} - \widetilde{D_{i-2}}\right) M_{\widetilde{d_{\mathrm{reg}}^{(i)}},\mathbf{1}}(i) M_{\widetilde{d_{\mathrm{reg}}^{(i)}},\mathbf{1}}(n)\right) \quad (4)$$

so that here again, working with quasi-homogeneous polynomials yields a gain or roughly P^3. Note that the exponent 3 (instead of the previous ω) is not really meaningful, because we assumed here that we were using the naive pivot algorithm to perform the Gauss reduction. However, if we assume $\omega = 3$ in the previous computations as well, we observe that our new bound is generally much better than the previous one: figure 1 shows a plot of data obtained both with algorithm 1 and with Matrix-F_5 through $\hom_W$, together with the different bounds we can compute.

Asymptotically, though, the gain does not look important, since the complexity is still $O(nD^3)$ where D is the degree of the ideal and $n \geq m$ the number of variables, or in $O(nd^{3n})$ where d is the greatest d_i.

Remark 3. One may also push the computations a bit further, and obtain an even more accurate bound, expressed in terms of the $b_{d,i}$ (these calculations are done in [2] for the homogeneous case, and can easily be transposed to the quasi-homogeneous case):

$$C_{\mathrm{F}_5} = O\left(\sum_{i=1}^{m-1} \sum_{d=0}^{\infty} \frac{b_{d+d_{i+1},i+1}}{P_{i+1} P_n} \cdot M_{d+d_{i+1}+S_{i+1}-i,\mathbf{1}}(i+1) \cdot M_{d+d_{i+1}+S_n-n+1,\mathbf{1}}(n)\right). \quad (5)$$

As an example, we computed that bound as well for a particular case, and included it in figure 1. As one can see, that bound is indeed better than the intermediate evaluation (3), but the difference is low enough to justify using the latter evaluation. Furthermore, the bound (3) expressed in terms of the D_i's is more useful in practice, since it has a closed formula using only the parameters of the system (n, m, d_i and w_i). That allows us to use it in complexity evaluations, in both theory and practice.

Remark 4. As one can see on figure 1, the number of operations needed by Matrix-F_5 on the homogenized system is not significantly higher than the number of operations needed by the quasi-homogeneous variant of Matrix-F_5. That is mostly true because the unmodified algorithm can make use of some of the structure of the quasi-homogeneous systems (for example, columns of zeroes in the matrices).

4. THE AFFINE CASE

We will now consider the case of input that do not necessarily consist of quasi-homogeneous polynomials. One of the methods to find a GREVLEX Gröbner basis of such a system is to apply F_5, considering at W-degree d the set of monomials having W-degree lower than or equal to d. This is equivalent to homogenizing the system, i.e. to adding a variable $X_1 > \cdots > X_n > H$, and applying the classical F_5 algorithm to this homogeneous system. The reverse transformation is done by evaluating each polynomial at $H = 1$.

However, this process makes it harder to compute the complexity of the F_5 algorithm. The main reason is that dehomogenizing does not necessarily preserve W-degree, and as a consequence, it is no longer true that running the Matrix-F_5 algorithm up to W-degree d provides us with a basis, truncated at W-degree d. What remains true though is that past some W-degree, we may obtain a Gröbner basis for the entire ideal.

Generally, we want to avoid *degree falls* in the run of F_5, that is, reductions where the W-degree of the reductee is less than the W-degrees of the polynomials forming the S-pair. This phenomenon is similar to reductions to zero in the quasi-homogeneous case. It can be ruled out by considering only systems which are *regular in the affine sense* (as found in [2] for gradings in total degree).

Definition 2. Let W be a system of weights, and $(f_1, \ldots, f_n)$ be a system of not-necessarily W-homogeneous polynomials. We denote

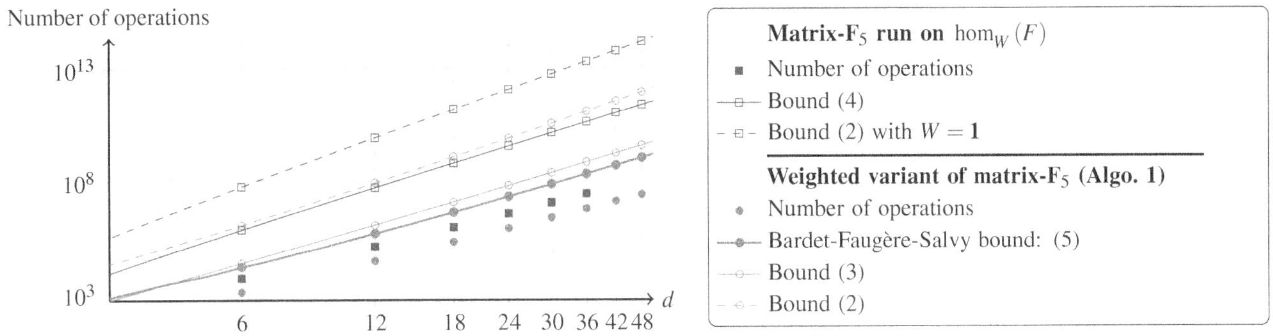

Figure 1: Bounds and values, on a log-log scale, for the number of arithmetic operations performed in Matrix-F_5 for a generic system with $W = (1,2,3)$ and $\mathbf{D} = (d,d,d)$

by h_i the quasi-homogeneous component of highest W-degree in f_i, for any $1 \leq i \leq n$. We say that the sequence (f_i) is *regular in the affine sense* when the sequence (h_i) is regular (in the quasi-homogeneous sense). We define the *degree of regularity* of the ideal $\langle f_i \rangle$ as the degree of regularity of the ideal $\langle h_i \rangle$.

Since a degree fall in a run of F_5 is precisely a reduction to zero in the highest W-degree quasi-homogeneous components of the system, we know that the F_5 criterion rules out all degree falls in a run of F_5 on such a regular system. In turns, it ensures that for such a system, running Matrix-F_5 up to degree d returns a d-Gröbner basis of F.

Hence we can study the complexity of F_5 by looking at a run of Matrix-F_5 on the homogenized system. As an example, we prove the following theorem:

THEOREM 11. *Let $W = (w_1, \dots, w_n)$ be a system of weights, and let $f_1, \dots, f_m$ be a generic system of polynomials of the form $f_i = g_i + \lambda_i$, with g_i W-homogeneous of W-degré d_i and $\lambda_i \in \mathbb{K}$. Let D be the degree of the system, d_{reg} its degree of regularity, and δ the gcd of the d_i's. We can compute a W-GREVLEX Gröbner basis of this system in time*

$$O\left(\frac{d_{\text{reg}}}{\delta^{\omega}} M_{d,W}(n)^{\omega}\right),$$

or in other words, we can divide the known complexity of the F_5 process on such a system by δ^{ω}.

PROOF. The idea is that when we homogenize the system, we can choose any suitable weight for H, not necessarily 1. More precisely, we can set the weight of H to be δ, so that the homogenized polynomials become $f_i^h = g_i + \lambda_i H^{d_i/\delta}$.

Thus, assuming the computations made at section 2.2 still hold, we have the same improvements on the bound on d_{reg} and on the size of matrices as before, and thus we have the wanted result.

Note that even if the initial system is generic, the homogenized system is not. However, one can check that if the initial system was regular in the affine sense, the homogenized system is still regular. Indeed, it's enough to check that no reduction to zero occur in a Matrix-F_5 run, but it is clear, since such a reduction would in particular be a degree fall. Also, the property of being in Noether position for the m first variables is clearly kept upon homogenizing.

As such, generically, our homogenized system is regular and in Noether position, so the previous computations indeed still hold. □

5. EXPERIMENTAL RESULTS

We have run some benchmarks[1], using the FGb library and the Magma algebra software. We present these results in Tables 1a and 1b. The examples are chosen with increasing n (number of variables and polynomials), two different classes of systems of weights W and systems of W-degrees D. With these conditions, we built a generic system of polynomials f_i in $\mathbb{F}_{65521}[\mathbf{X}]$, such that all monomials appearing in f_i have W-degree at most d_i. The last examples are systems arising in the study of the Discrete Logarithm Problem, when trying to compute the decompositions of points on an elliptic curve (see [17]). In both cases, we use a shortened notation for the systems of weights and the degrees, where for example $(2^3, 1^2)$ means $(2,2,2,1,1)$. The magma benchmarks were run on a machine with 128 GB RAM and 3 GHz CPU, running Magma v.2.17-1. The FGb benchmarks were run on a laptop with 16 GB RAM and 3 GHz CPU.

For each system, we compared our strategy ("qh") with the default strategy ("std"), for both steps. The algorithms used by the FGb library are F_5 and an implementation of FGLM taking advantage of the sparsity of the matrices ([13]). The algorithms used by Magma are F_4 and the classical FGLM. The complexity of sparse-FGLM depends on the number of solutions of the system and on the shape of the input basis, while the complexity of classical FGLM depends only on the number of solutions. This explains why we can see a speed-up on the FGLM step in FGb, even though the degree is unchanged.

Acknowledgments. This work was supported in part by the HPAC grant (ANR ANR-11-BS02-013) and by the EXACTA grant (ANR-09-BLAN-0371-01) of the French National Research Agency.

[1] All the systems we used are available online on `http://www-polsys.lip6.fr/~jcf/Software/benchsqhomog.html`.

6. REFERENCES

[1] G. Agnarsson. On the Sylvester denumerants for general restricted partitions. In *Proceedings of the Thirty-third Southeastern International Conference on Combinatorics, Graph Theory and Computing (Boca Raton, FL, 2002)*, volume 154, pages 49–60, 2002.

[2] M. Bardet. *Étude des systèmes algébriques surdéterminés. Applications aux codes correcteurs et à la cryptographie.* Thesis, Université Pierre et Marie Curie - Paris VI, Dec. 2004.

[3] M. Bardet, J.-C. Faugère, and B. Salvy. On the complexity of the F_5 Gröbner basis algorithm. Private communication, 2012.

[4] T. Becker and V. Weispfenning. *Gröbner bases*, volume 141 of *Graduate Texts in Mathematics*. Springer-Verlag, New York, 1993. A computational approach to commutative algebra, In cooperation with Heinz Kredel.

[5] W. Bosma, J. Cannon, and C. Playoust. The Magma algebra system. I. The user language. *J. Symbolic Comput.*, 24(3-4):235–265, 1997. Computational algebra and number theory (London, 1993).

System	deg(I)	t_{F_5} (qh)	t_{F_5} (std)	Speed-up for F_5	t_{FGLM} (qh)	t_{FGLM} (std)	Speed-up for FGLM
Generic $n=7$, $W=(1^4,2^3)$, $\mathbf{D}=(4^7)$	2048	2.7 s	3.4 s	1.2	0.4 s	1.1 s	2.6
Generic $n=8$, $W=(1^4,2^4)$, $\mathbf{D}=(4^8)$	4096	12.3 s	22.5 s	1.8	2.4 s	7.3 s	3.0
Generic $n=9$, $W=(1^5,2^4)$, $\mathbf{D}=(4^9)$	16384	314.9 s	778.5 s	2.5	119.6 s	327.8 s	2.7
Generic $n=7$, $W=(2^5,1^2)$, $\mathbf{D}=(4^7)$	512	0.1 s	0.3 s	3.2	0.1 s	0.1 s	1.7
Generic $n=8$, $W=(2^6,1^2)$, $\mathbf{D}=(4^8)$	1024	0.4 s	1.6 s	4.2	0.2 s	0.3 s	1.9
Generic $n=9$, $W=(2^7,1^2)$, $\mathbf{D}=(4^9)$	2048	1.6 s	8 s	4.9	0.6 s	1.2 s	2.0
Generic $n=10$, $W=(2^8,1^2)$, $\mathbf{D}=(4^{10})$	4096	7.5 s	40.4 s	5.4	2.4 s	6.2 s	2.6
Generic $n=11$, $W=(2^9,1^2)$, $\mathbf{D}=(4^{11})$	8192	33.3 s	213.5 s	6.4	17.5 s	41.2 s	2.4
Generic $n=12$, $W=(2^{10},1^2)$, $\mathbf{D}=(4^{12})$	16384	167.9 s	1135.6 s	6.8	115.8 s	246.7 s	2.1
Generic $n=13$, $W=(2^{11},1^2)$, $\mathbf{D}=(4^{13})$	32768	796.7 s	6700 s	8.4	782.7 s	1645.1 s	2.1
Generic $n=14$, $W=(2^{12},1^2)$, $\mathbf{D}=(4^{14})$	65536	5040.1 s	∞	∞	5602.3 s	NA	NA
DLP Edwards $n=4$, $W=(2^3,1)$, $\mathbf{D}=(8^4)$	512	0.1 s	0.1 s	1	0.1 s	0.1 s	1
DLP Edwards $n=5$, $W=(2^4,1)$, $\mathbf{D}=(16^5)$	65536	935.4 s	6461.2 s	6.9	2164.4 s	6935.6 s	3.2

(a) Benchmarks with FGb

System	deg(I)	t_{F_4} (qh)	t_{F_4} (std)	Speed-up for F_4	t_{FGLM} (qh)	t_{FGLM} (std)	Speed-up for FGLM
Generic $n=7$, $W=(1^4,2^3)$, $\mathbf{D}=(4^7)$	2048	7.9 s	14 s	1.7	214.2 s	222.7 s	1
Generic $n=8$, $W=(1^4,2^4)$, $\mathbf{D}=(4^8)$	4096	62.6 s	138.3 s	2.2	1774.7 s	1797.1 s	1
Generic $n=9$, $W=(1^5,2^4)$, $\mathbf{D}=(4^9)$	16384	3775.5 s	8830.5 s	2.3	∞	∞	NA
Generic $n=7$, $W=(2^5,1^2)$, $\mathbf{D}=(4^7)$	512	0.2 s	0.7 s	3.5	45.5 s	45.6 s	1
Generic $n=8$, $W=(2^6,1^2)$, $\mathbf{D}=(4^8)$	1024	1 s	6.2 s	6.2	512.3 s	515.6 s	1
Generic $n=9$, $W=(2^7,1^2)$, $\mathbf{D}=(4^9)$	2048	6 s	88.1 s	14.7	7965 s	8069.4 s	1
Generic $n=10$, $W=(2^8,1^2)$, $\mathbf{D}=(4^{10})$	4096	42.4 s	911.8 s	21.5	∞	∞	NA
Generic $n=11$, $W=(2^9,1^2)$, $\mathbf{D}=(4^{11})$	8192	292.5 s	12126.4 s	41.5	∞	∞	NA
Generic $n=12$, $W=(2^{10},1^2)$, $\mathbf{D}=(4^{12})$	16384	2463.2 s	146774.7 s	59.6	∞	∞	NA
Generic $n=13$, $W=(2^{11},1^2)$, $\mathbf{D}=(4^{13})$	32768	∞	∞	NA	∞	∞	NA
DLP Edwards $n=4$, $W=(2^3,1)$, $\mathbf{D}=(8^4)$	512	1 s	1 s	1	1 s	27 s	27
DLP Edwards $n=5$, $W=(2^4,1)$, $\mathbf{D}=(16^5)$	65536	6044 s	56105 s	9.3	∞	∞	NA

(b) Benchmarks with Magma

Table 1: Benchmarks with FGb and Magma for some affine systems

[6] B. Buchberger. A theoretical basis for the reduction of polynomials to canonical forms. *ACM SIGSAM Bull.*, 10(3):19–29, 1976.

[7] A. Dickenstein and I. Z. Emiris. Multihomogeneous resultant matrices. In *Proceedings of the 2002 International Symposium on Symbolic and Algebraic Computation*, pages 46–54, New York, 2002. ACM.

[8] D. Eisenbud. *Commutative algebra*, volume 150 of *Graduate Texts in Mathematics*. Springer-Verlag, New York, 1995. With a view toward algebraic geometry.

[9] J.-C. Faugére. A new efficient algorithm for computing Gröbner bases (F_4). *J. Pure Appl. Algebra*, 139(1-3):61–88, 1999. Effective methods in algebraic geometry (Saint-Malo, 1998).

[10] J.-C. Faugère. A new efficient algorithm for computing Gröbner bases without reduction to zero (F_5). In *Proceedings of the 2002 International Symposium on Symbolic and Algebraic Computation*, pages 75–83 (electronic) , New York, 2002. ACM.

[11] J. C. Faugère, P. Gianni, D. Lazard, and T. Mora. Efficient computation of zero-dimensional Gröbner bases by change of ordering. *J. Symbolic Comput.*, 16(4):329–344, 1993.

[12] J.-C. Faugère and A. Joux. Algebraic cryptanalysis of hidden field equation (HFE) cryptosystems using Gröbner bases. In *Advances in cryptology—CRYPTO 2003*, volume 2729 of *Lecture Notes in Comput. Sci.*, pages 44–60. Springer, Berlin, 2003.

[13] J.-C. Faugère and C. Mou. Sparse FGLM algorithms. Preprint available at `http://hal.inria.fr/hal-00807540`.

[14] J.-C. Faugère and S. Rahmany. Solving systems of polynomial equations with symmetries using SAGBI-Gröbner bases. In *ISSAC 2009—Proceedings of the 2009 International Symposium on Symbolic and Algebraic Computation*, pages 151–158. ACM, New York, 2009.

[15] J.-C. Faugère, M. Safey El Din, and P.-J. Spaenlehauer. Gröbner bases of bihomogeneous ideals generated by polynomials of bidegree $(1,1)$: algorithms and complexity. *J. Symbolic Comput.*, 46(4):406–437, 2011.

[16] J.-C. Faugère. FGb: A Library for Computing Gröbner Bases. In K. Fukuda, J. Hoeven, M. Joswig, and N. Takayama, editors, *Mathematical Software - ICMS 2010*, volume 6327 of *Lecture Notes in Computer Science*, pages 84–87, Berlin, Heidelberg, September 2010. Springer Berlin / Heidelberg.

[17] J.-C. Faugère, P. Gaudry, L. Huot, and G. Renault. Using symmetries in the index calculus for elliptic curves discrete logarithm. Cryptology ePrint Archive, Report 2012/199, 2012.

[18] R. Hartshorne. *Algebraic geometry*. Springer-Verlag, New York, 1977. Graduate Texts in Mathematics, No. 52.

[19] J. S. Milne. Algebraic geometry (v5.22), 2012. Available at `www.jmilne.org/math/`.

[20] L. Robbiano. On the theory of graded structures. *J. Symbolic Comput.*, 2(2):139–170, 1986.

[21] R. P. Stanley. Hilbert functions of graded algebras. *Advances in Math.*, 28(1):57–83, 1978.

Computing Rational Solutions of Linear Matrix Inequalities

Qingdong Guo
Key Laboratory of Mathematics Mechanization, Chinese Academy of Sciences
Beijing 100190, China
qdguo@mmrc.iss.ac.cn

Mohab Safey El Din
UPMC, Univ Paris 06
INRIA, Paris-Rocquencourt Center, POLSYS project-team
CNRS LIP6 UMR 7606
Institut Universitaire de France
Mohab.Safey@lip6.fr

Lihong Zhi
Key Laboratory of Mathematics Mechanization, Chinese Academy of Sciences
Beijing 100190, China
lzhi@mmrc.iss.ac.cn

ABSTRACT

Consider a $(D \times D)$ symmetric matrix A whose entries are linear forms in $\mathbb{Q}[X_1, \dots, X_k]$ with coefficients of bit size $\leq \tau$. We provide an algorithm which decides the existence of rational solutions to the linear matrix inequality $\mathsf{A} \succeq 0$ and outputs such a rational solution if it exists. This problem is of first importance: it can be used to compute algebraic certificates of positivity for multivariate polynomials. Our algorithm runs within $(k\tau)^{O(1)}2^{O(\min(k,D)D^2)}D^{O(D^2)}$ bit operations; the bit size of the output solution is dominated by $\tau^{O(1)}2^{O(\min(k,D)D^2)}$. These results are obtained by designing algorithmic variants of constructions introduced by Klep and Schweighofer. This leads to the best complexity bounds for deciding the existence of sums of squares with rational coefficients of a given polynomial. We have implemented the algorithm; it has been able to tackle Scheiderer's example of a multivariate polynomial that is a sum of squares over the reals but not over the rationals; providing the first computer validation of this counter-example to Sturmfels' conjecture.

Categories and Subject Descriptors

I.1.2 [**Symbolic and Algebraic Manipulation**]: Algorithms—*algebraic algorithms*; F.2.2 [**Analysis of Algorithms and Problem Complexity**]: Non numerical algorithms and problems—*complexity of proof procedures*

General Terms

Theory, Algorithms.

*The authors are supported by CNRS and INRIA through the ECCA project at LIAMA and by the EXACTA grant (ANR-09-BLAN-0371-01) of the French National Research Agency (ANR) and the Chinese National Natural Science Foundation under Grant 60911130369. The second author is also supported by the GeoLMI grant (ANR 2011 BS03 011 06) of the French National Research Agency (ANR) and the Institut Universitaire de France. The third author is also supported by NKBRPC 2011CB302400, the Chinese National Natural Science Foundation under Grants 91118001, 60821002/F02.

Keywords

effective real algebraic geometry, linear matrix inequality, rational sum of squares, semidefinite programming, convexity, complexity

1. INTRODUCTION

Motivation and problem statement. Let A be a symmetric $(D \times D)$-matrix whose entries are linear forms in $\mathbb{Q}[X_1, \dots, X_k]$ with coefficients of bit size τ. We consider the problem of computing a rational point $\mathbf{x} \in \mathbb{Q}^k$ which is a solution to the linear matrix inequality $\mathsf{A} \succeq 0$ (in other words $\mathsf{A}(\mathbf{x})$ is positive semi-definite, i.e. all its eigenvalues are non-negative).

This problem can be seen as a variant of integer linear programming or a diophantine version of semi-definite programming. It has become a topical question since semi-definite programming is used to compute sums-of-squares decompositions of polynomials which provide algebraic certificates of positivity [8, 9, 14, 5] and are used in polynomial optimization. In this framework, one issue is to get *rational* solutions to linear matrix inequalities.

This has been formalized through Sturmfels' conjecture asking whether all polynomials with coefficients in $\mathbb{Q}$ and which are sums of squares of polynomials with coefficients in $\mathbb{R}$ can be written as a sum of squares of polynomials with coefficients in $\mathbb{Q}$. More recently, Scheiderer gave an example showing that Sturmfels' conjecture is not true [21]. It is worth to remark that any algorithm designed for grabbing rational solutions to linear matrix inequalities may provide a computer proof to Scheiderer's example.

In [20], an algorithm is given to compute rational points in convex semi-algebraic sets. Linear Matrix Inequalities (LMIs) define convex semi-algebraic sets: these are defined by sign conditions on the coefficients of the characteristic polynomial [15]. As a by-product, [20] provides an algorithmic solution to the problem we consider here. Using the above notations, the algorithm in [20] applied to Linear Matrix Inequalities runs within $\tau^{O(1)}D^{O(k^3)}$.

However, recall that the original motivation for computing rational solutions to Linear Matrix Inequalities is sums of squares decompositions of polynomials. If $f \in \mathbb{Q}[Y_1, \dots, Y_n]$ has degree $2d$, the linear matrix inequality generated to decompose f as a sum of squares is such that $D = \binom{n+d}{n}$ and $k \leq \frac{1}{2}D(D+1) - \binom{n+2d}{n}$. Technical computations show that $\frac{1}{2}D(D+1) - \binom{n+2d}{n}$ lies in $O(\min(n^{2d}, d^{2n}))$ and $\binom{n+d}{n}$ lies in $O(\min(n^d, d^n))$. As a consequence, denoting $\min(n^d, d^n)$ by $\mathsf{M}(d,n)$ [20] yields a complexity $\tau^{O(1)}\mathsf{M}(d,n)^{\mathsf{M}(d,n)^6}$.

The goal of this paper is twofold:

1. improve the above complexity *by exploiting the special structure of LMIs;*
2. by designing an algorithm which is able to provide a "computer-proof" for the non-existence of a sum of squares decomposition over the rationals for Scheiderer's example [21].

Main results. Our study is more restrictive than the one in [20] since we do not consider general convex semi-algebraic sets but only those defined by Linear Matrix Inequalities. Consider a $(D \times D)$ symmetric matrix A whose entries are linear forms in $\mathbb{Q}[X_1, \ldots, X_k]$ with coefficients of bit size $\leq \tau$.

Our main results rely heavily on results obtained by Klep and Schweighofer in [12]. The algorithm we obtain can be seen as an effective variant of the results in [12] which provides some constructions of linear equations S and a linear matrix inequality $\widehat{\mathsf{A}} \succeq 0$ of size $(D-1, D-1)$ such that the set of common rational solutions of S and $\widehat{\mathsf{A}} \succeq 0$ is the same the set of rational solutions of $\mathsf{A} \succeq 0$.

Our algorithm is of recursive nature; it outputs a rational point $\mathbf{x} \in \mathbb{Q}^k$ at which A is positive semi-definite whenever such a point exists else it returns an empty list. It runs within $(k\tau)^{O(1)}2^{O(\min(k,D)D^2)}D^{O(D^2)}$ bit operations and in case of non-emptiness the output point has coordinates of bit size bounded by $\tau^{O(1)}2^{O(\min(k,D)D^2)}$ (see Theorem 4.1 below).

Note that on families of Linear Matrix Inequalities where $k \simeq D^2$ the obtained complexity bounds on runtime and size of output are better than the ones obtained in [20] (we get $k^{1.5}$ in the exponent instead of k^3). For the important application of sums of squares decompositions over the rationals of n-variate polynomials of degree $2d$, using the estimates on binomials which are given above, we obtain as a new bound for the runtime $\tau^{O(1)}2^{O(\mathsf{M}(d,n)^3)}\mathsf{M}(d,n)^{\mathsf{M}(d,n)^2}$ which lies in $\tau^{O(1)}2^{O(\mathsf{M}(d,n)^3)}$ and dramatically improves the one obtained from [20]. The same bound is obtained for the size of the output. This is summarized in the theorem below.

THEOREM 1.1. *Let $f \in \mathbb{Q}[X_1, \ldots, X_n]$ of degree $2d$ with coefficients of bit size $\leq \tau$. There exists an algorithm which decides the existence of a sum of squares decomposition of f over the rationals and computes such a decomposition whenever it exists within $\tau^{O(1)}2^{O(\mathsf{M}(d,n)^3)}$ bit operations where $\mathsf{M}(d,n) = \min(d^n, n^d)$. The bit size of the output is also dominated by $\tau^{O(1)}2^{O(\mathsf{M}(d,n)^3)}$.*

We also implemented our algorithm and ran it on several examples. In particular, our implementation, which uses routines provided by the RAGLIB package [17] has been able to provide the first *computer validation of Scheiderer's result.* The resulting Linear Matrix Inequality to solve over the rationals is rather small: there are 6 variables and the size of the matrix is 6×6. But, as far as we know, our implementation is the first one that can handle a non-trivial linear matrix inequality and solve it *over the rationals.*

Related works. Solving Linear Matrix Inequalities over the rationals has been mainly developed in [20]. It is worth to note that these algorithms are actually based on ideas derived by the ones in [10, 11] for solving Linear Matrix Inequalities over the *integers.* This paper mostly relies on results in [12] and it can be seen as an algorithmic variant exploiting some theoretical results in [12] (mostly Theorem 2.3.2 and Proposition 3.2.1). The algorithm makes use of several complexity results in real algebraic geometry mainly due to Basu, Pollack and Roy which are stated and proved in [2, 3]. Our implementation uses [17] which relies on variants of the algorithms presented in [4, 18]. As we explained previously, this paper is motivated by the global polynomial optimization for which computing algebraic certificates is of first importance. It is worth to note that several alternative approaches in Computer Algebra have been developed for polynomial optimizations, let us mention those based on the critical point method (see [1, 6, 19]) and those based on Cylindrical Algebraic Decomposition (see e.g. [1, 7]).

Structure of the paper.. We start in Section 2 with preliminaries, stating some properties of Linear Matrix Inequalities and complexity results in effective real algebraic geometry for solving polynomial systems over the reals. In Section 3, we design the main subroutines on which our main algorithm relies: the first one treats basic cases (univariate inequalities and those whose solution sets have non-empty interior); the other one constructs the aforementioned linear equations using [12]. Finally, we detail how our algorithm runs on Scheiderer's example; providing the first computer validation of the non-existence of rational sums-of-squares decompositions to polynomials with rational coefficients which are sums of squares with real coefficients.

2. PRELIMINARIES

Basic definitions and notations. A symmetric matrix M with entries in $\mathbb{R}$ is said to be positive definite (resp. positive semi-definite) when all its eigenvalues are positive (resp. non-negative). We will write respectively $\mathsf{M} \succ 0$ and $\mathsf{M} \succeq 0$.

In the sequel, given a matrix or a vector M with entries in a ring R, $\mathsf{M}^\star$ denotes the transposition of M.

Let $X_1, \ldots, X_k$ be indeterminates, $\mathsf{A}_0, \ldots, \mathsf{A}_k$ be $(D \times D)$ symmetric matrices with entries in $\mathbb{R}$ and A be the linear matrix $\mathsf{A}_0 + X_1\mathsf{A}_1 + \cdots + X_k\mathsf{A}_k$. For $\mathbf{x} = (\mathbf{x}_1, \ldots, \mathbf{x}_k) \in \mathbb{R}^k$, we write $\mathsf{A}(\mathbf{x})$ for $\mathsf{A}_0 + \mathbf{x}_1\mathsf{A}_1 + \cdots + \mathbf{x}_k\mathsf{A}_k$. We consider the linear matrix inequality

$$\mathsf{A} = \mathsf{A}_0 + X_1\mathsf{A}_1 + \cdots + X_k\mathsf{A}_k \succeq 0.$$

We denote by $\mathfrak{S}(\mathsf{A}) = \{\mathbf{x} \in \mathbb{R}^k \mid \mathsf{A}(\mathbf{x}) \succeq 0\}$ the feasible region of A. It is a closed convex semi-algebraic set lying in $\mathbb{R}^k$. We will say that the linear matrix inequality $\mathsf{A} \succeq 0$ is infeasible when $\mathfrak{S}(\mathsf{A}) = \emptyset$, else it is feasible.

Basic properties of Linear Matrix Inequalities. We start with immediate properties of Linear Matrix Inequalities.

LEMMA 2.1. *Let $\mathsf{A} = \mathsf{A}_0 + X_1\mathsf{A}_1 + \cdots + X_k\mathsf{A}_k$ where A_i is a $(D \times D)$ symmetric matrix with entries in $\mathbb{Q}$ (for $0 \leq i \leq k$) and $E \subset \mathbb{R}^k$ be an affine linear subspace defined by a set of linear equations. Assume that the entries of the i-th row and column of A vanish at all points in E and let $\widehat{\mathsf{A}}$ be the $(D-1, D-1)$-matrix obtained by removing the i-th row and column of A. Then $\mathfrak{S}(\widehat{\mathsf{A}}) \cap E = \mathfrak{S}(\mathsf{A}) \cap E$.*

LEMMA 2.2. *Let* $\mathsf{A} = \mathsf{A}_0 + X_1\mathsf{A}_1 + \cdots + X_k\mathsf{A}_k$ *where* A_i *is a* $(D \times D)$ *symmetric matrix with entries in* $\mathbb{Q}$ *(for* $0 \leq i \leq k$*),* P *be an invertible matrix and* $\mathsf{A}' = \mathsf{P}^\star\mathsf{A}\mathsf{P}$. *If* $\mathbf{x} \in \mathfrak{S}(\mathsf{A})$ *then,* $\mathbf{x} \in \mathfrak{S}(\mathsf{A}')$.

Encoding of real algebraic points. Our algorithm will manipulate points which are obtained by algorithms searching for real roots of semi-algebraic sets defined by polynomials with coefficients in $\mathbb{Q}$ (see e.g. [3, Chapter 13]).

The coordinates of these points are real algebraic numbers. As in [20], we will encode such a point $(\alpha_1, \ldots, \alpha_k)$ classically with a 0-dimensional parametrization

$$\mathscr{Q} = (q(T), q_0(T), q_1(T), \ldots, q_k(T))$$

where $q, q_0, \ldots, q_k$ lie in $\mathbb{Q}[T]$, $\mathsf{gcd}(q, q_0) = 1$, q is **irreducible** and such that for some root ϑ of q, $\alpha_i = q_i(\vartheta)/q_0(\vartheta)$ of its coordinates and a Thom-encoding Θ of ϑ (we refer to [3, Chapters 2 and 12] for details about Thom-encodings and univariate representations).

Given the encoding of a real algebraic point $\mathscr{Q}, \Theta$, we will consider the routines MinPol and Param which return respectively q and the vector $(\frac{q_1}{q_0}, \ldots, \frac{q_k}{q_0})$.

Now, consider a 0-dimensional parametrization $\mathscr{U} = (q, q_0, \mathscr{U}_1, \ldots, \mathscr{U}_D) \subset \mathbb{Q}[T]^{2+D\times D}$ of degree δ and a Thom-encoding Θ encoding a point $(\mathbf{u}_1, \ldots, \mathbf{u}_D)$ (with $\mathbf{u}_i \in \mathbb{R}^D$). We will use a routine ExtractFirstEntry$((\mathscr{U}, \Theta), D)$ which returns the encoding $((q, q_0, \mathscr{U}_1), \Theta)$ of $\mathbf{u}_1$.

Decision procedures over the reals. Let Φ be a quantifier-free formula involving s polynomials in k variables of degree $\leq \delta$ and with bit-size bounded by τ and let $\mathfrak{S} \subset \mathbb{R}^k$ be the semi-algebraic set defined by Φ. As in [20], we will consider a subroutine Decision which takes as input Φ and outputs a sample point in $\mathfrak{S}$ iff $\mathfrak{S} \neq \emptyset$ within $\tau s^{k+1}\delta^{O(k)}$ bit-operations, else it returns an empty list (see [3, Chapter 15] or [2]).

In the non-empty situation, such a real point encoded by $(\mathscr{Q}, \Theta)$ where $\mathscr{Q}$ is a 0-dimensional parametrization and Θ is a Thom-encoding of a real root of the minimal polynomial in $\mathscr{Q}$. Moreover, all polynomials in $\mathscr{Q}$ have degree bounded by $O(\delta^k)$ and the bit size of their coefficients is dominated by $\tau\delta^{O(k)}$. As in [20, Section 2.1], using factorization and Euclidean division on univariate polynomials, one can transform $\mathscr{Q}, \Theta$ to $\mathscr{Q}', \Theta'$ where $\mathscr{Q}'$ is a 0-dimensional parametrization and Θ' is a Thom-encoding which encode the same real point as $\mathscr{Q}, \Theta$ within a bit-complexity dominated by $\tau s^{k+1}\delta^{O(k)}$ and such that the minimal polynomial of $\mathscr{Q}'$ is *irreducible*.

We also recall that given a system of s strict inequalities in $\mathbb{Q}[X_1, \ldots, X_k]$ of degree δ and bit size $\leq \tau$ defining a semi-algebraic set $\mathfrak{S} \subset \mathbb{R}^k$, there exists a routine OpenDecision which computes a point with *rational* coordinates in $\mathfrak{S} \cap \mathbb{Q}^k$ within $\tau^{O(1)}s^{k+1}\delta^{O(k)}$ bit operations if and only if $\mathfrak{S} \cap \mathbb{Q}^k \neq \emptyset$ (else it returns an empty list); see [2, Proof of Theorem 4.1.2 pp. 1032]. In case of non-emptiness, the bit-size of the output is dominated by $\tau\delta^{O(k)}$.

For convenience, we summarize these complexity results in the Proposition below.

PROPOSITION 2.3. *[2] Let* Φ *be quantifier-free formula involving* s *polynomials in* k *variables of degree* $\leq \delta$ *and with bit-size bounded by* τ *and* $\mathfrak{S} \subset \mathbb{R}^k$ *be the semi-algebraic set defined by* Φ. *There exists an algorithm* Decision *which takes* Φ *as input and returns an encoding* $(\mathscr{Q}, \Theta)$ *of a point in* $\mathfrak{S}$ *if and only if* $\mathfrak{S} \neq \emptyset$ *else it returns* $\emptyset$ *within* $\tau^{O(1)}s^{k+1}\delta^{O(k)}$ *bit operations. The degrees of the polynomials in* $\mathscr{Q}$ *and the bit size of their coefficients are respectively bounded by* $O(\delta^k)$ *and* $\tau\delta^{O(k)}$.

When Φ *contains only strict inequalities, there exists an algorithm* OpenDecision *which returns a rational point in* $\mathfrak{S}$ *if and only if* $\mathfrak{S} \neq \emptyset$ *(else it returns an empty list) within* $\tau^{O(1)}s^{k+1}\delta^{O(k)}$ *bit operations. In case of non-emptiness, the bit-size of the output is dominated by* $\tau\delta^{O(k)}$.

Retrieving rational points. Below, we consider a real algebraic number $\vartheta \in \mathbb{R}$ and its minimal polynomial $q \in \mathbb{Q}[T]$ of degree δ. We also consider linear forms

$$\mathscr{L} = g_0(\vartheta) + g_1(\vartheta)X_1 + \cdots + g_k(\vartheta)X_k$$

such that $g_0, \ldots, g_k$ are rational fractions in $\mathbb{Q}(T)$ sharing the same denominator q_0 of degree $\leq \delta - 1$ and whose numerators $n_0, \ldots, n_k$ have also degree $\leq \delta - 1$. We assume that $\mathsf{gcd}(q_0, q) = 1$ and that there exists $0 \leq i \leq k$ such that $n_i \neq 0$.

Consider the linear forms $\ell_0, \ldots, \ell_{\delta-1}$ in $\mathbb{Q}[X_1, \ldots, X_k]$ which are the coefficients of $1, T, \ldots, T^{\delta-1}$ in the polynomial

$$n_0 + n_1X_1 + \cdots + n_kX_k.$$

Since, by assumption, there exists i such that $n_i \neq 0$, then there exists j such that $\ell_j \neq 0$. We denote by ExtractLinForms a routine which takes as input $\mathscr{L}, q$ and returns all linear forms ℓ_j such that $\ell_j \neq 0$.

The following Lemma is extracted from the correctness proof of the algorithm given in [20]. For clarity and completeness, we isolate this statement below and prove it.

LEMMA 2.4. *[20] Let* $\mathfrak{S} \subset \mathbb{R}^k$ *be a semi-algebraic set,* ϑ *be a real algebraic number of degree* δ*,* q *be its minimal polynomial,* τ *be a bound on the bit-size of the coefficients of* q *and* $\mathscr{L}$ *be a linear form in* $\mathbb{Q}(\vartheta)[X_1, \ldots, X_k]$. *Assume that* $\mathscr{L}$ *vanishes at all points in* $\mathfrak{S}$. *Then all linear forms in* $L =$ ExtractLinForms$(\mathscr{L}, q)$ *vanish at all points in* $\mathfrak{S} \cap \mathbb{Q}^k$ *and* L *is obtained within* $O(\tau k\delta^{O(1)})$ *bit operations. The bit size of the coefficients in the output is dominated by* $O(\tau)$.

PROOF. Take $\mathbf{x} = (\mathbf{x}_1, \ldots, \mathbf{x}_k) \in \mathfrak{S} \cap \mathbb{Q}^k$; by assumption $\mathscr{L}$ vanishes at $\mathbf{x}$; we conclude that $n_0 + n_1\mathbf{x}_1 + \cdots + n_k\mathbf{x}_k = 0$. Since ϑ is a real algebraic number of degree δ and the n_i's have degree $\leq \delta - 1$ we deduce that all the linear forms $\ell_0, \ldots, \ell_{\delta-1}$ vanish at $\mathbf{x}$. Runtime and bound on the bit size of the output are immediate. □

3. SUBROUTINES

Let $\mathsf{A}_0, \ldots, \mathsf{A}_k$ be symmetric matrices of size $D \times D$ and entries in $\mathbb{Q}$, A be the linear matrix $\mathsf{A}_0 + X_1\mathsf{A}_1 + \cdots + X_k\mathsf{A}_k$ and $\mathfrak{S}(\mathsf{A}) \subset \mathbb{R}^k$ be the feasible region of the linear matrix inequality $\mathsf{A} \succeq 0$. Recall that $\mathfrak{S}(\mathsf{A})$ is a convex semi-algebraic set. Our algorithm is based on a case distinction:

1. when $k = 1$, then $\mathfrak{S}(\mathsf{A})$ is either empty or a real point or an interval with non-empty interior;
2. when $D = 1$, then A is a linear form and unless $k = 0$ and $\mathsf{A} < 0$, $\mathfrak{S}(\mathsf{A})$ has non-empty interior;
3. when $\mathfrak{S}(\mathsf{A})$ is full dimensional, i.e. $\mathfrak{S}(\mathsf{A})$ has non-empty interior; in this case we say that the linear matrix inequality is strongly feasible;

These three cases will be tackled by a subroutine BasicCasesLMI;

4. when $\mathfrak{S}(A)$ is *not* full dimensional; if $\mathfrak{S}(A) = \emptyset$, we say that $A \succeq 0$ is infeasible, else we say that it is weakly feasible. In this latter case, according to [20, Lemma 3.4], there exists a hyperplane in $\mathbb{R}^k$ which contains $\mathfrak{S}(A)$.

   The routine WeakLMI below constructs linear forms whose coefficients are real algebraic numbers. From Lemma 2.4, this will allow us to deduce other linear forms S with coefficients in $\mathbb{Q}$ whose set of common solutions $\mathsf{Sols}(S) \in \mathbb{R}^k$ contains $\mathfrak{S}(A) \cap \mathbb{Q}^k$. It will also return a $(D-1, D-1)$ symmetric linear matrix $\widehat{A}$ such that $\mathfrak{S}(A) \cap \mathbb{Q}^k = \mathfrak{S}(\widehat{A}) \cap \mathsf{Sols}(S) \cap \mathbb{Q}^k$.

3.1 Subroutine BasicCasesLMI

Let $A = A_0 + X_1 A_1 + \cdots + X_k A_k$ where $A_0, \ldots, A_k$ are $(D \times D)$ symmetric matrices with entries in $\mathbb{Q}$ of bit size bounded by τ. We describe a subroutine BasicCasesLMI which takes as input $A, [X_1, \ldots, X_k]$ and

1. when $k = 1$, it returns a point with rational coordinates in $\mathfrak{S}(A)$ iff $\mathfrak{S}(A) \cap \mathbb{Q} \neq \emptyset$ else it returns an empty list;
2. a point with rational coordinates in $\mathfrak{S}(A)$ if $\mathfrak{S}(A)$ has a non-empty interior;

else it returns false.

Let $\chi(y) = y^D + m_{D-1} y^{D-1} + \cdots + m_0$ be the characteristic polynomial of A, we denote by Φ the following formula:

$$\Phi = \{(-1)^{(i+D)} m_i \geq 0,\ 0 \leq i \leq D-1\}$$

and by Ψ the following formula:

$$\Psi = \{(-1)^{(i+D)} m_i > 0,\ 0 \leq i \leq D-1\}.$$

By [15], the semi-algebraic set $\mathfrak{S}(A)$ is defined by Φ; the interior of $\mathfrak{S}(A)$ is defined by Ψ.

BasicCasesLMI($A, [X_1, \ldots, X_k]$)

1. If $k = 1$ and if there exists a linear factor $X - a$ of m_i (with $a \in \mathbb{Q}$) for some $0 \leq i \leq D-1$ such that $(-1)^{(j+D)} m_j(a) \geq 0$ for $j \neq i$ then return a.
2. $\mathscr{U} = \mathsf{OpenDecision}(\Psi)$.
3. If $\mathscr{U}$ is not empty or $k = 1$ then return $\mathscr{U}$ else return false.

PROPOSITION 3.1. *Let* $A = A_0 + X_1 A_1 + \cdots + X_k A_k$ *where* $A_0, \ldots, A_k$ *are* $(D \times D)$ *symmetric matrices with entries in* $\mathbb{Q}$ *of bit size bounded by* τ.

If $k = 1$ *and* $\mathfrak{S}(A) \cap \mathbb{Q} \neq \emptyset$, BasicCasesLMI($A, [X_1]$) *returns a rational point in* $\mathfrak{S}(A)$ *else it returns an empty list.*

If $\mathfrak{S}(A) \subset \mathbb{R}^k$ *is full-dimensional,* BasicCasesLMI($A, [X_1, \ldots, X_k]$) *returns a rational point in* $\mathfrak{S}(A) \cap \mathbb{Q}^k$ *else it returns* false.

It runs within $\tau^{O(1)} D^{O(k)}$ *bit operations, in case of non-empty output, it has bit size bounded by* $\tau^{O(1)} D^{O(k)}$.

PROOF. Assume that $\mathfrak{S}(A) \subset \mathbb{R}$ (Step 1). Then, since $\mathfrak{S}(A)$ is convex, it is either empty or a point or an interval with non-empty interior. Suppose that it is a point. Then, it is the unique solution of Φ since $\mathfrak{S}(A)$ is defined by Φ. By assumption, this solution is not a solution of Ψ (else $\mathfrak{S}(A)$ would have a non-empty interior). Then, it can be obtained as the root of linear factors of one m_i for some $0 \leq i \leq D-1$ at which the $(-1)^{j+D} m_j$'s (for $j \neq i$) are positive. Note that in the univariate case, the costs of factorizations and real root isolations are polynomial in τ and D and if a rational point is output at this step it has bit length bounded by $(D\tau)^{O(1)}$ (see [13]).

Now assume that $\mathfrak{S}(A)$ is full-dimensional; it has a non-empty interior. Then, by Proposition 2.3, Step 2 returns a rational point in $\mathfrak{S}(A) \cap \mathbb{Q}^k$ if and only if $\mathfrak{S}(A) \neq \emptyset$.

Suppose now that $\mathfrak{S}(A)$ is not full dimensional. If $k = 1$, we can conclude that $\mathfrak{S}(A)$ is empty and we are done. Now, assume that $k \geq 2$. Recall that the semi-algebraic set defined by the formula Ψ is the interior of $\mathfrak{S}(A)$; we deduce that it is empty. Then, by Proposition 2.3 $\mathscr{U}$ is empty and we return false.

By Proposition 2.3, runtime and bound on the output are immediate. □

3.2 Subroutine WeakLMI

Let $A = A_0 + X_1 A_1 + \cdots + X_k A_k$ where $A_0, \ldots, A_k$ are $(D \times D)$ symmetric matrices with entries in $\mathbb{Q}$ of bit size bounded by τ. We describe a routine WeakLMI which takes as input $A, [X_1, \ldots, X_k]$ such that the LMI $A \succeq 0$ is weakly feasible or infeasible and such that there does not exist $\mathbf{u} \in \mathbb{R}^D - \{\mathbf{0}\}$ such that $A\mathbf{u} = \mathbf{0}$. It returns $S, \widehat{A}$ such that

1. S is a sequence of linear forms in $\mathbb{Q}[X_1, \ldots, X_k]$ which vanish at all points in $\mathfrak{S}(A) \cap \mathbb{Q}^k$; we let $\mathsf{Sols}(S) \subset \mathbb{R}^k$ be the linear subspace of common solutions of S;
2. $\widehat{A}$ is a symmetric linear matrix of size $(D-1) \times (D-1)$ with entries in $\mathbb{Q}[X_1, \ldots, X_k]$ such that

$$\mathfrak{S}(\widehat{A}) \cap \mathsf{Sols}(S) \cap \mathbb{Q}^k = \mathfrak{S}(A) \cap \mathbb{Q}^k.$$

Recall that, as the feasible region of a linear matrix inequality, $\mathfrak{S}(A)$ is convex. Hence, by [20, Lemma 3.4], we already know that there exists a linear form $\mathscr{L} \in \mathbb{R}[X_1, \ldots, X_k]$ which vanishes at all points in $\mathfrak{S}(A)$. In [12, Prop. 3.3.1], Klep and Schweighofer exploited special properties of LMI's to prove that $A \succeq 0$ is a weak LMI (i.e. $\mathfrak{S}(A)$ has an empty interior) if and only if there exists a non-zero linear form $\mathscr{L} \in \mathbb{R}[X_1, \ldots, X_k]$ and a $(D \times D)$ matrix W with entries in $\mathbb{R}[X_1, \ldots, X_k]$ such that

$$\mathrm{Tr}(AW^\star W) = -\mathscr{L}^2.$$

As noticed in the first line of the proof of [12, Prop. 3.3.1], this implies that $\mathscr{L}$ vanishes at all points in $\mathfrak{S}(A)$. Another stronger result was given in the proof of [12, Lemma. 4.3.5]: there exist linear forms $\mathscr{L}_1, \ldots, \mathscr{L}_D$ in $\mathbb{R}[X_1, \ldots, X_k]$ and $(D \times D)$ matrices $W_1, \ldots, W_D$ such that

$$\mathrm{Tr}(AW_i^\star W_i) = -\mathscr{L}_i^2, \text{ for } 1 \leq i \leq D.$$

In the proof of [12, Lemma. 4.3.5], the matrices $W_1, \ldots, W_D$ are constructed from points in the semi-algebraic sets

$$G_1 = \{\mathbf{u} \in \mathbb{R}^D - \{\mathbf{0}\} \mid \mathbf{u}^\star A \mathbf{u} = 0\}$$

$$G_2 = \{(\mathbf{u}_1, \ldots, \mathbf{u}_D) \in \mathbb{R}^{D \times D} \mid \sum_{i=1}^{D} \mathbf{u}_i^\star A \mathbf{u}_i = 0, \mathbf{u}_1, \mathbf{u}_2 \neq \mathbf{0}\}$$

The algorithm described below can be seen as an effective counterpart and variant of the constructions in the proof of [12, Lemma. 4.3.5] to construct $\mathscr{L}_1, \ldots, \mathscr{L}_D$. These are obtained from the encodings of points in the aforementioned semi-algebraic sets. Finally, according to Lemma 2.4 (see also [20]), one can deduce from $\mathscr{L}_i$ linear equations with rational coefficients that must be satisfied by all elements in $\mathfrak{S}(\mathsf{A}) \cap \mathbb{Q}^k$.

We denote by ConstructFormula1, ConstructFormula2 routines that take as input A and return the following formulas defining respectively G_1, G_2:

$$||\mathbf{U}||^2 > 0, \mathbf{U}^\star \mathsf{A}_i \mathbf{U} = 0, \text{ for } 0 \leq i \leq k$$

$$||\mathbf{U}_1||^2 > 0, ||\mathbf{U}_2||^2 > 0, \sum_{i=1}^{D} \mathbf{U}_i^\star \mathsf{A}_j \mathbf{U}_i^\star = 0, \text{ for } 0 \leq j \leq k$$

where $\mathbf{U} = [U_1, \ldots, U_D]^\star$ is a vector of new indeterminates and $\mathbf{U}_1, \ldots, \mathbf{U}_D$ are vectors of new indeterminates $[U_{1,i}, \ldots, U_{D,i}]^\star$ for $1 \leq i \leq D$. We can now describe the algorithm WeakLMI.

WeakLMI(A, $[X_1, \ldots, X_k]$)

1. Let $\mathscr{U}$ = Decision(ConstructFormula1(A))
2. If $\mathscr{U}$ is not empty then
   (a) Let i be the smallest index of the non-null coordinates of the point encoded by $\mathscr{U}$
   (b) Let P be the matrix [Param($\mathscr{U}$), $(\mathbf{e}_j)_{1 \leq j \neq i \leq D}$] and $\mathsf{A}' = \mathsf{P}^\star \mathsf{A} \mathsf{P}$
   (c) Let $\mathscr{L}_1, \ldots, \mathscr{L}_D$ be the entries of element of $\mathsf{A}'\mathbf{e}_1$ and let $\widehat{\mathsf{A}}$ be the $(D-1, D-1)$ matrix obtained by removing the 1-st row and column in A'
   (d) Return (ExtractLinForms($\mathscr{L}_i$, MinPol($\mathscr{U}$)), $1 \leq i \leq D$), $\widehat{\mathsf{A}}$.
3. Let $\mathscr{V} = (\mathfrak{V}, \Theta)$ = Decision(ConstructFormula2(A))
4. Let $\mathscr{U}$ = ExtractFirstEntry($\mathscr{V}$, D)
   (a) Let i be the smallest index of the non-null coordinates of the point encoded by $\mathscr{U}$
   (b) Let P be the matrix [Param($\mathscr{U}$), $(\mathbf{e}_j)_{1 \leq j \neq i \leq D}$] and $\mathsf{A}' = \mathsf{P}^\star \mathsf{A} \mathsf{P}$
   (c) Let $\mathscr{L}_1, \ldots, \mathscr{L}_D$ be the entries of $\mathsf{A}'\mathbf{e}_1$ and let $\widehat{\mathsf{A}}$ be the $(D-1, D-1)$ matrix obtained by removing the 1-st row and column in A'
   (d) Return (ExtractLinForms($\mathscr{L}_i$, MinPol($\mathscr{U}$)), $1 \leq i \leq D$), $\widehat{\mathsf{A}}$.

PROPOSITION 3.2. *Let* $\mathsf{A} = \mathsf{A}_0 + X_1\mathsf{A}_1 + \cdots + X_k\mathsf{A}_k$ *where* $\mathsf{A}_0, \ldots, \mathsf{A}_k$ *are* $(D \times D)$ *symmetric matrices with entries in* $\mathbb{Q}$ *of bit size bounded by* τ. *Assume that* $\mathsf{A} \succeq 0$ *is either weakly feasible or infeasible and that there does not exist* $\mathbf{u} \in \mathbb{R}^D - \{\mathbf{0}\}$ *such that* $\mathsf{A}\mathbf{u} = \mathbf{0}$.

Then WeakLMI(A, $[X_1, \ldots, X_k]$) *returns* $S, \widehat{\mathsf{A}}$ *where* S *is a sequence of linear forms in* $\mathbb{Q}[X_1, \ldots, X_k]$ *which vanishes at all points in* $\mathfrak{S}(\mathsf{A}) \cap \mathbb{Q}^k$ *and* $\widehat{\mathsf{A}}$ *is a* $(D-1, D-1)$ *symmetric linear matrix with entries in* $\mathbb{Q}[X_1, \ldots, X_k]$ *such that* $\mathfrak{S}(\widehat{\mathsf{A}}) \cap \mathsf{Sols}(S) \cap \mathbb{Q}^k = \mathfrak{S}(\mathsf{A}) \cap \mathbb{Q}^k$.

It runs within $\tau^{O(1)}2^{O(D^2)}D^{O(D^2)}$ *bit operations; the bit size of the coefficients of forms in* S *is bounded by* $\tau 2^{O(D^2)}$ *and the bit size of coefficients of the entries of* $\widehat{\mathsf{A}}$ *is bounded by* $O(\tau)$.

Proof of correctness. We briefly sketch the construction in [12, Lemma. 4.3.5]. The construction in the proof of [12, Lemma. 4.3.5] is based on the following case distinction:

Case 1. Assume first that there exists a *non-null* vector $\mathbf{u} = (u_1, \ldots, u_D)^\star \in \mathbb{R}^D - \{0\}$ such that $\mathbf{u}^\star \mathsf{A} \mathbf{u} = 0$. Step 1 computes such a vector. In the proof of [12, Lemma. 4.3.5] it is shown that if $\mathbf{u} = \mathbf{e}_1$ then any non null element $\mathscr{L}$ of $\mathsf{A}\,\mathbf{e}_1$ vanishes at all points in $\mathfrak{S}(\mathsf{A})$; moreover since we have assumed that $\{\mathbf{u} \mid \mathsf{A}\mathbf{u} = \mathbf{0}\} = \{\mathbf{0}\}$ there exists such a non null element $\mathscr{L}$.

We denote by i the smallest index such that $u_i \neq 0$. Here, to retrieve a similar situation where $\mathbf{u} = \mathbf{e}_1$, Step 2 substitutes A by $\mathsf{A}' = \mathsf{P}^\star \mathsf{A} \mathsf{P}$ where P is the $(D \times D)$-matrix whose first column is $\mathbf{u}$ and next columns are the vectors $\mathbf{e}_j$ for $j \in \{1, \ldots, D\} - \{i\}$ (see Step 2b). Note that P is invertible and consider the matrix A' in Step 2b; Lemma 2.2 implies that $\mathfrak{S}(\mathsf{A}') = \mathfrak{S}(\mathsf{A})$. Moreover, following the proof of [12, Lemma. 4.3.5], all entries of $\mathsf{A}'\mathbf{e}_1$ (Step 2c) vanish at all points in $\mathfrak{S}(\mathsf{A}')$.

Let S be the set of linear forms obtained at Step 2d and $\mathsf{Sols}(S) \subset \mathbb{R}^k$ the affine linear subspace defined by their common solutions. Lemma 2.4 implies that

$$\mathfrak{S}(\mathsf{A}') \cap \mathbb{Q}^k = \mathfrak{S}(\mathsf{A}') \cap \mathsf{Sols}(S) \cap \mathbb{Q}^k.$$

Moreover, note that, by construction of S, the first row and column of A' is $\mathbf{0}$ at all points in $\mathsf{Sols}(S)$. Lemma 2.1 implies that $\mathfrak{S}(\widehat{\mathsf{A}}) \cap \mathsf{Sols}(S) = \mathfrak{S}(\mathsf{A}') \cap \mathsf{Sols}(S)$; we deduce that $\mathfrak{S}(\widehat{\mathsf{A}}) \cap \mathsf{Sols}(S) \cap \mathbb{Q}^k = \mathfrak{S}(\mathsf{A}) \cap \mathbb{Q}^k$ since we previously observed that $\mathfrak{S}(\mathsf{A}') = \mathfrak{S}(\mathsf{A})$.

Case 2. We assume that there doesn't exist $\mathbf{u} \in \mathbb{R}^D - \{\mathbf{0}\}$ such that $\mathbf{u}^\star \mathsf{A} \mathbf{u} = 0$. By [12], this implies that there exist vectors $\mathbf{u}_1, \ldots, \mathbf{u}_D$ in $\mathbb{R}^D$ such that $\sum_{i=1}^{D} \mathbf{u}_i^\star \mathsf{A} \mathbf{u}_i = 0$ and $\mathbf{u}_1 \neq \mathbf{0}, \mathbf{u}_2 \neq \mathbf{0}$. Step 3 computes $\mathscr{V}$ which encodes the concatenation of such vectors $\mathbf{u}_1, \ldots, \mathbf{u}_D$. Step 4 extracts from $\mathscr{V}$ the encoding $\mathscr{U}$ of the vector $\mathbf{u}_1 \in \mathbb{R}^D - \{\mathbf{0}\}$; note that $\mathbf{u}_1^\star \mathsf{A} \mathbf{u}_1 \neq 0$. In the proof of [12, Lemma. 4.3.5], when $\mathbf{u}_1 = \mathbf{e}_1$ it is proved that the entries of the first row and column of A vanish at all points in $\mathfrak{S}(\mathsf{A})$.

We retrieve such a situation when we substitute A with A' (Step 4b): here we have $\mathbf{e}_1^\star \mathsf{A}' \mathbf{e}_1 \neq 0$; note that $\mathfrak{S}(\mathsf{A}') = \mathfrak{S}(\mathsf{A})$ by Lemma 2.2. Also the linear forms $\mathscr{L}_1, \ldots, \mathscr{L}_D$ constructed at Step 4c correspond to the one constructed in the proof of [12, Lemma. 4.3.5] (Case 2) which vanish at all points in $\mathfrak{S}(\mathsf{A}') = \mathfrak{S}(\mathsf{A})$.

Now, consider

1. $\widehat{\mathsf{A}}$ is the matrix defined at Step 4c;
2. S be the set of linear forms obtained at Step 4d and $\mathsf{Sols}(S) \subset \mathbb{R}^k$ the affine linear subspace defined by their common solutions.

As in Case 1, note that by the construction of S, the first row and column of A' is $\mathbf{0}$ at $\mathsf{Sols}(S)$. Then, using the same reasoning based on Lemmas 2.1 and 2.4 as in Case 1, we conclude that

$$\mathfrak{S}(\widehat{\mathsf{A}}) \cap \mathsf{Sols}(S) = \mathfrak{S}(\mathsf{A}') \cap \mathsf{Sols}(S)$$

and that $\mathfrak{S}(\widehat{\mathsf{A}}) \cap \mathsf{Sols}(S) \cap \mathbb{Q}^k = \mathfrak{S}(\mathsf{A}) \cap \mathbb{Q}^k$ since we previously observed that $\mathfrak{S}(\mathsf{A}') = \mathfrak{S}(\mathsf{A})$. □

Complexity analysis. Proposition 2.3 implies that Step 1 requires $\tau^{O(1)}k^{O(D)}2^{O(D)}$ bit operations. Moreover, if $\mathscr{U}$ is not an empty list, it encodes a real point in G_1 using a 0-dimensional parametrization of degree $\leq O(2^D)$ and coefficients of bit size $\leq \tau 2^{O(D)}$.

Assume that $\mathscr{U}$ is not empty. Step 2a requires only gcd operations on univariate polynomials in this parametrization; the cost is polynomial in $\tau k^D 2^D$. Steps 2b-2c do not induce an extra cost. Finally, Step 2d is negligible in terms of bit operations and it returns linear forms of bit size $\leq \tau 2^{O(D)}$ (Lemma 2.4).

Assume now that $\mathscr{U}$ is empty. Then, Proposition 2.3 implies that Step 3 requires $\tau^{O(1)}D^{O(D^2)}2^{O(D^2)}$ bit operations; in case of non-emptiness, the output 0-dimensional parametrization has degree $\leq O(2^{D^2})$ and the bit size of its coefficients is bounded by $\tau 2^{O(D^2)}$. Step 4 runs within a negligible cost and as in the previous paragraph Steps 4a-4d do not induce extra cost and, by Lemma 2.4, the bit size of the output linear equations in Step 4d is bounded by $\tau 2^{O(D^2)}$.

Estimates on the bit size of the coefficients in the entries of the matrices $\widehat{\mathsf{A}}$ (Steps 2c and 4c) are immediate. □

4. MAIN ALGORITHM

Let $\mathsf{A} = \mathsf{A}_0 + X_1\mathsf{A}_1 + \cdots + X_k\mathsf{A}_k$ where $\mathsf{A}_0, \ldots, \mathsf{A}_k$ are $(D \times D)$ symmetric matrices with entries in $\mathbb{Q}$ of bit size bounded by τ. We describe now the main algorithm RationalLMI of this paper. It takes as input A, $[X_1, \ldots, X_k]$, and returns $\mathbf{x} = (\mathbf{x}_1, \ldots, \mathbf{x}_k) \in \mathfrak{S}(\mathsf{A}) \cap \mathbb{Q}^k$ encoded by the sequence $(X_1 - \mathbf{x}_1, \ldots, X_k - \mathbf{x}_k)$ if A has rational solutions; otherwise returns $\emptyset$.

At the beginning of the algorithm, we consider the following semi-algebraic set:

$$G = \{\mathbf{u} \in \mathbb{R}^D - \{\mathbf{0}\} \mid \mathsf{A}\mathbf{u} = \mathbf{0}\}.$$

We denote by ConstructFormula routine that takes as input A and returns the formula defining G.

We will also use several other basic subroutines:

1. LinearSolve: it takes a set of linear forms with coefficients in $\mathbb{Q}$ and returns a rational point in the set of their common solutions if it is not empty, else it returns an empty list.
2. GaussianElimination: it takes as input a set of linear forms in $\mathbb{Q}[X_1, \ldots, X_k]$ and performs Gaussian elimination to return $\mathcal{X}, \mathcal{H}, \mathcal{V}$ where $\mathcal{X}$ is a list of variables $X_{i_1}, \ldots, X_{i_\ell}$, $\mathcal{V}$ is the list of variables in $\{X_1, \ldots, X_k\} - \{X_{i_1}, \ldots, X_{i_\ell}\}$ and $\mathcal{H}$ is a list of linear forms $h_{i_1}, \ldots, h_{i_\ell}$ in $\mathbb{Q}[\mathcal{V}]$ such that the relations of the form $X_{i_r} = h_{i_r}(\mathcal{V})$ are generated by the input.
3. Substitute: it takes as input a list of variables $[X_1, \ldots, X_r]$, a list of linear forms $[h_1, \ldots, h_r]$ and a linear matrix A with entries depending on variables $X_{r+1}, \ldots, X_k$ and substitutes X_i by h_i in A for $1 \leq i \leq r$.
4. Evaluate: it takes as input a list of variables $\mathcal{X} = [X_1, \ldots, X_r]$, a list of polynomials $\mathcal{H} = [h_1, \ldots, h_r]$ in $\mathbb{Q}[Y_1, \ldots, Y_p]$ and a sequence of rational numbers $q = (q_1, \ldots, q_p)$; it returns the sequence $(X_i - h_i(q), 1 \leq i \leq r)$.

RationalLMI(A, $[X_1, \ldots, X_k]$)

1. $\mathscr{U}$ = BasicCasesLMI(A, $[X_1, \ldots, X_k]$)
2. If $\mathscr{U} \neq$ false is not empty then, denoting by $(\mathbf{x}_1, \ldots, \mathbf{x}_k)$ the point encoded by $\mathscr{U}$, return $X_1 - \mathbf{x}_1, \ldots, X_k - \mathbf{x}_k$
3. Let $\mathscr{U}$ = LinearSolve(ConstructFormula(A))
4. If $\mathscr{U}$ is not empty then
   (a) compute an invertible matrix P with entries in $\mathbb{Q}$ such that $\mathsf{P}\mathbf{e}_1 = \mathbf{u}$ and let $\mathsf{A}' = \mathsf{P}^\star\mathsf{A}\mathsf{P}$ and $\widehat{\mathsf{A}}$ be the $(D-1, D-1)$-matrix obtained by removing the first row and column from A'
   (b) return RationalLMI($\widehat{\mathsf{A}}$, $[X_1, \ldots, X_k]$)
5. $S, \widehat{\mathsf{A}}$ = WeakLMI(A, $[X_1, \ldots, X_k]$)
6. If LinearSolve(S) is empty then return $\emptyset$
7. $\mathcal{X}, \mathcal{H}, \mathcal{V}$ = GaussianElimination(S)
8. $\widetilde{\mathsf{A}}$ = Substitute($\mathcal{X}, \mathcal{H}, \widehat{\mathsf{A}}$) and R = RationalLMI($\widetilde{\mathsf{A}}, \mathcal{V}$)
9. If R is not empty then return R, Evaluate($\mathcal{X}, \mathcal{H}, R$) else return $\emptyset$

THEOREM 4.1. *Let* $\mathsf{A} = \mathsf{A}_0 + X_1\mathsf{A}_1 + \cdots + X_k\mathsf{A}_k$ *where* $\mathsf{A}_0, \ldots, \mathsf{A}_k$ *are* $(D\times D)$ *symmetric matrices with entries in* $\mathbb{Q}$ *of bit size bounded by* τ. *Then* RationalLMI(A, $[X_1, \ldots, X_k]$) *returns a point in* $\mathfrak{S}(\mathsf{A})\cap\mathbb{Q}^k$ *iff* $\mathfrak{S}(\mathsf{A})\cap\mathbb{Q}^k \neq \emptyset$ *else it returns an empty list. It runs within*

$$(k\tau)^{O(1)}2^{O(\min(k,D)D^2)}D^{O(D^2)}$$

bit operations and in case of non-emptiness the output point has coordinates of bit size bounded by $\tau^{O(1)}2^{O(\min(k,D)D^2)}$.

Proof of correctness. Assume for the moment that either $k = 1$ or $\mathfrak{S}(\mathsf{A})$ is full dimensional (note that if $D = 1$ and $k \geq 1$ $\mathfrak{S}(\mathsf{A})$ is full dimensional). Then, correctness follows from Proposition 3.1. The rest of the proof is by induction on D: our induction assumption is that for any linear symmetric matrix B of size $D-1$ with linear entries in $\mathbb{Q}[X_1, \ldots, X_p]$, RationalLMI(B, $[X_1, \ldots, X_p]$) outputs a rational point in $\mathfrak{S}(\mathsf{B})\cap\mathbb{Q}^p$ if and only if $\mathfrak{S}(\mathsf{B})\cap\mathbb{Q}^p$ is not empty.

Suppose first that there exists a vector $\mathbf{u} \in \mathbb{R}^D - \{0\}$ such that $\mathsf{A}.\mathbf{u} = \mathbf{0}$, then Step 3 computes such a vector. Lemma 2.2 ensures that $\mathfrak{S}(\mathsf{A}')$ (where A' is the symmetric matrix considered at Step 4a) equals $\mathfrak{S}(\mathsf{A})$. Moreover, by construction, we have $\mathsf{A}'\mathbf{e}_1 = \mathbf{0}$; then the first column (and row) of A' is $\mathbf{0}$. Then, by Lemmas 2.1 and 2.2, we conclude that $\mathfrak{S}(\widehat{\mathsf{A}}) = \mathfrak{S}(\mathsf{A}') = \mathfrak{S}(\mathsf{A})$. By the induction assumption applied to $\widehat{\mathsf{A}}$, we conclude that the call to RationalLMI at Step 4b outputs a rational point in $\mathfrak{S}(\mathsf{A})$ if $\mathfrak{S}(\mathsf{A}) \cap \mathbb{Q}^k \neq \emptyset$ else it returns $\emptyset$.

Now assume that there is no vector $\mathbf{u} \in \mathbb{R}^D - \{0\}$ such that $\mathsf{A}.\mathbf{u} = \mathbf{0}$; we are at Step 5. Proposition 3.2 implies that:

1. all linear forms in S vanish at all points in $\mathfrak{S}(\mathsf{A}) \cap \mathbb{Q}^k$ and at least one of them is not identically null; we denote by $\mathsf{Sols}(S) \subset \mathbb{R}^k$ the set of common solutions of the forms in S;
2. the $(D-1, D-1)$-matrix $\widehat{\mathsf{A}}$ is a symmetric linear matrix with entries in $\mathbb{Q}[X_1, \ldots, X_k]$ such that

$$\mathfrak{S}(\widehat{\mathsf{A}}) \cap \mathsf{Sols}(S) \cap \mathbb{Q}^k = \mathfrak{S}(\mathsf{A}) \cap \mathbb{Q}^k.$$

If S has no solution then $\mathfrak{S}(\mathsf{A})$ is empty (Step 6). In steps (7-8), the linear forms in S are used to eliminate variables in $\widehat{\mathsf{A}}$; this provides the symmetric linear matrix $\widetilde{\mathsf{A}}$ of size $(D-1, D-1)$ (Step 8). The induction assumption applied to $\widetilde{\mathsf{A}}$ allows us to conclude that, through the call to RationalLMI in Step 8, the last step returns a point in $\mathfrak{S}(\widehat{\mathsf{A}})\cap\mathsf{Sols}(S)\cap\mathbb{Q}^k$ if this set is non-empty else it returns $\emptyset$. Now recall that we previously observed that $\mathfrak{S}(\widehat{\mathsf{A}}) \cap \mathsf{Sols}(S) \cap \mathbb{Q}^k = \mathfrak{S}(\mathsf{A}) \cap \mathbb{Q}^k$; which concludes the proof. □

Complexity analysis. Let $\mathcal{A}$ be the set of symmetric linear matrices A of size D with entries in $\mathbb{Q}[X_1, \ldots, X_k]$ the coefficients of which have bit size bounded by τ. We denote by $\mathsf{C}(\tau, D, k)$ an upper bound on the runtime of the execution of RationalLMI for all possible inputs $\mathsf{A} \in \mathcal{A}$ and $[X_1, \ldots, X_k]$; we also denote by $\mathsf{T}(\tau, D, k)$ an upper bound on the bit size of the coordinates of the output of RationalLMI for all possible inputs $\mathsf{A} \in \mathcal{A}$.

By Propositions 3.1 and 3.2, there exist constants A and B large enough, independent of τ, D, k, such that

(A) Step 1 is performed within $A\tau^B D^{Bk}$ bit operations and if $\mathscr{U}$ is not empty then the returned point (Step 2) has coordinates of bit size bounded by $\tau^B D^{Bk}$;

(B) Step 3 (which consists of solving k linear systems of size $D \times D$ with coefficients of bit size $\leq \tau$), is performed within $A\tau^B D^B$ operations and the coefficients in $\mathscr{U}$ have bit size bounded by τD^B;

(C) Steps (4a-4b), which again consists of linear algebra operations, builds the matrix P in $A\tau^B D^B$ operations and the elements in the matrix P have bit size bounded by τD^B;

(D) Step 5 requires at most $\tau^B 2^{BD^2} D^{BD^2}$ bit operations and the bit size of the coefficients of forms in S is bounded by $\tau 2^{BD^2}$, and using elementary complexity results in linear algebra the bit size of coefficients of the entries of $\widetilde{\mathsf{A}}$ (in Step 8) is bounded by $\tau 2^{BD^2}$.

We let $m_{k,D} = \min(k, D)$ and prove below that

$$\begin{aligned}\mathsf{C}(\tau, D, k) &\leq Ak\tau^B 2^{B^2 m_{k,D} D^2} D^{B^2 D^2}\\ \mathsf{T}(\tau, D, k) &\leq A\tau^B 2^{B^2 m_{k,D} D^2}\end{aligned}$$

by decreasing induction on D and k. More precisely, we will assume that for $D' < D$ and $k' \leq k$

$$\begin{aligned}\mathsf{C}(\tau, D', k') &\leq Ak'\tau^B 2^{B^2 m_{k',D'} D'^2} D^{B^2 D'^2}\\ \mathsf{T}(\tau, D', k') &\leq A\tau^B 2^{B^2 m_{k',D'} D'^2}\end{aligned}$$

and that for $D' \leq D$ and $k' < k$

$$\begin{aligned}\mathsf{C}(\tau, D', k') &\leq Ak'\tau^B 2^{B^2 m_{k',D'} D'^2} D^{B^2 D'^2}\\ \mathsf{T}(\tau, D', k') &\leq A\tau^B 2^{B^2 m_{k',D'} D'^2}\end{aligned}$$

This induction is easily initialized for $k = 1$ or $D = 1$ at Steps 1-2 using Proposition 3.1.

We consider now the general case. Using the observations **(A)**, **(B)**, **(C)** and **(D)**, the worst case complexity is attained if $\mathscr{U}$ (computed in Step 1) is false and when the execution goes through Step 5. The cost of Step 3 is negligible compared to the cost of Step 1. In case of non-emptiness of $\mathscr{U}$ the recursive call at Step 4b requires to check that

$$\begin{aligned}\mathsf{C}(\tau, D, k) &\leq A\tau^B D^B + \mathsf{C}(\tau D^B, D-1, k)\\ &\leq A\tau^B D^B +\\ &\quad Ak\tau^B D^{B^2} 2^{B^2 m_{k,D-1}(D-1)^2} D^{B^2(D-1)^2}\\ &\leq Ak\tau^B 2^{B^2 m_{k,D} D^2} D^{B^2 D^2} \text{ by induction}\\ \mathsf{T}(\tau, D, k) &\leq \mathsf{T}(\tau D^B, D-1, k)\\ &\leq A\tau^B (D-1)^{B^2} 2^{B^2 m_{k,D-1}(D-1)^2} \text{ (induction)}\\ &\leq A\tau^B 2^{B^2 m_{k,D} D^2}\end{aligned}$$

Now, we need to check the worst case bound when the execution of $\mathsf{RationalLMI}$ goes through Step 5. By observation **(D)**, we get

$$\begin{aligned}\mathsf{C}(\tau, D, k) &\leq A\tau^B D^{BD^2} + \mathsf{C}(\tau 2^{BD^2}, D-1, k-1)\\ &\leq A\tau^B D^{BD^2} +\\ &\quad A(k-1)\tau^B 2^{B^2(D^2 + m_{k-1,D-1}(D-1)^2)} D^{B^2(D-1)^2}\\ &\leq Ak\tau^B 2^{B^2 m_{k,D} D^2} D^{BD^2} \text{ by induction}\\ \mathsf{T}(\tau, D, k) &\leq \mathsf{T}(\tau 2^{BD^2}, D-1, k-1)\\ &\leq A\tau^B 2^{B^2 D^2} 2^{B^2 m_{k-1,D-1}(D-1)^2} \text{ by induction}\\ &\leq A\tau^B 2^{B^2 m_{k,D} D^2}\end{aligned}$$

Finally, $\mathsf{C}(\tau, D, k)$ lies in $(k\tau)^{O(1)} 2^{O(m_{k,D} D^2)} D^{O(D^2)}$; we also have $\mathsf{T}(\tau, D, k)$ lies in $\tau^{O(1)} 2^{O(m_{k,D} D^2)}$. □

5. SCHEIDERER'S EXAMPLE

In [21], Scheiderer constructed explicit polynomials with rational coefficients which are sums of squares of polynomials with real coefficients, but not sums of squares of polynomials with rational coefficients.

By [21, Theorem 2.2], the polynomial

$$f = x^4 + x\,y^3 + y^4 - 3\,x^2\,y\,z - 4\,x\,y^2\,z + 2\,x^2\,z^2 + x\,z^3 + y\,z^3 + z^4$$

can be written as a sum of two polynomials with real coefficients but has no sos decomposition over rational numbers.

Suppose

$$f = [x^2, xy, y^2, xz, yz, z^2]\,\mathsf{A}\,[x^2, xy, y^2, xz, yz, z^2]^\star,$$

where the Gram matrix A is a 6×6 symmetric matrix

$$\begin{bmatrix} 1 & 0 & X_1 & 0 & -\frac{3}{2} - X_2 & X_3\\ 0 & -2\,X_1 & \frac{1}{2} & X_2 & -2 - X_4 & -X_5\\ X_1 & \frac{1}{2} & 1 & X_4 & 0 & X_6\\ 0 & X_2 & X_4 & -2\,X_3 + 2 & X_5 & \frac{1}{2}\\ -\frac{3}{2} - X_2 & -2 - X_4 & 0 & X_5 & -2\,X_6 & \frac{1}{2}\\ X_3 & -X_5 & X_6 & \frac{1}{2} & \frac{1}{2} & 1 \end{bmatrix}$$

with six variables: $X_1, X_2, X_3, X_4, X_5, X_6$ corresponding to seven symmetric matrices $\mathsf{A}_0, \mathsf{A}_1, \mathsf{A}_2, \mathsf{A}_3, \mathsf{A}_4, \mathsf{A}_5, \mathsf{A}_6$.

- Apply the routine $\mathsf{HasRealSolutions}$ in RAGLib [17] to compute

$$\mathscr{U} = \mathsf{OpenDecision}(\Psi).$$

The set $\mathscr{U}$ is empty, hence A is not strongly feasible.

- In step 5 of $\mathsf{RationalLMI}$, by $\mathsf{WeakLMI}(\mathsf{A}, [X_1, \cdots, X_6])$,

1. Using the routine $\mathsf{RationalUnivariateRepresentation}$ [16], we get an encoding of a real algebraic solution

$$\mathbf{u} = \begin{bmatrix} -1 + \frac{1}{2}\,\vartheta + \frac{1}{2}\,\vartheta^4\\ \frac{\vartheta^3}{2} + \frac{1}{2}\\ \vartheta^2\\ -2\,\vartheta + \frac{1}{2}\,\vartheta^2 + \frac{1}{2}\,\vartheta^5\\ \vartheta\\ 1 \end{bmatrix},$$

where ϑ is a real algebraic number satisfying

$$\vartheta^6 - 4\,\vartheta^2 - 1 = 0.$$

2. $\mathscr{U}$ is not empty then

(a) $i = 1$.

(b) $P = [\mathsf{Param}(\mathscr{U}), e_2, \ldots, e_6]$ and $\mathsf{A}' = \mathsf{P}^\star \mathsf{A} \mathsf{P}$.

(c) $\mathscr{L}_1, \ldots, \mathscr{L}_6$ in the first column of A' are

$$\begin{bmatrix} 0 \\ \frac{1}{2} X_2 \vartheta^5 - X_1 \vartheta^3 + \cdots - X_1 - X_5 \\ \frac{1}{2} X_4 \vartheta^5 + \frac{1}{2} X_1 \vartheta^4 + \cdots - X_1 + X_6 + \frac{1}{4} \\ (1 - X_3) \vartheta^5 + \frac{1}{2} X_2 \vartheta^3 + \cdots + \frac{1}{2} + \frac{1}{2} X_2 \\ \frac{1}{2} X_5 \vartheta^5 + \cdots + 1 + X_2 - \frac{1}{2} X_4 \\ \frac{1}{4} \vartheta^5 + \frac{1}{2} X_3 \vartheta^4 + \cdots - X_3 + 1 - \frac{1}{2} X_5 \end{bmatrix}$$

As $i = 1$, $\widehat{\mathsf{A}}$ is equivalent to the 5×5 matrix obtained by removing the 1-st row and column in A.

(d) The sequence of coefficients of $\vartheta^5, \ldots, \vartheta^1, 1$ in $\mathscr{L}_1, \ldots, \mathscr{L}_6$ is denoted as S, return $S, \widehat{\mathsf{A}}$.

- In step 6 of RationalLMI, the coefficient vector of ϑ^5 in $\mathscr{L}_1, \ldots, \mathscr{L}_6$ is

$$S_5 = \left[0, \frac{1}{2} X_2, \frac{1}{2} X_4, 1 - X_3, \frac{1}{2} X_5, \frac{1}{4}\right]^\star.$$

The last entry of S_5 is $\frac{1}{4}$, the linear system $S_5 = 0$ has no solutions. Therefore, LinearSolve(S) returns an empty set.

$\mathfrak{S}(\mathsf{A})$ has no rational solutions. The Maple worksheet can be downloaded from

http://www.mmrc.iss.ac.cn/~lzhi/Research/hybrid/RaLMI

6. REFERENCES

[1] H. Anai. Solving lmi and bmi problems by quantifier elimination. In R. Liska, editor, *The 4th International IMACS Conference on Applications of Computer Algebra*, volume 98 of *Proc. of IMACS-ACA*, 1998.

[2] S. Basu, R. Pollack, and M.-F. Roy. On the combinatorial and algebraic complexity of quantifier elimination. *J. ACM*, 43(6):1002–1045, November 1996.

[3] S. Basu, R. Pollack, and M.-F. Roy. *Algorithms in real algebraic geometry*, volume 10 of *Algorithms and Computation in Mathematics*. Springer-Verlag, second edition, 2006.

[4] J.C. Faugère, G. Moroz, F. Rouillier, and M. Safey El Din. Classification of the perspective-three-point problem, discriminant variety and real solving polynomial systems of inequalities. In D. Jeffrey, editor, *Proceedings of the twenty-first international symposium on Symbolic and algebraic computation*, ISSAC '08, pages 79–86, New York, NY, USA, 2008. ACM.

[5] A. Greuet, F Guo, M. Safey El Din, and L. Zhi. Global optimization of polynomials restricted to a smooth variety using sums of squares. *Journal of Symbolic Computation*, 47(5):503 – 518, 2012.

[6] A. Greuet and M. Safey El Din. Deciding reachability of the infimum of a multivariate polynomial. In É. Schost and I. Z. Emiris, editors, *Proceedings of the 36th international symposium on Symbolic and algebraic computation*, ISSAC '11, pages 131–138, New York, NY, USA, 2011. ACM.

[7] H. Iwane, H. Yanami, H. Anai, and K. Yokoyama. An effective implementation of a symbolic-numeric cylindrical algebraic decomposition for quantifier elimination. In H. Kai and H. Sekigawa, editors, *Proceedings of the 2009 conference on Symbolic numeric computation*, pages 55–64. ACM, 2009.

[8] E. Kaltofen, B. Li, Z. Yang, and L. Zhi. Exact certification of global optimality of approximate factorizations via rationalizing sums-of-squares with floating point scalars. In D. Jeffrey, editor, *Proceedings of the twenty-first international symposium on Symbolic and algebraic computation*, ISSAC '08, pages 155–164, New York, NY, USA, 2008. ACM.

[9] E. Kaltofen, B. Li, Z. Yang, and L. Zhi. Exact certification in global polynomial optimization via sums-of-squares of rational functions with rational coefficients. *J. Symb. Comput.*, 47(1):1–15, January 2012.

[10] L. Khachiyan and L. Porkolab. Computing integral points in convex semi-algebraic sets. In *Proceedings of the 38th Annual Symposium on Foundations of Computer Science*, FOCS '97, pages 162–171, Washington, DC, USA, 1997. IEEE Computer Society.

[11] L. Khachiyan and L. Porkolab. Integer optimization on convex semialgebraic sets. *Discrete and Computational Geometry*, 23(2):207–224, 2000.

[12] I. Klep and M. Schweighofer. An exact duality theory for semidefinite programming based on sums of squares. *ArXiv e-prints*, July 2012.

[13] A.K. Lenstra, H.W. Lenstra, and L. Lovàsz. Factoring polynomials with rational coefficients. *Mathematische Annalen*, 261:515–534, 1982.

[14] H. Peyrl and P. A. Parrilo. Computing sum of squares decompositions with rational coefficients. *Theor. Comput. Sci.*, 409(2):269–281, December 2008.

[15] V. Powers and T. Wörmann. An algorithm for sums of squares of real polynomials. *Journal of Pure and Applied Algebra*, 127:99–104, 1998.

[16] F. Rouillier. Solving zero-dimensional systems through the rational univariate representation. *Applicable Algebra in Engineering, Communication and Computing*, 9:433–461, 1999.

[17] M. Safey El Din. *RAGLib (Real Algebraic Geometry Library), Maple Package.* http://www-polsys.lip6.fr/~safey/RAGLib/.

[18] M. Safey El Din. Testing sign conditions on a multivariate polynomial and applications. *Mathematics in Computer Science*, 1(1):177–207, 2007.

[19] M. Safey El Din. Computing the global optimum of a multivariate polynomial over the reals. In D. Jeffrey, editor, *Proceedings of the twenty-first international symposium on Symbolic and algebraic computation*, ISSAC '08, pages 71–78, New York, NY, USA, 2008. ACM.

[20] M. Safey El Din and L. Zhi. Computing rational points in convex semialgebraic sets and sum of squares decompositions. *SIAM J. on Optimization*, 20(6):2876–2889, September 2010.

[21] C. Scheiderer. Descending the ground field in sums of squares representations. *eprint arXiv:1209.2976v2*, 09/2012.

Calculation of the Subgroups of a Trivial-Fitting Group

Alexander Hulpke
Department of Mathematics
Colorado State University
1874 Campus Delivery
Fort Collins, CO, 80523-1874, USA
hulpke@math.colostate.edu
http://www.math.colostate.edu/~hulpke

ABSTRACT

We describe an algorithm to determine representatives of the conjugacy classes of subgroups of a Trivial-Fitting group, this case being the one prior algorithms reduce to. As a subtask we describe an algorithm for determining conjugacy classes of complements to an arbitrary normal subgroup if the factor group is solvable.

Categories and Subject Descriptors

I.1.2 [**Symbolic and Algebraic Manipulation**]: Algebraic Algorithms

Keywords

Subgroups; Complement; Trivial-Fitting; Permutation group; Matrix group

1. INTRODUCTION

The question for determining the subgroups of a finite group is one of the earliest [Neu60] problems considered by computational group theory. It also is the task most frequently requested by users [Can11] which gives it extra prominence.

To save on memory this task is usually performed up to conjugacy in the group, that is the equivalence relation $U \sim U^g = g^{-1}Ug$ for $g \in G$. One thus determines for each such class (called a conjugacy class) a representative U, as well as its normalizer $N_G(U)$. A basic result about group actions then states that the elements of this class are in bijection with the cosets of $N_G(U)$ in G, in particular there are exactly $[G : N_G(U)]$ subgroups in the conjugacy class.

The method currently in use [CCH01] for determining the classes of subgroups uses the Trivial-Fitting paradigm: Let $R \lhd G$ be the largest solvable normal subgroup. First determine the classes of subgroups of G/R, then lift the result to G by complement computations. We shall concentrate on this first step, the subgroups of $H = G/R$. This group has no solvable normal subgroups and is thus called a Trivial-Fitting group.

The methods that have been suggested for determining the subgroups of H have been the use of pretabulated data, reduction to maximal subgroups, or the older method of cyclic extension; all of which are rather limited in scope. Instead we will describe a construction that uses the particular structure of such Trivial-Fitting groups as a subgroup of a direct product of wreath products. Section 2 summarizes this structure. Our construction has similarities with the enumeration of transitive subgroups of symmetric groups [Hul05], though of course here we do not impose any limiting conditions on the subgroups and face a more general situation. (In particular complements will arise for normal subgroups that are not solvable, see section 5.) The process also can be taken as a model for the problem of describing subgroups of a wreath product, a problem which sometimes arises in a combinatorial context.

Indeed the work in this paper has been motivated by the question of determining Möbius numbers for symmetric groups [Mon12, Sha97]; for settling the still open case of S_{18} a list of subgroups of $S_6 \wr S_3$ were desired.

2. TRIVIAL-FITTING GROUPS

Let H be a finite group with no solvable normal subgroup. Then the following well-known [BB99] facts hold:

Let $S = \mathrm{Soc}(H) \lhd H$ be the *socle* of H, that is the normal subgroup generated by all minimal normal subgroups. Then S is the direct product of simple nonabelian groups and H acts faithfully by conjugation on S, thus we can consider H as a subgroup of $\mathrm{Aut}(S)$. If the direct factors of S are the simple groups T_i ($i = 1, \ldots, k$), each arising with multiplicity n_i (that is $S = \bigtimes_i \bigtimes_{j=1}^{n_i} T_i$), then

$$\mathrm{Aut}(S) = \mathrm{Aut}(T_1) \wr S_{n_1} \times \cdots \times \mathrm{Aut}(T_k) \wr S_{n_k}$$

where $\wr$ denotes a permutational wreath product (that is $A \wr S_n = (\underbrace{A \times \cdots \times A}_{n \text{ copies}}) \rtimes S_n$ with S_n acting by permuting the copies of A. We shall assume that we have H represented as a subgroup of $\mathrm{Aut}(S)$ and thus can decompose according to this product structure. We then can embed

$$H/S \leq (\mathrm{Aut}(T_1)/T_1) \wr S_{n_1} \times \cdots \times (\mathrm{Aut}(T_k)/T_k) \wr S_{n_k}.$$

By the proof of Schreier's conjecture [Fei80], each $\mathrm{Aut}(T_i)/T_i$ is solvable and comparatively small. (If T_i is of Lie type defined over $\mathbb{F}$, then $\mathrm{Aut}(T_i)/T_i \leq \mathbb{F}^* \rtimes \mathrm{Gal}(\mathbb{F}/\mathbb{F}_p) \rtimes C_2$;

ISSAC'13, June 26–29, 2013, Boston, Massachusetts, USA.

otherwise $|\mathrm{Aut}(T_i)/T_i| \le 4$.) Thus if a is a bound for all $[\mathrm{Aut}(T_i){:}T_i]$, and we set $n = \sum n_i$ we have that

$$[H{:}S] \le a^n \cdot n! \tag{1}$$

In practice n has to be comparatively small (it is unlikely that n will be substantially larger than 5 as the total number of subgroup classes otherwise becomes infeasibly large for storage purposes), thus in practice this index $[H{:}S]$ will rarely be larger than a few 10000.

We thus see that S constitutes a large part of H and we shall use this fact in our construction: We first determine the conjugacy classes of subgroups of S, then fuse these to H-classes of subgroups of S and finally extend to classes of subgroups of H.

The structure of H as a subdirect product also means that it is feasible to represent it as a permutation group [Hular, Section 3]. In such a representation, we can determine subgroup normalizers, and test for subgroup conjugacy (constructively, that is finding a conjugating element) using standard backtrack methods [Hol91, Leo97, The97], [HEO05, Section 4.6].

3. SUBDIRECT PRODUCTS

The first step is to determine S-classes of subgroups of S. As S is a direct product of simple groups this is straightforward, using the concept of a subdirect product [Rem30]. Their computational construction is well understood (for example see [HP89] for an earlier application) and a description of how to enumerate their subgroups is found in [Hol10]. The content of this section thus is included only to make this paper self-contained. The section can easily be skipped by an experienced reader.

For simplicity we shall describe the case of a direct product of two factors. The case of multiple factors follows by induction by treating a product $F \times G \times H$ as $(F \times G) \times H$.

Consider a direct product $S = G \times H$. Then S has two projections $\alpha\colon S \to G$ and $\beta\colon S \to H$. A subgroup $U \le S$ then yields two images $A = U^\alpha \le G$ and $B = U^\beta \le H$. If we want to classify subgroups of S up to S-conjugacy, we clearly can assume that A is chosen up to G-conjugacy and B chosen up to H-conjugacy, conjugacy thus is reduced to $N_G(A) \times N_H(B)$.

To describe U further, consider the elements of U that project trivially on one of the components. Thus let $D = (\ker \beta_U)^\alpha \lhd A$ and $E = (\ker \alpha_U)^\beta \lhd B$ (where α_U means the restriction of α to U). If we denote the natural homomorphisms by $\rho\colon A \to A/D$ and $\sigma\colon B \to B/E$ then every element of A/D can be written as $(u^\alpha)^\rho$ for a suitable $u \in U$, similarly $\beta\sigma$ maps from U onto B/E. We thus can define a map $\zeta\colon A/D \to B/E$ by $(u^\alpha)^\rho \mapsto (u^\beta)^\sigma$. As $\ker \alpha \cap \ker \beta = \langle 1 \rangle$ this is well defined. It is easily seen that ζ is an isomorphism.

Clearly conjugates of D and E under $N_G(A)$ and $N_G(B)$ will lead to conjugates of U. When fixing D and E we thus can restrict conjugacy to $N \times M$ where $N = N_{N_G(A)}(D) = N_G(A) \cap N_G(D)$ and $M = N_{N_G(B)}(E) = N_G(B) \cap N_G(E)$.

The set of isomorphisms $A/D \to B/E$ is parameterized by $\mathrm{Aut}(A/D) \cong \mathrm{Aut}(B/E)$. Conjugation by N induces a subgroup $I_N \le \mathrm{Aut}(A/D)$, conjugation by M one of $\mathrm{Aut}(B/E)$. If we fix one isomorphism $\zeta_0\colon A/D$ (and also denote the induced isomorphism $\mathrm{Aut}(A/D) \to \mathrm{Aut}(B/E)$ by ζ_0), then conjugacy by $N \times M$ corresponds to a double coset in $I_N \setminus \mathrm{Aut}(A/D)/(I_M)^{\zeta_0^{-1}}$.

We have thus seen the following parameterization of S-classes of subgroups of the direct product S:

THEOREM 1. *[Hul96, Satz 32] For two finite group G, H let $\mathcal{A}$ be a set of representatives of the conjugacy classes of subgroups of G and $\mathcal{B}$ be a set of representatives of the conjugacy classes of H. Then a set of representatives of the conjugacy classes of subgroups of $G \times H$ is obtained in the following way:*

- *For each pair $A \in \mathcal{A}$, $B \in \mathcal{B}$ let $\mathcal{D}$ be a set of representatives of the $N_G(A)$-conjugacy classes of normal subgroups of A, $\mathcal{E}$ be a set of representatives of the $N_H(B)$-conjugacy classes of normal subgroups of B.*
- *For each pair $D \in \mathcal{D}$, $E \in \mathcal{E}$ (with natural homomorphisms $\rho\colon A \to A/D$, $\sigma\colon B \to B/E$) such that there is an isomorphism $\zeta_0\colon A/D \to B/E$ let $N = N_{N_G(A)}(D)$ and $M = N_{N_G(B)}(E)$. Let $\mathcal{Z}$ be a set of representatives of the double cosets $I_N \setminus Aut(A/D)/(I_M)^{\zeta_0^{-1}}$, where I_N are the automorphisms of A/D induced by conjugation with N, I_M the automorphisms of B/E induced by M.*
- *For each $\zeta \in \mathcal{Z}$ form the subgroup*

$$U := \left\{(a,b) \in G \times H \mid a \in A, b \in B, (a^\rho)^\zeta = b^\sigma\right\}$$

These subgroups U form representatives of the S-conjugacy classes of subgroups. The normalizer of U is the subgroup of $N \times M$ which stabilizes ζ via the induced action of I_N and I_M.

PROOF. The argument before the theorem statement yields the parameterization of subgroups. For such a subgroup U to be normalized, clearly the projections A and B, as well as the kernels E and F, must be stabilized by conjugation, so $N_S(U) \le N \times M$. If the isomorphism ζ defining U is stabilized by an element $x \in N \times M$ then $U^x \subset U$, and otherwise not, thus the stabilizer of ζ is exactly the normalizer of U. □

Note that this parameterization leads in an obvious way to an algorithm to enumerate representatives of the classes of subgroups of $G \times H$ and their normalizers. (We represent $\mathrm{Aut}(A/D)$ as permutation group to determine double cosets.) To find possibly pairings for isomorphic factor groups it can be useful to identify the isomorphism type of small factor groups first, using invariants obtained from the explicit list [BEO02].

We run this algorithm iteratively for a direct product of more than two factors. In the base case we need to obtain the subgroups of simple groups which we can do either with the traditional method of cyclic extension [Neu60], or simply use the large amount of theoretical information available, such as [Pfe97].

4. SUBGROUPS OF A TRIVIAL-FITTING GROUP

We now assume again that H is a Trivial-Fitting group with socle $S = \mathrm{Soc}(H) \lhd H$. We assume that we have determined the S-classes of subgroups of S, using the methods of the previous section. We aim to extend this list to a list of representatives of the H-classes of subgroups of H. (The process of extension shares some similarities with [NP12].)

First consider subgroups of S: The action of H can fuse S-classes of subgroups, as $S \lhd H$ the S-classes of subgroups form blocks for the action of H. If $U \leq S$ is such a subgroup then

$$b = \frac{[H:N_H(U)]}{[S:N_S(U)]} = \frac{[H:S]}{[N_H(U):N_S(U)]} \leq [H:S]$$

is a bound for the number of blocks. By computing $N_H(U)$ in a backtrack search we obtain b. If $b > 1$ the fusion of S-classes of subgroups to H-classes then can be done by an orbit algorithm [HEO05, Section 4.5.2] on blocks, given by representatives. In this, if U_i is a representative of an S-class, and $h \in H$, we determine the S-class of U_i^h and find $s \in S$ such that U_i^{hs} is the chosen representative of its S-class.

We thus obtain representatives of the H-classes of subgroups in S.

Now consider an arbitrary subgroup $V \leq H$. Then $U = V \cap S \lhd V$ and clearly $U \leq S$ and $V \leq N_H(U)$. Conjugates of V lead to conjugates of U, so it is sufficient to consider U up to H-conjugacy. We shall classify all subgroups V that intersect with S in the same subgroup U, up to conjugacy with $N_H(U)$:

Let $\varphi\colon N_H(U) \to N_H(U)/N_S(U)$. As φ is the restriction of the natural homomorphism $H \to H/S$ with domain restricted to $N_H(U)$ this can be constructed easily. Let $A := N_S(U) \cdot V$. Then $A^\varphi \leq N_H(U)^\varphi =: Q$. We obtain candidates for A up to conjugacy in $N_H(U)$ by determining representatives of the classes of subgroups of Q. This group Q in general is comparatively small (its order is again bounded by $[H:S]$) and often has a nontrivial radical (from the factors $\mathrm{Aut}(T_i)/T_i$, so the standard methods mentioned in the introduction work well.

We now shall classify all subgroups V that lead to the same U and the same A. As we permit conjugacy of $N_H(U)$, we shall consider A only up to $N_H(U)$ conjugacy. These classes of subgroups are in bijection with the Q-classes of subgroups of Q. Let $A/N_S(U)$ be one such subgroup and let $M = N_{N_H(U)}(A) = N_H(U) \cap N_H(A)$. Then (figure 1) we have that V/U is a complement to $N_S(U)/U$ in A/U, and every such complement gives a subgroup V with the desired properties.

We can obtain these classes of complements by first determining the classes of complements under A/U conjugacy and then fusing representatives under conjugacy by $N_H(U)$. (As in general $A \not\lhd N_H(U)$ this requires explicit conjugacy tests.) The representatives obtained are representatives of the desired subgroups V up to conjugacy in H.

The algorithm for all subgroups then consists of:

1. Take representatives U of the H classes of subgroups of H.
2. For each such U take representatives A of the subgroups of $N_H(U)$, containing $N_S(U)$.
3. For each pair U, A find subgroups V as complements, fused under the action of $N_H(U)$.

If $N_S(U)/U$ is solvable the complements can be obtained with cohomological methods, see [CNW90] and [HEO05,

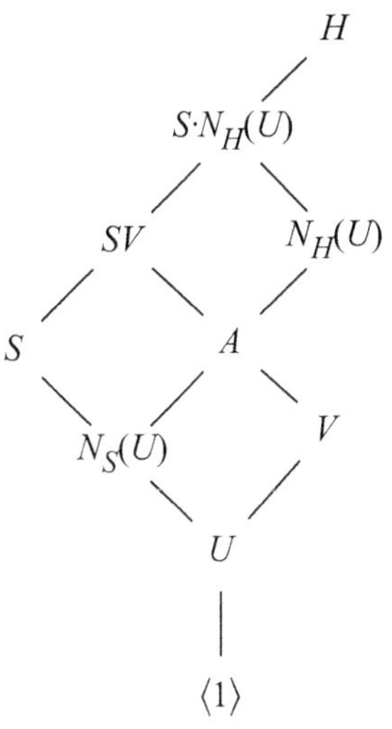

Figure 1: Subgroups in $N_H(U)$

Section 7.6.2], using a presentation for $A/N_S(U)$. (Such a presentation can be obtained for example using the methods of [BGK$^+$97].)

In the situation considered here the group $N_S(U)/U$ often is not solvable. An example of such a situation (which also indicates why it occurs frequently) is if U projects trivially on some direct factors of S. In this case all elements in thes (nonsolvable) direct factors then normalize U and thus contribute to $N_S(U)/U$.

In many cases however the factor group $N_H(U)/N_S(U)$ (and thus the group $A/N_S(U)$) is solvable. This is for example the case if U never has the same projection on more than four components (as S_4 is solvable) in particular if each $n_i \leq 4$.

We thus describe in section 5 a specific method to find classes of complements in the case of a solvable factor group. Section 6 then describes the general case.

5. COMPLEMENTS: SOLVABLE FACTOR GROUP

In this self-contained section we study the following situation (variables are not related to their use in the previous sections):

We are given a finite group G and $N \lhd G$ such that N is arbitrary but G/N is solvable. We want to determine, up to G-conjugacy, the subgroups $C \leq G$ such that $C \cap N = \langle 1 \rangle$ and $NC = G$.

As we actually need to determine complements in a factor group, we consider a slightly generalized situation (see figure 2):

Definition 2. *Given a group G and $N, M \lhd G$, we say that that a subgroup $M \leq C \leq G$ complements N in G modulo M if $NMC = G$ and $NM \cap C = M$.*

We assume that G/NM is solvable and want to determine these subgroups up to conjugacy by G.

Since we assumed that G/NM is solvable, there exists a normal subgroup $NM \leq A \lhd G$ such that A/NM is a p-group for some prime p. Clearly any complement C in G

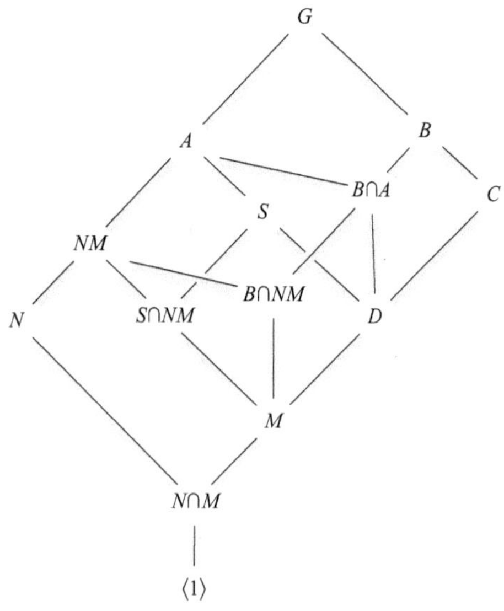

Figure 2: Complements in the factor group

must contain a normal subgroup $D = C \cap A$, this group D is a complement to NM in A modulo M. As $D/M \cong A/NM$ is a p-group, by the 2nd Sylow theorem we can assume – up to conjugacy – that any such complement D lies in S, where S/M is a particular p-Sylow subgroup of A/M. Thus D is a complement to $S \cap NM$ in S modulo M.

Given such a complement D, we known that $C \leq B := N_G(D)$ and thus that $N \cdot B = G$. Furthermore $C \cap A = D$, thus C is a complement to $B \cap A$ in B modulo D.

This motivates the following strategy, given G, N, M:

1. Determine a subgroup $A \geq N$, $A \lhd G$, such that A/NM is a p-group for some prime p. (See below on how to construct A.)

2. Determine (for example using [CCH97]) a p-Sylow subgroup $P \leq A$, let $S = MP$. (So $[S{:}M] = p^x$.)

3. Using cohomology [CNW90], compute representatives of the conjugacy classes of complements to $S \cap NM$ in S modulo M. Determine representatives of the G-classes of these complements. (This is done best by first fusing under the action of $N_G(S)$ which is an action on the cohomology group, followed by conjugacy tests in G.)

   The groups obtained in this step are representatives of the classes of complements to N in A modulo M, as every complement is G-conjugate to a complement in S.

4. For each such complement D determine $B := N_G(D)$ and check whether $N \cdot B = G$. If not discard D as its normalizer may not contain a complement to N in G.

5. Otherwise work recursively in B and compute B-classes of complements C to $B \cap A$ modulo D.

By the above argument clearly any complement to N modulo M is conjugate to one of the groups C in this classification. We can ensure that no duplicates are constructed if A/NM is characteristic (i.e. invariant under all automorphisms) in G/NM:

LEMMA 3. *Assume (notation as in the algorithm above) that A/NM is characteristic in G/NM. Let $\{D_i\}$ be a set of representatives of G-classes of complements to N in A modulo M such that $N \cdot N_G(D_i) = G$, and for each i let $\{C_{i,j}\}$ be a set of the $N_G(D_i)$-classes of complements to $N \cap N_G(D_i)$ modulo D_i. Then every subgroup C complementing N modulo M in G is conjugate to exactly one $C_{i,j}$.*

PROOF. The argument above already establishes that every subgroup C with the given properties is conjugate to at least one of the $C_{i,j}$.

Assume conversely that two of the $C_{i,j}$, call them C_1 and C_2 are conjugate: $C_1^g = C_2$, and that they were obtained from subgroups D_1, respectively D_2. The natural homomorphism $G/M \to G/NM$ induces isomorphisms $\phi_i\colon C_i/M \to G/NM$ such that $D_i/M^{\phi_i} = A/NM$.

If $D_1^g = D_2$ then $D_1 = D_2$ by the choice of the D_i and $g \in N_G(D_1)$, contradicting the choice of the $C_{1,j}$ to be representatives under the action of $N_G(D_1)$.

Otherwise consider the isomorphism $\phi\colon G/NM \to G/NM$ given by $x \mapsto ((x^{\phi_1^{-1}})^g)^{\phi_2}$. We then have that $A^\phi \neq A$, contradicting the choice of A being characteristic. □

5.1 Choice of normal subgroups

We finally just need to describe how to construct the subgroup A that leads to a characteristic factor A/NM. For this we are using standard methods for solvable groups.

DEFINITION 4 ([CELG04]). *Let G be a finite solvable group. The elementary abelian nilpotent-central (EANC) series*

$$G = G_1 \rhd G_2 \rhd \cdots \rhd G_{m+1} = \langle 1 \rangle$$

is obtained by iteratively refining normal series of G:

1. *Start with a series*

$$G = N_1 \rhd N_2 \rhd \cdots \rhd N_{m+1} = \langle 1 \rangle$$

   such that N_{i+1} is minimal normal in N_i such that N_i/N_{i+1} is nilpotent.

2. *Consider each nilpotent factor as a direct product of p-groups. For each direct factor take the lower exponent p-central series, take the direct products for all primes of subsequent factors of these series (thus the factors are direct products of elementary abelian p-groups).*

3. *Refine each of these factors by taking the series given by direct products of p-Sylow subgroups in order of ascending p.*

It is shown in [CELG04] that the subgroups in this series are characteristic and the subsequent factors are elementary abelian p-groups.

Efficient methods for computing this series are given in [CELG04] as well.

We thus choose A such that A/NM is the lowest nontrivial factor in the EANC series of G/NM. As we recurse only if $BNM = G$, the EANC series of $B/B \cap NM$ corresponds to the EANC series of G/NM. Furthermore (by the definition) we have that for a subgroup $G_i \lhd G$ in the EANC series we have that in the EANC series for G/G_i consists of the groups G_j/G_i for $j \leq i$.

Therefore this series needs to be computed only once and the subsequent factors can be used in the recursion.

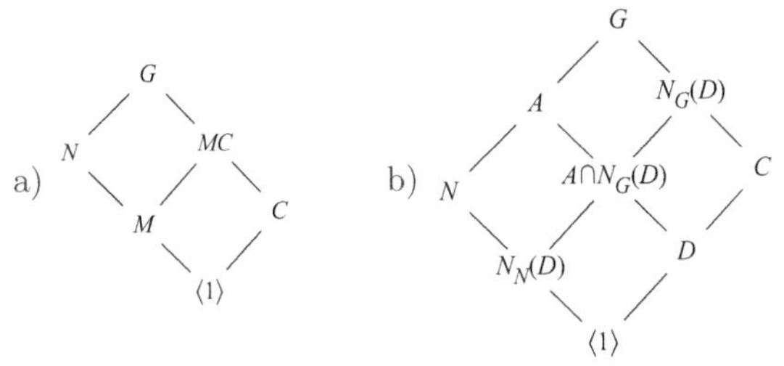

Figure 3: General Complement Situation

Group	Order	Classes	time	cyclic ext.
$S_5 \wr S_2$	28800	561	24	10
$S_5 \times S_6$	86400	3182	162	163
$S_5 \times S_7$	604800	5913	277	1441
$U_3(5).S_3$	756000	244	43	287
$S_6 \wr S_2$	1036800	8147	299	3745
$S_5 \wr S_3$	10368000	29155	1913	261561
$U_4(3).D_8$	26127360	3870	5245	319124
$M_{11} \wr S_2$	125452800	2048	367	-
$S_6 \wr S_3$	2239488000	7172632	680966	-

Table 1: Some run times of the algorithm

6. COMPLEMENTS: GENERAL CASE

We finally briefly describe the general situation of determining classes of complements to a normal subgroup $N \lhd G$. The methods from [CNW90] as well as from [CH04, Section 4] and from section 5 indicate the following two reductions (see figure 3) for the problem:

a) If there is $M \lhd G$, $M \leq N$, any complement C to N will have MC/M complement N/M in G/M. Furthermore C complements M in MC. Thus first find classes of complements to N/M in G/M and then for each subgroup $D \geq M$ such that D/M is representative of a class of complements find complements to M in D. Fuse under action of N.

b) If there is $A \lhd G$, $N \leq A$, then any complement C to N in G will contain a subgroup $D \lhd C$ complementing N in A and $C \leq N_G(D)$. Thus first find complements to N in A, fuse under the action of G. Then for each such complement D such that $N \cdot N_G(D) = G$, work in $N_G(D)$: Let $Z = A \cap N_G(D) = D \cdot N_N(D)$ and find complements to Z/D in $N_G(D)/D$. Every complement C arises this way, if A/N is characteristic in G/N the argument from lemma 3 shows that no further fusion under action of G can take place.

Thus the only remaining case in which neither reduction applies is applicable, nor normal subgroup nor factor group are solvable is that of N elementary nonabelian and G/N simple. The paper [CH04, Section 4] suggests in this case a reduction to maximal subgroups which it acknowledges as a not very efficient method.

In our situation however this means that for at least five projections T_i the quotient $N_S(U)/U$ must have a nonsolvable projection image and that these components are permuted transitively. This means that the whole group H already must be rather large and a calculation currently will likely already face obstacles of memory requirements for storing the subgroups. Thus this issue of complements, while being a theoretical problem, is it less of a practical concern here.

7. IMPLEMENTATION

The algorithm as described has been implemented by the author in GAP [GAP13] and is available in release 4.6 of the system. (The complement routine of section 5 is already available in release 4.5.) Table 1 shows the performance in comparison with the cyclic extension algorithm used by default. (For the base case of simple groups the new algorithm still used cyclic extension, and not table lookup, to get the classes of subgroups. Doing so would have given a further speedup.) All groups were represented by permutations in the smallest degree possible. Column "time" is the time taken by this new algorithm, column "cyclic extension" the timing for the default cyclic extension algorithm.

Runtimes are in seconds on an 2.3 GHz AMD Opteron 6276 under Linux with ample memory to allow storage of all subgroups. In the case of $M_{11} \wr S_2$ the cyclic extension algorithm was provided with a list of all perfect subgroups, obtained from the prior subdirect product calculation for $M_{11} \times M_{11}$, but still did not finish in weeks. The calculation for $S_6 \wr S_3$ by cyclic extension is absolutely hopeless.

It is easily seen that the new algorithm is an improvement except for the smallest cases. One reason for this is the large number of conjugacy tests the cyclic extension algorithm needs to do to eliminate duplicates.

8. ACKNOWLEDGMENTS

Experimental runs on large groups were aided substantially by the checkpointing facility of the DMTCP [AAC09] project that allowed to restart calculations after a reboot. The author's work has been supported in part by Simons Foundation Collaboration Grant 244502 which is gratefully acknowledged.

9. REFERENCES

[AAC09] Jason Ansel, Kapil Arya, and Gene Cooperman. DMTCP: Transparent checkpointing for cluster computations and the desktop. In *23rd IEEE International Parallel and Distributed Processing Symposium*, Rome, Italy, May 2009.

[BB99] László Babai and Robert Beals. A polynomial-time theory of black box groups. I. In C. M. Campbell, E. F. Robertson, N. Ruskuc, and G. C. Smith, editors, *Groups St Andrews 1997 in Bath*, volume 260/261 of *London Mathematical Society Lecture Note Series*, pages 30–64. Cambridge University Press, 1999.

[BEO02] Hans Ulrich Besche, Bettina Eick, and E. A. O'Brien. A millennium project: constructing small groups. *Internat. J. Algebra Comput.*, 12(5):623–644, 2002.

[BGK+97] László Babai, Albert J. Goodman, William M. Kantor, Eugene M. Luks, and Péter P. Pálfy. Short presentations for finite groups. *J. Algebra*, 194:97–112, 1997.

[Can11] John Cannon. Problems I would like to solve in

CGT. *Oberwolfach Reports*, 8(3):2127–2128, 2011.

[CCH97] John J. Cannon, Bruce C. Cox, and Derek F. Holt. Computing Sylow subgroups in permutation groups. *J. Symbolic Comput.*, 24(3-4):303–316, 1997. Computational algebra and number theory (London, 1993).

[CCH01] John Cannon, Bruce Cox, and Derek Holt. Computing the subgroup lattice of a permutation group. *J. Symbolic Comput.*, 31(1/2):149–161, 2001.

[CELG04] John J. Cannon, Bettina Eick, and Charles R. Leedham-Green. Special polycyclic generating sequences for finite soluble groups. *J. Symbolic Comput.*, 38(5):1445–1460, 2004.

[CH04] John Cannon and Derek Holt. Computing maximal subgroups of finite groups. *J. Symbolic Comput.*, 37(5):589–609, 2004.

[CNW90] Frank Celler, Joachim Neubuser, and Charles R. B. Wright. Some remarks on the computation of complements and normalizers in soluble groups. *Acta Appl. Math.*, 21:57–76, 1990.

[Fei80] Walter Feit. Some consequences of the classification of finite simple groups. In *The Santa Cruz Conference on Finite Groups (Univ. California, Santa Cruz, Calif., 1979)*, volume 37 of *Proc. Sympos. Pure Math.*, pages 175–181. Amer. Math. Soc., Providence, R.I., 1980.

[GAP13] The GAP Group, `http://www.gap-system.org`. *GAP – Groups, Algorithms, and Programming, Version 4.6.3*, 2013.

[HEO05] Derek F. Holt, Bettina Eick, and Eamonn A. O'Brien. *Handbook of Computational Group Theory*. Discrete Mathematics and its Applications. Chapman & Hall/CRC, Boca Raton, FL, 2005.

[Hol91] D. F. Holt. The computation of normalizers in permutation groups. *J. Symbolic Comput.*, 12(4-5):499–516, 1991. Computational group theory, Part 2.

[Hol10] Derek F. Holt. Enumerating subgroups of the symmetric group. In *Computational group theory and the theory of groups, II*, volume 511 of *Contemp. Math.*, pages 33–37. Amer. Math. Soc., Providence, RI, 2010.

[HP89] Derek F. Holt and W. Plesken. *Perfect groups*. Oxford University Press, 1989.

[Hul96] Alexander Hulpke. *Konstruktion transitiver Permutationsgruppen*. PhD thesis, Rheinisch-Westfalische Technische Hochschule, Aachen, Germany, 1996.

[Hul05] Alexander Hulpke. Constructing transitive permutation groups. *J. Symbolic Comput.*, 39(1):1–30, 2005.

[Hular] Alexander Hulpke. Computing conjugacy classes of elements in matrix groups. *J. Algebra*, to appear.

[Leo97] Jeffrey S. Leon. Partitions, refinements, and permutation group computation. In Larry Finkelstein and William M. Kantor, editors, *Proceedings of the 2nd DIMACS Workshop held at Rutgers University, New Brunswick, NJ, June 7–10, 1995*, volume 28 of *DIMACS: Series in Discrete Mathematics and Theoretical Computer Science*, pages 123–158. American Mathematical Society, Providence, RI, 1997.

[Mon12] Kenneth Monks. *The Moebius Number of the Symmetric Group*. PhD thesis, Colorado State University, Fort Collins, CO, 2012.

[Neu60] Joachim Neubüser. Untersuchungen des Untergruppenverbandes endlicher Gruppen auf einer programmgesteuerten elektronischen Dualmaschine. *Numer. Math.*, 2:280–292, 1960.

[NP12] L. Naughton and G. Pfeiffer. Computing the table of marks of a cyclic extension. *Math. Comp.*, 81(280):2419–2438, 2012.

[Pfe97] Götz Pfeiffer. The subgroups of M_{24}, or how to compute the table of marks of a finite group. *Experiment. Math.*, 6(3):247–270, 1997.

[Rem30] Robert Remak. Über die Darstellung der endlichen Gruppen als Untergruppen direkter Produkte. *J. Reine Angew. Math.*, 163:1–44, 1930.

[Sha97] John Shareshian. On the Möbius number of the subgroup lattice of the symmetric group. *J. Combin. Theory Ser. A*, 78(2):236–267, 1997.

[The97] Heiko Theißen. *Eine Methode zur Normalisatorberechnung in Permutationsgruppen mit Anwendungen in der Konstruktion primitiver Gruppen*. Dissertation, Rheinisch-Westfalische Technische Hochschule, Aachen, Germany, 1997.

Finding Hyperexponential Solutions of Linear ODEs by Numerical Evaluation

Fredrik Johansson*
RISC
Johannes Kepler University
4040 Linz, Austria
fjohanss@risc.jku.at

Manuel Kauers*
RISC
Johannes Kepler University
4040 Linz, Austria
mkauers@risc.jku.at

Marc Mezzarobba†
Inria, Univ. Lyon, AriC, LIP
ENS de Lyon, 46 allée d'Italie
69364 Lyon Cedex 07, France
marc@mezzarobba.net

ABSTRACT

We present a new algorithm for computing hyperexponential solutions of linear ordinary differential equations with polynomial coefficients. The algorithm relies on interpreting formal series solutions at the singular points as analytic functions and evaluating them numerically at some common ordinary point. The numerical data is used to determine a small number of combinations of the formal series that may give rise to hyperexponential solutions.

Categories and Subject Descriptors

I.1.2 [**Computing Methodologies**]: Symbolic and Algebraic Manipulation—*Algorithms*

General Terms

Algorithms

Keywords

Closed form solutions, D-finite equations, Effective analytic continuation

1. INTRODUCTION

We consider linear differential operators

$$P = p_r D^r + p_{r-1} D^{r-1} + \cdots + p_0$$

where $p_0, \dots, p_r$ are polynomials and D represents the standard derivation $\frac{d}{dx}$. Such operators act in a natural way on elements of a differential ring containing the polynomials. An object y is called a solution of the operator if P applied to y yields zero. We are interested in finding the hyperexponential solutions of a given operator. An object y is called hyperexponential if the quotient $D(y)/y$ can be identified with a rational function. Typical examples are rational functions (e.g. $(5x+3)/(3x+5)$), radicals (e.g. $\sqrt{x+1}$), exponentials (e.g. $\exp(3x^2-4)$ or $\exp(1/x)$), or combinations of these (e.g. $\sqrt{x+1}\exp(x^9/(x-1))$). Equivalently, y is called hyperexponential if there is some first order operator $q_1 D + q_0$ with q_0, q_1 polynomials which maps y to zero. If we regard differential operators as elements of an operator algebra $C(x)[D]$, then there is a one-to-one correspondence between the hyperexponential solutions y of an operator P and its first order right hand factors. In other words, if y is a hyperexponential term with $(q_1 D + q_0) \cdot y = 0$, then y is a solution of P if and only if there exist rational functions $u_0, \dots, u_{r-1}$ such that

$$P = (u_{r-1} D^{r-1} + u_{r-2} D^{r-2} + \cdots + u_0)(q_1 D + q_0).$$

Algorithms for finding the hyperexponential solutions of a linear differential equation (or equivalently, the first order right hand factors of the corresponding operators) are known since long. They are needed as subroutine in algorithms for factoring operators or for finding Liouvillian solutions. See Chapter 4 of [12] for details and references.

Classical algorithms first compute "local solutions" at singular points (cf. Section 2.3 below) and then test for each combination of local solutions whether it gives rise to a hyperexponential solution. This leads to a combinatorial explosion with exponential runtime. The situation is similar to classical algorithms for factoring polynomials over $\mathbb{Q}$, which first compute the irreducible factors modulo a prime and then test for each combination whether it gives rise to a factor in $\mathbb{Q}[x]$.

The algorithm of van Hoeij [13] avoids the combinatorial explosion as follows. It picks one local solution and considers the operator $Q = q_1 D + q_0$ with $q_1, q_0 \in C((x))$ which annihilates it. This operator is a right factor of P, though not with rational coefficients. The algorithm then constructs (if possible) a left multiple B of Q with rational coefficients of order at most $r-1$. This leads to a nontrivial factorization $P = AB$ in $C(x)[D]$. The procedure is then applied recursively to A and B until a complete factorization is found. The first order factors in this factorization give rise to at most r hyperexponential candidate solutions (possibly up to multiplication by a rational function). These are then checked in a second step. Van Hoeij's algorithm reminds of the polynomial factorization algorithm of Lenstra, Lenstra, Lovász [7, 16], which picks one modular factor and constructs (if possible) a multiple of this factor with integer coefficients but smaller

---

*Supported by the Austrian Science Fund (FWF) grant Y464-N18.
†UMR 5668 CNRS – ENS Lyon – Inria – UCBL

ISSAC'13, June 26–29, 2013, Boston, Massachusetts, USA.

degree than the original polynomial. This multiple is then a proper divisor in $\mathbb{Q}[x]$.

The algorithm we propose below avoids the combinatorial explosion in a different way. We start from the local solutions and regard them as asymptotic expansions of complex functions. By means of effective analytic continuation and arbitrary-precision numerical evaluation, we compute the values of these functions at some common ordinary reference point. Then a linear algebra algorithm is used to determine a small list of possible combinations of local solutions that may give rise to hyperexponential ones, possibly up to multiplication by a rational function. These are then checked in a second step. Our approach was motivated by van Hoeij's polynomial factorization algorithm [15], which associates to every modular factor a certain vector and then uses lattice reduction to determine a small list of combinations that may give rise to proper factors.

Although our algorithm avoids the combinatorial explosion problem, we do not claim that it runs in polynomial time. Indeed, no polynomial time algorithm can be expected because there are operators P which have hyperexponential solutions y that are exponentially larger than P. Also van Hoeij [13] makes no formal statement about the complexity of his algorithm. It is clear though that his algorithm is superior to the naive algorithm. Similarly, we believe that our algorithm has chances to outperform van Hoeij's algorithm, at least in examples that are not deliberately designed to exhibit worst case performance. The reason is partly that during the critical combination phase we only work with floating point numbers of moderate precision while van Hoeij's algorithm in general needs to do arithmetic in algebraic number fields whose degrees may grow during the computation. Another advantage of our algorithm is that it is conceptually simpler than van Hoeij's, at least if we take for granted that we can compute high-precision evaluations of D-finite functions.

In Section 2 we recall known results and definitions which we use. Section 3 contains a global description of the whole algorithm for finding hyperexponential solutions. Section 4 contains an abstract description of the algorithm for the combination phase, and in Section 5 we explain how this algorithm can be implemented using numerical evaluation. A detailed example is given in Section 6.

2. PRELIMINARIES

In this section, we recall some results from the literature and introduce notation that will be used in subsequent sections.

2.1 Differential Fields and Operator Algebras

A differential ring/field is a pair (K, D) where K is a ring/field and $D\colon K \to K$ is a derivation on K, i.e., a map satisfying $D(a+b) = D(a)+D(b)$ and $D(ab) = D(a)b+aD(b)$ for all $a, b \in K$. Throughout this paper, we consider the differential field $K = C(x)$, where C is some (computable) subfield of $\mathbb{C}$, together with the derivation $D\colon K \to K$ defined by $D(c) = 0$ for all $c \in C$ and $D(x) = 1$. For simplicity, we assume throughout that C is algebraically closed.

A differential ring/field E is called an extension of K if $K \subseteq E$, and the derivation of E restricted to K agrees with the derivation of K.

By $K[D]$ we denote the set of all polynomials in the indeterminate D with coefficients in K. Addition in $K[D]$ is defined in the usual way, and multiplication is defined subject to the commutation rule $Da = aD + D(a)$ for $a \in K$. The elements of $K[D]$ are called operators, and they act on the elements of some extension E of K in the obvious way: If $P = p_0+p_1D+\dots+p_rD^r$ is an operator of order r and $y \in E$, then $P\cdot y := \sum_{i=1}^{r} p_i D^i(y) \in E$. The noncommutative multiplication is compatible with operator application in the sense that we have $(PQ)\cdot y = P\cdot(Q\cdot y)$ for all $P, Q \in K[D]$ and all $y \in E$.

The elements $y \in E$ such that $P \cdot y = 0$ form a C-vector space V with $\dim V \leq r$. By making E sufficiently large it can always be assumed that $\dim V = r$.

2.2 Hyperexponential Terms

Let E be an extension of K. An element $h \in E \setminus \{0\}$ is called *hyperexponential* over K if $D(h)/h \in K$. Equivalently, h is hyperexponential if $Q \cdot h = 0$ for some nonzero first order operator $Q \in K[D]$.

Two hyperexponential terms h_1, h_2 are called *equivalent* if $h_1/h_2 \in K$. For example, the terms $\exp(3x^2 - x)$ and $(1 - 2x)^2 \exp(3x^2 - x)$ are equivalent, but $\exp(3x^2 - x)$ and $(1 - 2x)^{\sqrt{2}} \exp(3x^2 - x)$ are not. (Here and below, we use standard calculus notation to refer to elements of some extension E on which the derivation acts as the notation suggests, e.g. $D(\exp(3x^2 - x)) = (6x - 1)\exp(3x^2 - x)$.)

Every hyperexponential term can be written in the form $h = \exp(\int v)$, where v is a rational function. The additive constant of the integral amounts to a multiplicative constant for h, which is irrelevant in our context, because $P \cdot h = 0$ if and only if $P \cdot (ch) = 0$ for every $c \in C \setminus \{0\}$. If we consider the partial fraction decomposition of v and integrate it termwise, we obtain something of the form

$$g + \sum_{i=1}^{n} \gamma_i \log(p_i)$$

with $g \in K, \gamma_1, \dots, \gamma_n \in C$ and monic square free pairwise coprime polynomials $p_i \in C[x]$. In terms of this representation, two hyperexponential terms are equivalent if the difference of the corresponding rational functions g is a constant and any two corresponding coefficients γ_i differ by an integer.

The equivalence class of a hyperexponential term h is called the *exponential part* of h. The motivation for this terminology is that when we are searching for some hyperexponential solution h of P and we already know its equivalence class, then we can take an arbitrary element h_0 from this class and make an ansatz $h = uh_0$ for some rational function $u \in K$. The operator $\tilde{P} := P \otimes \big(D - \frac{D(1/h_0)}{1/h_0}\big) \in K[D]$ then has the property that u is a solution of $\tilde{P}$ if and only if uh_0 is a solution of P. This reduces the problem to finding rational solutions, which is well understood and will not be discussed here [1, 12].

2.3 Local Solutions

Consider an operator $P \in C(x)[D]$ of order r. By clearing denominators, if necessary, we may assume that $P \in C[x][D]$, say $P = p_rD^r + \dots + p_0$ with $p_r \neq 0$. A point $z \in \mathbb{C} \cup \{\infty\}$ is called *singular* if z is a root of p_r, or $z = \infty$. A point which is not singular is called *ordinary*. Note that there are only finitely many singular points, and that we include the "point at infinity" always among the singular points.

If $z = 0$ is an ordinary point then P admits r linearly independent power series solutions. If $z = 0$ is a singular point, it is still possible to find r linearly independent generalized series solutions of the form

$$x^{\alpha}\exp(u(x^{-1/s}))\sum_{k=0}^{m} b_k(x^{1/s})\log(x)^k \qquad (1)$$

where $\alpha \in C$, $u \in C[x]$ with $u(0)=0$, $s \in \mathbb{N}$, $m \in \mathbb{N}$ and $b_0,\dots,b_m \in C[[x]]$. These solutions are called the *local solutions* at 0. Their computation is well-known and will not be discussed here. (See [14, 12] and the references given there for details.)

Two series as in (1) are called equivalent if they have the same u and s and the difference of the respective values of α is in $\frac{1}{s}\mathbb{Z}$. The equivalence classes of generalized series under this equivalence relation are called the *exponential parts* of the series. Adopting van Hoeij's notation and defining $\operatorname{Exp}(e) := \exp(\int \frac{e}{x})$ for $e \in C[x^{-1/s}]$, we have that $\operatorname{Exp}(e_1)$ and $\operatorname{Exp}(e_2)$ are equivalent iff $e_1 - e_2 \in \frac{1}{s}\mathbb{Z}$. Note that if $m=0$ and $s=1$, two series are equivalent iff their quotient can be identified with a formal Laurent series. We will from now on make no notational distinction between $\operatorname{Exp}(e)$ and its equivalence class.

A point $z \neq 0$ can be moved to the origin by the change of variables $\tilde{x} = x - z$ (if $z \in C$) or $\tilde{x} = 1/x$ (if $z=\infty$). If $\tilde{P}$ is the operator obtained from P by replacing x by $\tilde{x}+z$ or $1/\tilde{x}$, then a local solution of $P \in C[x][D]$ at z is defined as the local solution of $\tilde{P} \in C[\tilde{x}][D]$ at 0.

Throughout the rest of this paper, we will use the following notation. P is some operator in $C[x][D]$ of order r, by $z_1,\dots,z_{n-1} \in C$ we denote its finite singular points, $z_n = \infty$. We write $\tilde{x}_i = x - z_i$ ($i=1,\dots,n-1$) and $\tilde{x}_n = 1/x$ for the variables with respect to which the singularities at z_i appear at the origin. For $i=1,\dots,n$, we consider the vector space V_i generated by all local solutions at z_i. There may be solutions with different exponential parts, say ℓ_i different parts $\operatorname{Exp}(e_{i,1}),\dots,\operatorname{Exp}(e_{i,\ell_i})$ for $e_{i,j} \in C[\tilde{x}_i^{-1/s_{i,j}}]$. By

$$V_{i,j} = V_i \cap \operatorname{Exp}(e_{i,j})C((\tilde{x}_i^{1/s_{i,j}}))[\log \tilde{x}_i]$$

we denote the vector space of all local solutions of P at z_i with exponential part (equivalent to) $\operatorname{Exp}(e_{i,j})$. Our $V_{i,j}$ are written $V_{e_{i,j}}(P)$ in van Hoeij's papers [14, 13].

The condition in the definition of equivalence that the difference of corresponding values of α be an integer (rather than, say, requiring exactly the same value of α) ensures that the $V_{i,j}$ are indeed vector spaces, because if some $V_{i,j}$ contains, for example, the two series

$$x^{\alpha}(1+x+x^2+\cdots) \quad \text{and} \quad x^{\alpha}(1+x+3x^2+\cdots)$$

then it must also contain their difference $x^{\alpha}(2x^2+\cdots) = x^{\alpha+2}(2+\cdots)$.

2.4 Analytic Solutions

It is classical that the formal power series solutions $\hat{y}$ of P at an ordinary point $z \in \mathbb{C}$ actually converge in a neighbourhood of z and thus give rise to analytic function solutions y of P. The correspondence is one-to-one. For any other ordinary point $z' \in \mathbb{C}$ and a path $z \rightsquigarrow z'$ avoiding singular points there exists a matrix $M_{z\rightsquigarrow z'} \in \mathbb{C}^{r\times r}$ such that

$$\left(D^j y(z')\right)_{j=0}^{r-1} = M_{z\rightsquigarrow z'}\left(D^j y(z)\right)_{j=0}^{r-1}$$

for every solution y analytic near z. There are algorithms [5, 9] for efficiently computing the entries of $M_{z\rightsquigarrow z'}$ for any given polygon path $z \rightsquigarrow z'$ with vertices in $\bar{\mathbb{Q}}$ to any desired precision. In other words, we can compute arbitrary precision approximations of y and its derivatives at every ordinary point ("effective analytic continuation").

Assume now that 0 is a singular point, and consider the case $s=1$ and $m=0$, i.e., let $\hat{y} = \operatorname{Exp}(e)b$ for some $e \in C[x^{-1}]$ and $b \in C[[x]]$ be a formal solution of P. To give an analytic meaning to $\operatorname{Exp}(e) = \exp(\int \frac{e}{x}) = \exp(u + \alpha\log x) = x^{\alpha}\exp(u)$ (for suitable $\alpha \in C$ and $u \in C[x^{-1}]$) amounts to making a choice for a branch of the logarithm. Every choice gives rise to the same function up to some multiplicative constant.

Since $\operatorname{Exp}(e)b$ is a solution of P iff b is a solution of the operator $P \otimes (D + \frac{e}{x})$, we may assume that $e=0$. Then the problem remains that the formal power series $\hat{y} = b$ may not be convergent if 0 is a singular point. However, by resummation theory [2, 3] it is still possible to associate to $\hat{y}$ an analytic function y defined on some sector

$$\Delta = \Delta(d,\varphi,\rho) := \{z \in \mathbb{C} : 0 < |z| \le \rho \wedge |d - \arg z| \le \varphi/2\}$$

(with $d \in [0,2\pi]$, $\rho,\varphi > 0$) such that $\hat{y}$ is the asymptotic expansion of y for $z \to 0$ in Δ.

The precise formulation of this result is technical and not really needed for our purpose (see [3, Chap. 6, 10, and 11] or [2, Chap. 5–7] for full details). It will be more than sufficient to know the following facts:

- For every $\boldsymbol{k} = (k_1,\dots,k_q) \in \mathbb{Q}^q$ with $k_1 > \dots > k_q$ and every $\boldsymbol{d} = (d_1,\dots,d_q) \in [0,2\pi]^q$ such that

  $$|d_{j+1} - d_j| \le (k_{j+1}^{-1} - k_j^{-1})\tfrac{\pi}{2}, \quad j = 1,\dots,q-1,$$

  one constructs [3, §10.2] a differential subring $\mathbb{C}\{x\}_{\boldsymbol{k},\boldsymbol{d}}$ of $\mathbb{C}[[x]]$ [3, Theorems 51 and 53] which contains the ring $\mathbb{C}\{x\}$ of all convergent power series.

- There is a differential ring homomorphism [3, Theorems 51 and 53] $\mathcal{S}_{\boldsymbol{k},\boldsymbol{d}}$ from $\mathbb{C}\{x\}_{\boldsymbol{k},\boldsymbol{d}}$ to the germs of analytic functions defined on sectors of the form $\Delta(d_1,\varphi,\rho)$ for suitable $\varphi,\rho > 0$, with the property that for every $\hat{y} \in \mathbb{C}\{x\}_{\boldsymbol{k},\boldsymbol{d}}$ the function $\mathcal{S}_{\boldsymbol{k},\boldsymbol{d}}(\hat{y})$ has $\hat{y}$ as its asymptotic expansion for $z \to 0$ [3, §10.2, Exercice 2]. The $\mathcal{S}_{\boldsymbol{k},\boldsymbol{d}}$ map convergent formal power series to their sum in the usual sense [3, Lemmas 8 and 20].

- For a given operator $P \in C[x][D]$ of order r, one can compute a tuple $\boldsymbol{k}$ and finite subsets $\mathcal{D}_1,\dots,\mathcal{D}_q$ of $[0,2\pi]$ such that any $\hat{y} \in \mathbb{C}[[x]]$ with $P\cdot\hat{y} = 0$ belongs to $\mathbb{C}\{x\}_{\boldsymbol{k},\boldsymbol{d}}$ for all $\boldsymbol{d}$ as above with $d_1 \notin \mathcal{D}_1,\dots,d_q \notin \mathcal{D}_q$. Additionally, given such a $\boldsymbol{d}$, one can compute $\varphi,\rho > 0$ such that each $\mathcal{S}_{\boldsymbol{k},\boldsymbol{d}}(\hat{y})$ is defined on $\Delta(d_1,\varphi,\rho)$.

- Furthermore, given a point $z \in \Delta(d_1,\varphi,\rho)$, a precision $\varepsilon > 0$, and $\hat{y} \in \mathbb{C}[[x]]$ with $P\cdot\hat{y} = 0$, one can efficiently compute an approximation Y_ε of the vector $Y(z) = (D^j\mathcal{S}_{\boldsymbol{k},\boldsymbol{d}}(\hat{y}))_{j=0}^{r-1}$ such that $\|Y(z) - Y_\varepsilon\| \le \varepsilon$.

The computational part of the last two items is a special case of Theorem 7 of van der Hoeven [11]. As an application, van der Hoeven [10] shows how to factor differential operators using numerical evaluation. Note that our k_j correspond to $1/k_j$ in van der Hoeven's articles, and the components of the tuples $\boldsymbol{k}$ and $\boldsymbol{d}$ appear in reverse order.

Also observe that in the last item, z is an ordinary point, so that from there we can use effective analytic continuation to compute values of $\mathcal{S}_{\boldsymbol{k},\boldsymbol{d}}(\hat{y})$ and its derivatives at any other ordinary point.

3. OUTLINE OF THE ALGORITHM

A hyperexponential term h can be expanded as a generalized series at every point $z \in \mathbb{C} \cup \{\infty\}$, in particular at its singularities. The resulting generalized series are local solutions of P if h is a solution of P. If $h = \exp(\int v)$ is a hyperexponential solution where $v \in \mathbb{C}(x)$, and if we write the partial fraction decomposition of v in the form

$$v = \frac{e_1}{x - z_1} + \frac{e_2}{x - z_2} + \cdots + \frac{e_n}{1/x},$$

where the e_i are polynomials in $\tilde{x}_i^{-1}$, then expanding this h at z_i yields a generalized series in $\tilde{x}_i$ whose exponential part matches $\mathrm{Exp}(e_i)$. The components e_i in the decomposition of v must hence show up among the exponential parts of the local solutions of P.

If $\mathrm{Exp}(e_{i,1}), \ldots, \mathrm{Exp}(e_{i,\ell_i})$ are (representatives of) the different exponential parts that appear among the local solutions at z_i, then any hyperexponential solution must be equivalent to the term $\exp(\int(\frac{e_{1,j_1}}{\tilde{x}_1} + \cdots + \frac{e_{n,j_n}}{\tilde{x}_n}))$ for some tuple $(j_1, \ldots, j_n)$. It then remains to check for each of these candidates whether some element of its equivalence class solves the given equation. The basic structure of the algorithm for finding hyperexponential solutions is thus as follows.

Algorithm 1. *Input: a linear differential operator* $P = p_0 + p_1 D + \cdots + p_r D^r$, $p_r \neq 0$, *with coefficients in* $C[x]$.
Output: all the hyperexponential terms h *with* $P \cdot h = 0$.

1. *Let* $z_1, \ldots, z_{n-1} \in \mathbb{C}$ *be the roots of* p_r *in* $\mathbb{C}$, *and let* $z_n = \infty$.
2. *For* $i = 1, \ldots, n$ *do*
3. *Find the exponential parts* $\mathrm{Exp}(e_{i,1}), \ldots, \mathrm{Exp}(e_{i,\ell_i})$ *of the local solutions of* P *at* z_i.
4. *Determine a set* $U \subseteq \{1, \ldots, \ell_1\} \times \cdots \times \{1, \ldots, \ell_n\}$ *s.t. for every hyperexponential solution* h *equivalent to* $\exp(\int \sum_{i=1}^n \frac{e_{i,j_i}}{\tilde{x}_i})$ *we have* $(j_1, \ldots, j_n) \in U$.
5. *For each* $(j_1, \ldots, j_n) \in U$ *do*
6. *Let* $h_0 := \exp(\int \sum_{i=1}^n \frac{e_{i,j_i}}{\tilde{x}_i})$, *and compute the operator* $\tilde{P} := P \otimes (D - \frac{D(1/h_0)}{1/h_0})$.
7. *Compute a basis* $\{u_1, \ldots, u_m\} \subseteq C(x)$ *of the vector space of all rational solutions of* $\tilde{P}$, *and output* $u_1 h_0, \ldots, u_m h_0$.

There is some freedom in step 4 of this algorithm. A naive approach would simply be to take all possible combinations, i.e., $U = \{1, \ldots, \ell_1\} \times \cdots \times \{1, \ldots, \ell_n\}$. This is a finite set, but its size is in general exponential in the number of singular points. For finding a smaller set U, Cluzeau and van Hoeij [6] use modular techniques to quickly discard unnecessary tuples. Our algorithm, explained in the following section, addresses the same issue. It computes a set U of at most r tuples.

4. THE COMBINATION PHASE

In general, the differential operator P may have several different solutions with the same exponential part, i.e., the dimension of the vector spaces $V_{i,j}$ might be greater than one. In this case, it might be that $V_{i,j}$ contains some series which is the expansion of a hyperexponential solution h at z_i as well as some other series which are not. If we compute some basis of $V_{i,j}$, we cannot expect it to contain the expansion of h. Instead, each basis element will in general be the linear combination of this series and some other one. Now, if the expansion of h at some other singular point $z_{i'}$ belongs to the space $V_{i',j'}$ (which possibly also has higher dimension), then, in some sense, h must belong to the intersection of the vector spaces $V_{i,j}$ and $V_{i',j'}$.

Our algorithm is based on testing which intersections are nontrivial. To make these intersections meaningful, we must first map the vector spaces we want to intersect into a common ambient space W. Let E be some differential ring containing $C(x)$ as well as all the hyperexponential solutions of P, and let $W \subseteq E$ be the C-vector space generated by solutions of P in E. For each i, let π_i be some vector space homomorphism

$$\bigoplus_{j=1}^{\ell_i} \mathrm{Exp}(e_{i,j})\mathbb{C}((\tilde{x}_i^{1/s_{i,j}}))[\log \tilde{x}_i] \supseteq V_i \xrightarrow{\pi_i} W$$

with the following properties:

1. The sum $\pi_i(V_{i,1}) + \cdots + \pi_i(V_{i,\ell_i})$ is direct.
2. If $h \in W$ is hyperexponential, then $\pi_i^{-1}(h)$ contains the formal series expansion $\hat{h}$ of h at z_i, possibly up to a multiplicative constant.

Define $W_{i,j} := \pi_i(V_{i,j})$. If h is some hyperexponential solution of P, say with exponential part

$$\exp\Bigl(\int\Bigl(\frac{e_{1,j_1}}{\tilde{x}_1} + \frac{e_{2,j_2}}{\tilde{x}_2} + \cdots + \frac{e_{n,j_n}}{\tilde{x}_n}\Bigr)\Bigr),$$

then $h \in W_{i,j_i}$ for all i, and hence the vector space $W_{\boldsymbol{j}} := W_{1,j_1} \cap \cdots \cap W_{n,j_n}$ is not the zero subspace (because it contains at least h). Our main observation is that there can be at most r tuples $\boldsymbol{j} = (j_1, \ldots, j_n)$ for which $W_{\boldsymbol{j}} \neq \{0\}$, and that they can be computed efficiently once we have bases of the $W_{i,j}$.

Postponing the discussion of making the π_i constructive to the next section, assume for the moment that W is some vector space over C, let $r = \dim W < \infty$ be its dimension, and suppose we are given n different decompositions of subspaces of W into direct sums:

$$\begin{gathered} W_{1,1} \oplus W_{1,2} \oplus \cdots \oplus W_{1,\ell_1} \subseteq W, \\ W_{2,1} \oplus W_{2,2} \oplus \cdots \oplus W_{2,\ell_2} \subseteq W, \\ \vdots \\ W_{n,1} \oplus W_{n,2} \oplus \cdots \oplus W_{n,\ell_n} \subseteq W. \end{gathered}$$

Without loss of generality, we may make the following assumptions:

- Each direct sum $\bigoplus_{j=1}^{\ell_i} W_{i,j}$ is in fact equal to W. If not, add one more vector space to the sum.
- $\ell_1 = \ell_2 = \cdots = \ell_n =: \ell$. If not, pad the sum with several copies of $\{0\}$.
- $\ell \leq r$. If not, then because the sums are supposed to be direct, each decomposition must contain at least $\ell - r$ copies of $\{0\}$, which can be dropped.

Lemma 2. *There are at most* $\dim W = r$ *different tuples*

$$\boldsymbol{j} = (j_1, \ldots, j_n) \in \{1, \ldots, \ell\}^n$$

such that $W_{\boldsymbol{j}} := W_{1,j_1} \cap W_{2,j_2} \cap \cdots \cap W_{n,j_n} \neq \{0\}$.

Proof. Induction on n. For $n = 1$, there are only $\ell \le r$ different tuples altogether: $(1), (2), \ldots, (\ell)$, so the claim is obviously true. Suppose now that the claim is shown for the case when $n-1$ decompositions of some vector space are given. Let $U \subset \{1, \ldots, \ell\}^n$ be a set of tuples $\boldsymbol{j}$ with $W_{\boldsymbol{j}} \neq \{0\}$. Partition the elements of U according to their first components,

$$U = U_1 \,\dot\cup\, U_2 \,\dot\cup \cdots \dot\cup\, U_\ell,$$

i.e., U_k is the set of all tuples $\boldsymbol{j}$ whose first component is k, for $k = 1, \ldots, \ell$.

For all $\boldsymbol{j} = (k, j_2, \ldots, j_n) \in U_k$ we have $\{0\} \neq W_{\boldsymbol{j}} \subseteq W_{1,k}$. Therefore, $(j_2, \ldots, j_n) \in \{1, \ldots, \ell\}^{n-1}$ is a valid solution tuple for the modified problem with $W'_{i,j} := W_{i+1,j} \cap W_{1,k}$ ($i = 1, \ldots, n-1$, $j = 1, \ldots, \ell$) in place of $W_{i,j}$ ($i = 1, \ldots, n$, $j = 1, \ldots, \ell$). By induction hypothesis, since the $W'_{i,j}$ form $n-1$ decompositions of the space $W_{1,k}$, there are at most $\dim W_{1,k}$ tuples $(j_2, \ldots, j_n)$ with $W_{(j_2,\ldots,j_m)} \neq \{0\}$. Consequently, there are altogether at most $\sum_{k=1}^{\ell} \dim W_{1,k} = \dim W = r$ different tuples for the original space W. ■

The desired index tuples can be computed efficiently using dynamic programming, as shown in the following algorithm.

Algorithm 3. *Input: a vector space W of dimension r, and a collection of subspaces $W_{i,j}$ ($i = 1, \ldots, n$; $j = 1, \ldots, \ell$) such that $W = \bigoplus_{j=1}^{\ell} W_{i,j}$ for $i = 1, \ldots, n$ and $\ell \le r$.*
Output: the set U of all tuples $\boldsymbol{j} = (j_1, \ldots, j_n)$ with the property $W_{\boldsymbol{j}} = \bigcap_{i=1}^{n} W_{i,j_i} \neq \{0\}$.

1. $U := \{\,(j) : W_{1,j} \neq \{0\}\,\}$
2. *For* $i = 2, \ldots, n$ *do*
3. $\quad U_{new} := \emptyset$
4. $\quad$ *For* $j = 1, \ldots, \ell$ *do*
5. $\quad\quad$ *For* $\boldsymbol{k} \in U$ *do*
6. $\quad\quad\quad$ *If* $W_{\boldsymbol{k}} \cap W_{i,j} \neq \{0\}$ *then*
7. $\quad\quad\quad\quad U_{new} := U_{new} \cup \{\text{append}(\boldsymbol{k}, j)\}$
8. $\quad U := U_{new}$
9. *Return* U

Proposition 4. *Algorithm 3 is correct and needs no more than $O(nr^5)$ operations in C, if the bases of the $W_{\boldsymbol{k}}$ are cached.*

Proof. Correctness is obvious by line 6 and the fact that whenever $\boldsymbol{k} = (k_1, \ldots, k_n)$ is such that $W_{\boldsymbol{k}} \neq \{0\}$ then we necessarily also have $W_{(k_1,\ldots,k_{n-1})} \neq \{0\}$.

For the complexity, observe that all the vector space intersections can be performed with a number of operations which is at most cubic in r. Taking also into account that we always have $|U| \le r$ by Lemma 2, and that $\ell \le r$, the bound $O(n\ell r^4) = O(nr^5)$ follows immediately. ■

By a more careful analysis it can be shown that the complexity is actually bounded by $8nr^4$. We omit the details of the corresponding argument as we will not need the tighter estimate in what follows.

5. NUMERICAL EVALUATION AT A REFERENCE POINT

We now turn to the question of how to construct the morphisms π_i. The basic idea is to choose a reference point z_0 that is an ordinary point of P, and let W be the space of analytic solutions of the equation in a neighborhood of z_0.

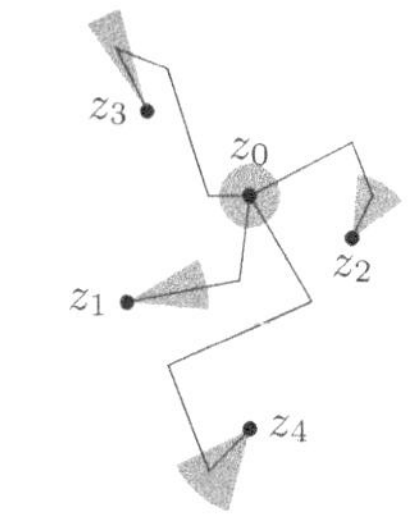

For each singular point z_i, let Δ_i be a sector rooted at z_i for which all formal power series appearing in the generalized series solutions of P at z_i admit an interpretation as analytic functions via some operator $\mathcal{S}_{\boldsymbol{k},\boldsymbol{d}}$ (depending on i, but not on the series), as described in Section 2.4. Such sectors exist and can be computed explicitly. Next, let γ_i ($i = 1, \ldots, n$) be polygonal paths from z_i to z_0 avoiding singular points and leaving the startpoint through Δ_i (meaning that for some $\varepsilon > 0$ all the points on γ_i with a distance to z_i less than ε should belong to Δ_i). Such paths exist. The analytic interpretations of the generalized series solutions at the singular points z_i defined in Δ_i admit a unique analytic continuation along the paths γ_i to the neighborhood of z_0.

We define $\pi_i \colon V_i \to W$ as follows. Let $V^0_{i,j}$ be the subspace of $V_{i,j}$ consisting of generalized series (1) with $s = 1$ and $m = 0$, and let $V'_{i,j}$ be a linear complement of $V^0_{i,j}$ in $V_{i,j}$. If $\hat y \in V^0_{i,j}$, i.e., if $\hat y = \operatorname{Exp}(e_{i,j}) b$ with $b \in \mathbb{C}[[\tilde x_i]]$, define $\pi_i(\hat y)$ to be the unique analytic continuation of the function $\mathrm{E}(e_{i,j}) \mathcal{S}_{\boldsymbol{k},\boldsymbol{d}}(b)$ along γ_i to z_0, where $\mathrm{E}(e_{i,j})$ refers to the function $z \mapsto \exp(\int_{z_0}^{z} e_{i,j}/\tilde x_i)$ with some arbitrary but fixed choice of the branch of the logarithm, and $\mathcal{S}_{\boldsymbol{k},\boldsymbol{d}}$ is as described in Section 2.4. Set $\pi_i(\hat y) = 0$ for $\hat y \in V'_{i,j}$, and then extend π_i to V_i by linearity. The precise values of $\pi_i(V_{i,j})$ depend on the choice of Δ_i and $\boldsymbol{d}$ (which is arbitrary, within the limits indicated in Section 2.4), but, as shown below, the properties of these spaces used in the algorithm do not.

Proposition 5. *The functions π_i defined above satisfy the two requirements imposed in Section 4: (1) $\pi_i(V_{i,1}) + \cdots + \pi_i(V_{i,\ell})$ is a direct sum; (2) if h is a hyperexponential term, then $\pi_i^{-1}(h)$ contains the formal series expansion of h at z_i, possibly up to a multiplicative constant.*

Proof. 1. Without loss of generality, we assume $z_i = 0$. Let $\hat y_j \in V_{i,j}$ ($j = 1, \ldots, \ell$) and consider $\hat y = \sum_{j=1}^{\ell} \hat y_j$. Write $\hat y_j = x^{\alpha_j} \exp(u_j) b_j + \hat y'_j$ where $\hat y'_j \in V'_{i,j}$, the (α_j, u_j) are pairwise distinct, $u_j(0) = 0$, and $b_j(0) \neq 0$ unless the series b_j is zero. Writing $u_j = \sum_k u_{j,k} x^{-k}$, choose a direction θ such that $\rho e^{i\theta} \in \Delta_i$ for small ρ and any two unequal $u_{j,k}^{1/k} e^{i\theta}$ have different real parts.

By changing x to $e^{-i\theta} x$, we can assume that $d = 0$. This tranforms u_j into $\sum_k (u_{j,k} e^{ik\theta}) x^{-k}$, so that the real parts of two polynomials u_j can be the same only if the u_j themselves are equal. Hence, we can reorder the nonzero terms in the expression of $\hat y$ by asymptotic growth rate, in such a way that the nonzero terms come first, $u_1 = \cdots = u_t$ and $\operatorname{Re}\alpha_1 = \cdots = \operatorname{Re}\alpha_t$, while

$$z^{\operatorname{Re}\alpha_1} e^{\operatorname{Re} u_1(z)} \gg z^{\operatorname{Re}\alpha_p} e^{\operatorname{Re} u_p(z)}, \quad z \to 0, z > 0$$

for all $p \ge t+1$ such that $y_p \neq 0$. Using the definition of π_i and the fact that $\mathcal{S}_{\boldsymbol{k},\boldsymbol{d}}(b_j)(z)$ tends to $b_j(0)$ as $z \to 0$ in the positive reals, it follows that

$$z^{-\operatorname{Re}\alpha_1}\exp(-u_1(z))\,\pi_i(\hat{y})(z) = \sum_{j=1}^{t} c_j b_j(0) z^{\mathrm{i}\operatorname{Im}\alpha_j} + \mathrm{o}(1)$$

(as $z \to 0$, $z > 0$) for some nonzero constants c_j. Since the (α_j, u_j) are pairwise distinct by assumption and the $(\operatorname{Re}\alpha_j, u_j)$ are equal for $j = 1, \dots, t$, the $\operatorname{Im}\alpha_j$ are pairwise distinct for $j = 1, \dots, t$.

Now assume that $\pi_i(\hat{y}) = 0$. Then, for all $\lambda > 0$, the expression $\sum_{j=1}^{t} c_j b_j(0)(\lambda z)^{\mathrm{i}\operatorname{Im}\alpha_j}$ tends to 0 as $z \to 0$, $z > 0$. Choosing $\lambda = \mathrm{e}^p$ for $p = 1, \dots, t$, it follows that if not all the $b_j(0)$ were zero, the $t \times t$ determinant

$$\det\Bigl((\mathrm{e}^p z)^{\mathrm{i}\operatorname{Im}\alpha_q}\Bigr)_{p,q} = z^{\mathrm{i}\operatorname{Im}(\alpha_1+\cdots+\alpha_t)}\det\Bigl((\mathrm{e}^{\mathrm{i}\operatorname{Im}\alpha_q})^p\Bigr)_{p,q}$$

would tend to zero as well, which however is not the case. Therefore $b_j(0) = 0$ for $j = 1, \dots, t$, and therefore $\hat{y}_j = 0$ for $j = 1, \dots, t$, and therefore $\hat{y}_j = 0$ for $j = 1, \dots, \ell$.

2. Let $h \in W$ be hyperexponential. Then the expansion $\hat{h}$ of h at z_i is clearly a local solution, so $\hat{h} \in V_{i,j}$ for some j. We show that $\pi_i(\hat{h}) = ch$ for some $c \in \mathbb{C}$. The map π_i is a differential homomorphism because $\mathcal{S}_{\boldsymbol{k},\boldsymbol{d}}$ is (as remarked in Section 2.4) and the (formal) exponential parts $\operatorname{Exp}(e_{i,j})$ are mapped to analytic functions satisfying the same differential equations. Since h is hyperexponential, it satisfies a first order linear differential equation. Since $\hat{h}$ is the expansion of h, it satisfies the same equations as h. Since π_i is a differential homomorphism, $\pi_i(\hat{h})$ satisfies the same equations as $\hat{h}$. Hence $\pi_i(\hat{h})$ and h satisfy the same first-order differential equation. The claim follows. ∎

The definition of the maps $\pi_1, \dots, \pi_m$ as outlined above relies on analytic continuation, a concept which is only available if $C = \mathbb{C}$. For actual computations, we must work in a computable coefficient domain. At this point, we use numerical approximations. By van der Hoeven's result quoted in Section 2.4, we are able to compute for every given $\hat{y} \in V_{i,j}$ and every given $\varepsilon > 0$ a vector $Y_\varepsilon \in \mathbb{Q}(\mathrm{i})^r$ with

$$\Bigl\|\bigl(D^k \pi_i(\hat{y})(z_0)\bigr)_{k=0}^{r-1} - Y_\varepsilon\Bigr\|_\infty < \varepsilon.$$

Using these approximations, the linear algebra parts of Algorithm 3 are then performed with ball arithmetic to keep track of accumulating errors during the calculations. The test in line 6 of this algorithm requires to check whether a certain matrix has full rank.

There are two possible outcomes: If during the Gaussian elimination we can find in every iteration an entry which is definitely different from zero, then the rank of the matrix is definitely maximal and the intersection of the vector spaces is definitely empty. We are then entitled to discard the possible extension of the partial tuple under consideration. On the other hand, if during the Gaussian elimination we encounter a column in which all the entries are balls that contain zero, this can either mean that the intersection is really nonempty, or that the accuracy of the approximation was insufficient. In this case, in order to be on the safe side, we must consider the intersection as nonempty and include the corresponding tuple.

Regardless of which initial accuracy ε is used, this variant of Algorithm 3 produces a set of tuples that is guaranteed to contain all correct ones, but may possibly contain additional ones. With sufficiently high precision, the number of tuples in the output that actually have an empty intersection will drop to zero.

We don't need to know in advance which precision is sufficient in this sense, because it is not dramatic to have some extra tuples in the output as long as they are not too many. Unfortunately, we don't have an implementation yet with which we could experimentally determine a good strategy for resolving this situation. An ad-hoc way to balance precision and output size might be to start the algorithm with some fixed precision ε and let it abort and restart with doubled precision whenever $|U|$ exceeds $2r$, say. Another idea, suggested by a referee, is to use our algorithm in combination with the modular approach from [6] for cross checking tuples for which the numerical data is not conclusive.

In any case, we want to stress that we use numerical approximations only to determine the (small) tuple set U, and not to somehow reconstruct the (long) exact symbolic hyperexponential solutions from it. We therefore don't expect to need very high precision in typical situations.

6. A DETAILED EXAMPLE

Consider the operator

$$P = p_0 + p_1 D + p_2 D^2 + p_3 D^3 \in \mathbb{Q}[x][D]$$

where

$$\begin{gathered} p_0 = -105x^{20} + 3570x^{19} - 58026x^{18} + 556216x^{17} - 3456830x^{16} + \\ 14810744x^{15} - 45667732x^{14} + 104614932x^{13} - 182764261x^{12} + \\ 249940430x^{11} - 276371642x^{10} + 257839924x^{9} - 211785148x^{8} + \\ 154714472x^{7} - 95675216x^{6} + 45214304x^{5} - 13863936x^{4} + \\ 1685888x^{3} + 424960x^{2} - 182784x + 20480, \end{gathered}$$

$$\begin{gathered} p_1 = (x-1)x(105x^{19} - 3150x^{18} + 51456x^{17} - 489796x^{16} + \\ 2938210x^{15} - 11903624x^{14} + 34247824x^{13} - 72603516x^{12} + \\ 116974957x^{11} - 148046826x^{10} + 153582952x^{9} - 137261696x^{8} + \\ 109046080x^{7} - 75250624x^{6} + 41559168x^{5} - 16084864x^{4} + \\ 3278080x^{3} + 163840x^{2} - 231424x + 32768), \end{gathered}$$

$$\begin{gathered} p_2 = -4(x-2)^2(x-1)^3x^2(30x^{15} - 693x^{14} + 7314x^{13} - 42905x^{12} + \\ 155930x^{11} - 378483x^{10} + 649718x^{9} - 828795x^{8} + 820160x^{7} - \\ 645092x^{6} + 398200x^{5} - 182384x^{4} + 54656x^{3} - 5696x^{2} - 2944x + 1024), \end{gathered}$$

$$\begin{gathered} p_3 = 4(x-2)^4(x-1)^5x^4(15x^{10} - 258x^{9} + 1492x^{8} - 4446x^{7} + \\ 8309x^{6} - 10972x^{5} + 10520x^{4} - 6456x^{3} + 1552x^{2} + 480x - 256). \end{gathered}$$

The leading coefficient p_3 has 13 distinct roots in $\mathbb{C}$, but those coming from the degree-10-factor turn out to be apparent singularities, so we can ignore them. (See [4] and the references there for more on apparent singularites.) It thus remains to study the singular points $z_1 := 0$, $z_2 := 1$, $z_3 := 2$, and $z_4 := \infty$.

For each singular point, we find three linearly independent generalized series solutions with two distinct exponential parts:

$$\begin{aligned} V_{1,1} &= \mathbb{C}\hat{y}_{1,1} & V_{1,2} &= \mathbb{C}\hat{y}_{1,2} + \mathbb{C}\hat{y}_{1,3}, \\ V_{2,1} &= \mathbb{C}\hat{y}_{2,1} & V_{2,2} &= \mathbb{C}\hat{y}_{2,2} + \mathbb{C}\hat{y}_{2,3}, \\ V_{3,1} &= \mathbb{C}\hat{y}_{3,1} & V_{3,2} &= \mathbb{C}\hat{y}_{3,2} + \mathbb{C}\hat{y}_{3,3}, \\ V_{4,1} &= \mathbb{C}\hat{y}_{4,1} & V_{4,2} &= \mathbb{C}\hat{y}_{4,2} + \mathbb{C}\hat{y}_{4,3} \end{aligned}$$

where

$$\hat{y}_{1,1} = \exp(\tfrac{1}{x})\Bigl(1 - \tfrac{4}{9}x + \tfrac{37}{32}x^2 + \tfrac{83}{384}x^3 + \cdots\Bigr),$$

$$\hat{y}_{1,2} = \sqrt{x}\Bigl(1 - x - \tfrac{25}{24}x^3 + \cdots\Bigr),$$

$$\hat{y}_{1,3} = \sqrt{x}\Bigl(x^2 - \tfrac{7}{4}x^3 + \tfrac{9}{32}x^4 + \cdots\Bigr),$$

$$\hat{y}_{2,1} = (x-1)^3 + (x-1)^5 - \tfrac{4}{3}(x-1)^6 + \cdots,$$

$$\hat{y}_{2,2} = \exp(\tfrac{1}{x-1})\Big(1 + \tfrac{1}{2}(x-1) + \tfrac{19}{120}(x-1)^3 + \cdots\Big),$$

$$\hat{y}_{2,3} = \exp(\tfrac{1}{x-1})\Big((x-1)^2 + \tfrac{23}{30}(x-1)^3 + \cdots\Big),$$

$$\hat{y}_{3,1} = 1 - \tfrac{3}{4}(x-2) + \tfrac{39}{32}(x-2)^2 - \tfrac{673}{384}(x-2)^3 + \cdots,$$

$$\hat{y}_{3,2} = \tfrac{1}{(x-2)^2}\exp(\tfrac{1}{x-2})\Big(1 + \tfrac{11}{4}(x-2) + \cdots\Big),$$

$$\hat{y}_{3,3} = \tfrac{1}{(x-2)^2}\exp(\tfrac{1}{x-2})\Big((x-2)^3 + \tfrac{1}{4}(x-2)^4 + \cdots\Big),$$

$$\hat{y}_{4,1} = x\Big(1 + 3x^{-1} + 9x^{-2} + \tfrac{79}{3}x^{-3} + 74x^{-4} + \cdots\Big),$$

$$\hat{y}_{4,2} = \sqrt{x}\Big(1 + x^{-1} + \tfrac{3}{2}x^{-2} + \tfrac{13}{6}x^{-3} + \cdots\Big),$$

$$\hat{y}_{4,2} = \sqrt{x}\Big(x^3 + x + \tfrac{19}{6}x^{-1} + \tfrac{283}{30}x^{-2} + \cdots\Big).$$

Let us choose $z_0 = 3$ as ordinary reference point and take the branch of the logarithm for which $\sqrt{x}$ is positive and real on the positive real axis. The example was chosen in such a way that all the power series are convergent in some neighborhood of the expansion point, so that we do not need to worry about sectors and resummation theory but can use the somewhat simpler algorithm for effective analytic continuation in the ordinary case to compute the values of the analytic functions $y_{i,j} := \pi_i(\hat{y}_{i,j})$ $(i = 1, \ldots, 4;\ j = 1, 2, 3)$.

The vectors $\big(y_{i,j}(z_0), Dy_{i,j}(z_0), D^2y_{i,j}(z_0)\big)$ to five decimal digits of accuracy are as follows.

$$W_{1,1} = \left[\begin{pmatrix}-200.15\\322.46\\-1184.8\end{pmatrix}\right],\quad W_{1,2} = \left[\begin{pmatrix}-70.513\\-46.308\\-101.17\end{pmatrix},\begin{pmatrix}-156.55\\-91.322\\-205.47\end{pmatrix}\right],$$

$$W_{2,1} = \left[\begin{pmatrix}30.349\\-48.896\\179.66\end{pmatrix}\right],\quad W_{2,2} = \left[\begin{pmatrix}12.494\\5.2891\\13.066\end{pmatrix},\begin{pmatrix}77.105\\44.216\\99.931\end{pmatrix}\right],$$

$$W_{3,1} = \left[\begin{pmatrix}.74285\\-.061904\\.14960\end{pmatrix}\right],\quad W_{3,2} = \left[\begin{pmatrix}15.580\\-31.307\\105.26\end{pmatrix},\begin{pmatrix}4.5433\\2.6503\\5.9631\end{pmatrix}\right],$$

$$W_{4,1} = \left[\begin{pmatrix}30.349\\-48.896\\179.66\end{pmatrix}\right],\quad W_{4,2} = \left[\begin{pmatrix}2.8557\\-.23797\\.57510\end{pmatrix},\begin{pmatrix}63.199\\41.308\\90.353\end{pmatrix}\right].$$

We now go through Algorithm 3. Start with the partial tuples (1) and (2) corresponding to the vector spaces $W_{1,1}$ and $W_{1,2}$, respectively. To compute the intersection of $W_{1,1}$ and $W_{2,1}$ we apply Gaussian elimination to the 3×2-matrix whose columns are the generators of $W_{1,1}$ and $W_{2,1}$:

$$\begin{pmatrix}-200.15 & 30.349\\322.46 & -48.896\\-1184.8 & 179.66\end{pmatrix} \longrightarrow \begin{pmatrix}-200.15 & 30.349\\ & 0.00\\ & 0.00\end{pmatrix}$$

The notation 0.00 refers to some complex number z with $|z| < 5 \cdot 10^{-3}$, which may or may not be zero, while the blank entries in the left column signify exact zeros that have been produced by the elimination. As the remaining submatrix does not contain any entry which is certainly nonzero, we regard the intersection as nonempty, which in this case means $W_{1,1} = W_{2,1}$. The partial tuple (1) is extended to (1, 1).

The intersections $W_{1,1} \cap W_{2,2}$ and $W_{1,2} \cap W_{2,1}$ turn out to be trivial, as they have to be if we really have $W_{1,1} = W_{1,2}$, because the sums $W_{1,1} \oplus W_{1,2}$ and $W_{2,1} \oplus W_{2,2}$ are direct.

It thus remains to consider the intersection $W_{1,2} \cap W_{2,2}$. Applying Gaussian elimination to the 3×4-matrix whose columns are the generators of $W_{1,2}$ and $W_{2,2}$, we find

$$\begin{pmatrix}-70.513 & -25.596 & 12.494 & 77.105\\-46.308 & 2.1330 & 5.2891 & 44.216\\-101.17 & -5.1548 & 13.066 & 99.931\end{pmatrix}$$

$$\longrightarrow \begin{pmatrix}-70.513 & -25.596 & 12.494 & 77.105\\ & -17.50 & 4.440 & 9.777\\ & & 0.00 & 0.00\end{pmatrix},$$

which suggests that we have $W_{1,2} = W_{2,2}$. We extend the partial tuple (2) to (2, 2). At the end of the first iteration, we have $U = \{(1,1), (2,2)\}$.

In the second iteration, we find $W_{(1,1)} \cap W_{3,1} = \{0\}$ and $W_{(1,1)} \subseteq W_{3,2}$, so we extend the partial tuple (1, 1) to (1, 1, 2) and record $W_{(1,1,2)} = W_{(1,1)} = W_{1,1}$. Furthermore we find $W_{3,1} \subseteq W_{(2,2)}$, so we extend (2, 2) to (2, 2, 1) and record $W_{(2,2,1)} = W_{3,1}$. Finally, there is a nontrivial intersection between $W_{(2,2)}$ and $W_{3,2}$:

$$\begin{pmatrix}12.494 & 77.105 & 15.580 & 4.5433\\5.2891 & 44.216 & -31.307 & 2.6503\\13.066 & 99.931 & 105.26 & 5.9631\end{pmatrix}$$

$$\longrightarrow \begin{pmatrix}12.494 & 77.105 & 15.580 & 4.5433\\ & -27.34 & 89.53 & -1.72\\ & & 216. & 0.00\end{pmatrix},$$

which suggests a common subspace of dimension 1 generated by the second listed generator of $W_{3,2}$. We therefore extend the partial tuple (2, 2) to (2, 2, 2) and record $W_{(2,2,2)} = [(4.5433, 2.6503, 5.9631)]$. At the end of the second iteration, we have $U = \{(1,1,2), (2,2,1), (2,2,2)\}$.

For the final iteration, we see by inspection that $W_{4,1} = W_{2,1} = W_{(1,1,2)}$, so we extend (1, 1, 2) to (1, 1, 2, 1). The other two partial tuples do not have a nontrivial intersection with $W_{4,1}$, nor is $W_{(1,1,2)} \cap W_{4,2}$ nontrivial. We do however have $W_{(2,2,1)} \subseteq W_{4,2}$ and $W_{(2,2,2)} \subseteq W_{4,2}$, so the algorithm terminates with the output

$$U = \{(1,1,2,1), (2,2,1,2), (2,2,2,2)\}.$$

At this point we know that every hyperexponential solution of the operator P must have one of the following three exponential parts:

$$\frac{1}{(x-2)^2}\exp\Big(\frac{1}{x} + \frac{1}{x-2}\Big) \qquad \text{from (1,1,2,1)},$$

$$\sqrt{x}\exp\Big(\frac{1}{x-1}\Big) \qquad \text{from (2,2,1,2)},$$

$$\sqrt{x}\exp\Big(\frac{1}{x-1} + \frac{1}{x-2}\Big) \qquad \text{from (2,2,2,2)}.$$

Following the steps of Algorithm 1, it remains to check if some rational function multiples of these terms are solutions of P. The important point is that we have to do this only for three different candidates, while the naive algorithm would have to go through all $2^4 = 16$ combinations. Indeed, it turns out that P has the following three hyperexponential solutions:

$$\frac{(x-1)^3}{(x-2)^2}\exp\Big(\frac{1}{x} + \frac{1}{x-2}\Big),\quad \sqrt{x}\exp\Big(\frac{1}{x-1}\Big),$$

$$(x-2)x^2\sqrt{x}\exp\Big(\frac{1}{x-1} + \frac{1}{x-2}\Big).$$

7. CONCLUDING REMARKS

Our algorithm as described above takes advantage of the fact that series expansions of hyperexponential terms cannot involve exponential terms with ramification ($s > 1$) or logarithms ($m > 0$), by letting the morphisms π_i map all these irrelevant series solutions to zero. As a result, we get smaller vector spaces $W_{i,j}$, which not only reduces the expected computation time per vector space intersection but also makes it somehow more likely for intersections to be empty, thus decreasing the chances of getting tuples that do not correspond to hyperexponential solutions.

As a further refinement in this direction, it would be desirable to exploit the fact that if $\hat{h} = \mathrm{Exp}(e)b$ is the expansion of some hyperexponential term h, then the formal power series b must be convergent in some neighborhood of the expansion point. Instead of the vector spaces $W_{i,j}$ used above, it would be sufficient to consider the subspaces $W'_{i,j} \subseteq W_{i,j}$ corresponding to generalized series solutions involving only convergent power series. Besides the advantage of having to work with even smaller vector spaces, an additional advantage would be that the numerical evaluation becomes simpler because algorithms for the regular case [5, 9] become applicable. Implementations of these algorithms are available [8], which to our knowledge is not yet the case for van der Hoeven's general algorithm for the divergent case [11]. Unfortunately however, it is not obvious how to compute from a given basis of $W_{i,j}$ a basis of the subspace $W'_{i,j}$. Miller's algorithm [17] numerically solves a similar problem, but so far we have not been able to turn the underlying convergence statements into explicit error bounds that would yield an algorithm producing output with certified precision.

Finally, it would of course be also interesting to see an analog of our algorithm for finding hypergeometric solutions of linear recurrence equations with polynomial coefficients. A translation is not immediate because there is no notion of local solution around a finite singularity in this case.

8. REFERENCES

[1] S.A. Abramov and K.Yu. Kvashenko. Fast algorithms to search for the rational solutions of linear differential equations with polynomial coefficients. In *Proceedings of ISSAC'91*, pages 267–270, 1991.

[2] Werner Balser. *From Divergent Power Series to Analytic Functions*, volume 1582 of *Lecture Notes in Mathematics*. Springer-Verlag, 1994.

[3] Werner Balser. *Formal power series and linear systems of meromorphic ordinary differential equations*. Springer, 2000.

[4] Shaoshi Chen, Maximilian Jaroschek, Manuel Kauers, and Michael F. Singer. Desingularization explains order-degree curves for univariate Ore operators. In *these ISSAC'13 proceedings*, 2013.

[5] David V. Chudnovsky and Gregory V. Chudnovsky. Computer algebra in the service of mathematical physics and number theory. In David V. Chudnovsky and Richard D. Jenks, editors, *Computers in Mathematics*, volume 125 of *Lecture Notes in Pure and Applied Mathematics*, pages 109–232, Stanford University, 1986. Dekker.

[6] Thomas Cluzeau and Mark van Hoeij. A modular algorithm to compute the exponential solutions of a linear differential operator. *Journal of Symbolic Computation*, 38:1043–1076, 2004.

[7] A. K. Lenstra, H. W. Lenstra, and L. Lovász. Factoring polynomials with rational coefficients. *Annals of Mathematics*, 126:515–534, 1982.

[8] Marc Mezzarobba. NumGfun: a package for numerical and analytic computation with D-finite functions. In *Proceedings of ISSAC'10*, 2010.

[9] Joris van der Hoeven. Fast evaluation of holonomic functions. *Theoretical Computer Science*, 210(1):199–216, 1999.

[10] Joris van der Hoeven. Around the numeric-symbolic computation of differential Galois groups. *Journal of Symbolic Computation*, 42:236–264, 2007.

[11] Joris van der Hoeven. Efficient accelero-summation of holonomic functions. *Journal of Symbolic Computation*, 42(4):389–428, 2007.

[12] Marius van der Put and Michael Singer. *Galois Theory of Linear Differential Equations*. Springer, 2003.

[13] Mark van Hoeij. Factorization of differential operators with rational functions coefficients. *Journal of Symbolic Computation*, 24:537–561, 1997.

[14] Mark van Hoeij. Formal solutions and factorization of differential operators with power series coefficients. *Journal of Symbolic Computation*, 24(1):1–30, 1997.

[15] Mark van Hoeij. Factoring polynomials and the knapsack problem. *Journal of Number Theory*, 95:167–189, 2002.

[16] Joachim von zur Gathen and Jürgen Gerhard. *Modern Computer Algebra*. Cambridge University Press, 1999.

[17] Jet Wimp. *Computing with Recurrence Relations*. Pitman Publishing Ltd., 1984.

Sparse Multivariate Function Recovery from Values with Noise and Outlier Errors*

Erich L. Kaltofen
Department of Mathematics
North Carolina State University
Raleigh, NC 27695-8205, USA
kaltofen@math.ncsu.edu www.kaltofen.us

Zhengfeng Yang
Shanghai Key Laboratory of Trustworthy Computing
East China Normal University
Shanghai 200062, China
zfyang@sei.ecnu.edu.cn

ABSTRACT

Error-correcting decoding is generalized to multivariate sparse rational function recovery from evaluations that can be numerically inaccurate and where several evaluations can have severe errors ("outliers"). The generalization of the Berlekamp-Welch decoder to exact Cauchy interpolation of univariate rational functions from values with faults is by Kaltofen and Pernet in 2012 [to be submitted]. We give a different univariate solution based on structured linear algebra that yields a stable decoder with floating point arithmetic. Our multivariate polynomial and rational function interpolation algorithm combines Zippel's symbolic sparse polynomial interpolation technique [Ph.D. Thesis MIT 1979] with the numeric algorithm by Kaltofen, Yang, and Zhi [Proc. SNC 2007], and removes outliers ("cleans up data") through techniques from error correcting codes. Our multivariate algorithm can build a sparse model from a number of evaluations that is linear in the sparsity of the model.

Categories and Subject Descriptors: I.1.2 [Symbolic and Algebraic Manipulation]: Algorithms; G.1.1 [Numerical Analysis]: Interpolation—smoothing

Keywords: error correcting coding, fault tolerance, Cauchy interpolation, rational function

1. INTRODUCTION

Reed-Solomon error correcting coding uses evaluation of a polynomial as the encoding device, and interpolation as the decoding device. The polynomial is oversampled, and $k \le E$ errors in the evaluations are corrected via an additional $2E$ sample points. Blahut's decoding algorithm [1], for evaluations at consecutive powers of roots of unity, locates the erroneous evaluations by sparse interpolation. Berlekamp/Welch decoding, for any set of distinct input arguments, reconstructs the error-corrected polynomial via Bezout coefficients in a polynomial extended Euclidean algorithm [23]. In [4] we address the situation when the polynomial is sparse: our decoding algorithm requires $2T(2E+1)$ evaluations for a polynomial with t non-zero terms, when bounds $T \ge t$ and $E \ge k$ are input. Here k is again the number of faulty evaluations, whose locations are unknown. We use the Prony/Blahut sparse interpolation algorithm and correct k errors in a linearly generated sequence of evaluations; thus we perform $T^{1+o(1)}E$ arithmetic operations. Our algorithm is deployed on numerical data [4, Section 6] via the floating-point versions of the Prony/Blahut algorithm [8, 9, 18].

In [14] we have generalized the Berlekamp/Welch procedure to reconstructing a rational number or a univariate rational function over a field from multiple residues or evaluations, under the assumption that some residues and values are faulty. Again, we use the extended Euclidean algorithm. Our algorithms are for exact arithmetic.

In [17] we generalize Zippel's interpolation algorithm for sparse multivariate polynomials [24] to sparse multivariate rational functions (cf. [6, Section 4]). We present a worst case analysis for exact arithmetic [17, Section 4.1], which for rational functions is more difficult than for polynomials, and implement the algorithm for noisy data with floating point arithmetic. The algorithm is numerically stable because 1. univariate rational recovery is accomplished by a structured total least norm algorithm on the original data, not by the extended Euclidean algorithm on derived data, and 2. multivariate recovery is performed on sparse candidates which constitute well-constrained models for loosely fitting data. As it turned out in [14], Berlekamp/Welch decoding is Cauchy interpolation, and the Euclidean algorithm computes an unreduced rational function.

We combine the insights from [17] and [14] and obtain the following:

1. a numerical, noise tolerant Berlekamp/Welch-like univariate polynomial and rational function interpolation algorithm that can remove outlier errors: our algorithm recovers the full coefficient vectors and for a sparse polynomial f with $t \le T$ terms requires $\deg(f) + 2E + 1$ evaluations compared to the $2T(2E+1)$ evaluations of the algorithm in [4, Section 6]. However, we can work with evaluations at arbitrary input arguments and can recover rational functions. Note that in [2, Section 3] we have shown stability for a numerical version of Blahut's error-correcting polynomial interpolation algorithm for $E = 1$.
2. an exact interpolation algorithm for sparse multivariate polynomials and rational functions à la Zippel which can correct errors in the evaluations: in Section 2 we give an analysis which allows for evaluations at poles of the

*This research was supported in part by the National Science Foundation under Grant CCF-1115772 (Kaltofen).

rational function. Errors in the evaluations may indicate false poles, and evaluations may falsely produce a value at a pole.

3. a numerical, noise tolerant interpolation algorithm for sparse multivariate polynomials and rational functions that can remove outlier errors. In least squares fitting of known models (note that our sparse models are computed by our algorithm), outliers can be identified by their leverage scores derived from the pseudo-inverse (projection matrix) of the normal equations. Our approach is entirely different, even for polynomial models. We locate those outliers via numerical error-correcting decoding, using structured linear algebra algorithms. Our computer experiments in Section 4 demonstrate that our approach is very feasible.

Remark 1.1. Reed-Solomon decoding reconstructs the coefficients of a polynomial f by interpolation where some of the evaluations are faulty. In fact, for d coefficients, i.e., $\deg(f) \le d-1$ and k errors, one needs $L = d+2E$ evaluations, where $E \ge k$ bounds the number of faults a-priori. In our algorithms, we will as substeps perform several interpolations, where each is designed to tolerate a given number E of errors. Since we view the acquisition of evaluations as probing a black box for the function f, the error rate of the black box can be related to E: Suppose the black box for any $L \ge L_{\min}$ evaluations produces faulty values for no more than $k \le L/q$ inputs, where $q > 2$. Here $1/q$ is the error rate, and $L_{\min}$ is a minimum on the number of each batch of evaluations: obviously, one cannot suppose that for $L = 1$ evaluation one always gets a correct answer. Then $E = \lfloor d/(q-2) \rfloor$ yields $L/q = (d+2E)/q \le (d+2d/(q-2))/q = d/(q-2)$, so $k \le E$ as required. For our multivariate algorithm, the situation is somewhat different, and the error rate of the black box for f/g cannot be too high: see Remark 2.6. □

Remark 1.2. Since our evaluations are numerically inaccurate, a question arises at what point a noisy value becomes an outlier. Outliers give rise to a common univariate polynomial factor, the error locator polynomial, in the sparse multivariate rational function reconstruction of the model (see the discussion after Assumption 5). If sufficiently large in magnitude, they markedly increase the numerical rank of the corresponding matrix (15). Their locations can be determined by their corresponding black box inputs being roots of the error locator polynomial factor (22), which must be present. See also Remark 3.1 below. □

As in [17], our multivariate algorithm takes advantage of multivariate sparsity, and a stable univariate algorithm for dense fractions, now with error correction. Cauchy interpolation [14] recovers the reduced fraction $(f/\mathrm{GCD}(f,g))/(g/\mathrm{GCD}(f,g))$. It was first observed in [19] that an unreduced fraction, e.g., $(x^d-y^d)/(x-y)$, can yield a much sparser model for the black box. Such models can be constructed by interpolation, both for exact and for numeric data. In [19] we give an exact univariate algorithm. In Kaltofen's ISSAC 2011 presentation at the FCRC in San Jose the use of [3] on numeric data was introduced. Example 4.1 shows the feasibility of that approach. Note that the number of evaluations in [19] depends logarithmically on the degrees.

2. ERROR-CORRECTING MULTIVARIATE RATIONAL FUNCTION INTERPOLATION

Here we generalize the analysis for exact arithmetic in [17, Section 4.1]. Consider the rational function $f/g \in \mathsf{K}(x_1,\ldots,x_n)$, where the numerator and denominator are represented as

$$f = \sum_{j=1}^{t_f} a_j \vec{x}^{\vec{d}_j},\ g = \sum_{m=1}^{t_g} b_m \vec{x}^{\vec{e}_m},\ a_j, b_m \in \mathsf{K} \setminus \{0\}, \qquad (1)$$

where K is an arbitrary field and the terms are denoted by $\vec{x}^{\vec{d}_j} = x_1^{d_{j,1}} \cdots x_n^{d_{j,n}}$ and $\vec{x}^{\vec{e}_m} = x_1^{e_{m,1}} \cdots x_n^{e_{m,n}}$. We analyze our variant of Zippel's sparse interpolation technique to recover the numerator and denominator. Zippel's technique [12, Section 4] determines the support of $f_i = f(x_1,\ldots,x_i,\alpha_{i+1},\ldots,\alpha_n)$ and $g_i = g(x_1,\ldots,x_i,\alpha_{i+1},\ldots,\alpha_n)$ iteratively from the support of f_{i-1} and g_{i-1}, where $\alpha_2,\ldots,\alpha_n \in \mathsf{K}$ is a random anchor point. We will use Zippel's probabilistic assumption.

Assumption 1. *Each term $x_1^{d_{j,1}} \cdots x_{i-1}^{d_{j,i-1}}$, where $1 \le j \le t_f$, and each term $x_1^{e_{m,1}} \cdots x_{i-1}^{e_{m,i-1}}$, where $1 \le m \le t_g$, has a non-zero coefficient in f_{i-1} and g_{i-1}.*

Note that for different j's and different m's one may have the same term prefix in $i-1$ variables. At this point we do not assume that f and g are relatively prime, but we will introduce relative primeness as Assumption 5 below for decoding; see also Remark 2.7.

We wish to recover f_i and g_i from the sparse supports of f_{i-1} and g_{i-1} and evaluations of $f_i(x_1,\ldots,x_i)/g_i(x_1,\ldots,x_i) = f(x_1,\ldots,x_i,\alpha_{i+1},\ldots,\alpha_n)/g(x_1,\ldots,x_i,\alpha_{i+1},\ldots,\alpha_n)$. We chose $\xi_1,\ldots,\xi_i \in \mathsf{K}$ and evaluate at powers $\xi_1^\ell,\ldots,\xi_i^\ell$, where $\ell = 0, 1, 2, \ldots$ We will obtain

$$\beta_{i,\ell} = \gamma_{i,\ell} + \gamma'_{i,\ell},\ \text{where } \gamma_{i,\ell} = \frac{f_i(\xi_1^\ell,\ldots,\xi_i^\ell)}{g_i(\xi_1^\ell,\ldots,\xi_i^\ell)} \in \mathsf{K} \cup \{\infty\} \qquad (2)$$

and where $\gamma'_{i,\lambda_\kappa} \neq 0$ exactly at the $k \le E$ unknown indices $0 \le \lambda_1 < \lambda_2 < \cdots < \lambda_k$ for ℓ, that is $\gamma'_{i,\ell} = 0$ for all $\ell \notin \{\lambda_1,\ldots,\lambda_k\}$.

Assumption 2. *We assume that we have the upper bound, E, on the number of erroneous evaluations, but not the actual count of errors, k, and not their locations λ_κ.*

If $g_i(\xi_1^\ell,\ldots,\xi_i^\ell) = 0$ we have $\gamma_{i,\ell} = \infty$, but $\beta_{i,\ell}$ can be erroneously $\in \mathsf{K}$. Similarly, if $g_i(\xi_1^\ell,\ldots,\xi_i^\ell) \neq 0$ we may erroneously have $\beta_{i,\ell} = \infty$.

Following the Berlekamp/Welch strategy, we attempt to recover $(f_i\Lambda_i)/(g_i\Lambda_i)$ where

$$\Lambda_i = (x_1 - \xi_1^{\lambda_1}) \cdots (x_1 - \xi_1^{\lambda_k}) \qquad (3)$$

is an error locator polynomial. The set of possible terms in $f_i\Lambda_i$ and $g_i\Lambda_i$ can now be restricted to

$$D_{f,i,E} = \{x_1^{d_{j,1}+\nu} x_2^{d_{j,2}} \cdots x_{i-1}^{d_{j,i-1}} x_i^{\delta_j} \mid 1 \le j \le t_f, 0 \le \nu \le E,$$
$$0 \le \delta_j \le \min(\deg(f) - d_{j,1} - \cdots - d_{j,i-1}, \deg_{x_i}(f))\}, \qquad (4)$$
$$D_{g,i,E} = \{x_1^{e_{m,1}+\nu} x_2^{e_{m,2}} \cdots x_{i-1}^{e_{m,i-1}} x_i^{\eta_m} \mid 1 \le m \le t_g, 0 \le \nu \le E,$$
$$0 \le \eta_m \le \min(\deg(g) - e_{m,1} - \cdots - e_{m,i-1}, \deg_{x_i}(g))\}. \qquad (5)$$

Note again that not all of the terms enumerated in (4) and/or (5) are distinct. See Remark 2.4 below for somewhat smaller candidate term sets. Here we make the assumption that the f_{i-1} and g_{i-1}, as said earlier, contain the full set of possible terms, that with high probability as we will inductively argue.

Assumption 3. *We assume that we know* $\deg(f)$, $\deg_{x_i}(f)$, $\deg(g)$ *and* $\deg_{x_i}(g)$.

Let $\vec{y}$ and $\vec{z}$ be the coefficient vectors of $f_i\Lambda_i$ and $g_i\Lambda_i$ for the (distinct) terms in $D_{f,i,E}$ and $D_{g,i,E}$. For any $\ell = 0, 1, 2, \ldots$

and any point $\xi_1, \ldots, \xi_i \in \mathsf{K}$ each value $\beta_{i,\ell}$ in (2) constitutes a linear equation for the coefficient vector,

$$\sum_{j,\nu,\delta} y_{j,\nu,\delta}(\xi_1^{d_{j,1}+\nu}\xi_2^{d_{j,2}}\cdots\xi_{i-1}^{d_{j,i-1}}\xi_i^{\delta})^{\ell} = \beta_{i,\ell}\sum_{m,\nu,\eta} z_{m,\nu,\eta}(\xi_1^{e_{m,1}+\nu}\xi_2^{e_{m,2}}\cdots\xi_{i-1}^{e_{m,i-1}}\xi_i^{\eta})^{\ell}. \quad (6)$$

Errors in the β_{i,λ_κ} are tolerated because $f_i\Lambda_i$ and $g_i\Lambda_i$ are both $= 0$ at $x_1 = \xi_1^{\lambda_\kappa}$. Note again that coefficients/indeterminates can be the same, $y_{j_1,\nu_1,\delta} = y_{j_2,\nu_2,\delta}$, for instance, if corresponding terms are the same.

With $\ell = 0, \ldots, L-1$, where L is yet to be determined, the equations (6) form a homogeneous linear system in the unknowns $y_{j,\nu,\delta}$ and $z_{m,\nu,\eta}$,

$$V_{i,L}(\xi_1, \ldots, \xi_i)\vec{\boldsymbol{y}}^T = \Gamma_{i,L}W_{i,L}(\xi_1, \ldots, \xi_i)\vec{\boldsymbol{z}}^T, \quad (7)$$

where $\Gamma_{i,L}$ is a diagonal matrix of rational function values $\beta_{i,\ell}$ and $V_{i,L}$ and $W_{i,L}$ are (transposed) Vandermonde matrices, with possible zero rows. If $\beta_{i,\ell} = \infty$ then the ℓ-th row in $V_{i,L}$ is set to a zero row, and $(\Gamma_{i,L})_{\ell,\ell}$ is set to 1.

Provided the term supports of f_{i-1} and g_{i-1} were correctly computed in the previous iterations, the coefficient vector $[\vec{\boldsymbol{y}}, \vec{\boldsymbol{z}}]$ of $f_i\Lambda_i$ and $g_i\Lambda_i$ solves (7). For the term sets in (4) and (5) let

$$D_{f,i,E}D_{g,i,E} = \{\tau_f \cdot \tau_g \mid \tau_f \in D_{f,i,E}, \tau_g \in D_{g,i,E}\}, \quad (8)$$

with $|D_{f,i,E}D_{g,i,E}| \le |D_{f,i,E}|\cdot|D_{g,i,E}|$. Now set $L \ge |D_{f,i,E} \times D_{g,i,E}|$ in (7). We argue that for random $\xi_1, \ldots, \xi_i$, the polynomials $\bar{f}$ and $\bar{g}$, which correspond to any non-zero solution vector $[\vec{\boldsymbol{y}}, \vec{\boldsymbol{z}}]$, respectively, of the linear system (7), with high probability satisfy $\bar{f}g_i\Lambda_i = \bar{g}f_i\Lambda_i$.

We shall first assume that the random choices for $\xi_1, \ldots, \xi_i \in S \subset \mathsf{K}$ are such that no two distinct terms in $D_{f,i,E}$ and no two distinct terms in $D_{g,i,E}$ evaluate at $x_\mu \leftarrow \xi_\mu$, $1 \le \mu \le i$, to the same element in K. Because

$$\forall \ell, 0 \le \ell \le L-1\colon \begin{cases} (f_i\Lambda_i)(\xi_1^\ell, \ldots, \xi_i^\ell) = \beta_{i,\ell}\,(g_i\Lambda_i)(\xi_1^\ell, \ldots, \xi_i^\ell), \\ \bar{f}(\xi_1^\ell, \ldots, \xi_i^\ell) = \beta_{i,\ell}\quad \bar{g}(\xi_1^\ell, \ldots, \xi_i^\ell), \end{cases}$$

we have

$$\forall \ell, 0 \le \ell \le L-1\colon (f_i\Lambda_i\bar{g})(\xi_1^\ell, \ldots, \xi_i^\ell) = (\bar{f}g_i\Lambda_i)(\xi_1^\ell, \ldots, \xi_i^\ell). \quad (9)$$

Note that if $\beta_{i,\ell} = 0$ then $(f_i\Lambda_i)(\xi_1^\ell, \ldots, \xi_i^\ell) = \bar{f}(\xi_1^\ell, \ldots, \xi_i^\ell) = 0$ and if $\beta_{i,\ell} = \infty$ then $(g_i\Lambda_i)(\xi_1^\ell, \ldots, \xi_i^\ell) = \bar{g}(\xi_1^\ell, \ldots, \xi_i^\ell) = 0$. The possibly occurring terms of the polynomial $f_i\Lambda_i\ \bar{g} - \bar{f}\ g_i\Lambda_i$ are contained in $D_{f,i,E}D_{g,i,E}$ of (8). Note that for $i = 1$ we have $L = |D_{f,1,E}D_{g,1,E}| = \deg_{x_1}(f) + \deg_{x_1}(g) + 1 + 2E$; see Remark 2.5 below.

Assumption 4. *Finally, we assume that the random choices for $\xi_1, \ldots, \xi_i \in S$ are such that no two terms in $D_{f,i,E}D_{g,i,E}$ evaluate to the same value, which subsumes our earlier assumption on the distinctness of term evaluations.*

For $L \ge |D_{f,i,E}E_{g,i,E}|$ we then must have

$$f_i\Lambda_i\ \bar{g} - \bar{f}\ g_i\Lambda_i = 0, \quad \bar{f} \ne 0, \bar{g} \ne 0, \quad (10)$$

because the coefficient vector of $f_i\Lambda_i\ \bar{g} - \bar{f}\ g_i\Lambda_i$ is by (9) a kernel vector in a square non-singular (transposed) Vandermonde matrix and therefore must be zero. Since $f_i \ne 0$, $g_i \ne 0$ and $\Lambda_i \ne 0$ and $[\vec{\boldsymbol{y}}, \vec{\boldsymbol{z}}] \ne 0$ both $\bar{f}$ and $\bar{g}$ are non-zero. We have concluded the analysis of the error correction property of our linear system. Next, we discuss the recovery of a sparse rational interpolant for f_i/g_i.

In [17] we have excluded 0 and ∞ from the evaluations $\gamma_{i,\ell}$ in (2), but here we show that those values are perfectly allowable. Our (generalized) Berlekamp/Welch decoding algorithm concludes as follows, at least for reduced fractions (1); see also Remark 2.7.

Assumption 5. $\mathrm{GCD}(f, g) = 1$ *in* $\mathsf{K}[x_1, \ldots, x_n]$ *in* (1).

Suppose now the f_i and g_i are relatively prime in $\mathsf{K}[x_1, \ldots, x_i]$. For random anchor points $\alpha_2, \ldots, \alpha_n$, this will be true with high probability. In fact, f_i and g_i are then random projections of the primitive parts of f and g after removing their contents in $\mathsf{K}[x_{i+1}, \ldots, x_n]$; see Remark 2.4. So we obtain from (10)

$$f_i/g_i = \bar{f}/\bar{g}, \quad (11)$$

by removing a common factor $h \in \mathsf{K}[x_1]$ of $\bar{f}$ and $\bar{g}$. Because of the degree constraints in the term supports in (4) and (5), no additional polynomial factors in the variables $x_2, \ldots, x_i$ are possible. We thus have $hf_i = \bar{f}$ and $hg_i = \bar{g}$. All $x_1 - \xi_1^{\lambda_\kappa}$ for all $\kappa = 1, \ldots, k$ divide $h(x_1)$, because by (11) $\bar{f}(\xi_1^{\lambda_\kappa}, \ldots, \xi_i^{\lambda_\kappa}) = \beta_{i,\lambda_\kappa}\bar{g}(\xi_1^{\lambda_\kappa}, \ldots, \xi_i^{\lambda_\kappa})$, but $f_i(\xi_1^{\lambda_\kappa}, \ldots, \xi_i^{\lambda_\kappa}) \ne \beta_{i,\lambda_\kappa}g_i(\xi_1^{\lambda_\kappa}, \ldots, \xi_i^{\lambda_\kappa})$. Because we estimate the number of errors by E in (4) and (5), $\bar{f}$ and $\bar{g}$ can have common factors in $\mathsf{K}[x_1]$ in addition to $\Lambda_i(x_1)$. Those common factors give the nullspace of (9) a dimension $1 + E - k$, and the kernel vectors corresponding to the lowest degree polynomials $\bar{f}$ and $\bar{g}$ in x_1 are the coefficient vectors of $\bar{f} = cf_i\Lambda_i$ and $\bar{g} = cg_i\Lambda_i$ for a $c \in \mathsf{K}$, $c \ne 0$.

Remark 2.1. Note that for $g = 1$ we obtain a sparse multivariate polynomial interpolation algorithm with error-correction, and Assumption 5 is satisfied. For the exact problem for multivariate polynomials, say with K a finite field, we also mention [22], where the minimum number of points is studied for unique recovery. There encoding/decoding is performed by combinatorial search for the erroneous evaluations. □

Remark 2.2. The linear system (7) of linear constraints (6) may yield f_i and g_i for smaller L. In [17] we have suggested to increment L until a 1-dimensional kernel is achieved. If $k = E$ such a strategy would work here. The system (7) has $M = |D_{f,i,E}| + |D_{g,i,E}|$ variables, so at least $L \ge M - 1$ equations are needed. In fact, from earlier iterations one has additional linear constraints for $\mu = 1, 2, \ldots, i-1$:

$$\sum_{j,\nu,\delta} y_{j,\nu,\delta}(\xi_1^{d_{j,1}+\nu}\xi_2^{d_{j,2}}\cdots\xi_\mu^{d_{j,\mu}})^{\ell}\alpha_{\mu+1}^{d_{j,\mu+1}}\cdots\alpha_i^{\delta} = \gamma_{\mu,\ell} \times \sum_{m,\nu,\eta} z_{m,\nu,\eta}(\xi_1^{e_{m,1}+\nu}\xi_2^{e_{m,2}}\cdots\xi_\mu^{e_{m,\mu}})^{\ell}\alpha_{\mu+1}^{e_{m,\mu+1}}\cdots\alpha_i^{\eta}. \quad (12)$$

Note that after f_μ and g_μ have been computed, the errors in the $\beta_{\mu,\lambda_\kappa}$ can be corrected.

If $k < E$, one may simultaneously grow $E = k = 0, 1, \ldots$ in $D_{f,i,E}$ and $D_{g,i,E}$ (see (4) and (5)), that is, add new columns to (7). The objective is to produce a single non-zero solution $\bar{f}$ and $\bar{g}$ with the property that the evaluations have k errors located by $h(x_1) = \Lambda_i(x_1)$. The constraints (12) guarantee that the corresponding $f_i = \bar{f}/\Lambda_i$ and $g_i = \bar{g}/\Lambda_i$ project to f_μ and g_μ for $\mu = 1, 2, \ldots, i-1$.

There is merit in not using some or all constraints (12). First, the system (7) retains its block Vandermonde structure and a fast solver can be deployed [21]. Second, if our assumptions hold, we must have $f_\mu = cf_i(x_1, \ldots, x_\mu, \alpha_{\mu+1}, \ldots, \alpha_i)$ and $g_\mu = cg_i(x_1, \ldots, x_\mu, \alpha_{\mu+1}, \ldots, \alpha_i)$ for some non-zero scalar $c \in \mathsf{K}$. Thus, unlucky anchor points $(\alpha_2, \ldots, \alpha_n)$ may be diagnosed via testing the projections.

In general, there is a trade-off of optimizing the number of evaluations against the arising cost of solving the linear systems. Note that instead of powers $(\xi_1^\ell, \ldots, \xi_i^\ell)$ in (6) one also could use fresh values $(\xi_{1,\ell},\dots,\xi_{i,\ell})$. The proof of property (10) then uses the idea that for symbolic values $\xi_\mu = v_\mu$ the property is true over the function field $\mathsf{K}(v_1,\dots,v_i)$. $\square$

Remark 2.3. At iteration i we have arbitrarily chosen the variable x_1 for our error-locator polynomial Λ_i. We could also have chosen x_2, or $x_3,\dots$, or x_i. Again, one may select that variable x_μ for which the sets $D_{f,i,E}$ and $D_{g,i,E}$ have the fewest elements. Clearly, if x_1 occurs sparsely and x_2 densely, x_2 is likely a better choice. Note that if x_i is chosen, one gets no overlap in the terms in $D_{f,i,E}$ or $D_{g,i,E}$.

The variable-by-variable interpolation depends on a variable order. Different orders may lead to a different number of evaluations. For numerical reasons, one should process the variables with smaller degrees first; see Remark 2.7.

For the record, we give an explicit worst case estimate for the exact algorithm. If we denote by $t_{f,i} = |D_{f,i,0}|$ and $t_{g,i} = |D_{g,i,0}|$, the number of terms in f_i and g_i respectively, with $t_{f,0} = t_{g,0} = 1$, and if we chose the variable x_i for Λ_i, one provably needs at most $\sum_{i=1}^{n} t_{f,i-1}t_{g,i-1}(\deg_{x_i}(f) + \deg_{x_i}(g) + 2E + 1) \le ((n-1)t_f t_g + 1)(\max_i\{\deg_{x_i}(f) + \deg_{x_i}(g)\} + 2E + 1)$ values $\beta_{i,\ell}$ (with high probability). $\square$

Remark 2.4. The term sets $D_{f,i,E}$ in (4) for $f_i\Lambda_i$ and $D_{g,i,E}$ in (5) for $g_i\Lambda_i$ should be as small as possible. One may restrict δ_j in (4) and η_m in (5) by $\delta_j \le \deg(f_i) - d_{j,1} - \cdots - d_{j,i-1}$, $\eta_m \le \deg(g_i) - e_{m,1} - \cdots - e_{m,i-1}$. This adds additional pairs of degrees $(\deg(f_i), \deg(g_i))$ to Assumption 3. Under Assumption 5, all degrees can be estimated by univariate rational recovery.

First, we show that f_i and g_i are relatively prime with high probability. We consider the substitutions $f_{i,u,x} = f(x_1, x_2 + u_2x_1, \dots, x_i + u_ix_1, x_{i+1}, \dots, x_n)$ and $g_{i,u,x} = g(x_1, x_2+u_2x_1, \dots, x_i+u_ix_1, x_{i+1}, \dots, x_n)$ over the function field $\mathsf{K}_u = \mathsf{K}(u_2,\dots,u_i)$. The map $x_\mu \mapsto x_\mu + u_\mu x_1$, were $2 \le \mu \le i$, constitutes a ring isomorphism on $\mathsf{K}_u[x_1,\dots,x_n]$. Therefore the polynomials $f_{i,u,x}$ and $g_{i,u,x}$ are relatively prime, because their pre-images f and g are relatively prime (extending to K_u cannot change this). Let $\rho_i \in \mathsf{K}_u[x_2,\dots,x_n]$ be the Sylvester resultant of $f_{i,u,x}$ and $g_{i,u,x}$ with respect to the variable x_1. We have $\rho_i \ne 0$ (relative primeness in $\mathsf{K}_u(x_2,\dots,x_n)[x_1]$), and if $\rho(x_2,\dots,x_i,\alpha_{i+1},\dots,\alpha_n) \ne 0$, the pair $f_{i,u} = f(x_1, x_2 + u_2x_1, \dots, x_i + u_ix_1, \alpha_{i+1}, \dots, \alpha_n)$ and $g_{i,u} = g(x_1, x_2+u_2x_1, \dots, x_i+u_ix_1, \alpha_{i+1}, \dots, \alpha_n)$ is relatively prime in $\mathsf{K}_u(x_2,\dots,x_i)[x_1]$. But the leading coefficients of $f_{i,u}$ and $g_{i,u}$ with respect to x_1 are in K_u, so relative primeness persists in $\mathsf{K}_u[x_1,\dots,x_i]$, and their pre-images f_i and g_i under the inverse isomorphism $x_\mu \mapsto x_\mu - u_\mu x_1$ are relatively prime over K_u, and also over the field K, which contains all coefficients of f_i and g_i.

We finally show how to compute $\deg(f_i)$ and $\deg(g_i)$ by randomization. Cauchy interpolation recovers $f_{i,u}$ and $g_{i,u}$ over $\mathsf{K}_u(x_2,\dots,x_i)$, and also with high probability the images under evaluation $x_\mu \leftarrow \phi_\mu$, $u_\mu \leftarrow \theta_\mu$ for random elements $\phi_\mu, \theta_\mu \in S \subseteq \mathsf{K}$ $(2 \le \mu \le i)$. The evaluations $\phi_2, \dots, \theta_i$ must preserve a non-zero leading subresultant coefficient.

We compute (with high probability) $\deg_{x_1}(f)$ as $\deg(f_1)$ and $\deg_{x_1}(g)$ as $\deg(g_1)$. Similarly, we compute (with high probability) $\deg_{x_i}(f)$ and $\deg_{x_i}(g)$ by rational function recovery (with outlier errors) of $f(\alpha_1,\dots,\alpha_{i-1},x_i,\alpha_{i+1},\dots,\alpha_n)/g(\alpha_1,\dots,\alpha_{i-1},x_i,\alpha_{i+1},\dots,\alpha_n)$, where $2 \le i \le n$. As a side consequence, we recover (with high probability) the term exponents of x_i in f and g in (1), namely $D_f^{[i]} = \{x_i^{d_{j,i}} \mid 1 \le j \le t_f\}$ and $D_g^{[i]} = \{x_i^{e_{m,i}} \mid 1 \le m \le t_g\}$. Thus, we can further restrict δ_j in (4) and η_m in (5) by $\delta_j \le \deg(f_i) - d_{j,1} - \cdots - d_{j,i-1}, x_i^{\delta_j} \in D_f^{[i]}$, $\eta_m \le \deg(g_i) - e_{m,1} - \cdots - e_{m,i-1}, x_i^{\eta_m} \in D_g^{[i]}$. $\square$

Remark 2.5. Our initialization for $i = 1$ uses for f_1 all terms x_1^δ, where $\delta = 0, 1, \dots, \deg_{x_1}(f)$, and for g_1 all terms x_1^η, where $\eta = 0, 1, \dots, \deg_{x_1}(g)$. By our definitions (4), (5) and (8) we evaluate the fraction at ξ_1^ℓ for $\ell = 0, 1, \dots, L-1$ with $L = \deg_{x_1}(f) + \deg_{x_1}(g) + 1 + 2E$. We suppose that $\xi^{\ell_1} \ne \xi^{\ell_2}$ in the range for ℓ. Our algorithm in the initialization phase essentially implements Berlekamp/Welch decoding at Blahut points for rational functions, and in the above we have proved that the errors are removed, that without appealing to the Euclidean algorithm. The linear system approach for Berlekamp/Welch decoding is also introduced in [21]. $\square$

Remark 2.6. As in Remark 1.1 for univariate interpolation, we can determine E from the error rate $1/q$ of the black box for f/g. For $L(E) \ge |D_{f,i,E}D_{g,i,E}|$ or, in practice, $L(E) \ge |D_{f,i,E}| + |D_{g,i,E}| + L_0$, which we use in Sections 3 and 4 with $L_0 = 10$, we must attain $k \le L(E)/q \le E$. Note that the latter may for $i \ge 2$ not have a solution for E if the rate $1/q$ is not sufficiently small. We have $|D_{f,i,E}| \le (E+1)|D_{f,i,0}|$ and $|D_{g,i,E}| \le (E+1)|D_{g,i,0}|$ in (4) and (5), so in practice in the worst case $L(E) \le (E+1)|D_{f,i,0}| + (E+1)|D_{g,i,0}| + L_0 \le qE \Longrightarrow q > |D_{f,i,0}| + |D_{g,i,0}|$. $\square$

Remark 2.7. The first algorithm for recovering a sparse rational function without Assumption 5 is described in [19]. As an example, the unreduced fraction $(x^d - y^d)/(x-y)$ is much sparser than the reduced polynomial. In fact, in [19] univariate fractions are recovered as sparse fractions, not using dense Cauchy interpolation; the number of evaluation points in the algorithms is proportional to $\log(\deg(f))$. Here we have followed the idea of lifting an unreduced fraction by delaying Assumption 5 until after establishing the key Berlekamp/Welch property (10), which is $f_i/g_i = \bar{f}/\bar{g}$. If f_i/g_i is unreduced, the error corrected $\bar{f}$ and $\bar{g}$ may not be equal to the sparse projections f_i and g_i. As we have supposed in [19], the sparsest possibly unreduced fraction f/g of lowest degree can be unique, hence liftable via $\bar{f}$ and $\bar{g}$. The initial sparse f_1 and g_1 can be also obtained by computing a sparse polynomial multiple [10]. Numerically, it may also be possible to recover a sparse unreduced fraction for $i = 1$ by optimizing the 1-norm of the solution vector via linear programming [3]. Example 4.1 below demonstrates such a recovery. Such sparse unreduced recovery is also useful when the evaluations at α_μ (see Remark 2.4) do not yield a numerically relatively prime univariate pair f_1, g_1.

In the exact case, there are other ways of determining the coefficients in $\mathsf{K}[x_i]$ of f_i and g_i, for example by interpolating or sparsely interpolating x_i, which yields a smaller linear system and possibly fewer evaluations (cf. [24]). One may also reconstruct the fraction using Strassen's removal of divisions approach: see [5] (cf. [11, end of Section 7] and [13, Section 4]). [5] recovers the sparse homogeneous parts from highest to lowest degree. Since their algorithm and Algorithm *Black Box Numerator and Denominator* in [15] are based on univariate Cauchy interpolation, any black box error rate $1/q < 1/2$ can be handled by those methods.

[5] does not address the problem of projections leading to a reducible univariate fraction. Especially in the numeric setting, approximate relative primeness of the projections is difficult to maintain throughout each univariate Cauchy

recovery (see [16, End of Section 6]). Our sparse system (7) is set up to avoid the reducedness requirement all together. The sparsity constraints numerically stabilize the algorithm, provided one starts with a correct term support for f_1 and g_1. By using more than one random anchor α_μ, where $2 \le \mu \le n$, one can improve the probability that no occurring term is falsely dropped from the term sets for f_i and g_i. □

Remark 2.8. After recovering the $\mathsf{K}[x_i]$ coefficients of f_i and g_i, one may sparsify those coefficients by shifting $x_i = x_i + \sigma_i$, where σ_i is either in K or algebraic over K. See [7] for computing such a sparsifying shift exactly, and [2] for an algorithm that tolerates numerical noise (and outlier). □

Remark 2.9. One may interpolate several sparse rational functions with a known common denominator (or numerator) simultaneously with fewer evaluations by the above method. An algorithm for the exact univariate dense recovery problem (without erroneous values) is in [20]. □

3. NUMERICAL INTERPOLATION WITH OUTLIER ERRORS

Based on the discussion in Section 2, we present a modified Zippel's sparse interpolation approach to recover sparse rational function from values with noise and outlier errors. In the approximate case, Θ is introduced to measure whether the evaluation is an outlier error, that is, we say the evaluation β at the point $(\zeta_1, \ldots, \zeta_n) \in \mathbb{C}^n$ is an outlier error, if $\beta = \gamma + \gamma'$, where $\gamma = f_i(\zeta_1, \ldots, \zeta_n)/g_i(\zeta_1, \ldots, \zeta_n) \in \mathbb{C}\cup\{\infty\}$, and $|\beta/\gamma| \ge \Theta$. Again false poles and non-poles are allowed; see explanation immediately after Assumption 2. Consider the rational function $f/g \in \mathbb{C}(x_1, \ldots, x_n)$, where f, g are represented as (1). Suppose a black box for f/g with noise and outlier errors at a known error rate is given. The upper bound on the number of erroneous evaluations E can be determined from the error rate; see Remark 1.1. In this Section, we at first present a method to interpolate a univariate rational function, and then discuss how to recover f_i and g_i when f_{i-1} and g_{i-1} are already computed.

Let $f^{[i]} = f(\alpha_1, \ldots, \alpha_{i-1}, x_i, \alpha_{i+1}, \ldots, \alpha_n) = \sum_{j=1}^{d_f} \psi_j^{[i]} x_i^j$, $g^{[i]} = g(\alpha_1, \ldots, \alpha_{i-1}, x_i, \alpha_{i+1}, \ldots, \alpha_n) = \sum_{m=1}^{d_g} \chi_m^{[i]} x_i^m$, and assume that (with high probability) the sets $D_f^{[i]}$ and $D_g^{[i]}$ at the end of Remark 2.4 are the corresponding nonzero terms of $f^{[i]}$ and $g^{[i]}$. Here we have a-priori total degree bounds $\bar{d}_f \ge \deg(f)$ and $\bar{d}_g \ge \deg(g)$. Now let us show how to compute those term supports $D_f^{[i]}$ and $D_g^{[i]}$ of the univariate polynomials $f^{[i]}$ and $g^{[i]}$ with respect to the variable x_i. Our discussion is for $i = 1$. Given a random root of unity $\zeta \in \mathbb{C}$, we compute the evaluations with outlier errors, that is, for $\ell = 0, 1, \ldots, \bar{d}_f + \bar{d}_g + 2E + 1$ we compute

$$\beta_\ell = \gamma_\ell + \gamma'_\ell, \text{ where } \gamma_\ell = \frac{f(\zeta^\ell, \alpha_2, \ldots, \alpha_n)}{g(\zeta^\ell, \alpha_2, \ldots, \alpha_n)} \in \mathbb{C}\cup\{\infty\}, \quad (13)$$

where γ'_ℓ denotes noise or possibly an outlier error. We have the upper bound of the number of erroneous evaluations E, which means that the number of ℓ, such that $|\beta_\ell/\gamma_\ell| \ge \Theta$, is $\le E$. Having (13), we construct the following linear equations for $\ell = 0, 1, \ldots, \bar{d}_f + \bar{d}_g + 2E$,

$$\sum_{j=0}^{\bar{d}_f+E} y_j \zeta^{\ell j} - \beta_\ell \sum_{m=0}^{\bar{d}_g+E} z_m \zeta^{\ell m} = 0, \quad (14)$$

The above equations form a linear system

$$G \begin{bmatrix} \vec{\boldsymbol{y}} & \vec{\boldsymbol{z}} \end{bmatrix}^T = [V_1, -\Gamma_1 W_1] \begin{bmatrix} \vec{\boldsymbol{y}} & \vec{\boldsymbol{z}} \end{bmatrix}^T = \mathbf{0}, \quad (15)$$

where $\Gamma_1 = \operatorname{diag}(\beta_0, \beta_1, \ldots, \beta_{\bar{d}_f+\bar{d}_g+2E})$, and where V_1, W_1 are Vandermonde matrices generated by the vectors $[1, \zeta, \ldots, \zeta^{\bar{d}_f+E}]^T$ and $[1, \zeta, \ldots, \zeta^{\bar{d}_g+E}]^T$. The numerical rank deficiency of G, denoted by ρ, can be computed by checking the number of small singular values of G or finding the largest gap among the singular values. Suppose

$$s = \min(\bar{d}_f - \deg_{x_1}(f), \bar{d}_g - \deg_{x_1}(g)).$$

According to the discussion following Assumption 5 in Section 2, we know that $\rho = 1 + E - k + s$. Having ρ, the linear equations (14) are transformed into the following reduced linear equations by removing some unknown coefficients of higher degree in (14), namely, for $\ell = 0, 1, \ldots, \bar{d}_f + \bar{d}_g + 2E$

$$\sum_{j=0}^{\bar{d}_f+E-\rho+1} y_j \zeta^{\ell j} - \beta_\ell \sum_{m=0}^{\bar{d}_g+E-\rho+1} z_m \zeta^{\ell m} = 0, \quad (16)$$

whose matrix form is

$$\widetilde{G} \begin{bmatrix} \vec{\boldsymbol{y}} & \vec{\boldsymbol{z}} \end{bmatrix}^T = [\widetilde{V}_1, -\Gamma_1 \widetilde{W}_1] \begin{bmatrix} \vec{\boldsymbol{y}} & \vec{\boldsymbol{z}} \end{bmatrix}^T = \mathbf{0}. \quad (17)$$

Note that the numerical rank deficiency of $\widetilde{G}$ is 1, since $\bar{d} + E + 1 - \rho = \bar{d} - s + k$. The coefficient vector $\vec{\boldsymbol{y}}^T$ of $f^{[1]}\Lambda_1$ and the coefficient vector of $g^{[1]}\Lambda_1$ are achieved from the last singular vector of $\widetilde{G}$. Note that Λ_1 should have the form $\Lambda_1 = (x_1 - \zeta_1^{\lambda_1}) \cdots (x_1 - \zeta_1^{\lambda_k})$. In that case, every root ζ^{λ_κ}, $1 \le \kappa \le k$, of Λ_1 can be detected by checking for $\ell = 0, 1, \ldots, \bar{d}_f + \bar{d}_g + 2E$ with a preset tolerance ϵ_{root}:

$$\ell \in \{\lambda_1, \ldots, \lambda_k\} \iff |(f^{[1]}\Lambda_1)(\zeta^\ell)| + |(g^{[1]}\Lambda_1)(\zeta^\ell)| \le \epsilon_{\text{root}}.$$

Having Λ_1, we obtain $f^{[1]}$ by applying the approximate univariate polynomial division technique between $f^{[1]}\Lambda_1$ with Λ_1. Similarly, $g^{[1]}$ can be obtained by approximate polynomial division. In the end, the actual supports $D_f^{[1]}$ and $D_g^{[1]}$ corresponding to $f^{[1]}$ and $g^{[1]}$ can be obtained by removing the terms whose coefficients are in absolute value $\le \epsilon_{\text{coeff}}$. Performing the above technique for each variable $x_i, 2 \le i \le n$, one may obtain all the nonzero terms $D_f^{[i]}$ and $D_g^{[i]}$ of $f^{[i]}$ and $g^{[i]}$.

Remark 3.1. The preset tolerance measures ϵ_{root} and ϵ_{coeff} require that the singular solution vector $[\vec{\boldsymbol{y}}, \vec{\boldsymbol{z}}]^T$ is normalized. We normalize the Euclidean 2-norm to 1. Because we oversample by $\bar{d}_f - \deg_{x_i}(f) + \bar{d}_g - \deg_{x_i}(g)$ evaluations in (16), noisy evaluations can be taken as extra outliers. The justification that $f^{[1]}(\zeta^\ell)$ and/or $g^{[1]}(\zeta^\ell)$ is separated from 0 for (almost) all $\ell \ne \lambda_\kappa$ is from [17, Section 3, Lemma 3.1]. As in [17], we use the same justification for correctly identifying non-zero terms via ϵ_{coeff}, but here an incorrectly dropped term cannot be reintroduced later. Therefore ϵ_{coeff} should be tight, and falsely kept terms will be removed later. □

Remark 3.2. The arising linear systems can be solved by structured linear solvers: e.g., the coefficient matrix in (17) is that in [21, Equ. (10)], provided $\beta_\ell \notin \{0, \infty\}$ for all ℓ. However, the values in Γ_1 are deformed by noise. In [17] we have used a structured total least norm (STLN) iteration to compute the optimal deformation of the diagonal of Γ_1 to achieve a rank deficiency of 1. The arising linear systems in the STLN iterations again have structure and are amenable to a displacement rank approach. How to deal with zeros and poles and the STLN iterations using structured solvers has yet to be worked out. □

We now turn to the main task, namely to interpolate f_i and g_i when f_{i-1} and g_{i-1} are computed. Suppose the actual supports of f_{i-1} and g_{i-1} are $D_{f,i-1}$ and $D_{g,i-1}$ (note Assumption 1). In this case, the possible terms in f_i, g_i are

$$\bar{D}_{f,i} = \{x_1^{d_{j,1}} \cdots x_{i-1}^{d_{j,i-1}} x_i^{\delta_j} \mid x_1^{d_{j,1}} \cdots x_{i-1}^{d_{j,i-1}} \in D_{f,i-1}, \quad x_i^{\delta_j} \in D_f^{[i]}, 0 \le \delta_j \le \bar{d}_f - d_{j,1} - \cdots - d_{j,i-1}\}, \tag{18}$$

$$\bar{D}_{g,i} = \{x_1^{e_{m,1}} \cdots x_{i-1}^{e_{m,i-1}} x_i^{\eta_m} \mid x_1^{e_{m,1}} \cdots x_{i-1}^{e_{m,i-1}} \in D_{g,i-1}, \quad x_i^{\eta_m} \in D_g^{[i]}, 0 \le \eta_m \le \bar{d}_g - e_{m,1} - \cdots - e_{m,i-1}\}. \tag{19}$$

Described in Remark 2.4, the new variable x_i is chosen among $x_i, \ldots, x_n$ such that the terms sets $D_{f,i,E}$ in (4) for $f_i\Lambda_i$ and $D_{g,i,E}$ in (5) for $g_i\Lambda_i$ are as small as possible. We designate the possible terms in $f_i\Lambda_i$ and $g_i\Lambda_i$, represented as (4) and (5), as $D_{f,i,E} = \{x_1^{\bar{d}_{j,1}} \cdots x_i^{\bar{d}_{j,i}} \mid j = 1, 2, \ldots, \bar{t}_{f,E}\}$ and $D_{g,i,E} = \{x_1^{\bar{e}_{m,1}} \cdots x_i^{\bar{e}_{m,i}} \mid m = 1, 2, \ldots, \bar{t}_{g,E}\}$. The unknown polynomials $f_i\Lambda_i$ and $g_i\Lambda_i$ are represented as $f_i\Lambda_i = \sum_{j=1}^{\bar{t}_{f,E}} y_j x_1^{\bar{d}_{j,1}} \cdots x_i^{\bar{d}_{j,i}}$, $g_i\Lambda_i = \sum_{m=1}^{\bar{t}_{g,E}} z_m x_1^{\bar{e}_{m,1}} \cdots x_i^{\bar{e}_{m,i}}$, where y_j and z_k are indeterminates.

Let $b_1, \ldots, b_i \in \mathbb{Z}_{>0}$ be sufficient large distinct prime numbers and s_j be random integers with $1 \le s_j < b_j$. We choose $\zeta_j = \exp(2\pi \boldsymbol{i}/b_j)^{s_j} \in \mathbb{C}$ for $1 \le j \le i$ (cf. [9]). In the exact case, discussed in Section 2 above, we know that the dimension of the nullspace of (7) is $1 + E - k$ for $L \ge \bar{t}_{f,E}\bar{t}_{g,E}$ evaluations. In fact, $\bar{t}_{f,E}\bar{t}_{g,E}$ is an upper bound which guarantees that the dimension of the nullspace of (7) is $1+E-k$. For the random examples shown in Table 1 and Table 2, our algorithm only needs $L = \bar{t}_{f,E} + \bar{t}_{g,E} + 10$ probes to obtain $f_i\Lambda_i$ and $g_i\Lambda_i$. In the noisy case, we start from the approximate evaluations for $\ell = 0, \ldots, L-1$,

$$\beta_{i,\ell}=\gamma_{i,\ell}+\gamma'_{i,\ell}, \text{ with } \gamma_{i,\ell}=f_i(\zeta_1^\ell,\ldots,\zeta_i^\ell)/g_i(\zeta_1^\ell,\ldots,\zeta_i^\ell), \tag{20}$$

where $\gamma'_{i,\ell}$ is noise or an outlier error. With y_j and z_m unknown, (20) yield the following linear system:

$$G\begin{bmatrix}\vec{\boldsymbol{y}}^T\\ \vec{\boldsymbol{z}}^T\end{bmatrix}=[V_{i,L}(\zeta_1,\ldots,\zeta_i), -\Gamma_{i,L}W_{i,L}(\zeta_1,\ldots,\zeta_i)]\begin{bmatrix}\vec{\boldsymbol{y}}^T\\ \vec{\boldsymbol{z}}^T\end{bmatrix}=\mathbf{0} \tag{21}$$

(cf. (7)), where $L = \bar{t}_{f,E} + \bar{t}_{g,E} + L_O$ with $L_O \ge 1$ constant, $V_{i,L}, W_{i,L}$ are Vandermonde matrices, and $\Gamma_{i,L} = \operatorname{diag}(\beta_{i,0}, \ldots, \beta_{i,L-1})$. One may estimate the numerical rank deficiency of G, denoted by ρ, by computing its SVD. In consequence, the actual count of errors $k = 1 + E - \rho$ is obtained.

Now let us show how to compute the coefficients of $f_i\Lambda_i$ and $g_i\Lambda_i$. Having the actual count of errors k, the possible terms in $f_i\Lambda_i$ and the possible terms in $g_i\Lambda_i$ are represented precisely by $D_{f,i,k}$ and $D_{g,i,k}$ instead of $D_{f,i,E}, D_{g,i,E}$. Furthermore, the numerical rank deficiency of $\widetilde{G}$, produced by (20) with fewer terms, is 1. In the sequel, the coefficient vector of $f_i\Lambda_i$ and $g_i\Lambda_i$ is achieved from the last singular vector of $\widetilde{G}$. Because all roots of Λ_i have the form of $\zeta_1^{\lambda_\kappa}$, $1 \le \kappa \le k$, similarly to the univariate case all roots can be identified by checking the evaluations $\ell \in \{\lambda_1, \ldots, \lambda_k\}$

$$\iff |(f_i\Lambda_i)(\zeta_1^\ell, \ldots, \zeta_i^\ell)| + |(g_i\Lambda_i)(\zeta_1^\ell, \ldots, \zeta_i^\ell)| \le \epsilon_{\text{root}}. \tag{22}$$

The remaining task is to compute f_i and g_i from the three polynomials $\Lambda_i, f_i\Lambda_i, g_i\Lambda_i$, which constitutes an approximate polynomial division problem. Since $\Lambda_i(x_1)$ is a univariate polynomial, Λ_i is the content of $f_i\Lambda_i$ and $g_i\Lambda_i$ w.r.t. the variables $x_2, \ldots, x_i$, and the corresponding primitive parts are f_i and g_i, respectively (note that f_i and g_i are assumed to be relatively prime—see Remark 2.4). Thus approximate univariate polynomial division can be employed to compute f_i and g_i. We then further get the exact supports of f_i and g_i by removing terms whose coefficients are in absolute value $\le \epsilon_{\text{coeff}}$. Remark 3.1 is relevant again, now with oversampling by $L_O = 10$.

Alternatively, one may compute f_i, g_i from Λ_i via the error locations. In fact, λ_κ is the location of an outlier error if $\zeta_i^{\lambda_\kappa}$ is a root of Λ_i. In this case, all error locations $\lambda_1, \ldots, \ldots \lambda_k$ can be determined by (22). By removing all the evaluations at $\lambda_1, \ldots, \lambda_k$, one gets the approximate evaluations without outlier errors, that is, for $\ell = 0, \ldots, L-1$, $\ell \notin \{\lambda_1, \ldots, \lambda_k\}$,

$$\beta_{i,\ell} \approx f_i(\zeta_1^\ell, \ldots, \zeta_i^\ell)/g_i(\zeta_1^\ell, \ldots, \zeta_i^\ell). \tag{23}$$

In this situation, the remaining problem can be transformed as the problem of interpolating f_i, g_i from their possible terms $\bar{D}_{f,i}$ and $\bar{D}_{g,i}$ and the approximate evaluations (23). One algorithm presented in [17], for interpolating sparse rational functions from noisy values, is applied to obtain f_i and g_i. More details will be found in [17].

Algorithm *Numerical Interpolation of Rational Functions with Outlier Errors*

Input: ▸ $\frac{f(x_1,\ldots,x_n)}{g(x_1,\ldots,x_n)} \in \mathbb{C}(x_1, \ldots, x_n)$ input as a black box with noise and outlier errors, the latter at a given rate (see Remark 1.1).

▸ $(x_1, \ldots, x_n)$: an ordered list of variables in f/g.

▸ $\bar{d}, \bar{e}$: total degree bounds $\bar{d} \ge \deg(f)$ and $\bar{e} \ge \deg(g)$.

▸ $\epsilon_{\text{coeff}} > 0$ (for "forcing underflow" of terms), $\epsilon_{\text{root}} > 0$ (for zero detection), the given tolerance.

Output: $f(x_1, \ldots, x_n)/c$ and $g(x_1, \ldots, x_n)/c$, where $c \in \mathbb{C}$.

1. Initialize the anchor points and the support of f and g: choose $\alpha_1, \alpha_2, \ldots, \alpha_n$ as random roots of unity, let $D_{f,0} = \{1\}$ and $D_{g,0} = \{1\}$.
2. For $i = 1, 2, \ldots, n$ do:
   Interpolate the univariate polynomials $f^{[i]}$ and $g^{[i]}$ and get their supports $D_f^{[i]}$ and $D_g^{[i]}$:

(a) Choose a random root of unity ζ and get the evaluations $\beta_{i,\ell}$ with the noise and the outlier errors as (13).

(b) Construct the matrix G in (15) from $\beta_{i,\ell}$ and ζ. Compute the SVD of G and find its numerical rank deficiency r. A relative tolerance ϵ_{rank} for a jump in the singular values can be provided as an additional input.

(c) Get the matrix $\widetilde{G}$ from the reduced linear system (16) with r, and then obtain $f^{[i]}\Lambda_i$ and $g^{[i]}\Lambda_i$ from the last singular vector of $\widetilde{G}$.

(d) Get the error locator polynomial Λ_i by checking (22).

(e) Obtain $f^{[i]}$ and $g^{[i]}$ by applying univariate polynomial division, and then get the actual support $D_f^{[i]}$ of $f^{[i]}$ by rounding coefficients that are absolutely $\le \epsilon_{\text{coeff}}$ to 0. Similarly, get the actual support $D_g^{[i]}$ of $g^{[i]}$.

3. Let $D_{f,1} = D_f^{[1]}$ and $D_{g,1} = D_g^{[1]}$. For $i = 2, \ldots, n$ do:
   Interpolate the polynomials f_i and g_i as follows:

(a) Choose the variable x_μ from $x_1, \ldots, x_i$ such that $D_{f,i,E}$ and $D_{g,i,E}$ have the fewest elments.

(b) Compute $f_i\Lambda_i$ and $g_i\Lambda_i$:

(b.1) Choose random roots of unity $\zeta_1, \ldots, \zeta_i$. For $\ell = 0, 1, 2, \ldots$, compute the evaluations $\beta_{i,\ell}$ with the noise and the outlier errors as (20).

(b.2) Construct the matrix G in (21) from $\beta_{i,\ell}$ and $D_{f,i,E}, D_{g,i,E}$.

(b.3) Compute the SVD of G and get the actual count of errors k from the numerical rank deficiency of G.

(b.4) Reconstruct the possible terms $D_{f,i,k}, D_{g,i,k}$ in $f_i\Lambda_i$ and $g_i\Lambda_i$.

(b.5) Get the shrunk matrix $\widetilde{G}$ from $\beta_{i,\ell}$ and $D_{f,i,k}, D_{f,i,k}$.

(b.6) Obtain $f_i\Lambda_i$, $g_i\Lambda_i$ from the last singular vector of $\widetilde{G}$.

(c) Obtain f_i and g_i

(c.1) Find the error location $\lambda_1, \ldots, \lambda_k$, for which $\zeta_\mu^{\lambda_\kappa}$ is the root of Λ_i.

(c.2) Get the approximate evaluations without outlier errors by removing ones correspond to λ_κ.

(c.3) Interpolate f_i and g_i by the structured total least norm technique presented in [17], and then get their actual supports $D_{f,i}, D_{g,i}$.

4. With the support of f_n and g_n, interpolate $f(x_1, \ldots, x_n)$ $/c$ and $g(x_1, \ldots, x_n)/c$ again to improve the accuracy of the coefficients:

(a) Construct the linear system from the approximate $\beta_{n,\ell}$ as (23) and the exact terms $D_{f,n}$ and $D_{f,n}$.

(b) Compute the refined solution $\vec{y}$ and $\vec{z}$ by use of STLN method.

(c) Obtain $f(x_1, \ldots, x_n)/c$ and $g(x_1, \ldots, x_n)/c$ from $\vec{y}, \vec{z}$ and $D_{f,n}$, $D_{g,n}$. □

4. EXPERIMENTS

Our algorithm has been implemented in Maple and the performance is reported in the following three tables. All examples in Table 1 and Table 2 are run in Maple 15 under Windows for *Digits*:=15. In Table 1 we exhibit the performance of our algorithm for recovering univariate rational functions from a black box that returns noisy values with outlier errors. For each example, we construct two relatively prime polynomials with random integer coefficients in the range $-5 \le c \le 5$. Here *Random Noise* denotes the range of relative noise randomly added to the black box evaluations of f/g; $\bar{d}_f \ge \deg(f)$ and $\bar{g}_g \ge \deg(g)$ denote the degree bound of the numerator and the denominator, respectively; t_f and t_g denote the number of terms of the numerator and denominator, respectively; $1/q$ denotes the error rate of the outlier error; *Rel. Error* is the relative error, namely $(\|c\tilde{f} - f\|_2^2 + \|c\tilde{g} - g\|_2^2)/(\|f\|_2^2 + \|g\|_2^2)$, where $\tilde{f}/\tilde{g}$ is the fraction computed by our algorithm and c is optimally chosen to minimize the error. For each example, the outlier error is the relative error of the evaluation, which is in the range of $0.01 \times [100, 200]$ or $0.01 \times [200, 300]$. Running times serve to give a rough idea on the efficiency, and are for SONY VAIO laptops with 8GB of memory and 2.67GHz and 2.80GHz Intel i7 processors.

Ex.	*Random Noise*	$\bar{d}_f, \bar{d}_g$	$\deg(f)$, $\deg(g)$	t_f, t_g	$1/q$	E	*Time* *(secs.)*	*Rel.* *Error*
1	$10^{-4} \sim 10^{-2}$	10, 10	3, 3	1, 3	1/3	37	5.9	6.0e–7
2	$10^{-5} \sim 10^{-3}$	6, 6	4, 5	2, 4	1/3	39	4.4	3.1e–6
3	$10^{-6} \sim 10^{-4}$	18, 13	8, 3	4, 3	1/4	26	1.5	2.3e–8
4	$10^{-5} \sim 10^{-3}$	20, 20	10, 10	4, 4	1/3	52	8.4	2.6e–4
5	$10^{-6} \sim 10^{-4}$	18, 30	3, 15	2, 6	1/4	30	9.9	8.8e–8
6	$10^{-6} \sim 10^{-4}$	40, 40	20, 20	5, 5	1/4	48	32	6.2e–9
7	$10^{-6} \sim 10^{-4}$	50, 30	30, 7	6, 3	1/5	34	24	8.6e–6
8	$10^{-7} \sim 10^{-5}$	30, 70	5, 40	4, 7	1/4	61	57	2.7e–9
9	$10^{-7} \sim 10^{-5}$	80, 80	50, 50	5, 5	1/5	52	29	7.2e–10
10	$10^{-7} \sim 10^{-5}$	80, 80	50, 50	51, 51	1/8	31	23	3.2e–7

Table 1: Performance: univariate case

Remark 4.1. In our tests, the k outlier errors are introduced in random locations after the bound $E \ge k$ is derived from the error rate $1/q$. However, our algorithm also makes random choices, the anchor points α and the random roots of unity ζ (see Step 2(a)). We perform 20 trials of the ζ's before giving up with recovery. Running times can fluctuate as a new random choice for ζ has a new number of outlier errors k in different places. For instance, Example 8 in Table 1 is a case where 10 trials are needed before success. Because our error rates are quite large, the tests cannot succeed simply because the batches have few outlier errors (see the column for E in Table 1). Example 10 in Table 1 is a dense rational function. Our algorithm currently fails to recover the fraction with an error rate of $1/q = 1/4$. □

Ex	*Random Noise*	$\bar{d}_f, \bar{d}_g$	$\deg(f)$, $\deg(g)$	t_f, t_g	n	$1/q$	E	N	*time secs.*	*Rel. Error*
1	$10^{-5} \sim 10^{-3}$	3, 3	1, 1	2, 2	2	1/10	12	403	6.2	7.3e–7
2	$10^{-5} \sim 10^{-3}$	5, 5	2, 2	3, 3	2	1/12	21	306	6.0	4.5e–8
3	$10^{-5} \sim 10^{-3}$	2, 5	1, 4	2, 4	3	1/15	13	561	13	4.7e–7
4	$10^{-6} \sim 10^{-4}$	8, 8	5, 2	10, 6	3	1/40	12	616	47	3.6e–6
5	$10^{-7} \sim 10^{-5}$	10, 10	7, 7	10,10	5	1/90	7	1508	197	5.1e–11
6	$10^{-7} \sim 10^{-5}$	15, 10	10, 3	15, 5	8	1/90	7	2423	273	7.4e–11
7	$10^{-7} \sim 10^{-5}$	10, 15	5, 13	4, 6	10	1/80	2	1289	24	8.1e–10
8	$10^{-7} \sim 10^{-5}$	25, 25	20, 20	7, 7	15	1/100	3	2890	137	2.9e–10
9	$10^{-8} \sim 10^{-6}$	35, 35	30, 30	6, 6	20	1/80	2	3881	230	5.5e–13
10	$10^{-8} \sim 10^{-6}$	45, 45	40, 40	6, 6	5	1/80	6	2080	219	3.7e–12
11	$10^{-8} \sim 10^{-6}$	85, 85	60, 60	7, 7	4	1/100	11	2787	1479	3.7e–13
12	$10^{-8} \sim 10^{-6}$	85, 85	80, 80	3, 3	5	1/30	4	1773	83	4.5e–12
13	$10^{-9} \sim 10^{-7}$	70, 0	40, 0	6, 1	15	1/70	2	2284	75	7.5e–18
14	$10^{-8} \sim 10^{-6}$	25, 25	20, 20	5, 5	102	1/80	1	10191	272	6.1e–12

Table 2: Performance: multivariate case

In Table 2 we exhibit the performance of our algorithm on multivariate inputs. For each example, we construct two relatively prime multivariate polynomials with random integer coefficients in the range $-5 \le c \le 5$. Here *Random Noise* denotes the noise in this range randomly added to the black box of f/g; $\bar{d}_f \ge \deg(f)$ and $\bar{g}_g \ge \deg(g)$ denote the degree bound of the numerator and the denominator, respectively; t_f and t_g denote the number of terms of the numerator and denominator, respectively; n denotes the number of variables of the rational functions; N denotes the number of the black box probes needed to interpolate the approximate multivariate rational function; E denotes the maximum number of outliers for each individual interpolation step, $1/q$ the resulting error rate of the outlier error; finally, *Rel. Error* denotes the relative backward error computed by our algorithm. About the setting of the outlier error, it is the same as the univariate case. Example 13 is one polynomial test which shows that our algorithm can also deal with the multivariate polynomial interpolation from values with noise and outlier errors.

Example 14 in Table 2 warrants further discussion, as the number of probes for a fraction with 5 terms in both the numerator and denominator takes over 10000 evaluations. There are $n = 102$ variables. Estimating the degree in each variable, using as upper bounds $\bar{d}_f$ and $\bar{d}_g$, consumes $(\bar{d}_f + \bar{d}_g + 2E + L_0) \cdot 102$ probes, about 5000. We then use x_i as the variable in the error locator polynomial Λ_i; see Remark 2.3. We have for each i the estimates $|D_{f,i,E}| = t_f(\deg_{x_i}(f) + 1 + E) = 5(3 + 1 + 1)$ and $|D_{g,i,E}| = t_g(\deg_{x_i}(g) + 1 + E) = 5(2 + 1 + 1)$, that is $45 + L_0$ evaluations, or about 5000 in total. Using sharper upper bounds for $\deg_{x_i}(f)$ and $\deg_{x_i}(g)$ one could reduce the number of probes to about 6000. The fact remains that 102 variables constitute a large recovery problem, with $(5 + 5) \times 102$ individual exponents $d_{j,\mu}$, $e_{m,\mu}$ to be determined.

Ex	*Random Noise*	*Rel. Outlier Error* Θ	deg(f), deg(g)	t_f,t_g	n	$1/q$	E	N	*time secs*	*Rel. Error*
1	10^{-4}~10^{-2}	1~2	5, 5	2, 3	1	1/4	24	94	0.8	8.1e–4
2	10^{-5}~10^{-3}	0.1~0.2	15, 15	3, 5	1	1/10	9	84	1.5	3.3e–4
3	10^{-7}~10^{-5}	0.001~0.002	20, 10	4, 3	1	1/15	5	74	0.6	9.3e–10
4	10^{-6}~10^{-4}	0.01~0.02	30, 25	4, 4	1	1/7	12	84	0.8	9.4e–9
5	10^{-7}~10^{-5}	0.001~0.002	50, 40	5, 4	1	1/40	3	288	3.0	8.1e–4
6	10^{-5}~10^{-3}	0.1~0.2	5, 8	1, 3	2	1/30	2	200	1.9	1.8e–2
7	10^{-6}~10^{-4}	0.01~0.02	10, 15	3, 3	4	1/40	2	860	8.2	9.5e–9
8	10^{-7}~10^{-5}	0.01~0.02	10, 10	3, 2	15	1/40	2	2433	18	3.0e–12
9	10^{-8}~10^{-6}	0.01~0.02	8, 8	4, 3	30	1/50	2	3236	35	1.2e–12
10	10^{-9}~10^{-7}	0.01~0.02	15, 15	3, 3	50	1/60	1	5299	50	2.5e–10

Table 3: Performance: small outliers

In Table 3 we give first tests with small outlier errors; see Remark 3.1. *Outlier Error* denotes the relative outlier error Θ, which is randomly selected in the given range.

Example 4.1. We now demonstrate the Candes-Tao recovery of unreduced sparse rational functions discussed in Remark 2.7, that on a small example with no outlier errors ($k = E = 0$): let $f = (x^{11}+1)(x-1) = x^{12} - x^{11} + x - 1$, $\bar{d}_f = 12$, and $g = (x+1)(x^5-1)$, $\bar{d}_g = 6$, $L = \bar{d}_f + \bar{d}_g + 1 = 19$. We have $f/g =$

$$\frac{x^{12}-x^{11}+x-1}{x^6+x^5-x-1} = \frac{x^{10}-x^9+x^8-x^7+x^6-x^5+x^4-x^3+x^2-x+1}{x^4+x^3+x^2+x+1}. \quad (24)$$

We compute $\beta_\ell = \gamma_\ell = (f/g)(\zeta^{\ell+1})$ for $\zeta = \exp(2\pi\boldsymbol{i}/31)$ and $\ell = 0, 1, \ldots, L-1$ in hardware precision complex floating point arithmetic. The shift in the exponent to $\ell + 1$ avoids $g(1) = 0$. Now we solve the linear system (15) for a real vector $[\vec{\boldsymbol{y}}, \vec{\boldsymbol{z}}]^T \in \mathbb{R}^{19}$ with the following constraint: $y_{12} = 1$, meaning the numerator polynomial f is monic of degree 12. By separating real and imaginary parts of the matrix G in (15) we have the linear equational constraints

$$\begin{bmatrix} \text{Realpart}(V_1), -\text{Realpart}(\Gamma_1 W_1) \\ \text{Imagpart}(V_1), -\text{Imagpart}(\Gamma_1 W_1) \end{bmatrix} \begin{bmatrix} \vec{\boldsymbol{y}}^T \\ \vec{\boldsymbol{z}}^T \end{bmatrix} = \mathbf{0}, y_{12} = 1. \quad (25)$$

The linear system (25) has a higher dimensional solution set because f and g are not relatively prime. We wish to discover a sparse solution by minimizing $\sum_j |y_j| + \sum_m |z_m| = \|[\vec{\boldsymbol{y}}, \vec{\boldsymbol{z}}]\|_1$ via Tshebysheff's linear programming formulation. In our case, Maple 16's call to `Optimization['LPSolve']` with `method = activeset` produces the solution $f = x^{12} - 1.0\,x^{11} - 1.78\times10^{-10}\,x^{10} + 1.72\times10^{-10}\,x^9 + 1.72\times10^{-10}\,x^7 - 1.78\times10^{-10}\,x^6 + 1.82\times10^{-10}\,x^5 - 1.88\times10^{-10}\,x^4 + 1.92\times10^{-10}\,x^3 - 1.88\times10^{-10}\,x^2 + 1.0\,x - 1.0$ and $g = 1.0\,x^6 + 1.0\,x^5 - 5.83\times10^{-12}\,x^3 - 2.07\times10^{-12}\,x^2 - 1.0\,x - 1.0$ with an objective value of 6.99999999999878. The rounded polynomials give the unreduced form in (24).

For some examples of lesser unreduced sparsity, the unreduced fraction can be recovered by oversampling at L_0 additional values. Without oversampling, the Maple 16 `LPSolve` call falsely reports infeasibility of the linear program.

We plan to study sparse unreduced recovery in the presence of outlier errors à la [19] and with Zippel lifting to several variables in the near future.

5. REFERENCES

[1] Blahut, R. E. A universal Reed-Solomon decoder. *IBM J. Res. Develop. 18*, 2 (Mar. 1984), 943–959.

[2] Boyer, B., Comer, M. T., and Kaltofen, E. L. Sparse polynomial interpolation by variable shift in the presence of noise and outliers in the evaluations. In *Electr. Proc. Tenth Asian Symposium on Computer Mathematics (ASCM 2012)* (2012). .

[3] Candes, E., and Tao, T. Decoding by linear programming. *IEEE Trans. Inf. Theory* IT-*51*, 12 (2005), 4203–4215.

[4] Comer, M. T., Kaltofen, E. L., and Pernet, C. Sparse polynomial interpolation and Berlekamp/Massey algorithms that correct outlier errors in input values. In *ISSAC 2012 Proc. 37th Internat. Symp. Symb. Alg. Comput.* (New York, N. Y., July 2012), J. van der Hoeven and M. van Hoeij, Eds., ACM, pp. 138–145. .

[5] Cuyt, A., and Lee, W. Sparse interpolation of multivariate rational functions. *Theoretical Comput. Sci. 412* (2011), 1445–1456.

[6] de Kleine, J., Monagan, M., and Wittkopf, A. Algorithms for the non-monic case of the sparse modular GCD algorithm. In *ISSAC'05 Proc. 2005 Internat. Symp. Symb. Alg. Comput.* (New York, N. Y., 2005), M. Kauers, Ed., ACM Press, pp. 124–131.

[7] Giesbrecht, M., Kaltofen, E., and Lee, W. Algorithms for computing sparsest shifts of polynomials in power, Chebychev, and Pochhammer bases. *J. Symbolic Comput. 36*, 3–4 (2003), 401–424. .

[8] Giesbrecht, M., Labahn, G., and Lee, W. Symbolic-numeric sparse interpolation of multivariate polynomials (Ext. Abstract). In *Proc. 9th Rhine Workshop Comput. Alg. (RWCA'04), Univ. Nijmegen, the Netherlands* (2004), pp. 127–139.

[9] Giesbrecht, M., Labahn, G., and Lee, W. Symbolic-numeric sparse interpolation of multivariate polynomials. *J. Symbolic Comput. 44* (2009), 943–959.

[10] Giesbrecht, M., Roche, D. S., and Tilak, H. Computing sparse multiples of polynomials. In *Proc. Internat. Symp. on Algorithms and Computation (ISAAC 2010)* (2010).

[11] Kaltofen, E. Greatest common divisors of polynomials given by straight-line programs. *J. ACM 35*, 1 (1988), 231–264. .

[12] Kaltofen, E. Factorization of polynomials given by straight-line programs. In *Randomness and Computation*, S. Micali, Ed., vol. 5 of *Advances in Computing Research*. JAI Press Inc., Greenwhich, Connecticut, 1989, pp. 375–412. .

[13] Kaltofen, E., and Koiran, P. Expressing a fraction of two determinants as a determinant. In *ISSAC 2008* (New York, N. Y., 2008), D. Jeffrey, Ed., ACM Press, pp. 141–146. .

[14] Kaltofen, E., and Pernet, C. Cauchy interpolation with errors in the values. Manuscript in preparation, Jan. 2013.

[15] Kaltofen, E., and Trager, B. Computing with polynomials given by black boxes for their evaluations: Greatest common divisors, factorization, separation of numerators and denominators. *J. Symbolic Comput. 9*, 3 (1990), 301–320. .

[16] Kaltofen, E., and Yang, Z. On exact and approximate interpolation of sparse rational functions. In *ISSAC 2007 Proc. 2007 Internat. Symp. Symb. Alg. Comput.* (New York, N. Y., 2007), C. W. Brown, Ed., ACM Press, pp. 203–210. .

[17] Kaltofen, E., Yang, Z., and Zhi, L. On probabilistic analysis of randomization in hybrid symbolic-numeric algorithms. In *SNC'07 Proc. 2007 Internat. Workshop Symb.-Numer. Comput.* (New York, N. Y., 2007), J. Verschelde and S. M. Watt, Eds., ACM Press, pp. 11–17. .

[18] Kaltofen, E. L., Lee, W., and Yang, Z. Fast estimates of Hankel matrix condition numbers and numeric sparse interpolation. In *SNC'11 Proc. 2011 Internat. Workshop Symb.-Numer. Comput.* (New York, N. Y., June 2011), M. Moreno Maza, Ed., ACM, pp. 130–136. .

[19] Kaltofen, E. L., and Nehring, M. Supersparse black box rational function interpolation. In *Proc. 2011 Internat. Symp. Symb. Alg. Comput. ISSAC 2011* (New York, N. Y., June 2011), A. Leykin, Ed., ACM, pp. 177–185. .

[20] Olesh, Z., and Storjohann, A. The vector rational function reconstruction problems. In *Proc. Waterloo Workshop on Computer Algebra: devoted to the 60th birthday of Sergei Abramov (WWCA)* (2007), pp. 137–149.

[21] Olshevsky, V., and Shokrollahi, M. A. A displacement approach to decoding algebraic codes. In *Algorithms for Structured Matrices: Theory and Applications*. AMS, Providence, RI, 2003, pp. 265–292. Contemp. Math., vol. 323.

[22] Saraf, S., and Yekhanin, S. Noisy interpolation of sparse polynomials, and applications. In *Proc. 26th Annual IEEE Conf. Comp. Complexity* (2011), IEEE Comp. Soc., pp. 86–92.

[23] Welch, L. R., and Berlekamp, E. R. Error correction of algebraic block codes. US Patent 4,633,470, 1986. Filed 1983; see http://patft.uspto.gov/.

[24] Zippel, R. Interpolating polynomials from their values. *J. Symbolic Comput. 9*, 3 (1990), 375–403.

Quantum Fourier Transform over Symmetric Groups

Yasuhito Kawano
NTT Communication Science Laboratories
3-1, Morinosato Wakamiya, Atsugi-shi
Kanagawa, 243-0198, Japan
kawano.yasuhito@lab.ntt.co.jp

Hiroshi Sekigawa*
Department of Math, Tokai University
4-1-1, Kitakaname, Hiratsuka-shi
Kanagawa, 259-1292, Japan
sekigawa@tokai-u.jp

ABSTRACT

This paper proposes an $O(n^4)$ quantum Fourier transform (QFT) algorithm over symmetric group S_n, the fastest QFT algorithm of its kind. We propose a fast Fourier transform algorithm over symmetric group S_n, which consists of $O(n^3)$ multiplications of unitary matrices, and then transform it into a quantum circuit form. The QFT algorithm can be applied to constructing the standard algorithm of the hidden subgroup problem.

Categories and Subject Descriptors

H.4 [**Information Systems Applications**]: Miscellaneous; D.2.8 [**Software Engineering**]: Metrics—*complexity measures, performance measures*

Keywords

Quantum Fourier Transform; Fast Fourier Transform; Symmetric Group; Representation Theory

1. INTRODUCTION

The quantum Fourier transform (QFT) plays an important role in many quantum algorithms exponentially faster than the classical counterparts. Shor's quantum algorithm [14] that efficiently solves the factoring problem applies the QFT over the cyclic groups. Quantum circuits for the QFT over the cyclic groups have been studied in detail, and many efficient QFT circuits over the cyclic groups have been proposed [5, 7, 4, 11].

Shor's algorithm can be naturally generalized to the standard algorithm for the hidden subgroup problem [12, §5.4.3]. An especially interesting application of the hidden subgroup problem is the graph isomorphism problem: Given two graphs, decide whether there exists an isomorphism map from one to the other. It is an open question whether there exists an efficient quantum algorithm that solves the graph isomorphism problem. The standard algorithm for the hidden subgroup problem has been thought to be one of the candidates that can efficiently solve the graph isomorphism problem or its subproblems [12, §5.4.4]. A standard algorithm that solves Simon's problem over non-abelian groups faster than classical algorithms has been found [1]. On the other hand, it is suggested that the graph isomorphism problem would not be solved using the standard algorithm (cf. [10]).

The standard algorithm for the graph isomorphism problem uses the QFT over symmetric groups instead of the QFT over cyclic groups. Efficient quantum circuits that perform the QFT over symmetric groups have thus been studied [2, 9]. An efficient QFT circuit over symmetric groups was first proposed by Beals [2]. Later, Moore, Rockmore, and Russell applied recent progress in the fast Fourier transform (FFT) algorithm, and proposed efficient QFT circuits over non-abelian groups, including symmetric groups [9]. However, their QFT circuits for symmetric groups sum amplitudes over cosets serially, where the sum-of-amplitudes is the most complex calculation in the QFT. On the other hand, since Coppersmith's well-known circuit [5, 12] for the QFT over the cyclic groups sums amplitudes in parallel, the time complexity of Coppersmith's circuit is much lower than the time complexities of circuits proposed in [2, 9]. It is known that Coppersmith's circuit is a quantum counterpart of the FFT algorithm over the cyclic groups.

The purpose of this paper is to propose an algorithm that performs QFT over symmetric groups more efficiently by calculating the sum-of-amplitudes in parallel. For this purpose, we propose an FFT (classical) algorithm over symmetric groups, which consists of a multiplication of sparse unitary matrices, and then transform it to a quantum circuit form. As a byproduct, we obtain an FFT algorithm over symmetric groups. A detailed discussion of applications can be found in [6, page 326].

This paper is constructed as follows. In section 2, we explain background notions and symbols of representation theory for symmetric groups. In section 3, the FFT algorithm is proposed. In section 4, the QFT algorithm is described. Finally, section 5 concludes the paper.

2. REPRESENTATION THEORY

Background notions and symbols of representation theory for symmetric groups are given in this section. More detailed explanations can be found in [15, 16, 8, 13].

*The current address of the author is: Department of Mathematical Information Science, Tokyo University of Science, 1-3 Kagurazaka, Shinjuku-ku, Tokyo, 162-8601, Japan. sekigawa@rs.tus.ac.jp

ISSAC'13, June 26–29, 2013, Boston, Massachusetts, USA.

2.1 Basic Notions

Let n be a positive integer. A permutation $i_1 \mapsto j_1, i_2 \mapsto j_2, \cdots, i_n \mapsto j_n$ over $\{1, 2, \cdots, n\}$ is denoted $\begin{pmatrix} i_1 & \cdots & i_n \\ j_1 & \cdots & j_n \end{pmatrix}$. When $i_k = j_k$ in the above permutation, the column of i_k and j_k is often abbreviated. A *multiplication* of permutations is defined by

$$\begin{pmatrix} j_1 & \cdots & j_n \\ k_1 & \cdots & k_n \end{pmatrix} \cdot \begin{pmatrix} i_1 & \cdots & i_n \\ j_1 & \cdots & j_n \end{pmatrix} = \begin{pmatrix} i_1 & \cdots & i_n \\ k_1 & \cdots & k_n \end{pmatrix}.$$

Define S_n as the group of the set of all permutations over $\{1, 2, \cdots, n\}$ with the multiplication. The S_n is called the *symmetric group* of order n. The number of elements in S_n, denoted $|S_n|$, is $n!$.

A permutation $\begin{pmatrix} i_1 & i_2 & \cdots & i_k \\ i_2 & i_3 & \cdots & i_1 \end{pmatrix}$ is called a *cyclic permutation.* It is denoted $c_{(i_1, i_2, \cdots, i_k)}$ or $i_1 \mapsto i_2 \mapsto \cdots \mapsto i_k \mapsto i_1$ in this paper. When $k = 2$, a permutation is called a *transposition.* An element in S_n can be decomposed into a multiplication of transpositions.

The quotient ring $\mathbb{Z}/n\mathbb{Z}$ is denoted $\mathbb{Z}_n$. For any element g in S_n, there is a unique $(i_1, i_2, \cdots, i_{n-1}) \in \mathbb{Z}_2 \times \mathbb{Z}_3 \times \cdots \times \mathbb{Z}_{n-1}$ such that

$$g = c_{(1,2,\cdots,n)}^{i_{n-1}} \cdots c_{(1,2,3)}^{i_2} c_{(1,2)}^{i_1}.$$

The map $g \mapsto (i_1, i_2, \cdots, i_{n-1})$ is called the *canonical coding* in this paper. The canonical coding will be used for coding an element in S_n on quantum states, i.e., an element g in S_n will be encoded as $|i_1, i_2, \cdots, i_{n-1}\rangle$ using qubit, qutrit, $\cdots$, and qu$n$it.

Let $g \mapsto (i_1, i_2, \cdots, i_{n-1})$ be the canonical coding. Define g_i, where $i = \sum_{j=1}^{n-1} i_j \cdot \frac{n!}{(j+1)!}$, as g. Then, $\{g_i | i = 0, 1, \cdots, n! - 1\}$ is an enumeration of elements in S_n, i.e., $S_n = \{g_0, g_1, \cdots, g_{n!-1}\}$. For example, $g_0 = id$ and $g_{n!/2} = c_{(1,2)}$. The S_3 is enumerated as id, $c_{(1,2,3)}$, $c_{(1,3,2)}$, $c_{(1,2)}$, $c_{(2,3)}$, and $c_{(1,3)}$. This order will be used for the column number of the Fourier transform matrix.

2.2 Irreducible Representations

Let V be a finite dimensional vector space over the field $\mathbb{C}$. Let $U(V)$ be the group defined by the set of unitary transforms from V to V. Given an orthonormal basis of V, each element in $U(V)$ is represented as an $n \times n$ unitary matrix, where n is the dimension of V.

If a function $\rho : S_n \to U(V)$ is homomorphic (i.e., $\rho(g_1 g_2) = \rho(g_1) \cdot \rho(g_2)$ is satisfied for all $g_1, g_2 \in S_n$) for a vector space V, then ρ is called a *representation* and V is called a *representation space.* If a subspace W of a representation space V satisfies $\rho(g)(W) \subseteq W$ for all $g \in S_n$, W is called an invariant subspace. Since V and $\{0\}$ are always invariant, they are called *trivial* invariant subspaces. If there is no non-trivial invariant space, then ρ is called an *irreducible* representation. The set of all irreducible representations of S_n is denoted Λ_n.

2.3 Young Diagrams

A Young diagram is a diagram with left-aligned and top-aligned square boxes. By enumerating the numbers of the boxes in the first, second, $\cdots$, and k-th rows, a Young diagram with n boxes is encoded as an ordered set of n numbers $(\lambda_1, \lambda_2, \cdots, \lambda_k, 0, \cdots, 0)$, where $n = \sum_{i=1}^{k} \lambda_i$ and $\lambda_1 \geq \lambda_2 \geq \cdots \geq \lambda_k > 0$. $(\lambda_1, \lambda_2, \cdots, \lambda_k, 0, \cdots, 0)$ is often written as $(\lambda_1, \lambda_2, \cdots, \lambda_k)$ by omitting zeros. Since a Young diagram of S_n is a non-increasing sequence of numbers such that the sum of them is n, it is sometimes called "a partition of n."

It is known that any irreducible representation for S_n corresponds to a Young diagram with n boxes, which is a partition of n. The set of all Young diagrams with n boxes (= the set of all partitions of n) can then be seen as the set of all irreducible representations of S_n. We will identify irreducible representations of S_n, the Young diagrams with n boxes, and partitions of n hereafter, and denote the set of them as Λ_n. For example, $(2, 1)$ is a partition of 3, a Young diagram with three boxes, and an irreducible representation of S_3. Hence, there is a vector space V such that $(2, 1) : S_3 \to U(V)$ is an irreducible representation.

2.4 Standard Young Tableaus

To calculate the dimension of the representation space, the notion of the standard Young tableau is introduced.

A *Young tableau* with n boxes is a diagram obtained from a Young diagram with n boxes by writing numbers from 1 to n into the boxes of the Young diagram. Here, different boxes must have different numbers. A Young tableau is standard if the number in each box is greater than both the number in the box above and number in the box to the left. It is known that the number of standard Young tableaus is equal to the dimension of the representation space.

The dimension of the representation space for $\lambda \in \Lambda_n$ is denoted by d_λ. Let $\rho_n = \bigoplus_{\lambda \in \Lambda_n} (I_{d_\lambda} \otimes \lambda)$ be the representation defined by $\rho_n(g) = \bigoplus_{\lambda \in \Lambda_n} (I_{d_\lambda} \otimes \lambda(g))$. Each $\rho_n(g)$ is then written as an $n! \times n!$ unitary matrix of $d_\lambda \times d_\lambda$ block matrices with duplication of d_λ for all $\lambda \in \Lambda_n$.

2.5 Bratteli diagram

While the column number of the Fourier transform is determined by the enumeration $\{g_0, g_1, \cdots, g_{n!-1}\}$ of the group elements in S_n, the row number is determined using the Bratteli diagram. The Bratteli diagram (Figure 1) is a directed acyclic graph with a root node such that

1. a Young diagram with n boxes is assigned on a node in the n-th row, and
2. μ, which is a Young diagram with $n+1$ boxes, is a child node of λ, which is a Young diagram with n boxes, if and only if μ is obtained by adding a box to λ.

We introduce a lexicographic order for nodes on each row of the Bratteli diagram, i.e., define that $(\lambda_1, \lambda_2, \cdots, \lambda_k) > (\lambda'_1, \lambda'_2, \cdots, \lambda'_{k'})$ if and only if $\lambda_i > \lambda'_i$ for the smallest i such that $\lambda_i \neq \lambda'_i$. The Bratteli diagram can then be drawn on a plane by drawing λ on the left-hand side of λ' if $\lambda > \lambda'$. Figure 1 shows the Bratteli diagram for $n \leq 6$. As already explained, an irreducible representation for S_n corresponds to a Young diagram with n boxes. Thus, all irreducible representations for S_n are enumerated in the n-th row of the Bratteli diagram. We will identify an irreducible representation of S_n, a partition of n, a Young diagram with n boxes, and a node in the n-th row of the Bratteli diagram hereafter. The symbol $\lambda \in \Lambda_n$, which is an irreducible representation of S_n, is often used to show the node of the Bratteli diagram corresponding to λ. Then, the number of paths from the root node to a node λ is equal to d_λ. Then, the following relation holds: $n! = \sum_{\lambda \in \Lambda_n} d_\lambda^2$.

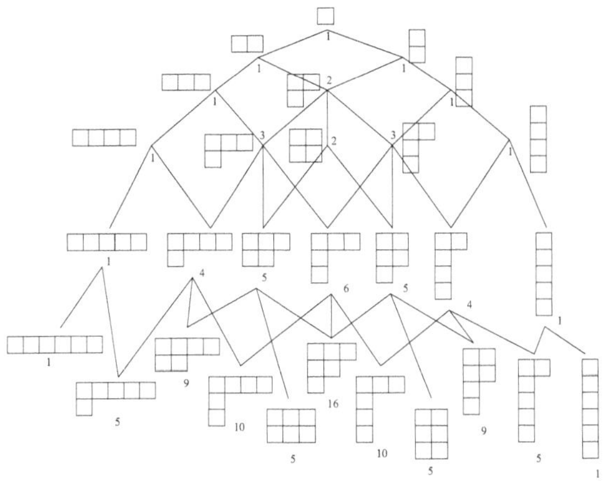

Figure 1: The Bratteli diagram of S_6. Each node of the nth column represents a Young diagram for S_n, which corresponds to an irreducible representation of S_n. The number written at each node shows the dimension of the representation space of the corresponding representation. Note that $n! = \sum_{\lambda \in \Lambda_n} d_\lambda^2$. For example, $4! = 1^2 + 3^2 + 2^2 + 3^2 + 1^2$.

Let $\mathcal{P}(\lambda)$ be the set of paths from the root node to node λ. Then, $|\mathcal{P}(\lambda)| = d_\lambda$. An element $p \in \mathcal{P}(\lambda)$ for $\lambda \in \Lambda_n$ can be encoded by an ordered set of $n-1$ numbers $(p_1, p_2, \cdots, p_{n-1})$ such that a new box is added to the (p_k+1)st row of the Young diagram λ at the kth stage. For example, $\mathcal{P}(2,1,1) = \{(0,1,2),(1,0,2),(1,2,0)\}$. It is easily seen that

$$(p_1, p_2, \cdots, p_{n-1}) \in \mathbb{Z}_2 \times \mathbb{Z}_3 \times \cdots \times \mathbb{Z}_n.$$

Define $<$ by $(p_1, p_2, \cdots, p_{n-1}) < (p'_1, p'_2, \cdots, p'_{n-1})$ if and only if $p_i < p'_i$ for the smallest i such that $p_i \neq p'_i$.

Since $n! = \sum_{\lambda \in \Lambda_n} d_\lambda^2$, $|\{|\lambda, p, q\rangle | \lambda \in \Lambda_n, p, q \in \mathcal{P}(\lambda)\}| = n!$. Define the order by $|\lambda, p, q\rangle < |\lambda', p', q'\rangle \Leftrightarrow \lambda > \lambda'$ or $\lambda = \lambda' \wedge p < p'$ or $\lambda = \lambda' \wedge p = p' \wedge q < q'$. (Notice that the direction of the inequality sign for λ is reversed) For each element in $\{|\lambda, p, q\rangle | \lambda \in \Lambda_n, p, q \in \mathcal{P}(\lambda)\}$, a number less than $n!$ is given according to the above order. It will be used as the row number of the Fourier transform.

2.6 Adapted Gel'fand-Tsetlin Bases

The $\lambda(g)$ $(g \in S_n, \lambda \in \Lambda_n)$ can be represented by a $d_\lambda \times d_\lambda$ matrix, but it depends on the selection of an orthonormal basis of the representation space. *Adapted Gel'fand-Tsetlin bases* are very useful for making an efficient Fourier transform algorithm on S_n $(n = 2,3,4,\cdots)$. Here, adapted Gel'fand-Tsetlin bases are series of orthonormal bases $\{v_{\lambda,p,q} | \lambda \in \Lambda_n, p, q \in \mathcal{P}(\lambda)\}_n$ on the representation spaces of irreducible representations for S_n $(n = 2,3,4,\cdots)$ that satisfies the following conditions.

Let λ be an irreducible representation of S_n. Then, λ corresponds to a Young diagram. (Hence, it is a node in the n-th row of the Bratteli diagram.) Let $\{\mu | \mu \searrow \lambda\}$ be irreducible representations which are parents of λ in the Bratteli diagram. From the properties of the Bratteli diagram, $d_\lambda = \sum_{\mu \searrow \lambda} d_\mu$.

For $\lambda \in \Lambda_n$, $\lambda(g)$ $(g \in S_n)$ is represented by a $d_\lambda \times d_\lambda$ matrix when we select an orthonormal basis of the representation space of λ. Denote this matrix as $\lambda(g)_{\mathcal{B}_\lambda}$, where $\mathcal{B}_\lambda$ is the basis. On the other hand, since μ's $(\mu \searrow \lambda)$ are irreducible representations for S_{n-1}, any $g \in S_{n-1}$ is represented by the direct sum of $d_\mu \times d_\mu$ matrices when we select orthonormal bases of the representation spaces of μ's $(\mu \searrow \lambda)$. Denote this matrix as $\oplus_{\mu \searrow \lambda} \mu(g)_{\mathcal{B}_\mu}$, where $\mathcal{B}_\mu$ $(\mu \searrow \lambda)$ are the bases. Generally, $\lambda(g)_{\mathcal{B}_\lambda}$ and $\oplus_{\mu \searrow \lambda} \mu(g)_{\mathcal{B}_\mu}$ are different because they depend on the selections of the bases. However, by selecting good bases,

$$\lambda(g)_{\mathcal{B}_\lambda} = \bigoplus_{\mu \searrow \lambda} \mu(g)_{\mathcal{B}_\mu} \quad (1)$$

holds for all $g \in S_{n-1}$. If (1) holds for all irreducible representations λ, the set of bases is called the adapted Gel'fand-Tsetlin bases. Elements in S_{n-1} are represented by block diagonal matrices in the $d_\lambda \times d_\lambda$ matrix that represents elements of S_n if we use the adapted Gel'fand-Tsetlin bases.

The adapted Gel'fand-Tsetlin bases are not unique. In this paper, we select a set of adapted Gel'fand-Tsetlin bases and use it consistently. Therefore, $\lambda(g)_{\mathcal{B}_\lambda}$ will be simply denoted $\lambda(g)$ if there's no confusion.

2.7 Specht Polynomial

Vandermonde determinant $\Delta(x_1, \ldots, x_n)$ is defined as

$$\Delta(x_1, \ldots, x_n) = \begin{vmatrix} x_1^{n-1} & x_2^{n-1} & \ldots & x_n^{n-1} \\ x_1^{n-2} & x_2^{n-2} & \ldots & x_n^{n-2} \\ \vdots & \vdots & \ddots & \vdots \\ x_1 & x_2 & \ldots & x_n \\ 1 & 1 & \ldots & 1 \end{vmatrix}.$$

Let $\mathcal{T}$ be an m-row 1-column Young tableau. Specht polynomial $\Delta(\mathcal{T})$ is defined by

$$\Delta(\mathcal{T}) = \Delta(x_1, \ldots, x_m).$$

For any Young tableau $\mathcal{T}$ with i columns, Specht polynomial $\Delta(\mathcal{T})$ is defined by

$$\Delta(\mathcal{T}) = \Delta(\mathcal{T}_1) \ldots \Delta(\mathcal{T}_i),$$

where $\mathcal{T}_j$ is the j-th column of $\mathcal{T}$. For example, $\mathcal{T}$ is a Young tableau such that it has three columns, the numbers of the first column are $1,4,5$, the second are $2,3$, and the third are $6,7$. Then,

$$\Delta(\mathcal{T}) = \Delta(x_1, x_4, x_5)\Delta(x_2, x_3)\Delta(x_6, x_7).$$

Let g be an element of S_n and $\mathcal{T}$ be a Young tableau. $g(\Delta(\mathcal{T}))$ is defined as $\Delta(g(\mathcal{T}))$, where $g(\mathcal{T})$ is the Young tableau obtained by replacing x_i by $x_{g(i)}$ for all $i = 1,2,\cdots,n$. Let V be the complex-number coefficient vector space on $\{g(\Delta(\mathcal{T})) | g \in S_n\}$. Generally, $g(\Delta(\mathcal{T}))$'s $(g \in S_n)$ are not linearly independent. It is known that Specht polynomials of standard Young tableaus are linearly independent. Since any $g \in S_n$ that maps $\Delta(\mathcal{T})$ to $g(\Delta(\mathcal{T}))$ defines a linear operation from V to V, a map from S_n to $U(V)$ is defined. This map is homomorphic, so it is a representation of S_n. In addition, the map is irreducible, and any irreducible representations of S_n can be enumerated by this method. However, the set of Specht polynomials for standard Young tableaus is not a set of adapted Gel'fand-Tsetlin bases. A method for making adapted Gel'fand-Tsetlin bases is proposed in what follows.

2.8 Construction of Adapted Gel'fand-Tsetlin Bases

In this subsection, we will give an algorithm that calculates unitary matrices that represent $\{\lambda(g)|\lambda \in \Lambda_n, g \in S_n\}$. The representation is not unique, but to make efficient QFT circuits later, those matrices should be chosen under the selection of special bases.

The following order between standard Young tableaus with n boxes is introduced.

DEFINITION 1. *For Young tableaus $\mathcal{T}_1$ and $\mathcal{T}_2$, $\mathcal{T}_1 < \mathcal{T}_2$ is defined as follows:*

> *There exists i $(1 \leq i \leq n)$ such that (1) for all j $(i < j \leq n)$ j belongs to the same columns of $\mathcal{T}_1$ and $\mathcal{T}_2$ and (2) the column that contains i in $\mathcal{T}_1$ is on the left-hand side of the column that contains i in $\mathcal{T}_2$.*

Then, "<" is the total order on the set of standard Young tableaus with n boxes.

Next, we introduce the Hermitian product on the vector space V defined from Specht polynomials that correspond to standard Young tableaus with n boxes. Since V is a subspace of the space W of linear combinations of monomials of $x_1, \ldots, x_n$, we can write $f, g \in W$ as

$$f = \sum_{\alpha} a_\alpha x^\alpha, \qquad g = \sum_{\alpha} b_\alpha x^\alpha,$$

where α is a multi-index. The Hermitian product on V is defined as the restriction of the Hermitian product

$$(f, g) = \sum_{\alpha} a_\alpha \overline{b_\alpha}$$

on W.

Adapted Gel'fand-Tsetlin bases on S_n are calculated as follows.

ALGORITHM 1 (ADAPTED GEL'FAND-TSETLIN BASES).

1. *Enumerate all Young diagrams with n boxes. The order is left-to-right of the n-th row of the Bratteli diagram.*
2. *Perform the following operations for each Young diagram according to the order determined in 1.*
   (a) *Enumerate all standard Young tableaus in the order of < of Definition 1. (Start from the smallest order and end at the largest).*
   (b) *From the end of the order of standard Young tableaus, orthonormalize Specht polynomials one by one. For example, on the scheme of the Gram-Schmidt orthonomalization, orthogonalize all Specht polynomials first, and then normalize the obtained polynomials at the end. (In this way, orthogonalizing can be performed by only four arithmetic operations.)*

The polynomials obtained by the above algorithm are adapted Gel'fand-Tsetlin bases. This comes from the following facts.

- Specht polynomials of standard Young tableaus for different Young diagrams are orthogonal.
- For a Young diagram λ with n boxes, divide the set of all standard Young tableaus obtained from λ into $\mathcal{A}_1 \cup \mathcal{A}_2 \cup \cdots \cup \mathcal{A}_q$ $(\mathcal{A}_i \neq \emptyset)$ such that $m < m'$ if and only if $\mu < \mu'$, where m and m' are the numbers of the columns that contain the boxes into which n is written for $\mathcal{T} \in \mathcal{A}_\mu$ and $\mathcal{T}' \in \mathcal{A}_{\mu'}$, respectively. Let V_i $(1 \leq i \leq q)$ be the subspace generated by the Specht polynomials $\Delta(\mathcal{T})$ $(\mathcal{T} \in \mathcal{A}_q \cup \mathcal{A}_{q-1} \cup \cdots \cup \mathcal{A}_{q-i+1})$. Then,

  $$V_1 \subset V_2 \subset \cdots \subset V_q$$

  is a sequence of invariant subspaces of S_{n-1}.

  In addition, inside V_{i+1}, V_i and $V_i^\perp$ (orthogonal complement of V_i) are invariant subspaces of S_{n-1}.
- An irreducible representation of S_n that corresponds to a Young diagram λ with n boxes is the direct sum of irreducible representations of Young diagrams obtained by subtracting a box from λ.

2.9 Fourier Transform

Let $f : S_n \to \mathbb{C}$ be a function. The Fourier transform $\hat{f}$ of f is defined as

$$\hat{f}(\lambda) = \sqrt{\frac{d_\lambda}{|S_n|}} \sum_{g \in S_n} f(g)\lambda(g)$$

for each $\lambda \in \Lambda_n$ (cf. [12, page 615]). The Fourier transform can be expressed as a matrix form

$$\mathfrak{F}_n = \sum_{\lambda \in \Lambda_n} \sum_{p,q \in \mathcal{P}(\lambda)} \sum_{g \in S_n} \sqrt{\frac{d_\lambda}{|S_n|}} [\lambda(g)]_{q,p} |\lambda, p, q\rangle\langle g|. \tag{2}$$

Here, $[\lambda(g)]_{q,p}$ is the (q, p) element of matrix representation of $\lambda(g)$. Notice that the index is not (p, q) but (q, p). Thanks to this definition, the algorithm proposed later becomes a little easier than that for the $\mathfrak{F}_n$ with $[\lambda(g)]_{p,q}$ as the coefficient.

The following are the concrete matrices of $\mathfrak{F}_2$ and $\mathfrak{F}_3$ calculated from the adapted Gel'fand-Tsetlin bases of the Specht polynomial defined above.

$$\mathfrak{F}_2 = \frac{1}{\sqrt{2}} \begin{pmatrix} 1 & 1 \\ 1 & -1 \end{pmatrix}$$

$$\mathfrak{F}_3 = \frac{1}{\sqrt{6}} \begin{pmatrix} 1 & 1 & 1 & 1 & 1 & 1 \\ \sqrt{2} & -\frac{1}{\sqrt{2}} & -\frac{1}{\sqrt{2}} & \sqrt{2} & -\frac{1}{\sqrt{2}} & -\frac{1}{\sqrt{2}} \\ 0 & \frac{\sqrt{3}}{\sqrt{2}} & -\frac{\sqrt{3}}{\sqrt{2}} & 0 & \frac{\sqrt{3}}{\sqrt{2}} & -\frac{\sqrt{3}}{\sqrt{2}} \\ 0 & -\frac{\sqrt{3}}{\sqrt{2}} & \frac{\sqrt{3}}{\sqrt{2}} & 0 & \frac{\sqrt{3}}{\sqrt{2}} & -\frac{\sqrt{3}}{\sqrt{2}} \\ \sqrt{2} & -\frac{1}{\sqrt{2}} & -\frac{1}{\sqrt{2}} & -\sqrt{2} & \frac{1}{\sqrt{2}} & \frac{1}{\sqrt{2}} \\ 1 & 1 & 1 & -1 & -1 & -1 \end{pmatrix}$$

It is easily seen that $\mathfrak{F}_2$ is the Hadamard matrix.

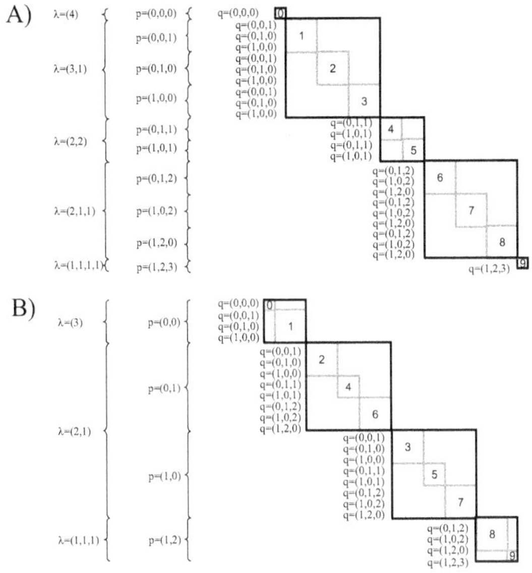

Figure 2: A) The form of $\rho_4(g)$, where $\rho_4 = \bigoplus_{\lambda \in \Lambda_4}(I_{d_\lambda} \otimes \lambda)$ and $g \in S_4$. It consists of one 1×1, three 3×3, two 2×2, three 3×3, and one 1×1 block diagonal matrices. The numbers 0-9 written in the submatrices are matrix numbers. Matrices 1, 2, and 3 are the same. Similarly, matrices 4 and 5 are the same, and matrices 6, 7, and 8 are the same. B) The form of $T_4^\dagger \rho_4(g) T_4$, which consists of a replacement of submatrices in panel A. The same matrix numbers in A and B mean the matrices are the same.

3. FFT ALGORITHM

3.1 Basis Transform T_n

A matrix for a basis transform, denoted T_n, is introduced before decomposing $\mathfrak{F}_n$ to simpler matrices. The T_n is an $n! \times n!$ matrix for a basis transform from the basis

$$\mathcal{B}_n = \{|\lambda, p, q\rangle | \lambda \in \Lambda_{n-1}, p \in \mathcal{P}(\lambda), \exists \mu(\lambda \searrow \mu \wedge q \in \mathcal{P}(\mu))\} \quad (3)$$

to the basis

$$\mathcal{B}'_n = \{|\lambda', p', q'\rangle | \lambda' \in \Lambda_n, p', q' \in \mathcal{P}(\lambda')\}.$$

It is defined as $T_n^\dagger |\lambda', p', q'\rangle = |\lambda, p, q\rangle$, where $q = q'$, $p = p'|_{n-2}$, and $p \in \mathcal{P}(\lambda)$. Here, $p'|_i$ is the restriction to the first i elements, i.e., $p'|_{n-2} = (p_1, p_2, \cdots, p_{n-2})$ when $p' = (p_1, p_2, \cdots, p_{n-1})$. It is easy to show that $T_n^\dagger$ is a one-to-one onto map from $\mathcal{B}'_n$ to $\mathcal{B}_n$. Hence, T_n is an $n! \times n!$ matrix with elements; zero or one.

Since Λ_n is the set of all irreducible representations of S_n, an element $g \in S_n$ can be written as $\bigoplus_{\lambda \in \Lambda_n} (I_{d_\lambda} \otimes \lambda(g))$. By the definition of T_n,

$$T_n^\dagger \left(\bigoplus_{\lambda \in \Lambda_n} (I_{d_\lambda} \otimes \lambda(g)) \right) T_n = \bigoplus_{\mu \in \Lambda_{n-1}} (I_{d_\mu} \otimes (\oplus_{\mu \searrow \lambda} \lambda(g))). \quad (4)$$

An example for $n = 4$ is given in Figure 2.

3.2 Inductive Decomposition of $\mathfrak{F}_n$

The H_n is defined by

$$H_n = T_n^\dagger \mathfrak{F}_n (\mathfrak{F}_{n-1} \otimes I_n)^\dagger. \quad (5)$$

Then, since $\mathfrak{F}_n = T_n H_n (\mathfrak{F}_{n-1} \otimes I_n)$,

$$\begin{aligned} \mathfrak{F}_n &= T_n H_n (\mathfrak{F}_{n-1} \otimes I_n) \\ &= T_n H_n (T_{n-1} H_{n-1} (\mathfrak{F}_{n-2} \otimes I_{n-1}) \otimes I_n) \\ &= \cdots. \end{aligned}$$

Hence, $\mathfrak{F}_n$ can be calculated from T_m, H_m $(m \le n)$. This is the existing FFT algorithm [3, 6].

It is proved that H_n is an $n! \times n!$ matrix consisting of $nd_\lambda \times nd_\lambda$ matrices with duplication of d_λ for $\lambda \in \Lambda_{n-1}$ on the basis $\mathcal{B}_n$. For example, for $n = 4$, the form of H_4 is the area inside the thick frame of Figure 2, panel B.

The sizes of submatrices of H_n, which are $nd_\lambda \times nd_\lambda$ for $\lambda \in \Lambda_n$, are smaller than the size of $\mathfrak{F}_n$, which is $n! \times n!$; however, nd_λ grows exponentially when n gets large. We will decompose H_n into a multiplication of $O(n^2)$ sparse matrices.

3.3 Inducing H'_{n-1} from H_{n-1}

We will introduce $n! \times n!$ matrices K_n, P_n, and A_n such that H_n is calculated by a multiplication of $\{K_n, P_n, A_n, T_n, H_{n-1}\}$. However, since H_n is an $n! \times n!$ matrix, the matrix sizes of H_n and H_{n-1} are different. To fix the problem, an $n! \times n!$ matrix H'_{n-1} is induced from H_{n-1}.

By (5), the input and output of H_{n-1} are defined on the bases $\mathcal{B}'_{n-2} \times \mathbb{Z}_{n-1}$ and $\mathcal{B}_{n-1}$, respectively. Since $\mathcal{B}_{n-1} \cong \mathcal{B}'_{n-2} \times \mathbb{Z}_{n-1}$, which is defined by the order of encoding, we can consider both the input and output of H_{n-1} to be defined on the basis $\mathcal{B}'_{n-2} \times \mathbb{Z}_{n-1}$, i.e., $H_{n-1} = \bigoplus_{\lambda \in \Lambda_{n-2}} (I_{d_\lambda} \otimes H_{n-1,\lambda})$, where $H_{n-1,\lambda}$ is an $(n-1)d_\lambda \times (n-1)d_\lambda$ matrix defined on the basis $\{|\lambda, p, q, i\rangle | q \in \mathcal{P}(\lambda), i < n-1\}$ for some $p \in \mathcal{P}(\lambda)$. Hence, there is a set of numbers $\{h_{\lambda,q,r,i,j} | q, r \in \mathcal{P}(\lambda), i, j < n-1\}$ such that

$$H_{n-1,\lambda} = \sum_{q,r \in \mathcal{P}(\lambda)} \sum_{i,j<n-1} h_{\lambda,q,r,i,j} |\lambda, p, q, i\rangle\langle\lambda, p, r, j|.$$

Define $H'_{n-1,\lambda}$ as

$$\sum_{q \in \mathcal{P}(\lambda)} |\lambda, p, q, 0\rangle\langle\lambda, p, q, 0| + \sum_{q,r \in \mathcal{P}(\lambda)} \sum_{i,j<n-1} h_{\lambda,q,r,i,j} |\lambda, p, q, i+1\rangle\langle\lambda, p, r, j+1|.$$

The $H'_{n-1,\lambda}$ is an $nd_\lambda \times nd_\lambda$ matrix, where $\lambda \in \Lambda_{n-2}$.

The H'_{n-1} is then defined as

$$\bigoplus_{\mu \in \Lambda_{n-1}} (I_{d_\mu} \otimes (\oplus_{\lambda \searrow \mu} H'_{n-1,\lambda})).$$

3.4 Basis Transform P_n

The matrix sizes of H_n and H'_{n-1} are the same; however, the output bases are different. i.e., the output bases of H_n and H'_{n-1} are defined on $\mathcal{B}_n$ and $\mathcal{B}'_{n-1} \times \mathbb{Z}_n$, respectively. To fix the difference, we introduce a basis transform P_n that changes the basis $\mathcal{B}'_{n-1} \times \mathbb{Z}_n$ to the basis $\mathcal{B}_n$.

The basis transform P_n is defined as a multiplication of two matrices $P_{n,1}$ and $P_{n,2}$, i.e., $P_n = P_{n,2} P_{n,1}$.

$P_{n,1}$ is $\bigoplus_{\lambda \in \Lambda_{n-1}} (I_{d_\lambda} \otimes P_{n,1,\lambda})$, where $P_{n,1,\lambda} = \sum_{x,y=0}^{nd_\lambda - 1} a_{x,y} |x\rangle\langle y|$ is the $nd_\lambda \times nd_\lambda$ matrix defined by

$$a_{x,y} = \begin{cases} 1 & \text{if } x < d_\lambda \wedge y = nx \\ 1 & \text{if } x \ge d_\lambda \wedge y = x - \lfloor \frac{nd_\lambda - x - 1}{n-1} \rfloor \\ 0 & \text{o.w.} \end{cases}.$$

$P_{n,2}$ is $\bigoplus_{\lambda\in\Lambda_{n-1}}(I_{d_\lambda}\otimes P_{n,2,\lambda})$ and $P_{n,2,\lambda}$ is determined as follows. Fix $\lambda\in\Lambda_{n-1}$ in the following argument. Enumerate the elements of $\{\mu\in\Lambda_n|\lambda\searrow\mu\}$ in the decreasing lexicographic order, i.e., from left to right in Figure 1. We denote the sequence as $\lambda_\downarrow$. For each element $\mu\in\lambda_\downarrow$, enumerate $\{\lambda'\in\Lambda_{n-1}|\lambda'\searrow\mu\}$ in the decreasing lexicographic order and denote it $\mu_\uparrow$. Then, we define $\lambda_{\downarrow\uparrow}$ as the sequence with repetition by concatenating $\mu_\uparrow$ for $\mu\in\lambda_\downarrow$. Similarly, we define $\lambda_{\uparrow\downarrow}$ as the sequence with repetition by concatenating $\nu_\downarrow$ for $\nu\in\lambda_\uparrow$. When we compare $\lambda_{\downarrow\uparrow}$ and $\lambda_{\uparrow\downarrow}$ as sets with duplication, it is easily proved that $\lambda_{\downarrow\uparrow}$ is $\lambda_{\uparrow\downarrow}$ plus λ. Hence, there are one-to-one maps from $\{\lambda_{\downarrow\uparrow}\}$ to $\{\lambda,\lambda_{\uparrow\downarrow}\}$. Select one of them and denote it $\mathfrak{f}_\lambda$, i.e., $\mathfrak{f}_\lambda(i)=j$ means that the ith element of $\{\lambda_{\downarrow\uparrow}\}$ is equal to the jth element of $\{\lambda,\lambda_{\uparrow\downarrow}\}$ for all $i=0,1,\cdots,|\lambda_{\downarrow\uparrow}|-1$. Then, define $P_{n,2,\lambda}=(b_{i,j})_{i,j=0}^{|\lambda_{\downarrow\uparrow}|-1}$ by

$$b_{i,j}=\begin{cases}I_{d_{\lambda_i}} & \text{if } \mathfrak{f}_\lambda(i)=j\\ 0 & \text{o.w.}\end{cases},$$

where $b_{i,j}$ is a $d_{\lambda_i}\times d_{\lambda_j}$ matrix such that λ_i is the ith element of $\lambda_{\downarrow\uparrow}$ and λ_j is the jth element of $\lambda_{\uparrow\downarrow}$.

P_n depends on the selection of $\mathfrak{f}_\lambda$; however, the key lemma (Lemma 1) holds for any selection of $\mathfrak{f}_\lambda$.

3.5 Controlled Cyclic Permutation K_n

The H_n and $P_nH'_{n-1}$ then have the same input and output bases. We can prove that for any $g\in S_{n-1}$, both $H_n^\dagger\cdot T_n^\dagger\rho_n(g)T_n\cdot H_n$ and $(P_nH'_{n-1})^\dagger\cdot T_n^\dagger\rho_n(g)T_n\cdot(P_nH'_{n-1})$ have non-zero elements at almost the same positions of the matrices. However, the values are different. To fix the difference, we introduce the following K_n defined as

$$K_n=\sum_{i=0}^{n-1}\left(\rho_{n-1}(c_{(i,i+1,\cdots,n-1)})\otimes|i\rangle\langle i|\right). \qquad (6)$$

Here, $c_{(i,i+1,\cdots,n-1)}$ is the cyclic permutation $i\mapsto i+1\mapsto\cdots\mapsto n-1\mapsto i$ when $0<i<n-1$ and is the identity if $i=0$ or $n-1$. Then, $H_n^\dagger\cdot T_n^\dagger\rho_n(g)T_n\cdot H_n$ and $(P_nH'_{n-1}K_n)^\dagger\cdot T_n^\dagger\rho_n(g)T_n\cdot(P_nH'_{n-1}K_n)$ are the same for any $g\in S_{n-1}$.

3.6 Residue A_n

Finally, A_n is defined as $H_nK_n^\dagger H'^\dagger_{n-1}P_n^\dagger$. Obviously, $H_n=A_nP_nH'_{n-1}K_n$. The following is the key lemma.

LEMMA 1. *For any $g\in S_{n-1}$, $T_n^\dagger\rho_n(g)T_n$ and A_n are commutative.*

Roughly speaking, Lemma 1 means that A_n is close to the identity matrix. The following theorem shows that the $n!$-dimensional Hilbert space can be separated into many small-dimensional invariant subspaces of A_n. Let e_λ be the number of children of λ in the Bratteli diagram. Obviously, $e_\lambda\le n$ for $n\ge 2$.

THEOREM 1. *For $\lambda\in\Lambda_{n-1}$ and $p,q\in\mathcal{P}(\lambda)$, let $W_{\lambda,p,q}$ be the e_λ-dimensional subspace spanned by $\{|\lambda,p,(q,x)\rangle\in\mathcal{B}_n|\lambda_p=\lambda_q\}$, where λ_p is the last node of p and (q,x) is $(q_1,\cdots,q_{n-2},x)$ when $q=(q_1,\cdots,q_{n-2})$. Then, $A_nW_{\lambda,p,q}\subseteq W_{\lambda,p,q}$ for any $\lambda\in\Lambda_{n-1}$ and $p,q\in\mathcal{P}(\lambda)$. Furthermore, for each $\lambda\in\Lambda_{n-1}$, there exists an $e_\lambda\times e_\lambda$ matrix $A'_{n,\lambda}$, independently of p and q, such that $A'_{n,\lambda}=A_n$ on $W_{\lambda,p,q}$. In addition, A_n performs as the identity on the subspace $\bigcap\{W^\perp_{\lambda,p,q}|\lambda\in\Lambda_{n-1},p,q\in\mathcal{P}(\lambda)\}$.*

The A_n was defined as $H_nK_n^\dagger H'^\dagger_{n-1}P_n^\dagger$; however, for calculating A_n, it is not necessary to multiply H_n, $K_n^\dagger$, $H'^\dagger_{n-1}$, and $P_n^\dagger$. By Theorem 1, A_n can be written as a direct sum of I_1's (the identity operator on a one-dimensional Hilbert space) and $e_\lambda\times e_\lambda$ matrices $A'_{n,\lambda}$ with duplication of d_λ^2 for all $\lambda\in\Lambda_{n-1}$. Theorem 2 shows that $A'_{n,\lambda}$ can be calculated easily from the diagonal elements of $T_n^\dagger\rho_n(c_{(n-1,n)})T_n$ and $\{d_\lambda\}_\lambda$. Note that $T_n^\dagger\rho_n(c_{(n-1,n)})T_n$ is $\bigoplus_{\lambda\in\Lambda_{n-1}}\left(I_{d_\lambda}\otimes\left(\oplus_{\lambda\searrow\mu}\mu(c_{(n-1,n)})\right)\right)$ and $\oplus_{\lambda\searrow\mu}\mu(c_{(n-1,n)})$ is

$$\sum_{\lambda\searrow\mu}\sum_{q,r\in\mathcal{P}(\mu)}[\mu(c_{(n-1,n)})]_{q,r}|\lambda,p,q\rangle\langle\lambda,p,r|$$

for some $p\in\mathcal{P}(\lambda)$.

THEOREM 2. *For (μ,ν) such that $\nu\searrow\lambda\searrow\mu$, define $q_{\mu,\nu}\in\mathcal{P}(\mu)$ as follows: $q_{\mu,\nu}|_{n-3}\in\mathcal{P}(\nu)$ and $q_{\mu,\nu}|_{n-2}\in\mathcal{P}(\lambda)$. Then, $A'_{n,\lambda}$ is equal to*

$$\sum_{\lambda\searrow\mu}\sum_{\nu\searrow\lambda}\sqrt{\frac{(n-1)d_\mu d_\nu}{nd_\lambda^2}}[\mu(c_{(n-1,n)})]_{q_{\mu,\nu},q_{\mu,\nu}}|\lambda,\mu\rangle\langle\lambda,\nu|$$
$$+\sum_{\lambda\searrow\mu}\sqrt{\frac{d_\mu}{nd_\lambda}}|\lambda,\mu\rangle\langle\lambda,\lambda|.$$

3.7 Algorithm and Complexity

We have the following relations.

(a) $H_2=\mathfrak{F}_2=H$ (the Hadamard matrix)

(b) $H_n=A_nP_nH'_{n-1}K_n$ for $n\ge 3$

(c) $\mathfrak{F}_n=T_nH_n(\mathfrak{F}_{n-1}\otimes I_n)$ for $n\ge 3$

$\mathfrak{F}_n$ for $n\ge 2$ can then be expressed as a multiplication of induced matrices from $\{A_m,P_m,K_m|m\le n\}$ and H as follows. By (b), $H_n=A_nP_nH'_{n-1}K_n=A_nP_n\,(A_{n-1}P_{n-1}H'_{n-2}\,K_{n-1})'\,K_n=\cdots=A_nP_nA^{(1)}_{n-1}P^{(1)}_{n-1}\cdots H^{(n-2)}\cdots K^{(1)}_{n-1}K_n$, where, e.g., $H^{(i)}=H''^{\cdots}$ (i times $'$). Substitute this H_n into (c), then H_n can be eliminated from (c). Since the obtained relation calculates $\mathfrak{F}_n$ from $\mathfrak{F}_{n-1}$, $\mathfrak{F}_n$ can be calculated inductively.

Suppose we are given $f:S_n\to\mathbb{C}$. Let $|f\rangle$ be the $n!$-dimensional vector corresponding to f. The Fourier transform $\mathfrak{F}_n|f\rangle$ is calculated by applying the matrices in the above matrix decomposition of $\mathfrak{F}_n$ to $|f\rangle$ one by one.

We evaluate the complexity of the algorithm. The complexity is counted as the number of multiplications and additions, which is the same rule as [6]. The total complexity will be shown as $O(n!n^3)$. The key is that all $A_m^{(n-m)}$'s, $P_m^{(n-m)}$'s, $K_m^{(n-m)}$'s, and $H^{(n-2)}$ are sparse matrices.

It suffices to show that the complexity for calculating $H_n|f\rangle$ from $|f\rangle$ is $O(n!n^2)$ because T_n is just an order change of elements of the vector, which can be performed by copying $n!$ values.

Since $c_{(i,i+1,\cdots,n-1)}=c_{(n-2,n-1)}\cdots c_{(i+1,i+2)}c_{(i,i+1)}$, K_n is calculated by a multiplication of $n-1$ matrices of adjacent transpositions. Each adjacent transposition is expressed as a direct sum of 1×1 and 2×2 matrices in Young's orthogonal representation. Hence, the complexity for calculating $K_n|f\rangle$ from $|f\rangle$ is $O(n!n)$. Similarly, the complexity for calculating $K_m^{(n-m)}|f\rangle$ ($m\le n$) from $|f\rangle$ is $O((n-1)!m^2)$. $H^{(n-2)}$ can

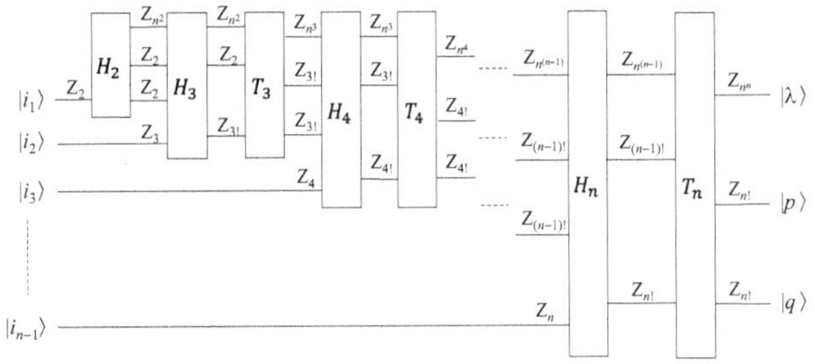

Figure 3: Circuit for $\mathfrak{F}_n$, where H_2, T_m, and H_m are given in Figures 4, 5, and 6, respectively. The input is $|i_1, \cdots, i_{n-1}\rangle$ such that $g = c_{(1,2,\cdots,n)}^{i_{n-1}} \cdots c_{(1,2,3)}^{i_2} c_{(1,2)}^{i_1}$ for $g \in S_n$. The output is $|\lambda, p, q\rangle$ such that $\lambda \in \mathbb{Z}_{n^n}$ and $p, q \in \mathbb{Z}_{n!}$, which are codes of $\lambda \in \Lambda_n$ and $p, q \in \mathcal{P}(\lambda)$, respectively.

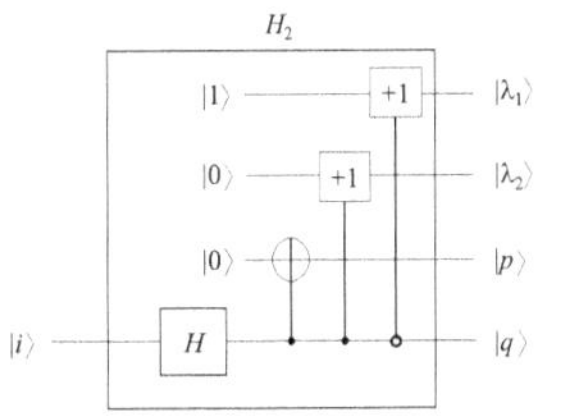

Figure 4: A quantum circuit for performing H_2. The input $|i\rangle$ is $|0\rangle$ or $|1\rangle$, which means the group element $c_{(1,2)}^0$ (the identity) or $c_{(1,2)}^1$ (the transposition $1 \leftrightarrow 2$) in S_2, respectively. The output is $\frac{1}{\sqrt{2}}(|(2,0),(0),(0)\rangle + |(1,1),(1),(1)\rangle$ or $\frac{1}{\sqrt{2}}(|(2,0),(0),(0)\rangle - |(1,1),(1),(1)\rangle$ according to the input $|i\rangle = |0\rangle$ or $|1\rangle$, respectively. The $+1$ gate performs an operation $|x\rangle \mapsto |x+1 \mod n\rangle$ on a qu$n$it.

be calculated with $O((n-1)!)$ operations. $P_m^{(n-m)}$ is an order change of elements of the vector, similar to $T_m^{(n-m)}$. Since $A_m^{(n-m)}$ is a direct sum of 1×1 and $e_\lambda \times e_\lambda$ matrices, where $\lambda \in \Lambda_m$ and $e_\lambda \leq m$, the complexity is $O(n!m)$. The complexity for calculating $H_n|f\rangle$ from $|f\rangle$ is then $O(n!n^2)$.

4. QFT ALGORITHM

The quantum counterpart of $\mathfrak{F}_n$ is called the quantum Fourier transform (QFT), which is defined the same as $\mathfrak{F}_n$ [See (2)]. We will use the same symbol $\mathfrak{F}_n$ since there's no confusion.

We propose a quantum circuit that performs $\mathfrak{F}_n$, shown in Figure 3. The circuit is $O(n^4)$-depth on $O(n \log n)$ qubits, where gates represented as a direct sum of 1×1 and 2×2 matrices whose elements are easily calculated in a classical computer are defined as elementary gates. Basic notions and symbols of quantum circuits are found in [12]. The words "qu$m$it" and "qu$n$it" will be used in the meaning of m-state and n-state quantum resources, respectively.

4.1 Algorithm

By the induction relation (c) in Subsection 3.7, $\mathfrak{F}_n$ can be calculated by $\mathfrak{F}_2$ $(= H_2)$, H_3, T_3, H_4, T_4, $\cdots$, H_n, and T_n. This can be depicted as the circuit in Figure 3.

The input of the $\mathfrak{F}_n$ circuit is $(i_1, i_2, \cdots, i_{n-1}) \in \mathbb{Z}_2 \times \mathbb{Z}_3 \times \cdots \times \mathbb{Z}_n$ such that $g = c_{(1,2,\cdots,n)}^{i_{n-1}} \cdots c_{(1,2,3)}^{i_2} c_{(1,2)}^{i_1}$. The output is (λ, p, q) such that λ is a partition of n, where p and q are paths from the root node to the node λ in the Bratteli diagram. A partition of n is encoded using n numbers less than n. Hence, λ is encoded as $(\lambda_1, \lambda_2, \cdots, \lambda_n) \in \mathbb{Z}_n \times \mathbb{Z}_n \times \cdots \times \mathbb{Z}_n$. In addition, p is encoded as $(p_1, p_2, \cdots, p_{n-1}) \in \mathbb{Z}_2 \times \mathbb{Z}_3 \times \cdots \times \mathbb{Z}_n$ and q is encoded as $(q_1, q_2, \cdots, q_{n-1}) \in \mathbb{Z}_2 \times \mathbb{Z}_3 \times \cdots \times \mathbb{Z}_n$. Therefore, $(\lambda, p, q) \in \mathbb{Z}_{n^n} \times \mathbb{Z}_{n!} \times \mathbb{Z}_{n!}$.

To compensate for the difference between the input and output, fresh quantum resources are added in the circuits for H_2, T_3, T_4, $\cdots$, T_n. The circuits for H_2, H_m, and T_m $(m \geq 3)$ are given below. The ordered set, e.g., (λ, p, q) will be denoted using the ket symbol $|\lambda, p, q\rangle$ hereafter.

4.2 Quantum Circuit for H_2

The H_2 is the unitary transform that performs

$$|0\rangle \mapsto \frac{1}{\sqrt{2}}(|(2,0),(0),(0)\rangle + |(1,1),(1),(1)\rangle,$$

$$|1\rangle \mapsto \frac{1}{\sqrt{2}}(|(2,0),(0),(0)\rangle - |(1,1),(1),(1)\rangle.$$

It can be performed by the circuit shown in Figure 4, where the first and second horizontal lines are qu$n$its and the third and fourth lines are qubits. Hence, two qu$n$its and one qubit are added as fresh quantum resources in the circuit.

4.3 Quantum Circuit for T_m

A quantum circuit that performs T_m consists of two parts. (See Figure 5.)

In the first part, a qu$m$it with the initial state $|0\rangle$ is added as p_{m-1} and changed to $q \backslash p$, where $q \backslash p$ means the setminus $\{q_1, q_2, \cdots, q_{m-1}\} \setminus \{p_1, p_2, \cdots, p_{m-2}\}$. By this operation, $(p_1, p_2, \cdots, p_{m-1})$ is in $\mathcal{P}(\lambda)$. The $q \setminus p$ is easily calculated by an addition circuit for $\sum_{i=1}^{m-1} q_i - \sum_{i=1}^{m-2} p_i \mod m$.

In the second part, a qu$n$it with the initial state $|0\rangle$ is added as λ_m and the number of $\lambda_{q \backslash p}$ is incremented by one. It can be performed using controlled addition circuits; $|\lambda_i\rangle$ is changed to $|\lambda_i + 1 \mod n\rangle$ when $p_{m-1} = i$.

4.4 Quantum Circuit for H_m

The circuit that performs H_m has an input $|\lambda, p, q, i\rangle$, where $\lambda = (\lambda_1, \cdots, \lambda_{m-1}) \in \mathbb{Z}_{n^{m-1}}$, $p = (p_1, \cdots, p_{m-2}) \in \mathbb{Z}_{(m-1)!}$, $q = (q_1, \cdots, q_{m-2}) \in \mathbb{Z}_{(m-1)!}$, $|i\rangle \in \mathbb{Z}_m$, and an output $|\lambda', p', q'\rangle$, where $\lambda' = (\lambda'_1, \cdots, \lambda'_{m-1}) \in \mathbb{Z}_{n^{m-1}}$, $p' = (p'_1, \cdots, p'_{m-2}) \in \mathbb{Z}_{(m-1)!}$, $q' = (q'_1, \cdots, q'_{m-1}) \in \mathbb{Z}_{m!}$.

Since $H_m = A_m P_m H'_{m-1} K_m$, the circuit is constructed as shown in Figure 6. Some notes are listed below.

K_m is the set of $\lambda(c_{(i,\cdots,m-1)})$ gates according to the value i of the fourth register. Hence, they are put in line in the first part of H_m.

Before performing H_{m-1}, the basis must be changed; $\lambda \mapsto (\lambda_1, \cdots, \lambda_{q_{m-2}} - 1, \cdots, \lambda_{m-1})$, $p \mapsto p|_{m-3}$, $q \mapsto q|_{m-3}$, and $i \mapsto i - 1 \mod m$. The change of λ is performed by the reverse operation of the second stage of T_{m-1}. The change of i is performed by an addition circuit. After performing H_{m-1}, the basis is restored to the previous point of the change.

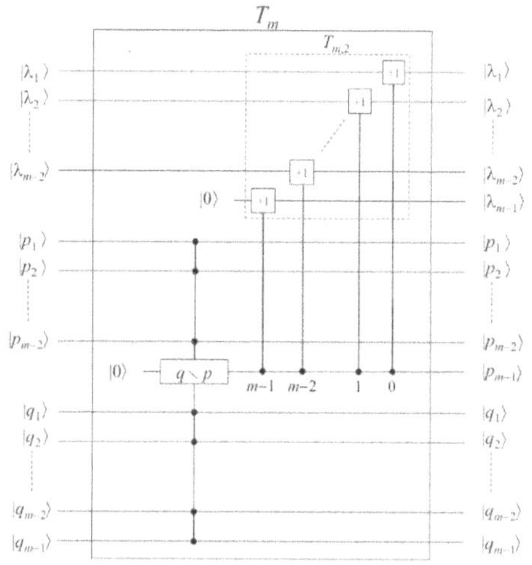

Figure 5: Quantum circuit for performing T_m. The $q \setminus p$ is the setminus $\{q_1, q_2, \cdots, q_{m-1}\} \setminus \{p_1, p_2, \cdots, p_{m-2}\}$, which can be calculated as $\sum_{i=1}^{m-1} q_i - \sum_{j=1}^{m-2} p_j \mod m$. The numbers on the p_{m-1} qu$m$it mean the conditions. The operation in the dashed line is named $T_{m,2}$.

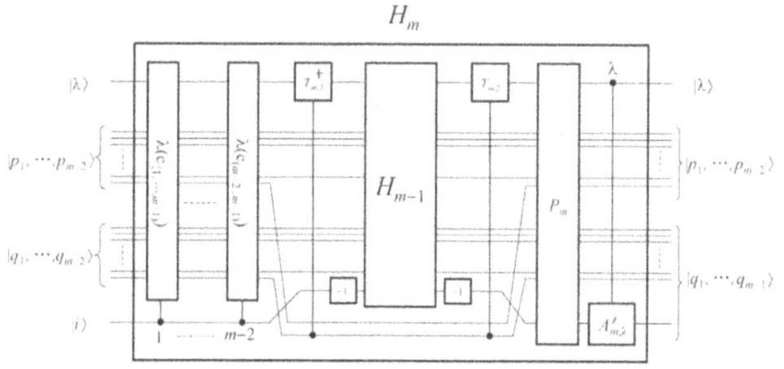

Figure 6: Quantum circuit for performing H_m, which contains H_{m-1} as a nested part. The gate $T_{m,2}$ is the operation in Figure 5. The -1 and $+1$ gates are $|i\rangle \mapsto |i-1 \mod m\rangle$ and $|i\rangle \mapsto |i+1 \mod m\rangle$, respectively.

P_n is a simple basis change, which can be performed in $O(n^2)$. A_m is the $A'_{m,\lambda}$ operation according to the value of $\lambda \in \Lambda_m$. Each $A'_{m,\lambda}$ is an $e_\lambda \times e_\lambda$ matrix, where $e_\lambda \leq m$. It is known that a $2^n \times 2^n$ unitary matrix can be performed using 4^n elementary gates. Thus, an $m \times m$ unitary matrix can be performed using m^2 elementary gates, which is $O(n^2)$.

4.5 Complexity

The circuit for H_n can be performed in $O(n^3)$. By Figure 3, the complexity of $\mathfrak{F}_n$ is then $O(n^4)$.

5. CONCLUSION

This paper proposed a QFT algorithm over symmetric groups. The time complexity of the algorithm is $O(n^4)$ for S_n. We also proposed an $O(n!n^3)$ FFT (classical) algorithm over symmetric groups. Estimating the time complexity of a QFT circuit using only elementary gates, such as single-qubit rotations and controlled-not gates, remains as further research. Another interesting problem is whether the new approach can be generalized to other groups.

6. ACKNOWLEDGEMENTS

This work was supported by JSPS KAKENHI Grant Number 25330021.

7. REFERENCES

[1] G. Alagic, C. Moore, and A. Russell. Quantum algorithms for Simon's problem over nonabelian groups. *ACM Transactions on Algorithms*, 6(1):No. 19, December 2009.

[2] R. Beals. Quantum computation of Fourier transforms over symmetric groups. In *Proceedings of the twenty-ninth annual ACM Symposium on the Theory of Computing (STOC)*, pages 48–53. ACM, May 1997.

[3] M. Clausen. Fast generalized Fourier transforms. *Theoret. Comput. Sci.*, 67:55–63, 1989.

[4] R. Cleve and J. Watrous. Fast parallel circuits for the quantum Fourier transform. In *Proceedings of the 41st Annual IEEE Symposium on Foundations of Computer Science (FOCS)*, pages 526–536. IEEE, 2000.

[5] D. Coppersmith. An approximate Fourier transform useful in quantum factoring. In *Technical Report RC19642*. IBM, 1994.

[6] P. Diaconis and D. Rockmore. Efficient computation of the Fourier transform on finite groups. *J. AMS*, 3(2):297–332, 1990.

[7] L. Hales and S. Hallgren. An improved quantum Fourier transform algorithm and applications. In *Proceedings of the 41st Annual IEEE Symposium on Foundations of Computer Science (FOCS)*, pages 515–525. IEEE, 2000.

[8] M. Hall. *The Theory of Groups*. Macmillan, New York, 1959.

[9] C. Moore, D. Rockmore, and A. Russell. Generic quantum Fourier transforms. *ACM Transactions on Algorithms*, 2(4):707–723, October 2006.

[10] C. Moore, A. Russell, and P. Sniady. On the impossibility of a quantum sieve algorithm for graph isomorphism: unconditional results. In *Proceedings of the thirty-ninth annual ACM symposium on Theory of computing (STOC)*, pages 536–545. ACM, 2007.

[11] M. Mosca and C. Zalka. Exact quantum Fourier transforms and discrete logarithm algorithms. *Int. J. Quant. Inf.*, 2(1):91–100, 2004.

[12] M. A. Nielsen and I. L. Chuang. *Quantum Computation and Quantum Information*. Cambridge Univ. Press, Cambridge, UK, 2000.

[13] J.-P. Serre. *Représentations Linéaires des Groupes Finis*. Hermann, Paris, 1971.

[14] P. W. Shor. Polynomial-time algorithms for prime factorization and discrete logarithms on a quantum computer. *SIAM J. Comput.*, 26(5):1484–1509, October 1997.

[15] H. Weyl. *The Classical Groups, their invariants and representations (2nd ed.)*. Princeton Univ. Press, Princeton, 1946.

[16] H. Weyl. *The Theory of Groups and Quantum Mechanics*. Dover, New York, 1950.

Second Order Differential Equations with Hypergeometric Solutions of Degree Three

Vijay Jung Kunwar*, Mark van Hoeij*
Department of Mathematics, Florida State University
Tallahassee, FL 32306-3027, USA
vkunwar@math.fsu.edu, hoeij@math.fsu.edu

ABSTRACT

Let L be a second order linear homogeneous differential equation with rational function coefficients. The goal in this paper is to solve L in terms of hypergeometric function ${}_2F_1(a, b; c \mid f)$ where f is a rational function of degree 3.

Categories and Subject Descriptors

I.1.2 [**Symbolic and Algebraic Manipulation**]: Algorithms; G.4 [**Mathematics of Computing**]: Mathematical Software

General Terms

Algorithms

Keywords

Symbolic Computation, Differential Equations, Closed Form Solutions

1. INTRODUCTION

A linear differential equation with rational function coefficients corresponds to a differential operator $L \in \mathbb{C}(x)[\partial]$ where $\partial = \frac{d}{dx}$. For example, if $L = a_2\partial^2 + a_1\partial + a_0$ is a differential operator with $a_2, a_1, a_0 \in \mathbb{C}(x)$, then the corresponding differential equation $L(y) = 0$ is $a_2y'' + a_1y' + a_0y = 0$. We assume that L has no Liouvillian solutions, otherwise L can be solved quickly using Kovacic's algorithm [7].

Definition 1.1. *If $S(x)$ is a special function that satisfies a differential operator L_S (called a base equation) of order n, then a function y is called a linear S-expression if there exist algebraic functions $f, r, r_0, r_1, \ldots$ such that $y =$*

$$\exp(\int r\,dx) \cdot \left(r_0S(f) + r_1S(f)' + \cdots + r_{n-1}S(f)^{(n-1)}\right). \quad (1)$$

More generally, we say that y can be expressed in terms of S if it can be written in terms of expressions of the form (1), using field operations and integrals.

Higher derivatives are not needed in (1) since they are linear combinations of $S(f), S(f)', \ldots, S(f)^{(n-1)}$. If $L_S \in \mathbb{C}(x)[\partial]$ is of order n and $k = \mathbb{C}(x, r, f, r_0, r_1, \ldots) \subseteq \overline{\mathbb{C}(x)}$ then y satisfies an equation $L \in k[\partial]$ of order $\leq n$.

If $L \in \mathbb{C}(x)[\partial]$ has order 3 or 4, and S is a special function that satisfies a second order equation, then the problem of solving L in terms of S can be reduced, with an algorithm and implementation [12], to the problem of solving second order equations. This reduction of order motivates a focus on second order equations.

If y and S satisfy second order operators, then products of (1) are not needed, and the form reduces to

$$y = \exp(\int r\,dx) \cdot \left(r_0S(f) + r_1S(f)'\right). \quad (2)$$

Although form (2) looks technical, it is the most natural form to consider, because it is closed under the known transformations that send irreducible second order operators in $\mathbb{C}(x)[\partial]$ to second order linear operators. Given an input operator L_{inp} of order 2, finding a solution of the form (2) corresponds to finding a sequence of transformations that sends L_S to L_{inp} (or a right hand factor of L_{inp}, but we assume L_{inp} to be irreducible):

(i) Change of variables: $y(x) \mapsto y(f)$
(ii) Gauge transformation: $y \mapsto r_0y + r_1y'$
(iii) Exponential product: $y \mapsto \exp(\int r\,dx)$

The function f in (i) above is called the *pullback* function. These transformations are denoted as $\xrightarrow{f}_C$, $\xrightarrow{r_0,r_1}_G$ and $\xrightarrow{r}_E$ respectively. They send expressions in terms of S to expressions in terms of S. So any solver for finding solutions in terms of S, if it is complete, then it must be able to deal with all three transformations. In other words, it must be able to find any solution of the form (2).

The goal in this paper is the following:

Given $L_{inp} \in \mathbb{C}(x)[\partial]$, irreducible, order 2, find, if it exists, a nonzero solution of form (2) where $S(x) = {}_2F_1(a, b; c \mid x)$, $f, r, r_0, r_1 \in \mathbb{C}(x)$ and f has degree 3.

Given L_{inp}, our task is to find:

$$L_S \xrightarrow{f}_C M \xrightarrow{r_0,r_1}_G \xrightarrow{r}_E L_{inp}.$$

*Supported by NSF grant 1017880

There are algorithms [2] to find the transformations $\xrightarrow{r_0,r_1}_G$ and $\xrightarrow{r}_E$ but to apply them we first need M (or equivalently, f and L_S). Thus the crucial part is to compute f.
We compute f from the singularities of M. Since we do not yet know M, the only singularities of M that we know are those singularities of L_{inp} that can not *disappear* (turn into regular points) under transformations $\xrightarrow{r_0,r_1}_G$ and $\xrightarrow{r}_E$.

Definition 1.2. *A singularity is called* non-removable *if it stays singular under any combination of* $\xrightarrow{r_0,r_1}_G$ *and* $\xrightarrow{r}_E$.

A singularity $x = p$ of L_{inp} that can become a regular point under $\xrightarrow{r_0,r_1}_G$ and/or $\xrightarrow{r}_E$ need not be a singularity of M. Such singularities (*removable* singularities) provide no information about f. They include *apparent* singularities (singularities p where all solutions are analytic at $x = p$, such singularities can disappear under $\xrightarrow{r_0,r_1}_G$). More generally, if there exist functions u, y_1, y_2 with y_1, y_2 analytic at $x = p$ such that uy_1, uy_2 is a basis of local solutions of L at $x = p$, then $x = p$ is removable (such p can be sent to an apparant singularity with $\xrightarrow{r}_E$).

1.1 Motivation

Equations with a ${}_2F_1$-type solution are common. We examined integer sequences $u(0), u(1), u(2), \ldots$ from the Online Encyclopedia of Integer Sequences (oeis.org) for which $y = \sum_n u(n)x^n \in \mathbb{Z}[x]$ is (a) convergent, and (b) holonomic, meaning that y satisfies a linear differential operator L. Among the L's obtained this way, all second order L's (including dozens that had no Liouvillian solutions) turned out to have ${}_2F_1$-type solutions. For third order operators we used order $3 \to 2$ reduction [12] to find solutions of the form $\exp(\int r dx)\left(r_0(S(f))^2 + r_1 S(f)S(f)' + r_2(S(f)')^2\right)$, where $S(x) = {}_2F_1(a, b; c \,|\, x)$.
The key step to find ${}_2F_1$-type solutions is to find the pullback function f, and the ${}_2F_1$-parameters a, b, c. Classifying all *rational* functions $f \in \mathbb{C}(x)$ that can occur as a pullback function for some L with d *non-removable* singular points is ongoing work, see [8] for $d = 4$ and [9] for $d = 5$ (with at least one logarithmic singularity). For a fixed d, a large table is needed to ensure that we can solve every second order L with d singularities that has a ${}_2F_1$-type solution. But the table can be greatly reduced by developing algorithms such as 2-*descent* [6]. If f is a rational function with a degree 2 decomposition, then we can apply the 2-descent algorithm to L to reduce the degree of f in half.
After trying 2-descent, the next case is to solve every L that has a ${}_2F_1$-type solution where f is a rational function of degree 3. This is useful in its own right because it solves many equations, but it also significantly reduces the tabulation work that is needed (many f's from [8],[9] have a decomposition factor of degree 2 or 3).

1.2 Hypergeometric solutions, an example

Consider the operator $L =$

$$2\left(2x^2-1\right)\left(8x^2-1\right)\partial^2 + 4x\left(24x^2-7\right)\partial + 24x^2 - 3. \quad (3)$$

L can be solved in terms of ${}_2F_1(a, b; c \,|\, f)$ where f is a rational function of degree 3. We give one such solution (a second independent solution looks similar):

$$sol_L = (1-2x\sqrt{2})^{-\frac{1}{3}}(1+x\sqrt{2})^{-\frac{1}{6}} \cdot {}_2F_1\Big(\frac{1}{6}, \frac{1}{3}; \frac{5}{6} \,|\, f\Big) \quad (4)$$

where $f = \frac{(2x-\sqrt{2})(4x+\sqrt{2})^2}{(2x+\sqrt{2})(4x-\sqrt{2})^2}$.
L is defined over $\mathbb{Q}$, i.e. $L \in \mathbb{Q}(x)[\partial]$ but $f \notin \mathbb{Q}(x)$; instead $f \in \mathbb{Q}(\sqrt{2}, x)$. Such a field extension can only occur when f is not unique (replacing $\sqrt{2}$ by $-\sqrt{2}$ gives another solution). The non-uniqueness of f in this example is explained by the fact that L has a symmetry $x \mapsto -x$ (The change of variables $x \mapsto -x$ produces an operator L_{-x} that equals L). The change of variables $x \mapsto \sqrt{x}$ produces an operator $L_{\sqrt{x}}$ that is still in $\mathbb{Q}(x)[\partial]$ (this is a trivial case of 2-descent).

$$L_{\sqrt{x}} = x(2x-1)(8x-1)\partial^2 + (32x^2 - 12x + \frac{1}{2})\partial + 3x - \frac{3}{8}. \quad (5)$$

Our program produces the following solution of $L_{\sqrt{x}}$:

$$sol_{L_{\sqrt{x}}} = {}_2F_1\Big(\frac{1}{12}, \frac{1}{4}; \frac{1}{2} \,|\, 2x\,(8x-3)^2\Big). \quad (6)$$

Applying $x \mapsto x^2$ to (6) produces another solution of L:

$$Sol_L = {}_2F_1\Big(\frac{1}{12}, \frac{1}{4}; \frac{1}{2} \,|\, 2x^2\,(8x^2-3)^2\Big). \quad (7)$$

The pullback function f in (7) has degree 6, but is nicer than the degree 3 function in (4). So we chose in our implementation to search only for f's which are defined over the field of constants specified in the input, and to apply 2-descent for the cases where f requires a field extension.
The following diagrams show the impact of the change of variables $x \mapsto f$.

Notation: (see section 2 for more details and definitions).
p: non-removable singularity with exponent-difference Δ_p.
$H^{a,b}_{c,x}$: hypergeometric differential operator.

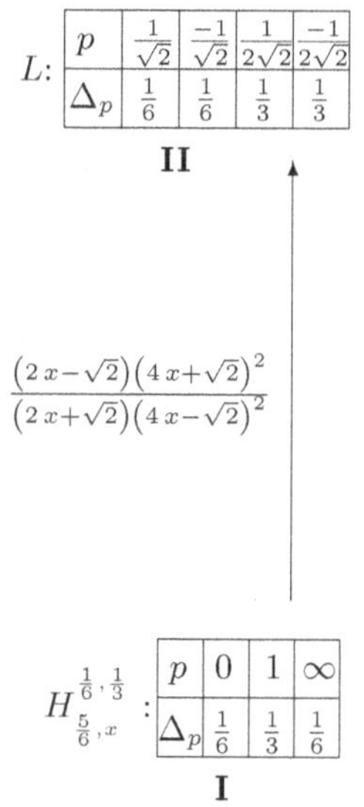

Diagram 1

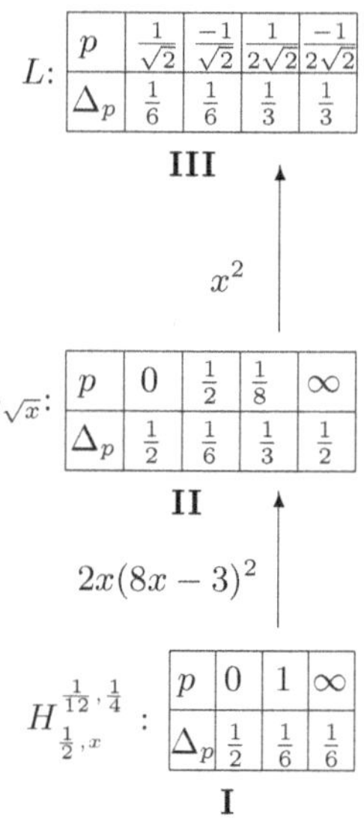

Diagram 2

The diagrams give the singularity structures (the p's and Δ_p's, see section 2 for definitions) of $L, L_{\sqrt{x}}$ and $H^{a,b}_{c,x}$. The hypergeometric function ${}_2F_1(a, b; c \,|\, x)$ is a solution of $H^{a,b}_{c,x}$ (see Section 2.2). Choosing $(a, b, c) = (\frac{1}{6}, \frac{1}{3}, \frac{5}{6})$ makes the exponent-differences of $H^{a,b}_{c,x}$ equal to $(\frac{1}{6}, \frac{1}{3}, \frac{1}{6})$. Then the pullback function $f = \frac{(2x-\sqrt{2})(4x+\sqrt{2})^2}{(2x+\sqrt{2})(4x-\sqrt{2})^2}$ in Diagram 1 sends the singularity structure of $H^{a,b}_{c,x}$ to that of L.

Diagram 2 has two changes of variables, $x \mapsto 2x(8x-3)^2$ produces the singularity structure of $L_{\sqrt{x}}$ from a hypergeometric equation with exponent-differences $(\frac{1}{2}, \frac{1}{6}, \frac{1}{6})$. After that, $x \mapsto x^2$ produces the singularity structure of L. Presenting (4) in the general form (2), we get $r = \frac{u'}{u} \in \mathbb{Q}(x, \sqrt{2})$ where $u = (1+2x\sqrt{2})^{-\frac{1}{3}}(1-x\sqrt{2})^{-\frac{1}{6}}$, $r_0 = 1$ and $r_1 = 0$. For (7), we also have $r_1 = 0$ (with $r = 0$, $r_0 = 1$). The case $r_1 \neq 0$ is more complicated as this involves integer shifts of the exponent-differences, see *section 2.4* for an example.

1.3 Goal of the paper

The examples in the above diagrams illustrate the effect of change of variables $x \mapsto f$ on the singularities and their exponent-differences; this paper will use that to reconstruct f whenever it has degree 3, a problem that turns out to consist of 18 cases. One of those cases (case sing5.2 in *section 4.2*) was treated previously in *section 2.6* in [4]. The goal in this paper is to cover all cases with f of degree 3. Combining this work with the 2-descent algorithm from [6] and the tables from [8, 9] leads to a solver that can find ${}_2F_1$-type solutions for many second order equations. The combined solver is very effective; it appears that closed form solutions exist for every second order L that has a non-zero convergent solution of the form $\sum_{n=0}^{\infty} u(n)x^n$ with $u(n) \in \mathbb{Z}$.

2. PRELIMINARIES AND NOTATIONS

This section gives a brief summary of prior results needed for this paper. Notations used throughout the paper:

C: a subfield of $\mathbb{C}$.
$L_{inp} \in C(x)[\partial]$: input differential operator.
$H^{a,b}_{c,x}$: Gauss hypergeometric differential operator.
$S(x) = {}_2F_1(a, b; c|x)$: hypergeometric function (a solution of $H^{a,b}_{c,x}$).
e_0, e_1, e_∞: exponent differences of $H^{a,b}_{c,x}$ at $0, 1, \infty$.
$f \in C(x)$: a rational function of degree 3.
$H^{a,b}_{c,f}$: obtained from $H^{a,b}_{c,x}$ by a change of variables $x \mapsto f$.
$S(f) = {}_2F_1(a, b; c\,|\,f)$: a solution of $H^{a,b}_{c,f}$.

2.1 Differential Operators and Singularities

A derivation $\partial = \frac{d}{dx}$ on $\mathbb{C}(x)$ produces a non-commutative ring $\mathbb{C}(x)[\partial]$. A differential operator of order n is an element $L \in \mathbb{C}(x)[\partial]$ of the form $L = \sum_{i=0}^{n} a_i \partial^i$, with $a_i \in \mathbb{C}(x)$ and $a_n \neq 0$. The solution space (set of all solutions in a *universal extension*) of L is denoted by $V(L) = \{y \,|\, L(y) = 0\}$.
After clearing denominators, we may assume $a_i \in \mathbb{C}[x]$. Then $p \in \mathbb{C}$ is called a *regular* (or *non-singular*) point when $a_n(p) \neq 0$. Otherwise it is called a *singular* point (or a *singularity*). The point $p = \infty$ is called regular if the change of variable $x \mapsto 1/x$ produces an operator $L_{1/x}$ with a regular point at $x = 0$. Given $p \in \mathbb{P}^1 = \mathbb{C} \cup \{\infty\}$, we define the local parameter as $t_p = \begin{cases} x - p & \text{if } x \neq \infty \\ \frac{1}{x} & \text{if } x = \infty. \end{cases}$

DEFINITION 2.1. *$L \in \mathbb{C}(x)[\partial]$ is called* Fuchsian *(or* regular singular*) if all its singularities are regular singularities. A singularity p is a* regular singularity *when:*
(1) (if $p \neq \infty$): $t_p^i \cdot \frac{a_{n-i}}{a_n}$ is analytic at p for $1 \leq i \leq n$.
(2) (if $p = \infty$): $L_{1/x}$ has a regular singularity at $x = 0$.
Otherwise p is called irregular singular.

This paper considers only Fuchsian operators, of order 2. For closed form solutions of non-Fuchsian equations (L having at least one irregular singularity) see [13].

THEOREM 2.2. *([10], Sections 2.1–2.4). Suppose L has order 2 and $p \in \mathbb{P}^1$. If $x = p$ is a regular singularity or a regular point of L, then there exists the following basis of $V(L)$ at $x = p$*

$y_1 = t_p^{e_1} \sum_{i=0}^{\infty} a_i t_p^i$, $a_0 \neq 0$ and

$y_2 = t_p^{e_2} \sum_{i=0}^{\infty} b_i t_p^i + c y_1 \log(t_p)$, $b_0 \neq 0$

where $e_1, e_2, a_i, b_i, c \in \mathbb{C}$ such that:
(i) If $e_1 = e_2$ then c must be non zero.
(ii) Conversely, if $c \neq 0$ then $e_1 - e_2$ must be in $\mathbb{Z}$.

NOTATION 2.3. *In Theorem 2.2:*

1. *If $c \neq 0$ then $x = p$ is called a* logarithmic *singularity.*
2. *e_1, e_2 are called the* exponents *of L at $x = p$ (they are computed as the roots of the* indicial equation*).*
3. *$\Delta_p(L) := \pm(e_1 - e_2)$ is the* exponent difference *at p.*
4. *If $\Delta_p(L_1) \equiv \Delta_p(L_2) \mod \mathbb{Z}$ then we say that $\Delta_p(L_1)$* matches *$\Delta_p(L_2)$.*

REMARK 2.4. *Logarithmic singularities are always non-removable (they stay logarithmic under the transformations in Definition 1.2). If $e_1 - e_2 \in \mathbb{Z}$ and $x = p$ is non logarithmic then $x = p$ is either a regular point or a removable singularity. Proofs and more details can be found in [5]. The relation with Theorem 2.2 is as follows:*
(1) $x = p$ is non-singular $\iff \{e_1, e_2\} = \{0, 1\}$ and $c = 0$.
(2) $x = p$ is non-removable $\iff c \neq 0$ or $e_1 - e_2 \notin \mathbb{Z}$.

DEFINITION 2.5. *The* singularity structure *of L is:*
$Sing(L) = \{(p, \Delta_p(L) \mod \mathbb{Z}) : p$ is non-removable$\}$.

It is often more convenient to express singularities in terms of monic irreducible polynomials.

DEFINITION 2.6. *Let C be a field of characteristic 0. places(C):= $\{f \in C[x] \,|\, f$ is monic and irreducible$\} \bigcup \{\infty\}$. The* degree *of a place p is 1 if $p = \infty$ and* $\deg(p)$ *otherwise.*

EXAMPLE 2.7. *Consider L in section 1.2.*

$$Sing(L) = \Big\{\Big(\frac{1}{\sqrt{2}}, \frac{1}{6}\Big), \Big(-\frac{1}{\sqrt{2}}, \frac{1}{6}\Big), \Big(\frac{1}{2\sqrt{2}}, \frac{1}{3}\Big), \Big(-\frac{1}{2\sqrt{2}}, \frac{1}{3}\Big)\Big\}.$$

In terms of places$(\mathbb{Q})$ it is written as:

$$Sing(L) = \Big\{\Big(x^2 - \frac{1}{2}, \frac{1}{6}\Big), \Big(x^2 - \frac{1}{8}, \frac{1}{3}\Big)\Big\}.$$

2.2 Gauss Hypergeometric Equation

The Gauss hypergeometric differential equation (GHE) is:

$$x(1-x)y'' + (c - (a+b+1)x)y' - aby = 0. \quad (8)$$

It has three regular singularities, with exponents $\{0, 1-c\}$ at $x = 0$, $\{0, c-a-b\}$ at $x = 1$ and $\{a, b\}$ at $x = \infty$. The corresponding differential operator is denoted by:

$$H^{a,b}_{c,x} = x(1-x)\partial^2 + (c - (a+b+1)x)\partial - ab. \quad (9)$$

One of the solutions at $x = 0$ is ${}_2F_1(a, b; c\,|\,x)$. Computing a ${}_2F_1$-type solution of a second order L_{inp} corresponds to computing transformations from $H^{a,b}_{c,x}$ to L_{inp}.

Remark 2.8. *The exponent differences of $H^{a,b}_{c,x}$ can be obtained from the parameters a, b, c and vice versa: $(e_0, e_1, e_\infty) = (1-c, c-a-b, b-a)$.*

Remark 2.9. *We assume that $H^{a,b}_{c,x}$ has no Liouvillian solutions. For such $H^{a,b}_{c,x}$, the points $0, 1, \infty$ are never non-singular or removable singularities. So if $H^{a,b}_{c,x}$ has $e_p \in \mathbb{Z}$ (with $p \in \{0,1,\infty\}$) then p is a logarithmic singularity.*

2.3 Properties of Transformations

For second order operators, we use the notation $L_1 \longrightarrow L_2$ if L_1 can be transformed to L_2 with any combination of the three transformations from *section 1*. If $L_1 \longrightarrow L_2$ then $L_1 \xrightarrow{f}_C \xrightarrow{r_0,r_1}_G \xrightarrow{r}_E L_2$. More details can be found in [1].

Remark 2.10.

1. $\xrightarrow{r_0,r_1}_G$ *and* $\xrightarrow{r}_E$ *are equivalence relations.*

2. *Δ_p remains same under $\xrightarrow{r}_E$ but may change by an integer under $\xrightarrow{r_0,r_1}_G$.*
   So if $L_1 \xrightarrow{f}_C M \xrightarrow{r_0,r_1}_G \xrightarrow{r}_E L_{inp}$ for some input L_{inp} with L_1, M unknown, then $\Delta_p(M)$ can be (mod $\mathbb{Z}$ and up to $\pm$) read from $\Delta_p(L_{inp})$,
   $$Sing(L_{inp}) = Sing(M).$$
   Hence L_1, f, M should be reconstructed from $Sing(L_{inp})$.

3. *If one of e_0, e_1, e_∞ is in $\frac{1}{2}+\mathbb{Z}$ then $H^{a,b}_{c,x}$ is determined, up to the equivalence relation $\xrightarrow{r_0,r_1}_G \xrightarrow{r}_E$, by the triple (e_0, e_1, e_∞) up to $\pm$ and* mod $\mathbb{Z}$.
   If $\{e_0, e_1, e_\infty\} \bigcap (\frac{1}{2}+\mathbb{Z}) = \emptyset$ then the triple leaves two separate cases for $H^{a,b}_{c,x}$ up to $\xrightarrow{r_0,r_1}_G \xrightarrow{r}_E$; we need to consider (e_0, e_1, e_∞) up to $\pm$ and mod $\mathbb{Z}$, *and (e_0+1, e_1, e_∞) up to $\pm$.* See Theorem 8, section 5.3 in [14] for details.

Because of the transformation $M \xrightarrow{r_0,r_1}_G \xrightarrow{r}_E L_{inp}$ in *remark 2.10* only non-removable singularities of L_{inp} provide usable data for M and f.

Definition 2.11. *Two operators L_1, L_2 are called projectively equivalent (notation: $L_1 \sim_p L_2$) if $L_1 \xrightarrow{r_0,r_1}_G \xrightarrow{r}_E L_2$.*

The following lemma gives the effect of $\xrightarrow{f}_C$ on the singularities and their exponent differences: (see [4] for more details)

Lemma 2.12. *Let e_0, e_1, e_∞ be the exponent differences of $H^{a,b}_{c,x}$ at $0, 1, \infty$. Let $H^{a,b}_{c,f}$ be the operator obtained from $H^{a,b}_{c,x}$ by applying $x \mapsto f$. Let $d = \Delta_p$ be the exponent difference of $H^{a,b}_{c,f}$ at $x = p$. Then:*

1. *If p is a root of f with multiplicity m, then $d = me_0$.*

2. *If p is a root of $1-f$ with multiplicity m, then $d = me_1$.*

3. *If p is a pole of f of order m, then $d = me_\infty$.*

2.4 An Example Involving All Three Transformations

Let $u(0) = 1,\ u(1) = 828,\ u(n+2) =$

$$\frac{4(592(n-1)^2 - 977)u(n+1) - 28^3(16n^2-9)u(n))}{(n+2)^2}. \quad (10)$$

This defines a sequence 1, 828, -121212, ... How to prove that this is an integer sequence?
Consider the following differential operator:

$$\tilde{L} = (x-37)\left(x^2+3\right)\partial^2 + \left(x^2+3\right)\partial - \frac{9}{16}(x+9). \quad (11)$$

Our implementation solves this equation. One solution is:

$$sol_{\tilde{L}} = s\left(g \cdot {}_2F_1\left(\frac{1}{12}, \frac{5}{12}; 1 \mid f\right) + h \cdot {}_2F_1\left(\frac{5}{12}, \frac{13}{12}; 1 \mid f\right)\right). \quad (12)$$

where $s = \frac{98^{\frac{1}{4}}}{126(3x-13)^{\frac{5}{4}}}$, $g = (3x+1)(3x-13)$, $h = 36x+40$ and $f = \frac{27(x-37)(x^2+3)}{(3x-13)^3}$.

One can convert between differential equations and recurrences (*see 'gfun' package in Maple*) and find:

$$sol_{\tilde{L}} = \sum_{n=0}^{\infty} u(n) \left(\frac{x-37}{2^7 \cdot 7^3}\right)^n \quad (13)$$

where $u(n)$ are given by the recurrence relation in (10).
The explicit expression (12) can be used to prove $u(n) \in \mathbb{Z}$ for $n = 0, 1, \ldots$ (it is not clear if there is a different way to prove that for this example).
The following diagram shows the effects of $\xrightarrow{f}_C$ and $\xrightarrow{r_0,r_1}_G$ on the set of non-removable singularities and their exponent differences ($\xrightarrow{r}_E$ does not affect them):

$\tilde{L}$:

p	37	$\sqrt{-3}$	$-\sqrt{-3}$	∞
Δ_p	0	1	1	$\frac{3}{2}$

↑ III: $y \mapsto r_0 y + r_1 y'$

$H^{\frac{1}{12},\frac{5}{12}}_{1,f}$:

p	37	$\sqrt{-3}$	$-\sqrt{-3}$	∞
Δ_p	0	0	0	$\frac{1}{2}$

↑ II: $f = \frac{27(x-37)(x^2+3)}{(3x-13)^3}$

$H^{\frac{1}{12},\frac{5}{12}}_{1,x}$:

p	0	1	∞
Δ_p	0	$\frac{1}{2}$	$\frac{1}{3}$

I

Diagram 3

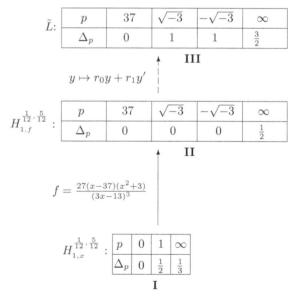

Suppose $y(x) = \sum_{n=0}^{\infty} u(n)x^n$ is convergent with $u(n) \in \mathbb{Z}$, $(n = 0, 1, 2, \ldots)$ and satisfies a second order differential operator $L \in \mathbb{Q}(x)[\partial]$. In all known examples such $y(x)$ is either algebraic or expressible in terms of ${}_2F_1$ hypergeometric functions. Hence algorithms for finding ${}_2F_1$-type solutions are useful for integer sequences.

3. PROBLEM DISCUSSION

Starting with a Fuchsian linear differential operator L_{inp}, of order 2, and no Liouvillian solutions, we want to find a solution of such L_{inp} in the form :

$$y = e^{\int r}(r_0 S(f) + r_1 S(f)') \qquad (14)$$

where $S(x) = {}_2F_1(a, b; c \,|\, x)$, $f, r, r_0, r_1 \in \mathbb{C}(x)$ and f has degree 3. There are two key steps:

- Compute f and (e_0, e_1, e_∞) such that

  $$Sing(H^{a,b}_{c,f}) = Sing(L_{inp}). \qquad (15)$$

  (see *remark 2.8* for the relation between (e_0, e_1, e_∞) and (a, b, c))
- Compute projective equivalence $\sim_p$ between $H^{a,b}_{c,f}$ and L_{inp} which sends solutions $S(f) = {}_2F_1(a, b; c \,|\, f)$ of $H^{a,b}_{c,f}$ to solutions of L_{inp} of the form (14).

If we find f then [2] takes care of the second step. Hence the crucial part is to compute f (as well as a, b, c, or equivalently, e_0, e_1, e_∞).
Let $f = A/B$ where $A, B \in \mathbb{C}[x]$ with $gcd(A, B) = 1$. The hypergeometric operator $H^{a,b}_{c,x}$ has singularities at $x = 0, 1, \infty$. So one might expect L_{inp} to have singularities whenever $f = 0, 1$ or ∞; i.e. at the roots of A, $A - B$ and B. If all roots of $A, A - B, B$ would appear among the singularities of L_{inp}, then it would be fairly easy to reconstruct $f = A/B$. However, that is not true in general (it is true for 8 out of the 18 cases in *Tab. 1* in *Section 4.2*). For example; if f has a root p with multiplicity 2 and e_0 is a half-integer (an odd integer divided by 2), then p will be a removable singularity or a non-singular point of $H^{a,b}_{c,f}$. Such p does not appear in $Sing(L_{inp})$.

4. COMPUTING PULLBACK FROM THE SINGULARITY STRUCTURE

4.1 Relating singularities to f

Let $[[a_1, \ldots, a_i], [b_1, \ldots, b_j], [c_1, \ldots, c_k]]$ denote the *branching pattern* of f. It contains the branching orders of f above 0, 1 and ∞ respectively (So f has i distinct roots with multiplicities $a_1, \ldots, a_i$. Likewise $1 - f$ and $\frac{1}{f}$ have j resp. k distinct roots). Using *lemma 2.12* and *remark 2.4*, the singularities of $H^{a,b}_{c,f}$ are as follows:

$$P_0 = \{x : f = 0 \text{ and } (e_0 \in \mathbb{Z} \text{ or } a_l\, e_0 \notin \mathbb{Z}) \text{ for } 1 \le l \le i\}$$

$$P_1 = \{x : 1 - f = 0 \text{ and } (e_1 \in \mathbb{Z} \text{ or } b_l\, e_1 \notin \mathbb{Z}) \text{ for } 1 \le l \le j\}$$

$$P_\infty = \{x : \tfrac{1}{f} = 0 \text{ and } (e_\infty \in \mathbb{Z} \text{ or } c_l\, e_\infty \notin \mathbb{Z}) \text{ for } 1 \le l \le k\}$$

where (e_0, e_1, e_∞) are the exponent differences of $H^{a,b}_{c,x}$ at $(0, 1, \infty)$ respectively. The union of P_0, P_1 and P_∞ are the non-removable singularities of $H^{a,b}_{c,f}$ (or L_{inp}, by eq. (15)). Since f has degree 3, L_{inp} could have at most 9 singularities. The least we could have is 2 when we choose the branching pattern of f as [[3],[1,2],[1,2]] and $(e_0, e_1, e_\infty) \equiv (\pm\frac{1}{3}, \frac{1}{2}, \frac{1}{2})$ mod $\mathbb{Z}$. But a hypergeometric equation with two exponent-differences in $\frac{1}{2} + \mathbb{Z}$ has Liouvillian solutions, so we do not treat this case here. If L_{inp} has 3 non-removable singularities, then we can transform these to $0, 1, \infty$ via a Möbius transformation and express the solution accordingly (with a rational function of degree 1). This case is already treated in [14]. So we exclude these cases (Liouvillian and 3 non-removable singularities) from our consideration.

4.2 Tabulating cases

Let d be the total number of non-removable singularities in L_{inp}. From *Section 4.1* we have $4 \le d \le 9$. The first step is to enumerate all possibilities for exponent differences e_0, e_1, e_∞ and branching patterns above $\{0, 1, \infty\}$ for each d. We express all such possibilities for degree 3 rational function f in the following table:

NOTATION 4.1.
d: number of non-removable singularities in L_{inp}.
$*$, E_1, E_2, E_3: elements of $\mathbb{C}$.
$\frac{*}{2}$: an element of $\frac{1}{2} + \mathbb{Z}$.
$\frac{*}{3}$: an element of $(\frac{1}{3} + \mathbb{Z}) \cup (\frac{2}{3} + \mathbb{Z})$.

d	Case	Exponent difference at 0, 1, ∞ resp.	Branching pattern above 0, 1, ∞ resp.
4	case4.1	$\frac{*}{2}, \frac{*}{3}, E_1$	[1,2], [3], [1,1,1]
	case4.2	$\neq \frac{*}{3}, \frac{*}{3}, E_1$	[3], [3], [1,1,1]
	case4.3	$\neq \frac{*}{2}, \neq \frac{*}{2}, \frac{*}{3}$	[1,2], [1,2], [3]
	case4.4	$\neq \frac{*}{3}, \neq \frac{*}{2}, \frac{*}{2}$	[3], [1,2], [1,2]
	Liouv	$\neq \frac{*}{2}, \frac{*}{2}, \frac{*}{2}$	[1,2], [1,2], [1,2]
5	case5.1	$\neq \frac{*}{3}, \neq \frac{*}{3}, E_1$	[3], [3], [1,1,1]
	Liouv	$\frac{*}{2}, \frac{*}{2}, E_1$	[1,2], [1,2], [1,1,1]
	case5.2	$\neq \frac{*}{2}, \frac{*}{3}, E_1$	[1,2], [3], [1,1,1]
	case5.3	$\frac{*}{2}, E_1, \neq \frac{*}{3}$	[1,2], [1,1,1], [3]
	case5.4	$\neq \frac{*}{2}, \neq \frac{*}{2}, \frac{*}{2}$	[1,2], [1,2], [1,2]
	case5.5	$\neq \frac{*}{3}, \neq \frac{*}{2}, \neq \frac{*}{2}$	[3], [1,2], [1,2]
6	case6.1	$\frac{*}{3}, E_1, E_2$	[3], [1,1,1], [1,1,1]
	case6.2	$\neq \frac{*}{2}, \frac{*}{2}, E_1$	[1,2], [1,2], [1,1,1]
	case6.3	$\neq \frac{*}{3}, \neq \frac{*}{2}, E_1$	[3], [1,2], [1,1,1]
	case6.4	$\neq \frac{*}{2}, \neq \frac{*}{2}, \neq \frac{*}{2}$	[1,2], [1,2], [1,2]
7	case7.1	$\neq \frac{*}{3}, E_1, E_2$	[3], [1,1,1], [1,1,1]
	case7.2	$\frac{*}{2}, E_1, E_2$	[1,2], [1,1,1], [1,1,1]
	case7.3	$\neq \frac{*}{2}, \neq \frac{*}{2}, E_1$	[1,2], [1,2], [1,1,1]
8	case8.1	$\neq \frac{*}{2}, E_1, E_2$	[1,2], [1,1,1], [1,1,1]
9	case9.1	E_1, E_2, E_3	[1,1,1], [1,1,1], [1,1,1]

Table 1: Cases for degree 3 pullback up to permutation of $0, 1, \infty$

Two cases (denoted *Liouv*) in *Tab.1* correspond to the hypergeometric equations with two singularities having a half-integer exponent difference. Such equations have Liouvillian solutions (this follows from Kovacic' algorithm and also from Theorem 8, section 5.3 in [14]). Now the main task is to compute f for the remaining 18 cases. Recall that non removable singularities of $H^{a,b}_{c,f}$ come from (form a subset of) the roots of f, $1 - f$ and poles of f. We will use the singularity structure of L_{inp} to recover f.

4.3 Treating one case

The main algorithm in *Section 5* takes as input C, L_{inp}, x where C is a field of characteristic 0, and $L_{inp} \in C(x)[\partial]$ has order 2 and no Liouvillian solutions. It computes $Sing(L_{inp})$ and d. Then it loops over the corresponding cases in *Tab.1*. For example; if $d = 4$ then it loops over cases 4.1 − 4.4 in *Tab.1*. Each case in *Tab.1* is a subprogram. Each of these subprograms takes C, $Sing(L_{inp})$ as input, checks if $Sing(L_{inp})$ is compatible with that particular case, and if so, returns a set of candidates for $f, (e_0, e_1, e_\infty)$ that are compatible with that particular case. We give details for only one case, namely Algorithm[5.3] (notation: casei, j is handled by Algorithm[i.j]). The other cases are treated by similar programs (details can be found in our implementation [11]).

Let $L_{inp} \in C(x)[\partial]$ be input differential operator with 5 non-removable singularities. In terms of places(C), there are 7 ways to end up with 5 points:

1. One place of degree 5 (note: a place of degree > 1 is always a monic irreducible polynomial of that degree. A place of degree 1 can be either ∞ or a monic polynomial of degree 1.)
2. Places of degrees 4, 1.
3. Places of degrees 3, 2.
4. Places of degrees 3, 1, 1.
5. Places of degrees 2, 2, 1.
6. Places of degrees 2, 1, 1, 1.
7. Places of degrees 1, 1, 1, 1, 1.

Algorithm[5.3]: Compute $f \in C(x)$ of degree 3 and exponent differences (e_0, e_1, e_∞) for $H^{a,b}_{c,x}$ corresponding to 'case5.3' in *Tab.1*.

Input: Field C and $Sing(L_{inp})$ in terms of *places*(C).

Output: A set of lists $[f, (e_0, e_1, e_\infty)]$ where $f \in C(x)$ has degree 3 and branching pattern [1,2], [1,1,1], [3] above $0, 1, \infty$ such that $Sing\left(H^{a,b}_{c,f}\right) = Sing(L_{inp})$ where (a, b, c) corresponds to (e_0, e_1, e_∞) by *remark 2.8* and (*see Tab.1*) $e_0 \in \frac{1}{2} + \mathbb{Z}$, e_1 is arbitrary, and $e_\infty \notin \pm\frac{1}{3} + \mathbb{Z}$.

1. Check if $Sing(L_{inp})$ is consistent with case 5.3 (if not, return the empty set and stop) as follows:

   The branching pattern [1,2] at $f = 0$ indicates that f has two roots $a_1, a_2 \in C \bigcup \{\infty\}$ with multiplicities 1 resp. 2. Then $x = a_1$ will have an exponent-difference $e_0 \in \frac{1}{2} + \mathbb{Z}$ but $x = a_2$ will be a regular point or a removable singularity, and so it does not appear in $Sing(L_{inp})$.

   The branching pattern [3] at $f = \infty$ indicates that f has precisely one pole $b \in C \bigcup \{\infty\}$, of order 3. Then $x = b$ will have an exponent-difference $\pm 3e_\infty$ mod $\mathbb{Z}$. In case 5.3 we have $e_\infty \notin \pm\frac{1}{3} + \mathbb{Z}$ and hence the point $x = b$ must be a non-removable singularity. Combined with $x = a_1$ we see that case 5.3 is only possible when $Sing(L_{inp})$ has at least two places of degree 1. So in the above listed 7 cases (5, 4+1, ...), we can exit Algorithm[5.3] immediately if we are not in case 4, 6, or 7.

   The branching pattern [1,1,1] at $f = 1$ indicates that $1 - f$ has three distinct roots, each of multiplicity 1. Thus there must be at least three distinct singularities that match the exponent-difference $\pm e_1$ mod $\mathbb{Z}$. If we can not find three singularities (one place of degree 3, or places of degrees 2 and 1, or three places of degree 1) whose exponent-differences match (up to $\pm$ and mod $\mathbb{Z}$) then Algorithm[5.3] stops. This condition determines e_1 (up to $\pm$ and mod $\mathbb{Z}$).

   We know from Kovacic' algorithm that if there are two $e_i \in \frac{1}{2} + \mathbb{Z}$ then $H^{a,b}_{c,x}$ will have Liouvillian solutions. Since we exclude Liouvillian cases, it follows that only e_0 is in $\frac{1}{2} + \mathbb{Z}$. We conclude that $Sing(L_{inp})$ must have either 1 or 2 singularities in $C \bigcup \{\infty\}$ with an exponent-difference in $\frac{1}{2} + \mathbb{Z}$ and that 2 such singularities can only occur when $e_\infty \in \pm\frac{1}{6} + \mathbb{Z}$. So if there are more than 2, then Algorithm[5.3] stops.

2. Set Cand $= \emptyset$ and write $f = k_1 \frac{(x-a_1)(x-a_2)^2}{(x-b)^3}$ where $a_1, a_2, b \in C \bigcup \{\infty\}$ and $k_1 \in C$. We replace any factor $x - \infty$ in f by 1 in the implementation. Compute the set of places with an exponent-difference in $\frac{1}{2} + \mathbb{Z}$. This set may only have 1 or 2 elements that must have degree 1. Now a_1 loops over this set, and e_0 is the exponent-difference at $x = a_1$.

3. Loop b over the places in $Sing(L_{inp})$ of degree 1, skipping a_1, and only considering a_1, b for which the remaining three singularities have matching exponent-differences. Let e_b be the exponent-difference at $x = b$. Now loop e_∞ over $\frac{e_b}{3}, \frac{(e_b-1)}{3}, \frac{(e_b+1)}{3}$. For e_1 one can take the exponent-difference at any of the 3 remaining singularities. The reason that there are three cases for e_∞ is because we have to determine e_∞ mod $\mathbb{Z}$. Now $3e_\infty = e_b$ but if a gauge transformation occurred, i.e. if the r_1 in the form (2) in *Section 1* is non-zero, then e_b is only known mod $\mathbb{Z}$, and this leaves in general three candidate values for e_∞ mod $\mathbb{Z}$ (it suffices to compute the e_i mod $\mathbb{Z}$, see section 5.3 in [14], summarized in Remark 2.10).

4. Among the remaining 3 singularities, let $P \in C[x]$ be the product of their places (replacing $x - \infty$ by 1 if that is among them). So P has degree 3 if ∞ is not among the 3 remaining singularities, and otherwise it has degree 2. In each loop, the a_1, b appearing in f are known, while k_1 and a_2 are unknown. Take the numerator of $1 - f$ and compute its remainder mod P. Equate the coefficients of this remainder to 0. This gives deg(P) equations for k_1, a_2. If deg(P) = 2 we obtain one more equation by setting $f(\infty) = 1$ (the resulting equation is $k_1 = 1$). Then we have 3 equations for 2 unknowns k_1, a_2. Compute the solutions $k_1 \in C$ and $a_2 \in C \bigcup \{\infty\}$. If any solution exists, then add the resulting $[f, (e_0, e_1, e_\infty)]$ to the set Cand.

5. Return the set Cand (which could be empty, but could also have one or more members).

EXAMPLE 4.2.
Take $C = \mathbb{Q}$. Let $Sing(L_{inp})$ in terms of places($\mathbb{Q}$) *be given by:*

$$Sing(L_{inp}) = \left\{[\infty, -\tfrac{1}{2}], [x, \tfrac{2}{7}], [x-2, \tfrac{1}{2}], [x^2 + 26\,x + 44, \tfrac{5}{7}]\right\}.$$

Our input is the following:

$Sing(L_{inp}) = \{[1, -\frac{1}{2}], [x, \frac{2}{7}], [x-2, \frac{1}{2}], [x^2 + 26\,x + 44, \frac{5}{7}]\}$. *Notations in the steps below come from* Algorithm[5.3]. *Write* $f(x) = k_1 \frac{(x-a_1)(x-a_2)^2}{(x-b)^3}$.

Step 1: *$Sing(L_{inp})$ satisfies the conditions for 'sing5.3';*
(1) $[1, -\frac{1}{2}]$ and $[x-2, \frac{1}{2}]$ have degree 1 and both have a half-integer exponent difference.
(2) The exponent differences in $[x, \frac{2}{7}]$ and $[x^2+26\,x+44, \frac{5}{7}]$ match, after all, we are working up to $\pm$ and mod $\mathbb{Z}$.

Step 2: *The candidates for $x - a_1$ are 1 and $x-2$. For the first case, we get $f = k_1 \frac{(x-a_2)^2}{(x-b)^3}$ and $e_0 = -\frac{1}{2}$ (note: we may equally well take $\frac{1}{2}$). For the second case, we get $f = k_1 \frac{(x-2)(x-a_2)^2}{(x-b)^3}$ and $e_0 = \frac{1}{2}$.*

Step 3: *For the first case, $x-b$ can only be $x-2$ and $e_b = \frac{1}{2}$ (because if we take $x - b = x$ then there would not remain three singularities with matching exponent-differences). Likewise, for the second case, $x - b$ can only be 1 and $e_b = -\frac{1}{2}$.*

First case: $f = k_1 \frac{(x-a_2)^2}{(x-2)^3}$ and $e_\infty = \frac{1}{6}$ (note: we should consider $e_\infty \in \{\frac{e_b}{3}, \frac{(e_b+1)}{3}, \frac{(e_b-1)}{3}\}$ since e_b is determined mod $\mathbb{Z}$, *and we have to determine e_∞* mod $\mathbb{Z}$. *However, $\frac{(e_b+1)}{3} = \frac{1}{2}$ is discarded since there should not be two $e_i's$ in $\frac{1}{2}+\mathbb{Z}$. And $\frac{(e_b-1)}{3} = -\frac{1}{6}$ but an exponent-difference $-\frac{1}{6}$ is equivalent to an exponent-difference $\frac{1}{6}$.)*

Second case: $f = k_1(x-2)(x-a_2)^2$ and $e_\infty = -\frac{1}{6}$.

Step 4: *In both cases $P = x \cdot (x^2 + 26x + 44)$ and $e_1 = \frac{2}{7}$ (we could equally well take $\frac{5}{7}$). Dividing the numerator of $1 - f$ by P produces equations in k_1 and a_2. In first case the equations have a solution; $\{k_1 = -32, a_2 = -\frac{1}{2}\}$, and in second case they do not.*

Step 5: *The output Cand has one element, namely* $\left\{[-32 \frac{(x+1/2)^2}{(x-2)^3}, (-\frac{1}{2}, \frac{2}{7}, \frac{1}{6})]\right\}$.

5. MAIN ALGORITHM

We have developed the algorithms to compute f's and possible exponent differences (e_0, e_1, e_∞) for $H^{a,b}_{c,x}$ corresponding to all 18 cases as given in *Tab.1*. Now we give our main algorithm:

Let $C \subseteq \mathbb{C}$ be a field and $L_{inp} \in C(x)[\partial]$ be the input differential operator. The main algorithm first computes the singularity structure of L_{inp} in terms of *places*(C). Suppose d is the total number of non-removable singularities of L_{inp}. Now we call all algorithms corresponding to d to produce a set of candidates for $f \in C(x)$ and the exponent differences $(e_0, e_1, e_\infty) = (1-c, c-a-b, b-a)$. For each member from that list we compute $H^{a,b}_{c,x}$, $H^{a,b}_{c,f}$ and apply *projective equivalence* [2] between $H^{a,b}_{c,f}$ and L_{inp} to find (if it exists) a nonzero map from $V\left(H^{a,b}_{c,f}\right)$ to $V(L_{inp})$ which sends solutions $S(f) = {}_2F_1(a, b; c\,|\, f)$ of $H^{a,b}_{c,f}$ to solutions $e^{\int r}(r_0 S(f) + r_1 S(f)')$ of L_{inp}.

Algorithm: Solve an irreducible second order linear differential operator $L_{inp} \in C(x)[\partial]$ in terms of ${}_2F_1(a, b; c\,|\, f)$, with $f \in C(x)$ of degree 3.

Input: A field C of characteristic 0, $L_{inp} \in C(x)[\partial]$ of order 2 which has no Liouvillian solutions, and a variable x.

Output: A non zero solution $y = e^{\int r}(r_0 S(f) + r_1 S(f)')$, if it exists, such that $L_{inp}(y) = 0$, where $S(f) = {}_2F_1(a, b; c\,|\, f)$, $f, r, r_0, r_1 \in C(x)$ and f has degree 3.

Step 1: Find the singularity structure of L_{inp} in terms of *places*(C). Let d be the total number of non-removable singularities.

Step 2: Let k be the total number of cases in *Tab.1* for d. For example; if $d = 6$ then $k = 4$.
Let *Candidates* $= \bigcup$ *Algorithm[n.a]*, where $a = \{1 \ldots k\}$. That produces a set of lists $[f, (e_0, e_1, e_\infty)]$ of all possible rational function $f \in C(x)$ of degree 3 and corresponding exponent differences (e_0, e_1, e_∞) for $H^{a,b}_{c,x}$.

Step 2.1 : $H^{a,b}_{c,x} = x(1-x)\partial^2 + (c - (a+b+1)x)\partial - ab$ where a, b, c come from the relation $(e_0, e_1, e_\infty) = (1-c, c-a-b, b-a)$.
For each element $[f, (e_0, e_1, e_\infty)]$ in *Candidates* (*Step 2*),
(i) If $\{e_0, e_1, e_\infty\} \bigcap \frac{1}{2} + \mathbb{Z} \neq \emptyset$ then Cand $:= \{[f, (e_0, e_1, e_\infty)]\}$ otherwise Cand $:= \{[f, (e_0, e_1, e_\infty)], [f, (e_0+1, e_1, e_\infty)]\}$ (That determines $H^{a,b}_{c,x}$ up to *projective equivalence, see remark 2.10*).
(ii) From each element in Cand above (a) compute a, b, c, (b) substitute the values of a, b, c in $H^{a,b}_{c,x}$ and (c) apply the change of variable $x \mapsto f$ on $H^{a,b}_{c,x}$. That produces a list of operators $H^{a,b}_{c,f}$.

Step 2.2 : Compute the projective equivalence [2] between each operator $H^{a,b}_{c,f}$ in *Step 2.1* and L_{inp}. If the output is zero, then go back to *Step 2.1* and take the next element from *Candidates*. Otherwise, we get a map of the form: $G = e^{\int r}(r_0 + r_1\partial)$, where $r, r_0, r_1 \in C(x)$ and $\partial = \frac{d}{dx}$.

Step 2.3: $S(f) = {}_2F_1(a, b; c\,|\, f)$ is a solution of $H^{a,b}_{c,f}$. Now compute $G(S(f))$. That gives a solution of L_{inp}.

Step 2.4: Repeat the same procedure for each element in *Candidates*. That gives us a list of solutions of L_{inp}.

Step 2.5: Choose the best solution (with shortest length) from the list (to obtain a second independent solution of L_{inp}, just use a second solution of $H^{a,b}_{c,x}$).

EXAMPLE 5.1. *Consider the operator in* Section 2.4;

$$L_{inp} = (x-37)\left(x^2+3\right)\partial^2 + \left(x^2+3\right)\partial - \tfrac{9}{16}(x+9).$$

Procedure to solve this equation is the following:

Step 1: *Read the file* hypergeomdeg3.txt *from the folder hypergeomdeg3 in* `www.math.fsu.edu/~vkunwar`.

Step 2: *$L_{inp} \in \mathbb{Q}(x)[\partial]$. We want the solution of L_{inp} in the base field $\mathbb{Q}$. Type hypergeomdeg3($\{\}, L_{inp}, x$). (in Maple $\{\}$ is the code for $\mathbb{Q}$)*

Step 3: *The program first finds the singularity structure;*

$Sing(L_{inp}) = \{[1, -\frac{3}{2}], [x-37, 0], [x^2+3, 1]\}$. *(our implementation uses "1" to encode a singularity at ∞, and polynomials to encode finite singularities).*

Step 4: *We get $d = 4$. The program loops over the four subprograms corresponding to case4.1,... case4.4 to compute f:*

1. Algorithm[4.1] *returns the following:* $F =$

$$\left\{[f, [-\frac{3}{2}, 0, \frac{1}{3}]], [f, [-\frac{3}{2}, 1, \frac{1}{3}]], [f, [-\frac{3}{2}, 0, \frac{2}{3}]], [f, [-\frac{3}{2}, 1, \frac{2}{3}]]\right\}$$

where $f = 8\frac{(9x+10)^2}{(3x-13)^3}$.
Note: this set contains $\sim_p$-duplicates, the four triples (e_0, e_1, e_∞) all give projectively equivalent $H^{a,b}_{c,x}$ so we could delete three and still find a solution (if it exists). The reason they were left in the current version of the implementation is because they may help to find a solution of smaller size. In the next version, we plan to make the code more efficient by removing $\sim_p$-duplicates, keeping only those for which the integer-differences between the exponent-differences of $H^{a,b}_{c,f}$ and L_{inp} are minimized (in this example, only the second element of F would be kept in this approach).

2. Algorithm[4.2] *returns NULL.*

3. Algorithm[4.3] *and* Algorithm[4.4] *require at least 3 linear polynomials in $\mathbb{Q}[x]$ for $Sing(L_{inp})$ which is not the case here. So $Sing(L_{inp})$ does not qualify the conditions for these algorithms.*

Hence F gives the Candidates. Note that we are in the case $\{e_0, e_1, e_\infty\} \bigcap \frac{1}{2} + \mathbb{Z} \neq \emptyset$.

Step 5: *Taking first element $i = [8\frac{(9x+10)^2}{(3x-13)^3}, [-\frac{3}{2}, 0, \frac{1}{3}]]$ in Candidates and applying* Step 2.1 *and* Step 2.2 *of the above main algorithm, we get $G = e^{\int r}(r_0 + r_1\partial)$ with*
$e^{\int r} = \frac{(\frac{9}{10}x+1)(\frac{1}{3}x^2+1)(-\frac{1}{37}x+1)}{(\frac{1}{12}x+1)(-\frac{3}{13}x+1)^{\frac{13}{4}}}$, $r_1 = 1 + \frac{90}{19}x - \frac{27}{19}x^2$ *and*
$r_0 = \frac{3}{38}\frac{729x^4-19845x^3-251919x^2+1114345x+239772}{(x-37)(3x-13)(9x+10)}$.

Step 6: *We have $S(f) = {}_2F_1(\frac{13}{12}, \frac{17}{12}; \frac{5}{2} \,|\, 8\frac{(9x+10)^2}{(3x-13)^3})$. Computing $G(S(f))$ we get $y = e^{\int r}(r_0 S(f) + r_1 S(f)')$ as a solution of L_{inp} where $e^{\int r}, r_0, r_1$ are given in* Step 5.

Step 7: *Taking second element $i = [8\frac{(9x+10)^2}{(3x-13)^3}, [-\frac{3}{2}, 1, \frac{1}{3}]]$ in Candidates we get another solution y with*
$e^{\int r} = \frac{(\frac{9}{10}x+1)}{(\frac{1}{12}x+1)(-\frac{3}{13}x+1)^{\frac{7}{4}}}$, $r_0 = \frac{(2187x^3+22284x^2-37813x+116484)}{98(13-3x)(9x+10)}$,
$r_1 = x^2 + 3$ *and* $S(f) = {}_2F_1(\frac{7}{12}, \frac{11}{12}; \frac{5}{2} \,|\, 8\frac{(9x+10)^2}{(3x-13)^3})$.

Steps 8 and 9: *Process the third and fourth element. Each produces a solution that looks quite similar to that given in Steps 6 and 7.*

Step 10: *The solution in* Step 7 *has the shortest length. So the implementation returns that as a solution of L_{inp}.*

After minor simplification this leads to the solution given in section 2.4.

6. REFERENCES

[1] R. Debeerst, M. van Hoeij, W. Koepf: *Solving Differential Equations in Terms of Bessel Functions*, ISSAC'08 Proceedings, 39-46 (2008).

[2] M. van Hoeij: *An implementation for finding equivalence map*, www.math.fsu.edu/~hoeij/files/equiv.

[3] Q. Yuan, M. van Hoeij: *Finding all Bessel type solutions for Linear Differential Equations with Rational Function Coefficients*, ISSAC'2010 Proceedings, 37-44 (2010).

[4] A. Bostan, F. Chyzak, M. van Hoeij, L. Pech: *Explicit formula for the generating series of diagonal 3D rook paths*, Seminaire Lotharingien de Combinatoire, B66a (2011).

[5] R. Vidunas: *Algebraic transformations of Gauss hypergeometric functions.* Funkcialaj Ekvacioj, Vol 52 (Aug 2009), 139-180.

[6] T. Fang, M. van Hoeij: *2-Descent for Second Order Linear Differential Equations*, ISSAC'2011 Proceedings, 107-114 (2011).

[7] J. Kovacic: *An algorithm for solving second order linear homogeneous equations*, J. Symbolic Computations, 2, 3-43 (1986).

[8] M. van Hoeij, R. Vidunas: *Belyi functions for hyperbolic hypergeometric-to-Heun transformations*, arXiv:1212.3803.

[9] M. van Hoeij, V. J. Kunwar: *Hypergeometric solutions of second order differential equations with 5 non-removable singularities*, (In Progress).

[10] Z. X. Wang, D. R. Guo: *Special Functions* World Scientific Publishing Co. Pte. Ltd, Singapore, 1989.

[11] V. J. Kunwar: *An implementation for hypergeometric solution of second order differential equation with degree 3 rational function*, www.math.fsu.edu/~vkunwar/hypergeomdeg3.

[12] M. van Hoeij: *Solving Third Order Linear Differential Equation in Terms of Second Order Equations*, ISSAC'07 Proceedings, 355-360, 2007. Implementation: www.math.fsu.edu/~hoeij/files/ReduceOrder.

[13] Q. Yuan: *Finding all Bessel type solutions for Linear Differential Equations with Rational Function Coefficients*, Ph.D thesis and implementation, available at www.math.fsu.edu/~qyuan (2012).

[14] T. Fang: *Solving Linear Differential Equations in Terms of Hypergeometric Functions by 2-Descent*, Ph.D thesis and implementation, available at www.math.fsu.edu/~tfang (2012).

Certified Symbolic Manipulation: Bivariate Simplicial Polynomials *

Laureano Lambán
Dept. of Mathematics and Computation
University of La Rioja
Calle Luis de Ulloa s/n.
26004 Logroño, Spain
lalamban@unirioja.es

Francisco J. Martín–Mateos
Dept. of Computer Science and Artificial Intelligence
University of Sevilla
Avda. Reina Mercedes, s/n.
41012 Sevilla, Spain
fjesus@us.es

Julio Rubio
Dept. of Mathematics and Computation
University of La Rioja
Calle Luis de Ulloa s/n.
26004 Logroño, Spain
julio.rubio@unirioja.es

José–Luis Ruiz–Reina
Dept. of Computer Science and Artificial Intelligence
University of Sevilla
Avda. Reina Mercedes, s/n.
41012 Sevilla, Spain
jruiz@us.es

ABSTRACT

Certified symbolic manipulation is an emerging new field where programs are accompanied by certificates that, suitably interpreted, ensure the correctness of the algorithms. In this paper, we focus on algebraic algorithms implemented in the proof assistant ACL2, which allows us to verify correctness in the same programming environment. The case study is that of *bivariate simplicial polynomials*, a data structure used to help the proof of properties in Simplicial Topology. Simplicial polynomials can be computationally interpreted in two ways. As symbolic expressions, they can be handled algorithmically, increasing the automation in ACL2 proofs. As representations of functional operators, they help proving properties of categorical morphisms. As an application of this second view, we present the definition in ACL2 of some morphisms involved in the Eilenberg-Zilber reduction, a central part of the *Kenzo* computer algebra system. We have proved the ACL2 implementations are correct and tested that they get the same results as *Kenzo* does.

*Partially supported by Ministerio de Ciencia e Innovación, project MTM2009-13842, and by European Union's 7th Framework Programme under grant agreement nr. 243847 (ForMath).

ISSAC'13, June 26–29, 2013, Boston, Massachusetts, USA.

Categories and Subject Descriptors

I.1.2 [**Computing Methodologies**]: Symbolic and Algebraic Manipulation—*Algebraic algorithms*; D.2.4 [**Software**]: Software Engineering—*Software/Program Verification*

General Terms

Verification

Keywords

Symbolic manipulation, ACL2, Simplicial topology

1. INTRODUCTION

A *certificate* is some datum added to an expression that can be used to verify (in an independent way) some property of the expression. The typical example is the certificate used to check the non-primality of a (big) natural number n: a pair of numbers whose product is equal to n. Many contributions have been made in this area, in particular in the area of *algebraic certificates* (see, for instance, [3]).

In an extreme case, we could identify a certificate with a complete proof, encoded in a concrete formalism, of the property under study. This is the point of view in the *proof-carrying code* paradigm [11], where a code *producer* attaches a proof of some property to the program; then a code *consumer* invokes a proof checker, that ensures a safety condition.

In this paper, we explore another path: programs, proofs, certificates and checking are generated in the same environment, the one of the theorem prover ACL2 [6]. In ACL2, once a proof has been mechanized, a certificate file is created, and the system only checks this file, without redoing the proof (which is, usually, a very time-consuming task) every time the mechanization is included in further developments.

Our area of application is *verified computer algebra*, and specifically algebraic computing in Algebraic Topology. Even more concretely, we are interested in applying formal methods to specify and analyze the behavior of *Kenzo* [5], a program to compute in Homological Algebra and Simplicial Topology. *Kenzo* is a Common Lisp program, created by F. Sergeraert, that can deal with infinite dimensional spaces, and is able to compute results that cannot be determined by any other means (theoretical or computational). In [13] a theorem corrected thanks to *Kenzo* is presented, together with other results computed with *Kenzo* which seem out of reach for any other method.

Due to these *Kenzo* features, a project was launched some years ago to formally study its correctness, trying to give to *Kenzo* results an status as close as possible to standard mathematical properties. To this aim, different methods and tools have been used. For instance, the proof assistants Isabelle and Coq have been used to model important algorithms appearing in *Kenzo* [2, 4].

Both Isabelle/HOL and Coq are very powerful tools (in particular, both are based on higher-order logic), but they are far from the *Kenzo* programming language: Common Lisp. So it is natural to use ACL2, a theorem prover intimately linked to this language. Even if ACL2 is not suitable to model all *Kenzo* features (*Kenzo* uses higher-order functional programming, while ACL2 is a first-order tool; this explains the role of Isabelle/HOL and Coq in the global project), it is superior to any other tool to formalize the *actual Kenzo* source code (for an example of such a formalization, see [9]).

In this area of ACL2 applications to Algebraic and Simplicial Topology, several contributions have been already made [1, 7, 9]. Our last development is a complete ACL2 proof of the so-called Eilenberg-Zilber theorem [8]. This is a central theorem in Computational Algebraic Topology. This fundamental aspect of Eilenberg-Zilber is reflected in its computational counterpart: experimental studies of *Kenzo* log files showed that most of the running time is devoted to Eilenberg-Zilber computations (and more concretely to the computation of the morphism which will be called *Shih* later). Thus, giving a mechanized proof of it seems a good challenge to demonstrate the usability of this kind of formal methods in computer algebra verification.

The formal proof of the Eilenberg-Zilber Theorem needed around 13000 lines of ACL2 code (326 definitions and 1368 lemmas and theorems whitout the generic development of polynomials and basic arithmetic properties), so being quite a big endeavour. In this paper, instead of explaining the main lines of the proof, we focus on some technical achivements, that we consider can be more interesting for a wider audience in symbolic computation. Specifically, we introduce the notion of *bivariate simplicial polynomial*, a data structure instrumental to getting a greater automation in the mechanized proofs. Let us remark that, when choosing the data structure representations, we have followed the most natural ideas (emulating, to a certain extent, the Kenzo way of working), guessing that they will produce the easiest proofs. Our data structures are inspired by those of Kenzo, but adapting them to the constraints imposed by ACL2. For instance, when in Kenzo an array is chosen to represent an entity, our ACL2 version would be a list.

In addition, simplicial polynomials can be executed, giving an experimental flavour to the deductive standard machinery, and can be evaluated (on concrete topological spaces), allowing an *automated testing* for some parts of the *Kenzo* system. To illustrate the way of working in ACL2, we chose an operation on simplicial polynomials called *derivative*, essential to define and prove properties about the above-mentioned *Shih* morphism.

The organization of the paper is as follows. Next section is devoted to state the mathematical problem, while Section 3 deals with the description of the ring of bivariate simplicial polynomials. We explain the notion of derivative in Section 4. Experimental results and computational aspects are presented in Section 5. The paper ends with conclusions and the bibliography.

2. MATHEMATICAL CONCEPTS AND APPLICATIONS

In this section, we introduce briefly the mathematical notions needed to understand the rest of the paper (for details and context about Simplicial Topology, see [10]).

DEFINITION 1. *A* simplicial set K *is a graded set* $\{K_n\}_{n\in\mathbb{N}}$ *together with functions:*

$$\begin{array}{llll} \partial_i^n : K_n \to K_{n-1}, & n>0, & i=0,\dots,n, \\ \eta_i^n : K_n \to K_{n+1}, & n\geq 0, & i=0,\dots,n, \end{array}$$

subject to the following equations:

$$\begin{array}{llllll} (1) & \partial_i^{n-1}\partial_j^n & = & \partial_j^{n-1}\partial_{i+1}^n & \textit{if} & i\geq j, \\ (2) & \eta_i^{n+1}\eta_j^n & = & \eta_{j+1}^{n+1}\eta_i^n & \textit{if} & i\leq j, \\ (3) & \partial_i^{n+1}\eta_j^n & = & \eta_{j-1}^{n-1}\partial_i^n & \textit{if} & i<j, \\ (4) & \partial_i^{n+1}\eta_j^n & = & \eta_j^{n-1}\partial_{i-1}^n & \textit{if} & i>j+1, \\ (5) & \partial_i^{n+1}\eta_i^n & = & \partial_{i+1}^{n+1}\eta_i^n & = & id^n, \end{array}$$

The elements of K_n are called simplices of *dimension* n, or simply n*-simplices.* The functions ∂ and η are called face and degeneracy operators, respectively. A simplex x is called *degenerate* if it can be written as $x = \eta_i y$ for some index i and some simplex y[1]. Otherwise, it is called *non-degenerate.* The set of non-degenerate n-simplices of K is denoted by K_n^{ND}.

With any simplicial set K we can associate an algebraic structure $C(K)$, a chain complex, in such a way that the homology of K is exactly the homology of $C(K)$:

DEFINITION 2. *A* chain complex *is a family of pairs* $C := \{(C_n, d_n)\}_{n\in\mathbb{Z}}$ *where each* C_n *is an abelian group, and each* d_n *is a homomorphism from* C_n *to* C_{n-1} *such that the boundary condition holds:* $d_n \circ d_{n+1} = 0$.

[Given a chain complex C, the boundary condition implies $\mathrm{Im}\ d_{n+1} \subseteq \mathrm{Ker}\ d_n$; then the *homology groups* of C are well-defined: $H_n(C) = \mathrm{Ker}\ d_n/\mathrm{Im}\ d_{n+1}$. These homology groups are the objects *Kenzo* finally computes.]

Let K be a simplicial set. For each $n \in \mathbb{N}$, let us consider $\mathbb{Z}[K_n^{ND}]$, the free abelian group generated by the non-degenerate n-simplices, denoted as $C_n(K)$. Then, the elements of such a group are formal linear combinations $\sum_{j=1}^{r} \lambda_j x_j$, where $\lambda_j \in \mathbb{Z}$ and $x_j \in K_n^{ND}, \forall j = 1,\dots,r$. These linear combinations are called *chains of simplices* or, in short, *chains.*

Now, given $n > 0$, we introduce the homomorphism $d_n : C_n(K) \to C_{n-1}(K)$, first defining it over each generator, and then extending it by linearity. Given $x \in K_n^{ND}$, define $d_n(x) = \sum_{i=0}^{n}(-1)^i\partial_i(x)$, where a term $\partial_i(x)$ is erased

[1] Note that, if the context is clear enough, the superindexes denoting dimension will be skipped.

when it is degenerate. It can be proved that the equations in the definition of simplicial set imply that $d_n \circ d_{n+1} = 0$, $\forall n \in \mathbb{N}$. That is to say, the family $\{d_n\}_{n\in\mathbb{N}}$ defines a *differential* (or boundary) homomorphism on the graded group $\{C_n(K)\}_{n\in\mathbb{N}}$, and then, the family of pairs $\{(C_n(K), d_n)\}_{n\in\mathbb{N}}$ is the *chain complex*[2] associated to the simplicial set K, denoted by $C(K)$.

An alternative definition can be given, by taking as generators *all* the simplices (degenerate and non-degenerate ones) in each dimension, and with the same expression for differentials: $\sum_{i=0}^{n}(-1)^i\partial_i(x)$ (now, it is not necessary to drop out any term). This (bigger) chain complex associated with a simplicial set K has homology groups canonically isomorphic to those of $C(K)$; more concretely, the so-called Normalization Theorem establishes a *reduction* between both chain complexes associated with a simplicial set (that reduction was programmed in ACL2 and proved correct in [7]).

Definition 3. *Given two chain complexes $C^1 := \{(C_n^1, d_n^1)\}_{n\in\mathbb{Z}}$ and $C^2 := \{(C_n^2, d_n^2)\}_{n\in\mathbb{Z}}$, a* reduction *between them is a triple (f, g, h) where $f : C^1 \to C^2$ and $g : C^2 \to C^1$ are chain morphisms (that is to say, they are families of homomorphisms $f_n : C_n^1 \to C_n^2$ and $g_n : C_n^2 \to C_n^1$ such that $f_{n-1} \circ d_n^1 = d_n^2 \circ f_n$ and $g_{n-1} \circ d_n^2 = d_n^1 \circ g_n$), and h is a family of homomorphisms (called* homotopy operator*) $h_n : C_n^1 \to C_{n+1}^1$ satisfying*

1. $f_n \circ g_n = id$,
2. $d_{n+1}^1 \circ h_n + h_{n-1} \circ d_n^1 + g_n \circ f_n = id$,
3. $f_{n+1} \circ h_n = 0$,
4. $h_n \circ g_n = 0$, *and*
5. $h_{n+1} \circ h_n = 0$.

We denote a reduction as $(f, g, h) : C^1 \Longrightarrow C^2$. The main property of a reduction is that it establishes a canonical isomorphism between the homology groups of C^1 and C^2. In fact, the components f and g are enough to determine such a canonical isomorphism, but the homotopy h is necessary to give stability to the concept and to construct reductions from other reductions (see the key instrument called *Basic Perturbation Lemma* in [2]).

To prove in ACL2 the Normalization Theorem, a data structure called *simplicial polynomial* was introduced in [7]. Before defining simplicial polynomials, let us start with an example. Consider $\partial_2^6\eta_3^5\partial_2^6\eta_5^5\eta_2^4\partial_1^5$, a composition of simplicial operators, defined on simplices of dimension 5. This defines a map from K_5 to K_5. First note that once we know the dimension on which it is applied, the superindexes are completely determined, so we can omit them. Second, note that if we apply the simplicial identities as rewriting rules, applied from left to right, we obtain an unique *canonical form* [1]: $\eta_4\eta_2\partial_1\partial_3$. In general, every composition of simplicial operators can be written as an equivalent expression consisting of a strictly decreasing sequence (w.r.t. its subindexes) of degeneracies and a strictly increasing sequence of faces. This canonical form is what we call a *simplicial term* and we represent it in ACL2 as a pair of lists of natural numbers, the first strictly decreasing and the second strictly increasing (in our example, `((4 2) (1 3))`).

[2]In our general definition of chain complex, the subindex ranges over $\mathbb{Z}$, so it is necessary to complete this definition with null groups and differentials in negative degrees.

In general, simplicial terms represent maps from K_n to K_m, where $m - n$ is called the *degree* of the term. The set of simplicial terms is endowed with a binary operation (composition): first, concatenate the terms and then reduce to canonical form. Thus, simplicial terms form a monoid.

Next, we can consider the *monoid ring* of the simplicial terms monoid over the integers $\mathbb{Z}$. This is the ring of *simplicial polynomials*. In ACL2, polynomials are stored in a *canonical form*: monomials have non-zero coefficients and they are sorted with respect to a total order on terms.[3] If all the terms in a simplicial polynomial represent maps from K_n to K_{n+r}, then the polynomial defines a linear map from $C_n(K)$ to $C_{n+r}(K)$, for any simplicial set K. This is the tool we used to prove in ACL2 the Normalization Theorem [7], by using polynomial expressions to describe the morphisms f, g and h in a reduction.

Our next observation was that the same tool could be useful in proving the correctness of other simplicial theorems, such as the Eilenberg-Zilber Theorem. We need the definitions of *Cartesian product* (of two simplicial sets) and of *tensor product* (of two chain complexes), in order to state that theorem.

Definition 4. *Given two simplicial sets K^1 and K^2, their* Cartesian product *is a new simplicial set, denoted by $K^1 \times K^2$, such that $(K^1 \times K^2)_n = K_n^1 \times K_n^2$ and faces and degeneracies, denoted as $\partial^\times$ and $\eta^\times$, are defined in a natural way: $\partial_i^\times(a, b) = (\partial_i a, \partial_i b)$ and $\eta_i^\times(a, b) = (\eta_i a, \eta_i b)$, respectively.*

The tensor product of two chain complexes can be defined in a general way, but in the following definition we focus on the tensor product of two *freely generated* chain complexes, the only case of application in our problem (because chain complexes associated to simplicial sets are freely generated).

Definition 5. *Given two freely generated chain complexes $C^1 := \{(C_n^1, d_n^1)\}_{n\in\mathbb{Z}}$ and $C^2 := \{(C_n^2, d_n^2)\}_{n\in\mathbb{Z}}$ (in other words, C_n^1 and C_n^2 are freely generated Abelian groups for all $n \in \mathbb{Z}$), the* tensor product *of C^1 and C^2, denoted by $C^1 \otimes C^2$, is the chain complex defined as follows. The groups $(C^1 \otimes C^2)_n$ are defined by the formula $(C^1 \otimes C^2)_n = \bigoplus_{p+q=n} C_p^1 \otimes C_q^2$, with $C_p^1 \otimes C_q^2$ the free abelian group generated by the pairs (x_p, y_q) (denoted $x_p \otimes y_q$), where x_p (y_q) ranges over the generators of C_p^1 (of C_q^2, respectively). Differentials are defined by $d_n^\otimes(x_p \otimes y_q) = d_p^1(x_p) \otimes y_q + (-1)^p x_p \otimes d_q^2(y_q)$ over generators*[4]*, and then extended linearly over elements of $(C^1 \otimes C^2)_n$.*

And, now, the statement:

Theorem 1 (Eilenberg-Zilber reduction). *Given two simplicial sets K^1 and K^2, there exists a reduction $C(K^1 \times K^2) \Longrightarrow C(K^1) \otimes C(K^2)$.*

Since we want to formalize, and execute, a proof of this theorem in ACL2, it will be necessarily a *constructive*

[3]Kenzo also stores combinations in a canonical form (ordered with respect to a total order over the set of generators in each dimension), in order to improve the efficiency of the operations among them.

[4]The operator $\otimes$ has been overloaded to denote its linear extension for combinations.

proof, providing explicitly the triple of the reduction[5]. Then, we need to consider mappings with the shape $t : \mathbb{Z}[K^1_p \times K^2_q] \to \mathbb{Z}[K^1_{p'} \times K^2_{q'}]$, where K^1 and K^2 are simplicial sets. Following the same ideas as above, we can represent such a transformation t as a polynomial over *pairs* of simplicial terms. We called it a *bivariate simplicial polynomial*, and we explain the formalization of that notion in ACL2 in the next section.

3. BIVARIATE SIMPLICIAL POLYNOMIALS

As we have said, the basic components of the reduction homomorphisms in the Eilenberg-Zilber theorem are mappings from $\mathbb{Z}[K^1_p \times K^2_q]$ to $\mathbb{Z}[K^1_{p'} \times K^2_{q'}]$. More concretely, these morphisms can be expressed as linear combinations of pairs of compositions of faces and degeneracies. To have a faithful formalization of the standard presentation of the theorem, these morphisms will have to be defined as functions in the ACL2 logic (and in fact that will be our approach in Subsection 5.2). Nevertheless, it turns out that most of the reasoning applied to prove the theorem is carried out viewing those pairs of compositions of simplicial operators (and their linear combinations) as symbolic expressions, and operating on them following certain rules derived from the simplicial identities. This is the point of view we adopt in this section, where we present what we call *bivariate simplicial polynomials*[6], linear combinations of pairs of compositions of simplicial operators; they are a representation of morphisms as symbolic expressions built using lists and natural numbers. In ACL2, the bivariate simplicial polynomials are stored in a canonical form: every component has to be a non-zero coefficient and they are sorted with respect to a total order on pairs of simplicial terms. We have defined the function `psp-p` that checks if an expression is a bivariate simplicial polynomial in this canonical form. For example, as the simplicial term $\eta_3\partial_1\partial_2$ is represented by the two element list `((3) (1 2))` and the simplicial term $\eta_1\partial_0$ by the list `((1) (0))`, then the pair of simplicial terms $(\eta_3\partial_1\partial_2, \eta_1\partial_0)$ is represented by the two element list `(((3) (1 2)) ((1) (0)))`, and the linear combination of pairs of simplicial terms $\boldsymbol{q_1} = 3\cdot(\eta_3\partial_1\partial_2, \eta_1\partial_0) - 2\cdot(\eta_4\eta_2\partial_3, \partial_0\partial_1)$ by the list `((3 (((3) (1 2)) ((1) (0)))) (-2 (((4 2) (3)) (() (0 1)))))`.

We define componentwise the composition of pairs of simplicial terms. Then, we also define on polynomials the operations of addition, composition and scalar (integer) product, formalizing the corresponding operations on the functions they represent. For example, the composition of $\boldsymbol{q_1}$ above and $2\cdot(\eta_2\partial_1, \eta_0\partial_1) - (\eta_4\eta_2, \partial_0\partial_1)$ is the polynomial $6\cdot(\eta_3\partial_1\partial_2, \eta_1\partial_1) - 3\cdot(\eta_3\eta_2\partial_1, \eta_1\partial_0\partial_1\partial_2) - 4\cdot(\eta_4\eta_2\partial_1, \partial_0\partial_1) + 2\cdot(\eta_5\eta_4\eta_2, \partial_0\partial_1\partial_2\partial_3)$, a result we obtain applying composition of pair of simplicial terms, distributing with respect to the sums and obtaining again a linear combination in canonical form. We defined in ACL2 three functions `add-psp-psp`, `cmp-psp-psp` and `scl-prd-psp`, respectively implementing addition, composition and scalar product on simplicial polynomials. We will denote these operations as $\boldsymbol{p_1} + \boldsymbol{p_2}$, $\boldsymbol{p_1}\cdot\boldsymbol{p_2}$ and $k\cdot\boldsymbol{p}$, respectively. We also denote $\mathbf{0}$ the zero polynomial (represented by the empty list in ACL2) and $\boldsymbol{id}$ the identity polynomial (that is a polynomial with only one pair of terms; each of these terms has empty lists of faces and degeneracies).

If we denote by $\mathcal{P}^\times$ the set of bivariate simplicial polynomials (that is, the set characterized by the function `psp-p`), we proved in ACL2 that $(\mathcal{P}^\times, +, \cdot)$ is a ring, with $\mathbf{0}$ being its identity with respect to addition and $\boldsymbol{id}$ the identity with respect to composition. For example, this is one of the properties proved, establishing right distributivity[7]:

THEOREM: `cmp-psp-psp-add-psp-psp-distributive-r`
$$(\boldsymbol{p}_1 \in \mathcal{P}^\times \wedge \boldsymbol{p}_2 \in \mathcal{P}^\times \wedge \boldsymbol{p}_3 \in \mathcal{P}^\times) \to \boldsymbol{p}_1\cdot(\boldsymbol{p}_2 + \boldsymbol{p}_3) = (\boldsymbol{p}_1\cdot\boldsymbol{p}_2) + (\boldsymbol{p}_1\cdot\boldsymbol{p}_3)$$

Some of these ring properties are not trivial to prove due to the fact that all these operations return its result in canonical form (see details in [8]). Nevertheless, note that the main advantage of requiring canonical forms is that we easily can check if two given polynomials represent the same function: just check if they are syntactically equal.

4. DERIVATIVES

We can use the bivariate simplicial polynomials to formalize the maps in the proof of the Eilenberg-Zilber Theorem. For instance, we can define the differential in the Cartesian product. First, let $\boldsymbol{\partial}^\times_i$ denote the pair of simplicial terms (∂_i, ∂_i) considered as a particular case of polynomial. Then, we can introduce the following function `cartesian-diff` that recursively defines $\boldsymbol{d}^\times_n$, the polynomial representing the differential in the Cartesian product[8]:

DEFINITION: $[\boldsymbol{d}^\times_n]$
`cartesian-diff`(n) :=
  if $n \notin \mathbb{N}^+$ **then** $\boldsymbol{\partial}^\times_0$
  else $(-1)^n \cdot \boldsymbol{\partial}^\times_n$ + `cartesian-diff`$(n-1)$

For example, $\boldsymbol{d}^\times_3$ is the polinomial $(\partial_0, \partial_0) - (\partial_1, \partial_1) + (\partial_2, \partial_2) - (\partial_3, \partial_3)$.

The most difficult morphism in the Eilenberg-Zilber Theorem is the one corresponding to h, the homotopy. Due to historical reasons this arrow is called *Shih* morphism [14]. To define polynomials representing *Shih*, we are going to define an operation on polynomials called *derivative* [12]. Given a simplicial term $\eta_{i_1}\ldots\eta_{i_k}\partial_{j_1}\ldots\partial_{j_l}$, its derivative is the simplicial term obtained inceasing by one the indexes of its operators, that is: $\eta_{i_1+1}\ldots\eta_{i_k+1}\partial_{j_1+1}\ldots\partial_{j_l+1}$. This operation is extended componentwise to pairs of simplicial terms, and by linearity, to polynomials. In our formalization, `derivative-psp`$(\boldsymbol{p})$ implements the derivative of a psp $\boldsymbol{p}$ (we will denote it as $\boldsymbol{p}'$). We can prove that the derivative is coherent regarding the operations of the polynomial ring, as stated by the following properties:

[5]There are more general versions of the Eilenberg-Zilber theorem (see [10]); but the formalization of those general versions seems harder than the task we undertake in our work

[6]Multivariate polynomials would appear if we extend the Eilenberg-Zilber reduction to any number of factors: $C(K^1 \times \ldots \times K^m) \Longrightarrow C(K^1) \otimes \ldots \otimes C(K^m)$. The generalization is straightforward, even if the formalization could become a bit cumbersome.

[7]In this and other statements, we adapt notations for the sake of readability, but they are the exact translation of ACL2 expressions; see [8].

[8]Note the expression between square brackets in the first line of the definition; in general, this will be the way we will show how a function will be denoted subsequently.

THEOREM: `derivative-psp-add-psp-psp`
$\boldsymbol{p}_1 \in \mathcal{P}^\times \wedge \boldsymbol{p}_2 \in \mathcal{P}^\times \rightarrow (\boldsymbol{p}_1 + \boldsymbol{p}_2)' = \boldsymbol{p}_1' + \boldsymbol{p}_2'$

THEOREM: `derivative-psp-cmp-psp-psp`
$\boldsymbol{p}_1 \in \mathcal{P}^\times \wedge \boldsymbol{p}_2 \in \mathcal{P}^\times \rightarrow (\boldsymbol{p}_1 \cdot \boldsymbol{p}_2)' = \boldsymbol{p}_1' \cdot \boldsymbol{p}_2'$

THEOREM: `derivative-psp-scl-prd-psp`
$\boldsymbol{p} \in \mathcal{P}^\times \wedge k \in \mathbb{Z} \rightarrow (k \cdot \boldsymbol{p})' = k \cdot \boldsymbol{p}'$

To prove these properties, we have to deal with how adding one to the subindexes of the simplicial operators affects the ring operations, taking into account that the polynomials are in canonical form. In particular, to prove the property `derivative-psp-cmp-psp-psp`, we needed two main lemmas.

First, we showed that the derivative distributes over composition of simplicial terms; recall that when we compose simplicial terms, they are returned in canonical form, and that canonical form is obtained exhaustively rewriting with the simplicial identities. Thus, the property on terms is a consequence of the fact that the simplicial identities are preserved if we add one to the subindexes. The second lemma deals with the ordering on terms needed for a polynomial to be considered in canonical form. We proved that this ordering is preserved by the derivative operation.

Note that these two lemmas were "suggested" by the ACL2 theorem prover, in a way we will explain now. Although the prover is automatic in the sense that there is no user interaction once a proof attempt starts, we can say that ACL2 is interactive in a wider sense. If a proof attempt fails, one can inspect the output, trying to guess in which point the prover digresses from the intended proof one has in mind. Usually this happens because it needs previously proved results that, used as simplification rewrite rules, would lead the prover to a successful proof.

In the case of the derivative of a composition, the two main lemmas mentioned above were suggested by inspecting failed proof attempts. In turn, to prove these two lemmas, several sublemmas were suggested about how derivatives affect to the canonical form of a simplicial term. Carrying out proofs in this way, the user provides the prover a collection of lemmas that finally let it to prove the intended lemma, only using induction and simplification. This is a standard way to interact with ACL2, called *The Method* [6] by the authors of the system. In the development of the formal proof of the Eilenberg-Zilber theorem, we followed *The Method*.

Finally, having defined derivatives, the function SH-pol(n) obtains the corresponding polynomial representing the *Shih* homomorphism in dimension n:

DEFINITION: $[\boldsymbol{SH}_n]$
SH-pol(n) :=
if $n \notin \mathbb{N}^+$ **then 0**
else $-1 \cdot ((\text{SH-pol}(n-1))' + (\sum_{i=0}^{n} \boldsymbol{EML}_{n-i,i} \cdot \boldsymbol{AW}_{n,i})' \cdot \boldsymbol{\eta}_0^\times)$

where $\boldsymbol{AW}$ and $\boldsymbol{EML}$ are the polynomials associated with the morphisms f and g, respectively, in the Eilenberg-Zilber reduction (see [14] for details), and $\boldsymbol{\eta}_0^\times$ is the polynomial (η_0, η_0).

5. EXECUTING AND EVALUATING POLYNOMIALS

Simplicial polynomials can be dealt with in two ways: as symbolic expressions, which can be handled by means of algorithms; and as codes for morphisms, which can be evaluated over chains of simplices. We will see now that both kinds of manipulations are useful for certified computing.

5.1 Simplifying and executing

The first point of view is very useful when proving properties, as we showed in the previous section. Identities between simplicial polynomials can be used as simplification rules for ACL2, in such a way that automation can be increased in proofs.

But we can use simplicial polynomials to help proofs in another way. Because ACL2 is also a programming environment, we can *execute* the recursive definitions, as that of $\boldsymbol{SH}$ in the previous section. Then we can also execute the expressions involved in statements. This could give us experimental insights about how proving the theorem, as well as guide us to the statement of new lemmas needed in the development. In fact, it has been the case in several of the most complicated parts of the mechanized proof of the Eilenberg-Zilber theorem. Let us illustrate these ideas with an example, only involving the *Shih* operator.

First, we can compute the explicit expression of $\boldsymbol{SH}$ for some (small) values of the dimension n.

```
ACL2 !>(sh-pol 4)
((-1 (((0) ()) ((4 3 2 1) (1 2 3))))
 (1 (((1) ()) ((4 3 2) (2 3))))
 (1 (((1 0) (4)) ((4 3 2) (1 2))))
 (-1 (((2) ()) ((4 3) (3))))
 (-1 (((2 0) (4)) ((4 3 1) (1 2))))
 (1 (((2 1) (4)) ((4 3) (2))))
 (-1 (((2 1 0) (3 4)) ((4 3) (1))))
 (1 (((3) ()) ((4) ())))
 (1 (((3 0) (4)) ((4 2 1) (1 2))))
 (-1 (((3 1) (4)) ((4 2) (2))))
 (1 (((3 1 0) (3 4)) ((4 2) (1))))
 (1 (((3 2) (4)) ((4) ())))
 (-1 (((3 2 0) (3 4)) ((4 1) (1))))
 (1 (((3 2 1) (3 4)) ((4) ())))
 (1 (((3 2 1 0) (2 3 4)) ((4) ())))
 (-1 (((4 0) (4)) ((3 2 1) (1 2))))
 (1 (((4 1) (4)) ((3 2) (2))))
 (-1 (((4 1 0) (3 4)) ((3 2) (1))))
 (-1 (((4 2) (4)) ((3) ())))
 (1 (((4 2 0) (3 4)) ((3 1) (1))))
 (-1 (((4 2 1) (3 4)) ((3) ())))
 (-1 (((4 2 1 0) (2 3 4)) ((3) ())))
 (1 (((4 3) (4)) ((3) ())))
 (-1 (((4 3 0) (3 4)) ((2 1) (1))))
 (1 (((4 3 1) (3 4)) ((2) ())))
 (1 (((4 3 1 0) (2 3 4)) ((2) ())))
 (-1 (((4 3 2) (3 4)) ((2) ())))
 (-1 (((4 3 2 0) (2 3 4)) ((1) ())))
 (1 (((4 3 2 1) (2 3 4)) ((1) ())))
 (-1 (((4 3 2 1 0) (1 2 3 4)) ((0) ()))))
```

Note that some of the pairs in the result are representing *degenerate* operators, that is, they produce degenerate simplices when they are evaluated over chains of simplices. In the Cartesian product, this means that they have a common index in the degeneracy list of each simplicial term in the pair. With our syntax, they are detected because there is a common integer in the first list of each simplicial term. For instance, in the last simplicial term `(((4 3 2 1 0) (1 2 3 4)) ((0) ()))` the index 0 is present in `(4 3 2 1 0)` and `(0)`. As bivariate simplicial polynomials, the degenerate pairs are kept, but they are eliminated later in a *normalization* process because the morphisms they represent always produce degenerate simplices.

By examining the previous expression, the ocurrence of *shuffles* can be expected, as foreseen in the formula presented in [14]. A (p,q)-shuffle $(\alpha_1,\ldots,\alpha_p,\beta_1,\ldots,\beta_q)$ is a permutation of the set $\{0,1,\ldots,p+q-1\}$ such that $\alpha_i < \alpha_{i+1}$ and $\beta_j < \beta_{j+1}$, when $i = 1,\ldots,p-1$ and $j = 1,\ldots,q-1$. Let us look for such shuffles in the two first monomials of the previous expression. The first monomial is `(-1 (((0) ()) ((4 3 2 1) (1 2 3))))`. After the coefficient -1, we have a pair of terms: the first one is `((0) ())`, denoting a term with only a degeneracy η_0 and no faces; and the second one is `((4 3 2 1) (1 2 3))` with $\eta_4\eta_3\eta_2\eta_1$ in the degenerate part. Here we find the $(1,4)$-shuffle $((0),(1,2,3,4))$. In the second monomial, `(1 (((1) ()) ((4 3 2) (2 3))))`, we find the *first derivative* (the level of derivation is determined by the smallest degenerate index in the first term) of the $(1,3)$-shuffle $((0),(1,2,3))$. By carefully examining the results of execution, new organizations of the formula presented in [14] can be found, that can be helpful when proving some properties of morphisms.

For instance, the morphism *Shih* must satisfy condition (5) in the definition of a reduction (that is to say, $h \circ h = 0$). Since the Eilenberg-Zilber theorem is stated in terms of the *normalized* chain complex, this means that the composition of $\boldsymbol{SH}$ with itself must give always degenerate terms. We can compute that composite for particular cases:

```
ACL2 !>(cmp-psp-psp (sh-pol 3) (sh-pol 2))
((-1 (((2 1 0) (2)) ((3 1) ())))
 (1 (((3 1 0) (2)) ((2 1) ())))
 (-1 (((3 1 0) (2)) ((3 2) ())))
 (-1 (((3 2 0) (2)) ((2 1) ())))
 (1 (((3 2 0) (2)) ((3 1) ())))
 (1 (((3 2 1) (2)) ((2 1) ())))
 (-1 (((3 2 1) (2)) ((3 1) ())))
 (1 (((3 2 1 0) (1 2)) ((1 0) ())))
 (-1 (((3 2 1 0) (1 2)) ((2 0) ())))
 (1 (((3 2 1 0) (1 2)) ((3 0) ()))))
```

It can be checked that, as foreseen, all the terms are denoting degenerate simplices in the Cartesian product, because there is always a common integer in the first list of each simplicial term.

Thus, by inspecting those expressions, we can *test* that the property holds, and even better, we can establish conjectures to organize the general proof of the property.

5.2 Evaluating

In order to interpret a simplicial polynomial as a morphism between chain complexes, we need to represent in ACL2 chain complexes (associated with simplicial sets) and then define how symplicial polynomials can be evaluated on chains of simplices. Finally, in order to get an executable function (corresponding to a simplicial polynomial) we need to instantiate generic simplicial sets (appearing in the proofs) to concrete ones.

Let us first deal with how we represent simplicial sets. Note that a simplicial set is characterized by a set K and families of functions (faces and degeneracies) satisfying certain properties. Since the Eilenberg-Zilber theorem is about any two simplicial sets, we have to introduce them in a completely generic way. Although in ACL2 the usual way to introduce functions in the logic is by the definition principle (using **`defun`**), it also provides the *encapsulation* principle (using **`encapsulate`**), which allows to introduce functions in the logic without defining them completely, only stating about them some assumed properties [6].

In our formalization, a generic simplicial set is defined by means of three functions `K`, `d` and `n`. The function `K` is a predicate of two arguments, with the intended meaning that `K`(n,x) holds when $x \in K_n$. Faces and degeneracies are represented, respectively, by the functions `d` and `n`, both with three arguments. The idea is that `d`(m,i,x) and `n`(m,i,x) respectively represent $\partial_i^m(x)$ and $\eta_i^m(x)$. These three functions are introduced using **`encapsulate`**, only assuming about them well-definedness and the simplicial identities. For example, the following are the assumptions corresponding respectively to the well-definedness of `d` and the first simplicial identity:

ASSUMPTION: **`d-well-defined`**

$$(x \in K_m \wedge m \in \mathbb{N}^+ \wedge i \in \mathbb{N} \wedge i \le m) \rightarrow \partial_i^m(x) \in K_{m-1}$$

ASSUMPTION: **`simplicial-id1`**

$$(x \in K_m \wedge m,i,j \in \mathbb{N} \wedge j \le i \wedge i < m \wedge 1 < m) \rightarrow \partial_i^{m-1}(\partial_j^m(x)) = \partial_j^{m-1}(\partial_{i+1}^m(x))$$

We omit here the rest of the assumptions (i.e., well-definedness of `n` and the rest of the simplicial identities), since they are stated in an analogous way.

As for the formalization of chains of simplices, since they are formal linear combinations of non-degenerate simplices, it is quite natural to represent them as lists of pairs of an integer coefficient and a non-degenerate simplex. As with polynomials, we consider chains in canonical form (as *Kenzo* does; see footnote 3): we do not allow zero coefficients and we require the pairs to be increasingly ordered with respect to a strict ordering on simplices. The following function `scn-p` defines chains in a given dimension n. It uses the auxiliary functions `ssn-p`, which recognizes pairs of a non-null integer and a non-degenerate simplex, and `ssn-<` which defines the lexicographic strict order between such pairs:

DEFINITION: $[c \in C_n(K)]$

scn-p(n,c) :=
  if endp(c) **then** c = **nil**
  elseif endp(rest(c))
    then ssn-p(n,first(c)) $\wedge$ rest(c) = **nil**
  else ssn-p(n,first(c)) $\wedge$
      ssn-<(n,first(c),second(c)) $\wedge$
      scn-p(n,rest(c))

We also define addition of chains, and the scalar product of an integer and a chain. These operations act on chains in the canonical form described above, and return chains also in canonical form. We proved that the set of chains of a given dimension is an Abelian group with respect to addition, where the identity is represented by the empty list.

Now, we have to formally specify the functional interpretation of a polynomial. That is, we define an ACL2 function such that given a polynomial and a chain of pairs of simplices of a given dimension, it computes the result of evaluating the function that the polynomial is supposed to represent, on the given chain.

First, we have to define some well-formedness conditions on polynomials. Think for example in the following simplicial term: $\eta_5\eta_1\partial_3$. This term cannot be interpreted as a function on $C_4(K)$, regardless of the simplicial set K, because in such case, η_5 would have to be applied to a simplex in $C_4(K)$, which is not possible. Nevertheless, it makes sense to apply it to any chain of dimension $n \geq 5$. We will say that a simplicial term is *valid for dimension m*, when interpreted as composition of simplicial operators, can be applied to any simplex of dimension m. Another notion to take into account is what we called previously the *degree* of a term: if a term is valid for n and it represents a function from K_n to K_m, its degree is $m - n$ (for example, the degree of the previous term is 1). Extending these concepts to pairs, we will say that a pair of simplicial terms (t_1, t_2) is valid for dimension (m_1, m_2) with degree (j_1, j_2) if t_i is valid for m_i and with degree j_i $(i = 1, 2)$. We say that a polynomial is *well-formed for dimension* (m_1, m_2) if all its terms are valid for that dimension and with the same degree. The *degree* of a polynomial is the common degree of its terms.

Well-formed polynomials for dimension (m_1, m_2) represent valid morphisms whose evaluation can be defined on $\mathbb{Z}[K^1_{m_1} \times K^2_{m_2}]$, where K^1 and K^2 are simplicial sets. We have defined in ACL2 a function eval-psp($\boldsymbol{p}$,m_1,m_2,c) that computes the result of evaluating a polynomial $\boldsymbol{p}$ on a linear combination c of pairs of simplices of dimension (m_1, m_2). This function does not remove degenerate simplices because this property is different if the result is in the Cartesian product or in the tensor product, but the evaluation is the same in both. We also proved that eval-psp is a homomorphism on the ring of polynomials. For example, under the corresponding well-formedness conditions, the evaluation of the composition of two polynomials is equal to the composition of the evaluations of the polynomials, and analogously for addition and scalar product. This is proved in a similar way as it is described in [7] for "univariate" simplicial polynomials.

Finally the definition of the SH function from the corresponding polynomial is as follows (SH_n is a function from $C_n(K^1 \times K^2)$ to $C_{n+1}(K^1 \times K^2)$). We can prove that the polynomial $\boldsymbol{SH}_n$ defined at the end of Section 4, is well-formed for dimension (n, n), with degree $(1, 1)$. So it is valid to define SH_n on a given chain, as first evaluating $\boldsymbol{SH}_n$ on the chain and then eliminate degenerate addends with respect to the Cartesian product (the function Fx-norm is in charge of eliminating these degenerate elements)

DEFINITION: $[SH_n(c)]$
SH(n,c) := Fx-norm($n+1$,eval-psp($\boldsymbol{SH}_n$,n,n,c))

Therefore, we have obtained a function SH from the simplicial polynomial associated with the morphism *Shih*. Nevertheless, it is not still executable, because proofs are carried out on *generic* simplicial sets, ensuring the properties are true for any two simplicial sets. In order to get running examples, we instantiate the previous generic theory over two concrete simplicial sets, proving all the assumptions that have been set on the generic simplical sets. As an example, we use the *standard simplex* Δ [10]. This simplicial set has some universal properties, since the simplicial identities are the unique constraints in it. In particular, any generic formula relating simplicial equalities will be faithfully drawn on Δ (see [10]). The definition of this simplicial set is the following: simplices in Δ are non-decreasing lists of natural numbers (lists of length $n + 1$ if we are in dimension n); a face of index i consists in erasing the element at position i; and a degeneracy of index i consists in repeating the element at position i in the list.

5.3 An example

Once the proof of the Eilenberg-Zilber theorem has been instantiated on the standard simplex Δ, we can *run* the different morphisms. We concentrate on the *Shih* morphism, being the more complex one. Furthermore, we can make computing *Kenzo* in the same examples, and then compare both results. Even if the data structures used in our implementation are different from that of *Kenzo*, and consequently the algorithms are also different, in essence both implementations are based on the same definition of the triple of the reduction.

The test is running over the Cartesian product $\Delta \times \Delta$, and then applied over the chain with only one component, with coefficient 1 and generator $((0, 1, \ldots, n), (0, 1, \ldots, n))$, belonging to $C_n(\Delta \times \Delta)$. Next we include the chain obtained by ACL2, in the case $n = 4$.

```
ACL2 !>(SH 4 (Delta1 4))
((1 ((0 0 0 0 0 1) (0 1 2 3 4 4)))
 (-1 ((0 0 0 0 1 1) (0 1 2 3 3 4)))
 (-1 ((0 0 0 0 1 2) (0 2 3 4 4 4)))
 (1 ((0 0 0 1 1 1) (0 1 2 2 3 4)))
 (1 ((0 0 0 1 1 2) (0 2 3 3 4 4)))
 (-1 ((0 0 0 1 2 2) (0 2 3 3 3 4)))
 (1 ((0 0 0 1 2 3) (0 3 4 4 4 4)))
 (-1 ((0 0 1 1 1 1) (0 1 1 2 3 4)))
 (-1 ((0 0 1 1 1 2) (0 2 2 3 4 4)))
 (1 ((0 0 1 1 2 2) (0 2 2 3 3 4)))
 (-1 ((0 0 1 1 2 3) (0 3 3 4 4 4)))
 (-1 ((0 0 1 2 2 2) (0 2 2 2 3 4)))
 (1 ((0 0 1 2 2 3) (0 3 3 3 4 4)))
 (-1 ((0 0 1 2 3 3) (0 3 3 3 3 4)))
 (-1 ((0 0 1 2 3 4) (0 4 4 4 4 4)))
 (1 ((0 1 1 1 1 2) (0 1 2 3 4 4)))
 (-1 ((0 1 1 1 2 2) (0 1 2 3 3 4)))
 (1 ((0 1 1 1 2 3) (0 1 3 4 4 4)))
 (1 ((0 1 1 2 2 2) (0 1 2 2 3 4)))
 (-1 ((0 1 1 2 2 3) (0 1 3 3 4 4)))
 (1 ((0 1 1 2 3 3) (0 1 3 3 3 4)))
 (1 ((0 1 1 2 3 4) (0 1 4 4 4 4)))
 (1 ((0 1 2 2 2 3) (0 1 2 3 4 4)))
 (-1 ((0 1 2 2 3 3) (0 1 2 3 3 4)))
 (-1 ((0 1 2 2 3 4) (0 1 2 4 4 4)))
 (1 ((0 1 2 3 3 4) (0 1 2 3 4 4))))
```

Let us interpret some of the components of this expression. The first one is (1 ((0 0 0 0 0 1) (0 1 2 3 4 4))). That means that the operator $(\eta_3\eta_2\eta_1\eta_0\partial_2\partial_3\partial_4, \eta_4)$ has been applied to ((0 1 2 3 4) (0 1 2 3 4)). And this corresponds to the bivariate simplicial monomial (1 (((3 2 1 0) (2 3 4)) ((4) ()))) which appears in the expression obtained by executing (sh-pol 4) in Subsection 5.1. It is the same for the rest of monomials, except for the degenerate ones, that, as announced in 5.1, have been cancelled during the normalization process. Of course, the results are also coherent in all dimensions in which the testing has been carried out.

The result of executing `(SH-n 4 (Delta1 4))` in ACL2 is also equal (up to combinations representation) to the one obtained with *Kenzo* (and it has been always the case in all the *automated* test cases we have checked). In addition, the implementation in *Kenzo* was based in the explicit formulas from [14]. As explained in Subsection 5.1, these formulas nicely correspond with the bivariate simplicial polynomials defined in Section 4. And it is from evaluating this polynomials how we have obtained the ACL2 executable morphisms. The coherence of these four components in this architecture gives a solid evidence that the formulas appearing in our constructive proof of the Eilenberg-Zilber theorem are the same implemented in *Kenzo*. As a consequence, the confidence in the correctness of *Kenzo* is reinforced.

6. CONCLUSIONS

In this paper, we have illustrated the role of symbolic manipulation in a context of certified computing. Even if the case study is taken from a rather specialized area (namely, Computational Simplicial Topology), we hope that the consequences of our methods can be useful for a range of researchers, working in automated reasoning, in algebraic computing or, at it is our case, in the frontier between both disciplines.

The main technical contribution of the paper is the introduction of *bivariate simplicial polynomials*, a data structure which allows us to enhance the ACL2 theorem prover with a kind of *algebraic* rewriting, greatly increasing the automation of the proofs.

Simplicial polynomials can be viewed from two different perspectives. In the first one, a simplicial polynomial is representing symbolically a family of natural transformations; interestingly enough, this allows us a kind of *logical* reduction: representing some higher-order constructs (i.e. natural transformations between functors) as first-order elements (lists of integers); recall that ACL2 is a first-order tool.

From the second point of view, a simplicial polynomial can be used to define morphisms between chain complexes, by means of a evaluation of the polynomial over chains of simplices.

In both perspectives executability is important (ACL2, as some other proof assistants, allows running definitions and expressions). In the first one, unfolding recursive definitions of polynomials was useful for conjecturing some lemmas which guided the proof of the main theorems. In the second perspective, executing morphisms (on concrete simplicial sets) permits us an *automated* testing of the *Kenzo* program: *Kenzo* results were confronted to their ACL2 verified counterpart.

Verifying technology is still too poor to undertake the complete correctness analysis of a software system as complex as *Kenzo*. So, we approach the problem with several tools and apply a separation of concerns strategy: sometimes we use proof assistants to verify the correctness of *algorithms* and in other occasions we concentrate on verifying actual running code. In this paper, we focused on a concrete result in Simplicial Topology: the Eilenberg-Zilber theorem, showing the feasibility and usefulness of our proposals.

As for further work, there is many room for extensions and improvements. Let us mention simply the possibility of using ACL2 tools (compilation, guards, single-threaded objects, and so on; see [6]) to speed up computation in the certified side; then the testing of *Kenzo* would be more complete, and so the confidence in its correctness would be still more reinforced.

7. REFERENCES

[1] M. Andrés, L. Lambán, J. Rubio, and J.-L. Ruiz-Reina. Formalizing Simplicial Topology in ACL2. *Proceedings ACL2 Workshop 2007*, 34–39, 2007.

[2] J. Aransay, C. Ballarin, and J. Rubio. A Mechanized Proof of the Basic Perturbation Lemma. *Journal of Automated Reasoning*, 40(4):271–292, 2008.

[3] F. Boudaoud, F. Caruso, and M.F. Roy. Certificates of Positivity in the Bernstein Basis. *Discrete and Computational Geometry*, 39(4):639–655, 2008.

[4] C. Domínguez, and J. Rubio. Effective homology of bicomplexes, formalized in Coq. *Theoretical Computer Science*, 412(11):962–970, 2011.

[5] X. Dousson, F. Sergeraert, and Y. Siret. The Kenzo program, Institut Fourier, Grenoble, 1999. `http://www-fourier.ujf-grenoble.fr/~sergerar/Kenzo/`

[6] M. Kaufmann, P. Manolios, and J S. Moore. *Computer-Aided Reasoning: An Approach.* Kluwer, 2010.

[7] L. Lambán, F.J. Martín-Mateos, J. Rubio, and J.-L. Ruiz-Reina. Formalization of a normalization theorem in simplicial topology. *Annals of Mathematics and Artificial Intelligence* 64(1):1–37, 2012.

[8] L. Lambán, F.J. Martín-Mateos, J. Rubio, and J.-L. Ruiz-Reina. Formalization of the Eilenberg-Zilber Theorem, 2012. `http://www.glc.us.es/fmartin/acl2/eztheorem`

[9] F.J. Martín-Mateos, J. Rubio, and J.-L. Ruiz-Reina. ACL2 verification of simplicial degeneracy programs in the Kenzo system, *Proceedings Calculemus 2009, Lecture Notes in Artificial Intelligence*, 5625:106–121, 2009.

[10] J. P. May. *Simplicial objects in Algebraic Topology.* Van Nostrand, 1967.

[11] G. C. Necula, and P. Lee. Proof-Carrying Code. Technical Report CMU-CS-96-165, 1996.

[12] P. Real. Homological Perturbation Theory and Associativity. *Homology, Homotopy and Applications* 2(5):51–88, 2000.

[13] A. Romero, and J. Rubio. Homotopy groups of suspended classifying spaces: an experimental approach. To appear in Mathematics of Computation, 2013.

[14] J. Rubio. Homologie effective des espaces de lacets itérés : un logiciel. Thèse, Institut Fourier, 1991. `http://dialnet.unirioja.es/servlet/tesis?codigo=1331`

On the Complexity of Solving Bivariate Systems: The Case of Non-singular Solutions

Romain Lebreton
LIRMM, UMR 5506 CNRS
Université de Montpellier II
Montpellier, France
lebreton@lirmm.fr

Esmaeil Mehrabi
Computer Science Dept.
Western University
London, ON, Canada
emehrab@uwo.ca

Éric Schost
Computer Science Dept.
Western University
London, ON, Canada
eschost@uwo.ca

ABSTRACT

We give an algorithm for solving bivariate polynomial systems over either $k(T)[X, Y]$ or $\mathbb{Q}[X, Y]$ using a combination of lifting and modular composition techniques.

Categories and Subject Descriptors

I.1.2 [**Computing Methodologies**]: Symbolic and Algebraic Manipulation—*Algebraic algorithms*

Keywords

Bivariate polynomial systems; complexity.

1. INTRODUCTION AND MAIN RESULTS

We investigate the complexity of solving bivariate polynomial systems. This question is interesting in its own right, but it also plays an important role in many higher-level algorithms, such as computing the topology of plane and space curves [13, 8] or solving general polynomial systems [18].

Many recent contributions on this question discuss computing real solutions of bivariate systems with integer or rational coefficients [15, 12, 30, 4, 14], by a combination of symbolic elimination and real root isolation techniques. Our interest here is on complexity of the "symbolic" component of such algorithms. One of our main results says that we can solve bivariate systems with integer coefficients in essentially optimal time, at least for non-singular solutions.

Geometric description. Let $\mathbb{A}$ be a domain, let $\mathbb{K}$ be its field of fractions and let $\overline{\mathbb{K}}$ be an algebraic closure of $\mathbb{K}$.

Let X, Y be the coordinates and let $Z \subset \overline{\mathbb{K}}^2$ be a finite set defined over $\mathbb{K}$ and of cardinality δ (so the defining ideal $I \subset \mathbb{K}[X, Y]$ of Z is generated by polynomials in $\mathbb{K}[X, Y]$). To describe Z, one may use a Gröbner basis of I, say for the lexicographic order $Y > X$. Such bases can however be unwieldy (they may involve a large number of polynomials, making modular computations difficult). Triangular decompositions are an alternative for which this issue is alleviated.

ISSAC'13, June 26–29, 2013, Boston, Massachusetts, USA.

Geometrically, performing a triangular decomposition of the defining ideal of Z amounts to writing Z as the disjoint union of finitely many *equiprojectable sets.* Let $\pi : \overline{\mathbb{K}}^2 \to \overline{\mathbb{K}}$ be the projection on the X-space given by $(x, y) \mapsto x$. To $p = (x, y)$ in Z, we associate the positive integer $N(Z, p)$ defined as the cardinality of the fiber $\pi^{-1}(x) \cap Z$: this is the number of points in Z lying above x. We say that Z is *equiprojectable* if there exists a positive integer n such that $N(Z, p) = n$ for all $p \in Z$ (see [10] for illustrations).

It is proved in [3] that Z is equiprojectable if and only if its defining ideal I admits a Gröbner basis for the lexicographic order $Y > X$ that is a *monic triangular set, i.e.* of the form $\mathbf{T} = (U(X), V(X, Y))$, with U and V monic in respectively X and Y and with coefficients in $\mathbb{K}$ (that result holds over a perfect field, so it applies over $\mathbb{K}$; the fact that I has generators in $\mathbb{K}[X, Y]$ implies that $\mathbf{T}$ has coefficients in $\mathbb{K}$). The degree $m = \deg(U, X)$ is the cardinality of $\pi(Z)$, and the equalities $n = \deg(V, Y)$ and $\delta = mn$ hold; we will say that $\mathbf{T}$ has *bidegree* (m, n).

When Z is not equiprojectable, it can be decomposed into equiprojectable sets, usually in a non-unique manner. The *equiprojectable decomposition* [10] is a canonical way to do so: it decomposes Z into subsets $Z_{n_1}, \dots, Z_{n_s}$, where for all $i \in \{1, \dots, s\}$, Z_{n_i} is the set of all $p \in Z$ for which $N(Z, p) = n_i$. This decomposition is implicit in the Cerlienco-Murredu description of the lexicographic Gröbner basis of the defining ideal of Z [7]; it can also be derived from Lazard's structure theorem for bivariate Gröbner bases [22].

If Z is defined over $\mathbb{K}$, then all Z_{n_i} are defined over $\mathbb{K}$ as well, so they can be represented by monic triangular sets

$$\mathbf{T}_1 \left| \begin{array}{l} V_1(X,Y) \\ U_1(X) \end{array} \right. \quad \cdots \quad \mathbf{T}_s \left| \begin{array}{l} V_s(X,Y) \\ U_s(X) \end{array} \right. \qquad (1)$$

with coefficients in $\mathbb{K}$. If we let $m_i = |\pi(Z_{n_i})|$, then $\mathbf{T}_i$ has bidegree (m_i, n_i) for all i, and $\sum_{i \le s} m_i n_i = \delta$.

By abuse of notation, we will call the family of monic triangular sets $\mathscr{T} = (\mathbf{T}_1, \dots, \mathbf{T}_s)$ the *equiprojectable decomposition* of Z. If I is a radical ideal of $\mathbb{K}[X, Y]$ that remains radical in $\overline{\mathbb{K}}[X, Y]$, its zero-set Z is defined over $\mathbb{K}$; then, we define the equiprojectable decomposition of I as that of Z.

Solving systems. Let now F and G be in $\mathbb{A}[X, Y]$. In this paper, we are interested in the set $Z(F, G)$ of *non-singular solutions* of the system $F = G = 0$, that is, the points (x, y) in $\overline{\mathbb{K}}^2$ such that $F(x, y) = G(x, y) = 0$ and $J(x, y) \neq 0$, where J is the Jacobian determinant of (F, G). Remark that $Z(F, G)$ is a finite set, defined over $\mathbb{K}$; if F and G have total degree at most d, then $Z(F, G)$ has cardinality $\delta \le d^2$.

For instance, for generic F and G, $Z(F,G)$ coincides with their whole zero-set $V(F,G)$, it is equiprojectable ($s = 1$), the corresponding triangular set $\mathbf{T} = \mathbf{T}_1$ takes the form $\mathbf{T} = (U(X), Y - \eta(X))$ and U is (up to a constant in $\mathbb{K}$) the resultant of F and G in Y.

Given F and G, our goal will be, up to a minor adjustment, to compute the triangular sets $\mathscr{T} = (\mathbf{T}_1, \ldots, \mathbf{T}_s)$ that define the equiprojectable decomposition of $Z(F,G)$.

Representing these polynomials requires $O(d^2)$ elements of $\mathbb{K}$. We will show below that one can compute them using $O\tilde{}(d^3)$ operations in $\mathbb{K}$, where $O\tilde{}(\,)$ indicates the omission of logarithmic factors. It is a major open problem to compute $\mathscr{T}$ in time $O\tilde{}(d^2)$, just like it is an open problem to compute the resultant of F and G in such a cost [16, Problem 11.10].

Size considerations. In this paper, we are mainly interested in a refinement of this situation to cases where $\mathbb{A}$ is endowed with a "length" function; in such cases, the cost analysis must take this length into account. Rather than giving an axiomatic treatment, we will assume that we are in one of the following situations:

- $\mathbb{A} = k[T]$ and $\mathbb{K} = k(T)$, for a field k, where we use the length function $\lambda(a) = \deg(a)$, for $a \in \mathbb{A} - \{0\}$;
- $\mathbb{A} = \mathbb{Z}$ and $\mathbb{K} = \mathbb{Q}$, where we use the length function $\lambda(a) = \log(|a|)$, for $a \in \mathbb{A} - \{0\}$.

In both cases, the length of $a \in \mathbb{A}$ represents the amount of storage needed to represent it, in terms of elements of k, resp. bits. It will be useful to introduce a notion of length for polynomials with coefficients in $\mathbb{K}$: if P is such a polynomial, $\lambda(P)$ denotes the maximum of the lengths $\lambda(n_i)$ and $\lambda(d_i)$, where n_i and d_i are the numerators and denominators of the coefficients of P, when written in reduced form using a common denominator.

When $\mathbb{A} = k[T]$, we are studying the intersection of two surfaces in a 3-dimensional space with coordinates T, X, Y; the output describes the solution curve for generic T.

In that case, write again $d = \max(\deg(F), \deg(G))$, as well as $\ell = \max(\lambda(F), \lambda(G))$. Then, the polynomials $U_1, \ldots, U_s$ in the equiprojectable decomposition (1) of $Z(F,G)$ are in $k(T)[X]$, and the sum of their degrees in X is at most d^2. These polynomials are all factors of the resultant $\mathrm{res}(F,G,Y)$, which implies that $\lambda(U_i)$ is at most $2d\ell$ for each i, so that representing them involves $O(d^3\ell)$ coefficients in k.

For the polynomials $V_1, \ldots, V_s$, however, the bounds are worse: [11] proves that $\lambda(V_i)$ only admits a weaker bound of order $d^3\ell + d^4$, so they involve $O(d^5\ell + d^6)$ coefficients in k. Practice shows that these bounds are realistic: the polynomials V_i are usually much larger than the polynomials U_i. In order to resolve this issue, we will use the polynomials $N_1, \ldots, N_s$ defined by $N_i = U_i' V_i \bmod U_i$ for all i. Then, Theorem 2 from [11] combined with the bi-homogeneous Bézout bound shows that $\lambda(N_i) \le 2d\ell + d^2$ for all i; thus, storing these polynomials uses $O(d^3\ell + d^4)$ coefficients in k.

Entirely similar considerations apply in the case $\mathbb{A} = \mathbb{Z}$; in that case, Theorem 1 from [11] and an arithmetic Bézout theorem [21] prove that $\lambda(U_i) \le 2d\ell + 24d^2$, and similarly for $\lambda(N_i)$, so $O(d^3\ell + d^4)$ bits are sufficient to store them.

We call *modified equiprojectable decomposition* of $Z(F,G)$ the set of polynomials $\mathscr{C} = (\mathbf{C}_1, \ldots, \mathbf{C}_s)$, with $\mathbf{C}_i = (U_i, N_i)$. These are not monic triangular sets anymore (N_i is not monic in Y), but *regular chains* [2]. In the particular case where $s = 1$ and $V = V_1$ has the form $V(X,Y) = Y - \eta(X)$, it coincides with the rational univariate representation [29].

Main results. Our main results are the following theorems, that give upper bounds on the cost of computing the modified equiprojectable decomposition. We start with the case $\mathbb{A} = k[T]$, where we count operations in k at unit cost. Our second result concerns the case $\mathbb{A} = \mathbb{Z}$; in this case, we measure the cost of our algorithm using bit operations.

In what follows, we let $\mathsf{M} : \mathbb{N} \to \mathbb{N}$ be such that over any ring, univariate polynomials of degree less than d can be multiplied in $\mathsf{M}(d)$ ring operations, under the super-linearity conditions of [16, Ch. 8]: using FFT techniques, we can take $\mathsf{M}(d) \in O(d\log(d)\log\log(d))$. We also let ω be such that we can multiply $n \times n$ matrices using $O(n^\omega)$ ring operations, over any ring. The best known bound is $\omega < 2.38$ [33].

THEOREM 1. *Let k be a field and let F, G be in $k[T][X,Y]$, with $d = \max(\deg(F), \deg(G))$ and $\ell = \max(\lambda(F), \lambda(G))$. If k has characteristic at least $4d^2(6d^2 + 9d\ell)$, one can compute the modified equiprojectable decomposition of $Z(F,G)$ over $k(T)[X,Y]$ by a probabilistic algorithm with probability of success at least $1/2$, using*

$$O\left(\mathsf{M}(d^2)\mathsf{M}(d\ell + d^2)d^{(\omega-1)/2}\log(d\ell)\right) \subset O\tilde{}\left(d^{3.69}\ell + d^{4.69}\right)$$

operations in k.

THEOREM 2. *Let $\varepsilon > 0$, let F, G be in $\mathbb{Z}[X,Y]$, and write $d = \max(\deg(F), \deg(G))$ and $\ell = \max(\lambda(F), \lambda(G))$. One can compute the modified equiprojectable decomposition of $Z(F,G)$ over $\mathbb{Q}[X,Y]$ by a probabilistic algorithm with probability of success at least $1/2$, using $O(d^{3+\varepsilon}\ell + d^{4+\varepsilon})$ bit operations.*

In both cases, one can easily obtain a cost of $O\tilde{}(d^4\ell + d^5)$ using modular methods: e.g., over $\mathbb{A} = k[T]$, solve the system at $O(d\ell + d^2)$ values of T, each of which costs $O\tilde{}(d^3)$ operations in k, and use rational function interpolation. Our main contribution is to show that this direct approach is sub-optimal; over $\mathbb{A} = \mathbb{Z}$, the cost of our algorithm almost matches the known upper bounds on the output size.

The structure of our algorithm is the same in both cases: we compute $Z(F,G)$ modulo an ideal $\mathfrak{m}$ of $\mathbb{A}$, lift the result modulo a high power of $\mathfrak{m}$ and reconstruct all rational function coefficients. This approach is similar to the algorithm of [10]; the key difference is in how we implement the lifting process. The result in [10] assumes that the input system is given by means of a straight-line program; here, we assume that the input is dense, and we rely on fast *modular composition techniques.*

Our results imply similar bounds for computing the resultant $R = \mathrm{res}(F,G,Y)$, at least for systems without singular roots: one can reconstruct R from $U_1, \ldots, U_s$, taking care if needed of the leading coefficients of F and G in Y; we leave the details to the reader. The main challenge is to handle systems with multiplicities without affecting the complexity. We expect that deflation techniques will make this possible.

After a section of preliminaries, we give (Section 3) an algorithm to compute $Z(F,G)$ over an arbitrary field in time $O\tilde{}(d^3)$. Section 4 is devoted to computing normal forms modulo triangular sets by means of modular composition techniques; this is the key ingredient of the main algorithm given in Section 5. Section 6 presents experimental results.

2. PRELIMINARIES

2.1 Notation and basic results

In the introduction, we defined monic triangular sets with coefficients in a field. We will actually allow coefficients in a ring $\mathbb{A}$; as in the introduction, all monic triangular sets will be bivariate, that is, in $\mathbb{A}[X, Y]$.

For positive integers m, n, $\mathbb{A}[X]_m$ denotes the set of all $F \in \mathbb{A}[X]$ such that $\deg(F) < m$, and $\mathbb{A}[X, Y]_{m,n}$ the set of all $F \in \mathbb{A}[X, Y]$ such that $\deg(F, X) < m$ and $\deg(F, Y) < n$. Using Kronecker's substitution, we can multiply polynomials in $\mathbb{A}[X, Y]_{m,n}$ in $O(\mathsf{M}(mn))$ operations in $\mathbb{A}$.

For a monic triangular set $\mathbf{T}$ in $\mathbb{A}[X, Y]$, the monicity assumption makes the notion of remainder modulo the ideal $\langle \mathbf{T} \rangle$ well-defined; if $\mathbf{T}$ has bidegree (m, n), then for any F in $\mathbb{A}[X, Y]$, the remainder $F \bmod \langle \mathbf{T} \rangle$ is in $\mathbb{A}[X, Y]_{m,n}$. In terms of complexity, we have the following result about computations with such a triangular set (see [25, 24]).

LEMMA 1. *Let $\mathbf{T}$ be a monic triangular set in $\mathbb{A}[X, Y]$ of bidegree (m, n). Then, given $F \in \mathbb{A}[X, Y]_{m',n'}$, with $m \le m'$ and $n \le n'$, one can compute $F \bmod \langle \mathbf{T} \rangle$ in $O(\mathsf{M}(m'n'))$ operations in $\mathbb{A}$. Additions, resp. multiplications modulo $\langle \mathbf{T} \rangle$ can be done in $O(mn)$, resp. $O(\mathsf{M}(mn))$ operations in $\mathbb{A}$.*

We continue with a result on polynomial matrix multiplication. The proof is the same as that of [5, Lemma 8], up to replacing univariate polynomials by bivariate ones. Remark that for such rectangular matrix multiplications, one could actually use an algorithm of Huang and Pan's [19], which features a slightly better exponent (for current values of ω).

LEMMA 2. *Let $\mathbf{M}_1, \mathbf{M}_2$ be matrices of sizes $(a \times b)$ and $(b \times c)$, with entries in $\mathbb{A}[X, Y]_{m,n}$. If $a = O(\ell^{1/2})$, $b = O(\ell^{1/2})$ and $c = O(\ell)$, one can compute $\mathbf{M} = \mathbf{M}_1\mathbf{M}_2$ using $O(\mathsf{M}(mn)\ell^{(\omega+1)/2})$ operations in $\mathbb{A}$.*

2.2 Chinese Remainder techniques

Let $\mathscr{T} = (\mathbf{T}_1, \ldots, \mathbf{T}_s)$ be a family of monic triangular sets in $\mathbb{A}[X, Y]$, where $\mathbb{A}$ is a ring. In such situations, we write $\langle \mathscr{T} \rangle = \langle \mathbf{T}_1 \rangle \cap \cdots \cap \langle \mathbf{T}_s \rangle$; if $\mathbb{A}$ is a field, we write $V(\mathscr{T}) = V(\mathbf{T}_1) \cup \cdots \cup V(\mathbf{T}_s)$, where $V(\mathbf{T})$ denotes the zero-set of $\mathbf{T}$ over the algebraic closure of $\mathbb{A}$.

Following [10], we say that $\mathscr{T}$ is *non-critical* if for i in $\{1, \ldots, s\}$, $F_i = U_1 \cdots U_{i-1}U_{i+1} \cdots U_s$ is invertible modulo U_i; if $\mathbb{A}$ is a field, this simply means that $U_1, \ldots, U_s$ are pairwise coprime. The family $\mathscr{T}$ is a *non-critical decomposition* of an ideal I if $\mathscr{T}$ is non-critical and $\langle \mathscr{T} \rangle = I$.

Let $\mathscr{T} = (\mathbf{T}_1, \ldots, \mathbf{T}_s)$ be a non-critical family of triangular sets, with $\mathbf{T}_i = (U_i(X), V_i(X, Y))$ of bidegree (m_i, n_i), and suppose that there exists n such that $n_i = n$ for all i; let also $m = m_1 + \cdots + m_s$. Under these conditions, the following lemma shows how to merge $\mathscr{T}$ into a single monic triangular set $\mathbf{T}$ of bidegree (m, n). Because $\mathbb{A}$ may not be a field, we assume that $\mathscr{R} = (R_1, \ldots, R_s)$ is part of the input, with $R_i = 1/F_i \bmod U_i$; we call them the *cofactors* of $\mathscr{T}$.

LEMMA 3. *Given a non-critical family $\mathscr{T}$ as above, under the assumption $n_i = n$ for all i, and given the cofactors $\mathscr{R}$, we can compute a monic triangular set $\mathbf{T}$ of bidegree (m, n) such that $\langle \mathbf{T} \rangle = \langle \mathscr{T} \rangle$ using $(n\mathsf{M}(m)\log(m))$ operations in $\mathbb{A}$.*

Given $F \in \mathbb{A}[X, Y]$ reduced modulo $\langle \mathbf{T} \rangle$, we can compute all polynomials $F_i = F \bmod \langle \mathbf{T}_i \rangle$ using $O(n\mathsf{M}(m)\log(m))$ operations in $\mathbb{A}$.

PROOF. For $i = 1, \ldots, s$, write $V_i = \sum_{j=0}^{n} v_{i,j}Y^j$, with all $v_{i,j}$ in $\mathbb{A}[X]$. Algorithm 10.22 in [16], where our polynomials R_i are written s_i, allows us to apply the Chinese Remainder Theorem, yielding $v_0, \ldots, v_n$ in $\mathbb{A}[X]$ such that $v_{i,j} = v_j \bmod U_i$ for all i, j. Since $v_{i,n} = 1$ for all i, $v_n = 1$ as well, so we let $U = U_1 \cdots U_s$, $V = \sum_{j=0}^{n} v_jY^j$ and $\mathbf{T} = (U, V)$. Computing U takes $O(\mathsf{M}(m)\log(m))$ by [16, Lemma 10.4] and computing V takes a total time of $O(n\mathsf{M}(m)\log(m))$ by [16, Coro. 10.23].

To prove the second point, write $F = \sum_{j=0}^{n-1} f_jY^j$, with all f_j in $\mathbb{A}[X]$. For $j = 0, \ldots, n-1$, we apply the modular reduction algorithm of [16, Algo. 10.16] to compute $f_{1,j}, \ldots, f_{s,j}$, with $f_{i,j} = f_j \bmod U_i$; we return $F_i = \sum_{j=0}^{n-1} f_{i,j}Y^j$, for $i = 1, \ldots, s$. The total time is n times the cost of modular reduction, that is, $O(n\mathsf{M}(m)\log(m))$. □

COROLLARY 1. *Let $\mathbb{K}$ be a field and let $\mathscr{T} = (\mathbf{T}_1, \ldots, \mathbf{T}_s)$ be a non-critical family of monic triangular sets in $\mathbb{K}[X, Y]$, with $\mathbf{T}_i = (U_i, V_i)$ of bidegree (m_i, n_i) for all i. Suppose that the ideal $\langle \mathscr{T} \rangle \cdot \overline{\mathbb{K}}[X, Y]$ is radical. Then one can compute the equiprojectable decomposition of the ideal $\langle \mathscr{T} \rangle$ using $O(\mathsf{M}(\delta)\log(\delta))$ operations in $\mathbb{K}$, with $\delta = \sum_{1 \le i \le s} m_in_i$.*

PROOF. Partition $\mathscr{T}$ in the classes of the equivalence relation where $(U, V) \equiv (U^\star, V^\star)$ if and only if $\deg(V, Y) = \deg(V^\star, Y)$. Let $\mathscr{T}_1, \ldots, \mathscr{T}_t$ be these classes; for $j \in \{1, \ldots, t\}$, let $\mu_j = \sum_{(U,V) \in \mathscr{T}_j} \deg(U, X)$ and let ν_j be the common value of $\deg(V, Y)$ for $(U, V) \in \mathscr{T}_j$; then, $\sum_{1 \le j \le t} \mu_j\nu_j = \delta$.

For $j = 1, \ldots, t$, let $\mathbf{T}_j^\star$ be the monic triangular set obtained by applying the previous lemma to $\mathscr{T}_j$. Since $\mathbb{K}$ is a field, the cofactors $\mathscr{R}_j$ are computed in time $O(\mathsf{M}(\mu_j)\log(\mu_j))$ using [16, Algo. 10.18], so the total time for any fixed j is $O(\nu_j\mathsf{M}(\mu_j)\log(\mu_j))$, which is $O(\mathsf{M}(\nu_j\mu_j)\log(\nu_j\mu_j))$. Summing over all j, the total cost is seen to be $O(\mathsf{M}(\delta)\log(\delta))$.

Since $\langle \mathscr{T} \rangle$ is radical in $\overline{\mathbb{K}}[X, Y]$, we deduce that for all $i \in \{1, \ldots, s\}$, the zero-set Z_i of $\mathbf{T}_i$ is equiprojectable, with fibers for the projection $\pi : \overline{\mathbb{K}}^2 \to \overline{\mathbb{K}}$ all having cardinality n_i. Thus, the triangular sets $\mathbf{T}_1^\star, \ldots, \mathbf{T}_t^\star$ form the equiprojectable decomposition of $\langle \mathscr{T} \rangle$. □

2.3 Specialization properties

Consider a domain $\mathbb{A}$, its fraction field $\mathbb{K}$, and a maximal ideal $\mathfrak{m} \subset \mathbb{A}$ with residual field $k = \mathbb{A}/\mathfrak{m}$. Given F and G in $\mathbb{A}[X, Y]$, our goal here is to relate the equiprojectable decomposition of $Z(F, G)$ to that of $Z(F \bmod \mathfrak{m}, G \bmod \mathfrak{m})$, where the former is defined over $\mathbb{K}$ and the latter over k.

The following results give quantitative estimates for ideals of "good reduction" in the two cases we are interested in, $\mathbb{A} = k[T]$ and $\mathbb{A} = \mathbb{Z}$; in both cases, we use the length function λ defined in the introduction. The case $\mathbb{A} = k[T]$, while not treated in [10], is actually the simpler, so we only sketch the proof; for $\mathbb{A} = \mathbb{Z}$, we can directly apply [10, Th. 1].

LEMMA 4. *Let F, G be in $k[T][X, Y]$ and let $\mathbf{T}_1, \ldots, \mathbf{T}_s \subset k(T)[X, Y]$ be the equiprojectable decomposition of $Z(F, G)$. If $d = \max(\deg(F), \deg(G))$ and $\ell = \max(\lambda(F), \lambda(G))$, there exist $A \in k[T] - \{0\}$ of degree at most $2d^2(6d^2 + 9d\ell)$ and with the following property.*

If an element $t_0 \in k$ does not cancel A, then none of the denominators of the coefficients of $\mathbf{T}_1, \ldots, \mathbf{T}_s$ vanishes at $T = t_0$ and their evaluation at $T = t_0$ forms the equiprojectable decomposition of $Z(F(t_0, X, Y),\ G(t_0, X, Y))$.

PROOF. The approach of [10, Section 3] still applies in this case, and shows that if an element $t_0 \in k$ satisfies three assumptions (denoted by H_1, H_2, H_3 in [10]), then the specialization property holds. These properties imply the existence of a non-zero $A \in k[T]$ as claimed in the lemma; its degree can be bounded using the results of [31, 11]. □

LEMMA 5. *Let F, G be in $\mathbb{Z}[X,Y]$ and let $\mathbf{T}_1,\dots,\mathbf{T}_s \subset \mathbb{Q}[X,Y]$ be the equiprojectable decomposition of $Z(F,G)$. If $d = \max(\deg(F), \deg(G))$ and $\ell = \max(\lambda(F), \lambda(G))$, there exists $A \in \mathbb{N} - \{0\}$, with $\lambda(A) \le 8d^5(3\ell + 10\log(d) + 22)$ and with the following property.*

If a prime $p \in \mathbb{N}$ does not divide A, then none of the denominators of the coefficients of $\mathbf{T}_1,\dots,\mathbf{T}_s$ vanishes modulo p, and their reduction modulo p forms the modified equiprojectable decomposition of $Z(F \bmod p, G \bmod p)$.

3. A DIRECT ALGORITHM

In this section, we work over a field $\mathbb{K}$. We give an algorithm that takes as input F, G in $\mathbb{K}[X,Y]$ and computes the equiprojectable decomposition $\mathbf{T}_1,\dots,\mathbf{T}_s$ of $Z(F,G)$. If F and G have degree at most d, the running time is $O\tilde{~}(d^3)$, that is, essentially the same as computing $\mathrm{res}(F,G,Y)$ (we count all operations in $\mathbb{K}$ at unit cost). This result is by no means surprising (a particular case appears in [23]) and certainly not enough to prove our main theorems. We will only use it as the initialization step of our lifting process.

The rest of this section is devoted to prove this proposition, following a few preliminaries.

PROPOSITION 1. *Let F, G be in $\mathbb{K}[X,Y]$, of total degree at most d. If the characteristic of $\mathbb{K}$ is greater than $2d^2 + d + 1$, one can compute the equiprojectable decomposition of $Z(F,G)$ using $O(\mathsf{M}(d)\mathsf{M}(d^2)\log(d)^2)$ operations in $\mathbb{K}$.*

Regular GCDs and quotients. Let R be a nonzero, squarefree polynomial in $\mathbb{K}[X]$, and let F, G be in $\mathbb{K}[X,Y]$. A *regular GCD* of (F,G) modulo R is a non-critical decomposition of the ideal $\langle R, F, G\rangle$; a *regular quotient* of F by G modulo R is a non-critical decomposition of the ideal $\langle R, F\rangle : G$. If $\mathscr{T} = (\mathbf{T}_1,\dots,\mathbf{T}_s)$ is a regular GCD of (F,G) modulo R, with $\mathbf{T}_i = (U_i, V_i)$ for all i, and if F is monic in Y, then $\mathscr{S} = (\S_1,\dots,\S_s)$, with $\S_i = (U_i, F/V_i \bmod U_i)$ for all i, is a regular quotient of F by G modulo R.

If F, G have degree at most d in Y, and R, F, G have degree at most m in X, then using the algorithm of [1], both operations can be done in time $O(\mathsf{M}(d)\mathsf{M}(m)\log(d)\log(m))$.

Radical computation. Regular quotients allow us to compute radicals. Let indeed $\mathbf{T} = (U, V)$ be a monic triangular set of bidegree (m, n) in $\mathbb{K}[X,Y]$, with U squarefree and with m and n less than the characteristic of $\mathbb{K}$; we prove that $I = \langle U, V\rangle : \partial V/\partial Y$ is the radical of the ideal $\langle \mathbf{T}\rangle$.

Let I' be the extension of I in $\overline{\mathbb{K}}[X,Y]$. Over $\overline{\mathbb{K}}$, the assumption on m ensures that U is still squarefree, so the ideal $\langle U, V\rangle$ is the intersection of primary ideals of the form $\mathfrak{p}_i = \langle X - x_i, (Y - y_i)^{e_i}\rangle$, where $(x_i, y_i)_{1\le i\le t}$ are the zeros of $\mathbf{T}$, and $e_i \in \mathbb{N}_{>0}$ is the multiplicity of the factor $Y - y_i$ in $V(x_i, Y)$. Then, I' is the intersection of the ideals $\mathfrak{p}_i : \partial V/\partial Y$, which we can rewrite as

$$I' = \bigcap_{1\le i\le t} \langle X - x_i, (Y - y_i)^{e_i}\rangle : (e_i(Y - y_i)^{e_i - 1}).$$

The assumption on n implies that $e_i \neq 0$ in $\mathbb{K}$, so that I' is the intersection of the maximal ideals $\langle X - x_i, Y - y_i\rangle$; our claim is proved. As a consequence, under the above assumption on $\mathbf{T}$, we can compute a non-critical decomposition of the radical of $\langle \mathbf{T}\rangle$ in time $O(\mathsf{M}(n)\mathsf{M}(m)\log(n)\log(m))$.

After these preliminaries, we can turn to the algorithm. In what follows, J is the Jacobian determinant of (F,G), $H = \gcd(F,G)$, $F^\star = F/H$ and $G^\star = G/H$. The idea is classical: we compute the resultant $R = \mathrm{res}(F^\star, G^\star, Y)$, then a regular GCD of $(F^\star, G^\star)$ modulo R; make the result radical and finally we remove all points where J vanishes.

Step 0. We compute H, $F^\star$, $G^\star$ as defined above. Corollary 11.9 in [16] gives an *expected* $O(d\mathsf{M}(d)\log(d))$ operations for computing H; we briefly explain how to make it deterministic, up to an acceptable increase in running time (this is routine; some details are left to the reader).

Choosing $2d^2 + d + 1$ values x_i in $\mathbb{K}$, we compute $H_i = \gcd(F(x_i, Y), G(x_i, Y))$, $F_i^\star = F(x_i, Y)/H_i$ and $G_i^\star = G(x_i, Y)/H_i$. Lucky values of x_i are those where the leading coefficient of (say) F in Y and the resultant of $(F^\star, G^\star)$ in Y are non-zero. We are sure to find at least $d^2 + 1$ of them; these will be those x_i's where H_i has maximal degree. These are enough to reconstruct H, $F^\star$ and $G^\star$ by interpolation, hence a total running time of $O(\mathsf{M}(d)\mathsf{M}(d^2)\log(d))$.

Step 1. Compute the (nonzero) resultant $R = \mathrm{res}(F^\star, G^\star, Y)$. Using Reischert's algorithm [28], this takes time $O(\mathsf{M}(d)\mathsf{M}(d^2)\log(d))$.

Step 2. Replace R by its squarefree part, which takes time $O(\mathsf{M}(d^2)\log(d))$. Note that $V(R, F^\star, G^\star) = V(F^\star, G^\star)$.

Step 3. Compute a regular GCD $\mathscr{T} = (\mathbf{T}_1,\dots,\mathbf{T}_s)$ of $(F^\star, G^\star)$ modulo R, in time $O(\mathsf{M}(d)\mathsf{M}(d^2)\log(d)^2)$. Note that $V(\mathscr{T})$ is equal to $V(R, F^\star, G^\star)$, that is, $V(F^\star, G^\star)$.

Step 4. For $i = 1,\dots,s$, writing $\mathbf{T}_i = (U_i, V_i)$, compute a regular quotient of V_i by $\partial V_i/\partial Y$ modulo U_i.

Letting (m_i, n_i) be the bidegree of $\mathbf{T}_i$, the cost for each i is $O(\mathsf{M}(d)\mathsf{M}(m_i)\log(d)\log(m_i))$. Using the super-linearity of M, the total is seen to be $O(\mathsf{M}(d)\mathsf{M}(d^2)\log(d)^2)$.

Let $\mathscr{S} = (\S_1,\dots,\S_t)$ be the union of all triangular sets obtained this way, with $\S_i = (P_i, Q_i)$. Since d^2 is less than the characteristic of $\mathbb{K}$, this is also the case for all m_i and n_i. As a result, by the discussion above, $\langle \mathscr{S}\rangle$ is the defining ideal of $V(F^\star, G^\star)$; in particular, it is radical in $\overline{\mathbb{K}}[X,Y]$.

Step 5. For $i = 1,\dots,t$, compute $J_i = J \bmod P_i$, where J is the Jacobian determinant of (F,G). Using fast simultaneous modular reduction, this costs $O(d\mathsf{M}(d^2)\log(d))$.

Step 6. For $i = 1,\dots,t$, compute a regular quotient of Q_i by J_i modulo P_i; again, the cost is $O(\mathsf{M}(d)\mathsf{M}(d^2)\log(d)^2)$. Let $\mathscr{U}$ be the union of all resulting triangular sets; then, $\langle \mathscr{U}\rangle$ is the defining ideal of $V(F^\star, G^\star) - V(J)$, and one verifies that the latter set is $Z(F,G)$.

Step 7. Finally, apply the algorithm of Corollary 1 to $\mathscr{U}$ to obtain the equiprojectable decomposition of $Z(F,G)$. Since $Z(F,G)$ has size at most d^2, the cost is $O(\mathsf{M}(d^2)\log(d))$.

4. NORMAL FORM ALGORITHMS

We consider now the problem of reducing $F \in \mathbb{A}[X,Y]$ modulo several triangular sets. Our input is as follows:

- $\mathscr{T} = (\mathbf{T}_1,\dots,\mathbf{T}_s)$ is a non-critical family of monic triangular sets in $\mathbb{A}[X,Y]$, where $\mathbf{T}_i = (U_i, V_i)$ of bidegree (m_i, n_i) for all i and $\mathbb{A}$ is a ring;

- $\mathscr{R} = (R_1, \ldots, R_s)$ is the family of cofactors associated to $\mathscr{T}$, as in Subsection 2.2;
- F is in $\mathbb{A}[X, Y]$, of total degree less than d.

We make the following assumptions:

$$\sum_{i=1,\ldots,s} m_i n_i \le d^2 \quad \text{and} \quad n_i \le d \text{ for all } i. \qquad \textbf{(H)}$$

Then, the size of input and output is $\Theta(d^2)$ elements of $\mathbb{A}$.

Already for $s = 1$, in which case we write (m, n) instead of (m_1, n_1), the difficulty of the problem can vary significantly: if both m and n are $O(d)$, Lemma 1 shows that the reduction can be done in optimal time $O\tilde{}(d^2)$; however, when $m \sim d^2$ and $n \simeq 1$, that same lemma gives a sub-optimal $O\tilde{}(d^3)$.

In this section, we prove the following propositions, which give algorithms with better exponents. The first one applies over any ring $\mathbb{A}$; it uses fast matrix multiplication to achieve an exponent $(\omega + 3)/2 \simeq 2.69$.

Proposition 2. *Under assumption* (H), *there exists an algorithm that takes as input polynomials $\mathscr{T}$, $\mathscr{R}$ and F as above and returns all $F \bmod \langle \mathbf{T}_i \rangle$, for i in $\{1, \ldots, s\}$, using $O(\mathsf{M}(d^2) d^{(\omega-1)/2} \log(d))$ operations in $\mathbb{A}$.*

When $\mathbb{A} = \mathbb{Z}/N\mathbb{Z}$, for some prime power N, better can be done in a *boolean* model: this second proposition shows that we can get arbitrarily close to linear time (in the boolean model, input and output sizes are $\Theta(d^2 \log(N))$).

Proposition 3. *Under assumption* (H), *for any $\varepsilon > 0$, there exists an algorithm that takes as input a prime power N, and polynomials $\mathscr{T}$, $\mathscr{R}$ and F as above, all with coefficients in $\mathbb{Z}/N\mathbb{Z}$, and returns all $F \bmod \langle \mathbf{T}_i \rangle$, for i in $\{1, \ldots, s\}$, using $d^{2+\varepsilon} O\tilde{}(\log(N))$ bit operations.*

4.1 Reduction modulo one triangular set

We first discuss a simplified version of the problem, where we reduce F modulo a single monic triangular set. In other words, we take $s = 1$; then, we simply write $\mathbf{T} = (U, V)$ instead of $\mathbf{T}_1$, and we denote its bidegree by (m, n) instead of (m_1, n_1). The polynomial F is in $\mathbb{A}[X, Y]$, of degree less than d; thus our assumptions are the following:

$$mn \le d^2 \quad \text{and} \quad n \le d. \qquad \textbf{(H}'\textbf{)}$$

In [27], Poteaux and Schost give two algorithms computing $F \bmod \langle \mathbf{T} \rangle$. The first one, originating from [26, Ths. 4-6], applies only in a boolean model, when $\mathbb{A} = \mathbb{Z}/p\mathbb{Z}$ for a prime p. Only a small change is needed to make it work modulo a prime power $N = p^\ell$. In both cases, when the base ring, or field, is too small, we need to enlarge it, by adding elements whose differences are invertible. In our case, we extend the basering $\mathbb{Z}/N\mathbb{Z}$ by a polynomial that is irreducible modulo p (since if $x - x'$ is a unit modulo p, it is a unit modulo N). The analysis of [26, Ths. 4-6] remains valid with this minor modification, and yields the following result.

Proposition 4. [26, 27] *Under assumption* (H′), *for any $\varepsilon > 0$, there exists an algorithm that takes as input a prime power N and F and $\mathbf{T}$ as above, with coefficients in $\mathbb{Z}/N\mathbb{Z}$, and returns $F \bmod \langle \mathbf{T} \rangle$ using $d^{2+\varepsilon} O\tilde{}\log(N))$ bit operations.*

Since in this case the input and output size is $\Theta(d^2 \log(N))$ bits, this algorithm is close to being optimal.

If we consider the question over an abstract ring $\mathbb{A}$, no quasi-optimal algorithm is known. Under assumption (H′), the second algorithm of [27] runs in time $O(d^{\omega+1})$; this is subquadratic in the size d^2 of the problem, but worse than $O\tilde{}(d^3)$. The following proposition gives an improved result.

Proposition 5. *Under assumption* (H′), *there exists an algorithm that takes as input F and $\mathbf{T}$ as above and returns $F \bmod \langle \mathbf{T} \rangle$ using $O(\mathsf{M}(d^2) d^{(\omega-1)/2})$ operations in $\mathbb{A}$.*

The rest of this subsection is devoted to prove this proposition. As in [27], we use a baby steps / giant steps approach inspired by Brent and Kung's algorithm [6], but with a slightly more refined subdivision scheme.

Let thus F be in $\mathbb{A}[X, Y]$, of total degree less than d, and let $\mathbf{T} = (U, V)$ be of bidegree (m, n). The steps of the algorithm are given below: they consist in computing some powers of Y modulo $\langle \mathbf{T} \rangle$ (baby steps, at Step 3), doing products of matrices with entries in $\mathbb{A}[X, Y]$ (Step 4), and concluding using Horner's scheme (giant steps, at Step 6).

Step 0. Replace F by $F \bmod U$; as a consequence, we can assume $F \in \mathbb{A}[X, Y]_{r,d}$, with $r = \min(d, m)$. For future use, note that $mn \le rd$: if $r = d$, this is because $mn \le d^2$. Else, $r = m$, and the claim follows from the fact that $n \le d$.

We do d reductions of polynomials of degree less than d by a polynomial of degree m in $\mathbb{A}[X]$; this is free if $d < m$ and costs $O(d\mathsf{M}(d))$ otherwise.

Step 1. Let $t = \lceil d/n \rceil - 1$ and write F as $F = F_0 + F_1 Y^n + \cdots + F_t Y^{nt}$, with all F_i in $\mathbb{A}[X, Y]_{r,n}$.

Step 2. Let $\rho = \lfloor d/n^{1/2} \rfloor$ and $\mu = \lceil (t+1)/\rho \rceil$; note that since $d \ge n$, $\rho \ge 1$ so μ is well-defined. Furthermore, both ρ and μ are $O(d/n^{1/2})$ and $\rho\mu \ge t + 1$. Set up the $(\mu \times \rho)$ matrix $\mathbf{M}_1 = [F_{i\rho+j}]_{0 \le i < \mu, 0 \le j < \rho}$ with entries in $\mathbb{A}[X, Y]_{r,n}$, where we set $F_k = 0$ for $k > t$.

Step 3. For $i = 0, \ldots, \rho$, compute $\sigma_i = Y^{ni} \bmod \langle \mathbf{T} \rangle$. Cost: since $\deg(V, Y) = n$, $\sigma_1 = (Y^n \bmod \langle \mathbf{T} \rangle)$ is equal to $Y^n - V$, so computing it takes time $O(mn)$. Then, it takes time $O(\rho \mathsf{M}(mn))$ to deduce all other σ_i's.

Step 4. Let $\nu = \lceil m/r \rceil - 1$; for $i = 0, \ldots, \rho - 1$, write $\sigma_i = \sigma_{i,0} + \sigma_{i,1} X^r + \cdots + \sigma_{i,\nu} X^{r\nu}$, with all $\sigma_{i,j}$ in $\mathbb{A}[X, Y]_{r,n}$. Build the $\rho \times (\nu + 1)$ matrix $\mathbf{M}_2 = [\sigma_{i,j}]_{0 \le i < \rho, 0 \le j \le \nu}$ and compute $\mathbf{M} = \mathbf{M}_1 \mathbf{M}_2$.

Cost: we have seen that $mn \le rd$, so that $m/r \le d/n$ and thus $\nu = O(d/n)$. Using the bounds on ρ, μ, ν and Lemma 2, we deduce that the cost is $O(\mathsf{M}(rn)(e/n)^{(\omega+1)/2})$.

Step 5. Let $[m_{i,j}]_{0 \le i < \mu, 0 \le j \le \nu}$ be the entries of $\mathbf{M}$, which are in $\mathbb{A}[X, Y]_{2r-1, 2n-1}$. For $i = 0, \ldots, \mu - 1$, compute $G_i = m_{i,0} + m_{i,1} X^r + \cdots + m_{i,\nu} X^{r\nu}$ and $H_i = G_i \bmod \langle \mathbf{T} \rangle$. Because $m_{i,j} = F_{i\rho}\sigma_{0,j} + F_{i\rho+1}\sigma_{1,j} + \cdots + F_{(i+1)\rho-1}\sigma_{\rho-1,j}$, we deduce that $G_i = F_{i\rho}\sigma_0 + F_{i\rho+1}\sigma_1 + \cdots + F_{(i+1)\rho-1}\sigma_{\rho-1}$. Since $\sigma_j = Y^{nj} \bmod \langle \mathbf{T} \rangle$ for all j, this proves that $H_i = F_{i\rho} + F_{i\rho+1} Y^n + \cdots + F_{(i+1)\rho-1} Y^{n(\rho-1)} \bmod \langle \mathbf{T} \rangle$.

Computing a single G_i takes $O(rn\nu)$ additions in $\mathbb{A}$, which is $O(mn)$ since $r\nu = O(m)$. By construction, G_i is in $\mathbb{A}[X, Y]_{r(\nu+2)-1, 2n-1}$; since $r(\nu + 2) = O(m)$, Lemma 1 implies that reducing G_i to obtain H_i takes time $O(\mathsf{M}(mn))$. The total for all G_i's is $O(\mu \mathsf{M}(mn))$.

Step 6. Compute $H = H_0 + H_1 \sigma_\rho + \cdots + H_{\mu-1} \sigma_\rho^{\mu-1} \bmod \langle \mathbf{T} \rangle$ using Horner's scheme; the expression given above for the polynomials H_i implies that $H = F \bmod \langle \mathbf{T} \rangle$. Cost: $O(\rho)$ additions and multiplications modulo $\mathbf{T}$, each of which costs $O(\mathsf{M}(mn))$ operations in $\mathbb{A}$.

Summary. Summing all contributions, we obtain

$$O\left(\mathsf{M}(d)d + \mathsf{M}(mn)(d/n)^{1/2} + \mathsf{M}(rn)(d/n)^{(\omega+1)/2}\right).$$

The first two terms are easily seen to be $O(\mathsf{M}(d^2)d^{1/2})$. To deal with the last term, note that $r \le d$ implies $\mathsf{M}(rn) \le \mathsf{M}(dn)$, and the super-linearity of M implies that $\mathsf{M}(dn) \le \mathsf{M}(d^2)n/d$. Thus, the third term is $O(\mathsf{M}(d^2)(d/n)^{(\omega-1)/2})$, which is $O(\mathsf{M}(d^2)d^{(\omega-1)/2})$. Proposition 5 is proved.

4.2 Reduction modulo several triangular sets

We now prove Proposition 2 and 3. To simplify the presentation, we give details for the first result (in the algebraic model); the proof in the boolean model requires no notable modification (just use Proposition 4 instead of 5 below).

Let thus $\mathscr{T} = (\mathbf{T}_1, \dots, \mathbf{T}_s)$ be monic triangular sets of the form $\mathbf{T}_i = (U_i, V_i)$, with coefficients in $\mathbb{A}$ and bidegrees (m_i, n_i). We also assume that the cofactors $\mathscr{R} = (R_1, \dots, R_s)$ are given. Given F in $\mathbb{A}[X, Y]$ of degree less than d, we consider the question of reducing F modulo all $\langle \mathbf{T}_i \rangle$, under assumption (H). Our proof distinguishes three cases, from the most particular to the general case.

Identical n_i's. Assume first that there exists n such that $n_i = n$ for all i. Writing $m = m_1 + \cdots + m_s$, Lemma 3 shows that we can build a monic triangular set $\mathbf{T}$ in $\mathbb{A}[X, Y]$ of bidegree (m, n), such that the ideal $\langle \mathbf{T} \rangle$ is the intersection of all $\langle \mathbf{T}_i \rangle$, in time $O(n\mathsf{M}(m)\log(m))$

To compute all $F \bmod \langle \mathbf{T}_i \rangle$, because (H) implies $mn \le d^2$, we first compute $H = F \bmod \langle \mathbf{T} \rangle$ using Proposition 5, in time $O(\mathsf{M}(d^2)d^{(\omega-1)/2})$. Then, we use Lemma 3 to reduce H modulo all $\langle \mathbf{T}_i \rangle$ in time $O(n\mathsf{M}(m)\log(m))$. Since $n\mathsf{M}(m)$ is $O(\mathsf{M}(d^2))$, the cost of this step is negligible.

Similar n_i's. We now relax the assumption that all n_i's are the same; instead, we assume that there exists n such that $n_i \in \{n, \dots, 2n-1\}$ for all i; as above, we write $m = m_1 + \cdots + m_s$, and we introduce $n' = 2n - 1$.

For $i = 1, \dots, s$, define $V_i^\star = Y^{n'-n_i} V_i$ and $\mathbf{T}_i^\star = (U_i, V_i^\star)$, so that the $\mathbf{T}_i^\star$'s are monic triangular sets of bidegrees (m_i, n'). These new triangular sets and F may not satisfy (H) anymore, but they will, provided we replace d by $d' = 2d$. Indeed, notice that the inequality $n' \le 2n_i$ holds for all i; using (H), this yields

$$\sum_{i=1,\dots,s} m_i n' \le 2 \sum_{i=1,\dots,s} m_i n_i \le 2d^2 \le d'^2,$$

and similarly $n' \le d'$. The algorithm in the previous paragraph then allows us to compute all $H_i = F \bmod \langle \mathbf{T}_i^\star \rangle$, still in time $O(\mathsf{M}(d^2)d^{(\omega-1)/2})$.

Then, for all i, we compute the remainder $H_i \bmod \langle \mathbf{T}_i \rangle$. Using Lemma 1, this can be done in time $O(\mathsf{M}(m_i n))$ for each i, for a negligible total cost of $O(\mathsf{M}(mn)) \subset O(\mathsf{M}(d^2))$.

Arbitrary degrees. Finally, we drop all assumptions on the degrees n_i. Instead, we partition the set $S = \{1, \dots, s\}$ into $S_0, \dots, S_\kappa$ such that i is in S_ℓ if and only if n_i is in $\{2^\ell, \dots, 2^{\ell+1} - 1\}$. Because all n_i satisfy $n_i \le d$, κ is in $O(\log(d))$. We write as usual $m = m_1 + \cdots + m_s$.

We are going to apply the algorithm of the previous paragraph to all S_ℓ independently. Remark that if all $\mathbf{T}_i$ and F satisfy assumption (H), the subset $\{\mathbf{T}_i \mid i \in S_\ell\}$ and F still satisfy this assumption. Let us thus fix $\ell \in \{0, \dots, \kappa\}$. The only thing that we need to take care of are the cofactors required for Chinese Remaindering. As input, we assumed that we know all $R_i = 1/(U_1 \cdots U_{i-1} U_{i+1} \cdots U_s) \bmod U_i$; what we need are the inverses $R_{\ell,i} = 1/\prod_{i' \in S_\ell, i' \ne i} U_{i'} \bmod U_i$ for $i \in S_\ell$. These polynomials are computed easily: first, we form the product $B_\ell = \prod_{i \notin S_\ell} U_i$; using [16, Lemma 10.4], this takes $O(\mathsf{M}(m)\log(m))$ operations in $\mathbb{A}$. Then, we reduce B_ℓ modulo all U_i, for $i \in S_\ell$, for the same amount of time as above. Finally, we obtain all $R_{\ell,i}$ as $R_{\ell,i} = R_i B_\ell \bmod U_i$; the time needed for these products is $O(\mathsf{M}(m))$.

Once the polynomials $R_{\ell,i}$ are known, the algorithm above gives us $F \bmod \langle \mathbf{T}_i \rangle$, for $i \in S_\ell$, in time $O(\mathsf{M}(d^2)d^{(\omega-1)/2})$; this dominates the cost of computing the polynomials $R_{\ell,i}$. Summing over ℓ finishes the proof of Prop. 2.

5. PROOF OF THE MAIN RESULTS

We assume here that we are one of the cases $\mathbb{A} = k[T]$ or $\mathbb{A} = \mathbb{Z}$ and we prove Theorems 1 and 2. Given F, G in $\mathbb{A}[X, Y]$ and writing as before $\mathscr{T} = (\mathbf{T}_1, \dots, \mathbf{T}_s)$ for the equiprojectable decomposition of $Z(F, G)$ and $\mathscr{C} = (\mathbf{C}_1, \dots, \mathbf{C}_s)$ for its modified version, we show here how to compute the latter.

The algorithm follows the template given in [10]: compute the equiprojectable decomposition modulo a randomly chosen maximal ideal $\mathfrak{m}$ of $\mathbb{A}$, lift it modulo $\mathfrak{m}^N$, for N large enough, and reconstruct all rational fractions that appear as coefficients in $\mathscr{C}$ from their expansion modulo $\mathfrak{m}^N$.

We suppose that $\lambda(F), \lambda(G) \le \ell$ and $\deg(F), \deg(G) \le d$, where λ is the length function defined in the introduction. For $i \le s$, we write (m_i, n_i) for the bidegree of $\mathbf{T}_i$; then, we have the upper bound $\sum_{i \le s} m_i n_i \le d^2$; besides, each n_i is at most d, so assumption (H) of Section 4 holds.

5.1 One lifting step

Here, $\mathfrak{m}$ is a maximal ideal of $\mathbb{A}$; we assume that none of the denominators of the coefficients of the polynomials in $\mathscr{T}$ vanishes modulo $\mathfrak{m}$. Thus, for $N \ge 1$, we can define $\mathscr{T}_N = \mathscr{T} \bmod \mathfrak{m}^N$ by reducing all coefficients of $\mathscr{T} \bmod \mathfrak{m}^N$. Given $\mathscr{T}_N$, we show how to compute $\mathscr{T}_{2N}$; this will be the core of our main algorithm. We start by describing some basic operations in $\mathbb{A}_N = \mathbb{A}/\mathfrak{m}^N$ (when $N = 1$, we also use the notation k to denote the residual field $\mathbb{A}/\mathfrak{m}$).

For complexity analyzes, we assume that $\mathbb{A} = k[T]$ and that $\mathfrak{m}$ has the form $\langle T - t_0 \rangle$, for some t_0 in k; we discuss the case $\mathbb{A} = \mathbb{Z}$ afterwards. Remark in particular that operations $(+, -, \times)$ in $\mathbb{A}_N$ can be done in $O(\mathsf{M}(N))$ operations in k.

Univariate inversion. Given Q monic of degree m in $\mathbb{A}_N[X]$ and $F \in \mathbb{A}_N[X]$ of degree less than m, consider the problem of computing $1/F$ in $\mathbb{A}_N[X]/\langle Q \rangle$, if it exists.

This is done by computing the inverse modulo $\mathfrak{m}$ (*i.e.*, over $k[X]$) by an extended GCD algorithm and lifting it by Newton iteration [16, Ch. 9]. The first step uses $O(\mathsf{M}(m)\log(m))$ operations in k, the second one $O(\mathsf{M}(m)\mathsf{M}(N))$.

Bivariate inversion. Given a monic triangular set $\mathbf{T}$ in $\mathbb{A}_N[X, Y]$ of bidegree (m, n) and $F \in \mathbb{A}_N[X, Y]_{m,n}$, consider the computation of $1/F$ in $\mathbb{A}_N[X, Y]/\langle \mathbf{T} \rangle$, assuming it exists.

We use the same approach as above, but with bivariate computations. For inversion modulo $\mathfrak{m}$, we use [9, Prop. 6], which gives a cost $O(\mathsf{M}(m)\mathsf{M}(n)\log(m)^3\log(n)^3)$. The lifting now takes $O(\mathsf{M}(mn)\mathsf{M}(N))$.

Lifting $\mathscr{T}_N$. We can now explain the main algorithm, called Lift in the next subsection. In what follows, we write $\mathscr{T}_N = (\mathbf{T}_{N,1}, \dots, \mathbf{T}_{N,s})$; all computations take place over $\mathbb{A}_{2N}$.

Step 0. First, as in the proof of Corollary 1, we compute the cofactors $\mathscr{R}_N$ associated to $\mathscr{T}_N$ using [16, Algo. 10.18]; this time, though, we work over the ring $\mathbb{A}_{2N}$. Steps 1 and 2 of that algorithm work over any ring; Step 3, which computes inverses modulo the polynomials $\mathbf{T}_{N,j}$, is dealt with using the remarks made above on univariate inversion. Because $\mathbf{T}_{N,j}$ has bidegree (m_j, n_j) for all j, with $\sum_{j\leq s} m_j n_j \leq d^2$, the total cost is $O(\mathsf{M}(d^2)\mathsf{M}(N)\log(d))$ operations in k.

Step 1. We will use formulas from [31] to lift from $\mathscr{T}_N$ to $\mathscr{T}_{2N}$. First, we reduce the polynomials F, G and the entries of their Jacobian matrix J modulo $\mathfrak{m}^{2N}$; as a result, we will now see these polynomials as elements of $\mathbb{A}_{2N}[X,Y]$.

Over $\mathbb{A} = k[T]$, the assumption that $\lambda(F), \lambda(G) \leq \ell$ means that F and G have degree at most ℓ in T; we are reducing them modulo the polynomial $(T - t_0)^{2N}$. The time for one coefficient reduction is $O(\mathsf{M}(\ell))$, since when $2N > \ell$, no work is needed. The total time is $O(d^2\mathsf{M}(\ell))$.

Step 2. We compute $F_{N,j} = F \bmod \langle \mathbf{T}_{N,j}\rangle$ over $\mathbb{A}_{2N}[X,Y]$ for all $j \in \{1,\dots,s\}$, as well as $G_{N,j} = G \bmod \langle \mathbf{T}_{N,j}\rangle$ and $J_{N,j} = J \bmod \langle \mathbf{T}_{N,j}\rangle$. This is the most costly part of the algorithm: because we know the cofactors $\mathscr{R}_N$ associated to $\mathscr{T}_N$, and because assumption (H) of Section 4 is satisfied, Proposition 2 shows that one can compute all $F_{N,j}$ using $O(\mathsf{M}(d^2)\mathsf{M}(N)d^{(\omega-1)/2}\log(d))$ operations in k. The same holds for all $G_{N,j}$ and $J_{N,j}$.

Step 3. Finally, for all j, we compute the (2×2) Jacobian matrix $M_{N,j}$ of $\mathbf{T}_{N,j}$ in $\mathbb{A}_{2N}[X,Y]$ and the vector

$$\delta_{N,j} = M_{N,j} J_{N,j}^{-1} \begin{bmatrix} F_{N,j} \\ G_{N,j} \end{bmatrix} \quad \text{over } \mathbb{A}_{2N}[X,Y]/\langle \mathbf{T}_{N,j}\rangle.$$

Proposition 4 in [31] then proves that $\mathbf{T}_{2N,j} = \mathbf{T}_{N,j} + \delta^\star_{N,j}$, where $\delta^\star_{N,j}$ is the canonical preimage of $\delta_{N,j}$ over $\mathbb{A}_{2N}[X,Y]$.

The dominant cost is the inversion of the matrices $J_{N,j}$. By the remark above, the cost for a given j is $O(\mathsf{M}(m_j)\mathsf{M}(n_j)\log(m_j)^3\log(n_j)^3 + \mathsf{M}(m_j n_j)\mathsf{M}(N))$; summing over j, this step is negligible compared to Step 2.

Summary. When $\mathbb{A} = k[T]$, the cost of deducing $\mathscr{T}_{2N}$ from $\mathscr{T}_N$ is $O(d^2\mathsf{M}(\ell) + \mathsf{M}(d^2)\mathsf{M}(N)d^{(\omega-1)/2}\log(d))$ operations in k, which is $O\tilde{~}(d^2\ell + d^{(\omega+3)/2}N)$.

When $\mathbb{A} = \mathbb{Z}$ and $\mathfrak{m} = \langle p\rangle$, for a prime p, the algorithm does not change, but the complexity analysis does. Using the fact that computations modulo p^r can be done in $O\tilde{~}(\log(p^r))$ bit operations, and using Proposition 3, we obtain a cost of $d^{2+\varepsilon}O\tilde{~}(\ell + N\log(p))$ bit operations, for any $\varepsilon > 0$.

5.2 Main algorithm

We will now analyze the main steps of the following algorithm, proving our main theorems. For simplicity, we suppose that $\mathbb{A} = k[T]$; the modifications for $\mathbb{A} = \mathbb{Z}$ follow.

Input: F, G in $\mathbb{A}[X,Y]$, $\mathfrak{m} \subset \mathbb{A}$, $\ell \in \mathbb{N}$, $d \in \mathbb{N}$
Output: $\mathscr{C} = (\mathbf{C}_1,\dots,\mathbf{C}_s)$
(1) $\mathscr{T}_1 \leftarrow Z(F \bmod \mathfrak{m}, G \bmod \mathfrak{m})$
(2) $i \leftarrow 1$
(3) **while** $\lambda(\mathfrak{m}^{2^i}) < 4d\ell + 48d^2$ **do**
  (3.a) $\mathscr{T}_{2^i} \leftarrow \mathsf{Lift}(\mathscr{T}_{2^{i-1}}, F, G)$
  (3.b) $i \leftarrow i+1$
end
(4) $\mathscr{C}_{2^{i-1}} \leftarrow \mathsf{Convert}(\mathscr{T}_{2^{i-1}})$
(5) **return** $\mathsf{RationalReconstruction}(\mathscr{C}_{2^{i-1}})$

Step 1. Over $\mathbb{A} = k[T]$, the maximal ideal $\mathfrak{m}$ has the form $\mathfrak{m} = \langle T - t_0\rangle$, for some $t_0 \in k$. Reducing F and G modulo $\mathfrak{m}$ takes $O(\ell d^2)$ operations in k by the plain algorithm.

We assume that t_0 is not a root of the polynomial A defined in Lemma 4. By assumption, the cardinality of k is at least twice the degree of A, so choosing t_0 at random, our assumption is satisfied with probability at least $1/2$.

We use the algorithm of Section 3 over k to compute the equiprojectable decomposition $\mathscr{T}_1$ of $Z(F \bmod \mathfrak{m}, G \bmod \mathfrak{m})$; under our assumption on t_0, $\mathscr{T}_1$ coincides with $\mathscr{T} \bmod \mathfrak{m}$. This step takes $O(\mathsf{M}(d)\mathsf{M}(d^2)\log(d)^2)$ operations in k.

Step 3. We saw in the introduction that over either $\mathbb{A} = k[T]$ or $\mathbb{A} = \mathbb{Z}$, all polynomials U_i and N_i in $\mathscr{C}$ satisfy $\lambda(U_i), \lambda(N_i) \leq 2d\ell + 24d^2$. In order to reconstruct the coefficients of these polynomials from their expansion modulo $\mathfrak{m}^N$, it is thus enough to ensure that $2(2d\ell + 24d^2) \leq \lambda(\mathfrak{m}^N)$; this accounts for the bound in the **while** loop. If we wanted to compute $\mathscr{T}$ instead, the bound would be of order $d^3\ell + d^4$.

Step 3.a. For each value of i, we call the algorithm described in the previous subsection; we saw that the cost is $O(d^2\mathsf{M}(\ell) + \mathsf{M}(d^2)\mathsf{M}(2^i)d^{(\omega-1)/2}\log(d))$ operations in k. The last value i_0 of the loop index is such that $2^{i_0} < 4d\ell + 48d^2 \leq 2^{i_0+1}$. We deduce the total running time:

$$O\left(d^2\mathsf{M}(\ell)\log(\ell) + \mathsf{M}(d^2)\mathsf{M}(d\ell + d^2)d^{(\omega-1)/2}\log(d)\right).$$

Step 4. We obtain $\mathscr{C} \bmod \mathfrak{m}^{2^{i_0}}$ from $\mathscr{T} \bmod \mathfrak{m}^{2^{i_0}}$ by applying subroutine $\mathsf{Convert}$, which does the following: for all $i \leq s$, $\mathbf{T}_i$ has the form (U_i, V_i) and $\mathbf{C}_i = (U_i, N_i)$, with $N_i = V_i U_i' \bmod U_i$, over the ring $\mathbb{A}_{2^{i_0}}[X,Y]$. The cost is negligible compared to that of the lifting.

Step 5. Finally, $\mathsf{RationalReconstruction}$ recovers the rational coefficients appearing in $\mathscr{C}$ from their expansion modulo $\mathfrak{m}^{2^{i_0}}$ (the index i_0 was chosen such that this precision is sufficient). There are $O(d^2)$ coefficients, each of them having numerator and denominator of degree $O(d\ell + d^2)$, so the total time is $O(d^2\mathsf{M}(d\ell + d^2)\log(d\ell))$ operations in k.

Summary. Summing all previous costs, we see that the total time admits the upper bound claimed in Theorem 1,

$$O\left(\mathsf{M}(d^2)\mathsf{M}(d\ell + d^2)d^{(\omega-1)/2}\log(d\ell)\right).$$

Over $\mathbb{A} = \mathbb{Z}$, $\mathfrak{m}$ is of the form $\langle p\rangle$, for a suitable p chosen as follows: let $B = 6 \cdot 8d^5(3\ell + 10\log(d) + 22)$. Using [16, Th. 18.10], we can compute in time $O\tilde{~}(\log(d\ell))$ an integer $p \in [B+1,\dots,2B]$ such that with probability at least $1/2$, p is prime and does not divide the integer A of Lemma 5. We apply the same algorithm as above (in particular, since $p \geq B$, the computation modulo p will satisfy the requirement on the characteristic of the field $k = \mathbb{Z}/p\mathbb{Z}$ of Proposition 1).

Using the analysis in the previous subsection and the bounds on the bit-size of the output, it is straightforward to derive an upper bound of $d^{2+\varepsilon}O\tilde{~}(d\ell + d^2)$ bit operations, for any $\varepsilon > 0$. Up to doubling ε, the polylogarithmic terms can be discarded, and we get the result of Theorem 2.

6. EXPERIMENTAL RESULTS

We report here on preliminary results obtained with an experimental implementation of our main algorithm in the case $\mathbb{A} = \mathbb{Z}$, based on Shoup's NTL [32]. Although Theorem 2 features the best complexity, it relies ultimately on an idea of Kedlaya and Umans' [20], and we are not aware of an efficient implementation of it, nor do we know how to derive one. Instead, we used the baby steps / giant steps idea underlying Theorem 1, which applies over any ring.

Our prototype is limited to inputs with word-size coefficients, and handles only the generic case described in the introduction, with only one triangular set of the form $U(X), Y - \eta(X)$ in $\mathscr{T}$. We did implement some classical optimizations not described above in the lifting process, such as halving the precision needed for the Jacobian matrix [18, § 4.4]. In the size ranges below, we choose our prime p of about 50 bits (this agrees with the bound given in the previous section; also, in this generic case, it is easy to verify that such a prime is "lucky"). Our implementation does polynomial matrix multiplication with exponent $\omega = 3$. Nevertheless, this step was carefully implemented, using FFT techniques for evaluation / interpolation and fast multiplication of matrices modulo small primes.

We compare our results to a Chinese Remainder approach that computes the resultant and the last subresultant modulo many primes. NTL only computes resultants, so we used an implementation of the fast subresultant algorithm already used in [17] that mimics NTL's built-in resultant implementation. We give timings for the two kinds of modular arithmetic supported by NTL, `ZZ_p` and `lzz_p`, for respectively "large" primes and word-size primes. The latter is usually faster, as confirmed below, but the former allows us to choose fewer but larger primes for modular computations, which may be advantageous.

The following table shows timings needed to compute the output modulo p^N, where p is a 50 bit prime, and N is a power of 2, using these various approaches; inputs are random dense polynomials, and correctness was verified by comparing that the results of all approaches agreed. On these examples, our lifting algorithm does better than our CRT-based resultant implementation. The next step in our implementation will be to confirm whether this is still the case when we lift the general position assumption.

degree	precision	Lifting	CRT, `ZZ_p`	CRT, `lzz_p`
100	32	295.67	1474.88	899.48
100	64	558.75	2949.76	1798.96
100	128	1241.4	5899.52	3597.93
120	32	421.78	2711.36	1990.40
120	64	774.14	5422.72	3980.80
120	128	1728.1	10845.44	7961.60
140	32	818.97	4902.24	2671.89
140	64	1486.35	9804.48	5343.79
140	128	3045.91	19608.96	10687.59
160	32	1072.1	7610.6	5293.64
160	64	1896.64	15221.2	10587.28
160	128	3958.17	30442.4	21174.56
180	32	1394.61	11121.48	6541.90
180	64	2399.61	22242.96	13097.57
180	128	4951.37	44485.92	26195.15

Acknowledgements. We acknowledge support from NSERC, the CRC program and ANR grant HPAC (ANR-11-BS02-013). We thank all reviewers for their remarks.

7. REFERENCES

[1] C. J. Accettella, G. M. D. Corso, and G. Manzini. Inversion of two level circulant matrices over $\mathbb{Z}_p$. *Lin. Alg. Appl.*, 366:5 – 23, 2003.

[2] P. Aubry, D. Lazard, and M. M. Maza. On the theories of triangular sets. *JSC*, 28(1,2):45–124, 1999.

[3] P. Aubry and A. Valibouze. Using Galois ideals for computing relative resolvents. *JSC*, 30(6):635–651, 2000.

[4] E. Berberich, P. Emeliyanenko, and M. Sagraloff. An elimination method for solving bivariate polynomial systems: Eliminating the usual drawbacks. In *ALENEX*, pages 35–47. SIAM, 2011.

[5] A. Bostan, C.-P. Jeannerod, and E. Schost. Solving structured linear systems with large displacement rank. *Theor. Comput. Sci.*, 407(1-3):155–181, 2008.

[6] R. P. Brent and H. T. Kung. Fast algorithms for manipulating formal power series. *J. ACM*, 25(4):581–595, 1978.

[7] L. Cerlienco and M. Mureddu. From algebraic sets to monomial linear bases by means of combinatorial algorithms. *Discrete Math.*, 139:73–87, 1995.

[8] J. Cheng, S. Lazard, L. M. Peñaranda, M. Pouget, F. Rouillier, and E. P. Tsigaridas. On the topology of real algebraic plane curves. *Mathematics in Computer Science*, 4(1):113–137, 2010.

[9] X. Dahan, X. Jin, M. Moreno Maza, and É. Schost. Change of order for regular chains in positive dimension. *Theor. Comput. Sci.*, 392(1–3):37–65, 2008.

[10] X. Dahan, M. Moreno Maza, É. Schost, W. Wu, and Y. Xie. Lifting techniques for triangular decompositions. In *ISSAC'05*, pages 108–115. ACM, 2005.

[11] X. Dahan and É. Schost. Sharp estimates for triangular sets. In *ISSAC*, pages 103–110. ACM, 2004.

[12] D. Diochnos, I. Emiris, and E. Tsigaridas. On the asymptotic and practical complexity of solving bivariate systems over the reals. *JSC*, 44(7):818–835, 2009.

[13] M. El Kahoui. Topology of real algebraic space curves. *J. Symb. Comput.*, 43(4):235–258, 2008.

[14] P. Emeliyanenko and M. Sagraloff. On the complexity of solving a bivariate polynomial system. In *ISSAC'12*. ACM, 2012.

[15] I. Z. Emiris and E. P. Tsigaridas. Real solving of bivariate polynomial systems. In *CASC*, pages 150–161. Springer, 2005.

[16] J. von zur Gathen and J. Gerhard. *Modern Computer Algebra*. Cambridge University Press, 1999.

[17] P. Gaudry and É. Schost. Construction of secure random curves of genus 2 over prime fields. In *Eurocrypt'04*, volume 3027 of *LNCS*, pages 239–256. Springer, 2004.

[18] M. Giusti, G. Lecerf, and B. Salvy. A Gröbner free alternative for polynomial system solving. *J. Comp.*, 17(1):154–211, 2001.

[19] X. Huang and V. Pan. Fast rectangular matrix multiplication and applications. *J. Complexity*, 14(2):257–299, 1998.

[20] K. S. Kedlaya and C. Umans. Fast polynomial factorization and modular composition. *SICOMP*, 40(6):1767–1802, 2011.

[21] T. Krick, L. M. Pardo, and M. Sombra. Sharp estimates for the arithmetic Nullstellensatz. *Duke Math. J.*, 109:521–598, 2001.

[22] D. Lazard. Ideal bases and primary decomposition: case of two variables. *J. Symbolic Comput.*, 1(3):261–270, 1985.

[23] X. Li, M. Moreno Maza, and W. Pan. Computations modulo regular chains. In *ISSAC*, pages 239–246. ACM, 2009.

[24] X. Li, M. Moreno Maza, and É. Schost. Fast arithmetic for triangular sets: from theory to practice. *JSC*, 44:891–907, 2009.

[25] C. Pascal and E. Schost. Change of order for bivariate triangular sets. In *ISSAC'06*, pages 277–284. ACM, 2006.

[26] A. Poteaux and É. Schost. Modular composition modulo triangular sets and applications. Comput. Comp. (to appear).

[27] A. Poteaux and É. Schost. On the complexity of computing with zero-dimensional triangular sets. JSC (to appear).

[28] D. Reischert. Asymptotically fast computation of subresultants. In *ISSAC'97*, pages 233–240. ACM, 1997.

[29] F. Rouillier. Solving zero-dimensional systems through the rational univariate representation. *Appl. Algebra Engrg. Comm. Comput.*, 9(5):433–461, 1999.

[30] F. Rouillier. On solving systems of bivariate polynomials. In *ICMS*, volume 6327 of *LNCS*, pages 100–104. Springer, 2010.

[31] É. Schost. Computing parametric geometric resolutions. *Appl. Algebra Engrg. Comm. Comput.*, 13(5):349–393, 2003.

[32] V. Shoup. A new polynomial factorization algorithm and its implementation. *JSC*, 20(4):363–397, 1995.

[33] V. Vassilevska Williams. Multiplying matrices faster than Coppersmith-Winograd. STOC '12, pages 887–898. ACM, 2012.

Enhanced Computations of Gröbner Bases in Free Algebras as a New Application of the Letterplace Paradigm

Viktor Levandovskyy
RWTH Aachen University
Templergraben 64
52062 Aachen, Germany
viktor.levandovskyy@math.rwth-aachen.de

Grischa Studzinski*
RWTH Aachen University
Templergraben 64
52062 Aachen, Germany
grischa.studzinski@rwth-aachen.de

Benjamin Schnitzler†
RWTH Aachen University
Templergraben 64
52062 Aachen, Germany
benjamin.schnitzler@rwth-aachen.de

ABSTRACT

Recently, the notion of "letterplace correspondence" between ideals in the free associative algebra $K\langle \mathbf{X}\rangle$ and certain ideals in the so-called letterplace ring $K[\mathbf{X} \mid \mathbb{N}]$ has evolved. We continue this research direction, started by La Scala and Levandovskyy, and present novel ideas, supported by the implementation, for effective computations with general non-graded ideals in the free algebra by utilizing the generalized letterplace correspondance. In particular, we provide a direct algorithm to compute Gröbner bases of non-graded ideals. Surprizingly we realize its behavior as "homogenizing without a homogenization variable". Moreover, we develop new shift-invariant data structures for this family of algorithms and discuss about them.

Furthermore we generalize the famous criteria of Gebauer-Möller to the non-commutative setting and show the benefits for the computation by allowing to skip unnecessary critical pairs. The methods are implemented in the computer algebra system SINGULAR. We present a comparison of performance of our implementation with the corresponding implementations in the systems MAGMA [BCP97] and GAP [GAP13] on the representative set of nontrivial examples.

Categories and Subject Discriptors:
G.4 [**Mathematical Software**]: Algorithm design and analysis; Certification and testing;
I.1.2 [**Algorithms**]

General Terms:
Algorithms, Theory

Keywords:
non-commutative Groebner Bases, Free Algebra, Gebauer-Moeller, non-graded Ideals, Letterplace

*Supported by DFG priority project of SPP 1489, Project LE 2697/2-1
†Partially supported by DFG priority project SPP 1489, Project LE 2697/2-1

1. INTRODUCTION

Let K be an arbitrary field and $F = K\langle \mathbf{X}\rangle$ be the free associative algebra. Consider $P = K[\mathbf{X} \mid \mathbb{N}]$ the corresponding letterplace ring as introduced in [LL09]. Hereby the set of variables or *letters* $\mathbf{X}$ is blended with another structure, so-called *places* from $\mathbb{N}$. We denote the generators by $x_i(k)$ with x_i resp. k being the letter resp. the place. The *letterplace monoid* $[\mathbf{X} \mid \mathbb{N}]$ contains the neutral element 1 and all finite products $x_{i_1}(k_1)\ldots x_{i_d}(k_d)$ for $x_{i_j} \in X$ and $k_j \in \mathbb{N}$. The corresponding monoid ring is called the *letterplace ring* $K[\mathbf{X} \mid \mathbb{N}]$. It is an infinitely-generated commutative K-algebra, generated by the set $\{x_i(j) : x_i \in X, j \in \mathbb{N}\}$. Note, that $K[\mathbf{X} \mid \mathbb{N}]$ is not Noetherian and hence its ideals have usually infinite generating sets.

The monoid $\mathbb{N}$ acts on P by shifting the places, thus providing an important additional structure. We call the number $s(m) = \min\{k_i \mid m = x_{i_1}(k_1)\ldots x_{i_d}(k_d)\} - 1$ the *shift* of the monomial m. For each $n \in \mathbb{N}$ and $m = x_{i_1}(k_1)\ldots x_{i_d}(k_d) \in [\mathbf{X} \mid \mathbb{N}]$ n acts on m as $n \cdot m =: x_{i_1}(k_1+n)\ldots x_{i_d}(k_d+n)$. From now on $\cdot$ will be used to denote this action. We call the *place support* of $m \in [\mathbf{X} \mid \mathbb{N}]$ the set of places, occuring in m. A monomial m is called *place-multilinear*, if each number from the place support of m appears at most once.

Indeed, there is an embedding of K-vector spaces $\iota : F \to P$, $\iota(x_{i_1} \cdot \ldots \cdot x_{i_d}) = x_{i_1}(1)\ldots x_{i_d}(d)$ for all monomials $m = x_{i_1}\ldots x_{i_d} \in \langle \mathbf{X}\rangle$. However, ι is not a ring homomorphism. Denote by $V := im(\iota)$, then V is spanned by all monomials of $K[\mathbf{X} \mid \mathbb{N}]$, whose place support is of the form $1,\ldots,d$ for some $d \in \mathbb{N}$. In particular, such elements have shift 0 and are place-multilinear.

We aim at studying infinite structures, which are *invariant under the action of the shift* or shortly *shift-invariant.* Let $J \subset K[\mathbf{X} \mid \mathbb{N}]$ be an ideal, then it is *shift-invariant* if $\forall s \in \mathbb{N}$ holds $s \cdot J \subseteq J$. Observe, that for any ideal $L \subset K[\mathbf{X} \mid \mathbb{N}]$ the ideal $\cup_{s\in\mathbb{N}} s \cdot L$ is shift-invariant by construction. Let us define a shift-invariant vector space $V' = \cup_{s\in\mathbb{N}} s \cdot V$.

DEFINITION 1.1. *For a graded ideal $J \subset K[\mathbf{X} \mid \mathbb{N}]$, where the ring is graded with respect to the total degree function of letters* $\mathbf{deg}()$ *we denote by J_d the graded component of total degree d. We call J a letterplace ideal if J is generated by $\bigcup_{s,d\in\mathbb{N}} s \cdot (J_d \cap V)$. In this case, J is shift-invariant and place-multigraded.*

The following theorem is the key result proven by Levandovskyy and La Scala in [LL09].

THEOREM 1.2. *Let $I \trianglelefteq K\langle \boldsymbol{X}\rangle$ be a graded ideal and set $J = \langle \iota(I)\rangle \subseteq K[\boldsymbol{X} \mid \mathbb{N}]$. Then J is a letterplace ideal of $K[\boldsymbol{X} \mid \mathbb{N}]$. Conversely, let $J \trianglelefteq K[\boldsymbol{X} \mid \mathbb{N}]$ be a letterplace ideal and set $I = \iota^{-1}(J \cap V)$. Then I is a graded ideal of $K\langle \boldsymbol{X}\rangle$. The mappings $I \to J$ and $J \to I$ define a bijective correspondence between graded ideals of $K\langle \boldsymbol{X}\rangle$ and letterplace ideals of $K[\boldsymbol{X} \mid \mathbb{N}]$.*

With this correspondence in mind we would like to introduce the main idea behind the computation of Gröbner bases via the letterplace approach: one adds all shifts of the elements of a generating set to the set and computes a commutative Gröbner basis. After removing superfluous elements one is then left with a generating system corresponding to a Gröbner basis.

DEFINITION 1.3.

- *Let J be a letterplace ideal of $K[X|P]$ and $H \subset K[X|P]$. We say that H is a letterplace basis of J if $H \subset \bigcup_{d\in\mathbb{N}} J_d \cap V$ and $\bigcup_{s\in\mathbb{N}} s \cdot H$ is a generating set of the ideal J.*
- *Let J be an ideal of $K[X|P]$ and $H \subset J$. Then H is called a (Gröbner) shift-basis of J if $\bigcup_{s\in\mathbb{N}} s \cdot H$ is a (Gröbner) basis of J.*

The following theorem from [LL09] derscribes the connection between Gröbner shift-basis and the original Gröbner basis for the ideal of the free algebra.

THEOREM 1.4.

- *Let I be a graded two-sided ideal of $K\langle \boldsymbol{X}\rangle$ and put $J = \langle \bigcup_{s\in\mathbb{N}} s \cdot \iota(I)\rangle$. Moreover, let $G \subset \bigcup_{d\in\mathbb{N}} I_d$ and define $H = \iota(G) \subset \bigcup_{d\in\mathbb{N}} J_d \cap V$. Then G is a generating set of I as a two-sided ideal if and only if H is a letterplace basis of J.*
- *Let $I \trianglelefteq K\langle \boldsymbol{X}\rangle$ be a graded two-sided ideal and put $J = \langle \bigcup_{s\in\mathbb{N}} s \cdot \iota(I)\rangle$. Moreover, let H be a Gröbner letterplace basis of J and put $G = \iota^{-1}(H \cap V) \subset \bigcup_{d\in\mathbb{N}} I_d$. Then G is a two-sided Gröbner basis of I.*

This allows one to use commutative methods in infinite dimension "up to shifting". Since $K\langle \mathbf{X}\rangle$ is not Noetherian as well, one expects a finite output rather rarely. Instead, one passes to the semi-algorithms, that is the computations are performed up to given degree bound.

ALGORITHM 1.5.

Input: *G_0, a generating set for an graded ideal $I \trianglelefteq K\langle \boldsymbol{X}\rangle$*
Output: *G, a Gröbner basis for I*
$H := \iota(G_0 \setminus \{0\})$;
$P = \{(f, s\cdot g) \mid f, g \in H, s \in \mathbb{N}, f \neq s\cdot g, \mathbf{gcd}(\mathbf{lm}(f), \mathbf{lm}(s\cdot g)) \neq 1, \mathbf{lcm}(\mathbf{lm}(f), \mathbf{lm}(s \cdot g)) \in V\}$;
while $P \neq \emptyset$ ***do***
  Choose $(f, s \cdot g) \in P$;
  $P = P \setminus (f, s \cdot g)$;
  $h := \text{REDUCE}(S(f, s \cdot g), \bigcup_{t\in\mathbb{N}} t \cdot H)$;
  if $h \neq 0$ ***then***
    $P := P \cup \{(h, s \cdot g) \mid g \in H, s \in \mathbb{N}, \mathbf{gcd}(\mathbf{lm}(h), \mathbf{lm}(s \cdot g)) \neq 1, \mathbf{lcm}(\mathbf{lm}(h), \mathbf{lm}(s \cdot g)) \in V\}$;
    $P := P \cup \{(g, s \cdot h) \mid g \in H, s \in \mathbb{N}, \mathbf{gcd}(\mathbf{lm}(g), \mathbf{lm}(s \cdot h)) \neq 1, \mathbf{lcm}(\mathbf{lm}(g), \mathbf{lm}(s \cdot h)) \in V\}$;
    $H := H \cup \{h\}$;
  end if
end while;
$G := \iota^{-1}(H)$;
return G;

As we can see, the algorithm resembles Buchberger's algorithm and in this formulation can be seen as a member of "critical pair and completion" family of algorithms. The crucial novelty is the presence of an additional structure via shift.

Analysis shows, that indeed the set P in the algorithm can be shortened to contain exactly the pairs corresponding to overlaps of leading monomials of the participating elements. This is due to the fact that the condition $\mathbf{lcm}(\mathbf{lm}(f), \mathbf{lm}(s \cdot g)) \in V$ gets rid of superfluous elements produced by senseless shifting. Nevertheless, in order to establish the set P as well as for the reduction step, a large amount of shifted polynomials will be generated.

The algorithm above has been implemented in SINGULAR in an almost verbatim way in order to check the new idea with letterplace and shifting. Nevertheless the naive implementation performed nicely [LL09, SL13]. In this article we report on the work done for optimizing and generalizing this algorithm.

For simplicity we will only consider graded monomial orderings, that is $m > n$ if $deg(m) > deg(n) \quad \forall m, n \in K\langle \mathbf{X}\rangle$.

2. A SEPARATING INVARIANT

It is easy to see that the letterplace Gröbner basis algorithm depends heavily on the computation of shifts and a large set of shifted polynomials is generated in each step. However, only a few of them are needed. Luckily there is better way to search for critical pairs. Therefore we start with a closer investigation of the shift action.

In commutative computer algebra one often uses *exponent vectors* in order to determine if a monomial divides another.

EXAMPLE 2.1. *Consider $m_1 = x_1^{a_1} \cdots x_n^{a_n}$, $m_2 = x_1^{b_1} \cdots x_n^{b_n} \in K[x_1, \ldots, x_n]$. Then $m_1 | m_2 \Leftrightarrow \forall i\ a_i \leq b_i$.*

For the free algebra there has been no direct way to attach exponent vectors from $\mathbb{N}^k$ to monomials of $\langle \mathbf{X}\rangle$. However, the letterplace ring is commutative, so there is a way to use the exponent vectors, since the support of a monomial is always finite.

EXAMPLE 2.2. *Consider $K[x_1, x_2, x_3 \mid \mathbb{N}]$ and take $p = x_1(1)x_3(2)x_2(3) + x_2(1)x_3(2)$. We order the variables by the lowest place first, that is $x_i(k) < x_i(l)$ if $k > l$. So for $x_1(1)x_3(2)x_2(3)$ we have the exponent vector $(1,0,0,\ 0,0,1,\ 0,1,0)$ and for $x_2(1)x_3(2)$ we have $(0,1,0,\ 0,0,1)$. In other words, one of the natural ways to order the variables of $K[\boldsymbol{X} \mid \mathbb{N}]$ is to use blocks of original variables; in the example it is*

$$x_1(1), x_2(1), x_3(1), x_1(2), x_2(2), x_3(2), x_1(3), x_2(3), x_3(3), \ldots$$

REMARK 2.3. *Take $m \in V' \subset K[\boldsymbol{X} \mid \mathbb{N}]$ of total degree d and shift s with exponent vector $e \in \mathbb{N}^k$.*
Then $k = d + s$ and $e = (e_1, \ldots, e_{d+s})$, where e_i is a block of length n for all $1 \leq i < d$. Then:

- *For $1 \leq i < s$:* e_i *contains only zeros.*
- *For $s \geq i$:* e_i *contains exactly one 1 and $n-1$ zeros. Note that the one is on position j, if and only if $(x_j \mid i) | m$.*

LEMMA 2.4. *Take two monomials $m, m' \in V'$ such that $m' = s \cdot m$ for some $s \in \mathbb{N}$ and construct their exponent vectors e and e', respectively, as explained above. Suppose that $\mathbf{deg}(m) = d$. By setting $\tilde{e} := (e'_s, \ldots, e'_{s+d})$ we obtain $\tilde{e} = e$.*

PROOF. Denote by $\tilde{m}$ the monomial corresponding to $\tilde{e}$. Since $shift(m') \geq s$ we have $e'_i = (0, \ldots, 0)\ \forall i < s$. We conclude that $shift(\tilde{m}) = shift(m') - s$ and $s \cdot \tilde{m} = m'$, which already implies $\tilde{m} = m$ and thus $\tilde{e} = e$. □

DEFINITION 2.5. *Let $m \in V'$ with shift s and $\mathbf{deg}(m) = d > 0$ and assume the exponent vector e has been constructed as before. Set $\tilde{e} = (e_s, \ldots, e_{d+s})$. Construct an integer vector D as follows: The first entry of D is the unique position of 1 in e_s. For $1 < i \leq d$ denote by z the number of zeros between the two occurring 1 in the vector (e_{s+i}, e_{s+i-1}). Then we set the $i - th$ entry of D to $z + 1$. We call D a distance vector and denote by dv the map that assigns to each monomial $m \in V'$ its distance vector. Moreover, we postulate $dv(1) := 0 \in \mathbb{N}^1$.*

EXAMPLE 2.6. *As above consider $p = x_1(1)x_3(2)x_2(3) + x_2(1)x_3(2)$. The distance vector for $x_1(1)x_3(2)x_2(3)$ is given by $(1, 5, 2)$ and for $x_2(1)x_3(2)$ we have $(2, 4)$.*

PROPOSITION 2.7. *The map dv is invariant with respect to the shift action, that is it separates the orbits. Moreover, for all $m, m' \in V'$ we have $dv(m) = dv(m') \Leftrightarrow m' = s \cdot m$ or $m = s \cdot m'$ for some $s \in \mathbb{N}$.*

In particular, this leads to the fact recognition of shifted monomials in $K[\mathbf{X} \mid \mathbb{N}]$ and at the same time – via letterplace encoding – to the divisibility check $m|m'$ for monomials $m, m' \in K\langle \mathbf{X} \rangle$.

DEFINITION 2.8. *For two distance vectors d and d' we say that d is contained in d', if $d = 0$ or $size(d') \geq size(d)$ and there exists i such that $d[1] = d'[i] + in - \sum_{1}^{i-1} d'[i]$ and $d[j] = d'[i + j - 1]$ for $1 < j \leq size(d)$.*

LEMMA 2.9. *Let $m, m' \in V'$ be two monomials. Then $m|m'$ if and only if $dv(m)$ is contained in $dv(m')$.*

COROLLARY 2.10.
Take two monomials $w, w' \in \langle X \rangle$. Then $w|w' \Leftrightarrow dv(\iota(w))$ is contained in $dv(\iota(w'))$.

REMARK 2.11.
The closer analysis of Algorithm 1.5 shows, that one creates new critical pairs in the pairset P with shifts of elements from the would-be-Gröbner-basis H. In addition, the reduction of a polynomial takes place with respect to all shifts of H. In the practical but still naive version of the algorithm, running with a given degree bound, one possibility is to store all shifts of elements of H and add to the source of the pairset all shifts of new polynomials as well. By this approach one can use the usual commutative divisibility on monomials and also use classical monomial orderings for comparisons.

REMARK 2.12. *Now we follow a different way: by using distance vectors we can make a fast division test for monomials in V' respectively V. In particular, the shift of a monomial can be read off and stored during the computation of its distance vector.*

Thus keeping the distance vector of a leading monomial of a polynomial adjoint to the polynomial data in the Algorithm 1.5 directly improves the algorithm. Namely, the overlaps-based computation of critical pairs is more effective and one can directly use special optimized algorithms for the shift-divisibility and shift-reductions, vital for the performance of the Algorithm.

3. NON-GRADED IDEALS

Although the theoretical aspect of the correspondence between non-graded ideals and their homogenized counterparts is technically involved, the basic idea is similar to the classical homogenization in the commutative case. While being algorithmically feasible in the Noetherian case, the computation of a Gröbner basis of a non-graded ideal in the non-Noetherian case has the following problem: A non-graded ideal $I \in K\langle \mathbf{X} \rangle$ has a finite Gröbner basis, while the homogenized set of generators leads to an infinite Gröbner basis.

As mentioned before in [Sca12] La Scala proposed a generalization of the letterplace approach to the non graded case by introducing another variable to homogenize the generators of an ideal and translating the process of non commutative homogenization to the letterplace ring. While showing good results this approach does not solve the general problem.

Since there are concrete examples of this behavior, communicated to us by Victor Ufnarovskij, we are looking for direct Gröbner basis theory for non-graded ideals of $K\langle \mathbf{X} \rangle$.

3.1 Place grading, or homogenization without a homogenization variable

To improve the method of La Scala our first step is to show that introducing a new variable is superfluous. In fact one can use the structure given by the letterplace ring quite successfully.

DEFINITION 3.1.

- *Denote by $W' \subset K[\boldsymbol{X} \mid \mathbb{N}]$ the vector space, spanned by all place-multilinear monomials, that is monomials of $[X \mid \mathbb{N}]$, whose place support is irredundant as a set. Let $W \subset W'$ be spanned by all place-multilinear monomials of shift zero.*
- *For a monomial $m \in W$, define the* ***place-degree*** $\mathbf{pdeg}(m)$ *to be the highest place occurring in the place-support of m and we set $\mathbf{pdeg}(m) = 0$ for $m \in K$ by convention. For a polynomial $p \in W \setminus \{0\}$ we set $\mathbf{pdeg}(p) = \max_i\{\mathbf{pdeg}(m_i) | p = \sum a_i m_i, a_i \in \mathbb{K} \setminus \{0\}\}$.*
- *If there is $1 \leq k \leq \mathbf{pdeg}(m)$, such that k is not in the place-support, we say m has a hole at place k. The number of holes between the first occurring variable and the last one is called the* ***place defect*** *of m.*
- *Let $\cdot_{lp}$ be the* ***letterplace multiplication*** *on $K[\boldsymbol{X} \mid \mathbb{N}]$ [LL09], that is $m_1 \cdot_{lp} m_2 = m_1(\mathbf{pdeg}(m_1) \cdot m_2)$ for monomials $m_1, m_2 \in [X \mid \mathbb{N}]$.*
- *Define $W_k = \{w \in W \mid \mathbf{pdeg}(w) = k\} \subseteq W$.*

PROPOSITION 3.2. *The following holds:*

1. $W' = \bigcup_{s \in \mathbb{N}_0} s \cdot W$.
2. $\mathbf{pdeg}(m_1 \cdot_{lp} m_2) = \mathbf{pdeg}(m_1) + \mathbf{pdeg}(m_2) = \mathbf{pdeg}(m_2 \cdot_{lp} m_1)$ *and thus* $W_l \cdot_{lp} W_k \subseteq W_{l+k}\ \forall l, k \in \mathbb{N}_0$.
3. $\forall m \in W$: $\mathbf{pdeg}(m) = \mathbf{deg}(m) +$ *shift*$(m) +$ *place-defect*(m).
4. $W = \bigoplus_{k \in \mathbb{N}_0} W_k$ *is graded with respect to place degree.*
5. $V_0 = W_0 = K$, $V_1 = W_1 = \oplus_{i=1}^{n} K x_i(1)$ *and* $V_k \subsetneq W_k\ \forall k \geq 2$. *Thus* $V \subsetneq W$ *and* $V' \subsetneq W'$.
6. *Place grading respects shifts, that is* $s \cdot W_k \subset W_{k+s}$ $\forall k, s \in \mathbb{N}$ *holds.*

EXAMPLE 3.3. *To get a deeper insight into the structure of* W *a look into the graded parts is useful. It is easy to see that* $W_0 = K \cdot \{1\} = V_0$ *and* $W_1 = K \cdot \{x_i \mid 1 \leq i \leq n\} = V_1$. *According to the definition,* $W_2 = V_2 \oplus (1 \cdot V_1)$. *Further on, we find* $W_3 = V_3 \oplus (1 \cdot W_2) \oplus (W_1 \times 2 \cdot W_1)$, *where* $W_1 \times 2 \cdot W_1 = \{w(2 \cdot \tilde{w}) \mid w, \tilde{w} \in W_1\}$. *Substituting previous expressions we obtain* $W_3 = V_3 \oplus (1 \cdot V_2) \oplus (2 \cdot V_1) \oplus (V_1 \times 2 \cdot V_1)$.

Recall that the letterplace ring is equipped with the shift action, which also defines an equivalence relation: $m_1 \simeq m_2$ if either $m_1 = s \cdot m_2$ or $m_2 = s \cdot m_1$ for some $s \in \mathbb{N}$. Using this relation we can identify a monomial of the free algebra with the orbit of monomials from $K[\mathbf{X} \mid \mathbb{N}]$ under the shift action. A natural choice for the representative of an orbit is $\iota(m) \in V$.

DEFINITION 3.4. *Let* $G \subset K\langle \boldsymbol{X} \rangle$ *be a set of polynomials and put* $\tilde{G} = \iota(G)$. *Then for each* $\tilde{p} = \sum_i a_i \tilde{m}_i \in \tilde{G}$ *with* $a_i \in K, \tilde{m}_i \in [X \mid \mathbb{N}]$ *we set*

$$p_h = \sum_i a_i(\mathbf{pdeg}(\tilde{p}) - \mathbf{pdeg}(m_i)) \cdot m_i \in K[\boldsymbol{X} \mid \mathbb{N}].$$

Then p_h *is graded with respect to* **pdeg** *(or place-homogeneous). We call* p_h *the place-homogenization of* $\tilde{p}$.

So instead of adding a new variable to the free algebra one is able to use places to homogenize polynomials. Note that normally one adds commutators to the ideal in order to allow the new variable to be commuted. This is also possible with the use of holes in view of the fact, that any representative of an orbit can be chosen, which implies that any polynomial in $K[\mathbf{X} \mid \mathbb{N}]$ can be viewed as place-homogenized.

Using this new place-homogenization the algorithm presented in [LL09] can be applied, since the homogenization is shift-invariant. However, dehomogenization is not as simple, since monomials with positive place-defect are not in an orbit with an element of V under the shift-action. However, if a s-polynomial is computed or one does reduction it is easy to see that polynomials with positive place-defect will occur.

DEFINITION 3.5. *Define a* K*-linear map* $\mathbf{shrink}: W' \to V$ *as follows:* $\mathbf{shrink}(m) = m$ *for* $m \in V$. *Now suppose that* $m \in W' \setminus V$ *with* $1 \leq \mathbf{deg}(m) = d$ *and* $\mathbf{pdeg}(m) = d' > d$, *then there exist* $s_1, \ldots, s_d \in \mathbb{N}_0$ *such that* $m = s_1 \cdot x_{i_1}(1) \ldots s_d \cdot x_{i_d}(d)$, *where* $s_d = d' - d \geq 1$. *We put* $\mathbf{shrink}(m) := x_{i_1}(1) \cdots x_{i_d}(d) \in V$.

DEFINITION 3.6.

- *Define an equivalence relation on* W' *respectively on* W *by* $m_1 \sim m_2 \Leftrightarrow \mathbf{shrink}(m_1) = \mathbf{shrink}(m_2)$. *We set* $\sim: W' \to W'/\sim$ *to be the natural surjection.*
- *Define a map* $\star : V \times V \to W'/\sim$ *as follows:* $v_1 \star v_2 := [v_1(\mathbf{pdeg}(v_1) \cdot v_2)]$.

LEMMA 3.7. *Define a* K*-linear map*

$$\eta : V \to W'/\sim, \quad f = \sum_i a_i m_i \mapsto [f].$$

Then η *is an isomorphism of vector spaces.*

PROOF. Since V is stable under shrinking, η is injective. Let $w \in W'$ be place-multilinear, then $\mathbf{shrink}(w) \in V$ belongs to $[w]$ and thus η is surjective. □

Note that the inverse map for η is given by **shrink**. If we identify a residue class $[w] \in W'/\sim$ with the unique element $v \in V$ contained in this class we can think of $\star$ as a multiplication on V, which respects the total degree of polynomials, thus giving V an K-algebra structure.

LEMMA 3.8. *Define the map* $\star : V \times V \to V$, $(v_1, v_2) \mapsto \mathbf{shrink}(v_1(\mathbf{pdeg}(v_1) \cdot v_2))$.

1. $\star$ *is bilinear.*
2. *We have* $p \star q = 0 \Leftrightarrow p = 0 \vee q = 0$.
3. $\star$ *is associative.*

PROOF.

1. Recall that **pdeg** of a polynomial is the highest occurring place in any monomial with non-zero coefficient. Moreover, we have $\mathbf{shrink}(v_1(s \cdot v_2)) = \mathbf{shrink}(v_1(s' \cdot v_2))\ \forall v_1, v_2 \in V$ if $s, s' \geq \mathbf{pdeg}(v_1)$ holds. The claim follows by the linearity of **shrink**, shift and the multiplication on $K[\mathbf{X} \mid \mathbb{N}]$ as a ring.
2. Because we have $p \star q = (\sum_d^D p_d) \star (\sum_e^E q_e) = \sum_{k=0}^{D+E} (\sum_{d=0}^{k} p_d q_{k-d})$, where p_d, q_e are **pdeg**–graded components of p and q respectively we need to proof the claim for **deg**-graded components $\sum_d p_d q_{k-d}$ only. Because $p, q \in V$ we have that $\sum_d p_d q_{k-d}$ is homogeneous and the claim follows by the fact, that we can use the letterplace multiplication for the graded case.
3. Routine computation. □

Indeed the map $\star$ can be extended to a map $K[\mathbf{X} \mid \mathbb{N}] \times K[\mathbf{X} \mid \mathbb{N}] \to V$, which enjoys similar properties.

THEOREM 3.9. *Define a K-linear map*

$$\vartheta : (K\langle \boldsymbol{X}\rangle, \cdot) \to (V, \star),\ p = \sum_{c_j \in K\setminus\{0\}} c_j m_j \mapsto \sum c_j \iota(m_j).$$

Then ϑ is a K-algebra isomorphism.

PROOF. By definition ϑ is K-linear and we have $\vartheta(p+q) = \vartheta(p) + \vartheta(q)\ \forall p, q \in K\langle \mathbf{X}\rangle$. We need to show that $\vartheta(p\cdot q) = \vartheta(p) \star \vartheta(q)\ \forall p, q \in K\langle \mathbf{X}\rangle$. We have $\vartheta(p) \star \vartheta(q) = \vartheta(\sum_i a_i p_i) \star \vartheta(\sum_j b_j q_j) = \sum_{i,j} a_i b_i \vartheta(p_i) \star \vartheta(q_j)$. Because $\vartheta(p_i) \star \vartheta(q_j) = \iota(p_i) \star \iota(q_j) = \mathbf{shrink}(\iota(p_i)(\mathbf{deg}(p)\cdot \iota(q_j))) = \iota(p_i)(\mathbf{deg}(p_i)\cdot \iota(q_j)) = \iota(p_i q_j)$ where we used the remark given in the proof of 3.8. So we have $\vartheta(p) \star \vartheta(q) = \sum_{i,j} a_i b_i \iota(p_i q_j) = \sum_{i,j} a_i b_i \vartheta(p_i q_j) = \vartheta(p\cdot q)$ using linearity of ϑ. Because of 3.8 (2) ϑ is injective. Since V is defined as the image of ι we have that ϑ is also surjective, which completes the proof. □

We have revealed that one can interpret holes in letterplace monomials as traces of the appearance of homogenization variable. The results of La Scala for correspondence of ideals and generating systems, especially Gröbner bases can be used for the correspondence here, thereby proving the correctness of our method.

3.2 Saturation on the fly

Knowing about the problem behind homogenization mentioned earlier there are two steps one can take in order to avoid it. The first one is to apply an ordering which allows one to simplify the homogenization in each step and the second one is an alternative for homogenization which natural occurs on the letterplace ring, namely the use of distance vectors.

In our opinion, by using graded techniques on the homogenized ideal, our aim is not to compute the trusted homogenized ideal, but to come as directly as possible to the non-graded Gröbner basis, which is usually obtained via the post-computation of the saturation.

As the first step let us recall the homogenization.

DEFINITION 3.10. *Consider the free algebra $K\langle \boldsymbol{X}\rangle$ and let h be a new variable commuting with all $x_i \in \boldsymbol{X}$. Define $\overline{\boldsymbol{X}} = \boldsymbol{X} \cup \{h\}$ and $F = K\langle \overline{\boldsymbol{X}}\rangle$. Then each $p \in K\langle \boldsymbol{X}\rangle$ is the image of some homogeneous element $\overline{p} \in K\langle \overline{\boldsymbol{X}}\rangle$ under the algebra homomorphism Φ defined via $\Phi(x_i) = x_i$, $\Phi(h) = 1$. More precisely, if we have $f = \sum_{k=0}^{d} p_k$ with $p_k \in K\langle \boldsymbol{X}\rangle_k$, $p_d \neq 0$, then $\tilde{p} = \sum_{k=0}^{d} p_k h^{d-k}$ is a homogeneous element, satisfying $\Phi(\tilde{p}) = p$.*

REMARK 3.11. *In order to compute a Gröbner basis via classical homogenization one has to employ an ordering that has the following property: $h^k \mid \mathbf{lm}(\tilde{p})$ then h^k divides each term occurring in $\tilde{p}$ with non-zero coefficient. An example for such orderings can be found in [Li12] and in [BB98] as well as in [Mor88] there is a full introduction to this topic. In particular, note that for a homogenized polynomial $\tilde{p}$ we always have $h \nmid \mathbf{lm}(p)$ with respect to such an ordering.*

After one has computed the Gröbner basis of a homogenized ideal, a saturation of the result with respect to h must be computed. If we introduce the commutators to the homogenized ideal one is always able to move the homogenization variable to the end of each monomial using reduction if needed. Indeed, for each computed s-polynomial p, such that $h^k | \mathbf{lm}(p)$, one can replace p with the polynomial p/h^k. This procedure is called saturation on the fly, because p/h^k belongs to the saturated homogenized ideal. This allows one to reduce significantly the total degree of considered polynomials during the computation. Note, that the somewhat analogous effect in the commutative case can be achieved by using the notion of ecart.

Recognizing holes as traces of the homogenization, one can apply the method presented by La Scala rather effectively. The big advantage hereby is that one does not need to introduce an extra variable and in each step of the algorithm a sort of saturation-on-the-fly is applied. Also, it is not necessary to choose a special ordering for the homogenization variable. In the following we present the full algorithm.

Note that the classical operations with polynomials (creation of s-polynomials, reductions etc.) usually produces holes in the monomials of inhomogeneous input. Hence the new reduction routine SHRINK-REDUCE is introduced, which applies shrinking after each elementary reduction step $f = f - c_i m_i h$, where h is an appropriately shifted reductor, $c_i \in K \setminus \{0\}$ and $m_i \in V$.

ALGORITHM 3.12.
Input: *G_0, a generating set for an ideal $I \trianglelefteq K\langle \boldsymbol{X}\rangle$*
Output: *G, a Gröbner basis for I*
$H := \iota(G_0 \setminus \{0\})$;
$P = \{(f, s\cdot g) \mid f, g \in H, s \in \mathbb{N}, f \neq s\cdot g, \mathbf{gcd}(\mathbf{lm}(f), \mathbf{lm}(s\cdot g)) \neq 1, \mathbf{lcm}(\mathbf{lm}(f), \mathbf{lm}(s\cdot g)) \in V\}$;
while $P \neq \emptyset$ ***do***
Choose $(f, s\cdot g) \in P$*;*
$P = P \setminus (f, s\cdot g)$*;*
$h :=$ SHRINK-REDUCE$(\mathbf{shrink}(S(f, s\cdot g)), \bigcup_{t\in\mathbb{N}} t\cdot H)$*;*
if $h \neq 0$ ***then***
$P := P \cup \{(h, s\cdot g) \mid g \in H, s \in \mathbb{N}, \mathbf{gcd}(\mathbf{lm}(h), \mathbf{lm}(s\cdot g)) \neq 1, \mathbf{lcm}(\mathbf{lm}(h), \mathbf{lm}(s\cdot g)) \in V\}$;
$P := P \cup \{(g, s\cdot h) \mid g \in H, s \in \mathbb{N}, \mathbf{gcd}(\mathbf{lm}(g), \mathbf{lm}(s\cdot h)) \neq 1, \mathbf{lcm}(\mathbf{lm}(g), \mathbf{lm}(s\cdot h)) \in V\}$;
$H := H \cup \{h\}$;
end if
end while*;*
$G := \iota^{-1}(H)$;
return *G;*

THEOREM 3.13. *If the algorithm above terminates it returns a reduced Gröbner basis for the ideal I.*

PROOF. As explained before H can be viewed as a set of homogenized generators, where holes were added at the end of each monomial. Since leading monomials are not affected by the homogenization P clearly contains all critical pairs, as shown in the proof for graded ideals.
So the only thing to prove is the correctness of the computation of H, which is clear by the correspondence given in the previous section. □

3.3 Applying the new data structure and the new homogenization

Equipped with the knowledge that we can compute Gröbner bases of non-graded ideals using the letterplace approach without introducing direct homogenization one can ask if there is a better way, because applying shrinking to each new s-polynomial can be very inefficient. Luckily there is better way.

As in the section before the methods of distance vectors can be applied to reconstruct Buchberger's original algorithm. It is important to note that the ideal and generating set correspondence still holds even without explicit homogenization. In addition to that, distance vectors can be used to represent monomials in the shift invariant way. By switching to this new representation one can multiply monomials more effectively.

PROPOSITION 3.14. *Denote by* $\mathbf{lg}$ *the size of a distance vector. For two monomials* $m_1, m_2 \in \langle X \rangle$ *set* $\tilde{m}_1 := \iota(m_1)$, $\tilde{m}_2 := \iota(m_2)$, $dm_1 := dv(\tilde{m})$, $dm_2 := dv(\tilde{m}_2)$.
Define a new vector d *by setting* $d[1 \ldots \mathbf{lg}(dm_1)] = dm_1$, $d[\mathbf{lg}(dm_1)+1] = \mathbf{lg}(dm_1)n - (\sum_{k=1}^{\mathbf{lg}(dm_1)} dm_1[k]) + dm_2[1]$, $d[(\mathbf{lg}(dm_1)+2) \ldots (\mathbf{lg}(dm_1)+\mathbf{lg}(dm_2))] = dm_2[2 \ldots \mathbf{lg}(dm_2)]$. *Then* $dv(\iota(m_1 m_2)) = d$.

PROOF. To see that the claim is correct one only needs to notice that the entry $d[\mathbf{lg}(dm_1)+1]$ is exactly the gap in the exponent vector of $\tilde{m}_1 \mathbf{lg}(dm_1) \cdot \tilde{m}_2$ between the last variable of $\tilde{m}_1$ and the first of $\mathbf{lg}(dm_1) \cdot \tilde{m}_2$. □

REMARK 3.15. *Using this multiplication allows one to completely eliminate the need for shrinking. Since the shift is not needed either, one has a sparse representation of the orbit under the shift action on a monomial. Since the procedure is directly inherited from the methods of homogenization, its correctness is granted.*

4. GEBAUER-MÖLLER'S CRITERION

In commutative as well as in non-commutative Gröbner basis theory it is well-known that the practical use of criteria to reduce the set of critical pairs has very effective impact on the performance. Out of several criteria, first formulated by Buchberger, the product criterion in the case of free algebras is naturally applied during the consideration of overlaps of polynomials. The chain criterion applies as well, but it can be refined further, following the work of Gebauer and Möller [GM88] in the commutative case. Gebauer-Möller's criterion has been generalized to the setup of modules in [KR00] and [KR05], while in the non-commutative case Mora gave a detailed presentation of superfluos pairs in [Mor94], which was adapted to fit practical computations, as for example in [Xiu12].
Here we will present the theoretical layout as well as the practical use of the criterion in the letterplace framework.
We want to point out that our research was done simultaneously and independently, comparing to the recent work [KX13].

For this section we will assume that each set $P \subset K\langle \mathbf{X} \rangle$ is *interreduced*, meaning $\forall p, q \in P, p \neq q : \mathbf{lm}(p) \nmid \mathbf{lm}(q)$ and that each $p \in P$ is monic.

4.1 The non-commutative theory

In the non-commutative version of Buchberger's algorithm one constructs s-polynomials from so-called *obstructions*, that is a six-tuple $(l, p, r; \lambda, q, \rho)$ with $l, r, \lambda, \rho \in K\langle \mathbf{X} \rangle$, $p, q \in P$ and $\mathbf{lm}(lpr) = \mathbf{lm}(\lambda q \rho)$.
The classical "product criterion theorem" states that only those pairs need to be considered, leading monomials of which involve an *overlap*, that is $\mathbf{lm}(p) = ab$ and $\mathbf{lm}(q) = bc$ for some monomials a, b, c. Therefore one only has to consider pairs $\pi = (1, p_i, r; \lambda, p_j, 1)$, such that $\mathbf{lm}(p_i r) = \mathbf{lm}(\lambda p_j)$.

DEFINITION 4.1. *For an obstruction* $\pi = (1, p_i, r; \lambda, p_j, 1)$ *we denote by* $\mathbf{cm}(\pi) := \mathbf{lm}(p_i r) = \mathbf{lm}(p_i) r = \lambda \mathbf{lm}(p_j)$ *the common multiple of* p_i *and* p_j *with respect to the overlap considered in* π.

Let us consider a set of polynomials P and construct the set of all critical pairs $\pi(P)$ by searching for overlaps in the leading monomials. We want to apply the criteria to $\pi(P)$ to reduce its size.

THEOREM 4.2.
Suppose that we are given a set of polynomials P*, its set of critical pairs* $\pi(P)$ *and a pair* $\pi = (1, p_i, r_i; \lambda_k, p_k, 1) \in \pi(P)$.

1. *If there exist two pairs* $\pi_1 = (1, p_i, r_i'; \lambda_j, p_j, 1), \pi_2 = (1, p_j, r_j; \lambda_k', p_k, 1) \in \pi(P) \setminus \{\pi\}$*, such that* $\mathbf{lm}(p_j) | \mathbf{cm}(\pi)$*, then the s-polynomial* $s(\pi)$ *of* π *will reduce to zero.*

2. *If there exists a pair* $\pi_1 = (1, p_j, r_j; \lambda_k', p_k, 1) \in \pi(P) \setminus \{\pi\}$*, such that* $\mathbf{cm}(\pi_1)$ *divides* $\mathbf{cm}(\pi)$ *from the right, then the s-polynomial* $s(\pi)$ *of* π *will reduce to zero.*

PROOF. 1. Because of the assumptions we have $\mathbf{lm}(p_j) = abc$, $\mathbf{lm}(p_k) = bct_k$ and $\mathbf{lm}(p_i) = t_i ab$ for some monomials a, b, c, t_i, t_k. Since P is interreduced, none of the leading monomials can divide the overlap cofactors. This implies $\lambda_k = t_i a$ and $r_i = ct_k$. Moreover, the existence of π_1 and π_2 and the form of the leading monomials imply that there exist pairs $\pi_1' = (1, p_i, c; t_i, p_j, 1)$ and $\pi_2' = (1, p_j, t_k; a, p_k, 1)$. Then $s(\pi) = p_i ct_k - t_i a p_k = t_i abct_k + \mathbf{tail}(p_i) ct_k - t_i abct_k - t_i a \mathbf{tail}(p_k) \to -t_i \mathbf{tail}(p_j) t_k + \mathbf{tail}(p_i) ct_k + t_i \mathbf{tail}(p_j) t_k - t_i a \mathbf{tail}(p_k) = -s(\pi_1') t_k - t_i s(\pi_s') \to 0$.
Note that the reductions used are performed according to the fixed monomial ordering.

2. We first note that $\mathbf{lm}(p_j) r_j = \mathbf{lm}(p_j r_j) = \mathbf{lm}(\lambda_k' p_k) = \lambda_k' \mathbf{lm}(p_k)$ and $\tilde{l} \mathbf{lm}(p_j) r_j = \tilde{\lambda} \lambda_k' \mathbf{lm}(p_k) = \lambda_k \mathbf{lm}(p_k) = \mathbf{lm}(p_i) r_i$ for some monomials $\tilde{l}, \tilde{\lambda}$. This already implies $\tilde{l} = \tilde{\lambda}$ and $\tilde{\lambda} \lambda' = \lambda_k$. Moreover, $\tilde{l} \mathbf{lm}(p_j) r_j = \mathbf{lm}(p_i) r_i$ implies that one of the following holds:

   - $\tilde{l} \mathbf{lm}(p_j) | \mathbf{lm}(p_i)$. Then the set of polynomials is not interreduced, which leads to a contradiction.
   - There exists $\hat{r}_i$ such that $r_j = \hat{r}_i r_i$. This implies the existence of a pair $(1, p_i, \hat{r}_i; \tilde{l}, p_j, 1)$ and the claim follows from the first case. □

REMARK 4.3. *One can apply these criteria in a straight forward way: If the set of critical pairs during some step of Buchberger's algorithm has been constructed, then one can just check the pairs and search for redundant ones. However, to decide if a monomial divides another is not as cheap and easy as in the commutative case. So, in this situation the usage of distance vectors leads to much more effective computations.*

4.2 Translation to letterplace

Our final goal now is to translate the criteria into the letterplace realm. Notably, in this criterion there is no distinction between graded and non-graded cases.

THEOREM 4.4. *Let P be the set of critical pairs. Suppose it contains a pair $\pi = (p_i, s \cdot p_k)$ for $p_i, p_k \in W \subset K[\boldsymbol{X} \mid \mathbb{N}]$ and $s \in \mathbb{N}$.*

1. *If there exist two pairs $\pi_1 = (p_i, s' \cdot p_j)$ and $\pi_2 = (p_j, s'' \cdot p_k)$, such that $\mathbf{lm}(s' \cdot p_j) | \mathbf{lcm}(p_i, s \cdot p_k)$, then the s-polynomial $s(\pi)$ of π will reduce to zero.*

2. *If there exists a pair $\pi_1 = (p_j, s \cdot p_k) \neq \pi$, such that $\mathbf{lcm}(p_j, s \cdot p_k)$ divides $\mathbf{lcm}(p_i, s \cdot p_k)$, then the s-polynomial $s(\pi)$ of π will reduce to zero.*

REMARK 4.5. *Ad 1.: We have $s'' = s - s'$. This follows immediately from the non-commutative proof and the form of the overlap. In the case, when the shifts are already known, commutative methods can be used to check the divisibility. For condition 1. this is especially easy, since from the concrete pair we check its shift is known.*
Ad 2.: Since we assume that the shift of p_k is the same for π and π_1, the condition, that $\mathbf{cm}(\pi_1)$ divides $\mathbf{cm}(\pi)$ from the right, is always satisfied.

5. IMPLEMENTATION AND TIMINGS

The new methods have been implemented into the kernel part of the computer algebra system SINGULAR. As mentioned before, the implementation of the letterplace structure is discussed in [LL09], so we will not discuss this any further. For an introduction to SINGULAR we refer to the online-manual [DGPS12].

We will now present some important examples and compare our timings with those given by the implementation of letterplace Gröbner bases by Viktor Levandovskyy in the current distribution of SINGULAR, as well as with the implementations in GAP and MAGMA. We must mention that the older implementation in SINGULAR [LL09] has been released for graded ideals; its functionality with non-graded ideals is experimental.

Note that the implementation of the LETTERPLACE:DVEC algorithm is not yet distributed with SINGULAR. The merge of our development branch with the main branch of SINGULAR will be done soon.

All tests were performed on a PC equipped with two Intel Core i7 Quadcore Processor (8×2933 MHz) with 16GB RAM running Linux.

We used MAGMA V2.18-12 [BCP97], GAP Version 4.5.6 [GAP13] with the package GBNP, version 1.0.1 and SINGULAR version 3-1-6.

Testing methodology. In order to make the tests reproducible, we used the new SDEVALV2 framework, created by Albert Heinle of the SYMBOLICDATA project ([BG00]) for our benchmarking. It means that the input polynomials have been out into the system SYMBOLICDATA. Then, for each computer algebra system the files to be executed were generated by the SYMBOLICDATA using scripts, written by ourselves for this purpose. With the help of SDEVALV2 the *computing task* was formed, put to the compute server, executed and evaluated. The functions of SYMBOLICDATA as well as the data are free to use. In such a way our comparison is easily and trustfully reproducible by any other person. Note, that among other the function, which is used to measure the time, can be customized within this approach.

5.1 Examples

Many of the examples are explained in detail in [LL09] or [Stu10] and we use the same notation. In the following we explain only the new ones.

One-relator quotients

In [CHN] the authors present a list of 48 examples of one relator quotients of the modular group. All these examples were considered with a degree bound by the total degree of the maximal generator. The enumeration is chosen according to the paper and the examples are denoted by H.

LS

The examples *LS_5d9* and *LS_6d10* were presented to us during discussions with Roberto La Scala and are connected to Clifford algebras. Infinite Gröbner bases are expected from this generating sets, therefore degree bounds are employed. The first number denotes the number of generators, while the number following the d denotes the degree bound.

Example	Sing 1	Sing 2	Magma	GAP
2tri_4v7d	4.10	1.75	1.40	31.67
3nilp_d6	0.41	0.29	0.96	4.76
3nilp_d10	2410.15†	36.65	2.89	31.08
4nilp_d8	380.23†	747.95	10.25	1133.82
Braid3_11	273.40†	15.73	1.52	185.39
Braid4_11	51.82	3.10	1.14	31.97
plBraid3d_6	0.18	0.08	0.91	926.80
lp1_10	31.31	2.33	1.00	11.10
lv2d10	0.23	0.15	0.78	3.29
s_e6d10	10.56	1.84	1.12	12.45
s_e6d13	976.32	44.74	7.81	274.63
s_eha112d10	1.12	0.26	0.96	6.20
s_eha112d12	462.36	4.19	1.40	62.40
s_f4_d10	4.35	0.58	0.97	5.35
s_f4_d15	1103.33 †	147.31	13.54	2241.62
s_ha11_d10	2.18	0.32	0.81	3.51
LS_5d9	23.46	2.49	0.79	2.90
LS_6d10	411.33 †	704.97	16.86	372.06
C_4_1_7W	3.23	1.19	0.91	5.76
C_4_1_7Y	0.09	0.09	0.91	2.91
H_5	0.62	0.24	0.62	2.90
H_19	0.88	0.32	0.62	2.99
H_37	0.86	0.32	0.68	2.89
H_40	0.98	0.29	0.62	2.89
H_48	0.88	0.31	0.62	2.91

5.2 Timings

In [LL09] we used external time measuring for the whole computation via `/usr/bin/time` command. This included the initializing of a computer algebra system as well as the loading of standard libraries. In this paper we use IEEE standard for measuring (POSIX.2) and present the timings for the *system* record of the time output.

In the following tables the resulting timings are presented. Sing 1 refers to the implementation by Viktor Levandovskyy, currently distributed with SINGULAR, while Sing 2 is the new implementation of the authors using distance vectors. Results are presented in seconds. By † we denote the situation when the computation run out of memory after the indicated time.
While MAGMA is still faster in some cases the timings show that the new letterplace implementation can compete with other systems and since the test were done with a first implementation there is still space for improvements.

Acknowledgements

We would like to thank Roberto La Scala for numerous discussions on the topics, connected to this article and for examples he suggested to us. The expertise and friendliness of Martin Kreuzer and Victor Ufnarovskij was a big help in our research. Special thanks go to Hans Schönemann and Olexander Motsak from the SINGULAR team, whose advices and help in technical questions greatly contributed to the software development. We are grateful to Simon King and to David Green for the motivation and inspiration via concrete problems. With the help of Albert Heinle and the package SDEVALV2 [HLN13] for SYMBOLICDATA [BG00] the testing became comfortable, reproduceable and transparent.

6. REFERENCES

[BB98] Miguel Angel Borges and Mijail Borges. Gröbner bases property on elimination ideal in the noncommutative case. In B. Buchberger and F. Winkler, editors, *Gröbner bases and applications*, pages 323–337. Cambridge University Press, 1998.

[BCP97] Wieb Bosma, John Cannon, and Catherine Playoust. The Magma algebra system. I. The user language. *J. Symbolic Comput.*, 24(3-4):235–265, 1997. http://dx.doi.org/10.1006/jsco.1996.0125.

[BG00] Olaf Bachmann and Hans-Gert Gräbe. The SYMBOLICDATA project. In *Reports on Computer Algebra*, volume 27. 2000. Available from: http://www.mathematik.uni-kl.de/~zca.

[CHN] Marston Conder, George Havas, and M.F. Newman. On one-relator quotients of the modular group. 2011. Cambridge University Press. London Mathematical Society Lecture Note Series 387, 183-197.

[DGPS12] W. Decker, G.-M. Greuel, G. Pfister, and H. Schönemann. SINGULAR 3-1-6 — A computer algebra system for polynomial computations. 2012. Available from: http://www.singular.uni-kl.de.

[GAP13] The GAP Group. *GAP – Groups, Algorithms, and Programming, Version 4.6.3*, 2013. Available from: http://www.gap-system.org.

[GM88] Rüdiger Gebauer and H. Michael Möller. On an installation of Buchberger's algorithm. *J. Symb. Comput.*, 6(2-3):275–286, 1988. doi:10.1016/S0747-7171(88)80048-8.

[HLN13] Albert Heinle, Viktor Levandovskyy and Andreas Nareike. Symbolicdata:sdeval - benchmarking for everyone. 2013. submitted.

[KR00] Martin Kreuzer and Lorenzo Robbiano. *Computational commutative algebra. 1.* Springer-Verlag, Berlin, 2000.

[KR05] Martin Kreuzer and Lorenzo Robbiano. *Computational commutative algebra. 2.* Springer-Verlag, Berlin, 2005.

[KX13] Martin Kreuzer and Xingqiang Xiu. Non-commutative Gebauer-Moeller criteria. 2013. arXiv:arXiv/1302.3805.

[Li12] Huishi Li. *Gröbner Bases in Ring Theory.* World Scientific Publishing, 2012.

[LL09] R. La Scala and V. Levandovskyy. Letterplace ideals and non-commutative Gröbner bases. *J. Symbolic Computation*, 44(10):1374–1393, 2009. doi:doi:10.1016/j.jsc.2009.03.002.

[Mor88] Teo Mora. Gröbner bases in non-commutative algebras. *Proceedings of ISSAC conference, 358:150–161, 1988.* Lecture Notes in Computer Science. Springer-Verlag.

[Mor94] Teo Mora. An introduction to commutative and non-commutative Gröbner bases. *Theoretical Computer Science*, 134:131–173, 1994.

[Sca12] Roberto La Scala. Extended letterplace correspondence for nongraded noncommutative ideals and related algorithms. 2012. arxiv.org/abs/1206.6027.

[SL13] Roberto La Scala and Viktor Levandovskyy. Skew polynomial rings, Gröbner bases and the letterplace embedding of the free associative algebra. *Journal of Symbolic Computation*, 48:110 – 131, 2013. doi:10.1016/j.jsc.2012.05.003.

[Stu10] Grischa Studzinski. Algorithmic computations for factor algebras. Diploma thesis, RWTH Aachen, 2010. Available from: http://www.math.rwth-aachen.de/~Grischa.Studzinski/DA.pdf.

[Xiu12] Xingqiang Xiu. *Non-Commutative Gröbner Bases and Applications.* PhD thesis, University of Passau, 2012. Available from: http://www.opus-bayern.de/uni-passau/volltexte/2012/2682/.

Multivariate Difference-Differential Dimension Polynomials and New Invariants of Difference-Differential Field Extensions

Alexander Levin
The Catholic University of America
Washington, D. C. 20064
levin@cua.edu

ABSTRACT

We introduce a method of characteristic sets with respect to several term orderings for difference-differential polynomials. Using this technique, we obtain a method of computation of multivariate dimension polynomials of finitely generated difference-differential field extensions. Furthermore, we find new invariants of such extensions and show how the computation of multivariate difference-differential polynomials is applied to the equivalence problem for systems of algebraic difference-differential equations.

Categories and Subject Descriptors

I.1.2 [**Symbolic and Algebraic Manipulation**]: Algorithms—*algebraic algorithms*

General Terms

Theory, Algorithms

Keywords

Difference-differential field, dimension polynomial, reduction, characteristic set

1. INTRODUCTION

The role of Hilbert polynomials in commutative and homological algebra, as well as in algebraic geometry, is well known. A similar role in differential algebra is played by differential dimension polynomials, which describe in exact terms the freedom degree of a dynamic system, as well as the number of arbitrary constants in the general solution of a system of algebraic differential equations.

The notion of a differential dimension polynomial was introduced by E. Kolchin [4] for a finitely generated differential field extension $L = K\langle\eta_1, \dots, \eta_n\rangle$ (Char $K = 0$). He proved that there is a polynomial $\omega_{\eta|K}(t)$ associated with the set of generators $\eta = \{\eta_1, \dots, \eta_n\}$ such that for all sufficiently large $r \in \mathbf{Z}$, $\omega_{\eta|K}(r)$ is the transcendence degree of the field extension of K generated by all derivations of η_i $(1 \le i \le n)$ of order $\le r$. $\omega_{\eta|K}(t)$ is called the *differential dimension polynomial* of the extension L/K associated with the set of differential generators η.

If P is a prime differential ideal of a finitely generated differential algebra $R = K\{\zeta, \dots, \zeta_n\}$ over a differential field K, then the quotient field of R/P is a differential field extension of K generated by the images of ζ_i $(1 \le i \le n)$ in R/P. The corresponding differential dimension polynomial, therefore, characterizes the ideal P; it is denoted by $\omega_P(t)$. Assigning such polynomials to prime differential ideals has led to a number of new results on the Krull-type dimension of differential algebras and dimension of differential varieties (see, for example, [2] and [3]).

Furthermore, as it was shown in [11], one can naturally assign a differential dimension polynomial to a system of algebraic differential equations and this polynomial expresses the A. Einstein's strength of the system (see [1]). Methods of computation of (univariate) differential dimension polynomials and the strength of systems of differential equations via the Ritt-Kolchin technique of characteristic sets can be found, for example, in [12] and [6, Chapters 5, 9]. Note also, that there are quite many works on computation of dimension polynomials of differential, difference and difference-differential modules with the use of various generalizations of the Gröbner basis method (see, for example, [6, Chapters V - XI], [7], [8], [9], [10, Chapter 3], and [13]). This method, however, does not work for non-linear difference-differential polynomial ideals, which, generally speaking, do not have finite Gröbner bases.

In this paper, we develop a method of characteristic sets with respect to several orderings for algebras of difference-differential polynomials over a difference-differential fields whose basic set of derivations is partied into several disjoint subsets. We apply this method to prove the existence, outline a method of computation, and determine invariants of a multivariate dimension polynomial associated with a finite system of generators of a difference-differential field extension (and a partition of the basic sets of derivations). We also show that most of these invariants are not carried by univariate dimension polynomials and show how the consideration of the new invariants can be applied to the isomorphism problem for difference-differential field extensions and equivalence problem for systems of algebraic difference-differential equations.

ISSAC'13, June 26–29, 2013, Boston, Massachusetts, USA.

2. PRELIMINARIES

Throughout the paper, $\mathbf{N}, \mathbf{Z}, \mathbf{Q}$, and $\mathbf{R}$ denote the sets of all non-negative integers, integers, rational numbers, and real numbers, respectively. $\mathbf{Q}[t_1, \dots, t_k]$ will denote the ring of polynomials in variables $t_1, \dots, t_k$ over $\mathbf{Q}$.

By a *difference-differential ring* we mean a commutative ring R together with finite sets $\Delta = \{\delta_1, \dots, \delta_m\}$ and $\sigma = \{\alpha_1, \dots, \alpha_n\}$ of derivations and automorphisms of R, respectively, such that any two mappings of the set $\Delta \cup \sigma$ commute. The set $\Delta \cup \sigma$ is called the *basic set* of the difference-differential ring R, which is also called a Δ-σ-ring. If R is a field, it is called a *difference-differential field* or a Δ-σ-field. Furthermore, in what follows, we denote the set

$$\{\alpha_1, \dots, \alpha_n, \alpha_1^{-1}, \dots, \alpha_n^{-1}\}$$

by σ^*. If R is a Δ-σ-ring, then Λ will denote the free commutative semigroup of all power products of the form

$$\lambda = \delta_1^{k_1} \dots \delta_m^{k_m} \alpha_1^{l_1} \dots \alpha_n^{l_n}$$

where $k_i \in \mathbf{N}$, $l_j \in \mathbf{Z}$ $(1 \le i \le m,\ 1 \le j \le n)$. For any such an element λ, we set

$$\lambda_\Delta = \delta_1^{k_1} \dots \delta_m^{k_m}, \quad \lambda_\sigma = \alpha_1^{l_1} \dots \alpha_n^{l_n},$$

and denote by Λ_Δ and Λ_σ the commutative semigroup of power products $\delta_1^{k_1} \dots \delta_m^{k_m}$ and the commutative group of elements of the form $\alpha_1^{l_1} \dots \alpha_n^{l_n}$, respectively. The *order* of λ is defined as

$$\operatorname{ord} \lambda = \sum_{i=1}^{m} k_i + \sum_{j=1}^{n} |l_j|,$$

and for every $r \in \mathbf{N}$, we set $\Lambda(r) = \{\lambda \in \Lambda \mid \operatorname{ord} \lambda \le r\}$ $(r \in \mathbf{N})$.

A subring (ideal) R_0 of a Δ-σ-ring R is called a difference-differential (or Δ-σ-) subring of R (respectively, difference-differential (or Δ-σ-) ideal of R) if R_0 is closed with respect to the action of any operator of $\Delta \cup \sigma^*$. If a prime ideal P of R is closed with respect to the action of $\Delta \cup \sigma^*$, it is called a *prime* difference-differential (or Δ-σ-) *ideal* of R.

If R is a Δ-σ-field and R_0 a subfield of R which is also a Δ-σ-subring of R, then R_0 is said to be a Δ-σ-subfield of R; R, in turn, is called a difference-differential (or Δ-σ-) field extension or a Δ-σ-overfield of R_0. In this case we also say that we have a Δ-σ-field extension R/R_0.

If R is a Δ-σ-ring and $\Sigma \subseteq R$, then the intersection of all Δ-ideals of R containing the set Σ is the smallest Δ-σ-ideal of R containing Σ; it is denoted by $[\Sigma]$. (Clearly, $[\Sigma]$ is generated, as an ideal, by the set $\{\lambda\xi \mid \xi \in \Sigma,\ \lambda \in \Lambda\}$). If the set Σ is finite, $\Sigma = \{\xi_1, \dots, \xi_q\}$, we say that the Δ-ideal I is finitely generated (we write this as $I = [\xi_1, \dots, \xi_q]$) and call $\xi_1, \dots, \xi_q$ difference-differential (or Δ-σ-)generators of I.

If K_0 is a Δ-σ-subfield of the Δ-σ-field K and $\Sigma \subseteq K$, then the smallest Δ-σ-subfield of K containing K_0 is denoted by $K_0\langle\Sigma\rangle$. If $K = K_0\langle\Sigma\rangle$ and the set Σ is finite, $\Sigma = \{\eta_1, \dots, \eta_s\}$, then K is said to be a finitely generated Δ-σ-extension of K_0 with the set of Δ-σ-generators $\{\eta_1, \dots, \eta_s\}$. In this case we write $K = K_0\langle\eta_1, \dots, \eta_s\rangle$. It is easy to see that the field $K_0\langle\eta_1, \dots, \eta_s\rangle$ coincides with the field $K_0(\{\lambda\eta_i \mid \lambda \in \Lambda, 1 \le i \le s\})$.

A ring homomorphism of Δ-σ-rings $\phi : R \longrightarrow S$ is called a *difference-differential* (or Δ-σ-) *homomorphism* if $\phi(\tau a) = \tau\phi(a)$ for any $\tau \in \Delta \cup \sigma$, $a \in R$.

If K is a Δ-σ-field and $Y = \{y_1, \dots, y_s\}$ is a finite set of symbols, then one can consider the countable set of symbols

$$\Lambda Y = \{\lambda y_j \mid \lambda \in \Lambda, 1 \le j \le s\}$$

and the polynomial ring

$$R = K[\{\lambda y_j \mid \lambda \in \Lambda, 1 \le j \le s\}]$$

in the set of indeterminates ΛY over the field K. This polynomial ring is naturally viewed as a Δ-σ-ring where

$$\tau(\lambda y_j) = (\tau\lambda) y_j$$

for any $\tau \in \Delta \cup \sigma$, $\lambda \in \Lambda$, $1 \le j \le s$, and the elements of $\Delta \cup \sigma$ act on the coefficients of the polynomials of R as they act in the field K. The ring R is called a *ring of difference-differential* (or Δ-σ-) *polynomials* in the set of differential (Δ-σ-)indeterminates $y_1, \dots, y_s$ over K. This ring is denoted by $K\{y_1, \dots, y_s\}$ and its elements are called difference-differential (or Δ-σ-) polynomials.

Let $L = K\langle\eta_1, \dots, \eta_s\rangle$ be a difference-differential field extension of K generated by a finite set $\eta = \{\eta_1, \dots, \eta_s\}$. As a field, $L = K(\{\lambda\eta_j \mid \lambda \in \Lambda, 1 \le j \le s\})$.

The following is a unified version of E. Kolchin's theorem on differential dimension polynomial and the author's theorem on the dimension polynomial of a difference field extension (see [7] or [10, Theorem 4.2.5]).

THEOREM 2.1. *With the above notation, there exists a polynomial $\phi_{\eta|K}(t) \in \mathbf{Q}[t]$ such that*

(i) $\phi_{\eta|K}(r) = \operatorname{tr.deg}_K K(\{\lambda\eta_j \mid \lambda \in \Lambda(r), 1 \le j \le s\})$ *for all sufficiently large* $r \in \mathbf{Z}$;

(ii) $\deg \phi_{\eta|K} \le m + n$ *and* $\phi_{\eta|K}(t)$ *can be written as*

$$\phi_{\eta|K}(t) = \sum_{i=0}^{m+n} a_i \binom{t+i}{i}, \text{ where } a_i \in \mathbf{Z} \text{ and } 2^n | a_{m+n}.$$

(iii) $d = \deg \phi_{\eta|K}$, a_{m+n} *and* a_d *do not depend on the set of difference-differential generators* η *of* L/K ($a_d \ne a_{m+n}$ *if and only if* $d < m + n$). *Moreover,* $\frac{a_{m+n}}{2^n}$ *is equal to the difference-differential transcendence degree of* L *over* K *(denoted by* Δ-σ-$\operatorname{tr.deg}_K L$*), that is, to the maximal number of elements* $\xi_1, \dots, \xi_k \in L$ *such that the family* $\{\lambda\xi_i \mid \lambda \in \Lambda, 1 \le i \le k\}$ *is algebraically independent over* K.

The polynomial whose existence is established by this theorem is called a *univariate difference-differential* (or Δ-σ-) *dimension polynomial* of the extension L/K associated with the system of difference-differential generators η.

3. PARTITION OF THE SET Δ. FORMULATION OF THE MAIN THEOREM

Let K be a difference-differential field of zero characteristic with basic sets $\Delta = \{\delta_1, \dots, \delta_m\}$ and $\sigma = \{\alpha_1, \dots, \alpha_n\}$ of derivations and automorphisms, respectively. Suppose that the set of derivations is represented as the union of p disjoint subsets $(p \ge 1)$:

$$\Delta = \Delta_1 \cup \dots \cup \Delta_p \tag{1}$$

where $\Delta_1 = \{\delta_1, \dots, \delta_{m_1}\}$, $\Delta_2 = \{\delta_{m_1+1}, \dots, \delta_{m_1+m_2}\}$, $\dots$,

$$\Delta_p = \{\delta_{m_1+\dots+m_{p-1}+1}, \dots, \delta_m\} \ (m_1 + \dots + m_p = m).$$

If

$$\lambda = \delta_1^{k_1} \dots \delta_m^{k_m} \alpha_1^{l_1} \dots \alpha_n^{l_n} \in \Lambda$$

($k_i \in \mathbf{N}$, $l_j \in \mathbf{Z}$), then the orders of λ with respect to Δ_i ($1 \le i \le p$) and σ are defined as

$$\operatorname{ord}_i \lambda = \sum_{\nu=m_1+\dots+m_{i-1}+1}^{m_1+\dots+m_i} k_\nu \quad \text{and} \quad \operatorname{ord}_\sigma \lambda = \sum_{j=1}^{n} |l_j|,$$

respectively. (If $i = 1$, then ν changes from 1 to m_1 in the first sum.) For any $r_1, \dots, r_{p+1} \in \mathbf{N}$, the set

$$\{\lambda \in \Lambda \,|\, \operatorname{ord}_i \lambda \le r_i \, (i = 1, \dots, p), \operatorname{ord}_\sigma \lambda \le r_{p+1}\}$$

will be denoted by $\Lambda(r_1, \dots, r_{p+1})$.

In what follows, for any permutation $(j_1, \dots, j_{p+1})$ of the set $\{1, \dots, p+1\}$, $<_{j_1,\dots,j_{p+1}}$ will denote the lexicographic order on $\mathbf{N}^{p+1}$ such that

$$(r_1, \dots, r_{p+1}) <_{j_1,\dots,j_{p+1}} (s_1, \dots, s_{p+1})$$

if and only if either $r_{j_1} < s_{j_1}$ or there exists $k \in \mathbf{N}$, $1 \le k \le p$, such that $r_{j_\nu} = s_{j_\nu}$ for $\nu = 1, \dots, k$ and $r_{j_{k+1}} < s_{j_{k+1}}$.

If $\Sigma \subseteq \mathbf{N}^{p+1}$, then Σ' will denote the set of all $e \in \Sigma$ that are maximal elements of this set with respect to one of the $(p+1)!$ orders $<_{j_1,\dots,j_{p+1}}$.

The following statement is the main result of this paper.

THEOREM 3.1. *Let $L = K\langle\eta_1, \dots, \eta_s\rangle$ be a Δ-σ-field extension generated by a set $\eta = \{\eta_1, \dots, \eta_s\}$. Then there exists a polynomial $\Phi_\eta \in \mathbf{Q}[t_1, \dots, t_{p+1}]$ such that*

(i) $\Phi_\eta(r_1, \dots, r_{p+1}) = \operatorname{tr.deg}_K K(\bigcup_{j=1}^{s} \Lambda(r_1, \dots, r_{p+1})\eta_j)$

for all sufficiently large $(r_1, \dots, r_{p+1}) \in \mathbf{N}^{p+1}$ (it means that there exist $s_1, \dots, s_{p+1} \in \mathbf{N}$ such that the equality holds for all $(r_1, \dots, r_{p+1}) \in \mathbf{N}^{p+1}$ with $r_1 \ge s_1, \dots, r_{p+1} \ge s_{p+1}$);

(ii) *$\deg_{t_i} \Phi_\eta \le m_i$ ($1 \le i \le p$), $\deg_{t_{p+1}} \Phi_\eta \le n$ and $\Phi_\eta(t_1, \dots, t_{p+1})$ can be represented as*

$$\Phi_\eta = \sum_{i_1=0}^{m_1} \dots \sum_{i_p=0}^{m_p} \sum_{i_{p+1}=0}^{n} a_{i_1 \dots i_{p+1}} \binom{t_1 + i_1}{i_1} \dots \binom{t_{p+1} + i_{p+1}}{i_{p+1}}$$

where $a_{i_1 \dots i_{p+1}} \in \mathbf{Z}$ and $2^n \,|\, a_{m_1 \dots m_p n}$.

(iii) *Let $E_\eta = \{(i_1, \dots, i_{p+1}) \in \mathbf{N}^{p+1} \,|\, 0 \le i_k \le m_k$ for $k = 1, \dots, p$, $0 \le i_{p+1} \le n$, and $a_{i_1 \dots i_{p+1}} \ne 0\}$. Then $d = \deg \Phi_\eta$, $a_{m_1 \dots m_{p+1}}$, elements $(k_1, \dots, k_{p+1}) \in E'_\eta$, the corresponding coefficients $a_{k_1 \dots k_{p+1}}$ and the coefficients of the terms of total degree d do not depend on the choice of the system of Δ-σ-generators η.*

DEFINITION 3.2. *$\Phi_\eta(t_1, \dots, t_{p+1})$ is called the difference-differential* (or Δ-σ-) *dimension polynomial of the Δ-σ-field extension L/K associated with the set of Δ-σ-generators η and partition* (1) *of the basic set of derivations.*

The Δ-σ-dimension polynomial associated with partition (1) has the following interpretation as the strength of a system of difference-differential equations.

Consider a system of difference-differential equations

$$A_i(f_1, \dots, f_s) = 0 \qquad (i = 1, \dots, q) \tag{2}$$

over a field of functions of m real variables $x_1, \dots, x_m$ (f_i are unknown functions of $x_1, \dots, x_m$). Suppose that $\Delta = \{\delta_1, \dots, \delta_m\}$ where δ_i is the partial differentiation $\partial/\partial x_i$ and the basic set of automorphisms $\sigma = \{\alpha_1, \dots, \alpha_m\}$ where

$$\alpha_i : f(x_1, \dots, x_m) \mapsto f(x_1, \dots, x_{i-1}, x_i + h_i, x_{i+1}, \dots, x_m)$$

($h_1, \dots, h_m \in \mathbf{R}$). Thus, we assume that the left-hand sides of the equations in (2) contain unknown functions f_i, their partial derivatives, their images under the shifts α_j, and various compositions of such shifts and partial derivations. Furthermore, we suppose that system (2) is algebraic, that is, all $A_i(y_1, \dots, y_s)$ are elements of a ring of Δ-σ-polynomials $K\{y_1, \dots, y_s\}$ over some functional Δ-σ-field K.

Let us consider a grid with equal cells of dimension $h_1 \times \dots \times h_m$ that fills $\mathbf{R}^m$. We fix some node $\mathcal{P}$ and say that *a node $\mathcal{Q}$ has order i* if the shortest path from $\mathcal{P}$ to $\mathcal{Q}$ along the edges of the grid consists of i steps (by a step we mean a path from a node of the grid to a neighbor node along the edge between them). We also fix partition (1) of the set of basic derivations Δ (such a partition can be, for example, a natural separation of (all or some) derivations with respect to coordinates and the derivation with respect to time).

For any $r_1, \dots, r_{p+1} \in \mathbf{N}$, let us consider the values of the unknown functions $f_1, \dots, f_s$ and their partial derivatives, whose order with respect to Δ_i does not exceed r_i ($1 \le i \le p$), at the nodes whose order does not exceed r_{p+1}. If $f_1, \dots, f_s$ should not satisfy any system of equations (or any other condition), these values can be chosen arbitrarily. Because of the system (and equations obtained from the equations of the system by partial differentiations and transformations of the form

$$f_j(x_1, \dots, x_m) \mapsto f_j(x_1 + k_1 h_1, \dots, x_m + k_m h_m)$$

with $k_1, \dots, k_m \in \mathbf{Z}$, $1 \le j \le s$), the number of independent values of the functions $f_1, \dots, f_s$ and their partial derivatives whose ith order does not exceed r_i ($1 \le i \le p$) at the nodes of order $\le r_{p+1}$ decreases. This number, which is a function of $p+1$ variables $r_1, \dots, r_{p+1}$, is the "measure of strength" of the system in the sense of A. Einstein. We denote it by $S_{r_1,\dots,r_{p+1}}$.

Suppose that the Δ-σ-ideal J of $K\{y_1, \dots, y_s\}$ generated by the Δ-σ-polynomials $A_1, \dots, A_q$ is prime (e. g., the polynomials are linear). Then the field of fractions L of the Δ-σ-integral domain $K\{y_1, \dots, y_s\}/J$ is a Δ-σ-field extension of K generated by the finite set $\eta = \{\eta_1, \dots, \eta_s\}$ where η_i is the canonical image of y_i in $K\{y_1, \dots, y_s\}/J$ ($1 \le i \le s$). It is easy to see that the Δ-σ-dimension polynomial $\Phi_\eta(t_1, \dots, t_{p+1})$ of the extension L/K associated with the system of Δ-σ-generators η has the property that

$$\Phi_\eta(r_1, \dots, r_{p+1}) = S_{r_1,\dots,r_{p+1}}$$

for all sufficiently large $(r_1, \dots, r_{p+q}) \in \mathbf{N}^{p+1}$, so this dimension polynomial is the measure of strength of the system of difference-differential equations (2) in the sense of A. Einstein.

4. NUMERICAL POLYNOMIALS

DEFINITION 4.1. *A polynomial $f(t_1, \dots, t_p) \in \mathbf{Q}[t_1, \dots, t_p]$ is called* numerical *if $f(r_1, \dots, r_p) \in \mathbf{Z}$ for all sufficiently large $(r_1, \dots, r_p) \in \mathbf{Z}^p$.*

The following theorem proved in [6, Chapter 2] gives the "canonical" representation of a numerical polynomial. (As usual, $\binom{t}{k} = \frac{t(t-1)\dots(t-k+1)}{k!}$.)

THEOREM 4.2. *Let $f(t_1, \dots, t_p)$ be a numerical polynomial in p variables and let $\deg_{t_i} f = m_i$ ($m_1, \dots, m_p \in \mathbf{N}$).*

Then $f(t_1,\ldots,t_p)$ *can be represented as*

$$f(t_1,\ldots t_p)=\sum_{i_1=0}^{m_1}\cdots\sum_{i_p=0}^{m_p}a_{i_1\ldots i_p}\binom{t_1+i_1}{i_1}\cdots\binom{t_p+i_p}{i_p}$$

with uniquely defined integer coefficients $a_{i_1\ldots i_p}$.

In what follows, we deal with subsets of $\mathbf{N}^m\times\mathbf{Z}^n$ $(m,n\geq 1)$ and a fixed partition of the set $\mathbf{N}_m=\{1,\ldots,m\}$ into p disjoint subsets $(p\geq 1)$:

$$\mathbf{N}_m=N_1\cup\cdots\cup N_p \qquad (3)$$

where $N_1=\{1,\ldots,m_1\},\ldots,\ N_p=\{m_1+\cdots+m_{p-1}+1,\ldots,m\}$ $(m_1+\cdots+m_p=m)$.

If $a=(a_1,\ldots,a_{m+n})\in\mathbf{N}^m\times\mathbf{Z}^n$ we set

$$\text{ord}_i\, a=\sum\nolimits_{j\in N_i}a_j,\quad 1\leq i\leq p$$

and

$$\text{ord}_{p+1}\, a=\sum_{i=m+1}^{m+n}|a_i|.$$

Furthermore, we consider the set $\mathbf{Z}^n$ as a union

$$\mathbf{Z}^n=\bigcup\nolimits_{1\leq j\leq 2^n}\mathbf{Z}_j^{(n)} \qquad (4)$$

where $\mathbf{Z}_1^{(n)},\ldots,\mathbf{Z}_{2^n}^{(n)}$ are all different Cartesian products of n sets each of which is either $\mathbf{N}$ or $\mathbf{Z}_-=\{a\in\mathbf{Z}|a\leq 0\}$. We assume that $\mathbf{Z}_1^{(n)}=\mathbf{N}^n$ and call $\mathbf{Z}_j^{(n)}$) the *jth orthant* of $\mathbf{Z}^n$. The set $\mathbf{N}^m\times\mathbf{Z}^n$ will be treated as a partially ordered set with the order $\trianglelefteq$ such that

$$(e_1,\ldots,e_m,f_1,\ldots,f_n)\trianglelefteq(e_1',\ldots,e_m',f_1',\ldots,f_n')$$

if and only if the n-tuples $(f_1,\ldots,f_n)$ and $(f_1',\ldots,f_n')$ lie in the same orthant $\mathbf{Z}_k^{(n)}$ and

$$(e_1,\ldots,e_m,|f_1|,\ldots,|f_n|)<_P(e_1',\ldots,e_m',|f_1'|,\ldots,|f_n'|)$$

where $<_P$ is the product order on $\mathbf{N}^{m+n}$. (Recall that the product order on $\mathbf{N}^k$ is a partial order $<_P$ such that

$$c=(c_1,\ldots,c_k)<_P c'=(c_1',\ldots,c_k')$$

if and only if $c_i<c_i'$ for $i=1,\ldots,k$. We write $c\leq_P c'$ if $c<_P c'$ or $c=c'$).

If A is a subset of $\mathbf{N}^m\times\mathbf{Z}^n$, then W_A will denote the set of all elements $w\in\mathbf{N}^m\times\mathbf{Z}^n$ such that there is no $a\in A$ with $a\trianglelefteq w$. Furthermore, for any $r_1,\ldots r_{p+1}\in\mathbf{N}$, $A(r_1,\ldots r_{p+1})$ denotes the set

$$\{x=(x_1,\ldots,x_m,x_1',\ldots,x_n')\in A\,|\,\text{ord}_i\, x\leq r_i, 1\leq i\leq p{+}1\}.$$

If $E\subseteq\mathbf{N}^m$ and $s_1,\ldots,s_p\in\mathbf{N}$, then $E(s_1,\ldots,s_p)$ will denote the set $\{(e_1,\ldots,e_m)\in E\,|\,\text{ord}_i(e_1,\ldots,e_m,0,\ldots,0)\leq s_i$ for $i=1,\ldots,p\}$ ($(e_1,\ldots,e_m,0,\ldots,0)$ ends with n zeros; it is treated as a point in $\mathbf{N}^m\times\mathbf{Z}^n$.) Furthermore V_E will denote the set of all m-tuples $v=(v_1,\ldots,v_m)\in\mathbf{N}$ which are not greater than or equal to any m-tuple from E with respect to the product order on $\mathbf{N}^m$. Clearly, $v=(v_1,\ldots,v_m)\in V_E$ if and only if for any element $(e_1,\ldots,e_m)\in E$, there exists $i\in\mathbf{N}, 1\leq i\leq m$, such that $e_i>v_i$.

The following two theorems are proved in [6, Chapter 2].

THEOREM 4.3. *Let* E *be a subset of* $\mathbf{N}^m$ *where* $m=m_1+\cdots+m_p$ *for some* $m_1,\ldots,m_p\in\mathbf{N}$ $(p\geq 1)$. *Then there exists a numerical polynomial* $\omega_E(t_1,\ldots,t_p)$ *such that*

(i) $\omega_E(r_1,\ldots,r_p)=\text{Card}\, V_E(r_1,\ldots,r_p)$ *for all sufficiently large* $(r_1,\ldots,r_p)\in\mathbf{N}^p$. (As usual, Card M denotes the number of elements of a finite set M).

(ii) $\deg_{t_i}\omega_E\leq m_i$ *for all* $i=1,\ldots,p$.

(iii) $\deg\omega_E=m$ *if and only if* $E=\emptyset$. *In this case*

$$\omega_E(t_1,\ldots,t_p)=\prod_{i=1}^{p}\binom{t_i+m_i}{m_i}.$$

DEFINITION 4.4. *The polynomial* $\omega_E(t_1,\ldots,t_p)$ *is called the dimension polynomial of the set* $E\subseteq\mathbf{N}^m$ *associated with the partition* $(m_1,\ldots,m_p)$ *of* m.

THEOREM 4.5. *Let* $E=\{e_1,\ldots,e_q\}$ $(q\geq 1)$ *be a finite subset of* $\mathbf{N}^m$ *and let partition* (3) *of* $\mathbf{N}_m$ *be fixed. Let* $e_i=(e_{i1},\ldots,e_{im})$ $(1\leq i\leq q)$ *and for any* $l\in\mathbf{N}$, $0\leq l\leq q$, *let* $\Gamma(l,q)$ *denote the set of all* l*-element subsets of the set* $\mathbf{N}_q=\{1,\ldots,q\}$. *Furthermore, for any* $\sigma\in\Gamma(l,q)$, *let* $\bar{e}_{\emptyset j}=0$, $\bar{e}_{\sigma j}=\max\{e_{ij}\,|\,i\in\sigma\}$ *if* $\sigma\neq\emptyset$ $(1\leq j\leq m)$, *and* $b_{\sigma k}=\sum_{h\in N_k}\bar{e}_{\sigma h}$ $(k=1,\ldots,p)$. *Then*

$$\omega_E(t_1,\ldots,t_p)=\sum_{l=0}^{q}(-1)^l\sum_{\sigma\in\Gamma(l,\,q)}\prod_{j=1}^{p}\binom{t_j+m_j-b_{\sigma j}}{m_j} \qquad (5)$$

Remark. It is clear that if $E\subseteq\mathbf{N}^m$ and E^* is the set of all minimal elements of the set E with respect to the product order on $\mathbf{N}^m$, then the set E^* is finite and $\omega_E(t_1,\ldots,t_p)=\omega_{E^*}(t_1,\ldots,t_p)$. Thus, Theorem 4.5 gives an algorithm that allows one to find a numerical polynomial associated with any subset of $\mathbf{N}^m$ (and with a given partition of the set $\{1,\ldots,m\}$): one should first find the set of all minimal points of the subset and then apply Theorem 4.5.

The following result can be obtained by mimicking the proof of [8, Theorem 3.4].

THEOREM 4.6. *Let* $A\subseteq\mathbf{N}^m\times\mathbf{Z}^n$ *and let partition* (3) *of* $\mathbf{N}_m$ *be fixed. Then there exists a numerical polynomial* $\phi_A(t_1,\ldots,t_{p+1})$ *such that*

(i) $\phi_A(r_1,\ldots,r_{p+1})=\text{Card}\, W_A(r_1,\ldots,r_{p+1})$ *for all sufficiently large* $(r_1,\ldots,r_{p+1})\in\mathbf{N}^{p+1}$.

(ii) $deg_{t_i}\phi_A\leq m_i$ $(1\leq i\leq p)$, $deg_{t_{p+1}}\phi_A\leq n$ *and the coefficient of* $t_1^{m_1}\ldots t_p^{m_p}t_{p+1}^n$ *in* ϕ_A *is of the form*

$$\frac{2^n a}{m_1!\ldots m_p!n!},$$

with $a\in\mathbf{Z}$.

(iii) *Let us consider a mapping*

$$\rho:\mathbf{N}^m\times\mathbf{Z}^n\longrightarrow\mathbf{N}^{m+2n}$$

such that $\rho((e_1,\ldots,e_{m+n})=(e_1,\ldots,e_m,\max\{e_{m+1},0\},\ldots,\max\{e_{m+n},0\},\max\{-e_{m+1},0\},\ldots,\max\{-e_{m+n},0\})$.

Let

$$B=\rho(A)\cup\{\bar{e}_1,\ldots,\bar{e}_n\},$$

where $\bar{e}_i$ $(1\leq i\leq n)$ *is a* $(m+2n)$*-tuple in* $\mathbf{N}^{m+2n}$ *whose* $(m+i)$*th and* $(m+n+i)$*th coordinates are equal to 1 and all other coordinates are equal to 0. Then*

$$\phi_A(t_1,\ldots,t_{p+1})=\omega_B(t_1,\ldots,t_{p+1})$$

where $\omega_B(t_1,\ldots,t_{p+1})$ *is the dimension polynomial of the set* B *(see Definition 4.4) associated with the partition* $\mathbf{N}_{m+2n}=\{1,\ldots,m_1\}\cup\{m_1+1,\ldots,m_1+m_2\}\cup\cdots\cup\{m_1+\cdots+m_{p-1}+1,\ldots,m\}\cup\{m+1,\ldots,m+2n\}$ *of the set* $\mathbf{N}_{m+2n}$.

(iv) *If $A = \emptyset$, then*

$$\phi_A = \prod_{j=1}^{p} \binom{t_j + m_j}{m_j} \sum_{i=0}^{n} (-1)^{n-i} 2^i \binom{n}{i} \binom{t_{p+1} + i}{i}. \quad (6)$$

The polynomial $\phi_A(t_1, \dots, t_{p+1})$ is called the *dimension polynomial* of the set $A \subseteq \mathbf{N}^m \times \mathbf{Z}^n$ associated with partition (3) of $\mathbf{N}_m$.

5. PROOF OF THE MAIN THEOREM

In this section, we prove Theorem 3.1 and give a method of computation of difference-differential dimension polynomials of Δ-σ-field extensions based on constructing a characteristic set of the defining prime Δ-σ-ideal of the extension.

In what follows we use the conventions of section 3. In particular, we assume that partition (1) of the set of basic derivations $\Delta = \{\delta_1, \dots, \delta_m\}$ is fixed.

Let us consider total orderings $<_1, \dots, <_p, <_\sigma$ of the set of power products Λ such that

$$\lambda = \delta_1^{k_1} \dots \delta_m^{k_m} \alpha_1^{l_1} \dots \alpha_n^{l_n} <_i \lambda' = \delta_1^{k'_1} \dots \delta_m^{k'_m} \alpha_1^{l'_1} \dots \alpha_n^{l'_n}$$

$(1 \le i \le p)$ if and only if $(\mathrm{ord}_i\, \lambda, \mathrm{ord}\, \lambda, \mathrm{ord}_1\, \lambda, \dots, \mathrm{ord}_{i-1}\, \lambda, \mathrm{ord}_{i+1}\, \lambda, \dots, \mathrm{ord}_p\, \lambda,$ $\mathrm{ord}_\sigma\, \lambda, k_{m_1+\dots+m_{i-1}+1}, \dots, k_{m_1+\dots+m_i}, k_1, \dots, k_{m_1+\dots+m_{i-1}},$ $k_{m_1+\dots+m_i+1}, \dots, k_m, |l_1|, \dots, |l_n|, l_1, \dots, l_n)$ is less than the corresponding $(m + 2n + p + 2)$-tuple for λ' with respect to the lexicographic order on $\mathbf{N}^{m+2n+p+2}$.

Similarly, $\lambda <_\sigma \lambda'$ if and only if $(\mathrm{ord}_\sigma\, \lambda, \mathrm{ord}\, \lambda, \mathrm{ord}_1\, \lambda, \dots,$ $\mathrm{ord}_p\, \lambda, |l_1|, \dots, |l_n|, l_1, \dots, l_n, k_1, \dots, k_m)$ is less than the corresponding $(m + 2n + p + 2)$-tuple for λ' with respect to the lexicographic order on $\mathbf{N}^{m+2n+p+2}$.

Let $\lambda_1 = \delta_1^{k_1} \dots \delta_m^{k_m} \alpha_1^{l_1} \dots \alpha_n^{l_n}$, $\lambda_2 = \delta_1^{r_1} \dots \delta_m^{r_m} \alpha_1^{s_1} \dots \alpha_n^{s_n}$ be elements of Λ. They are called *similar*, if $(l_1, \dots, l_n)$ and $(s_1, \dots, s_n)$ lie in the same orthant of $\mathbf{Z}^n$ (see (4)). In this case we write $\lambda_1 \sim \lambda_2$. We say that λ_1 *divides* λ_2 (or λ_2 is a *multiple* of λ_1) and write $\lambda_1 | \lambda_2$ if $\lambda_1 \sim \lambda_2$ and there exists $\lambda \in \Lambda$ such that $\lambda \sim \lambda_1$, $\lambda \sim \lambda_2$ and $\lambda_2 = \lambda\lambda_1$.

Let K be a Δ-σ-field (Char $K = 0$) and let partition (1) of the set Δ be fixed. Let $K\{y_1, \dots, y_s\}$ be the ring of Δ-σ-polynomials over K and let ΛY denote the set of all elements λy_i $(\lambda \in \Lambda, 1 \le i \le s)$ called *terms*. Note that as a ring,

$$K\{y_1, \dots, y_s\} = K[\Lambda Y].$$

Two terms $u = \lambda y_i$ and $v = \lambda' y_j$ are called *similar* if λ and λ' are similar; in this case we write $u \sim v$. If $u = \lambda y_i$ is a term and $\lambda' \in \Lambda$, we say that u is similar to λ' and write $u \sim \lambda'$ if $\lambda \sim \lambda'$. Furthermore, if $u, v \in \Lambda Y$, we say that u *divides* v or v *is a multiple of* u, if $u = \lambda' y_i$, $v = \lambda'' y_i$ for some y_i and $\lambda' | \lambda''$. (If $\lambda'' = \lambda\lambda'$ for some $\lambda \in \Lambda$, $\lambda \sim \lambda'$, we write $\frac{v}{u}$ for λ.)

Let us consider $p+1$ orders $<_1, \dots, <_p, <_\sigma$ on the set ΛY that correspond to the orders on Λ (we use the same symbols for the orders on Λ and ΛY). These orders are defined as follows: $\lambda y_j <_i$ (or $<_\sigma$) $\lambda' y_k$ if and only if $\lambda <_i$ (respectively, $<_\sigma$)λ' in Λ or $\lambda = \lambda'$ and $j < k$ $(1 \le i \le p, 1 \le j, k \le s)$.

The order of a term $u = \lambda y_k$ and its orders with respect to the sets Δ_i $(1 \le i \le p)$ and σ are defined as the corresponding orders of λ (we use the same notation $\mathrm{ord}\, u$, $\mathrm{ord}_i\, u$, and $\mathrm{ord}_\sigma\, u$ for the corresponding orders).

If $A \in K\{y_1, \dots, y_s\} \setminus K$ and $1 \le k \le p$, then the highest with respect to $<_k$ term that appears in A is called the *k-leader* of A. It is denoted by $u_A^{(k)}$. The highest term of A with respect to $<_\sigma$ is called the *σ-leader* of A; it is denoted by v_A. If A is written as a polynomial in v_A,

$$A = I_d (v_A)^d + I_{d-1} (v_A)^{d-1} + \dots + I_0,$$

where all terms of $I_0, \dots, I_d$ are less than v_A with respect to $<_\sigma$, then I_d is called the *initial* of A. The partial derivative

$$\partial A / \partial v_A = d I_d (v_A)^{d-1} + (d-1) I_{d-1} (v_A)^{d-2} + \dots + I_1$$

is called the *separant* of A. The initial and the separant of a Δ-σ-polynomial A are denoted by I_A and S_A, respectively.

If $A, B \in K\{y_1, \dots, y_s\}$, then A is said to have lower rank than B (we write $\mathrm{rk}\, A < \mathrm{rk}\, B$) if either $A \in K$, $B \notin K$, or

$$(v_A, \deg_{v_A} A, \mathrm{ord}_1\, u_A^{(1)}, \dots, \mathrm{ord}_p\, u_A^{(p)})$$

is less than

$$(v_B, \deg_{v_B} B, \mathrm{ord}_1\, u_B^{(1)}, \dots, \mathrm{ord}_p\, u_B^{(p)})$$

with respect to the lexicographic order (v_A and v_B are compared with respect to $<_\sigma$). If the vectors are equal (or $A, B \in K$) we say that A and B are of the same rank and write $\mathrm{rk}\, A = \mathrm{rk}\, B$.

DEFINITION 5.1. *If $A, B \in K\{y_1, \dots, y_s\}$, then B is said to be reduced with respect to A if*

(i) B does not contain terms λv_A such that $\lambda \sim v_A$, $\lambda_\Delta \neq 1$, and

$$\mathrm{ord}_i (\lambda u_A^{(i)}) \le \mathrm{ord}_i\, u_B^{(i)}, \quad i = 1, \dots, p.$$

(ii) If B contains a term λv_A, where $\lambda \sim v_A$ and $\lambda_\Delta = 1$, then either there exists j, $1 \le j \le p$, such that

$$\mathrm{ord}_j\, u_B^{(j)} < \mathrm{ord}_j \left(\lambda u_A^{(j)}\right) \quad or \quad \mathrm{ord}_j \left(\lambda u_A^{(j)}\right) \le \mathrm{ord}_j\, u_B^{(j)}$$

for all $j = 1, \dots, p$ and $\deg_{\lambda v_A} B < \deg_{v_A} A$.

If $B \in K\{y_1, \dots, y_s\}$, then B is said to be *reduced with respect to a set* $\Sigma \subseteq K\{y_1, \dots, y_s\}$ if B is reduced with respect to every element of Σ.

A set $\Sigma \subseteq K\{y_1, \dots, y_s\}$ is called *autoreduced* if $\Sigma \cap K = \emptyset$ and every element of Σ is reduced with respect to any other element of this set.

The proof of the following lemma can be found in [5, Chapter 0, Section 17].

LEMMA 5.2. *Let A be any infinite subset of $\mathbf{N}^m \times \mathbf{N}_n$ $(n \ge 1)$. Then there exists an infinite sequence of elements of A, strictly increasing relative to the product order, in which every element has the same projection on $\mathbf{N}_n$.*

As a consequence, we obtain the following statement.

LEMMA 5.3. *Let S be any infinite set of terms λy_j $(\lambda \in \Lambda, 1 \le j \le s)$ in $K\{y_1, \dots, y_s\}$. Then there exists an index j and an infinite sequence of terms $\lambda_1 y_j, \lambda_2 y_j, \dots, \lambda_k y_j, \dots$ such that $\lambda_k | \lambda_{k+1}$ for every $k = 1, 2, \dots$.*

PROPOSITION 5.4. *Every autoreduced set is finite.*

PROOF. Suppose that Σ is an infinite autoreduced subset of $K\{y_1, \dots, y_s\}$. Then Σ must contain an infinite set Σ' whose Δ-σ-polynomials have different σ-leaders similar to each other. Indeed, if it is not so, then Σ contains an infinite set Σ_1 whose Δ-σ-polynomials have the same σ-leader v. By Lemma 5.2, the infinite set

$$\left\{ \left(\mathrm{ord}_1\, u_A^{(1)}, \dots, \mathrm{ord}_p\, u_A^{(p)} \right) \mid A \in \Sigma_1 \right\}$$

contains a nondecreasing infinite sequence

$$(\mathrm{ord}_1\, u_{A_1}^{(1)}, \dots, \mathrm{ord}_p\, u_{A_1}^{(p)}) \le_P (\mathrm{ord}_1\, u_{A_2}^{(1)}, \dots, \mathrm{ord}_p\, u_{A_2}^{(p)}) \le_P \dots$$

($A_1, A_2, \dots \in \Sigma_1$ and $\le_P$ denotes the product order on $\mathbf{N}^p$). Since the sequence

$$\{\deg_{v_{A_i}} A_i | i = 1, 2, \dots\}$$

cannot be strictly decreasing, there are two indices i and j such that $i < j$ and $\deg_{v_{A_i}} A_i \le \deg_{v_{A_j}} A_j$. We see that A_j is not reduced with respect to A_i that contradicts the fact that Σ is an autoreduced set.

Thus, we can assume that all Δ-σ-polynomials in Σ have distinct σ-leaders similar to each other. Then (see Lemma 5.3) there exists an infinite sequence $B_1, B_2, \dots$ of elements of Σ such that $v_{B_i} | v_{B_{i+1}}$ and $\left(\frac{v_{B_{i+1}}}{v_{B_i}}\right)_\Delta \ne 1$ $(i = 1, 2, \dots)$. Let

$$k_{ij} = \mathrm{ord}_j\, v_{B_i} \quad \text{and} \quad l_{ij} = \mathrm{ord}_j\, u_{B_i}^{(j)}, \quad 1 \le j \le p.$$

Then $l_{ij} \ge k_{ij}$, so that

$$\{(l_{i1} - k_{i1}, \dots, l_{ip} - k_{ip}) | i = 1, 2, \dots\} \subseteq \mathbf{N}^p.$$

By Lemma 5.2, there exists an infinite sequence of indices $i_1 < i_2 < \dots$ such that $(l_{i_1 1} - k_{i_1 1}, \dots, l_{i_1 p} - k_{i_1 p}) \le_P (l_{i_2 1} - k_{i_2 1}, \dots, l_{i_2 p} - k_{i_2 p}) \le_P \dots$. Then for any $j = 1, \dots, p$, we have

$$\mathrm{ord}_j \left(\frac{v_{B_{i_2}}}{v_{B_{i_1}}} u_{B_{i_1}}^{(j)} \right) = k_{i_2 j} - k_{i_1 j} + l_{i_1 j} \le l_{i_2 j} = \mathrm{ord}_j\, u_{B_{i_2}}^{(j)},$$

so that B_{i_2} contains a term $\lambda v_{B_{i_1}} = v_{B_{i_2}}$ with $\lambda_\Delta \ne 1$ and

$$\mathrm{ord}_j \left(\lambda u_{B_{i_1}}^{(j)}\right) \le \mathrm{ord}_j\, u_{B_{i_2}}^{(j)}, \quad j = 1, \dots, p.$$

Thus, B_{i_2} is reduced with respect to B_{i_1} that contradicts the fact that Σ is an autoreduced set. □

Throughout the rest of the paper, while considering and autoreduced set $\Sigma = \{A_1, \dots, A_d\}$ in $K\{y_1, \dots, y_s\}$, we always assume that $\mathrm{rk}\, A_1 < \dots < \mathrm{rk}\, A_d$.

The proof of the following statement is similar to the proof of Theorem 3.5.27 in [6].

Proposition 5.5. *Let $\Sigma = \{A_1, \dots, A_d\}$ be an autoreduced set in $K\{y_1, \dots, y_s\}$ and let I_k and S_k denote the initial and separant of A_k, respectively. Let $I(\Sigma) = \{X \in K\{y_1, \dots, y_s\} \,|\, X = 1$ or X is a product of finitely many elements of the form $\gamma(I_k)$ and $\gamma'(S_k)$ where $\gamma, \gamma' \in \Lambda_\sigma\}$. Then for any Δ-σ-polynomial B, there exist $B_0 \in K\{y_1, \dots, y_s\}$ and $J \in I(\Sigma)$ such that B_0 is reduced with respect to Σ and $JB \equiv B_0 \mod [\Sigma]$ (that is, $JB - B_0 \in [\Sigma]$).*

With the notation of the last proposition, we say that the Δ-σ-polynomial B *reduces to* B_0 modulo Σ.

Definition 5.6. *Let $\Sigma = \{A_1, \dots, A_d\}$ and $\Sigma' = \{B_1, \dots, B_e\}$ be two autoreduced sets in $K\{y_1, \dots, y_s\}$. Then Σ is said to have lower rank than Σ' if one of the following two cases holds:*

(i) There exists $k \in \mathbf{N}$ such that $k \le \min\{d, e\}$, $\mathrm{rk}\, A_i = \mathrm{rk}\, B_i$ for $i = 1, \dots, k-1$ and $\mathrm{rk}\, A_k < \mathrm{rk}\, B_k$.

(ii) $d > e$ and $\mathrm{rk}\, A_i = \mathrm{rk}\, B_i$ for $i = 1, \dots, e$.

If $d = e$ and $\mathrm{rk}\, A_i = \mathrm{rk}\, B_i$ for $i = 1, \dots, d$, then Σ is said to have the same rank as Σ'.

As in [5, Chapter I, Section 9], we obtain the following proposition.

Proposition 5.7. *In every nonempty family of autoreduced sets of difference-differential polynomials there exists an autoreduced set of lowest rank.*

Let J be any ideal of $K\{y_1, \dots, y_s\}$. Since the set of all autoreduced subsets of J is not empty (if $A \in J$, then $\{A\}$ is an autoreduced subset of J), the last statement shows that J contains an autoreduced subset of lowest rank. Such an autoreduced set is called a *characteristic set* of the ideal J. The following statement can be obtained by mimicking the proof of Lemma 8 in [5, Chapter I].

Proposition 5.8. *Let $\Sigma = \{A_1, \dots, A_d\}$ be a characteristic set of a Δ-σ-ideal J of the ring $R = K\{y_1, \dots, y_s\}$. Then an element $B \in R$ is reduced with respect to the set Σ if and only if $B = 0$.*

Since for any $A \in K\{y_1, \dots, y_s\}$ and $\gamma \in \Lambda_\sigma$, $\mathrm{ord}_i(\gamma A) = \mathrm{ord}_i\, A$ for $i = 1, \dots, p$, one can introduce the concept of a coherent autoreduced set of a linear Δ-σ-ideal of $K\{y_1, \dots, y_s\}$ (that is, a Δ-σ-ideal generated by a finite set of linear Δ-σ-polynomials) in the same way as it is defined in the case of difference polynomials (see [6, Section 6.5]): an autoreduced set $\Sigma = \{A_1, \dots, A_d\} \subseteq K\{y_1, \dots, y_s\}$ consisting of linear Δ-σ-polynomials is called *coherent* if it satisfies the following two conditions:

(i) λA_i reduces to zero modulo Σ for any $\lambda \in \Lambda$, $1 \le i \le d$.

(ii) If $v_{A_i} \sim v_{A_j}$ and $w = \lambda v_{A_i} = \lambda' v_{A_j}$, where $\lambda \sim \lambda' \sim v_{A_i} \sim v_{A_j}$, then the Δ-σ-polynomial

$$(\lambda' I_{A_j})(\lambda A_i) - (\lambda I_{A_i})(\lambda' A_j)$$

reduces to zero modulo Σ.

The following two propositions can be proved precisely in the same way as the corresponding statements for difference polynomials, see [6, Theorem 6.5.3 and Corollary 6.5.4]).

Proposition 5.9. *Any characteristic set of a linear Δ-σ-ideal of $K\{y_1, \dots, y_s\}$ is a coherent autoreduced set. Conversely, if Σ is a coherent autoreduced set in $K\{y_1, \dots, y_s\}$ consisting of linear Δ-σ-polynomials, then Σ is a characteristic set of the linear Δ-σ-ideal $[\Sigma]$.*

Proposition 5.10. *Let us consider a partial order $\preccurlyeq$ on $K\{y_1, \dots, y_s\}$ such that $A \preccurlyeq B$ if and only if $v_A | v_B$. Let A be a linear Δ-σ-polynomial in $K\{y_1, \dots, y_s\}$, $A \notin K$. Then the set of all minimal with respect to $\preccurlyeq$ elements of the set $\{\lambda A \,|\, \lambda \in \Lambda\}$ is a characteristic set of the Δ-σ-ideal $[A]$.*

Now we are ready to prove Theorem 3.1.

Proof. Let $L = K\langle \eta_1, \dots, \eta_s \rangle$ be a Δ-σ-field extension of K generated by a finite set $\eta = \{\eta_1, \dots, \eta_s\}$. Then there exists a natural Δ-σ-homomorphism Υ_η of the ring of Δ-σ-polynomials $K\{y_1, \dots, y_s\}$ onto $K\{\eta_1, \dots, \eta_s\}$ such that $\Upsilon_\eta(a) = a$ for any $a \in K$ and $\Upsilon_\eta(y_j) = \eta_j$ for $j = 1, \dots, s$. (If $A \in K\{y_1, \dots, y_s\}$, then $\Upsilon_\eta(A)$ is called the *value* of A at η; it is denoted by $A(\eta)$.) Obviously, the kernel P of the Δ-σ-homomorphism Υ_η is a prime Δ-σ-ideal of $K\{y_1, \dots, y_s\}$. This ideal is called the *defining* ideal of η over K or the defining ideal of the extension $L = K\langle \eta_1, \dots, \eta_s \rangle$.

It is easy to see that the quotient Δ-σ-field of the factor ring $\bar{R} = K\{y_1, \dots, y_s\}/P$ is naturally Δ-σ-isomorphic to the field L. The corresponding isomorphism is identity on K and maps the images of the Δ-σ-indeterminates $y_1, \dots, y_s$ in the factor ring $\bar{R}$ to the elements $\eta_1, \dots, \eta_s$, respectively.

Let $\Sigma = \{A_1, \dots, A_d\}$ be a characteristic set of P and for any $r_1, \dots, r_{p+1} \in \mathbf{N}$, let

$U_{r_1 \dots r_{p+1}} = \{u \in \Lambda Y \mid \operatorname{ord}_i u \leq r_i$ for $i = 1, \dots, p$, $\operatorname{ord}_\sigma u \leq r_{p+1}$, and either u is not a multiple of any v_{A_i} or for every $\lambda \in \Lambda, A \in \Sigma$ such that $u = \lambda v_A$ and $\lambda \sim v_A$, there exists $j \in \{1, \dots, p\}$ such that $\operatorname{ord}_j(\lambda u_A^{(j)}) > r_j\}$.

Applying the arguments of the proof of Theorem 6 in [5, Chapter II], we obtain that the set

$$\bar{U}_{r_1 \dots r_{p+1}} = \{u(\eta) \mid u \in U_{r_1 \dots r_{p+1}}\}$$

is a transcendence basis of $K(\bigcup_{j=1}^{n} \Lambda(r_1, \dots, r_{p+1})\eta_j)$ over K.

Let $U^{(1)}_{r_1 \dots r_{p+1}} = \{u \in \Lambda Y \mid \operatorname{ord}_i u \leq r_i$ for $i = 1, \dots, p$, $\operatorname{ord}_\sigma u \leq r_{p+1}$, and u is not a multiple of any v_{A_j}, $j = 1, \dots, d\}$ and let

$U^{(2)}_{r_1 \dots r_{p+1}} = \{u \in \Lambda Y \mid \operatorname{ord}_i u \leq r_i$, $\operatorname{ord}_\sigma u \leq r_{p+1}$ for $i = 1, \dots, p$ and there exists at least one pair i, j $(1 \leq i \leq p, 1 \leq j \leq d)$ such that $u = \lambda v_{A_j}$, $\lambda \sim v_{A_j}$, and $\operatorname{ord}_i(\lambda u_{A_j}^{(i)}) > r_i\}$.

Clearly,

$$U_{r_1 \dots r_{p+1}} = U^{(1)}_{r_1 \dots r_{p+1}} \cup U^{(2)}_{r_1 \dots r_{p+1}}$$

and

$$U^{(1)}_{r_1 \dots r_{p+1}} \cap U^{(2)}_{r_1 \dots r_{p+1}} = \emptyset.$$

By Theorem 4.6, there exists a numerical polynomial $\phi \in \mathbf{Q}[t_1, \dots, t_{p+1}]$ such that

$$\phi(r_1, \dots, r_{p+1}) = \operatorname{Card} U^{(1)}_{r_1 \dots r_{p+1}}$$

for all sufficiently large $(r_1, \dots, r_{p+1}) \in \mathbf{N}^{p+1}$, $\deg_{t_i} \phi \leq m_i$ $(1 \leq i \leq p)$, and $\deg_{t_{p+1}} \phi \leq n$. Repeating the arguments of the proof of Theorem 4.1 of [9], we obtain that there is a linear combination $\psi(t_1, \dots, t_{p+1})$ of polynomials of the form (6) such that

$$\psi(r_1, \dots, r_{p+1}) = \operatorname{Card} U^{(2)}_{r_1 \dots r_{p+1}}$$

for all sufficiently large $(r_1, \dots, r_{p+1}) \in \mathbf{N}^{p+1}$. Then the polynomial

$$\Phi_\eta(t_1, \dots, t_{p+1}) = \phi(t_1, \dots, t_{p+1}) + \psi(t_1, \dots, t_{p+1})$$

satisfies conditions (i) and (ii) of Theorem 3.1.

In order to prove the last part of the theorem, suppose that $\zeta = \{\zeta_1, \dots, \zeta_q\}$ is another system of Δ-σ-generators of L/K, that is, $L = K\langle \eta_1, \dots, \eta_s \rangle = K\langle \zeta_1, \dots, \zeta_q \rangle$. Let

$$\Phi_\zeta = \sum_{i_1=0}^{m_1} \cdots \sum_{i_p=0}^{m_p} \sum_{i_{p+1}=0}^{n} b_{i_1 \dots i_{p+1}} \binom{t_1 + i_1}{i_1} \cdots \binom{t_{p+1} + i_{p+1}}{i_{p+1}}$$

be the dimension polynomial of L/K associated with the system of generators ζ. Then there exist positive integers $h_1, \dots, h_{p+1}$ such that

$$\eta_i \in K\left(\bigcup_{j=1}^{q} \Lambda(h_1, \dots, h_{p+1})\zeta_j\right)$$

and

$$\zeta_k \in K\left(\bigcup_{j=1}^{s} \Lambda(h_1, \dots, h_{p+1})\eta_j\right)$$

for any $i = 1, \dots, s$ and $k = 1, \dots, q$, whence

$$\Phi_\eta(r_1, \dots, r_{p+1}) \leq \Phi_\zeta(r_1 + h_1, \dots, r_{p+1} + h_{p+1})$$

and

$$\Phi_\zeta(r_1, \dots, r_{p+1}) \leq \Phi_\eta(r_1 + h_1, \dots, r_{p+1} + h_{p+1})$$

for all sufficiently large $(r_1, \dots, r_{p+1}) \in \mathbf{N}^{p+1}$. Now the statement of the third part of Theorem 3.1 follows from the fact that for any element $(k_1, \dots, k_{p+1}) \in E'_\eta$, the term $\binom{t_1+k_1}{k_1} \cdots \binom{t_{p+1}+k_{p+1}}{k_{p+1}}$ appears in Φ_η and Φ_ζ with the same coefficient $a_{k_1 \dots k_{p+1}}$. The equality of the coefficients of the corresponding terms of total degree $d = \deg \Phi_\eta$ in Φ_η and Φ_ζ can be shown as in the proof of [10, Theorem 3.3.21]. $\square$

The result of Theorem 3.1 can be generalized to the case when both sets of basic operators Δ and σ are represented as unions of disjoint subsets. The proof is, however, essentially longer; the author will present the generalized version of Theorem 3.1 in his forthcoming paper.

EXAMPLE 5.11. Let us find the Δ-σ-dimension polynomial that expresses the strength of the difference-differential equation

$$\frac{\partial^2 y(x_1, x_2)}{\partial x_1^2} + \frac{\partial^2 y(x_1, x_2)}{\partial x_2^2} + y(x_1 + h, x_2) + a(x_1, x_2) = 0 \quad (7)$$

over some Δ-σ-field of functions of two real variables K, where the basic set of derivations $\Delta = \{\delta_1 = \frac{\partial}{\partial x_1}, \delta_2 = \frac{\partial}{\partial x_2}\}$ has the partition $\Delta = \{\delta_1\} \cup \{\delta_1\}$ and σ consists of one automorphisms $\alpha : f(x_1, x_2) \mapsto f(x_1 + h, x_2)$ $(h \in \mathbf{R})$.

In this case, the associated Δ-σ-extension $K\langle \eta \rangle / K$ is Δ-σ-isomorphic to the field of fractions of

$$K\{y\}/[\alpha y + \delta_1^2 y + \delta_2^2 y + a]$$

($a \in K$ corresponds to the function $a(x_1, x_2)$). Applying Proposition 5.10, we obtain that the characteristic set of the defining ideal of $K\langle \eta \rangle / K$ consists of the Δ-σ-polynomials

$$g_1 = \alpha y + \delta_1^2 y + \delta_2^2 y + a$$

and

$$g_2 = \alpha^{-1} g_1 = \alpha^{-1} \delta_1^2 y + \alpha^{-1} \delta_2^2 y + y + \alpha^{-1}(a).$$

With the notation of the proof of Theorem 3.1, the application of the procedure described in this proof, Theorem 4.6(iii), and formula (5) leads to the following expressions for the numbers of elements of the sets $U^{(1)}_{r_1 r_2 r_3}$ and $U^{(2)}_{r_1 r_2 r_3}$:

$$\operatorname{Card} U^{(1)}_{r_1 r_2 r_3} = r_1 r_2 + 2 r_2 r_3 + r_1 + r_2 + 2 r_3 + 1$$

and

$$\operatorname{Card} U^{(2)}_{r_1 r_2 r_3} = 4 r_1 r_3 + 2 r_2 r_3 - 2 r_3$$

for all sufficiently large $(r_1, r_2, r_3) \in \mathbf{N}^3$. Thus, the strength of equation (7) corresponding to the given partition of the basic set of derivations is expressed by the Δ-σ-polynomial

$$\Phi_\eta(t_1, t_2, t_3) = t_1 t_2 + 4 t_1 t_3 + 4 t_2 t_3 + t_1 + t_2 + 1.$$

EXAMPLE 5.12. Let K be a Δ-σ-field where the basic set of derivations $\Delta = \{\delta_1, \delta_2\}$ is considered together with its partition $\Delta = \{\delta_1\} \cup \{\delta_2\}$ and $\sigma = \{\alpha\}$ for some automorphism α of K. Let $L = K\langle\eta\rangle$ be a Δ-σ-field extension with the defining equation

$$\delta_1^a\delta_2^b\alpha^c\eta + \delta_1^a\delta_2^b\alpha^{-c}\eta + \delta_1^a\delta_2^{b+c}\eta + \delta_1^{a+c}\delta_2^b\eta = 0 \qquad (8)$$

where a, b, and c are positive integers. Let $\Phi_\eta(t_1, t_2, t_3)$ denote the corresponding difference-differential dimension polynomial (which expresses the strength of equation (8) with respect to the given partition of Δ). In order to compute Φ_η, notice, first , that the defining Δ-σ-ideal P of the extension L/K is the linear Δ-σ-ideal of $K\{y\}$ generated by the Δ-σ-polynomial

$$f = \delta_1^a\delta_2^b\alpha^c y + \delta_1^a\delta_2^b\alpha^{-c}y + \delta_1^a\delta_2^{b+c}y + \delta_1^{a+c}\delta_2^b y.$$

By Proposition 5.10, the characteristic set of P consists of f and $\alpha^{-1}f = \alpha^{-(c+1)}\delta_1^a\delta_2^b y + \delta_1^a\delta_2^b\alpha^{c-1}y + \delta_1^a\delta_2^{b+c}\alpha^{-1}y + \delta_1^{a+c}\delta_2^b\alpha^{-1}y$. The procedure described in the proof of Theorem 3.1 shows that

$$\operatorname{Card} U^{(1)}_{r_1r_2r_3} = \phi_A(r_1, r_2, r_3)$$

for all sufficiently large $(r_1, r_2, r_3) \in \mathbf{N}^3$, where ϕ_A is the dimension polynomial of the set

$$A = \{(a, b, c), (a, b, -(c+1))\} \subseteq \mathbf{N}^2 \times \mathbf{Z}.$$

Applying Theorem 4.6(iii), and formula (5) we obtain that $\phi_A(t_1, t_2, t_3) = 2ct_1t_2 + 2bt_1t_3 + 2at_2t_3 + (b + 2c - 2bc)t_1 + (a + 2c - 2ac)t_2 + (2a + 2b - 2ab)t_3 + a + b + 2c - ab - 2ac - 2bc + 2abc$. The computation of $\operatorname{Card} U^{(2)}_{r_1r_2r_3}$ with the use of the method of inclusion and exclusion described in the proof of Theorem 3.1 yields the following:

$$\operatorname{Card} U^{(2)}_{r_1r_2r_3} = (2r_3 - 2c + 1)[c(r_2 - b + 1) + c(r_1 - a + 1) - c^2]$$

for all sufficiently large $(r_1, r_2, r_3) \in \mathbf{N}^3$. Therefore, the Δ-σ-dimension polynomial that expresses the strength of equation (8), is as follows.

$$\Phi_\eta = 2ct_1t_2 + 2(b+c)t_1t_3 + 2(a+c)t_2t_3 + (b+3c-2bc-c^2)t_1$$

$$+(2a+2b+4c-2ab-2ac-2bc-2c^2)t_3 + a+b+4c-ab-3ac-3bc$$

$$+(a + 3c - 2ac - 2c^2)t_2 + +2abc + 2ac^2 + 2bc^2 + 2c^3 - 5c^2.$$

The computation of the univariate Δ-σ-dimension polynomial (see Theorem 2.1) via the method of Kähler differentials described in [6, Section 6.5] (by mimicking Example 6.5.6 of [6]) leads to the following result:

$$\phi_{\eta|K}(t) = \frac{D}{2}t^2 - \frac{D(D-2)}{2}t + \frac{D(D-1)(D-2)}{6} \qquad (9)$$

where $D = a + b + c$. In this case the polynomial $\phi_{\eta|K}(t)$ carries just one invariant $a+b+c$ of the extension L/K while $\Phi_\eta(t_1, t_2, t_3)$ determines three such invariants: c, $b + c$, and $a + c$ (see Theorem 3.1(iii)), that is, Φ_η determines all three parameters a, b, c of the defining equation while $\phi_\eta(t)$ gives just the sum of these parameters.

The extension $K\langle\zeta\rangle/K$ with a Δ-σ-generator ζ, the same basic set $\Delta \cup \sigma$ ($\Delta = \{\delta_1, \delta_2\}$, $\sigma = \{\alpha\}$), the same partition of Δ and defining equation

$$\delta_1^{a+b}\alpha^c\zeta + \delta_2^{a+b}\alpha^{-c}\zeta = 0 \qquad (10)$$

has the same univariate difference-dimension polynomial (9). However, its Δ-σ-dimension polynomial is not only different, but also has different invariants described in part (iii) of Theorem 3.1:

$$\Phi_\zeta = 2ct_1t_2 + 2(a+b)t_1t_3 + 2(a+b)t_2t_3 + At_1 + Bt_2 + Ct_3 + E$$

where $A = B = (a+b)(1-2c) + 2c$, $C = 2[1 - (a+b-1)^2]$, and $E = 1 + 2c(a + b - 1)^2$.

Two systems of algebraic difference-differential (Δ-σ-) equations with coefficients in a Δ-σ-field K are said to be *equivalent* if there is a Δ-σ-isomorphism between the Δ-σ-field extensions of K with these defining equations, which is identity on K. Our example shows that using a partition of the basic set of derivations and the computation of the corresponding multivariate Δ-σ-dimension polynomials, one can determine that two systems of Δ-σ-equations (see systems (8) and (10)) are not equivalent, even though they have the same univariate difference-dimension polynomial.

6. ACKNOWLEDGMENTS

This research was supported by the NSF Grant CCF-1016608

7. REFERENCES

[1] A. Einstein. *The Meaning of Relativity. Appendix II (Generalization of gravitation theory)*, pages 153–165. Princeton University Press, Princeton, NJ, 1953.

[2] J. L. Johnson. Kähler differentials and differential algebra. *Ann. of Math.*, 89(2):92–98, 1969.

[3] J. L. Johnson. A notion on Krull dimension for differential rings. *Comm. Math. Helv.*, 44:207–216, 1969.

[4] E. R. Kolchin. The notion of dimension in the theory of algebraic differential equations. *Bull. Amer. Math. Soc.*, 70:570–573, 1964.

[5] E. R. Kolchin. *Differential Algebra and Algebraic Groups.* Academic Press, 1973.

[6] M. V. Kondrateva, A. B. Levin, A. V. Mikhalev, and E. V. Pankratev. *Differential and Difference Dimension Polynomials.* Kluwer Academic Publishers, 1999.

[7] A. B. Levin. Characteristic polynomials of filtered difference modules and difference field extensions. *Russian Mathematical Surveys*, 33(3):165–166, 1978.

[8] A. B. Levin. Reduced Gröbner bases, free difference-differential modules and difference-differential dimension polynomials. *J. Symb. Comput.*, 29:1–26, 2000.

[9] A. B. Levin. Gröbner bases with respect to several orderings and multivariable dimension polynomials. *J. Symb. Comput.*, 42(5):561–578, 2007.

[10] A. B. Levin. *Difference Algebra.* Springer, 2008.

[11] A. V. Mikhalev and E. V. Pankratev. Differential dimension polynomial of a system of differential equations. In *Algebra*, pages 57–67. Moscow State University Press, 1980.

[12] A. V. Mikhalev and E. V. Pankratev. *Computer Algebra. Calculations in Differential and Difference Algebra.* Moscow State University Press, 1989.

[13] M. Zhou and F. Winkler. Computing difference-differential dimension polynomials by relative Gröbner bases in difference-differential modules. *J. Symb. Comput.*, 43(10):726–745, 2008.

Sparse Difference Resultant*

Wei Li, Chun-Ming Yuan, Xiao-Shan Gao
KLMM, Institute of Systems Science, AMSS, Chinese Academy of Sciences, Beijing 100190, China
{liwei,cmyuan,xgao}@mmrc.iss.ac.cn

ABSTRACT

In this paper, the concept of sparse difference resultant for a Laurent transformally essential system of Laurent difference polynomials is introduced and its properties are proved. In particular, order and degree bounds for the sparse difference resultant are given. Based on these bounds, an algorithm to compute the sparse difference resultant is proposed, which is single exponential in terms of the number of variables, the Jacobi number, and the size of the system. Also, the precise order, degree, a determinant representation, and a Poisson-type product formula for the difference resultant are given.

Categories and Subject Descriptors

I.1.2 [**Computing Methodologies**]: Symbolic and Algebraic Manipulation - Algebraic algorithms

General Terms

Algorithms, Theory

Keywords

Sparse difference resultant, difference resultant, Laurent transformally essential system, Jacobi number, single exponential algorithm.

1. INTRODUCTION

The resultant, which gives conditions for an overdetermined system of polynomial equations to have common solutions, is a basic concept in algebraic geometry and a powerful tool in elimination theory [3, 6, 8, 16, 17, 26]. The concept of sparse resultant originated from the work of Gelfand, Kapranov and Zelevinsky on generalized hypergeometric functions, where the central concept of $\mathcal{A}$-discriminant is studied [15]. Kapranov, Sturmfels and Zelevinsky introduced the concept of $\mathcal{A}$-resultant [18]. Sturmfels further introduced the general mixed sparse resultant and gave a single exponential algorithm to compute the sparse resultant [26, 27]. Canny and Emiris showed that the sparse resultant is a factor of the determinant of a Macaulay-style matrix and gave an efficient algorithm to compute sparse resultants based on this matrix representation [10]. A determinant representation for the sparse resultant was finally given by D'Andrea [7]. Li [23] studied resultants and subresultants for linear differential and linear difference polynomials. Recently, in [13], a rigorous definition for the differential resultant of $n+1$ generic differential polynomials in n variables was presented [13] and also the theory of sparse differential resultants for Laurent differentially essential systems was developed [20, 21]. It is meaningful to generalize the theory of sparse resultant to difference polynomial systems.

In this paper, the concept of sparse difference resultant for a Laurent transformally essential system consisting of $n+1$ Laurent difference polynomials in n difference variables is introduced and its basic properties are proved. In particular, we give order and degree bounds for the sparse difference resultant. Based on these bounds, we give an algorithm to compute the sparse difference resultant. The complexity of the algorithm in the worst case is single exponential of the form $O(m^{O(nlJ^2)}(nJ)^{O(lJ)})$, where n, m, J, and l are the number of variables, the degree, the Jacobi number, and the size of the Laurent transformally essential system respectively. Besides these, the difference resultant, which is non-sparse, is introduced and its basic properties are given, such as its precise order, degree, determinant representation, and Poisson-type product formula.

Although most properties for sparse difference resultants and difference resultants are similar to their differential counterparts given in [20, 21, 13], some of them are quite different in terms of descriptions and proofs. Firstly, the definition for difference resultant is more subtle than the differential case as illustrated in section 7. Secondly, the criterion for transformally essential systems given in Section 3.3 is quite different and much simpler than its differential counterpart given in [21]. Also, a determinant representation for the difference resultant is given in Section 6, but such a representation is still not known for differential resultants [28, 25]. Finally, some properties are more difficult in the difference case. For instance, we can only show that the vanishing of the difference resultant is a necessary condition for the corresponding difference polynomial system to have a common nonzero solution. However, the sufficient condition part is still open. Also, there does not exist a definition for homo-

* Partially supported by a National Key Basic Research Project of China (2011CB302400) and by grants from NSFC (60821002,11101411).

ISSAC'13, June 26–29, 2013, Boston, Massachusetts, USA.

geneous difference polynomials, and the definition we give in this paper is different from its differential counterpart.

The rest of the paper is organized as follows. In Section 2, we prove some preliminary results. In Section 3, we first introduce the concepts of Laurent difference polynomials and Laurent transformally essential systems, and then define the sparse difference resultant for Laurent transformally essential systems. Then basic properties of sparse difference resultant are proved in Section 4. And in Section 5, we present an algorithm to compute the sparse difference resultant. Then we introduce the notion of difference resultant and prove its basic properties in section 6. In Section 7, we conclude the paper by proposing several problems for future research.

2. PRELIMINARIES

In this section, some basic notations and preliminary results in difference algebra will be given. For more details about difference algebra, please refer to [5, 19].

Let $\mathcal{F}$ be an ordinary difference field with a transforming operator σ. For each $a \in \mathcal{F}$ and $n \in \mathbb{N}_0$, we denote $\sigma^n(a)$ by $a^{(n)}$, and by $a^{[n]}$ we mean the set $\{a, a^{(1)}, \ldots, a^{(n)}\}$. A typical example of difference field is $\mathbb{Q}(x)$ with $\sigma(f(x)) = f(x+1)$. Throughout this paper, we shall often use the prefix "σ-" to replace "difference" or "transformally".

Let $\mathcal{G}$ be a σ-extension field of $\mathcal{F}$. A subset $\mathcal{S}$ of $\mathcal{G}$ is said to be *σ-independent* (resp. *σ-dependent*) over $\mathcal{F}$ or a set of difference indeterminates if the set $\{\sigma^k a | a \in \mathcal{S}, k \geq 0\}$ is algebraically independent (resp. dependent) over $\mathcal{F}$. We use $\Delta\text{tr.deg}\,\mathcal{G}/\mathcal{F}$ and $\text{tr.deg}\,\mathcal{G}/\mathcal{F}$ to denote the *σ-transcendence degree* and the algebraic transcendence degree of $\mathcal{G}$ over $\mathcal{F}$ respectively. The following property will be needed.

Lemma 2.1 *Let $P_i(\mathbb{U}, \mathbb{Y}) \in \mathcal{F}\langle\mathbb{Y}\rangle\{\mathbb{U}\}$ $(i = 1, \ldots, m)$ where $\mathbb{U} = (u_1, \ldots, u_r)$ and $\mathbb{Y}$ are sets of σ-indeterminates. If $P_i(\mathbb{U}, \mathbb{Y})$ are σ-dependent over $\mathcal{F}\langle\mathbb{U}\rangle$, then for any $\overline{\mathbb{U}} \in \mathcal{F}^r$, $P_i(\overline{\mathbb{U}}, \mathbb{Y})$ are σ-dependent over $\mathcal{F}$.*

Proof: It suffices to show the case $r = 1$. We denote $u = u_1$. Since $P_i(u, \mathbb{Y})$ are σ-dependent over $\mathcal{F}\langle u\rangle$, there exist s and l such that $\mathbb{P}_i^{(k)}(u, \mathbb{Y})\,(k \leq s)$ are algebraically dependent over $\mathcal{F}(u^{(k)}|k \leq s+l)$. By [16, p.168], $\mathbb{P}_i^{(k)}(\bar{u}, \mathbb{Y})\,(k \leq s)$ are algebraically dependent over $\mathcal{F}$ and the lemma follows. □

Let $\mathcal{F}\{\mathbb{Y}\} = \mathcal{F}\{y_1, \ldots, y_n\}$ be a σ-polynomial ring and $\mathscr{R}$ a ranking endowed on it. For a σ-polynomial $f \in \mathcal{F}\{\mathbb{Y}\}$, the greatest $y_j^{(k)}$ w.r.t. $\mathscr{R}$ which appears effectively in f is called the *leader* of f, denoted by $\text{ld}(f)$ and correspondingly y_j is called the *leading variable* of f, denoted by $\text{lvar}(f) = y_j$. The leading coefficient of f as a univariate polynomial in $\text{ld}(f)$ is called the *initial* of f and is denoted by I_f.

For each subset $S \subset \mathcal{F}\{\mathbb{Y}\}$, we use (S) and $[S]$ to denote the algebraic ideal and the σ-ideal in $\mathcal{F}\{\mathbb{Y}\}$ generated by S. A σ-ideal $\mathcal{I} \subset \mathcal{F}\{\mathbb{Y}\}$ is called *reflexive* if $a^{(1)} \in \mathcal{I} \Longrightarrow a \in \mathcal{I}$. An n-tuple over $\mathcal{F}$ is of the form $(a_1, \ldots, a_n)$ where the a_i are in some σ-overfield of $\mathcal{F}$. An n-tuple η is called a *generic zero* of a σ-ideal $\mathcal{I} \subset \mathcal{F}\{\mathbb{Y}\}$ if for each $P \in \mathcal{F}\{\mathbb{Y}\}$ we have $P(\eta) = 0 \Leftrightarrow P \in \mathcal{I}$. It is well known that

Lemma 2.2 *[5, p.77] A σ-ideal possesses a generic zero if and only if it is a reflexive prime σ-ideal other than [1].*

Let $\mathcal{I}$ be a reflexive prime σ-ideal and η a generic zero of $\mathcal{I}$. The *dimension* of $\mathcal{I}$ is defined to be $\Delta\text{tr.deg}\mathcal{F}\langle\eta\rangle/\mathcal{F}$. There is another description of dimension in terms of characteristic set. Let $\mathcal{A}$ be a characteristic set of a reflexive prime σ-ideal $\mathcal{I}$ w.r.t. some ranking $\mathscr{R}$. We rewrite $\mathcal{A}$ in the following form [14]

$$\mathcal{A} = \begin{cases} A_{11}, \ldots, A_{1k_1} \\ \cdots \\ A_{p1}, \ldots, A_{pk_p} \end{cases}$$

where $\text{lvar}(A_{ij}) = y_{c_i}$ and $\text{ord}(A_{ij}, y_{c_i}) < \text{ord}(A_{i,j+l}, y_{c_i})$. Then p is equal to the *codimension* of $\mathcal{I}$, that is $n - \dim(\mathcal{I})$. Unlike the differential case, here even though $\mathcal{I}$ is of codimension one, it may happen that $k_1 > 1$. Below, we will show a property of uniqueness still exists. Before this, we list several algebraic results about regular chains [2].

Let $\mathcal{B} = B_1, \ldots, B_m$ be an algebraic triangular set in $\mathcal{F}[\mathbb{Y}]$ with $\text{lvar}(B_i) = y_i$ and $U = \mathbb{Y}\backslash\{y_1, \ldots, y_m\}$. A polynomial f is said to be invertible w.r.t. $\mathcal{B}$ if either $f \in \mathcal{F}[U]$ or $(f, B_1, \ldots, B_s) \cap \mathcal{F}[U] \neq \{0\}$ where $\text{lvar}(f) = \text{lvar}(B_s)$. We call $\mathcal{B}$ a *regular chain* if for each $i > 1$, the initial of B_i is invertible w.r.t. $B_1, \ldots, B_{i-1}$. By $\text{asat}(\mathcal{B})$, we mean the algebraic saturation ideal $(\mathcal{B}) : \text{I}_{\mathcal{B}}^{\infty}$. For a regular chain $\mathcal{B}$, a polynomial f is said to be invertible w.r.t. $\text{asat}(\mathcal{B})$ if $(f, \text{asat}(\mathcal{B})) \cap \mathcal{F}[U] \neq \{0\}$.

Lemma 2.3 *Let $\mathcal{B} \subset \mathcal{F}[\mathbb{Y}]$ be a regular chain. If $\sqrt{\text{asat}(\mathcal{B})} = \bigcap_{i=1}^m \mathcal{P}_i$ is a minimal prime decomposition, then $f \in \mathcal{F}[\mathbb{Y}]$ is invertible w.r.t. $\text{asat}(\mathcal{B})$ if and only if $f \notin \mathcal{P}_i$ for all i.*

Proof: By [12], the parametric set of $\mathcal{B}$ is that of $\mathcal{P}_i$ for each i. The lemma follows from the fact that for prime ideals $\mathcal{P}_i$, $f \notin \mathcal{P}_i$ if and only if $(f, \mathcal{P}_i) \cap \mathcal{F}[U] \neq \{0\}$. □

Lemma 2.4 *[2] Let $\mathcal{B}$ be a regular chain in $\mathcal{F}[U, \mathbb{Y}]$, $L \neq 0$ invertible w.r.t $\mathcal{B}$, and $Lf \in (\mathcal{B})$. Then $f \in \text{asat}(\mathcal{B})$.*

Lemma 2.5 *Let $A \in \mathcal{F}\{\mathbb{Y}\}$ be irreducible with $\deg(A, y_{i_0}) > 0$ for some i_0. If f is invertible w.r.t. $A^{[k]}$ when $A^{[k]}$ is treated as an algebraic triangular set, then $\sigma(f)$ is invertible w.r.t. $A^{[k+1]}$. In particular, $A^{[k]}$ is a regular triangular set for any $k \geq 0$.*

Proof: It is a direct consequence of [14, Theorem 4.2]. □

The following fact is needed to define sparse σ-resultant.

Lemma 2.6 *Let $\mathcal{I}$ be a reflexive prime σ-ideal of codimension one in $\mathcal{F}\{\mathbb{Y}\}$. The first element in any characteristic set of $\mathcal{I}$ w.r.t. any ranking, when taken irreducible, is unique up to a factor in $\mathcal{F}$.*

Proof: Let $\mathcal{A} = A_1, \ldots, A_m$ be a characteristic set of $\mathcal{I}$ w.r.t. some ranking $\mathscr{R}$ with A_1 irreducible. Suppose $\text{lvar}(\mathcal{A}) = y_1$. Given another characteristic set $\mathcal{B} = B_1, \ldots, B_l$ of $\mathcal{I}$ w.r.t. some other ranking $\mathscr{R}'$ (B_1 is irreducible), we will show there exists $c \in \mathcal{F}$ s.t. $B_1 = c \cdot A_1$. It suffices to consider the case $\text{lvar}(\mathcal{B}) \neq y_1$. Suppose $\text{lvar}(B_1) = y_2$. Clearly, y_2 appears effectively in A_1 for $\mathcal{B}$ reduces A_1 to 0. And since $\mathcal{I}$ is reflexive, there exists some i_0 such that $\deg(A_1, y_{i_0}) > 0$.

Suppose $\text{ord}(A_1, y_2) = o_2$. Take another ranking under which $y_2^{(o_2)}$ is the leader of A_1 and we use $\tilde{A}_1$ to distinguish it from A_1 under $\mathscr{R}$. By Lemma 2.5, for each k, $A_1^{[k]}$ and $\tilde{A}_1^{[k]}$ are regular triangular sets.

Now we claim that $\text{asat}(A_1^{[k]}) = \text{asat}(\tilde{A}_1^{[k]})$. Suppose $f \in \text{asat}(A_1^{[k]})$, then $(\prod_{i=0}^{k} \sigma^i(\mathrm{I}_{A_1}))^a f \in (A_1^{[k]})$. Since I_{A_1} is invertible w.r.t. $\tilde{A}_1$, by Lemma 2.5, $(\prod_{i=0}^{k} \sigma^i(\mathrm{I}_{A_1}))^a$ is invertible w.r.t. $\tilde{A}_1^{[k]}$. So by Lemma 2.4, $f \in \text{asat}(\tilde{A}_1^{[k]})$ and $\text{asat}(A_1^{[k]}) \subseteq \text{asat}(\tilde{A}_1^{[k]})$ follows. Similarly, $\text{asat}(\tilde{A}_1^{[k]}) \subseteq \text{asat}(A_1^{[k]})$. Thus, $\text{asat}(A_1^{[k]}) = \text{asat}(\tilde{A}_1^{[k]})$.

Suppose $\text{ord}(B_1, y_2) = o_2'$. Clearly, $o_2 \geq o_2'$. Assume $o_2 > o_2'$. Then B_1 is invertible w.r.t. $\text{asat}(\tilde{A}_1^{[k]})$. Since $\text{asat}(A_1^{[k]}) = \text{asat}(\tilde{A}_1^{[k]})$, by Lemma 2.3, B_1 is invertible w.r.t. $\text{asat}(A_1^{[k]})$. Thus, there exists a nonzero polynomial H with $\text{ord}(H, y_1) < \text{ord}(A_1, y_1)$ s.t. $H \in (B_1, \text{asat}(A_1^{[k]})) \subset \mathcal{I}$, which is a contradiction. Thus, $o_2 = o_2'$. Since $\mathcal{B}$ reduces A_1 to zero and A_1 is irreducible, there exists $c \in \mathcal{F}$ such that $B_1 = c \cdot A_1$. □

3. SPARSE DIFFERENCE RESULTANT

In this section, the concepts of Laurent σ-polynomials and Laurent σ-essential systems are first introduced, and then the sparse σ-resultant for Laurent σ-essential systems is defined. And we also give a criterion for Laurent σ-essential systems in terms of the support of the given system.

3.1 Laurent difference polynomial

Similar to [21], before defining sparse σ- resultant, we first introduce the concept of Laurent σ-polynomials.

Definition 3.1 *A Laurent σ-monomial of order s is of the form $\prod_{i=1}^{n} \prod_{k=0}^{s} (y_i^{(k)})^{d_{ik}}$ where d_{ik} are integers which can be negative. A* Laurent σ-polynomial *over $\mathcal{F}$ is a finite linear combination of Laurent σ-monomials with coefficients in $\mathcal{F}$.*

Clearly, the collections of all Laurent σ-polynomials form a commutative σ-ring. We denote the σ-ring of Laurent σ-polynomials with coefficients in $\mathcal{F}$ by $\mathcal{F}\{\mathbb{Y}, \mathbb{Y}^{-1}\}$.

Definition 3.2 *For each $F \in \mathcal{F}\{\mathbb{Y}, \mathbb{Y}^{-1}\}$, the* normal form *of F, denoted by $\mathrm{N}(F)$, is defined to be the σ-polynomial in $\mathcal{F}\{\mathbb{Y}\}$ obtained by clearing denominators from F. The order and degree of F is defined to be the* order *and* degree *of $\mathrm{N}(F)$, denoted by $\text{ord}(F)$ and $\deg(F)$.*

Definition 3.3 *Let $F \in \mathcal{F}\{\mathbb{Y}, \mathbb{Y}^{-1}\}$. An n-tuple $(a_1, \ldots, a_n)$ over $\mathcal{F}$ with each $a_i \neq 0$ is called a* nonzero σ-zero *of F if $F(a_1, \ldots, a_n) = 0$.*

3.2 Def nition of sparse difference resultant

In this section, the definition of the sparse σ-resultant will be given. We first define sparse σ-resultants for Laurent σ-polynomials whose coefficients are σ-indeterminates.

Suppose $\mathcal{A}_i = \{M_{i0}, \ldots, M_{il_i}\}$ $(i = 0, \ldots, n)$ are finite sets of Laurent σ-monomials in $\mathbb{Y}$. Consider $n+1$ *generic Laurent σ-polynomials* defined over $\mathcal{A}_0, \ldots, \mathcal{A}_n$:

$$\mathbb{P}_i = \sum_{k=0}^{l_i} u_{ik} M_{ik} \ (i = 0, \ldots, n), \tag{1}$$

where all the u_{ik} are σ-independent over $\mathbb{Q}$. Denote

$$\mathbf{u}_i = (u_{i0}, u_{i1}, \ldots, u_{in}) \text{ and } \mathbf{u} = \bigcup_{i=0}^{n} \mathbf{u}_i \backslash \{u_{i0}\}. \tag{2}$$

The number $l_i + 1$ is called the *size* of $\mathbb{P}_i$. To avoid the triviality, $l_i \geq 1$ $(i = 0, \ldots, n)$ are always assumed.

Definition 3.4 *A set of Laurent σ-polynomials of the form (1) is called* Laurent σ-essential *if there exist k_i $(i = 0, \ldots, n)$ with $1 \leq k_i \leq l_i$ s.t. $\Delta\text{tr.deg}\, \mathbb{Q}\langle \frac{M_{0k_0}}{M_{00}}, \ldots, \frac{M_{nk_n}}{M_{n0}} \rangle / \mathbb{Q} = n$. In this case, $\mathcal{A}_0, \ldots, \mathcal{A}_n$ are also called Laurent σ-essential.*

Although M_{i0} are used as denominators in the above definition, the σ-essential condition does not depend on the choices of M_{i0}. Let $\mathcal{I}_{\mathbb{Y},\mathbf{u}} = ([\mathrm{N}(\mathbb{P}_0), \ldots, \mathrm{N}(\mathbb{P}_n)] : \mathbf{m}) \subset \mathbb{Q}\{\mathbb{Y}, \mathbf{u}_0, \ldots, \mathbf{u}_n\}$ where $\mathbf{m}$ is the set of all σ-monomials in $\mathbb{Y}$. The following result is a foundation for defining sparse σ-resultants.

Theorem 3.5 *Let $\mathbb{P}_0, \ldots, \mathbb{P}_n$ be the Laurent σ-polynomials defined in (1). Then the following assertions hold.*

1. *$\mathcal{I}_{\mathbb{Y},\mathbf{u}}$ is a reflexive prime σ-ideal in $\mathbb{Q}\{\mathbb{Y}, \mathbf{u}_0, \ldots, \mathbf{u}_n\}$.*
2. *$\mathcal{I}_{\mathbb{Y},\mathbf{u}} \cap \mathbb{Q}\{\mathbf{u}_0, \ldots, \mathbf{u}_n\}$ is of codimension one if and only if $\mathbb{P}_0, \ldots, \mathbb{P}_n$ form a Laurent σ-essential system.*

Proof: Let $\eta = (\eta_1, \ldots, \eta_n)$ be a sequence of σ-independent elements over $\mathbb{Q}\langle \mathbf{u} \rangle$, where $\mathbf{u}$ is defined in (2). Let

$$\zeta_i = -\sum_{k=1}^{l_i} u_{ik} \frac{M_{ik}(\eta)}{M_{i0}(\eta)} \ (i = 0, 1, \ldots, n), \tag{3}$$

and $\zeta = (\zeta_0, u_{01}, \ldots, u_{0l_0}; \ldots; \zeta_n, u_{n1}, \ldots, u_{nl_n})$. It is easy to show that $(\eta; \zeta)$ is a generic zero of $\mathcal{I}_{\mathbb{Y},\mathbf{u}}$, and by Lemma 2.2, $\mathcal{I}_{\mathbb{Y},\mathbf{u}}$ is a reflexive prime σ-ideal.

Consequently, $\mathcal{I}_{\mathbf{u}} = \mathcal{I}_{\mathbb{Y},\mathbf{u}} \cap \mathbb{Q}\{\mathbf{u}_0, \ldots, \mathbf{u}_n\}$ is a reflexive prime σ-ideal with a generic zero ζ. So $\mathcal{I}_{\mathbf{u}}$ is of codimension one $\Leftrightarrow$ $\Delta\text{tr.deg}\, \mathbb{Q}\langle \zeta \rangle / \mathbb{Q} = \sum_{i=0}^{n} l_i + n$ $\Leftrightarrow$ there exist distinct $i_1, \ldots, i_n$ s.t. $\zeta_{i_1}, \ldots, \zeta_{i_n}$ are σ-independent over $\mathbb{Q}\langle \mathbf{u} \rangle$ $\Leftrightarrow$ $\mathbb{P}_0, \ldots, \mathbb{P}_n$ form a Laurent σ-essential system. And the last "$\Leftrightarrow$" follows from Lemma 2.1. □

Let $[\mathbb{P}_0, \ldots, \mathbb{P}_n]$ be the σ-ideal in $\mathbb{Q}\{\mathbb{Y}, \mathbb{Y}^{-1}; \mathbf{u}_0, \ldots, \mathbf{u}_n\}$ generated by $\mathbb{P}_i$. Then we have

Corollary 3.6 *$\mathcal{I}_{\mathbf{u}} = [\mathbb{P}_0, \ldots, \mathbb{P}_n] \cap \mathbb{Q}\{\mathbf{u}_0, \ldots, \mathbf{u}_n\}$ is a reflexive prime σ-ideal of codimension one if and only if $\{\mathbb{P}_i : i = 0, \ldots, n\}$ is a Laurent σ-essential system.*

Proof: It is a direct consequence of Theorem 3.5 and the fact that $[\mathbb{P}_0, \ldots, \mathbb{P}_n] \cap \mathbb{Q}\{\mathbf{u}_0, \ldots, \mathbf{u}_n\} = \mathcal{I}_{\mathbb{Y},\mathbf{u}} \cap \mathbb{Q}\{\mathbf{u}_0, \ldots, \mathbf{u}_n\}$. □

Now suppose $\{\mathbb{P}_0, \ldots, \mathbb{P}_n\}$ is a Laurent σ-essential system. Since $\mathcal{I}_{\mathbf{u}}$ is a reflexive prime σ-ideal of codimension one, by Lemma 2.6, there exists a unique irreducible σ-polynomial $\mathbf{R}(\mathbf{u}_0, \ldots, \mathbf{u}_n) \in \mathbb{Q}\{\mathbf{u}_0, \ldots, \mathbf{u}_n\}$ such that $\mathbf{R}$ can serve as the first polynomial in each characteristic set of $\mathcal{I}_{\mathbf{u}}$ w.r.t. any ranking endowed on $\mathbf{u}_0, \ldots, \mathbf{u}_n$. That is,

Lemma 3.7 *Among all the polynomials in $\mathbb{Q}\{\mathbf{u}_0, \ldots, \mathbf{u}_n\}$ vanishing at $(\mathbf{u}; \zeta_0, \ldots, \zeta_n)$, $\mathbf{R}$ is of minimal order and degree in each u_{i0} $(i = 0, \ldots, n)$.*

Now the definition of sparse σ-resultant is given as follows:

Definition 3.8 *The above $\mathbf{R} \in \mathbb{Q}\{\mathbf{u}_0, \ldots, \mathbf{u}_n\}$ is defined to be the* sparse difference resultant *of the Laurent σ-essential system $\mathbb{P}_0, \ldots, \mathbb{P}_n$, denoted by $\text{Res}_{\mathcal{A}_0, \ldots, \mathcal{A}_n}$ or $\text{Res}_{\mathbb{P}_0, \ldots, \mathbb{P}_n}$.*

We give an example to show that for a Laurent σ-essential system, $\mathbf{R}$ may not involve the coefficients of some $\mathbb{P}_i$.

Example 3.9 *Let $n = 2$ and $\mathbb{P}_i$ has the form $\mathbb{P}_0 = u_{00} + u_{01}y_1y_2$, $\mathbb{P}_1 = u_{10}+u_{11}y_1^{(1)}y_2^{(1)}$, $\mathbb{P}_2 = u_{20}+u_{21}y_1^{(1)}y_2$. Clearly, $\mathbb{P}_0, \mathbb{P}_1, \mathbb{P}_2$ form a Laurent σ-essential system. The sparse σ-resultant of $\mathbb{P}_0, \mathbb{P}_1, \mathbb{P}_2$ is $\mathbf{R} = u_{00}^{(1)}u_{11} - u_{01}^{(1)}u_{10}$, which is free from the coefficients of $\mathbb{P}_2$.*

Example 3.9 can be used to illustrate the difference between the differential and difference cases. If $\mathbb{P}_0, \mathbb{P}_1, \mathbb{P}_2$ are differential polynomials, then the sparse differential resultant is $u_{01}u_{11}u_{20}u_{21}u_{00}' - u_{11}u_{20}u_{00}u_{21}u_{01}' - u_{11}u_{20}^2u_{01}^2 - u_{01}u_{00}u_{21}^2u_{10}$, which contains all the coefficients of $\mathbb{P}_0, \mathbb{P}_1, \mathbb{P}_2$.

We now define the sparse σ-resultant for specific Laurent σ-polynomials. For any finite set $\mathcal{A}$ of Laurent σ-monomials in $\mathbb{Y}$, let $\mathcal{L}(\mathcal{A}) = \{\sum_{M\in\mathcal{A}} a_M M\}$ where the a_M are in some σ-extension field of $\mathbb{Q}$. Then $\mathcal{L}(\mathcal{A})$ can be considered as the set of all l-tuples over $\mathbb{Q}$ where $l = |\mathcal{A}|$.

Definition 3.10 *Let $\mathcal{A}_i = \{M_{i0}, \ldots, M_{il_i}\}$ $(i = 0, \ldots, n)$ be a Laurent σ-essential system. Consider $n+1$ Laurent σ-polynomials $(F_0, \ldots, F_n) \in \prod_{i=0}^n \mathcal{L}(\mathcal{A}_i)$. The sparse σ-resultant of $F_0, \ldots, F_n$ is obtained by replacing $\mathbf{u}_i$ by the corresponding coefficient vector of F_i in $\mathrm{Res}_{\mathcal{A}_0,\ldots,\mathcal{A}_n}(\mathbf{u}_0, \ldots, \mathbf{u}_n)$.*

The following lemma shows that the sparse σ-resultant gives a necessary condition for a system to have common nonzero solutions.

Lemma 3.11 *Suppose $(F_0, \ldots, F_n) \in \prod_{i=0}^n \mathcal{L}(\mathcal{A}_i)$ have common nonzero solutions. Then $\mathrm{Res}_{F_0,\ldots,F_n} = 0$.*

Proof: By Definition 3.8, $\mathrm{Res}_{\mathcal{A}_0,\ldots,\mathcal{A}_n} \in [\mathbb{P}_0, \ldots, \mathbb{P}_n]$. If the F_i have a common nonzero solution, clearly, $\mathrm{Res}_{\mathcal{A}_0,\ldots,\mathcal{A}_n}$ vanishes at the coefficients of F_i. □

3.3 Criterion for Laurent transformally essential systems in terms of the supports

Let $\mathcal{A}_i$ $(i = 0, \ldots, n)$ be finite sets of Laurent σ-monomials. According to Definition 3.4, in order to check whether they are Laurent σ-essential, we need to check whether there exist $M_{ik_i} \in \mathcal{A}_i$ s.t. $\Delta\mathrm{tr.deg}\,\mathbb{Q}\langle M_{0k_0}/M_{00}, \ldots, M_{nk_n}/M_{n0}\rangle/\mathbb{Q} = n$. This can be done with the characteristic set method [14]. In this section, we will give a conceptually and computationally simpler criterion which is based on linear algebra.

Let $B_i = \prod_{j=1}^n \prod_{k\geq 0}^s (y_j^{(k)})^{d_{ijk}}$ $(i = 1, \ldots, m)$ be Laurent σ-monomials. We now introduce a new algebraic indeterminate x and let $d_{ij} = \sum_{k=0}^s d_{ijk}x^k$ $(i = 1, \ldots, m; j = 1, \ldots, n)$ be univariate polynomials in $\mathbb{Z}[x]$. If y_j and its transforms do not occur in B_i, then set $d_{ij} = 0$. The vector $(d_{i1}, d_{i2}, \ldots, d_{in})$ is called the *symbolic support vector* of B_i. The matrix $M = (d_{ij})_{m\times n}$ is called the *symbolic support matrix* of $B_1, \ldots, B_m$.

Definition 3.12 *A matrix $M = (d_{ij})_{m\times n}$ over $\mathbb{Q}(x)$ is called* normal upper-triangular of rank r *if for each $i \leq r$, $d_{ii} \neq 0$ and $d_{i,i-k} = 0\,(1 \leq k \leq i-1)$, and the last $m - r$ rows are zero vectors.*

Definition 3.13 *A set of Laurent σ-monomials $B_1, \ldots, B_m$ is said to be in* r-upper-triangular *form if its symbolic support matrix is a normal upper-triangular matrix of rank r.*

The following lemma shows that it is easy to compute the σ-transcendence degree of a set of Laurent σ-monomials in upper-triangular form.

Lemma 3.14 *Let $B_1, \ldots, B_m$ be an r-upper-triangular set. Then $\Delta\mathrm{tr.deg}\,\mathbb{Q}\langle B_1, \ldots, B_m\rangle/\mathbb{Q} = r$.*

Proof: Clearly, for each $i \leq r$, $B_i = \prod_{j=i}^n \prod_{k\geq 0}(y_j^{(k)})^{d_{ijk}}$ with $\mathrm{ord}(B_i, y_i) \geq 0$, and $B_{r+k} = 1\,(k > 0)$. Let $B_i' = \prod_{j=i}^r \prod_{k\geq 0}(y_j^{(k)})^{d_{ijk}}$. Then $r \geq \Delta\mathrm{tr.deg}\,\mathbb{Q}\langle B_1, \ldots, B_m\rangle/\mathbb{Q} \geq \Delta\mathrm{tr.deg}\,\mathbb{Q}\langle B_1', \ldots, B_r'\rangle/\mathbb{Q}$. So it suffices to prove that $\Delta\mathrm{tr.deg}\,\mathbb{Q}\langle B_1', \ldots, B_r'\rangle/\mathbb{Q} = r$.

It is clear for $r = 1$. Suppose it holds for $r - 1$. Let $B_i'' = \prod_{j=i}^{r-1} \prod_{k\geq 0}(y_j^{(k)})^{d_{ijk}}$, then $\Delta\mathrm{tr.deg}\,\mathbb{Q}\langle B_1'', \ldots, B_{r-1}''\rangle/\mathbb{Q} = r-1$. Thus, $\Delta\mathrm{tr.deg}\,\mathbb{Q}\langle B_1', \ldots, B_r'\rangle/\mathbb{Q} = \Delta\mathrm{tr.deg}\,\mathbb{Q}\langle B_r'\rangle/\mathbb{Q} + \Delta\mathrm{tr.deg}\,\mathbb{Q}\langle B_1', \ldots, B_r'\rangle/\mathbb{Q}\langle B_r'\rangle \geq 1 + \Delta\mathrm{tr.deg}\,\mathbb{Q}\langle B_1'', \ldots, B_{r-1}''\rangle/\mathbb{Q} = r$. So $\Delta\mathrm{tr.deg}\mathbb{Q}\langle B_1, \ldots, B_m\rangle/\mathbb{Q} = r$ follows. □

In the following, we will show that each set of Laurent difference monomials can be transformed to an upper-triangular set with the same σ-transcendence degree. Here we use three types of elementary matrix transformations. For a matrix M over $\mathbb{Q}[x]$, Type 1 (resp. Type 3) operations consist of interchanging two rows (resp. columns) of M; and Type 2 operations consist of adding an $f(x)$-multiple of the j-th row to the i-th row, where $f(x) \in \mathbb{Q}[x]$. Note that these operations correspond to certain transformations of the σ-monomials. For example, multiplying the i-th row of M by a polynomial $f(x) = a_dx^d + \cdots + a_0$ and adding the result to the j-th row means changing B_j to $\prod_{k=0}^d (\sigma^k B_i)^{a_k} B_j$.

Lemma 3.15 *Let $C_1, \ldots, C_m$ be obtained from $B_1, \ldots, B_m$ by performing a series of transformations. Then* $\Delta\mathrm{tr.deg}\,\mathbb{Q}\langle B_1, \ldots, B_m\rangle/\mathbb{Q} = \Delta\mathrm{tr.deg}\,\mathbb{Q}\langle C_1, \ldots, C_m\rangle/\mathbb{Q}$.

Proof: It suffices to show Type 2 operations keep the σ-transcendence degree. Indeed, given $\sum_{i=0}^d \frac{p_i}{q}x^i \in \mathbb{Q}[x]$ with $p_i, q \in \mathbb{Z}^*$, $\Delta\mathrm{tr.deg}\,\mathbb{Q}\langle B_1, \prod_{k=0}^d (B_1^{(k)})^{a_k}B_2\rangle/\mathbb{Q} = \Delta\mathrm{tr.deg}\,\mathbb{Q}\langle \prod_{k=0}^d (B_1^{(k)})^{p_k}, \prod_{k=0}^d (B_1^{(k)})^{p_k}B_2^q\rangle/\mathbb{Q} = \Delta\mathrm{tr.deg}\,\mathbb{Q}\langle B_1, B_2\rangle/\mathbb{Q}$. □

Theorem 3.16 *Suppose M is the symbolic support matrix of $B_1, \ldots, B_m$. Then $\Delta\mathrm{tr.deg}\,\mathbb{Q}\langle B_1, \ldots, B_m\rangle/\mathbb{Q} = \mathrm{rk}(M)$.*

Proof: Since $\mathbb{Q}[x]$ is an Euclidean domain, it is clear that each matrix M can be reduced to a normal upper-triangular matrix by performing a series of elementary transformations. By Lemma 3.14 and Lemma 3.15, the theorem follows. □

Example 3.17 *Let $B_1 = y_1y_2$ and $B_2 = y_1^{(a)}y_2^{(b)}$. Then $M = \begin{pmatrix} 1 & 1 \\ x^a & x^b \end{pmatrix}$ and $\mathrm{rk}(M) = \begin{cases} 1 & \text{if } a = b \\ 2 & \text{if } a \neq b. \end{cases}$ Thus, by Theorem 3.16, if $a \neq b$, B_1 and B_2 are σ-independent over $\mathbb{Q}$. Otherwise, they are σ-dependent over $\mathbb{Q}$.*

Consider the set of generic Laurent σ-polynomials defined in (1). Let β_{ik} be the symbolic support vector of M_{ik}/M_{i0}. Then the vector $w_i = \sum_{k=0}^{l_i} u_{ik}\beta_{ik}$ is called the *symbolic support vector* of $\mathbb{P}_i$ and the matrix $M_{\mathbb{P}}$ whose rows are $w_0, \ldots, w_n$ is called the *symbolic support matrix* of $\mathbb{P}_0, \ldots, \mathbb{P}_n$.

Now, we give the main theorem in this section.

Theorem 3.18 *$\mathbb{P}_0, \ldots, \mathbb{P}_n$ form a Laurent σ-essential system if and only if* $\text{rk}(M_\mathbb{P}) = n$.

Proof: By Theorem 3.16 and Definition 3.4, $\mathbb{P}_0, \ldots, \mathbb{P}_n$ are Laurent σ-essential iff there exist M_{ik_i} with $1 \le k_i \le l_i$ s.t. the symbolic support matrix of $M_{0k_0}/M_{00}, \ldots, M_{nk_n}/M_{n0}$ is of rank n. And the latter is equivalent to $\text{rk}(M_\mathbb{P}) = n$. □

We will end this section by introducing a new notion, namely super-essential systems. Let $\mathbb{T} \subset \{0, 1, \ldots, n\}$. We denote by $\mathbb{P}_\mathbb{T}$ the Laurent σ-polynomial set consisting of $\mathbb{P}_i\,(i \in \mathbb{T})$, and $M_{\mathbb{P}_\mathbb{T}}$ its symbolic support matrix.

Definition 3.19 *Let $\mathbb{T} \subset \{0, 1, \ldots, n\}$. Then we call $\mathbb{T}$ or* $\mathbb{P}_\mathbb{T}$ super-essential *if* $\text{card}(\mathbb{T}) - \text{rk}(M_{\mathbb{P}_\mathbb{T}}) = 1$ *and for each* $\mathbb{J} \subsetneq \mathbb{T}$, $\text{card}(\mathbb{J}) = \text{rk}(M_{\mathbb{P}_\mathbb{J}})$.

Note that super-essential systems are the difference analogue of essential systems introduced in [27] and also that of rank essential systems introduced in [21].

Theorem 3.20 *If $\{\mathbb{P}_0, \ldots, \mathbb{P}_n\}$ is a Laurent σ-essential system, then for any $\mathbb{T} \subset \{0, 1, \ldots, n\}$,* $\text{card}(\mathbb{T}) - \text{rk}(M_{\mathbb{P}_\mathbb{T}}) \le 1$ *and there exists a unique $\mathbb{T}$ which is super-essential. In this case, the sparse σ-resultant of $\mathbb{P}_0, \ldots, \mathbb{P}_n$ involves only the coefficients of $\mathbb{P}_i\,(i \in \mathbb{T})$.*

Proof: By [22, Theorem 3.24], the theorem follows. □

Using this property, one can determine which polynomial is needed for computing the sparse σ-resultant, which will eventually reduce the computation complexity.

Example 3.21 *Continue from Example 3.9. It is easy to show that $\mathbb{P}_0, \mathbb{P}_1$ constitute a super-essential system. Recall that the sparse σ-resultant is free from the coefficients of $\mathbb{P}_2$.*

4. BASIC PROPERTIES OF SPARSE DIFFERENCE RESULTANT

In this section, we will prove some basic properties for the sparse σ-resultant $\mathbf{R}(\mathbf{u}_0, \ldots, \mathbf{u}_n)$.

4.1 Sparse difference resultant is transformally homogeneous

We introduce the concept of σ-homogeneous polynomials.

Definition 4.1 *A σ-polynomial $f \in \mathcal{F}\{y_0, \ldots, y_n\}$ is called* transformally homogeneous *if for a new σ-indeterminate λ, there exists a σ-monomial $M(\lambda)$ such that $f(\lambda y_0, \ldots, \lambda y_n) = M(\lambda) f(y_0, \ldots, y_n)$. And f is called σ-homogeneous of degree m if* $\deg(M(\lambda)) = m$.

The difference analogue of Euler's theorem related to homogeneous polynomials is valid.

Lemma 4.2 *$f \in \mathcal{F}\{y_0, y_1, \ldots, y_n\}$ is σ-homogeneous if and only if for each $r \in \mathbb{N}_0$, there exists $m_r \in \mathbb{N}_0$ such that*

$$\sum_{i=0}^{n} y_i^{(r)} \frac{\partial f(y_0, \ldots, y_n)}{\partial y_i^{(r)}} = m_r f.$$

Sparse σ-resultants have the following property.

Theorem 4.3 *The sparse σ-resultant is σ-homogeneous in each $\mathbf{u}_i$ which is the coefficient set of $\mathbb{P}_i$.*

Proof: Suppose $\text{ord}(\mathbf{R}, \mathbf{u}_i) = h_i \ge 0$. Follow the notations used in Theorem 3.5. By Lemma 3.7, $\mathbf{R}(\mathbf{u}; \zeta_0, \ldots, \zeta_n) = 0$. Differentiating this identity w.r.t. $u_{ij}^{(k)}$ $(j = 1, \ldots, l_i)$ respectively, we have

$$\overline{\frac{\partial \mathbf{R}}{\partial u_{ij}^{(k)}}} + \overline{\frac{\partial \mathbf{R}}{\partial u_{i0}^{(k)}}}\Big(-\frac{M_{ij}(\eta)}{M_{i0}(\eta)}\Big)^{(k)} = 0. \qquad (4)$$

In the above equations, $\overline{\frac{\partial \mathbf{R}}{\partial u_{ij}^{(k)}}} = \frac{\partial \mathbf{R}}{\partial u_{ij}^{(k)}}\big|_{(u_{00}, \ldots, u_{n0}) = (\zeta_0, \ldots, \zeta_n)}$.

Multiplying (4) by $u_{ij}^{(k)}$ and for j from 1 to l_i, adding them together, we get $\overline{\frac{\partial \mathbf{R}}{\partial u_{i0}^{(k)}}}\zeta_i^{(k)} + \sum_{j=1}^{l_i} u_{ij}^{(k)} \overline{\frac{\partial \mathbf{R}}{\partial u_{ij}^{(k)}}} = 0$. Thus, $f_k = \sum_{j=0}^{l_i} u_{ij}^{(k)} \frac{\partial \mathbf{R}}{\partial u_{ij}^{(k)}}$ vanishes at $(\zeta_0, \ldots, \zeta_n)$. Since $\text{ord}(f_k, u_{i0}) \le \text{ord}(\mathbf{R}, u_{i0})$ and $\deg(f_k) = \deg(\mathbf{R})$, by Lemma 3.7, there exists an $m_k \in \mathbb{Z}$ such that $f_k = m_k \mathbf{R}$. Thus, by Lemma 4.2, $\mathbf{R}$ is σ-homogeneous in $\mathbf{u}_i$. □

4.2 Order bound in terms of Jacobi number

In this section, we will give an order bound for the sparse σ-resultant in terms of the Jacobi number of the given system.

Consider a generic Laurent σ-essential system $\{\mathbb{P}_0, \ldots, \mathbb{P}_n\}$ defined in (1) with $\mathbf{u}_i$ being the coefficient vector of $\mathbb{P}_i$. Suppose $\mathbf{R}$ is the sparse σ-resultant of $\mathbb{P}_0, \ldots, \mathbb{P}_n$. Denote $\text{ord}(\mathbf{R}, \mathbf{u}_i) = \max_k \text{ord}(\mathbf{R}, u_{ik})$. If $\mathbf{u}_i$ does not occur in $\mathbf{R}$, then set $\text{ord}(\mathbf{R}, \mathbf{u}_i) = -\infty$.

Lemma 4.4 *For fixed i and s, if there exists a k_0 such that $\deg(\mathbf{R}, u_{ik_0}^{(s)}) > 0$, then for all $k \in \{0, \ldots, l_i\}$, $\deg(\mathbf{R}, u_{ik}^{(s)}) > 0$. In particular, $\text{ord}(\mathbf{R}, u_{ik}) = \text{ord}(\mathbf{R}, \mathbf{u}_i)\,(k = 0, \ldots, l_i)$.*

Proof: By (4) and lemma 3.7, the lemma follows. □

Let $s_{ij} = \text{ord}(\text{N}(\mathbb{P}_i), y_j)$ and $s_i = \text{ord}(\text{N}(\mathbb{P}_i))$. We call the $(n+1) \times n$ matrix $A = (s_{ij})$ the *order matrix* of $\mathbb{P}_0, \ldots, \mathbb{P}_n$. By $A_{\hat{i}}$, we mean the submatrix of A obtained by deleting the $(i+1)$-th row from A. We use $\mathbb{P}$ to denote the set $\{\text{N}(\mathbb{P}_0), \ldots, \text{N}(\mathbb{P}_n)\}$ and by $\mathbb{P}_{\hat{i}}$, we mean the set $\mathbb{P} \backslash \{\text{N}(\mathbb{P}_i)\}$. For a matrix $B = (a_{ij})_{n \times n}$, the *Jacobi number* is $\text{Jac}(B) = \max_{\tau \in S_n}\{a_{1\tau(1)} + \cdots + a_{n\tau(n)}\}$, where S_n is the set of all permutations of $\{1, \ldots, n\}$. We call $J_i = \text{Jac}(A_{\hat{i}})$ the *Jacobi number* of the system $\mathbb{P}_{\hat{i}}$, also denoted by $\text{Jac}(\mathbb{P}_{\hat{i}})$.

The following theorem shows that Jacobi numbers are order bounds for sparse σ-resultants.

Theorem 4.5 *Let $\mathbb{P}$ be a Laurent σ-essential system and $\mathbf{R}$ the sparse σ-resultant of $\mathbb{P}$. Then*

$$\text{ord}(\mathbf{R}, \mathbf{u}_i) = \begin{cases} -\infty & \text{if } J_i = -\infty, \\ h_i \le J_i & \text{if } J_i \ge 0. \end{cases}$$

Proof: For details, see [22, Theorem 4.14]. □

Example 4.6 *Let $n = 2$ and $\mathbb{P}_i$ have the form $\mathbb{P}_0 = u_{00} + u_{01} y_1 y_1^{(1)}$, $\mathbb{P}_1 = u_{10} + u_{11} y_1$, $\mathbb{P}_2 = u_{10} + u_{11} y_2^{(1)}$. In this example, $J_0 = 1, J_1 = 2, J_2 = -\infty$. And $\text{ord}(\mathbf{R}, \mathbf{u}_0) = 0 < J_0, \text{ord}(\mathbf{R}, \mathbf{u}_1) = 1 < J_1, \text{ord}(\mathbf{R}, \mathbf{u}_2) = -\infty$.*

5. SINGLE EXPONENTIAL ALGORITHM

In this section, we give an algorithm to compute the sparse σ-resultant for a Laurent σ-essential system with single exponential complexity. The idea is to estimate the degree bounds for the resultant and then to use linear algebra to find the coefficients of the resultant.

The following result gives an upper bound for the degree of the sparse σ-resultant.

Theorem 5.1 *Let $\mathbb{P}_0, \ldots, \mathbb{P}_n$ be a Laurent σ-essential system of form (1) with $\mathrm{ord}(\mathrm{N}(\mathbb{P}_i)) = s_i$ and $\deg(\mathrm{N}(\mathbb{P}_i), \mathbb{Y}) = m_i$. Suppose $\mathrm{N}(\mathbb{P}_i) = \sum_{k=0}^{l_i} u_{ik} N_{ik}$ and $J_i = \mathrm{Jac}(\mathbb{P}_{\hat{i}})$. Denote $m = \max_i\{m_i\}$. Let $\mathbf{R}$ be the sparse difference resultant of $\mathbb{P}_i\,(i = 0, \ldots, n)$. Suppose $\mathrm{ord}(\mathbf{R}, \mathbf{u}_i) = h_i$ for each i. Then*

1) $\deg(\mathbf{R}) \le \prod_{i=0}^n (m_i + 1)^{h_i+1} \le (m+1)^{\sum_{i=0}^n (J_i+1)}$.

2) *$\mathbf{R}$ has a representation*

$$\prod_{i=0}^{n}\prod_{k=0}^{n} (N_{i0}^{(k)})^{\deg(\mathbf{R})} \cdot \mathbf{R} = \sum_{i=0}^{n}\sum_{k=0}^{h_i} G_{ik}\mathrm{N}(\mathbb{P}_i)^{(k)} \qquad (5)$$

where $G_{ik} \in \mathbb{Q}[\mathbf{u}_0^{[h_0]}, \ldots, \mathbf{u}_n^{[h_n]}, \mathbb{Y}^{[h]}]$ and $h = \max\{h_i + e_i\}$ such that $\deg(G_{ik}\mathrm{N}(\mathbb{P}_i)^{(k)}) \le [m + 1 + \sum_{i=0}^n (h_i + 1)\deg(N_{i0})]\deg(\mathbf{R})$.

Proof: We sketch the proof here. In $\mathbf{R}$, replace u_{i0} by $\big(\mathrm{N}(\mathbb{P}_i) - \sum_{k=1}^{l_i} u_{ik}N_{ik}\big)/N_{i0}$ for each $i = 0, \ldots, n$ and let $\mathbf{R}$ be expanded as a σ-polynomial in $\mathrm{N}(\mathbb{P}_i)$ and their transforms. (5) follows from the fact that $\mathcal{I} \cap \mathbb{Q}\{\mathbf{u}, \mathbb{Y}\} = \{0\}$. To obtain the degree bound of $\mathbf{R}$, we consider the algebraic ideal $\mathcal{J} = \big(\mathrm{N}(\mathbb{P}_0)^{[h_0]}, \ldots, \mathrm{N}(\mathbb{P}_n)^{[h_n]}\big) : \mathbf{m}^{[h]}$. By Bézout theorem, $\deg(\mathcal{J}) \le \prod_{i=0}^n (m_i+1)^{h_i+1}$. Since $\mathcal{J} \cap \mathbb{Q}[\mathbf{u}_0^{[h_0]}, \ldots, \mathbf{u}_n^{[h_n]}] = (\mathbf{R})$, by [20, Theorem 2.1], $\deg(\mathbf{R}) \le \deg(\mathcal{J})$ and 1) follows. In order to estimate $\deg(G_{ik}\mathrm{N}(\mathbb{P}_i)^{(k)})$ and a_{ik}, it suffices to consider each monomial M in $\mathbf{R}$ and substitute u_{i0} by $\big(\mathrm{N}(\mathbb{P}_i) - \sum_{k=1}^{l_i} u_{ik}N_{ik}\big)/N_{i0}$ into M and then expand it. □

With given order and degree bounds, we can give the algorithm **SDResultant** to compute sparse σ-resultants based on linear algebra techniques.

Theorem 5.2 *Let $\mathbb{P}_0, \ldots, \mathbb{P}_n$ be a Laurent σ-essential system of form (1). Denote $\mathbb{P} = \{\mathrm{N}(\mathbb{P}_0), \ldots, \mathrm{N}(\mathbb{P}_n)\}$, $J_i = \mathrm{Jac}(\mathbb{P}_{\hat{i}})$, $J = \max_i J_i$ and $m = \max_{i=0}^n \deg(\mathbb{P}_i, \mathbb{Y})$. Algorithm* **SDResultant** *computes sparse σ-resultant $\mathbf{R}$ of $\mathbb{P}_0, \ldots, \mathbb{P}_n$ with the following complexities:*

1) In terms of a given degree bound D of $\mathbf{R}$, the algorithm needs at most $O(D^{O(lJ)}(nJ)^{O(lJ)})$ $\mathbb{Q}$-arithmetic operations, where $l = \sum_{i=0}^n (l_i + 1)$ is the size of all $\mathbb{P}_i$.

2) The algorithm needs at most $O(m^{O(nlJ^2)}(nJ)^{O(lJ)})$ $\mathbb{Q}$-arithmetic operations.

Proof: In each loop of Step 3, the complexity is clearly dominated by Step 3.1.2, where we need to solve a system of linear equations $\mathcal{P} = 0$ over $\mathbb{Q}$ in $\mathbf{c}_0$ and $\mathbf{c}_{ij}$. $\mathcal{P} = 0$ is a linear system with $N = \binom{d+L-1}{L-1} + \sum_{i=0}^n (h_i+1)\binom{d_1-m_i-1+L+n(h+1)}{L+n(h+1)}$ variables and $M = \binom{d_1+L+n(h+1)}{L+n(h+1)}$ equations, where $L = \sum_{i=0}^n (h_i+1)(l_i+1)$. So we need at most $(\max\{M, N\})^{\omega}$ arithmetic operations over $\mathbb{Q}$ to solve it, where ω is the matrix multiplication exponent and the currently best known ω is 2.376.

Algorithm 1 — SDResultant($\mathbb{P}_0, \ldots, \mathbb{P}_n$)

Input: A generic Laurent σ-essential system $\mathbb{P}_0, \ldots, \mathbb{P}_n$.
Output: The sparse σ-resultant $\mathbf{R}$ of $\mathbb{P}_0, \ldots, \mathbb{P}_n$.

1. Set $\mathrm{N}(\mathbb{P}_i) = \sum_{k=0}^{l_i} u_{ik}N_{ik}$ with $\deg(N_{i0}) \le \deg(N_{ik})$.
   Set $m_i = \deg(\mathrm{N}(\mathbb{P}_i))$, $m_{i0} = \deg(N_{i0})$, $\mathbf{u}_i = \mathrm{coeff}(\mathbb{P}_i)$.
   Set $s_{ij} = \mathrm{ord}(\mathrm{N}(\mathbb{P}_i), y_j)$, $A = (s_{ij})$, $J_i = \mathrm{Jac}(A_{\hat{i}})$.
2. Set $\mathbf{R} = 0$, $o = 0$, $m = \max_i\{m_i\}$.
3. While $\mathbf{R} = 0$ do
   - 3.1. For each $(h_0, \ldots, h_n) \in \mathbb{N}_0^{n+1}$ with $\sum_{i=0}^n h_i = o$ and $h_i \le J_i$ do
     - 3.1.1. $U = \cup_{i=0}^n \mathbf{u}_i^{[h_i]}$, $h = \max_i\{h_i + e_i\}$, $d = 1$.
     - 3.1.2. While $\mathbf{R} = 0$ and $d \le \prod_{i=0}^n (m_i+1)^{h_i+1}$ do
       - 3.1.2.1. Set $\mathbf{R}_0$ to be a GHPol of degree d in U.
       - 3.1.2.2. Set $\mathbf{c}_0 = \mathrm{coeff}(\mathbf{R}_0, U)$.
       - 3.1.2.3. Set H_{ij} to be GPols in $\mathbb{Y}^{[h]}, U$ of degree $[m + 1 + \sum_{i=0}^n (h_i+1)m_{i0}]d - m_i - 1$.
       - 3.1.2.4. Set $\mathbf{c}_{ij} = \mathrm{coeff}(H_{ij}, \mathbb{Y}^{[h]} \cup U)$.
       - 3.1.2.5. Set $\mathcal{P}$ to be the set of coefficients of $\prod_{i=0}^n \prod_{k=0}^{h_i} (N_{i0}^{(k)})^d \mathbf{R}_0 - \sum_{i=0}^n \sum_{j=0}^{h_i} H_{ij}(\mathrm{N}(\mathbb{P}_i))^{(j)}$ as a polynomial in $\mathbb{Y}^{[h]}, U$.
       - 3.1.2.6. Solve linear equations $\mathcal{P}(\mathbf{c}_0, \mathbf{c}_{ij}) = 0$.
       - 3.1.2.7. If $\mathbf{c}_0$ has a nonzero solution $\bar{\mathbf{c}}_0$, then $\mathbf{R} = \mathbf{R}_0(\bar{\mathbf{c}}_0)$ and goto 4., else $\mathbf{R} = 0$.
       - 3.1.2.8. d:=d+1.
   - 3.2. o:=o+1.
4. Return $\mathbf{R}$.

/*/ GPol(GHpol) stands for generic (homogenous) poly.
/*/ coeff(P, V) returns coefficients of P in variables V.

The iteration in Step 3.1.2 may go through 1 to $\prod_{i=0}^n (m_i + 1)^{h_i+1}$, and the iteration in Step 3.1 at most will repeat $\prod_{i=0}^n (J_i+1)$ times. And by Theorem 5.1, Step 3 may loop from $o = 0$ to $\sum_{i=0}^n (J_i+1)$. Thus, the complexity follows. □

6. DIFFERENCE RESULTANT

In this section, we introduce the notion of σ-resultant and prove its basic properties.

Definition 6.1 *Let $\mathbf{m}_{s,r}$ be the set of all σ-monomials in $\mathbb{Y}$ of order $\le s$ and degree $\le r$. Let $\mathbf{u} = \{u_M\}_{M \in \mathbf{m}_{s,r}}$ be a set of σ-indeterminates over $\mathbb{Q}$. Then, $\mathbb{P} = \sum_{M \in \mathbf{m}_{s,r}} u_M M$ is called* a generic σ-polynomial *of order s and degree r.*

Throughout this section, a generic σ-polynomial is assumed to be of degree greater than zero. Let

$$\mathbb{P}_i = u_{i0} + \sum_{\substack{\alpha \in \mathbb{Z}_{\ge 0}^{n(s_i+1)} \\ 1 \le |\alpha| \le m_i}} u_{i\alpha}(\mathbb{Y}^{[s_i]})^{\alpha} \; (i = 0, \ldots, n) \qquad (6)$$

be generic σ-polynomials of order s_i, degree m_i, and coefficients $\mathbf{u}_i$ respectively. Clearly, they form a super-essential system. We define the sparse σ-resultant $\mathrm{Res}_{\mathbb{P}_0,\ldots,\mathbb{P}_n}$ to be the *difference resultant* of $\mathbb{P}_0, \ldots, \mathbb{P}_n$. That is, The *difference resultant* of $\mathbb{P}_0, \ldots, \mathbb{P}_n$ is defined as the irreducible σ-polynomial of minimal order in each $\mathbf{u}_i$ which is contained in $[\mathbb{P}_0, \ldots, \mathbb{P}_n] \cap \mathbb{Q}\{\mathbf{u}_0, \ldots, \mathbf{u}_n\}$.

Difference resultants satisfy all the properties we have proved for sparse σ-resultants in previous sections. Apart from these, in the following, we will show σ-resultants possess other better properties. Firstly, we will give the precise degree for the σ-resultant, which is of BKK-type [1, 6]. Here, we need results about algebraic sparse resultants from [27]. For what needed here, please refer to [22, Sec. 6].

Theorem 6.2 *Let $\mathbf{R}(\mathbf{u}_0,\ldots,\mathbf{u}_n)$ be the σ-resultant of the $n+1$ generic σ-polynomials $\mathbb{P}_0,\ldots,\mathbb{P}_n$ in (6). Denote $s=\sum_{i=0}^n s_i$. Then $\mathrm{ord}(\mathbf{R},\mathbf{u}_i)=s-s_i$ and $\mathbf{R}(\mathbf{u}_0,\ldots,\mathbf{u}_n)$ is also the algebraic sparse resultant of $\mathbb{P}_0^{[s-s_0]},\ldots,\mathbb{P}_n^{[s-s_n]}$ as polynomials in $\mathbb{Y}^{[s]}$. In particular, for each i and k, $\deg(\mathbf{R},\mathbf{u}_i^{(k)})$ is equal to the mixed volume of the* Newton *polytopes of* $\bigcup_{j=0}^n \mathbb{P}_j^{[s-s_j]}\backslash\{\mathbb{P}_i^{(k)}\}$.

Proof: Regard $\mathbb{P}_i^{(k)}$ as polynomials in the $n(s+1)$ variables $\mathbb{Y}^{[s]}$, and we denote its support by $\mathcal{B}_{ik}$. Since the coefficients of $\mathbb{P}_i^{(k)}$ are algebraic indeterminates, $\mathbb{P}_i^{(k)}$ are generic sparse polynomials with supports $\mathcal{B}_{ik}$ respectively. We claim that:

C1) $\overline{\mathcal{B}}=\{\mathcal{B}_{ik}:0\le i\le n; 0\le k\le s-s_i\}$ is essential.

C2) $\overline{\mathcal{B}}$ jointly spans the affine lattice $\mathbb{Z}^{n(s+1)}$.

Note that $|\overline{\mathcal{B}}|=n(s+1)+1$. To prove C1), it suffices to show that any $n(s+1)$ distinct $\mathbb{P}_i^{(k)}$ are algebraically independent. Without loss of generality, we prove that for a fixed $l\in\{0,\ldots,s-s_0\}$,

$$S_l=\{(\mathbb{P}_i^{[s-s_i]})_{1\le i\le n},\mathbb{P}_0,\ldots,\mathbb{P}_0^{(l-1)},\mathbb{P}_0^{(l+1)},\ldots,\mathbb{P}_0^{(s-s_0)}\}$$

is algebraically independent. Clearly, $\{y_j^{(k)},\ldots,y_j^{(s_i+k)}|j=1,\ldots,n\}$ is a subset of the support of $\mathbb{P}_i^{(k)}$. Now we choose a monomial from each $\mathbb{P}_i^{(k)}$ and denote it by $m(\mathbb{P}_i^{(k)})$. Let

$$m(\mathbb{P}_0^{(k)})=\begin{cases} y_1^{(k)} & 0\le k\le l-1\\ y_1^{(s_0+k)} & l+1\le k\le s-s_0\end{cases}\quad\text{and}$$

$$m(\mathbb{P}_1^{(k)})=\begin{cases} y_1^{(l+k)} & 0\le k\le s_0\\ y_2^{(s_1+k)} & s_0+1\le k\le s-s_1\end{cases}.$$

For each $i\in\{2,\ldots,n\}$, let

$$m(\mathbb{P}_i^{(k)})=\begin{cases} y_i^{(k)} & 0\le k\le \sum_{j=0}^{i-1}s_j\\ y_{i+1}^{(s_i+k)} & \sum_{j=0}^{i-1}s_j+1\le k\le s-s_i\end{cases}.$$

So $m(S_l)$ is equal to $\{y_j^{[s]}:1\le j\le n\}$, which are algebraically independent over $\mathbb{Q}$. Thus, by the algebraic version of Lemma 2.1,the $n(s+1)$ members of S_l are algebraically independent over $\mathbb{Q}$ and claim C1) is proved. Claim C2) follows from the fact that 1 and $\mathbb{Y}^{[s]}$ are contained in the support of $\mathbb{P}_0^{[s-s_0]}$.

By C1) and C2), the sparse resultant of $(\mathbb{P}_i^{[s-s_i]})_{0\le i\le n}$ exists and we denote it by G. Then $\mathrm{ord}(G,\mathbf{u}_i)=s-s_i$ and $G\in[\mathbb{P}_0,\ldots,\mathbb{P}_n]$. By Lemma 3.7 and C1) again, $\mathbf{R}=c\cdot G$ for some $c\in\mathbb{Q}$. Thus, $\mathbf{R}$ is equal to the algebraic sparse resultant of $\mathbb{P}_0^{[s-s_0]},\ldots,\mathbb{P}_n^{[s-s_n]}$ and the theorem follows. □

As a direct consequence of the above theorem and the determinant representation for algebraic sparse resultant given by D'Andrea [7], we have the following result.

Corollary 6.3 *The σ-resultant of $\mathbb{P}_i$ can be written as the form $\det(M_1)/\det(M_0)$ where M_1 and M_0 are matrices whose elements are coefficients of $\mathbb{P}_i$ and their transforms up to the order $s-s_i$ and M_0 is a minor of M_1.*

Based on the matrix representation given as above, the single exponential algorithms given by Canny, Emiris, and Pan [10, 11] can be used to compute σ-resultants.

Now, we proceed to give a Poisson-type product formula for σ-resultant. Let $\tilde{\mathbf{u}}=\cup_{i=0}^n\mathbf{u}_i\setminus\{u_{00}\}$ and $\mathbb{Q}\langle\tilde{\mathbf{u}}\rangle$ be the σ-transcendental extension of $\mathbb{Q}$ in the usual sense. Let $\mathbb{Q}_0=\mathbb{Q}\langle\tilde{\mathbf{u}}\rangle(u_{00},\ldots,u_{00}^{(s-s_0-1)})$. Here, $\mathbb{Q}_0$ is not necessarily a σ-overfield of $\mathbb{Q}$, for the transforms of u_{00} are not defined. Consider $\mathbf{R}$ as an irreducible polynomial $r(u_{00}^{(s-s_0)})$ in $\mathbb{Q}_0[u_{00}^{(s-s_0)}]$. In a suitable algebraic extension field of $\mathbb{Q}_0$, $\mathbf{R}(\mathbf{u}_0,\mathbf{u}_1,\ldots,\mathbf{u}_n)=A\prod_{\tau=1}^{t_0}(u_{00}^{(s-s_0)}-\gamma_\tau)$, where $A\in\mathbb{Q}_0$. Let $\mathcal{I}_{\mathbf{u}}=[\mathbb{P}_0,\ldots,\mathbb{P}_n]\cap\mathbb{Q}\{\mathbf{u}_0,\ldots,\mathbf{u}_n\}$. Then by the definition of difference resultant, $\mathcal{I}_{\mathbf{u}}$ is an essential reflexive prime σ-ideal in the decomposition of $\{\mathbf{R}\}$ which is not held by any σ-polynomial of order less than $s-s_0$ in u_{00}. Suppose $\mathbf{R},\mathbf{R}_1,\mathbf{R}_2,\ldots$ is a basic sequence[1] of $\mathbf{R}$ corresponding to $\mathcal{I}_{\mathbf{u}}$. That is, $\mathcal{I}_{\mathbf{u}}=\bigcup_{k\ge0}\mathrm{asat}(\mathbf{R},\mathbf{R}_1,\ldots,\mathbf{R}_k)$. Let $(\gamma_\tau,\gamma_{\tau1},\ldots,\gamma_{\tau k})$ be a generic zero of $\mathrm{asat}(\mathbf{R},\mathbf{R}_1,\ldots,\mathbf{R}_k)$ for each k. Let $\mathcal{G}_\tau=\mathbb{Q}_0(\gamma_\tau,\gamma_{\tau1},\ldots)$. Then $\mathcal{G}_\tau$ is isomorphic to the quotient field of $\mathbb{Q}\{\mathbf{u}_0,\ldots,\mathbf{u}_n\}/\mathcal{I}_{\mathbf{u}}$, which is also a σ-field. So we can introduce a transforming operator σ_τ into $\mathcal{G}_\tau$ to make it a difference field such that the isomorphism becomes a difference one. That is, $\sigma_\tau|_{\mathbb{Q}_0}=\sigma|_{\mathbb{Q}_0}$ and

$$\sigma_\tau^k(u_{00})=\begin{cases} u_{00}^{(k)} & 0\le k\le s-s_0-1\\ \gamma_{\tau,k-s-s_0} & k\ge s-s_0\end{cases}$$

In this way, $(\mathcal{G}_\tau,\sigma_\tau)$ is a difference field.

Let $F\in\mathbb{Q}\{\mathbf{u}_0,\ldots,\mathbf{u}_n\}$. By saying F vanishes at $u_{00}^{(s-s_0)}=\gamma_\tau$, we mean $F|_{u_{00}^{(s-s_0+k)}=\gamma_{\tau k},k\ge0}=0$. The following lemma is a direct consequence of the above discussion.

Lemma 6.4 *$F\in\mathcal{I}_{\mathbf{u}}$ iff F vanishes at $u_{00}^{(s-s_0)}=\gamma_\tau$.*

Proof: Since $\mathcal{I}_{\mathbf{u}}=\bigcup_{k\ge0}\mathrm{asat}(\mathbf{R},\mathbf{R}_1,\ldots,\mathbf{R}_k)$ and $u_{00}^{(s-s_0+i)}=\gamma_{\tau i}\ (0\le i\le k)$ is a generic point of $\mathrm{asat}(\mathbf{R},\mathbf{R}_1,\ldots,\mathbf{R}_k)$, the lemma follows. □

Difference resultants have a Poisson-type product formula which is similar to their algebraic and differential analogues.

Theorem 6.5 *Let $\mathbf{R}$ be the σ-resultant of $\mathbb{P}_0,\ldots,\mathbb{P}_n$. Let $\deg(\mathbf{R},u_{00}^{(s-s_0)})=t_0$. Then there exist $\xi_{\tau k}\ (\tau=1,\ldots,t_0;k=1,\ldots,n)$ in overfields $(\mathcal{G}_\tau,\sigma_\tau)$ of $(\mathbb{Q}\langle\tilde{\mathbf{u}}\rangle,\sigma)$ such that*

$$\mathbf{R}=A\prod_{\tau=1}^{t_0}\mathbb{P}_0(\xi_{\tau1},\ldots,\xi_{\tau n})^{(s-s_0)},\qquad(7)$$

and the points $\xi_\tau=(\xi_{\tau1},\ldots,\xi_{\tau n})\ (\tau=1,\ldots,t_0)$ in (7) are generic zeroes of the σ-ideal $[\mathbb{P}_1,\ldots,\mathbb{P}_n]\subset\mathbb{Q}\langle\mathbf{u}_1,\ldots,\mathbf{u}_n\rangle\{\mathbb{Y}\}$. Note that (7) is formal and should be understood in the following precise meaning: $\mathbb{P}_0(\xi_{\tau1},\ldots,\xi_{\tau n})^{(s-s_0)}\triangleq\sigma^{s-s_0}u_{00}+\sigma_\tau^{s-s_0}(\sum_\alpha u_{0\alpha}(\xi_\tau^{[s-s_0]})^\alpha)$.

[1] For the rigorous definition of *basic sequence*, please refer to [4]. Here, we list its basic properties: i) For each $k\ge0$, $\mathrm{ord}(\mathbf{R}_k,u_{00})=s-s_0+k$ and $\mathbf{R},\mathbf{R}_1,\ldots,\mathbf{R}_k$ is an irreducible algebraic ascending chain, and ii) $\bigcup_{k\ge0}\mathrm{asat}(\mathbf{R},\mathbf{R}_1,\ldots,\mathbf{R}_k)$ is a reflexive prime σ-ideal.

Proof: By Theorem 4.3, there exists $m \in \mathbb{N}$ s.t. $u_{00}\frac{\partial \mathbf{R}}{\partial u_{00}} + \sum_\alpha u_{0\alpha}\frac{\partial \mathbf{R}}{\partial u_{0\alpha}} = m\mathbf{R}$. Let $\xi_{\tau\alpha} = (\frac{\partial \mathbf{R}}{\partial u_{0\alpha}}/\frac{\partial \mathbf{R}}{\partial u_{00}})\big|_{u_{00}^{(s-s_0)}=\gamma_\tau}$. Then $u_{00} = -\sum_\alpha u_{0\alpha}\xi_{\tau\alpha}$ with $u_{00}^{(s-s_0)} = \gamma_\tau$. That is, $\gamma_\tau = -\sigma_\tau^{s-s_0}(\sum_\alpha u_{0\alpha}\xi_{\tau\alpha}) = -(\sum_\alpha u_{0\alpha}\xi_{\tau\alpha})^{(s-s_0)}$. Thus, we have $\mathbf{R} = A\prod_{\tau=1}^{t_0}(u_{00} + \sum_\alpha u_{0\alpha}\xi_{\tau\alpha})^{(s-s_0)}$. For $j = 1, \ldots, n$, let $\xi_{\tau j} = (\frac{\partial \mathbf{R}}{\partial u_{0j0}}/\frac{\partial \mathbf{R}}{\partial u_{00}})\big|_{u_{00}^{(s-s_0)}=\gamma_\tau}$, where u_{0j0} is the coefficient of y_j in $\mathbb{P}_0$. Let $\xi_\tau = (\xi_{\tau 1}, \ldots, \xi_{\tau n})$. Similar to the proof of [13, Theorem 6.4] and by Lemma 6.4, we can show that $\xi_{\tau\alpha} = (\xi_\tau^{[s_0]})^\alpha$. Thus, (7) follows. And by Lemma 6.4, it is easy to show that ξ_τ are generic zeroes of the σ-ideal $[\mathbb{P}_1, \ldots, \mathbb{P}_n] \subset \mathbb{Q}\langle \mathbf{u}_1, \ldots, \mathbf{u}_n\rangle\{\mathbb{Y}\}$. □

7. CONCLUSION AND PROBLEM

In this paper, we first introduce the concept of Laurent σ-essential systems and give a criterion for Laurent σ-essential systems in terms of their supports. Then the sparse σ-resultant for a Laurent σ-essential system is defined and its basic properties are proved. In particular, order and degree bounds are given. Based on these bounds, an algorithm to compute the sparse σ-resultant is proposed, which is single exponential in terms of the order, the number of variables, and the size of the system. Besides these, the σ-resultant is introduced and its precise order, degree, determinant representation and the Poisson-type product formula are given.

Below, we propose several questions for further study.

It is useful to represent the sparse σ-resultant as the quotient of two determinants, as done in [7, 10] in the algebraic case. In the difference case, Theorem 6.2 shows that σ-resultant has such a matrix formula, but for sparse σ-resultant, we do not have such a formula yet.

The degree of the algebraic sparse resultant is equal to the mixed volume of the Newton polytopes of certain polynomials [24] or [15, p.255]. A similar degree bound is given [21, Theorem 1.3] for the differential resultant. And Theorem 6.2 shows that the degree of σ-resultants is exactly of such BKK-type. We conjecture that sparse σ-resultant has such degree bounds.

There exist very efficient algorithms to compute algebraic sparse resultants [9, 10, 11, 7] based on matrix representations. How to apply the principles behind these algorithms to compute sparse σ-resultants is an important problem.

8. REFERENCES

[1] D. N. Bernshtein. The Number of Roots of a System of Equations. *Funct. Anal. Appl.*, 9(3), 183-185, 1975.

[2] D. Bouziane, A. Kandri Rody, and H. Maârouf. Unmixed-dimensional Decomposition of a Finitely Generated Perfect Differential Ideal. *J. Symb. Comput.*, 31(6), 631-649, 2001.

[3] J. F. Canny. Generalized Characteristic Polynomials. *J. Symb. Comput.*, 9, 241-250, 1990.

[4] R. M. Cohn. Manifolds of Difference Polynomials. *Trans. Amer. Math. Soc.*, 64(1), 1948.

[5] R. M. Cohn. *Difference Algebra.* Interscience Publishers, New York, 1965.

[6] D. Cox, J. Little, D. O'Shea. *Using Algebraic Geometry.* Springer, 1998.

[7] C. D'Andrea. Macaulay Style Formulas for Sparse Resultants. *Trans. Amer. Math. Soc.*, 354(7), 2595-2629, 2002.

[8] D. Eisenbud, F. O. Schreyer, and J. Weyman. Resultants and Chow Forms via Exterior Syzygies. *J. Amer. Math. Soc.*, 16(3), 537-579, 2004.

[9] I. Z. Emiris. On the Complexity of Sparse Elimination. *J. Complexity*, 12, 134-166, 1996.

[10] I. Z. Emiris and J. F. Canny. Efficient Incremental Algorithms for the Sparse Resultant and the Mixed Volume. *J. Symb. Comput.*, 20(2), 117-149, 1995.

[11] I. Z. Emiris and V. Y. Pan. Improved Algorithms for Computing Determinants and Resultants. *J. Complexity*, 21, 43-71, 2005.

[12] X. S. Gao and S. C. Chou. On the Dimension for Arbitrary Ascending Chains. *Chinese Bull. of Sciences,* vol. 38, 396-399, 1993.

[13] X. S. Gao, W. Li, C. M. Yuan. Intersection Theory in Differential Algebraic Geometry: Generic Intersections and the Differential Chow Form. Accepted by *Trans. of Amer. Math. Soc.*, http://dx.doi.org/10.1090/S0002-9947-2013-05633-4.

[14] X. S. Gao, Y. Luo, C. M. Yuan. A Characteristic Set Method for Ordinary difference Polynomial Systems. *J. Symb. Comput.*, 44(3), 242-260, 2009.

[15] I. M. Gelfand, M. Kapranov, A. Zelevinsky. *Discriminants, Resultants and Multidimensional Determinants.* Boston, Birkhäuser, 1994.

[16] W.V.D. Hodge and D. Pedoe. *Methods of Algebraic Geometry, Volume I.* Cambridge Univ. Press, 1968.

[17] J. P. Jouanolou. Le formalisme du rèsultant. *Advances in Mathematics*, 90(2), 117-263, 1991.

[18] M. Kapranov, B. Sturmfels, and A Zelevinsky. Chow Polytopes and General Resultants. *Duke Math. J.*, 67, 189-218, 1992.

[19] A. Levin. *Difference Algebra.* Springer, 2008.

[20] W. Li, X. S. Gao, C. M. Yuan. Sparse Differential Resultant. *Proc. ISSAC 2011*, 225-232, ACM Press, New York, 2011.

[21] W. Li, C. M. Yuan, X. S. Gao. Sparse Differential Resultant for Laurent Differential Polynomials. *ArXiv:1111.1084v3,* 1-70, 2012.

[22] W. Li, C. M. Yuan, X. S. Gao. Sparse Difference Resultant. *ArXiv:1212.3090v1,* 1-34, 2012.

[23] Z.M. Li. A Subresultant Theory for Linear Differential, Linear Difference and Ore Polynomials, with Applications. PhD thesis, Johannes Kepler University, Linz, 1996.

[24] P. Pedersen and B. Sturmfels. Product Formulas for Resultants and Chow Forms. *Mathematische Zeitschrift*, 214(1), 377-396, 1993.

[25] S. L. Rueda. Linear Sparse Differential Resultant Formulas. *Linear Algebra and its Applications*, 438(11), 4296-4321, 2013.

[26] B. Sturmfels. Sparse Elimination Theory. In *Computational Algebraic Geometry and Commutative Algebra*, Eisenbud, D., Robbiano, L. eds. 264-298, Cambridge University Press, 1993.

[27] B. Sturmfels. On The Newton Polytope of the Resultant. *J. Algebraic Comb.*, 3, 207-236, 1994.

[28] Z. Y. Zhang, C. M. Yuan, X. S. Gao. Matrix Formula of Differential Resultant for First Order Generic Ordinary Differential Polynomials. ArXiv:1204.3773, 2012.

From Approximate Factorization to Root Isolation

Kurt Mehlhorn
Max Planck Institute for Informatics
mehlhorn@mpi-inf.mpg.de

Michael Sagraloff
Max Planck Institute for Informatics
msagralo@mpi-inf.mpg.de

Pengming Wang
Max Planck Institute for Informatics
s9pewang@stud.uni-saarland.de

ABSTRACT

We present an algorithm for isolating all roots of an arbitrary complex polynomial p which also works in the presence of multiple roots provided that arbitrary good approximations of the coefficients of p and the number of distinct roots are given. Its output consists of pairwise disjoint disks each containing one of the distinct roots of p, and its multiplicity. The algorithm uses approximate factorization as a subroutine. For the case, where Pan's algorithm [16] is used for the factorization, we derive complexity bounds for the problems of isolating and refining all roots which are stated in terms of the geometric locations of the roots only. Specializing the latter bounds to a polynomial of degree d and with integer coefficients of bitsize less than τ, we show that $\tilde{O}(d^3+d^2\tau+d\kappa)$ bit operations are sufficient to compute isolating disks of size less than $2^{-\kappa}$ for all roots of p, where κ is an arbitrary positive integer.

Our new algorithm has an interesting consequence on the complexity of computing the topology of a real algebraic curve specified as the zero set of a bivariate integer polynomial and for isolating the real solutions of a bivariate system. For input polynomials of degree n and bitsize τ, the currently best running time improves from $\tilde{O}(n^9\tau+n^8\tau^2)$ (deterministic) to $\tilde{O}(n^6+n^5\tau)$ (randomized) for topology computation and from $\tilde{O}(n^8+n^7\tau)$ (deterministic) to $\tilde{O}(n^6+n^5\tau)$ (randomized) for solving bivariate systems.

Categories and Subject Descriptors

G.1.5 [**Roots of Nonlinear Equations**]: Polynomials, methods for

Keywords

Root Isolation, Root Refinement, Bit Complexity, Numerical Algorithms, Certified Algorithms, Cylindrical Algebraic Decomposition

1. INTRODUCTION

Root isolation is a fundamental problem of computational algebra. Given a univariate polynomial p with complex coefficients and possibly multiple roots, the goal is to compute disjoint disks in the complex plane each containing exactly one root. We assume the existence of an oracle that can be asked for rational approximations of the coefficients of arbitrary precision. In particular, non-rational coefficients can never be learned exactly in finite time.

In this generality, the problem is unsolvable. Consider $p(x)=(x-\sqrt{2})^2(x+1)=x^3+(-2\sqrt{2}-1)x^2+(2+2\sqrt{2})x+2$ and assume that an algorithm terminates after having received approximations 1, α, β, and 2 for the coefficients, where α and β are rational and the polynomial $x^3+\alpha x^2+\beta x+2$ has three distinct roots. If the algorithm outputs two disks, the adversary commits to this α and β as the two middle coefficients. If the algorithm outputs three disks, the adversary commits to p.

The example shows that the problem needs to be restricted. In addition to our assumption that the coefficients of our input polynomial p are provided by coefficient oracles, we further assume that *the number k of distinct roots* is also given. Root isolation is a key ingredient in the computation of a CAD (cylindrical algebraic decomposition) for a given set of multivariate polynomials and, in particular, for computing the topology of algebraic curves and surfaces. In these applications, one has to deal with polynomials with multiple roots and algebraic coefficients which can be approximated to an arbitrary precision. In addition, the number of distinct roots is readily available from an algebraic decomposition of the input. At this point, we refer to some recent symbolic-numeric algorithms [20, 6, 3, 2] which combine structural information derived from symbolic computation and the use of numerical root finding methods to isolate the roots of polynomials with algebraic coefficients for which only approximations are given.

We now give a short overview of our algorithm and our results. Let $p(x)=\sum_{i=0}^{n}p_ix^i$ be a polynomial with k distinct roots $z_1,\dots,z_k$. For $i=1,\dots,k$, let $m_i:=\operatorname{mult}(z_i,p)$ be the *multiplicity* of z_i, and let $\sigma_i:=\sigma(z_i,p):=\min_{j\neq i}|z_i-z_j|$ be the *separation* of z_i from the other roots of p. Then, our algorithm outputs isolating disks $\Delta_i=\Delta(\tilde{z}_i,R_i)$ for the roots z_i and the corresponding multiplicities m_i. The radii satisfy $R_i<\frac{\sigma_i}{64n}$, thus the center $\tilde{z}_i$ of Δ_i approximates z_i to an error of less than $\frac{\sigma_i}{64n}$. If the number of distinct roots of p differs from k, we make no claims about termination and output.

The coefficients of p are provided by oracles. That is, on input L, the oracle essentially returns binary fraction approximations $\tilde{p}_i$ of the coefficients p_i such that $\left\|p-\sum_{i=0}^{n}\tilde{p}_ix^i\right\|\leq 2^{-L}\|p\|$. Here, $\|p\|:=\|p\|_1=|p_0|+\ldots+|p_n|$ denotes the *one-norm* of p. The details are given in Section 2.

Our algorithm has a simple structure. We first use any algorithm (e.g. [4, 19, 16, 22]) for approximately factorizing the input polynomial. It is required that it can be run with different levels of precision, and that, for any given integer b, it returns approximations $\hat{z}_1$ to $\hat{z}_n$ for the roots of p such that

$$\left\|p-p_n\prod_{1\leq j\leq n}(x-\hat{z}_j)\right\|\leq 2^{-b}\|p\|. \tag{1}$$

In a second step, we partition the root approximations $\hat{z}_1$ to $\hat{z}_n$ into

ISSAC'13, June 26–29, 2013, Boston, Massachusetts, USA.

k clusters $C_1, \ldots, C_k$ based on geometric vicinity. We enclose each cluster C_i in a disk $D_i = \Delta(\tilde{z}_i, r_i)$ and make sure that the disks are pairwise disjoint and that the radii r_i are not "too small" compared to the pairwise distances of the centers $\tilde{z}_i$.[1] In a third step, we verify that the n-times enlarged disks $\Delta_i = \Delta(\tilde{z}_i, R_i) = \Delta(\tilde{z}_i, n \cdot r_i)$ are disjoint and that each of them contains exactly the same number of approximations as roots of p counted with multiplicity. If the clustering and the verification succeed, we return the disks $\Delta_1, \ldots, \Delta_k$ and the number of approximations $\hat{z} \in \{\hat{z}_1, \ldots, \hat{z}_n\}$ in the disk as the multiplicity of the root isolated by the disk. If either clustering or verification does not succeed, we repeat with a higher precision. Strzebonski [20] has previously described a similar approach. The main difference is that he used a heuristic for the clustering step and hence could neither prove completeness of his approach nor analyze its complexity. He reports that his algorithm does very well in the context of CAD computation.

In the example above, we would have the additional information that p has exactly two distinct roots. We ask the oracle for an L-approximation of p for sufficiently large L and approximately factor it. Assume, we obtain approximations -1.01, 1.4, and 1.5 of the roots, and let $\hat{p} = (x+1.01)(x-1.4)(x-1.5)$. The clustering step may then put the first approximation into a singleton cluster and the other two approximations into a cluster of size two. It also computes disjoint enclosing disks. The verification step tries to certify that p and $\hat{p}$ contain the same number of roots in both disks. If L and b are sufficiently large, clustering and verification succeed.

If Pan's algorithm [16] is used for the approximate factorization step, then the overall algorithm has bit complexity[2]

$$\tilde{O}\left(n^3 + n^2 \sum_{i=1}^{k} \log M(z_i) + n \sum_{i=1}^{k} \log\left(M(\sigma_i^{-m_i}) \cdot M(P_i^{-1})\right)\right) \quad (2)$$

where $P_i := \prod_{j \neq i}(z_i - z_j)^{m_j} = \frac{p^{(m_i)}(z_i)}{m_i! p_n}$, and $M(x) := \max(1, |x|)$. Observe that our algorithm is adaptive in a very strong sense, namely, the above bound exclusively depends on the actual multiplicities and the geometry (i.e. the actual modulus of the roots and their distances to each other) of the roots.

Our algorithm can also be used to further refine the isolating disks to a size of $2^{-\kappa}$ or less, where κ is a given integer. The bit complexity for the refinement is given by the bound in (2) plus an additional term $\tilde{O}(n \cdot \kappa \cdot \max_i m_i)$. In particular for square-free polynomials, the amortized cost per root and bit of precision is one showing that the method is optimal up to polylogarithmic factors.

For the benchmark problem of isolating all roots of a polynomial p with *integer* coefficients of absolute value bounded by 2^τ, the bound in (2) becomes $\tilde{O}(n^3 + n^2\tau)$. The bound for the refinement becomes $\tilde{O}(n^3 + n^2\tau + n\kappa)$, even if there exist multiple roots.

For a square-free integer polynomial p, we are aware of only one method [8, Theorem 3.1] that achieves a comparable complexity bound for the benchmark problem. That is, based on the gap theorem from Mahler, one can compute a theoretical worst case bound b_0 of size $\Theta(n\tau)$ with the property that if n points $\hat{z}_j \in \mathbb{C}$ fulfill the inequality (1) for a $b \geq b_0$, then they approximate the corresponding roots z_j to an error less than $\sigma_j/(2n)$; cf. Lemma 4 for an adaptive version. Hence, for $b \geq b_0$, Pan's factorization algorithm also yields isolating disks for the roots of p using $\tilde{O}(n^2\tau)$ bit operations. Note that this approach achieves a good worst case complexity, however, for the price of running the factorization algorithm with $b = \Theta(n\tau)$, even if the roots are well conditioned. In contrast, our algorithm turns Pan's factorization algorithm into a highly adaptive method for isolating and approximating the roots of a general polynomial. Also, for general polynomials, there exist theoretical worst case bounds [19, Section 19] for the distance between the roots of p and corresponding approximations fulfilling (1). They are optimal for roots of multiplicity $\Omega(n)$ but they constitute strong overestimations if all roots have considerably smaller multiplicities. For the task of root refinement, the bit complexity of our method (i.e. $\tilde{O}(n \max_i m_i \cdot \kappa)$ for κ dominating) adapts to the highest occurring multiplicity, whereas this is not given for the currently best methods [12, 16, 18] which achieve the bound $\tilde{O}(n^2\kappa)$.

[1] This will turn out to be crucial to control the cost for the final verification step. For details, we refer to Sections 3.2 and 3.3.

[2] $\tilde{O}$ indicates that we omit logarithmic factors.

We would also like to remark that we are aware of only two previous root isolation algorithms [20, 14] that can cope with multiple roots. The latter algorithm can cope with at most one multiple root and needs to know the number of distinct complex roots as well as the number of distinct real roots. The former algorithm has the same applicability as our algorithm, but it has heuristic steps.

Our new root isolation algorithms has an interesting consequence on the complexity of computing the topology of a real planar algebraic curve specified as the zero set of an integer polynomial and for isolating the real solutions of a bivariate polynomial system. Both problems are well-studied [1, 10, 11, 20, 6, 5, 2, 7, 13]. More specifically, in an extended version [15] of this paper, we apply our root isolation method to a recent randomized algorithm [2] for computing the topology of a planar algebraic curve. This yields bounds on the *expected* number of bit operations which improve the currently best (which are both deterministic) bounds [7, 13] from $\tilde{O}(n^9\tau + n^8\tau^2)$ to $\tilde{O}(n^6 + n^5\tau)$ for topology computation and from $\tilde{O}(n^8 + n^7\tau)$ to $\tilde{O}(n^6 + n^5\tau)$ for solving bivariate systems, where n and τ are upper bounds on the degree and the bitsize of the input polynomials, respectively.

2. BASIC PROPERTIES

We consider a polynomial $p(x) = p_n x^n + \ldots + p_0 \in \mathbb{C}[x]$ of degree $n \geq 2$, where $p_n \neq 0$. In addition to our notations from the introduction, we fix the following definitions:

- τ_p denotes the minimal non-negative integer with $\frac{|p_i|}{|p_n|} \leq 2^{\tau_p}$ for all $i = 0, \ldots, n-1$,
- $\Gamma_p := M(\max_i \log |z_i|)$ the *logarithmic root bound* of p,
- $\mathrm{Mea}(p) := |p_n| \cdot \prod_{i=1}^{k} M(z_i)^{m_i}$ the *Mahler Measure* of p.

A straight forward argument shows that the quantities τ_p, Γ_p, $|p_n|$ and $\mathrm{Mea}(p)$ are closely related; see [15] for details.

LEMMA 1. *$\Gamma_p \leq 1 + \tau_p$ and $\tau_p - n - 1 \leq \log \frac{\mathrm{Mea}(p)}{|p_n|} \leq n\Gamma_p$.*

We assume the existence of an oracle which provides arbitrary good approximations of the polynomial p. Let $L \geq 1$ be an integer. We call a polynomial $\tilde{p} = \tilde{p}_n x^n + \ldots + \tilde{p}_0$, with $\tilde{p}_i = s_i \cdot 2^{-\ell}$ and $s_i, \ell \in \mathbb{Z}$, an *approximation of precision L* of p if $|\tilde{p}_i - p_i| \leq 2^{-L-\log(n+1)} \|p\|$, $\ell \leq L + \lceil \log(n+1) \rceil - \lfloor \log \|p\| \rfloor$, and $\log |s_i| \leq L + \lceil \log(n+1) \rceil + 1$ for all i. When considering p_i as infinite bitstring $p_i = \mathrm{sgn}(p_i) \cdot \sum_{k=-\infty}^{+\infty} b_k 2^k$, $b_k \in \{0,1\}$, then we can obtain $\tilde{p}_i$ from the partial string which starts at index $k_1 = \lfloor \log \|p\| \rfloor$ and ends at index $k_2 = \lfloor \log \|p\| \rfloor - L - \lceil \log(n+1) \rceil$, that is, $s_i := 2^l \cdot \mathrm{sgn}(p_i) \cdot \sum_{k=k_2}^{k_1} b_k 2^k$, and $l = L + \lceil \log(n+1) \rceil - \lfloor \log \|p\| \rfloor$. *We assume that we can ask for an approximation of precision L of p at cost $O(n(L + \log n)) = \tilde{O}(nL)$.* This is the cost of reading the approximation of precision L. The next Lemma summarizes some elementary properties of approximations of precision L. Again, we refer to the extended version [15] of this paper for its simple proof.

LEMMA 2. *If $\tilde{p}$ is an approximation of precision L of p, then*

- $\|\tilde{p}\|/2 \le \|p\| \le 2\|\tilde{p}\|$.
- *If $L \ge \tau_p + 4$, then $2^{-L-\log(n+1)}\|\tilde{p}\| \le |\tilde{p}_n|/4$.*
- *If $2^{-L-\log(n+1)}\|\tilde{p}\| \le |\tilde{p}_n|/4$, then $|\tilde{p}_n|/2 \le |p_n| \le 2|\tilde{p}_n|$.*

The above lemma suggests an efficient method for estimating p_n. We ask for approximations $\tilde{p}$ of precision L of p for $L = 1,2,4,\dots$ until $2^{-L-\log(n+1)}\|\tilde{p}\| \le |\tilde{p}_n|/4$. Then, $|\tilde{p}_n|/2 \le |p_n| \le 2|\tilde{p}_n|$ by part 3 of the Lemma. Also $L \le 2(\tau_p + 4)$ by part 2 of the above Lemma. The cost is $\tilde{O}(n\tau_p) = \tilde{O}(n^2\Gamma_p)$ bit operations, where we used the upper bound for τ_p from Lemma 1. Observe that this bound depends only on the geometry of the roots (i.e. the actual root bound Γ_p) and the degree but not (directly) on the size of the coefficients of p. We remark that a "good" integer approximation Γ of Γ_p can also be computed with $\tilde{O}(n^2\Gamma_p)$ bit operations. The proof (see [15, Theorem 1]) of the latter fact is almost identical to the one given in [17, Section 6.1], however, a small modification (essentially, we replaced linear search by exponential binary search) yields an improvement from $\tilde{O}(n^2\Gamma^2)$ to $\tilde{O}(n^2\Gamma)$.

THEOREM 1. *An integer $\Gamma \in \mathbb{N}$ with*

$$\Gamma_p \le \Gamma < 8\log n + \Gamma_p \tag{3}$$

can be computed with $\tilde{O}(n^2\Gamma_p)$ bit operations. The computation uses an approximation of precision L of p with $L = O(n\Gamma_p)$.

3. ALGORITHM

We present an algorithm for isolating the roots of a polynomial $p(x) = \sum_{i=0}^n p_i x^i = p_n \prod_{i=1}^k (x - z_i)^{m_i}$, where the coefficients p_i are given as described in the previous section. The algorithm uses an arbitrary polynomial factorization algorithm to produce approximations for the roots $z_1,\dots,z_k$, and then performs a clustering and certification step to verify that the candidates are of sufficient quality. For concreteness, we pick Pan's factorization algorithm [16] for the factorization step, which also currently offers the best worst case bit complexity.[3] If the candidates do not pass the verification step, we reapply the factorization algorithm with a higher precision. For a given positive integer b denoting the desired precision, the factorization algorithm computes n root approximations $\hat{z}_1,\dots,\hat{z}_n$. The quality of approximation and the bit complexity are as follows:

THEOREM 2. *For an arbitrary integer $b \ge n\log n$, complex numbers $\hat{z}_1,\dots,\hat{z}_n$ can be computed such that*

$$\left\|p - p_n \prod_{i=1}^n (x - \hat{z}_i)\right\| \le 2^{-b}\|p\|$$

using $\tilde{O}(n^2\Gamma + bn)$ bit-operations. We write $\hat{p} := p_n \prod_{i=1}^n (x - \hat{z}_i)$. The algorithm returns the real and imaginary part of the $\hat{z}_i$'s as dyadic fractions of the form $A \cdot 2^{-B}$ with $A \in \mathbb{Z}$, $B \in \mathbb{N}$ and $B = O(b + n\Gamma_p)$. All fractions have the same denominator.

PROOF. If all roots of p have absolute value less than 1, then we can use Pan's Algorithm to obtain the above result; see [16, Theorem 2.1.1]. For general polynomials, we first scale p such that the roots of the scaled polynomial are contained in the unit disk $\Delta(0,1)$. For this purpose, we compute a Γ as in Theorem 1, and then consider the polynomial $f(x) := p(s \cdot x) = \sum_{i=0}^n f_i x^i$ with $s := 2^\Gamma$. Then, $f(x)$ has roots $\xi_i = z_i/s \in \Delta(0,1)$, and thus we can use Pan's algorithm with $b' := n\Gamma + b$ to compute an approximate factorization $\hat{f}(x) := \sum_{i=0}^n \hat{f}_i x^i := f_n \prod_{i=1}^n (x - \hat{\xi}_i)$ such that $\|f - \hat{f}\| < 2^{-b'}\|f\|$. According to Theorem 1, the cost for computing Γ is bounded by $\tilde{O}(n^2\Gamma)$ bit operations. The cost for running Pan's Algorithm is bounded by $\tilde{O}(n^2\Gamma) + \tilde{O}(nb') = \tilde{O}(n^2\Gamma + nb)$ bit operation. Furthermore, we need an approximation of precision b' of f, and thus an approximation of precision L of p with $L = O(ns + b) = \tilde{O}(n\Gamma_p + b)$. Again, the cost is bounded by $\tilde{O}(n^2\Gamma + nb)$.

Let $\hat{z}_i := s \cdot \hat{\xi}_i$ for all i and $\hat{p}(x) := p_n \cdot \prod_{i=1}^n (x - \hat{z}_i) = \hat{f}(x/s) = \sum_{i=0}^n \hat{f}_i / s^i x^i$. Then, $\hat{p}$ has the desired property. Namely, $\|\hat{p} - p\| \le \sum_{i=0}^n |f_i - \hat{f}_i| \le 2^{-b'} \sum_{i=0}^n |f_i| \le 2^{-b'} s^n \sum_{i=0}^n |f_i/s^i| = 2^{-b}\|p\|$. □

We now examine how far the approximations $\hat{z}_1,\dots,\hat{z}_n$ can deviate from the actual roots for a given value of b. Let $\Delta(z,r)$ be the disk with center z and radius r and let $\mathrm{bd}\,\Delta(z,r)$ be its boundary. We further define $P_i := \prod_{j \ne i}(z_i - z_j)^{m_j}$. Then, $p^{(m_i)}(z_i) = m_i! p_n P_i$.

LEMMA 3. *If $r \le \frac{\sigma_i}{n}$, then $|p(x)| > r^{m_i} \cdot \frac{|p_n P_i|}{4}$ for all $x \in \Delta(z_i, r)$.*

PROOF. We have

$$\begin{aligned}
|p(x)| &= |p_n|\,|x - z_i|^{m_i} \prod_{j \ne i} |x - z_j|^{m_j} \\
&\ge |p_n|\,|x - z_i|^{m_i} \prod_{j \ne i} |z_i - z_j|^{m_j} \cdot (1 - |x - z_i|/|z_i - z_j|)^{m_j} \\
&\ge r^{m_i}(1 - 1/n)^{n - m_i} |p_n| \prod_{j \ne i} |z_i - z_j|^{m_j} > r^{m_i}|p_n P_i|/4.
\end{aligned}$$

□

Based on the above Lemma, we can now use Rouché's theorem to show that, for sufficiently large b, the disk $\Delta(z_i, 2^{-b/(2m_i)})$ contains exactly m_i root approximations.

LEMMA 4. *Let $\hat{p}$ be such that $\|p - \hat{p}\| \le 2^{-b}\|p\|$. If*

$$b \ge \max(8n, n\log(n)), \text{ and } b \text{ is a power of two} \tag{4}$$

$$2^{-b/(2m_i)} \le \min(1/(2n^2), \sigma_i/(2n)), \text{ and} \tag{5}$$

$$2^{-b/2} \le \frac{|P_i|}{16(n+1)2^{\tau_p} M(z_i)^n} \tag{6}$$

for all i, the disk $\Delta(z_i, 2^{-b/(2m_i)})$ contains exactly m_i root approximations. For $i \ne j$, let $\hat{z}_i$ and $\hat{z}_j$ be arbitrary approximations in the disks $\Delta(z_i, 2^{-b/(2m_i)})$ and $\Delta(z_j, 2^{-b/(2m_j)})$, respectively. Then,

$$(1 - 1/n)|z_i - z_j| \le |\hat{z}_i - \hat{z}_j| \le (1 + 1/n)|z_i - z_j|.$$

PROOF. Let $\delta_i := \left(16 \cdot (n+1) \cdot 2^{-b} 2^{\tau_p} |P_i|^{-1} M(z_i)^n\right)^{1/m_i}$. It is easy to verify that $\delta_i \le 2^{-b/(2m_i)} \le \min(1, \sigma_i)/(2n)$ from (6) and (5). Note that to show that $\Delta(z_i, \delta_i)$ contains m_i approximations, it suffices to show that $|(p - \hat{p})(x)| < |p(x)|$ for all x on the boundary of $\Delta(z_i, \delta_i)$. Then, Rouché's theorem guarantees that $\Delta(z_i, \delta_i)$ contains the same number of roots of p and $\hat{p}$ counted with multiplicity. Since z_i is of multiplicity m_i and $\delta_i < \sigma_i/n$, the disk contains exactly m_i roots of p counted with multiplicity. We have (note that $M(x) \le (1 + 1/(2n^2)) \cdot M(z_i)$ for $x \in \mathrm{bd}\,\Delta(z_i, \delta_i)$)

$$\begin{aligned}
|(p - \hat{p})(x)| &\le \|p - \hat{p}\| \cdot M(x)^n < 2^{-b}\|p\| M(x)^n \\
&\le 2^{-b}\|p\| \cdot (1 + 1/(2n^2))^n \cdot M(z_i)^n \\
&\le 4 \cdot 2^{-b} \cdot 2^{\tau_p} |p_n| \cdot (n+1) \cdot M(z_i)^n \\
&\le \delta_i^{m_i} |p_n P_i|/4 < |p(x)|,
\end{aligned}$$

where the inequality in line three follows from $\|p\| \le (n+1)|p_n|2^{\tau_p}$, the first one in line four follows from the definition of δ_i, and the

[3] In practice, one might consider a numerical root finder [4] based on the Aberth-Ehrlich method instead. There is empirical evidence that such methods achieve comparable complexity bounds.

last inequality follows from Lemma 3. It follows that $\Delta(z_i, 2^{-b/(2m_i)})$ contains exactly m_i approximations. Furthermore, since $\delta_i \le \sigma_i/(2n)$ for all i, the disks $\Delta(z_i, \delta_i)$, $1 \le i \le k$, are pairwise disjoint.

For the second claim, we observe that $|\hat{z}_\ell - z_\ell| \le 2^{-b/(2m_\ell)} \le \sigma_\ell/(2n) \le |z_i - z_j|/(2n)$ for $\ell = i, j$ and hence $|\hat{z}_i - z_i| + |\hat{z}_j - z_j| \le |z_i - z_j|/n$. The claim now follows from the triangle inequality. □

We have now established that the disks $\Delta(z_i, 2^{-b/(2m_i)})$, $1 \le i \le k$, are pairwise disjoint and that the i-th disk contains exactly m_i root approximations provided that b satisfies (4) to (6). We want to stress that the radii $2^{-b/(2m_i)}$, $1 \le i \le k$, are vastly different. Assume $b = 40$. For a one-fold root ($m = 1$), the radius is 2^{-20}, for a double root ($m = 2$) the radius is 2^{-10}, for a four-fold root ($m = 4$) the radius is 2^{-5}, and for a twenty-fold root ($m = 20$), the radius is as large as $1/2$; see Figure 1 for an illustration.

Unfortunately, the conditions on b are stated in terms of the unknown quantities m_i, σ_i and $|P_i|$, as well as the center z_i. In the remainder of the section, we will show how to cluster root approximations and to certify them. We will need the following more stringent properties for the clustering and certification step.

$$2^{-b/(2m_i)} < \min\left(\left(\frac{\sigma_i}{4n}\right)^8, \frac{\sigma_i}{1024n^2}\right) \tag{7}$$

$$2^{-b/8} < \min(1/16, |P_i|/((n+1)\cdot 2^{2n\Gamma_p + 8n})) \tag{8}$$

Let b_0 be the smallest integer satisfying (4) to (8) for all i. Then,

$$b_0 = O(n\log n + n\Gamma_p + \max_i(m_i \log M(\sigma_i^{-1})) + \max_i \log M(P_i^{-1})).$$

3.1 Overview of the Algorithm

On input p and the number k of distinct roots, the algorithm outputs isolating disks $\Delta_i = \Delta(\tilde{z}_i, R_i)$ for the roots of p as well as the corresponding multiplicities m_i. The radii satisfy $R_i < \sigma_i/(64n)$.

The algorithm uses the factorization step with an increasing precision until the result can be certified. If either the clustering step or the certification step fails, we simply double the precision. There are a couple of technical safeguards to ensure that we do not waste time on iterations with an insufficiently large precision (Steps 2, 5, and 6); also recall that we need to scale our initial polynomial.

1. Compute an integer bound Γ for Γ_p that fulfills inequality (3).
2. Compute a 2-approximation $\lambda = 2^{l_\lambda}$, $l_\lambda \in \mathbb{Z}$, of $\|p\|/|p_n|$.
3. Scale p, that is, $f(x) := p(s \cdot x)$, with $s := 2^\Gamma$, to ensure that the roots $\xi_i = z_i/S$, $i = 1, \ldots, k$, of f are contained in the unit disk. Let b be the smallest integer satisfying (4)
4. Run Pan's algorithm on input f with parameter $b' := b + n\Gamma$ to produce approximations $\hat{\xi}_1, \ldots, \hat{\xi}_n$ for the roots of f. Then, $\hat{z}_i := s \cdot \hat{\xi}_i$ are approximations of the roots of p, and $\|\hat{p} - p\| < 2^{-b}\|p\|$, where $\hat{p}(x) := p_n \prod_{i=1}^n (x - \hat{z}_i)$.
5. If there is a $\hat{z}_i$ with $\hat{z}_i \ge 2^{\Gamma+1}$, return to Step 4 with $b := 2b$.
6. If $\prod_{i=1}^n M(\hat{z}_i) > 8\lambda$, return to Step 4 with $b := 2b$.
7. Partition $\hat{z}_1, \ldots, \hat{z}_n$ into k clusters $C_1, \ldots, C_k$. Compute (well separated) enclosing disks $D_1, \ldots, D_k$ for the clusters. If the clustering fails to find k clusters and corresponding disks, return to Step 4 with $b := 2b$.
8. For each i, let Δ_i denote the disk with the same center as D_i but with an n-times larger radius. We now verify the existence of $|C_i|$ roots (counted with multiplicity) of p in Δ_i. If the verification fails, return to Step 4 with $b := 2b$.
9. If the verification succeeds, output the disks Δ_i and report the number $|C_i|$ of root approximations $\hat{z} \in \{\hat{z}_1, \ldots, \hat{z}_n\}$ contained in the disks as the corresponding multiplicities.

Note that Steps 5 and 6 ensure that $\log M(\hat{z}_i) = O(\Gamma_p + \log n)$ for all i, and $\log \prod_{i=1}^n M(\hat{z}_i) = O(\log(\|p\|/|p_n|)) = O(\log n + \tau_p) = \tilde{O}(n\Gamma_p)$. The following Lemma guarantees that the algorithm passes these steps if $b \ge b_0$.

LEMMA 5. *For any $b \ge b_0$, it holds that $|\hat{z}_i| < 2^{\Gamma+1}$ for all i, and $\prod_{i=1}^n M(\hat{z}_i) < 8\lambda$.*

PROOF. In the proof of Lemma 4, we have already shown that $|\hat{z}_i| \le (1 + 1/(2n^2)) \cdot M(z_i)$ for all i. Hence, it follows that $|\hat{z}_i| \le (1 + 1/(2n^2)) \cdot 2^{\Gamma_p} < 2 \cdot 2^{\Gamma_p} \le 2^{\Gamma_p + 1}$, and (note that $(1 + \frac{1}{2n^2})^n < 4$)

$$\prod_{i=1}^n M(\hat{z}_i) \le 4 \cdot \prod_{i=1}^k M(z_i)^{m_i} < \frac{4\,\mathrm{Mea}(p)}{|p_n|} \le \frac{4\|p\|_2}{|p_n|} \le \frac{4\|p\|}{|p_n|} < 8\lambda.$$

□

3.2 Clustering

After candidate approximations $\hat{z}_1, \ldots, \hat{z}_n$ are computed using a fixed precision parameter b, we perform a partitioning of these approximations into k clusters $C_1, \ldots, C_k$. Our clustering algorithm works in phases. In each phase, it attempts to form a cluster based on some unclustered approximation as seed. After all approximations have been clustered, ideally, each of the clusters now corresponds to a distinct root of p. The algorithm satisfies the following properties: (1) For $b < b_0$, the algorithm may or may not succeed in finding k clusters. (2) For $b \ge b_0$, the clustering always succeeds. Whenever the clustering succeeds, the cluster C_i with seed $\tilde{z}_i$ is contained in the disk $D_i := \Delta(\tilde{z}_i, r_i)$, where $r_i \approx \min(\frac{1}{n^2}, \frac{\tilde{\sigma}_i}{256n^2})$, and $\tilde{\sigma}_i = \min_{j \ne i} |\tilde{z}_i - \tilde{z}_j|$. Furthermore, for $b \ge b_0$, D_i contains the root z_i (under suitable numbering) and exactly m_i approximations.

Before we describe our clustering method, we discuss two evident approaches that do not work for any b of size comparable to b_0 or smaller. A clustering with a fixed grid does not work as root approximations coming from roots with different multiplicities may move by vastly distinct amounts. As a consequence, we can only succeed if $b > (\max_i m_i) \cdot \log(\min_i \sigma_i)^{-1}$ which can be considerably larger than b_0, see Figure 1. A clustering based on Gershgorin disks does not work as well because very good approximations of a multiple root lead to large disks which then fail to separate approximations of distinct roots. In particular, if approximations are identical, the corresponding Gershgorin disks have infinite radius.

For our clustering, we use the fact that the factorization algorithm provides approximations $\hat{z}$ of the root z_i with distance less than $2^{-b/(2m_i)}$ (for $b \ge b_0$). Thus, we aim to determine clusters C of maximal size such that the pairwise distance between two elements in the same cluster is less than $2 \cdot 2^{-b/(2|C|)}$. We give details.

1. Initialize $\mathscr{C}$ to the empty set (of clusters).
2. Initialize C to the set of all unclustered approximations and choose $\hat{z} \in C$ arbitrarily. Let $a := 2^{\lfloor \log n \rfloor + 2}$ and $\delta := 2^{-b/4}$.
3. Update C to the set of points $q \in C$ satisfying $|\hat{z} - q| \le 2\sqrt[a/2]{\delta}$.
4. If $|C| \ge a/2$, add C to $\mathscr{C}$. Otherwise, set $a := a/2$ and continue with step 3.
5. If there are still unclustered approximations, continue with step 2.
6. If the number of clusters in $\mathscr{C}$ is different from k, report failure, and return to the factorization step with $b := 2b$.

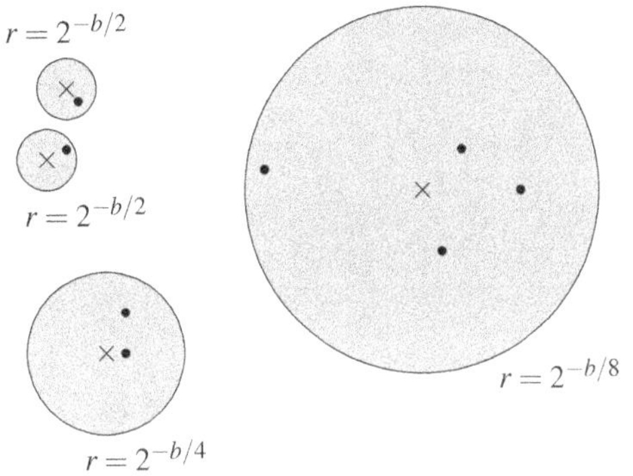

Figure 1: Example of a polynomial with four distinct roots with multiplicities 1, 1, 2, and 4. Crosses are roots of the polynomial, dots represent the approximations. The disk around a root shows the potential locations of its approximations. Note that the straightforward approach to cluster with a fixed distance threshold fails for all b with $b < (\max_i m_i) \cdot \log(\min_i \sigma_i)^{-1}$: For each such b, one can not choose any threshold that allows detecting the simple roots without splitting the four-fold root.

Note that, for $b \geq b_0$, the disks $\Delta(z_i, 2^{-b/(2m_i)})$ are disjoint. Let Z_i denote the set of root approximations in $\Delta(z_i, 2^{-b/(2m_i)})$. Then, $|Z_i| = m_i$ according to Lemma 4. We show that, for $b \geq b_0$, the clustering algorithm terminates with $C = Z_i$ if called with an approximation $\hat{z} \in Z_i$.

LEMMA 6. *Assume $b \geq b_0$, $\hat{z}_i \in Z_i$, $\hat{z}_j \in Z_j$, and $i \neq j$. Then,*

$$|\hat{z}_i - \hat{z}_j| \geq 2 \cdot (2^{-b/(16m_i)} + 2^{-b/(16m_j)}).$$

PROOF. Since $b \geq b_0$, we have $2^{-b/(2m_\ell)} \leq \sigma_\ell$ for $\ell = i, j$ by (5) and $2^{-b/(16m_\ell)} = (2^{-b/(2m_\ell)})^{1/8} \leq \sigma_\ell/(4n) \leq \sigma_\ell/8$ by (7). Thus,

$$|\hat{z}_i - \hat{z}_j| \geq \max(\sigma_i, \sigma_j) - 2^{-\frac{b}{2m_i}} - 2^{-\frac{b}{2m_j}} \geq \frac{\sigma_i}{2} + \frac{\sigma_j}{2} - \frac{\sigma_i}{4} - \frac{\sigma_j}{4}$$
$$\geq 2 \cdot (2^{-b/(16m_i)} + 2^{-b/(16m_j)}).$$

□

LEMMA 7. *If $b \geq b_0$, the clustering algorithm produces clusters C_1 to C_k such that $C_i = Z_i$ for all i (under suitable numbering). Let $\tilde{z}_i$ be the seed of C_i and let $\tilde{\sigma}_i = \min_{j \neq i} |\tilde{z}_i - \tilde{z}_j|$. Then, $(1 - 1/n)\sigma_i \leq \tilde{\sigma}_i \leq (1 + 1/n)\sigma_i$ and C_i as well as the root z_i is contained in $\Delta(\tilde{z}_i, \min(\frac{1}{n^2}, \frac{\tilde{\sigma}_i}{256n^2}))$.*

PROOF. Assume that the algorithm has already produced Z_1 to Z_{i-1} and is now run with a seed $\hat{z} \in Z_i$. We prove that it terminates with $C = Z_i$. Let ℓ be a power of two such that $\ell \leq m_i < 2\ell$. The proof consists of two parts. We first assume that steps 2 and 3 are executed for $a = 2\ell$. For this case, we show that the algorithm will terminate with $C = Z_i$. In the second part of the proof, we show that the algorithm does not terminate as long as $a > 2\ell$.

Assume the algorithm reaches steps 2 and 3 with $a/2 = \ell$, i.e. $a/2 \leq m_i < a$. For any approximation $q \in Z_i$, we have $|\hat{z} - q| \leq 2 \cdot 2^{-b/(2m_i)} = 2 \sqrt[m_i/2]{\delta} \leq 2 \sqrt[a/2]{\delta}$. Thus, $Z_i \subseteq C$. Conversely, consider any approximation $q \notin Z_i$. Then, $|\hat{z} - q| \geq 2 \cdot 2^{-b/(16m_i)} > 2 \sqrt[4m_i]{\delta} \geq 2 \sqrt[2a]{\delta}$, and thus no such approximation is contained in C. This shows that $C = Z_i$. Since $|C| \geq a/2$, the algorithm terminates and returns Z_i.

It is left to argue that the algorithm does not terminate before $a/2 = \ell$. Since ℓ and a are powers of two, assume we terminate with $a/2 \geq 2\ell$, and let C be the cluster returned. Then, $m_i < a/2 \leq |C| < a$ and Z_i is a proper subset of C. Consider any approximation $q \in C \setminus Z_i$, say $q \in Z_j$ with $j \neq i$. Since $q \notin Z_i$, we have $|q - \hat{z}| \geq 2 \cdot (2^{-b/(16m_i)} + 2^{-b/(16m_j)}) > 2 \cdot 2^{-b/(16m_i)} > 2 \sqrt[4m_i]{\delta}$. And since $q \in C$, we have $|q - \hat{z}| \leq 2 \sqrt[a/2]{\delta}$. Thus, $4m_i \leq a/2$ and, hence, there are at least $3a/8$ many approximations in $C \setminus Z_i$. Furthermore, $|z_i - z_j| \leq |z_i - \hat{z}| + |\hat{z} - q| + |q - z_j| \leq 2^{-b/(2m_i)} + 2 \sqrt[a/2]{\delta} + 2^{-b/(2m_j)} \leq 2^{-b/(16m_i)} + 2 \sqrt[a/2]{\delta} + 2^{-b/(16m_j)} \leq 3 \sqrt[a/2]{\delta}$. Consequently, there are at least $3a/8$ roots $z_j \neq z_i$ counted with multiplicity within distance $3 \sqrt[a/2]{\delta}$ to z_i. This observation allows us to upper bound the value of $|P_i|$, namely

$$|P_i| = \prod_{j \neq i} |z_i - z_j|^{m_j} \leq (3 \sqrt[a/2]{\delta})^{3a/8} 2^{(n - m_i - 3a/8)\Gamma_p}$$
$$< 3^n \delta^{3/4} 2^{n\Gamma_p} \leq 3^n 2^{-3b/16} 2^{n\Gamma_p} < 3^n 2^{-b/8} \cdot 2^{n\Gamma_p},$$

a contradiction to (8).

We now come to the claims about $\tilde{\sigma}_i$ and the disks defined in terms of it. The relation between σ_i and $\tilde{\sigma}_i$ follows from the second part of Lemma 4. All points in $C_i = Z_i$ have distance at most $2 \cdot 2^{-b/(2m_i)}$ from $\tilde{z}_i$. Also, by (5) and (7),

$$2 \cdot 2^{-b/(2m_i)} < \min(1/n^2, \sigma_i/(512n^2)) \leq \min(1/n^2, \tilde{\sigma}_i/(256n^2))$$

Hence, C_i as well as z_i is contained in $\Delta(\tilde{z}_i, \min(1/n^2, \tilde{\sigma}_i/(256n^2)))$. □

LEMMA 8. *For a fixed precision b, computing a complete clustering needs $\tilde{O}(nb + n^2\Gamma_p)$ bit operations.*

PROOF. For each approximation, we examine the number of distance computations we need to perform. Recall that b (property (4)) and a are powers of two, $a \leq 4n$ by definition, and $b \geq 8n \geq 2a$ by property (4). Then, $\sqrt[a/2]{\delta} = 2^{-b/(2a)} \in 2^{-\mathbb{N}}$. Thus, the number $\sqrt[a/2]{\delta}$ has a very simple format in binary notation. There is a single one, and this one is $b/(2a)$ positions after the binary point. In addition, all approximations $\hat{z}$ have absolute value less than $2 \cdot 2^{\Gamma}$ due to Step 5 in the overall algorithm. Thus, each evaluation of the form $|\hat{z} - q| \leq 2 \sqrt[a/2]{\delta}$ can be done with

$$O(\Gamma + \log \delta^{-2/a}) = O((b/a) + \Gamma) = O((b/a) + \Gamma_p + \log n)$$

bit operations.

For a fixed seed $\hat{z}$, in the i-th iteration of step 2, we have at most $a \leq n/2^{i-2}$ many unclustered approximations left in C, since otherwise we would have terminated in an earlier iteration. Hence, we perform at most a evaluations of the form $|\hat{z} - q| \leq 2 \sqrt[a/2]{\delta}$, resulting in an overall number of bit operations of $a \cdot O((b/a) + \Gamma) = O(b + a\Gamma)$ for a fixed iteration. As we halve a in each iteration, we have at most $\log n + 2$ iterations for a fixed $\hat{z}$, leading to a bit complexity of $O(b \log n + n\Gamma) = \tilde{O}(b + n\Gamma) = \tilde{O}(b + n\Gamma_p)$.

In total, performing a complete clustering has a bit complexity of at most $\tilde{O}(nb + n^2\Gamma_p)$. □

When the clustering succeeds, we have k clusters C_1 to C_k and corresponding seeds $\tilde{z}_1, \ldots, \tilde{z}_k \subseteq \{\hat{z}_1, \ldots, \hat{z}_n\}$. For $i = 1, \ldots, k$, we define $D_i := \Delta(\tilde{z}_i, r_i)$, where $\tilde{z}_i$ is the seed for the cluster C_i and

$$r_i := \min(2^{-\lceil 2\log n \rceil}, 2^{\lceil \log \tilde{\sigma}_i/(256n^2) \rceil}) \geq \min(\frac{1}{2n^2}, \frac{\tilde{\sigma}_i}{256n^2}). \quad (9)$$

In particular, r_i is a 2-approximation of $\min(1/n^2, \tilde{\sigma}_i/(256n^2))$. Notice that the cost for computing the separations $\tilde{\sigma}_i$ is bounded by $\tilde{O}(nb + n^2\Gamma_p)$ bit operations since we can compute the nearest neighbor graph of the points $\tilde{z}_i$ (and thus the values $\tilde{\sigma}_i$) in $O(n \log n)$ steps [9] with a precision of $O(b + n\Gamma)$.

Now, suppose that $b \geq b_0$, Then, according to Lemma 7, the cluster C_i is contained in the disk D_i. Furthermore, D_i contains exactly one root z_i of p (under suitable numbering of the roots), and it holds that $m_i = \text{mult}(z_i, p) = |C_i|$ and $\min(1/(2n^2), \sigma_i/(512n^2)) \leq r_i \leq \min(1/n^2, \sigma_i/(64n^2))$. Hence, before we proceed, we verify that each disk D_i actually contains the cluster C_i. If this is not the case, we report a failure, and return to the factorization with $b := 2b$.

In the next step, we aim to show that each of the enlarged disks $\Delta_i := \Delta(\tilde{z}_i, R_i) := \Delta(\tilde{z}_i, nr_i)$, $i = 1, \ldots, k$, contains exactly one root z_i of p, and that the number of elements in $C_i \subseteq \Delta_i$ equals the multiplicity of z_i. Notice that, from the definition of r_i and Δ_i, it obvious that the disks Δ_i are pairwise disjoint and that $C_i \subseteq D_i \subseteq \Delta_i$.

3.3 Certification

In order to show that Δ_i contains exactly one root of p with multiplicity $|C_i|$, we show that each Δ_i contains the same number of roots of p and $\hat{p}$ counted with multiplicity. For the latter, we compute a lower bound for $|\hat{p}(z)|$ on the boundary $\text{bd}\,\Delta_i$ of Δ_i, and check whether this bound is larger than $|(\hat{p}-p)(z)|$ for all points $z \in \text{bd}\,\Delta_i$. If this is the case, then we are done according to Rouché's theorem. Otherwise, we start over the factorization algorithm with $b = 2b$. We now come to the details:

1. Let $\lambda = 2^{l_\lambda}$ be the 2-approximation of $\|p\|/|p_n|$ as defined in step 2 of the overall algorithm.

2. For $i = 1, \ldots, k$, let $z_i^* := \tilde{z}_i + n \cdot r_i \in \Delta_i$. Note that $|z_i^*| \leq (1+1/n) \cdot M(\tilde{z}_i)$ since $nr_i \leq 1/n$.

3. We try to establish the inequality
$$|\hat{p}(z_i^*)/p_n| > E_i := 32 \cdot 2^{-b} \lambda M(\tilde{z}_i)^n \qquad (10)$$
for all i. If this check is satisfied, we know that each disk Δ_i contains exactly one root z_i of p and that its multiplicity equals the number $|C_i|$ of approximations in Δ_i (Lemma 10). In order to establish the inequality, we consider $\rho = 1, 2, 4, \ldots$ and compute $|\hat{p}(z_i^*)/p_n|$ to an absolute error less than $2^{-\rho}$. If, for all $\rho \leq b$, we fail to show that $|p(z_i^*)/p_n| > E_i$, we again return to the factorization step with $b := 2b$. Otherwise, let ρ_i be the smallest ρ for which we are successful.

4. If, at any stage of the algorithm, $\sum_i \rho_i > b$, we also report a failure and go back to the factorization step with $b := 2b$.

5. If we can verify that $|\hat{p}(\tilde{z}_i + nr_i)/p_n| > E_i$ for all i, we return the disks Δ_i and the multiplicities $m_i = |C_i|$.

LEMMA 9. *For any i, we can compute $|\hat{p}(z_i^*)/p_n|$ to an absolute error less than $2^{-\rho}$ with a number of bit operations less than*
$$\tilde{O}(n(n+\rho+n\log M(\tilde{z}_i)+\tau_p)).$$
For a fixed b, the total cost for all evaluations in the above certification step is bounded by $\tilde{O}(nb+n^2\tau_p+n^3)$.

PROOF. Consider an arbitrary subset $S \subseteq \{\hat{z}_1, \ldots, \hat{z}_n\}$. We first derive an upper bound for $\prod_{\hat{z} \in S} |z_i^* - \hat{z}|$. For that, consider the polynomial $\hat{p}_S(x) := \prod_{\hat{z} \in S}(x - \hat{z})$. The i-th coefficient of $\hat{p}_S$ is bounded by $\binom{|S|}{i} \cdot \prod_{\hat{z} \in S} M(\hat{z}) \leq 2^n \prod_{i=1}^n M(\hat{z}_i) \leq 8\lambda \cdot 2^n$ due to Step 6 in the overall algorithm. It follows that
$$\prod_{\hat{z} \in S} |z_i^* - \hat{z}| = |\hat{p}_S(z_i^*)| \leq (n+1) M(z_i^*)^n \cdot 8\lambda \cdot 2^n < 2^{4n+\tau_p+6} M(\tilde{z}_i)^n,$$
where we used that $M(z_i^*) < (1+1/n) \cdot M(\tilde{z}_i)$ and $\lambda < 2\|p\|/|p_n| < 2^{\tau_p+1}(n+1)$. In order to evaluate $|\hat{p}(z_i^*)/p_n| = \prod_{j=1}^n |z_i^* - \hat{z}_j|$, we use approximate interval evaluation with an absolute precision $K = 1, 2, 4, \ldots$. More precisely, we compute the distance of z_i^* to each of the points $\hat{z}_j$, $j = 1, \ldots, n$, up to an absolute error of 2^{-K}, and then take the product over all distances using a fixed point precision of K bits after the binary point.[4] We stop when the resulting interval has size less than $2^{-\rho}$. The above consideration shows that all intermediate results have at most $O(n + \tau_p + n\log M(\tilde{z}_i))$ bits before the binary point. Thus, we eventually succeed for a $K = O(\rho + \tau_p + n + n\log M(\tilde{z}_i))$. Since we perform n subtractions and n multiplications, the cost is bounded by $\tilde{O}(nK)$ bit operations for each K. Hence, the bound for evaluating $|\hat{p}(z_i^*)/p_n|$ follows.

We come to the second claim. Since we double ρ in each iteration and consider at most $\log b$ iterations, the cost for the evaluation of $|\hat{p}(z_i^*)/p_n|$ are bounded by $\tilde{O}(n(n+\rho_i+n\log M(\tilde{z}_i)+\tau_p))$. Since we ensure that $\sum_i \rho_i \leq b$, it follows that the total cost is bounded by $\tilde{O}(nb + n^2\tau_p + n^3 + n^2 \log(\prod_{i=1}^k M(\tilde{z}_i)))$. The last summand is smaller than $n^2 \cdot 8\lambda$ according to Step 6 in the overall algorithm, and $\lambda < 2\|p\|/|p_n| < 2(n+1)2^{\tau_p}$. This shows the claim. □

We now show that inequality (10) implies that the disk Δ_i contains the same number of roots of the polynomials $\hat{p}$ and p.

LEMMA 10. *1. If inequality (10) holds for all i, then Δ_i isolates a root of z_i of p of multiplicity $m_i = \text{mult}(z_i, p) = |C_i|$.*

2. If $b \geq b_0$, then
$$\frac{|\hat{p}(z_i^*)|}{|p_n|} > \left(\frac{\min(256, \sigma_i)}{1024n}\right)^{m_i} \cdot \frac{|P_i|}{8} \geq 64 \cdot 2^{-b} \lambda M(\tilde{z}_i)^n$$

PROOF. We first show that $|\hat{p}(x)| \geq \frac{1}{4}|\hat{p}(z_i^*)|$ for all $x \in \text{bd}\,\Delta_i$. We fix an approximation $\hat{z}$ in some disk D_j. Suppose that x is the farthest point on $\text{bd}\,\Delta_i$ from $\hat{z}$, and y the nearest. Then, for $i \neq j$, we have $|x - \hat{z}| \leq |x - \tilde{z}_i| + |\tilde{z}_i - \tilde{z}_j| + |\tilde{z}_j - \hat{z}| \leq (1+1/n)|\tilde{z}_i - \tilde{z}_j|$, and $|y - \hat{z}| \geq |\tilde{z}_i - \tilde{z}_j| - |y - \tilde{z}_i| - |\tilde{z}_j - \hat{z}| \geq (1-1/n)|\tilde{z}_i - \tilde{z}_j|$. Similarly, for $i = j$, it holds that $|x - \hat{z}| \leq |x - \tilde{z}_i| + |\tilde{z}_i - \hat{z}| \leq (1+1/n)nr_i$ and $|y - \hat{z}| \geq |y - \tilde{z}_i| - |\tilde{z}_i - \hat{z}| \geq (1-1/n)nr_i$. Hence, for arbitrary points $x, y \in \text{bd}\,\Delta_i$ and an arbitrary approximation $\hat{z}$, it follows that
$$(1-1/n)|y - \hat{z}| \leq |x - \hat{z}| \leq (1+1/n)|y - \hat{z}|\,.$$
We conclude that $(1-1/n)^n|\hat{p}(z_i^*)| \leq |\hat{p}(x)| \leq (1+1/n)^n|\hat{p}(z_i^*)|$ for all $x \in \text{bd}\,\Delta_i$. This shows the above claim.

We can now prove Part 1 of the lemma. We have $|x| < (1+1/n)M(\tilde{z}_i)$ for all $x \in \text{bd}\,\Delta_i$ since $nr_i < 1/n$,. Now, if $|\hat{p}(z_i^*)|/|p_n| > 32 \cdot 2^{-b}\lambda M(\tilde{z}_i)^n$, then
$$|\hat{p}(x)| > |\hat{p}(z_i^*)|/4 > 8|p_n|\lambda 2^{-b}(1-1/n)^n M(x)^n$$
$$> \|p\| 2^{-b} M(x)^n \geq \|\hat{p} - p\| M(x)^n \geq |\hat{p}(x) - p(x)|.$$
Hence, according to Rouché's theorem. Δ_i contains the same number (namely, $|C_i|$) of roots of p and $\hat{p}$. If this holds for all disks Δ_i, then each of the disks must contain exactly one root since p has k distinct roots. In addition, the multiplicity of each root equals the number $|C_i|$ of approximations within Δ_i.

It remains to show the third claim. Since $b \geq b_0$, it follows that $\min(1/(2n^2), \sigma_i/(512n^2)) \leq r_i \leq \min(1/n^2, \sigma_i/(64n^2))$ and $|\tilde{z}_i - z_i| < r_i$; cf. the remark following the definition of r_i in (9). Thus,
$$\begin{aligned}
|\hat{p}(z_i^*)| &\geq |p(z_i^*)| - 2^{-b}\|p\| \cdot M(z_i^*)^n \\
&= |p(z_i + (\tilde{z}_i - z_i + nr_i))| - 2^{-b}\|p\| \cdot M(z_i^*)^n \\
&\geq ((n-1)r_i)^{m_i}|p_n P_i|/4 - 4 \cdot 2^{-b}\|p\| M(z_i)^n \\
&\geq \left(\frac{(n-1)\min(256, \sigma_i)}{512n^2}\right)^{m_i} |p_n P_i|/4 - 4 \cdot 2^{-b}\|p\| M(z_i)^n \\
&\geq \left(\frac{\min(256, \sigma_i)}{1024n}\right)^{m_i} |p_n P_i|/4 - 4 \cdot 2^{-b}\|p\| M(z_i)^n,
\end{aligned}$$

[4]In fact, we compute an interval I_j of size less than 2^{-K} such that $|z_i^* - \hat{z}_j| \in I_j$, and then consider the product $\prod_j I_j$.

where the first inequality is due to $|(p-\hat{p})(x)| < 2^{-b}\|p\| \cdot M(x)^n$, the second inequality follows from $|\tilde{z}_i - z_i + nr_i| \le (n+1)r_i \le \sigma_i/n$, Lemma 3 and $M(z_i^*) < (1+1/n)\cdot M(z_i)$, and the third inequality follows from $r_i \ge \min(\frac{1}{2n^2}, \frac{\sigma_i}{512n^2})$. In addition, we have

$$2^{-b}\|p\| M(z_i)^n \le \left(\frac{\min(256,\sigma_i)}{1024n}\right)^{m_i} \cdot \frac{|p_n P_i|}{4096}, \qquad (11)$$

since

$$\begin{aligned} 2^{-b}&\|p\| M(z_i)^n \\ &\le 2^{-b/8}\cdot 2^{-b/2}\cdot 2^{\tau_p}|p_n|\cdot(n+1)\cdot M(z_i)^n \\ &\le \frac{|P_i|}{(n+1)2^{2n\Gamma_p+8n}}\left(\frac{\min(256,\sigma_i)}{1024n}\right)^{m_i} 2^{\tau_p}|p_n|(n+1)M(z_i)^n \\ &\le \left(\frac{\min(256,\sigma_i)}{1024n}\right)^{m_i}\cdot\frac{|p_n P_i|}{2^{7n-1}} \le \left(\frac{\min(256,\sigma_i)}{1024n}\right)^{m_i}\cdot\frac{|p_n P_i|}{4096}, \end{aligned}$$

where the second inequality follows from (8), (7), and (5) [5], and the third inequality follows from $\tau_p \le n\Gamma_p + n + 1$ (Lemma 1) and $M(z_i)^n \le 2^{n\Gamma_p}$. Finally,

$$\begin{aligned} \frac{|\hat{p}(z_i^*)|}{|p_n|} &> \left(\frac{\min(256,\sigma_i)}{1024n}\right)^{m_i}\cdot\frac{|P_i|}{8} \ge 512\cdot 2^{-b}\frac{\|p\|}{|p_n|}M(z_i)^n \\ &\ge 64\cdot 2^{-b}\lambda M(\hat{z}_i)^n, \end{aligned}$$

where the first and the second inequality follow from (11) and the third inequality holds since λ is a 2-approximation of $\|p\|/|p_n|$ and $|z_i|^n \le (1+1/n)^n|\tilde{z}_i|^n \le 4|\tilde{z}_i|^n$. □

LEMMA 11. *There exists a $b^* \ge b_0$ bounded by*

$$O\left(n\log n + n\Gamma_p + \sum_{i=1}^k\left(\log M(P_i^{-1}) + m_i \log M(\sigma_i^{-1})\right)\right)$$

such that the certification step succeeds for any $b > b^*$. The total cost in the certification algorithm (i.e. for all iterations until we eventually succeed) is bounded by

$$\tilde{O}\left(n^3 + n^2\tau_p + n\cdot\sum_{i=1}^k\left(\log M(P_i^{-1}) + m_i\log M(\sigma_i^{-1})\right)\right)$$

bit operations.

PROOF. Due to Lemma 10, $|\hat{p}(z_i^*)/p_n| > \left(\frac{\min(256,\sigma_i)}{1024n}\right)^{m_i}\cdot\frac{|P_i|}{8} > 64\cdot 2^{-b_0}\lambda M(\hat{z}_i)^n$. Thus, in order to verify inequality (10), it suffices to evaluate $|\hat{p}(z_i^*)/p_n|$ to an error of less than $|\hat{p}(z_i^*)/2p_n|$. It follows that we succeed for some ρ_i with

$$\rho_i = O(m_i\log n + m_i\max(1,\log\sigma_i^{-1}) + \log\max(1,|P_i|^{-1})).$$

In Step 3 of the certification algorithm, we require that the sum over all ρ_i does not exceed b. Hence, we eventually succeed in verifying the inequality (10) for all i if b is larger than some b^* with

$$\begin{aligned} b^* &= O(b_0 + \textstyle\sum_i m_i\log n + \sum_i(\log M(P_i^{-1}) + m_i\log M(\sigma_i^{-1}))) \\ &= O(n\log n + n\Gamma_p + \textstyle\sum_i(\log M(P_i^{-1}) + m_i\log M(\sigma_i^{-1}))). \end{aligned}$$

For the bound for the overall cost, we remark that, for each b, the certification algorithm needs $\tilde{O}(n^3 + nb + n^2\tau_p)$ bit operations due to Lemma 9. Thus, the above bound follows from the fact that that we double b in each step and that the certification algorithm succeeds under guarantee for all $b > b^*$. □

[5] Observe $2^{-b/(2m_i)} \le \min(\frac{1}{2n^2}, \frac{\sigma_i}{1024n^2}) \le \frac{\min(256,\sigma_i)}{1024n}$.

4. COMPLEXITY ANALYSIS

We now turn to the complexity analysis of the root isolation algorithm. In the first step, we provide a bound for general polynomials p with complex coefficients. In the second step, we give a simplified bound for the special case, where p has integer coefficients. We also give bounds for the number of bit operations that is needed to refine the isolating disks to a size less than $2^{-\kappa}$, with κ an arbitrary positive integer.

THEOREM 3. *Let $p(x) \in \mathbb{C}[x]$ be a polynomial as defined in Section 2. We assume that the number k of distinct roots of p is given. Then, for all $i = 1,\dots,k$, the algorithm from Section 3 returns an isolating disk $\Delta(\tilde{z}_i, R_i)$ for the root z_i and the corresponding multiplicity m_i, and it holds that $R_i < \frac{\sigma_i}{64n}$.*
For that, it uses a number of bit operations bounded by

$$\tilde{O}\left(n^3 + n^2\tau_p + n\cdot\sum_{i=1}^k\left(\log M(P_i^{-1}) + m_i\log M(\sigma_i^{-1})\right)\right) \qquad (12)$$

The algorithm needs an approximation of precision L of p, with

$$L = O\left(n\Gamma_p + \sum_{i=1}^k\left(\log M(P_i^{-1}) + m_i\log M(\sigma_i^{-1})\right)\right). \qquad (13)$$

PROOF. For a fixed b, let us consider the cost for each of the steps in the algorithm: Steps 1-3, 5 and 6 do not use more than $\tilde{O}(n^2\Gamma_p + nb)$ bit operations (Theorem 1 and Lemma 2). The Steps 4 and 7 do not use more than $\tilde{O}(n^2\Gamma_p + nb)$ bit operations (Corollary 2 and Lemma 8), and the Steps 8 and 9 use a number of bit operations bounded by (12) (Lemma 11).

Furthermore, the oracle must provide an approximation of precision $O(n\Gamma_p + b)$ of p in order to compute the bound Γ for Γ_p, to compute the 2-approximation λ of $\|p\|/|p_n|$, and to run Pan's algorithm. The algorithm succeeds in computing isolating disks if $b > b^*$ with a b^* as in Lemma 11. Since we double b in each step, we need at most $\log b^*$ iterations and the total cost for each iteration is bounded by (12). This shows the complexity result.

It remains to prove the bound for R_i. When the clustering succeeds, it returns disks $D_i = \Delta(\tilde{z}_i, r_i)$ with $\min(\frac{1}{2n^2}, \frac{\tilde{\sigma}_i}{256n^2}) \le r_i \le \min(\frac{1}{n^2}, \frac{\tilde{\sigma}_i}{128n^2})$ for all $i = 1,\dots,m$. It follows that $R_i = n\cdot r_i \le \frac{\tilde{\sigma}_i}{128n}$, and thus $|z_i - z_j| \ge |\tilde{z}_i - \tilde{z}_j| - |z_i - \tilde{z}_i| - |z_j - \tilde{z}_j| > |\tilde{z}_i - \tilde{z}_j|\cdot(1 - 1/(64n)) > |\tilde{z}_i - \tilde{z}_j|/2$ for all i,j with $i \ne j$. We conclude that $\sigma_i > \tilde{\sigma}_i/2 \ge 64nR_i$. □

We remark that the bound (12) can be reformulated in terms of values that exclusively depend on the degree n and the geometry of the roots (i.e. their absolute values and their separations). Namely, due to Lemma 1, $\tau_p \le n + 1 + \log\frac{\mathrm{Mea}(p)}{|p_n|}$, and the latter expression only involves the degree and the absolute values of the roots of p. This yields the bound (2) that we stated in the introduction.

Next, we show that combining our algorithm with Pan's factorization algorithm also yields a very efficient method to further refine the isolating disks.

THEOREM 4. *Let $p(x)$ be a polynomial as in Theorem 3, and κ be a given positive integer. We can compute isolating disks $\Delta_i(\tilde{z}_i, R_i)$ with radius $R_i < 2^{-\kappa}$ in a number of bit operations bounded by*

$$\mathcal{B} + \tilde{O}(n\kappa\cdot\max_{1\le i\le k} m_i), \qquad (14)$$

where $\mathcal{B}$ is bounded by (12). For that, we need an approximation of precision L of p with $L = \mathcal{L} + \tilde{O}(n\kappa\cdot\max_{1\le i\le k} m_i)$, where $\mathcal{L}$ is bounded by (13).

PROOF. As a first step, we use the algorithm from Section 3 to compute isolating disks $\Delta_i = \Delta(\tilde{z}_i, R_i)$ with $R_i \le \sigma_i/(64n)$. Each disk Δ_i contains the root z_i, $m_i = \text{mult}(z_i, p)$ approximations $\hat{z} \in \{\hat{z}_1, \ldots, \hat{z}_n\}$ of z_i, and it holds that $\sigma_i/2 < \tilde{\sigma}_i < 2\sigma_i$. We further define $\hat{P}_i := \prod_{j:\hat{z}_j \notin \Delta_i}(\tilde{z}_i - \hat{z}_j)$. Since $|\tilde{z}_i - z_i| < \sigma_i/(64n)$ for all i, we have $(1 - \frac{1}{64n})|z_i - z_j| \leq |\tilde{z}_i - \hat{z}| \leq (1 + \frac{1}{64n})|z_i - z_j|$ for all $j \neq i$ and $\hat{z} \in \Delta_j$. Thus, $|\hat{P}_i|$ is a 2-approximation of $|P_i|$, that is, $1/2|P_i| < |\hat{P}_i| < 2|P_i|$. Similar as in the certification step, we now use approximate interval arithmetic to compute a 2-approximation μ_i of $|\hat{P}_i|$, and thus a 4-approximation of $|P_i|$. A completely similar argument as in the proofs of Lemma 9 and Lemma 11 then shows that we can compute such μ_i's with less than $\tilde{O}(n^3 + n^2\tau_p + n\sum_i \log M(P_i^{-1}))$ bit operations. Now, from the 2- and 4-approximations of σ_i and $|P_i|$, we can determine a b_κ such that (A) the properties (4) to (6) are fulfilled, and, in addition, (B) the inequality $2^{-b/(2m_i)} < 2^{-\kappa}$ holds. Then, from Corollary 2 and Lemma 4, we conclude that Pan's factorization algorithm (if run with $b \geq b_\kappa$) returns, for all i, m_i approximations $\hat{z}$ of z_i with $|\hat{z} - z_i| < 2^{-b/(2m_i)} < 2^{-\kappa}$. Thus, for each i, we can simply choose an arbitrary approximation $\hat{z} \in \Delta_i$ and return the disk $\Delta(\hat{z}, 2^{-\kappa})$ which isolates z_i. The total cost splits into the cost for the initial root isolation and the cost for running Pan's Algorithm with $b = b_\kappa$. Since the latter cost is bounded by $\tilde{O}(nb_\kappa + n^2\Gamma_p)$, the bound (14) follows. □

Finally, we apply the above results to the important special case, where p is a polynomial with integer coefficients.

THEOREM 5. *Let $p(x) \in \mathbb{Z}[x]$ be a polynomial of degree n with integer coefficients of size less than 2^τ. Then, we can compute isolating disks $\Delta(\tilde{z}_i, R_i)$, with $R_i < \frac{\sigma_i}{64n}$, for all roots z_i together with the corresponding multiplicities m_i using*

$$\tilde{O}(n^3 + n^2\tau) \tag{15}$$

bit operations. For a given positive integer κ, we can further refine all disks Δ_i to a size of less than $2^{-\kappa}$ with a number of bit operations bounded by $\tilde{O}(n^3 + n^2\tau + n\kappa)$.

PROOF. In a first step, we compute the square-free part p^* of p. According to [21, §11.2], we need $\tilde{O}(n^2\tau)$ bit operations for this step, and p^* has integer coefficients of bitsize $O(n+\tau)$. The degree of p^* yields the number k of distinct roots of p. Thus, we can directly apply our algorithm from Section 3 to the polynomial p. In order to derive the bound in (15), we have to reformulate the bound from (12) in terms of the degree n and the bitsize τ of p. From [7, Theorem 2], we conclude that $\sum_i m_i \log\max(1, \sigma_i^{-1}) = \tilde{O}(n^2 + n\tau)$. Furthermore, we have $\tau_p \leq \tau$. Finally, we can bound $\sum_{i=1}^k \log M(P_i^{-1})$ by $\tilde{O}(n^3 + n^2\tau)$. For that, we use a similar approach as in the proof of [7, Theorem 2]. That is, we consider a square-free factorization $p(x) = \prod_{l=1}^n (Q_l(x))^l$ of p and write each partial product $\prod_{i:Q_l(z_i)=0} P_i$ in terms of the leading coefficients of Q_l and the resultant $\text{res}(Q_l, Q_l') \in \mathbb{Z}$ of Q_l and its derivative Q_l'. For more details, we refer to [15] and [7].

We turn to the proof of the bound for the cost of refining the isolating disks $\Delta_i(\tilde{z}_i, R_i)$ to a size of less than $2^{-\kappa}$. For the refinement, we consider the square-free part p^*. Note that the disks Δ_i obtained in the first step are also isolating for the roots of p^* (p and p^* have exactly the same distinct roots) and that $R_i < \sigma(z_i, p)/(64n) = \sigma(z_i, p^*)/(64n) \leq \sigma(z_i, p^*)/(64 \deg p^*)$. Thus, proceeding in completely analogous manner as in the proof of Theorem 4 (with the square-free part p^* instead of p) shows that we need $\tilde{O}(n^3 + n^2\tau + n\kappa)$ bit operations for the refinement. □

5. REFERENCES

[1] D. S. Arnon, G. E. Collins, and S. McCallum. Cylindrical Algebraic Decomposition I. *SIAM Journal of Computing*, 13(4):865–889, 1984.

[2] E. Berberich, P. Emeliyanenko, A. Kobel, and M. Sagraloff. Exact Symbolic-Numeric Computation of Planar Algebraic Curves. *CoRR*, abs/1201.1548, 2012.

[3] E. Berberich, M. Kerber, and M. Sagraloff. An efficient algorithm for the stratification and triangulation of an algebraic surface. *Comput. Geom.*, 43(3):257–278, 2010.

[4] D. Bini and G. Fiorentino. Design, Analysis, and Implementation of a Multiprecision Polynomial Rootfinder. *Numerical Algorithms*, 23:127–173, 2000.

[5] D. I. Diochnos, I. Z. Emiris, and E. P. Tsigaridas. On the Asymptotic and Practical Complexity of Solving Bivariate Systems Over the Reals. *J. Symb. Comput.*, 44(7):818–835, 2009.

[6] A. Eigenwillig, M. Kerber, and N. Wolpert. Fast and Exact Analysis of Real Algebraic Plane Curves. In *ISSAC*, pages 151–158, New York, NY, USA, 2007. ACM.

[7] P. Emeliyanenko and M. Sagraloff. On the Complexity of Solving a Bivariate Polynomial System. In *ISSAC*, pages 154–161, New York, NY, USA, 2012. ACM.

[8] I. Z. Emiris, V. Y. Pan, and E. P. Tsigaridas. Algebraic Algorithms. available at `tr.cs.gc.cuny.edu/tr/files/TR-2012001.pdf`, 2012.

[9] D. Eppstein, M. Paterson, and F. Yao. On Nearest-Neighbor Graphs. *Discrete & Comput. Geometry*, 17:263–282, 1997.

[10] L. Gonzalez-Vega and M. E. Kahoui. An Improved Upper Complexity Bound for the Topology Computation of a Real Algebraic Plane Curve. *J. Complexity*, 12(4):527–544, 1996.

[11] H. Hong. An Efficient Method for Analyzing the Topology of Plane Real Algebraic Curves. *Mathematics and Computers in Simulation*, 42(4-6):571–582, 1996.

[12] M. Kerber and M. Sagraloff. Efficient Real Root Approximation. In *ISSAC*, pages 209–216, New York, NY, USA, 2011. ACM.

[13] M. Kerber and M. Sagraloff. A Worst-case Bound for Topology Computation of Algebraic Curves. *J. Symb. Comput.*, 47(3):239–258, 2012.

[14] K. Mehlhorn and M. Sagraloff. A Deterministic Descartes Algorithm for Real Polynomials. *J. Symb. Comput.*, 46(1):70 – 90, 2011. A preliminary version appeared in ISSAC 2009.

[15] K. Mehlhorn, M. Sagraloff, and P. Wang. From Approximate Factorization to Root Isolation with Application to Cylindrical Algebraic Decomposition. *CoRR*, abs:1301.4870, 2013. `http://arxiv.org/abs/1301.4870`.

[16] V. Pan. Univariate Polynomials: Nearly Optimal Algorithms for Numerical Factorization and Root Finding. *J. Symb. Comput.*, 33(5):701–733, 2002.

[17] M. Sagraloff. On the Complexity of Real Root Isolation. *CoRR*, abs/1011.0344, 2010.

[18] M. Sagraloff. When newton meets descartes: a simple and fast algorithm to isolate the real roots of a polynomial. In *ISSAC*, pages 297–304, 2012.

[19] A. Schönhage. The Fundamental Theorem of Algebra in Terms of Computational Complexity. Technical report, Math. Inst. Univ. Tübingen, 1982.

[20] A. Strzebonski. Cylindrical Algebraic Decomposition Using Validated Numerics. *J. Symb. Comp*, 41:1021–1038, 2006.

[21] J. von zur Gathen and J. Gerhard. *Modern Computer Algebra*. Cambridge University Press, Cambridge, 1999.

[22] C. K. Yap and M. Sagraloff. A Simple but Exact and Efficient Algorithm for Complex Root Isolation. In *ISSAC*, pages 353–360, New York, NY, USA, 2011. ACM.

The Termination of the F5 Algorithm Revisited

Senshan Pan Yupu Hu Baocang Wang
State Key Laboratory of Integrated Service Networks
Xidian University, Xi'an, China 710071
pansenshan@gmail.com, yphu@mail.xidian.edu.cn, bcwang79@yahoo.com.cn

ABSTRACT

The F5 algorithm [8] is generally believed as one of the fastest algorithms for computing Gröbner bases. However, its termination problem is still unclear. The crux lies in the non-determinacy of the F5 in selecting which from the critical pairs of the same degree. In this paper, we construct a generalized algorithm F5GEN which contain the F5 as its concrete implementation. Then we prove the correct termination of the F5GEN algorithm. That is to say, for any finite set of homogeneous polynomials, the F5 terminates correctly.

Categories and Subject Descriptors

I.1.2 [**Symbolic and Algebraic Manipulation**]: Algorithms – Analysis of algorithms

Keywords

Gröbner basis; termination; F5; F5GEN

1. INTRODUCTION

In 1965 Buchberger's [3] thesis he described the appropriate framework for the study of polynomial ideals, with the introduction of Gröbner bases. Since then, Gröbner basis has become a fundamental tool of computational algebra and it has found countless applications in coding theory, cryptography and even directions of Physics, Biology and other sciences.

Although Buchberger presented several improvements to his algorithm for computing Gröbner bases in [4], the efficiency is not so good. Recent years have seen a surge in the number of algorithms in computer algebra research, but efficient ones are few. Faugère [8] proposed the idea of signatures and utilized two powerful criteria to avoid useless computation in the F5 algorithm. Faugère and Joux broke the first Hidden Field Equation (HFE) Cryptosystem Challenge (80 bits) by using the F5 algorithm in [9]. The proof of the termination in [8] is based on the hypothesis that the input polynomials are homogeneous and regular, which was labeled as a conjecture in [15]. Gash [12] pointed out that Theorem 2 in [8] is false and he proposed another conjecture for the termination of the F5. The proof under that conjecture, we will show in Section 6.4, can be viewed as the proof for a possible implementation of the F5. In [1], the authors did an inspiring work by constructing a simpler algorithm and proving its termination. [13] gave a generalized TRB algorithm and proposed the "compatible" concept that sheds light on the sufficient and necessary conditions for the termination of the TRB. The author claimed to have proved the termination of the F5, but it can also be viewed as a partial proof of the original F5. Besides, his proofs for the correctness are hard to understand due to mistakes. Though the F5 algorithm seems to terminate for any homogeneous polynomial ideals, the proof of it has been admitted as an open problem in [16, 7, 5]. After our preliminary paper appeared on the arXiv we have learned that independently of our work here, Vasily Galkin tried to give a direct proof of the F5 in [10] without any modifications. His proofs are different from ours, but they are slightly too long to understand for us.

In this paper, we show that the reason why "compatible" property is implicitly satisfied between the monomial order and the module order in almost all signature-based algorithms. We propose the F5GEN algorithm (F5 algorithm with a generalized insertion strategy) to cover the behavior of original F5 of [8]. Then we prove that the F5GEN terminates correctly, which, on the other hand, shows the correct termination of the F5.

The paper is organized as follows. We start by settling basic notations in Section 2. In Section 3, we present the strict definition of the admissible module order. Then two admissible orders and their connection are described in Section 4 and under the "compatible" condition, the top-reduced S-Gröbner basis for a polynomial ideal is proved finite. After that, the F5GEN algorithm is described and its correct termination is proved in Section 5. We compare the F5 of [8] with the F5GEN, and show that the F5 implements the F5GEN in Section 6. In Section 7, we conclude this paper with an open problem.

2. PRELIMINARIES

Let $R = K[x_1, \dots, x_n]$ be the polynomial ring in n variables over the field K. An **admissible monomial order** $\leq_m$ on the monoid $\mathcal{M} = \{\Pi_{i=1}^n x_i^{a_i} \mid a_i \in \mathbb{N}\}$ is a linear order (i.e. a connex, reflexive, antisymmetric and transitive order) such that (i) $1 \leq_m s, \forall s \in \mathcal{M}$, (ii) $m_1 \leq_m m_2 \Rightarrow m_1 \cdot s \leq_m m_2 \cdot s$, $\forall s, m_1, m_2 \in \mathcal{M}$.

ISSAC'13, June 26–29, 2013, Boston, Massachusetts, USA.

It can be seen that the admissible order $\leq_m$ is a well-order on $\mathcal{M}$. Sometimes we write $\leq$ for $\leq_m$ for brevity. For any $p \in R$, without confusion, we denote $HM(p)$ (resp. $HT(p)$, $HC(p)$) for the head monomial (resp. head term, head coefficient) of p.

Let $\mathcal{I}$ be the ideal generated by the set $F = \{f_1, \ldots, f_d\} \in R$, i.e.,

$$\mathcal{I} =< f_1, \ldots, f_d >= \{p_1 f_1 + \cdots + p_d f_d \,|\, p_1, \ldots, p_d \in R\}.$$

Consider the following R-submodule of $R^d \times R$:

$$\mathcal{P} = \{(\mathbf{u}, p) \in R^d \times R \,|\, \mathbf{u} \cdot \mathbf{f} = p\}, \qquad (1)$$

where $\mathbf{f} = (f_1, \ldots, f_d) \in R^d$ and $\mathbf{e}_i$ is ith unit vector of R^d such that the free R-module R^d is generated by the set $\Sigma = \{\mathbf{e}_1, \ldots, \mathbf{e}_d\}$. The element $\alpha = (\mathbf{u}, p)$ in $\mathcal{P}$ we call a **sig-polynomial** and $poly(\alpha)$ is its polynomial part p. A subset $Syz = \{(\mathbf{u}, 0) \in \mathcal{P}\}$ is defined by the **syzygy submodule** for $\mathcal{P}$, and $NS = \mathcal{P} \setminus Syz$ is called the set of **non-syzygy sig-polynomials**. Let $(\mathbf{u}_1, p_1)$ and $(\mathbf{u}_2, p_2)$ be two sig-polynomials in NS. The module generated by syzygies of the form $(p_2\mathbf{u}_1 - p_1\mathbf{u}_2, 0)$ is called a **principal syzygy submodule** PS.

Other basic concepts not presented here can be found in [2].

3. THE ADMISSIBLE MODULE ORDER

Let $\preceq$ be a quasi-order (i.e. a reflexive and transitive order) on M and $N \subseteq M$. Then a subset B of N is called a **Dickson basis** of N w.r.t. $\preceq$ if for every $a \in N$ there exists some $b \in B$ with $b \preceq a$. We say that $\preceq$ has the **Dickson property**, or is a **Dickson quasi-order**, if every subset N of M has a finite Dickson basis w.r.t. $\preceq$.

If $\preceq$ is a (Dickson) quasi-order on M, then we call $(M, \preceq)$ a (Dickson) quasi-ordered set. For example, the natural order $\leq$ on $\mathbb{N}$ has the Dickson property. Similarly, the divisibility relation on $\{x^a \,|\, a \in \mathbb{N}\}$ is a Dickson quasi-order since the map $a \mapsto x^a$ is an isomorphism between $(\mathbb{N}, \leq)$ and $(\{x^a \,|\, a \in \mathbb{N}\}, |)$. Let now $(M, \preceq)$ and $(N, \preceq')$ be quasi-ordered sets, then a quasi-order $\preceq''$ on Cartesian product $M \times N$ is defined as follows:

$$(a, b) \preceq'' (c, d) \;\Leftrightarrow\; a \preceq b \text{ and } c \preceq' d,$$

$\forall (a, b), (c, d) \in M \times N$. The **direct product** of the quasi-order sets $(M, \preceq)$ and $(N, \preceq')$ is denoted by $(M \times N, \preceq'')$. The Dickson property can be derived as follows.

Lemma 1. *[2] Let $(M, \preceq)$ and $(N, \preceq')$ be Dickson quasi-ordered sets, $(M \times N, \preceq'')$ their direct product. Then $(M \times N, \preceq'')$ is a Dickson quasi-ordered set.*

The immediate corollary is that $(\mathbb{N}^n, \leq')$, the direct product of n copies of the natural numbers $(\mathbb{N}, \leq)$ with their natural ordering, is a Dickson quasi-ordered set. This is Dickson's lemma, and another version of which is given below by an isomorphism: $(a_1, \ldots, a_n) \mapsto \prod_{i=1}^n x_i^{a_i}$, where $(a_1, \ldots, a_n) \in \mathbb{N}^n$.

Corollary 1. *[2] The divisibility relation $|$ is a Dickson quasi-order on $\mathcal{M} = \{\prod_{i=1}^n x_i^{a_i} \,|\, a_i \in \mathbb{N}\}$. More explicitly, every non-empty subset S of $\mathcal{M}$ has a finite subset B such that for all $s \in S$, there exists $t \in B$ with $t \,|\, s$.*

Let $\mathcal{M}_d = \{m\mathbf{e}_i \,|\, m \in \mathcal{M}, i \in [1, d]\} \in R^d$ be the $\mathcal{M}$-monomodule. The definition of the divisibility relation $|$ on $\mathcal{M}_d$ is

$$m_1\mathbf{e}_i \,|\, m_2\mathbf{e}_j \Leftrightarrow m_1 | m_2 \text{ and } i = j \in [1, d]. \qquad (2)$$

Since $(\mathcal{M}, |)$ is a Dickson quasi-ordered set, by Lemma 1, $(\mathcal{M}_d, |)$ is also a Dickson quasi-ordered set. On $\mathcal{M}_d$, we will define the admissible order similarly.

Definition 1. An **admissible module order** $\leq_s$ is a linear order on $\mathcal{M}_d$ that satisfies the following restrictions.

1. $\mathbf{e}_i \leq_s m\mathbf{e}_i, \forall m\mathbf{e}_i \in \mathcal{M}_d$,
2. $m_1\mathbf{e}_i \leq_s m_2\mathbf{e}_j$ implies $t \cdot m_1\mathbf{e}_i \leq_s t \cdot m_2\mathbf{e}_j$, $\forall t \in \mathcal{M}, \forall m_1\mathbf{e}_i, m_2\mathbf{e}_j \in \mathcal{M}_d$.

As any two elements of $\mathcal{M}_d$ are comparable w.r.t. the linear order $\leq_s$, in this paper we always assume that $\mathbf{e}_1 <_s \cdots <_s \mathbf{e}_d$ without loss of generality.

If no misunderstanding occurs, $\leq_s$ is replaced by $\leq$. By Dickson's lemma and the definition above, the admissible module order $\leq_s$ has the following properties.

Proposition 1. *The admissible $\leq_s$ is a well-order (i.e. a well-founded connex order) on $\mathcal{M}_d$, and it extends the order $|$ on $\mathcal{M}_d$, which means $m_1\mathbf{e}_i \,|\, m_2\mathbf{e}_i$ implies $m_1\mathbf{e}_i \leq_s m_2\mathbf{e}_i$, for all $m_1\mathbf{e}_i, m_2\mathbf{e}_i \in \mathcal{M}_d, i \in [1, d]$.*

It should be noticed that $\leq_s$ may or may not be related to $\leq_m$. The **compatible** property [14] between $\leq_m$ and $\leq_s$ was used by [13] for the proof of termination for the F5 and GVW algorithms: $\sigma\mathbf{e}_j \leq_s \mu\mathbf{e}_j$ if and only if $\sigma \leq_m \mu$. The following section will show that this relation is sufficient for the proof of finiteness.

For any $\alpha = (\mathbf{u}, p) \in \mathcal{P}$, let $\mathcal{S}(\alpha)(resp.\mathcal{S}(\mathbf{u}))$ be the **signature** of α (resp. $\mathbf{u}$), $HM(\alpha) = HM(p)$ the head monomial of α. We call $idx(\alpha) = k$ the **index** of α if $\mathcal{S}(\alpha) = \mu\mathbf{e}_k$.

4. PROPERTIES OF SIG-POLYNOMIALS

Without loss of generality, assume that the $poly(\alpha)$ is always monic for each non-syzygy $\alpha \in NS$. We use the map

$$\begin{aligned} \vartheta : NS \mapsto \quad & \mathcal{M}_d \times \mathcal{M} \\ \alpha \mapsto & (\mathcal{S}(\alpha), HM(\alpha)) \end{aligned}$$

in [1] and call the image $\vartheta(\alpha)$ a **head pair**.

Let α be a non-syzygy one in NS. α is called **top-reducible** by B, if there exists a sig-polynomial $\beta \in B$ satisfying one of the following conditions,

1. $HM(t\beta) = HM(\alpha)$ and $\mathcal{S}(t\beta) <_s \mathcal{S}(\alpha)$,
2. $\mathcal{S}(t\beta) = \mathcal{S}(\alpha)$ and $HM(t\beta) <_m HM(\alpha)$,
3. $HM(t\beta) = HM(\alpha)$ and $\mathcal{S}(t\beta) = \mathcal{S}(\alpha)$, $t \in \mathcal{M}$;

otherwise, α is top-irreducible[1] by B.

The process $\alpha - t\beta$ is called an **$\mathcal{S}$-reduction** (resp. **$\mathcal{M}$-reduction**[2], **super top-reduction**), if item 1 (resp. 2, 3)

[1]By the following lemma, we will see that this notion is the same as that of [11] and the "primitive $\mathcal{S}$-irreducible" of [1].

[2]The term "$\mathcal{M}$-reducible" has a similar meaning as the "M-pair" in [17]. It serves as one type of rewritten criterion which can be traced back in [1, 13, 11]. Defining so because our version of rewritten criterion is unrelated to this term.

is satisfied. β or $t\beta$ of item 1 (resp. 2) is called an $\mathcal{S}$-reducer (resp. $\mathcal{M}$-reducer). Let $\gamma = \alpha - t\beta$, we denote $\alpha \underset{B}{\longrightarrow} \gamma$. $\underset{B}{\overset{*}{\rightarrow}}$ is the reflexive-transitive closure of $\underset{B}{\longrightarrow}$.

LEMMA 2. *Let α be a non-syzygy one in NS. α is $\mathcal{M}$-reducible by $\mathcal{P}^*$ if and only if it is $\mathcal{S}$-reducible by $\mathcal{P}^*$.*

PROOF. If α is $\mathcal{M}$-reducible, let $t\beta$ be its $\mathcal{M}$-reducer and $\gamma = \alpha - t\beta$. It can be verified readily that $\gamma \in \mathcal{P}^*$ can $\mathcal{S}$-reduce α. The other direction can be proved similarly. □

A set $\mathcal{G} \subset \mathcal{P}$ is called an **S-Gröbner basis** for the module $\mathcal{P}$, if any non-syzygy $\alpha \in NS$ is top-reducible by $\mathcal{G}$. By Lemma 2, each nonzero polynomial in $\mathcal{I}$ can be reduced by $\{poly(\alpha) \mid \alpha \in \mathcal{G}, poly(\alpha) \neq 0\}$, a Gröbner basis of $\mathcal{I}$. So the "S-Gröbner basis" is in fact a term in [16], which is a simpler version of "strong Gröbner basis" in [11].

All non-syzygy top-irreducible sig-polynomials form a special kind of S-Gröbner basis called the **top-reduced S-Gröbner basis** $\mathcal{H}$ for $\mathcal{P}$. The signature of a top-irreducible sig-polynomial is called a **top-irreducible signature** w.r.t. $\mathcal{P}$. Besides, we call two sig-polynomials α and β **equivalent** if $\beta \neq \alpha$ and $\vartheta(\beta) = \vartheta(\alpha)$. If we store only one for all equivalent sig-polynomials in $\mathcal{H}$, for fixed orders $\leq_m$ and $\leq_s$, the top-reduced S-Gröbner basis $\mathcal{H}$ is uniquely determined by the module $\mathcal{P}$ up to equivalence. Those top-reducible sig-polynomials in $\mathcal{P} \setminus \mathcal{H}$ are also called **redundant sig-polynomials**.

Since $(\mathcal{M}, |)$ and $(\mathcal{M}_d, |)$ are Dickson quasi-ordered sets, by Lemma 1, we have $(\mathcal{M}_d \times \mathcal{M}, |^*)$ is also a Dickson quasi-ordered set of which the order $|^*$ is defined as follows:

$$\vartheta(\alpha) \,|^*\, \vartheta(\beta) \Leftrightarrow \mathcal{S}(\alpha) \mid \mathcal{S}(\beta) \text{ and } HM(\alpha) \mid HM(\beta), \qquad (3)$$

where $\alpha, \beta \in \mathcal{P}^*$.

The proofs of [1, Prop. 14] and [7, Lem. 15] implicitly use the "compatible" property. Here, we reprove the following results to show that "compatible" property is a sufficient condition for the finiteness of the top-reduced S-Gröbner basis.

LEMMA 3. *Let α and β be two arbitrary sig-polynomials in $\mathcal{P}^*$ such that $\vartheta(\alpha) \,|^*\, \vartheta(\beta)$. If the admissible monomial order $\leq_m$ and the admissible module order $\leq_s$ are compatible, then β is top-reducible by α.*

PROOF. Let s and t be two monomials in $\mathcal{M}$ such that $s = \mathcal{S}(\beta)/\mathcal{S}(\alpha)$ and $t = HM(\beta)/HM(\alpha)$. There are three cases as follows.

1. If $s = t$, then β is super top-reducible by α.

2. If $s <_m t$, then $sHM(\alpha) <_m HM(\beta)$, i.e., β is $\mathcal{M}$-reducible by α.

3. If $s >_m t$, as $\leq_m$ and $\leq_s$ are compatible, $t\mathcal{S}(\alpha) <_s \mathcal{S}(\beta)$, i.e., β is $\mathcal{S}$-reducible by α.

□

By the above lemma and the Dickson property of $|^*$, we have the following truth.

THEOREM 1. *The top-reduced S-Gröbner basis for $\mathcal{P}$ is finite.*

It can be seen that the "compatible" property is a sufficient condition for the finiteness of the top-reduced S-Gröbner basis $\mathcal{H}$. Below we provide an example by extracting the proof of [13, Th. 13] to show that we may get an infinite $\mathcal{H}$ if the "compatible" property is not satisfied.

Example 1. Let $\leq_{m^*} = \leq_{invlex}$ be the inverse lexicographical order[3] on $\mathcal{M}$ and let $\leq_{s^*}$ be the order on $\mathcal{M}_d$ with the following definition: $m\mathbf{e}_i <_{s^*} m'\mathbf{e}_j$ if

1. $i > j$,

2. $i = j$, $m <_{lex} m'$,

where $\leq_{lex}$ is the lexicographical order on $\mathcal{M}$. Particularly, we have $m\mathbf{e}_i = m'\mathbf{e}_j$, if $i = j$ and $m = m'$.

Readers can verify that $\leq_{m^*}$ (resp. $\leq_{s^*}$) is an admissible monomial (resp. module) order and $\leq_{m^*}$ and $\leq_{s^*}$ are not compatible. Let the polynomials to be computed are $f_1 = x_1$ and $f_2 = x_3x_2 - x_3x_1$, the initialized sig-polynomials $\alpha_1 = (\mathbf{e}_1, x_1)$ and $\alpha_2 = (\mathbf{e}_2, x_3x_2 - x_3x_1)$. We have

$$\alpha_3 = x_2x_3\alpha_1 - x_1\alpha_2 = (x_2x_3\mathbf{e}_1, x_3x_1^2),$$

and α_3 cannot be top-reduced by α_1 and α_2. It can be inferred that α_3 is top-irreducible and we can generate infinite top-irreducible sig-polynomials

$$\alpha_k = x_2\alpha_{k-1} - x_1^{k-2}\alpha_2 = (x_2^{k-2}x_3\mathbf{e}_1, x_3x_1^{k-1}),$$

for $k \geq 3$. As they are not pairwise equivalent (i.e., their head pairs are unequal), $\mathcal{H}$ is an infinite sequence $\{\alpha_\ell\}_{\ell \in \mathbb{N}^*}$.

Hence, in the remaining sections of this paper we will assume that the admissible monomial order $\leq_m$ and the admissible module order $\leq_s$ are compatible. Suppose there are two sig-polynomials $\alpha, \beta \in NS$. Denote by $\Gamma_{\alpha\beta}$ the least common multiple $lcm(HM(\alpha), HM(\beta))$. Let $m_\alpha = \frac{\Gamma_{\alpha\beta}}{HM(\alpha)}$ and $m_\beta = \frac{\Gamma_{\alpha\beta}}{HM(\beta)}$. If $m_\alpha\mathcal{S}(\alpha) >_s m_\beta\mathcal{S}(\beta)$, then $m_\alpha\alpha$ is called the (**first**) **J-pair** (cf. [11]) of α and β; $m_\beta\beta$ is called the **second J-pair** w.r.t. $m_\alpha\alpha$; α (resp. β) is called the **first** (resp. **second**) **component** of $m_\alpha\alpha$; m_α (resp. m_β) is called the **multiplier** of α (resp. β).

5. THE F5GEN ALGORITHM

5.1 Pseudo code

Without loss of generality, we assume that inter-reducing the input $\{f_1, \ldots, f_d\}$ (i.e. reducing each polynomial by the rest) does not lead to zero cancellation. It can be deduced $\mathbf{e}_1, \ldots, \mathbf{e}_d$ are top-irreducible signatures as in [17]. Let α be a non-syzygy one in NS of signature $\mathbf{e}_i$, $i \in [1, d]$. If α is $\mathcal{M}$-reducible by $\mathcal{P}^*$, a repeated $\mathcal{S}$-reduction of α by $\mathcal{P}^*$ would generate an $\mathcal{S}$-irreducible sig-polynomial β. By Lemma 2, β is also $\mathcal{M}$-irreducible. As β cannot be super top-reduced by $\mathcal{P}^*$, β is a top-irreducible sig-polynomial and $\mathbf{e}_i$ is a top-irreducible signature.

Note that in the computation process of the F5GEN, there may exist syzygies in $\mathcal{G}$. Let S be a set of polynomials (resp. sig-polynomials), sort(S, $\leq_m$ (resp. $\leq_s$)) means that we arrange S by ascending head monomials (resp. signatures) of polynomials (resp. sig-polynomials) with respect to the order $\leq_m$ (resp. $\leq_s$).

[3] $\mu \leq_{invlex} \sigma$ if μ has smaller degree in the last variable for which they differ.

Algorithm 1 The F5GEN Algorithm (F5 algorithm with a generalized insertion strategy)

1: **inputs:**
$F = \{f_1, \ldots, f_d\} \in R$, a list of polynomials
$\leq_m$ an admissible monomial order on $\mathcal{M}$
$\leq_s$, an admissible module order on $\mathcal{M}_d$ such that $\leq_s$ is compatible with $\leq_m$ and $\mathbf{e}_1 <_s \cdots <_s \mathbf{e}_d$
2: **outputs:**
a Gröbner basis of $\mathcal{I} =< f_1, \ldots, f_d >$
3: inter-reduce F and F :=sort($\{f_1, \ldots, f_d\}$, $\leq_m$), $F_i = (\mathbf{e}_i, f_i)$ for $i \in [1, d]$
4: **init:**
$CP := \{J-pair[F_i, F_j] \,|\, 1 \leq i < j \leq d\}$,
$\mathcal{G} = \{F_i \,|\, i \in [1, d]\}$
5: **while** $CP \neq \emptyset$ **do**
6: $\quad\gamma$:=select_F5(CP) and $CP := CP \backslash \{\gamma\}$
7: $\quad$**if** $\mathcal{S}(\gamma) \notin \mathcal{S}(PS)$ and γ is not rewritable by $\mathcal{G}$ **then**
8: $\quad\quad\gamma \xrightarrow[\mathcal{G}]{*} \alpha$
9: $\quad\quad\mathcal{G}$:=insert_F5GEN(α, $\mathcal{G}$, γ)
10: $\quad\quad$**if** $poly(\alpha) \neq 0$ **then**
11: $\quad\quad\quad CP := CP \cup \{J\text{-}pair(\alpha, \beta) \,|\, \forall \beta \in \mathcal{G} \backslash Syz, \beta \neq \alpha\}$
12: **return** $\{poly(\alpha) \,|\, \alpha \in \mathcal{G} \setminus Syz\}$

Algorithm 2 select_F5

1: **inputs:**
CP, a set of J-pairs
2: $i := min\{idx(\delta) | \delta \in CP\}$ and
$CP^i := \{\delta \in CP \,|\, idx(\delta) = i\}$
3: $D := min\{deg(\mathcal{S}(\delta)) | \delta \in CP^i\}$ and
$CP^i_D := \{\delta \in CP^i \,|\, deg(\mathcal{S}(\delta)) = D\}$
4: $\gamma :=$ any J-pair in CP^i_D
5: **return** γ

The select_F5 function seems to be too cumbersome for applications, but it simulates fairly well the process of selecting critical pairs in the F5 algorithm of [8]. In fact, if γ is replaced by the $\leq_s$-minimal J-pair at line 6, we can also obtain another version of the F5GEN and its proof is similar to that in Subsection 5.3. This paper is tailored to deal with the F5 algorithm, so we do not consider other selection strategies here. Call a J-pair γ being considering if the F5GEN algorithm is executing the while-loop where γ is selected at line 6. If the algorithm has completed that while-loop, γ has been considered, otherwise γ has not been considered.

Algorithm 3 rewritable

1: **inputs:**
$m\mathcal{G}(k)$, a J-pair
$\mathcal{G} := \{\mathcal{G}(1), \ldots, \mathcal{G}(r)\}$
2: **outputs:**
true if $\mathcal{S}(m\mathcal{G}(k))$ is a multiple of another sig-polynomial appearing later than $\mathcal{G}(k)$ in $\mathcal{G}$
3: find the first position j_b and the last position j_e in $\mathcal{G}$ such that $idx(\mathcal{G}(k)) = idx(\mathcal{G}(j_b)) = idx(\mathcal{G}(j_e))$
4: **for** $i = j_e$ **to** j_b **do**
5: $\quad$**if** $\mathcal{S}(\mathcal{G}(i)) \,|\, \mathcal{S}(m\mathcal{G}(k))$ **then**
6: $\quad\quad$**return** $i \neq k$
7: **return false**

Algorithm 4 insert_F5GEN

1: **inputs:**
α, a sig-polynomial
$\mathcal{G} := \{\mathcal{G}(1), \ldots, \mathcal{G}(r)\}$
$\gamma = m\mathcal{G}(k)$, the J-pair which is $\mathcal{S}$-reduced to α
2: find the first position j_b and the last position j_e in $\mathcal{G}$ such that

$$idx(\mathcal{G}(j_b)) = idx(\mathcal{G}(j_e)) = idx(\alpha) = idx(\gamma)$$

3: insert α into $\mathcal{G}$ after $\mathcal{G}(i)$, where $j_b - 1 \leq i \leq j_e$, such that α appears later in $\mathcal{G}$ than $\mathcal{G}(k)$ (i.e., $k \leq i \leq j_e$)
4: **return**

It is important to note that after the execution of line 3 in the insert_F5GEN function, α is in $\mathcal{G}$: α in $\mathcal{G}$ may be of the form $\mathcal{G}(k+1)$ or $\mathcal{G}(k+100)$ ($k + 100 \leq j_e + 1$). We will show that the original F5 algorithm of [8], which did not specify an order when adding critical pairs of the same degree into $Rule[idx(\alpha)]$, can be viewed as a restricted case of the insertion strategy.

5.2 Simplifications compared with F5

If $\mathbf{s}$ is the signature (not necessary top-irreducible) in $\mathcal{S}(\mathcal{P}^*)$, we denote by $\mathcal{P}_{\leq_s(\mathbf{s})}$ the subset of sig-polynomials in $\mathcal{P}$ of which the signatures are smaller than or equal to $\mathbf{s}$ and denote by $\mathcal{G}_{\leq_s(\mathbf{s})}$ the S-Gröbner basis for $\mathcal{P}_{\leq_s(\mathbf{s})}$. That is, any non-syzygy sig-polynomial in $\mathcal{P}_{\leq_s(\mathbf{s})}$ is top-reducible by $\mathcal{G}_{\leq_s(\mathbf{s})}$. $\mathcal{P}_{\leq_s(\mathbf{s})}$ and $\mathcal{G}_{<_s(\mathbf{s})}$ are defined similarly.

We denote by **F5-reduction** the process of checking the **F5-criteria** (i.e. line 7 of the F5GEN algorithm) for the $\mathcal{S}$-reducer.[4] We think that omitting the F5-criteria check for

[4]The terms F5-reduction and F5-criteria are summarized from the pseudo code of the F5 algorithm of [8].

the $\mathcal{S}$-reducers at line 8 as in [7] would not affect the correct termination of this algorithm. However, Galkin pointed out that the simplifications above are not as obvious as we imaged. Using the idea of [10], we have the following lemma to confirm this point but our result is slightly different from Corollary 35 of [10].

LEMMA 4. *Assume the F5GEN has computed correctly up to signature* $\mathbf{s}$ *(i.e.,* $\mathcal{G} = \mathcal{G}_{<_s(\mathbf{s})}$ *and the F5GEN has not considered any sig-polynomial of signature* $\mathbf{s}$*). Let* $\zeta \in NS$ *be a non-syzygy one of signature* $\mathbf{s}$*. If* ζ *can be* $\mathcal{S}$*-reduced by* $\mathcal{G}$*, it would also be F5-reduced by the updated* $\mathcal{G}$ *after finite runs through the while-loop.*

The detailed proof can be seen in the appendix.

Therefore, in the remaining part of this paper the sole reduction used will be $\mathcal{S}$-reduction and the F5-criteria check for the second J-pair will be omitted too because the second J-pair is also an $\mathcal{S}$-reducer.

5.3 Proof of F5GEN

Define an order $\preceq_p$ on $\mathcal{P}^*$ in the following way:

$$\alpha \preceq_p \beta \Leftrightarrow HM(\alpha)\mathcal{S}(\beta) \leq_s HM(\beta)\mathcal{S}(\alpha)$$

This order are closely related to the terms defined earlier. If α is $\mathcal{S}$-reducible by β, then $(HM(\alpha)/HM(\beta))\mathcal{S}(\beta) <_s \mathcal{S}(\alpha)$ and $\alpha \prec_p \beta$. If β is $\mathcal{M}$-reducible by α, then

$$(\mathcal{S}(\beta)/\mathcal{S}(\alpha))HM(\alpha) <_m HM(\beta)$$

and $idx(\alpha) = idx(\beta)$. As $<_s$ and $<_m$ are compatible, $\alpha \prec_p \beta$.

It can be deduced that $\preceq_p$ is a well-order on $\mathcal{P}^*$ as $\leq_s$ is the well-order and the $\preceq_p$-minimal elements on $\mathcal{P}^*$ are $\{(\mathbf{u}, g) \in \mathcal{P}^* \mid g = 0\}$.

From the idea of [13, Th. 13], we derive a crucial lemma for the termination.

LEMMA 5. *Let* $\alpha \in NS$ *be the non-syzygy top-irreducible one of the maximum signature. After finite while-loops, assume that the F5GEN algorithm has computed correctly up to signature* $\mathbf{s}$*, where* $\mathbf{s} >_s \mathcal{S}(\alpha)$*. Then the F5GEN would terminate.*

PROOF. By the assumption the S-Gröbner basis $\mathcal{G}$ is a finite set, we denote by $\{\mathcal{G}(k_1), \ldots, \mathcal{G}(k_w)\}$ all non-syzygy ones in $\mathcal{G}$. Then we use a permutation ς of $\{1, \ldots, w\}$ to ensure that

$$\mathcal{G}(k_{\varsigma(1)}) \succeq_p \cdots \succeq_p \mathcal{G}(k_{\varsigma(w)})$$

since $\preceq_p$ is a well-order. All the remainders in CP can be counted in a vector $N_{CP} = (n_1, \ldots, n_w) \in \mathbb{N}^w$ as follows. n_j stands for the number of J-pairs $\preceq_p$-equal to $\mathcal{G}(k_{\varsigma(j)})$, $j \in [1, w]$.

Consider the relation between N_{CP} and $N_{CP'}$ after two consecutive runs through the while-loop w.r.t the lexicographical order, denoted by $<_{lex}$. Let

$$N_{CP} = (n_1, \ldots, n_w) \text{ and } N_{CP'} = (n'_1, \ldots, n'_w),$$

we have $N_{CP} >_{lex} N_{CP'}$ whenever the leftmost non-zero component, say $n_a - n'_a$, is positive, where $a \in [1, w]$. During an execution of the loop, a J-pair γ is extracted from CP, which will be either discarded or $\mathcal{S}$-reduced to an $\mathcal{S}$-irreducible sig-polynomial β. If $\beta \neq 0$, new J-pairs η would be generated at line 11. Whether the first components of η would be β or not, $\eta \preceq_p \beta \prec_p \gamma$ always holds. Because η can be super top-reduced by $\mathcal{G}$, $N_{CP} >_{lex} N_{CP'}$ and the algorithm would terminate because $<_{lex}$ on $\mathbb{N}^w$ is well-founded. □

The following lemma shows under what condition a top-irreducible sig-polynomial can be computed by the F5GEN algorithm.

LEMMA 6. *Let* α *be a non-syzygy top-irreducible one in* NS*. After finite steps, assume that the F5GEN has computed correctly up to a top-irreducible signature* $\mathcal{S}(\alpha)$*.* $\mathcal{G}$ *would be* $\mathcal{G}_{\leq_s(\mathcal{S}(\alpha))}$ *after finite runs of the loop.*

PROOF. Suppose $\mathcal{G}$ does not contain any sig-polynomial of signature equivalent to α since otherwise $\mathcal{G} = \mathcal{G}_{\leq_s(\mathbf{s})}$. As $\mathbf{e}_1, \ldots, \mathbf{e}_d$ are top-irreducible signatures, we first prove that there is a J-pair in CP can pass F5-criteria by distinguishing between two cases.

1. If $\mathcal{S}(\alpha) = \mathbf{e}_v$ for some $v \in [1, d]$, there is at least one sig-polynomial of signature $\mathbf{e}_v$ in $\mathcal{G}$ (e.g., $(\mathbf{e}_v, f_v)$), so we can choose one of signature $\mathbf{e}_v$ in the maximum position of $\mathcal{G}_{\leq_s(\mathbf{s}_i)}$, denoted by γ. Since γ is $\mathcal{M}$-reducible by α, by Lemma 2 it must be $\mathcal{S}$-reducible by $\mathcal{G}$. Then γ is a J-pair in CP and can pass the F5-criteria check at line 7 of the F5GEN algorithm.

2. If $\mathcal{S}(\alpha) \neq \mathbf{e}_v$ for all $v \in [1, d]$, as $\mathcal{G}$ is a finite set, there exists a non-syzygy $\beta \in \mathcal{G}$ and a monomial $m \geq 1$ such that $\mathcal{S}(m\beta) = \mathcal{S}(\alpha) \notin \mathcal{S}(PS)$ and $m\beta$ is not rewritable by $\mathcal{G}$.[5] Since $m\beta$ can be $\mathcal{M}$-reduced by α, by Lemma 2 it can also be $\mathcal{S}$-reduced by some non-syzygy top-irreducible one in $\mathcal{G}$, say δ. If $m = 1$, then β would be a J-pair in CP and pass the F5-criteria check.

   If $m \neq 1$, we denote by $m_\beta\beta$ the J-pair of β and δ, where $m_\beta \mid m$. Assume for a contradiction that m_β properly divides m. As the F5GEN has considered $\mathcal{S}(m_\beta\beta)$ and β rewrites α, either $\mathcal{S}(m_\beta\beta) = \mathcal{S}(PS)$ or $m_\beta\beta$ is rewritable by $\mathcal{G}$.

   If $\mathcal{S}(m_\beta\beta) = \mathcal{S}(PS)$, then $\mathcal{S}(m\beta)$ would be in $\mathcal{S}(PS)$, a contradiction. If $m_\beta\beta$ is rewritable by $\mathcal{G}$, let $\eta \in \mathcal{G}$ be the sig-polynomial rewriting $m_\beta\beta$. As η appears later in $\mathcal{G}$ than β and

   $$\mathcal{S}(\eta) \mid \mathcal{S}(m_\beta\beta) \mid \mathcal{S}(m\beta),$$

   $m\beta$ is rewritable by η, a contradiction.

   Therefore, $m_\beta = m$, i.e., $\gamma = m\beta$ is the J-pair of two non-syzygy β and δ such that $\mathcal{S}(\alpha) = \mathcal{S}(\gamma) \notin \mathcal{S}(PS)$ and γ is not rewritable by $\mathcal{G}$.

Then we prove that such a J-pair γ can be extracted after finite runs of the while-loop. If the F5GEN has not considered the sig-polynomial of index $idx(\alpha)$, since $\mathcal{G} = \mathcal{G}_{<_s(\mathcal{S}(\alpha))}$ and the F5GEN is incremental, the F5GEN would consider sig-polynomial of index $idx(\alpha)$ after finite runs of the loop by Lemma 5. If the F5GEN is considering the sig-polynomial of index $idx(\alpha)$, from the pseudo code of the select_F5 function we know that γ would be extracted after finite steps. As γ can pass the F5-criteria, $\mathcal{G}$ would be $\mathcal{G}_{\leq_s(\mathcal{S}(\alpha))}$. □

[5] We choose the sig-polynomial in the maximum position of $\mathcal{G}$ that its signature divides $\mathcal{S}(\alpha)$.

THEOREM 2. *For any finite subset $F = \{f_1, \ldots, f_d\}$ of polynomials in R, the F5GEN algorithm terminates correctly.*

PROOF. We proceed by induction on top irreducible sig-polynomials. The base case is vacuously true. As the induction step is guaranteed by Lemma 6 and top-reduced S-Gröbner basis is finite by Theorem 1, we would get an S-Gröbner basis after finite while-loops of the F5GEN. Then we can prove the termination of the F5GEN by Lemma 5. □

6. COMPARISON WITH F5

6.1 Orderings for rewritten

In the original F5 algorithm of [8], the admissible module order, denoted by $\leq_{s_0}$[6], is compatible with the admissible monomial order $\leq_m$. Let α and β be two arbitrary sig-polynomials in NS and $m_\alpha = \frac{\Gamma_{\alpha\beta}}{HM(\alpha)}, m_\beta = \frac{\Gamma_{\alpha\beta}}{HM(\beta)}$. Because the input polynomials $F = \{f_1, \ldots, f_d\}$ of the F5 algorithm are homogeneous, F5-reducing a sig-polynomial does not lead to a decrease in the degree of its polynomial part and

$$deg(HM(\alpha)) = deg(\mathcal{S}(\alpha)) + deg(HM(f_{idx(\alpha)}))$$

always holds. Assume that $m_\alpha \mathcal{S}(\alpha) >_s m_\beta \mathcal{S}(\beta)$. During the ith run through the **F5** function of [8], sorting critical pairs by the degrees of $\Gamma_{\alpha\beta}$ equals sorting them by the degrees of $m_\alpha \mathcal{S}(\alpha)$ because these critical pairs share the index i. Therefore, line 2 and 3 of the select_F5 function in this paper do the same thing as the **F5** function did.

The crux of proving the termination of the F5 algorithm of [8] is the author did not specify that critical pairs of the smallest degree are chosen in what sequence. This causes the non-determinacy of the ordering in *Rule*[i] and the answer to the question whether a sig-polynomial is rewritable. Any proof by breaking ties using a concrete ordering would be a proof for one implementation of the F5. For example, the TRB-F5 algorithm of [13] selects the $\leq_{s_0}$-minimal one from CP_D^i[7] at the beginning of each run through the while-loop, if there are more than one J-pairs in CP_D^i. In addition, sig-polynomials are added at the head of *Rule*[i] in [13]. We will give an example to show that, upon input a set of homogeneous polynomials, the original F5 with different concrete orderings (on critical pairs of the same degree) may output different results. In the insert_F5GEN function of this paper, we generalize the insertion strategy in $\mathcal{G}$ which covers all possible orderings in *Rule*[i] for all i.

6.2 Criteria

Instead of sorting sig-polynomials in $\mathcal{G}$ in the F5GEN algorithm, the original F5 algorithm uses an array *Rule*[i] to store sig-polynomials of index i, for each $i \in [1, d]$. The sig-polynomial of F5-reduction from a J-pair γ appears earlier in *Rule*[i] than the first component of γ. So the insertion strategy of *Rule*[i] of [8] satisfies (inversely) the description of the insert_F5GEN function. Besides, if α appears in *Rule*[i] later than $\beta \in$ *Rule*[i], $deg(\mathcal{S}(\alpha)) \geq deg(\mathcal{S}(\beta))$ always holds, whereas in $\mathcal{G}$ it may not hold. That means the rewritten criterion here is more generalized than the one in [8].

[6] That is, $\mu \mathbf{e}_i <_{s_0} \sigma \mathbf{e}_j$ if $i > j$ or $i = j$ and $\mu <_m \sigma$.

[7] CP_D^i is not a notation in [13], but is used for summarizing the operations in the TRB-F5.

In [8], checking whether a sig-polynomial of index i is a normal form of itself w.r.t. the computed S-Gröbner basis of index greater than i can be viewed as a relaxation of principal syzygy check because two sig-polynomials of index i can also generate a principal syzygy sig-polynomial. Therefore, the criteria of the F5 is an implementation of the criteria of the F5GEN algorithm.

6.3 Orderings for reduction

It is important to note that the sig-polynomial selected at each time for the **TopReduction** function is determinate: the $\mathcal{S}$-minimal J-pair[8] in the minimal degree J-pairs of the current index, say i, is always selected by the F5. In our F5GEN algorithm, any J-pair in CP_D^i can be selected for $\mathcal{S}$-reduction, a relaxation of the F5. In addition, by the discussion in Section 5.2, we can add the check for the $\mathcal{S}$-reducers and the second J-pairs w.r.t. the J-pairs in CP without affecting the correct termination of the F5GEN. Hence the F5 algorithm of [8] is one implementation of the F5GEN.

6.4 Example

For the original F5 algorithm, [12] made a conjecture on which the termination is based. That is, there will not be a sig-polynomial added in $\mathcal{G}$ such that it is super top-reducible by a sig-polynomial already in $\mathcal{G}$. We doubt the truth of the conjecture and prove Lemma 5 without that conjecture. The example given in [12] can also serve as a counterexample to show that the sig-polynomials generated by the F5 have one element super top-reducible by another one.

Example 2. The admissible monomial order $\leq_m$ is the degree reverse lexicographical order ($x >_m y >_m z >_m t$) and the input polynomials are $(x^2y - z^2t, xz^2 - y^2t, yz^3 - x^2t^2)$. The admissible module order $\leq_{s_0}$ is automatically specified. The head pairs of the S-Gröbner basis generated in [12] are

$$\begin{aligned} \vartheta(r_3) &= (\mathbf{e}_3, x^2y), & \vartheta(r_2) &= (\mathbf{e}_2, xz^2), \\ \vartheta(r_4) &= (xy\mathbf{e}_3, xy^3t), & \vartheta(r_5) &= (xyz^2\mathbf{e}_2, z^6t), \\ \vartheta(r_1) &= (\mathbf{e}_1, yz^3), & \vartheta(r_6) &= (x\mathbf{e}_1, y^3zt), \\ \vartheta(r_7) &= (x^2\mathbf{e}_1, z^5t), & \vartheta(r_9) &= (x^2z\mathbf{e}_1, y^5t^2), \\ \vartheta(r_8) &= (x^3\mathbf{e}_1, x^5t^2), & \vartheta(r_{10}) &= (z^3t\mathbf{e}_1, y^6t^2), \\ \vartheta(r_{11}) &= (x^3z\mathbf{e}_1, x^5zt^2). \end{aligned}$$

Comparing the S-Gröbner basis in [8], only one more sig-polynomial r_{11} of head pair $\vartheta(r_{11}) = (x^3z\mathbf{e}_1, x^5zt^2)$ is generated. We can see that r_{11} can be super top-reduced by r_8.

The reason for two different S-Gröbner bases computed by the same algorithm lies in that the positions of r_9 and r_8 in *Rule*[1] are different and r_{11} is the sig-polynomial F5-reduced from the J-pair $x \cdot r_9$. The original F5 did not specify that the critical pairs of the same degree should be added into *Rule*[i] in what sequence, so the case of r_9 appearing earlier in *Rule*[1] than r_8[9] is possible as the degrees of their signatures are the same. In that case, r_{11} is kept in contrast to the example in [8]. Therefore, if the conjecture is assumed to be correct, the proof of the correct termination would be only a partial proof.

[8] It makes no difference to consider J-pairs instead of the result of Spoly in [8] when the signature and degree of a sig-polynomial are taken into consideration.

[9] We assume that new elements are added at the beginning of *Rule*[i].

7. CONCLUSION

This paper presents a proof of the correct termination of the F5GEN algorithms under the condition that the admissible monomial order and the admissible module order are compatible. And the F5GEN is a generalization of the F5 algorithm, so the termination of the F5 is solved.

In fact, our original goal was to prove the F5GEN with the following generalization.

CONJECTURE 1. *At line 6 of the F5GEN algorithm, assume that any J-pair can be selected from CP. Then the F5GEN terminates correctly.*

We tend to think the above is true because with $\preceq_s$-minimal J-pair in CP instead of the cumbersome select_F5 function, the correct termination of the F5GEN would be proved similarly. Unfortunately, we face hurdles in proving the corresponding Lemma 4 (to be specific, Lemma 7) and Lemma 5 for that conjecture. Therefore, we leave that conjecture as an **open problem**.

8. ACKNOWLEDGMENTS

We would like to thank Christian Eder, John Perry and Vasily Galkin for valuable feedback and discussions on this work. Comments by Yao Sun, Dingkang Wang and Xiaodong Ma greatly improved this paper. We are indebted to the anonymous referee for a careful reading of this paper and many useful suggestions. Yupu Hu is partially supported by the National Natural Science Foundation of China (No. 61173151). Baocang Wang would like to thank partial supported of the National Natural Science Foundation of China (No. 61173152).

9. REFERENCES

[1] A. Arri and J. Perry. The F5 criterion revised. *Journal of Symbolic Computation*, 46(9):1017–1029, 2011.

[2] T. Becker, H. Kredel, and V. Weispfenning. *Gröbner bases: a computational approach to commutative algebra*. Springer-Verlag, London, UK, April 1993.

[3] B. Buchberger. *Ein Algorithmus zum Auffinden der Basiselemente des Restklassenrings nach einem nulldimensionalen Polynomideal*. PhD thesis, Universität Innsbruck, Austria, 1965.

[4] B. Buchberger. A criterion for detecting unnecessary reductions in the construction of Gröbner-bases. In *Symbolic and Algebraic Computation*, volume 72 of *Lecture Notes in Computer Science*, pages 3–21. Springer Berlin Heidelberg, 1979.

[5] C. Eder, J. Gash, and J. Perry. Modifying Faugère's F5 algorithm to ensure termination. *ACM Communications in Computer Algebra*, 45(1/2):70–89, July 2011.

[6] C. Eder and J. Perry. F5C: A variant of Faugère's F5 algorithm with reduced Gröbner bases. *Journal of Symbolic Computation*, 45(12):1442–1458, 2010.

[7] C. Eder and J. Perry. Signature-based algorithms to compute Gröbner bases. In *Proceedings of the 36th international symposium on Symbolic and algebraic computation*, ISSAC '11, pages 99–106, New York, USA, 2011. ACM.

[8] J.-C. Faugère. A new efficient algorithm for computing Gröbner bases without reduction to zero (F5). In *Proceedings of the 27th international symposium on Symbolic and algebraic computation*, ISSAC '02, pages 75–83, New York, USA, 2002. ACM.

[9] J.-C. Faugère and A. Joux. Algebraic cryptanalysis of hidden field equation (HFE) cryptosystems using Gröbner bases. In *Advances in Cryptology - CRYPTO 2003*, volume 2729 of *Lecture Notes in Computer Science*, pages 44–60. Springer Berlin Heidelberg, 2003.

[10] V. Galkin. Termination of original F5. Preprint, arXiv: 1203.2402 [math.AC], March 2012.

[11] S. Gao, F. Volny, and M. Wang. A new algorithm for computing Gröbner bases. Cryptology ePrint Archive, Report 2010/641, December 2010.

[12] J. Gash. *On efficient computation of Gröbner bases*. PhD thesis, Indiana University, USA, 2009.

[13] L. Huang. A new conception for computing Gröbner basis and its applications. Preprint, arXiv: 1012.5425 [cs.SC], December 2010.

[14] M. Kreuzer and L. Robbiano. *Computational Commutative Algebra 1*. Computational Commutative Algebra. Springer, 2000.

[15] T. Stegers. Faugère's F5 algorithm revisited. Master's thesis, Technische Universität Darmstadt, 2005.

[16] Y. Sun and D. Wang. A generalized criterion for signature related Gröbner basis algorithms. In *Proceedings of the 36th international symposium on Symbolic and algebraic computation*, ISSAC '11, pages 337–344, New York, USA, 2011. ACM.

[17] F. Volny. *New Algorithms for Computing Gröbner Bases*. PhD thesis, Clemson University, 2011.

APPENDIX

During an execution of the while-loop of the F5GEN algorithm, assume that $\mathcal{G}$ computed is the set of finite sig-polynomials. Let $\mathcal{G}(j)$ and $\mathcal{G}(k)$ be two sig-polynomials in $\mathcal{G}$. We adopt the order $\lessdot$ in [10], which is defined as follows. $m_j\mathcal{G}(j) \lessdot m_k\mathcal{G}(k)$ if

1. $m_j\mathcal{S}(\mathcal{G}(j)) <_s m_k\mathcal{S}(\mathcal{G}(k))$,
2. $m_j\mathcal{S}(\mathcal{G}(j)) = m_k\mathcal{S}(\mathcal{G}(k))$ and $j > k$ (i.e., $\mathcal{G}(j)$ appears later in $\mathcal{G}$ than $\mathcal{G}(k)$),

where m_j and m_k are two monomials in $\mathcal{M}$. Particularly, $m_j\mathcal{G}(j)$ is $m_k\mathcal{G}(k)$ if $m_j\mathcal{S}(\mathcal{G}(j)) = m_k\mathcal{S}(\mathcal{G}(k))$ and $j = k$. It can be seen that $\lessdot$ is a well-order on $\mathcal{M} \times \mathcal{G}$, because $\preceq_s$ is a well-order and $\mathcal{G}$ is finite.

Below we will introduce representations that are inspired by the ideas of [6, 10]. Let α be a sig-polynomial, if

$$poly(\alpha) = \sum_{k=1}^{\ell} p_k \cdot poly(\mathcal{G}(i_k)), \mathcal{G}(i_k) \in \mathcal{G}, \tag{4}$$

$\sum_{k=1}^{\ell} p_k\mathcal{G}(i_k)$ is called an ($\mathcal{G}$-)**representation** of α. Each $m_{k_v}\mathcal{G}(i_k)$ is called an **element** of the representation, where $p_k = \sum_v c_{k_v} m_{k_v}$, $0 \neq c_{k_v} \in K$ and $m_{k_v} \in \mathcal{M}$. Of course, we can store all elements of a representation in an array $A \in (\mathcal{M} \times \mathcal{G})^v$ of size v such that $A_1 \gtrdot \cdots \gtrdot A_v$. Let another $\mathcal{G}$-representation of α be $\sum_{k=1}^{\ell'} p'_k\mathcal{G}(j_k)$ and its array of elements be $A' \in (\mathcal{M} \times \mathcal{G})^{v'}$. The representation A' is called $\lessdot$-smaller than A if A' has $\lessdot$-smaller element at the leftmost position for which they differ[10]. If $\mathcal{S}(p_k\mathcal{G}(i_k)) \leq_s \mathcal{S}(\alpha), \forall k \in [1,\ell]$, the $\mathcal{G}$-representation is called **$\mathcal{S}$-safe** (i.e. the term "signature-safe" in [10]). Let t be a monomial in $\mathcal{M}$. The $\mathcal{S}$-safe $\mathcal{G}$-representation is called a **t-representation** if $HM(p_k\mathcal{G}(i_k)) \leq_m t, \forall k \in [1,\ell]$. We can give few examples for those representations.

Example 3. Let $\mathcal{G}(j) = (\mathbf{u}, p) \in \mathcal{G}$ be a sig-polynomial and m be a monomial.

- $m\mathcal{G}(j)$ itself is an $\mathcal{S}$-safe representation of $m\mathcal{G}(j)$, called a **trivial representation**.
- Assume that $\mathcal{G}(j) \notin \{F_1, \ldots, F_d\}$. As $\mathcal{G}$ is initialized with $\{F_1, \ldots, F_d\}$, $\sum_{k=1}^d m \cdot u_k F_k$ is an $\mathcal{S}$-safe $\mathcal{G}$-representation of $m\mathcal{G}(j)$, where $(u_1, \ldots, u_d) = \mathbf{u} \in R^d$. It is called a **signature representation**[11]. It is important to note that we have
$$\mathcal{S}(mHM(u_k)F_k) <_s \mathcal{S}(m\mathcal{G}(j)), \forall k \in [1,d],$$
except one, say w, such that
$$\mathcal{S}(mHM(u_w)F_w) = \mathcal{S}(m\mathcal{G}(j)).$$
Clearly, $\mathcal{G}(j)$ appears later than F_w by the insertion strategy of the F5GEN, i.e., $m\mathcal{G}(j) \lessdot mHM(u_w)F_w$. So the trivial representation $m\mathcal{G}(j)$ is $\lessdot$-smaller than the signature representation $\sum_{k=1}^d m \cdot u_k F_k$.
- Let $\alpha = (\mathbf{u}, 0)$ be a principal syzygy sig-polynomial generated by $\mathcal{G}(i)$ and $\mathcal{G}(j)$ in $\mathcal{G}$. That is,
$$\alpha = poly(\mathcal{G}(i))\mathcal{G}(j) - poly(\mathcal{G}(j))\mathcal{G}(i),$$
which is an $\mathcal{S}$-safe $\mathcal{G}$-representation.

Then Lemma 4 is proved as follows.

PROOF. Suppose ζ cannot be F5-reduced by the present $\mathcal{G}$. Let $m\mathcal{G}(k)$ be a non-syzygy sig-polynomial $\mathcal{S}$-reducing ζ. Either $\mathcal{S}(m\mathcal{G}(k)) \in \mathcal{S}(PS)$ or $m\mathcal{G}(k)$ is rewritable by $\mathcal{G}$.

1. If $\mathcal{S}(m\mathcal{G}(k))$ is in $\mathcal{S}(PS)$, there is a principal syzygy sig-polynomial $(\mathbf{u}', 0)$, such that $m'\mathcal{S}(\mathbf{u}') = m\mathcal{S}(\mathcal{G}(k))$. Adding $(c \cdot m\mathbf{u}', 0)$ to $m\mathcal{G}(k)$, we can obtain a sig-polynomial $(\mathbf{u}^*, m \cdot poly(\mathcal{G}(k)))$ such that $\mathcal{S}(\mathbf{u}^*) <_s m\mathcal{S}(\mathcal{G}(k))$. The signature representation $\sum_t u_t^* F_t$ of $(\mathbf{u}^*, m \cdot poly(\mathcal{G}(k)))$ is a representation of $m\mathcal{G}(k)$ $\lessdot$-smaller than the trivial representation $m\mathcal{G}(k)$. If the signature of an element $m_{t_v} F_t \in \sum_t u_t^* F_t$ is in $\mathcal{S}(PS)$, we can find an $\mathcal{S}$-safe $\mathcal{G}$-representation $\lessdot$-smaller than $\sum_t u_t^* F_t$ by the same method discussed above.
2. If $m\mathcal{G}(k)$ is rewritable by $\mathcal{G}$, let $\mathcal{G}(j)$ be the one rewriting $m\mathcal{G}(k)$. So $j > k$ and $\mathcal{S}(m'\mathcal{G}(j)) = \mathcal{S}(m\mathcal{G}(k))$, i.e., $m'\mathcal{G}(j) \lessdot m\mathcal{G}(k)$. As $m\mathcal{G}(k)$ and $m'\mathcal{G}(j)$ share a common $\lessdot$-maximum element, say $m^* F_w$ with
$$w = idx(m\mathcal{G}(k)) = idx(m'\mathcal{G}(j)),$$
by canceling $m^* F_w$, $m\mathcal{G}(k)$ has an $\mathcal{S}$-safe representation of the form
$$m'\mathcal{G}(j) + \sum_t p_t F_t, \tag{5}$$
which is $\lessdot$-smaller than the trivial $\mathcal{G}$-representation $m(SG)'(k)$. Similarly, if $m_{t_v} F_t$ in representation (5) can be rewrite by $\mathcal{G}_{\leq_s(\mathbf{s}_i)}$, we would obtain a $\lessdot$-smaller $\mathcal{S}$-safe $\mathcal{G}$-representation.

Because $\lessdot$ is a well-order on $\mathcal{M} \times \mathcal{G}$, we can ensure that there is an $\mathcal{S}$-safe $\mathcal{G}$-representation of $m\mathcal{G}(k)$ such that each element is neither in PS nor rewritable by $\mathcal{G}$. Together with the following lemma, we would have an $\mathcal{S}$-safe $HM(m\mathcal{G}(k))$-representation such that each element would be neither in PS nor rewritable by updated $\mathcal{G}$ after finite while-loops. Therefore, there exists a sig-polynomial in updated $\mathcal{G}$ F5-reducing $m\mathcal{G}(k)$. □

Making use of the proof for Lemma 32 of [10], we can obtain a short lemma below.

LEMMA 7. *Assume the F5GEN has computed correctly up to signature* $\mathbf{s}$. *Let* $\sum_t p_t\mathcal{G}(j_t)$ *be an* $\mathcal{S}$*-safe* $\mathcal{G}$*-representation of* $m\mathcal{G}(k)$, *where* $m\mathcal{S}(\mathcal{G}(k)) <_s \mathbf{s}$. *We assume that each element* $m_{t_y}\mathcal{G}(j_t)$ *is neither in* PS *nor rewritable by* $\mathcal{G}$. *If there exists an element* $m_{t_v}\mathcal{G}(j_t)$ *such that* $HM(m_{t_v}\mathcal{G}(j_t)) > HM(m\mathcal{G}(k))$, *where* $m_{t_v} \in p_t$, $m\mathcal{G}(k)$ *would have an* $\mathcal{S}$*-safe* $\mathcal{G}$*-representation* $\lessdot$*-smaller than* $\sum_t p_t\mathcal{G}(j_t)$ *after finite runs through the while-loop.*

PROOF. Because $poly(m\mathcal{G}(k)) = \sum_t p_t \cdot poly(\mathcal{G}(j_t))$ and $HM(m_{t_v}\mathcal{G}(j_t)) > HM(m\mathcal{G}(k))$, there exists another element $m_{r_w}\mathcal{G}(j_r)$ such that
$$HM(m_{r_w}\mathcal{G}(j_r)) = HM(m_{t_v}\mathcal{G}(j_t)).$$
Let m' and m'' be two monomials such that
$$HM(m'\mathcal{G}(j_t)) = HM(m''\mathcal{G}(j_r)) = lcm(\mathcal{G}(j_t), \mathcal{G}(j_r))$$
and $m'\mathcal{S}(\mathcal{G}(j_t)) > m''\mathcal{S}(\mathcal{G}(j_r))$. If $m'\mathcal{G}(j_t)$ had been considered by the F5GEN algorithm, there would be a sig-polynomial in current $\mathcal{G}$ rewriting $m'\mathcal{G}(j_t)$, a contradiction. From the selection strategy of the F5GEN, we know that $deg(m'\mathcal{S}(\mathcal{G}(j_t)) = deg(\mathbf{s})$ and $m'\mathcal{G}(j_t)$ has not been considered. Obviously, the F5GEN cannot generate infinite J-pairs such that their signatures are of a fixed degree. So $m'\mathcal{G}(j_t)$ would be output at line 6 of the algorithm after finite runs through the while-loop. By the assumption of the lemma, the J-pair $m'\mathcal{G}(j_t)$ can pass the F5-criteria of the F5GEN algorithm and would be F5-reduced to a new sig-polynomial $\mathcal{G}(\ell)$ stored into $\mathcal{G}$. Hence
$$\mathcal{S}(\mathcal{G}(\ell)) = m'\mathcal{S}(\mathcal{G}(j_t)) \text{ and } \ell > j_t,$$
i.e., $\mathcal{G}(\ell) \lessdot m'\mathcal{G}(j_t)$ and $\mathcal{G}(\ell)$ can rewrite the element $m'\mathcal{G}(j_t)$. By the method of case (2) in the above lemma, we can find an $\mathcal{S}$-safe $\mathcal{G}$-representation $\lessdot$-smaller than $\sum_t p_t\mathcal{G}(j_t)$. □

[10] The comparison is proceeded by padding with zeros at the right of the shorter array if A and A' are not of equal length.

[11] Different from the input-representation defined in [10], the signature representation of a fixed sig-polynomial is unique.

On the Boolean Complexity of Real Root Refinement

Victor Y. Pan
Depts. of Mathematics and Computer Science
Lehman College and Graduate Center
of the City University of New York
Bronx, NY 10468 USA
victor.pan@lehman.cuny.edu
http://comet.lehman.cuny.edu/vpan/

Elias P. Tsigaridas
INRIA, Paris-Rocquencourt Center,
PolSys Project
UPMC, Univ Paris 06, LIP6
CNRS, UMR 7606, LIP6
Paris, France
elias.tsigaridas@inria.fr

ABSTRACT

We assume that a real square-free polynomial A has a degree d, a maximum coefficient bitsize τ and a real root lying in an isolating interval and having no nonreal roots nearby (we quantify this assumption). Then, we combine the *Double Exponential Sieve* algorithm (also called the *Bisection of the Exponents*), the bisection, and Newton iteration to decrease the width of this inclusion interval by a factor of $t = 2^{-L}$. The algorithm has Boolean complexity $\widetilde{\mathcal{O}}_B(d^2\tau + dL)$. Our algorithms support the same complexity bound for the refinement of r roots, for any $r \leq d$.

Categories and Subject Descriptors:
F.2 [Theory of Computation]: Analysis of Algorithms and Problem Complexity; I.1 [Computing Methodology]: Symbolic and algebraic manipulation: Algorithms

General Terms
Algorithms, Experimentation, Theory

Keywords
real root refinement; polynomial; real root problem, Boolean complexity

1. INTRODUCTION

Given a polynomial A, of degree d and maximum coefficient bitsize τ, and an interval with rational endpoints that contains one of its real roots (isolating interval), we devise an algorithm that refines this *inclusion interval* to decrease its width by a factor $t = 2^{-L}$, for some positive integer L.

The problem of real root refinement appears very often as an important ingredient of various algorithms in computer algebra and nonlinear computational geometry, for example in algorithms for computing the topology of real plane algebraic curves [11, 22], solving systems of polynomial equations [24, 14, 17], isolating the real roots of polynomials with coefficients in an extension field [38, 20], cylindrical algebraic decomposition [2, 13], and many others.

For the complexity of approximating (all) the roots of a polynomial we refer the reader to [29], see also [27], [25, Chapter 15] where a Boolean complexity of $\widetilde{\mathcal{O}}_B(d^3 + d^2L)$ is proved, which is "within polylogarithmic factors from the optimum", provided $\tau = \mathcal{O}(L)$.

The problem of refinement is also an important ingredient for algorithms that tackle the real root problem, that is the problem of isolating and approximating the real roots of a polynomial that has only real roots. In this context we refer to the work of Ben-Or and Tiwari [3] that introduced interlacing polynomials and *Double Exponential Sieve*. Pan and Linzer [30] and Bini and Pan [8] in a sequence of works, see also [6, 7], modified the approach of [3] (they called it *Bisection of the Exponents*) to approximate the eigenvalues of a real symmetric tridiagonal matrix by using Courant-Fischer minimax characterization theorem. In [32] a variant of the refinement algorithms in [30, 8] is used, for approximating all the real roots of a polynomial.

Collins and Krandick [12] presented a variant of Newton's algorithm where all the evaluations involve only dyadic numbers, as well as a comparison with the case where operations are performed with rationals of arbitrary size. Quadratic convergence of Newton's iterations is guaranteed by point estimates and α-theory of Smale, e.g. [9]. For robust approximation of zeros based on bigfloats operations we refer to [37]. A very interesting and efficient algorithm that combines bisection and Newton iterations is the *Quadratic Interval Refinement* (QIR) by Abbott [1]. For a detailed analysis of the Boolean complexity of QIR we refer the reader to [21]. Kerber and Sagraloff [22] modify QIR to use interval arithmetic and approximations and they achieve a bound of $\widetilde{\mathcal{O}}_B(d^3\tau^2 + dL)$. A factor of τ could be saved if we use fast algorithms for root isolation of univariate polynomials, e.g. [35], [29, 36]. We should also mention [34] that is based on Kantorovich point estimates which is efficient in practice but of unknown complexity.

We revisit the approach of [3], [30, 8] to devise our *Real Root Refinement* ($\mathtt{R_3}$) algorithm and present a detailed analysis in the bit complexity model, based on exact operations with rationals (Thm. 12). We also introduce an approximate variant ($\alpha\mathtt{R_3}$) based on interval arithmetic, Sec. 2.1, where we use multi-precision floating point numbers for computations and to represent the endpoints of intervals, and we estimate in advance the maximum precision needed. For this we use tools from Kerber and Sagraloff [22] for evaluating a polynomial at a rational number using interval arithmetic. We also study the Newton operator both from an exact and approximate point of view (Sec. 2.3). We provide Boolean complexity bounds for approximate variants of Double Exponential Sieve (Lem. 4) and Newton iteration (Lem. 10).

ISSAC'13, June 26–29, 2013, Boston, Massachusetts, USA.

The Boolean complexity of $\mathtt{R}_3$ and $\alpha\mathtt{R}_3$ is $\widetilde{\mathcal{O}}_B(d^2\tau + dL)$ (Theorem 12). The same algorithms support the bound $\widetilde{\mathcal{O}}_B(d^2\tau + d^2L)$ and $\widetilde{\mathcal{O}}_B(d^2\tau + dL)$, respectively, for the refinement of up to d roots (Section 2.4). We assume that there is no complex root of the polynomial in the complex disc that has the isolating interval as diameter. Such an interval could be the outcome of root-finding algorithms. We detail this in Section 2.5.

The rest of the paper is structured as follows. First we introduce our notation. Section 2 presents a high level description of the real root refinement algorithm. We detail its three steps, in Sec. 2.1, Sec. 2.2, and Sec. 2.3. Section 3 estimates the expected number of steps of DES and αDES when the input polynomial is random of type Weyl, Sec. 3.1, or $SO(2)$, Sec. 3.2. Finally, in Section 4 we conclude and suggest directions for furhter study.

Notation. In what follows $\mathcal{O}_B$, resp. $\mathcal{O}$, means bit, resp. arithmetic, complexity and the $\widetilde{\mathcal{O}}_B$, resp. $\widetilde{\mathcal{O}}$, notation means that we are ignoring logarithmic factors. For a polynomial $A = \sum_{i=0}^{d} a_i\, x^i \in \mathbb{Z}[x]$, $\mathsf{deg}(A) = d$ denotes its degree and $\mathcal{L}(A) = \tau$ the maximum bitsize of its coefficients, including a bit for the sign. For $a \in \mathbb{Q}$, $\mathcal{L}(a) \geq 1$ is the maximum bitsize of the numerator and the denominator. $\mathsf{M}(\tau)$ denotes the bit complexity of multiplying two integers of size τ; we have $\mathsf{M}(\tau) = \widetilde{\mathcal{O}}_B(\tau)$. 2^{Γ} is an upper bound on the magnitude of the roots of A. We write $\Delta_\alpha(A)$ or just Δ_α to denote the minimum distance between a root α of a polynomial A and any other root, we call this quantity *local separation bound*. If we are considering the i-th root, α_i, then we also write Δ_i instead of Δ_{α_i}. $\Delta(A) = \min_\alpha \Delta_\alpha(A)$ or just Δ denotes the *separation bound*, that is the minimum distance between all the roots of A. The Mahler bound (or measure) of A is $\mathcal{M}(A) = a_d \prod_{|\alpha|\geq 1} |\alpha|$, where α runs through the complex roots of A, e.g. [26, 40]. If $A \in \mathbb{Z}[x]$ and $\mathcal{L}(A) = \tau$, then $\mathcal{M}(A) \leq \|A\|_2 \leq \sqrt{d+1}\|A\|_\infty = 2^\tau\sqrt{d+1}$. If we evaluate a function F (e.g. $F = A$) at a number c using interval arithmetic, then we denote the resulting interval by $[F(c)]$, provided that we fix the evaluation algorithm and the precision of computing. Let $D(c, r) = \{x : |x - c| \leq r\}$.

2. THE $\mathtt{R}_3$ AND $\alpha\mathtt{R}_3$ ALGORITHMS

In what follows $A = \sum_{i=0}^{d} a_i x^i \in \mathbb{Z}[x]$ with $\mathcal{L}(A) = \tau$. Let α_1 be the real root of A that lies in an (isolating) interval $I = [a \,..\, b]$. The width of I, $w = (b - a)/2$, has bitsize $\mathcal{L}(w) = \mathcal{O}(\lg \Delta(A)) = \mathcal{O}(d\tau)$, in the worst case (see also Prop. 1). We write $|I| = b - a$ and $m = (a + b)/2$. We wish to refine I to include α_1 into a subinterval of the width $t = 2^{-L}w$. We define the isolation ratio of a real isolating interval I of a root α of A as $\mathtt{ir}(I) = 2\,|m - \alpha_c|/|I|$, where α_c is the root of A that is the closest to α.

The high-level description that we present follows [3] and [30]. For details and various improvements we refer the reader to [30, 6, 8, 32]. The algorithm for refining the isolating interval of a real root α_1 consists of three steps: Double Exponential Sieve (DES), Bisection (BIS) and Newton iteration (NEWTON). We denote the approximate variants by αDES, αBIS, and αNEWTON, respectively. The three procedures are as follows:

(1) DES or αDES achieves an isolation ratio at least 3.

(2) Sufficiently many bisections (BIS or αBIS), but $\widetilde{\mathcal{O}}(\lg(d))$, increase the isolation ratio.

(3) Then Newton iteration, NEWTON or αNEWTON, converges quadratically and yields an inclusion interval of the desired width.

In the sequel we describe in detail and analyze the complexity of the three sub-algorithms.

The following proposition estimates the separation bounds for a univariate (integer) polynomial. For variants and proof techniques we refer the reader to e.g. [13, 15, 26, 20]. We use a variant from [23, Thm. 4].

Proposition 1. *Suppose $A \in \mathbb{Z}[x]$ is a square-free univariate polynomial of degree d, $\mathcal{L}(A) = \tau$, and $\alpha_1, \ldots, \alpha_d$, are the d distinct roots of A. Let $\mathbf{SR}_0(A, A')$ be the resultant of A and its derivative, A', w.r.t. x, and $\Delta_i = \Delta_i(A) = \Delta_{\alpha_i}(A)$, where $i = 1, \ldots, d$. Then it holds*

$$
\begin{aligned}
|\alpha_i| &\leq 2^{\Gamma} \leq 2\|A\|_\infty \leq 2^{\tau+1}\,, && (1)\\
-\sum_i \lg \Delta_i(A) &\leq 30\,d\,\lg \mathcal{M}(A) + 3\,\mathcal{L}\left(\mathbf{SR}_o(A, A')\right) && (2)\\
&\leq 36\,d\,\tau + 42\,d \lg d\,.
\end{aligned}
$$

Lemma 2. *[22] Suppose we evaluate A at c, where $|c| \leq 2^{\Gamma+2}$, and suppose we use a working precision (or fixed precision arithmetic) ρ. Then*

$$\mathtt{width}[A(c)] \leq 2^{-\rho+1}(d+1)^2 2^{\tau+d(\Gamma+2)}\,.$$

The following lemma generalizes [22, Lemma 4].

Lemma 3. *Let x_0 be such that $|x_0 - \alpha_i| \geq \Delta_i/c$ for all* real *α_i such that $i \neq 1$ and $c \geq 2$. Then*

$$|A(x_0)| > |a_d|\,|x_0 - \alpha_1|\,c^{1-d}\,\mathcal{M}(A)^{-1}\,2^{\lg \prod_i \Delta_i - 1}\,.$$

Proof: Let $\Im(\alpha)$ be the imaginary part of $\alpha \in \mathbb{C}$. It holds $|x_0 - \alpha_i| \geq |\Im(\alpha_i)| \geq \Delta_i/2 \geq \Delta_i/c$ and so $|x_0 - \alpha_i| \geq \Delta_i/c$ is true for all the roots of A. Now

$$
\begin{aligned}
|A(x_0)| &= |a_d| \prod_{i=1}^{d} |x_0 - \alpha_i| = |a_d||x_0 - \alpha_1| \prod_{i=2}^{d} |x_0 - \alpha_i|\\
&\geq |a_d||x_0 - \alpha_1| \frac{c}{\Delta_1} \prod_{i=1}^{d} \frac{\Delta_i}{c}\\
&\geq |a_d||x_0 - \alpha_1| c^{1-d} \mathcal{M}(A)^{-1}\, 2^{\lg \prod_i \Delta_i - 1}\,.
\end{aligned}
$$

For the last inequality we use $\Delta_1 \leq 2\,\mathcal{M}(A)$, that in turn relies on $\Delta_1 = |\alpha_1 - \alpha_{c_1}| \leq |\alpha_1| + |\alpha_{c_1}| \leq 2\,\mathcal{M}(A)$, where α_{c_1} is a the root closest to α_1. □

2.1 Double Exponential Sieve

In this subsection we follow [3], [8] and [32] to compute an interval that contains the real root and has endpoints "far away" from the endpoints of the initial interval. The difficult case is when the real root is very close to one of the endpoints of I. Next we outline this procedure referring the reader to [32] for its detailed treatment and efficient implementation.

Initially let $\alpha_1 \in I = [a \,..\, b]$ for $a < b$. We compute a new interval $\bar{I} = [\bar{a} \,..\, \bar{b}]$ containing α_1 and such that either $0 \leq \bar{b} - \bar{a} \leq 2\,t$ or $\mathtt{ir}(\bar{I}) \geq 3$. In the first case the midpoint of $\bar{I}$, $\bar{m} = (\bar{a}+\bar{b})/2$, approximates α_1 within a desired error bound t and hence we return either $[\bar{a} \,..\, \bar{m}]$ or $[\bar{m} \,..\, \bar{b}]$, depending on the sign of $A(\bar{m})$. In the second case we can apply bisections to increase the isolation ratio to the level supporting Newton's iteration. We present the analysis of the bisection iteration in the next subsection.

During the first step of DES, we decide whether α_1 lies in $[a \,..\, \frac{a+b}{2}]$ or $[\frac{a+b}{2} \,..\, b]$. W.l.o.g. assume that $a = 0$ and $b = 2$

because our claims are invariant in the shifts and scaling the variable x. Furthermore assume that $\alpha_1 \in [0..1]$ for the other case is treated similarly. Now the bound $\mathtt{ir}(\bar{I}) \geq 3$ follows where

$$\bar{b} \leq 2\bar{a} \ . \tag{3}$$

Next, we write $a_0 = 0$, $b_0 = 1$, $I_0 = [0..1]$ and evaluate A at the sequence of points $c_k = a_0 + 2w_0/2^{2^k}$, $k = 1, \ldots, g_1$, where $2w_0 = b_0 - a_0$ and $g_1 - 1$ is the maximum index such that $\alpha_1 \in [0..c_{g_1}]$. If $g_1 = 1$, then we write $a' = c_{g_1}$, $b' = 1$, obtain that $\alpha_1 \in [a'..b']$ and $\mathtt{ir}([a'..b']) \geq 5/3$, and yield an interval $\bar{I}$ with $\mathtt{ir}(\bar{I}) \geq 3$ in at most two bisections. Otherwise (if $g_1 > 1$) we reapply the DES procedure to the interval $[c_{g_1}..1]$, denote by g_2 the number of evaluations of $A(x)$ with $x < \alpha_1$ in this process and ensure (3) unless $g_2 < g_1$. Recursively we obtain a strictly decreasing sequence of intervals I_i, each defined by means of g_i evaluations of $A(x)$ where the sequence $g_1, g_2, \ldots$ strictly decreases. This means that the overall number of evaluations of $A(x)$ in the DES procedure is at most $1 + \sum_{i=1}^{g} g_i \leq 1 + (g_1 + 1)g_1/2$ for $u \leq g_1$. Moreover, $g_1 = \lceil \lg(\lg(w) + L - 1) \rceil = \mathcal{O}(\lg(\tau + L))$ because otherwise we would have $0 \leq \bar{b} - \bar{a} \leq 2t$.

The next lemma provides an approximate variant of the algorithm where at each step of the procedure we use exactly the number of bits needed. We call this variant αDES, from *approximate* Double Exponential Sieve.

Lemma 4 (αDES). *The procedure αDES compresses the isolating interval I to an interval J such that $|J| \leq 1/2^L$ or $\mathtt{ir}(J) \geq 3$ using a working precision and time*

$$\mathcal{O}(-g_1 \lg w + 2^{g_1} + g_1(\lg \mathcal{M}(A) + \tau + d\Gamma + \lg \prod_i \Delta_i)) \quad \textit{bits},$$

$$\widetilde{\mathcal{O}}_B(d^2\tau g_1^2 - d g_1^2 \lg w + d g_1 2^{g_1} + d g_1^2(\lg \mathcal{M}(A) + \tau + d\Gamma + \lg \prod_i \Delta_i))$$

or $\widetilde{\mathcal{O}}(d\tau + L)$ and $\widetilde{\mathcal{O}}_B(d^2\tau + dL)$, respectively.

Proof: Initially α_1 lies in the interval $I = I_0 = [a..b]$. Let $w = w_0 = |b - a|$ be its width. We want to compute the maximum integer g_1 such that $\alpha_1 \in (a..a + w/2^{2^{g_1}})$. For this we need to evaluate A on $a + w/2^{2^k}$, for $k = 1, \ldots, g_1$. It might happen that the evaluation of A at one of these numbers is zero. To avoid this, at each step, we evaluate A at two points, instead of one. This multiple evaluation is borrowed from [22].

For each step k we define m_1 and m_2 such that

$$a < m_1 = a + w/2^{2^k+1} < m_2 = a + w/2^{2^k} < b$$

and evaluate A over them. At least one of them is not a zero of A. Let $j \in \{1, 2\}$, then for all $i \neq 1$ it holds $|m_j - \alpha_i| \geq w/2^{2^k+1} \Rightarrow w \leq 2^{2^k+1}|m_j - \alpha_i|$ and $|\alpha_1 - m_j| \leq w/2^{2^k}$. Then

$$\begin{aligned}\Delta_i &\leq |\alpha_1 - \alpha_i| \leq |\alpha_1 - m_j| + |m_j - \alpha_i| \\ &\leq w/2^{2^k} + |m_j - \alpha_i| \leq |m_j - \alpha_i| 2^{2^k+1}/2^{2^k} + |m_j - \alpha_i| \\ &\leq 3|m_j - \alpha_i| \ .\end{aligned}$$

Using Lemma 3 with $c = 3$ we get

$$|A(m_j)| > |m_j - \alpha_1||a_d| 3^{1-d} \mathcal{M}(A)^{-1} 2^{\lg \prod_i \Delta_i - 1} \ .$$

For at least one of m_j's it holds that $|m_j - \alpha_1| > w/2^{2^k+2}$ (actually this is the half the distance between m_1 and m_2) and hence

$$\begin{aligned}|A(m_j)| &> |a_d| \frac{w}{2^{2^k+2}} 3^{1-d} \mathcal{M}(A)^{-1} 2^{\lg \prod_i \Delta_i - 1} \\ &\geq w\, 3^{1-d} 2^{-2^{g_1} - 3 + \lg \prod_i \Delta_i - \lg \mathcal{M}(A) + \lg |a_d|} \ .\end{aligned}$$

Using Lemma 2 the precision needed for this step, ρ_1, satisfies the equation

$$\begin{aligned}&2^{-\rho_1+1}(d+1)^2 2^{\tau + d(\Gamma+2)} < \\ &\quad w\, 3^{1-d} 2^{-2^{g_1} - 3 + \lg \prod_i \Delta_i - \lg \mathcal{M}(A) + \lg |a_d|} \ ,\end{aligned}$$

and thus

$$\rho_1 = \widetilde{\mathcal{O}}(-\lg w + 2^{g_1} + \lg \mathcal{M}(A) + \tau + d\Gamma - \lg \prod_i \Delta_i) \ .$$

To support our computation of the desired interval J we double the precision of computing at every DES step. We perform $\lg(\rho_1)$ steps overall; each is essentially an evaluation of A. By applying Horner's rule we yield the cost bound $\widetilde{\mathcal{O}}_B(d(d\tau + \rho_1))$. Similarly, at the i-th step of αDES we perform $g_i \lg(\rho_i)$ evaluations, each at the cost of $\widetilde{\mathcal{O}}_B(d(d\tau + \rho_i))$, where

$$\rho_i = \widetilde{\mathcal{O}}(-\lg w_{i-1} + 2^{g_i} + \lg \mathcal{M}(A) + \tau + d\Gamma - \lg \prod_i \Delta_i) \ .$$

Summarizing, the overall cost of performing v steps is bounded by

$$\widetilde{\mathcal{O}}_B\left(\sum_{i=1}^{v} d(d\tau + \rho_i) g_i \lg \rho_i\right) \ .$$

The sequence of g_i's is strictly decreasing, that is $g_i > g_{i+1}$, and so $v < g_1^2$. The produced intervals I_i have widths $w_i < w/2^{2^i}$, and so

$$\sum_{i=0}^{v-1} \lg w_i = g_1 \lg w - 2^{g_1} + 1 \ ,$$

$$\sum_i \rho_i = \mathcal{O}(-g_1 \lg w + 2^{g_1} + g_1(\lg \mathcal{M}(A) + \tau + d\Gamma - \lg \prod_i \Delta_i)) \ ,$$

and the overall cost is bounded by

$$\widetilde{\mathcal{O}}_B(d^2\tau g_1^2 - d g_1^2 \lg w + d g_1 2^{g_1} + d g_1^2(\lg \mathcal{M}(A) + \tau + d\Gamma - \lg \prod_i \Delta_i)) \ .$$

By noticing that $|a_d| \geq 1$, $\lg w = \mathcal{O}(d\tau)$, $\Gamma = \mathcal{O}(\tau)$, $\mathcal{O}(\lg \mathcal{M}(A)) = \mathcal{O}(\tau + \lg d)$ and using Prop. 1 to bound $\lg \prod_i \Delta_i$, we get that the maximum precision needed is $\widetilde{\mathcal{O}}(d\tau + L)$. Finally, the complexity of αDES is $\widetilde{\mathcal{O}}_B(d^2\tau + dL)$. □

Remark 5 (DES). *We call this procedure DES if it uses only exact arithmetic with rational numbers. Then, in the worst case, we perform $g_1^2 = \mathcal{O}(\lg^2(\tau + L))$ evaluations of A at numbers of bitsize $\mathcal{O}(L)$.*

Using Horner's rule, each evaluation costs $\widetilde{\mathcal{O}}_B(d^2(\tau + L))$. Hence, the overall complexity is $\widetilde{\mathcal{O}}_B(d^2(\tau + L)\lg^2(\tau + L)) = \widetilde{\mathcal{O}}_B(d^2(\tau + L))$. This bound is greater by a factor of d than the one supported by αDES.

However, since we are working with exact arithmetic in the bit complexity model, Horner's arithmetic is not optimal. To see this, notice that the output of the evaluation is of bitsize $\widetilde{\mathcal{O}}(d(\tau + L))$. If we use the divide and conquer approach [10, 19], each evaluation costs $\widetilde{\mathcal{O}}_B(d(\tau + L))$ and the overall *complexity is $\widetilde{\mathcal{O}}_B(d(\tau+L)\lg^2(\tau+L)) = \widetilde{\mathcal{O}}_B(d(\tau+L))$. This bound is the same as the one supported by αDES.*

Nevertheless, this approach has some drawbacks. First, we are forced to work with full precision right from the beginning. Even though this does not affect the worst case bound it is a serious disadvantage for implementations. Second, this approach does not scale well, in the case where we want to refine all the real roots of A. Then, we have to multiply the bound by d, which is not the case for the approximate algorithm. For further details we refer to Section 2.4.

2.2 Bisection(s)

This section covers the second step of the refinement algorithm. Recall that the isolation ratio of a real isolating interval I of a root α_1 is defined as $\mathtt{ir}(I) = 2|m - \alpha_{c_1}|/|I|$, where α_{c_1} is the root of A closest to α [30]. Our goal is, using bisections, to achieve an isolation ratio of $\mathtt{ir}(I) \geq 5d^2$, which ensures the quadratic convergence of Newton iteration right from the start [33, Corollary 4.5].

If $\mathtt{ir}(I) = 1 + \delta$ for an interval $I = [a\mathinner{..}b]$, then after k bisection steps, we obtain an interval I_k, such that $\mathtt{ir}(I_k) \geq 1 + 2^k\delta$. Details of the algorithm appear in Alg. 2. We increase the isolation ratio by applying bisections, for which we need to evaluate A. We set dir, equal to -1, 0, or 1 to indicate the search direction. If the initial isolation ratio is $\mathtt{ir}(I) = 1+2\delta$, then after k bisections, in the worst case, we increase the isolation ratio to $\mathtt{ir}(I_k) = 1 + 2^k\delta$, where I_k is the new, refined, isolating interval. The variable dir takes the values `left` or `right` and specifies if the roots is on the left or on the right of m. If we know in advance whether the closest root of A not belonging to the interval I lies to the left or the right of the midpoint of I then we may set dir accordingly. Otherwise we set $dir = \emptyset$. In the latter case, after the first bisection, we are sure about the value of dir. Even if this observation does not affect the asymptotic behavior of the algorithm, it can save us a constant number of bisections, which might be important in practice.

If the initial isolation ratio is $r = 1 + 2 \cdot 2^{\lg(r-1)-1}$, then after k steps it becomes $1 + 2^k \cdot 2^{\lg(r-1)-1}$. If our goal is to achieve an isolation ratio R, then

$$1 + 2^k \cdot 2^{\lg(r-1)-1} \geq R \Leftrightarrow 2^k \cdot 2^{\lg(r-1)-1} \geq R - 1 \Leftrightarrow$$
$$k + \lg(r-1) - 1 \geq \lg(R-1) \Leftrightarrow k \geq 1 + \lg\frac{R-1}{r-1} \ .$$

In our case $R = 5d^2$. From the previous step αDES guarantees an isolation ratio at least 3 and thus we need to perform $k = \mathcal{O}(\lg(d))$ bisections.

Each bisection consists of an evaluation of A over the midpoint of the corresponding interval and setting dir accordingly. We will perform this in an approximate way using the algorithm from [22]. We need the following lemma for the approximate variant, αBIS.

Lemma 6. *[22] The approximate bisection for a root $\alpha_1 \in I = [a\mathinner{..}b]$ of A requires a working precision of $\rho = \widetilde{\mathcal{O}}(-\lg w + \tau + d\Gamma + \lg\prod_i \Delta_i))$ bits and has bit complexity $\widetilde{\mathcal{O}}_B(d(-\lg w + \tau + d\Gamma + \lg\prod_i \Delta_i))$, where w is the width of I.*

A single bisection halves an interval of width w, k bisections decrease the width to $w_k = w/2^k$.

In the worst case the interval has the width $w\,2^{-L}$, and so the number of bits needed in the worst case is $\rho = \widetilde{\mathcal{O}}(-\lg w + L + \tau + d\Gamma + \lg\prod_i \Delta_i)$. We perform $\mathcal{O}(\lg(d))$ bisections, and so the overall cost is $\widetilde{\mathcal{O}}_B(d(-\lg w + L + \tau + d\Gamma + \lg\prod_i \Delta_i))$.

Recall that $-\lg w = -\mathcal{O}(\lg \Delta(A)) = \widetilde{\mathcal{O}}(d\tau)$.

Lemma 7 (αBIS). *The cost of αBIS is $\widetilde{\mathcal{O}}_B(d(-\lg w + L + \tau + d\Gamma + \lg\prod_i \Delta_i))$ or $\widetilde{\mathcal{O}}_B(d^2\tau + dL)$.*

Remark 8 (BIS). *If we perform the bisection step using only exact arithmetic with rational numbers, then we perform $\mathcal{O}(\lg(d))$ evaluations of A over numbers of bitsize $\mathcal{O}(L - \lg w)$, in the worst case. Each evaluation costs $\widetilde{\mathcal{O}}_B(d(\tau - \lg w + L))$, using the divide and conquer scheme [19, 10]. Hence, the overall complexity is $\widetilde{\mathcal{O}}_B(d(\tau - \lg w + L)\lg(d)) = \widetilde{\mathcal{O}}_B(d(d\tau + L)) = \widetilde{\mathcal{O}}_B(d^2\tau + dL)$.*

2.3 Bounding the Newton operator

The last step of the refinement algorithm consists in performing a suitable number of Newton iterations to refine the isolating interval up to the required width. The bisections of the previous step ensure that the interval is small enough that Newton iteration converges quadratically, right from the beginning. Actually it satisfies the conditions of the following theorem [33, Corollary 4.5].

Theorem 9. *Suppose both discs $D(m,r)$ and $D(m,r/s)$ for $s \geq 5d^2$ contain a single simple root of a polynomial $A = A(x)$ of degree d. Then Newton's iteration*

$$x_{k+1} = x_k - A(x_k)/A'(x_k), k = 0, 1, \ldots \qquad (4)$$

converges quadratically to the root α right from the start provided $x_0 = m$.

First, we estimate the precision needed at each of Newton iteration. Given an interval $[a_k\mathinner{..}b_k]$, let $m_k = (a_k + b_k)/2$ be its middle point, where we apply the Newton operator, that is

$$NA(m_k) = m_k - \frac{A(m_k)}{A'(m_k)} \ ,$$

where $A_k = A(m_k)$, $A'_k = A'(m_k)$, and A' is the derivative of A. We assume that $A_k > 0$ and $A'_k < 0$. The other sign combinations could be treated similarly. Suppose we compute A_k and A'_k using interval arithmetic and a working precision ρ to be specified later. We can assume that their interval representation is $[A_k] = [A_k - \epsilon\mathinner{..}A_k + \epsilon]$ and $[A'_k] = [A'_k - \epsilon\mathinner{..}A'_k + \epsilon]$, both having the same width, ϵ.

The interval evaluation of Newton operator, using the same working precision ρ, results in the interval

$$[NA(m_k)] = \left[m_k - \epsilon - \frac{A_k - \epsilon}{A'_k - \epsilon}\mathinner{..}m_k + \epsilon - \frac{A_k + \epsilon}{A'_k + \epsilon}\right] \ .$$

The width of $[NA(m_k)]$ is $2\epsilon - \frac{A_k+\epsilon}{A'_k+\epsilon} + \frac{A_k-\epsilon}{A'_k-\epsilon}$, and now we ensure its upper bound $t = 2^{-L}w$.

$$2\epsilon - \frac{A_k + \epsilon}{A'_k + \epsilon} + \frac{A_k - \epsilon}{A'_k - \epsilon} \leq t \Rightarrow$$
$$P(\epsilon) = 2\epsilon^3 - t\epsilon^2 - 2(A_k - A'_k + {A'_k}^2)\epsilon + t{A'_k}^2 \geq 0 \ .$$

The coefficient list of P has 2 sign variations and hence from Descartes' rule of signs it follows that P has at most two positive real roots. If there exists such a pair of roots, let them be $\epsilon_1 < \epsilon_2$ and assume that P is positive between 0 and ϵ_1. For the width of $[NA(m_k)]$ to be smaller than $t = 2^{-L}w$, it suffices ϵ to satisfy $0 < \epsilon \leq \min\{1, \epsilon_1\}$. To guarantee this we estimate a (positive) lower bound on the roots of the P and require ϵ to be smaller than it. Combine Lemma 2 and Cauchy's bound, e.g. [40, 26] to obtain

$$\epsilon \leq \frac{t{A'_k}^2}{t+2(A_k - A'_k + {A'_k}^2)} .$$

by working with a precision ρ that satisfies

$$2^{-\rho+1}(d+1)^2 2^{\tau+d(\Gamma+2)} \leq \epsilon \leq \frac{t{A'_k}^2}{t+2(A_k - A'_k + {A'_k}^2)} . \quad (5)$$

Hence, we can express ρ as a function of the desired width.

It remains to bound the evaluations A_k and A'_k. At the k-th step, given an interval I_k, we apply Newton operator on its midpoint, m_k, and deduce that $\Delta_i < 2|m_k - \alpha_i|$ for all $i \neq 1$. Indeed for $m_k \leq \alpha_1 \leq \alpha_i$, it holds $|m_k - \alpha_i| \geq |\alpha_1 - \alpha_i| \geq \Delta_i$, whereas for $\alpha_1 \leq m_k \leq \alpha_i$ it holds $|\alpha_1 - m_k| \leq |m_k - \alpha_i|$, because α_i lies outside the isolating interval. Therefore $\Delta_i \leq |\alpha_i - \alpha_1| \leq |\alpha_i - m_k| + |\alpha_1 - m_k| \leq 2|m_k - \alpha_i|$.

So using Lemma 3 we get

$$|A_k| \geq |m_k - \alpha_1||a_d| \, 3^{1-d} \mathcal{M}(A)^{-1} 2^{\lg \prod_i \Delta_i - 1} . \quad (6)$$

For the approximations achieved by Newton iterations [32, 8], when the convergence is quadratic, we have

$$|m_k - \alpha_1| = 2^{4-2^k} |m_0 - \alpha_1| . \quad (7)$$

Obviously

$$|m_0 - \alpha_1| \geq t = 2^{-L} w ,$$

since the required width is not achieved from the initial interval, and so

$$|m_k - \alpha_1| = 2^{4-2^k} |m_0 - \alpha_1| \geq 2^{4-2^k - L + \lg w} , \quad (8)$$

which leads to

$$|A_k| \geq |a_d| \, 2^{4-2^k-L+\lg w} \, 3^{1-d} \mathcal{M}(A)^{-1} 2^{\lg \prod_i \Delta_i - 1} . \quad (9)$$

We need a similar bound for $|A'_k|$. Let α'_i be the roots of A'. We assume that A' is square-free. This is no loss of generality since we can estimate the required quantities using the square-free part of A'.

Let the two roots of A' that are closer to α_1 be α'_1 and α'_2, which are located to the left and to the right of α_1, respectively. Let α' denote any other root of A'. Then it holds that $|m_k - \alpha'_i| \geq |\alpha' - \alpha'_i| \geq \Delta'_i$, where Δ'_i is the local separation bound of α'_i and where α' is either α'_1 or α'_2 depending on which side from m_k the root α'_i lies. To see this assume that α'_i lies on the right of α'_2. Then it holds $m_k \leq \alpha'_2 \leq \alpha'_i$, and so $|m_k - \alpha'_2| \geq |\alpha'_2 - \alpha'_i| \geq \Delta'_i$. A similar argument holds when α'_i lies on the left of α'_1. Therefore

$$\begin{aligned} |A'_k| = |A'(m_k)| &= |d\,a_d| \prod_{i=1}^{d-1} |m_k - \alpha'_i| = \\ &= |d\,a_d||m_k - \alpha'_1||m_k - \alpha'_2| \prod_{i=3}^{d} |m_k - \alpha'_i| \\ &\geq |d\,a_d||m_k - \alpha'_1||m_k - \alpha'_2| \frac{1}{\Delta'_1 \Delta'_2} \prod_{i=1}^{d} \Delta'_i \\ &\geq |d\,a_d||m_k - \alpha'_1||m_k - \alpha'_2| \, \mathcal{M}(A')^{-2} 2^{-2} 2^{\lg \prod_i \Delta'_i} , \end{aligned}$$

where we used the inequality $\Delta'_i \leq 2\mathcal{M}(A')$.

We bound $|m_k - \alpha'_1|$ and $|m_k - \alpha'_2|$ using the width of I_k. We notice that m_k is closer to α_1 than to α'_1 and α'_2. This is so because both α'_1 and α'_2 lie outside the interval of interest, which holds because of the quadratic convergence of Newton operator.

The complex disc with diameter the interval I_0 satisfies the assumptions of Thm. 9, and this is also the case for all discs with the centers at m_k and radii $|I_k|/2$. In other words, the roots α'_1 and α'_2 lie outside the $5d^2$-dilations of these discs, that is the centers m_k lie much closer to α_1 than to α'_1 and α'_2.

So using Eq. (8) we obtain $|m_k - \alpha'_1| \geq |m_k - \alpha_1| \geq 2^{4-2^k-L+\lg w}$. Hence

$$|A'_k| \geq |d\,a_d| \, 2^{8-2^{k+1}-2L+2\lg w} \mathcal{M}(A')^{-2} 2^{-2+\lg \prod_i \Delta'_i} . \quad (10)$$

Now we can bound the right-hand side of inequality (5). First, we bound the numerator. For all m_k it holds that $|m_k| \leq 2^{\Gamma} \leq 2^{\tau+1}$, see Eq. (1). Thus

$$|A'_k| = |A'(m_k)| \leq \sum_{i=1}^{d} i\,|a_i|\, 2^{(i-1)\Gamma} \leq d\, 2^{\tau} 2^{d\Gamma} \leq 2^{2d\tau + \lg d} ,$$

and so

$$t{A'_k}^2 \leq 2^{-L+\lg w + 6d\tau + 2\lg d} . \quad (11)$$

Next, we bound the denominator of the upper bound in (5). To simplify the notation let $|A_k| \geq 2^{-\ell_1} > 0$ and $|A'_k| \geq 2^{-\ell_2} > 0$. For the exact bounds we refer the reader to equations (9) and (10). Recall that we assume A'_k to be negative and hence $-A'_k > 0$. Then

$$\begin{aligned} t + 2(A_k - A'_k + {A'_k}^2) &\geq 2^{\min\{-\ell_1, -2\ell_2, -L+\lg w\}} \\ &\geq 2^{-\max\{\ell_1, 2\ell_2, L-\lg w\}} , \end{aligned}$$

and so

$$\frac{1}{t + 2(A_k - A'_k + {A'_k}^2)} \leq 2^{\max\{\ell_1, 2\ell_2, L-\lg w\}} . \quad (12)$$

Combining (9) and (10) with (11), (12) and (5) we deduce that the required precision is

$$\widetilde{\mathcal{O}}\left(L - \lg w + 2^k + \tau + \lg\left(\mathcal{M}(A)\,\mathcal{M}(A')\right) - \lg \prod_i \Delta_i \prod_i \Delta'_i\right) .$$

We want to achieve $|m_k - \alpha_1| \leq w/2^L$. Using Eq. (7) we get

$$|m_k - \alpha_1| \leq 2^{4-2^k} \frac{w}{2} \leq \frac{w}{2^L} \Rightarrow k \geq \lg(L+3) .$$

Hence, we need to apply the Newton operator $k = \mathcal{O}(\lg(L))$ times, to refine the interval by a factor of 2^{-L}. So the overall bit complexity of this step is

$$\widetilde{\mathcal{O}}_B\left(d\left(L - \lg w + d\tau + \lg\left(\mathcal{M}(A)\,\mathcal{M}(A')\right) - \lg \prod_i \Delta_i \prod_i \Delta'_i\right)\right) .$$

Note that $|a_d| \geq 1$, $\mathcal{O}(\lg \mathcal{M}(A)) = \mathcal{O}(\tau + \lg d)$, $\mathcal{O}(\lg \mathcal{M}(A')) = \mathcal{O}(\tau \lg d)$, $-\lg w = \mathcal{O}(d\tau)$, use Prop. 1 to bound $\lg \prod_i \Delta_i$ and $\lg \prod_i \Delta'_i$, and obtain that the precision $\widetilde{\mathcal{O}}(d\tau + L)$ is sufficient for us. So the Boolean complexity of αDES is $\widetilde{\mathcal{O}}_B(d^2\tau + dL)$.

Lemma 10 (αNEWTON). *The maximum number of bits needed by Newton iterations is $\widetilde{\mathcal{O}}(d\tau + L)$ and the total complexity of the Newton step is $\widetilde{\mathcal{O}}_B(d^2\tau + dL)$.*

We should mention that there is no need to realize the Newton iteration using interval arithmetic. However, it is easier to estimate theoretically the working precision needed using the formalization of interval arithmetic.

Remark 11 (NEWTON). *We can also estimate the complexity of Newton iterations in the case where only exact arithmetic with rational number is used. We need to perform $\mathcal{O}(\lg(L))$ Newton iterations, each of which consists of an evaluation of A and its derivative over numbers of bitsize $\mathcal{O}(\tau + L - \lg w)$, in the worst case. The cost of the evaluations is $\widetilde{\mathcal{O}}_B(d(\tau + L - \lg w))$. Hence the overall complexity is $\widetilde{\mathcal{O}}_B(d(\tau + L - \lg w)\lg(L)) = \widetilde{\mathcal{O}}_B(d^2\tau + dL))$.*

2.4 Overall complexity of $\mathtt{R}_3$ and $\alpha\mathtt{R}_3$

Theorem 12 ($\alpha\mathtt{R}_3$). *We can refine an isolating interval of a real root of A to decrease its width by a factor of 2^{-L} by using $\alpha\mathtt{R}_3$ or $\mathtt{R}_3$ with Boolean complexity $\widetilde{\mathcal{O}}_B(d^2\tau + dL)$.*

The bound of Th. 12 for $\alpha\mathtt{R}_3$ holds even if we want to refine all the real roots of A. The main operation needed is the evaluation of a polynomial at, at most, r rational numbers, where $r \leq d$ is the number of real roots that we need refine. We can evaluate a polynomial at d points by using $\widetilde{\mathcal{O}}(d)$ arithmetic (field) operations, e.g. [5, 39]. However, these fast multipoint evaluation algorithms are numerically unstable, see, e.g. [4]. From Lemmata 4, 6 and 10 we deduce that the maximum working precision needed by the three subroutines of $\alpha\mathtt{R}_3$ is $\widetilde{\mathcal{O}}(d\tau + L)$. Supported also by the multipoint evaluation techniques of [31] this should provide an overall complexity in $\widetilde{\mathcal{O}}_B(d^2\tau + dL)$ for refining all the roots. We refer the reader to the journal version of our work and to [31] for a detailed presentation.

In the exact version, $\mathtt{R}_3$, we perform evaluations at numbers of bitsize L. The output of such an evaluation results in rational numbers of bitsize $\widetilde{\mathcal{O}}(d(\tau + L))$. When we isolate all the r real roots, the bitsize of the output is $\widetilde{\mathcal{O}}(r\,d(\tau + L)) = \widetilde{\mathcal{O}}(d^2(\tau + L))$, which is also a lower bound on the Boolean complexity of the refinement process. This exceeds the bound for $\alpha\mathtt{R}_3$ by a factor of d.

2.5 Requirements for the isolating intervals

Our algorithms support the complexity bound of Thm. 12 provided that we are given a real m and a positive r such that the *root-isolation disc* $D(m, r) = \{x : |x - m| < r\}$ contains a single simple real root α of A, and no other roots of A, and furthermore α is not very close to the boundary circle of the disc, namely

$$|\alpha - m|(1 + c'/d^c) \leq r \ , \qquad (13)$$

for two real constants $c' > 0$ and c. Our argument in Section 2.2 shows that under the latter assumption it is sufficient to apply $\mathcal{O}(\log d)$ bisections to strengthen bound (13) to the level $5d^2|\alpha - m| \leq r$. Then we can apply Theorem 9 to ensure quadratic convergence of Newton's iteration, and then complete our algorithms and proofs.

Is it simple to ensure bound (13) at a low cost? For the worst case input this is not simpler than to approximate the root α very closely. Indeed the divide-and-conquer algorithms (cf. [29], [36]) can compute a real isolation interval for a single simple root, but produce such intervals already well isolated from all other roots, and then our construction is not needed. On the other hand root-finders working on the real line such as the subdivision algorithms produce such intervals independently of the distribution of nonreal roots on the complex plane. In this case we cannot exclude any unfavorable distribution of them. On the average input, however, violation of the isolation assumption (13) seems to be rather pathological (see also the next section).

Algorithm 1: ISOLATION_DISCS(A, $I_1, \ldots, I_r$)

Input: $A \in \mathbb{Z}[X]$, and isolating intervals $I_1, \ldots, I_r$
Output: Isolation ratio for convergence of Newton.

1. Apply a fixed number of bisections as well as bisections of exponenent to all r input intervals $I_1, \ldots, I_r$, which transforms them into subintervals $\bar{I}_1, \ldots, \bar{I}_r$.
2. Recall that the Möbius map $z = \frac{x+\sqrt{-1}}{x-\sqrt{-1}}$ maps the real line $\{x : \Im(x) = 0\}$ into the unit circle $C_1 = \{z : |z| = 1\}$. Note that $x = \sqrt{-1}\frac{z-1}{z+1}$ and compute the coefficients of the polynomial $B(z) = (z+1)^n A(\sqrt{-1}\frac{z-1}{z+1})$.
3. Apply the root-radii algorithm (cf. [29], [36]) to approximate the root radii of this polynomial. In particular this defines an annulus about the circle C_1, which contains the images of the r real roots of the polynomial $A(x)$ and no images of its nonreal roots.
4. Compute the boundary circles of the image of this annulus in the converse map $x = \sqrt{-1}\frac{z-1}{z+1}$.
5. For all subintervals $\bar{I}_1, \ldots, \bar{I}_r$ compute at first the distances $\bar{d}_1, \ldots, \bar{d}_r$ from their midpoints to these two boundary circles and then the ratios $2d_1/|\bar{I}_1, \ldots, 2d_r/|\bar{I}_r|$.
6. RETURN the minimum of the ratios as the quaranteed isolation ratio for the input intervals.

A natural question arises: How can we test whether this assumption holds for a given polynomial A and a real interval I containing its single simple root α? In fact very easily: we can just apply our algorithms. They compute a sequence of real inclusion intervals $(a_h \mathinner{..} b_h)$, for $h = 0, 1, \ldots,$ where $(a_0 \mathinner{..} b_0) = I$ and $b_h > a_h$ for all h. We verify the inclusion property by checking whether $A(a_h)A(b_h) < 0$ and either observe that h bisection steps decrease the width of the isolating interval by a factor of 2^h or otherwise conclude that the assumption (13) is certainly violated. This *test by action* requires negligible extra cost.

Alternatively, given m and r, we can test whether the disc $D(m, r)$ contains only one root by applying the Schur–Conn test, partial inverses of Descartes' rule of sign, e.g. [16] or the root-radii algorithms of [36] (cf. [28, Section 4]), which approximate the distances from m to all roots of A within, say 1% error. These a priori tests, however, have no advantage versus the test by action and have a little greater cost.

Remark 13. *Suppose we have r real intervals $I_1, \ldots, I_r$, each containing a single simple root of A. In this case our algorithm* ISOLATION_DISCS *in Alg. 1 is a single a priori test of the existence of all the r root-isolation discs. Our present paper does not use this algorithm, but it may be of independent interest, and our next research plan includes estimation of its Boolean complexity, which seems to be dominated at Stage 3. This stage is quite inexpensive, according to the estimates in [36] and [29].*

3. AVERAGE ANALYSIS OF DES AND αDES

The most time consuming part of R_3 and αR_3 is the αDES procedure. It requires, in the worst case,

$$g_1^2 = \lceil \lg(\lg(w) + L - 1) \rceil^2 = \mathcal{O}(\lg^2(\tau + L))$$

evaluations of our input polynomial A. This occurs where (other) roots of A lie very close to the endpoints of the initial interval I. However, in practice this behavior is rare, if it occurs at all. To explain this phenomenon, Pan and Linzer [30] estimated the average number of steps of DES under the assumption that a real root is uniformly distributed in an interval and concluded that in this case R_3, and hence αR_3, needed a constant number of steps, with a high probability.

Even though the assumption on the equidistribution of the real roots in [30] is plausible, we are not aware of any distribution on the coefficients that results in such a behavior for the roots. We consider an average case analysis in the case where the root we approximate is a real root of a random Weyl or $SO(2)$ polynomial [18]. The density function of the real roots is considerably different in these cases. We will also arrive to the same conclusion, that is R_3 and αR_3 perform a constant number of steps with high probability.

3.1 Weyl polynomials

Weyl polynomials, are random polynomials of the form $A = \sum_{i=0}^{d} a_i x^i / \sqrt{i!}$, where the coefficients a_i are independent standard normals. Alternatively, we could consider them as $A = \sum_{i=0}^{d} a_i x^i$, where a_i are normals of mean zero and variance $1/\sqrt{i!}$. As the degree grows, the real roots, except a constant number of them, are in the interval $[-\sqrt{d} \,..\, \sqrt{d}]$. Their (asymptotic) probability density function is $f(t) = \frac{1}{2\sqrt{d}}$.

The probability that $\alpha_1 \in [a \,..\, a + w/2^{2^k}]$ is

$$\Pr[\alpha_1 \in [a \,..\, a{+}w/2^{2^k}]] = \Pr[g_1 > k] = \int_a^{a+w/2^{2^k}} f(t)dt = \frac{w}{2^{2^k+1}\sqrt{d}} \, ,$$

which rapidly goes to zero as k, but also as d, grows.

3.2 $SO(2)$ polynomials

Random polynomials of the form $A = \sum_{i=0}^{d} a_i x^i$, where the coefficients are i.d. normals with mean zero and variances $\binom{d}{i}$, where $0 \leq i \leq d$ are called $SO(2)$ polynomials. Alternatively, we could consider them as $A = \sum_{i=0}^{d} \sqrt{\binom{d}{i}}\, a_i\, x^i$, where a_i are i.i.d. standard normals. They are called $SO(2)$ because the joint probability distribution of their zeros is $SO(2)$ invariant, after homogenization.

The (asymptotic) probability density function of the real roots of $SO(2)$ random polynomials is $f(t) = \frac{1}{\pi(1+t^2)}$.

Assume that the isolating interval $I = [a \,..\, b]$ is a subset of $(0 \,..\, 1)$, and let its width be w. We can treat the case where I is subset of $(1 \,..\, \infty)$ similarly. The probability that $\alpha_1 \in [a \,..\, a + w/2^{2^k}]$ is

$$\Pr\left[\alpha_1 \in [a \,..\, a + w/2^{2^k}]\right] = \Pr[g_1 > k]$$

$$= \int_a^{a+w/2^{2^k}} f(t)dt = \arctan(a) - \arctan(a + w/2^{2^k})$$

$$= \arctan \frac{w/2^{2^k}}{1 + a^2\, w/2^{2^k}} \leq \frac{w}{2^{2^k} + a^2 w} - \frac{1}{3}\left(\frac{w}{2^{2^k} + a^2 w}\right)^3 \, ,$$

which rapidly goes to zero as in the case of Weyl polynomials.

However, now there is no dependence on the degree but only on the endpoints of the (initial) isolating interval.

4. CONCLUSIONS AND FUTURE WORK

We present an approximate variant of a real root refinement algorithm that is based on the *Bisection of the Exponents*, or *Double Exponential Sieve* algorithm, bisection and Newton operator. The complexity of the algorithm is $\widetilde{\mathcal{O}}_B(d^2\tau + dL)$. We are currently implementing the presented algorithm for real root refinement and the first results are quite encouraging. We plan to report a detailed experimental analysis in the near future.

Can we combine αDES with the approximate version of QIR in [22] to provide an alternative method to guarantee quadratic behavior? We believe that this is a very interesting approach to explore.

For random polynomials, it is reasonable to assume that we can derive an even faster algorithm for real root refinement that takes advantage of the distribution of the roots.

Acknowledgments

VP is supported by NSF Grant CCF–1116736 and PSC CUNY Awards 64512–0042 and 65792–0043. ET is partially supported by the EXACTA grant of the National Science Foundation of China (NSFC 60911130369) and the French National Research Agency (ANR-09-BLAN-0371-01), GeoLMI (ANR 2011 BS03 011 06), HPAC (ANR ANR-11-BS02-013) and an FP7 Marie Curie Career Integration Grant. Both authors are also grateful to Michael Sagraloff for his helpful comment to section 2.4.

5. REFERENCES

[1] J. Abbott. Quadratic interval refinement for real roots. In *ISSAC 2006, poster presentation*, 2006. http://www.dima.unige.it/~abbott/.

[2] S. Basu, R. Pollack, and M-F.Roy. *Algorithms in Real Algebraic Geometry*, volume 10 of *Algorithms and Computation in Mathematics*. Springer-Verlag, 2nd edition, 2006.

[3] M. Ben-Or and P. Tiwari. Simple algorithms for approximating all roots of a polynomial with real roots. *Journal of Complexity*, 6(4):417–442, 1990.

[4] D. Bini and G. Fiorentino. Design, analysis, and implementation of a multiprecision polynomial rootfinder. *Numerical Algorithms*, pages 127–173, 2000.

[5] D. Bini and V. Pan. *Polynomial and Matrix Computations*, volume 1: Fundamental Algorithms. Birkhäuser, Boston, 1994.

[6] D. Bini and V. Y. Pan. Parallel complexity of tridiagonal symmetric eigenvalue problem. In *Proc. 2nd Annual ACM-SIAM Sympos. Discrete Algorithms (SODA)*, pages 384–393, 1991.

[7] D. Bini and V. Y. Pan. Practical improvement of the divide-and-conquer eigenvalue algorithms. *Computing*, 48:109–123, 1992.

[8] D. Bini and V. Y. Pan. Computing matrix eigenvalues and polynomial zeros where the output is real. *SIAM J. Comput.*, 27(4):1099–1115, 1998.

[9] L. Blum, F. Cucker, M. Shub, and S. Smale. *Complexity and Real Computation*. Springer-Verlag, 1998.

[10] M. Bodrato and A. Zanoni. Long integers and polynomial evaluation with Estrin's scheme. In *Proc. 11th Int'l Symp. on Symbolic and Numeric Algorithms for Scientific Computing (SYNASC)*, pages 39–46. IEEE, 2011.

[11] J. Cheng, S. Lazard, L. M. Peñaranda, M. Pouget, F. Rouillier, and E. P. Tsigaridas. On the topology of planar algebraic curves. In J. Hershberger and E. Fogel, editors, *Proc. 25th Annual ACM Symp. Comput. Geom. (SoCG)*, pages 361–370, Århus, Denmark, 2009.

[12] G. E. Collins and W. Krandick. A hybrid method for high precision calculation of polynomial real roots. In *Proc. ACM*

Int'l Symp. on Symbolic & Algebraic Comp. (ISSAC), pages 47–52, 1993.

[13] J. H. Davenport. Cylindrical algebraic decomposition. Technical Report 88–10, School of Mathematical Sciences, University of Bath, England, available at: http://www.bath.ac.uk/masjhd/, 1988.

[14] D. I. Diochnos, I. Z. Emiris, and E. P. Tsigaridas. On the asymptotic and practical complexity of solving bivariate systems over the reals. *J. Symbolic Computation*, 44(7):818–835, 2009. (Special issue on ISSAC 2007).

[15] Z. Du, V. Sharma, and C. K. Yap. Amortized bound for root isolation via Sturm sequences. In D. Wang and L. Zhi, editors, *Int. Workshop on Symbolic Numeric Computing*, pages 113–129, School of Science, Beihang University, Beijing, China, 2005. Birkhauser.

[16] A. Eigenwillig. *Real root isolation for exact and approximate polynomials using Descartes' rule of signs*. PhD thesis, Universität des Saarlandes, 2008.

[17] P. Emeliyanenko and M. Sagraloff. On the complexity of solving a bivariate polynomial system. In *Proc. 37th ACM Int'l Symp. on Symbolic & Algebraic Comp. (ISSAC)*, pages 154–161, Grenoble, France, July 2012. ACM.

[18] I. Z. Emiris, A. Galligo, and E. P. Tsigaridas. Random polynomials and expected complexity of bisection methods for real solving. In S. Watt, editor, *Proc. 35th ACM Int'l Symp. on Symbolic & Algebraic Comp. (ISSAC)*, pages 235–242, Munich, Germany, July 2010. ACM.

[19] W. Hart and A. Novocin. Practical divide-and-conquer algorithms for polynomial arithmetic. In *Proc. CASC*, volume 6885 of *LNCS*, pages 200–214. Springer, 2011.

[20] J. R. Johnson. *Algorithms for Polynomial Real Root Isolation*. PhD thesis, The Ohio State University, 1991.

[21] M. Kerber. On the complexity of reliable root approximation. In *In Proc. 11th Int'l Wkshp on Computer Algebra in Scientific Computing (CASC)*, volume 5743 of *LNCS*, pages 155–167, 2009.

[22] M. Kerber and M. Sagraloff. Efficient real root approximation. In *Proc. 36th ACM Int'l Symp. on Symbolic & Algebraic Comp. (ISSAC)*, pages 209–216, San Jose, CA, USA, June 2011. ACM.

[23] M. Kerber and M. Sagraloff. A worst-case bound for topology computation of algebraic curves. *J. Symb. Comput.*, 47(3):239–258, 2012.

[24] A. Mantzaflaris, B. Mourrain, and E. P. Tsigaridas. On continued fraction expansion of real roots of polynomial systems, complexity and condition numbers. *Theor. Comput. Sci.*, 412(22):2312–2330, 2011.

[25] J. M. McNamee and V. Y. Pan. *Numerical methods for roots of polynomials (II)*, chapter 15. Elsevier, 2013.

[26] M. Mignotte. *Mathematics for Computer Algebra*. Springer-Verlag, New York, 1991.

[27] V. Pan. Optimal and nearly optimal algorithms for approximating polynomial zeros. *Comp. and Math. (with Appl.)*, 31:97–138, 1996.

[28] V. Pan. Approximating complex polynomial zeros: modified Weyl's quadtree construction and improved Newton's iteration. *J. of Complexity*, 16(1):213–264, 2000.

[29] V. Pan. Univariate polynomials: Nearly optimal algorithms for numerical factorization and rootfinding. *J. Symbolic Computation*, 33(5):701–733, 2002.

[30] V. Pan and E. Linzer. Bisection acceleration for the symmetric tridiagonal eigenvalue problem. *Numerical Algorithms*, 22(1):13–39, 1999.

[31] V. Y. Pan. Transformations of matrix structures work again. Fast approximate multipoint polynomial evaluation and interpolation by means of transformation of matrix structures. *Tech. Reports TR 2013004 & 2013005, PhD Program in Comp. Sci., Graduate Center, CUNY*, 2013. Available at http://www.cs.gc.cuny.edu/tr/techreport.php?id=446.

[32] V. Y. Pan, B. Murphy, R. E. Rosholt, G. Qian, and Y. Tang. Real root-finding. In S. M. Watt and J. Verschelde, editors, *Proc. 2nd ACM Int'l Work. Symbolic Numeric Computation (SNC)*, pages 161–169, 2007.

[33] J. Renegar. On the worst-case arithmetic complexity of approximating zeros of polynomials. *Journal of Complexity*, 3(2):90–113, 1987.

Algorithm 2: INCRISOLATIONRATIO_BY_SUBDIV(A, I, r, R, dir)

Input: $A \in \mathbb{Z}[X], I = [a, b],\ L \in \mathbb{Z}$
Output: (r, I), where $r \geq R$
Data: Initially $r < R$. The direction of the closest root, dir, may or may not be given.

if $dir = \emptyset$ **then**
  $m \leftarrow \frac{a+b}{2}$;
  $s_m \leftarrow \text{sgn}(A(m))$;
  if $s = 0$ **then** $I = [m..m]$; RETURN $(r = \infty, I)$;
  if $s_l \cdot s_m < 0$ **then** $I \leftarrow [a..m]$; $dir = left$;
  if $s_r \cdot s_m < 0$ **then** $I \leftarrow [m..b]$; $dir = right$;
  $r \leftarrow 2(r-1)+1$;
while $r < R$ **do**
  $m \leftarrow \frac{a+b}{2}$;
  $s_m \leftarrow \text{sgn}(A(m))$;
  if $s = 0$ **then** $I = [m..m]$; RETURN $(r = \infty, I)$;
  if $s_l \cdot s_m < 0$ **then** $I \leftarrow [a..m]$; $cdir = left$;
  if $s_r \cdot s_m < 0$ **then** $I \leftarrow [m..b]$; $cdir = right$;
  if $cdir \neq dir$ **then**
    $r \leftarrow 2(r-1)+1$
  else
    $r \leftarrow 2(r+1)-1$
RETURN (r, I)

Algorithm 3: NEWTON(A, I, t)

Input: $A \in \mathbb{Z}[X], I = [a, b],\ t \in \mathbb{Z}$
Output: Interval J such that $w(J) < 2^{-L}w$
Data: It holds that $\mathtt{ir}(I) > 20d^2$

$x_1 \leftarrow \mathtt{m}(I)$;
$x_0 \leftarrow \infty$;
while $|x_1 - x_0| > 2^{-L}w$ **do**
  SWAP(x_1, x_0) ;
  $x_1 \leftarrow x_0 - A(x_0)/A'(x_0)$;
RETURN $[x_0..x_1]$

[34] F. Rouillier. On solving systems of bivariate polynomials. *Mathematical Software-ICMS 2010*, 6327:100–104, 2010.

[35] M. Sagraloff. When Newton meets Descartes: A simple and fast algorithm to isolate the real roots of a polynomial. In *Proc. 37th ACM Int'l Symp. on Symbolic & Algebraic Comp. (ISSAC)*, pages 297–304, Grenoble, France, July 2012. ACM.

[36] A. Schönhage. The fundamental theorem of algebra in terms of computational complexity. Manuscript. Univ. of Tübingen, Germany, 1982. URL: http://www.iai.uni-bonn.de/~schoe/fdthmrep.ps.gz.

[37] V. Sharma, Z. Du, and C. Yap. Robust approximate zeros. *Proc. European Symposium of Algorithms (ESA)*, pages 874–886, 2005.

[38] A. Strzeboński and E. P. Tsigaridas. Univariate real root isolation in an extension field. In A. Leykin, editor, *Proc. 36th ACM Int'l Symp. on Symbolic & Algebraic Comp. (ISSAC)*, pages 321–328, San Jose, CA, USA, June 2011. ACM.

[39] J. von zur Gathen and J. Gerhard. *Modern Computer Algebra*. Cambridge Univ. Press, Cambridge, U.K., 2nd edition, 2003.

[40] C. Yap. *Fundamental Problems of Algorithmic Algebra*. Oxford University Press, New York, 2000.

Computing the Invariant Structure of Integer Matrices: Fast Algorithms into Practice

Colton Pauderis
cpauderi@uwaterloo.ca

Arne Storjohann
astorjoh@uwaterloo.ca

David R. Cheriton School of Computer Science
University of Waterloo, Ontario, Canada N2L 3G1

ABSTRACT

We present a new heuristic algorithm for computing the determinant of a nonsingular $n \times n$ integer matrix. Extensive empirical results from a highly optimized implementation show the running time grows approximately as $n^3 \log n$, even for input matrices with a highly nontrivial Smith invariant structure. We extend the algorithm to compute the Hermite form of the input matrix. Both the determinant and Hermite form algorithm certify correctness of the computed results.

Categories and Subject Descriptors

I.1.2 [**Symbolic and Algebraic Manipulation**]: Algorithms; G.4 [**Mathematical Software**]: Algorithm Design and Analysis; F.2.1 [**Analysis of Algorithms and Problem Complexity**]: Numerical Algorithms and Problems

Keywords

Integer matrix; determinant; Hermite normal form

1. INTRODUCTION

Computing the exact integer determinant of a nonsingular integer matrix $A \in \mathbb{Z}^{n\times n}$ is a classical problem. The problem provides a "canonical" example of a key feature of symbolic computation: the growth in bitlength of numbers in the output compared to those in the input. On the one hand, if we let $||A|| = \max_{ij} |A_{ij}|$, then considering a diagonal input matrix we see that we may have $\log |\det A| \geq n \log ||A||$. On the other hand, Hadamard's bound gives that $\log |\det A| \leq (n/2)\log n + n \log ||A||$. Thus, the bitlength of the determinant can be up to n times that of entries in the input matrix. Textbooks on computational mathematics often use the problem of computing the determinant to illustrate the technique of homomorphic imaging and Chinese remaindering: this gives a deterministic algorithm to compute $\det A$ using $O(n^4(\log n + \log ||A||)^2)$ bit operations, even assuming standard, quadratic, integer arithmetic.

A lot of effort has been devoted to obtaining improved upper bounds for the complexity of computing $\det A$. We will not give a complete survey here but refer to [9, 10, 17]. An initial breakthrough was Kaltofen's 1992 [8] division free algorithm to compute the determinant of a matrix over a ring: the algorithm can be adapted to compute $\det A$ in $O\tilde{}(n^{3.5} \log ||A||)$ bit operations. High-order lifting and integrality certification [17] can be used to compute the determinant in about the same time (asymptotically, up to logarithmic factors) as required to multiply together two matrices having the same dimension and size of entries as A, thus in $O\tilde{}(n^3 \log ||A||)$ bit operations assuming standard matrix multiplication and pseudo-linear integer arithmetic. While the algorithm in [17] does achieve the important so called "reduction to matrix multiplication" goal, it does not make an attractive candidate for implementation as it is presented because the constant suppressed by the $O\tilde{}$ notation seems very large. For example, the algorithm begins by embedding the input matrix into a matrix of more than twice the dimension which has many entries chosen randomly. The algorithm we describe in this paper also relies on high-order lifting but does not require an increase in the dimension.

The heuristic determinant algorithm of Abbot, Bronstein & Mulders [1] is based on the well-known phenomenon that the largest invariant factor of the matrix (the smallest positive integer s_n such that $s_n A^{-1}$ is integral) is a factor of the determinant that is often very large. For randomly chosen $v_1, v_2 \in \mathbb{Z}^{n\times 2}$, the minimal $s \in \mathbb{Z}_{>0}$ such that both $sA^{-1}v_1$ and $sA^{-1}v_2$ are integral is likely to be equal to s_n, or at least a large factor, thus decreasing the number of images of the determinant that need to be computed using the classical Chinese remainder based determinant algorithm mentioned above. Consider the following nonsingular input matrix.

$$A = \begin{bmatrix} 33 & 8 & -50 & 45 & -38 \\ -20 & 62 & 39 & 11 & -79 \\ 13 & -82 & -52 & -65 & -37 \\ -35 & -81 & 3 & 114 & 7 \\ -100 & 14 & -114 & -22 & -10 \end{bmatrix}. \qquad (1)$$

If we choose

$$\begin{bmatrix} v_1 & v_2 \end{bmatrix} = \begin{bmatrix} 10 & 45 \\ -16 & -81 \\ -9 & -38 \\ -50 & -18 \\ -22 & 87 \end{bmatrix}$$

ISSAC'13, June 26–29, 2013, Boston, Massachusetts, USA.

then

$$A^{-1}\left[\begin{array}{cc} v_1 & v_2 \end{array}\right] = \left[\begin{array}{cc} \frac{1428470455}{3313087328} & \frac{43150207}{161614016} \\ \frac{673936589}{2484815496} & \frac{66351701}{121210512} \\ -\frac{1462901509}{9939261984} & -\frac{516047293}{484842048} \\ -\frac{1221838091}{9939261984} & \frac{138504781}{484842048} \\ \frac{89642859}{414135916} & \frac{18215255}{20201752} \end{array}\right].$$

The denominators of $A^{-1}\left[\begin{array}{cc} v_1 & v_2 \end{array}\right]$ have least common multiple $s = 19878523968$, which for this example is actually equal to $-\det A$. But even if the heuristic finds a large factor or even the complete determinant, the running time of the method is still quartic in n, even for a random matrix, because the expected bitlength of the gap between $|\det A|$ and Hadamard's bound (and thus the bitlength of the images needed by the homomorphic imaging scheme to guarantee to compute the correct determinant) is $\Theta(n)$ [1, Section 3].

Eberly, Giesbrecht & Villard [5] relax the requirement that the determinant should be certified correct, and use additive preconditioners combined with binary search to get a Monte Carlo algorithm requiring

$$O\tilde{}(n^3(\log ||A||)^2\sqrt{\log|\det A|}) \tag{2}$$

bit operations to recover the Smith form (and thus also the determinant) of A. Recall that the Smith form of A is a diagonal matrix $S = \mathrm{Diag}(s_1, s_2, \ldots, s_n)$ such that $S = UAV$ for unimodular matrices U and V. Note that $|\det A| = s_1 s_2 \ldots s_n$. Using Hadamard's bound for $|\det A|$ in (2) gives $O\tilde{}(n^{3.5}(\log ||A||)^{2.5})$ in the worst case. This last cost estimate is quite pessimistic in the average case because the algorithm is highly sensitive to the Smith invariant structure of A: for an input matrix with only $O(1)$ nontrivial invariant factors the running time is only $O(n^3(\log n + \log ||A||)^2(\log n))$ bit operations. Moreover, a careful analysis in [5] shows that an integer matrix with entries chosen uniformly and randomly from an interval

$$\Lambda = \{a, a+1, \ldots, a+\lambda-1\} \tag{3}$$

for any $a \in \mathbb{Z}$ and $\lambda \in \Omega(n)$ is expected to have $O(1)$ nontrivial invariant factors.

In this paper we describe a new heuristic algorithm for computing $\det A$ that has the following features:

- The algorithm certifies correctness of the computed determinant.

- The algorithm is especially fast for propitious inputs (e.g., matrices with few nontrivial invariant factors) but still seems to work very effectively for matrices with a highly nontrivial invariant structure.

We can illustrate the main idea of our algorithm by using the input example in (1). Consider the first random projection $A^{-1}v_1$. The minimal $d_1 \in \mathbb{Z}_{>0}$ such that $d_1 A^{-1} v_1$ is integral is $d_1 = 9939261984$, giving us a large factor of $\det A$. But instead of only using the denominator of $A^{-1}v_1$, we also use the vector of numerators to produce an upper triangular basis T_1 of a superlattice of the integer lattice generated by the rows of A. In particular, in Section 2 we give an algorithm to produce a nonsingular and upper triangular matrix T_1 with minimal magnitude determinant such that $T_1 A^{-1} v_1$ is integral. For this example we obtain

$$T_1 = \left[\begin{array}{ccccc} 1 & & & 15 & 183835840 \\ & 1 & & 4 & 294625615 \\ & & 1 & 1 & 159758078 \\ & & & 24 & 300295265 \\ & & & & 414135916 \end{array}\right].$$

The matrix T_1 has positive diagonal entries and the off-diagonal entries in each column are nonnegative and strictly less than the diagonal entries in the same column. Recall that such matrices are said to be in Hermite form, a canonical presentation of row lattices. Continuing with the example, we can use the second projection $A^{-1}v_2$ to compute a second triangular factor T_2, the minimal triangular denominator of

$$T_1 A^{-1} v_2 = \left[\begin{array}{c} \frac{165758732}{1} \\ \frac{531308447}{2} \\ \frac{144048601}{1} \\ \frac{541532731}{2} \\ \frac{746825455}{2} \end{array}\right]. \tag{4}$$

We obtain

$$T_2 = \left[\begin{array}{ccccc} 1 & & & & 0 \\ & 1 & & & 1 \\ & & 1 & & 0 \\ & & & 1 & 1 \\ & & & & 2 \end{array}\right].$$

Since the rows of T_1 generate a superlattice of the lattice generated by the rows of A, we can remove T_1 from A by computing

$$B_1 = AT_1^{-1} = \left[\begin{array}{ccccc} 33 & 8 & -50 & -18 & 12 \\ -20 & 62 & 39 & 1 & -51 \\ 13 & -82 & -52 & 5 & 69 \\ -35 & -81 & 3 & 40 & 43 \\ -100 & 14 & -114 & 64 & 32 \end{array}\right].$$

The factor T_2 can also be removed to obtain

$$B_2 = B_1 T_2^{-1} = \left[\begin{array}{ccccc} 33 & 8 & -50 & -18 & 11 \\ -20 & 62 & 39 & 1 & -57 \\ 13 & -82 & -52 & 5 & 73 \\ -35 & -81 & 3 & 40 & 42 \\ -100 & 14 & -114 & 64 & -23 \end{array}\right].$$

Fast unimodularity certification [12] can now be used to certify that B_2 is unimodular (i.e., with determinant ± 1). If B_2 is unimodular then $(\det T_1)(\det T_2)$ must be, up to sign, equal to the determinant of A. (Note that if B_2 is determined to be unimodular, the sign of the determinant can be recovered by computing $\det B_2 \bmod p$ for single odd prime p.) If B_2 had not been unimodular then further projections $v_3, v_4, \ldots$ can be computed and used to find further triangular matrices $T_3, T_4, \ldots$ which can be factored from the work matrix. On the one hand, if k is the number of nontrivial invariant factors in the Smith form of A, then at least k projections will be required to capture the complete determinant. On the other hand, from the lattice conditioning analysis of [3] we know $k + O(1)$ projections will be sufficient with high probability. We remark that in [18, 14] a "bonus idea" method is developed that uses the numerators of the projections to recover the penultimate invariant factor in addition to just the last invariant.

A main computational task used in the approach just described is to solve nonsingular linear systems to compute projections $A^{-1}v$ for some $v \in \mathbb{Z}^{n\times m}$. The most efficient algorithms for nonsingular linear system solving are based on linear p-adic lifting [3, 4, 11]. The simplest variant of linear p-adic lifting has two phases. The first phase computes a low precision inverse $C := A^{-1} \bmod p$ for a prime p with $\log p \in \Theta(\log n + \log ||A||)$. This first phase thus has cost $O(n^3(\log n + \log ||A||)^2)$ bit operations assuming standard, quadratic integer arithmetic. The second phase computes a truncated p-adic expansion

$$A^{-1}v \equiv c_0 + c_1 p + c_2 p^2 + \cdots + c_{k-1} p^{\ell-1} \bmod p^{\ell},$$

each $c_i \in \mathbb{Z}^{n\times m}$ with entries reduced modulo p, from which the rational solution vector can be recovered using rational number reconstruction provided the precision ℓ is high enough. The required precision depends on the size of numerators and denomininators if $A^{-1}v$. Computing one of the terms in the p-adic expansion requires a constant number of premultiplications with C and A of matrices of dimension $n \times m$ filled with entries of bitlength $O(\log n + \log ||A||)$. Thus, the second phase has cost $O(\ell m \times n^2(\log n + \log ||A||)^2)$. Note that if the required precision k and the number of projections m satisfy $\ell m \in O(n)$, the overall cost of producing $A^{-1}v$ is bounded by $O(n^3(\log n + \log ||A||)^2)$ bit operations.

We claimed above that our algorithm is also highly effective for matrices with nontrivial invariant structure, even with $k \in \Omega(n)$ nontrivial Smith invariant factors. However, the method as sketched above would solve $\Omega(n)$ nonsingular rational systems at full precision, equivalent in cost to computing the exact inverse of A. Instead of computing the projection $T_1 A^{-1} v_2$ by first computing $A^{-1}v_2$ and then premultiplying by T_1, consider computing $B_1^{-1}v_2$, equal to the vector in (4). Since the first factor T_1 has captured a large factor of the det A, the denominator of $B_1^{-1}v_2$ is very small. It would be desirable to reduce the number of p-adic lifting steps (the precision ℓ) to be proportional to the bitlength of information actually contained in the projection, namely the bitlength of the denominator. We can accomplish this by first precondtioning the projection with a so called high-order residue [12]. A high-order residue R of B_1 can be computed in $O(n^3(\log n + \log ||B_1||)^2(\log n))$ bit operations, and has the property that $B_1^{-1}Rv_2$ will be nearly proper. For this example, one example of such a high-order residue is

$$R = \begin{bmatrix} -15 & -15 & -15 & -15 & 1 \\ 31 & 31 & 31 & 31 & 6 \\ -47 & -47 & -47 & -47 & -4 \\ -30 & -30 & -30 & -30 & 1 \\ -104 & -104 & -104 & -104 & 55 \end{bmatrix}$$

with

$$B_1^{-1}Rv_2 = \begin{bmatrix} -\frac{92}{1} \\ -\frac{97}{2} \\ -\frac{92}{1} \\ -\frac{97}{2} \\ -\frac{97}{2} \end{bmatrix}.$$

The above projection will yield the same triangular factor T_2. Thus, our implementation can effectively exploit the same bitlength versus dimension paradigm we used for our worst case determinant algorithm [17]. For example, compute one projection at full precision, two at about half the precision, four at (at most) a quarter of the precision, etc. In particular, the p-adic lifting precision ℓ and the number of projections m will satisfy $\ell m \in O(n)$ at each phase.

For various reasons, we don't give a rigorous cost analysis of our algorithm in this paper. First, a reduction to matrix multiplication for the problem of computing the determinant has already been given in [17]. Second, a detailed analysis of most of the key subroutines is already available. (For nonsingular solving we refer to [4, 11, 17]. For the high order residue computation see [12].) Third, for the worst case of the problem, when the Hermite form has many nontrivial columns, there remain some technical challenges to overcome to arrive at an algorithm that has expected running time provably cubic in n, see Section 6. We can, though, state the following rigorous cost estimate for an important class of matrices. For an input matrix $A \in \mathbb{Z}^{n\times n}$ that has a Hermite with only $k \in O(1)$ nontrivial columns, the entire Hermite form can be captured with high probability using a single random projection of column dimension $O(1)$. The dominant cost of our algorithm in this case is a single high-order residue computation to certify correctness. The expected running time of our algorithm in this case is $O(n^3(\log n + \log ||A||)^2(\log n))$ bit operations assuming standard integer arithmetic. Moreover, it follows from the proof of [5, Lemma 6.1] that if entries in $A \in \mathbb{Z}^{n\times n}$ are chosen uniformly and randomly from an interval

$$\Lambda = \{a, a+1, \ldots, a+\lambda-1\} \qquad (5)$$

for any $a \in \mathbb{Z}$ and $\lambda \in \Omega(n)$, then the Hermite form of A has the shape

$$\left[\begin{array}{c|c} I_{n-k} & * \\ \hline & * \end{array}\right]$$

where the expected value of k is $O(1)$. Our algorithm is thus very likely to be effective for random inputs.

The rest of this paper is organised as follows. Section 2 details our procedure for computing the minimal triangular denominator. Sections 3 and 4 give our algorithms for the determinant and Hermite normal form, respectively. In Section 5, we compare our implementation against implementations of alternative heuristic Hermite form algorithms [13]. Section 6 concludes.

2. AN ALGORITHM FOR MINIMAL TRIANGULAR DENOMINATOR

Let $v \in \mathbb{Q}^{n\times 1}$. In this section we show how to compute a nonsingular upper triangular matrix $T \in \mathbb{Z}^{n\times n}$ with minimal magnitude determinant such that Tv is integral; we call such a T a *minimal triangular denominator* of v. Our algorithm is based on the following lemma.

LEMMA 1. *Let $v \in \mathbb{Q}^{n\times 1}$ and $d \in \mathbb{Z}_{>0}$ be such that $w := dv$ is integral. Write the Hermite form of*

$$B := \left[\begin{array}{c|c} d & \\ \hline w & I_n \end{array}\right] \in \mathbb{Z}^{(n+1)\times(n+1)} \qquad (6)$$

as

$$\left[\begin{array}{c|c} * & * \\ \hline & H \end{array}\right] \in \mathbb{Z}^{(n+1)\times(n+1)}. \qquad (7)$$

Then $H \in \mathbb{Z}^{n\times n}$ is a minimal triangular denominator of v.

PROOF. The unique unimodular matrix U that transforms B to Hermite form is given by

$$\begin{aligned} U &= \left[\begin{array}{c|c} * & * \\ \hline & H \end{array}\right]\left[\begin{array}{c|c} d & \\ \hline w & I_n \end{array}\right]^{-1} \\ &= \left[\begin{array}{c|c} *d^{-1} & * \\ \hline -Hv & H \end{array}\right] \in \mathbb{Z}^{(n+1)\times(n+1)}. \end{aligned} \tag{8}$$

Since U is integral the submatrix $-Hv$ shown in (8) is integral. Now, suppose that T is a minimal triangular denominator of v. Then we could replace H in the definition of U with T and still arrive at an integral matrix in (8). Since U is unimodular, with determinant ± 1, we must have $|\det H| = |\det T|$. □

Let $v \in \mathbb{Q}^{n\times 1}$ and $d \in \mathbb{Z}_{>0}$ be such that $w := dv$ is integral, as in Lemma 1. Let us define the entries of our input vector w and target H as

$$w = \left[\begin{array}{c} w_1 \\ w_2 \\ \vdots \\ w_n \end{array}\right] \quad \text{and} \quad H = \left[\begin{array}{cccc} h_1 & h_{12} & \cdots & h_{1n} \\ & h_2 & \cdots & h_{2n} \\ & & \ddots & \vdots \\ & & & h_n \end{array}\right]. \tag{9}$$

The obvious approach to compute a minimal triangular denominator (not necessarily in Hermite form) of v is to simply apply unimodular row operations to triangularize the input matrix in (6). To this end, define Gcdex to be the operation that takes as input $a, b \in \mathbb{Z}$ and returns as output s, t, v, h, g such that

$$\left[\begin{array}{cc} s & t \\ v & h \end{array}\right]\left[\begin{array}{c} a \\ b \end{array}\right] = \left[\begin{array}{c} g \\ 0 \end{array}\right]$$

with $sh - tv = \pm 1$ and g a greatest common divisor of a and b. We further specify that $h > 0$ and $0 \leq t < h$.

Now, if we compute $s_n, t_n, v_n, h_n, g_n = \text{Gcdex}(d, w_n)$, then the following unimodular transformation of the input matrix zeroes out the entry occupied by w_n.

$$\left[\begin{array}{c|cccc} s_n & & & & t_n \\ \hline & 1 & & & \\ & & \ddots & & \\ & & & 1 & \\ v_n & & & & h_n \end{array}\right]\left[\begin{array}{c|cccc} d & & & & \\ \hline w_1 & 1 & & & \\ \vdots & & \ddots & & \\ w_{n-1} & & & 1 & \\ w_n & & & & 1 \end{array}\right] = \left[\begin{array}{c|cccc} g_n & & & & t_n \\ \hline w_1 & 1 & & & \\ \vdots & & \ddots & & \\ w_{n-1} & & & 1 & \\ & & & & h_n \end{array}\right]$$

If we initialize $B \in \mathbb{Z}^{(n+1)\times(n+1)}$ to be the input matrix in (6), the following loop applies similar transformations to zero out the entries occupied by $w_n, w_{n-1}, \ldots, w_1$. After the loop completes the work matrix B will be upper triangular.

$g_{n+1} := d$;
for $i = n$ **downto** 1 **do**
  $s_i, t_i, v_i, h_i, g_i := \text{Gcdex}(g_{i+1}, w_i)$;
  $\left[\begin{array}{c} B[1,*] \\ B[i+1,*] \end{array}\right] := \left[\begin{array}{cc} s_i & t_i \\ v_i & h_i \end{array}\right]\left[\begin{array}{c} B[1,*] \\ B[i+1,*] \end{array}\right]$
od

The problem with this approach is that the top row of the work matrix will fill in and the subsequent application of the unimodular transformations will become too expensive. Indeed, at the start of iteration $i-1$ the work matrix has the shape

$$B = \left[\begin{array}{c|ccc|ccc} g_i & & & & t_i & \cdots & * \\ \hline w_1 & 1 & & & & & \\ \vdots & & \ddots & & & & \\ w_{i-1} & & & 1 & & & \\ \hline & & & & h_i & \cdots & * \\ & & & & & \ddots & \vdots \\ & & & & & & h_n \end{array}\right] \tag{10}$$

Instead, we perform the gcd operations only to determine the diagonal entries h_i, for $i = n, n-1, \ldots, 1$, omitting the application of the unimodular transformation to the work matrix. Once the diagonal entries have been precomputed, the off-diagonal entries $h_{1,i}, \ldots, h_{i-1,i}$ in column i of H, now in increasing order $i = 1, 2, \ldots, n$, are computed by appealing to the definition of H as the minimal denominator of the vector w/d.

We begin by initializing $w^{(0)}$ to be the the first column of the matrix in (6). For some $i \geq 1$, suppose the first $i-1$ columns of H have been computed. Then considering (10), we see that the vector

$$\left[\begin{array}{c} d \\ \hline \bar{w}_1 \\ \vdots \\ \bar{w}_{i-1} \\ \hline w_i \\ \vdots \\ w_n \end{array}\right] := \left[\begin{array}{c|ccc|ccc} 1 & & & & & & \\ \hline & h_1 & \cdots & h_{1,i-1} & & & \\ & & \ddots & \vdots & & & \\ & & & h_i & & & \\ \hline & & & & 1 & & \\ & & & & & \ddots & \\ & & & & & & 1 \end{array}\right]\left[\begin{array}{c} d \\ \hline w_1 \\ \vdots \\ w_{i-1} \\ \hline w_i \\ \vdots \\ w_n \end{array}\right]$$

will have all entries divisible by $g_i = d/(h_i h_{i+1} \cdots h_n)$, which is a multiple of $h_1 h_2 \cdots h_{i-1}$. At iteration i the algorithm works with the vector

$$w^{(i-1)} := \left[\begin{array}{c} d \\ \hline \bar{w}_1 \\ \vdots \\ \bar{w}_{i-1} \\ \hline w_i \\ \vdots \\ w_n \end{array}\right] \frac{1}{h_1 h_2 \cdots h_{i-1}}. \tag{11}$$

We now compute off-diagonal entries $h_{1,i}, h_{2,i}, \ldots, h_{i-1,i}$ to be the unique integers in the range $[0, h_i)$ such that the vector

$$\left[\begin{array}{c|ccc|ccc} 1 & & & & & & \\ \hline & 1 & & & h_{1,i} & & \\ & & \ddots & & \vdots & & \\ & & & 1 & h_{i-1,i} & & \\ \hline & & & & h_i & & \\ & & & & & \ddots & \\ & & & & & & 1 \end{array}\right] w^{(i-1)} \tag{12}$$

has all entries divisible by h_i.

Algorithm `hcol`, shown in Figure 1, implements the above method to compute H. Given a nonzero integer b, operation $\text{Rem}(a, b)$ computes the unique integer in the range $0 \leq$

```
hcol(w, d)
Input: A vector w ∈ Z^{n×1} and d ∈ Z.
Output: H ∈ Z^{n×n}, a minimal triangular denominator
of w/d.

  1. [1. Diagonal entries]
     g := d;
     for i = n downto 1 do
        *, t_i, *, h_i, g := Gcdex(g, w_i)
     od;

  2. [2. Off-diagonal entries]
     for i = 1 to n do
        assert: [d; w] = w^{(i-1)}
        if h_i = 1 then next fi;
        for k = 1 to i - 1 do
           h_{k,i} := Rem(-t_i w_k, h_i);
           w_k := Rem(w_k + h_{k,i} w_i, d)
        od;
        w_i := h_i w_i;
        d, w := d/h_i, w/h_i
     od;
     return H as in (9)
```

Figure 1: Algorithm hcol

$\mathrm{Rem}(a, b) < |b|$ that is congruent to a modulo b. Phase 1 computes the diagonal entries of H. Phase 2 computes the off-diagonal entries in column i of H for $i = 1, \ldots, n$. The vector w and modulus d are updated in place. At the start of loop iteration i the algorithm works with the $n+1$ integers in $w^{(i-1)}$ as shown in (12). The inner loop and first line after the inner loop computes the off-diagonal entries in column i of H and performs the update shown in (12). The last line removes the common factor h_i to arrive at $w^{(i)}$.

To analyse the cost of algorithm hcol we need to introduce a cost model. The number of bits in the binary representation of an integer a is given by

$$\lg a = \begin{cases} 1, & \text{if } a = 0 \\ 1 + \lfloor \log_2 |a| \rfloor, & \text{if } a > 0. \end{cases}$$

Using standard integer arithmetic [2, Chapter 3], a and b can be multiplied together in $O((\lg a)(\lg b))$ bit operations. Operation Gcdex also costs $O((\lg a)(\lg b))$ bit operations. For an integer a and nonzero integer b, we can express $a = qb + r$ with $0 \le r < |b|$ using $O((\lg q)(\lg b))$ bit operations.

THEOREM 2. *If entries in $w \in \mathbb{Z}^{n\times 1}$ are reduced modulo d, then algorithm* hcol *computes the minimal triangular denominator (in Hermite form) of w/d using $O(n(\log d)^2)$ bit operations.*

PROOF. Correctness of the algorithm follows from the previous discussion.

For the cost analysis, let D denote the initial value of d as passed into the algorithm. Assume without loss of generality that $D > 1$. Phase 1 performs n extended gcd computations with numbers bounded in magnitude by D. This has cost bounded by $O(n(\log D)^2)$ bit operations.

Now consider phase 2. At the start of each loop iterations all entries in w are reduced modulo d, a divisor of D. Using the fact that $|t_i| < h_i$ and $|h_{k,i}| < h_i$, each individual operation from $\{+, -, \times, /, \mathrm{Rem}\}$ performed during iteration i of the loop uses $O((\lg D)(\lg h_i))$ bit operations. Since the number of operations from $\{+, -, \times, /, \mathrm{Rem}\}$ performed during a single iteration of the outer loop is bounded by $O(n)$, there exists an absolute constant c such that loop iteration i has cost bounded by $cn(\lg D)(\lg h_i)$ bit operations. Let $L = \{i \mid h_i > 1\}$. Then the total cost of phase 2 is bounded by

$$\begin{aligned} \sum_{i\in L} cn(\lg D)(\lg h_i) &\le cn(1 + \log_2 D)\sum_{i\in L}(1 + \log_2 h_i) \\ &\le cn(2\log_2 D)(2\sum_{i\in L}\log_2 h_i) \\ &= 4cn(\log_2 D)(\log_2 h_1 h_2 \cdots h_n) \\ &\le 4cn(\log_2 D)(\log_2 D). \end{aligned}$$

□

We remark that our determinant algorithm will call hcol with a sequence of vectors that have denominator $d_1, \ldots, d_k$ with $d_1 \cdots d_k = |\det A|$. The total cost of all calls to hcol will thus be bounded by $O(n(\log|\det A|)^2)$ bit operations, which becomes $O(n^3(\log n + \log||A||)^2)$ in the worst case using Hadamard's bound for $|\det A|$.

3. THE PROJECTION METHOD FOR DETERMINANT

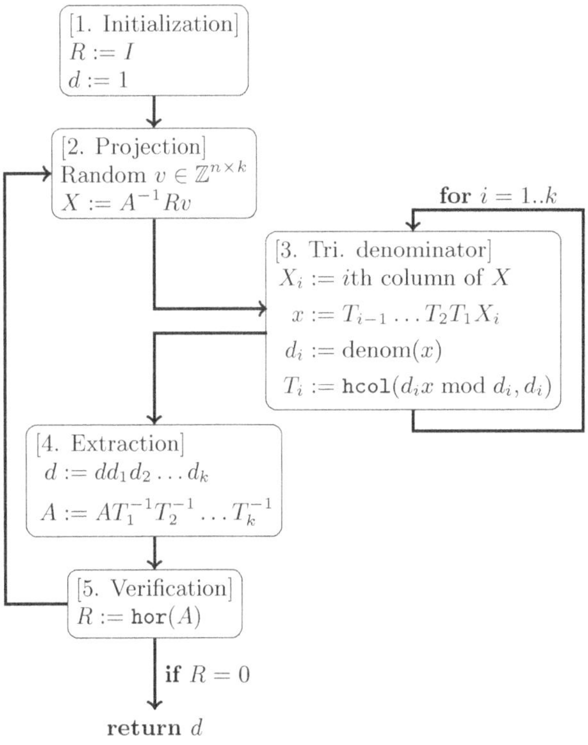

Figure 2: Overview of determinant algorithm.

Figure 2 gives a high-level overview of our determinant algorithm. The first phase uses a highly-optimized implementation of p-adic linear system solving [3] to compute

$X = A^{-1}v$, a projection of the inverse A^{-1} for a random block of vectors $v \in \mathbb{Z}^{n \times k}$. The procedure described in the previous section computes minimal triangular denominators T_i such that $T_i T_{i-1} \cdots T_1$ is a minimal triangular denominator of the first i columns of X, $i = 1, 2, \ldots, k$. The factor $T_k \cdots T_2 T_1$ is next extracted from the input by updating A as $A := AT_1^{-1}T_2^{-1} \cdots T_k^{-1}$. Finally, to check the completeness of the determinant extracted thus far, a high-order residue R is computed by the method of double-plus-one lifting [12] (denoted `hor` in Figure 2). If this check fails, a new random block of vectors is chosen and the process repeats.

We note that the above algorithm is randomized in the Las Vegas sense: the output is always correct, but the number of iterations required to produce that output may vary. In the case of generic matrices, the computation of the high order residue (the "Verification" phase in Figure 2) serves only to verify the correctness of the result. Omitting this verification step, then, yields a faster Monte Carlo randomized algorithm: one that computes the determinant from a single projection with high probability.

4. EXTENSION TO HERMITE FORM

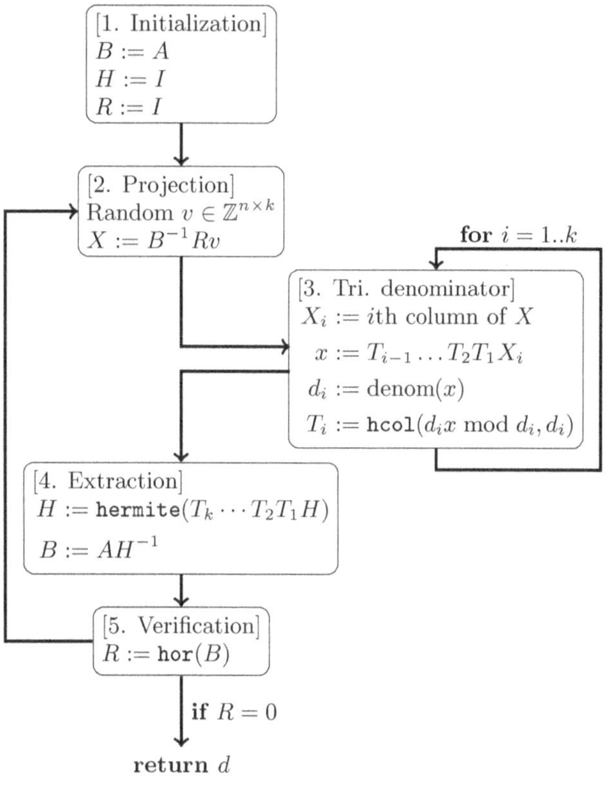

Figure 3: Overview of Hermite form algorithm.

A conceptually straightforward extension of the projection method allows the recovery of the entire Hermite form. Figure 3 gives an overview. The Hermite form algorithm differs from the preceding determinant algorithm only in the "extraction" phase.

In the determinant algorithm, the triangular denominators are repeatedly extracted from the same matrix in place (i.e., $A := AT_1^{-1} \ldots T_k^{-1}$) and then discarded. Here, each set of T_i is combined into a single work matrix H (i.e., $H := T_k \ldots T_1 H$) which eventually contains the Hermite form of A. This process may cause growth in the off-diagonal entries of H. A special Hermite normal form algorithm [16] (denoted `hermite` in Figure 3) for triangular matrices provides an efficient scheme for appropriately reducing the off-diagonal entries.

Additionally, H must be extracted from the original input matrix — not a work matrix — at each stage. That is, A is not updated in-place; rather, the subsequent projection and verification phases operate on $B := AH^{-1}$.

The additional cost required to combine the minimal triangular denominators and extract them from the input matrix complicates attempts to obtain a strong result for the asymptotic complexity of the algorithm. Even if there are very few invariant factors, the Hermite form may have many non-trivial columns in the worst case.

5. IMPLEMENTATION

Although the Hermite normal form algorithm of the previous section is not asymptotically optimal, empirical tests with a careful C implementation bear out its effectiveness in practice.

Our implementation relies on several existing, highly efficient libraries. Nonsingular system solving is provided by the Integer Matrix Library (IML) [3]. Additional routines not available elsewhere (algorithm `hcol`, for one) are implemented in terms of the integer arithmetic routines of the GNU Multi-Precision Arithmetic (GMP) library [6]. The algorithm and implementation presented in [12] gives a routine for computing the high-order residue.

The projection-based method for extracting the invariant structure is sensitive to the number of non-trivial invariant factors. A matrix with many non-trivial invariant factors requires the computation of many projections. However, as generic matrices are expected to have very few non-trivial invariant factors - and often only one - the practical performance of the algorithm in the common case is very good. Typically, only a single projection is required to extract the entirety of the invariant structure.

Yet, while the projection method performs most dramatically on generic matrices, the algorithm can also be effectively applied to matrices specifically constructed to have many non-trivial invariant factors and, in turn, highly non-trivial Hermite and Smith forms. This implementation is robust in its ability to handle all input matrices without making undue concessions to either the common or exceptional cases. No special manual tuning is used to better handle any particular case.

We discuss some specific implementation concerns below.

Projection size.

The column dimension k of the random $v \in \mathbb{Z}^{n \times k}$ used to compute the projection $x := A^{-1}v$ (cf. the "Projection" phase in figure 3) may be varied from one iteration to the next. Choosing a value for k, the size of the projection, is an inexact process driven mostly by empirical observation. A reality of the lifting-based linear system solving algorithms, like those in IML, is that the cost of initialization can meet or surpass the cost of the lifting steps themselves. Indeed, to cover a wide range of inputs, IML has been tuned to balance the cost of initialization with the cost of lifting. The relevant consequence here, then, is that the cost of computing a projection of multiple columns is negligibly more than the cost of working with a single column.

Preliminary observations suggest that an initial projection of eight columns works well. Eight columns are sufficient to capture all invariant factors in the case of a random matrix without being prohibitively more costly than working with a single column.

Subsequent iterations can use much larger projections as the largest invariant factors will have already been extracted and, consequently, the system can be solved at a much lower precision. The scheme used in this implementation is somewhat coarse, but effective. The second iteration uses a projection of $n/10$ columns; the third iteration uses a projection of n columns. Additionally, rather than a randomly chosen matrix, the third iteration uses the identity matrix. This guarantees that only three iterations are ever required and obviates a final high-order residue computation to certify the result.

It is perhaps possible to choose the size of projections adaptively, based on the size of the denominator of the previous projection (or, even better, based on the expected size of the next one). If the previous denominator is larger than expected, the next denominator may be relatively small and, thus, the next projection could be made larger without incurring much additional cost. An adaptive scheme of this sort would require quantifying, perhaps only in a heuristic sense, the expected size of the invariant factors extracted at each iteration.

Combining slices.

As each projection yields a portion of the Hermite form corresponding to only a few invariant factors, combining these "slices" involves operations on highly structured matrices. Generally, the non-zero elements of these matrices are confined to only a few columns. Thus, storing only the non-trivial columns immediately improves both the time and space requirements. Two operations — multiplication and reduction to Hermite form — are implemented for matrices in this packed representation. Both operations use existing algorithms modified to concern themselves with only the non-trivial columns: the former is based on classical matrix multiplication while the latter uses an algorithm for the Hermite form for triangular matrices [16].

Experimental results.

Two classes of matrices are used to illustrate the performance of the implementation at both extremes of the spectrum of input matrices.

Firstly, to test the implementation on the generic case, we use matrices with random 8-bit entries; these results are shown in Table 1. Each of the random matrices used in Table 1 had three or fewer non-trivial elements on the diagonal of the Hermite form. As expected, only a single projection was required in each case.

Following the example of Jäger and Wagner [7], we use the following class of matrices to test performance on inputs with many non-trivial invariant factors:

$$A_n = [a_{i,j}] \text{ with } a_{i,j} = (i-1)^{j-1} \bmod n \text{ for } 1 \le i,j \le n$$

If n is prime, A_n is nonsingular, typically has more than $n/2$ non-trivial invariant factors, and s_n is very large relative to n. For instance, A_{113} has 72 non-trivial invariant factors, the largest of which is 253 bits in length. In addition to having the desired structural properties, A_n can be quickly and straightforwardly constructed, allowing for consistent comparisons between implementations. Results for the A_n matrices are shown in Table 2.

The following tables summarize the experimental results. These timings were made on an "M1 Medium" Amazon EC2 instance with 3.75 GiB RAM and a 64-bit Intel Xeon E5-2650 at 2GHz. The software was compiled with GCC 4.6.3 and linked against IML 1.0.3, ATLAS 3.10.1, and GMP 5.1.1.

Sage 5.5 [15], compiled from source and run on the same machine, provides a point of comparison for our implementation. For matrices of the size considered here, Sage uses a modular algorithm [13] most effective in the random case. Our implementation compares favourably with Sage.

n	time (s)		n	time (s)	
	iherm	Sage		iherm	Sage
100	0.09	0.635	125	0.14	0.844
200	0.42	2.25	250	0.75	3.83
400	3.08	12.2	500	5.66	24.7
800	22.5	81.1	1000	42.0	152
1600	171	681	2000	348	1365
3200	1625		4000	3214	

Table 1: Time to compute Hermite form of random $n \times n$ matrix with 8-bit entries.

For random matrices (Table 1), the computation time grows roughly as $n^3 \log n$. That is, doubling the input dimension increases the cost by a factor of slightly less than nine. The fit is not perfect, however: smaller inputs slightly outperform expectations while larger inputs are slightly underperforming. For instance, the ratio between results for $n = 4000$ and $n = 2000$ is greater than nine, but is near seven between $n = 1000$ and $n = 500$. Smaller input matrices may be taking advantage of some beneficial machine-specific cache effects.

n	k	time (s)	
		iherm	Sage
101	56	0.52	1.13
211	118	2.91	19.4
401	266	18.6	531
809	503	118	
1601	1060	831	
n	k	time (s)	
		iherm	Sage
127	77	0.83	2.24
251	132	5.90	44.6
503	252	32.0	1520
1009	663	221	
2003	1041	1410	

Table 2: Time to compute Hermite form of A_n with k non-trivial invariant factors.

For inputs with many non-trivial invariant factors (Table 2), the empirical timings again grow roughly as $n^3 \log n$. Although the algorithm runs much faster overall on generic inputs, the rate of growth exhibited by the timings is the same for both types of inputs.

6. CONCLUSIONS AND FUTURE WORK

Although experiments demonstrate excellent performance in practice, we do not provide a worst-case cost analysis for our Hermite normal form algorithm. There are two essential roadblocks. Firstly, repeatedly combining the results of `hcol` may, in the worst-case, be costly; there is the potential for expressions swell, and while each minimal triangular denominator has few non-trivial columns in practice, this need not always be the case. Secondly, preliminary analysis suggests that in the worst case the entries in AH^{-1} can have n more bits compared to the entries in A. However, this growth has not been observed in practice, even for matrices with highly non-trivial Hermite forms.

The algorithm for Hermite normal form given here can be extended to one for Smith normal form. An efficient algorithm for finding the Smith normal form of a triangular input matrix is given in [16]; this algorithm can be directly applied to the result of our Hermite form computation.

7. REFERENCES

[1] J. Abbott, M. Bronstein, and T. Mulders. Fast deterministic computation of determinants of dense matrices. In S. Dooley, editor, *Proc. Int'l. Symp. on Symbolic and Algebraic Computation: ISSAC'99*, pages 197–204. ACM Press, New York, 1999.

[2] E. Bach and J. Shallit. *Algorithmic Number Theory*, volume 1 : Efficient Algorithms. MIT Press, 1996.

[3] Z. Chen and A. Storjohann. A BLAS based C library for exact linear algebra on integer matrices. In M. Kauers, editor, *Proc. Int'l. Symp. on Symbolic and Algebraic Computation: ISSAC'05*, pages 92–99. ACM Press, New York, 2005.

[4] J. D. Dixon. Exact solution of linear equations using p-adic expansions. *Numer. Math.*, 40:137–141, 1982.

[5] W. Eberly, M. Giesbrecht, and G. Villard. Computing the determinant and Smith form of an integer matrix. In *Proc. 31st Ann. IEEE Symp. Foundations of Computer Science*, pages 675–685, 2000.

[6] T. Granlund and the GMP development team. Gnu mp: The GNU multiple precision arithmetic library, 2011. Edition 5.0.2. `http://gmplib.org`.

[7] G. Jäger and C. Wagner. Efficient parallelizations of hermite and smith normal form algorithms. *Parallel Comput.*, 35(6):345–357, 2009.

[8] E. Kaltofen. On computing determinants of matrices without divisions. In P. S. Wang, editor, *Proc. Int'l. Symp. on Symbolic and Algebraic Computation: ISSAC'92*, pages 342–349. ACM Press, New York, 1992.

[9] E. Kaltofen and G. Villard. Computing the sign or the value of the determinant of an integer matrix, a complexity survey. *J. Computational Applied Math.*, 162(1):133–146, Jan. 2004. Special issue: Proceedings of the International Conference on Linear Algebra and Arithmetic 2001, held in Rabat, Morocco, 28–31 May 2001, S. El Hajji, N. Revol, P. Van Dooren (guest eds.).

[10] E. Kaltofen and G. Villard. On the complexity of computing determinants. *Computational Complexity*, 13(3–4):91–130, 2004.

[11] T. Mulders and A. Storjohann. Diophantine linear system solving. In S. Dooley, editor, *Proc. Int'l. Symp. on Symbolic and Algebraic Computation: ISSAC'99*, pages 281–288. ACM Press, New York, 1999.

[12] C. Pauderis and A. Storjohann. Deterministic unimodularity certification. In J. van der Hoeven and M. van Hoeij, editors, *Proc. Int'l. Symp. on Symbolic and Algebraic Computation: ISSAC'12*, pages 281–288. ACM Press, New York, 2012.

[13] C. Pernet and W. Stein. Fast computation of Hermite normal forms of integer matrices. *Journal of Number Theory*, 130(7), 2010.

[14] D. Saunders and Z. Wan. Smith normal form of dense integer matrices, fast algorithms into practice. In J. Gutierrez, editor, *Proc. Int'l. Symp. on Symbolic and Algebraic Computation: ISSAC'04*, pages 274–281. ACM Press, New York, 2004.

[15] W. Stein et al. *Sage Mathematics Software (Version 5.5.0)*. The Sage Development Team, 2012. `http://www.sagemath.org`.

[16] A. Storjohann. Computing Hermite and Smith normal forms of triangular integer matrices. *Linear Algebra and its Applications*, 282:25–45, 1998.

[17] A. Storjohann. The shifted number system for fast linear algebra on integer matrices. *Journal of Complexity*, 21(4):609–650, 2005. Festschrift for the 70th Birthday of Arnold Schönhage.

[18] Z. Wan. *Computing the Smith Forms of Integer Matrices and Solving Related Problems*. PhD thesis, University of Deleware, 2005.

Termination Conditions for Positivity Proving Procedures

Veronika Pillwein*
RISC-Linz
Johannes Kepler University
4040 Linz (Austria)
vpillwei@risc.jku.at

ABSTRACT

Proving positivity of a sequence given by a linear recurrence with polynomial coefficients (P-finite recurrence) is a non-trivial task for both humans and computers. Algorithms dealing with this task are rare or non-existent. One method that was introduced in the last decade by Gerhold and Kauers succeeds on many examples, but termination of this procedure has been proven so far only up to order three for special cases. Here we present an analysis that extends the previously known termination results on recurrences of order three, and also provides termination conditions for recurrences of higher order.

Categories and Subject Descriptors

I.1.2 [**Computing Methodologies**]: Symbolic and Algebraic Manipulation—*Algorithms*; G.2.1 [**Discrete Mathematics**]: Combinatorics—*Recurrences and difference equations*

General Terms

Algorithms

Keywords

P-finite Sequences, Positivity, Cylindrical decomposition

1. INTRODUCTION

Special functions are interesting mathematical objects, in particular from the point of view of symbolic computation. They are often defined by linear recurrences, differential equations, or mixed difference-differential equations which makes them a suitable input for several computer algebra methods. Nowadays, a multitude of algorithms is available that can automatically prove or derive transformations, identities or closed forms for special functions.

*Supported by the Austrian Science Fund (FWF) grant P22748-N18.

The situation is quite different for special functions inequalities, which arise in many applications in mathematics and physics. These are still a serious challenge for both humans and computers. Even for special cases such as sequences satisfying linear recurrence relations with constant coefficients, deciding whether the given sequence is positive leads to hard number theoretic questions to which no solutions are known today [12, 13].

Despite these difficulties, some progress has been made in the last decade. Only a few years ago, Mezzarobba and Salvy have given an algorithm for effectively computing tight upper bounds for sequences defined by linear recurrences with polynomial coefficients [22]. Recently, Cha [6] has extended a method for finding closed form solutions of difference equations introduced by Cha, van Hoeij, and Levy [7] to a method that determines closed form solutions that are linear combinations of sums of squares. In the lucky case, it suffices to prove positivity of the rational function coefficients using, e.g., Cylindrical Algebraic Decomposition [2, 9, 10, 5] (CAD), to conclude positivity of the given sequence.

In 2005, Gerhold and Kauers [14] proposed a method that is applicable to proving inequalities concerning sequences that satisfy recurrence equations of a very general type. Their method consists of constructing a sequence of polynomial sufficient conditions that would imply the non-polynomial inequality under consideration. The truth of these conditions can be detected for instance using CAD. If the inequality does not hold, then the method terminates after a finite number of steps and returns a counterexample. If the inequality holds, then either the program terminates and returns True or it may fail to detect this and run forever.

Despite its simplicity, the method has been quite successful in applications. Not only did it provide the first computer proofs of some special function inequalities from the literature [14, 15, 18, 19], but it even helped to resolve some open conjectures [1, 20, 19, 23]. At the same time, the lack of termination conditions was unsatisfactory from a computational point of view. First results on a priori conditions, for some restricted classes, guaranteeing that the method (or some variation of it) will succeed were determined only recently [21] for sequences defined by linear recurrence equations with polynomial coefficients (P-finite sequences) up to order three.

The aim of this paper is to extend these known termination results for the original proving procedure introduced by Gerhold and Kauers (Algorithm 1 below) for recurrences of order three and to obtain new termination conditions for recurrences of higher order. As auxiliary tool, we propose a new variation of the method (Algorithm 3 below) for proving positivity of P-finite sequences. If Algorithm 3 terminates on a given input, then Algorithm 1 also terminates on this input. The new method, however, allows for a different analysis than the initial one which leads to an extension of the known termination conditions (Theorem 2 and Corollary 1). Some parts of the analysis are not affected by the order and thus also termination conditions for higher orders are obtained (Corollary 2 and Theorem 3). An interesting aspect of the analysis is that algorithms for real quantifier elimination are not only used in the execution of Algorithm 1 and 3, but also enter in the proof of the termination theorems.

2. PRELIMINARIES

A sequence $f\colon \mathbb{N} \to K := \mathbb{R} \cap \overline{\mathbb{Q}}$ is called *P-finite* (or *holonomic*) if there exist polynomials $p_0, \dots, p_d \in K[x]$, not all zero, such that

$$p_0(n)f(n) + p_1(n)f(n+1) + \cdots + p_d(n)f(n+d) = 0$$

for all $n \in \mathbb{N}$. Such an equation is called a (P-finite) *recurrence*, and d is called its order if p_0 and p_d are not vanishing entirely. In the special case when all coefficients in the recurrence are from the ground field K we call it a *C-finite* recurrence. If $p_d(n) \neq 0$ for all $n \in \mathbb{N}$, then the infinite sequence f is uniquely determined by the recurrence and d *initial values* $f(0), f(1), \dots, f(d-1)$. In this case, we may also rewrite the recurrence with rational function coefficients as

$$f(n+d) = r_{d-1}(n)f(n+d-1) + \cdots + r_0(n)f(n),$$

with $r_k(n) = -\frac{p_k(n)}{p_d(n)}$. The assumption $p_d(n) \neq 0$ for all $n \in \mathbb{N}$ can be adopted without loss of generality, because we can substitute $g(n) = f(n+N)$ for some N larger than the biggest integer root of p_d and then consider g instead of f and check non-negativity of the finitely many terms $f(0), \dots, f(N-1)$ by inspection.

A P-finite recurrence is called *balanced* if $\deg p_0 = \deg p_d$ and $\deg p_i \leq \deg p_0$ $(i = 1, \dots, d)$. The *characteristic polynomial* of a balanced recurrence is defined as

$$\chi(x) = \operatorname{lc}_y\big(p_0(y) + p_1(y)x + p_2(y)x^2 + \cdots + p_d(y)x^d\big) \in \mathbb{Q}[x].$$

Its roots $\alpha_1, \dots, \alpha_d \in \mathbb{C}$ are called the *eigenvalues* of the recurrence. (The α_i are not necessarily distinct.) Note that for a balanced recurrence, the characteristic polynomial has always degree d and it has never zero as a root.

An eigenvalue α_i is called *dominant* if $|\alpha_j| \leq |\alpha_i|$ for all $j = 1, \dots, d$. If none of the dominant eigenvalues α_i is real and positive, then it is clear that f will be ultimately oscillating [3] (if the recurrence is of minimal order), and so $f(n) \geq 0$ cannot possibly be true for all n. This case can be sorted out trivially beforehand, and we may therefore assume that there is a real and positive dominant eigenvalue. In that case, if $\alpha_i \neq 1$, we consider the P-finite sequence $g(n) = f(n)/\alpha_i^n$ instead, whose dominant eigenvalues are of modulus 1. Since $f(n) \geq 0 \Leftrightarrow g(n) \geq 0$ it suffices to consider the case when $\alpha_i = 1$ for some i and $|\alpha_j| \leq |\alpha_i|$ for all $j \neq i$. In the analysis of the terminating region we assume that for the given sequence there is a unique dominant real eigenvalue, $\alpha_i = 1$, in order to exploit that this eigenvalue governs its asymptotic behaviour.

3. INDUCTION BASED PROVING PROCEDURES

Let $f(n)$ be a P-finite sequence of order d given in terms of its defining recurrence relation and initial values. Based on this information it is to be decided whether the sequence is non-negative for all $n \geq 0$. Here and in what follows we stick to the terminology used in [21] and refer to the positivity proving procedures as *algorithms*, even though identifying conditions that imply termination is precisely the topic of this note.

3.1 Summary of previous algorithms

The initial version of the algorithm as introduced by Gerhold and Kauers [14] proceeds by induction:

$$\begin{aligned}&f(n) \geq 0 \wedge \cdots \wedge f(n+d-1) \geq 0 \Longrightarrow\\ &r_{d-1}(n)f(n+d-1) + \cdots + r_1(n)f(n+1) + r_0(n)f(n) \geq 0\end{aligned}$$

A sufficient condition for this to hold for all $n \in \mathbb{N}$ is that the *induction step formula*

$$\begin{aligned}&\forall\, y_0, y_1, \dots, y_{d-1} \in \mathbb{R}\ \forall\, x \in \mathbb{R}:\\ &\quad (x \geq 0 \wedge y_0 \geq 0 \wedge \cdots \wedge y_{d-1} \geq 0)\\ &\qquad \Longrightarrow r_{d-1}(x)y_{d-1} + \cdots + r_0(x)y_0 \geq 0.\end{aligned}$$

is true, and this can be decided by a quantifier elimination algorithm. If it is true, the induction step is established and f is non-negative everywhere if and only if it is non-negative for $n = 0, \dots, d-1$, which can be checked. Passing from discrete values of n to a real variable x (and possibly ignoring other relevant information passing from $f(n+i)$ to real variables y_i), the induction step formula may be false even though the given sequence is non-negative. If the formula is false, then in the next step the induction hypothesis is extended and the formulas

$$f(n) \geq 0 \wedge \cdots \wedge f(n+D-1) \geq 0 \Longrightarrow f(n+D) \geq 0$$

for $D > d$ are constructed. Using the recurrence, each term $f(n+i)$ can be rewritten as a linear combination of $f(n)$, $\dots$, $f(n+d-1)$ with rational function coefficients, and using this rewriting, *refined induction step formulas* are built :

$$\begin{aligned}\Phi(D) := {}&\forall\, y_0, y_1, \dots, y_{d-1} \in \mathbb{R}\ \forall\, x \in \mathbb{R}:\\ &(x \geq 0 \wedge y_0 \geq 0 \wedge \cdots \wedge y_{d-1} \geq 0\\ &\wedge R_0(d,x)y_0 + \cdots + R_{d-1}(d,x)y_{d-1} \geq\\ &\wedge R_0(d+1,x)y_0 + \cdots + R_{d-1}(d+1,x)y_{d-1} \geq 0\\ &\vdots\\ &\wedge R_0(D-1,x)y_0 + \cdots + R_{d-1}(D-1,x)y_{d-1} \geq 0)\\ &\quad \Rightarrow R_0(D,x)y_0 + \cdots + R_{d-1}(D,x)y_{d-1} \geq 0,\end{aligned}$$

where the $R_i(j,\cdot)$ are some rational functions. The full method then reads as follows.

ALGORITHM 1.
Input: A P-finite recurrence of order d and a vector of initial values defining a sequence $f\colon \mathbb{N} \to \mathbb{Q}$.
Output: True if $f(n) \geq 0$ for all $n \in \mathbb{N}$, False if $f(n) < 0$ for some $n \in \mathbb{N}$, possibly no output at all.

1 **for** $n = 0$ **to** $d-1$ **do**
2 **if** $f(n) < 0$ **then return** False
3 **for** $n = d, d+1, d+2, d+3, \dots$ **do**

if $\Phi(n)$ **then return** True
if $f(n) < 0$ **then return** False

In [21] a variation of Algorithm 1 was introduced that covers some of its non-terminating cases. The basic idea is to determine a constant factor $M > 0$ such that the sequence $M^n f(n)$ is increasing and such that this fact can be proven by an application of Algorithm 1 without extending the induction hypothesis. We refer to this algorithm as Algorithm 2. Termination for both variants has been investigated up to recurrences of order three [21]. These results are summarized in the figure below in terms of the coefficients of the characteristic polynomial u_1, u_0, where

$$\chi(x) = (x-1)(x^2 + u_1 x + u_0).$$

The outer triangle corresponds to the area where the roots are in the interior of the complex unit disk, the gray shaded area is the termination region of Algorithm 2, and the dashed areas are terminating regions for Algorithm 1. For the latter, the parts I_D correspond to induction hypotheses of length D.

In particular, the region where Algorithm 1 terminates in the first step is $I_3 = \{(u_1, u_0) \mid 0 < u_0 < u_1 < 1\}$. This corresponds to the trivial case when all coefficients in the recurrence eventually are positive.

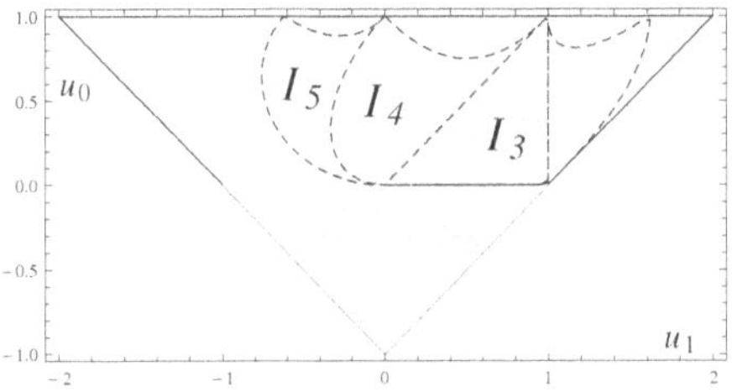

For an induction hypothesis of length three, i.e., the case when the induction succeeds in the first step, the algorithm terminates only in the case when all the coefficients of the characteristic polynomial are non-negative [21]. Induction hypotheses of length six and more as well as higher order cases have not been discussed in the previous paper because of computational limitations.

3.2 Algorithm 3

Our aim is to extend the known termination results for Algorithm 1 and our tool in the analysis is a variant of it. This new method, Algorithm 3, has the property that if it terminates then also the original algorithm terminates, but it allows to obtain better statements about the terminating region. The efficiency of Algorithm 3 is not investigated.

Let $f(n)$ be a given P-finite sequence of order d. Instead of proceeding by induction on the sequence itself, we consider the shifted subsequences $f(n+m)$, for some fixed natural number $m \geq d$, and aim at proving inductively that non-negativity of d successive sequence elements implies non-negativity of one such shifted sequence. That is, we prove that

$$f(n) \geq 0 \wedge f(n+1) \geq 0 \wedge \cdots \wedge f(n+d-1) \geq 0 \Rightarrow f(n+m) \geq 0,$$

for $n \geq n_0$, for some lower bound $n_0 \geq 0$. Using the recurrence relation, $f(n+m)$ can be expressed as linear combination of $f(n), \ldots, f(n+d-1)$ with rational function coefficients. Checking initial values $f(0) \geq 0, \ldots, f(n_0+m-1) \geq 0$ completes the proof. The (modified) refined induction step formula for this variation reads as

$$\widetilde{\Psi}(m) := \exists \xi \, \forall y_0, y_1, \ldots, y_{d-1} \in \mathbb{R} \; \forall x \in \mathbb{R}: \\ (x \geq \xi \geq 0 \wedge y_0 \geq 0 \wedge \cdots \wedge y_{d-1} \geq 0) \\ \Longrightarrow R_0(m,x) y_0 + \cdots + R_{d-1}(m,x) y_{d-1} \geq 0,$$

where the $R_i(m, \cdot)$ are some rational functions. In order to determine the lower bound ξ we perform quantifier elimination on the formula where we drop the first quantifier in $\widetilde{\Psi}(m)$. The full method then reads as follows.

ALGORITHM 3.
Input: A P-finite recurrence of order d and a vector of initial values defining a sequence $f\colon \mathbb{N} \to \mathbb{Q}$.
Output: True if $f(n) \geq 0$ for all $n \in \mathbb{N}$, False if $f(n) < 0$ for some $n \in \mathbb{N}$, possibly no output at all.

1 **for** $n = 0$ **to** $d-1$ **do**
2 **if** $f(n) < 0$ **then return** False
3 **for** $m = d, d+1, d+2, d+3, \ldots$ **do**
4 Determine a quantifier free formula $\Psi(m, \xi)$ equivalent to

$$\xi > 0 \wedge \forall x, y_0, y_1, \ldots, y_{d-1} \in \mathbb{R}: \\ (x \geq \xi \wedge y_0 \geq 0 \wedge \cdots \wedge y_{d-1} \geq 0) \\ \Rightarrow R_0(m,x) y_0 + \cdots + R_{d-1}(m,x) y_{d-1} \geq 0,$$

5 **if** $\exists \xi_0 \colon \Psi(m, \xi_0)$ **then**
6 **for** $n = m, \ldots, m + \lceil \xi_0 \rceil - 1$ **do**
7 **if** $f(n) < 0$ **then return** False
8 **return** True
9 **if** $f(m) < 0$ **then return** False

Essentially correctness of this algorithm follows from the correctness of Algorithm 1.

THEOREM 1. *Algorithm 3 is correct.*

PROOF. Correctness is obvious whenever the algorithm returns False, because this happens only when an explicit point n with $f(n) < 0$ has been found. Suppose now that the algorithm returns True at some value m of the iteration index of the for loop. Then $f(0), \ldots, f(m-1)$ are all non-negative (this has been verified for the initial values in line 2 and in each iteration of the loop in line 9).

It remains to verify non-negativity for the initial values that have not been checked yet, i.e., that $f(n) \geq 0$ for $m \leq n \leq m + n_0 - 1$. This is carried out in line 7. Then, for any d consecutive, non-negative terms of the sequence $f(n), \ldots, f(n+d-1)$ with $n \geq n_0$, it follows by the definition of Ψ (line 5) and the choice of ξ_0 that $f(n+m) \geq 0$, and thus inductively that $f(n) \geq 0$ for all n. □

In Algorithm 3 the length of the induction hypothesis in the refined induction step formula remains constantly equal to the order of the recurrence, whereas in Algorithm 1 additional requirements (or additional informations) are added to the hypothesis. In both algorithms the conclusion in the induction step formula, however, is the same. Since $(A \Rightarrow B) \Rightarrow ((A \wedge C) \Rightarrow B)$, termination of Algorithm 3 implies termination of Algorithm 1.

In Algorithm 1 the quantifier free part in the induction step formula is of the form

$$\mathcal{H}_d(x, y_0, \ldots, y_{d-1}) \quad \Longrightarrow \quad \mathcal{C}(x, y_0, \ldots, y_{d-1}).$$

Both sides, the hypothesis and the conclusion, define semi-algebraic sets and from the geometric point of view it is checked whether the set defined by $\mathcal{H}$ is contained in the set defined by $\mathcal{C}$. If this is not the case, then a further restriction is added to the hypothesis defining a new set that can, if anything, only decrease in size. At the same time the conclusion is altered in order to determine a set large enough to fit the hypothesis. The process is repeated until (in the lucky case) both sides are made to fit. On the other hand, Algorithm 3 searches for an appropriate superset to fit the hypothesis.

EXAMPLE 1. *Let $f\colon \mathbb{N} \to K$ be defined by*

$$(20n+1)f(n+3) = 3(5n+1)f(n+2) - (13n+1)f(n+1) + (18n+7)f(n),$$

and $f(0) = f(1) = 1, f(2) = 3$. The characteristic polynomial of this recurrence has the coefficients $u_0 = \frac{9}{10}, u_1 = \frac{1}{4}$ and eigenvalues $1, -\frac{1}{8} \pm \mathbf{i}\frac{\sqrt{1415}}{40}$. These values are chosen outside the previously proven terminating regions.

An application [17] of Algorithm 1 succeeds proving positivity by extending the induction hypothesis up to length nine and also Algorithm 3 succeeds proving positivity for the shifted sequence $f(n+9)$ for $n \geq 0$. In general, there might be a large gap between the length of the induction hypothesis needed for Algorithm 1 and the shift length, hence we do not advertise Algorithm 3 as an alternative for real world problems.

Identifying the terminating region of Algorithm 3 corresponds to identifying the region for which the recurrence coefficients $R_i(m,\cdot)$ are in the interior of the terminating region I_m of Algorithm 1 for some $m \geq d$. This is made more precise in the next section.

4. ORDER THREE

Let $f\colon \mathbb{N} \to K$ be a sequence defined by a balanced, P-finite recurrence of order d, where we assume that the leading coefficient has no positive root and we consider the recurrence

$$f(n+d) = r_{d-1}(n)f(n+d-1) + \cdots + r_0(n)f(n), \quad (1)$$

with rational coefficients $r_k(n) = -\frac{p_k(n)}{p_d(n)}$. The main results in this section are on recurrences of order three, i.e., $d = 3$. We assume that 1 is the dominant eigenvalue, and let $\alpha_k \in K$ with $|\alpha_k| < 1$, $k = 1, \ldots, d-1$, be such that

$$\chi(x) = (x-1)(x-\alpha_1)\cdots(x-\alpha_{d-1})$$

is the characteristic polynomial of the recurrence. Depending on the context we also use the following alternative representations of the characteristic polynomial

$$\begin{aligned}\chi(x) &= x^d - c_{d-1}x^{d-1} \cdots - c_1 x - c_0\\ &= (x-1)(x^{d-1} + u_{d-2}x^{d-2} + \cdots + u_0).\end{aligned}$$

The latter allows for direct comparison of the order three results as obtained previously [21]. These three representations are related and we may freely switch between them. For $d = 3$, e.g., we have

$$\begin{aligned}u_0 &= c_0 = \alpha_1\alpha_2,\\ u_1 &= c_0 + c_1 = -\alpha_1 - \alpha_2,\end{aligned}$$

and in general, because of $\chi(1) = 0$, that

$$1 = c_0 + c_1 + \cdots + c_{d-1}.$$

Repeated application of the recurrence (1) allows us to compute $f(n+m)$ (for some $m \geq d$) from $d-1$ consecutive sequence elements $f(n), f(n+1), \ldots, f(n+d-1)$,

$$\begin{aligned}f(n+m) = R_{d-1}(m,n)f(n+d-1) + \cdots\\ \cdots + R_1(m,n)f(n+1) + R_0(m,n)f(n),\end{aligned}$$

where the coefficients $R_k(m,\cdot)$ are rational functions. The repeated application of the recurrence acts also on the coefficients $r_k(n)$ and it is easy to verify that for $k = 0, \ldots, d-1$ they satisfy the recurrence

$$\begin{aligned}R_k(m+d,n) = r_{d-1}(n+m)R_k(m+d-1,n) + \cdots\\ \cdots + r_0(n+m)R_k(m,n), \quad m \geq 0,\end{aligned} \quad (2)$$

with initial values $R_k(j,n) = \delta_{k,j}$ for all $j = 0, \ldots, d-1$ and k, where $\delta_{k,j}$ denotes the Kronecker delta. Note that $R_k(d,n) = r_k(n)$. In particular, for the C-finite recurrence

$$y(n+d) = c_{d-1}y(n+d-1) + \cdots + c_1 y(n+1) + c_0 y(n),$$

with c_k being the coefficients of the characteristic polynomial $\chi(x)$, we have that

$$\begin{aligned}y(n+m) = \gamma_{d-1}(m)y(n+d-1) + \cdots\\ \cdots + \gamma_0(m)y(n), \quad m \geq 0,\end{aligned}$$

where the $\gamma_k(m)$ satisfy the recurrences

$$\gamma_k(m+d) = c_{d-1}\gamma_k(m+d-1) + \cdots + c_0\gamma_k(m), \quad (3)$$

with initial values $\gamma_k(j) = \delta_{k,j}$ for $j,k = 0, \ldots, d-1$. Also these recurrences have characteristic polynomial $\chi(x)$ with largest root 1 and roots α_k in the interior of the complex unit disk. Because (3) is a C-finite recurrence, it can be solved explicitly. The general solution is a linear combination of the sequences $(1)_{m\geq 0}$ and sequences of the form

$$(\alpha^m)_{m\geq 0}, (m\alpha^m)_{m\geq 0}, \ldots, (m^{e-1}\alpha^m)_{m\geq 0},$$

where $\alpha \in \{\alpha_1, \ldots, \alpha_d\}$ and e denotes its multiplicity. Since all the roots are strictly less than one in absolute value, the limit for m tending to infinity is given by the coefficient of the eigenvalue 1.

For the coefficients $r_k(n)$ of the defining recurrence of $f(n)$, we have, using the characteristic polynomial, that

$$\lim_{n\to\infty} r_k(n) = c_k, \quad k = 0, \ldots, d-1.$$

The recurrences (2) and (3) satisfied by the iterated coefficient sequences $R_k(m,n)$ and $c_k(m)$ are structurally of the same form. This, in combination with the limit relation above, yields that for every *fixed* $m \geq 0$ we have

$$\lim_{n\to\infty} R_k(m,n) = \gamma_k(m).$$

With these preliminary considerations at hand, we are in the position to prove that Algorithm 3 terminates, if all the eigenvalues (except 1) have negative real part. Note that until now we have not put any restrictions on the order d. For the first termination result given below, we consider recurrences of order $d = 3$. This statement is formulated in terms of the coefficients u_0, u_1 in order to better compare with the previous results.

THEOREM 2. *Algorithm 3 terminates, if $0 < u_1 < 2$ and $\max(0, u_1 - 1) < u_0 < 1$.*

PROOF. The algorithm terminates if for some $m \geq d$ the coefficients $R_k(m,n)$ in the iterated recurrence

$$f(n+m) = R_2(m,n)f(n+2) + R_1(m,n)f(n+1) + R_0(m,n)f(n),$$

are positive which corresponds to the terminating region I_3 of Algorithm 3. Next, we define the C-finite sequences $\gamma_k(m)$, by

$$\gamma_k(m+3) = c_2\gamma_k(m+2) + c_1\gamma_k(m+1) + c_0\gamma_k(m), \quad (4)$$

with initial values $\gamma_k(j) = \delta_{k,j}$, $k, j = 0, 1, 2$, and denote their respective limits by $\zeta_k = \lim_{m\to\infty} \gamma_k(m)$. Then, for any $\epsilon > 0$ there is an m_0 such that for all $m \geq m_0$,

$$\left|\gamma_k(m) - \zeta_k\right| < \frac{\epsilon}{2}.$$

On the other hand, for every *fixed* m_0, the (iterated coefficient) sequences $R_k(m_0, n)$ tend to $\gamma_k(m_0)$ as $n \to \infty$, i.e., there exists an n_0 such that for all $n \geq n_0$,

$$\left|R_k(m_0, n) - \gamma_k(m_0)\right| < \frac{\epsilon}{2}.$$

Combining these two estimates using the triangle inequality yields that for any $\epsilon > 0$, there exist m_0 and n_0 such that for all $n \geq n_0$,

$$\left|R_k(m_0, n) - \zeta_k\right| < \epsilon.$$

The proof is completed once we can show that the ζ_k are positive under the assumptions of the theorem.

The C-finite recurrence (4) has the eigenvalues $1, \alpha_1, \alpha_2$, where either there are two simple roots $\alpha_1 \neq \alpha_2$ (real or complex conjugate), or there is a double root $\alpha = \alpha_1 = \alpha_2$. In the first case the general ansatz for the closed form of the sequence is

$$\gamma_k(m) = \nu_k(0) + \nu_k(1)\alpha_1^m + \nu_k(2)\alpha_2^m,$$

in the second case the ansatz is

$$\gamma_k(n) = \nu_k(0) + \big(\nu_k(1) + m\nu_k(2)\big)\alpha^m.$$

The constants $\nu_k(0), \nu_k(1), \nu_k(2)$ are obtained using the initial values $\gamma_k(j) = \delta_{k,j}$ and solving the resulting linear system. The limits are then given by $\zeta_k = \nu_k(0)$ and in terms of u_1, u_0 for either of the cases (distinct or double root) the limits turn out to be

$$\zeta_0 = \frac{u_0}{u_1 + u_0 + 1}, \; \zeta_1 = \frac{u_1}{u_1 + u_0 + 1}, \; \zeta_2 = \frac{1}{u_1 + u_0 + 1}.$$

It can be shown by CAD-computations that under the given assumptions on u_1, u_0 these limits are positive. Since ζ_k can be computed a priori from the given recurrence, ϵ above is chosen such that $\min_k(\zeta_k - \epsilon) > 0$. The algorithm terminates no later than at iteration $n_0 + m_0$. $\square$

REMARK 1. *All CAD computations carried out in this paper were performed with Mathematica's built-in implementation of CAD [24, 25]. The computation time is negligible for all of them and could certainly be carried out with other implementations such as, e.g., [11, 4, 8] as well.*

The strategies of Algorithms 1 and 3 can be combined to prove that $f(n+m) \geq 0$, for some $m \geq D$, with an extended induction hypothesis of length D. The proof of termination for these variations is essentially the same as the proof of Theorem 2, merely the choice of ϵ (influencing n_0 and m_0) has to be adapted so that all the $R_k(m_0, n)$ appearing are in an appropriate neighbourhood of the limits ζ_k. The terminating region I_4 for Algorithm 1 with an induction hypothesis of length four is given by [21]

$$u_1 < 1 \wedge u_0 > 0 \wedge 1 - u_1 + u_1^2 - u_0 > 0$$
$$\wedge \big(u_1 > 0 \vee u_1^2 - u_0 - u_1u_0 + u_0^2 < 0\big).$$

The variant of Algorithm 3 with an extended induction hypothesis of length four extends the terminating region of Theorem 2 by the semialgebraic set defined via

$$-1 < u_1 < 0 \;\wedge\; u_1^2 < u_0 < -\frac{1 + u_1 + u_1^2}{u_1}. \quad (5)$$

The computational effort for doing quantifier elimination for analyzing variants extending the induction hypothesis beyond length five becomes prohibitive. However, also termination of the variation with induction hypothesis of length four implies termination of Algorithm 1 and hence as a consequence we have extended the terminating region of Algorithm 1.

COROLLARY 1. *Algorithm 1 terminates, if either $-1 < u_1 < 0$ and $u_1^2 < u_0 < -\frac{1+u_1+u_1^2}{u_1}$, or $0 < u_1 < 2$ and $\max(0, u_1 - 1) < u_0 < 1$.*

The eigenvalues corresponding to this extension are all complex and of the form $\alpha_1 = z_1 + \mathbf{i}z_2 = \overline{\alpha_2}$ inside the unit disk with either negative real part or

$$0 < z_1 < \tfrac{1}{2} \;\wedge\; \sqrt{3}z_1 < |z_2| < \sqrt{1 - z_1^2}.$$

The results obtained in this section are summarized in the figure below. The dark gray area shows the terminating region of Algorithm 3 extending the induction hypothesis up to length four. For better comparison, the previously known terminating region of Algorithm 1 with induction hypothesis up to length four is indicated by dashed lines. The remaining light gray area are the non overlapping parts of I_5 and the terminating region of Algorithm 2.

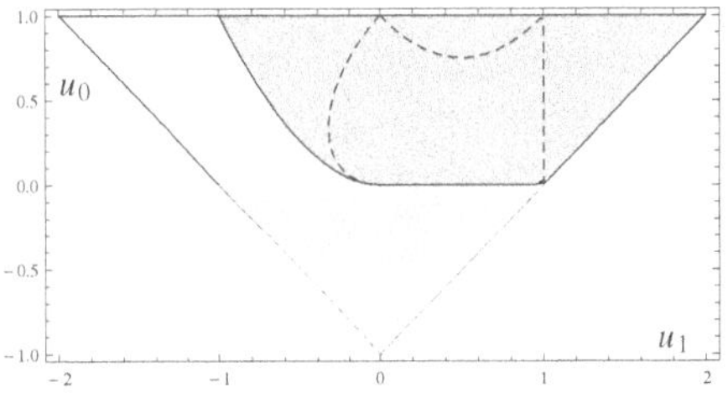

5. ORDER FOUR AND HIGHER

The proof of the termination result of Theorem 2 carries over immediately to higher orders. The induction step formula for a given P-finite sequence of order d with a hypothesis of length d,

$$\forall\, y_0, y_1, \ldots, y_{d-1} \in \mathbb{R}\; \forall\, x \in \mathbb{R}:$$
$$\big(x \geq 0 \wedge y_0 \geq 0 \wedge \cdots \wedge y_{d-1} \geq 0\big)$$
$$\Longrightarrow r_{d-1}(x)y_{d-1} + \cdots + r_0(x)y_0 \geq 0,$$

is true, if the coefficients in the recurrence are eventually positive. Hence, following the previous discussion, to obtain a terminating region it suffices to determine the cases when the recurrence coefficients of the shifted sequence become eventually positive.

Again, we assume that 1 ist the unique dominant eigenvalue and let $\alpha_1, \ldots, \alpha_\ell$, $1 \leq \ell \leq d-1$, be the distinct eigenvalues of the given recurrence and

$$\chi(x) = (x-1)(x^{d-1} + u_{d-2}x^{d-2} + \cdots + u_0) = (x-1)U(x).$$

its characteristic polynomial. Let $\gamma_k(m)$ be the associated C-finite sequences defined by the recurrence (3). A closed form solution can be computed using the inverse of the generalized Vandermonde matrix. For the further reasoning only the limits $\zeta_k = \lim_{n\to\infty} \gamma_k(n)$ are needed and so just the coefficients of the simple eigenvalue 1 for $k = 0, \ldots, d-1$ have to be computed. These values turn out to be

$$\zeta_k = \frac{u_k}{U(1)}, \quad k = 0, \ldots, d-1.$$

The coefficients of $U(x)$ are given by the elementary symmetric polynomials $e_j(\alpha_1, \ldots, \alpha_d)$ (where the eigenvalues are written with multiplicity) as

$$u_{d-1-j} = (-1)^j e_j(\alpha_1, \ldots, \alpha_d).$$

If all the eigenvalues have negative real part, then all the limits (and thus eventually all the recurrence coefficients) are trivially positive. Note that in our setting complex roots may only appear in pairs of complex conjugates, since for proving positivity the sequence can only assume real values.

Corollary 2. *Algorithm 3 (and thus Algorithm 1) terminates, if all eigenvalues α_k inside the interior of the unit disk have negative real part.*

Combining the strategies of Algorithms 1 and 3 is in theory also possible for higher orders. In practice the analysis cannot be carried out because of computational limitations. Under the assumption that the terminating region for order d is contained in the terminating region of order $d+1$ at least a guess for sufficient conditions can be computed. For recurrences of order four with an induction hypothesis of length five, such a guess could be found in reasonable time and verified easily using CAD. For this guess and prove-strategy the representation

$$\chi(x) = (x-1)(x^3 + u_2x^2 + u_1x + u_0)$$

for the characteristic polynomial was used, because the involved inequalities as well as the output are simpler and lower in degree than using the eigenvalues. The following statement concerns only the extension beyond the cases covered in the corollary.

Theorem 3. *If $0 < u_2 < 1$ and $\min\left(-\frac{1}{3}u_2, U\right) < u_1 < \max\left(0, U\right)$ and $V_1 < u_0 < \min\left(u_2(1+u_1-u_2), V_2\right)$, where*

$$U = \frac{u_2^2(u_2-1)}{u_2^2 - u_2 + 1},$$

$$V_{1,2} = \frac{u_1+u_2}{2} \mp \frac{\sqrt{(3u_1+u_2)(u_2-u_1)}}{2},$$

then Algorithm 1 terminates.

Proof. By the limit arguments used earlier it is sufficient to prove that the induction step formula for the limiting case of the recurrence coefficients holds. Let $\Omega \subset \mathbb{R}^3$ denote the set described by the conditions in the theorem. The eigenvalues of a recurrence with coefficients taken from Ω are a pair of complex conjugated roots and a simple real root. For recurrences of order four the limits in terms of u_0, u_1, u_2 are given by

$$\zeta_k = \frac{u_k}{1+u_2+u_1+u_0}, \quad k = 0, 1, 2,$$

and $\zeta_3 = 1 - \zeta_2 - \zeta_1 - \zeta_0$. A CAD-computation confirms that

$$\begin{aligned}
&\forall\,(u_2,u_1,u_0) \in \Omega\ \forall y_0,y_1,y_2,y_3 \in \mathbb{R}:\\
&\quad y_0 > 0 \wedge y_1 > 0 \wedge y_2 > 0 \wedge y_3 > 0\\
&\qquad \wedge\ (1-u_2)y_3 + (u_2-u_1)y_2 + (u_1-u_0)y_1 + u_0y_0 > 0\\
&\qquad \Longrightarrow \zeta_3y_3 + \zeta_2y_2 + \zeta_1y_1 + \zeta_0y_0 > 0.
\end{aligned}$$

Hence, as soon as the recurrence coefficients of the sequence and of the shifted sequence are in a neighbourhood close enough to $(1-u_2), \ldots, u_0$ and $\zeta_3, \ldots, \zeta_0$, respectively, the algorithm terminates. □

The terminating region was determined assuming that the complex conjugate roots are in the terminating region (5) and applying quantifier elimination using CAD to the induction step formula above with the first quantifier dropped. A further iteration of this idea extending to recurrences of order five is very space and time consuming and different strategies are needed.

Besides being hard to analyze, approaches based on CAD also (usually) do not provide insight in why the given sequence is non-negative and there is no human readable output that can be verified independently. Future directions may build on recent results on sums-of-squares representations [16] either to verify the final induction step formula or in combination with recent methods to determine a representation of a given P-finite sequence as linear combination of squares [6].

6. REFERENCES

[1] Horst Alzer, Stefan Gerhold, Manuel Kauers, and Alexandru Lupaş. On Turán's inequality for Legendre polynomials. *Expositiones Mathematicae*, 25(2):181–186, 2007.

[2] Saugata Basu, Richard Pollack, and Marie-Françoise Roy. *Algorithms in Real Algebraic Geometry*, volume 10 of *Algorithms and Computation in Mathematics*. Springer, 2nd edition, 2006.

[3] Jason P. Bell and Stefan Gerhold. On the positivity set of a linear recurrence sequence. *Israel J. Math.*, 157:333–345, 2007.

[4] Chris W. Brown. QEPCAD B – a program for computing with semi-algebraic sets. *Sigsam Bulletin*, 37(4):97–108, 2003.

[5] Bob F. Caviness and Jeremy R. Johnson, editors. *Quantifier Elimination and Cylindrical Algebraic Decomposition*, Texts and Monographs in Symbolic Computation. Springer, 1998.

[6] Yongjae Cha. Closed form solutions of linear difference equations in terms of symmetric products. Submitted.

[7] Yongjae Cha, Mark van Hoeij, and Giles Levy. Solving recurrence relations using local invariants. In Stephen Watt, editor, *Proceedings of the 2010 International Symposium on Symbolic and Algebraic Computation*, ISSAC '10, pages 303–309, New York, NY, USA, 2010. ACM.

[8] Changbo Chen, Marc Moreno Maza, Bican Xia, and Lu Yang. Computing cylindrical algebraic decomposition via triangular decomposition. In *Proceedings of ISSAC'09*, pages 95–102, New York, NY, USA, 2009. ACM.

[9] George E. Collins. Quantifier elimination for the elementary theory of real closed fields by cylindrical algebraic decomposition. *Lecture Notes in Computer Science*, 33:134–183, 1975.

[10] George E. Collins and Hoon Hong. Partial cylindrical algebraic decomposition for quantifier elimination. *Journal of Symbolic Computation*, 12(3):299–328, 1991.

[11] Andreas Dolzmann and Thomas Sturm. Guarded expressions in practice. In *Proceedings of ISSAC'97*, 1997.

[12] Graham Everest, Alf van der Poorten, Igor Shparlinski, and Thomas Ward. *Recurrence Sequences*, volume 104 of *Mathematical Surveys and Monographs*. American Mathematical Society, 2003.

[13] Stefan Gerhold. *Combinatorial Sequences: Non-Holonomicity and Inequalities*. PhD thesis, RISC-Linz, Johannes Kepler Universität Linz, 2005.

[14] Stefan Gerhold and Manuel Kauers. A procedure for proving special function inequalities involving a discrete parameter. In Manuel Kauers, editor, *Proceedings of ISSAC'05*, pages 156–162, 2005.

[15] Stefan Gerhold and Manuel Kauers. A computer proof of Turán's inequality. *Journal of Inequalities in Pure and Applied Mathematics*, 7(2):#42, 2006.

[16] Erich L. Kaltofen, Bin Li, Zhengfeng Yang, and Lihong Zhi. Exact certification in global polynomial optimization via sums-of-squares of rational functions with rational coefficients. *Journal of Symbolic Computation*, 47(1):1–15, 2012.

[17] Manuel Kauers. SumCracker – A Package for Manipulating Symbolic Sums and Related Objects. *Journal of Symbolic Computation*, 41(9):1039–1057, 2006.

[18] Manuel Kauers. Computer algebra and power series with positive coefficients. In *Proceedings of FPSAC'07*, 2007.

[19] Manuel Kauers. Computer algebra and special function inequalities. In Tewodros Amdeberhan and Victor H. Moll, editors, *Tapas in Experimental Mathematics*, volume 457 of *Contemporary Mathematics*, pages 215–235. AMS, 2008.

[20] Manuel Kauers and Peter Paule. A computer proof of Moll's log-concavity conjecture. *Proceedings of the AMS*, 135(12):3847–3856, 2007.

[21] Manuel Kauers and Veronika Pillwein. When can we detect that a P-finite sequence is positive? In Stephen Watt, editor, *Proceedings of ISSAC'10*, pages 195–202, 2010.

[22] Marc Mezzarobba and Bruno Salvy. Effective bounds for P-recursive sequences. *J. Symbolic Comput.*, 45(10):1075–1096, 2010.

[23] Veronika Pillwein. Positivity of certain sums over Jacobi kernel polynomials. *Advances in Applied Mathematics*, 41(3):365–377, 2008.

[24] Adam Strzeboński. Solving systems of strict polynomial inequalities. *Journal of Symbolic Computation*, 29:471–480, 2000.

[25] Adam Strzeboński. Cylindrical algebraic decomposition using validated numerics. *Journal of Symbolic Computation*, 41(9):1021–1038, 2006.

Integration of Unspecified Functions and Families of Iterated Integrals

Clemens G. Raab
Deutsches Elektronen-Synchrotron (DESY)
15738 Zeuthen, Germany
clemens.raab@desy.de

ABSTRACT

An algorithm for parametric elementary integration over differential fields constructed by a differentially transcendental extension is given. It extends current versions of Risch's algorithm to this setting and is based on some first ideas of Graham H. Campbell transferring his method to more formal grounds and making it parametric, which allows to compute relations among definite integrals. Apart from differentially transcendental functions, such as the gamma function or the zeta function, also unspecified functions and certain families of iterated integrals such as the polylogarithms can be modeled in such differential fields.

Categories and Subject Descriptors

I.1.2 [**Symbolic and Algebraic Manipulation**]: Algorithms

General Terms

Algorithms, Theory

Keywords

Symbolic Integration; Parametric Elementary Integration; Differentially Transcendental Functions; Differential Fields

1. INTRODUCTION

Among the many approaches to compute integrals in closed from we take the differential algebra approach using differential fields. The first complete algorithm for finding elementary integrals of a class of transcendental functions was published by Risch [10] and deals with integration of elementary functions. Later, this was generalized to other types of integrands, see [2, 9] and references therein. Given an element of a suitable differential field these algorithms look for an antiderivative in elementary extensions of that field. Often such algorithms consider an integrand depending linearly on parameters as in the following problem.

ISSAC'13, June 26–29, 2013, Boston, Massachusetts, USA.

PROBLEM 1 (PARAMETRIC ELEMENTARY INTEGRATION). *Given a differential field (F, D) and $f_0, \dots, f_m \in F$, compute $\mathbf{c}_1, \dots, \mathbf{c}_n \in C^{m+1}$, where $C := \mathrm{Const}(F)$, and corresponding $g_1, \dots, g_n$ from some elementary extension of (F, D) such that*

$$Dg_j = (f_0, \dots, f_m) \cdot \mathbf{c}_j$$

for all j and $\{\mathbf{c}_1, \dots, \mathbf{c}_n\}$ is a basis of the C-vector space of all $\mathbf{c} \in C^{m+1}$ for which $(f_0, \dots, f_m) \cdot \mathbf{c}$ has an elementary integral over (F, D).

Parametric integration is important for definite integration as on an interval (a, b) a relation of the form

$$c_0 f_0(x) + \dots + c_m f_m(x) = g'(x),$$

where the c_i do not depend on x, yields a linear relation for the corresponding definite integrals

$$c_0 \int_a^b f_0(x)\,dx + \dots + c_m \int_a^b f_m(x)\,dx = g(b) - g(a).$$

For more on definite integration based on this principle see e.g. [1, 9] and references therein. Note that above formulation of the problem requires finding *all* choices for the parameters giving rise to an elementary integral.

Typically the field (F, D) is generated by functions satisfying first-order differential equations of certain forms, mostly linear but not exclusively. In strong contrast to this, in this paper we consider differential fields $(F, D) = (K\langle t\rangle, D)$ generated from some underlying differential field (K, D) by adjoining a generator t such that all derivatives $t, Dt, D^2t, \dots$ are algebraically independent over K, i.e., t is *differentially transcendental* over (K, D). As simple example we could choose $(K, D) = (\mathbb{Q}(x), \frac{d}{dx})$, then t differentially transcendental over (K, D) can represent, for example, the gamma function $\Gamma(x)$, the digamma function $\psi(x)$, or the Riemann zeta function $\zeta(x)$ as those functions do not satisfy any algebraic differential equation with rational function coefficients. For other functions with this property see [8] for example. A differentially transcendental t may also be regarded as differential indeterminate or unspecified function, which is what was informally considered by Campbell [3] for a single integrand, i.e. in the non-parametric case. We also note that certain families of iterated integrals, such as the polylogarithms $\mathrm{Li}_2(x), \mathrm{Li}_3(x), \dots$, are such that the generic element of the family is differentially transcendental. Hence an integral like

$$\int \frac{\mathrm{Li}_{n-2}(x)\mathrm{Li}_n(x)}{x\mathrm{Li}_{n-1}(x)^2}\,dx = \ln(x) - \frac{\mathrm{Li}_n(x)}{\mathrm{Li}_{n-1}(x)}$$

with symbolic n can be obtained in the same framework, for example.

We carefully analyze the problem in the language of differential fields which leads to a new subproblem, see Problem 11, that we solve. Our algorithm, summarized as Algorithm 3, solves the parametric elementary integration problem over differential fields $(K\langle t\rangle, D)$, where t is differentially transcendental over (K, D). Note that an algorithm for this type of field extensions can also be used to at least heuristically treat any given function to which no other algorithm applies, even if the function satisfies some algebraic differential equation with coefficients from K, it just may not find all solutions in that case.

An analogous problem for summation of unspecified sequences was treated by Kauers and Schneider [5, 6].

In Section 2 we recall the definitions used in this paper, give some properties of the notions used, and introduce a flexible way of representing elements of the fields dealt with. Then, in Section 3 we prove a refined version of Liouville's theorem, which will be crucial for the algorithm in Section 4. Section 5 deals with representing families of iterated integrals and a simple sufficient condition is given for algebraic independence of their members. Finally, Section 6 presents some examples of various applications of the algorithm.

Parts of the material presented in this paper were already included in the author's PhD thesis [9]. All fields are implicitly understood to be of characteristic 0.

2. DEFINITIONS AND BASIC PROPERTIES

We will write $\operatorname{coeff}(p, x_1^{i_1}\cdots x_n^{i_n})$, explicitly keeping factors of the form x_j^0 in case $i_j = 0$, for the coefficient of $x_1^{i_1}\cdots x_n^{i_n}$ of p as a polynomial in $x_1, \dots, x_n$. Let a and b be univariate polynomials, then we write $a \div b$ and $a \bmod b$ for their quotient and remainder (the variable will be clear from the context). We sometimes write linear combinations as products of vectors $(f_0, \dots, f_m)\cdot \mathbf{c} = \sum_{i=0}^m c_i f_i$. By $(\mathbf{v}_1, \dots, \mathbf{v}_n)$ we denote a matrix given by its column vectors. When extending a differential field (K, D) by adjoining new elements $x_1, \dots, x_n$ we need to distinguish between the field $K(x_1, \dots, x_n)$ they generate and the differential field $K\langle x_1, \dots, x_n\rangle = K(x_1, \dots, x_n, Dx_1, \dots, Dx_n, D^2x_1, \dots)$ generated by those elements.

2.1 Elementary Extensions

We briefly recall the precise definition of the type of integrals we want to compute.

DEFINITION 2. *Let (F, D) be a differential field and let $(E, D) = (F(s_1, \dots, s_n), D)$ be a differential field extension. Then (E, D) is called an* elementary *extension of (F, D), if each s_i is elementary over $(E_{i-1}, D) := (F(s_1, \dots, s_{i-1}), D)$, or more explicitly if each s_i*

1. *is a* logarithm *over (E_{i-1}, D), i.e. there exists $a \in E_{i-1}$ such that $Ds_i = \frac{Da}{a}$, or*
2. *is an* exponential *over (E_{i-1}, D), i.e. there exists $a \in E_{i-1}$ such that $\frac{Ds_i}{s_i} = Da$, or*
3. *is algebraic over E_{i-1}.*

DEFINITION 3. *Let (F, D) be a differential field and $f \in F$. Then we say that f has an* elementary integral *over (F, D) if there exist an elementary extension (E, D) of (F, D) and $g \in E$ such that $Dg = f$.*

2.2 Differentially Transcendental Extensions

DEFINITION 4. *Let (F, D) be a differential field, (K, D) a differential subfield, and $t \in F$. If all derivatives $t, Dt, D^2t, \dots$ are algebraically independent over K, then t is* differentially transcendental *over (K, D).*

We will, however, consider a more general way of representing the elements of the field $K\langle t\rangle = K(t, Dt, D^2t, \dots)$ by choosing $t_0, t_1, \dots \in K\langle t\rangle$ such that

$$t_0 = t \tag{1}$$
$$Dt_n = a_n t_{n+1} + b_n \tag{2}$$

with $a_n, b_n \in K(t_0, \dots, t_n)$ for $n \in \mathbb{N}$. Then, the following lemma shows that automatically $a_n \neq 0$ and the equality $K(t, Dt, \dots, D^n t) = K(t_0, \dots, t_n)$ are satisfied for all $n \in \mathbb{N}$. So we will mainly consider $K\langle t\rangle$ as $K(t_0, t_1, \dots)$. For instance, the flexibility in the representation introduced by (2) allows to represent the polylogarithms $\mathrm{Li}_m(x)$ for symbolic m in a convenient way. If $(K, D) = (C(x), \frac{d}{dx})$ and $a_n = \frac{1}{x}$ and $b_n = 0$ in (2), then t_n corresponds to $\mathrm{Li}_{m-n}(x)$ and functions involving $\mathrm{Li}_m(x), \mathrm{Li}_{m-1}(x), \dots$ are directly represented in terms of these functions instead of $\mathrm{Li}_m(x), \mathrm{Li}_m'(x), \dots$, see also Section 5.

LEMMA 5. *Let t be differentially transcendental over (K, D) and let $t_0, t_1, \dots \in K\langle t\rangle$ such that (1) and (2). Then*

1. *$a_n \neq 0$ for all $n \in \mathbb{N}$ and*
2. *for all $n \in \mathbb{N}$ there are $\tilde{a}_n, \tilde{b}_n \in K(t, Dt, \dots, D^n t)$ such that $t_{n+1} = \tilde{a}_n D^{n+1}t + \tilde{b}_n$ and $\tilde{a}_n \neq 0$.*

PROOF. We prove both statements in parallel by induction, for which we artificially include the case $n = -1$. For $n = -1$ we define $a_{-1} := 1$, $\tilde{a}_{-1} := 1$, and $\tilde{b}_{-1} := 0$, then trivially $a_{-1}, \tilde{a}_{-1} \in K^*$ and $t_0 = \tilde{a}_{-1}t + \tilde{b}_{-1}$ by definition. For $n \in \mathbb{N}$ we assume that for all $i \in \{-1, 0, \dots, n-1\}$ there are $\tilde{a}_i, \tilde{b}_i \in K(t, Dt, \dots, D^i t)$ such that $t_{i+1} = \tilde{a}_i D^{i+1}t + \tilde{b}_i$ and $\tilde{a}_i \neq 0$. Then, from the assumptions above we obtain that $a_n t_{n+1} = Dt_n - b_n = D(\tilde{a}_{n-1}D^n t + \tilde{b}_{n-1}) - b_n = \tilde{a}_{n-1}D^{n+1}t + (D\tilde{a}_{n-1})D^n t + D\tilde{b}_{n-1} - b_n$. If we had $a_n = 0$, then this and $\tilde{a}_{n-1} \neq 0$ would imply that $D^{n+1}t$ equals $\frac{(D\tilde{a}_{n-1})D^n t + D\tilde{b}_{n-1} - b_n}{-\tilde{a}_{n-1}}$. The latter is in $K(t, \dots, D^n t)$ by induction hypothesis, which would be in contradiction to the algebraic independence of $t, \dots, D^{n+1}t$ over K. Hence, $a_n \neq 0$ and we set $\tilde{a}_n := \frac{\tilde{a}_{n-1}}{a_n}$ and $\tilde{b}_n := \frac{(D\tilde{a}_{n-1})D^n t + D\tilde{b}_{n-1} - b_n}{a_n}$, which both are in $K(t, \dots, D^n t)$ by the induction hypothesis. □

The second statement of the lemma above has some important immediate consequences, which we emphasize by stating the following corollary. The proof is trivial and so we omit it.

COROLLARY 6. *Let t be differentially transcendental over (K, D) and let $t_0, t_1, \dots \in K\langle t\rangle$ such that (1) and (2). Then*

1. *$t_0, t_1, \dots$ are algebraically independent over K, and*
2. *$K(t, Dt, \dots, D^n t) = K(t_0, \dots, t_n)$ for all $n \in \mathbb{N}$.*

In the following we will formalize the ideas of Campbell [3] in this framework. In this context we define the coefficient lifting $\kappa_D : K[t_0, t_1, \ldots] \to K[t_0, t_1, \ldots]$ of D by $\kappa_D(\sum_\alpha f_\alpha t^\alpha) := \sum_\alpha (Df_\alpha)t^\alpha$, where we use multiindex notation for brevity, and extend it to a derivation κ_D on $K(t_0, t_1, \ldots)$ in the natural way by the quotient rule. It is easy to see that

$$Df = \kappa_D f + \sum_{k=0}^{\infty} \frac{\partial f}{\partial t_k} Dt_k. \tag{3}$$

Note that the sum contains only finitely many nonzero summands since $\frac{\partial f}{\partial t_k} = 0$ from some point on. Generalizing the definition of κ_D above, for each $n \in \mathbb{N}$ we define the derivation $\kappa_{D,n}$ on $K(t_0, t_1, \ldots)$ by

$$\kappa_{D,n} f := \kappa_D f + \sum_{k=0}^{n-1} \frac{\partial f}{\partial t_k} Dt_k. \tag{4}$$

These derivations obviously obey $\kappa_{D,n} f + \frac{\partial f}{\partial t_n} Dt_n = \kappa_{D,n+1} f$ for $f \in K(t_0, t_1, \ldots)$ with $\kappa_{D,0} = \kappa_D$. For $f \in K(t_0, \ldots, t_{n-1})$ we have in particular $\kappa_{D,n} f = Df$. An important measure on the elements of $K(t_0, t_1, \ldots)$ is the highest index of any of the generators needed to represent the particular element of the field.

Definition 7. *Let t be differentially transcendental over (K, D), then we define the* differential degree *of $f \in K\langle t \rangle$ by*

$$\operatorname{ddeg}_t(f) := \begin{cases} \min\{k \in \mathbb{N} \mid f \in K(t, \ldots, D^k t)\} & \text{if } f \notin K \\ -\infty & \text{if } f \in K. \end{cases}$$

Note that for $t_0, t_1, \ldots \in K\langle t \rangle$ with (1) and (2) the above corollary implies

$$\operatorname{ddeg}_t(f) = \begin{cases} \min\{k \in \mathbb{N} \mid f \in K(t_0, \ldots, t_k)\} & \text{if } f \notin K \\ -\infty & \text{if } f \in K. \end{cases}$$

So we can say that $\frac{\partial f}{\partial t_k} = 0$ for $k > \operatorname{ddeg}_t(f)$ in (3). The differential degree obeys the following properties with respect to the operations of a differential field. For $f, g \in K(t_0, t_1, \ldots)^*$ we have

$$\begin{aligned} \operatorname{ddeg}_t(f+g) &\le \max(\operatorname{ddeg}_t(f), \operatorname{ddeg}_t(g)), \\ \operatorname{ddeg}_t(fg) &\le \max(\operatorname{ddeg}_t(f), \operatorname{ddeg}_t(g)), \\ \operatorname{ddeg}_t(1/f) &= \operatorname{ddeg}_t(f), \\ \operatorname{ddeg}_t(Df) &= \operatorname{ddeg}_t(f) + 1 \end{aligned}$$

with equality in the first two relations if $\operatorname{ddeg}_t(f) \neq \operatorname{ddeg}_t(g)$. In particular, the last property implies that $\operatorname{Const}_D(K\langle t \rangle) = \operatorname{Const}_D(K)$. Furthermore, the following corollary highlights important properties implied by (2) and (3).

Corollary 8. *Let t be differentially transcendental over (K, D), let $t_0, t_1, \ldots \in K\langle t \rangle$ such that (1) and (2), and let $F := K(t_0, t_1, \ldots)$. Then $\operatorname{Const}_D(F) = \operatorname{Const}_D(K)$ and for all $f \in F$ and any $k \in \mathbb{N}$ with $k \geq \operatorname{ddeg}_t(f)$ there exist $a, b \in K(t_0, \ldots, t_k)$ with $a = a_k \frac{\partial f}{\partial t_k}$ and*

$$Df = at_{k+1} + b.$$

3. LIOUVILLE'S THEOREM

Based on the properties stated in the previous section we are now ready to prove a refinement of Liouville's theorem for this situation. It was not made explicit in [3] which type of expressions are dealt with exactly. Above we gave one interpretation in terms of differential fields and the following results present the corresponding precise details of the ideas from [3].

Theorem 9. *Let t be differentially transcendental over (K, D) and let $t_0, t_1, \ldots \in K\langle t \rangle$ such that (1) and (2). Let $f \in F := K(t_0, t_1, \ldots)$ such that f has an elementary integral over (F, D) and let $k := \operatorname{ddeg}_t(f)$. Then there are $v \in K(t_0, \ldots, t_{k-1})$, $c_1, \ldots, c_n \in \operatorname{Const}_D(K)$, and $u_1, \ldots, u_n \in K(c_1, \ldots, c_n, t_0, \ldots, t_{k-1})^*$ such that*

$$f = Dv + \sum_{i=1}^{n} c_i \frac{Du_i}{u_i}. \tag{5}$$

If $k \geq 1$, we can also write this as

$$f = a_{k-1} \left(\frac{\partial v}{\partial t_{k-1}} + \sum_{i=1}^{n} c_i \frac{\frac{\partial u_i}{\partial t_{k-1}}}{u_i} \right) t_k + b$$

for some $b \in K(t_0, \ldots, t_{k-1})$.

Proof. By Liouville's theorem (e.g. [2, Thm 5.5.3]) we know that there are $v \in F$, $c_1, \ldots, c_n \in \operatorname{Const}_D(F)$, and $u_1, \ldots, u_n \in F(c_1, \ldots, c_n)^*$ such that (5) and by Corollary 8 we deduce $c_1, \ldots, c_n \in \operatorname{Const}_D(K)$. Define

$$m := \max(\operatorname{ddeg}_t(v), \operatorname{ddeg}_t(u_1), \ldots, \operatorname{ddeg}_t(u_n)).$$

If $m < 0$, then $f \in K$ and the statement is trivially fulfilled. So assume $m \geq 0$ now and assume without loss of generality that $v, c_1, \ldots, c_n, u_1, \ldots, u_n$ are chosen such that $u_1, \ldots, u_n$ are pairwise relatively prime polynomials from $K(c_1, \ldots, c_n, t_0, \ldots, t_{m-1})[t_m]$. Then, applying Corollary 8 to each summand in (5) implies that

$$f = a_m \left(\frac{\partial v}{\partial t_m} + \sum_{i=1}^{n} c_i \frac{\frac{\partial u_i}{\partial t_m}}{u_i} \right) t_{m+1} + b$$

for some $b \in K(c_1, \ldots, c_n, t_0, \ldots, t_m)$ and by comparing the differential degree of both sides we obtain $k \leq m+1$. Since by Corollary 6 $t_0, t_1, \ldots$ are algebraically independent over K they are also algebraically independent over $K(c_1, \ldots, c_n)$. So we even have $b \in K(t_0, \ldots, t_m)$ by comparing the coefficient of t_{m+1}^0. Next, Lemma 5 implies that $a_m \neq 0$. If we had $m > k-1$, then by comparing the coefficient of t_{m+1} we could conclude $\tilde{f} := \frac{\partial v}{\partial t_m} + \sum_{i=1}^{n} c_i \frac{\frac{\partial u_i}{\partial t_m}}{u_i} = 0$. From this we would obtain $\max(\operatorname{ddeg}_t(u_1), \ldots, \operatorname{ddeg}_t(u_n)) < m$ by applying Lemma 5.6.1 from [2] in the differential field $(K(t_0, \ldots, t_m), \frac{\partial}{\partial t_m})$ and noting that an irreducible polynomial $p \in K(c_1, \ldots, c_n, t_0, \ldots, t_{m-1})[t_m]$ can divide at most one of $u_1, \ldots, u_n$ as they are pairwise relatively prime. Therefore, the definitions of m and $\tilde{f}$ would imply $\operatorname{ddeg}_t(v) = m$ and $\tilde{f} = \frac{\partial v}{\partial t_m}$, respectively, which would give $\tilde{f} \neq 0$ altogether in contradiction to $\tilde{f} = 0$. Hence we have $m = k-1$. □

In particular, for the special case $k \leq 0$ this theorem contains analogs of Corollary 5.11.1 from [2] and Theorem 3.15 from [9], which we make explicit in the following corollary.

Corollary 10. *Let t be differentially transcendental over (K, D), let $t_0, t_1, \ldots \in K\langle t \rangle$ with (1) and (2) and let $F := K(t_0, t_1, \ldots)$. If $f \in K(t_0)$ has an elementary integral over (F, D), then $f \in K$. If $f \in K$ has an elementary integral over (F, D), then it has an elementary integral over (K, D).*

4. ALGORITHM

In the following we will assume that the differential field (K, D) is computable, i.e., that we can effectively compute the basic arithmetic operations as well as derivation and zero-testing. Furthermore, with $C := \mathrm{Const}_D(K)$ we will need to compute C-vector space bases of the constant solutions $\{\mathbf{c} \in C^n \mid A \cdot \mathbf{c} = 0\}$ of linear systems with coefficients in K, i.e., $\ker(A) \cap C^n$ for $A \in K^{m \times n}$. This task can be reduced to the solution of linear systems with coefficients in C by an algorithm of Bronstein [2, Lemma 7.1.2], which computes $B \in C^{\tilde{m} \times n}$ such that $\ker(A) \cap C^n = \ker(B)$.

4.1 Computing the Logarithmic Part

Theorem 9 suggests that in order to compute elementary integrals over $(K(t_0, t_1, \ldots), D)$ we should look at the following new subproblem, which was not part of [3]. The difficulty of this problem is related to the computation of the logarithmic part of the integral and will be taken care of by Algorithm 1. Based on this algorithm it is straightforward to solve the full subproblem, Algorithm 2 shows how this can be done.

PROBLEM 11. *Given: t differentially transcendental over (K, D), $t_0, t_1, \ldots \in K\langle t \rangle$ with (1) and (2), $k \in \mathbb{N}^+$, and $f_0, \ldots, f_m \in K(t_0, \ldots, t_{k-1})$.*

Find: a basis $\mathbf{c}_1, \ldots, \mathbf{c}_n \in C^{m+1}$, where $C := \mathrm{Const}_D(K)$, of the C-vector space of all $\mathbf{c} \in C^{m+1}$ such that there exist $v \in K(t_0, \ldots, t_{k-1})$, $d_1, \ldots, d_l \in \overline{C}$, and $u_1, \ldots, u_l \in K(d_1, \ldots, d_l, t_0, \ldots, t_{k-1})^*$ with

$$(f_0, \ldots, f_m) \cdot \mathbf{c} = \frac{\partial v}{\partial t_{k-1}} + \sum_{i=1}^{l} d_i \frac{\frac{\partial u_i}{\partial t_{k-1}}}{u_i}$$

as well as corresponding $v_j \in K(t_0, \ldots, t_{k-1})$, $d_{j,1}, \ldots, d_{j,l_j} \in \overline{C}$, and $u_{j,1}, \ldots, u_{j,l_j} \in K(d_{j,1}, \ldots, d_{j,l_j}, t_0, \ldots, t_{k-1})^*$ for each $j \in \{1, \ldots, n\}$.

At first glance this problem may look like it was just parametric elementary integration over $(K(t_0, \ldots, t_{k-1}), \frac{\partial}{\partial t_{k-1}})$ and we could solve it easily by well-known algorithms, but there is a subtle difference. Observe that the above formulation requires linear combinations with coefficients belonging to $C = \mathrm{Const}_D(K)$ instead of $\mathrm{Const}_{\frac{\partial}{\partial t_{k-1}}}(K(t_0, \ldots, t_{k-1})) = K(t_0, \ldots, t_{k-2})$. Furthermore, instead of allowing residues $d_i \in \overline{K(t_0, \ldots, t_{k-2})}$ they are restricted to $d_i \in \overline{\mathrm{Const}_D(K)}$ above. Because of these requirements Problem 11 cannot be solved directly by standard algorithms. However, it can be solved algorithmically as follows by adapting the ideas of Theorem 3.9 from [9] in order to make use of both derivations D as well as $\frac{\partial}{\partial t_{k-1}}$. In the proof, we will use the following formula, implied by Lemma 3.2.2 from [2], where $p \in K(t_0, \ldots, t_n)[z]$ and f is in some differential field extension of $(K(t_0, t_1, \ldots), D)$.

$$D(p(f)) = \sum_{i=0}^{\deg_z(p)} (\kappa_{D,n+1} \operatorname{coeff}(p, z^i)) f^i + \frac{\partial p}{\partial z}(f) Df \quad (6)$$

THEOREM 12. *Let t be differentially transcendental over (K, D), let $t_0, t_1, \ldots \in K\langle t \rangle$ such that (1) and (2), and let $C := \mathrm{Const}_D(K)$. Let $k \in \mathbb{N}^+$, let $\tilde{K} := K(t_0, \ldots, t_{k-2})$, let $a_0, \ldots, a_m, b \in \tilde{K}[t_{k-1}]$ with $b \neq 0$ and $\gcd(b, \frac{\partial b}{\partial t_{k-1}}) = 1$, and let z be an indeterminate. Then by Algorithm 1 we can compute linear independent $\mathbf{c}_1, \ldots, \mathbf{c}_n \in C^{m+1}$ such that:*

1. *If $\mathbf{c} \in C^{m+1}$ is such that there exist $v \in \tilde{K}(t_{k-1})$, $d_1, \ldots, d_l \in \overline{C}$, and $u_1, \ldots, u_l \in \tilde{K}(d_1, \ldots, d_l, t_{k-1})^*$ with*
$$\frac{\partial v}{\partial t_{k-1}} + \sum_{i=1}^{l} d_i \frac{\frac{\partial u_i}{\partial t_{k-1}}}{u_i} = \frac{(a_0, \ldots, a_m) \cdot \mathbf{c}}{b},$$
then $\mathbf{c} \in \mathrm{span}_C\{\mathbf{c}_1, \ldots, \mathbf{c}_n\}$.

2. *For all $j \in \{1, \ldots, n\}$ there exists $r \in C[z]$ such that*
$$\frac{(a_0, \ldots, a_m) \cdot \mathbf{c}_j}{b} - \sum_{r(\alpha)=0} \alpha \frac{\frac{\partial g_\alpha}{\partial t_{k-1}}}{g_\alpha} \in \tilde{K}[t_{k-1}],$$
where for all $\alpha \in \overline{C}$ with $r(\alpha) = 0$ we define $g_\alpha := \gcd((a_0, \ldots, a_m) \cdot \mathbf{c}_j - \alpha \frac{\partial b}{\partial t_{k-1}}, b)) \in \tilde{K}[t_{k-1}]$.

Algorithm 1 Restrict residues as specified in Theorem 12

---

1. Let $\tilde{a}_0, \ldots, \tilde{a}_m, \tilde{b} \in \tilde{K}[z]$ such that $\tilde{a}_i(t_{k-1}) = a_i$ and $\tilde{b}(t_{k-1}) = b$
2. Let $\tilde{b}_j := \operatorname{coeff}(\tilde{b}, z^j)$ for all $j \in \{0, \ldots, \deg_z(\tilde{b})\}$
3. For all $i \in \{0, \ldots, m\}$ compute $p_i \in \tilde{K}[z]$ with $\deg_z(p_i) < \deg_z(\tilde{b})$ such that
$$\tilde{a}_i \equiv p_i \frac{\partial \tilde{b}}{\partial z} \pmod{\tilde{b}}$$
4. For all $i \in \{0, \ldots, m\}$ and $l \in \{0, 1\}$ compute $\tilde{p}_{i,l} \in \tilde{K}[z]$ with $\deg_z(\tilde{p}_{i,l}) < \deg_z(\tilde{b})$ such that
$$\frac{\partial p_i}{\partial z} \cdot \sum_{j=0}^{\deg_z(\tilde{b})} \operatorname{coeff}(\kappa_{D,k-1} \tilde{b}_j, t_{k-1}^l) z^j \equiv \tilde{p}_{i,l} \frac{\partial \tilde{b}}{\partial z} \pmod{\tilde{b}}$$
5. For all $i \in \{0, \ldots, m\}$ compute
$$q_i := \sum_{j=0}^{\deg_z(p_i)} (\kappa_{D,k-1} \operatorname{coeff}(p_i, z^j)) z^j - (\tilde{p}_{i,1} t_{k-1} + \tilde{p}_{i,0})$$
6. Construct a matrix $A \in \tilde{K}^{2 \deg_z(\tilde{b}) \times (m+1)}$ by
$$A := \begin{pmatrix} \operatorname{coeff}(q_i, t_{k-1}^0 z^j))_{j,i} \\ \operatorname{coeff}(q_i, t_{k-1}^1 z^j))_{j,i} \end{pmatrix},$$
where $j \in \{0, \ldots, \deg(\tilde{b}) - 1\}$ and $i \in \{0, \ldots, m\}$
7. Compute a C-vector space basis $\mathbf{c}_1, \ldots, \mathbf{c}_n \in C^{m+1}$ of $\ker(A) \cap C^{m+1}$

PROOF. First, we prove that the steps of Algorithm 1 can indeed be executed. Since $\gcd(b, \frac{\partial b}{\partial t_{k-1}}) = 1$ implies $\gcd(\tilde{b}, \frac{\partial \tilde{b}}{\partial z}) = 1$ the p_i and $\tilde{p}_{i,l}$ defined in Steps 3 and 4 exist and we apply the half-extended (i.e. computing one Bézout coefficient but not both) Euclidean algorithm in $\tilde{K}[z]$ for computing them. Step 7 can be reduced to computing a basis of the nullspace of a matrix with entries in C, either by exploiting the differential structure provided by D as done in Lemma 7.1.2 from [2], or by exploiting knowledge of the generators which generate the field $\tilde{K}$.

Next, we want prove that $q_0, \ldots, q_m \in \tilde{K}[t_{k-1}][z]$ satisfy $\deg_z(q_i) < \deg_{t_{k-1}}(b)$ and

$$\forall \beta \in \overline{\tilde{K}}, \tilde{b}(\beta) = 0 : q_i(\beta) = D\left(\frac{\tilde{a}_i(\beta)}{\frac{\partial \tilde{b}}{\partial z}(\beta)}\right) \tag{7}$$

for all $i \in \{0, \ldots, m\}$. By construction we have $\deg_z(q_i) < \deg_z(\tilde{b}) = \deg_{t_{k-1}}(b)$. For verifying (7) we take $\beta \in \overline{\tilde{K}}$ such that $\tilde{b}(\beta) = 0$ and obtain

$$p_i(\beta) = \frac{\tilde{a}_i(\beta)}{\frac{\partial \tilde{b}}{\partial z}(\beta)}$$

from the definition of p_i. From the definition of $\tilde{p}_{i,0}$ and $\tilde{p}_{i,1}$ by (4) and (6) we also get

$$\begin{aligned}\tilde{p}_{i,1}(\beta)t_{k-1} + \tilde{p}_{i,0}(\beta) &= \frac{\frac{\partial p_i}{\partial z}(\beta)\cdot\sum_{j=0}^{\deg_z(\tilde{b})}(\kappa_{D,k-1}\tilde{b}_j)\beta^j}{\frac{\partial \tilde{b}}{\partial z}(\beta)}\\ &= \frac{\partial p_i}{\partial z}(\beta)\frac{D(\tilde{b}(\beta)) - \frac{\partial \tilde{b}}{\partial z}(\beta)\cdot D\beta}{\frac{\partial \tilde{b}}{\partial z}(\beta)} = -\frac{\partial p_i}{\partial z}(\beta)\cdot D\beta.\end{aligned}$$

Therefore, using (4) and (6) again we obtain

$$\begin{aligned}q_i(\beta) = \sum_{j=0}^{\deg_z(p_i)}(\kappa_{D,k-1}\operatorname{coeff}(p_i, z^j))\beta^j + \frac{\partial p_i}{\partial z}(\beta)\cdot D\beta\\ = D(p_i(\beta)) = D\left(\frac{\tilde{a}_i(\beta)}{\frac{\partial \tilde{b}}{\partial z}(\beta)}\right).\end{aligned}$$

Now let $\mathbf{c} \in C^{m+1}$ be fixed and define $q := (q_0, \ldots, q_m)\cdot\mathbf{c} \in \tilde{K}[t_{k-1}][z]$. Then, by construction $\deg_z(q) < \deg_z(b)$. The roots of $r := \operatorname{res}_{t_{k-1}}((a_0, \ldots, a_m)\cdot\mathbf{c} - z\frac{\partial b}{\partial t_{k-1}}, b) \in \tilde{K}[z]$ are those $\alpha \in \overline{\tilde{K}}$ such that there exists a $\beta \in \overline{\tilde{K}}$ with $\tilde{b}(\beta) = 0$ and $(\tilde{a}_0(\beta), \ldots, \tilde{a}_m(\beta))\cdot\mathbf{c} - \alpha\cdot\frac{\partial \tilde{b}}{\partial z}(\beta) = 0$. Hence if $\beta \in \overline{\tilde{K}}$ ranges over the roots of b then $\alpha = \frac{(a_0(\beta), \ldots, a_m(\beta))\cdot\mathbf{c}}{\frac{\partial \tilde{b}}{\partial z}(\beta)}$ ranges over the roots of r. By (7) this implies

$$\{q(\beta) \mid \beta \in \overline{\tilde{K}}, b(\beta) = 0\} = \{D\alpha \mid \alpha \in \overline{\tilde{K}}, r(\alpha) = 0\}. \tag{8}$$

For verifying the first part of the statement of the theorem assume that there exist $v \in \tilde{K}(t_{k-1})$, $d_1, \ldots, d_l \in \overline{C}$, and $u_1, \ldots, u_l \in \tilde{K}(d_1, \ldots, d_l, t_{k-1})^*$ with $\frac{\partial v}{\partial t_{k-1}} + \sum_{i=1}^l d_i\frac{\frac{\partial u_i}{\partial t_{k-1}}}{u_i} = \frac{(a_0, \ldots, a_m)\cdot\mathbf{c}}{b}$. Let $\alpha \in \overline{\tilde{K}}$ be such that $r(\alpha) = 0$. By Lemma 3.7 from [9] applied in $\tilde{K}(d_1, \ldots, d_l)[t_{k-1}]$ there exists an irreducible $s \in \tilde{K}(d_1, \ldots, d_l)[t_{k-1}]$ such that the residue satisfies $\operatorname{res}_s(\frac{(a_0, \ldots, a_m)\cdot\mathbf{c}}{b}) = \pi_s(\frac{(a_0, \ldots, a_m)\cdot\mathbf{c}}{\frac{\partial b}{\partial t_{k-1}}}) = \alpha$. Here π_s denotes the canonical projection onto the residue field of the valuation ring which is associated to s via the valuation $\nu_s(f) = \sup\{\nu \in \mathbb{Z} \mid \gcd(\operatorname{den}(fs^{-\nu}), s) = 1\}$. Hence by Lemma 5.6.1 from [2] we obtain $\alpha = \operatorname{res}_s(\frac{(a_0, \ldots, a_m)\cdot\mathbf{c}}{b}) = \sum_i d_i\nu_s(u_i) \in \overline{C}$. Therefore, we have that $\alpha \in \overline{C}$ for all roots of r, i.e., $q(\beta) = 0$ for all roots $\beta \in \overline{\tilde{K}}$ of $\tilde{b}$ by (8). Since $\tilde{b}$ is squarefree it has $\deg_z(\tilde{b})$ distinct roots in $\overline{\tilde{K}}$ and it follows that $q = 0$. Consequently, by definition we have $A\cdot\mathbf{c} = 0$, i.e., $\mathbf{c} \in \operatorname{span}_C\{\mathbf{c}_1, \ldots, \mathbf{c}_n\}$ as required. For verifying the second part of the statement we fix some $j \in \{1, \ldots, n\}$ and assume $\mathbf{c} = \mathbf{c}_j$. Then $q = (1, z, \ldots, z^{\deg_z(\tilde{b})-1}, t_{k-1}, t_{k-1}z, \ldots, t_{k-1}z^{\deg_z(\tilde{b})-1})\cdot A\cdot\mathbf{c}_j = 0$. So by (8) all roots $\alpha \in \overline{K}$ of r lie in $\overline{C}$. Therefore $\frac{r}{\operatorname{lc}_z(r)} \in C[z]$ and it fulfils the statement by Theorem 3.8.1 from [9]. □

COROLLARY 13. *We can solve Problem 11.*

Algorithm 2 Solve Problem 11

Abbreviate $\tilde{K} := K(t_0, \ldots, t_{k-2})$

1. For all $i \in \{0, \ldots, m\}$ compute $g_i \in \tilde{K}(t_{k-1})$ such that $h_i := f_i - \frac{\partial g_i}{\partial t_{k-1}} \in \tilde{K}(t_{k-1})$ has squarefree denominator (e.g. by Hermite reduction [2])
2. $b := \operatorname{lcm}_{t_{k-1}}(\operatorname{den}_{t_{k-1}}(h_0), \ldots, \operatorname{den}_{t_{k-1}}(h_m)) \subset \tilde{K}[t_{k-1}]$
3. Apply Algorithm 1 to k and $h_0b, \ldots, h_mb, b \in \tilde{K}[t_{k-1}]$ to obtain $\mathbf{c}_1, \ldots, \mathbf{c}_n \in C^{m+1}$
4. For all $j \in \{1, \ldots, n\}$ compute $d_{j,1}, \ldots, d_{j,l_j} \in \overline{C}$, $u_{j,1}, \ldots, u_{j,l_j} \in \tilde{K}(d_{j,1}, \ldots, d_{j,l_j}, t_{k-1})^*$, $p_j \in \tilde{K}[t_{k-1}]$ such that
$$(h_0, \ldots, h_m)\cdot\mathbf{c}_j = \frac{\partial p_j}{\partial t_{k-1}} + \sum_{i=1}^l d_i\frac{\frac{\partial u_i}{\partial t_{k-1}}}{u_i}$$
(see [7, 4] for example)
5. For all $j \in \{1, \ldots, n\}$ compute $v_j := (g_0, \ldots, g_m)\cdot\mathbf{c}_j + p_j$

4.2 Main Algorithm

The following theorem is the main result of this paper showing that we can do parametric elementary integration over $(K(t_0, t_1, \ldots), D)$ provided we can do parametric elementary integration over (K, D). The corresponding algorithm resembles the one stated in [3] and extends it to the parametric version of the integration problem. Moreover, by incorporating Algorithm 2, motivated by our Theorem 9, non-integrability is detected at an earlier stage in some situations.

THEOREM 14. *Let t be differentially transcendental over (K, D), let $t_0, t_1, \ldots \in K\langle t\rangle$ with (1) and (2) and let $F := K(t_0, t_1, \ldots)$ and $C := \operatorname{Const}_D(F)$. Assume we can solve the parametric elementary integration problem over (K, D). Then we can solve the parametric elementary integration problem over (F, D) by Algorithm 3.*

PROOF. Note that $t_0, t_1, \ldots$ are algebraically independent over K by Corollary 6 and $C = \operatorname{Const}(K)$ by Corollary 8.

First, we prove that the steps of Algorithm 3 can indeed be executed. Step 3 can be computed by assumption. Steps 5 and 12 can be done by clearing the denominator

$$b := \operatorname{lcm}_{t_k}(\operatorname{den}_{t_k}(f_0), \ldots, \operatorname{den}_{t_k}(f_m))$$

and constructing the rows of A by coefficient extraction

$$(\operatorname{coeff}(f_ib \div b, t_k^j))_{i=0,\ldots,m}$$

for $j \in \{\min(k+1, 2), \ldots, \max_i(\deg_{t_0}(f_ib)) - \deg_{t_0}(b)\}$ and

$$(\operatorname{coeff}(f_ib \bmod b, t_k^j))_{i=0,\ldots,m}$$

for $j \in \{0, \ldots, \deg_{t_0}(b) - 1\}$. Alternatively, we can construct a matrix A based on partial fraction decomposition instead of computing b. In Step 12 the entries of a matrix generated that way are in $\tilde{K}$ and a matrix with entries in K can be

Algorithm 3 Parametric elementary integration over differentially transcendental extensions

Require: t differentially transcendental over (K, D), $t_0, t_1, \ldots \in K\langle t\rangle$ with (1) and (2), $F := K(t_0, t_1, \ldots)$, $C := \mathrm{Const}_D(F)$, and $f_0, \ldots, f_m \in F$

Ensure: $\mathbf{c}_1, \ldots, \mathbf{c}_n \in \mathrm{Const}(K)^{m+1}$ and $g_1, \ldots, g_n$ from some elementary extension of (F, D) such that:

1. If $(f_0, \ldots, f_m)\cdot\mathbf{c} \in F$ has an elementary integral over (F, D) for $\mathbf{c} \in C^{m+1}$, then $\mathbf{c} \in \mathrm{span}_C\{\mathbf{c}_1, \ldots, \mathbf{c}_n\}$.
2. $\forall j \in \{1, \ldots, n\} : Dg_j = (f_0, \ldots, f_m)\cdot\mathbf{c}_j$.

1. $k := \max_i(\mathrm{ddeg}_t(f_i))$
2. **if** $k < 0$ **then**
3. Solve the parametric elementary integration problem over (K, D) with $f_0, \ldots, f_m \in K$
4. **else if** $k = 0$ **then**
5. Compute a matrix $A \in K^{l\times(m+1)}$ such that $A\cdot\mathbf{c} = 0$ is equivalent to $(f_0, \ldots, f_m)\cdot\mathbf{c} \in K$ for all $\mathbf{c} \in C^{m+1}$
6. Compute a C-vector space basis $\bar{\mathbf{c}}_1, \ldots, \bar{\mathbf{c}}_{\bar n} \in C^{m+1}$ of $\ker(A) \cap C^{m+1}$
7. For all $j \in \{1, \ldots, \bar n\}$ set $\bar f_j := (f_0, \ldots, f_m)\cdot\bar{\mathbf{c}}_j \in K$
8. Apply Algorithm 3 recursively to $\bar f_1, \ldots, \bar f_{\bar n} \in K$ to obtain some $\tilde{\mathbf{c}}_1, \ldots, \tilde{\mathbf{c}}_n \in C^{\bar n}$ and $g_1, \ldots, g_n$ from some elementary extension of (K, D)
9. Compute $(\mathbf{c}_1, \ldots, \mathbf{c}_n) := (\bar{\mathbf{c}}_1, \ldots, \bar{\mathbf{c}}_{\bar n})\cdot(\tilde{\mathbf{c}}_1, \ldots, \tilde{\mathbf{c}}_n)$
10. **else** {i.e. $k > 0$}
11. Abbreviate $\tilde K := K(t_0, \ldots, t_{k-1})$
12. Compute a matrix $A \in K^{l\times(m+1)}$ such that $A\cdot\mathbf{c} = 0$ is equivalent to $(f_0, \ldots, f_m)\cdot\mathbf{c} \in \tilde K[t_k]$ with $\deg_{t_k} \leq 1$ for all $\mathbf{c} \in K^{m+1}$
13. Compute a C-vector space basis $\bar{\mathbf{c}}_1, \ldots, \bar{\mathbf{c}}_{\bar n} \in C^{m+1}$ of $\ker(A) \cap C^{m+1}$
14. Set $\tilde f_{j,1}t_k + \tilde f_{j,0} := (f_0, \ldots, f_m)\cdot\bar{\mathbf{c}}_j$ with $\tilde f_{j,0}, \tilde f_{j,1} \in \tilde K$ for all $j \in \{1, \ldots, \bar n\}$
15. Apply Algorithm 2 to k and $\frac{\tilde f_{1,1}}{a_{k-1}}, \ldots, \frac{\tilde f_{\bar n,1}}{a_{k-1}} \in \tilde K$ to obtain $\tilde{\mathbf{c}}_1, \ldots, \tilde{\mathbf{c}}_{\tilde n} \in C^{\bar n}$ and corresponding $v_j, d_{j,i}, u_{j,i}$ for $j \in \{1, \ldots, \tilde n\}$
16. For $j \in \{1, \ldots, \tilde n\}$ set $\tilde g_j := v_j + \sum_{i=1}^{l_j} d_{j,i}\log(u_{j,i})$ and $\bar f_j := (\tilde f_{1,1}t_k + \tilde f_{1,0}, \ldots, \tilde f_{\bar n,1}t_k + \tilde f_{\bar n,0})\cdot\tilde{\mathbf{c}}_j - D\tilde g_j$
17. Apply Algorithm 3 recursively to $\bar f_1, \ldots, \bar f_{\tilde n} \in \tilde K$ to obtain some $\hat{\mathbf{c}}_1, \ldots, \hat{\mathbf{c}}_n \in C^{\tilde n}$ and $\hat g_1, \ldots, \hat g_n$ from some elementary extension of (F, D)
18. For $j \in \{1, \ldots, n\}$ set $g_j := (\tilde g_1, \ldots, \tilde g_{\tilde n})\cdot\hat{\mathbf{c}}_j + \hat g_j$ and $(\mathbf{c}_1, \ldots, \mathbf{c}_n) := (\bar{\mathbf{c}}_1, \ldots, \bar{\mathbf{c}}_{\bar n})\cdot(\tilde{\mathbf{c}}_1, \ldots, \tilde{\mathbf{c}}_{\tilde n})\cdot(\hat{\mathbf{c}}_1, \ldots, \hat{\mathbf{c}}_n)$
19. **end if**

constructed from it by extracting appropriate coefficients. The definition in Step 5 implies $\bar f_j \in K$ in Step 7, similarly we infer the existence of $\tilde f_{j,0}, \tilde f_{j,1} \in \tilde K$ in Step 14.

Showing the algorithm terminates requires to show $\bar f_j \in \tilde K$ in Step 17. With $D\tilde g_j = Dv_j + \sum_i d_{j,i}\frac{Du_{j,i}}{u_{j,i}}$ in Step 16 and by Corollary 8 we can write $\bar f_j$ as

$$\left((\tilde f_{1,1}, \ldots, \tilde f_{\bar n,1})\cdot\tilde{\mathbf{c}}_j - a_{k-1}\left(\frac{\partial v_j}{\partial t_{k-1}} + \sum_i d_{j,i}\frac{\frac{\partial u_{j,i}}{\partial t_{k-1}}}{u_{j,i}}\right)\right)t_k + h_j$$

for some $h_j \in \tilde K$. Since we have $\frac{\partial v_j}{\partial t_{k-1}} + \sum_i d_{j,i}\frac{\frac{\partial u_{j,i}}{\partial t_{k-1}}}{u_{j,i}} = \frac{(\tilde f_{1,1}, \ldots, \tilde f_{\bar n,1})\cdot\tilde{\mathbf{c}}_j}{a_{k-1}}$ by Step 15, this implies $\bar f_j = h_j \in \tilde K$.

We prove correctness by induction on $k = \max_i(\mathrm{ddeg}_t(f_i))$. If $k < 0$, the $\mathbf{c}_1, \ldots, \mathbf{c}_n \in C^{m+1}$ and $g_1, \ldots, g_n$ obtained satisfy the two properties by Corollary 10.
$k = 0$: For showing the first property we fix a $\mathbf{c} \in C^{m+1}$ such that $f := (f_0, \ldots, f_m)\cdot\mathbf{c} \in K(t_0)$ has an elementary integral over (F, D). By Corollary 10 we have $f \in K$ and hence by construction of $\bar{\mathbf{c}}_1, \ldots, \bar{\mathbf{c}}_{\bar n}$ there exists a $\bar{\mathbf{c}} \in C^{\bar n}$ such that $\mathbf{c} = (\bar{\mathbf{c}}_1, \ldots, \bar{\mathbf{c}}_{\bar n})\cdot\bar{\mathbf{c}}$. Now, by invoking the case $k < 0$ we get $\bar{\mathbf{c}} \in \mathrm{span}_C\{\tilde{\mathbf{c}}_1, \ldots, \tilde{\mathbf{c}}_n\}$ and therefore $\mathbf{c} \in \mathrm{span}_C\{\mathbf{c}_1, \ldots, \mathbf{c}_n\}$. The second property is verified easily by just plugging in the definitions of g_j and $\mathbf{c}_j$.
$k > 0$: In order to prove the first property we fix a $\mathbf{c} \in C^{m+1}$ such that $f := (f_0, \ldots, f_m)\cdot\mathbf{c} \in \tilde K(t_k)$ has an elementary integral over (F, D). By Theorem 9 there are $v, b \in \tilde K$, $d_1, \ldots, d_N \in \overline{C}$, and $u_1, \ldots, u_N \in \tilde K(d_1, \ldots, d_N)$ such that $f = a_{k-1}\left(\frac{\partial v}{\partial t_{k-1}} + \sum_{i=1}^{N} d_i\frac{\frac{\partial u_i}{\partial t_{k-1}}}{u_i}\right)t_k + b$. Hence by construction of $\bar{\mathbf{c}}_1, \ldots, \bar{\mathbf{c}}_{\bar n}$ and $\tilde{\mathbf{c}}_1, \ldots, \tilde{\mathbf{c}}_{\tilde n}$ there is a $\hat{\mathbf{c}} \in C^{\tilde n}$ such that $\mathbf{c} = (\bar{\mathbf{c}}_1, \ldots, \bar{\mathbf{c}}_{\bar n})\cdot(\tilde{\mathbf{c}}_1, \ldots, \tilde{\mathbf{c}}_{\tilde n})\cdot\hat{\mathbf{c}}$. Now, by $\bar f_j \in \tilde K$ we have $f - D((\tilde g_1, \ldots, \tilde g_{\tilde n})\cdot\hat{\mathbf{c}}) = (\bar f_1, \ldots, \bar f_{\tilde n})\cdot\hat{\mathbf{c}} \in \tilde K$. So by construction of $\hat{\mathbf{c}}_1, \ldots, \hat{\mathbf{c}}_n$ we obtain $\mathbf{c} \in \mathrm{span}_C\{\mathbf{c}_1, \ldots, \mathbf{c}_n\}$. The second property is easily verified based on the construction. □

5. FAMILIES OF ITERATED INTEGRALS

Now we will be concerned with representing functions depending on an additional (discrete) variable. Let $(\tilde F, D)$ be a differential field and let $(s_k)_{k=1,2,\ldots}$ be a sequence in $\tilde F$. For representing s_n for symbolic n we adapt the definitions from difference fields, see [11, Section 2.5], to our setting. Let (F, D) be a differential field containing the indeterminate n in its constant field. Then elements $t_0, t_1, \ldots \in F$ represent $s_n, s_{n-1}, \ldots$ for symbolic n, if there exists a differential field monomorphism $\sigma : (F, D) \to (F, D)$ and maps $\varphi_1, \varphi_2, \ldots$ from F to $\tilde F$ with the following properties. One should think of φ_j as evaluation at $n = j$.

1. $\sigma(n) = n - 1$ and $\sigma(t_k) = t_{k+1}$ for all $k \in \mathbb{N}$.
2. $\varphi_j(n) = j$ and $\varphi_j(t_k) = s_{j-k}$ for all $j, k \in \mathbb{N}$ s.t. $k < j$.
3. For all $f, g \in F$ there exists $j_0 \in \mathbb{N}^+$ such that for all $j \geq j_0$ we have
$$\varphi_j(f + g) = \varphi_j(f) + \varphi_j(g), \quad \varphi_j(fg) = \varphi_j(f)\varphi_j(g),$$
$$\varphi_j(Df) = D\varphi_j(f), \quad \text{and} \quad \varphi_{j+1}(\sigma(f)) = \varphi_j(f).$$

These conditions imply that $f \mapsto (\varphi_1(f), \ldots)$ is both a differential ring monomorphism and a difference ring monomorphism from F into the ring of sequences $\tilde F^{\mathbb{N}}/\!\sim$. The latter is defined by considering two sequences equivalent iff they differ only at finitely many entries and it is equipped with componentwise addition, multiplication, and derivation as well as with $\sigma(f_1, f_2, \ldots) := (0, f_1, f_2, \ldots)$. Note that the above properties in particular imply that $\varphi_j(0) = 0$ and $\varphi_j(1) = 1$ from some point on. Since F is a field they imply further that for all $f \in F \setminus \{0\}$ there exists j_0 such that $\varphi_j(f)\varphi_j(\frac{1}{f}) = 1$ and hence $\varphi_j(f) \neq 0$ for all $j \geq j_0$. In other words, any relation among $t_0, t_1, \ldots$ corresponds to a relation among $s_n, s_{n-1}, \ldots$ that is valid for all n above some lower bound and vice versa.

Now, let $(\tilde{K}, D)$ be a differential subfield of $(\tilde{F}, D)$ and assume in particular that $s_1, s_2, \ldots$ are such that

$$Ds_1 = \tilde{a}_0 \quad (9)$$
$$Ds_k = \tilde{a}_{k-1}s_{k-1} \quad (10)$$

for $k \geq 2$ and some $\tilde{a}_0, \tilde{a}_1, \ldots \in \tilde{K}^*$. We want to represent s_n for symbolic n by a differentially transcendental t over (K, D). Therefore we set $(F, D) = (K\langle t\rangle, D)$ and require $\varphi_j(K) \subseteq \tilde{K}$. Following the considerations above, it is also necessary that there is no algebraic relation among $s_n, s_{n-1}, \ldots$ over K that is valid for all but possibly finitely many values of n. This needs to be verified for a particular sequence $(s_k)_{k=1,2,\ldots}$ in some way. One way to do so is to check the stronger condition that there is no algebraic relation among $s_1, s_2, \ldots$ over $\tilde{K}$. This is by no means necessary. Nevertheless, in the following we will give a criterion to check the stronger condition provided the following additional assumption holds. Assume that for all $i \in \mathbb{N}^+$ there exist $b_{i,1}, \ldots, b_{i,i} \in \tilde{K}$ such that

$$b_{i,i} = 1 \quad (11)$$
$$Db_{i,j} = -\tilde{a}_j b_{i,j+1}. \quad (12)$$

This assumption seems rather restrictive, but it covers several relevant cases as we will see below. First, it is important to note that we have

$$b_{i,1}\tilde{a}_0 = D\left(\sum_{j=1}^{i} b_{i,j}s_j\right). \quad (13)$$

Next, we make use of the following theorem from differential algebra in order to obtain a criterion on the algebraic independence of $s_1, s_2, \ldots$ over $\tilde{K}$.

THEOREM 15. *Let (K, D) be a differential field and define $C := \mathrm{Const}(K)$. Let $w_1, \ldots, w_n \in K$ such that no non-trivial C-linear combination of them has an integral in (K, D). Then any $t_1, \ldots, t_n$ with $Dt_i = w_i$ are algebraically independent over K and $\mathrm{Const}(K(t_1, \ldots, t_n)) = \mathrm{Const}(K)$.*

Letting $w_i := b_{i,1}\tilde{a}_0$ and relying on (13) we obtain the following corollary on the sequence $(s_k)_{k=1,2,\ldots}$.

COROLLARY 16. *Let $(\tilde{K}, D)$ and $s_k, \tilde{a}_k, b_{i,j} \in \tilde{K}$ be as above in (9)-(12) and define $\tilde{C} := \mathrm{Const}(\tilde{K})$. If no non-trivial $\tilde{C}$-linear combination of $b_{1,1}\tilde{a}_0, b_{2,1}\tilde{a}_0, \ldots$ has an integral in $(\tilde{K}, D)$, then $s_1, s_2, \ldots$ are algebraically independent over $\tilde{K}$ and $\mathrm{Const}(\tilde{K}(s_1, s_2, \ldots)) = \mathrm{Const}(\tilde{K})$.*

Once we verified for a particular choice of $s_1, s_2, \ldots$ that they are algebraically independent over $\tilde{K}$ we can represent s_n for symbolic n in terms of a differentially transcendental extension t. If for some k the derivatives $t, Dt, \ldots, D^k t$ (representing $s_n, Ds_n, \ldots, D^k s_n$) were algebraically dependent over K, then as outlined above the evaluation maps $\varphi_1, \varphi_2, \ldots$ would translate this to an algebraic dependence of $s_n, Ds_n, \ldots, D^k s_n$, or equivalently of $s_n, s_{n-1}, \ldots, s_{n-k}$, over $\tilde{K}$ for all specific n sufficiently large.

Polylogarithms.

The polylogarithms are defined as $\mathrm{Li}_2(x) = -\int_0^x \frac{\ln(1-t)}{t}\,dt$ and $\mathrm{Li}_{k+1}(x) = \int_0^x \frac{1}{t}\mathrm{Li}_k(t)\,dt$ for $k \geq 2$. Let $(\tilde{K}, D) = (\tilde{C}(x, \ln(x), \ln(1-x)), \frac{d}{dx})$ and $(\tilde{F}, D) = (K(s_1, s_2, \ldots), D)$, then s_k represents $\mathrm{Li}_{k+1}(x)$ if we choose $\tilde{a}_0 := -\frac{\ln(1-x)}{x}$ and $\tilde{a}_k := \frac{1}{x}$, $k \geq 1$. In addition, it is easily verified that $b_{i,j} := \frac{(-\ln(x))^{i-j}}{(i-j)!}$ satisfy (11) and (12). Now, $b_{k,1}\tilde{a}_0 = \frac{(-1)^k}{(k-1)!}\frac{\ln(x)^{k-1}\ln(1-x)}{x}$ and no non-trivial $\tilde{C}$-linear combination of them has an integral in (K, D) as can be verified based on the fact that there are no $g \in \tilde{C}(x)$ and $c \in \tilde{C}$ such that $Dg = \frac{1}{1-x} + \frac{c}{x}$, for details on how the verification proceeds in general we refer to [2]. Hence by Corollary 16 we conclude that $\mathrm{Li}_2(x), \mathrm{Li}_3(x), \ldots$ are algebraically independent over $\tilde{K}$. For symbolic n we deduce as above that $\mathrm{Li}_n(x)$ is differentially transcendental over $(K, D) = C(n)(x, \ln(x), \ln(1-x)), \frac{d}{dx})$.

Repeated antidifferentiation.

For some function $f(x)$ we consider the iterated antiderivatives $\int f(x)\,dx, \int\int f(x)\,dx\,dx, \ldots$, so let $(\tilde{K}, D)$ be a differential field extension of $(\tilde{C}(x), \frac{d}{dx})$ such that $\tilde{a}_0 \in \tilde{K}$ represents $f(x)$. Then s_k represents the k-fold antiderivative if we choose $\tilde{a}_k := 1$ for $k \in \mathbb{N}^+$. We have that $b_{i,j} := \frac{(-x)^{i-j}}{(i-j)!}$ satisfy (11) and (12), so for applying Corollary 16 we need to check whether there exists a nonzero polynomial $p \in \tilde{C}[x]$ such that $p\tilde{a}_0$ has an integral in $(\tilde{K}, D)$. Depending on $\tilde{a}_0$ this may or may not be the case.

6. EXAMPLES

In this section we give some applications of the algorithm with various kinds of integrands. First, we look at some indefinite integrals. Then, we show how the parametric nature of the algorithm can be exploited in the context of definite integrals.

Example 1. We consider the polylogarithms $\mathrm{Li}_n(x)$ for symbolic n and want to compute the integral

$$\int \frac{\mathrm{Li}_{n-2}(x)\mathrm{Li}_n(x)}{x\mathrm{Li}_{n-1}(x)^2}\,dx.$$

For applying our algorithm we use $C := \mathbb{Q}(n)$ and $(K, D) := (C(x), \frac{d}{dx})$. We set $a_k = \frac{1}{x}$ and $b_k = 0$ in (2), so t_k from $(F, D) := (C(x, t_0, t_1, \ldots), D)$ corresponds to $\mathrm{Li}_{n-k}(x)$ as detailed in Section 5. The integrand is represented by

$$f := \frac{t_0 t_2}{x t_1^2},$$

which has $\mathrm{ddeg}_t(f) = 2$ and even is of the form $\tilde{f}_1 t_2 + \tilde{f}_0$ with $\tilde{f}_1 = \frac{t_0}{x t_1^2} \in K(t_0, t_1)$ and $\tilde{f}_0 = 0$. So by applying Algorithm 3 we just need to solve Problem 11 for $x\tilde{f}_1$ and $k = 2$ (Step 15). We obtain $v = -\frac{t_0}{t_1}$ and $f - Dv = \frac{1}{x}$, which is easily dealt with in Step 17 resp. Step 3. Altogether, we obtain the following elementary integral of f over (F, D).

$$-\frac{t_0}{t_1} + \log(x)$$

Translating back yields

$$\int \frac{\mathrm{Li}_{n-2}(x)\mathrm{Li}_n(x)}{x\mathrm{Li}_{n-1}(x)^2}\,dx = \ln(x) - \frac{\mathrm{Li}_n(x)}{\mathrm{Li}_{n-1}(x)}.$$

Note that *Mathematica* and *Maple* in their current versions do not find the integral even for specific $n \in \{3, 4, \dots\}$. $\square$

With the following example we illustrate that the algorithm also works heuristically in cases where the assumption of t being differentially transcendental over (K, D) does not reflect the true situation. An integral computed that way will still be valid provided it does not represent a situation where division by zero occurs. More explicitly, the denominator of an expression in $K(t_0, t_1, \dots)$ should not reduce to zero after applying the true relations among $t_0, t_1, \dots$. This is completely analogous to specializing an unspecified function to a specific function.

Example 2. The Weierstraß elliptic function $\wp(x)$ fulfills the relation $\wp'(x)^2 = 4\wp(x)^3 - g_2\wp(x) - g_3$ for some constants g_2 and g_3, but in the following computation we will ignore this relation and treat $\wp(x)$ and its derivatives as algebraically independent. Let the invariants g_2 and g_3 be fixed. In addition to $\wp(x)$ we also consider the Weierstraß zeta and sigma functions $\zeta(x)$ and $\sigma(x)$, where $\zeta'(x) = -\wp(x)$ and $\sigma'(x) = \zeta(x)\sigma(x)$. We will compute a closed form of

$$\int (x - \sigma(x))\wp(x) + \sigma(x)\zeta(x)^2\, dx.$$

To this end, we set $C := \mathbb{Q}(g_2, g_3)$, $(K, D) := (C(x), \frac{d}{dx})$, and $F := K(t_0, \dots)$ where we choose $a_0 = t_0$, $a_1 = -1$, and $a_n = 1$ for $n \geq 2$ as well as $b_n = 0$ for all $n \in \mathbb{N}$ in (2). Hence we have

$$Dt_0 = t_0t_1 \qquad Dt_1 = -t_2 \qquad Dt_2 = t_3$$

and $\wp(x), \wp'(x), \zeta(x), \sigma(x)$ are represented by t_2, t_3, t_1, t_0, respectively. Then the integrand is represented by

$$f := (x - t_0)t_2 + t_0t_1^2,$$

which has $\operatorname{ddeg}_t(f) = 2$ and even is the form $\tilde{f}_1t_2 + \tilde{f}_0$ with $\tilde{f}_0, \tilde{f}_1 \in K(t_0, t_1)$. Following Algorithm 3 we first need to solve Problem 11 for $\frac{\tilde{f}_1}{a_1} = t_0 - x$ and $k = 2$, which gives $v = (t_0 - x)t_1$. Proceeding recursively with the remaining integrand $f - Dv = t_1$, which obviously has $\operatorname{ddeg}_t(t_1) = 1$, we solve Problem 11 for $\frac{1}{a_0} = \frac{1}{t_0}$ and $k = 1$, giving $\frac{1}{t_0} = \frac{\partial}{\partial t_0}\log(t_0)$. Since $t_1 - D\log(t_0) = 0$ in Step 16 we are done and arrived at

$$(t_0 - x)t_1 + \log(t_0)$$

for the elementary integral of f over (F, D). When translated back to

$$\int (x - \sigma(x))\wp(x) + \sigma(x)\zeta(x)^2\, dx = (\sigma(x) - x)\zeta(x) + \ln(\sigma(x))$$

we verify that this remains valid and we successfully computed a closed form of the integral. $\square$

With the next example we illustrate the ability of our algorithm to find such relations of definite integrals.

Example 3. Consider the Laplace transform of an unspecified function $f(x)$. For the function $f(x)$ we merely assume that it is sufficiently regular, e.g. that $f(x)$ has a derivative on $\mathbb{R}_0^+$ that is continuous and bounded. Using our algorithm we want to relate the Laplace transform of $f'(x)$ to that of $f(x)$. For the differential fields we may choose $C := \mathbb{Q}(s)$, $(K, D) := (C(e^{-sx}), \frac{d}{dx})$, as well as $F := K(t_0, \dots)$ with $a_n = 1$ and $b_n = 0$ in (2), so that t_n represents $f^{(n)}(x)$.

In order to relate $\int_0^\infty e^{-sx}f(x)\,dx$ and $\int_0^\infty e^{-sx}f'(x)\,dx$ we set $f_0 := e^{-sx}t_0$ and $f_1 := e^{-sx}t_1$. Then Algorithm 3 easily computes the relation

$$-sf_0 + f_1 = D(e^{-sx}t_0)$$

in (F, D). Translating back and integrating gives

$$-s\int_0^\infty e^{-sx}f(x)\,dx + \int_0^\infty e^{-sx}f'(x)\,dx = e^{-sx}f(x)\Big|_{x=0}^\infty$$

where the right hand side evaluates to $0 - f(0)$ since $f(x)$ is assumed not to grow too fast. In other words, we automatically discovered the identity

$$\mathcal{L}_x(f'(x))(s) = s\mathcal{L}_x(f(x))(s) - f(0)$$

satisfied by the Laplace transform. $\square$

7. ACKNOWLEDGMENTS

The author would like to thank the reviewers for their careful reading and for their suggestions contributing to the improvement of the presentation of the material. The author was supported by the Austrian Science Fund (FWF), grant no. W1214-N15 project DK6, by the strategic program "Innovatives OÖ 2010 plus" of the Upper Austrian Government, and by the Research Executive Agency (REA) of the European Union under the Grant Agreement number PITN-GA-2010-264564 (LHCPhenoNet).

8. REFERENCES

[1] Gert E. T. Almkvist, Doron Zeilberger, *The method of differentiating under the integral sign*, J. Symbolic Computation 10, pp. 571–591, 1990.

[2] Manuel Bronstein, *Symbolic Integration I – Transcendental Functions*, Springer, Heidelberg, 1997.

[3] Graham H. Campbell, *Symbolic integration of expressions involving unspecified functions*, ACM SIGSAM Bulletin 22, pp. 25–27, 1988.

[4] Günter Czichowski, *A Note on Gröbner Bases and Integration of Rational Functions*, J. Symbolic Computation 20, pp. 163–167, 1995.

[5] Manuel Kauers, Carsten Schneider, *Indefinite summation with unspecified summands*, Discrete Math. 306, pp. 2073-2083, 2006.

[6] Manuel Kauers, Carsten Schneider, *Application of Unspecified Sequences in Symbolic Summation*, Proceedings of ISSAC'06, pp. 177-183, 2006.

[7] Daniel Lazard, Renaud Rioboo, *Integration of Rational Functions: Rational Computation of the Logarithmic Part*, J. Symbolic Computation 9, pp. 113–115, 1990.

[8] Žarko Mijajlović, Branko Malešević, *Differentially transcendental functions*, Bull. Belg. Math. Soc. Simon Stevin 15, pp. 193–201, 2008.

[9] Clemens G. Raab, *Definite Integration in Differential Fields*, PhD Thesis, JKU Linz, 2012.

[10] Robert H. Risch, *The problem of integration in finite terms*, Trans. Amer. Math. Soc. 139, pp. 167–189, 1969.

[11] Carsten Schneider, *Symbolic Summation in Difference Fields*, PhD Thesis, JKU Linz, 2001.

Signature Rewriting in Gröbner Basis Computation

Christian Eder
INRIA, Paris-Rocquencourt Center, PolSys Project
UPMC, Univ. Paris 06, LIP6
CNRS, UMR 7606, LIP6
UFR Ingénierie 919, LIP6
Case 169, 4, Place Jussieu, F-75252 Paris
Christian.Eder@inria.fr

Bjarke Hammersholt Roune
Department of Mathematics
University of Kaiserslautern
Postbox 3049
67653 Kaiserslautern, Germany
www.broune.com

ABSTRACT

We introduce the **RB** algorithm for Gröbner basis computation, a simpler yet equivalent algorithm to **F5GEN**. **RB** contains the original unmodified $\mathbf{F_5}$ algorithm as a special case, so it is possible to study and understand $\mathbf{F_5}$ by considering the simpler **RB**. We present simple yet complete proofs of this fact and of $\mathbf{F_5}$'s termination and correctness.

RB is parametrized by a rewrite order and it contains many published algorithms as special cases, including **SB**. We prove that **SB** is the best possible instantiation of **RB** in the following sense. Let **X** be any instantiation of **RB** (such as $\mathbf{F_5}$). Then the S-pairs reduced by **SB** are always a subset of the S-pairs reduced by **X** and the basis computed by **SB** is always a subset of the basis computed by **X**.

Categories and Subject Descriptors

G.4 [**Mathematical Software**]: Algorithm design and analysis; I.1.2 [**Algorithms**]: Algebraic algorithms

Keywords

Gröbner basis, Syzygy Gröbner basis, Signature Gröbner basis, Signature rewriting, F5

1. INTRODUCTION

The $\mathbf{F_5}$ algorithm [5] is a significant development in Gröbner basis computation and the difficulty of understanding it is a widely cited concern. We address this problem in this paper by introducing **RB**, an algorithm simpler than yet equivalent to **F5GEN** [8]. **RB** contains the original unmodified $\mathbf{F_5}$ as a special case [8], so we can understand and prove statements about $\mathbf{F_5}$ by considering the simpler **RB**. In this way we present simplified proofs of the termination and correctness of $\mathbf{F_5}$. Much of the paper is concerned with showing exactly how to derive $\mathbf{F_5}$ as a special case of **RB**.

There are many publications that present signature-based algorithms for Gröbner basis computation. Seeing as these algorithms are based on signatures, they are all somehow related to $\mathbf{F_5}$, yet it is often not clear what the exact relationship is — is $\mathbf{F_5}$ a special case of the new algorithm? If not, what are the exact differences? Pan, Hu and Wang [8] (PHW) addressed this problem by introducing the **F5GEN** algorithm. **F5GEN** contains many published algorithms as special cases which makes it possible to compare algorithms theoretically. PHW pose the open problem of which instantiation of **F5GEN** is faster:

> Moreover, with this proved F5GEN algorithm, researchers can shift their focus on the different variants of the F5GEN algorithm and find out the fastest one. — Pan, Hu, Wang [8]

We answer this open problem by proving that the **SB** algorithm [9] is the best possible instantiation of **RB** (and hence **F5GEN**) in the following sense. Let **X** be any instantiation of **RB** (such as $\mathbf{F_5}$). Then the S-pairs reduced by **SB** are always a subset of the S-pairs reduced by **X** up to signature and the basis computed by **SB** is always a subset of the basis computed by **X** up to sig-lead pairs.

Roune and Stillman's algorithm **SB** [9] is an improvement of Gao, Volny and Wang's algorithm **GVW** [7]. **SB** is equivalent to Arri and Perry's algorithm **AP** [2] and also to the modification **GVWHS** [11] of **GVW**, so all statements made about **SB** also apply to **AP** and **GVWHS**.

Section 2 introduces terminology and **SB**. Section 3 introduces **RB** and proves that **SB** is the best possible instantiation of **RB**. Section 4 introduces $\mathbf{RB_5}$, a special case of **RB** that is more similar to $\mathbf{F_5}$. Section 5 proves that $\mathbf{F_5}$ is a special case of $\mathbf{RB_5}$ (and hence **RB**). Section 6 proves termination of **RB** (and hence $\mathbf{F_5}$).

REMARK 1. *The $\mathbf{F_5}$ paper [5] requires that the input polynomials form a homogeneous regular sequence. Faugère must have known of an algorithm without these restrictions since the $\mathbf{F_5}$ paper contains benchmarks on ideals that do not meet the restrictions. In this paper, the term $\mathbf{F_5}$ refers exclusively to $\mathbf{F_5}$ exactly as described in the $\mathbf{F_5}$ paper [5] including the restriction to homogeneous regular sequences.*

2. THE SIGNATURE BASIS ALGORITHM

We briefly introduce **SB** in this section. We refer to [9] for a full treatment of **SB** with proofs. If we consider a polynomial ring R then the main differences between **SB** and the classic Buchberger algorithm is that everything is lifted from R to R^m, that polynomial reduction is constrained in that some reduction steps are not allowed and that the S-pair elimination criteria are more powerful. It is not actually necessary to represent polynomials in R^m when implementing **SB** on a computer (see Remark 2).

2.1 Notation and Terminology

Let R be a polynomial ring over a field κ. All polynomials $f \in R$ can be uniquely written as a finite sum $f = \sum_{cx^v \in M} cx^v$ where $c \in \kappa$, $x^v := \prod_i x_i^{v_i}$ and M is minimal. The elements of M are the *terms* of f. A *monomial* is a polynomial with exactly one term. A monomial with a coefficient of 1 is *monic*. Neither monomials nor terms of polynomials are necessarily monic. We write $f \simeq g$ for $f, g \in R$ if there exists a non-zero $s \in \kappa$ such that $f = sg$.

Let R^m be a free R-module and let $\boldsymbol{e}_1, \ldots, \boldsymbol{e}_m$ be the standard basis of unit vectors in R^m. All module elements $\alpha \in R^m$ can be uniquely written as a finite sum $\alpha = \sum_{a\boldsymbol{e}_i \in M} a\boldsymbol{e}_i$ where the a are monomials and M is minimal. The elements of M are the *terms* of α. A *module monomial* is an element of R^m with exactly one term. A module monomial with a coefficient of 1 is *monic*. Neither module monomials nor terms of module elements are necessarily monic. Let $\alpha \simeq \beta$ for $\alpha, \beta \in R^m$ if $\alpha = s\beta$ for some non-zero $s \in \kappa$.

Let $\leq$ denote two different orders – one for R^m and one for R. The order for R is a monomial order, which means that it is a well-order on the set of monomials in R such that $a \leq b$ implies $ca \leq cb$ for all monomials $a, b, c \in R$. The order for R^m is a module monomial order which means that it is a well-order on the set of module monomials in R^m such that $L \leq T$ implies $cL \leq cT$ for all module monomials $L, T \in R^m$ and monomials $c \in R$. We require the two orders to be compatible in the sense that $a \leq b$ if and only if $a\boldsymbol{e}_i \leq b\boldsymbol{e}_i$ for all monomials $a, b \in R$ and $i = 1, \ldots, m$.

Consider a finite sequence of polynomials $g_1, \ldots, g_m \in R$ that we call the *input polynomials*. Define the homomorphism $\alpha \mapsto \overline{\alpha}$ from R^m to R by $\overline{\alpha} := \sum_{i=1}^m \alpha_i g_i$. $\alpha \in R^m$ is a *syzygy* if $\overline{\alpha} = 0$.

Let the *lead term* $\operatorname{lt}(f)$ be the $\leq$-maximal term of $f \in R \setminus \{0\}$. Let the *signature* $\mathfrak{s}(\alpha)$ be the $\leq$-maximal term of $\alpha \in R^m \setminus \{0\}$. In this way every non-syzygy module element $\alpha \in R^m$ has two main associated characteristics – the signature $\mathfrak{s}(\alpha) \in R^m$ and the lead term $\operatorname{lt}(\overline{\alpha}) \in R$ of its image $\overline{\alpha}$ in R. Lead terms and signatures include a coefficient for mathematical convenience, though an implementation of **SB** need not store the signature coefficients. If $a\boldsymbol{e}_i = \mathfrak{s}(\alpha)$ then $\operatorname{ind}(\alpha) := i$ is the *index of signature* of α.

Let $\mathcal{G} \subseteq R^m$ be finite and assume that $\mathfrak{s}(\alpha) \simeq \mathfrak{s}(\beta) \Rightarrow \alpha = \beta$ for $\alpha, \beta \in \mathcal{G}$. $\alpha, \beta \in R^m$ are *equal up to sig-poly pairs* if $\mathfrak{s}(\alpha) = \mathfrak{s}(s\beta)$ and $\overline{\alpha} = \overline{s\beta}$ for some non-zero $s \in \kappa$.

Consider the unique extension of the module monomial order $<$ to module monomials that may have negative exponents such that $a\boldsymbol{e}_i < b\boldsymbol{e}_j \Leftrightarrow ca\boldsymbol{e}_i < cb\boldsymbol{e}_j$ for all monomials $a, b, c \in R$. The *sig-lead ratio* of $\alpha \in R^m$ is $r_\alpha := \frac{\mathfrak{s}(\alpha)}{\operatorname{lt}(\overline{\alpha})}$.

This notation is the **SB** notation [10], which is improved from the notation of the preceding paper [9]. In particular, there is no module R^n and no function $\phi \colon R^n \to R^m$.

REMARK 2. *The* **SB** *notation using module elements differs from the standard notation which considers sig-poly pairs* $(\mathfrak{s}(\alpha), \overline{\alpha})$ *in place of* $\alpha \in R^m$. **SB** *can also use sig-poly pairs (see Section 5.1). We believe that the* **SB** *notation makes signature algorithms easier to understand and reason about.*

2.2 Reduction With Signatures

Both **SB** and the classic Buchberger algorithm are based on reducing S-pairs. We first describe classic polynomial reduction and then describe **SB**'s reduction in similar terms.

Classic polynomial reduction

Let $f \in R$ and let t be a term of f. Then we can *reduce* t by $g \in R$ if $\operatorname{lt}(g) | t$ or equivalently if

- there exists a monomial b such that $\operatorname{lt}(bg) = t$.

The outcome of the reduction step is then $f - bg$ and g is the *reducer*. It holds that $\operatorname{lt}(bg) \leq \operatorname{lt}(f)$, but that is not listed as a requirement since it is implied. If $\operatorname{lt}(bg) \simeq \operatorname{lt}(f)$ then the reduction step is a *top reduction step* and otherwise it is a *tail reduction step*.

The result of classic polynomial reduction of $f \in R$ is a polynomial $h \in R$ that has been calculated from f through a sequence of reduction steps such that h cannot be further reduced. The reduction is a *tail reduction* if only tail reduction steps are allowed and it is a *top reduction* if only top reduction steps are allowed.

The implied condition that $\operatorname{lt}(bg) \leq \operatorname{lt}(f)$ is equivalent to $\operatorname{lt}(f - bg) \leq \operatorname{lt}(f)$ so during classic polynomial reduction it is not allowed to increase the leading term. For tail reduction we perform only those reduction steps that do not change the leading term at all.

Signature reduction

Let $\alpha \in R^m$ and let t be a term of $\overline{\alpha}$. Then we can $\mathfrak{s}$-*reduce* t by $\beta \in R^m$ if

- there exists a monomial b such that $\operatorname{lt}(\overline{b\beta}) = t$ and
- $\mathfrak{s}(b\beta) \leq \mathfrak{s}(\alpha)$.

The outcome of the $\mathfrak{s}$-reduction step is then $\alpha - b\beta$ and β is the $\mathfrak{s}$-*reducer*. The second condition is analogous to the implied condition $\operatorname{lt}(bg) \leq \operatorname{lt}(f)$ from classic polynomial reduction, the condition is just lifted to R^m so that it involves signatures. When β $\mathfrak{s}$-reduces t we also say for convenience that $b\beta$ $\mathfrak{s}$-reduces α. That way b is introduced implicitly instead of having to repeat the equation $\operatorname{lt}(\overline{b\beta}) = t$.

Just as for classic polynomial reduction, if $\operatorname{lt}(\overline{b\beta}) \simeq \operatorname{lt}(\overline{\alpha})$ then the $\mathfrak{s}$-reduction step is a *top* $\mathfrak{s}$-*reduction step* and otherwise it is a *tail* $\mathfrak{s}$-*reduction step*. We need words for the analogous distinction for signatures, so if $\mathfrak{s}(b\beta) \simeq \mathfrak{s}(\alpha)$ then the reduction step is a *singular* $\mathfrak{s}$-*reduction step* and otherwise it is a *regular* $\mathfrak{s}$-*reduction step*.

The result of $\mathfrak{s}$-reduction of $\alpha \in R^m$ is a $\gamma \in R^m$ that has been calculated from α through a sequence of $\mathfrak{s}$-reduction steps such that γ cannot be further $\mathfrak{s}$-reduced. The reduction is a *tail* $\mathfrak{s}$-*reduction* if only tail $\mathfrak{s}$-reduction steps are allowed and it is a *top* $\mathfrak{s}$-*reduction* if only top $\mathfrak{s}$-reduction steps are allowed. The reduction is a *regular* $\mathfrak{s}$-*reduction* if only regular $\mathfrak{s}$-reduction steps are allowed. A module element $\alpha \in R^m$ is $\mathfrak{s}$-*reducible* if it can be $\mathfrak{s}$-reduced. If α $\mathfrak{s}$-reduces to γ and γ is a syzygy then we say that α $\mathfrak{s}$-*reduces to zero* even if $\gamma \neq 0$.

As for classic polynomial reduction, the implied condition $\operatorname{lt}(\overline{b\beta}) \leq \operatorname{lt}(\overline{\alpha})$ is equivalent to $\operatorname{lt}(\overline{\alpha - b\beta}) \leq \operatorname{lt}(\overline{\alpha})$, so during $\mathfrak{s}$-reduction it is not allowed to increase the leading term. For tail $\mathfrak{s}$-reduction we perform only those $\mathfrak{s}$-reduction steps that do not change the leading term at all. Analogously, the condition $\mathfrak{s}(b\beta) \leq \mathfrak{s}(\alpha)$ is equivalent to $\mathfrak{s}(\alpha - b\beta) \leq \mathfrak{s}(\alpha)$, so during $\mathfrak{s}$-reduction it is not allowed to increase the signature. For regular $\mathfrak{s}$-reduction, we perform only those $\mathfrak{s}$-reduction steps that do not change the signature at all.

Classic reduction is always with respect to a finite *basis* $B \subseteq R$. The reducers in classic polynomial reduction are chosen from the basis B. Analogously, $\mathfrak{s}$-reduction is always with respect to a finite *basis* $\mathcal{G} \subseteq R^m$. The $\mathfrak{s}$-reducers in $\mathfrak{s}$-reduction are chosen from the basis $\mathcal{G}$.

$\mathcal{G}$ is a *signature Gröbner basis in signature* T if all $\alpha \in R^m$ with $\mathfrak{s}(\alpha) = T$ $\mathfrak{s}$-reduce to zero. $\mathcal{G}$ is a *signature Gröbner basis up to signature* T if $\mathcal{G}$ is a signature Gröbner basis in all signatures L such that $L < T$. $\mathcal{G}$ is a *signature Gröbner basis* if it is a signature Gröbner basis in all signatures.

SB computes a signature Gröbner basis. If $\mathcal{G}$ is a signature Gröbner basis then $\{\overline{\alpha} \mid \alpha \in \mathcal{G}\}$ is a Gröbner basis of $\langle g_1, \ldots, g_m \rangle$, so we can use **SB** to compute Gröbner bases.

2.3 S-pairs and Syzygies

Let $f, g \in R$ and let $c := \gcd(\operatorname{lt}(f), \operatorname{lt}(g))$ be the monic greatest common divisor of $\operatorname{lt}(f)$ and $\operatorname{lt}(g)$. In the classic Buchberger algorithm the *S-polynomial* between f and g is

$$\frac{\operatorname{lt}(g)}{c} f - \frac{\operatorname{lt}(f)}{c} g.$$

The classic Buchberger algorithm proceeds by reducing S-polynomials. If an S-polynomial reduces to $h \neq 0$ then h is added to the basis so that the S-polynomial now reduces to zero by this larger basis. The classic Buchberger algorithm terminates once all S-polynomials between elements of the basis reduce to zero. The **SB** algorithm works the same way except that these computations are lifted from R to R^m in the following way.

Let $\alpha, \beta \in R^m$ and let $c := \gcd(\operatorname{lt}(\overline{\alpha}), \operatorname{lt}(\overline{\beta}))$ be the monic greatest common divisor of $\operatorname{lt}(\overline{\alpha})$ and $\operatorname{lt}(\overline{\beta})$. The *S-pair* between α and β is

$$\mathcal{S}(\alpha, \beta) := \frac{\operatorname{lt}(\overline{\beta})}{c}\alpha - \frac{\operatorname{lt}(\overline{\alpha})}{c}\beta.$$

If $\mathfrak{s}\left(\frac{\operatorname{lt}(\overline{\beta})}{c}\alpha\right) \simeq \mathfrak{s}\left(\frac{\operatorname{lt}(\overline{\alpha})}{c}\beta\right)$ then the S-pair is *singular* and otherwise it is *regular*. By "S-pair" we always mean "regular S-pair". Observe that $\mathcal{S}(\alpha, \beta) \in R^m$ and that $\overline{\mathcal{S}(\alpha, \beta)}$ is the S-polynomial between $\overline{\alpha}$ and $\overline{\beta}$.

SB proceeds by $\mathfrak{s}$-reducing S-pairs. If an S-pair $\mathfrak{s}$-reduces to γ and $\overline{\gamma}$ is not zero then γ is added to the basis. Theorem 3 implies that if all S-pairs and all $\boldsymbol{e}_i$ $\mathfrak{s}$-reduce to zero then the basis is a signature Gröbner basis.

THEOREM 3. *Let T be a module monomial of R^m and let $\mathcal{G} \subseteq R^m$ be a finite basis. Assume that all S-pairs $p := \mathcal{S}(\alpha, \beta)$ with $\alpha, \beta \in \mathcal{G}$ and $\mathfrak{s}(p) < T$ $\mathfrak{s}$-reduce to zero and all $\boldsymbol{e}_i$ with $\boldsymbol{e}_i < T$ $\mathfrak{s}$-reduce to zero. Then $\mathcal{G}$ is a signature Gröbner basis up to signature T.*

The outcome of classic polynomial reduction depends on the choice of reducer, so the choice of reducer can change what the intermediate bases are in the classic Buchberger algorithm. Lemma 4 implies that all S-pairs with the same signature yield the same regular $\mathfrak{s}$-reduced result as long as we process S-pairs in order of increasing signature.

LEMMA 4. *Let $\alpha, \beta \in R^m$ and let $\mathcal{G}$ be a signature Gröbner basis up to signature $\mathfrak{s}(\alpha) = \mathfrak{s}(\beta)$. If α and β are both regular top $\mathfrak{s}$-reduced then $\operatorname{lt}(\overline{\alpha}) = \operatorname{lt}(\overline{\beta})$ or $\overline{\alpha} = \overline{\beta} = 0$. If α and β are both regular $\mathfrak{s}$-reduced then $\overline{\alpha} = \overline{\beta}$.*

```
SimpleSignatureBasisAlgorithm({g_1, ..., g_m} ⊆ R)
  G ← ∅  (G will be the signature Gröbner basis)
  P ← {e_1, ..., e_m}  (P is the set of pending reductions)
  H ← ⟨0⟩ ⊆ R^m  (H will be the initial module of syzygies)
  while P ≠ ∅ do
    p ← an element of P with ≤-minimal signature
    P ← P \ {p}
    p' ← result of regular s-reducing p
    if p̄' = 0 then
      H ← H + ⟨s(p')⟩
    else if p' is not singular top reducible then
      P ← P ∪ {S(α, p') | α ∈ G and S(α, p') is regular}
      G ← G ∪ {p'}
  return (G, H)
```

Figure 1: Pseudo code for a simple version of SB.

A signature Gröbner basis is *minimal* if no basis element top $\mathfrak{s}$-reduces any other basis element. Theorem 5 implies that the minimal signature Gröbner basis is unique and is contained in all signature Gröbner bases up to sig-lead pairs. **SB** computes a minimal signature Gröbner basis.

THEOREM 5. *Let A be a minimal signature Gröbner basis and let B be a signature Gröbner basis of $g_1, \ldots, g_m$. Then it holds for all $\alpha \in A$ that there exists a non-zero scalar $c \in \kappa$ and a $\beta \in B$ such that $\mathfrak{s}(\alpha) = c\,\mathfrak{s}(\beta)$ and $\operatorname{lt}(\overline{\alpha}) = c \operatorname{lt}(\overline{\beta})$.*

We can extract a Gröbner basis from a signature Gröbner basis, but that is not the only reason to be interested in signature Gröbner bases. It is also possible to extract a Gröbner basis of the syzygy module of the input basis $\{g_1, \ldots, g_m\}$ from a signature Gröbner basis. Note that this is the syzygy module of the *input* basis rather than the syzygy module of a Gröbner basis of the same ideal. The former is in general much harder to compute than the latter.

The key to computing the syzygy module is Theorem 6 which implies that we can determine the initial module of the module of syzygies from looking at those S-pairs and $\boldsymbol{e}_i$ that regular $\mathfrak{s}$-reduce to zero. If we are computing with full elements of R^m (as opposed to sig-poly pairs) and we have stored the syzygies that were computed (as opposed to storing just their signatures) then those syzygies will form the minimal Gröbner basis of the syzygy module.

THEOREM 6. *Let $\alpha \in R^m$ be a syzygy and let $\mathcal{G}$ be a signature Gröbner basis up to signature $\mathfrak{s}(\alpha)$. Then there exists a $\beta \in R^m$ with $\mathfrak{s}(\beta) \mid \mathfrak{s}(\alpha)$ such that β is an S-pair or has the form $\boldsymbol{e}_i$ and such that β regular $\mathfrak{s}$-reduces to zero.*

Figure 1 contains pseudo code for a simple version of **SB** that returns a signature Gröbner basis $\mathcal{G}$ and and also the initial submodule H of the syzygy module of $g_1, \ldots, g_m$. This pseudo code is intended to succinctly state the essence of **SB** without getting bogged down in any of the complexities of implementation on a computer. Among many other things, a reasonable implementation would use criteria to eliminate S-pairs (see Section 2.4) and use the sig-poly pair optimization (see Remark 2).

We stated that **SB** proceeds by $\mathfrak{s}$-reducing S-pairs and then adding the $\mathfrak{s}$-reduced result to the basis if it is not a syzygy. In Figure 1 we *regular* $\mathfrak{s}$-reduce the S-pair p and then add the *regular* $\mathfrak{s}$-reduced result p' to the basis if it is not a syzygy and not singular $\mathfrak{s}$-reducible. The reason for this is that if p' is singular $\mathfrak{s}$-reducible then $\mathfrak{s}$-reduction of p' is going to $\mathfrak{s}$-reduce p' to zero anyway so we might as well not spend the time on doing that $\mathfrak{s}$-reduction.

2.4 S-Pair Elimination

Three things can happen when **SB** regular $\mathfrak{s}$-reduces an S-pair in signature T and gets a result $\gamma \in R^m$.

Syzygy If γ is a syzygy then T is recorded in H.

Singular If γ is singular top $\mathfrak{s}$-reducible then γ $\mathfrak{s}$-reduces to zero so γ is discarded.

Basis Otherwise γ is recorded in $\mathcal{G}$ as a new basis element.

For these three cases T is respectively a *syzygy*, *singular* or *basis* signature. This is well defined due to Lemma 4.

If L is a syzygy signature and $L|T$ then T is also a syzygy signature. **SB** eliminates an S-pair p by the *signature criterion* if $\mathfrak{s}(p) \in H$ since then $\mathfrak{s}(p)$ is syzygy.

The *Koszul syzygy* between $\alpha, \beta \in \mathcal{G}$ is $\mathcal{K}(\alpha, \beta) := \overline{\beta}\alpha - \overline{\alpha}\beta$. If $\mathfrak{s}(\overline{\beta}\alpha) \not\simeq \mathfrak{s}(\overline{\alpha}\beta)$ then the Koszul syzygy is *regular*. By "Koszul syzygy" we always mean "regular Koszul syzygy". **SB** eliminates an S-pair p by the *Koszul criterion* if there exists a Koszul syzygy σ such that $\mathfrak{s}(p) = \mathfrak{s}(\sigma)$. In this case $\mathfrak{s}(p)$ is recorded in H since $\mathfrak{s}(p)$ is syzygy.

A signature T is *predictably syzygy* if $\mathfrak{s}(p) = \mathfrak{s}(\sigma)$ for a Koszul syzygy σ or if there exists a syzygy $\sigma \in R^m$ such that $\mathfrak{s}(\sigma) < T$ and $\mathfrak{s}(\sigma)|T$. The combined effect of the signature criterion and the Koszul criterion is to eliminate all S-pairs in predictably syzygy signatures.

If there are two or more S-pairs in the same signature T, then we only have to regular $\mathfrak{s}$-reduce one of them as they all regular $\mathfrak{s}$-reduce to the same thing by Lemma 4. Since $\mathfrak{s}$-reduction proceeds by decreasing the lead term, we can speed up the process by choosing an S-pair p in signature T whose lead term $\operatorname{lt}(\overline{p})$ is minimal. If $\mathfrak{s}(\mathcal{S}(\alpha,\beta)) = \mathfrak{s}(a\alpha)$, then we get the same result from regular $\mathfrak{s}$-reducing $\mathcal{S}(\alpha,\beta)$ as for regular $\mathfrak{s}$-reducing $a\alpha$. So we should choose the $a\alpha \in M$ with minimal lead term $\operatorname{lt}(\overline{a\alpha})$, where M is the set

$$\{a\alpha \,|\, \alpha \in \mathcal{G},\ a \text{ is a monomial and } \mathfrak{s}(a\alpha) = T\}.$$

Note that α might not be involved in any S-pair in signature T. If $a\alpha$ is not regular top $\mathfrak{s}$-reducible, then T is a singular signature. In this case **RB** eliminates all the S-pairs in signature T by the *singular criterion*. The effect of the singular criterion is to eliminate all S-pairs in singular signatures.

3. THE REWRITE BASIS ALGORITHM

In this section we introduce the **RB** algorithm, which is simpler than yet equivalent to **F5GEN** [8]. Our motivation for studying **RB** is that it contains many published algorithms as special cases including **SB** and $\mathbf{F_5}$.

The difference between **RB** and **SB** is that **RB** uses the concept of *rewriting* instead of the concept of singular $\mathfrak{s}$-reduction. This has two main consequences. First, **RB** uses a different criterion for eliminating S-pairs. Second, **RB** adds a regular $\mathfrak{s}$-reduced $\alpha \in R^m$ to the basis even if α is singular top $\mathfrak{s}$-reducible. **SB** does not add such α to the basis because such α $\mathfrak{s}$-reduce to zero anyway. **RB** must add such α to the basis because rewriting requires it.

```
SimpleRewriteBasisAlgorithm({g1, ..., gm} ⊆ R)
  G ← ∅  (G will be the rewrite basis)
  P ← {e1, ..., em}  (P is the set of pending reductions)
  H ← ⟨0⟩ ⊆ R^m  (H will be the initial module of syzygies)
  while P ≠ ∅ do
    p ← an element of P with ≤-minimal signature
    P ← P \ {p}
    if not RBEliminated(p) then
      p' ← result of regular s-reducing p
      if p̄' = 0 then
        H ← H + ⟨s(p')⟩
      else
        P ← P ∪ {S(α, p') | α ∈ G and S(α, p') is regular}
        G ← G ∪ {p'} (insert p' into the basis even when p'
        is singular top s-reducible)
  return (G, H)

RBEliminated(S-pair aα − bβ ∈ R^m)
  if aα is rewritable or aα is predictably syzygy then
    return true (eliminate S-pair)
  if bβ is rewritable or bβ is predictably syzygy then
    return true (eliminate S-pair)
  return false (do not eliminate S-pair)
```

Figure 2: Pseudo code for a simple version of RB.

RB is parametrized by a *rewrite order*. Despite the differences between **RB** and **SB**, PHW [8] show that **GVWHS** (and hence **SB**) is an instantiation of **F5GEN** (and hence **RB**). We simplify their proof in the setting of **SB** and **RB** and extend the result to say that **SB** is the best possible instantiation of **RB** in the sense of Theorem 13.

Figure 2 contains pseudo code for a simple version of **RB**.

3.1 Rewriting and S-pair Elimination

Let $\preceq$ be a *rewrite order*, which means that $\preceq$ is a total order on $\mathcal{G}$ such that $\mathfrak{s}(\alpha) \,|\, \mathfrak{s}(\beta) \Rightarrow \alpha \preceq \beta$. Such an order always exists due to our assumption that $\mathfrak{s}(\alpha) \simeq \mathfrak{s}(\beta) \Rightarrow \alpha = \beta$. A basis element $\alpha \in \mathcal{G}$ is a *rewriter in signature* T if $\mathfrak{s}(\alpha)\,|T$. If $\mathfrak{s}(a\alpha) = T$ for a monomial a we also say for convenience that $a\alpha$ is a rewriter of T. The $\preceq$-maximal rewriter in signature T is the *canonical rewriter*. A basis element multiple $a\alpha$ is *rewritable* if α is not the canonical rewriter of $\mathfrak{s}(a\alpha)$. An S-pair $a\alpha - b\beta$ is eliminated by **RB** if $a\alpha$ is predictably syzygy or rewritable, or if $b\beta$ is predictably syzygy or rewritable.

Note that **RB**'s S-pair elimination criterion applies equally to both components $a\alpha$ and $b\beta$ of an S-pair $a\alpha - b\beta$. No criterion in **SB** can eliminate an S-pair based on $\mathfrak{s}(b\beta)$ where $\mathfrak{s}(a\alpha) > \mathfrak{s}(b\beta)$. Can **RB** eliminate some S-pairs that **RB** cannot due to this difference? Since both **SB** and **RB** regular $\mathfrak{s}$-reduce at most one S-pair in any given signature, what matters is the ability to eliminate *all* S-pairs in a signature. We will prove that **SB** has a stronger S-pair elimination criterion than **RB** does in the sense that if **SB** must regular $\mathfrak{s}$-reduce an S-pair in a signature then so must **RB**.

3.2 Rewrite bases

$\mathcal{G}$ is a *rewrite basis in signature* T if the canonical rewriter in signature T is not regular top $\mathfrak{s}$-reducible or if T is a syzygy signature. $\mathcal{G}$ is a *rewrite basis up to signature* T if $\mathcal{G}$ is a rewrite basis for all signatures L such that $L < T$. $\mathcal{G}$ is a *rewrite basis* if it is a rewrite basis in all signatures. We prove Theorem 7 later in this section.

THEOREM 7. **RB** *computes a rewrite basis.*

LEMMA 8. *If $\mathcal{G}$ is a rewrite basis up to signature T then $\mathcal{G}$ is also a signature Gröbner basis up to signature T.*

PROOF. Suppose to get a contradiction that $\mathcal{G}$ is a rewrite basis up to signature T but that it is not a signature Gröbner basis up to signature T. Since the monomial order is a well-order there exists an $\alpha \in R^m$ with minimal signature $\mathfrak{s}(\alpha) < T$ such that α does not $\mathfrak{s}$-reduce to zero. Then $\mathcal{G}$ is a signature basis up to signature $\mathfrak{s}(\alpha)$ and it is a rewrite basis in signature $\mathfrak{s}(\alpha)$.

Let β be the result of regular $\mathfrak{s}$-reducing α and let $c\gamma$ be the canonical rewriter in signature $\mathfrak{s}(\alpha)$. Then $\mathfrak{s}(c\gamma) = \mathfrak{s}(\beta)$ and both $c\gamma$ and β are not regular top $\mathfrak{s}$-reducible, so $\operatorname{lt}(\overline{c\gamma}) = \operatorname{lt}(\overline{\beta})$ by Lemma 4. Perform the singular $\mathfrak{s}$-reduction step $\beta - c\gamma$. Since $\mathfrak{s}(\beta - c\gamma) < \mathfrak{s}(\alpha)$ and $\mathcal{G}$ is a signature Gröbner basis up to signature $\mathfrak{s}(\alpha)$ we then get that $\beta - c\gamma$ $\mathfrak{s}$-reduces to zero. Hence α $\mathfrak{s}$-reduces to zero which is a contradiction. □

SB computes a minimal signature Gröbner basis so the **SB** basis is a subset of the **RB** basis up to sig-poly pairs. Both **SB** and **RB** have to perform a regular $\mathfrak{s}$-reduction in all syzygy signatures that are not predictably syzygy, so there are no differences between the two algorithms in terms of how many S-pairs are reduced to zero. For non-syzygy signatures T, both algorithms will add an element to the basis with signature T if and only if they reduce an S-pair in signature T.[1] This implies that if **SB** regular $\mathfrak{s}$-reduces an S-pair in signature T then so does **RB**. In other words, the S-pairs reduced by **SB** form a subset of the S-pairs reduced by **RB** up to signature. If A is the basis computed by **SB** and B is the basis computed by **RB**, then **RB** regular $\mathfrak{s}$-reduced $|B| - |A| \geq 0$ more S-pairs than **SB** did. If $|A| = |B|$ then **SB** and **RB** reduced the same S-pairs up to signature.

Our proof of Theorem 7 is based on the following series of lemmas. Lemma 10 connects S-pairs to rewrite bases. Lemma 11 is an important technical lemma that we use here to construct S-pairs and which we use again in Section 4.1. Lemma 12 gives a precise criterion for when **RB** regular $\mathfrak{s}$-reduces an S-pair in a signature.

LEMMA 9. *Let $\mathcal{G}$ be a rewrite basis up to signature T. Let $a\alpha$ be the canonical rewriter in signature T and let $b\beta$ be a regular top $\mathfrak{s}$-reducer of $a\alpha$. Then $\mathcal{S}(\alpha, \beta) = a\alpha - b\beta$ and $\mathfrak{s}(\mathcal{S}(\alpha, \beta)) = T$.*

PROOF. If $g := \gcd(a, b)$ then $a\alpha - b\beta = g\mathcal{S}(\alpha, \beta)$. Suppose to get a contradiction that $g \neq 1$. Let $c\gamma$ be the canonical rewriter in signature $\frac{T}{g}$. Then $\gamma \succeq \alpha$ since $\frac{a}{g}\alpha$ is a rewriter in signature $\frac{T}{g}$. Also $\gamma \preceq \alpha$ since $gc\gamma$ is a rewriter in signature T. Hence $\alpha = \gamma$. This is a contradiction since then $\frac{b}{g}\beta$ is a regular top $\mathfrak{s}$-reducer of $\frac{a}{g}\alpha = c\gamma$ but $c\gamma$ is not regular top $\mathfrak{s}$-reducible. □

LEMMA 10. *$\mathcal{G}$ is a rewrite basis up to signature T if $\mathcal{G}$ is a rewrite basis in all signatures $\mathfrak{s}(p) < T$ where p is an S-pair or $p = \mathbf{e}_i$.*

PROOF. Suppose to get a contradiction that $\mathcal{G}$ is not a rewrite basis up to signature T. Since the module monomial order $<$ is a well-order there exists a minimal non-syzygy signature $L < T$ such that $\mathcal{G}$ is not a rewrite basis in signature L. Then $\mathcal{G}$ is a rewrite basis up to signature L. Then there exists an S-pair with signature L by Lemma 9 so $\mathcal{G}$ is a rewrite basis in signature L which is a contradiction. □

LEMMA 11. *Let $\alpha \in R^m$, let $\mathcal{G}$ be a rewrite basis up to signature $\mathfrak{s}(\alpha)$ and let t be a regular $\mathfrak{s}$-reducible term of $\overline{\alpha}$. Let M be the set of $c\gamma$ that regular $\mathfrak{s}$-reduce t. Let $b\beta$ be the canonical rewriter in signature $L := \min_{c\gamma \in M} \mathfrak{s}(c\gamma)$. Then*

- *$b\beta$ is a regular $\mathfrak{s}$-reducer of t,*
- *$b\beta$ is not regular top $\mathfrak{s}$-reducible,*
- *$b\beta$ is not rewritable and*
- *$\mathfrak{s}(b\beta)$ is not syzygy.*

PROOF. $b\beta$ is the canonical rewriter in signature $L < \mathfrak{s}(\alpha)$ and $\mathcal{G}$ is a rewrite basis up to signature $\mathfrak{s}(\alpha)$, which establishes the two middle statements.

***$b\beta \in M$*:** Let $c\gamma \in M$ such that $\mathfrak{s}(c\gamma) = L$. Suppose to get a contradiction that $d\delta$ regular top $\mathfrak{s}$-reduces $c\gamma$. Then $\operatorname{lt}(\overline{d\delta}) = \operatorname{lt}(\overline{c\gamma}) = t$ and $\mathfrak{s}(d\delta) < \mathfrak{s}(c\gamma) < \mathfrak{s}(\alpha)$, so $d\delta \in M$ and $\mathfrak{s}(d\delta) < L$ which is a contradiction. Then $\operatorname{lt}(\overline{b\beta}) = \operatorname{lt}(\overline{c\gamma}) = t$ by Lemma 4 so $b\beta \in M$.

$\mathfrak{s}(b\beta)$ is not syzygy: Suppose to get a contradiction that there exists a syzygy σ such that $\mathfrak{s}(\sigma) = \mathfrak{s}(b\beta)$. Since $\mathfrak{s}(b\beta - \sigma) < \mathfrak{s}(\alpha)$ there exists a top $\mathfrak{s}$-reducer $c\gamma$ of $b\beta - \sigma$. Then $\mathfrak{s}(c\gamma) \leq \mathfrak{s}(b\beta - \sigma) < \mathfrak{s}(b\beta)$ so $c\gamma$ is a regular top $\mathfrak{s}$-reducer of $b\beta$ but $b\beta$ is not regular top $\mathfrak{s}$-reducible. □

LEMMA 12. *Let $\mathcal{G}$ be a rewrite basis up to signature T. Let $a\alpha$ be the canonical rewriter in signature T. Then* **RB** *$\mathfrak{s}$-reduces an S-pair in signature T if and only if $a\alpha$ is regular top $\mathfrak{s}$-reducible and T is not predictably syzygy.*

PROOF. **if:** Let $b\beta$ be the regular top $\mathfrak{s}$-reducer of $a\alpha$ from Lemma 11 so that $b\beta$ is not rewritable and $\mathfrak{s}(b\beta)$ is not predictably syzygy. Then $\mathcal{S}(\alpha, \beta) = a\alpha - b\beta$ by Lemma 9 and this S-pair is not eliminated by **RB**.

only if: Let $\mathcal{S}(\alpha, \beta) = a\alpha - b\beta$ such that $T = \mathfrak{s}(a\alpha) > \mathfrak{s}(b\beta)$ and such that **RB** does not eliminate $\mathcal{S}(\alpha, \beta)$. The latter implies that $a\alpha$ is the canonical rewriter in signature T and that T is not predictably syzygy. Observe that $a\alpha$ is regular top $\mathfrak{s}$-reducible since $b\beta$ regular top $\mathfrak{s}$-reduces it. □

PROOF OF THEOREM 7. Suppose that **RB** has terminated with the basis $\mathcal{G}$. Suppose to get a contradiction that p is an S-pair such that $\mathcal{G}$ is not a rewrite basis in signature $\mathfrak{s}(p)$. Let $a\alpha$ be the canonical rewriter in signature $\mathfrak{s}(p)$. Then $a\alpha$ is regular top $\mathfrak{s}$-reducible and $\mathfrak{s}(a\alpha) = \mathfrak{s}(p)$ is not syzygy, so **RB** has reduced an S-pair in signature $\mathfrak{s}(p)$ by Lemma 12. Then **RB** has also added a basis element β with signature $\mathfrak{s}(p)$ whereby $\beta = a\alpha$ is the canonical reducer in signature $\mathfrak{s}(p)$. This is a contradiction since β is not regular top $\mathfrak{s}$-reducible. Then $\mathcal{G}$ is a rewrite basis by Lemma 10. □

3.3 SB is the Best Possible Instantiation of RB

Let the *sig-lead ratio order* $\preceq_r$ be the order on $\mathcal{G}$ such that $\alpha \preceq_r \beta$ if $r_\alpha \leq r_\beta$ or if $r_\alpha = r_\beta$ and $\mathfrak{s}(\alpha) \leq \mathfrak{s}(\beta)$.

THEOREM 13. **RB** *is equivalent to* **SB** *when using the sig-lead ratio rewrite order. If* **RB** *uses any other rewrite*

[1] If an S-pair p regular $\mathfrak{s}$-reduces to p' then it is true that **SB** will not add p' to the basis if p' is singular top $\mathfrak{s}$-reducible. However, the singular criterion always eliminates such S-pairs p before the regular $\mathfrak{s}$-reduction happens.

order, then **SB** *computes a basis that is a subset of the basis computed by* **RB** *up to sig-poly pairs and* **SB** *regular* $\mathfrak{s}$*-reduces a subset of the S-pairs regular* $\mathfrak{s}$*-reduced by* **RB** *up to signature.*

PROOF. Let **RB** use $\preceq_r$ for rewriting and let $\mathcal{G}$ be the minimal signature Gröbner basis. We will prove that $\preceq_r$ is a rewrite order and that $\mathcal{G}$ is a rewrite basis. Then **RB** computes $\mathcal{G}$ as it only regular $\mathfrak{s}$-reduces an S-pair in a non-syzygy signature T if the basis is not already a rewrite basis in signature T. Thus **SB** and **RB** compute the same basis and then the result follows from the statements in Section 3.2.

$\preceq_r$ is a rewrite order: If $\mathfrak{s}(\alpha) \mid \mathfrak{s}(\beta)$ then $r_\alpha \le r_\beta$ by Lemma 14 and $\mathfrak{s}(\alpha) \le \mathfrak{s}(\beta)$ so $\alpha \preceq_r \beta$.

$\mathcal{G}$ is a rewrite basis: Let $a\alpha$ be a canonical rewriter. We need to prove that $a\alpha$ is not regular top $\mathfrak{s}$-reducible. Let γ be the result of regular $\mathfrak{s}$-reducing $a\alpha$. $\mathcal{G}$ is a signature Gröbner basis, so there exists a $b\beta$ that singular top $\mathfrak{s}$-reduces γ. Then $r_\alpha \le r_\beta$ by Lemma 14 so $\alpha = \beta$. □

LEMMA 14. *Let $\mathcal{G}$ be a signature Gröbner basis up to signature $\mathfrak{s}(a\alpha) = \mathfrak{s}(b\beta)$ where $\alpha, \beta \in \mathcal{G}$ and a, b are monomials. If $b\beta$ is not regular top $\mathfrak{s}$-reducible then $r_\alpha \le r_\beta$.*

PROOF. Let γ be the result of regular $\mathfrak{s}$-reducing $a\alpha$. $b\beta$ and γ are not regular top $\mathfrak{s}$-reducible so $\operatorname{lt}(\overline{a\alpha}) \ge \operatorname{lt}(\overline{\gamma}) = \operatorname{lt}(\overline{b\beta})$ by Lemma 4. Also $\mathfrak{s}(\gamma) = \mathfrak{s}(a\alpha) = \mathfrak{s}(b\beta)$, so

$$r_\alpha = \frac{\mathfrak{s}(a\alpha)}{\operatorname{lt}(\overline{a\alpha})} \le \frac{\mathfrak{s}(\gamma)}{\operatorname{lt}(\overline{\gamma})} = \frac{\mathfrak{s}(b\beta)}{\operatorname{lt}(\overline{b\beta})} = r_\beta. \quad \square$$

4. RB₅ — BRINGING RB CLOSER TO F₅

In this section we define $\mathbf{RB_5}$, which is a special case of **RB**. $\mathbf{RB_5}$ is more similar to $\mathbf{F_5}$ than **RB** is. We will show why the differences between $\mathbf{RB_5}$ and **RB** are not significant, except for the fact that **RB** is more general. In Section 5 we resolve the differences between $\mathbf{RB_5}$ and $\mathbf{F_5}$, which completes our proof that $\mathbf{F_5}$ is a special case of **RB**.

$\mathbf{RB_5}$ is specialized from **RB** in three main ways. First, the module monomial order must be position-first, which means that if $i < j$ then $a\mathbf{e}_i < b\mathbf{e}_j$ for all monomials $a, b \in R$. Second, the input basis elements $g_1, \dots, g_m$ must form a regular sequence. Third, the rewrite order must be the $\mathbf{F_5}$ *rewrite order* $\preceq_5$. Let $a\mathbf{e}_i := \mathfrak{s}(\alpha)$ and $\mathfrak{s}(b\mathbf{e}_j) := \beta$. Then $\alpha \preceq_5 \beta$ if $i < j$ or if $i = j$ and the total degree of a is strictly smaller than the total degree of b. Break ties arbitrarily.

4.1 Reduction in RB₅

Let $\alpha \in R^m$ and let t be a term of $\overline{\alpha}$. We can $\mathfrak{r}$*-reduce* t by a basis element $\beta \in \mathcal{G}$ if

- there exists a monomial b such that $\operatorname{lt}(\overline{b\beta}) = t$,
- $\mathfrak{s}(b\beta) < \mathfrak{s}(\alpha)$,
- $\mathfrak{s}(b\beta)$ is not syzygy and
- $b\beta$ is not rewritable.

The outcome of the $\mathfrak{r}$-reduction step is then $\alpha - b\beta$ and β is the $\mathfrak{r}$*-reducer*. When β $\mathfrak{r}$-reduces a term of $\overline{\alpha}$ we also say for convenience that $b\beta$ $\mathfrak{r}$-reduces α. If $t = \operatorname{lt}(\overline{\alpha})$ then the $\mathfrak{r}$-reduction step is a *top $\mathfrak{r}$-reduction step.*

The first condition on $\mathfrak{r}$-reducers is the same as for $\mathfrak{s}$-reduction. The second condition requires $<$ where $\mathfrak{s}$-reduction requires only $\le$, so $\mathfrak{r}$-reduction is implicitly regular. The final two conditions are not used by **RB**. In Section 5.3 we motivate these extra conditions in terms of S-pair elimination. All $\mathfrak{r}$-reducers are regular $\mathfrak{s}$-reducers but not vice versa.

LEMMA 15. *Let $\alpha \in R^m$, let t be a term of $\overline{\alpha}$ and let $\mathcal{G}$ be a rewrite basis up to signature $\mathfrak{s}(\alpha)$. Then t is regular $\mathfrak{s}$-reducible if and only if t is $\mathfrak{r}$-reducible*

PROOF. If t is regular $\mathfrak{s}$-reducible then by Lemma 11 there exists a regular $\mathfrak{s}$-reducer $b\beta$ of t such that $\mathfrak{s}(b\beta)$ is not syzygy and $b\beta$ is not rewritable. So $b\beta$ is an $\mathfrak{r}$-reducer. □

During regular $\mathfrak{s}$-reduction there can be a choice of which regular $\mathfrak{s}$-reducer to use. Lemma 15 shows that the effect of the extra conditions imposed on $\mathfrak{r}$-reducers is to exclude some of the $\mathfrak{s}$-reducers from consideration, but it is never the case that all of the $\mathfrak{s}$-reducers are excluded. The outcome of regular $\mathfrak{s}$-reduction does not depend on the choice of $\mathfrak{s}$-reducer up to sig-poly pairs, which proves Corollary 16.

COROLLARY 16. *Let $\alpha \in R^m$ and let $\mathcal{G}$ be a rewrite basis up to signature $\mathfrak{s}(\alpha)$. Let $\alpha_\mathfrak{s}$ be the result of regular $\mathfrak{s}$-reducing α and let $\alpha_\mathfrak{r}$ be the result of $\mathfrak{r}$-reducing α. Then $\mathfrak{s}(\alpha_\mathfrak{s}) = \mathfrak{s}(\alpha_\mathfrak{r})$ and $\overline{\alpha_\mathfrak{s}} = \overline{\alpha_\mathfrak{r}}$.*

4.2 S-pair elimination in RB₅

Due to the extra assumptions on the input, $\mathbf{RB_5}$ can eliminate all S-pairs that reduce to zero by considering only the Koszul syzygies. This is stated already in the $\mathbf{F_5}$ paper [5]. **RB** uses condition (2) of Theorem 17 while $\mathbf{RB_5}$ uses condition (4). Theorem 17 implies that these two are equivalent.

THEOREM 17. *Let $\mathcal{G}$ be a signature Gröbner basis up to signature $\mathbf{e}_i$. Assume that the input basis element g_i is not a zero divisor of $R/\langle g_1, \dots, g_{i-1}\rangle$. Assume that the module monomial order $<$ on R^m has the property that $b\mathbf{e}_j < a\mathbf{e}_i < c\mathbf{e}_k$ for all j, k such that $j < i < k$ and for all monomials $a, b, c \in R$. Then the following statements are equivalent for all monomials $a \in R$.*

1. *The signature $a\mathbf{e}_i$ is syzygy.*
2. *The signature $a\mathbf{e}_i$ is predictably syzygy.*
3. *$\exists \alpha \in \mathcal{G}$: $\operatorname{ind}(\alpha) < i$ and $\mathfrak{s}(\mathcal{K}(\mathbf{e}_i, \alpha)) \mid a\mathbf{e}_i$.*
4. *$\exists \alpha \in \mathcal{G}$: $\operatorname{ind}(\alpha) < i$ and $\operatorname{lt}(\overline{\alpha}) \mid a$.*

PROOF. Let $G := \{\alpha \in \mathcal{G} \mid \operatorname{ind}(\alpha) < i\}$, $\overline{G} := \{\overline{\alpha} \mid \alpha \in G\}$ and $F := \langle g_1, \dots, g_{i-1}\rangle$. That (3) ⇒ (2) ⇒ (1) is immediate.

$\overline{G}$ is a Gröbner basis of F: If $\alpha \in G$ then $\alpha_j = 0$ for $j \ge i$ so it is immediate that $\langle \overline{G} \rangle \subseteq \langle g_1, \dots, g_{i-1}\rangle$. To prove the other inclusion, let $f \in F$. We need to prove that f reduces to zero on classic polynomial reduction by $\overline{G}$.

There exists an $\alpha \in R^m$ such that $f = \overline{\alpha}$ and $\alpha_j = 0$ for $j \ge i$. Then $\operatorname{ind}(\alpha) < i$ so $\mathfrak{s}(\alpha) < \mathbf{e}_i$ whereby α $\mathfrak{s}$-reduces to zero when $\mathfrak{s}$-reducing by $\mathcal{G}$. All the $\mathfrak{s}$-reducers $\gamma \in \mathcal{G}$ in that $\mathfrak{s}$-reduction must have $\mathfrak{s}(\gamma) \le \mathfrak{s}(\alpha)$ whereby $\operatorname{ind}(\gamma) < i$. Thus α also $\mathfrak{s}$-reduces to zero when $\mathfrak{s}$-reducing by G. Then $\overline{\alpha} = f$ reduces to zero on classic polynomial reduction by $\overline{G}$.

(1) ⇒ (4): Let $\beta \in R^m$ be a syzygy such that $\mathfrak{s}(\beta) = a\mathbf{e}_i$. Then $\beta_j = 0$ for all $j > i$ whereby $\beta_i g_i \in F$ since

$$0 = \overline{\beta} = \sum_{j=1}^{m} \beta_j g_j = \beta_i g_i + \sum_{j=1}^{i-1} \beta_j g_j.$$

$\phi_i \in \mathcal{G}$	reduced from	$\operatorname{lt}(\overline{\phi_i})$	$\mathfrak{s}(\phi_i)$
ϕ_1	$\mathbf{e}_1$	y^3	$\mathbf{e}_1$
ϕ_2	$\mathbf{e}_2$	xyz	$\mathbf{e}_2$
ϕ_3	$y^2\phi_2 - xz\phi_1 = \mathcal{S}(\phi_2, \phi_1)$	x^3z^2	$y^2\mathbf{e}_2$
ϕ_4	$\mathbf{e}_3$	yz^2	$\mathbf{e}_3$
ϕ_5	$x\phi_3 - z\phi_2 = \mathcal{S}(\phi_3, \phi_2)$	xz^3	$x\mathbf{e}_3$
ϕ_6	$y^2\phi_3 - z^2\phi_1 = \mathcal{S}(\phi_3, \phi_1)$	x^2z^3	$y^2\mathbf{e}_3$
ϕ_7	$y\phi_5 - z^2\phi_2 = \mathcal{S}(\phi_5, \phi_2)$	x^2y^2t	$xy\mathbf{e}_3$
ϕ_8	$x\phi_5 - \phi_6 = \mathcal{S}(\phi_5, \phi_6)$	z^5	$x^2\mathbf{e}_3$
ϕ_9	$x\phi_6 - z\phi_3 = \mathcal{S}(\phi_6, \phi_3)$	x^4zt	$xy^2\mathbf{e}_3$
ϕ_{10}	$y\phi_8 - z^3\phi_4 = \mathcal{S}(\phi_8, \phi_4)$	x^3y^2t	$x^2y\mathbf{e}_3$
ϕ_{11}	$x^3\phi_4 - y\phi_3 = \mathcal{S}(\phi_4, \phi_3)$	x^4yt	$x^3\mathbf{e}_3$
ϕ_{12}	$z\phi_{11} - x^3\phi_2 = \mathcal{S}(\phi_{11}, \phi_2)$	x^3zt^3	$x^3z\mathbf{e}_3$
ϕ_{13}	$y\phi_{10} - x^3\phi_1 = \mathcal{S}(\phi_{10}, \phi_1)$	x^5zt	$x^2y^2\mathbf{e}_3$
ϕ_{14}	$x\phi_{12} - \phi_9 = \mathcal{S}(\phi_{12}, \phi_9)$	x^4t^4	$x^4z\mathbf{e}_3$

Figure 3: Computations for $\mathbf{RB_5}$ in Example 19.

Hence $\beta_i \in F$ as g_i is not a zero divisor. As $\overline{G}$ is a Gröbner basis of F there exists an $\alpha \in G$ such that $\operatorname{lt}(\overline{\alpha}) \mid \operatorname{lt}(\beta_i) = a$.

(4) $\Rightarrow$ (3): Let $\alpha \in \mathcal{G}$ such that $\operatorname{ind}(\alpha) < i$ and $\operatorname{lt}(\overline{\alpha}) | a$. By definition $\mathcal{K}(\mathbf{e}_i, \alpha) = \overline{\alpha}\mathbf{e}_i - g_i\alpha$. As $\operatorname{lt}(\overline{\alpha})\,\mathbf{e}_i > \operatorname{lt}(g_i)\,\mathfrak{s}(\alpha)$ we see that $\mathfrak{s}(\mathcal{K}(\mathbf{e}_i, \alpha)) = \operatorname{lt}(\overline{\alpha})\,\mathbf{e}_i | a\mathbf{e}_i$. □

4.3 Examples

Example 18 shows a homogeneous ideal where $\mathbf{RB_5}$ (and hence $\mathbf{F_5}$) computes a larger basis than **SB** does. Example 19 shows details of how $\mathbf{RB_5}$ computes a basis.

EXAMPLE 18. *Consider the homogenized* `Eco-6` *ideal using the graded reverse lexicographic monomial order. Here* $\mathbf{RB_5}$ *computes a basis with* 100 *elements while* **SB** *computes a basis with* 87 *elements. Both* $\mathbf{RB_5}$ *and* **SB** *compute a basis element* α *with* $\mathfrak{s}(\alpha) = x_0x_2\mathbf{e}_5$. $\mathbf{RB_5}$ *additionally computes basis elements* β *and* γ *with* $\mathfrak{s}(\beta) = x_0x_2x_3^2\mathbf{e}_5$ *and* $\mathfrak{s}(\gamma) = x_0x_2x_3x_4\mathbf{e}_5$. *As* β *and* γ *are not necessary to have a signature Gröbner basis, we can replacing them with a single element with signature* $x_0x_2x_3\mathbf{e}_5$ *and thus get a rewrite basis that contains fewer elements than the basis computed by* $\mathbf{RB_5}$. *Furthermore, the basis computed by* **SB** *is not a rewrite basis with respect to the* $\mathbf{F_5}$ *rewrite order.*

EXAMPLE 19. *Let* κ *be the finite field with 13 elements and let* $R := \kappa[x, y, z, t]$. *Let* $<$ *be the graded reverse lexicographic monomial order. Consider the three input elements*

$$g_1 := -2y^3 - x^2z - 2x^2t - 3y^2t, \quad g_2 := 3xyz + 2xyt,$$

$$g_3 := 2xyz - 2yz^2 + 2z^3 + 4yzt.$$

Figure 3 shows the $\mathfrak{r}$*-reductions performed by* $\mathbf{RB_5}$ *to compute the rewrite basis* $\{\phi_1, \ldots, \phi_{14}\}$. **SB** *regular* $\mathfrak{s}$*-reduce the same S-pairs except for the ones that lead to* ϕ_{10} *and* ϕ_{13}, *so the basis computed by* **SB** *contains 12 elements. The S-pair* $\mathcal{S}(\phi_8, \phi_4)$ *leads to* ϕ_{10} *and it is eliminated by the singular criterion since* $\mathfrak{s}(x\phi_7) = \mathfrak{s}(\mathcal{S}(\phi_8, \phi_4))$ *and* $x\phi_7$ *is not regular top* $\mathfrak{s}$*-reducible. The S-pair* $\mathcal{S}(\phi_{10}, \phi_1)$ *leads to* ϕ_{13} *and it is not considered by* **SB** *since* ϕ_{10} *is not part of* **SB**'*s basis.*

5. $\mathbf{F_5}$ IS EQUIVALENT TO $\mathbf{RB_5}$

The $\mathbf{F_5}$ paper [5] assumes that the input basis elements $g_1, \ldots, g_m$ are homogeneous. In this section we show that $\mathbf{F_5}$ and $\mathbf{RB_5}$ become the same algorithm with this assumption.

5.1 The Sig-Poly Pair Optimization

The notation for $\mathbf{RB_5}$ concerns elements $\alpha \in R^m$ while the $\mathbf{F_5}$ paper uses notation based on sig-poly pairs $(\mathfrak{s}(\alpha), \overline{\alpha})$. We believe that the **SB** notation makes signature computations easier to understand and reason about. However, the cost in time and space for computations on elements of R^m is higher than for sig-poly pairs.

Consider that $\mathbf{RB_5}$ never performs singular $\mathfrak{s}$-reduction steps, so we never need to know any of the terms in R^m other than the leading one — the signature. The singular $\mathfrak{s}$-reduction steps appear only in theorems and proofs. So it is possible to apply the *sig-poly pair optimization* when implementing $\mathbf{RB_5}$ (and **RB** and **SB**), which involves replacing $\alpha \in R^m$ with the sig-poly pair $(\mathfrak{s}(\alpha), \overline{\alpha})$. In $\mathbf{F_5}$ this idea is used in the notation while in $\mathbf{RB_5}$ we present it as an optimization. The outcome is the same.

5.2 $\mathbf{F_5}$ maintains a list of rewriting rules

The $\mathbf{F_5}$ paper [5] does not contain the phrases "canonical rewriter" and "rewrite order". Instead, there is a pseudo code function named `Rewritten` that takes a signature as a parameter and returns the canonical rewriter in that signature. The concept of the canonical rewriter is thus represented implicitly in the $\mathbf{F_5}$ paper as that thing which `Rewritten` returns. We do not specify how the $\mathbf{F_5}$ rewrite order breaks ties since $\mathbf{F_5}$'s implicit rule for this seems more an arbitrary outcome of how the pseudo code was written than an intentional choice (it is not a function of signature).

In $\mathbf{F_5}$ there is a list of *rewriting rules*. A rewriting rule is added to the front of the list when an S-polynomial is calculated. When `Rewritten` is called, it goes through the rewriting rules and returns the first rewriter it finds. The first rewriter found is also the rewriter that was most recently added to the basis. S-polynomials are only calculated for S-pairs whose signature has the current index and the current total degree, so the rewriting rules are added to the front of the list first in order of index of the signature and then in order of total degree of the signature. This proves that $\mathbf{F_5}$ is indeed implicitly using the $\mathbf{F_5}$ rewrite order.

The only exception to this description is that $\mathbf{F_5}$ does not record $\alpha \in \mathcal{G}$ into the list of rewriters if $\mathfrak{s}(\alpha) \simeq \mathbf{e}_i$. However, this does not change whether a basis element multiple is rewritable, so this difference does not change the algorithm.

5.3 $\mathbf{F_5}$ is written to resemble $\mathbf{F_4}$ reduction

$\mathbf{F_4}$ is an algorithm that speeds up classic polynomial reduction by using a symbolic preprocessing step to turn the reduction of many polynomials into row reduction of a single matrix [4]. The more reductions in the same homogeneous degree that can be done at the same time, the larger the matrix becomes and the more of a speed up there is from using $\mathbf{F_4}$ over using classic polynomial reduction. Albrecht and Perry describe a variation of $\mathbf{F_5}$ that uses $\mathbf{F_4}$ reduction [1].

The pseudo code in the $\mathbf{F_5}$ paper is written without $\mathbf{F_4}$ reduction. However, it is still written in a way that heavily suggests how to adapt the symbolic preprocessing step from $\mathbf{F_4}$ to $\mathbf{F_5}$. For example, the $\mathbf{F_5}$ pseudo code is written to process all S-pairs in a homogeneous degree at the same time, which is important when using $\mathbf{F_4}$. This is done by having a set P of all S-pairs in the current homogeneous degree. In $\mathbf{F_4}$ $\mathfrak{r}$-reduction of all the S-pairs in P would then be carried out simultaneously within one big matrix.

The above scheme for $\mathfrak{r}$-reducing all S-pairs in a homogeneous degree does not succeed in getting *all* of the S-pairs in that homogeneous degree. When reducing an S-pair to a new basis element α, it is possible that α can form an S-pair in the same homogeneous degree with another basis element. Since $\mathbf{F_5}$ requires the input ideal to be homogeneous, this happens if and only if there is another basis element β such that $b\beta$ would top $\mathfrak{r}$-reduce α except that $\mathfrak{s}(b\beta) > \mathfrak{s}(\alpha)$. Call such an S-pair $\mathcal{S}(\beta, \alpha)$ a *late S-pair*. The late S-pairs will not initially be included in P since they only appear after some $\mathfrak{r}$-reductions have already been carried out. This is unfortunate when using $\mathbf{F_4}$ $\mathfrak{r}$-reduction because for $\mathbf{F_4}$ $\mathfrak{r}$-reduction we want to identify *all* of the S-pairs in each homogeneous degree up front — including the late ones.

In $\mathbf{F_5}$ there is a mechanism for discovering late S-pairs while $\mathfrak{r}$-reducing another S-pair. The mechanism is that $\mathbf{F_5}$ allows reducers to increase the signature of the reduction. In that case $\mathbf{F_5}$ will split the $\mathfrak{r}$-reduction into two $\mathfrak{r}$-reductions — one $\mathfrak{r}$-reduction in the old signature that continues to be carried out and another $\mathfrak{r}$-reduction in the new signature that is recorded in P. These higher signature $\mathfrak{r}$-reductions are precisely the late S-pairs and this mechanism would allow an $\mathbf{F_4}$ version of $\mathbf{F_5}$ to get all the S-pairs in a degree before doing any $\mathfrak{r}$-reduction. Note that it is possible for such a variant of $\mathbf{F_5}$ to detect more late S-pairs than actually exist because the symbolic preprocessing step does not take account of cancellations.

We have seen that the only effect of $\mathbf{F_5}$'s higher signature reducers is to indicate how to translate $\mathbf{F_5}$ into a variant that uses $\mathbf{F_4}$ reduction. The pseudo code for reduction in $\mathbf{F_5}$ looks different from $\mathfrak{r}$-reduction because that pseudo code is not just doing $\mathfrak{r}$-reduction — it is also constructing late S-pairs. This also explains the purpose of the extra conditions on $\mathfrak{r}$-reducers — they are S-pair elimination criteria. So what $\mathbf{F_5}$ is doing remains equivalent to $\mathfrak{r}$-reduction, which proves that $\mathbf{F_5}$ is a special case of $\mathbf{RB_5}$ despite $\mathbf{F_5}$'s higher-signature reducers and its multiple results of reduction.

6. TERMINATION OF RB AND HENCE $\mathbf{F_5}$

The proof of termination in the $\mathbf{F_5}$ paper [5] is incorrect. Proving termination of $\mathbf{F_5}$ had been an open problem for a decade before it was settled by Galkin [6]. PHW later proved that **F5GEN** (and hence **RB**) terminates [8], which implies that $\mathbf{F_5}$ terminates. Our proof of termination is similar to but significantly simpler than Galkin's and PHW's.

THEOREM 20. *The* **RB** *algorithm terminates.*

PROOF. Let $\phi_1, \phi_2, \ldots$ be the sequence of basis elements computed by **RB** and let $\mathcal{G} := \{\phi_1, \phi_2, \ldots\}$. **RB** processes S-pairs in increasing order of signature so $\mathfrak{s}(\phi_1) < \mathfrak{s}(\phi_2) < \cdots$. Partition $\mathcal{G}$ into sets $R_r := \{\phi_i \,|\, r_{\phi_i} = r\}$. We prove that there are only finitely many non-empty sets R_r and that each R_r is finite. Hence $\mathcal{G}$ is finite whereby **RB** terminates.

Only finitely many $\boldsymbol{R_r}$ are non-empty: If $\alpha \in \mathcal{G}$ then $\mathcal{G}_\alpha := \{\beta \in \mathcal{G} \,|\, \mathfrak{s}(\beta) < \mathfrak{s}(\alpha)\}$ is a signature Gröbner basis up to signature $\mathfrak{s}(\alpha)$. Call $\alpha \in \mathcal{G}$ *minimal* if there is no other $\beta \in \mathcal{G}$ such that $\mathfrak{s}(\beta) \,|\, \mathfrak{s}(\alpha)$ and $\operatorname{lt}(\overline{\beta}) \,|\, \operatorname{lt}(\overline{\alpha})$. It follows from Lemma 21 that a non-minimal $\alpha \in \mathcal{G}$ is top $\mathfrak{s}$-reducible by $\mathcal{G}_\alpha$. No basis element in **RB** is regular top $\mathfrak{s}$-reducible, so if $\alpha \in \mathcal{G}$ is non-minimal, then α must be singular top $\mathfrak{s}$-reducible by $\mathcal{G}_\alpha$. Thus there exists a $\beta \in \mathcal{G}_\alpha$ and a monomial $m \neq 1$ such that $\mathfrak{s}(m\beta) = \mathfrak{s}(\alpha)$ and $\operatorname{lt}(\overline{m\beta}) = \operatorname{lt}(\overline{\alpha})$ whereby α and β lie in the same set R_r. This shows that there are exactly as many non-empty sets R_r as there are minimal basis elements in $\mathcal{G}$. Both R and R^m are Noetherian, so there are only finitely many minimal basis elements. Hence there are also only finitely many non-empty sets R_r.

Each $\boldsymbol{R_r}$ is finite: We prove by induction on the finitely many non-empty sets R_r that each R_r is finite. Assume by induction that all sets $R_{r'}$ with $r' < r$ are finite. We need to prove that R_r is finite. The base case is immediate.

Let $\gamma \in R_r$ and let $\mathcal{S}(\alpha, \beta) = a\alpha - b\beta$ be the S-pair that **RB** regular $\mathfrak{s}$-reduced to get γ where $\mathfrak{s}(a\alpha) > \mathfrak{s}(b\beta)$. Let c be the monic greatest common divisor of $\operatorname{lt}(\overline{\alpha})$ and $\operatorname{lt}(\overline{\beta})$. Then $\frac{1}{c}\operatorname{lt}(\overline{\beta})\,\mathfrak{s}(\alpha) = \mathfrak{s}(a\alpha) > \mathfrak{s}(b\beta) = \frac{1}{c}\operatorname{lt}(\overline{\alpha})\,\mathfrak{s}(\beta)$ so $r_\alpha > r_\beta$. Also $\operatorname{lt}(\overline{\gamma}) < \operatorname{lt}(\overline{a\alpha})$ and $\mathfrak{s}(\gamma) = \mathfrak{s}(a\alpha)$ so

$$r = \frac{\mathfrak{s}(\gamma)}{\operatorname{lt}(\overline{\gamma})} > \frac{\mathfrak{s}(a\alpha)}{\operatorname{lt}(\overline{a\alpha})} = \frac{\mathfrak{s}(\alpha)}{\operatorname{lt}(\overline{\alpha})} > \frac{\mathfrak{s}(\beta)}{\operatorname{lt}(\overline{\beta})}. \qquad (1)$$

Hence $\alpha \in R_{r'}$ and $\beta \in R_{r''}$ where $r', r'' < r$, so R_r contains at most as many elements as there are pairs of elements in the set $\cup_{r'<r} R_{r'}$ which is finite by induction. There may also be one more element with signature $\boldsymbol{e}_i$ since those elements do not come from S-pairs. So by induction R_r is finite. □

LEMMA 21 (EDER AND PERRY [3]). *Let $\alpha \in R^m$ and let $\mathcal{G}$ be a signature Gröbner basis up to signature $\mathfrak{s}(\alpha)$. If there exists a $\beta \in \mathcal{G}$ such that $\operatorname{lt}(\overline{\beta}) \,|\, \operatorname{lt}(\overline{\alpha})$ and $\mathfrak{s}(\beta) \,|\, \mathfrak{s}(\alpha)$ then α is top $\mathfrak{s}$-reducible by $\mathcal{G}$.*

7. REFERENCES

[1] Albrecht, M. and Perry, J. F4/5. 2010. `http://arxiv.org/abs/1006.4933`.

[2] Arri, A. and Perry, J. The F5 Criterion revised. *Journal of Symbolic Computation*, 46(2):1017–1029, June 2011.

[3] Eder, C. and Perry, J. Signature-based Algorithms to Compute Gröbner Bases. In *Proceedings of the 2011 international symposium on Symbolic and algebraic computation*, pages 99–106, 2011.

[4] Faugère, J.-C. A new efficient algorithm for computing Gröbner bases (F4). *Journal of Pure and Applied Algebra*, 139(1–3):61–88, June 1999.

[5] Faugère, J.-C. A new efficient algorithm for computing Gröbner bases without reduction to zero F5. In *Proceedings of the 2002 international symposium on Symbolic and algebraic computation*, pages 75–82, July 2002. Revised version from `http://fgbrs.lip6.fr/jcf/Publications/index.html`.

[6] Galkin, V. Termination of original F5. `http://arxiv.org/abs/1203.2402`, 2012.

[7] Gao, S., Volny IV, F., and Wang, D. A new algorithm for computing Groebner bases. `http://eprint.iacr.org/2010/641`, 2010.

[8] Pan, S., Hu, Y., and Wang, B. The Termination of Algorithms for Computing Gröbner Bases. 2012. `http://arxiv.org/abs/1202.3524`.

[9] Roune, B. H. and Stillman, M. Practical Gröbner basis computation. In *Proceedings of the 2011 international symposium on Symbolic and algebraic computation*.

[10] Roune, B. H. and Stillman, M. Practical signature Gröbner basis computation. In preparation.

[11] Frank Volny IV. *New Algorithms for Computing Gröbner Bases.* PhD thesis, Clemson University, 2011. `http://etd.lib.clemson.edu/documents/1306872881/Volny_clemson_0050D_11180.pdf`.

Finding Points on Real Solution Components and Applications to Differential Polynomial Systems

Wenyuan Wu*
Chongqing Inst. of Green and Intelligent Techn.
Chinese Academy of Sciences
wuwenyuan@cigit.ac.cn

Greg Reid†
Applied Mathematics Dept.
Western University, Canada
reid@uwo.ca

ABSTRACT

In this paper we extend complex homotopy methods to finding witness points on the irreducible components of real varieties. In particular we construct such witness points as the isolated real solutions of a constrained optimization problem.

First a random hyperplane characterized by its random normal vector is chosen. Witness points are computed by a polyhedral homotopy method. Some of them are at the intersection of this hyperplane with the components. Other witness points are the local critical points of the distance from the plane to components. A method is also given for constructing regular witness points on components, when the critical points are singular.

The method is applicable to systems satisfying certain regularity conditions. Illustrative examples are given. We show that the method can be used in the consistent initialization phase of a popular method due to Pryce and Pantelides for preprocessing differential algebraic equations for numerical solution.

Categories and Subject Descriptors: G.1.8 **General Terms:** algorithms, design

Keywords: numerical algebraic geometry, real algebraic geometry, homotopy continuation, singular critical points, witness points, differential algebraic equations.

*The work is partly supported by the projects NSFC-11001040, West Light Foundation of CAS from China (Y21Z010C10).

†Work partly funded by Reid's NSERC grant (government of Canada).

ISSAC'13, June 26–29, 2013, Boston, Massachusetts, USA.

1. INTRODUCTION

This article is a contribution to the development of numerical algorithms for real computational algebraic geometry, and the extension of such methods to systems of real differential polynomials.

It is motivated by recent progress and numerical algorithms for complex algebraic geometry, and in particular the blossoming area of numerical algebraic geometry pioneered by Sommese, Wampler, Verschelde and others. See the book [22] and also [2] for references. Recent progress in extending such methods to the real case are by Lu [17] and Besana et al. [3]. Of particular interest and closest to this paper is the recent work of Hauenstein [11]. Our work is also motivated by progress in symbolic methods for determining features of real solutions of general real polynomial systems (e.g. see [19]).

In this paper we give a numerical method for computing solution (witness) points on each real connected component for real polynomial systems satisfying certain regularity conditions. Given a random vector $\mathbf{n}$ the method computes local critical points of the distance of a hyperplane to the component in the direction $\mathbf{n}$. A method for regularizing singular critical points is also given. This method takes advantage of the availability of efficient homotopy solvers which exploit sparsity and structure of the polynomial system [14].

The real solving method we describe in this paper is applied to the consistent initialization step of the Pryce-Pantelides method [25, 18] for preprocessing differential algebraic equations (DAE) for numerical solution. The Pryce-Pantelides method is successful under certain regularity conditions (in particular that certain system Jacobians are nonsingular). Though not guaranteed the success rate is high enough that it has been implemented in a number of problem solving environments. For example it is the first method of choice in Maple's DAE solving environment MapleSim (see [25] for references).

The Pryce-Pantelides method takes as input a square system of DAE (i.e. the number of equations and unknowns are equal) and consists of a differentiation (prolongation) step and a consistent initialization step.

The prolongation step involves solving an optimization problem. The results are used to determine which higher order derivatives of equations in the system should be appended to the original DAE system, to form an prolonged system of DAE, that implicitly includes its missing constraints. For background references and interpretation in terms of differential elimination theory see our paper [25]. For example as shown in [25] the optimization corresponds to the choice of a partial ranking minimizing a differential Hilbert function of the DAE.

The consistent initialization step requires the determination of initial conditions lying on the variety of the DAE in the space where its unknowns and derivatives are regarded as indeterminates. We will apply our real solving method to this step, and exploit the fact that the required regularity conditions are exactly the same as those needed by Pryce-Pantelides. When these steps are successful standard index one numerical DAE solvers can be applied to the output. Indeed there is even a complexity analysis available showing the low (polynomial) cost of this numerical solution [13].

1.1 Numerical algebraic geometry

The methods of numerical algebraic geometry compute approximate complex points on all irreducible solution components of multivariate complex systems of polynomial equations. Such witness points on components of each possible dimension are obtained by slicing with random planes of equal co-dimension. The points are computed with efficient homotopy methods.

Consider the set $\mathbb{C}[x_1, x_2, ..., x_n]$ of multivariate polynomials with complex coefficients in the complex variables $x = (x_1, x_2, ..., x_n) \in \mathbb{C}^n$. Then $\mathbb{C}[x_1, x_2, ..., x_n]$ is a ring and a system of m polynomials $p_1(x)$, $p_2(x)$, ... , $p_m(x)$ in $\mathbb{C}[x_1, x_2, ..., x_n]$ yields a system of m multi-variate polynomial equations $p(x) = (p_1(x), p_2(x), ..., p_m(x)) = 0$. Its solution set or variety is $V(p) = \{x \in \mathbb{C}^n : p(x) = 0\}$. The regular points of $V(p)$, or reg$(V(p))$, are points at which $V(p)$ is a local complex Euclidean manifold of possible dimension: 0 (points), 1 (curves), ... , $n-1$ (hyper-surfaces). Then reg$(V(p))$ can be partitioned into a disjoint union of subsets, under the equivalence relation of connectedness. Finally closure in the Euclidean topology of these sets yields the irreducible components of a complex polynomial system.

1.2 Real algebraic geometry

Finding real solution components, is usually the case of interest in applications. Naive extension of the complex approach to the real case fails, since such random planes may not intersect some (e.g. compact) components.

Suppose that $x = (x_1, x_2, ..., x_n) \in \mathbb{R}^n$ and consider a system of k multivariate polynomials $p_1(x)$, $p_2(x)$, ... , $p_k(x)$ in the polynomial ring $\mathbb{R}[x_1, x_2, ..., x_n]$. Its solution set or variety is

$$V_{\mathbb{R}}(p_1, ..., p_k) = \{x \in \mathbb{R}^n : p_j(x) = 0,\ 1 \le j \le k\} \quad (1)$$

Real algebraic geometry is a vast subject with many applications. For a modern text with many references on computational real algebraic geometry see [1].

Sturm's ancient method on counting real roots of a polynomial in an interval is central to Tarski's real quantifier elimination [23] and was further developed by Seidenberg [21]. One of the most important algorithms of real algebraic geometry is cylindrical algebraic decomposition. CAD was introduced by Collins [7] and improved by Hong [12] who made Tarski's quantifier elimination algorithmic. This algorithm decomposes $\mathbb{R}^n$ into cells on which each polynomial of a given system has constant sign. The projections of two cells in $\mathbb{R}^n$ to $\mathbb{R}^k$ with $k < n$ either don't intersect or are equal. The double exponential cost of this algorithm [8], is a major barrier to its application. See [6] and [5] for modern improvements using triangular decompositions.

A paper for obtaining witness points for the real positive dimensional case, closely related to our approach, is [19] (also see [9]). Homotopy methods are used in [17] and [3] for real algebraic geometry. Lasserre et al [15] uses semi-definite programming and interestingly that approach is related to the prolongation-projection method used in geometrical completion of differential systems (also see Wu and Zhi [26]).

2. COMPUTING WITNESS POINTS ON REAL COMPONENTS

We consider systems

$$f = (f_1, f_2, ..., f_k) = 0 \quad (2)$$

of k polynomials from $\mathbb{R}[x_1, ..., x_n]$ which satisfy the following assumptions:

A_1: $V_{\mathbb{R}}(f_1, f_2, ..., f_i)$ has dimension $n - i$ for $1 \le i \le k$.

A_2: the ideal $I_i = \langle f_1, f_2, ..., f_i \rangle$ is radical for $1 \le i \le k$.

These assumptions mean that the Jacobian of the system $\{f_1, f_2, ..., f_i\}$ has full row rank at generic points of $V_{\mathbb{R}}(f_1, f_2, ..., f_i)$ for $1 \le i \le k$. If a system f satisfies these two assumptions, we say f is *regular.*

The application of the Pryce-Pantelides method to DAE in Section 4 requires the assumptions A_1 and A_2. Implementations and many successful applications of this method in problem solving environments such as MapleSim, Mathematica, gPROMS, Modelica and EMSO attest to these assumptions being often satisfied in practise.

In the longer term we aim to also develop methods that don't require such assumptions. However they lead to very efficient and fast algorithms. Our approach is to treat well-conditioned problems first before addressing more singular and ill-conditioned problems.

We outline the main steps of our method, postponing details, such as singular cases to later sections.

To begin with, we choose a random point $\hat{x}$ and a random vector $\mathbf{n}$ in $\mathbb{R}^n$. Consider the random hyperplane H in $\mathbb{R}^n$ through $\hat{x}$ with normal $\mathbf{n}$.

For illustration of some of the main ideas we use a simple example in $\mathbb{R}^2$ which has variety given in Figure 1. Many methods are known for this case, and also for the case of real hyper-surfaces in $\mathbb{R}^3$. The novelty of our methods, is primarily for the case of $k > 1$ polynomials in $\mathbb{R}^n$, defining equi-dimensional varieties of dimension $n-k$. Our use of the co-dimension one case and Example 2.1 is purely illustrative.

EXAMPLE 2.1. *Figure 1 displays the variety of a single polynomial $\mathbb{R}^2$ given by:*

$$f = (x^2 + y^2 - 1) \cdot (x - 3 - y^2) \cdot (x + (y+2)^3) \cdot \left((2y-4)^2 - (2x - x^2)^3\right) = 0$$

First we compute the intersection of f and a random line H, that lie distance 0 from the variety, and obtain the point P_1. Note that the resulting system $f = 0$ together with $H = 0$ is one dimension less, and belongs to the case of $k+1$ polynomials. Since we have intersected with a random hyperplane the resulting system of $k+1$ polynomials will also satisfy the regularity assumptions A_1 and A_2.

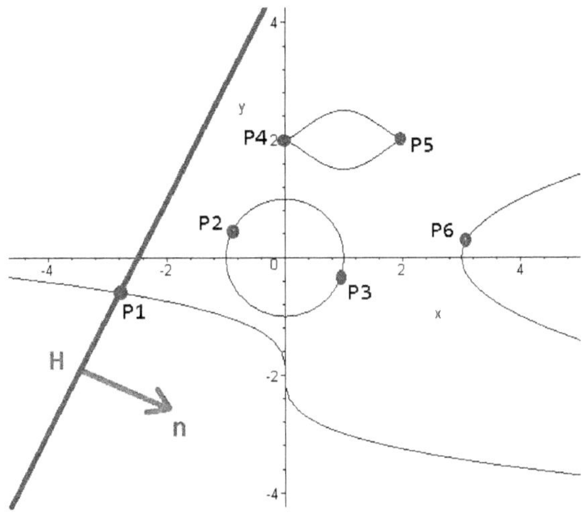

Figure 1: H is a random line through the random point $\hat{x}$ with random normal n. The variety consists of 4 one dimensional components. Point P_1 is the only point at which the variety intersects H at distance 0 from H. The regular local critical points of the normal distance from H are given by P_2, P_3, P_6. The singular critical points are P_4 and P_5.

In the simple case in $\mathbb{R}$ above it would amount to finding the isolated zeros of a zero dimensional system defined by H and $f = 0$.

The closure of any component that does not intersect H must (with probability one) contain critical points of the normal distance to H. If not, then the component would become asymptotically close to H, without touching H. However the randomness of the defining normal of H, and finiteness of the number of such asymptotes, implies that this can only happen on a set of lower measure.

Returning to our illustrative example, it shows the critical points P_2, P_3, P_4, P_5 and P_6 on their corresponding varieties. The above description is essentially a geometric one. However any algorithm that computes it, must use the equations defining the variety and results in well-known and nontrivial difficulties for higher multiplicities and over-determined systems. Thus we invoke the assumptions A_1 and A_2 to yield enough regularity so that our algorithms are well-conditioned, and the encountered Jacobians have full rank.

Regular critical points (e.g. P_2, P_3, P_6) will have $\lambda \nabla f = \mathbf{n}$ in $\mathbb{R}^2$. Generally, in $\mathbb{R}^n$

$$\mathbf{n} \in \operatorname{span}\{\nabla f\} = \operatorname{span}\{\nabla f_1, \cdots, \nabla f_k\} \tag{3}$$

Consequently these critical points are the solutions of the Lagrange optimization problem

$$f = 0, \qquad \sum_{i=1}^{k} \lambda_i \nabla f_i = \mathbf{n} \tag{4}$$

Here $\mathbf{n}$ is a random vector in $\mathbb{R}^n$ and (4) has $n+k$ equations and $n+k$ unknowns $(x, \lambda) = (x_1, ..., x_n, \lambda_1, ..., \lambda_k)$.

Remark 2.1. *The point-distance formulation to obtain real points as the critical points of the distance from a component to a random point can be found in [19, 11]. The main difference between that previous work and our paper is that we find critical points of the distance to a random hyperplane rather than to a random point. If the random point in the point-distance formulation is far away from a component, the corresponding system may have poor condition number. However, in our plane-distance approach the random hyperplane or random normal vector is invariant under translation. So the distance to the component does not affect the conditioning. This usually leads to a square system with lower mixed volume the solution of which is more numerically stable and efficient.*

2.1 The case of regular critical points

We first consider the case of regular local critical points. The singular cases will be considered in Section 2.2. For the regular case we use homotopy continuation methods to solve the new system and find the real critical points which are real. Homotopy continuation methods can determine all isolated complex roots. To apply the methods, we need to show the critical points are isolated in complex space.

Proposition 2.2. *Suppose (x^0, λ^0) is a real solution of (4) with random vector $\mathbf{n}$ and also that the Jacobian of f at this point is of full row rank. Then (x^0, λ^0) is an isolated root of (4) in $\mathbb{C}^{n+k}$ with probability 1.*

Proof. Since the Jacobian of f at (x^0, λ^0) has full row rank k, by the implicit function theorem [10] there are k variables that can be locally expressed as smooth functions of the other variables. Without loss of generality, we write $x_1 = x_1(z), ..., x_k = x_k(z)$, where $z = (x_{k+1}, ..., x_n)$. Substituting into the second equation of (4), that is into $\sum_{i=1}^{k} \lambda_i \nabla f_i = \mathbf{n}$, yields a square system of n equations, which we denote by $g(z, \lambda) = \mathbf{n}$.

Thus, g is a smooth mapping from $\mathbb{R}^n$ to $\mathbb{R}^n$. By Sard's Theorem [20], [10], for almost all $\mathbf{n}$, the Jacobian of g at (z^0, λ^0) has rank n. Consequently, the Jacobian of the full system of $n+k$ equations (4) has full rank $n+k$ at $(x^0, \lambda^0) \in \mathbb{R}^{n+k}$. The rank is the same at this point in $\mathbb{C}^{n+k}$. So (x^0, λ^0) is an isolated root in $\mathbb{C}^{n+k}$. □

Example 2.2. *Let $f = x^2 + y^2 - 1$. Let H be the line through $\hat{x} = (0, -3)$ with $\mathbf{n} = (2, -1)^t$, that is the line $y = 2x - 3$. Then (4) is*

$$f = x^2 + y^2 - 1 = 0, \quad \lambda \begin{pmatrix} 2x \\ 2y \end{pmatrix} = \begin{pmatrix} -2 \\ 1 \end{pmatrix} \tag{5}$$

Simplifying the system, we have $x^2 + y^2 - 1 = 0, x = -2y$. This leads to two real roots: $P_1 = (-2\sqrt{5}/5, \sqrt{5}/5)$ and $P_2 = (2\sqrt{5}/5, -\sqrt{5}/5)$ corresponding to the minimum distance point and the maximum distance point respectively as shown in Figure 2.

Here f satisfies the full rank Jacobian condition and the critical points are regular. In Example 2.3 however the critical points on a connected component are singular.

Example 2.3. *Let $f = y^2 - (x - x^2)^3$ with graph shown in Figure 3 and consider H given by $y = kx + b$.*

Consider the case when the absolute value of the slope of H, i.e. $|k|$, is larger than the maximum slope of the tangent line to the curve $y^2 - (x - x^2)^3 = 0$. Then when H is $y = 2x + 3$, the critical points are the singular points $(0, 0)$ and $(1, 0)$.

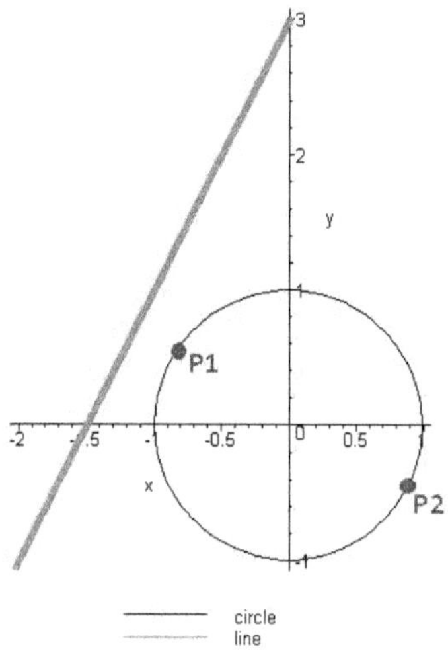

Figure 2: The circle $x^2 + y^2 = 1$ and the line $y = 2x + 3$. The normal is $(2, -1)^t$ and the critical points of the normal distance from the line to the circle are displayed as P_1 and P_2.

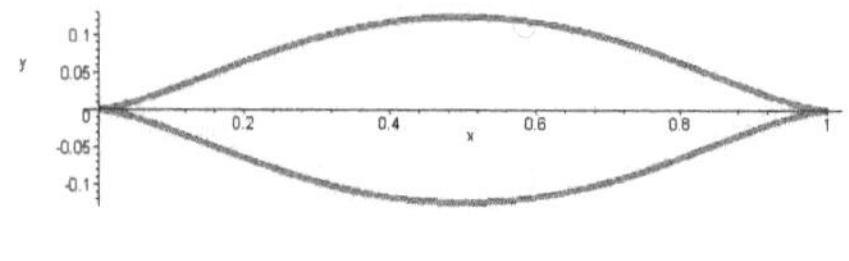

Figure 3: $y^2 - (x - x^2)^3 = 0$

Then the corresponding optimization problem

$$f = y^2 - (x - x^2)^3 = 0, \quad \lambda \begin{pmatrix} -3(x - x^2)^2(1 - 2x) \\ 2y \end{pmatrix} = \begin{pmatrix} -k \\ 1 \end{pmatrix} \tag{6}$$

has no real solutions. In fact, the probability of hitting the singular points can be quite large. We illustrate this by calculating this probability for the curve $t\,y^2 - (x - x^2)^3 = 0$ *where* t *is a parameter. For the solution branch,* $y = (x - x^2)^{3/2}/\sqrt{t}$ *we have*

$$\frac{d^2y}{dx^2} = \frac{3}{4} \frac{(1 - 8x + 8x^2)}{\sqrt{x(1 - x)}\sqrt{t}}.$$

So when $x = \frac{2 \pm \sqrt{2}}{4}$ *the slope of the curve* $\frac{dy}{dx}$ *attains the critical values* $\frac{\mp 3}{8\sqrt{t}}$. *Thus (6) has no real solutions if and only if* $k \in (-\infty, -\frac{8}{3}\sqrt{t}] \cup [\frac{8}{3}\sqrt{t}, \infty)$.

Let H *have normal* $\mathbf{n} = (sin(\theta), cos(\theta))$. *Assume the random variable* θ *is uniformly distributed in* $[0, 2\pi]$. *Then the probability of hitting the singular points* $(0, 0)$ *and* $(1, 0)$ *is* $\frac{2}{\pi}\arctan(\frac{8}{3}\sqrt{t})$. *When* $t = 1$, *the probability is* 0.772 *and if* $t = 10$ *the probability increases to* 0.925. *This fact indicates that avoiding hitting singular points can be difficult for such curves.*

2.2 Singular critical point

As is usual in numerical investigations, we first considered the regular cases before considering singular cases. In this section we consider the case illustrated in Figure 1, where some of the critical points are singular.

The key is to study the local properties of a perturbed system. We use the following notation. For any point $p \in \mathbb{R}^n$, $B_p(r) = \{x \in \mathbb{R}^n \mid \|x - p\| \le r\}$. If $p \in V_{\mathbb{R}}$, we define $N_p = V_{\mathbb{R}} \cap B_p$.

Let f be a regular system of k equations and consider a regular point $p \in V_{\mathbb{R}}(f)$. By appending linear equations $L(x) = L(p)$ to f, where L is orthogonal to the Jacobian of f at p, we obtain a square system:

$$G = \begin{pmatrix} f \\ L \end{pmatrix} \tag{7}$$

with $\mathcal{J} = \frac{\partial G}{\partial x}$. A perturbation $\epsilon \in \mathbb{R}^k$ is added to f to yield the perturbed system

$$\tilde{G} = \begin{pmatrix} f + \epsilon \\ L \end{pmatrix} \tag{8}$$

The variety of a perturbed system with perturbation $\epsilon \in \mathbb{R}^k$ is denoted by $V_{\mathbb{R}}^{\epsilon}$. Similarly $N_p^{\epsilon} = V_{\mathbb{R}}^{\epsilon} \cap B_p$.

Consider the linear homotopy connecting these two systems given by $H(x, t) = t\tilde{G}(x) + (1 - t)G(x) = 0$. We follow a curve $x(t)$ starting from $x(0) = p$ satisfying $H(x(t), t) = 0$ by solving a differential algebraic equation:

$$\frac{dH}{dt} = \frac{d}{dt}\begin{pmatrix} f + t\epsilon \\ L \end{pmatrix} = \mathcal{J} \cdot \frac{dx}{dt} + \begin{pmatrix} \epsilon \\ 0 \end{pmatrix} = 0 \tag{9}$$

with initial condition $x(0) = p$. The solution $q = x(1)$ of (9) is uniquely determined if the perturbation is small enough so that the Jacobian $\mathcal{J}$ is invertible for $t \le 1$.

PROPOSITION 2.3. *Suppose* $\mathcal{J}$ *is invertible in* $B_p(r)$ *for some* $r > 0$. *Let* $\delta = \max\{\|\mathcal{J}^{-1}(x)\| \mid x \in B_p(r)\}$. *If* $\|\epsilon\| < r/\delta$, *then* $\|p - q\| \le \|\epsilon\|\delta$. *Moreover, the local dimension of the perturbed variety at* q *is* $n - k$ *and* $dist(p, N_p^{\epsilon}) \le \|\epsilon\|\delta$.

PROOF. Since

$$\left\| \int_0^1 \mathcal{J}^{-1} \cdot \begin{pmatrix} \epsilon \\ 0 \end{pmatrix} dt \right\| \le \|\epsilon\| \cdot \int_0^1 \|\mathcal{J}^{-1}\|\, dt \le \|\epsilon\|\delta$$

the distance between p and q is at most $\|\epsilon\|\delta$. Thus $dist(p, N_p^{\epsilon}) \le dist(p, q) \le \|\epsilon\|\delta$. Because $\mathcal{J}$ is invertible in $B_p(r)$, the Jacobian of $\tilde{G}$ is also invertible at q. Thus the local dimension of $f + \epsilon$ at q is $n - k$. □

THEOREM 2.4 (REGULARIZATION THEOREM). *If* $f = \{f_1, ..., f_k\}$ *is regular and* c *is sufficiently small, then for almost all* $\|\epsilon\| < c$, $f + \epsilon$ *has dimension* $n - k$ *and there are no singular points on* $V_{\mathbb{R}}(f + \epsilon)$.

PROOF. Consider f as a smooth mapping from $\mathbb{R}^n$ to $\mathbb{R}^k$ with $k < n$. The image of the critical set, denoted by S has Lebesgue measure zero in $\mathbb{R}^k$ by Sard's theorem.

By Proposition 2.3, when the perturbation is small enough, $V_{\mathbb{R}}(f + \epsilon) \ne \emptyset$. Thus, for almost all $\|\epsilon\| < c$, $f + \epsilon$ has no singular points and $\dim V_{\mathbb{R}}(f + \epsilon) = n - k$. □

EXAMPLE 2.4. *Consider* $f = \{y^2 + z^2 - (2x - x^2)^3, z - y^2\}$. *Figure 4 clearly shows a singular point at the origin. To show the intersection explicitly, we solve* y *and* z *in terms of* x *obtaining two solution branches:*

$$y = \sqrt{z}, z = (-1 + \sqrt{1 - 32x^3 - 48x^4 + 24x^5 - 4x^6})/2$$

and

$$y = -\sqrt{z}, z = (-1 + \sqrt{1 - 32x^3 - 48x^4 + 24x^5 - 4x^6})/2.$$

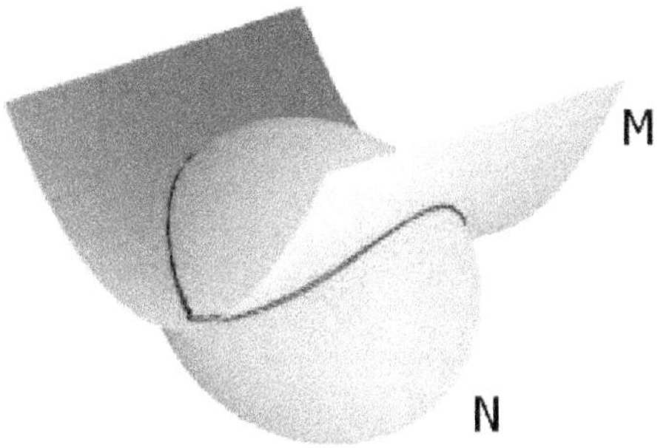

Figure 4: Two surfaces M and N and their intersection. Here M: $z-y^2=0$ and N: $y^2+z^2-(2x-x^2)^3=0$.

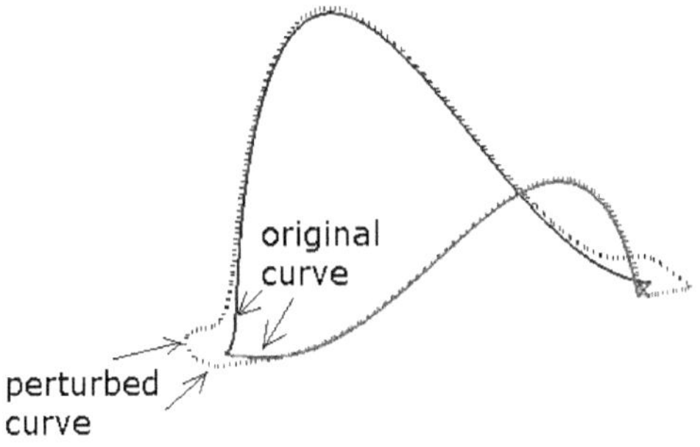

Figure 5: Perturbed and original real varieties

Consider the perturbed system $f+\epsilon=\{y^2+z^2-(2\,x-x^2)^3+\epsilon_1, z-y^2+\epsilon_2\}$. *Consequently,*

$$z=(-1+\sqrt{1-32\,x^3-48\,x^4+24\,x^5-4\,x^6-4\epsilon_1-4\epsilon_2})/2.$$

Suppose $\epsilon_1=-0.04, \epsilon_2=0.015$. *The perturbed and unperturbed curves are shown in Figure 5.*

Set $\mathbf{n}=(-1,0,0.2)$ *in (4). We solve the system* $\{f+\epsilon=0, \lambda\cdot\nabla(f+\epsilon)=\mathbf{n}\}$ *by Hom4Ps2 and obtain* 26 *complex roots, including the following two real roots*

$$(-0.108289277,0,0.01), \quad (2.108289277,0,0.01).$$

Let us consider $p=(-0.108289277,0,0.01)$. *To address the numerical difficulties encountered near the singularity, we project onto the original curve. After* 20 *Newton iterations, we obtain the point* $(-0.000179,0,0)$. *The smallest singular value of the Jacobian at this point is* 0.769×10^{-6}.

One possible way to overcome the numerical difficulty is to solve $G=\{f=0,\lambda\cdot\nabla f=\mathbf{n}\}$ *by using deflation methods [16]. However it is easily seen that the critical point* $(0,0,0)$ *does not satisfy* G.

Here we present alternative way to avoid the numerically difficult region. Firstly, we move p along the perturbed curve to another point $q=(0.101,-0.135,0.0283)$ *where the singular values of the Jacobian are* $(1.04,0.34)$. *Secondly, we project the point q to the original curve and find* $q'=(q_1',q_2',q_3')$ *where* $q_1'=0.122267538554529$, $q_2'=-0.109354334436423$ *and* $q_3'=0.0119583704597381$. *At* q' *the residue is* $(0.59\times10^{-12},-0.29\times10^{-12})$ *and the singular values are* $(1.02,0.353)$. *So* q' *is a regular point on the target curve with high accuracy. Now we use this example to verify Proposition 2.3. By the definition* $\delta\le1/0.34$ *and* $\|\epsilon\|=0.0427$, *and we have* $dist(q,q')=0.0372<\|\epsilon\|\delta$.

REMARK 2.5. *In contrast to the approach in Hauenstein [11], our critical points are defined differently, ie. by distance to a hyperplane, rather than distance to a point. In [11] the author applied an endgame and adaptive precision tracking technique to deal with singular cases. Such cases can occur with high probability especially when the components have "cusps". In the case where there are nearby smooth real points, we give a regularization method for the singular critical points without extending the hardware precision.*

3. IMPLEMENTATION

To address the potential singularities of a given system f, we first perturb the input to yield a nearby system $\tilde{F}$. Then a random linear equation $L=\mathbf{n}\cdot x+1$ is defined. There are two systems to analyze:
(1) System $G=\{\tilde{F},\lambda\cdot\nabla F=\mathbf{n}\}$ (2) System $\{\tilde{F},L\}$.

We solve the square system $\tilde{G}$ of (8) in $\mathbb{C}$ by a Homotopy continuation package, e.g. HOM4PS2 by T.Y. Li et al [14]. We choose only the real roots and discard any imaginary roots. For each real point, we need to project to the variety $V_{\mathbb{R}}(F)$ by the algorithm PROJ2MANIFOLD described below. If the Jacobian is near rank deficiency, then we apply the algorithm FOLLOWCURVE to move the point along $V_{\mathbb{R}}(\tilde{F})$ until the condition of the Jacobian is tolerable. Finally, we compute the projection from this new point onto $V_{\mathbb{R}}(F)$.

For the second system of lower dimension, we can consider it as a new input and solve it recursively by the method introduced above.

Now we describe our algorithms.

Algorithm PROJ2MANIFOLD

Input: System$f=\{f_1,...,f_k\}\subset\mathbb{R}[x_1,\ldots,x_n]$,point p

- Solve N the right null space of $\mathcal{J}_p(F)$ by SVD
- Construct the linear system $L:\ N\cdot x=N\cdot p$
- Let $\bar{F}=\{F,L\}$ which is a square system
- Apply Newton iteration to p to yield q

Output: A regular point q on $V_{\mathbb{R}}(f)$

Algorithm FOLLOWCURVE

Input: $f+\epsilon$: a perturbed system given by $\{f_1+\epsilon_1,...,f_k+\epsilon_k\}\subset\mathbb{R}[x_1,\ldots,x_n]$
$p=$ an approximate solution of $f+\epsilon=0$
$K=$ control condition ♯ used in curve tracking

- Solve N the right null space of $\mathcal{J}_p(f+\epsilon)$ by SVD
- Construct an $(n-k-1)\times n$ linear system $L:\ A\cdot x=A\cdot p$, where A consists of the last $n-k-1$ vectors of N to produce a curve to follow to a regular point.
- Let $\bar{F}=\{f+\epsilon,L\}$ which is $(n-1)\times n$ system and $\bar{F}(p)=0$
- Track the curve from p to q by prediction-projection method until $\|\mathcal{J}_q^+\|<K$, where $\mathcal{J}^+$ is the pseudo-inverse of $\mathcal{J}$.

Output: A regular point q on $V_{\mathbb{R}}(f+\epsilon)$

Algorithm RealWitnessPoint

Input: $f = \{f_1, ..., f_k\} \subset \mathbb{R}[x_1, \ldots, x_n]$ satisfying regularity assumptions A_1 and A_2

c = upper bound on the perturbation

K = control condition $\sharp$ used in curve tracking

- Choose a random real hyperplane $H: \mathbf{n} \cdot x = 1$
- Construct $G = \{f + \epsilon, \lambda \cdot \nabla f = \mathbf{n}\}$, where $\epsilon \in \mathbb{R}^k$ and $\|\epsilon\| < c$
- Let S be the set of real roots of $G = 0$ by Hom4Ps2
- For each point $p \in S$,
  - $*$ if $\|\mathcal{J}_p^+\| > K$ (i.e. condition poor) then
    $q = \text{FollowCurve}(f + \epsilon, p, K)$
    $q' = \text{Project2Manifold}(f, q)$
  - $*$ else $q' = \text{Project2Manifold}(f, p)$
- replace p by q' in S

Output: The real witness points of $f = 0$:
$S \cup \text{RealWitnessPoints}(\{f, H\})$

Usually the poor conditioning region where $\|\mathcal{J}_p^+\| > K$ appears close to a singular component or point (e.g. at the intersection of two irreducible components). Since the dimension of the singular set is lower than the dimension of regular set, the likelihood of leaving this poor conditioning region is quite large.

We implemented the algorithms in Maple 16 together with a Maple interface to Hom4Ps2. Although we can not verify the regularity assumption in advance, it can be detected if the perturbed system has no real solutions or the Jacobian is always near rank-deficiency during path tracking.

4. APPLICATION TO DAE

In this section we show how our real solving method can be applied to the consistent initialization of DAE in the Pryce-Pantelides method.

As shown in [25] that method is equivalent implicitly to a Riquier Basis (an object which could be computed for exact input by symbolic differential elimination algorithms). In this section we consider a crane control example. We compare the application of symbolic differential elimination with Pryce-Pantelides (coupled with our real solving method).

If successful it is very efficient since the prolongation step can be solved in polynomial time and an efficient polynomial cost method can be used to numerically solve the prolonged DAE [13]. The reader may wish to look ahead at the Table 1 below, where Pryce's method partnered by our real-solving method is much more efficient than a standard differential elimination method on a class of DAE. The reader should view this comparison cautiously as symbolic differential elimination algorithms yield more theoretically complete results, since they follow cases (and a radical differential membership result is available). Also they can apply to over-determined DAE and systems with multiplicities.

In fact as shown in our paper [25] the Pryce-Pantelides method produces an implicit form of a Riquier Basis without making potentially costly symbolic eliminations. Also the nonsingular Jacobians are precisely the conditions for the implicit function theorem, to transform Pryce's system into a Riquier Basis (see Theorem 6.12 in [25]). Those conditions are equivalent to A_1 and A_2.

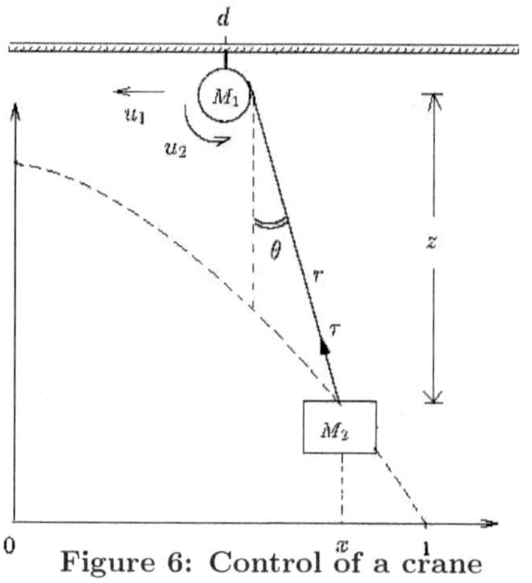

Figure 6: Control of a crane

Example 4.1 (Index 5 DAE for a crane).
This model which is illustrated in Figure 6, is discussed in [25]. The problem is to determine the horizontal velocity $u_1(t)$ and the angular velocity $u_2(t)$ of a winch of mass M_1, so that the attached load M_2 moves along a prescribed path – the dashed curve in Figure 6.

The equations of motion are given by Visconti [24] with unknowns $\{x, x', z, z', d, d', r, r', \theta, \tau, u_1, u_2\}$:

$$x_t - x' = 0, \quad z_t - z' = 0, \quad d_t - d' = 0, \quad r_t - r' = 0$$
$$M_2\, x'_t + \tau \sin(\theta) = 0, \quad M_1\, d'_t + C_1\, d_t - u_1 - \tau \sin(\theta) = 0$$
$$M_2\, z'_t + \tau \cos(\theta) - mg = 0, \quad J\, r'_t + C_2\, r_t + C_3\, u_2 - C_3{}^2 \tau = 0$$
$$r \sin(\theta) + d - x = 0, \quad r \cos(\theta) - z = 0$$
$$H_1(x, z, t) = 0, \quad H_2(x, z, t) = 0.$$

The prescribed path of the mass M_2 is described by an algebraic equations $\{H_1 = 0, H_2 = 0\}$. The winch has moment of inertia J and is attached with a cable of length $r(t)$, making an angle $\theta(t)$ to the vertical. Substituting $sin(\theta) = s(t)$ and $cos(\theta) = c(t)$ and appending $s(t)^2 + c(t)^2 = 1$ converts the DAE to a system of differential polynomials. Applying the Pryce-Pantelides method [25], we obtain 13 ODE and 39 algebraic constraints.

To illustrate how to find a real initial point for this DAE, we reuse the polynomial system of Example 2.4 with $\{H_1 = z(t)^2 - t = 0,\ H_2 = z(t)^2 + t^2 - \left(2\, x(t) - x(t)^2\right)^3 = 0\}$. Then the total Bézout degree of the constraints becomes 21233664, but it has a block triangular structure enabling its solutions by bottom up substitution.

Choosing the initial time t randomly as in [25], say $t = 4$ and applying the homotopy method Hom4Ps2 *yields 24 solutions. But all of them are complex. Since the real variety of the bottom block $H_1 = H_2 = 0$ is a bounded curve in the (x, z, t)-space as shown in Figure 5, it leads to a large chance of missing the curve when we apply a random real slicing used in our previous paper [25].*

The method we have presented in the current paper, however, can find real initial points for this example, with good condition as we explained already in Example 2.4. Further the Block structure of the system, enables us to easy verify that all the relevant Jacobians for the success of the method are non-singular, by efficient bottom up substitution. Equivalently the variety defined by the DAE satisfies our assumptions A_1 and A_2.

j	diff-elim (sec) (rifsimp)	fast prolongation + real solving (sec)	# real points
1	0.31	0.063+0.12	4
2	0.69	0.063+0.22	16
3	2.39	0.063+0.17	10
4	22.48	0.063+0.39	12
5	> 3 hr	0.063+0.28	10

Table 1: Times for the crane problem by symbolic differential elimination (rifsimp) and the Pryce method. Here $H_1 = x(t)^j + z(t)^j - x(t)z(t) - t - j, H_2 = z(t)^j + x(t)z(t) - t^j + j$ for $j = 1, 2, 3, 4, 5$. Exsecuted in Maple16 on a PC under Windows 7 with 4GB of RAM, I5 cpu at 2.5 GHz.

Moreover, the computational difficulty of this problem for the symbolic differential elimination algorithm Rifsimp explosively increases with the degree d of H_1, H_2 in comparison with the numerical method as shown in the Table 1.

5. CONCLUSIONS

In this paper we give a numerical method for computing witness points on real connected components for real polynomial systems satisfying the regularity assumptions A_1 and A_2. Given a random vector $\mathbf{n}$ the method computes local critical points of the distance of a hyperplane to the component in the direction $\mathbf{n}$. A method for regularizing singular critical points is also given.

This method takes advantage of the availability of efficient homotopy solvers which exploit sparsity and structure of the polynomial system to potentially significantly reduce the number of paths following in homotopy solving. The method is pleasingly parallelisable, since homotopy paths can be followed independently on different processors. Once a witness point is determined, additional points on the component, can be further generated by other homotopies.

We demonstrated the usefulness of our plane-distance method in the consistent initialization step of the Pryce-Pantelides method. In particular its regularity conditions are the same as those of Pryce-Pantelides.

Theoretically, if we take a point at infinity, Hauenstein's distance-point method [11] is equivalent to the plane-distance method. But our plane-distance method has the advantage that it is translation invariant. Thus its conditioning does not depend on the distance, as does the most closely related method, the distance-point method of Hauenstein. In addition, in contrast to the approach in [11] for singular critical points, we present a way to move away from a singularity by stepping along a perturbed component to obtain a regular point with hardware precision rather than tracking at singular endpoints with higher precision.

Future research includes loosening the assumptions A_1 and A_2, to address more degenerate cases. Such research includes methods for deflating higher multiplicity components, and also addressing problems with equations which are sums of squares. We will explore the relations with the closest method to ours, that of Hauenstein [11], especially as regards the effect on sparsity of the equation systems for these methods.

Acknowledgements

We thank professor T.Y. Li and Tsung-Lin Lee for their helpful suggestions and the work on Hom4ps2-Maple interface. We also especially thank one of the reviewers for their valuable comments.

6. REFERENCES

[1] S. Basu, R. Pollack and M-F Roy, Algorithms in real algebraic geometry. 2nd edition. Algorithms and Computation in Math., 10. Springer-Verlag, 2006.

[2] D.J. Bates, J.D. Hauenstein, A.J. Sommese, and C.W. Wampler, Numerically Solving Polynomial Systems with the Software Package Bertini. In preparation. To be published by SIAM, 2013.

[3] G.M Besana, S. DiRocco, J.D. Hauenstein, A.J. Sommese, and C.W. Wampler. Cell decomposition of almost smooth real algebraic surfaces. Num. Algorithms, DOI:10.1007/s11075-012-9646-y, (published online Sept 28, 2012).

[4] B. Buchberger, An Algorithm for Finding the Basis Elements of the Residue Class Ring of a Zero Dimensional Polynomial Ideal. Ph.D. dissertation, University of Innsbruck, 1965.

[5] C. Chen, J.H. Davenport, J.P. May, M.M. Maza, B. Xia and R. Xiao. *Triangular decomposition of semi-algebraic systems* Journal of Symbolic Computation Vol49-0, pp 3-26, 2013.

[6] C. Chen, M. Moreno Maza, B. Xia, L. Yang, Computing Cylindrical Algebraic Decomposition via Triangular Decomposition. Proc. of ISSAC 2009, pages 95-102, ACM Press, New York, 2009.

[7] G. Collins, Quantifier elimination for real closed fields by cylindrical algebraic decomposition. Springer Lec. Notes Comp. Sci. v33. 515-532, 1975.

[8] J.H. Davenport and J. Heintz, Real quantifier elimination is doubly exponential, J. Symbolic Comp., 5:29-35, 1988.

[9] D. Grigorév and N. Vorobjov. Counting connected components of a semialgebraic set in subexponential time. Comput. Complexity, 2(2), 133-186, 1992.

[10] Gunning, R. and Rossi, H. Analytic functions in several complex variables. Eaglewood Cliffs N.J., Prentice Hall, 1965.

[11] J. Hauenstein, Numerically computing real points on algebraic sets. Acta Appl. Math. DOI:10.1007/s10440-012-9782-3, (published online September 27, 2012).

[12] H. Hong. Improvement in CAD-Based Quantifer Elimination. Ph.D. thesis, the Ohio State University, Columbus, Ohio, 1990.

[13] S. Ilie, R.M. Corless, G. Reid, Numerical solutions of index-1 differential algebraic equations can be computed in polynomial time, Numerical Algorithms, 41(2), 161–171, 2006.

[14] T.Y. Li and Tsung-Lin Lee. HOMotopy method for solving Polynomial Systems, software at

`http://www.math.msu.edu/~li/Software.htm`

[15] J.B. Lasserre, M. Laurent and P. Rostalski. A prolongation-projection algorithm for computing the finite real variety of an ideal. Theoret. Comput. Sci., 410(27-29), 2685-2700, 2009.

[16] A. Leykin, J. Verschelde, and A. Zhao. Higher-order deflation for polynomial systems with isolated singular solutions. In Algorithms in algebraic geometry, IMA Volume 146, pages 79-97. Springer, New York, 2008.

[17] Y. Lu, Finding all real solutions of polynomial systems, Ph.D Thesis, University of Notre Dame, 2006. Results of this thesis appear in: (with D.J. Bates, A.J. Sommese, and C.W. Wampler), Finding all real points of a complex curve, Contemporary Mathematics 448 (2007), 183–205.

[18] C. Pantelides. The Consistent Initialization of Differential-Algebraic Systems, SIAM J. Sci. and Stat. Comput. Volume 9, Issue 2, pp. 213-231, 1988.

[19] F. Rouillier, M.-F. Roy, and M. Safey El Din. Finding at least one point in each connected component of a real algebraic set defined by a single equation. J. Complexity, 16 (4), 716-750, 2000.

[20] A. Sard, The measure of the critical values of differentiable maps, Bulletin of the American Mathematical Society 48 (12): 883-890, 1942.

[21] A. Seidenberg, A new decision method for elementary algebra, Ann. of Math. 60, 365-374, 1954.

[22] A.J. Sommese and C.W. Wampler. *The Numerical solution of systems of polynomials arising in engineering and science.* World Scientific Press, 2005.

[23] A. Tarski, A decision method for elementary algebra and geometry, Fund. Math. 17, 210-39, 1931.

[24] J. Visconti. *Numerical Solution of Differential Algebraic Equations, Global Error Estimation and Symbolic Index Reduction.* Ph.D. Thesis. Laboratoire de Modélisation et Calcul. Grenoble. 1999.

[25] W. Wu, G. Reid and S. Ilie, *Implicit Riquier Bases for PDAE and their semi-discretizations.* Journal of Symbolic Computation, Volume 44 Issue 7, Pages 923-941, 2009.

[26] X. Wu and L. Zhi. Computing the multiplicity structure from geometric involutive form, Journal of Symbolic Computation, 47(3): 227-238, 2012.

Gröbner Bases of Ideals Invariant under a Commutative Group: The Non-Modular Case

Jean-Charles Faugère, Jules Svartz
INRIA, Paris-Rocquencourt Center, PolSys Project
UPMC, Univ Paris 06, LIP6
CNRS, UMR 7606, LIP6
{Jean-Charles.Faugere,Jules.Svartz}@lip6.fr

ABSTRACT

We propose efficient algorithms to compute the Gröbner basis of an ideal $I \subset k[x_1, \ldots, x_n]$ globally invariant under the action of a commutative matrix group G, in the non-modular case (where $char(k)$ doesn't divide $|G|$). The idea is to simultaneously diagonalize the matrices in G, and apply a linear change of variables on I corresponding to the base-change matrix of this diagonalization. We can now suppose that the matrices acting on I are diagonal. This action induces a grading on the ring $R = k[x_1, \ldots, x_n]$, compatible with the degree, indexed by a group related to G, that we call G-degree. The next step is the observation that this grading is maintained during a Gröbner basis computation or even a change of ordering, which allows us to split the Macaulay matrices into $|G|$ submatrices of roughly the same size. In the same way, we are able to split the canonical basis of R/I (the staircase) if I is a zero-dimensional ideal. Therefore, we derive *abelian* versions of the classical algorithms F_4, F_5 or FGLM. Moreover, this new variant of F_4/F_5 allows complete parallelization of the linear algebra steps, which has been successfully implemented. On instances coming from applications (NTRU crypto-system or the Cyclic-n problem), a speed-up of more than 400 can be obtained. For example, a Gröbner basis of the Cyclic-11 problem can be solved in less than 8 hours with this variant of F_4. Moreover, using this method, we can identify new classes of polynomial systems that can be solved in polynomial time.

Categories and Subject Descriptors

I.1.2 [**Computing Methodologies**]: Symbolic and Algebraic Manipulation—*Algorithms*

Keywords

Gröbner Basis, Invariant Ideals, Group Action

1. INTRODUCTION

Solving multivariate polynomial systems is a fundamental problem in Computer Algebra, since algebraic systems can arise from many applications (cryptology, robotics, biology, physics, coding theory, etc...). One method to solve such systems is based on Gröbner basis theory. Efficient algorithms to compute Gröbner bases have been proposed, for instance Buchberger's algorithm [1] and Faugère's F_4 or F_5 [4, 5]. If the system has only a finite number of solutions the usual strategy is to compute a Gröbner basis for the DRL ordering, and then perform a change of ordering to obtain a Gröbner basis for the lexicographic ordering with the FGLM algorithm [10]. However, problems coming from applications are often highly structured: in several algebraic problems the set of solutions (the algebraic variety) is invariant under the action of a finite group. The underlying algebraic problem is to compute the variety $V(I)$ associated to an ideal $I \subseteq k[x_1, \ldots, x_n]$ that is *globally* stable under a finite matrix group $G \subset \mathrm{GL}_n(k)$, which means that $\forall f \in I \quad \forall A \in G \quad f^A \in I$. If all the equations are invariant under the action of the group, several approaches have been proposed to solve the system while taking the symmetries into account. In [2] Colin proposes to use invariants [19] to solve the system. This method is very efficient if the Hironaka Decomposition of the ring of invariants is simple, but for the Cyclic-n problem [12] for example, it seems better to use a second method based on SAGBI Gröbner Basis techniques [7]. However, it remains an open issue to solve efficiently the system in the general case. In the biology problem [6] or in the physics problem [9], an approach has been proposed if the group G is the symmetric group or copies of the symmetric group (elements of the form $(\sigma, \ldots, \sigma) \in \mathfrak{S}_m^j$ with $mj = n$.)

MAIN RESULTS. We present efficient algorithms together with complexity analysis to solve polynomial systems which are *globally invariant* under the action of any *commutative* group G. The algorithms are based on three main ideas: first, since the group G is commutative, it is possible to diagonalize the group G, assuming that the characteristic of the field k and $|G|$ are coprime. Thus, up to some linear change of variables, we obtain an ideal $I_{\mathscr{D}}$ invariant under a diagonal group $G_{\mathscr{D}}$ isomorphic to G.

The second idea is to introduce a grading on $R = k[x_1, \ldots, x_n]$ given by the group $G_{\mathscr{D}}$. This grading exists for every finite group H and is indexed on $X(H)$, the set of irreducible linear representations of the group H. The decomposition $R = \bigoplus_{\chi \in X(H)} R_\chi$ is known as the decomposition of R into *isotypic components* (see [17]). In our case, since $G_{\mathscr{D}}$ is diagonal, the set $X(G_{\mathscr{D}})$ is isomorphic to $G_{\mathscr{D}}$ and the isotypic components are generated by monomials. Therefore, we introduce the notion of $G_{\mathscr{D}}$-degree of a polynomial: assuming that $G_{\mathscr{D}}$ is generated by diagonal matrices $Diag(\beta_{i,1}, \ldots, \beta_{i,n})$ of order q_i with $q_1|q_2|\ldots|q_\ell = e$ and that β is a primitive e-root of 1, we say that a polynomial $f \in k[x_1, \ldots, x_n]$ is $G_{\mathscr{D}}$-homogeneous of $G_{\mathscr{D}}$-degree $(d_1, \ldots, d_\ell) \in \mathbb{Z}_{q_1} \times \cdots \times \mathbb{Z}_{q_\ell}$ if $f(\beta_{i,1}x_1, \cdots, \beta_{i,n}x_n) = \beta^{d_i \frac{e}{q_i}} f(x_1, \ldots, x_n)$ for all i. Notice that the action of diagonal groups on polynomials has been used in invariant theory or to speed up Gröbner basis computation in [17, 19, 18, 13]. However, to the best of our knowledge, the impact of such a grading on the complexity of Gröbner bases has not been studied.

Taking into account that the operation of taking the S-polynomial preserves this grading, the final idea is to observe that this can be used to speed up the Gröbner basis computation. More precisely, Macaulay matrix can be decomposed into $|G_{\mathscr{D}}|$ smaller *in-*

ISSAC'13, June 26–29, 2013, Boston, Massachusetts, USA.

dependent matrices, being roughly the same size. In particular, this allows us to split the matrices arising in classical Gröbner basis algorithms based on linear algebra like Macaulay/Lazard algorithm [15], F_4 [4] or F_5 [5]. Therefore, the complexity (in time and in memory) of computing Gröbner bases of such invariant ideals can be decreased in both, theory and practice. In the same way, in the case of a zero-dimensional ideal $I_{\mathscr{D}}$, the canonical basis of the ring $R/I_{\mathscr{D}}$ can also be decomposed in monomials having same $G_{\mathscr{D}}$-degree and thus we are able to split the multiplication matrices arising in FGLM.

In addition, this grading can be used to transform very easily a globally invariant problem into a problem for which all the equations are $G_{\mathscr{D}}$-homogeneous: we show that for each original equation f we can take the $G_{\mathscr{D}}$-homogeneous components of f.

We have implemented, in the computer algebra system Magma, "abelian" versions of the F_5 and $FGLM$ algorithms that run several times faster, compared to the same implementation of these classical algorithms. For example, applying FGLM on the Cyclic-10 problem (a system with 34940 solutions), instead of computing 10 multiplication matrices of size 34940, our algorithm computes 900 quasi-square matrices of size at most 348.

In order to compare similar implementations, we have implemented an "abelian" version of F_4 [4] in FGb (C language): computing a Gröbner basis of the Cyclic-10 problem is about 410 times faster with the new approach. Moreover, a grevlex Gröbner basis for the Cyclic-11 problem (184756 solutions) can be computed in less than 8 hours. We also demonstrate that our approach has a significant impact in other fields: NTRU is a well known cryptosystem and the underlying problem can easily be modeled by quadratic equations which are left globally invariant by the action of a cyclic group. We observe a factor of 250 in favor of the new approach for small size problems and more importantly we can solve previously untractable problems. Surprisingly, during these experiments, the linear algebra parts (that is building the matrices and the gaussian elimination parts) can sometimes be so accelerated that the management of the list of critical pairs becomes the most time-consuming part whereas it is usually negligible.

More generally, the algorithms given in this paper can also be used for other kinds of structured polynomial systems like quasi-homogeneous or multi-homogeneous polynomials. Hence we now have a systematic and uniform approach to solve those structured problems. Several further developments can be made on the subject: the Abelian-F_5 and Abelian-FGLM algorithms have to be implemented in C, and it seems possible to obtain a parallelized version of the Abelian-FGLM algorithm. We have already identified new classes of invariant problems which can be solved in polynomial time; for other classes of problems the degree reached during the Gröbner basis computation is much lower than expected and it would be very useful to compute explicitly the Hilbert Series of ideals invariant under a diagonal group.

The organization of the paper is as follows: in section 2, we recall classical notations and explain the relations between the ideals I and $I_{\mathscr{D}}$, and the matrix groups G and $G_{\mathscr{D}}$. In section 3, we explain the grading induced by the diagonal matrix group $G_{\mathscr{D}}$, and introduce the notion of $G_{\mathscr{D}}$-degree of monomials and polynomials. The vector space generated by all monomials having same $G_{\mathscr{D}}$-degree is nothing else than an *isotypic component* ([17]) but since the formulation is simpler in the case of a diagonal group, we introduce the notion of $G_{\mathscr{D}}$-degree of monomials and $G_{\mathscr{D}}$-homogeneous polynomials. Sections 4 and 5 provide variants of the F_5 and FGLM algorithms. The complexity questions are answered in section 6, and benchmarks are made in section 7.

2. LINEAR CHANGE OF VARIABLES

2.1 Frequently used notations

From now on we assume that G is a finite commutative subgroup of $GL_n(k)$, the set of square matrices with coefficients in a field k of characteristic 0 or p such that p and $|G|$ are coprime. $G_{\mathscr{D}}$ will be used to denote a diagonal matrix group, conjugated to G. $R_k = k[x_1,\dots,x_n]$ is the ring of polynomials with coefficients in k. In the following, we will have to consider a finite simple extension of k that will be denoted $K = k(\xi)$. The set of monomials of R_k (or R_K) will be denoted $\mathscr{M}$. We fix an admissible monomial ordering $\preceq$ on the set of monomials (only admissible orderings are allowed, for a precise definition, we refer to [3] p. 53). For a given degree d, $\mathscr{M}_d$ will be the set of all monomials in R of degree d. For a polynomial in R, $\mathsf{LC}(f)$ (resp $\mathsf{LM}(f)$, $\mathsf{LT}(f)$) denotes the leading coefficient (resp leading monomial, leading term) in f. We have the relation $\mathsf{LT}(f) = \mathsf{LC}(f)\mathsf{LM}(f)$.

2.2 Action of $GL_n(k)$ on $k[x_1,\dots,x_n]$. Invariant rings.

This subsection describes the basic properties of the action of $GL_n(k)$ on polynomials. We recall that G is a finite subgroup of $GL_n(k)$. Let X be the column vector whose entries are $x_1,\dots,x_n$. For f a polynomial in R and $A \in G$, let f^A be the polynomial obtained by substituting the components of $A.X$ to $x_1,\dots,x_n$. Since $(f^A)^B = f^{AB}$, we obtain an action of G on R. Let R_d be the vector space of all homogeneous polynomials of degree d. Then $R = \bigoplus_{d=0}^{\infty} R_d$ and we observe that the action of G preserves the homogeneous components.

Definition 1 *We denote by R^G the set of* invariant polynomials, *that means polynomials invariant under the action of G : $f^A = f$ for every A in G.*

Although we won't work exclusively in the ring R^G of invariant polynomials, we will use several known properties of this set, especially in the complexity section.

Example 1 *The symmetric group $\mathfrak{S}_n$ can be embedded in $GL_n(k)$, and $R^{\mathfrak{S}_n}$ is nothing else than the set of the so called* symmetric polynomials. *Let C_n be the subgroup of $\mathfrak{S}_n$ generated by the n-cycle $\sigma = (12\dots n)$. C_n is a cyclic group of order n, embedded in $GL_n(k)$ and generated by M_σ.*

$$M_\sigma = \begin{pmatrix} 0 & 1 & 0 & \cdots & 0 \\ 0 & 0 & 1 & \cdots & 0 \\ \vdots & \vdots & \vdots & \ddots & \vdots \\ 0 & 0 & 0 & \cdots & 1 \\ 1 & 0 & 0 & \cdots & 0 \end{pmatrix}$$

For example if $n = 3$ then $x_1^2x_2 + x_2^2x_3 + x_3^2x_1$ belongs to $R^{C_n} \backslash R^{\mathfrak{S}_n}$.

2.3 From commutative group to diagonal group

This subsection presents one of the main ideas of the paper, although it is very simple. We recall some well known facts about commutative matrix groups.

Theorem 1 *Any finite commutative group is uniquely isomorphic to a product $\mathbb{Z}/q_1\mathbb{Z} \times \cdots \times \mathbb{Z}/q_\ell\mathbb{Z}$ with $q_1 | \dots | q_\ell$.*

Definition-Proposition 1 *Following the notations of the previous theorem, the integer $e = q_\ell$ is called the* exponent *of the group and is the lowest common multiple of the orders of the elements of the group.*

Theorem 2 *Let G be a finite commutative matrix group, and e be its exponent. Let ξ be a primitive e-th root of 1, in an extension of k and $K = k(\xi)$. The subgroup G is diagonalizable over K, meaning that there exists a matrix P in $GL_n(K)$, such that the group $G_{\mathscr{D}} = P^{-1}GP = \{P^{-1}AP \mid A \in G\}$ is a diagonal group.*

PROOF. Every matrix $A \in G$ satisfies the polynomial $X^e - 1$, which fully splits in K and has simple roots since $char(k) \not| |G|$, so every matrix of G is diagonalizable, and it is well known that a commutative set of diagonalizable matrices is codiagonalizable. □

Example 2 *Let k be any field of characteristic 0 or coprime with n. Then if we denote $K = k(\xi)$ where ξ is a primitive n-root of 1 in an extension of k, then the cyclic group C_n defined in example 1 is diagonalizable with the base-change matrix $P = (\xi^{ij})_{i,j \in \{1,\dots,n\}}$. The matrix associated to the cycle $(1\dots n)$ becomes the diagonal matrix $D_\sigma = diag(\xi,\dots,\xi^{n-1},1)$.*

Definition 2 *Let I be an ideal in $R_k = k[x_1,\dots,x_n]$. I is said to be* stable *under the action of G (G-*stable*) if: $\forall f \in I, \forall A \in G \quad f^A \in I$*

Proposition 1 *Let I be a G-stable ideal, and let $G_{\mathscr{D}}$ and P be the diagonal group and the base-change matrix obtained in theorem 2. Then $I_{\mathscr{D}} = K \bigotimes_k \{f^P, f \in I\}$ is an ideal of R_K stable under $G_{\mathscr{D}}$. If $I = \langle f_1,\dots,f_m\rangle_{R_k}$, then $I_{\mathscr{D}} = \langle f_1^P,\dots,f_m^P\rangle_{R_K}$.*

Example 3 *To illustrate the definition, we will use the well known* Cyclic-n problem. *The ideal I of R_k is generated by:*

$$\begin{cases} h_1 = x_1 + \cdots + x_n \\ h_2 = x_1x_2 + x_2x_3 + \cdots + x_nx_1 \\ \vdots \\ h_{n-1} = x_1x_2\dots x_{n-1} + x_2\dots x_nx_1 + \cdots + x_nx_1\dots x_{n-2} \\ h_n = x_1x_2\dots x_{n-1}x_n - 1 \end{cases}$$

The ideal I is obviously invariant under the cyclic group C_n, since each h_i satisfies $h_i^{M_\sigma} = h_i$ and is also stable under the scalar matrix ξI_n with ξ a primitive n-root of 1, since $h_i^{\xi I_n} = \xi^i h_i$. The group G is generated by M_σ and ξI_n. With P the matrix given in example 2, $G_{\mathscr{D}} = P^{-1}GP$, generated by D_σ and ξI_n, is a diagonal group isomorphic to $\mathbb{Z}/n\mathbb{Z} \times \mathbb{Z}/n\mathbb{Z}$. We denote by f_i the polynomials h_i^P, which generate $I_{\mathscr{D}}$: for instance, $f_1 = 3x_3, f_2 = -3x_1x_2 + 3x_3^2, f_3 = x_1^3 + x_2^3 + 3x_1x_2x_3 + x_3^3 - 1$ when $n = 3$. It is easy to prove that for the Cyclic-n problem, the polynomial f_1 is always equal to nx_n.

3. GRADING INDUCED BY A DIAGONAL MATRIX GROUP

In this section, we define the $G_{\mathscr{D}}$-degree of a monomial where $G_{\mathscr{D}}$ is a diagonal matrix group. This $G_{\mathscr{D}}$-degree induces a grading of R_K given by the isomorphism $G_{\mathscr{D}} \simeq \prod \mathbb{Z}/q_i\mathbb{Z}$.

3.1 $G_{\mathscr{D}}$-degree of monomials

Let $G_{\mathscr{D}}$ be a diagonal group of $GL_n(K)$, with diagonal coefficients in $\mathbb{U}_e = \{\xi^0, \xi^1, \dots, \xi^{e-1}\}$, with e the exponent of G and ξ a primitive e-root of 1, as defined in the previous section. Let ϕ be an isomorphism

$$\phi : \begin{pmatrix} G_{\mathscr{D}} & \longrightarrow & \mathbb{Z}/q_1\mathbb{Z} \times \cdots \times \mathbb{Z}/q_\ell\mathbb{Z} \\ D & \longmapsto & \phi(D) \end{pmatrix}$$

and let D_i be the preimage of $(0,\dots,0,\underset{i}{1},0,\dots,0)$, so D_i generates a subgroup of $G_{\mathscr{D}}$ of cardinality $|q_i|$.

Example 4 *With $G_{\mathscr{D}}$ the group arising in the previous example 3, we take ϕ such that $\phi(D_\sigma) = (1,0) \in \mathbb{Z}/n\mathbb{Z} \times \mathbb{Z}/n\mathbb{Z}$ and $\phi(\xi I_n) = (0,1)$.*

Proposition 2 *For every monomial $m \in \mathscr{M}$ and for each i, there exists a unique $\mu_i \in \{0,\dots,q_i-1\}$ such that $m^{D_i} = \xi^{\frac{e}{q_i}\mu_i} m$.*

PROOF. Let $m = \prod x_j^{\alpha_j}$ and $D_i = Diag(\beta_1,\dots,\beta_n)$. Since D_i has order q_i, the coefficients β_j are q_i-roots of 1, so can be denoted $\xi^{\ell_j \frac{e}{q_i}}$. Then

$$m_i^D = (\beta_1 x_1)^{\alpha_1} \times \cdots \times (\beta_n x_n)^{\alpha_n} = \left(\prod \beta_j^{\alpha_j}\right) m = \xi^{\frac{e}{q_i}\sum \ell_j\alpha_j} m$$

Then we can take $\mu_i = \sum \ell_j \alpha_j \bmod q_i$. Since ξ has order e, $\xi^{\frac{e}{q_i}}$ has order q_i and the unicity of μ_i is clear. □

Instead of considering μ_i in $\{0,\dots,q_i-1\}$, we take μ_i in $\mathbb{Z}/q_i\mathbb{Z}$, which makes sense since $\xi^{\frac{e}{q_i}}$ has order q_i.

Definition 3 *The k-tuple $(\mu_1,\dots,\mu_k) \in \prod \mathbb{Z}/q_i\mathbb{Z}$ is said to be the* $G_{\mathscr{D}}$-degree *of m and is denoted $\mathsf{deg}_{G_{\mathscr{D}}}(m)$, although it depends on the choice of the matrices D_i (more exactly, the choice of ϕ). We denote by $\hat{G} = \prod \mathbb{Z}/q_i\mathbb{Z}$ the set of all $G_{\mathscr{D}}$-degrees.*

Remark 1 *It is yet unclear that every $\mu \in \hat{G}$ is the $G_{\mathscr{D}}$-degree of some monomial. This will be proved in the complexity section.*

Proposition 3 *Since $\mathsf{deg}_{G_{\mathscr{D}}}(m) + \mathsf{deg}_{G_{\mathscr{D}}}(m') = \mathsf{deg}_{G_{\mathscr{D}}}(mm')$ for all $m, m' \in \mathscr{M}$, R can be graded by $R = \bigoplus_{g\in\hat{G}} Vect(\mathscr{M}_g)$, where $\mathscr{M}_g$ is the set of monomials of $G_{\mathscr{D}}$-degree g.*

PROOF. Let $i \in \{1,\dots,k\}$, $m, m' \in \mathscr{M}$ and μ_i, μ_i' such that $m^{D_i} = \xi^{\frac{e}{q_i}\mu_i} m$ and $m'^{D_i} = \xi^{\frac{e}{q_i}\mu_i'} m'$. Then $(mm')^{D_i} = \xi^{\frac{e}{q_i}(\mu_i+\mu_i')} mm'$. It follows that the $G_{\mathscr{D}}$-degree verifies $\mathsf{deg}_{G_{\mathscr{D}}}(mm') = \mathsf{deg}_{G_{\mathscr{D}}}(m) + \mathsf{deg}_{G_{\mathscr{D}}}(m')$ for all monomials $m, m' \in \mathscr{M}$ and $\mathsf{deg}_{G_{\mathscr{D}}}$ is a monoid morphism between $\mathscr{M}$ and $\hat{G}$ since $\mathsf{deg}_{G_{\mathscr{D}}}(1) = (0,\dots,0)$. □

Remark 2 *If we denote by $\mathscr{M}_{d,g}$ the set of monomials of degree d and $G_{\mathscr{D}}$-degree g, $\mathscr{M}_{d,g}\mathscr{M}_{d',g'} \subseteq \mathscr{M}_{d+d',g+g'}$ for all d, d', g, g'. Therefore $R = \bigoplus_{d\in\mathbb{N}, g\in\hat{G}} Vect(\mathscr{M}_{d,g})$.*

Notice that for computing $\mathsf{deg}_{G_{\mathscr{D}}}(m)$ with $m = \prod x_i^{\alpha_i}$, we just have to know $\mathsf{deg}_{G_{\mathscr{D}}}(x_i)$ since $\mathsf{deg}_{G_{\mathscr{D}}}(m) = \sum \alpha_i \mathsf{deg}_{G_{\mathscr{D}}}(x_i)$. This grading will be used to reduce the sizes of the matrices in the Diagonal-F_5 algorithm.

Example 5 *Let $G_{\mathscr{D}}$ be the matrix group generated by the diagonal matrix $D_\sigma = Diag(\xi, \xi^2, 1)$ where ξ is a primitive third root of 1. Each x_i has $G_{\mathscr{D}}$-degree i mod 3, so $m = \prod x_j^{\alpha_j}$ has $G_{\mathscr{D}}$-degree $\sum j\alpha_j$ mod 3. Hence, $x_1x_2x_3$ (resp. $x_1x_2^2$) has $G_{\mathscr{D}}$-degree 0 (resp. 2).*

Example 6 *(cont. of example 3) The $G_{\mathscr{D}}$-degree of x_i is $(i,1)$.*

3.2 $G_{\mathscr{D}}$-homogeneous polynomials

In this subsection, we define the notion of $G_{\mathscr{D}}$-homogeneity. The cornerstone of the Abelian-F_5 algorithm (section 4) is that the S-polynomial of two $G_{\mathscr{D}}$-homogeneous polynomials is $G_{\mathscr{D}}$-homogeneous, which will be proved in theorem 3.

Definition 4 *A polynomial f in R_K is said to be $G_{\mathscr{D}}$-*homogeneous *if all monomials of f share the same $G_{\mathscr{D}}$-degree $(\mu_1,\dots,\mu_k) \in \hat{G}$. In this case, we set $\mathsf{deg}_{G_{\mathscr{D}}}(f) = \mathsf{deg}_{G_{\mathscr{D}}}(\mathsf{LM}(f))$.*

Proposition 4 *If f is $G_{\mathscr{D}}$-homogeneous and m is a monomial, then mf is $G_{\mathscr{D}}$-homogeneous. Moreover, $\mathsf{deg}_{G_{\mathscr{D}}}(mf) = \mathsf{deg}_{G_{\mathscr{D}}}(m) + \mathsf{deg}_{G_{\mathscr{D}}}(f)$.*

PROOF. For any monomial $\tilde{m}$ of f, $\mathsf{deg}_{G_{\mathscr{D}}}(\tilde{m}m) = \mathsf{deg}_{G_{\mathscr{D}}}(\tilde{m}) + \mathsf{deg}_{G_{\mathscr{D}}}(m) = \mathsf{deg}_{G_{\mathscr{D}}}(f) + \mathsf{deg}_{G_{\mathscr{D}}}(m)$, so all monomials of mf share the same $G_{\mathscr{D}}$-degree $\mathsf{deg}_{G_{\mathscr{D}}}(f) + \mathsf{deg}_{G_{\mathscr{D}}}(m) = \mathsf{deg}_{G_{\mathscr{D}}}(mf)$. □

Theorem 3 *Let f, g be two $G_{\mathscr{D}}$-homogeneous polynomials of R_K. The S-polynomial of (f,g), defined by*

$$S(f,g) = \frac{\mathsf{LM}(f) \vee \mathsf{LM}(g)}{\mathsf{LM}(f)} f - \frac{\mathsf{LM}(f) \vee \mathsf{LM}(g)}{\mathsf{LM}(g)} \frac{\mathsf{LC}(f)}{\mathsf{LC}(g)} g$$

is $G_{\mathscr{D}}$-homogeneous of $G_{\mathscr{D}}$-degree $\mathsf{deg}_{G_{\mathscr{D}}}(\mathsf{LM}(f) \vee \mathsf{LM}(g))$. ($\mathsf{LM}(f) \vee \mathsf{LM}(g)$ denotes the lowest common multiple of $\mathsf{LM}(f)$ and $\mathsf{LM}(g)$.)

PROOF. Since $\mathsf{LM}(f)$ and $\mathsf{LM}(g)$ divide $\mathsf{LM}(f) \vee \mathsf{LM}(g)$, both fractions $\frac{\mathsf{LM}(f)\vee\mathsf{LM}(g)}{\mathsf{LM}(f)}$ and $\frac{\mathsf{LM}(f)\vee\mathsf{LM}(g)}{\mathsf{LM}(g)}$ are monomials, therefore by previous proposition, $\frac{\mathsf{LM}(f)\vee\mathsf{LM}(g)}{\mathsf{LM}(g)} \frac{\mathsf{LC}(f)}{\mathsf{LC}(g)} g$ and $\frac{\mathsf{LM}(f)\vee\mathsf{LM}(g)}{\mathsf{LM}(f)} f$ are $G_{\mathscr{D}}$-homogeneous. Moreover, they share the same leading monomial, so they have same $G_{\mathscr{D}}$-degree, which is the $G_{\mathscr{D}}$-degree of $S(f,g)$. We actually proved that $\mathsf{deg}_{G_{\mathscr{D}}}(S(f,g)) = \mathsf{deg}_{G_{\mathscr{D}}}(\mathsf{LM}(f) \vee \mathsf{LM}(g))$. □

Example 7 *Following example 3, it appears that each f_i has $G_{\mathscr{D}}$-degree $(0,i) \in \mathbb{Z}/n\mathbb{Z} \times \mathbb{Z}/n\mathbb{Z}$ under $G_{\mathscr{D}}$ generated by D_σ and ξI_n.*

3.3 $G_{\mathscr{D}}$-homogeneous ideals

In this subsection, $G_{\mathscr{D}}$ is a diagonal group, and $I_{\mathscr{D}}$ is a $G_{\mathscr{D}}$-stable ideal generated by $f_1, \ldots, f_m$. A Gröbner basis computation preserves the $G_{\mathscr{D}}$-degree, but the polynomials f_i are not necessarily $G_{\mathscr{D}}$-homogeneous. Our aim here is to prove that the $G_{\mathscr{D}}$-homogeneous components of the f_i are in $I_{\mathscr{D}}$, and so to compute a Gröbner basis of $I_{\mathscr{D}}$, we take the $G_{\mathscr{D}}$-homogeneous components of generators of $I_{\mathscr{D}}$ as inputs. This operation has a negligible cost since at each degree d, the abelian-F_5 algorithm (presented in the next section) separates the set $\mathscr{M}_d$ into subsets $\mathscr{M}_{d,g}$ of same $G_{\mathscr{D}}$-degree g.

Definition 5 *An ideal J of R_K is said to be* $G_{\mathscr{D}}$-homogeneous *iff for any polynomial $f \in J$, its $G_{\mathscr{D}}$-homogeneous components are also in J.*

Theorem 4 *An ideal is $G_{\mathscr{D}}$-homogeneous if and only if it is $G_{\mathscr{D}}$-stable.*

It is obvious that a $G_{\mathscr{D}}$-homogeneous ideal is $G_{\mathscr{D}}$-stable. To prove the other implication, we will first prove a lemma.

Lemma 1 *Let $f \in I_{\mathscr{D}}$, and $D \in G_{\mathscr{D}}$, then the $\langle D \rangle$-homogeneous components of f are in $I_{\mathscr{D}}$, where $\langle D \rangle$ is the subgroup generated by D.*

PROOF. Let q be the order of D in $G_{\mathscr{D}}$, and $\xi_D = \xi^{\frac{e}{q}}$. Then, all diagonal coefficients of D are powers of ξ_D. f can be written $\sum_{j=0}^{q-1} f_j$, with $f_j^D = \xi_D^j f_j$; in other words, the f_j are the $\langle D \rangle$-homogeneous components of f. Let $X_f = {}^t(f_0, f_1, \ldots, f_{q-1})$, $V = (\xi_D^{ij})_{0 \leq i,j \leq q-1}$, and $Y_f = V X_f$. Since $f_j^{D^i} = \xi_D^{ij} f_j$, the column vector Y_f is equal to ${}^t(f, f^D, \ldots, f^{D^{q-1}})$. Since $f \in I_{\mathscr{D}}$ and $I_{\mathscr{D}}$ is $G_{\mathscr{D}}$-stable, all components of Y_f belong to $I_{\mathscr{D}}$. But V is a VanDerMonde invertible matrix, so the components of X_f are obtained from Y_f by linear combinations, and the f_j belong to $I_{\mathscr{D}}$. □

PROOF. We now prove theorem 4 by induction on ℓ where $G \simeq \mathbb{Z}/q_1\mathbb{Z} \times \cdots \times \mathbb{Z}/q_\ell\mathbb{Z}$: the case $\ell = 1$ is the lemma. Now assume that $\ell \geq 2$ and let D_i be the matrices generating $G_{\mathscr{D}}$ as defined in section 2. Let $f \in I_{\mathscr{D}}$. By the lemma, the $\langle D_\ell \rangle$-homogeneous components of f are in $I_{\mathscr{D}}$. Denote by $\tilde{G_{\mathscr{D}}}$ the subgroup of $G_{\mathscr{D}}$ generated by $D_1, \ldots, D_{\ell-1}$, then $\tilde{G_{\mathscr{D}}} \simeq \mathbb{Z}/q_1\mathbb{Z} \times \cdots \times \mathbb{Z}/q_{\ell-1}\mathbb{Z}$, and $I_{\mathscr{D}}$ is also $\tilde{G_{\mathscr{D}}}$-stable, and by induction the $\tilde{G_{\mathscr{D}}}$-homogeneous components of each $\langle D_\ell \rangle$-homogeneous component of f are in $I_{\mathscr{D}}$, but they are exactly the $G_{\mathscr{D}}$-homogeneous components of f. □

Remark 3 *In representation theory, the $G_{\mathscr{D}}$-homogeneous components of a polynomial in $I_{\mathscr{D}}$ are the images of the projections onto each isotypic component. [17]*

Example 8 *Let $G_{\mathscr{D}}$ be the diagonal group of order 2 generated by the matrix $diag(-1,1)$, acting on $R = k[x_1, x_2]$. Suppose that $x_1^3x_2 + x_1^2x_2^2 - x_1 + 1 \in I_{\mathscr{D}}$, with $I_{\mathscr{D}}$ a $G_{\mathscr{D}}$-stable ideal. Then since $\deg_{G_{\mathscr{D}}}(x_i) = i \bmod 2$, $\deg_{G_{\mathscr{D}}}(x_1^3x_2) = \deg_{G_{\mathscr{D}}}(x_1) = 1$ and $\deg_{G_{\mathscr{D}}}(1) = \deg_{G_{\mathscr{D}}}(x_1^2x_2^2) = 0$, so $x_1^3x_2 - x_1$ and $x_1^2x_2^2 + 1$ belong to $I_{\mathscr{D}}$.*

4. ABELIAN-F_5 ALGORITHM

Now, we are able to describe the Abelian-F_5 algorithm, which is a variant of F_5 that takes advantage of the action of the abelian group $G_{\mathscr{D}}$. As usual, $I_{\mathscr{D}}$ is a $G_{\mathscr{D}}$-stable ideal, with $G_{\mathscr{D}}$ a diagonal group isomorphic to $\hat{G}$, the set of $G_{\mathscr{D}}$-degrees. Let $f_1, \ldots, f_m$ be $G_{\mathscr{D}}$-homogeneous polynomials generating $I_{\mathscr{D}}$ (according to theorem 4). All computation of the reduced Gröbner basis of $I_{\mathscr{D}}$ would implicitly use the grading $R = \bigoplus_{g \in \hat{G}} R_g$ since it computes S-polynomials. There exist several versions of the F_5-algorithm (see [7, 5]), we present here a variant of the matrix version. The F_5-algorithm constructs matrices degree by degree. At a fixed degree d, it constructs m matrices of the form $M_{d,i}$ for each i between 1 and m, and performs row reduction on them to obtain $\tilde{M}_{d,i}$. In the homogeneous case, $\tilde{m}_1, \ldots, \tilde{m}_\nu$ are all monomials of degree d, whereas in the affine case, they are all monomials of degrees between 0 and d. For the sake of simplicity, we assume that all polynomials f_i are homogeneous. The rows are indexed by couples of the form $m_\mu f_j$, the matrix $M_{d,i}$ is deduced from the matrix $\tilde{M}_{d,i-1}$ by adding all rows $m_\mu f_i$ with m_μ describing the set of monomials of degree $d - \deg(f_i)$, except monomials removed by the F_5-criterion (see [7, 5]).

$$
M_{d,i} = \begin{array}{c} \\ m_1 f_1 \\ \vdots \\ m_\mu f_j \\ \vdots \\ m_\gamma f_i \end{array}
\begin{array}{c} \begin{array}{cccc} \tilde{m}_1 & \tilde{m}_2 & \ldots & \tilde{m}_\nu \end{array} \\
\begin{pmatrix} \cdots & \cdots & \cdots & \cdots \\ & & & \\ \cdots & \cdots & \cdots & \cdots \\ \cdots & \cdots & \cdots & \cdots \\ & & & \\ \cdots & \cdots & \cdots & \cdots \\ \cdots & \cdots & \cdots & \cdots \end{pmatrix} \end{array}
$$

The key of the Abelian-F_5 algorithm is the following : the polynomials f_i are $G_{\mathscr{D}}$-homogeneous, and also the polynomials $m_\mu f_i$. Therefore, the only non-zero coefficients of the row indexed by $m_\mu f_i$ are on columns indexed by monomials having same $G_{\mathscr{D}}$-degree. So, instead of building one Macaulay matrix $M_{d,i}$, we will construct $|G_{\mathscr{D}}|$ matrices $M_{d,i,g}$, for all $g \in \hat{G}$.

Abelian-F_5 (homogeneous-case)

Input: The set $\hat{G}$ of $G_{\mathscr{D}}$-degrees, homogeneous and $G_{\mathscr{D}}$-homogeneous polynomials $(f_1, \ldots, f_m)$ with degrees $d_1 \leq \ldots \leq d_m$ and a maximal degree D.
Output: the elements of degree at most D of a Gröbner basis of $(f_1, \ldots, f_i)$ for $i = 1, \ldots, m$.

for i **from** 1 **to** m **do** $\mathscr{G}_i := \emptyset$ **end for**
for d **from** d_1 **to** D **do**
  for g **in** $\hat{G}$ **do**
    $M_{d,0,g} := \emptyset$, $\tilde{M}_{d,0,g} := \emptyset$
    for i **from** 1 **to** m **do**
      case
        $d < d_i$) $M_{d,i,g} := \tilde{M}_{d,i-1,g}$
        $d = d_i$) **if** $g = \deg_{G_{\mathscr{D}}}(f_i)$ **then**
          $M_{d,i,g} :=$ add new row f_i to $\tilde{M}_{d,i-1,g}$ with index $(i,1)$
          else
          $M_{d,i,g} := \tilde{M}_{d,i-1,g}$
          end if
        $d > d_i$) $M_{d,i,g}$:=add new row $m.f_i$ for all monomials m of degree $d - d_i$ with $\deg_{G_{\mathscr{D}}}(m) = g - \deg_{G_{\mathscr{D}}}(f_i)$ that do not appear as leading monomials in the matrix $\tilde{M}_{d-d_i,i-1,u-\deg_{G_{\mathscr{D}}}(f_i)}$ to $\tilde{M}_{d,i-1,g}$ with index (i,m).
      end case
      Compute $\tilde{M}_{d,i,g}$ by Gaussian elimination from $M_{d,i,g}$.
      Add to $\mathscr{G}_i$ all rows of $\tilde{M}_{d,i,g}$ not reducible by $\mathsf{LM}(\mathscr{G}_i)$.
    end for
  end for
end for
return $\mathscr{G}_1, \cdots, \mathscr{G}_m$

Notice that all the loops on $g \in \hat{G}$ are independent, so at each degree d, it is possible to parallelize on $|G|$ different processors to speed up the computations. Assuming that the degrees of the *primary invariants* are relatively prime, we will see in the complexity section 6 that the number of monomials of $\mathscr{M}_d$ having same $G_{\mathscr{D}}$-degree is almost the same for all g. In the affine case, we will prove without any assumption that the monomials of degree between 0 and d are evenly distributed on $\hat{G}$. These considerations allow us to bound the complexity of the computation of a Gröbner basis on such ideals, and we will verify that in practice they make an improvement on the timings (see section 7).

5. ABELIAN-FGLM ALGORITHM

In this section, we explain how to take advantage of the $G_{\mathscr{D}}$-grading to speed up the change of ordering, using a variant of the classical FGLM algorithm [10]. We suppose that $\dim(I_{\mathscr{D}}) = 0$, and that a Gröbner basis $\mathscr{G}_{\preceq_1}$ for an ordering $\preceq_1$ (for instance the DRL ordering) of the ideal $I_{\mathscr{D}} \subset R_K$ has already been computed, and we are interested in computing the Gröbner basis of $I_{\mathscr{D}}$ for an other ordering $\preceq_2$ (for example, the lexicographical ordering). In this section, $\mathbf{Deg}(I_{\mathscr{D}})$ will denote the degree of $I_{\mathscr{D}}$, defined by the dimension of $R/I_{\mathscr{D}}$. The idea of both FGLM and Abelian-FGLM algorithms is to pick up monomials m in $\mathscr{M}$ by increasing order for $\preceq_2$, and look for linear combinations in $R/I_{\mathscr{D}}$ between the Normal Forms $\mathsf{NF}(m, \mathscr{G}_{\preceq_1})$. To this end, these algorithms use linear algebra: we first compute the staircase

$$\mathscr{E} = \{m \in \mathscr{M} \mid m \text{ not reducible by } \mathsf{LM}(\mathscr{G}_{\preceq_1})\}$$

The elements of $\mathscr{E}$ form a basis of $R/I_{\mathscr{D}}$, which has dimension $\mathbf{Deg}(I_{\mathscr{D}})$, the degree of $I_{\mathscr{D}}$. Since $I_{\mathscr{D}}$ is $G_{\mathscr{D}}$-homogeneous, this staircase can be splitted in $|G_{\mathscr{D}}|$ parts, and we will denote by $\mathscr{E}_g$ the set of monomials in $\mathscr{E}$ having $G_{\mathscr{D}}$-degree g. As in the FGLM-algorithm, we use the linear maps given by the multiplication by one variable x_i in $R/I_{\mathscr{D}}$. The main difference lies in the following proposition:

Proposition 5 *Let f be a $G_{\mathscr{D}}$-homogeneous polynomial, and $\mathscr{G}_{\preceq_1}$ be the Gröbner basis of the ideal $I_{\mathscr{D}}$ for $\preceq_1$. Then $\mathsf{NF}(f, \mathscr{G}_{\preceq_1})$ is $G_{\mathscr{D}}$-homogeneous and has same $G_{\mathscr{D}}$-degree as f.*

PROOF. We have seen that being $G_{\mathscr{D}}$-homogeneous is a property stable under S-polynomials operations, so $\mathsf{NF}(f, \mathscr{G}_{\preceq_1})$ is $G_{\mathscr{D}}$-homogeneous. Moreover the only operations used in a Normal-Form computation are of the form $\tilde{f} \leftarrow f - \lambda m h$ with $h \in \mathscr{G}_{\preceq_1}$, $\lambda \in K$ and m a monomial such that $\mathsf{LM}(h) \times m$ is equal to some monomial in f, so $\mathsf{deg}_{G_{\mathscr{D}}}(f) = \mathsf{deg}_{G_{\mathscr{D}}}(\tilde{f})$. □

Diagonal-FGLM algorithm

Input: Multiplication matrices $M_{i,g}$, the sub-staircases $\mathscr{E}_g$, an ordering $\preceq_2$
Output: The Gröbner basis of $I_{\mathscr{D}}$ for $\preceq_2$.

```
L := [(1,0̂,n),(1,0̂,n−1),...,(1,0̂,1)]   // list of 3-uples (j,g,i) symbolizing the monomials S_g[j] × x_i, ordered by increasing order.
S_g := [] for g ∈ Ĝ\{0̂} and S_0̂ = [1].   // subsets of the staircase 𝒮 for the ordering ≼_2 having same G_𝒟-degree.
V_g := [] for g ∈ Ĝ\{0̂} and V_0̂ = [ᵗ(1,0,...,0)].  // V_g contains the expressions of NF(S_g[j],𝒢_≼1) in ℰ_g, each vector in V_g has n_g components.
G := []  // The Gröbner basis for ≼_2
Q_g := I_{n_g} for all g ∈ Ĝ.
 Do
   m := first(L) and remove m from L.
   j := m[1]; g' := m[2]; i := m[3]; g := g' + deg_{G_𝒟}(x_i)
   v := M_{i,g'} V_{g'}[j]  // components of NF(x_i S_g[j],𝒢_≼1) in ℰ_g
   s := #S_g  // number of elements in S_g.
   λ = ᵗ(λ_1,...,λ_{n_g}) := Q_g v
     if λ_{s+1} = ··· = λ_{n_g} = 0 then
        G := G ∪ [m − Σ_{j=1}^{s} λ_j · S_g[j]]
     else
        S_g := S_g ∪ [S_{g'}[j] × x_i]
        V_g := V_g ∪ [v]
        L := Sort(L ∪ [(s+1,g,i) | i = 1,...,n], ≼_2)
        Remove duplicates from L
        Update(Q_g,λ,v)  // Now Q_g v = ᵗ(0,...,0, 1_{s+1},0,...,0)
     end if
   Remove from L all multiples of LM_≼2(G)
     if L = ∅ then return G end if
```

Therefore, if m has $G_{\mathscr{D}}$-degree g, $\mathsf{deg}_{G_{\mathscr{D}}}(x_i m) = \mathsf{deg}_{G_{\mathscr{D}}}(x_i) + g$ and $\mathsf{NF}(x_i m, \mathscr{G}_{\preceq_1})$ is of same $G_{\mathscr{D}}$-degree. The map of multiplication by x_i in $\mathsf{Vect}(\mathscr{E})$ can be splitted into the following maps:

$$\begin{array}{rccc} M_{i,g}: & \mathsf{Vect}(\mathscr{E}_g) & \longrightarrow & \mathsf{Vect}(\mathscr{E}_{g+\mathsf{deg}_{G_{\mathscr{D}}}(x_i)}) \\ & f & \longmapsto & \mathsf{NF}(x_i f, \mathscr{G}_{\preceq_1}) \end{array}$$

The Diagonal-FGLM algorithm needs the matrices of multiplication $M_{i,g}$ and proceeds just like FGLM-algorithm:a new monomial to consider (except 1) is of the form $m = x_i m'$, with $m' \preceq_2 m$. Assume that $\mathsf{deg}_{G_{\mathscr{D}}}(m') = g'$, so we already know the expression of $\mathsf{NF}(m', \mathscr{G}_{\preceq_1})$ in terms of $\mathscr{E}_{g'}$, which is a vector V'. It follows that $\mathsf{NF}(m, \mathscr{G}_{\preceq_1})$ is computed by the product $V = M_{i,g'} V'$. Then we have to decide if m belongs to the new staircase in construction $\mathscr{S}$ or if it is the leading monomial of a polynomial of the Gröbner basis for $\preceq_2$. To this end, we use base-change matrices Q_g between $\mathscr{E}_g$ an $\mathscr{S}_g$, the subsets of the staircases having same $G_{\mathscr{D}}$-degree g. If s is the number of elements of the staircase $\mathscr{S}_g = \{u_1 \preceq_2 \cdots \preceq_2 u_s\}$ at the current point of the algorithm, and V_i the vectors corresponding to $\mathsf{NF}(u_i, \mathscr{G}_{\preceq_1})$, then $Q_g V_i$ is equal to the i-th vector of the canonical basis. Since the matrix Q_g is invertible, if all the components but the s first ones of QV are zero, then we deduce a new element of the Gröbner basis $\mathscr{G}_{\preceq_2}$, otherwise m is a new element of $\mathscr{S}_g$ and we have to update Q_g, to map V on the $(i+1)$-th element of the canonical basis. We can now give the pseudocode of the Diagonal-FGLM algorithm, here $\hat{0}$ means the $G_{\mathscr{D}}$-degree $(0, \ldots, 0)$. We suppose that $x_1, \ldots, x_n$ is the set of variables, with $x_n \preceq_2 x_{n-1} \preceq_2 \cdots \preceq_2 x_1$, and denote by n_g the number of elements in $\mathscr{E}_g$. Notice that with $\mathsf{deg}_{G_{\mathscr{D}}}(x_i) = \hat{0}$ for each i, we recover the standard FGLM algorithm.

Remark 4 *According to a point of view of representation theory, the sub-staircases $\mathscr{E}_g$ and $\mathscr{S}_g$ can be seen as two distinct bases of an isotypic component of the representation $R/I_{\mathscr{D}}$.*

6. COMPLEXITY QUESTIONS

In this section, we discuss the arithmetic complexity of the algorithms presented before. This complexity will be counted in terms of operations in $K = k(\xi)$.

Remark 5 *[18] A very interesting case is when ξ belongs to k, so $K = k$. Assume that k is the finite group $\mathbb{F}_p$ with p prime. Then*
$\xi \in k \iff X^e - 1$ splits on $k \iff \mathbb{Z}/e\mathbb{Z} \subseteq \mathbb{Z}/(p-1)\mathbb{Z} \iff p \equiv 1[e]$
By Dirichlet's theorem, there are infinitely many such primes and the distribution of such primes is $1/\varphi(e)$, where φ is the Euler's totient function. To compute the Gröbner basis of an ideal over $\mathbb{Q}$, it is more efficient to compute modulo some such primes and use modular methods to recover the original Gröbner basis.

We start by giving without proof a bound on the cost of the two first linear steps:

Proposition 6 *The cost of the diagonalization of the matrix group G is bounded by $O((q_1 + \cdots + q_k)n^{\omega})$, with ω the constant of linear algebra. With m polynomials f_i of degree less or equal than d, the cost of computing the f_i^P is bounded by $O(\binom{n+d}{d} ndm \log d \log \log d)$.*

6.1 Dimensions of the subspaces $R_{d,g}$

6.1.1 General facts about the ring of invariants

The first object we are interested in is the ring of invariants $R^{G_{\mathscr{D}}}$, with $G_{\mathscr{D}}$ the diagonal matrix group. Notice that we consider the invariants in a theoretical point of view to obtain complexity bounds, so we don't have to compute them. In this paragraph, we recall some well known facts about R^G, without any assumption on G, excepted that G is a finite matrix group of $GL_n(K)$, $charK$ doesn't divide $|G|$, and G is diagonalizable on K. We follow the presentation of [19]. Although Sturmfels works on $\mathbb{C}$, the results can be easily extended since the characteristic polynomials of matrices in G fully split on K.

Theorem 5 *[19] The invariant ring R_K^G is Cohen-Macaulay : there exist a set of n homogeneous polynomials $\theta_1, .., \theta_n$ and t other invariant polynomials $\eta_1, .., \eta_t$ such that $R_K^G = \bigoplus_{i=1}^{t} \eta_i \mathbb{K}[\theta_1, \ldots, \theta_n]$.*

The set of polynomials θ_i is called a set of *primary invariants* of G and the set of η_j a set of *secondary invariants* of G. A consequence of the previous theorem is the following proposition

Proposition 7 *[19] The Hilbert (Molien) series of the ring R_K^G is*

$$H(R_K^G, z) = \sum_{d=0}^{\infty} z^d \dim(R_{K,d}^G) = \frac{\sum_{i=0}^{t} z^{\mathsf{deg}(\eta_i)}}{\prod_{j=1}^{n}(1 - z^{\mathsf{deg}(\theta_j)})}$$

Proposition 8 *[19] The set of secondary invariants depends on the chosen set of primary invariants, moreover the degrees of the primary invariants and the number of secondary invariants are related by the formula :* $t = \prod_j deg(\theta_j)/|G|$

Now, we want to give an estimation of the size of R_d^G (set of invariant polynomials of degree d) compared to R_d. To give an estimation of the complexities of Abelian-F_5 and Abelian-FGLM algorithms, we are interested in two quantities.

Definition 6 *We define the* density *of R_d^G in R_d and the density of R^G in R by*

$$\delta(R_d^G) = \frac{\dim(R_d^G)}{\dim(R_d)} \qquad \text{and} \qquad \delta(R^G) = \lim_{D\to+\infty} \frac{\sum_{d=0}^D \dim(R_d^G)}{\sum_{d=0}^D \dim(R_d)}$$

The goal of this subsection is to prove the following theorem:

Theorem 6 *The density $\delta(R^G)$ is well defined and is equal to $1/|G|$. If a set of primary invariants of R^G can be chosen such that their degrees are relatively prime, the density $\delta(R_d^G)$ has the limit $\delta(R^G) = 1/|G|$ as d tends to infinity.*

PROOF. Denote by α_i the degree of θ_i, and by α the greatest common divisors of the α_i. We are interested in an asymptotic estimation of the coefficient in z^d in the Hilbert series of R_K^G. For now, denote by $f(z)$ the power series $1/(\prod_{j=1}^n (1-z^{\alpha_j}))$, and $[z^d]f(z)$ the coefficient in z^d in the expansion of f. Clearly, $[z^d]f(z) = 0$ if α doesn't divide d. Then, if $\alpha|d$, $[z^d]f(z) = [z^{d/\alpha}]\frac{1}{\prod_{j=1}^n (1-z^{\alpha_j/\alpha})}$.
Since the integers α_i/α have no common factor, it follows that 1 is the unique pole of multiplicity n in the previous rational function, the other poles having a smaller multiplicity. Following the idea of [11] Theorem 4.9, p.256, we obtain that

$$[z^{d/\alpha}]\frac{1}{\prod_{j=1}^n (1-z^{\alpha_j/\alpha})} = [z^{d/\alpha}]\frac{1}{(1-z)^n \prod_{j=1}^n (\sum_{\ell=0}^{(\alpha_j/\alpha)-1} z^\ell)} = \gamma\binom{d/\alpha+n-1}{n-1}$$

with γ the coefficient of $\frac{1}{1-z^n}$ in the partial fraction expansion:
$1/\gamma = \lim_{z\to 1}\prod_{j=1}^n (\sum_{\ell=0}^{(\alpha_j/\alpha)-1} z^\ell) = \prod_{j=1}^n \alpha_j/\alpha^n$
Since $\binom{d/\alpha+n-1}{n-1} \underset{d\to+\infty}{\sim} (d/\alpha)^{n-1}$, we have obtained that:

$$[z^d]f(z) = \begin{cases} 0 & \text{if } \alpha \nmid d \\ \frac{\alpha d^{n-1}}{\prod_j \alpha_j} + o(d^{n-1}) & \text{if } \alpha \mid d \end{cases}$$

We are now able to give the density of R^G :
$\sum_{d=0}^D \dim R_d^G \underset{D\to+\infty}{\sim} t\sum_{0\le d\le D, \alpha|d} \frac{\alpha d^{n-1}}{\prod_j \alpha_j} \underset{D\to+\infty}{\sim} \frac{t}{\prod_j \alpha_j}\sum_{d=0}^D d^{n-1}$
But $\sum_{d=0}^D \dim R_d \underset{D\to+\infty}{\sim} \sum_{d=0}^D d^{n-1}$, and by applying proposition 8, we conclude that $\delta(R_K^G) = 1/|G|$. Assume now that $\alpha = 1$, then $[z^d]f(z) = \frac{d^{n-1}}{\prod \alpha_i} + o(d^{n-1})$, so $[z^d]H(R^G,z) = \frac{td^{n-1}}{\prod \alpha_i} + o(d^{n-1})$, and the second part of the theorem follows. □

Remark 6 *If the degrees of the primary invariants have a common factor, the second part of the theorem is false. The following (trivial) example illustrates this fact.*

Example 9 *Let $G = \{Diag(\pm1,\pm1)\}$ Then $\mathbb{K}[x,y]^G = \mathbb{K}[x^2,y^2]$, and all the densities $\delta(R_d^G)$ are zero for odd d.*

6.1.2 Application to diagonal groups

Now we go back to the situation where G is a diagonal group isomorphic to $\prod_{i=1}^k \mathbb{Z}/q_i\mathbb{Z}$. Recall that $\mathscr{M}_{d,g}$ is the set of monomials of degree d and $G_{\mathscr{D}}$-degree g. We denote by $\hat{0} = (0,\ldots,0)$ the $G_{\mathscr{D}}$-degree of 1. Then $R_d^G = \mathrm{Vect}_K(\mathscr{M}_{d,\hat{0}})$, and $\dim(R_d^G) = |\mathscr{M}_{d,\hat{0}}|$.

Definition 7 *Following definition 6, we define the densities $\delta(R_g)$ and $\delta(R_{d,g})$ for any $g \in \hat{G}$ as*

$$\delta(R_{d,g}) = \frac{\dim(R_{d,g})}{\dim(R_d)} = \frac{|\mathscr{M}_{d,g}|}{|\mathscr{M}_d|} \quad \text{and} \quad \delta(R_g) = \lim_{D\to+\infty}\frac{\sum_{d=0}^D |\mathscr{M}_{d,g}|}{\sum_{d=0}^D |\mathscr{M}_d|}$$

Theorem 7 *The density $\delta(R_g)$ is well defined and is equal to $1/|G|$. If a set of primary invariants of R^G can be chosen such that their degrees are relatively prime, the density $\delta(R_{d,g})$ has limit $\delta(R^G) = 1/|G|$ as d tends to infinity.*

PROOF. First of all assume that all the sets $\mathscr{M}_g$ are non-empty, and let $m_g \in \mathscr{M}_g$ for all $g \in \hat{G}$. Denote by d_{m_g} its degree. Then $\mathscr{M}_g$ can be written $\mathscr{M}_g = m_g \mathscr{M}_{\hat{0}} \sqcup \{m \in \mathscr{M}_g \mid m_g \nmid m\}$. Therefore, for d big enough, $\mathscr{M}_{d,g} = m_g \mathscr{M}_{d-d_{m_g},\hat{0}} \sqcup \{m \in \mathscr{M}_{d,g} \mid m_g \nmid m\}$.
Assuming the condition of the degrees of the primary invariants, we obtain by theorem 6

$$\frac{|\mathscr{M}_{d,g}|}{|\mathscr{M}_d|} = \underbrace{\frac{|\mathscr{M}_{d-d_{m_g},\hat{0}}|}{|\mathscr{M}_{d-d_{m_g}}|}\frac{|\mathscr{M}_{d-d_{m_g}}|}{|\mathscr{M}_d|}}_{\underset{d\to\infty}{\to} 1/|G|} + \underbrace{\frac{|\{m \in \mathscr{M}_{d,g} \mid m_g \nmid m\}|}{|\mathscr{M}_d|}}_{\underset{d\to\infty}{\to} 0}$$

and the second part of the theorem is proved. In the same way, we conclude by sketching the proof of theorem 6 that

$$\delta(R_g) = \begin{cases} 1/|G| & \text{if } \mathscr{M}_g \neq \emptyset \\ 0 & \text{if } \mathscr{M}_g = \emptyset \end{cases}$$

But by definition, $\sum \delta(R_g) = 1$, so we proved that every set $\mathscr{M}_g$ is non-empty and $\delta(R_g) = 1/|G|$. □

Remark 7 *We have seen that* asymptotically, *the sets $\mathscr{M}_{d,g}$ have roughly the same size (with the assumption on the degrees of the primary invariants) and that the same result holds without assumption on the sets $\cup_{d=0}^D \mathscr{M}_{d,g}$, and the sizes of these sets correspond to the number of columns in the matrices of the abelian-F_5 algorithm, in the homogeneous or affine case. Actually, these sets are very fast evenly distributed, as we will see in section 7. To perform a complexity analysis, we will suppose that this is the case.*

6.2 Application to the complexity of abelian-F_5 and abelian-FGLM algoritms

6.2.1 Abelian-F_5 algorithm

To analyse the efficiency of our algorithm to compute a Gröbner basis of $I_{\mathscr{D}}$, we have to compare the complexity of the classical F_5 algorithm on I and $I_{\mathscr{D}}$ and the abelian-F_5 algorithm on $I_{\mathscr{D}}$. In order to bound the complexity of F_5 we bound the complexity of the so called Macaulay/Lazard algorithm [15], consisting in building a row echelon form of the Macaulay's matrix; this computation can be seen as a redundant variant of the F_5 algorithm. Since the base-change matrix P defined in section 2 induces an isomorphism between the homogenous components of same degree of I and $I_{\mathscr{D}}$, assuming they are homogeneous, so these ideals have same Hilbert series. Therefore, the index of regularity (homogeneous case) or the degree of regularity (affine case) are the same. For a good introduction to these notions, see [16]. From the Lazard algorithm [15] it is possible to derive a complexity bound of the computation of a Gröbner basis of zero dimensional homogeneous system.

Theorem 8 *[16] Let $\mathbf{F} = (f_1,\ldots,f_m) \in R^m$ be a family of homogeneous polynomials generating a zero-dimensional ideal. The complexity of computing a Gröbner basis for the DRL ordering of the ideal $\langle \mathbf{F}\rangle$ is bounded by* $O\left(m\binom{n+d_{reg}(\mathbf{F})}{d_{reg}(\mathbf{F})}^\omega\right)$
where ω is the constant of linear algebra.

The proof of the previous theorem is obtained by analyzing size and rank of the Macaulay's matrix, and by the fact that a row echelon form of a matrix of size (ℓ,c) and rank r can be computed in times $O(\ell c r^{\omega-2})$. In the case of an ideal $\mathbf{F}$ invariant under a diagonal group $G_{\mathscr{D}}$, we have seen that such a matrix can be slitted into $|G_{\mathscr{D}}|$ parts, and previous analysis of the size of the sets $\mathscr{M}_{d,g}$ in theorem 7 proves that, under parallelization on the computations of row echelon form of the $|G_{\mathscr{D}}|$ submatrices, the following theorem holds:

Theorem 9 *Let $\mathbf{F} = (f_1, \ldots, f_m) \in R^m$ be a family of homogeneous polynomials generating a 0-dimensional ideal, invariant under a diagonal group $G_{\mathscr{D}}$ such that a set of primary invariants of $G_{\mathscr{D}}$ can be chosen with degrees relatively prime. The complexity of computing a Gröbner basis for the DRL ordering of the ideal $\langle \mathbf{F} \rangle$ is bounded by*

$$O\left(\frac{m}{|G_{\mathscr{D}}|^{\omega}} \binom{n + d_{reg}(\mathbf{F})}{d_{reg}(\mathbf{F})}^{\omega}\right)$$

Remark 8 *In the affine case, it seems that a bound similar to theorem 8 could be obtained (see [16], page 53), therefore we could obtain a similar improvement than in theorem 9.*

6.2.2 Abelian-FGLM algorithm

Let $I_{\mathscr{D}}$ be a zero-dimensional ideal invariant under the diagonal group $G_{\mathscr{D}}$. We have to consider the two parts of the algorithm to give a complexity estimation : the construction of the multiplication's matrices $M_{i,g}$ and the loop in FGLM. We denote by $\mathbf{Deg}(I_{\mathscr{D}})$ the degree of the ideal $I_{\mathscr{D}}$.

Theorem 10 *Under the hypothesis that the monomials of $\mathscr{E}$ are evenly distributed over the staircases $\mathscr{E}_g$ (which is verified in practice), it is possible to obtain the reduced Gröbner basis $\mathscr{G}_{\preceq_2}$ from $\mathscr{G}_{\preceq_1}$ of $I_{\mathscr{D}}$ with $O(\frac{n}{|G_{\mathscr{D}}|^2} \mathbf{Deg}(I_{\mathscr{D}})^3)$ arithmetics operations in K.*

PROOF. We follow the notations of [10].

- To compute the multiplication matrices, we have to compute the normal forms $NF(m, \mathscr{G}_{\preceq_1})$ for all $m \in B(\mathscr{G}_{\preceq_1}) \cup M(\mathscr{G}_{\preceq_1})$. For at most $n\mathbf{Deg}(I_{\mathscr{D}})$ of these monomials, arithmetic computations are needed and since the staircases $\mathscr{E}_g$ have size about $\mathbf{Deg}(I_{\mathscr{D}})/G_{\mathscr{D}}$, each of these normal forms can be computed with $O((\frac{\mathbf{Deg}(I_{\mathscr{D}})}{|G_{\mathscr{D}}|})^2)$ arithmetic operations in K.
- In the same way, the loop in the FGLM algorithm presented in section 5 has to be done at most $n\mathbf{Deg}(I_{\mathscr{D}})$ times. The cost of the linear operations was $O(\mathbf{Deg}(I_{\mathscr{D}})^2)$ in the original FGLM algorithm [10] but it is reduced to $O(\mathbf{Deg}(I_{\mathscr{D}})^2/|G_{\mathscr{D}}|^2)$ here since the square matrices have a number of lines and columns divided by about $|G_{\mathscr{D}}|$. □

6.3 Polynomial complexity

Suppose that $g_1, \ldots, g_m$ are affine polynomials of R of degree 2, which are individually invariant under the cyclic-n group. Usually, computing a Gröbner basis of $I = \langle g_1, \ldots, g_m \rangle$ is exponential, but we will see that we can obtain a Gröbner basis of $I_{\mathscr{D}}$ in polynomial time in n and m. With $P = (\xi^{ij})$, and $f_i = g_i^P$, each f_i is invariant under $D_\sigma = diag(\xi, \xi^2, \ldots, \xi^{n-1}, 1)$ and f_i has $G_{\mathscr{D}}$-degree 0.

Lemma 2 *The support of each f_i is contained in $\{1, x_n, x_n^2\} \cup \{x_i x_{n-i}, \mid 1 \leq i \leq \lfloor (n-1)/2 \rfloor\}$.*

PROOF. Each x_i has $G_{\mathscr{D}}$-degree i mod n, so $\deg_{G_{\mathscr{D}}}(x_i x_j) = i + j$ mod n, and the only monomials of degree 2 having $G_{\mathscr{D}}$-degree 0 are $x_i x_{n-i}$. The only monomial of degree 1 and $G_{\mathscr{D}}$-degree 0 is x_n, and 1 is also of $G_{\mathscr{D}}$-degree 0. □

Theorem 11 *A Gröbner Basis for every monomial ordering of a system of m equations invariant under $D_\sigma = diag(\xi, \ldots, \xi^{n-1}, 1)$ can be computed in polynomial time in $n+m$.*

PROOF. We set $y_i = x_i x_{n-i}$ for each $i \in \{0, \ldots, \lfloor (n-1)/2 \rfloor\}$ to linearize the equations, and perform a Gauss elimination on the equations. The result is a Gröbner Basis since the leading monomials of any pair of the obtained polynomials are coprime. The matrix we have to reduce has m lines and $\lfloor (n+5)/2 \rfloor$ columns, and the complexity is polynomial in $n+m$. □

Remark 9 *Similar results can be obtained for other groups and systems. This will be discussed in an extended version of this paper.*

7. EXPERIMENTS

In this section, we report some experiments that show the improvements given by our approach on the computation of Gröbner bases of ideals invariant under a commutative group. We first present sizes of the sets $\mathscr{M}_{d,g}$ and $\mathscr{E}_g$, and then give timings obtained with an implantation of the algorithm Abelian-F_4. A web page has been made for other softwares and benchmarks, see [8].

7.1 Sizes of the sets $\mathscr{M}_g$ or $\mathscr{E}_g$

In this subsection, we suppose that G is the cyclic group generated by the matrix M_σ presented in example 1. Therefore $G_{\mathscr{D}}$ is the group generated by the diagonal matrix with diagonal coefficients $(\xi, \xi^2, \ldots, \xi^{n-1}, 1)$. We want to compare the size of $\mathscr{M}_{d,g}$ with $|\mathscr{M}_d|/n$ (recall that n is the order of $G_{\mathscr{D}}$). To this end we compute the relative standard deviation of the sets $|\mathscr{M}_{d,g}|$ to $|\mathscr{M}_d|/n$, for several n and d. The formula is $\sigma_{d,n} = \frac{\sqrt{\frac{1}{n}\sum_{g \in G_{\mathscr{D}}}(|\mathscr{M}_{d,g}| - |\mathscr{M}_d|/n)^2}}{|\mathscr{M}_d|/n}$. The following table presents some values of $\sigma_{d,n}$. We see that the monomials are very fast evenly distributed over $g \in \hat{G}$. In the same way,

d/n	2	3	4	5	10	15
3	0.00	0.14	0.00	0.09	0.00	0.01
4	0.20	0.00	0.10	0.09	0.02	0.01
5	0.00	0.09	0.00	0.02	0.00	0.00
10	0.09	0.00	0.02	0.00	0.00	0.00

Table 1: Repartition of the monomials under $G_{\mathscr{D}}$

the stairs $\mathscr{E}_g$ that appear in the abelian-FGLM algorithm have about same size. Table 2 presents some zero dimensional ideals together with the size of the group and the size of the stairs. The final column is the relative standard deviation between $|\mathscr{E}_g|$ and $|\mathscr{E}|/|G_{\mathscr{D}}|$.

n	$\lvert\mathscr{E}\rvert$	$\lvert G_{\mathscr{D}}\rvert$	$\lvert\mathscr{E}_g\rvert/\lvert G_{\mathscr{D}}\rvert$	Max $\lvert\mathscr{E}_g\rvert$	$\sigma_{\mathscr{E}}$
5	70	25	2.80	6	0.286
6	156	36	4.33	6	0.133
7	924	49	18.86	24	0.045
10	34940	100	349.40	354	0.0043
11	184756	121	1526.91	1536	0.00060

Table 2: Cyclic-n: Repartition of the monomials into $\mathscr{E}_g$

From the experimental side, applying the F_4 algorithm on the cyclic 9 problem we obtain, in degree 15, a matrix of size 72558×93917; applying the abelian-F_4 algorithm we obtain 9 independent matrices of roughly the same size: $8340 \times 10703, 8180 \times 10544, 8122 \times 10484, 7804 \times 10171, 7993 \times 10358, 8042 \times 10404, 7796 \times 10162, 7967 \times 10369$ and 8314×10722.

7.2 Abelian-F_4 implementation

A first implementation of the Abelian-F_4-algorithm [4] has been made. The algorithm constructs $|G_{\mathscr{D}}|$ matrices at each degree, using the usual strategy of F_4. Notice that only the construction of the matrices and the operations of row-reduction on them have been parallelized, the handle of the list of critical pairs is still sequential. Surprisingly, the linear algebra can sometimes be so accelerated that this handling can become the most time-consuming part whereas it is usually negligible. Therefore we report in the following tables two timings or ratios in each column: the timings are related to $F_4^{A,n}$, which is the new abelian algorithm parallelized on n cores. The first one is the total timing and the second one is only the parallelized part (that is to say, building the matrices and the linear algebra parts). The other columns contain the ratios between

F_4^A or F_4 and $F_4^{A,n}$. F_4 means the standard F_4 applied on I and F_4^A the standard F_4 applied on $I_{\mathscr{D}}$. F_4^M is the F_4 of Magma, and there is only the ratio for the total timing. In each case except table 7, the group G acting on I is the cyclic group C_n generated by the matrix M_σ defined in example 1, and $G_{\mathscr{D}}$ is the group generated by the diagonal matrix $\mathrm{diag}(\xi,\xi^2,\ldots,1)$. Notice that we have to reach big-sized problems to have a significant impact. In table 3, we consider n randomized equations of degree 3 stable under C_n, which give rise to equations of $G_{\mathscr{D}}$-degree 0 in $I_{\mathscr{D}}$. Table 4 presents n equations of degree 2, half of these equations in $I_{\mathscr{D}}$ are of $G_{\mathscr{D}}$-degree 0, and half of $G_{\mathscr{D}}$-degree 1. In this case, the computation on $I_{\mathscr{D}}$ becomes polynomial in n and the handling of the critical pairs is the most time-consuming part. All computations have been made on a computer with 4 Intel(R) Xeon(R) CPU E5-4620 0 @ 2.20GHz with 387 GB of RAM, on a field where $X^{|G|}-1$ fully splits (most of the time $\mathbb{F}_{65521}$), according to remark 5.

n	$F_4^{A,n}$ total; // part	$F_4^A/F_4^{A,n}$ tot;p.p	$F_4/F_4^{A,n}$ tot;p.p	$F_4^{\mathrm{M}}/F_4^{A,n}$ tot
8	3.46s;2.48s	2.2;2.7	33.0;45.4	22
9	77.04s;64.21s	7.3;8.6	67.8;81.0	50
10	762s;672s	10.0;11.3	160.9;182.1	134
11	22162s;20425s	13.0;14.0	∞	∞

Table 3: n cubic equations of $G_{\mathscr{D}}$-degree 0

n	$F_4^{A,n}$ total; // part	$F_4^A/F_4^{A,n}$ tot;p.p	$F_4/F_4^{A,n}$ tot;p.p
25	0.25s;0.06s	1.9;4.5	56.60;230.0
30	0.58s;0.11s	1.5;4.6	80.79;415.1
35	0.86s;0.11s	1.9;8.5	228.5;1755
40	1.55s;0.21s	2.0;8.5	300.6;2174
50	3.96s;0.45s	2.6;13.3	753.8;6504
60	10.85s;0.96s	2.8;17.2	1294;14330

Table 4: n quadratic equations of $G_{\mathscr{D}}$-degree 0 or 1

Table 5 presents equations coming from a cryptographic application : the cryptosystem NTRU [14]. The underlying problem is the following: given $h \in \mathbb{F}_p[x]$, we are looking for a polynomial $f \in \mathbb{F}_p[x]$ of degree $n-1$ and coefficients in $\{0,1\}$ such that $g = fh \bmod x^n - 1$ has also its coefficients in $\{0,1\}$. Denote $f = \sum_{i=0}^{n-1} f_i x^i$ and $g = \sum_{i=0}^{n-1} g_i x^i$, then the g_i's are linear forms in the f_i's verifying $g_i^{M_\sigma} = g_{\sigma(i)}$. Since the conditions of f_i and g_i to be in $\{0,1\}$ can be written $f_i^2 - f_i = g_i^2 - g_i = 0$, the system consists of $2n$ quadratic equations in the f_i's generating an ideal globally stable under the action of C_n. The speed-up between F_4 and $F_4^{A,n}$ is roughly 250 with 24 variables, and the use of $F_4^{A,n}$ has a significant impact since we can achieve bigger problems. In this case the handling of the critical pairs is also the most time-consuming part.

n	$F_4^{A,n}$ total; // part	$F_4^A/F_4^{A,n}$ tot;p.p	$F_4/F_4^{A,n}$ tot;p.p
21	4.52s;1.21s	4.0;11.9	90.15;334.0
23	11.16s;1.87s	3.3;17.2	115.2;686.1
24	128s;14.3s	5.2;36.5	241.1;2149.
25	218s;31.0s	5.8;32.5	∞
28	1214s;192s	7.1;36.1	∞

Table 5: NTRU equations

Table 6 presents timings on the Cyclic-n problem, we see that Cyclic-11 could be solved in less than 8 hours although it is untractable with F_4. Table 7 is an example of ideals generating by random polynomials of degree 3 invariant under the group $C_{k_1} \times C_{k_2}$, each subgroup C_k acting on k variables. We see that the algorithm is more efficient where $k_1 = k_2$, which makes sense since the size of the group is $k_1 k_2$.

Acknowledgments. This work was supported in part by the HPAC grant (ANR ANR-11-BS02-013) and by the EXACTA grant (ANR-09-BLAN-0371-01) of the French National Research Agency.

n	$F_4^{A,n}$ total; // part	$F_4^A/F_4^{A,n}$ tot;p.p	$F_4/F_4^{A,n}$ tot;p.p	$F_4^{\mathrm{M}}/F_4^{A,n}$ tot
8	0.50s;0.40s	2.5;2.7	7.8;9.3	6.0
9	10.21s;7.71s	4.3;5.4	37.0;48.4	30.5
10	334s;290s	13.2;14.8	411.0;472.3	207
11	27539s;25454s	∞	∞	∞

Table 6: The Cyclic-n problem

k_1,k_2	F_4^{A,k_1k_2} tot; // p.p	$F_4^A/F_4^{A,k_1k_2}$ tot;p.p	$F_4/F_4^{A,k_1k_2}$ tot;p.p	$F_4^{\mathrm{M}}/F_4^{A,k_1k_2}$ tot
4,4	2.0s;1.3s	2.4;3.2	61.8;94.6	37
6,2	2.9s;2.4s	2.2;2.5	76.4;91.4	44
5,5	70s;43s	11.8;16.2	∞	∞
6,4	92s;76s	17.7;19.8	∞	∞
8,2	107s;100s	12.1;12.3	∞	∞

Table 7: $n = k_1 + k_2$ cubic equations invariant under $C_{k_1} \times C_{k_2}$

8. REFERENCES

[1] B. Buchberger. Bruno buchberger's phd thesis 1965: An algorithm for finding the basis elements of the residue class ring of a zero dimensional polynomial ideal. *Journal of Symbolic Computation*, 41(3-4):475 – 511, 2006. Logic, Mathematics and Computer Science: Interactions in honor of Bruno Buchberger (60th birthday).

[2] A. Colin. Solving a system of algebraic equations with symmetries. *J. Pure Appl. Algebra*, 117/118:195–215, 1997. Algorithms for algebra (Eindhoven, 1996).

[3] D. Cox, J. Little, and D. O'Shea. *Ideals, varieties, and algorithms*. Undergraduate Texts in Mathematics. Springer, New York, third edition, 2007. An introduction to computational algebraic geometry and commutative algebra.

[4] J.-C. Faugère. A new efficient algorithm for computing Gröbner bases (F4). *Journal of Pure and Applied Algebra*, 139(1–3):61–88, June 1999.

[5] J.-C. Faugère. A new efficient algorithm for computing Gröbner bases without reduction to zero (F5). In *Proceedings of the 2002 international symposium on Symbolic and algebraic computation*, ISSAC '02, pages 75–83, New York, NY, USA, 2002. ACM.

[6] J.-C. Faugère, M. Hering, and J. Phan. The membrane inclusions curvature equations. *Advances in Applied Mathematics*, 31(4):643–658, June 2003.

[7] J.-C. Faugère and S. Rahmany. Solving systems of polynomial equations with symmetries using SAGBI-Gröbner bases. In *ISSAC '09: Proceedings of the 2009 international symposium on Symbolic and algebraic computation*, ISSAC '09, pages 151–158, New York, NY, USA, 2009. ACM.

[8] J.-C. Faugère and J. Svartz. Software and benchmarks. http://www-polsys.lip6.fr/~jcf/Software/benchssym.html.

[9] J.-C. Faugère and J. Svartz. Solving polynomial systems globally invariant under an action of the symmetric group and application to the equilibria of n vortices in the plane. In *ISSAC '12: Proceedings of the 2012 International Symposium on Symbolic and Algebraic Computation*, ISSAC '12, pages 170–178, New York, NY, USA, 2012. ACM. accepted.

[10] J.C. Faugère, P. Gianni, D. Lazard, and T. Mora. Efficient Computation of Zero-dimensional Gröbner Bases by Change of Ordering. *Journal of Symbolic Computation*, 16(4):329–344, 1993.

[11] P. Flajolet and R. Sedgewick. *Analytic combinatorics*. Cambridge University Press, 2009.

[12] G. Björck. Functions of modulus 1 on Zn , whose Fourier transforms have constant modulus, and "cyclic n-roots". *NATO, Adv. Sci. Inst. Ser. C, Math. Phys. Sci.*, 315:131–140, 1990. Recent Advances in Fourier Analysis and its applica tions.

[13] K. Gatemann. Symbolic solution polynomial equation systems with symmetry. In *Proceedings of the international symposium on Symbolic and algebraic computation*, ISSAC '90, pages 112–119, New York, NY, USA, 1990. ACM.

[14] J. Hoffstein, J. Pipher, and J.H. Silverman. Ntru: a ring-based public key cryptosystem. In *Algorithmic number theory (Portland, OR, 1998)*, volume 1423 of *Lecture Notes in Comput. Sci.*, pages 267–288. Springer, Berlin, 1998.

[15] D. Lazard. Gröbner bases, Gaussian elimination and resolution of systems of algebraic equations. In *Computer Algebra, EUROCAL'83*, volume 162 of *LNCS*, pages 146–156. Springer, 1983.

[16] P-J. Spaenlehauer. *Solving multi-homogeneous and determinantal systems. Algorithms - Complexity - Applications*. PhD thesis, Université Paris 6, 2012.

[17] R. P. Stanley. Invariants of finite groups and their applications to combinatorics. *Bull. Amer. Math. Soc. (new series)*, 1:475–511, 1979.

[18] S. Steidel. Gröbner bases of symmetric ideals. *Journal of Symbolic Computation*, 54(0):72 – 86, 2013.

[19] B. Sturmfels. *Algorithms in Invariant Theory*. Texts and Monographs in Symbolic Computation. SpringerWienNewYork, Vienna, second edition, 2008.

Structured FFT and TFT: Symmetric and Lattice Polynomials*

Joris van der Hoeven
Laboratoire d'informatique
UMR 7161 CNRS
École Polytechnique
91128 Palaiseau Cedex, France
vdhoeven@lix.polytechnique.fr

Romain Lebreton
LIRMM
UMR 5506 CNRS
Université de Montpellier II
Montpellier, France
lebreton@lirmm.fr

Éric Schost
Computer Science Department
Western University
London, Ontario
Canada
eschost@uwo.ca

ABSTRACT

In this paper, we consider the problem of efficient computations with structured polynomials. We provide complexity results for computing Fourier Transform and Truncated Fourier Transform of symmetric polynomials, and for multiplying polynomials supported on a lattice.

Categories and Subject Descriptors

F.2.2 [**ANALYSIS OF ALGORITHMS AND PROBLEM COMPLEXITY**]: Numerical Algorithms and Problems–*Computations on polynomials*

Keywords

symmetric FFT; lattice FFT; TFT

1. INTRODUCTION

Fast computations with multivariate polynomials and power series have been of fundamental importance since the early ages of computer algebra. The representation is an important issue which conditions the performance in an intrinsic way; see [24, 33, 8] for some historical references.

It is customary to distinguish three main types of representations: dense, sparse, and functional. A *dense representation* is made of a compact description of the support of the polynomial and the sequence of its coefficients. The main example concerns *block supports* – it suffices to store the coordinates of two opposite vertices. In a dense representation all the coefficients of the considered support are stored, even if they are zero. If a polynomial has only a few non-zero terms in its bounding block, we shall prefer to use a *sparse representation* which stores only the sequence of the non-zero terms as pairs of monomials and coefficients. Finally, a *functional representation* stores a function that can produce values of the polynomials at any given point. This can be a pure blackbox (which means that its internal structure is not supposed to be known) or a specific data structure such as *straight-line programs* (see Chapter 4 of [4], for instance, and [11] for a concrete library for the manipulation of polynomials with a functional representation).

For dense representations with block supports, it is classical that the algorithms used for the univariate case can be naturally extended: the naive algorithm, Karatsuba's algorithm, and even Fast Fourier Transforms [7, 32, 6, 30, 17] can be applied recursively in each variable, with good performance. Another classical approach is the Kronecker substitution which reduces the multivariate product to one variable only; for all these questions, we refer the reader to classical books such as [29, 14]. When the number of variables is fixed and the partial degrees tend to infinity, these techniques lead to softly linear costs.

After the discovery of sparse interpolation [2, 25, 26, 27, 12], probabilistic algorithms with a quasi-linear complexity have been developed for sparse polynomial multiplication [5]. It has recently be shown that such asymptotically fast algorithms may indeed become more efficient than naive sparse multiplication [20].

In practice however, it frequently happens that multivariate polynomials with a dense flavor do not admit a block support. For instance, it is common to consider polynomials of a given total degree. In a recent series of works [16, 18, 19, 22, 21], we have studied the complexity of polynomial multiplication in this "semi-dense" setting; see also [10, 31]. In the case when the supports of the polynomials are initial segments for the partial ordering on $\mathbb{N}^n$, the truncated Fourier transform is a useful device for the design of efficient algorithms.

Besides polynomials with supports of a special kind, we may also consider what will call "structured polynomials". By analogy with linear algebra, such polynomials carry a special structure which might be exploited for the design of more efficient algorithms. In this paper, we turn our attention to a first important example of this kind: polynomials which are invariant under the action of certain matrix groups. We consider only two special cases: finite subgroups of $\mathfrak{S}_n$ and finite groups of diagonal matrices. These cases are already sufficient to address questions raised *e.g.* in celestial mechanics [13]; it is hoped that more general groups can be dealt with using similar ideas.

In the limited scope of this paper, our main objective is to prove complexity results that demonstrate the savings induced by a proper use of the symmetries. Our complexity analyses take into account the number of arithmetic operations in the base field, and as often, we consider that operations with groups and lattices take a constant number of operations. A serious implementation of our algorithms would require an improved study of these aspects.

* This work has been partly supported by the DIGITEO 2009-36HD grant of the Région Ile-de-France, the ANR grant HPAC (ANR-11-BS02-013), NSERC and the CRC program.

ISSAC'13, June 26–29, 2013, Boston, Massachusetts, USA.

Of course, there already exists an abundant body of work on some of these questions. *Crystallographic* FFT algorithms date back to [34], with contributions as recent as [28], but are dedicated to crystallographic symmetries. A more general framework due to [1] was recently revisited under the point of view of high-performance code generation [23]; our treatment of permutation groups is in a similar spirit, but to our understanding, these previous papers do not prove results such as those we give below (and they only consider the FFT, not its truncated version).

Also our results on diagonal groups, which fit in the general context of FFTs over lattices, use similar techniques as in a series of papers initiated by [15] and continued as recently as [35, 3], but the actual results we prove are not in those references.

2. THE CLASSICAL FFT

2.1 Notation

We will work over an effective base field (or ring) $\mathbb{K}$ with sufficiently many roots of unity; the main objects are polynomials $\mathbb{K}[\boldsymbol{x}] := \mathbb{K}[x_1, \ldots, x_n]$ in n variables over $\mathbb{K}$. For any $\boldsymbol{k} = (\boldsymbol{k}_1, \ldots, \boldsymbol{k}_n) \in \mathbb{N}^n$ and $P \in \mathbb{K}[\boldsymbol{x}]$, we denote by $\boldsymbol{x}^{\boldsymbol{k}}$ the monomial $x_1^{\boldsymbol{k}_1} \cdots x_n^{\boldsymbol{k}_n}$ and by $P_{\boldsymbol{k}}$ the coefficient of P in $\boldsymbol{x}^{\boldsymbol{k}}$. The support $\operatorname{supp}(P)$ of a polynomial $P \in \mathbb{K}[\boldsymbol{x}]$ is the set of exponents $\boldsymbol{k} \in \mathbb{N}^n$ such that $P_{\boldsymbol{k}} \neq 0$.

For any subset S of $\mathbb{N}^n$, we define $\mathbb{K}[\boldsymbol{x}]_S$ as the polynomials with support included in S. As an important special case, when $S = \mathbb{N}_d^n := \{0, \ldots, d-1\}^n$, we denote by $\mathbb{K}[\boldsymbol{x}]_d := \mathbb{K}[\boldsymbol{x}]_S$ the set of polynomials with partial degree less than d in all variables.

Let $\omega \in \mathbb{K}$ be a primitive d-th root of unity (in Section 4, it will be convenient to write this root $\mathrm{e}^{2\pi \mathrm{i}/d}$). For any $\boldsymbol{k} = (\boldsymbol{k}_1, \ldots, \boldsymbol{k}_n) \in \mathbb{N}^n$, we define $\omega^{\boldsymbol{k}} = (\omega^{\boldsymbol{k}_1}, \ldots, \omega^{\boldsymbol{k}_n})$. One of our aims is to compute efficiently the map

$$\mathrm{FFT}_\omega : \left\{ \begin{array}{ccc} \mathbb{K}[\boldsymbol{x}]_d & \longrightarrow & \mathbb{K}^{\mathbb{N}_d^n} \\ P & \longmapsto & (P(\omega^{\boldsymbol{i}}))_{\boldsymbol{i} \in \mathbb{N}_d^n} \end{array} \right.$$

when P is a "structured polynomial". Here, $\mathbb{K}^{\mathbb{N}_d^n}$ denotes the set of vectors with entries in $\mathbb{K}$ indexed by $\mathbb{N}_d^n$; in other words, $\boldsymbol{v} \in \mathbb{K}^{\mathbb{N}_d^n}$ implies that $\boldsymbol{v}_{\boldsymbol{i}} \in \mathbb{K}$ for all $\boldsymbol{i} \in \mathbb{N}_d^n$. Likewise, $\mathbb{K}^{\mathbb{N}_d^n \times \mathbb{N}_d^n}$ denotes the set of matrices with indices in $\mathbb{N}_d^n$, that is, $\mathbf{M} \in \mathbb{K}^{\mathbb{N}_d^n \times \mathbb{N}_d^n}$ implies that $\mathbf{M}_{\boldsymbol{i},\boldsymbol{j}} \in \mathbb{K}$ for all $\boldsymbol{i}, \boldsymbol{j} \in \mathbb{N}_d^n$.

Most of the time, we will take $d = 2^\ell$ with $\ell \in \mathbb{N}$, although we will also need more general d of mixed radices $d = p_1 \cdots p_\ell$.

We denote by $\boldsymbol{i} \cdot \boldsymbol{j}$ the inner product of two vectors in $\mathbb{N}^n$. We also let $\boldsymbol{e}_k := \left(0, \overset{k-1}{\ldots}, 0, 1, 0, \ldots, 0\right)$, for $1 \leqslant k \leqslant n$, be the k-th element of the canonical basis of $\mathbb{K}^n$. At last, for $\ell \in \mathbb{N}$ and $k \in \mathbb{N}_{2^\ell}$ we denote by $[k]_\ell$ the bit reversal of k in length ℓ and we extend this notation to vectors by $[\boldsymbol{k}]_s = ([\boldsymbol{k}_1]_s, \ldots, [\boldsymbol{k}_n]_s)$.

2.2 The classical multivariate FFT

Let us first consider the in-place computation of a full n-dimensional FFT of length $d = 2^\ell$ in each variable. We first recall the notations and operations of the decimation in time variant of the FFT, and we refer the reader to [9] for more details. In what follows, ω is a d-th primitive root of unity.

We start with the FFT of a univariate polynomial $P \in \mathbb{K}[x]_d$. Decimation in time amounts to decomposing P into its even and odd parts, and proceeding recursively, by means of ℓ decimations applied to the variable x. For $0 \leqslant k < d$, we will write $c_k^0 = P_k$ for the input and denote by $\boldsymbol{c}^s = (c_k^s)_{0 \leqslant k < d}$ the result after s decimation steps, for $s \in \{1, \ldots, \ell\}$.

At stage s, the decimation is computed using butterflies of span $\delta := 2^{\ell - s}$. If $i \in \mathbb{N}_{2^s}$ is even and j belongs to $\mathbb{N}_\delta$ then these butterflies are given by

$$\begin{pmatrix} c_{i\delta+j}^s \\ c_{(i+1)\delta+j}^s \end{pmatrix} = \begin{pmatrix} 1 & \omega^{[i]_s \delta} \\ 1 & -\omega^{[i]_s \delta} \end{pmatrix} \begin{pmatrix} c_{i\delta+j}^{s-1} \\ c_{(i+1)\delta+j}^{s-1} \end{pmatrix}.$$

Putting the coefficients of all these linear relations in a matrix $\mathbf{B}^s \in \mathbb{K}^{\mathbb{N}_d \times \mathbb{N}_d}$, we get $\boldsymbol{c}^s = \mathbf{B}^s \boldsymbol{c}^{s-1}$. The matrix $\mathbf{B}^s$ is sparse, with at most two non-zero coefficients on each row and each column; up to permutation, it is block-diagonal, with blocks of size 2. After ℓ stages, we get the evaluations of P in the bit reversed order: $c_k^\ell = P(\omega^{[k]_\ell})$.

We now adapt this notation to the multivariate case. The computation is still divided in ℓ stages, each stage doing one decimation in every variable $x_1, \ldots, x_n$. Therefore, we will denote by $c_{\boldsymbol{k}}^0 := P_{\boldsymbol{k}}$ the coefficients of the input for $\boldsymbol{k} \in \mathbb{N}_d$ and by $c_{\boldsymbol{k}}^{s,t}$ the coefficients obtained during the s-th stage, after the decimations in $x_1, \ldots, x_t$ are done, with $s \in \{1, \ldots, \ell\}$ and $t \in \{0, \ldots, n\}$, so that $c_{\boldsymbol{k}}^{s-1,n} = c_{\boldsymbol{k}}^{s,0}$. We abbreviate $c_{\boldsymbol{k}}^{s-1} := c_{\boldsymbol{k}}^{s,0}$ for the coefficients after $s-1$ stages.

The intermediate coefficients $c_{\boldsymbol{k}}^s$ can be seen as evaluations of intermediate polynomials: for every $s \in \{0, \ldots, \ell\}$, $\boldsymbol{i} \in \mathbb{N}_{2^s}^n$ and $\boldsymbol{j} \in \mathbb{N}_\delta^n$, one has

$$c_{\boldsymbol{i}\delta+\boldsymbol{j}}^s = P_{\boldsymbol{j}}^s((\omega^\delta)^{[\boldsymbol{i}]_s}) \qquad (1)$$

where $P_{\boldsymbol{j}}^s = \sum_{\boldsymbol{i}' \in \mathbb{N}_{2^s}^n} P_{\boldsymbol{i}'\delta+\boldsymbol{j}} \boldsymbol{x}^{\boldsymbol{i}'}$ are obtained through an s-fold decimation of P. Equivalently, the coefficients $c_{\boldsymbol{k}}^s$ satisfy

$$c_{\boldsymbol{i}\delta+\boldsymbol{j}}^s = \sum_{\boldsymbol{i}' \in \mathbb{N}_{2^s}^n} P_{\boldsymbol{i}'\delta+\boldsymbol{j}} \, \omega^{[\boldsymbol{i}]_s \cdot \boldsymbol{i}' \delta}. \qquad (2)$$

Thus, $c_{\boldsymbol{j}}^\ell = P(\omega^{[\boldsymbol{j}]_\ell})$ yields $\mathrm{FFT}_\omega(P)$ in bit reversed order.

For the concrete computation of the coefficients $c_{\boldsymbol{k}}^s$ at stage s from the coefficients $c_{\boldsymbol{k}}^{s-1}$ at stage $s-1$, we use n so called "elementary transforms" with respect to each of the variables $x_1, \ldots, x_n$. For any $t \in \{1, \ldots, n\}$, the coefficients $c_{\boldsymbol{k}}^{s,t}$ are obtained from the coefficients $c_{\boldsymbol{k}}^{s,t-1}$ through butterflies of span $\delta = 2^{\ell-s}$ with respect to the variable x_t; this can be rewritten by means of the formula

$$\begin{pmatrix} c_{\boldsymbol{i}\delta+\boldsymbol{j}}^{s,t} \\ c_{(\boldsymbol{i}+\boldsymbol{e}_t)\delta+\boldsymbol{j}}^{s,t} \end{pmatrix} = \begin{pmatrix} 1 & \omega^{[\boldsymbol{i}_t]_s \delta} \\ 1 & -\omega^{[\boldsymbol{i}_t]_s \delta} \end{pmatrix} \begin{pmatrix} c_{\boldsymbol{i}\delta+\boldsymbol{j}}^{s,t-1} \\ c_{(\boldsymbol{i}+\boldsymbol{e}_t)\delta+\boldsymbol{j}}^{s,t-1} \end{pmatrix} \qquad (3)$$

for any $\boldsymbol{i} \in \mathbb{N}_{2^s}^n$ with even t-th coordinate $\boldsymbol{i}_t$ and $\boldsymbol{j} \in \mathbb{N}_\delta^n$. Equation (3) can be rewritten more compactly in matrix form $\boldsymbol{c}^{s,t} = \mathbf{B}^{s,t} \boldsymbol{c}^{s,t-1}$ where $\mathbf{B}^{s,t}$ is a sparse matrix in $\mathbb{K}^{\mathbb{N}_d^n \times \mathbb{N}_d^n}$ which is naturally indexed by pairs of multi-indices in $\mathbb{N}_d^n$. Setting $\mathbf{B}^s = \mathbf{B}^{s,n} \cdots \mathbf{B}^{s,1} \in \mathbb{K}^{\mathbb{N}_d^n \times \mathbb{N}_d^n}$, we also obtain a short formula for $\boldsymbol{c}^s$ as a function of $\boldsymbol{c}^{s-1}$:

$$\boldsymbol{c}^s = \mathbf{B}^s \boldsymbol{c}^{s-1} \qquad (4)$$

Remark that the matrices $\mathbf{B}^{s,1}, \ldots, \mathbf{B}^{s,n}$ commute pairwise. Notice also that each of the rows and columns of $\mathbf{B}^{s,t}$ has at most 2 non zero entries and consequently those of $\mathbf{B}^s$ contains at most 2^n non zero entries. For this reason, we can apply the matrices $\mathbf{B}^{s,t}$ (resp. $\mathbf{B}^s$) within $\mathcal{O}(|\mathbb{N}_d^n|) = \mathcal{O}(d^n)$ (resp. $\mathcal{O}(n\, d^n)$) operations in $\mathbb{K}$. Finally, the full n-dimensional FFT of length $d = 2^\ell$ costs $\mathsf{F}(d, n) := 3/2\, \ell\, n\, d^n = 3/2\, d^n \log(d^n)$ operations in $\mathbb{K}$ (see [14, Section 8.2]).

Example 1. Let us make these matrices explicit for polynomials in $n=2$ variables and degree less than $d=4$, so that $\ell=2$. We start with the first decimation in x_1 whose butterflies of span δ are captured by $\mathbf{B}^{s,t}$ with $\delta = 2^{\ell-s} = 2^1$, $s=1$ and $t=1$. It takes as input $\boldsymbol{c}_{\boldsymbol{k}}^0 = \boldsymbol{c}_{\boldsymbol{k}}^{1,0} := P_{\boldsymbol{k}}$ and outputs $\boldsymbol{c}_{\boldsymbol{k}}^{1,1}$ for $\boldsymbol{k} \in \mathbb{N}_4^2 = \{0, \ldots, 3\}^2$. For any $\boldsymbol{j}_1 \in \mathbb{N}_2 = \{0, 1\}$ and $\boldsymbol{k}_2 \in \{0, \ldots, 3\}$, we let

$$\begin{pmatrix} \boldsymbol{c}_{(\boldsymbol{j}_1,\boldsymbol{k}_2)}^{1,1} \\ \boldsymbol{c}_{(2+\boldsymbol{j}_1,\boldsymbol{k}_2)}^{1,1} \end{pmatrix} = \begin{pmatrix} 1 & 1 \\ 1 & -1 \end{pmatrix} \begin{pmatrix} \boldsymbol{c}_{(\boldsymbol{j}_1,\boldsymbol{k}_2)}^{1,0} \\ \boldsymbol{c}_{(2+\boldsymbol{j}_1,\boldsymbol{k}_2)}^{1,0} \end{pmatrix}.$$

So, $\mathbf{B}^{1,1}$ is a 16 by 16 matrix, which is made of diagonal blocks of $\left(\begin{smallmatrix} 1 & 1 \\ 1 & -1 \end{smallmatrix}\right)$ in a suitable basis. The decimation in x_2 during stage 1 is similar :

$$\begin{pmatrix} \boldsymbol{c}_{(\boldsymbol{k}_1,\boldsymbol{j}_2)}^{1,2} \\ \boldsymbol{c}_{(\boldsymbol{k}_1,2+\boldsymbol{j}_2)}^{1,2} \end{pmatrix} = \begin{pmatrix} 1 & 1 \\ 1 & -1 \end{pmatrix} \begin{pmatrix} \boldsymbol{c}_{(\boldsymbol{k}_1,\boldsymbol{j}_2)}^{1,1} \\ \boldsymbol{c}_{(\boldsymbol{k}_1,2+\boldsymbol{j}_2)}^{1,1} \end{pmatrix}$$

for $\boldsymbol{j}_2 \in \mathbb{N}_2$ and $\boldsymbol{k}_1 \in \mathbb{N}_4$. Consequently, $\mathbf{B}^{1,2}$ is made of diagonal blocks $\left(\begin{smallmatrix} 1 & 1 \\ 1 & -1 \end{smallmatrix}\right)$ (in another basis than $\mathbf{B}^{1,1}$). Their product $\mathbf{B}^1 := \mathbf{B}^{1,2}\mathbf{B}^{1,1}$ corresponds to the operations

$$\begin{pmatrix} \boldsymbol{c}_{(\boldsymbol{j}_1,\boldsymbol{j}_2)}^1 \\ \boldsymbol{c}_{(2+\boldsymbol{j}_1,\boldsymbol{j}_2)}^1 \\ \boldsymbol{c}_{(\boldsymbol{j}_1,2+\boldsymbol{j}_2)}^1 \\ \boldsymbol{c}_{(2+\boldsymbol{j}_1,2+\boldsymbol{j}_2)}^1 \end{pmatrix} = \begin{pmatrix} 1 & 1 & 1 & 1 \\ 1 & -1 & 1 & -1 \\ 1 & 1 & -1 & -1 \\ 1 & -1 & -1 & 1 \end{pmatrix} \begin{pmatrix} \boldsymbol{c}_{(\boldsymbol{j}_1,\boldsymbol{j}_2)}^0 \\ \boldsymbol{c}_{(2+\boldsymbol{j}_1,\boldsymbol{j}_2)}^0 \\ \boldsymbol{c}_{(\boldsymbol{j}_1,2+\boldsymbol{j}_2)}^0 \\ \boldsymbol{c}_{(2+\boldsymbol{j}_1,2+\boldsymbol{j}_2)}^0 \end{pmatrix}$$

for $\boldsymbol{j}_1, \boldsymbol{j}_2 \in \mathbb{N}_2$. Thus $\mathbf{B}^1$ is made of such 4 by 4 matrices on the diagonal (in yet another basis). Note that in general, the matrices $\mathbf{B}^{s,t}$ and $\mathbf{B}^s$ are still made of diagonal blocks, but these blocks vary along the diagonal.

We can sum it all up in the two following algorithms.

Algorithm Butterfly
INPUT: n, degree d, stage s, index $(2\boldsymbol{i}', \boldsymbol{j})$ of the butterfly with $\boldsymbol{i}' \in \mathbb{N}_{2^{s-1}}^n$, $\boldsymbol{j} \in \mathbb{N}_\delta^n$, root of unity ω and coefficients $\boldsymbol{c} \in \mathbb{K}^{\mathbb{N}_2^n}$ for the butterfly ($\boldsymbol{c}_{\boldsymbol{b}} := \boldsymbol{c}_{(2\boldsymbol{i}'+\boldsymbol{b})\delta+\boldsymbol{j}}^{s-1}$ for $\boldsymbol{b} \in \mathbb{N}_2^n$).
OUTPUT: the output of the butterfly $\boldsymbol{d} \in \mathbb{K}^{\mathbb{N}_2^n}$

For $t=1, \ldots, n$ do //decimation in x_t
  For $\boldsymbol{b} \in \mathbb{N}_2^n = \{0,1\}^n$ such that $\boldsymbol{b}_t = 0$ do
    $r = \omega^{[2\boldsymbol{i}_t']_s \delta} \boldsymbol{c}_{\boldsymbol{b}+\boldsymbol{e}_t}$ //butterfly in one variable
    $\boldsymbol{d}_{\boldsymbol{b}} = \boldsymbol{c}_{\boldsymbol{b}} + r$
    $\boldsymbol{d}_{\boldsymbol{b}+\boldsymbol{e}_t} = \boldsymbol{c}_{\boldsymbol{b}} - r$
Return $\boldsymbol{d}_{\boldsymbol{b}}$

Algorithm FFT
INPUT: n, degree d, ω and coefficients $\boldsymbol{c}^0 \in \mathbb{K}^{\mathbb{N}_d^n}$ of P
OUTPUT: coefficients $\boldsymbol{c}^\ell \in \mathbb{K}^{\mathbb{N}_d^n}$ in bit reversed order

For $s=1, \ldots, \ell$ do //stage s
  $\delta := d/2^s$
  For $\boldsymbol{i}' \in \mathbb{N}_{2^{s-1}}^n$, $\boldsymbol{j} \in \mathbb{N}_{2^{\ell-s}}^n$ do //pick a butterfly
    For $\boldsymbol{b} \in \mathbb{N}_2^n$ do $\boldsymbol{c}_{\boldsymbol{b}} := \boldsymbol{c}_{(2\boldsymbol{i}'+\boldsymbol{b})\delta+\boldsymbol{j}}^{s-1}$
    $\boldsymbol{d}_{\boldsymbol{b}} = \mathbf{Butterfly}(n, d, s, \boldsymbol{i}', \boldsymbol{j}, \boldsymbol{c}_{\boldsymbol{b}})$
    For $\boldsymbol{b} \in \mathbb{N}_2^n$ do $\boldsymbol{c}_{(2\boldsymbol{i}'+\boldsymbol{b})\delta+\boldsymbol{j}}^s := \boldsymbol{d}_{\boldsymbol{b}}$
Return $\boldsymbol{c}^\ell$

In the last section, we will also use the ring isomorphism

$$\begin{array}{ccc} \mathbb{K}[\boldsymbol{x}]_d/(x_1^d - 1, \ldots, x_n^d - 1) & \to & [\mathbb{K}[\boldsymbol{x}]_\delta/(x_1^\delta - 1, \ldots, x_n^\delta - 1)]^{2^{ns}} \\ P & \mapsto & Q^s = (Q_{\boldsymbol{i}}^s(\boldsymbol{x}))_{\boldsymbol{i} \in \mathbb{N}_{2^s}^n} \end{array}$$

with $Q_{\boldsymbol{i}}^s(\boldsymbol{x}) := \sum_{\boldsymbol{j} \in \mathbb{N}_\delta^n} \left(\sum_{\boldsymbol{i}' \in \mathbb{N}_{2^s}^n} P_{\boldsymbol{i}'\delta+\boldsymbol{j}} \, \boldsymbol{\omega}^{\boldsymbol{i}\cdot\boldsymbol{i}'\delta} \right) (\boldsymbol{\omega}^{\boldsymbol{i}} \, \boldsymbol{x})^{\boldsymbol{j}}$ for $\boldsymbol{i} \in \mathbb{N}_{2^s}^n$, and $(\boldsymbol{\omega}^{\boldsymbol{i}} \, \boldsymbol{x})^{\boldsymbol{j}} = (\omega^{\boldsymbol{i}_1} x_1)^{\boldsymbol{j}_1} \cdots (\omega^{\boldsymbol{i}_n} x_n)^{\boldsymbol{j}_n}$. These polynomials generalize the decomposition of a univariate polynomials P into $P \bmod (x^{d/2} - 1)$ and $P \bmod (x^{d/2} + 1)$. We could obtain them through a decimation in frequency, but it is enough to remark that we can reconstruct them from the coefficients $\boldsymbol{c}^s$ thanks to the formula

$$Q_{\boldsymbol{i}}^s(\boldsymbol{x}) = \sum_{\boldsymbol{j} \in \mathbb{N}_\delta^n} \boldsymbol{c}_{[\boldsymbol{i}]_s\delta+\boldsymbol{j}}^s \, (\boldsymbol{\omega}^{\boldsymbol{i}} \, \boldsymbol{x})^{\boldsymbol{j}}. \qquad (5)$$

3. THE SYMMETRIC FFT

In this section, we let $G \subseteq \mathfrak{S}_n$ be a permutation group. The group G acts on the polynomials $\mathbb{K}[\boldsymbol{x}]$ *via* the map

$$\varphi_g : \left\{ \begin{array}{ccc} \mathbb{K}[\boldsymbol{x}] & \longrightarrow & \mathbb{K}[\boldsymbol{x}] \\ P(x_1, \ldots, x_n) & \longmapsto & P^g(\boldsymbol{x}) := P(x_{g(1)}, \ldots, x_{g(n)}) \end{array} \right. .$$

We denote by $\mathbb{K}[\boldsymbol{x}]^G := \mathrm{stab}_G \, \mathbb{K}[\boldsymbol{x}]$ the set of polynomials invariant under the action of G. Our main result here is that one can save a factor (roughly) $|G|$ when computing the FFT of an invariant polynomial. Our approach is in the same spirit as the one in [1], where similar statements on the savings induced by (very general) symmetries can be found. However, we are not aware of published results similar to Theorem 7 below.

3.1 The bivariate case

Let $P \in \mathbb{K}[x_1, x_2]$ be a symmetric bivariate polynomial of partial degrees less than $d=8$, so that $\ell=3$; let also $\omega \in \mathbb{K}$ be a primitive eighth root of unity. We detail on this easy example the methods used to exploit the symmetries to decrease the cost of the FFT.

The coefficients of P are placed on a 8×8 grid. The bivariate classical FFT consists in the application of butterflies of size $\delta := 2^{\ell-s}$ for s from 1 to 3, as in Figure 1. When some butterflies overlap, we draw only the shades of all but one of them. The result of stage $s=3$ is the set of evaluations $(P(\omega^{\boldsymbol{i}_1}, \omega^{\boldsymbol{i}_2}))_{\boldsymbol{i} \in \mathbb{N}_8^2}$ in bit reversed order.

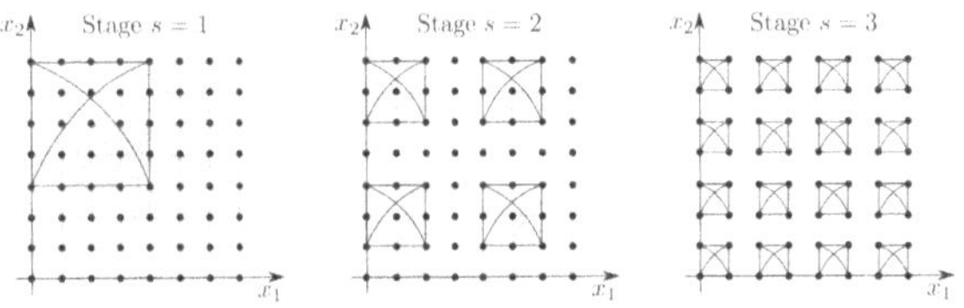

Figure 1. Classical bivariate FFT

We will remark below that each stage s preserves the symmetry of the input. In particular, since our polynomial P is symmetric, the coefficients $\boldsymbol{c}_{\boldsymbol{k}}^1$ at stage 1 are symmetric too, so we only need to compute the coefficients $\boldsymbol{c}_{(\boldsymbol{k}_1,\boldsymbol{k}_2)}^1$ for $(\boldsymbol{k}_1, \boldsymbol{k}_2)$ in the *fundamental domain* (sometimes called asymmetric unit) $0 \leqslant k_2 \leqslant k_1 < 8$. We choose to compute at stage s only the butterflies which involves at least an element of F; the set of indices that are included in a butterfly of span $\delta = 2^{\ell-s}$ that meets F will be called the extension F_δ of F.

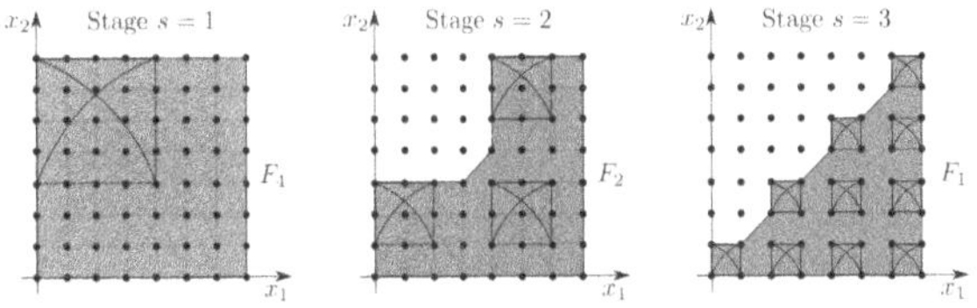

Figure 2. Symmetric bivariate FFT

Every butterfly of stage 1 meets the fundamental domain, so we do not save anything there. However, we save 4 out of 16 butterflies at stage 2 and 6 out of 16 butterflies at the third stage. Asymptotically in d, we gain a factor two in the number of arithmetic operations in $\mathbb{K}$ compared to the classical FFT; this corresponds to the order of the group.

3.2 Notation

The group G also acts on $\mathbb{N}^n$ with $g(\boldsymbol{i}) = (\boldsymbol{i}_{g(1)}, \ldots, \boldsymbol{i}_{g(n)})$ for any $\boldsymbol{i} \in \mathbb{N}^n$. This action is consistent with the action on polynomials since $\varphi_g(\boldsymbol{x}^{\boldsymbol{k}}) = \boldsymbol{x}^{g^{-1}(\boldsymbol{k})}$. Because G acts on polynomials, it acts on the vector of its coefficients. More generally, if $I \subseteq \mathbb{N}^n$ and $\boldsymbol{v} \in \mathbb{K}^{G(I)}$, we denote by $\boldsymbol{v}^g \in \mathbb{K}^I$ the vector defined by $\boldsymbol{v}^g_{\boldsymbol{i}} = \boldsymbol{v}_{g(\boldsymbol{i})}$ for all $\boldsymbol{i} \in I$. If $I \subseteq \mathbb{N}^n$ is stable by G, *i.e.* $G(I) \subseteq I$, we define the set of invariant vectors $\mathbb{K}^{I,G}$ by $\mathbb{K}^{I,G} := \{\boldsymbol{v} \in \mathbb{K}^I : \forall g \in G, \boldsymbol{v}^g = \boldsymbol{v}\}$.

For any two sets I, J satisfying $J \subseteq I \subseteq \mathbb{N}^n$, we define the restriction $\boldsymbol{v}_J \in \mathbb{K}^J$ of $\boldsymbol{v} \in \mathbb{K}^I$ by $(\boldsymbol{v}_J)_{\boldsymbol{j}} = \boldsymbol{v}_{\boldsymbol{j}}$ for all $\boldsymbol{j} \in J$. Recall the definition of the lexicographical order $\leqslant_{\text{lex}}$: for all $\boldsymbol{i}, \boldsymbol{j} \in \mathbb{N}^n$, $\boldsymbol{i} <_{\text{lex}} \boldsymbol{j}$ if there exists $k \in \{0, \ldots, n-1\}$ such that $\boldsymbol{i}_t = \boldsymbol{j}_t$ for any $1 \leqslant t \leqslant k$ and $\boldsymbol{i}_{k+1} < \boldsymbol{j}_{k+1}$. Finally for any subset $S \subseteq \mathbb{N}^n$ stable by G, we define a fundamental domain S^G of the action of G on S by $S^G := \{\boldsymbol{i} \in S : \forall g \in G, \boldsymbol{i} \geqslant_{\text{lex}} g(\boldsymbol{i})\}$ together with the projection $\pi^G : \mathbb{N}^n \rightarrow \mathbb{N}^n$ such as $\{\pi^G(\boldsymbol{i})\} := G(\{\boldsymbol{i}\}) \cap (\mathbb{N}^n)^G$.

3.3 Fundamental domains

Let $F = F^{[d]} := (\mathbb{N}^n_d)^G$ be the fundamental domain associated to the action of G on $\mathbb{N}^n_d$. Any G-symmetric vector $\boldsymbol{c} \in \mathbb{K}^{\mathbb{N}^n_d}$ can be reconstructed from its restriction $\boldsymbol{c}_F \in \mathbb{K}^F$ using the formula $\boldsymbol{c}_{\boldsymbol{i}} = (\boldsymbol{c}_F)_{\pi^G(\boldsymbol{i})}$. As it turns out, the in-place FFT algorithm from Section 2.2 admits the important property that the coefficients at all stage are still G-symmetric.

LEMMA 2. *Let $P \in \mathbb{K}[\boldsymbol{x}]_d$ and the vectors $\boldsymbol{c}^0, \ldots, \boldsymbol{c}^\ell \in \mathbb{K}^{\mathbb{N}^n_d}$ be as in Section 2.2. Then if $P \in \mathbb{K}[\boldsymbol{x}]^G_d$, $\boldsymbol{c}^0, \ldots, \boldsymbol{c}^\ell \in \mathbb{K}^{\mathbb{N}^n_d, G}$.*

PROOF. Given $P \in \mathbb{K}[\boldsymbol{x}]^G_d$, $\boldsymbol{k} \in \mathbb{N}^n_d$ and $g \in G$, we clearly have $\boldsymbol{c}^0_{g(\boldsymbol{k})} = \boldsymbol{c}^0_{\boldsymbol{k}}$. For any $s \in \{0, \ldots, \ell\}$, $\boldsymbol{i}, \boldsymbol{i}' \in \mathbb{N}^n_{2^s}$, $\boldsymbol{j} \in \mathbb{N}^n_{2^{\ell-s}}$ and $g \in G \subseteq \mathfrak{S}_n$, we also notice that $g(\boldsymbol{i}\, 2^{\ell-s} + \boldsymbol{j}) = g(\boldsymbol{i})\, 2^{\ell-s} + g(\boldsymbol{j})$, $[g(\boldsymbol{i})]_s = g([\boldsymbol{i}]_s)$ and $g(\boldsymbol{i}) \cdot g(\boldsymbol{i}') = \boldsymbol{i} \cdot \boldsymbol{i}'$. Hence, using Equation (2), we get

$$\begin{aligned}
\boldsymbol{c}^s_{g(\boldsymbol{i}2^{\ell-s}+\boldsymbol{j})} &= \boldsymbol{c}^s_{g(\boldsymbol{i})2^{\ell-s}+g(\boldsymbol{j})} \\
&= \sum_{\boldsymbol{i}' \in \mathbb{N}^n_{2^s}} P_{\boldsymbol{i}'2^{\ell-s}+g(\boldsymbol{j})}\, \omega^{[g(\boldsymbol{i})]_s \cdot \boldsymbol{i}' 2^{\ell-s}} \\
&= \sum_{\boldsymbol{i}' \in \mathbb{N}^n_{2^s}} P_{g(\boldsymbol{i}')2^{\ell-s}+g(\boldsymbol{j})}\, \omega^{[g(\boldsymbol{i})]_s \cdot g(\boldsymbol{i}')2^{\ell-s}} \\
&= \sum_{\boldsymbol{i}' \in \mathbb{N}^n_{2^s}} P_{g(\boldsymbol{i}'2^{\ell-s}+\boldsymbol{j})}\, \omega^{g([\boldsymbol{i}]_s) \cdot g(\boldsymbol{i}')2^{\ell-s}} \\
&= \sum_{\boldsymbol{i}' \in \mathbb{N}^n_{2^s}} P_{\boldsymbol{i}'2^{\ell-s}+\boldsymbol{j}}\, \omega^{[\boldsymbol{i}]_s \cdot \boldsymbol{i}'2^{\ell-s}} \\
&= \boldsymbol{c}^s_{\boldsymbol{i}2^{\ell-s}+\boldsymbol{j}}.
\end{aligned}$$

Thus, $\boldsymbol{c}^s_{g(\boldsymbol{k})} = \boldsymbol{c}^s_{\boldsymbol{k}}$ for all $\boldsymbol{k} \in \mathbb{N}^n_d$, whence $\boldsymbol{c}^0, \ldots, \boldsymbol{c}^\ell \in \mathbb{K}^{\mathbb{N}^n_d, G}$. □

This lemma implies that for the computation of the FFT of a G-symmetric polynomial P, it suffices to compute the projections $\boldsymbol{c}^0_F, \ldots, \boldsymbol{c}^\ell_F$.

In order to apply formula (4) for the computation of $\boldsymbol{c}^s_F$ as a function of $\boldsymbol{c}^{s-1}_F$, it is not necessary to completely reconstruct $\boldsymbol{c}^{s-1}$, due to the sparsity of the matrix $\mathbf{B}^s$. Instead, we define the δ*-expansion* of the set F by $F_\delta = \{\boldsymbol{k} \boxplus \boldsymbol{\epsilon}\, \delta : \boldsymbol{k} \in F, \boldsymbol{\epsilon} \in \mathbb{N}^n_2\}$, where $\boldsymbol{i} \boxplus \boldsymbol{j}$ stands for the "bitwise exclusive or" of $\boldsymbol{i}$ and $\boldsymbol{j}$. For any $\boldsymbol{k} \in \mathbb{N}^n_d$, the set $\{\boldsymbol{k} \boxplus \boldsymbol{\epsilon}\, \delta : \boldsymbol{\epsilon} \in \mathbb{N}^n_2\}$ describes the vertices of the butterfly of span δ that includes the point $\boldsymbol{k}$. Thus, F_δ is indeed the set of indices of $\boldsymbol{c}^{s-1}$ that are involved in the computation of $\boldsymbol{c}^s$ *via* the formula $\boldsymbol{c}^s = \mathbf{B}^s \boldsymbol{c}^{s-1}$.

LEMMA 3. *For any $\delta \in \{1, 2, 4, \ldots, 2^{\ell-1}\}$, let $F_{[\delta]} = \{\boldsymbol{k} \boxplus \boldsymbol{\epsilon} : \boldsymbol{k} \in F, \boldsymbol{\epsilon} \in \mathbb{N}^n_\delta\}$, so that $F_\delta \subseteq F_{[2\delta]}$. Then we have $F_{[\delta]} = \delta\, F^{[d/\delta]} + \mathbb{N}^n_\delta$.*

PROOF. Assume that $\boldsymbol{i} \in F^{[d/\delta]}$. Then clearly, $\delta\, \boldsymbol{i} \in F$, whence $\delta\, \boldsymbol{i} + \mathbb{N}^n_\delta \subseteq F_{[\delta]}$. Conversely, if $\boldsymbol{i} \in F_{[\delta]}$, then there exists a $\boldsymbol{j} \in F$ with $\boldsymbol{j} = \boldsymbol{i} \boxplus \boldsymbol{\epsilon}$ and $\boldsymbol{\epsilon} \in \mathbb{N}^n_\delta$. Let $\boldsymbol{k} = \lfloor \boldsymbol{j}/\delta \rfloor$. For any $g \in G$, we have $g(\boldsymbol{k}) = \lfloor g(\boldsymbol{j})/\delta \rfloor \leqslant_{\text{lex}} \lfloor \boldsymbol{j}/\delta \rfloor = \boldsymbol{k}$. Consequently, $\lfloor \boldsymbol{j}/\delta \rfloor \in F^{[d/\delta]}$, whence $\boldsymbol{i} \in \delta\, \lfloor \boldsymbol{j}/\delta \rfloor + \mathbb{N}^n_\delta$. □

For the proof of the next lemma, for $(j, k) \in \{1, \ldots, n\}^2$, define $\Delta_{j,k} = \{\boldsymbol{i} \in \mathbb{N}^n : \boldsymbol{i}_j = \boldsymbol{i}_k\}$ and $\Upsilon = \mathbb{N}^n_d \setminus \bigcup_{k \neq j} \Delta_{j,k}$.

LEMMA 4. *There exists a constant C (depending on n) such that $\left| |F^{[d]}| - \frac{d^n}{|G|} \right| \leqslant C\, d^{n-1}$.*

PROOF. With the notation above, we have $|\Delta_{j,k} \cap \mathbb{N}^n_d| = d^{n-1}$. Taking $C = \binom{n}{2}$, it follows that $\left| \left(\bigcup_{k \neq j} \Delta_{j,k} \right) \cap \mathbb{N}^n_d \right| \leqslant C\, d^{n-1}$. On the other hand, the orbit of any element in Υ under G contains exactly $|G|$ elements and one element in F. In other words, $|F \cap \Upsilon| = |\Upsilon|/|G|$ and so $0 \leqslant |F| - |\Upsilon|/|G| \leqslant C\, d^{n-1}$. Finally, $0 \leqslant (d^n - |\Upsilon|)/|G| \leqslant C/|G|\, d^{n-1}$ which implies $-C/|G|\, d^{n-1} \leqslant |F| - d^n/|G| \leqslant C\, d^{n-1}$. □

3.4 The symmetric FFT

Let $\boldsymbol{c}^{s-1}_{F_\delta}$ be the restriction of $\boldsymbol{c}^{s-1}$ to F_δ and notice that $\mathbb{K}^{F_\delta}$ is stable under $\mathbf{B}^s$, *i.e.* $\mathbf{B}^s(\mathbb{K}^{F_\delta}) \subseteq \mathbb{K}^{F_\delta}$. Therefore the restriction $\mathbf{B}^s_{\mathbb{K}^{F_\delta}}$ of the map $\mathbf{B}^s : \mathbb{K}^{\mathbb{N}^n_d} \rightarrow \mathbb{K}^{\mathbb{N}^n_d}$ to vectors in $\mathbb{K}^{F_\delta}$ is a $\mathbb{K}$-algebra morphism from $\mathbb{K}^{F_\delta}$ to $\mathbb{K}^{F_\delta}$. By construction, we now have $\boldsymbol{c}^s_{F_\delta} = \mathbf{B}^s_{\mathbb{K}^{F_\delta}} \boldsymbol{c}^{s-1}_{F_\delta}$. This allows us to compute $\boldsymbol{c}^s_F$ as a function of $\boldsymbol{c}^{s-1}_F$ using

$$\boldsymbol{c}^s_F \;=\; (\mathbf{B}^s_{\mathbb{K}^{F_\delta}} \xi_\delta(\boldsymbol{c}^{s-1}_F))_F \qquad (6)$$

where $\xi_\delta(\boldsymbol{c}^{s-1}_F)$ denotes the δ-expansion $\boldsymbol{c}^{s-1}_{F_\delta}$ of $\boldsymbol{c}^{s-1}_F$.

Remark 5. We could have saved a few more operations by only computing the coefficients $\boldsymbol{c}^s_F$. To do so, we would have performed just the operations of a butterfly corresponding to vertices inside F. However, the potential gain of complexity is only linear in the numbers of monomials d^n.

The formula (6) yields a straightforward way to compute the direct FFT of a G-symmetric polynomial:

Algorithm Symmetric-FFT
INPUT: n, d and coefficients $\boldsymbol{c}^0_F \in \mathbb{K}^{\mathbb{N}^n_d}$ of $P \in \mathbb{K}[\boldsymbol{x}]^G_d$ in F
OUTPUT: $\boldsymbol{c}^\ell_F \in \mathbb{K}^{\mathbb{N}^n_d}$ in bit reversed order

For $s = 1, \ldots, \ell$ do
  $\delta := d/2^s$
  For $\boldsymbol{i}' \in \mathbb{N}^n_{2^{s-1}}$, $\boldsymbol{j} \in \mathbb{N}^n_{2^{\ell-s}}$ s.t. $2\boldsymbol{i}'\delta + \boldsymbol{j} \in F_\delta$ do
    For $\boldsymbol{b} \in \mathbb{N}^n_2$ do $\boldsymbol{c}_{\boldsymbol{b}} := \boldsymbol{c}^{s-1}_{\pi^G((2\boldsymbol{i}'+\boldsymbol{b})\delta+\boldsymbol{j})}$
    $\boldsymbol{d}_{\boldsymbol{b}} =$ **Butterfly**$(n, d, s, \boldsymbol{i}', \boldsymbol{j}, \boldsymbol{c}_{\boldsymbol{b}})$
    For $\boldsymbol{b} \in \mathbb{N}^n_2$ do $\boldsymbol{c}^s_{\pi^G((2\boldsymbol{i}'+\boldsymbol{b})\delta+\boldsymbol{j})} := \boldsymbol{d}_{\boldsymbol{b}}$
Return $\boldsymbol{c}^\ell$

Remark 6. The main challenge for an actual implementation of our algorithms consists in finding a way to iterate on sets like F_δ without too much overhead. We give a hint on how to iterate over F in Lemma 8 (see also [23] for similar considerations).

The inverse FFT can be computed classically by unrolling the loops in the inverse order and inverting the operations of the butterflies. The main result of this section is then the following theorem, where $\mathsf{F}(d,n)$ denotes the cost of a full n-dimensional FFT of multi-degree d.

THEOREM 7. *For fixed n and G, and for $d\to\infty$, the direct (resp. inverse) symmetric FFT can be computed in time*

$$\mathsf{T}(d,n) \;=\; \tfrac{1}{|G|}\,\mathsf{F}(d,n)+\mathcal{O}(d^n).$$

PROOF. At stage s of the computation, the ratio between the computation of $\boldsymbol{c}_F^s$ as a function of $\boldsymbol{c}_F^{s-1}$ and $\boldsymbol{c}^s$ as a function of $\boldsymbol{c}^{s-1}$ is $|F_{2^{\ell-s}}|/|\mathbb{N}_d^n|$, so that $\frac{\mathsf{T}(d,n)}{\mathsf{F}(d,n)}\leqslant\frac{|F_{2^{\ell-1}}|+|F_{2^{\ell-2}}|+\cdots+|F_2|}{\ell\,|\mathbb{N}_d^n|}$. Lemmas 3 and 4 thus imply $|F_\delta|\leqslant |F_{[2\delta]}|\leqslant|F^{[d/(2\delta)]}|\,(2\,\delta)^n\leqslant d^n/|G|+2\,C\,\delta\,d^{n-1}$, whence $\frac{\mathsf{T}(d,n)}{\mathsf{F}(d,n)}\leqslant\frac{\ell\,d^n/|G|+2\,C\,d^n}{\ell\,d^n}=\frac{1}{|G|}+\frac{2\,C}{\ell}$. The result follows for $d\to\infty$. □

Finally, the following lemma gives a description of an "open" subset of F, characterized only by simple inequalities. Although we do not describe implementation questions in detail here, we point out that this simple characterization would naturally be used in order to iterate over the sets F_δ, possibly using tree-based techniques as in [18].

LEMMA 8. *Let Π be the set of $(j,k)\in\{1,\ldots,n\}^2$ such that $j<k$ and $\exists g\in G$ such that $g(i)=i$ for $i\leqslant j$ and $g(j)=k$. For $(j,k)\in\{1,\ldots,n\}^2$, define $H_{j,k}=\{\boldsymbol{i}\in\mathbb{N}^n:\boldsymbol{i}_j>\boldsymbol{i}_k\}$. Then*

$$F\cap\Upsilon \;=\; \bigcap_{(j,k)\in\Pi} H_{j,k}\cap\mathbb{N}_d^n.$$

PROOF. Assume that $\boldsymbol{i}\in\Upsilon$ does not lie in F. Then $g(\boldsymbol{i})>_{\text{lex}}\boldsymbol{i}$ for some $g\in G$. Let $j\in\{1,\ldots,n\}$ be minimal with $k=g(j)\neq j$, whence $k>j$ and $(j,k)\in\Pi$. Since $\boldsymbol{i}_j\neq\boldsymbol{i}_k$, it follows that $\boldsymbol{i}_k>\boldsymbol{i}_j$, so $\boldsymbol{i}\notin H_{j,k}$. Inversely, assume that $\boldsymbol{i}\in\mathbb{N}_d^n$ does not lie in $\bigcap_{(j,k)\in\Pi}H_{j,k}$, so that there exists $(j,k)\in\Pi$ with $\boldsymbol{i}\notin H_{j,k}$. Let $g\in G$ be such that $g(i)=i$ for $i\leqslant j$ and $g(j)=k$. Then, $g(\boldsymbol{i})>_{\text{lex}}\boldsymbol{i}$. □

4. THE LATTICE FFT

In this section, we deal with polynomials supported on lattices; our main result is an algorithm for the multiplication of such polynomials using Fourier transforms.

The first subsection introduces the main objects we will need (a lattice Λ and its dual Γ); then, we will give an algorithm, based on Smith normal form computation, for the FFT on what we will call a *basic domain*, and we will finally deduce a multiplication algorithm for the general case.

The techniques we use, based on Smith normal form computation, can be found in several papers, originating from [15]; in particular, the results in Section 4.2 are essentially in [35] (in a more general form).

4.1 Lattice polynomials

Assume that $\mathbb{K}$ admits a primitive k-th root of unity for any order $k>1$, which we will denote by $\mathrm{e}^{2\pi\mathrm{i}/k}$. Let Λ be a free $\mathbb{Z}$-submodule of $\mathbb{Z}^n$ of rank n, generated by the vectors $\boldsymbol{\lambda}_1,\ldots,\boldsymbol{\lambda}_n\in\mathbb{N}^n$. Then $\mathbb{K}[\boldsymbol{x}]_\Lambda=\{P\in\mathbb{K}[\boldsymbol{x}]:\operatorname{supp}P\subseteq\Lambda\}$ is a subring of $\mathbb{K}[\boldsymbol{x}]$, and we will call elements of $\mathbb{K}[\boldsymbol{x}]_\Lambda$ *lattice polynomials*. First, we show that these polynomials are the invariant polynomial ring for a diagonal matrix group.

The set $\mathbb{Q}^n$ acts on $\mathbb{K}[\boldsymbol{x}]$ *via*, for any $\boldsymbol{\gamma}=(\gamma_1,\ldots,\gamma_n)\in\mathbb{Q}^n$, the isomorphism $\varphi_{\boldsymbol{\gamma}}$ of $\mathbb{K}[\boldsymbol{x}]$ given by $\varphi_{\boldsymbol{\gamma}}$: $P(x_1,\ldots,x_n)\mapsto P(\mathrm{e}^{2\pi\mathrm{i}\gamma_1}x_1,\ldots,\mathrm{e}^{2\pi\mathrm{i}\gamma_n}x_n)$. Note that $\varphi_{\boldsymbol{\gamma}+\boldsymbol{v}}=\varphi_{\boldsymbol{\gamma}}$ for any $\boldsymbol{v}\in\mathbb{Z}^n$. The action of $\varphi_{\boldsymbol{\gamma}}$ on the monomial $\boldsymbol{x}^{\boldsymbol{\lambda}_i}$ is given by $\varphi_{\boldsymbol{\gamma}}(\boldsymbol{x}^{\boldsymbol{\lambda}_i})=\mathrm{e}^{2\pi\mathrm{i}(\boldsymbol{\lambda}_i\cdot\boldsymbol{\gamma})}\boldsymbol{x}^{\boldsymbol{\lambda}_i}$.

In particular, all elements of $\mathbb{K}[\boldsymbol{x}]_\Lambda$ are invariants under the action of $\varphi_{\boldsymbol{\gamma}}$ if and only if

$$^{t}\boldsymbol{\Lambda}\,\boldsymbol{\gamma}\;\in\;\mathbb{Z}^n, \tag{7}$$

where $\boldsymbol{\Lambda}\in\mathbb{N}^{n\times n}$ is the matrix with columns $\boldsymbol{\lambda}_1,\ldots,\boldsymbol{\lambda}_n$, that is $\boldsymbol{\Lambda}_{i,j}=(\boldsymbol{\lambda}_j)_i$. Let Γ be the dual (or reciprocal) lattice of Λ, that is the set of all $\boldsymbol{\gamma}\in\mathbb{Q}^n$ satisfying Equation (7). A basis of Γ is given by the columns of $\boldsymbol{\Gamma}={}^t\boldsymbol{\Lambda}^{-1}\in\mathbb{Q}^{n\times n}$.

Let G be the group of actions $\{\varphi_{\boldsymbol{\gamma}}:\boldsymbol{\gamma}\in\boldsymbol{\Gamma}\}$. It is generated by $\{\varphi_{\boldsymbol{\gamma}_i}\}_{1\leqslant i\leqslant n}$, where $\boldsymbol{\gamma}_i$ are the columns of $\boldsymbol{\Gamma}$. From Equation (7), we deduce that G is the diagonal matrix group of all actions $\varphi_{\boldsymbol{\gamma}}$ which leave $\mathbb{K}[\boldsymbol{x}]_\Lambda$ invariant. Conversely, because monomials are mapped to monomials by elements of G, the ring of invariants is spanned by the monomials $x^{\boldsymbol{\lambda}}$ with ${}^t\boldsymbol{\lambda}\,\boldsymbol{\Gamma}\in\mathbb{Z}^n$, *i.e.* $\boldsymbol{\lambda}$ belongs to the dual of Γ. Since the dual of the dual of a lattice is the lattice itself, we deduce that $\boldsymbol{\lambda}\in\Lambda$ and $\mathbb{K}[\boldsymbol{x}]_\Lambda=\mathbb{K}[\boldsymbol{x}]^G$. Note that only $\boldsymbol{\Gamma}$ modulo $\mathbb{Z}^{n\times n}$ matters to determine the group G.

Example 9. Consider the lattice Λ generated by $\boldsymbol{\lambda}_1=(2,0)$ and $\boldsymbol{\lambda}_2=(1,1)$, and let $\boldsymbol{\Lambda}=\begin{pmatrix}2&1\\0&1\end{pmatrix}$. The lattice polynomials $\mathbb{K}[\boldsymbol{x}]_\Lambda$ are the polynomials $\mathbb{K}[x_1^2,x_1\,x_2,x_2^2]$. We have $\boldsymbol{\Gamma}=\begin{pmatrix}1/2&0\\-1/2&1\end{pmatrix}$, so G is the group generated by $\varphi_{\boldsymbol{\gamma}_1}$: $P(x_1,x_2)\mapsto P(-x_1,-x_2)$ and $\varphi_{\boldsymbol{\gamma}_2}=\mathrm{Id}$. The lattice polynomials $\mathbb{K}[\boldsymbol{x}]_\Lambda=\mathbb{K}[x_1^2,x_1x_2,x_2^2]$ are those polynomials invariant under the symmetry $P(x_1,x_2)\mapsto P(-x_1,-x_2)$.

4.2 The lattice FFT on a basic domain

Given $\boldsymbol{d}=(\boldsymbol{d}_1,\ldots,\boldsymbol{d}_n)$, we define

$$\begin{aligned}\mathbb{K}[\boldsymbol{x}]_{\boldsymbol{d}} &= \{P\in\mathbb{K}[\boldsymbol{x}]:\deg_{x_1}(P)<\boldsymbol{d}_1,\ldots,\deg_{x_n}(P)<\boldsymbol{d}_n\}\\ \mathbb{K}[\boldsymbol{x}]_{\Lambda,\boldsymbol{d}} &= \mathbb{K}[\boldsymbol{x}]_\Lambda\cap\mathbb{K}[\boldsymbol{x}]_{\boldsymbol{d}}.\end{aligned}$$

For $i\in\{1,\ldots,n\}$, let $p_i>0$ be minimal such that $x_i^{p_i}\in\mathbb{K}[\boldsymbol{x}]_\Lambda$. We call the block $\mathbb{N}_{\boldsymbol{p}}:=\mathbb{N}_{p_1}\times\cdots\times\mathbb{N}_{p_n}$ a *basic domain* for Λ. In this subsection, we consider the computation of the FFT at order $\boldsymbol{p}$ of a polynomial $P\in\mathbb{K}[\boldsymbol{x}]_{\Lambda,\boldsymbol{p}}$.

In what follows, we set $\omega_i=\mathrm{e}^{2\pi\mathrm{i}/p_i}$ for each i, so that we have to show how to compute the set of evaluations $P(\boldsymbol{\omega}^{\boldsymbol{k}})=P(\omega_1^{\boldsymbol{k}_1},\ldots,\omega_n^{\boldsymbol{k}_n})$ for $\boldsymbol{k}\in\mathbb{N}_{\boldsymbol{p}}$. If we proceeded directly, we would compute more evaluations than the number of monomials of P, which is $|\Lambda\cap\mathbb{N}_{\boldsymbol{p}}|$. We show how to reduce the problem to computing exactly $|\Lambda\cap\mathbb{N}_{\boldsymbol{p}}|$ evaluations.

For any $\boldsymbol{\gamma}=(\gamma_1,\ldots,\gamma_n)\in\Gamma$, notice that

$$\begin{aligned}P(\boldsymbol{\omega}^{\boldsymbol{k}}) &= P(\mathrm{e}^{2\pi\mathrm{i}\boldsymbol{k}_1/p_1},\ldots,\mathrm{e}^{2\pi\mathrm{i}\boldsymbol{k}_n/p_n})\\ &= P(\mathrm{e}^{2\pi\mathrm{i}(\boldsymbol{k}_1/p_1+\gamma_1)},\ldots,\mathrm{e}^{2\pi\mathrm{i}(\boldsymbol{k}_n/p_n+\gamma_n)}).\end{aligned}$$

Therefore we would only need to consider the evaluations with multi-indices in $(1/p_1\mathbb{Z}, \ldots, 1/p_n\mathbb{Z})$ modulo Γ. There are only $|\Lambda \cap \mathbb{N}_{\boldsymbol{p}}|$ such evaluations but, as in Section 3.4, we would have to expand the fundamental domain of evaluations at each stage of the FFT to compute the butterflies.

Instead, we propose a more direct method with no domain expansion. We show that, regarding the evaluations, polynomials P in $\mathbb{K}[\boldsymbol{x}]_\Lambda$ can be written $Q(\boldsymbol{x}^{\boldsymbol{\lambda}_1'}, \ldots, \boldsymbol{x}^{\boldsymbol{\lambda}_n'})$ where $\boldsymbol{\lambda}_1'$, ..., $\boldsymbol{\lambda}_n'$ is a basis of Λ and the evaluation can be done directly in this rewritten form.

We introduce the notation $\overline{k}^p \in \{0, \ldots, p-1\}$ for the remainder of $k \in \mathbb{Z}$ by $p \in \mathbb{N}^*$. If $\boldsymbol{k} = (\boldsymbol{k}_1, \ldots, \boldsymbol{k}_n) \in \mathbb{Z}^n$ and $\boldsymbol{p} = (\boldsymbol{p}_1, \ldots, \boldsymbol{p}_n) \in (\mathbb{N}^*)^n$, we let $\overline{\boldsymbol{k}}^{\boldsymbol{p}} := (\overline{\boldsymbol{k}_1}^{p_1}, \ldots, \overline{\boldsymbol{k}_n}^{p_n})$. This notation is motivated by the remark that $\boldsymbol{x}^{\boldsymbol{\lambda}}(\boldsymbol{\omega}^{\boldsymbol{k}})$ depends only on the class of $\boldsymbol{\lambda}$ and $\boldsymbol{k}$ modulo $(p_1\mathbb{Z}, \ldots, p_n\mathbb{Z})$.

LEMMA 10. *There exists a basis $(\boldsymbol{\lambda}_1', \ldots, \boldsymbol{\lambda}_n')$ of Λ and a basis $(\boldsymbol{k}_1, \ldots, \boldsymbol{k}_n)$ of $\mathbb{Z}^n$ such that*

$$\boldsymbol{y}_i(\boldsymbol{\omega}^{\boldsymbol{k}_j}) = \mathrm{e}^{2\pi \mathrm{i} \delta_{i,j}/q_i} \tag{8}$$

where $\boldsymbol{y}_i = \boldsymbol{x}^{\overline{\boldsymbol{\lambda}_i'}^{\boldsymbol{p}}}$, the q_i's are positive integers satisfying $q_1 \mid q_2 \mid \ldots \mid q_n$ and $\delta_{i,j} = 1$ if $i = j$ and 0 otherwise.

PROOF. Let us consider the lattice of the exponents of $(\boldsymbol{x}^{\boldsymbol{\lambda}_i}(\boldsymbol{\omega}^{\boldsymbol{k}}))_{1 \leqslant i \leqslant n}$ for $\boldsymbol{k} \in \mathbb{N}^n$. If $\boldsymbol{k} = (\boldsymbol{k}_1, \ldots, \boldsymbol{k}_n) \in \mathbb{N}^n$, then

$$\boldsymbol{x}^{\boldsymbol{\lambda}_i}(\boldsymbol{\omega}^{\boldsymbol{k}}) = \mathrm{e}^{2\pi \mathrm{i}[(\boldsymbol{k}_1/p_1, \ldots, \boldsymbol{k}_n/p_n)\cdot \boldsymbol{\lambda}_i]} = \mathrm{e}^{2\pi \mathrm{i}[\boldsymbol{k}\cdot((\boldsymbol{\lambda}_i)_1/p_1, \ldots, (\boldsymbol{\lambda}_i)_n/p_n)]}.$$

We define the lattice Δ spanned by the columns of $\boldsymbol{\Delta} = {}^t\boldsymbol{\Lambda}\begin{pmatrix} 1/p_1 & 0 & 0 \\ 0 & \ddots & 0 \\ 0 & 0 & 1/p_n \end{pmatrix}$, that is $\boldsymbol{\Delta}_{i,j} = (\boldsymbol{\lambda}_i)_j/p_j$. We want to take the Smith normal form of $\boldsymbol{\Delta}$ but the coefficients of $\boldsymbol{\Delta}$ are not necessarily integers. So we multiply the matrix by $\ell := \mathrm{LCM}(p_1, \ldots, p_n)$, take the Smith normal form and divide by ℓ. Therefore there exists ${}^t\boldsymbol{H}, \boldsymbol{K} \in \mathrm{GL}_n(\mathbb{Z})$ and integers $d_n \mid \ldots \mid d_2 \mid d_1$ such that

$${}^t\boldsymbol{H}\,\boldsymbol{\Delta}\,\boldsymbol{K} = \begin{pmatrix} d_1/\ell & 0 & 0 \\ 0 & \ddots & 0 \\ 0 & 0 & d_n/\ell \end{pmatrix}. \tag{9}$$

Let us prove that $\mathbb{Z}^n \subseteq \Delta$. By definition of $p_1, \ldots, p_n$, there exists $\boldsymbol{S} \in \mathbb{Z}^{n \times n}$ such that $\boldsymbol{\Lambda}\,\boldsymbol{S} = \begin{pmatrix} p_1 & 0 & 0 \\ 0 & \ddots & 0 \\ 0 & 0 & p_n \end{pmatrix}$. Thus we have ${}^t\boldsymbol{S}\,{}^t\boldsymbol{\Lambda}\begin{pmatrix} 1/p_1 & 0 & 0 \\ 0 & \ddots & 0 \\ 0 & 0 & 1/p_n \end{pmatrix} = \mathrm{Id}$, implying that $\boldsymbol{\Delta}\,{}^t\boldsymbol{S} = \mathrm{Id}$ and our result. Because $\mathbb{Z}^n \subseteq \Delta$, we have $d_i \mid \ell$ and by setting $q_i := \ell/d_i \in \mathbb{N}^*$, we get $q_1 \mid q_2 \mid \ldots \mid q_n$.

With a geometrical point of view, the equality of Equation (9) gives the existence of the two required bases. The columns of $\boldsymbol{K}$ give the basis $(\boldsymbol{k}_1, \ldots, \boldsymbol{k}_n)$ of $\mathbb{Z}^n$ and the columns of $\boldsymbol{\Lambda}' := \boldsymbol{\Lambda}\,\boldsymbol{H}$ give the basis $(\boldsymbol{\lambda}_1', \ldots, \boldsymbol{\lambda}_n')$ of Λ. To sum up, we have

$${}^t\boldsymbol{\Lambda}'\begin{pmatrix} 1/p_1 & 0 & 0 \\ 0 & \ddots & 0 \\ 0 & 0 & 1/p_n \end{pmatrix}\boldsymbol{K} = \begin{pmatrix} 1/q_1 & 0 & 0 \\ 0 & \ddots & 0 \\ 0 & 0 & 1/q_n \end{pmatrix} \tag{10}$$

which is the matricial form of Equation (8). □

PROPOSITION 11. *The following ring morphism Φ*

$$\begin{array}{ccc} \mathbb{K}[\boldsymbol{y}]/(y_1^{q_1}-1, \ldots, y_n^{q_n}-1) & \longrightarrow & \mathbb{K}[\boldsymbol{x}]/(x_1^{p_1}-1, \ldots, x_n^{p_n}-1) \\ y_i & \longmapsto & \boldsymbol{x}^{\overline{\boldsymbol{\lambda}_i'}^{\boldsymbol{p}}} \end{array}$$

is well-defined, injective and its image is $\mathbb{K}[\boldsymbol{x}]_{\Lambda,\boldsymbol{p}}$. Moreover, if $P = \Phi(Q)$ then

$$P(\boldsymbol{\omega}^{\ell_1\boldsymbol{k}_1 + \cdots + \ell_n\boldsymbol{k}_n}) = Q(\mathrm{e}^{2\pi\mathrm{i}\ell_1/q_1}, \ldots, \mathrm{e}^{2\pi\mathrm{i}\ell_n/q_n}). \tag{11}$$

PROOF. First, we prove that Φ is well defined. For this matter we have to check that $\Phi(y_i^{q_i}-1) = \boldsymbol{x}^{q_i\overline{\boldsymbol{\lambda}_i'}^{\boldsymbol{p}}} - 1 = 0$ modulo $(x_1^{p_1}-1, \ldots, x_n^{p_n}-1)$. It is sufficient to prove that $q_i\,\overline{\boldsymbol{\lambda}_i'}^{\boldsymbol{p}} \in (p_1\mathbb{Z}, \ldots, p_n\mathbb{Z})$, which follows from Equation (10).

Then we prove Equation (11). Let $Q \in \mathbb{K}[\boldsymbol{y}]$ and $P(\boldsymbol{x}) := \Phi(Q)(\boldsymbol{x})$. As a consequence of Lemma 10, one has

$$P(\boldsymbol{\omega}^{\ell_1\boldsymbol{k}_1 + \cdots + \ell_n\boldsymbol{k}_n}) = Q(\mathrm{e}^{2\pi\mathrm{i}\ell_1/q_1}, \ldots, \mathrm{e}^{2\pi\mathrm{i}\ell_n/q_n}).$$

Now, we prove that Φ is injective. Indeed if $Q \in \mathbb{K}[\boldsymbol{y}]$ satisfies $\Phi(Q) = 0$ then for all $\boldsymbol{k} \in \mathbb{N}^n$, one has $\Phi(Q)(\boldsymbol{\omega}^{\boldsymbol{k}}) = 0$. Using Equation (11), we get that for all $\ell_1, \ldots, \ell_n \in \mathbb{Z}$, $Q(\mathrm{e}^{2\pi\mathrm{i}\ell_1/q_1}, \ldots, \mathrm{e}^{2\pi\mathrm{i}\ell_n/q_n}) = 0$. As a result, Q belongs to the ideal generated by $(y_1^{q_1}-1, \ldots, y_n^{q_n}-1)$.

Finally it is trivial to see that the image of Φ is included in $\mathbb{K}[\boldsymbol{x}]_{\Lambda,\boldsymbol{p}}$. Reciprocally, let $\boldsymbol{x}^{\boldsymbol{\lambda}} \in \mathbb{K}[\boldsymbol{x}]_{\Lambda,\boldsymbol{p}}$. We write $\boldsymbol{\lambda} = \sum_{i=1}^n \ell_i\,\overline{\boldsymbol{\lambda}_i'}^{\boldsymbol{p}}$ with $\ell_i \in \mathbb{Z}$ and define $\bar{\boldsymbol{\lambda}} = \sum_{i=1}^n \overline{\ell_i}^{q_i}\,\overline{\boldsymbol{\lambda}_i'}^{\boldsymbol{p}}$. Now because the lattice $q_1\,\boldsymbol{\lambda}_1'\,\mathbb{Z} \oplus \cdots \oplus q_n\,\boldsymbol{\lambda}_n'\,\mathbb{Z}$ is included in $(p_1\mathbb{Z}, \ldots, p_n\mathbb{Z})$, we have that $\Phi(\boldsymbol{y}^{(\overline{\ell_1}^{q_1}, \ldots, \overline{\ell_n}^{q_n})}) = x^{\bar{\boldsymbol{\lambda}}} = x^{\boldsymbol{\lambda}}$ modulo $(x_1^{p_1}-1, \ldots, x_n^{p_n}-1)$. □

In particular, the FFT of P is uniquely determined by its restriction to evaluations of the form $P(\boldsymbol{\omega}^{\boldsymbol{k}})$ with $\boldsymbol{k} = \ell_1\,\boldsymbol{k}_1 + \cdots + \ell_n\,\boldsymbol{k}_n$ and $\boldsymbol{\ell} = (\ell_1, \ldots, \ell_n) \in \mathbb{N}_{\boldsymbol{q}}$. Notice that Proposition 11 implies that the number $q_1 \cdots q_n$ of evaluations equals the numbers $|\Lambda \cap \mathbb{N}_{\boldsymbol{p}}|$ of monomials in $\mathbb{K}[\boldsymbol{x}]_{\Lambda,\boldsymbol{p}}$.

We define the *lattice FFT* of P to be the vector of values $(P(\boldsymbol{\omega}^{\boldsymbol{\ell}_1\boldsymbol{k}_1 + \cdots + \boldsymbol{\ell}_n\boldsymbol{k}_n}))_{\boldsymbol{\ell} \in \mathbb{N}_{\boldsymbol{q}}}$. We have thus shown that the computation of the lattice FFT of P reduces to the computation of an ordinary multivariate FFT of order $\boldsymbol{q}$.

4.3 The general lattice FFT multiplication

Assume that $\boldsymbol{d} \in (\mathbb{N}^*)^n$ is such that $p_i \mid d_i$ for each i and consider the computation of the FFT at order $\boldsymbol{d}$ of a lattice polynomial $P \in \mathbb{K}[\boldsymbol{x}]_{\Lambda,\boldsymbol{d}}$. In practice, one often has $\boldsymbol{d}_i = p_i\,2^{\ell_i}$, and this is what we will assume from now on. For simplicity, we also suppose that $\ell_1 = \cdots = \ell_n$ and denote by ℓ this common value. The results are valid in general but we would have to do more decimations in some variables than in others, which would not match the notations of Section 2.

We give here an algorithm for the multiplication of two polynomials $P_1, P_2 \in \mathbb{K}[\boldsymbol{x}]_{\Lambda,\boldsymbol{d}/2}$. We start by doing ℓ FFT stages. These stages preserve the lattice Λ, because the butterflies have a span which belongs to Λ. Therefore we compute only the butterflies whose vertices are in Λ. After these ℓ stages, we twist the coefficients thanks to Formula 5 and obtain the polynomials $Q^\ell = (Q_{\boldsymbol{i}}^\ell)$, for $\boldsymbol{i} \in \mathbb{N}_{2^\ell}^n$. As explained in Section 2.2, up to a minor modification (here, the partial degrees are not all the same), these polynomials reduce our multiplication to $2^{n\ell}$ multiplications in $\mathbb{K}[\boldsymbol{x}]_{\Lambda,\boldsymbol{p}}/(x_1^{p_1}-1, \ldots, x_n^{p_n}-1)$. Thanks to Proposition 11, we are able to perform these latter multiplications as multiplications in $\mathbb{K}[\boldsymbol{y}]_{\boldsymbol{q}}/(y_1^{q_1}-1, \ldots, y_n^{q_n}-1)$.

Algorithm Lattice-partial-FFT
INPUT: n, $\boldsymbol{d}$, ω and coefficients $\boldsymbol{c}^0 \in \mathbb{K}^{\mathbb{N}_d^n}$ of $P \in \mathbb{K}[\boldsymbol{x}]_{\Lambda,d}$
OUTPUT: the coefficients $\boldsymbol{c}^\ell \in \mathbb{K}^{\mathbb{N}_d^n}$

For $s = 1, \ldots, \ell$ do   //ℓ decimation steps in each variable
  $\delta := d/2^s$
  For $\boldsymbol{i}' \in \mathbb{N}_{2^{s-1}}^n$, $\boldsymbol{j} \in \Lambda \cap \mathbb{N}_{d/2^s}^n$ do   //a butterfly in Λ
    For $\boldsymbol{b} \in \mathbb{N}_2^n$ do $\boldsymbol{c}_{\boldsymbol{b}} := \boldsymbol{c}_{(2\boldsymbol{i}'+\boldsymbol{b})\delta+\boldsymbol{j}}^{s-1}$
    $\boldsymbol{d}_{\boldsymbol{b}} = \mathbf{Butterfly}(n, d, s, \boldsymbol{i}', \boldsymbol{j}, \boldsymbol{c}_{\boldsymbol{b}})$
    For $\boldsymbol{b} \in \mathbb{N}_2^n$ do $\boldsymbol{c}_{(2\boldsymbol{i}'+\boldsymbol{b})\delta+\boldsymbol{j}}^s := \boldsymbol{d}_{\boldsymbol{b}}$
Return $\boldsymbol{c}^\ell$

Algorithm Lattice-FFT-multiplication
INPUT: n, $\boldsymbol{d}$, ω and $P_1, P_2 \in \mathbb{K}[\boldsymbol{x}]_{\Lambda,d}$
OUTPUT: the coefficients product $P_1 P_2$

For $i = 1, 2$ do
  $c^\ell = $ **Lattice-partial-FFT**$(n, 2\boldsymbol{d}, \omega, ((P_i)_{\boldsymbol{k}})_{\boldsymbol{k}\in\mathbb{N}^n_{2d}})$
  Let $Q_i := Q^\ell \in (\mathbb{K}[\boldsymbol{x}]_{\Lambda,\boldsymbol{p}})^{\mathbb{N}^n_{2^\ell}}$ obtained from c^ℓ
  Transform Q_i as a vector of $\mathbb{K}[\boldsymbol{y}]_{\boldsymbol{q}}$ using Proposition 11
Multiply $Q = Q_1 Q_2$ in $(\mathbb{K}[\boldsymbol{y}]/(y_1^{q_1}-1, \ldots, y_n^{q_n}-1))^{\mathbb{N}^n_{2^\ell}}$
Transform Q as a vector of $\mathbb{K}[\boldsymbol{x}]_{\Lambda,\boldsymbol{p}}$ using Proposition 11
Recover c^ℓ from $Q \in (\mathbb{K}[\boldsymbol{x}]_{\Lambda,\boldsymbol{p}})^{\mathbb{N}^n_{2^\ell}}$
$c^0 = $ **Lattice-partial-FFT**$^{-1}(n, 2\boldsymbol{d}, \omega, (c^\ell_{[\boldsymbol{k}]_\ell})_{\boldsymbol{k}\in\mathbb{N}^n_{2d}})$
Return c^0

As for symmetric polynomials, the inverse algorithm **Lattice-partial-FFT**$^{-1}$ is obtained by reversing the loops and inverting the butterflies of **Lattice-partial-FFT**.

Let us analyze the cost of this algorithm. It starts with ℓ stages of classical n-dimensional FFT, computing only the butterflies whose vertices are in Λ. This amounts to $\frac{3}{2} n\, \ell\, |\Lambda \cap \mathbb{N}^n_d|$ arithmetic operations in $\mathbb{K}$. The second part is made of the transformations of Equation 5 and Proposition 11. Since we consider that operations with the lattice Λ take time $\mathcal{O}(1)$, these transformations take time $\mathcal{O}(|\Lambda \cap \mathbb{N}^n_d|)$. Finally, the componentwise multiplication of $(\mathbb{K}[\boldsymbol{y}]/(y_1^{q_1}-1, \ldots, y_n^{q_n}-1))^{\mathbb{N}^n_{2^\ell}}$ costs $2^{n\ell}\, \mathsf{M}(\boldsymbol{q}_1, \ldots, \boldsymbol{q}_n)$ where $\mathsf{M}(\boldsymbol{d})$ stands for the arithmetic complexity of polynomials multiplication in $\mathbb{K}[\boldsymbol{x}]_{\boldsymbol{d}}$. So our multiplication algorithms cost $\mathsf{T}(\boldsymbol{d}) := 3/2\; n\; \ell\; |\Lambda \cap \mathbb{N}^n_d| + 2^{n\ell}\, \mathsf{M}(\boldsymbol{q}) + \mathcal{O}(|\Lambda \cap \mathbb{N}^n_d|)$. By contrast, a symmetry-oblivious approach would consist in doing ℓ stages of classical n-dimensional FFT and then $2^{n\ell}$ multiplications in $\mathbb{K}[\boldsymbol{x}]/(x_1^{p_1}-1, \ldots, x_n^{p_n}-1)$. The cost analysis is similar to the one before and the classical approach's cost is $\mathsf{F}(\boldsymbol{d}) := 3/2\, n\, \ell\, d^n + 2^{n\ell}\, \mathsf{M}(\boldsymbol{p}) + \mathcal{O}(d^n)$.

The ratio $d^n / |\Lambda \cap \mathbb{N}^n_d|$ is exactly the volume $\mathrm{vol}(\Lambda)$ of the lattice, defined by $\mathrm{vol}(\Lambda) := \det(\boldsymbol{\Lambda}) \in \mathbb{Z}$. Under the superlinearity assumption for the function M, we get that $\mathrm{vol}(\Lambda)\, \mathsf{M}(\boldsymbol{q}) \leqslant \mathsf{M}(\boldsymbol{p})$ and we deduce the following theorem.

THEOREM 12. *For a fixed lattice Λ and $d \to \infty$, the direct (resp. inverse) lattice FFT can be computed in time*

$$\mathsf{T}(\boldsymbol{d}) \;=\; \frac{1}{\mathrm{vol}(\Lambda)} \mathsf{F}(\boldsymbol{d}) + \mathcal{O}(d^n).$$

5. THE SYMMETRIC TFT

To conclude this paper, we briefly discuss the extensions of the previous results to the Truncated Fourier Transform (TFT). With the notation of Section 2, let $I \subseteq \mathbb{N}^n_d$ be an initial segment subset of $\mathbb{N}^n_d$: for $\boldsymbol{i}, \boldsymbol{j} \in \mathbb{N}^n_d$, if $\boldsymbol{i} \in I$ and $\boldsymbol{j} \leqslant \boldsymbol{i}$, then $\boldsymbol{j} \in I$. Given $P \in \mathbb{K}[\boldsymbol{x}]_I$, the TFT of P is defined to be the vector $\hat{P} \in \mathbb{K}^I$ defined by $\hat{P}_{\boldsymbol{i}} = P(\omega^{[\boldsymbol{i}]_\ell})$, where ω is a root of unity of order d.

In [17, 18], van der Hoeven described fast algorithms for computing the TFT and its inverse. For a fixed dimension n, the cost of these algorithms is bounded by $\mathcal{O}(|I| \log |I| + d^n)$ instead of $\mathcal{O}(d^n \log d^n)$. In this section we will outline how a further acceleration can be achieved for symmetric polynomials of the types studied in Sections 3 and 4.

5.1 The symmetric TFT

For each $\delta \in \{1, \ldots, 2^\ell\}$, let $I_{[\delta]} = \{\boldsymbol{i} \boxplus \boldsymbol{\epsilon} : \boldsymbol{i} \in I, \boldsymbol{\epsilon} \in \mathbb{N}^n_\delta\}$. The entire computation of the TFT and its inverse can be represented schematically by a graph Γ. The vertices of the graph are pairs $(s, \boldsymbol{i})$ with $s \in \{0, \ldots, \ell\}$ and $\boldsymbol{i} \in I_{[2^{\ell-s}]}$. The edges are between vertices $(s, \boldsymbol{i})$ and $(s+1, \boldsymbol{j})$ with $\boldsymbol{i} = \boldsymbol{j} \boxplus (2^{\ell-s-1}\, \boldsymbol{\epsilon})$ and $\boldsymbol{\epsilon} \in \mathbb{N}^n_2$. The edge is labeled by a constant that we will simply write $c_{s,\boldsymbol{i},\boldsymbol{j}}$ such that

$$P^{s+1}_{\boldsymbol{j}} \;=\; \sum_{\boldsymbol{i}} c_{s,\boldsymbol{i},\boldsymbol{j}}\, P^s_{\boldsymbol{i}}. \tag{12}$$

For a direct TFT, we are given $P^0_{\boldsymbol{i}}$ on input and "fill out" the remaining values $P^s_{\boldsymbol{i}}$ for increasing values of s using (12). In the case of an inverse TFT, we are given the $P^\ell_{\boldsymbol{i}}$ with $\boldsymbol{i} \in I$ on input, as well as the coefficients $P^0_{\boldsymbol{i}} = 0$ with $\boldsymbol{i} \in \mathbb{N}^n_d \setminus I$. We next "fill out" the remaining values $P^s_{\boldsymbol{i}}$ using the special algorithm described in [18].

Now let G be as in Section 3 and assume that I is stable under G. Then each of the $I_{[\delta]}$ is also stable under G. Furthermore, any $\boldsymbol{i} \in I$ lies in the orbit of some element of $I \cap F$ under the action of G. Let $(I \cap F)_{[\delta]} = \{\boldsymbol{i} \boxplus \boldsymbol{\epsilon} : \boldsymbol{i} \in I \cap F, \boldsymbol{\epsilon} \in \mathbb{N}^n_\delta\}$. Given a G-symmetric input polynomial on input, the idea behind the symmetric TFT is to use the restriction Γ' of the above graph Γ by keeping only those vertices $(s, \boldsymbol{i})$ such that $\boldsymbol{i} \in (F \cap I)_{[2^{\ell-s}]}$. The symmetric TFT and its inverse can then be computed by intertwining the above "filling out" process with steps in which we compute $P^s_{g(\boldsymbol{i})} = P^s_{\boldsymbol{i}}$ for all $\boldsymbol{i} \in \mathbb{N}^n_d$ and $g \in G$ such that $P^s_{\boldsymbol{i}}$ is known but not $P^s_{g(\boldsymbol{i})}$.

The computational complexities of the symmetric TFT and its inverse are both proportional to the size $|\Gamma'|$ of Γ'. For many initial domains I of interest, it can be shown that $|\Gamma'| \sim |\Gamma| / |G|$, as soon as d gets large. This is for instance the case for $I = \Sigma_k = \{\boldsymbol{i} \in \mathbb{N}^n_d : i_1 + \cdots + i_n \leqslant k\}$, when $k \to \infty$, and where $\ell = \lceil \log_2 k \rceil$. Note that Σ_k is stable under any permutation group. Indeed, using similar techniques as in the proof of Theorem 7, we first show that $|(\Sigma_k \cap F)_{[\delta]} - (\Sigma_k)_{[\delta]} / |G|| = \mathcal{O}(\delta\, d^{n-1})$, and then conclude in a similar way.

5.2 The lattice TFT

Let us now consider the case of a polynomial $P \in \mathbb{K}[\boldsymbol{x}]_{\Lambda \cap I}$, with Λ as in Section 4. In order to design a fast "lattice TFT", the idea is simply to replace I by a slightly larger set J which preserves the fundamental domains. More precisely, with $p_1, \ldots, p_n$ as in Section 4, we take

$$\begin{aligned} J &= \{(p_1 i_1 + j_1, \ldots, p_n i_n + j_n) : \boldsymbol{i} \in K, \boldsymbol{j} \in \mathbb{N}_{\boldsymbol{p}}\} \\ K &= \{(\lfloor i_1/p_1 \rfloor, \ldots, \lfloor i_n/p_n \rfloor) : \boldsymbol{i} \in I\}. \end{aligned}$$

A lattice TFT of order $(p_1\, 2^{\ell_1}, \ldots, p_n\, 2^{\ell_n})$ can then be regarded as $|\Lambda \cap \mathbb{N}_{\boldsymbol{p}}|$ TFTs at order $(2^{\ell_1}, \ldots, 2^{\ell_n})$ and initial segment K, followed by $|K|$ TFTs lattice FFTs on a fundamental domain. Asymptotically speaking, we thus gain a factor $|\mathbb{N}_{\boldsymbol{p}}| / |\Lambda \cap \mathbb{N}_{\boldsymbol{p}}|$ with respect to a usual TFT with initial segment J. In the case when $I = \Sigma_k = \{\boldsymbol{i} \in \mathbb{N}^n_d : i_1 + \cdots + i_n \leqslant k\}$, we finally notice that $|J| \sim |I|$ for $k \to \infty$.

6. CONCLUSION

Let us quickly outline possible generalizations of the results of this paper.

It seems that the most general kinds of finite groups for which the techniques in this paper work are finite subgroups G of $\Pi_n\, U_n$, where Π_n is the group of $n \times n$ permutation matrices and U_n the group of diagonal matrices whose entries are all roots of unity. Indeed, any such group G both acts on the sets $\mathbb{K}[\boldsymbol{x}]$ and on the torus $\mathbb{U}^n$, where $\mathbb{U} = \{e^{2\pi i/n} : i, n \in \mathbb{N}\}$. Even more generally, the results may still hold for closed algebraic subgroups G generated by an infinite number of elements of $\Pi_n U_n$.

Many other interesting groups can be obtained as conjugates $G' = T^{-1}\, G\, T$ of groups G of the above kind. For certain applications, such as the integration of dynamical systems, the change of coordinates T can be done "once and for all" on the initial differential equations, after which the results of this paper again apply.

It is classical that the FFT of a polynomial with real coefficients can be computed twice as fast (roughly speaking) as a polynomial with complex coefficients. Real polynomials can be considered as symmetric polynomials for complex conjugation: $P(\bar{z}) = \overline{P(z)}$. Some further extensions of our setting are possible by including this kind of symmetries.

On some simple examples, we have verified that the ideas of this paper generalize to other evaluation-interpolation models for polynomial multiplication, such as Karatsuba multiplication and Toom-Cook multiplication.

We intend to study the above generalizations in more detail in a forthcoming paper.

7. REFERENCES

[1] L. Auslander, J. R. Johnson, and R. W. Johnson. An equivariant Fast Fourier Transform algorithm. Drexel University Technical Report DU-MCS-96-02, 1996.

[2] M. Ben-Or and P. Tiwari. A deterministic algorithm for sparse multivariate polynomial interpolation. In *STOC '88: Proceedings of the twentieth annual ACM symposium on Theory of computing*, pages 301–309, New York, NY, USA, 1988. ACM Press.

[3] R. Bergmann. The fast Fourier transform and fast wavelet transform for patterns on the torus. *Applied and Computational Harmonic Analysis*, 2012. In Press.

[4] P. Bürgisser, M. Clausen, and M. A. Shokrollahi. *Algebraic complexity theory*. Springer-Verlag, 1997.

[5] J. Canny, E. Kaltofen, and Y. Lakshman. Solving systems of non-linear polynomial equations faster. In *Proc. ISSAC '89*, pages 121–128, Portland, Oregon, A.C.M., New York, 1989. ACM Press.

[6] D.G. Cantor and E. Kaltofen. On fast multiplication of polynomials over arbitrary algebras. *Acta Informatica*, 28:693–701, 1991.

[7] J.W. Cooley and J.W. Tukey. An algorithm for the machine calculation of complex Fourier series. *Math. Computat.*, 19:297–301, 1965.

[8] S. Czapor, K. Geddes, and G. Labahn. *Algorithms for Computer Algebra*. Kluwer Academic Publishers, 1992.

[9] P. Duhamel and M. Vetterli. Fast Fourier transforms: a tutorial review and a state of the art. *Signal Process.*, 19(4):259–299, April 1990.

[10] R.J. Fateman. What's it worth to write a short program for polynomial multiplication. http://www.cs.berkeley.edu/~fateman/papers/shortprog.tex, 2010.

[11] T. S. Freeman, G. M. Imirzian, E. Kaltofen, and Y. Lakshman. DAGWOOD a system for manipulating polynomials given bystraight-line programs. *ACM Trans. Math. Software*, 14:218–240, 1988.

[12] S. Garg and É. Schost. Interpolation of polynomials given by straight-line programs. *Theoretical Computer Science*, 410(27-29):2659–2662, 2009.

[13] M. Gastineau and J. Laskar. TRIP 1.2.26. Trip reference manual, IMCCE, 2012. http://www.imcce.fr/trip/.

[14] J. von zur Gathen and J. Gerhard. *Modern Computer Algebra*. Cambridge University Press, 2-nd edition, 2002.

[15] A. Guessoum and R. Mersereau. Fast algorithms for the multi-dimensional discrete Fourier transform. *IEEE Transactions on Acoustics, Speech and Signal Processing*, 34(4):937 – 943, 1986.

[16] J. van der Hoeven. Relax, but don't be too lazy. *JSC*, 34:479–542, 2002.

[17] J. van der Hoeven. The truncated Fourier transform and applications. In J. Gutierrez, editor, *Proc. ISSAC 2004*, pages 290–296, Univ. of Cantabria, Santander, Spain, July 4–7 2004.

[18] J. van der Hoeven. Notes on the Truncated Fourier Transform. Technical Report 2005-5, Université Paris-Sud, Orsay, France, 2005.

[19] J. van der Hoeven. Newton's method and FFT trading. *JSC*, 45(8):857–878, 2010.

[20] J. van der Hoeven and G. Lecerf. On the bit-complexity of sparse polynomial multiplication. Technical report, HAL, 2010. http://hal.archives-ouvertes.fr/hal-00476223, accepted for publication in JSC.

[21] J. van der Hoeven and G. Lecerf. On the complexity of blockwise polynomial multiplication. In *Proc. ISSAC '12*, pages 211–218, Grenoble, France, July 2012.

[22] J. van der Hoeven and É. Schost. Multi-point evaluation in higher dimensions. Technical report, HAL, 2010. http://hal.archives-ouvertes.fr/hal-00477658, accepted for publication in AAECC.

[23] J. Johnson and X. Xu. Generating symmetric DFTs and equivariant FFT algorithms. In *ISSAC'07*, pages 195–202. ACM, 2007.

[24] S. C. Johnson. Sparse polynomial arithmetic. *SIGSAM Bull.*, 8(3):63–71, 1974.

[25] E. Kaltofen and Y. N. Lakshman. Improved sparse multivariate polynomial interpolation algorithms. In *ISSAC '88: Proceedings of the international symposium on Symbolic and algebraic computation*, pages 467–474. Springer Verlag, 1988.

[26] E. Kaltofen, Y. N. Lakshman, and J.-M. Wiley. Modular rational sparse multivariate polynomial interpolation. In *ISSAC '90: Proceedings of the international symposium on Symbolic and algebraic computation*, pages 135–139, New York, NY, USA, 1990. ACM Press.

[27] E. Kaltofen, W. Lee, and A. A. Lobo. Early termination in Ben-Or/Tiwari sparse interpolation and a hybrid of Zippel's algorithm. In *ISSAC '00: Proceedings of the 2000 international symposium on Symbolic and algebraic computation*, pages 192–201, New York, NY, USA, 2000. ACM Press.

[28] A Kudlicki, M. Rowicka, and Z. Otwinowski. The crystallographic Fast Fourier Transform. recursive symmetry reduction. *Acta Cryst.*, A63:465–480, 2007.

[29] V. Pan and D. Bini. *Polynomial and matrix computations*. Birkhauser, 1994.

[30] V. Y. Pan. Simple multivariate polynomial multiplication. *J. Symb. Comput.*, 18(3):183–186, 1994.

[31] D. S. Roche. Chunky and equal-spaced polynomial multiplication. *J. Symbolic Comput.*, 46(7):791–806, 2011.

[32] A. Schönhage and V. Strassen. Schnelle Multiplikation grosser Zahlen. *Computing*, 7:281–292, 1971.

[33] D. R. Stoutemyer. Which polynomial representation is best? In *Proceedings of the 1984 MACSYMA Users' Conference: Schenectady, New York, July 23–25, 1984*, pages 221–243, 1984.

[34] L. F. Ten Eyck. Crystallographic Fast Fourier Transform. *Acta Cryst.*, A29:183–191, 1973.

[35] A. Vince and X. Zheng. Computing the Discrete Fourier Transform on a hexagonal lattice. *Journal of Mathematical Imaging and Vision*, 28:125–133, 2007.

Interfacing Mathemagix with C++

Joris van der Hoeven
Laboratoire d'Informatique, UMR 7161 CNRS
Campus de l'École polytechnique
91128 Palaiseau Cedex, France
vdhoeven@lix.polytechnique.fr

Grégoire Lecerf
Laboratoire d'Informatique, UMR 7161 CNRS
Campus de l'École polytechnique
91128 Palaiseau Cedex, France
gregoire.lecerf@math.cnrs.fr

ABSTRACT

In this paper, we give a detailed description of the interface between the MATHEMAGIX language and C++. In particular, we describe the mechanism which allows us to import a C++ template library (which only permits static instantiation) as a fully generic MATHEMAGIX template library.

Categories and Subject Descriptors

D.2.11 [**SOFTWARE ENGINEERING**]: Software Architectures–*Languages*; D.3.4 [**PROGRAMMING LANGUAGES**]: Processors–*Translator writing systems and compiler generators*

General Terms

Languages, Performance

Keywords

Mathemagix; C++; generic programming; template library

1. INTRODUCTION

1.1 Motivation behind Mathemagix

Until the mid nineties, the development of computer algebra systems tended to exploit advances in the area of programming languages, and sometimes even influenced the design of new languages. The FORMAC system [2] was developed shortly after the introduction of FORTRAN. Symbolic algebra was an important branch of the artificial intelligence project at MIT during the sixties. During a while, the MACSYMA system [22, 24, 26] was the largest program written in LISP, and motivated the development of better LISP compilers.

The MODLISP [17] and SCRATCHPAD II [12] systems were at the origin of yet another interesting family of computer algebra systems, especially after the introduction of domains as function return values and dependent types [18, 20, 30]. These developments were at the forefront of language design and type theory [10, 23, 25]. SCRATCHPAD II later evolved into the AXIOM system [1, 19]. In the A# project [32, 33], later renamed into ALDOR, the language and compiler were redesigned from scratch and further purified.

After this initial period, computer algebra systems have been less keen on exploiting new ideas in language design. One important reason is that a good language for computer algebra is more important for developers than for end users. Indeed, typical end users tend to use computer algebra systems as enhanced pocket calculators, and rarely write programs of substantial complexity themselves. Another reason is specific to the family of systems that grew out of SCRATCHPAD II: after IBM's decision to no longer support the development, there has been a long period of uncertainty for developers and users on how the system would evolve. This has discouraged many of the programmers who did care about the novel programming language concepts in these systems.

In our opinion, this has led to an unpleasant current situation in computer algebra: there is a dramatic lack of a modern, sound and fast general purpose programming language. The major systems MATHEMATICA™ [34], MAPLE™ [21] and MAGMA [3] are interpreted, weakly typed (even though the dynamic types of MAGMA do admit some expressive power), besides being proprietary and expensive. The SAGE system [28] relies on PYTHON and merely contents itself to glue together various existing libraries and other software components.

The absence of modern languages for computer algebra is even more critical whenever performance is required. Nowadays, many important computer algebra libraries (such as GMP [11], MPFR [7], FLINT [13], FGB [6], etc.) are directly written in C or C++. Performance issues are also important whenever computer algebra is used in combination with numerical algorithms. We would like to emphasize that high level ideas can be important even for traditionally low level applications. For instance, in a suitable high level language it should be easy to operate on SIMD vectors of, say, 256 bit floating point numbers. Unfortunately, MPFR would have to be completely redesigned in order to make such a thing possible.

For these reasons, we have started the design of a new software system, MATHEMAGIX [15, 16], based on a compiled and strongly typed language, featuring signatures, dependent types, and overloading. MATHEMAGIX is intended as a general purpose language, which supports both functional and imperative programming styles. Although the design has greatly been influenced by SCRATCHPAD II and its successors AXIOM and ALDOR, the type system of MATH-

ISSAC'13, June 26–29, 2013, Boston, Massachusetts, USA.

EMAGIX contains several novel aspects, as described in [14]. MATHEMAGIX is also a free software, which can be downloaded from http://www.mathemagix.org.

1.2 Interfacing Mathemagix with C++

One major design goal of the MATHEMAGIX compiler is to admit a good compatibility with existing programming languages. For the moment, we have focussed on C and C++. Indeed, on the one hand, in parallel with the development of the compiler, we have written several high performance C++ template libraries for various basic mathematical structures (polynomials, matrices, series, etc.). On the other hand, the compiler currently generates C++ code.

We already stated that MATHEMAGIX was inspired by AXIOM and ALDOR in many respects. Some early work on interfacing with C++ was done in the context of ALDOR [5, 8]. There are two major differences between C++ and ALDOR which are important in this context.

On the one hand, ALDOR provides support for genuine generic functional programming: not only functions, but also data types can be passed as function arguments. For instance, one may write a routine which takes a ring R and an integer n on input and which returns the ring $R[X_1]\cdots[X_n]$. The language also provides support for dependent types: a function may very well take a ring R together with an instance x of R as its arguments or return value.

On the other hand, C++ provides support for templates. We may write a routine `cube` which takes an instance `x` of an arbitrary type `R` on input and returns `x*x*x`. However, and even though there is some work in this direction [9, 27], it is currently not possible to add the requirement that `R` must be a ring when declaring the template `cube`. Hence, the correctness of the template body `x*x*x` can only be checked at the moment when the template is instantiated for a particular type `R`. Furthermore, types and templates being all fully determined at compile time, only a finite number of these instantiations can occur in a program or library, and template parameters cannot be passed to functions as objects.

In ALDOR, there is no direct equivalent of templates. Nevertheless, it is possible to implement a function `cube` which takes a ring `R` and an instance `x` of `R` on input, and which returns `x*x*x`. It thus makes sense to consider importing C++ template libraries into ALDOR. Although [5, 8] contain a precise strategy for realizing such interfacing, part of the interface still had to be written by hand.

MATHEMAGIX features two main novelties with respect to the previous work which was done in the context of AXIOM and ALDOR. First of all, the language itself admits full support for templates with typed parameters; see our paper [14] on the type system for more details. Secondly, C++ template libraries can be imported into MATHEMAGIX in a straightforward way, without the need to write any non trivial parts of the interface by hand.

The ability to transform a C++ template library which only permits static instantiation into a fully generic template library is not straightforward. Our strategy mainly relies on the specification of the interface itself. Indeed, the interface should in particular provide the missing type information about the parameters of the C++ templates. In this paper, we will describe in more details how this mechanism works. We think that similar techniques can be applied for the generic importation of C++ templates into other languages such as ALDOR or OCAML (in particular, the fact that the MATHEMAGIX compiler generates C++ code is not essential for our mechanism to work). It might also be useful for future extensions of C++ itself.

The paper is organized as follows. In Section 2, we describe how to import and export non templated classes and functions from and to C++. In Section 3, we briefly recall how genericity works in MATHEMAGIX, and we describe what kind of C++ code is generated by the compiler for generic classes and functions. The core of the paper is Section 4, where we explain how C++ templates are imported into MATHEMAGIX. In Section 5 we summarize the main C++ libraries that have been interfaced to MATHEMAGIX, and Section 6 contains a conclusion and several ideas for future extensions.

2. INTERFACE PRINCIPLES TO C++

2.1 Preparing imports from C++

Different programming languages have different conventions for compiling programs, organizing projects into libraries, and mechanisms for separate compilation.

C++ is particularly complex, since the language does not provide any direct support for the management of big projects. Instead, this task is delegated to separate configuration and Makefile systems, which are responsible for the detection and specification of external and internal dependencies, and the determination of the correct compilation flags. Although these tasks may be facilitated up to a certain extent when using integrated development environments such as ECLIPSE™, XCODE™, or C++ BUILDER™, they usually remain non trivial for projects of a certain size.

MATHEMAGIX uses a different philosophy for managing big projects. Roughly speaking, any source file contains all information which is necessary for building the corresponding binary. Consequently, there is no need for external configuration or Makefile systems.

Whenever we import functionality from C++ into MATHEMAGIX, our design philosophy implies that we have to specify the necessary instructions for compiling and/or linking the imported code. To this effect, MATHEMAGIX provides special primitives `cpp_flags`, `cpp_libs` and `cpp_include` for specifying the compilation and linking flags, and C++ header files to be included.

For instance, the **numerix** library of MATHEMAGIX contains implementation for various numerical types. In particular, it contains wrappers for the GMP and MPFR libraries [7, 11] with implementations of arbitrary precision integers, rational numbers and floating point numbers. The MATHEMAGIX interface to import the wrapper for arbitrary precision integers starts as follows:

```
foreign cpp import {
  cpp_flags    "`numerix-config --cppflags`";
  cpp_libs     "`numerix-config --libs`";
  cpp_include "numerix/integer.hpp";
  ... }
```

On installation of the **numerix** library, a special script **numerix-config** is installed in the user's path. In the above example, we use this script in order to retrieve the compilation and linking flags. Notice also that **numerix/integer.hpp** is the C++ header file for basic arbitrary precision integer arithmetic.

2.2 Importing simple classes and functions

Ideally speaking, the bulk of an interface between MATHEMAGIX and a foreign language is simply a dictionary which specifies how concepts in one system should be mapped into the other one. For ordinary classes, functions and constants, there is a direct correspondence between MATHEMAGIX and C++, so the interface is very simple in this case.

Assume for instance that we want to map the C++ class `integer` from `integer.hpp` into the MATHEMAGIX class `Integer`, and import the basic constructors and arithmetic operations on integers. This is done by completing the previous example into:

```
foreign cpp import {
  cpp_flags   "`numerix-config --cppflags`";
  cpp_libs    "`numerix-config --libs`";
  cpp_include "numerix/integer.hpp";

  class Integer == integer;
  literal_integer: Literal -> Integer ==
    make_literal_integer;

  prefix -: Integer -> Integer == prefix -;
  infix +: (Integer, Integer) -> Integer == infix +;
  infix -: (Integer, Integer) -> Integer == infix -;
  infix *: (Integer, Integer) -> Integer == infix *;
  ... }
```

The special constructor `literal_integer` allows us to write literal integer constants such as `12345678987654321` using the traditional notation. This literal constructor corresponds to the C++ routine

```
integer make_literal_integer (const literal&);
```

where `literal` is a special C++ class for string symbols.

2.3 Syntactic sugar

The syntax of C++ is quite rigid and often directly related to implementation details. For instance, in C++ the notation `p.x` necessarily presupposes the definition of a class or structure with a field `x`. In MATHEMAGIX, the operator `postfix .x` can be defined anywhere. More generally, the language provides a lot of syntactic sugar which allows for a flexible mapping of C++ functionality to MATHEMAGIX.

Another example is type inheritance. In C++, type inheritance can only be done at the level of class definitions. Furthermore, type inheritance induces a specific low level representation in memory for the corresponding class instances. In MATHEMAGIX, we may declare any type `T` to inherit from a type `U` by defining an operator `downgrade: T -> U` as an ordinary function. This operator really acts as a converter with the special property that for any second converter `X -> T`, MATHEMAGIX automatically generates the converter `X -> U`. This allows for a more high level view of type inheritance.

MATHEMAGIX also provides a few built-in type constructors: `Alias T` provides a direct equivalent for the C++ reference types, the type `Tuple T` can be used for writing functions with an arbitrary number of arguments of the same type `T`, and `Generator T` corresponds to a stream of coefficients of type `T`. The built-in types `Alias T`, `Tuple T` and `Generator T` are automatically mapped to the C++ types `T&`, `mmx::vector<T>` and `mmx::iterator<T>` in definitions of foreign interfaces. The containers `mmx::vector<T>` and `mmx::iterator<T>` are defined in the C++ support library `basix` for MATHEMAGIX, where `mmx` represents the MATHEMAGIX namespace.

2.4 Compulsory functions

When importing a C++ class `T` into MATHEMAGIX, we finally notice that the user should implement a few compulsory operators on `T`. These operators have fixed named in C++ and in MATHEMAGIX, so it is not necessary to explicitly specify them in foreign interfaces.

The first compulsory operator is `flatten: T -> Syntactic`, which converts instances of type `T` into syntactic expression trees which can then be printed in several formats (ASCII, LISP, $\text{T}_\text{E}\text{X}_{\text{MACS}}$, etc.). The other compulsory operators are three types of equality (and inequality) tests and the corresponding hash functions. Indeed, MATHEMAGIX distinguishes between "semantic" equality, exact "syntactic" equality and "hard" pointer equality. Finally, any C++ type should provide a default constructor with no arguments.

2.5 Exporting basic functionality to C++

Simple MATHEMAGIX classes and functions can be exported to C++ in a similar way as C++ classes and functions are imported. Assume for instance that we wrote a MATHEMAGIX class `Point` with a constructor, accessors, and a few operations on points. Then we may export this functionality to C++ as follows:

```
foreign cpp export {
  class Point == point;
  point: (Double, Double) -> Point ==
    keyword constructor;
  postfix .x: Point -> Double == get_x;
  postfix .y: Point -> Double == get_y;
  middle: (Point, Point) -> Point == middle; }
```

3. CATEGORIES AND GENERICITY

Before we discuss the importation of C++ template libraries into MATHEMAGIX, let us first describe how to define generic classes and functions in MATHEMAGIX, and how such generic declarations are reflected on the C++ side. For a description of the MATHEMAGIX type system we refer the reader to [14].

MATHEMAGIX provides the `forall` construct for the declaration of generic functions. For instance, a simple generic function for the computation of a cube is the following:

```
forall (M: Monoid) cube (x: M): M == x*x*x;
```

This function can be applied to any element `x` whose type `M` is a monoid. For instance, we may write

```
c: Int == cube 3;
```

The parameters of generic functions are necessarily typed. In our example, the parameter `M` is a type itself and the type of `M` a *category*. The category `Monoid` specifies the requirements which are made upon the type `M`, and a typical declaration would be the following:

```
category Monoid == { infix *: (This, This) -> This; }
```

Hence, a type `M` is considered to have the structure of a `Monoid` in a given context, as soon as the function `infix *: (M, M) -> M` is defined in this context. Notice that the compiler does not provide any means for checking mathematical axioms that are usually satisfied, such as associativity.

Already on this simple example, we notice several important differences with the C++ "counterpart" of the declaration of `cube`:

```
template<typename M> M
cube (const M& x) { return x*x*x; }
```

First of all, C++ does not provide a means for checking that `M` admits the structure of a monoid. Consequently, the correctness of the body `return x*x*x` can only be checked for actual instantiations of the template. In particular, it is not possible to compile a truly generic version of `cube`.

By default, MATHEMAGIX always compiles functions prefixed with the `forall` keyword, such as `cube`, in a generic way. Let us briefly describe how this is implemented. First of all (and similarly to [5, 8]), the definition of the category `Monoid` gives rise to a corresponding abstract base class on the C++ side:

```
class Monoid_rep: public rep_struct {
  inline Monoid_rep ();
  virtual inline ~Monoid_rep ();
  virtual generic mul (const generic&,
                       const generic&) const = 0;
  ... };
```

A concrete monoid is a "managed pointer" (i.e. the objects to which they point are reference counted) to a derived class of `Monoid_rep` with an actual implementation of the multiplication `mul`. Instances of the MATHEMAGIX type `generic` correspond to managed pointers to objects of arbitrary types. The declaration of `cube` gives rise to the following code on the C++ side:

```
generic
cube (const Monoid& M, const generic& x) {
  // x is assumed to contain an object "of type M"
  return M->mul (x, M->mul (x, x)); }
```

The declaration `c: Int == cube 3;` gives rise to the automatic generation of a class `Int_Monoid_rep` which corresponds to the class `Int` with the structure of a `Monoid`:

```
struct Int_Ring_rep: public Ring_rep {
  ...
  generic
  mul (const generic& x, const generic& y) const {
    return as_generic<int> (from_generic<int> (x) *
                            from_generic<int> (y)); }
  ... };
```

The declaration itself corresponds to the following C++ code:

```
Monoid Int_Ring= new Int_Ring_rep ();
int c= from_generic<int>
        (cube (Int_Ring, as_generic<int> (3)));
```

Notice that we did not generate any specific instantiation of `cube` for the `Int` type. This may lead to significantly smaller executables with respect to C++ when the function `cube` is applied to objects of many different types. Indeed, in the case of C++, a separate instantiation of the function needs to be generated for each of these types during the compilation stage.

Remark 1. Of course, for very low level types such as `Int`, the use of generic functions does imply a non trivial overhead. Nevertheless, since the type `generic` is essentially a `void*`, the overhead is kept as small as possible. In particular, the overhead is guaranteed to be bounded by a fixed constant. We also notice that MATHEMAGIX provides an experimental keyword `specialize` which allows for the explicit instantiation of a generic function, thus achieving a fine compromise between speed and compiler code size.

Remark 2. Although generic functions such as `cube` are not instantiated by default, our example shows that we do have to generate special code for converting the type parameter `Int` to a `Monoid`. Although this code is essentially trivial, it may become quite voluminous when there are many different types and categories. We are still investigating how to reduce this size as much as possible while keeping the performance overhead small.

MATHEMAGIX also allows for the declaration of generic container classes; the user simply has to specify the typed parameters when declaring the class:

```
class Complex (R: Ring) == {
  re: R;
  im: R;
  constructor complex (r: R, i: R) == {
    re == r;
    im == i; } }
```

Again, only the generic version of this class is compiled by default. In particular, the internal representation of the corresponding C++ class is simply a class with two fields `re` and `im` of type `generic`.

Regarding functions and templates, a few other features of MATHEMAGIX extend those in C++:

1. General non mutual dependencies are allowed between function and template parameters and return values, as in the following example:

```
forall (R: Ring, M: Module R)
infix * (c: R, v: Vector M): Vector M ==
  [ c * x | x: M in v ];
```

2. Template parameters can be arbitrary types or (not necessarily constant) instances. For instance, one may define a container `Vec (R: Ring, n: Int)` for vectors with a fixed size.

3. Functions can be used as arguments and as values:

```
compose (f: Int -> Int, g: Int -> Int)
        (x: Int): Int == f g x;
```

Notice that AXIOM and ALDOR admit the same advantages with respect to C++.

4. IMPORTING C++ TEMPLATES

One of the most interesting aspects of our interface between MATHEMAGIX and C++ is its ability to import C++ template classes and functions. This makes it possible to provide a fully generic MATHEMAGIX interface on top of an existing C++ template library. We notice that the interface between ALDOR and C++ [5, 8] also provided a strategy for importing templates. However, the bulk of the actual work still had to be done by hand.

4.1 Example of a generic C++ import

Before coming to the technical details, let us first give a small example of how to import part of the univariate polynomial arithmetic from the C++ template library `algebramix`, which is shipped with MATHEMAGIX:

```
foreign cpp import {
  ...
  class Pol (R: Ring) == polynomial R;

  forall (R: Ring) {
    pol: Tuple R -> Pol R == keyword constructor;
    upgrade: R -> Pol R == keyword constructor;

    deg: Pol R -> Int == deg;
    postfix []: (Pol R, Int) -> R == postfix [];

    prefix -: Pol R -> Pol R == prefix -;
    infix +: (Pol R, Pol R) -> Pol R == infix +;
    infix -: (Pol R, Pol R) -> Pol R == infix -;
    infix *: (Pol R, Pol R) -> Pol R == infix *;
    ... } }
```

As is clear from this example, the actual syntax for template imports is a straightforward extension of the syntax of usual imports and the syntax of generic declarations on the MATHEMAGIX side.

Actually, the above code is still incomplete: in order to make it work, we also have to specify how the ring operations on R should be interpreted on the C++ side. This is done by *exporting* the category Ring to C++:

```
foreign cpp export
category Ring == {
  convert: Int -> This == keyword constructor;
  prefix -: This -> This == prefix -;
  infix +: (This, This) -> This == infix +;
  infix -: (This, This) -> This == infix -;
  infix *: (This, This) -> This == infix *; }
```

This means that the ring operations in C++ are the constructor from int and the usual operators +, - and *. The programmer should make sure that the C++ implementations of the imported templates only rely on these ring operations.

4.2 Generation of generic instance classes

The first thing the compiler does with the above C++ export of Ring is the creation of a C++ class capable of representing generic instances of arbitrary ring types. Any mechanism for doing this has two components: we should not only store the actual ring elements, but also the rings themselves to which they belong. This can actually be done in two ways.

The most straightforward idea is to represent an instance of a generic ring by a pair (R, x), where R is the actual ring (similar to the example of the C++ counterpart of a monoid in Section 3) and x an actual element of R. This approach has the advantage of being purely functional, but it requires non trivial modifications on the C++ side.

Indeed, whenever a function returns a ring object, we should be able to determine the underlying ring R from the input arguments. In the case of a function such as postfix []: (Pol R, Int) -> R, this means that R has to be read off from the coefficients of the input polynomial. But the most straightforward implementation of the zero polynomial does not have any coefficients! In principle, it is possible to tweak all C++ containers so as to guarantee the ability to determine the underlying generic parameters from actual instances. We have actually implemented this idea, but it required a lot of work, it involved long compilation times from the C++ side, and it violated the principle that writing a MATHEMAGIX interface for a C++ template library should essentially be trivial.

The second approach, which is less functional but more efficient, is to store the ring R in a global variable, whose value will frequently be changed in the course of actual computations. In fact, certain templates might carry more than one parameter of type Ring, in which case we need more than one global ring. For this reason, we chose to implement a container instance<Cat,Nr> for generic instances of a type of category Cat, with an additional integer parameter Nr for distinguishing between various parameters of the same category Cat. The container instance<Cat,Nr> is really a wrapper for generic:

```
template<typename Cat, int Nr>
class instance {
public:
  generic rep;
  static Cat Cur;
  inline instance (const instance& prg2):
    rep (prg2.rep) {}
  inline instance (const generic& prg):
    rep (prg) {}
  instance ();
  template<typename C1> instance (const C1& c1);
  ... };
```

For instance, objects of type instance<Ring,2> are instances of the second generic Ring parameter of templates. The corresponding underlying ring is stored in the global static variable instance<Ring,2>::Cur.

When exporting the Ring category to C++, the MATHEMAGIX compiler automatically generates generic C++ counterparts for the ring operations. For instance, the following multiplication is generated for instance<Ring,Nr>:

```
template<int Nr> inline instance<Ring,Nr>
operator * (const instance<Ring,Nr>& a1,
            const instance<Ring,Nr>& a2) {
  typedef instance<Ring,Nr> Inst;
  return Inst (Inst::Cur->mul (a1.rep, a2.rep)); }
```

Since the C++ language does not allow to directly specialize constructors of instance<Cat,Nr>, we provide a general default constructor of instance<Cat,Nr> from an arbitrary type T, which relies on the in place routine

```
void set_as (instance<Ring,Nr>&, const T&);
```

This routine can be specialized for particular categories. For instance, the converter convert: Int -> This from Ring gives rise to following routine, which induces a constructor for instance<Ring,Nr> from int:

```
template<int Nr> inline void
set_as (instance<Ring,Nr>& ret, const int& a1) {
  typedef instance<Ring,Nr> Inst;
  ret = Inst (Inst::Cur->cast (a1)); }
```

In this example, Inst::Cur->cast represents the function that sends an int into an element of the current ring.

4.3 Importing C++ templates

Now that we have a way to represent arbitrary MATHEMAGIX classes R with the structure of a Ring by a C++ type instance<Ring,Nr>, we are in a position to import arbitrary C++ templates with Ring parameters. This mechanism is best explained on an example. Consider importing the routine

```
forall (R: Ring)
infix *: (Pol R, Pol R) -> Pol R;
```

The compiler essentially generates the following C++ code for this import (the actual code is slightly more complicated and follows the *resource acquisition is initialization* idiom [29] in order to be exception safe):

```
polynomial<generic>
mul (const Ring &R,
     const polynomial<generic>& p1,
     const polynomial<generic>& p2) {
  typedef instance<Ring,1> Inst;
  Ring old_R= Inst::Cur;
  Inst::Cur= R;
  polynomial<Inst> P1= as<polynomial<Inst> > (p1);
  polynomial<Inst> P2= as<polynomial<Inst> > (p2);
  polynomial<Inst> Q = P1 * P2;
  polynomial<generic> q=
    as<polynomial<generic> > (Q);
  Inst::Cur= old_R;
  return q; }
```

There are two things to be observed in this code. First of all, for the computation of the actual product `P1 * P2`, we have made sure that `Inst::Cur` contains the ring `R` corresponding to the coefficients of the generic coefficients of the inputs `p1` and `p2`. Moreover, the old value of `Inst::Cur` is restored on exit. Secondly, we notice that `polynomial<Inst>` and `polynomial<generic>` have exactly the same internal representation. The template `as` simply casts between these two representations. In the actual code generated by the compiler, these casts are done statically without any cost, directly on pointers.

The above mechanism provides us with a fully generic way to import C++ templates. However, as long as the template parameters are themselves types which were imported from C++, it is usually more efficient to shortcut the above mechanism and directly specialize the templates on the C++ side. For instance, the MATHEMAGIX program

```
p: Pol Integer == ...;
q: Pol Integer == p * p;
```

is simply compiled into the following C++ code:

```
polynomial<integer> p= ...;
polynomial<integer> q= p * p;
```

5. INTERFACED C++ LIBRARIES

Currently, most of the mathematical features available in MATHEMAGIX are imported from C++ libraries, either of our own or external [16]. In this section, we briefly describe what these libraries provide, and the main issues we encountered.

5.1 Mathemagix libraries

C++ libraries of the MATHEMAGIX project provide the user with usual data types and mathematical objects. We have already mentioned the **basix** library which is devoted to vectors, iterators, lists, hash tables, generic objects, parsers, pretty printers, system commands, and the $\mathrm{T_EX_{MACS}}$ interface. The **numerix** library is dedicated to numerical types including integers, modular integers, rational numbers, floating point numbers, complex numbers, intervals, and balls. Univariate polynomials, power series, fraction fields, algebraic numbers, and matrices are provided by the **algebramix** library, completed by **analyziz** for when working with numerical coefficient types. Multivariate polynomials, jets, and power series and gathered in the **multimix** library. Finally **continewz** implements analytic functions and numerical homotopy continuation for polynomial system solving.

The MATHEMAGIX compiler is itself written in MATHEMAGIX on the top of the **basix** library. In order to produce a first binary for this compiler, we designed a mechanism for producing standalone C++ sources from its MATHEMAGIX sources (namely the **mmcompileregg** package). This mechanism is made available to the user via the option **--keep-cpp** of the **mmc** compiler command.

In the following example we illustrate simple calculations with analytic functions. We use the notation ==> for macro definitions. We first construct the polynomial indeterminate x of $\mathbb{C}[x]$, and convert it into the analytic function indeterminate z. We display $\exp z$, $\exp 1$, and $\exp(z+1)$ on the standard output `mmout`. Internal computations are performed up to 256 bits of precision, but printing is restricted to 5 decimal digits. Analytic functions are displayed as their underlying power series at the origin, for which we set the output order to 5.

```
include "basix/fundamental.mmx";
include "numerix/floating.mmx";
include "numerix/complex.mmx";
include "continewz/analytic.mmx";

R ==> Floating;
C ==> Complex R;
Pol ==> Polynomial C;
Afun ==> Analytic (R, C);
bit_precision := 256;

x: Pol == polynomial (complex (0.0 :> R),
                      complex (1.0 :> R));
z: Afun == x :> Pol;
f: Afun == exp z;

significant_digits := 5;
set_output_order (x :> (Series C), 5);

mmout << "f= " << f << lf;
mmout << "f (1)= " << f (1.0 :> C) << lf;
mmout << "f (1 + z)= " << move (f, 1.0 :> C) << lf;
```

Compiling and running this program in a textual terminal yields:

```
f= 1.0000 + 1.0000 * z + 0.50000 * z^2
  + 0.16667 * z^3 + 0.041667 * z^4 + O (z^5)
f (1)= 2.7183
f (1 + z)= 2.7183 + 2.7183 * z + 1.3591 * z^2
  + 0.45305 * z^3 + 0.11326 * z^4 + O (z^5)
```

5.2 External libraries

Importing a library that is completely external to the MATHEMAGIX project involves several issues. First of all, as mentioned in Section 2.4, all the data types to be imported should satisfy mild conditions in order to be properly usable from MATHEMAGIX. Usually, these conditions can easily be satisfied by writing a C++ wrapper whenever necessary.

However, when introducing new types and functions, one usually wants them to interact naturally with other librairies. For instance, if several libraries have their own arbitrarily long integer type, straightforward interfaces introduce several MATHEMAGIX types of such integers, leaving to the user the responsibility of the conversions in order to use functions of different librairies within a single program.

For C and C++ libraries not involving tempates, we prefer to design lower level interfaces to the C++ libraries of MATHEMAGIX. In this way, we focus on writing efficient converters between external and C++ MATHEMAGIX objects, and then on interfacing new functions at the MATHEMAGIX language level. This is the way we did for instance with lattice reduction of the FPLLL library [4], where we mainly had to write converters for integer matrices. Similarly the interface with FGB [6] mainly consists in converters between different representations of multivariate polynomials.

When libraries contain many data types, functions, and have their own memory management, the interface quickly becomes tedious. This situation happened with the PARI library [31]. We first created a C++ wrapper of generic PARI objects, so that wrapped objects are reference counted and have memory space allocated by MATHEMAGIX. Before calling a PARI function, the arguments are copied onto the PARI stack. Once the function has terminated, the result from the stack is wrapped into a MATHEMAGIX object. Of course converters for the different representations of integers, rationals, polynomials and matrices were needed. The following example calls the PARI function `nfbasis` to compute an integral basis of the number field defined by $x^2 + x - 1001$:

```
include "basix/fundamental.mmx";
include "mpari/pari.mmx";
Pol ==> Polynomial Integer;
x: Pol == polynomial (0 :> Integer, 1 :> Integer);
p: Pol == x^2 + x - 1001;
mmout << pari_nf_basis p << lf;
```

```
[1, 1 / 3 * x - 1 / 3]
```

6. FUTURE EXTENSIONS

The current mechanism for importing C++ template libraries has been tested for the standard mathematical libraries which are shipped with MATHEMAGIX. For this purpose, it has turned out to be very user friendly, flexible and robust. We think that other languages may develop facilities for the importation of C++ template libraries along similar lines. In the future, our approach may even be useful for adding more genericity to C++ itself. A few points deserve to be developed further:

Exporting MATHEMAGIX *containers and templates.* So far, we have focussed on the importation of C++ containers and templates, and MATHEMAGIX only allows for the exportation of simple, non generic functions and non parameterized classes. Nevertheless, one could add support for the more general exportation of generic functions and parameterized classes. Of course, information on the categories of the template parameters would be lost in this process, and the resulting templates will only allow for static instantiation.

Multi-threading. The main disadvantage of relying on global variables for storing the current values of template parameters is that this strategy is not thread-safe. In order to allow generic code to be run simultaneously by several threads, the global variables have to be replaced by fast lookup tables which determine the current values of template parameters as a function of the current thread.

Non class parameters. The current interface only allows for the importation of C++ templates with type parameters. This is not a big limitation, because templates with value parameters are only supported for built-in types and they can only be instantiated for constant values. Nevertheless, it is possible to define auxiliary classes for storing mutable static variables, and use these instead as our template parameters; notice that this is exactly the purpose of the `instance<Cat,Nr>` template. In MATHEMAGIX, we also use this mechanism for the implementation of modular arithmetic, with a modulus that can be changed during the execution. After fixing a standard convention for the creation of auxiliary classes, our implementation could be extended to the importation of C++ with "value parameters" of this kind.

Interfacing more libraries. Interfacing libraries often involves portability issues, and also create dependencies that have a risk to be broken in case the library stops being maintained. In the MATHEMAGIX project we considered that functionalities imported from an external library should be implemented even naively directly in MATHEMAGIX (excepted for GMP and MPFR). This represents a certain amount of work (for lattice reduction, Gröbner basis, finite fields, etc), but this eases testing the interfaces and allows the whole software to run on platforms where some libraries are not available.

Acknowledgments

This work has been partly supported by the DIGITEO 2009-36HD grant of the Région Ile-de-France. We would like to thank Jean-Charles Faugère for helping us in the interface with FGB, and also Karim Belabas and Bill Allombert for their precious advices in the design of our interface with the PARI library.

7. REFERENCES

[1] Axiom computer algebra system. Software available from http://wiki.axiom-developer.org.

[2] E. Bond, M. Auslander, S. Grisoff, R. Kenney, M. Myszewski, J. Sammet, R. Tobey, and S. Zilles. FORMAC an experimental formula manipulation compiler. In *Proceedings of the 1964 19th ACM national conference*, ACM '64, pages 112.101–112.1019, New York, NY, USA, 1964. ACM.

[3] W. Bosma, J. Cannon, and C. Playoust. The Magma algebra system. I. The user language. *J. Symbolic Comput.*, 24(3-4):235–265, 1997. Computational algebra and number theory (London, 1993).

[4] D. Cade, X. Pujol, and D. Stehlé. Fplll, library for LLL-reduction of Euclidean lattices, 1998. Software available from http://perso.ens-lyon.fr/damien.stehle/fplll.

[5] Y. Chicha, F. Defaix, and S. M. Watt. Automation of the Aldor/C++ interface: User's guide. Technical Report Research Report D2.2.2c, FRISCO Consoritum, 1999. Available from

http://www.csd.uwo.ca/~watt/pub/reprints/1999-frisco-aldorcpp-ug.pdf.

[6] J.-C. Faugère. FGb: A Library for Computing Gröbner Bases. In K. Fukuda, J. van der Hoeven, M. Joswig, and N. Takayama, editors, *Mathematical Software - ICMS 2010, Third International Congress on Mathematical Software, Kobe, Japan, September 13-17, 2010*, volume 6327 of *Lecture Notes in Computer Science*, pages 84–87. Springer Berlin / Heidelberg, 2010.

[7] L. Fousse, G. Hanrot, V. Lefèvre, P. Pélissier, and P. Zimmermann. MPFR: A multiple-precision binary floating-point library with correct rounding. *ACM Transactions on Mathematical Software*, 33(2), 2007. Software available from http://www.mpfr.org.

[8] M. Gaëtano and S. M. Watt. An object model correspondence for Aldor and C++. Technical Report Research Report D2.2.1, FRISCO Consortium, 1997. Available from http://www.csd.uwo.ca/~watt/pub/reprints/1997-frisco-aldorcppobs.pdf.

[9] R. Garcia, J. Järvi, A. Lumsdaine, J. G. Siek, and J. Willcock. A comparative study of language support for generic programming. In *Proceedings of the 2003 ACM SIGPLAN conference on Object-oriented programming, systems, languages, and applications (OOPSLA'03)*, October 2003.

[10] J. Y. Girard. Une extension de l'interprétation de Gödel à l'analyse, et son application à l'élimination de coupures dans l'analyse et la théorie des types. In J. E. Fenstad, editor, *Proceedings of the Second Scandinavian Logic Symposium*, pages 63–92. North-Holland Publishing Co., 1971.

[11] T. Granlund and the GMP development team. GNU MP: The GNU Multiple Precision Arithmetic Library, 1991. Software available from http://gmplib.org.

[12] J. H. Griesmer, R. D. Jenks, and D. Y. Y. Yun. *SCRATCHPAD User's Manual*. Computer Science Department monograph series. IBM Research Division, 1975.

[13] W. Hart. An introduction to Flint. In K. Fukuda, J. van der Hoeven, M. Joswig, and N. Takayama, editors, *Mathematical Software - ICMS 2010, Third International Congress on Mathematical Software, Kobe, Japan, September 13-17, 2010*, volume 6327 of *Lecture Notes in Computer Science*, pages 88–91. Springer Berlin / Heidelberg, 2010.

[14] J. van der Hoeven. Overview of the Mathemagix type system. In *Electronic proc. ASCM '12*, Beijing, China, October 2012. Available from http://hal.archives-ouvertes.fr/hal-00702634.

[15] J. van der Hoeven, G. Lecerf, B. Mourrain, et al. Mathemagix, 2002. Software available from http://www.mathemagix.org.

[16] J. van der Hoeven, G. Lecerf, B. Mourrain, Ph. Trébuchet, J. Berthomieu, D. Diatta, and A. Manzaflaris. Mathemagix, the quest of modularity and efficiency for symbolic and certified numeric computation. *ACM Commun. Comput. Algebra*, 45(3/4):186–188, 2012.

[17] R. D. Jenks. The SCRATCHPAD language. *SIGPLAN Not.*, 9(4):101–111, 1974.

[18] R. D. Jenks. MODLISP – an introduction (invited). In *Proceedings of the International Symposium on Symbolic and Algebraic Computation*, EUROSAM '79, pages 466–480, London, UK, 1979. Springer-Verlag.

[19] R. D. Jenks and R. Sutor. *AXIOM: the scientific computation system*. Springer-Verlag, New York, NY, USA, 1992.

[20] R. D. Jenks and B. M. Trager. A language for computational algebra. *SIGPLAN Not.*, 16(11):22–29, 1981.

[21] Maple user manual. Toronto: Maplesoft, a division of Waterloo Maple Inc., 2005–2012. Maple is a trademark of Waterloo Maple Inc. http://www.maplesoft.com/products/maple.

[22] W. A. Martin and R. J. Fateman. The MACSYMA system. In *Proceedings of the second ACM symposium on symbolic and algebraic manipulation*, SYMSAC '71, pages 59–75, New York, NY, USA, 1971. ACM.

[23] P. Martin-Löf. Constructive mathematics and computer programming. *Logic, Methodology and Philosophy of Science VI*, pages 153–175, 1979.

[24] Maxima, a computer algebra system (free version). Software available from http://maxima.sourceforge.net, 2011.

[25] R. Milner. A theory of type polymorphism in programming. *Journal of Computer and System Sciences*, 17:348–375, 1978.

[26] J. Moses. Macsyma: A personal history. *Journal of Symbolic Computation*, 47(2):123–130, 2012.

[27] G. Dos Reis and B. Stroustrup. Specifying C++ concepts. *SIGPLAN Not.*, 41(1):295–308, 2006.

[28] W. A. Stein et al. *Sage Mathematics Software*. The Sage Development Team, 2004. Software available from http://www.sagemath.org.

[29] B. Stroustrup. *The C++ Programming Language*. Addison-Wesley Longman Publishing Co., Inc., Boston, MA, USA, 3rd edition, 2000.

[30] R. S. Sutor and R. D. Jenks. The type inference and coercion facilities in the Scratchpad II interpreter. *SIGPLAN Not.*, 22(7):56–63, 1987.

[31] The PARI Group, Bordeaux. *PARI/GP*, 2012. Software available from http://pari.math.u-bordeaux.fr.

[32] S. Watt, P. A. Broadbery, S. S. Dooley, P. Iglio, S. C. Morrison, J. M. Steinbach, and R. S. Sutor. A first report on the A# compiler. In *Proceedings of the international symposium on symbolic and algebraic computation*, ISSAC '94, pages 25–31, New York, NY, USA, 1994. ACM.

[33] S. Watt et al. Aldor programming language. Software available from http://www.aldor.org, 1994.

[34] S. Wolfram. *Mathematica: A System for Doing Mathematics by Computer*. Addison-Wesley, second edition, 1991. Mathematica is a trademark of Wolfram Research, Inc. http://www.wolfram.com/mathematica.

Verified Error Bounds for Real Solutions of Positive-dimensional Polynomial Systems *

Zhengfeng Yang
Shanghai Key Laboratory of Trustworthy Computing
East China Normal University, Shanghai 200062, China
zfyang@sei.ecnu.edu.cn

Lihong Zhi
Key Laboratory of Mathematics Mechanization
Chinese Academy of Sciences
Beijing 100190, China
lzhi@mmrc.iss.ac.cn

Yijun Zhu
Shanghai Key Laboratory of Trustworthy Computing
East China Normal University, Shanghai 200062, China
yjzhu@ecnu.cn

ABSTRACT

In this paper, we propose two algorithms for verifying the existence of real solutions of positive-dimensional polynomial systems. The first one is based on the critical point method and the homotopy continuation method. It targets for verifying the existence of real roots on each connected component of an algebraic variety $V \cap \mathbb{R}^n$ defined by polynomial equations. The second one is based on the low-rank moment matrix completion method and aims for verifying the existence of at least one real roots on $V \cap \mathbb{R}^n$. Combined both algorithms with the verification algorithms for zero-dimensional polynomial systems, we are able to find verified real solutions of positive-dimensional polynomial systems very efficiently for a large set of examples.

Categories and Subject Descriptors: G.4 [Mathematics of computing]: Mathematical Software; I.1.2 [Symbolic and Algebraic Manipulation]: Algorithms;
General Terms: Algorithms, experimentation
Keywords: positive-dimensional polynomial systems, real solutions, verification, error bounds.

1. INTRODUCTION

Let $F(\mathbf{x}) = [f_1, \ldots, f_m]^T$ be a polynomial system in $\mathbb{Q}[\mathbf{x}] = \mathbb{Q}[x_1, \ldots, x_n]$, and $V \subset \mathbb{C}^n$ be the algebraic variety defined by:

$$f_1(x_1, \ldots, x_n) = \cdots = f_m(x_1, \ldots, x_n) = 0. \tag{1}$$

We are interested in verifying the existence of real solutions on $V \cap \mathbb{R}^n$.

Suppose $I = \langle f_1, \ldots, f_m \rangle$ is a radical ideal and V is equidimensional, i.e., the irreducible components of V have same dimensions, then a point $\hat{\mathbf{x}} \in V$ is called a *regular point* of V, or V is called smooth at $\hat{\mathbf{x}}$ if and only if the rank of the Jacobian matrix $F_{\mathbf{x}}(\hat{\mathbf{x}})$ satisfies

$$\dim V = n - \text{rank}(F_{\mathbf{x}}(\hat{\mathbf{x}})). \tag{2}$$

The set V_{reg} of regular points of V is called the *regular locus* of V. A point $\hat{\mathbf{x}}$ is called *singular* at V if and only if

$$\text{rank}(F_{\mathbf{x}}(\hat{\mathbf{x}})) < n - \dim V. \tag{3}$$

The set $V_{sing} := V \backslash V_{reg}$ is called the *singular locus* of V. If all points on V are regular, then V is called *smooth*.

Remark 1 If $I = \langle f_1, \ldots, f_m \rangle$ is not a radical ideal, then a point $\hat{\mathbf{x}} \in V$ is called a *regular point* of V if and only if the rank of the Jacobian matrix $G_{\mathbf{x}}(\hat{\mathbf{x}})$ satisfies $\dim V = n - \text{rank}(G_{\mathbf{x}}(\hat{\mathbf{x}}))$, where $G(\mathbf{x}) = [g_1, \ldots, g_s]^T$ is a polynomial basis of $\sqrt{I}$.

Computing real roots of a polynomial system is a fundamental problem of computational real algebraic geometry. There are symbolic methods based on Cylindrical Algebraic Decomposition [12] and critical point methods [5, 8, 16, 15, 19, 31, 38] for finding real points on the variety $V \cap \mathbb{R}^n$. Algorithms proposed in [1, 3, 4, 33, 37] find at least one real point on each connected component of $V \cap \mathbb{R}^n$. Recent work for computing verified real roots based on homotopy methods include certified homotopy-tracking method in [6], certifying solutions to polynomial systems using Smale's α-theorem [18].

A square zero-dimensional polynomial system . Suppose $F(\mathbf{x})$ is a square and zero-dimensional polynomial system, i.e., $m = n$. Standard verification methods for nonlinear square systems are based on the following theorem [21, 30, 34].

Theorem 1 *Let $F(\mathbf{x}) : \mathbb{R}^n \to \mathbb{R}^n$ be a polynomial system, and $\tilde{\mathbf{x}} \in \mathbb{R}^n$. Let $\mathbb{IR}$ be the set of real intervals, and $\mathbb{IR}^n$ and $\mathbb{IR}^{n \times n}$ be the set of real interval vectors and real interval matrices, respectively. Given $\mathbf{X} \in \mathbb{IR}^n$ with $\mathbf{0} \in \mathbf{X}$ and $M \in \mathbb{IR}^{n \times n}$ satisfies $\nabla f_i(\tilde{\mathbf{x}} + \mathbf{X}) \subseteq M_{i,:}$, for $i = 1, \ldots, n$. Denote by I_n the $n \times n$ identity matrix and assume*

$$-F_{\mathbf{x}}^{-1}(\tilde{\mathbf{x}})F(\tilde{\mathbf{x}}) + (I_n - F_{\mathbf{x}}^{-1}(\tilde{\mathbf{x}})M)\mathbf{X} \subseteq \text{int}(\mathbf{X}), \tag{4}$$

where $F_{\mathbf{x}}(\tilde{\mathbf{x}})$ is the Jacobian matrix of $F(\mathbf{x})$ at $\tilde{\mathbf{x}}$. Then there is a unique $\hat{\mathbf{x}} \in \mathbf{X}$ with $F(\hat{\mathbf{x}}) = 0$. Moreover, every matrix

*This material is based on work supported by a NKBRPC 2011CB302400, the Chinese National Natural Science Foundation under Grants 91118001, 60821002/F02, 60911130369(Zhi). This material is supported in part by the National Natural Science Foundation of China under Grants 91118007 and 61021004(Yang, Zhu).

$\widetilde{M} \in M$ *is nonsingular. In particular, the Jacobian matrix* $F_{\mathbf{x}}(\hat{\mathbf{x}})$ *is nonsingular.*

The non-singularity of the Jacobian matrix $F_{\mathbf{x}}(\hat{\mathbf{x}})$ restricts the application of Theorem 1 to regular solutions of a square polynomial system. If $F_{\mathbf{x}}(\hat{\mathbf{x}})$ is singular and $\mathbf{x}$ is an isolated singular solution of $F(\mathbf{x})$, in [26, 27, 36], by adding smoothing parameters properly to $F(\mathbf{x})$, an extended regular and square polynomial system is generated for computing verified error bounds, such that a slightly perturbed polynomial system of $F(\mathbf{x})$ is guaranteed to possess an isolated singular solution within the computed bounds. The method in [29] can also be used to verify the isolated singular solutions.

Remark 2 There are two functions verifynlss and verifynlss2 in the INTLAB package implemented by Rump in Matlab [35]. The procedure verifynlss can be used to verify the existence of a simple root of a square and regular zero-dimensional polynomial system and verifynlss2 can be used to verify the existence of a double root of a slightly perturbed polynomial system of $F(\mathbf{x})$. If the polynomial system $F(\mathbf{x})$ has an isolated singular root with multiplicity larger than 2, then the function viss designed in [26, 27] and implemented by Li and Zhu in Matlab can be applied to obtain verified error bounds such that a slightly perturbed polynomial system of $F(\mathbf{x})$ is guaranteed to possess an isolated singular solution within the computed bounds.

An overdetermined zero-dimensional polynomial system . Suppose $F(\mathbf{x})$ is an overdetermined zero-dimensional polynomial system, i.e., $m > n$. A natural procedure for obtaining a square polynomial system from $F(\mathbf{x})$ is to pick up a full rank random matrix $A \in \mathbb{Q}^{n \times m}$ and form a square polynomial system $A \cdot F(\mathbf{x})$. According to [42, Theorem 13.5.1], we have the following theorem.

Theorem 2 *There is a nonempty Zariski open subset* $\mathcal{A} \in \mathbb{C}^{n \times m}$ *such that for every* $A \in \mathcal{A}$, *a solution of* $F(\mathbf{x})$ *is regular if and only if it is a nonsingular solution of the square system* $A \cdot F(\mathbf{x})$. *Moreover, if* $F(\mathbf{x})$ *is a zero-dimensional system, then* $A \cdot F(\mathbf{x})$ *is also a zero-dimensional system.*

According to Theorem 2, we can apply Theorem 1 to regular solutions of the square polynomial system $A \cdot F(\mathbf{x})$ and check whether the verified solution of $A \cdot F(\mathbf{x})$ is a solution of $F(\mathbf{x})$ by computing the residual of $F(\hat{\mathbf{x}})$ as an additional test, see also [18, Lemma 3.1]. If $F(\hat{\mathbf{x}})$ is small, with high probability, the verified real solution of $A \cdot F(\mathbf{x})$ is a real solution of $F(\mathbf{x})$.

A positive-dimensional polynomial system . Suppose $F(\mathbf{x})$ is a positive-dimensional polynomial system. It is clear that an underdetermined system $F(\mathbf{x})$ is a positive-dimensional system whose dimension is at least $n - m \geq 1$. A square polynomial system and an overdetermined system can also be positive-dimensional. In [9, 10], the authors transformed an underdetermined system into a regular square system by choosing m independent variables and setting $n - m$ remaining variables to be anchors, then they used a Krawczyk-type interval operator to verify the existence of the solutions of the transformed regular and square system. It is very impressive that they can verify a solution of a polynomial system with more than 10000 variables and 20000 equations with degrees as high as 100. More general methods using linear slices to reduce the underdetermined system to a square system were proposed in [40, 41, 42]. We notice that it is very important to choose independent variables and initial values for the dependent variables or linear slices. Especially, we might have a big chance to miss the real points because of the bad choice for values of some variables.

Example 1 Consider the polynomial *Vor2*, which appears in a problem studying Voronoi Diagram of three lines in $\mathbb{R}^3$ [13]. *Vor2* is a polynomial in five variables with degree 18. It has an infinite number of real solutions. Let us set four variables as rational numbers chosen in the range $[-\frac{3000}{1000}, \frac{3000}{1000}]$, e.g.

$$\hat{x}_2 = \frac{177}{500},\ \hat{x}_3 = \frac{423}{1000},\ \hat{x}_4 = \frac{209}{1000},\ \hat{x}_5 = \frac{143}{50},$$

the univariate polynomial $V(x_1) = \mathit{Vor2}\,(x_1, \hat{x}_2, \hat{x}_3, \hat{x}_4, \hat{x}_5) \in \mathbb{Q}[x_1]$ has no real solutions.

Remark 3 If there is only one polynomial $f(x_1, \ldots, x_n)$ and the degree of f with respect to the variable x_i is odd, the univariate polynomial $f(\hat{x}_1, \ldots, \hat{x}_{i-1}, x_i, \hat{x}_{i+1}, \ldots, \hat{x}_n)$ will always have a real root $\hat{x}_i \in \mathbb{R}$ for arbitrary fixed values $\hat{x}_j \in \mathbb{Q}, 1 \leq j \leq n, j \neq i$. Hence, it is easy to verify that $(\hat{x}_1, \ldots, \hat{x}_{i-1}, \hat{x}_i, \hat{x}_{i+1}, \ldots, \hat{x}_n)$ is the real root of $f(\mathbf{x})$.

The main task of this paper is to construct a square and zero-dimensional polynomial system for computing verified real solutions of positive-dimensional polynomial systems. Let $I = \langle f_1, \ldots, f_m \rangle$ and V be an algebraic variety defined by $\{f_1 = 0, \ldots, f_m = 0\}$. We propose below different strategies for computing verified solutions on $V \cap \mathbb{R}^n$.

1. If the ideal I is radical and contains regular real solutions, we propose two algorithms for computing verified real solutions on $V \cap \mathbb{R}^n$:

   a. We use theoretical results developed in real algebraic geometry for finding one point on each connected component of $V \cap \mathbb{R}^n$ to construct a square and regular zero-dimensional polynomial system [1, 3, 4, 33, 37], then use the homotopy continuation solver HOM4PS-2.0 [24] to find its approximate real solutions. Finally, we apply verifynlss in the INTLAB package [35] to verify the existence of real solutions in the neighborhood of the computed approximate real solutions on connected components of $V \cap \mathbb{R}^n$.

   b. We compute an approximate real solution $\tilde{\mathbf{x}}$ of $F(\mathbf{x})$ by the low-rank moment matrix completion method in [28]. If the Jacobian matrix $F_{\mathbf{x}}(\tilde{\mathbf{x}})$ is singular, we compute a normalized null vector $\mathbf{v}$ of $F_{\mathbf{x}}(\tilde{\mathbf{x}})$ and add new polynomials $\sum_{j=1}^{m} \mathbf{v}_i \frac{\partial f_j(\mathbf{x})}{\partial x_i}$ for $1 \leq i \leq n$ to $F(\mathbf{x})$. Otherwise, we choose a normalized random vector λ and add polynomials $F_{\mathbf{x}}(\mathbf{x})\lambda - F_{\mathbf{x}}(\tilde{\mathbf{x}})\lambda$ to $F(\mathbf{x})$. Finally, we apply verifynlss to verify the existence of a real solution $\hat{\mathbf{x}}$ in the neighborhood of $\tilde{\mathbf{x}}$ on $V \cap \mathbb{R}^n$.

2. If the ideal I is not radical, we add tiny perturbations to the polynomial system $F(\mathbf{x})$ and modify above two algorithms accordingly.

c. The critical variety for the perturbed system is a zero-dimensional polynomial system containing not only regular solutions but also approximate singular solutions. For approximate singular solutions, we apply the verification algorithms verifynlss2 in [35] or viss in [26, 27] to compute verified error bounds, such that a slightly perturbed polynomial system of $F(\mathbf{x})$ possesses a real solution within the computed error bounds.

d. The real solutions computed by the method in [28] can be approximate singular solutions. We need to apply verification algorithms verifynlss2 or viss to compute verified error bounds of a slightly perturbed polynomial system.

Structure of the paper. In Section 2, we introduce theoretical results and methods for computing verified real solutions for positive-dimensional polynomial systems. In Section 3, we present three routines: verifyrealroot0 computes verified real solutions for zero-dimensional polynomial systems; verifyrealrootpc aims for computing verified real solutions on each connected components of $V \cap \mathbb{R}^n$; verifyrealrootpm is designed for computing at least one verified real solution for positive-dimensional polynomial systems. In Section 4, we demonstrate the effectiveness of the algorithms for computing verified real roots of a set of benchmark systems.

2. POSITIVE-DIMENSIONAL POLYNOMIAL SYSTEMS

2.1 The Radical Ideal Case

Let us consider the case where the ideal I generated by $f_1(\mathbf{x}), \ldots, f_m(\mathbf{x})$ is radical and V is of dimension d and contains a regular point in $\mathbb{R}^n$.

The critical point method

Theorem 3 *[4, Lemma 1]Let C be a connected component of the real variety V containing a regular point. Then, with respect to the Euclidean topology, there exists a non-empty open subset U_C of $\mathbb{R}^n \backslash V$ that satisfies the following condition: Let $\mathbf{u}$ be an arbitrary point of U_C and let $\hat{\mathbf{x}}$ be any point of V that minimizes the Euclidean distance to $\mathbf{u}$ with respect to V. Then $\hat{\mathbf{x}}$ is a regular point belonging to C.*

According to Theorem 3, one can compute a regular real sample point on V by computing its critical points of a distance function to a generic point restricted to V. This method was proposed in [1, 32, 33], see also [2, 4, 7] for some recent results when $F(\mathbf{x})$ has real singular solutions. Let us briefly introduce the method in [1].

Definition 1 *[1, Notation 2.4] For an arbitrary point $\mathbf{u} = (u_1, \ldots, u_n) \in \mathbb{R}^n$, let $g = \frac{1}{2}(x_1 - u_1)^2 + \cdots + \frac{1}{2}(x_n - u_n)^2$ and*

$$J_g(F) = \begin{bmatrix} \frac{\partial f_1}{\partial x_1} & \cdots & \frac{\partial f_m}{\partial x_1} & \frac{\partial g}{\partial x_1} \\ \vdots & & \vdots & \vdots \\ \frac{\partial f_1}{\partial x_n} & \cdots & \frac{\partial f_m}{\partial x_n} & \frac{\partial g}{\partial x_n} \end{bmatrix}. \quad (5)$$

We define the algebraic set:

$$C(V, \mathbf{u}) = \{\hat{\mathbf{x}} \in V, \operatorname{rank}(J_g(F(\hat{\mathbf{x}}))) \le n - d\}. \quad (6)$$

Let $\Delta_{\mathbf{u},d}(F)$ be the set of all the minors of order $n-d+1$ in the matrix $J_g(F)$ such that their last column contains the entries in the last column of $J_g(F)$.

Theorem 4 *[1, Theorem 2.3] Let V be an algebraic variety of dimension d and I be a radical equidimensional ideal. If D is a large enough positive integer, there exists at least one point $\mathbf{u}$ in $\{1, \ldots, D\}^n$ such that:*

1. *$C(V, \mathbf{u})$ meets every semi-algebraically connected component of $V \cap \mathbb{R}^n$;*

2. *$C(V, \mathbf{u}) = V_{sing} \cap V_{0,\mathbf{u}}$, where $V_{0,\mathbf{u}}$ is a finite set of points in $\mathbb{C}^n$ and V_{sing} are singular points on V whose Jacobian matrix have rank less than $n-d$.*

Moreover,

$$\dim(C(V, \mathbf{u})) < \dim(V). \quad (7)$$

According to Theorem 4, for almost all $\mathbf{u}$, the dimension of the algebraic variety $C(V, \mathbf{u})$ of $\Delta_{\mathbf{u},d}(F) \cup F(\mathbf{x})$ will be strict less than the dimension of V. Therefore, inductively, we will obtain a zero-dimensional polynomial system which can be used to verify the existence of regular real solutions on V. As stated in [1, 32], the main bottleneck for the critical points method is the computation of $\Delta_{\mathbf{u},d}$ since the number of elements in $\Delta_{\mathbf{u},d}$ is equal to $\binom{m}{n-d}\binom{n}{n-d+1}$ and the polynomials in $\Delta_{\mathbf{u},d}$ are usually dense and have large coefficients. An alternative way to avoid the computation of the minors is to introduce extra variables $\lambda_0, \ldots, \lambda_{n-d}$ and pick up randomly $n-d$ real numbers $a_0, \ldots, a_{n-d}$ and polynomials in $F(\mathbf{x})$ such as $f_1, \ldots, f_{n-d}$, and replace the minors in $\Delta_{\mathbf{u},d}$ by polynomials defined below

$$p_i = \lambda_0 \frac{\partial g}{\partial x_i} + \lambda_1 \frac{\partial f_1}{\partial x_i} + \ldots + \lambda_{n-d} \frac{\partial f_{n-d}}{\partial x_i}, \quad \text{for } 1 \le i \le n,$$
$$p_{n+1} = a_0\lambda_0 + \ldots + a_{n-d}\lambda_{n-d} - 1.$$

This is the way used in [17, Theorem 5] to generate solution paths leading to real solutions on V using the homotopy continuation method.

If V is compact and smooth, and the variables $x_1, \ldots, x_n$ are in a generic position with respect to $f_1, \ldots, f_m$, then as shown in [3, Theorem 10], one can change the distance function g to a coordinate function $g = x_i, 1 \le i \le n$ such that the dimension of the real variety of $\Delta_{\mathbf{u},d}(F) \cup F(\mathbf{x})$ will be zero and contains at least one real point on each connected component of $V \cap \mathbb{R}^n$. Moreover, in [37], Safey El Din and Schost extended the result in [3] to deal with the case where $V \cap \mathbb{R}^n$ is non-compact.

The low-rank moment matrix completion method. Recently, there is also an arising interest in using numerical semidefinite programming (SDP) based method [11, 20, 23] for characterizing and computing the real solutions of polynomial systems. As pointed out in [23], the great benefit of using SDP techniques is that it exploits the real algebraic nature of the problem right from the beginning and avoids the computation of complex components. For example, if $V \cap \mathbb{R}^n$ is zero-dimensional, then the moment-matrix algorithm in [23] can compute all real solutions of $F(\mathbf{x})$ by solving a sequence of SDP problems.

If the polynomial system $F(\mathbf{x})$ has an infinite number of real solutions, then the algorithm in [23] can not be used.

Hence, in [20, 22], they replaced the constant object function by the trace of the moment matrix and showed that their software GloptiPoly is very efficient for finding a partial set of real solutions for a large set of polynomial systems [22, Table 6.3, 6.4]. Since the trace of a semidefinite moment matrix is equal to its nuclear norm defined as the sum of its singular values, the optimization problem can be transformed to the following nuclear norm minimization problem:

$$\begin{cases} \min & \|M_t(y)\|_* \\ \text{s. t.} & y_0 = 1, \\ & M_t(y) \succeq 0, \\ & M_{t-d_j}(f_j\, y) = 0, \quad j = 1, \ldots, m, \end{cases} \tag{8}$$

In [28], a new algorithm based on accelerated fixed point continuation method and alternating direction method was presented to solve the minimization problem (8) for finding real solutions of $F(\mathbf{x})$ even when its real variety $V \cap \mathbb{R}^n$ is positive-dimensional. Although the method based on function values and gradient evaluations cannot yield as high accuracy as interior point methods, much larger problems can be solved since no second-order information needs to be computed and stored.

Encouraged by the results shown in [28, Table1] and noted that the main bottleneck for the critical point method is the computation of $\Delta_{\mathbf{u},d}$, we explain below how to avoid the computation of minors by constructing a zero-dimensional polynomial system based on the approximate real solution $\tilde{\mathbf{x}}$ computed by the algorithm MMCRSolver in [28] for verifying the existence of real solutions in $V \cap \mathbb{R}^n$ in the neighborhood of $\tilde{\mathbf{x}}$ when V is positive-dimensional.

Suppose $\tilde{\mathbf{x}}$ is an approximate real root of $F(\mathbf{x})$ computed by MMCRSolver. If the rank of the Jacobian matrix $F_{\mathbf{x}}(\tilde{\mathbf{x}})$ is less than $n-d$, then $\tilde{\mathbf{x}}$ is a singular point on V. Stimulated by the deflation method used in [25] for constructing extended regular polynomial systems, we compute a normalized null vector $\mathbf{v}$ ($|\mathbf{v}|_2 = 1$) of $F_{\mathbf{x}}(\tilde{\mathbf{x}})$ and generate a new polynomial system $\widetilde{F}(\mathbf{x}) = F(\mathbf{x}) \cup F_{\mathbf{x}}(\mathbf{x})\mathbf{v}$. It is clear that $\tilde{\mathbf{x}}$ is a real solution of

$$\begin{cases} F(\mathbf{x}) & = & \mathbf{0}, \\ F_{\mathbf{x}}(\mathbf{x})\mathbf{v} & = & \mathbf{0}. \end{cases} \tag{9}$$

It is possible that $\tilde{\mathbf{x}}$ is still a singular solution of $\widetilde{F}$, then we can perform the similar deflations to the system $\widetilde{F}(\mathbf{x})$ again.

If the approximate solution $\tilde{\mathbf{x}}$ is not a singular point on the variety $V \cap \mathbb{R}^n$, i.e., the rank of the Jacobian matrix $F_{\mathbf{x}}(\tilde{\mathbf{x}})$ is $n-d$, then we choose a normalized random vector λ and construct a new polynomial system $\widetilde{F}(\mathbf{x}) = F(\mathbf{x}) \cup \{F_{\mathbf{x}}(\mathbf{x})\lambda - F_{\mathbf{x}}(\tilde{\mathbf{x}})\lambda\}$. It is clear that $\tilde{\mathbf{x}}$ is a solution of

$$\begin{cases} F(\mathbf{x}) & = & \mathbf{0}, \\ F_{\mathbf{x}}(\mathbf{x})\lambda - F_{\mathbf{x}}(\tilde{\mathbf{x}})\lambda & = & \mathbf{0}. \end{cases} \tag{10}$$

Suppose we obtain a zero-dimensional regular system $\widetilde{F}(\mathbf{x})$ after above steps, then we apply verifynlss [35] to verify the existence of a real solution $\hat{\mathbf{x}}$ in the neighborhood of $\tilde{\mathbf{x}}$ on $V \cap \mathbb{R}^n$. However, it is not guaranteed that the variety of the new polynomial system $\widetilde{F}(\mathbf{x})$ generated above will be a zero-dimensional regular system. In order to obtain a zero-dimensional regular system, we may need to add more random polynomials vanishing at $\tilde{\mathbf{x}}$.

The verification algorithm based on using the null vector of $F_{\mathbf{x}}(\tilde{\mathbf{x}})$ or a random vector to construct new polynomial systems can be more efficient since it avoids the computations of minors or the introduction of new variables. However, since we only use local information about the approximate real root $\tilde{\mathbf{x}}$ of $F(\mathbf{x})$ in order to construct the new extended system, it is limited to verify the existence of a real root $\hat{\mathbf{x}}$ in the neighborhood of $\tilde{\mathbf{x}}$. For some interesting applications, it is enough to verify the existence of one real solution, e.g., for deciding reachability of the infimum of a multivariate polynomial [14], if we can verify the existence of one real solution for $f - f^*$, then we prove that f^* is a minimum which can be attained.

Example 2 [7, Example 4] To illustrate the above method, consider the polynomial $f(x_1, x_2) = x_1^2 - x_2\,(x_2+1)(x_2+2)$. MMCRSolver yields one approximate real solution

$$\tilde{\mathbf{x}} = [3.671518 \times 10^{-8}, -0.999902]^T.$$

Since the approximate solution $\tilde{\mathbf{x}}$ is not a singular solution of $f(\mathbf{x})$, we choose a random vector $\lambda = [0.715927, -0.328489]^T$ and construct a new square polynomial system by adding one more polynomial g defined by λ and $\tilde{\mathbf{x}}$ in (10). The curves of f and g are displayed in the following figure with dot and dash line styles respectively.

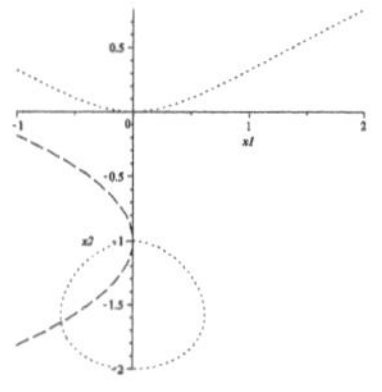

We run the algorithm verifynlss and prove that $f(x_1, x_2)$ has a verified real solution within the inclusion

x_1	x_2
$4.3211387 \times 10^{-8} \pm 2.7 \times 10^{-15}$	$-1 \pm 2.2 \times 10^{-15}$

2.2 The Non-radical Ideal Case

If the ideal I generated by polynomials in $F(\mathbf{x})$ is not radical, as pointed out in [1], the inequality (7) in Theorem 4 is not true. It is difficult to verify the exact existence of real points on singular locus V_{sing} which might have the same dimension as V.

Example 3 [43, 5.15] Consider the system $F(\mathbf{x})$ containing polynomials $f_1 = x_1^3 x_3^2 + x_3, f_2 = x_1^2 x_2 + x_3$.

The ideal I generated by polynomials f_1, f_2 is not radical. The real algebraic variety $V \cap \mathbb{R}^n$ defined by $\{f_1 = 0, f_2 = 0\}$ contains three one-dimensional solutions $V_1 = \{x_1 = 0, x_3 = 0\}, V_2 = \{x_2 = 0, x_3 = 0\}, V_3 = \{x_1^5 x_2 - 1 = 0, x_3 x_1^3 + 1 = 0\}$.

Since the rank of the Jacobian matrix at all points on the variety V_1 is 1, we know that the variety $C(V, \mathbf{u})$ defined in (6) for an arbitrary chosen point $\mathbf{u}$ contains the one-dimensional variety V_1. Hence $\dim(C(V, \mathbf{u})) = \dim(V) = 1$, the inequality (7) in Theorem 4 is not true for this example.

Let us choose $\mathbf{u}$ as

$$\{u_1 = 1, u_2 = 2, u_3 = 3\}.$$

The set $\Delta_{\mathbf{u},1}(F)$ consists of the determinant of $J_g(F)$ defined by (5) in Theorem 4. Applying the homotopy solver HOM4PS-2.0 to the polynomial system $F(\mathbf{x}) \cup \Delta_{\mathbf{u},1}(F)$, we obtain 5 real approximate solutions of $C(V, \mathbf{u})$.

- The real solution $\{x_1 = 0, x_2 = 0, x_3 = 0\}$ is on $V_1 \cap C(V, \mathbf{u})$. It is not an isolated singular solution. Therefore, it can not be verified by verifynlss2 or viss.

  However, it is interesting to notice that there is another real root computed by HOM4PS-2.0 which is very near to V_1. Run the algorithms viss, we are able to compute the verified error bound, such that the slightly perturbed system (within 4.16×10^{-15}) of $F(\mathbf{x})$ has a verified real solution within the inclusion

x_1	x_2	x_3
0	$2 \pm 4.44 \times 10^{-16}$	0

- Applying the algorithm verifynlss to other three approximate real roots computed by HOM4PS-2.0, we obtain:

  - two verified regular real solutions within inclusions on the component V_3,

x_1	x_2	x_3
$1.7 \pm 2 \times 10^{-15}$	$0.07 \pm 5 \times 10^{-15}$	$-0.2 \pm 8 \times 10^{-16}$
$-1.1 \pm 7 \times 10^{-16}$	$-0.57 \pm 1 \times 10^{-15}$	$0.71 \pm 1 \times 10^{-15}$

  - one verified regular real solution within the inclusion on the component V_2,

x_1	x_2	x_3
$1 \pm 4.4440892 \times 10^{-16}$	0	0

If I is not radical, a well-known method to get a smooth algebraic variety is to add one or more infinitesimal deformations to polynomials in $F(\mathbf{x})$ and work over a non-archimedean real closed extension of the ground field [5, 33]. The computation could be quite expensive. Therefore, instead of proving the exact existence of real roots on the variety defined by a non-radical ideal, we perturb the system by a tiny real number and show the existence of real roots of this slightly perturbed polynomial system.

Theorem 5 *[17, Lemma 4] Suppose G consists of $n - d$ polynomials and $V(G)$ is a pure d-dimensional variety. There is a nonempty Zariski open set $Z \subset \mathbb{C}^{n-d}$ such that, for every $z \in Z$, $V(G - z)$ is a smooth algebraic set of dimension d.*

For $m = 1$, it is a well known consequence of Sard theorem, see [33, Lemma 3.5].

Let us add a small perturbation 10^{-25} to f_1 above and run the homotopy solver HOM4PS-2.0 for the perturbed system $\{f_1 + 10^{-25}, f_2\} \cup \Delta_{\mathbf{u},1}(F)$, we also obtain 5 approximate real solutions on $C(V, \mathbf{u})$. The algorithm verifynlss computes inclusions of three real solutions near to V_2 and V_3:

x_1	x_2	x_3
$1 \pm 3 \times 10^{-16}$	0	0
$1.7 \pm 2 \times 10^{-15}$	$0.07 \pm 5 \times 10^{-15}$	$-0.2 \pm 8 \times 10^{-16}$
$-1.1 \pm 7 \times 10^{-16}$	$-0.57 \pm 1 \times 10^{-15}$	$0.71 \pm 1 \times 10^{-15}$

Applying the algorithm viss, we obtain inclusions for another two real solutions near to V_1:

x_1	x_2	x_3
$7.4 \times 10^{-9} \pm 3 \times 10^{-24}$	$1.8 \times 10^{-9} \pm 2 \times 10^{-24}$	0
0	$2 \pm 9 \times 10^{-16}$	0.

Remark 4 Notice here, the perturbed system is smooth, $C(V, \mathbf{u})$ is a zero-dimensional variety. However, it contains approximate singular solutions [29]. Hence, it is necessary to apply the algorithm viss to verify the existence of real singular solutions of a slightly perturbed system. Moreover, since computations in Matlab have limited precisions, with or without tiny perturbations, we may get similar results.

We can also apply MMCRSolver to obtain an approximate real solution of $F(\mathbf{x})$ near to $(0, 0, 0)$. Running the algorithms verifyrealrootpm and verifynlss2, we prove that the slightly perturbed system (within 10^{-58}) of $F(\mathbf{x})$ has a verified real solution within the inclusion

x_1	x_2	x_3
$1.35 \times 10^{-15} \pm 2 \times 10^{-30}$	$6.77 \times 10^{-15} \pm 9 \times 10^{-29}$	0

3. ALGORITHMS FOR COMPUTING VERIFIED SOLUTIONS OF POLYNOMIAL SYSTEMS

Based on discussions in above sections, we present three procedures: verifyrealroot0 is based on the verification algorithms verifynlss, verifynlss2 [35] and viss [26, 27], and computes verified real solutions for zero-dimensional polynomial systems; verifyrealrootpc is based on the critical point method and the homotopy continuation method, and aims for computing at least one verified real solution on each connected component of $V \cap \mathbb{R}^n$; verifyrealrootpm is based on the low-rank moment matrix completion method in [28], and aims for computing at least one verified real solution for positive-dimensional polynomial systems. Before we show the algorithms, we would like to point out that unlike symbolic methods [1, 32, 33, 37], our algorithms can not be used to verify the nonexistence of real solutions on $V \cap \mathbb{R}^n$, i.e., the failure of our algorithms does not mean there exist no real solutions on $V \cap \mathbb{R}^n$.

Remark 5 According to Theorem 4, suppose I is a radical equidimensional ideal, then the variety $C(V, \mathbf{u})$ meets every semi-algebraic connected component of $V \cap \mathbb{R}^n$. Furthermore, applying Theorem 4 recursively, one is able to obtain a zero-dimensional overdetermined system, which contains a regular real root on $V \cap \mathbb{R}^n$ if it is not empty. It should be pointed out that we do not perform the equidimensional decomposition of the ideal $I = \langle f_1, \ldots, f_m \rangle$. The function verifyrealrootpc does not guarantee to verify real roots on each connected component of $V \cap \mathbb{R}^n$ if V is not equidimensional.

Remark 6 If I is not radical, it is still possible to verify the existence of regular real solutions. However, as observed from the Example 3, the variety $C(V, \mathbf{u})$ may have singular locus with the same dimension as the variety V. Hence, we could only verify the existence of a singular real root near to the slightly perturbed polynomial system. Another possibility would be to perturb the polynomial system $F(\mathbf{x})$ at the beginning by a tiny number. We notice in the non-radical case, it is necessary to call verifynlss2 or viss for verifying the existence of a singular solution of a slightly perturbed polynomial system.

verifyrealroot0

Input: A zero-dimensional polynomial system $F(\mathbf{x}) = [f_1, \ldots, f_m]^T$ in $\mathbb{Q}[x_1, \ldots, x_n]$, a given small tolerance $\epsilon \in \mathbb{R}$.

Output: A set L of real root inclusions X.

1. If $m > n$, choose a random matrix $A \in \mathbb{Q}^{n \times m}$; otherwise, set $A = I_n$.
2. Set $\widetilde{F} = A \cdot F(\mathbf{x})$, apply MMCRSolver or HMO4PS-2.0 to obtain approximate real roots of $\widetilde{F}(\mathbf{x})$, denoted by $\tilde{\mathbf{x}}_1, \ldots, \tilde{\mathbf{x}}_k$.
3. Set $L = \{\}$. For $i = 1, \ldots, k$ do:
   (a) If the Jacobian matrix $\widetilde{F}_{\mathbf{x}}(\tilde{\mathbf{x}}_i)$ is regular
   call verifynlss to obtain the real root inclusion X of $\widetilde{F}(\mathbf{x})$, set $b = 0$.
   Otherwise
   call verifynlss2 or viss to obtain the verified error bound b and the real root inclusion X.
   (b) Compute the residue $\tau = F(X)$. If $\tau < \epsilon$ and $b < \epsilon$, set $L = L \cup \{X\}$.
4. Output L.

Figure 1: The Verification Algorithm for Zero-Dimensional Polynomial Systems.

verifyrealrootpc

Input: A positive-dimensional polynomial system $F(\mathbf{x}) = [f_1, \ldots, f_m]^T$ in $\mathbb{Q}[x_1, \ldots, x_n]$, a given small tolerance $\epsilon \in \mathbb{R}$.

Output: A set L of real root inclusions X.

1. Construct a zero-dimensional overdetermined system, denoted by $\widetilde{F}(\mathbf{x})$ via the critical point method.
2. Suppose the number of polynomials in $\widetilde{F}(\mathbf{x})$ is s, choose a random matrix $A = \mathbb{Q}^{n \times s}$ and update $\widetilde{F}(\mathbf{x})$ to be $A \cdot \widetilde{F}(\mathbf{x})$.
3. Apply HOM4PS-2.0 to obtain approximate real roots of $\widetilde{F}(\mathbf{x})$, denoted by $\tilde{\mathbf{x}}_1, \ldots, \tilde{\mathbf{x}}_k$.
4. Run Step 3 of verifyrealroot0 for $\widetilde{F}(\mathbf{x})$ and $\tilde{\mathbf{x}}_1, \ldots, \tilde{\mathbf{x}}_k$.
5. Output L.

Figure 2: The Verification Algorithm for Positive-Dimensional Polynomial Systems Based on the Critical Point Method and the Homotopy Continuation Method

Remark 7 The polynomial system $\widetilde{F}(\mathbf{x})$ generated in Step 2(a) of verifyrealrootpm is not guaranteed to be of zero dimension. If $\widetilde{F}(\mathbf{x})$ is still of positive dimension and Step 2(c) fails for $\tilde{\mathbf{x}}_i$, then we add more polynomials vanishing at $\tilde{\mathbf{x}}_i$ to $\widetilde{F}(\mathbf{x})$.

verifyrealrootpm

Input: A positive-dimensional polynomial system $F(\mathbf{x}) = [f_1, \ldots, f_m]^T$ in $\mathbb{Q}[x_1, \ldots, x_n]$, a given small tolerance $\epsilon \in \mathbb{R}$.

Output: A set L of real root inclusions X.

1. Apply MMCRSolver to obtain approximate real roots of $F(\mathbf{x})$, denoted by $\tilde{\mathbf{x}}_1, \ldots, \tilde{\mathbf{x}}_k$.
2. Set $L = \{\}$. For $i = 1, \ldots, k$ do:
   (a) i. If $F_{\mathbf{x}}(\tilde{\mathbf{x}}_i)$ is singular
   compute a normalized null vector $\mathbf{v}$ of $F_{\mathbf{x}}(\tilde{\mathbf{x}}_i)$
   $\widetilde{F}(\mathbf{x}) = \left\{\sum_{j=1}^{m} \mathbf{v}_i \frac{\partial f_j(\mathbf{x})}{\partial x_i},\ 1 \le i \le n\right\} \cup F(\mathbf{x})$.
   ii. Otherwise
   compute a normalized random vector λ
   $\widetilde{F}(\mathbf{x}) = F(\mathbf{x}) \cup \{F_{\mathbf{x}}(\mathbf{x})\lambda - F_{\mathbf{x}}(\tilde{\mathbf{x}}_i)\lambda\}$
   (b) Choose a random matrix $A = \mathbb{Q}^{n \times (m+n)}$, update $\widetilde{F}$ to be $A \cdot \widetilde{F}$.
   (c) Run Step 3(a)(b) of verifyrealroot0 for $\widetilde{F}(\mathbf{x})$ and $\tilde{\mathbf{x}}_i$.
3. Output L.

Figure 3: The Verification Algorithm for Positive-dimensional Polynomial Systems Based on Low-rank Moment Matrix Completion Method.

Remark 8 In verifyrealrootpc and verifyrealrootpm, if the polynomial system $F(\mathbf{x})$ is underdetermined, i.e., $m < n$, our first choice of the random matrix A will always have the block structure $\begin{bmatrix} I_m & 0 \\ 0 & A_{sub} \end{bmatrix}$, where A_{sub} is chosen randomly. In this case, we do not need to compute the residue. The verified solution of $\widetilde{F}$ will be a verified solution of $F(\mathbf{x})$.

4. EXPERIMENTS

Our algorithms have been implemented in Matlab (2011R) and the performance is reported in the following tables. All examples are run on Intel(R) Core(TM) at 2.6GHz under Windows. We also translate the Maple codes of MMCRSolver [28] and viss [26, 27] into Matlab codes. The codes can be downloaded from `http://www.mmrc.iss.ac.cn/~lzhi/Research/hybrid/VerifyRealRoots/`

In Table 1, we exhibit the performance of the algorithm verifyrealroot0 for computing verified real solutions of zero-dimensional polynomial systems. All problems are taken from the homepage of Jan Verschelde `http://www.math.uic.edu/~jan/`. Here *var* and *deg* denote the number of the variables and the highest degree of polynomials; *ctrs* denotes the number of the equations; verifyrealroot0(M) and verifyrealroot0(H) refer to the two methods based on the low-rank moment matrix completion method and the homotopy method respectively for computing approximate roots in verifyrealroot0; *sol* denotes the number of the verified solutions; *time* is given in seconds for computing verified real solutions; whereas *width* denotes the largest of widths of all verified solutions computed by our algorithms.

In Table 2, Table 3 and Table 4, we exhibit the performance of our algorithms on positive dimensional polynomial

problem	*var*	*deg*	verifyrealroot0(M)			verifyrealroot0(H)		
			time	*sol*	*width*	*time*	*sol*	*width*
cohn2	4	6	10.8	1	6.3e-29	20.1	3	6.5e-12
cohn3	4	6	24.7	1	2.4e-26	137	5	2.9e-9
comb3000	10	3	1.56	1	2.0e-20	1.38	4	2.7e-20
d1	12	3	52.3	2	1.7e-14	6.24	16	1.8e-14
boon	6	4	27.6	1	5.1e-15	1.98	8	2.9e-15
des22_24	10	2	1.79	1	2.5e-14	1.73	10	1.1e-8
discret3	8	2	51.5	1	1.3e-13	107	102	1.5e-14
geneig	6	3	6.53	2	6.7e-15	4.63	10	2.7e-13
heart	8	4	24.9	2	5.3e-15	1.40	2	4.9e-15
i1	10	3	1.23	1	1.7e-16	11.0	16	9.1e-08
katsura5	6	2	1.35	1	2.2e-16	3.26	12	1.8e-15
kin1	12	3	52.3	2	1.8e-14	5.91	16	1.8e-14
ku10	10	2	37.8	1	4.7e-14	0.96	2	6.7e-14
noon3	3	3	1.88	1	1.6e-16	11.7	8	1.6e-15
noon4	4	3	9.70	1	3.6e-15	30.2	22	3.9e-15
puma	8	2	5.85	2	2.9e-14	3.99	16	1.8e-13
quadfor2	4	4	1.48	2	5.6e-16	0.71	2	2.2e-16
rbp1	6	3	5.59	1	2.6e-15	23.2	4	8.4e-14
redeco5	5	2	0.95	1	8.3e-17	1.07	4	1.3e-15
reimer5	5	6	26.7	3	8.4e-14	5.83	24	3.2e-13

Table 1: Algorithm Performance on Zero-dimensional Polynomial Systems

systems. The symbol $\triangle$ denotes the singular solutions verified by verifynlss2 or viss; the symbol $*$ denotes that the verified solution is a real solution of the original polynomial system with high probability. *curve0-5* are examples from [7]; *ex4* and *ex5* are cited from [33]; *Vor2* is from [13]; the remaining examples are taken from the homepage of Jan Verschelde and the polynomial test suite of D. Bini and B. Mourrain `http://www-sop.inria.fr/saga/POL/`.

problem	*var*	*ctrs*	*deg*	verifyrealrootpm			verifyrealrootpc		
				time	*sol*	width	*time*	*sol*	*width*
curve1	2	1	6	3.43	1	1.1e-14	4.13	9	5.0e-14
curve2	2	1	12	8.87	1	9.5e-20	160	27, $11_\triangle$	5.2e-10
curve3	2	1	6	2.02	1	8.7e-15	20.0	8	2.1e-14
curve4	2	1	3	1.26	1	8.7e-15	3.96	3	3.3e-15
curve5	2	1	6	6.11	$1_\triangle$	4.9e-16	12.9	4	4.7e-11
ex4	3	1	5	4.72	1	5.0e-14	46.6	10	6.0e-12
ex5	4	1	4	5.13	1	1.4e-26	122	46, $2_\triangle$	2.2e-9
adjmin22e4	6	2	2	9.86	1	9.0e-30	234	14, $22_\triangle$	1.8e-12
butcher	4	2	3	3.41	1	8.9e-15	319	30	1.7e-12
gerdt2	5	3	4	4.82	1	1.6e-15	506	31	1.2e-10

Table 2: Algorithm Performance on Positive-dimensional Polynomial Systems

The algorithm verifyrealrootpc is designed for computing the verified solutions on each connected component on V by adding all minors in $\Delta_{n,d}$. However, it is well-known that polynomials in $\Delta_{n,d}$ are usually dense and have large coefficients. It is difficult for HOM4PS-2.0 to handle large polynomials in Matlab. Therefore, verifyrealrootpc can only find successfully verified real solutions for polynomial systems in Table 2. In order to apply verifyrealrootpc to polynomial systems in Table 3, we use the most possible canonical projections, i.e., fixing as many variables as possible, to construct the zero-dimensional polynomial system. The modified version of verifyrealrootpc is denoted as verifyrealrootpc*. Therefore, in Table 3, both algorithms are aiming only for verifying the existence of at least one real root of polynomial systems.

problem	*var*	*ctrs*	*deg*	verifyrealrootpm			verifyrealrootpc*		
				time	*sol*	width	*time*	*sol*	*width*
vor2	5	1	18	19.9	$1_\triangle$	3.2e-11	587	$1_\triangle$	1.7e-6
curve0	2	1	12	9.28	$3_\triangle$	3.9e-15	10.8	$4_\triangle$	4.4e-16
birkhoff	4	1	10	127	$1_\triangle$	2.2e-26	7.72	7	1.0e-14
adjmin23e5	8	3	2	1.24	1	2.3e-28	1.09	1	7.8e-16
adjmin24e6	10	4	2	1.68	1	4.8e-28	1.46	1	1.1e-15
adjmin25e7	12	5	2	6.19	1	3.7e-27	1.68	1	6.2e-15
adjmin26e8	14	6	2	4.05	1	3.3e-29	2.32	1	3.1e-15
adjmin27e9	16	7	2	3.51	1	1.0e-29	1.98	1	2.7e-15
adjmin28eA	18	8	2	26.6	1	3.9e-29	3.29	1	3.3e-15
adjmin29eB	20	9	2	6.39	1	2.3e-29	9.22	1	4.0e-15
geddes2	5	4	6	18.9	1	5.8e-14	5.43	11	3.6e-11
geddes3	11	2	3	2.58	1	5.5e-28	1.26	1	7.1e-15
geddes4	12	3	3	3.05	1	1.3-27	1.34	1	7.1e-15
hairer1	8	6	3	2.06	1	1.2e-14	1.25	1	5.8e-15
hairer2	9	7	4	244	3	1.3e-12	17.7	6	9.3e-12
lanconelli	8	2	3	5.38	1	6.7e-15	1.48	2	4.9-13
bronestein2	4	3	4	14.7	$1_\triangle$	1.3e-25	3.18	2	3.8e-15
hawes1	5	4	9	16.1	$1_\triangle$	4.5e-19	2.09	1	3.6e-14
raksanyi	8	4	3	2.47	1	1.4e-19	1.69	2	1.2e-15
spatburmel	6	5	2	11.9	1	5.9e-15	3.92	2	1.6e-12

Table 3: Algorithm Performance on Positive-dimensional Polynomial Systems

In Table 4 we show the performance of our algorithms on non-radical polynomial systems cited from [40, 43, 44]. Let *pert.* denote the real number added to the original polynomial system. It should be noticed that only a very limited number of small non-radical polynomial systems are tested above. We are working on providing a more reliable algorithm for certifying real roots of non-radical polynomial systems.

Ex	*var*	*ctrs*	*deg*	verifyrealrootpm				verifyrealrootpc			
				pert.	*time*	*sol*	width	*pert.*	*time*	*sol*	*width*
Ex.1	3	2	4	0	18.9	$1_\triangle$	2.5e-16	0	18.7	1	1.4e-15
				1e-15	1.48	1	5.6e-17	1e-14	8.84	1, $1_\triangle$	1.6e-15
Ex.2	3	3	2	0	3.33	1*	4.2e-27	0	7.12	3*	2.2e-11
				1e-15	0.99	1	8.5e-21	1e-15	0.48	1	9.9e-31
Ex.3	3	3	2	0	3.90	1*	4.4e-9	0	3.13	1	3.3e-16
				1e-10	22.2	1	8.9e-15	1e-14	2.63	1	8.5e-9
Ex.4	2	2	5	0	3.51	1*	2.3e-19	0	22.9	2*, $1^*_\triangle$	5.6e-13
				1e-20	8.01	1	2.0e-31	1e-15	0.655	1	1.9e-11
Ex.5	3	2	5	0	8.92	$1_\triangle$	4.2e-17	0	32.3	3	2.0e-15
				1e-14	11.5	$1_\triangle$	5.7e-18	1e-15	6.98	5	2.2e-15
Ex.6	2	2	8	0	92.9	$1_\triangle$	6.6e-16	0	15.2	3	2.6e-15
				1e-9	43.6	1	3.6e-12	1e-8	11.3	5	2.9e-12

Table 4: Algorithm Performance on Nonradical Positive-dimensional Polynomial Systems

5. REFERENCES

[1] Aubry, P., Rouillier, F., and Safey El Din, M. Real solving for positive dimensional systems. *J. Symb. Comput. 34*, 6 (2002), 543–560.

[2] Bank, B., Giusti, M., Heintz, J., Lehmann, L., and Pardo, L. M. Algorithms of intrinsic complexity for point searching in compact real singular hypersurfaces. *Foundations of Computational Mathematics 12* (2012), 75–122.

[3] Bank, B., Giusti, M., Heintz, J., and Mbakop, G. M. Polar varieties and efficient real elimination. *MATHEMATISCHE ZEITSCHRIFT 238* (2000), 2001.

[4] Bank, B., Giusti, M., Heintz, J., Safey El Din, M., and Schost, E. On the geometry of polar varieties. *Applicable Algebra in Engineering, Communication and Computing 21* (2010), 33–83.

[5] Basu, S., Pollack, R., and Roy, M.-F. On the combinatorial and algebraic complexity of quantifier elimination. *J. ACM 43*, 6 (Nov. 1996), 1002–1045.

[6] Beltrán, C., and Leykin, A. Certified Numerical Homotopy Tracking. *Experimental Mathematics 21* (2012), 69–83.

[7] Camilla, M. H., and Ragni, P. Polars of real singular plane curves. In *Alorithm in Algebraic Geometry* (2008), Springer, pp. 99–115.

[8] Canny, J. Computing roadmaps of general semi-algebraic sets. In *Applied Algebra, Algebraic Algorithms and Error-Correcting Codes*, H. Mattson, T. Mora, and T. Rao, Eds., vol. 539 of *Lecture Notes in Computer Science*. Springer Berlin Heidelberg, 1991, pp. 94–107.

[9] Chen, X., Frommer, A., and Lang, B. Computational existence proofs for spherical t-designs. *Numerische Mathematik 117* (2011), 289–305.

[10] Chen, X., and Womersley, R. S. Existence of solutions to systems of underdetermined equations and spherical designs. *SIAM J. NUMER. ANAL. 44*, 6 (2006), 2326–2341.

[11] Chesi, G., Garulli, A., Tesi, A., and Vicino, A. Characterizing the solution set of polynomial systems in terms of homogeneous forms: an LMI approach. *International Journal of Robust and Nonlinear Control 13*, 13 (2003), 1239–1257.

[12] Collins, G. Quantifier elimination for real closed fields by cylindrical algebraic decompostion. vol. 33 of *Lecture Notes in Computer Science*. 1975, pp. 134–183.

[13] Everett, H., Lazard, S., Lazard, D., and Safey El Din, M. The voronoi diagram of three lines. In *Proceedings of the twenty-third annual symposium on Computational geometry* (2007), SCG '07, ACM, pp. 255–264.

[14] Greuet, A., and Safey El Din, M. Deciding reachability of the infimum of a multivariate polynomial. In *Proceedings of the 36th international symposium on Symbolic and algebraic computation* (New York, NY, USA, 2011), ISSAC '11, ACM, pp. 131–138.

[15] Grigor'ev, D., and Vorobjov, N.N., J. Counting connected components of a semialgebraic set in subexponential time. *computational complexity 2* (1992), 133–186.

[16] Grigor'ev, D. Y., and Jr, N. V. Solving systems of polynomial inequalities in subexponential time. *Journal of Symbolic Computation 5*, 1ĺC2 (1988), 37 - 64.

[17] Hauenstein, J. D. Numerically computing real points on algebraic sets. *ArXiv e-prints* (May 2011).

[18] Hauenstein, J. D., and Sottile, F. Algorithm 921: alphacertified: Certifying solutions to polynomial systems. *ACM Trans. Math. Softw. 38*, 4 (Aug. 2012), 28:1–28:20.

[19] Heintz, J., Roy, M.-F., and SolernÃş, P. Description of the connected components of a semialgebraic set in single exponential time. *Discrete & Computational Geometry 11* (1994), 121–140.

[20] Henrion, D., and Lasserre, J. Detecting global optimality and extracting solutions in GloptiPoly. In *Positive polynomials in control*, vol. 312 of *Lecture Notes in Control and Inform. Sci.* Springer, Berlin, 2005, pp. 293–310.

[21] Krawczyk, R. Newton-algorithmen zur bestimmung von nullstellen mit fehlerschranken. *Computing* (1969), 187–201.

[22] Lasserre, J. *Moments, Positive Polynomials and Their Applications.* Imperial College Press, 2009.

[23] Lasserre, J., Laurent, M., and Rostalski, P. Semidefinite characterization and computation of zero-dimensional real radical ideals. *Foundations of Computational Mathematics 8* (2008), 607–647.

[24] Lee, T.-L., Li, T.-Y., and Tsai, C.-H. Hom4ps-2.0: a software package for solving polynomial systems by the polyhedral homotopy continuation method. *Computing 83*, 2-3 (2008), 109–133.

[25] Leykin, A., Verschelde, J., and Zhao, A. Newton's method with deflation for isolated singularities of polynomial systems. *Theoretical Computer Science 359*, 1 (2006), 111–122.

[26] Li, N., and Zhi, L. Verified error bounds for isolated singular solutions of polynomial systems. Preprint, arxiv.org/pdf/1201.3443.

[27] Li, N., and Zhi, L. Verified error bounds for isolated singular solutions of polynomial systems: case of breadth one. To appear in Theoretical Computer Science, DOI: 10.1016/j.tcs.2012.10.028.

[28] Ma, Y., and Zhi. Computing real solutions of polynomial systems via low-rank moment matrix completion. In *ISSAC* (2012), ACM, pp. 249–256.

[29] Mantzaflaris, A., and Mourrain, B. Deflation and certified isolation of singular zeros of polynomial systems. In *Proceedings of the 36th international symposium on Symbolic and algebraic computation* (New York, NY, USA, 2011), A. Leykin, Ed., ISSAC '11, ACM, pp. 249–256.

[30] Moore, R. E. A test for existence of solutions to nonlinear systems. *SIAM Journal on Numerical Analysis 14*, 4 (1977), pp. 611–615.

[31] Renegar, J. On the computational complexity and geometry of the first-order theory of the reals. part i: Introduction. preliminaries. the geometry of semi-algebraic sets. the decision problem for the existential theory of the reals. *Journal of Symbolic Computation 13*, 3 (1992), 255 – 299.

[32] Rouillier, F. Efficient algorithms based on critical points method. In *Algorithmic and Quantitative Aspects of Real Algebraic Geometry in Mathematics and Computer Science* (2001), S. Basu and L. Gonzĺćlez-Vega, Eds., American Mathematical Society, pp. 123–138.

[33] Rouillier, F., Roy, M.-F., and Safey El Din, M. Finding at least one point in each connected component of a real algebraic set defined by a single equation. *J. Complexity 16*, 4 (2000), 716–750.

[34] Rump, S. Solving algebraic problems with high accuracy. In *Proc. of the symposium on A new approach to scientific computation* (San Diego, CA, USA, 1983), Academic Press Professional, Inc., pp. 51–120.

[35] Rump, S. INTLAB - INTerval LABoratory. In *Developments in Reliable Computing* (1999), T. Csendes, Ed., Kluwer Academic Publishers, Dordrecht, pp. 77–104.

[36] Rump, S., and Graillat, S. Verified error bounds for multiple roots of systems of nonlinear equations. *Numerical Algorithms 54*, 3 (2009), 359–377.

[37] Safey El Din, M., and Schost, É. Polar varieties and computation of one point in each connected component of a smooth real algebraic set. In *ISSAC* (2003), ACM, pp. 224–231.

[38] Seidenberg, A. A new decision method for elementary algebra. *Annals Math. 60* (1954), 365–374.

[39] Shen, F., Wu, W., and Xia, B. Real Root Isolation of Polynomial Equations Based on Hybrid Computation. *ArXiv e-prints*, July 2012.

[40] Sommese, A. J., and Verschelde, J. Numerical homotopies to compute generic points on positive dimensional algebraic sets. *J. Complex. 16*, 3 (Sept. 2000), 572–602.

[41] Sommese, A. J., and Wampler, C. W. Numerical algebraic geometry. In *The mathematics of numerical analysis (Park City, UT, 1995)* (1996), S. S. J. Renegar, M. Shub, Ed., Lectures in Appl. Math.,32, Amer. Math. Soc., Providence, RI, pp. 749–763.

[42] Sommese, A. J., and Wampler, C. W. *The numerical solution of systems of polynomials - arising in engineering and science.* World Scientific, 2005.

[43] Spang, S. J. *On the computation of the real radical.* Thesis, Technische Universität Kaiserslautern, 2007.

[44] Stetter, H. *Numerical Polynomial Algebra.* SIAM, 2004.

[45] Verschelde, J. Algorithm 795: Phcpack: a general-purpose solver for polynomial systems by homotopy continuation. *ACM Trans. Math. Softw. 25*, 2 (1999), 251–276.

Computing Column Bases of Polynomial Matrices

Wei Zhou and George Labahn
Cheriton School of Computer Science
University of Waterloo,
Waterloo, Ontario, Canada
{w2zhou,glabahn}@uwaterloo.ca

ABSTRACT

Given a matrix of univariate polynomials over a field $\mathbb{K}$, its columns generate a $\mathbb{K}[x]$-module. We call any basis of this module a column basis of the given matrix. Matrix gcds and matrix normal forms are examples of such module bases. In this paper we present a deterministic algorithm for the computation of a column basis of an $m \times n$ input matrix with $m \leq n$. If s is the average column degree of the input matrix, this algorithm computes a column basis with a cost of $O^{\sim}\left(nm^{\omega-1}s\right)$ field operations in $\mathbb{K}$. Here the soft-O notation is Big-O with log factors removed while ω is the exponent of matrix multiplication. Note that the average column degree s is bounded by the commonly used matrix degree that is also the maximum column degree of the input matrix.

Categories and Subject Descriptors: I.1.2 [Symbolic and Algebraic Manipulation]: Algorithms; F.2.2 [Analysis of Algorithms and Problem Complexity]: Nonnumerical Algorithms and Problems

General Terms: Algorithms, Theory

Keywords: Order Basis, Kernel basis, Nullspace basis, Column Basis

1. INTRODUCTION

In this paper, we consider the problem of efficiently computing a column basis of a polynomial matrix $\mathbf{F} \in \mathbb{K}[x]^{m\times n}$ with $n \geq m$. A column basis of $\mathbf{F}$ is a basis for the $\mathbb{K}[x]$-module

$$\{\mathbf{F}\mathbf{p} \mid \mathbf{p} \in \mathbb{K}[x]^n \ \} \ .$$

Such a basis can be represented as a full rank matrix $\mathbf{T} \in \mathbb{K}[x]^{m\times r}$ whose columns are the basis elements. A column basis is not unique and indeed any column basis right multiplied by a unimodular polynomial matrix gives another column basis. As a result, a column basis can have arbitrarily high degree. In this paper, the computed column basis has column degrees bounded by the largest column degrees of the input matrix.

ISSAC'13, June 26–29, 2013, Boston, Massachusetts, USA.

Column bases are fundamental constructions in polynomial matrix algebra. As an example, when the row dimension is one (i.e. $m = 1$), then finding a column basis coincides with finding a greatest common divisor (GCD) of all the polynomials in the matrix. Similarly, the nonzero columns of column reduced forms, Popov normal forms, and Hermite normal forms are all column bases satisfying additional degree constraints. A column reduced form gives a special column basis whose column degrees are the smallest possible, while Popov and Hermite forms are special column reduced or shifted column reduced forms satisfying additional conditions that make them unique. Efficient column basis computation is thus useful for fast computation for such core procedures as determining matrix GCDs [4], column reduced forms [7] and Popov forms [15] of $\mathbf{F}$ with any dimension and rank. Column basis computation also provides a deterministic alternative to randomized lattice compression [11, 14].

Column bases are produced by column reduced, Popov and Hermite forms and considerable work has been done on computing such forms, for example [1, 8, 9, 12, 13]. However most of these existing algorithms require that the input matrices be square nonsingular and so start with existing column bases. It is however pointed out in [12, 13] that randomization can be used to relax the square nonsingular requirement.

To compute a column basis, we know from [3] that any matrix polynomial $\mathbf{F} \in \mathbb{K}[x]^{m\times n}$ can be unimodularly transformed to a column basis by repeatedly working with the leading column coefficient matrices. However this method of computing a column basis can be expensive. Indeed one needs to work with up to $\sum \vec{s}$ such coefficient matrices, which could involve up to $\sum \vec{s}$ polynomial matrix multiplications, where a sum without index denotes the sum of all entries.

In this paper we give a fast, deterministic algorithm for the computation of a column basis for $\mathbf{F}$ having complexity $O^{\sim}\left(nm^{\omega-1}s\right)$ field operations in $\mathbb{K}$ with s being the average column degree of $\mathbf{F}$. Here the soft-O notation is Big-O with log factors removed while ω is the exponent of matrix multiplication. Our algorithm works for both rectangular and non-full column rank matrices. Our method depends on kernel basis computation of $\mathbf{F}$ along with finding a factorization of the input matrix polynomial into a column basis and a left kernel basis of a right kernel basis of $\mathbf{F}$. Finding the right and left kernel basis then makes use of the fast kernel basis and order basis algorithms from [20] and [18, 19], respectively.

The remainder of this paper is as follows. Basic definitions and preliminary results on both kernel and order bases are given in the next section. Section 3 provides the matrix factorization form of our input polynomial matrix that forms the core of our procedure, with a column basis being the left factor, and the right factor is a left kernel basis of a right kernel basis of the input matrix. Section 4 provides an algorithm for fast computation of a left kernel basis making use of order bases computation with unbalanced shift. The column basis algorithm is given in Section 5 with the following section giving details on how the methods can be improved when the number of columns is significantly larger than the number of rows. The paper ends with a conclusion along with topics for future research.

2. PRELIMINARIES

In this paper computational cost is analyzed by bounding the number of arithmetic operations in the coefficient field $\mathbb{K}$ on an algebraic random access machine. We assume the cost of multiplying two polynomial matrices with dimension n and degree d is $O^{\sim}(n^{\omega}d)$ field operations, where the multiplication exponent ω is assumed to satisfy $2 < \omega \leq 3$. We refer to the book by [16] for more details and references about the cost of polynomial multiplication and matrix multiplication.

In this section we first describe the notations used in this paper, and then give the basic definitions and properties of *shifted degree*, *order basis* and *kernel basis* for a matrix of polynomials. These will be the building blocks used in our algorithm.

2.1 Notations

For convenience we adopt the following notations in this paper.

Comparing Unordered Lists For two lists $\vec{a} \in \mathbb{Z}^n$ and $\vec{b} \in \mathbb{Z}^n$, let $\bar{a} = [\bar{a}_1, \ldots, \bar{a}_n] \in \mathbb{Z}^n$ and $\bar{b} = [\bar{b}_1, \ldots, \bar{b}_n] \in \mathbb{Z}^n$ be the lists consists of the entries of $\vec{a}$ and $\vec{b}$ but sorted in increasing order.

$$\begin{cases} \vec{a} \geq \vec{b} & \text{if } \bar{a}_i \geq \bar{b}_i \text{ for all } i \in [1, \ldots n] \\ \vec{a} \leq \vec{b} & \text{if } \bar{a}_i \leq \bar{b}_i \text{ for all } i \in [1, \ldots n] \\ \vec{a} > \vec{b} & \text{if } \vec{a} \geq \vec{b} \text{ and } \bar{a}_j > \bar{b}_j \text{ for at least one } j \in [1, \ldots n] \\ \vec{a} < \vec{b} & \text{if } \vec{a} \leq \vec{b} \text{ and } \bar{a}_j < \bar{b}_j \text{ for at least one } j \in [1, \ldots n]. \end{cases}$$

Uniformly Shift a List For a list $\vec{a} = [a_1, \ldots, a_n] \in \mathbb{Z}^n$ and $c \in \mathbb{Z}$, we write $\vec{a} + c$ to denote $\vec{a} + [c, \ldots, c] = [a_1 + c, \ldots, a_n + c]$, with subtraction handled similarly.

Compare a List with a Integer For $\vec{a} = [a_1, \ldots, a_n] \in \mathbb{Z}^n$ and $c \in \mathbb{Z}$, we write $\vec{a} < c$ to denote $\vec{a} < [c, \ldots, c]$, and similarly for $>, \leq, \geq, =$.

2.2 Shifted Degrees

Our methods depend extensively on the concept of *shifted* degrees of polynomial matrices [5]. For a column vector $\mathbf{p} = [p_1, \ldots, p_n]^T$ of univariate polynomials over a field $\mathbb{K}$, its column degree, denoted by $\text{cdeg}\,\mathbf{p}$, is the maximum of the degrees of the entries of $\mathbf{p}$, that is,

$$\text{cdeg } \mathbf{p} = \max_{1 \leq i \leq n} \deg p_i.$$

The *shifted column degree* generalizes this standard column degree by taking the maximum after shifting the degrees by a given integer vector that is known as a *shift*. More specifically, the shifted column degree of $\mathbf{p}$ with respect to a shift $\vec{s} = [s_1, \ldots, s_n] \in \mathbb{Z}^n$, or the *$\vec{s}$-column degree* of $\mathbf{p}$ is

$$\text{cdeg}_{\vec{s}}\, \mathbf{p} = \max_{1 \leq i \leq n} [\deg p_i + s_i] = \deg(x^{\vec{s}} \cdot \mathbf{p}),$$

where $x^{\vec{s}} = \text{diag}\,(x^{s_1}, x^{s_2}, \ldots, x^{s_n})$. For a matrix $\mathbf{P}$, we use $\text{cdeg}\,\mathbf{P}$ and $\text{cdeg}_{\vec{s}}\mathbf{P}$ to denote respectively the list of its column degrees and the list of its shifted $\vec{s}$-column degrees. When $\vec{s} = [0, \ldots, 0]$, the shifted column degree specializes to the standard column degree. The shifted row degree of a row vector $\mathbf{q} = [q_1, \ldots, q_n]$ is defined similarly as

$$\text{rdeg}_{\vec{s}}\mathbf{q} = \max_{1 \leq i \leq n} [\deg q_i + s_i] = \deg(\mathbf{q} \cdot x^{\vec{s}}).$$

Shifted degrees have been used previously in polynomial matrix computations and in generalizations of some matrix normal forms [6].

The usefulness of the shifted degrees can be seen from their applications in polynomial matrix computation problems [19, 20]. One of its uses is illustrated by the following lemma from [17, Chapter 2], which can be viewed as a stronger version of the predictable-degree property [10].

LEMMA 2.1. *Let $\mathbf{A} \in \mathbb{K}[x]^{m \times n}$ be a $\vec{u}$-column reduced matrix with no zero columns and with $\text{cdeg}_{\vec{u}}\mathbf{A} = \vec{v}$. Then a matrix $\mathbf{B} \in \mathbb{K}[x]^{n \times k}$ has $\vec{v}$-column degrees $\text{cdeg}_{\vec{v}}\mathbf{B} = \vec{w}$ if and only if $\text{cdeg}_{\vec{u}}(\mathbf{AB}) = \vec{w}$.*

The following lemma from [17, Chapter 2] describes a relationship between shifted column degrees and shifted row degrees.

LEMMA 2.2. *A matrix $\mathbf{A} \in \mathbb{K}[x]^{m \times n}$ has $\vec{u}$-column degrees bounded by $\vec{v}$ if and only if its $-\vec{v}$-row degrees are bounded by $-\vec{u}$.*

Another essential fact needed in our algorithm, also based on the use of the shifted degrees, is the efficient multiplication of matrices with unbalanced degrees [20, Theorem 3.7].

THEOREM 2.3. *Let $\mathbf{A} \in \mathbb{K}[x]^{m \times n}$ with $m \leq n$, $\vec{s} \in \mathbb{Z}^n$ a shift with entries bounding the column degrees of $\mathbf{A}$ and ξ, a bound on the sum of the entries of $\vec{s}$. Let $\mathbf{B} \in \mathbb{K}[x]^{n \times k}$ with $k \in O(m)$ and the sum θ of its $\vec{s}$-column degrees satisfying $\theta \in O(\xi)$. Then we can multiply $\mathbf{A}$ and $\mathbf{B}$ with a cost of $O^{\sim}(n^2 m^{\omega-2} s) \subset O^{\sim}(n^{\omega} s)$, where $s = \xi/n$ is the average of the entries of $\vec{s}$.*

2.3 Order Basis

Let $\mathbb{K}$ be a field, $\mathbf{F} \in \mathbb{K}[[x]]^{m \times n}$ a matrix of power series and $\vec{\sigma} = [\sigma_1, \ldots, \sigma_m]$ a vector of non-negative integers.

DEFINITION 2.4. *A vector of polynomials $\mathbf{p} \in \mathbb{K}[x]^{n \times 1}$ has* order *$(\mathbf{F}, \vec{\sigma})$ (or* order *$\vec{\sigma}$ with respect to $\mathbf{F}$) if $\mathbf{F} \cdot \mathbf{p} \equiv \mathbf{0} \bmod x^{\vec{\sigma}}$, that is,*

$$\mathbf{F} \cdot \mathbf{p} = x^{\vec{\sigma}} \mathbf{r}$$

for some $\mathbf{r} \in \mathbb{K}[[x]]^{m \times 1}$. If $\vec{\sigma} = [\sigma, \ldots, \sigma]$ has entries uniformly equal to σ, then we say that $\mathbf{p}$ has order $(\mathbf{F}, \sigma)$. The set of all order $(\mathbf{F}, \vec{\sigma})$ vectors is a free $\mathbb{K}[x]$-module denoted by $\langle(\mathbf{F}, \vec{\sigma})\rangle$.

An order basis for $\mathbf{F}$ and $\vec{\sigma}$ is simply a basis for the $\mathbb{K}[x]$-module $\langle(\mathbf{F}, \vec{\sigma})\rangle$. We again represent order bases using matrices, whose columns are the basis elements. We only work with those order bases having minimal or shifted minimal degrees (also referred to as a reduced order basis in [3]), that is, their column degrees or shifted column degrees are the smallest possible among all bases of the module.

An order basis [2, 3] $\mathbf{P}$ of $\mathbf{F}$ with order $\vec{\sigma}$ and shift $\vec{s}$, or simply an $(\mathbf{F},\vec{\sigma},\vec{s})$-basis, is a basis for the module $\langle(\mathbf{F},\vec{\sigma})\rangle$ having minimal $\vec{s}$-column degrees. If $\vec{\sigma}=[\sigma,\ldots,\sigma]$ is uniform then we simply write $(\mathbf{F},\sigma,\vec{s})$-basis. The precise definition of an $(\mathbf{F},\vec{\sigma},\vec{s})$-basis is as follows.

DEFINITION 2.5. *A polynomial matrix $\mathbf{P}$ is an order basis of $\mathbf{F}$ of order $\vec{\sigma}$ and shift $\vec{s}$, denoted by $(\mathbf{F},\vec{\sigma},\vec{s})$-basis, if the following properties hold:*

1. *$\mathbf{P}$ is a nonsingular matrix of dimension n and is $\vec{s}$-column reduced.*
2. *$\mathbf{P}$ has order $(\mathbf{F},\vec{\sigma})$ (or equivalently, each column of $\mathbf{P}$ is in $\langle(\mathbf{F},\vec{\sigma})\rangle$).*
3. *Any $\mathbf{q}\in\langle(\mathbf{F},\vec{\sigma})\rangle$ can be expressed as a linear combination of the columns of $\mathbf{P}$, given by $\mathbf{P}^{-1}\mathbf{q}$.*

In this paper, a $(\mathbf{F},\vec{\sigma},\vec{s})$-basis is also called a $(\mathbf{F},\vec{\sigma},\vec{s})$-order basis to further distinguish it from the kernel basis notation given in the following subsection.

Note that any pair of $(\mathbf{F},\vec{\sigma},\vec{s})$-order bases $\mathbf{P}$ and $\mathbf{Q}$ are column bases of each other and are unimodularly equivalent.

We will need to compute order bases with unbalanced shifts using Algorithm 2 from [18]. This computation can be done efficiently as given by the following result from [20].

THEOREM 2.6. *If $\vec{s}$ satisfies $\vec{s}\le 0$ and $-\sum\vec{s}\le m\sigma$, then a $(\mathbf{F},\sigma,\vec{s})$-basis can be computed with a cost of $O^{\sim}(n^{\omega}d)$ field operations, where $d=m\sigma/n$.*

2.4 Kernel Bases

The kernel of $\mathbf{F}\in\mathbb{K}[x]^{m\times n}$ is the $\mathbb{F}[x]$-module

$$\{\mathbf{p}\in\mathbb{K}[x]^{n}\mid\mathbf{F}\mathbf{p}=0\}$$

with a kernel basis of $\mathbf{F}$ being a basis of this module. Kernel bases are closely related to order bases, as can be seen from the following definitions.

DEFINITION 2.7. *Given $\mathbf{F}\in\mathbb{K}[x]^{m\times n}$, a polynomial matrix $\mathbf{N}\in\mathbb{K}[x]^{n\times *}$ is a (right) kernel basis of $\mathbf{F}$ if the following properties hold:*

1. *$\mathbf{N}$ is full-rank.*
2. *$\mathbf{N}$ satisfies $\mathbf{F}\cdot\mathbf{N}=0$.*
3. *Any $\mathbf{q}\in\mathbb{K}[x]^{n}$ satisfying $\mathbf{F}\mathbf{q}=0$ can be expressed as a linear combination of the columns of $\mathbf{N}$, that is, there exists some polynomial vector $\mathbf{p}$ such that $\mathbf{q}=\mathbf{N}\mathbf{p}$.*

Any pair of kernel bases $\mathbf{N}$ and $\mathbf{M}$ of $\mathbf{F}$ are column bases of each other and are unimodularly equivalent.

A $\vec{s}$-minimal kernel basis of $\mathbf{F}$ is just a kernel basis that is $\vec{s}$-column reduced.

DEFINITION 2.8. *Given $\mathbf{F}\in\mathbb{K}[x]^{m\times n}$, a polynomial matrix $\mathbf{N}\in\mathbb{K}[x]^{n\times *}$ is a $\vec{s}$-minimal (right) kernel basis of $\mathbf{F}$ if $\mathbf{N}$ is a kernel basis of $\mathbf{F}$ and $\mathbf{N}$ is $\vec{s}$-column reduced. We also call a $\vec{s}$-minimal (right) kernel basis of $\mathbf{F}$ a $(\mathbf{F},\vec{s})$-kernel basis.*

In our earlier paper [20], a minimal kernel basis is called a minimal nullspace basis. In this paper, the term kernel basis is now used to emphasize the fact that we compute a basis for a $\mathbb{F}[x]$-module instead of a basis for a $\mathbb{F}(x)$-vector space, since the term nullspace basis usually refers to a basis of some vector space as in [14].

We will need the following result from [20] to bound the sizes of kernel bases.

THEOREM 2.9. *Suppose $\mathbf{F}\in\mathbb{K}[x]^{m\times n}$ and $\vec{s}\in\mathbb{Z}_{\ge 0}^{n}$ is a shift with entries bounding the corresponding column degrees of $\mathbf{F}$. Then the sum of the $\vec{s}$-column degrees of any $\vec{s}$-minimal kernel basis of $\mathbf{F}$ is bounded by $\sum\vec{s}$.*

We will also need the following result from [20] to compute kernel bases by rows.

THEOREM 2.10. *Let $\mathbf{G}=\left[\mathbf{G}_1^T,\mathbf{G}_2^T\right]^T\in\mathbb{K}[x]^{m\times n}$ and $\vec{t}\in\mathbb{Z}^{n}$ a shift vector. If $\mathbf{N}_1$ is a $\left(\mathbf{G}_1,\vec{t}\right)$-kernel basis with $\vec{t}$-column degrees $\vec{u}$, and $\mathbf{N}_2$ is a $(\mathbf{G}_2\mathbf{N}_1,\vec{u})$-kernel basis with $\vec{u}$-column degrees $\vec{v}$, then $\mathbf{N}_1\mathbf{N}_2$ is a $\left(\mathbf{G},\vec{t}\right)$-kernel basis with $\vec{t}$-column degrees $\vec{v}$.*

Also recall the cost of kernel basis computation from [20].

THEOREM 2.11. *Given an input matrix $\mathbf{F}\in\mathbb{K}[x]^{m\times n}$. Let $\vec{s}=\operatorname{cdeg}\mathbf{F}$ and $s=\sum\vec{s}/n$ be the average column degree of $\mathbf{F}$. Then a $(\mathbf{F},\vec{s})$-kernel basis can be computed with a cost of $O^{\sim}(n^{\omega}s)$ field operations.*

3. COLUMN BASIS VIA FACTORIZATION

In this section we reduce the problem of determining a column basis of a polynomial matrix into three separate processes. For this reduction it turns out to be useful to look at following relationship between column basis, kernel basis, and unimodular matrices.

LEMMA 3.1. *Let $\mathbf{F}\in\mathbb{K}[x]^{m\times n}$ and suppose $\mathbf{U}\in\mathbb{K}[x]^{n\times n}$ is a unimodular matrix such that $\mathbf{F}\mathbf{U}=[0,\mathbf{T}]$ with $\mathbf{T}$ of full column rank. Partition $\mathbf{U}=[\mathbf{U}_L,\mathbf{U}_R]$ such that $\mathbf{F}\cdot\mathbf{U}_L=0$ and $\mathbf{F}\mathbf{U}_R=\mathbf{T}$. Then*

1. *$\mathbf{U}_L$ is a kernel basis of $\mathbf{F}$ and $\mathbf{T}$ is a column basis of $\mathbf{F}$.*
2. *If $\mathbf{N}$ is any other kernel basis of $\mathbf{F}$, then $\mathbf{U}^*=[\mathbf{N},\ \mathbf{U}_R]$ is also unimodular and also unimodularly transforms $\mathbf{F}$ to $[0,\mathbf{T}]$.*

PROOF. Since $\mathbf{F}$ and $[0,\mathbf{T}]$ are unimodularly equivalent with $\mathbf{T}$ having full column rank we have that $\mathbf{T}$ is a column basis of $\mathbf{F}$. It remains to show that $\mathbf{U}_L$ is a kernel basis of $\mathbf{F}$. Since $\mathbf{F}\mathbf{U}_L=0$, $\mathbf{U}_L$ is generated by any kernel basis $\mathbf{N}$, that is, $\mathbf{U}_L=\mathbf{N}\mathbf{C}$ for some polynomial matrix $\mathbf{C}$. Let r be the rank of $\mathbf{F}$, which is also the column dimension of $\mathbf{T}$ and $\mathbf{U}_R$. Then both $\mathbf{N}$ and $\mathbf{U}_L$ have column dimension $n-r$. Hence $\mathbf{C}$ is a square $(n-r)\times(n-r)$ matrix. The unimodular matrix $\mathbf{U}$ can be factored as

$$\mathbf{U}=[\mathbf{N}\mathbf{C},\mathbf{U}_R]=[\mathbf{N},\mathbf{U}_R]\begin{bmatrix}\mathbf{C}&0\\0&I\end{bmatrix},$$

implying that both factors $[\mathbf{N},\mathbf{U}_R]$ and $\begin{bmatrix}\mathbf{C}&0\\0&I\end{bmatrix}$ are unimodular. Therefore, $\mathbf{C}$ is unimodular and $\mathbf{U}_L=\mathbf{N}\mathbf{C}$ is also a kernel basis. Notice that the unimodular matrix $[\mathbf{N},\mathbf{U}_R]$ also transforms $\mathbf{F}$ to $[0,\mathbf{T}]$. □

REMARK 3.2. *It is interesting to see what Lemma 3.1 implies in the case of unimodular matrices. Let* $\mathbf{U} \in \mathbb{K}[x]^{n\times n}$ *be a unimodular matrix with inverse* $\mathbf{V}$*, which, for a given* k*, are partitioned as* $\mathbf{U} = [\mathbf{U}_L, \mathbf{U}_R]$ *and* $\mathbf{V} = \begin{bmatrix} \mathbf{V}_U \\ \mathbf{V}_D \end{bmatrix}$ *with* $\mathbf{U}_L \in \mathbb{K}[x]^{n\times k}$ *and* $\mathbf{V}_U \in \mathbb{K}[x]^{k\times n}$*. Since* $\mathbf{U}$ *and* $\mathbf{V}$ *are inverses of each other we have the identities*

$$\mathbf{VU} = \begin{bmatrix} \mathbf{V}_U\mathbf{U}_L & \mathbf{V}_U\mathbf{U}_R \\ \mathbf{V}_D\mathbf{U}_L & \mathbf{V}_D\mathbf{U}_R \end{bmatrix} = \begin{bmatrix} I_k & 0 \\ 0 & I_{n-k} \end{bmatrix}. \tag{1}$$

Lemma 3.1 then gives:

1. I_k *is a column basis of* $\mathbf{V}_U$ *and a row basis of* $\mathbf{U}_L$*,*
2. I_{n-k} *is a column basis of* $\mathbf{V}_D$ *and a row basis of* $\mathbf{U}_R$*,*
3. $\mathbf{V}_D$ *and* $\mathbf{U}_L$ *are kernel bases of each other,*
4. $\mathbf{V}_U$ *and* $\mathbf{U}_R$ *are kernel bases of each other.*

LEMMA 3.3. *Let* $\mathbf{F} \in \mathbb{K}[x]^{m\times n}$ *with rank* r*. Suppose* $\mathbf{N} \in \mathbb{K}[x]^{n\times(n-r)}$ *is a right kernel basis of* $\mathbf{F}$ *and* $\mathbf{G} \in \mathbb{K}[x]^{r\times n}$ *is a left kernel basis of* $\mathbf{N}$*. Then* $\mathbf{F} = \mathbf{T}\cdot\mathbf{G}$ *with* $\mathbf{T} \in \mathbb{K}[x]^{m\times r}$ *a column basis of* $\mathbf{F}$*.*

PROOF. Let $\mathbf{U} = [\mathbf{U}_L, \mathbf{U}_R]$ be a unimodular matrix with inverse $\mathbf{V} = \begin{bmatrix} \mathbf{V}_U \\ \mathbf{V}_D \end{bmatrix}$ partitioned as in equation (1) and satisfying $\mathbf{F}\cdot\mathbf{U} = [0, \mathbf{B}]$ with $\mathbf{B} \in \mathbb{K}[x]^{m\times r}$ a column basis of $\mathbf{F}$. Then $\mathbf{F} = [0, \mathbf{B}]\,\mathbf{U}^{-1} = \mathbf{B}\,[0, I]\,\mathbf{V} = \mathbf{B}\mathbf{V}_D$. Since $\mathbf{V}_D$ is a left kernel basis of $\mathbf{U}_L$, any other left kernel basis $\mathbf{G}$ of $\mathbf{U}_L$ is unimodularly equivalent to $\mathbf{V}_D$, that is, $\mathbf{V}_D = \mathbf{W}\cdot\mathbf{G}$ for some unimodular matrix $\mathbf{W}$. Thus $\mathbf{F} = \mathbf{B}\cdot\mathbf{W}\cdot\mathbf{G}$. Then $\mathbf{T} = \mathbf{B}\cdot\mathbf{W}$ is a column basis of $\mathbf{F}$ since it is unimodularly equivalent to the column basis $\mathbf{B}$. □

Lemma 3.3 outlines a procedure for computing a column basis of $\mathbf{F}$ with three main steps. The first step is to compute a right kernel basis $\mathbf{N}$ of $\mathbf{F}$, something which can be efficiently done using the kernel basis algorithm of [20]. The second step, computing a left kernel basis $\mathbf{G}$ for $\mathbf{N}$ and the third step, computing the column basis $\mathbf{T}$ from $\mathbf{F}$ and $\mathbf{G}$, will still require additional work for efficient computation. Note that, while Lemma 3.3 does not require the bases computed to be minimal, working with minimal kernel bases keeps the degrees well controlled, an important consideration for efficient computation.

EXAMPLE 3.4. *Let*

$$\mathbf{F} = \begin{bmatrix} x^2 & x^2 & x+x^2 & 1+x^2 \\ 1+x+x^2 & x^2 & 1+x^2 & 1+x^2 \end{bmatrix}$$

be a matrix over $\mathbb{Z}_2[x]$*. Then the matrix*

$$\mathbf{N} = \begin{bmatrix} x & 1 \\ 1 & x \\ x & 1 \\ 0 & x \end{bmatrix}$$

is a right kernel basis of $\mathbf{F}$ *and the matrix*

$$\mathbf{G} = \begin{bmatrix} 1 & 0 & 1 & 0 \\ x & x^2 & 0 & 1+x^2 \end{bmatrix}$$

is a left kernel basis of $\mathbf{N}$*. Finally the matrix*

$$\mathbf{T} = \begin{bmatrix} x+x^2 & 1 \\ 1+x^2 & 1 \end{bmatrix}$$

satisfies $\mathbf{F} = \mathbf{TG}$*, and is a column basis of* $\mathbf{F}$*.* □

4. COMPUTING A RIGHT FACTOR

Let $\mathbf{N}$ be an $(\mathbf{F}, \vec{s})$-kernel basis computed using the existing algorithm from [20]. Consider now the problem of computing a left $-\vec{s}$-minimal kernel basis $\mathbf{G}$ for $\mathbf{N}$, or equivalently, a right $(\mathbf{N}^T, -\vec{s})$-kernel basis $\mathbf{G}^T$. For this problem, the kernel basis algorithm of [20] cannot be applied directly, since the input matrix $\mathbf{N}^T$ has nonuniform row degrees and negative shift. Comparing to the earlier problem of computing a $\vec{s}$-minimal kernel basis $\mathbf{N}$ for $\mathbf{F}$, it is interesting to note that the original output $\mathbf{N}$ now becomes the new input matrix $\mathbf{N}^T$, while the new output matrix $\mathbf{G}$ has size bounded by the size of $\mathbf{F}$. In other words, the new input has degrees that match the original output, while the new output has degrees bounded by the original input. It is therefore reasonable to expect that the new problem can be computed efficiently. However, we need to find some way to work with the more complicated input degree structure. On the other hand, the simpler output degree structure makes it easier to apply order basis computation in order to compute a $(\mathbf{N}^T, -\vec{s})$-kernel basis.

4.1 Kernel Bases via Order Bases

In order to see how order basis computations can be applied here, let us first recall the following result (Lemma 3.3 [20]) on a relationship between order bases and kernel bases.

LEMMA 4.1. *Let* $\mathbf{P} = [\mathbf{P}_L, \mathbf{P}_R]$ *be any* $(\mathbf{F}, \sigma, \vec{s})$*-order basis and* $\mathbf{N} = [\mathbf{N}_L, \mathbf{N}_R]$ *be any* $\vec{s}$*-minimal kernel basis of* $\mathbf{F}$*, where* $\mathbf{P}_L$ *and* $\mathbf{N}_L$ *contain all columns from* $\mathbf{P}$ *and* $\mathbf{N}$*, respectively, whose* $\vec{s}$*-column degrees are less than* σ*. Then* $[\mathbf{P}_L, \mathbf{N}_R]$ *is a* $\vec{s}$*-minimal kernel basis of* $\mathbf{F}$*, and* $[\mathbf{N}_L, \mathbf{P}_R]$ *is a* $(\mathbf{F}, \sigma, \vec{s})$*-order basis.*

It is not difficult to extend this result to the following lemma to accommodate our situation here.

LEMMA 4.2. *Given a matrix* $\mathbf{A} \in \mathbb{K}[x]^{m\times n}$ *and some integer lists* $\vec{u} \in \mathbb{Z}^n$ *and* $\vec{v} \in \mathbb{Z}^m$ *such that* $\operatorname{rdeg}_{\vec{u}}\mathbf{A} \le \vec{v}$*, or equivalently,* $\operatorname{cdeg}_{-\vec{v}}\mathbf{A} \le -\vec{u}$*. Let* $\mathbf{P}$ *be a* $(\mathbf{A}, \vec{v}+1, -\vec{u})$*-order basis and* $\mathbf{Q}$ *be any* $(\mathbf{A}, -\vec{u})$*-kernel basis. Partition* $\mathbf{P} = [\mathbf{P}_L, \mathbf{P}_R]$ *and* $\mathbf{Q} = [\mathbf{Q}_L, \mathbf{Q}_R]$ *where* $\mathbf{P}_L$ *and* $\mathbf{Q}_L$ *contain all the columns from* $\mathbf{P}$ *and* $\mathbf{Q}$*, respectively, whose* $-\vec{u}$*-column degrees are no more than* 0*. Then*

(i) $[\mathbf{P}_L, \mathbf{Q}_R]$ *is an* $(\mathbf{A}, -\vec{u})$*-kernel basis, and*

(ii) $[\mathbf{Q}_L, \mathbf{P}_R]$ *is an* $(\mathbf{A}, \vec{v}+1, -\vec{u})$*-order basis.*

PROOF. We can use the same proof from Lemma 3.3 in [20]. We know $\operatorname{cdeg}_{-\vec{v}}\mathbf{A}\mathbf{P}_L \le \operatorname{cdeg}_{-\vec{u}}\mathbf{P}_L \le 0$, or equivalently, $\operatorname{rdeg}\mathbf{A}\mathbf{P}_L \le \vec{v}$. However it also has order greater than $\vec{v}$ and hence $\mathbf{A}\mathbf{P}_L = 0$. Thus $\mathbf{P}_L$ is generated by the kernel basis $\mathbf{Q}_L$, that is, $\mathbf{P}_L = \mathbf{Q}_L\mathbf{U}$ for some polynomial matrix $\mathbf{U}$. On the other hand, $\mathbf{Q}_L$ certainly has order $(\mathbf{A}, \vec{v}+1)$ and therefore is generated by $\mathbf{P}_L$, that is, $\mathbf{Q}_L = \mathbf{P}_L\mathbf{V}$ for some polynomial matrix $\mathbf{V}$. We now have $\mathbf{P}_L = \mathbf{P}_L\mathbf{VU}$ and $\mathbf{Q}_L = \mathbf{Q}_L\mathbf{UV}$, implying both $\mathbf{U}$ and $\mathbf{V}$ are unimodular. The result then follows from the unimodular equivalence of $\mathbf{P}_L$ and $\mathbf{Q}_L$ and the fact that they are $-\vec{u}$-column reduced. □

With the help of Lemma 4.2 we can return to the problem of efficiently computing a $(\mathbf{N}^T, -\vec{s})$-kernel basis. In fact, we just need to use a special case of Lemma 4.2, where all the elements of the kernel basis have shifted degrees bounded by 0, thereby making the partial kernel basis be a complete kernel basis.

LEMMA 4.3. *Let* $\mathbf{N}$ *be a* $(\mathbf{F}, \vec{s})$*-kernel basis with* $\mathrm{cdeg}_{\vec{s}}\,\mathbf{N} = \vec{b}$. *Then any* $(\mathbf{N}^T, -\vec{s})$*-kernel basis* $\mathbf{G}^T$ *satisfies* $\mathrm{cdeg}_{-\vec{s}}\mathbf{G}^T \le 0$. *Let* $\mathbf{P} = [\mathbf{P}_L, \mathbf{P}_R]$ *be a* $\left(\mathbf{N}^T, \vec{b}+1, -\vec{s}\right)$*-order basis, where* $\mathbf{P}_L$ *consists of all columns* $\mathbf{p}$ *satisfying* $\mathrm{cdeg}_{-\vec{s}}\,\mathbf{p} \le 0$. *Then* $\mathbf{P}_L$ *is a* $(\mathbf{N}^T, -\vec{s})$*-kernel basis.*

PROOF. The column dimension of any $(\mathbf{N}^T, -\vec{s})$-kernel basis $\mathbf{G}^T$ equals the rank r of $\mathbf{F}$. Since both $\mathbf{F}$ and $\mathbf{G}$ are in the left kernel of $\mathbf{N}$, we know $\mathbf{F}$ is generated by $\mathbf{G}$, and the $-\vec{s}$-minimality of $\mathbf{G}$ ensures that the $-\vec{s}$-row degrees of $\mathbf{G}$ are bounded by the corresponding r largest $-\vec{s}$-row degrees of $\mathbf{F}$. These are in turn bounded by 0 since $\mathrm{cdeg}\,\mathbf{F} \le \vec{s}$. Therefore, any $(\mathbf{N}^T, -\vec{s})$-kernel basis $\mathbf{G}^T$ satisfies $\mathrm{cdeg}_{-\vec{s}}\mathbf{G}^T \le 0$. The result follows from Lemma 4.2. □

While Lemma 4.3 shows that a complete $(\mathbf{N}^T, -\vec{s})$-kernel basis can be computed by computing a $\left(\mathbf{N}^T, \vec{b}+1, -\vec{s}\right)$-order basis, in fact we do not compute such a order basis, as the computational efficiency can be improved by using Theorem 2.10 to compute a $(\mathbf{N}^T, -\vec{s})$-kernel basis by rows. More specifically, we can partition $\mathbf{N}$ into $[\mathbf{N}_1, \mathbf{N}_2]$ with $\vec{s}$-column degrees $\vec{b}_1$, $\vec{b}_2$ respectively, compute a $(\mathbf{N}_1^T, -\vec{s})$-kernel basis $\mathbf{Q}_1$ with $-\vec{s}$-column degrees $-\vec{s}_2$, and then compute a $(\mathbf{N}_2^T\mathbf{Q}_1, -\vec{s}_2)$-kernel basis $\mathbf{Q}_2$, then $\mathbf{Q}_1\mathbf{Q}_2$ is a $(\mathbf{N}^T, -\vec{s})$-kernel basis. In order to compute the kernel bases $\mathbf{Q}_1$ and $\mathbf{Q}_2$, we still use order basis computations but work with subsets of rows rather than the whole matrix $\mathbf{N}^T$. We now need to make sure that the order bases computed from subsets of rows contain these kernel bases.

LEMMA 4.4. *Let* $\mathbf{N}$ *be partitioned as* $[\mathbf{N}_1, \mathbf{N}_2]$, *with* $\vec{s}$*-column degrees* $\vec{b}_1$, $\vec{b}_2$, *respectively. Then we have the following:*

1. *A* $\left(\mathbf{N}_1^T, \vec{b}_1+1, -\vec{s}\right)$*-order basis contains a* $(\mathbf{N}_1^T, -\vec{s})$*-kernel basis whose* $-\vec{s}$*-column degrees are bounded by* 0.
2. *If* $\mathbf{Q}_1$ *is this* $(\mathbf{N}_1^T, -\vec{s})$*-kernel basis from above and* $-\vec{s}_2 = \mathrm{cdeg}_{-\vec{s}}\mathbf{Q}_1$, *then a* $\left(\mathbf{N}_2^T\mathbf{Q}_1, \vec{b}_2+1, -\vec{s}_2\right)$*-basis contains a* $(\mathbf{N}_2^T\mathbf{Q}_1, -\vec{s}_2)$*-kernel basis,* $\mathbf{Q}_2$, *whose* $-\vec{s}$*-column degrees are bounded by* 0.
3. *The product* $\mathbf{Q}_1\mathbf{Q}_2$ *is a* $(\mathbf{N}^T, -\vec{s})$*-kernel basis.*

PROOF. To see that a $\left(\mathbf{N}_1^T, \vec{b}_1+1, -\vec{s}\right)$-basis contains a $(\mathbf{N}_1^T, -\vec{s})$-kernel basis whose $-\vec{s}$-column degrees are bounded by 0, we just need to show that $\mathrm{cdeg}_{-\vec{s}}\bar{\mathbf{Q}}_1 \le 0$ for any $(\mathbf{N}_1^T, -\vec{s})$-kernel basis $\bar{\mathbf{Q}}_1$ and then apply Lemma 4.2. Note that there exists a polynomial matrix $\bar{\mathbf{Q}}_2$ such that $\bar{\mathbf{Q}}_1\bar{\mathbf{Q}}_2 = \bar{\mathbf{G}}$ for any $(\mathbf{N}^T, -\vec{s})$-kernel basis $\bar{\mathbf{G}}$, as $\bar{\mathbf{G}}$ satisfies $\mathbf{N}_1^T\bar{\mathbf{G}} = 0$ and is therefore generated by the $(\mathbf{N}_1^T, -\vec{s})$-kernel basis $\bar{\mathbf{Q}}_1$. If $\mathrm{cdeg}_{-\vec{s}}\bar{\mathbf{Q}}_1 \not\le 0$, then Lemma 2.1 forces $\mathrm{cdeg}_{-\vec{s}}\left(\bar{\mathbf{Q}}_1\bar{\mathbf{Q}}_2\right) = \mathrm{cdeg}_{-\vec{s}}\bar{\mathbf{G}} \not\le 0$,a contradiction since we know from Lemma 4.3 that $\mathrm{cdeg}_{-\vec{s}}\bar{\mathbf{G}} \le 0$.

As before, to see that a $\left(\mathbf{N}_2^T\mathbf{Q}_1, \vec{b}_2+1, -\vec{s}_2\right)$-basis contains a $(\mathbf{N}_2^T\mathbf{Q}_1, -\vec{s}_2)$-kernel basis whose $-\vec{s}$-column degrees are no more than 0, we can just show $\mathrm{cdeg}_{-\vec{s}_2}\hat{\mathbf{Q}}_2 \le 0$ for any $(\mathbf{N}_2^T\mathbf{Q}_1, -\vec{s}_2)$-kernel basis $\hat{\mathbf{Q}}_2$ and then apply Lemma 4.2. Since $\mathrm{cdeg}_{\vec{s}}\mathbf{N}_2 = \vec{b}_2$, we have $\mathrm{rdeg}_{-\vec{b}_2}\mathbf{N}_2 \le -\vec{s}$ or equivalently, $\mathrm{cdeg}_{-\vec{b}_2}\mathbf{N}_2^T \le -\vec{s}$. Then combining this with $\mathrm{cdeg}_{-\vec{s}}\mathbf{Q}_1 = -\vec{s}_2$ we get $\mathrm{cdeg}_{-\vec{b}_2}\mathbf{N}_2^T\mathbf{Q}_1 \le -\vec{s}_2$ using Lemma 2.1. Let $\hat{\mathbf{G}} = \mathbf{Q}_1\hat{\mathbf{Q}}_2$, which is a $(\mathbf{N}^T, -\vec{s})$-kernel basis by Theorem 2.10. Note that $\mathrm{cdeg}_{-\vec{s}_2}\hat{\mathbf{Q}}_2 = \mathrm{cdeg}_{-\vec{s}}\mathbf{Q}_1\hat{\mathbf{Q}}_2 = \mathrm{cdeg}_{-\vec{s}}\hat{\mathbf{G}} \le 0$. □

Algorithm 1 MinimalKernelBasisReversed$(\mathbf{M}, \vec{s}, \xi)$

(Kernel basis computation with reversed degree structure)

Input: $\mathbf{M} \in \mathbb{K}[x]^{k\times n}$ and $\vec{s} \in \mathbb{Z}^n_{\ge 0}$ such that $\sum \mathrm{rdeg}_{\vec{s}}\mathbf{M} \le \xi$, $\sum \vec{s} \le \xi$, and any $(\mathbf{M}, -\vec{s})$-kernel basis having row degrees bounded by $\vec{s}$ (equivalently, having $-\vec{s}$-column degrees bounded by 0).

Output: $\mathbf{G} \in \mathbb{K}[x]^{n\times *}$, a $(\mathbf{M}, -\vec{s})$-kernel basis.

$\left[\mathbf{M}_1^T, \mathbf{M}_2^T, \cdots, \mathbf{M}_{\lceil \log k\rceil -1}^T, \mathbf{M}_{\lceil \log k\rceil}^T\right] := \mathbf{M}^T$, with $\mathbf{M}_{\lceil \log k\rceil}, \mathbf{M}_{\lceil \log k\rceil-1}, \cdots, \mathbf{M}_2, \mathbf{M}_1$ having $\vec{s}$-row degrees in the range $\left[0, \frac{2\xi}{k}\right], (\frac{2\xi}{k}, \frac{4\xi}{k}], ..., (\frac{\xi}{4}, \frac{\xi}{2}], (\frac{\xi}{2}, \xi]$.
for i **from** 1 **to** $\lceil \log k\rceil$ **do**
  $\sigma_i := \left\lceil \frac{\xi}{2^{i-1}} \right\rceil + 1; \vec{\sigma}_i := [\sigma_i, \ldots, \sigma_i]$, number of entries matching the row dimension of $\mathbf{M}_i$;
end for
$\vec{\sigma} := \left[\vec{\sigma}_1, \vec{\sigma}_2, \ldots, \vec{\sigma}_{\lceil \log k\rceil}\right]$;
$\hat{\mathbf{N}} := x^{\vec{\sigma}-\vec{b}-1}\mathbf{M}$;
$\mathbf{G}_0 := I_n$; $\tilde{\mathbf{G}}_0 := I_n$;
for i **from** 1 **to** $\lceil \log k\rceil$ **do**
  $\vec{s}_i := -\mathrm{cdeg}_{-\vec{s}}\mathbf{G}_{i-1}$; (note $\vec{s}_1 = \vec{s}$)
  $\mathbf{P}_i := \mathrm{UnbalancedFastOrderBasis}\left(\hat{\mathbf{N}}_i\tilde{\mathbf{G}}_{i-1}, \sigma_i, -\vec{s}_i\right)$;
  $[\mathbf{G}_i, \mathbf{Q}_i] := \mathbf{P}_i$, where $\mathbf{G}_i$ is a $\left(\hat{\mathbf{M}}_i, -\vec{s}_i\right)$-kernel basis;
  $\tilde{\mathbf{G}}_i := \tilde{\mathbf{G}}_{i-1} \cdot \mathbf{G}_i$;
end for
return $\tilde{\mathbf{G}}_i$

4.2 Efficient Computation of Kernel Bases

Now that we can correctly compute a $(\mathbf{N}^T, -\vec{s})$-kernel basis by rows with the help of order basis computation using Lemma 4.4, we need to look at how to do this efficiently. One major difficulty is that the order $\vec{b}+1$, or equivalently, the $\vec{s}$-row degrees of $\mathbf{N}^T$ may be unbalanced and can have degree as large as $\sum \vec{s}$. Note that the existing kernel basis algorithm from [20] handles input matrices with unbalanced column degrees, but not unbalanced row degrees. For example, in the simpler special case of $\vec{s} = [s, \ldots, s]$ having uniformly equal entries, the sum of the row degrees is $O(ns)$, but the sum of column degrees can be $\Theta\left(n^2 s\right)$, which puts an extra factor n to the cost if the algorithm from [20] is used. To overcome this problem with unbalanced $\vec{s}$-row degrees, we separate the rows of $\mathbf{N}^T$ into blocks according to their $\vec{s}$-row degrees, and then work with these blocks one by one successively using Lemma 4.4.

Let k be the column dimension of $\mathbf{N}$ and ξ be an upper bound of $\sum \vec{s}$. Since

$$\sum \mathrm{cdeg}_{\vec{s}}\mathbf{N} = \sum \vec{b} \le \sum \vec{s} \le \xi$$

by Theorem 2.9, at most $\frac{k}{c}$ columns of $\mathbf{N}$ have $\vec{s}$-column degrees greater than or equal to $\frac{c\,\xi}{k}$ for any $c \ge 1$. Without loss of generality we can assume that the rows of $\mathbf{N}^T$ are arranged in decreasing $\vec{s}$-row degrees. We divide $\mathbf{N}^T$ into $\lceil \log k\rceil$ row blocks according to the $\vec{s}$-row degrees of its rows, or equivalently, divide $\mathbf{N}$ into blocks of columns according to the $\vec{s}$-column degrees. Let

$$\mathbf{N} = \left[\mathbf{N}_1, \mathbf{N}_2, \cdots, \mathbf{N}_{\lceil \log k \rceil - 1}, \mathbf{N}_{\lceil \log k \rceil}\right]$$

with $\mathbf{N}_{\lceil \log k \rceil}, \mathbf{N}_{\lceil \log k \rceil - 1}, \ldots, \mathbf{N}_2, \mathbf{N}_1$ having $\vec{s}$-column degrees in the range $[0, 2\xi/k]$, $(2\xi/k, 4\xi/k]$, $(4\xi/k, 8\xi/k]$, ..., $(\xi/4, \xi/2]$, $(\xi/2, \xi]$, respectively. Let $\sigma_i = \lceil \xi/2^{i-1} \rceil + 1$ and $\vec{\sigma}_i = [\sigma_i, \ldots, \sigma_i]$ with the same dimension as the row dimension of $\mathbf{N}_i$ and $\vec{\sigma} = \left[\vec{\sigma}_1, \vec{\sigma}_2, \ldots, \vec{\sigma}_{\lceil \log k \rceil}\right]$ be the orders in the order basis computation.

To further simplify our task, we also make the order of our problem in each block uniform. Rather than of using $\mathbf{N}^T$ as the input matrix, we instead use

$$\hat{\mathbf{N}} = \begin{bmatrix} \hat{\mathbf{N}}_1 \\ \vdots \\ \hat{\mathbf{N}}_{\lceil \log k \rceil} \end{bmatrix} = x^{\vec{\sigma} - \vec{b} - 1} \begin{bmatrix} \mathbf{N}_1^T \\ \vdots \\ \mathbf{N}_{\lceil \log k \rceil}^T \end{bmatrix} = x^{\vec{\sigma} - \vec{b} - 1} \mathbf{N}^T$$

so that a $\left(\hat{\mathbf{N}}, \vec{\sigma}, -\vec{s}\right)$-order basis is a $\left(\mathbf{N}^T, \vec{b} + 1, -\vec{s}\right)$-order basis.

In order to compute a $\left(\mathbf{N}^T, -\vec{s}\right)$-kernel basis we determine a series of kernel bases via a series of order basis computations as follows:

1. Let $\vec{s}_1 = \vec{s}$. Compute an $\left(\hat{\mathbf{N}}_1, \vec{\sigma}_1, -\vec{s}_1\right)$-order basis $\mathbf{P}_1$ using Algorithm 2 from [19] for order basis computation with unbalanced shift. Note that here the order $\vec{\sigma}_1 = [\sigma_1, \ldots, \sigma_1]$ is uniform, an $\left(\hat{\mathbf{N}}_1, \vec{\sigma}_1, -\vec{s}_1\right)$ order basis is also $\left(\hat{\mathbf{N}}_1, \sigma_1, -\vec{s}_1\right)$-order basis. Partition $\mathbf{P}_1$ as $\mathbf{P}_1 = [\mathbf{G}_1, \mathbf{Q}_1]$, where $\mathbf{G}_1$ is a $\left(\hat{\mathbf{N}}_1, -\vec{s}_1\right)$-kernel basis by Lemma 4.4. Set $\tilde{\mathbf{G}}_1 = \mathbf{G}_1$ and $\vec{s}_2 = -\mathrm{cdeg}_{-\vec{s}}\mathbf{G}_1$.
2. Compute an $\left(\hat{\mathbf{N}}_2 \tilde{\mathbf{G}}_1, \sigma_2, -\vec{s}_2\right)$-order basis $\mathbf{P}_2$ and partition $\mathbf{P}_2 = [\mathbf{G}_2, \mathbf{Q}_2]$ with $\mathbf{G}_2$ a $\left(\hat{\mathbf{N}}_2, -\vec{s}_2\right)$-kernel basis. Set $\vec{s}_3 = -\mathrm{cdeg}_{-\vec{s}_2}\mathbf{G}_2$ and $\tilde{\mathbf{G}}_2 = \tilde{\mathbf{G}}_1 \mathbf{G}_2$.
3. Continuing this process, at each step i we compute a $\left(\hat{\mathbf{N}}_i \tilde{\mathbf{G}}_{i-1}, \sigma_i, -\vec{s}_i\right)$-order basis $\mathbf{P}_i$ and then partition $\mathbf{P}_i = [\mathbf{G}_i, \mathbf{Q}_i]$ with $\mathbf{G}_i$ a $\left(\hat{\mathbf{N}}_i \tilde{\mathbf{G}}_{i-1}, -\vec{s}_i\right)$-kernel basis. Let $\tilde{\mathbf{G}}_i = \prod_{j=1}^{i} \mathbf{G}_i = \tilde{\mathbf{G}}_{i-1} \mathbf{G}_i$.
4. Return $\tilde{\mathbf{G}}_{\lceil \log k \rceil}$, a $\left(\mathbf{N}^T, -\vec{s}\right)$-kernel basis.

This process of computing a $\left(\mathbf{N}^T, -\vec{s}\right)$-kernel basis is formally given in Algorithm 1.

4.3 Cost of Left Kernel Basis Computation

The cost of Algorithm 1 is dominated by the order basis computations and the multiplications $\hat{\mathbf{N}}_i \tilde{\mathbf{G}}_{i-1}$ and $\tilde{\mathbf{G}}_{i-1} \mathbf{G}_i$. Let $s = \xi/n$.

LEMMA 4.5. *An $\left(\hat{\mathbf{N}}_i \tilde{\mathbf{G}}_{i-1}, \sigma_i, -\vec{s}_i\right)$-order basis can be computed with a cost of $O^\sim(n^\omega s)$.*

PROOF. Note that $\mathbf{N}_i$ has less than 2^i columns. Otherwise, since $\mathrm{cdeg}_{\vec{s}}\mathbf{N}_i > \xi/2^i$, we have $\sum \mathrm{cdeg}_{\vec{s}}\mathbf{N}_i > 2^i \xi/2^i = \xi$, contradicting with $\sum \mathrm{cdeg}_{\vec{s}}\mathbf{N} = \sum \vec{b} \le \sum \vec{s} \le \xi$. It follows that $\hat{\mathbf{N}}_i$, and therefore $\hat{\mathbf{N}}_i \tilde{\mathbf{G}}_{i-1}$, also have less than 2^i rows. We also have $\sigma_i = \lceil \xi/2^{i-1} \rceil + 1 \in \Theta(\xi/2^i)$. Therefore, Algorithm 2 from [19] for order basis computation with unbalanced shift can be used with a cost of $O^\sim(n^\omega s)$. □

LEMMA 4.6. *The multiplications $\hat{\mathbf{N}}_i \tilde{\mathbf{G}}_{i-1}$ can be done with a cost of $O^\sim(n^\omega s)$.*

PROOF. The dimension of $\hat{\mathbf{N}}_i$ is bounded by $2^i \times n$ and $\sum \mathrm{rdeg}_{\vec{s}}\hat{\mathbf{N}}_i \le 2^i \cdot \xi/2^{i-1} \in O(\xi)$. We also have $\mathrm{cdeg}_{-\vec{s}}\tilde{\mathbf{G}}_{i-1} \le 0$, or equivalently, $\mathrm{rdeg}\,\tilde{\mathbf{G}}_{i-1} \le \vec{s}$. We can now use Theorem 2.3 to multiply $\tilde{\mathbf{G}}_{i-1}^T$ and $\hat{\mathbf{N}}_i^T$ with a cost of $O^\sim(n^\omega s)$. □

LEMMA 4.7. *The multiplication $\tilde{\mathbf{G}}_{i-1}\mathbf{G}_i$ can be done with a cost of $O^\sim(n^\omega s)$.*

PROOF. We know $\mathrm{cdeg}_{-\vec{s}}\tilde{\mathbf{G}}_{i-1} = -\vec{s}_i$, and $\mathrm{cdeg}_{-\vec{s}_i}\mathbf{G}_i = -\vec{s}_{i+1} \le 0$. In other words, $\mathrm{rdeg}\,\mathbf{G}_i \le \vec{s}_i$, and $\mathrm{rdeg}_{\vec{s}_i}\tilde{\mathbf{G}}_{i-1} \le \vec{s}$, hence we can again use Theorem 2.3 to multiply $\mathbf{G}_i^T$ and $\tilde{\mathbf{G}}_{i-1}^T$ with a cost of $O^\sim(n^\omega s)$. □

LEMMA 4.8. *Given an input matrix $\mathbf{M} \in \mathbb{K}[x]^{k \times n}$, a shift $\vec{s} \in \mathbb{Z}^n$, and an upper bound $\xi \in \mathbb{Z}$ such that*

(i) $\sum \mathrm{rdeg}_{\vec{s}}\mathbf{M} \le \xi$,

(ii) $\sum \vec{s} \le \xi$,

(iii) any $(\mathbf{M}, -\vec{s})$-kernel basis having row degrees bounded by $\vec{s}$, or equivalently, $-\vec{s}$-column degrees bounded by 0.

Then Algorithm 1 costs $O^\sim(n^\omega s)$ field operations to compute a $(\mathbf{M}, -\vec{s})$-kernel basis.

Note that while the upper bound ξ can be simply replaced by $\sum \vec{s}$ in Lemma 4.8 and Algorithm 1 for computing a right factor in this section, keeping it separate makes the algorithm more general and allows it to be reused in the next section.

It may also be informative to note again the correspondence between Lemma 4.8 and Theorem 2.11, on the reversal of the degree structures of the input matrices and the output kernel bases.

THEOREM 4.9. *A right factor $\mathbf{G}$ satisfying $\mathbf{F} = \mathbf{TG}$ for a column basis $\mathbf{T}$ can be computed with a cost of $O^\sim(n^\omega s)$.*

5. COMPUTING A COLUMN BASIS

Once a right factor $\mathbf{G}$ of $\mathbf{F}$ has been computed, we are in a position to determine a column basis $\mathbf{T}$ using the equation $\mathbf{F} = \mathbf{TG}$. In order to do so efficiently, however, the degree of $\mathbf{T}$ cannot be too large. We see that this is the case from the following lemma.

LEMMA 5.1. *Let $\mathbf{F}$ and $\mathbf{G}$ be as before and $\vec{t} = -\mathrm{rdeg}_{-\vec{s}}\mathbf{G}$. Then*

(i) the column degrees of $\mathbf{T}$ are bounded by the corresponding entries of $\vec{t}$;

(ii) if $\vec{t}$ has r entries and $\vec{s}\,'$ is the list of the r largest entries of $\vec{s}$, then $\vec{t} \le \vec{s}\,'$.

PROOF. Since $\mathbf{G}$ is $-\vec{s}$-row reduced, and $\mathrm{rdeg}_{-\vec{s}}\mathbf{F} \le 0$, by Lemma 2.1 $\mathrm{rdeg}_{-\vec{t}}\mathbf{T} \le 0$, or equivalently, $\mathbf{T}$ has column degrees bounded by $\vec{t}$.

Let $\mathbf{G}'$ be the $-\vec{s}$-row Popov form of $\mathbf{G}$ and the square matrix $\mathbf{G}''$ consist of only the columns of $\mathbf{G}'$ that contains pivot entries, and has the rows permuted so the pivots are in the diagonal. Let $\vec{s}\,''$ be the list of the entries in $\vec{s}$ that correspond to the columns of $\mathbf{G}''$ in $\mathbf{G}'$. Note that $\mathrm{rdeg}_{-\vec{s}\,''}\mathbf{G}'' = -\vec{t}\,''$ is just a permutation of $-\vec{t}$ with the same entries. By the definition of shifted row degree, $-\vec{t}\,''$ is the sum of $-\vec{s}\,''$ and the list of the diagonal pivot degrees, which are nonnegative. Therefore, $-\vec{t}\,'' \geq -\vec{s}\,''$. The result then follows as $\vec{t}$ is a permutation of $\vec{t}\,''$ and $\vec{s}\,'$ consists of the largest entries of $\vec{s}$. □

Having determined a bound on the column degrees of $\mathbf{T}$, we are now ready to compute $\mathbf{T}$. This is done again by computing a kernel basis using an order basis computation as before.

LEMMA 5.2. *Let $\vec{t}^* = [0, \ldots, 0, \vec{t}] \in \mathbb{Z}^{m+r}$. Then any $\left(\left[\mathbf{F}^T, \mathbf{G}^T\right], -\vec{t}^*\right)$-kernel basis has the form $\begin{bmatrix} V \\ \bar{\mathbf{T}} \end{bmatrix}$, where $V \in \mathbb{K}^{m \times m}$ is a unimodular matrix and $\left(\bar{\mathbf{T}}V^{-1}\right)^T$ is a column basis of $\mathbf{F}$.*

PROOF. Note first that the matrix $\begin{bmatrix} -I \\ \mathbf{T}^T \end{bmatrix}$ is a kernel basis of $\left[\mathbf{F}^T, \mathbf{G}^T\right]$ and is therefore unimodularly equivalent to any other kernel basis. Hence any other kernel basis has the form $\begin{bmatrix} -I \\ \mathbf{T}^T \end{bmatrix} U = \begin{bmatrix} V \\ \bar{\mathbf{T}} \end{bmatrix}$, with U and $V = -U$ unimodular. Thus $\mathbf{T} = \left(\bar{\mathbf{T}}V^{-1}\right)^T$. Also note that the $-\vec{t}^*$ minimality forces the unimodular matrix V in any $\left(\left[\mathbf{F}^T, \mathbf{G}^T\right], -\vec{t}^*\right)$-kernel basis to be of degree 0, the same degree as I. □

EXAMPLE 5.3. *Let*

$$\mathbf{F} = \begin{bmatrix} x^2 & x^2 & x + x^2 & 1 + x^2 \\ 1 + x + x^2 & x^2 & 1 + x^2 & 1 + x^2 \end{bmatrix},$$

a matrix over $\mathbb{Z}_2[x]$, and

$$\mathbf{G} = \begin{bmatrix} 1 & 0 & 1 & 0 \\ x & x^2 & 0 & 1 + x^2 \end{bmatrix},$$

a minimal left kernel basis of a right kernel basis of $\mathbf{F}$. In order to compute the column basis $\mathbf{T}$ satisfying $\mathbf{F} = \mathbf{T}\mathbf{G}$, first we can determine $\operatorname{cdeg} \mathbf{T} \leq \vec{t} = [2, 0]$ *from Lemma 5.1. Then we can compute a $\left[0, 0, -\vec{t}\right]$-minimal left kernel basis of $\begin{bmatrix} \mathbf{F} \\ \mathbf{G} \end{bmatrix}$. The matrix*

$$\left[V^T, \bar{\mathbf{T}}^T\right] = \begin{bmatrix} V \\ \bar{\mathbf{T}} \end{bmatrix}^T = \begin{bmatrix} 1 & 0 & x + x^2 & 1 \\ 1 & 1 & 1 + x & 0 \end{bmatrix}$$

is such a left kernel basis. A column basis can then be computed as

$$\mathbf{T} = \left(V^T\right)^{-1} \bar{\mathbf{T}}^T = \begin{bmatrix} x + x^2 & 1 \\ 1 + x^2 & 1 \end{bmatrix}.$$

□

In order to compute a $\left(\left[\mathbf{F}^T, \mathbf{G}^T\right], -\vec{t}^*\right)$-kernel basis, we can again use order basis computation as before, as we again have an order basis that contains a $\left(\left[\mathbf{F}^T, \mathbf{G}^T\right], -\vec{t}^*\right)$-kernel basis.

LEMMA 5.4. *Any $\left(\left[\mathbf{F}^T, \mathbf{G}^T\right], \vec{s} + 1, -\vec{t}^*\right)$-order basis contains a $\left(\left[\mathbf{F}^T, \mathbf{G}^T\right], -\vec{t}^*\right)$-kernel basis whose $-\vec{t}^*$-row degrees are bounded by 0.*

PROOF. As before, Lemma 4.2 can be used here. We just need to show that a $\left(\left[\mathbf{F}^T, \mathbf{G}^T\right], -\vec{t}^*\right)$-kernel basis has $-\vec{t}^*$-row degrees no more than 0. This follows from the fact that $\operatorname{rdeg}_{-\vec{t}^*} \begin{bmatrix} I \\ \mathbf{T}^T \end{bmatrix} \leq 0$. □

Algorithm 2 ColumnBasis($\mathbf{F}$)

Input: $\mathbf{F} \in \mathbb{K}[x]^{m \times n}$.
Output: a column basis of $\mathbf{F}$.

$\vec{s}$:= cdeg $\mathbf{F}$;
$\mathbf{N}$:= MinimalKernelBasis $(\mathbf{F}, \vec{s})$;
$\mathbf{G}$:= $\left(\text{MinimalKernelBasisReversed}(\mathbf{N}^T, \vec{s}, \sum \vec{s})\right)^T$;
$\vec{t}^*$:= $[0, \ldots, 0, -\operatorname{rdeg}_{-\vec{s}} \mathbf{G}]$, with rowDimension($\mathbf{G}$) number of 0's ;
$\begin{bmatrix} V \\ \bar{\mathbf{T}} \end{bmatrix}$:= MinimalKernelBasisReversed($\left[\mathbf{F}^T, \mathbf{G}^T\right], \vec{t}^*, \sum \vec{s}$) with a square V;
$\mathbf{T}$ = $\left(\bar{\mathbf{T}}V^{-1}\right)^T$;
return $\mathbf{T}$;

In order to compute a $\left(\left[\mathbf{F}^T, \mathbf{G}^T\right], -\vec{t}^*\right)$-kernel basis efficiently, we notice that we have the same type of problem as in Section 4.2 and hence we can again use Algorithm 1.

LEMMA 5.5. *A $\left(\left[\mathbf{F}^T, \mathbf{G}^T\right], -\vec{t}^*\right)$-kernel basis can be computed using Algorithm 1 with a cost of $O^\sim(n^\omega s)$, where $s = \xi/n$ is the average column degree of $\mathbf{F}$ as before.*

PROOF. Just use the algorithm with input $\left(\left[\mathbf{F}^T, \mathbf{G}^T\right], \vec{t}^*, \xi\right)$. We can verify the conditions on the input are satisfied.

- To see that $\sum \operatorname{rdeg}_{\vec{t}^*} \left[\mathbf{F}^T, \mathbf{G}^T\right] \leq \xi$, note that from $\vec{t} = -\operatorname{rdeg}_{-\vec{s}} \mathbf{G}$ and Lemma 2.2 that $\operatorname{cdeg}_{\vec{t}} \mathbf{G} \leq \vec{s}$, or equivalently, $\operatorname{rdeg}_{\vec{t}} \mathbf{G}^T \leq \vec{s}$. Since we also have $\operatorname{rdeg} \mathbf{F}^T \leq \vec{s}$, it follows that $\operatorname{rdeg}_{\vec{t}^*} \left[\mathbf{F}^T, \mathbf{G}^T\right] \leq \vec{s}$.
- The condition $\sum \vec{t}^* \leq \xi$ follows from Lemma 5.1.
- The third condition holds since $\begin{bmatrix} -I \\ \mathbf{T}^T \end{bmatrix}$ is a kernel basis with row degrees bounded by $\vec{t}^*$.

□

With a $\left(\left[\mathbf{F}^T, \mathbf{G}^T\right], -\vec{t}^*\right)$-kernel basis $\left[V^T, \bar{\mathbf{T}}^T\right]^T$ computed, a column basis is then given by $\mathbf{T} = \left(\bar{\mathbf{T}}V^{-1}\right)^T$.

The complete algorithm for computing a column basis is then given in Algorithm 2.

THEOREM 5.6. *A column basis $\mathbf{T}$ of $\mathbf{F}$ can be computed with a cost of $O^\sim(n^\omega s)$, where $s = \xi/n$ is the average column degree of $\mathbf{F}$ as before.*

PROOF. The cost is dominated by the cost of the three kernel basis computations in the algorithm. The first one is handled by the algorithm from [20] and Theorem 2.11, while the remaining two are handled by Algorithm 1, Lemma 4.8 and Lemma 5.5. □

6. A SIMPLE IMPROVEMENT

When the input matrix $\mathbf{F}$ has column dimension n much larger than the row dimension m, then we can separate $\mathbf{F} = [\mathbf{F}_1, \mathbf{F}_2, \ldots, \mathbf{F}_{n/m}]$ into n/m blocks, each with dimension $m \times m$, assuming without loss of generality n is a multiple of m, and the columns are arranged in increasing degrees. We then do a series of column basis computations. First we compute a column basis $\mathbf{T}_1$ of $[\mathbf{F}_1, \mathbf{F}_2]$. Then compute a column basis $\mathbf{T}_2$ of $[\mathbf{T}_1, \mathbf{F}_3]$. Repeating this process, at step i, we compute a column basis $\mathbf{T}_i$ of $[\mathbf{T}_{i-1}, \mathbf{F}_{i+1}]$, until $i = n/m - 1$, when a column basis of $\mathbf{F}$ is computed.

Lemma 6.1. *Let $\bar{s}_i = \left(\sum \text{cdeg}\, \mathbf{F}_i\right)/m$. Then at step i, computing a column basis $\mathbf{T}_i$ of $[\mathbf{T}_{i-1}, \mathbf{F}_{i+1}]$ can be done with a cost of $O^{\sim}\left(m^{\omega}(\bar{s}_i + \bar{s}_{i+1})/2\right)$ field operations.*

Proof. From Lemma 5.1, the column basis $\mathbf{T}_{i-1}$ of $[\mathbf{F}_1, \ldots, \mathbf{F}_i]$ has column degrees bounded by the largest column degrees of $\mathbf{F}_i$, hence $\sum \text{cdeg}\, \mathbf{T}_{i-1} \leq \sum \text{cdeg}\, \mathbf{F}_i$. The lemma then follows by combining this with the result from Theorem 5.6 that a column basis $\mathbf{T}_i$ of $[\mathbf{T}_{i-1}, \mathbf{F}_{i+1}]$ can be computed with a cost of $O^{\sim}(m^{\omega}\hat{s}_i)$, where

$$\hat{s}_i = \left(\sum \text{cdeg}\, \mathbf{T}_{i-1} + \sum \text{cdeg}\, \mathbf{F}_{i+1}\right)/2m \leq \frac{(\bar{s}_i + \bar{s}_{i+1})}{2}.$$

□

Theorem 6.2. *If $s = \left(\sum \text{cdeg}\, \mathbf{F}\right)/n$, then a column basis of $\mathbf{F}$ can be computed with a cost of $O^{\sim}(m^{\omega}s)$.*

Proof. Summing up the cost of all the column basis computations,

$$\sum_{i=1}^{n/m-1} O^{\sim}\left(m^{\omega}\left(\bar{s}_i + \bar{s}_{i+1}\right)/2\right) \subset O^{\sim}\left(m^{\omega}\left(\sum_{i=1}^{n/m} \bar{s}_i\right)\right) = O^{\sim}\left(nm^{\omega-1}s\right),$$

since $\sum \text{cdeg}\, \mathbf{F} = \sum_{i=1}^{n/m}(m\bar{s}_i) = ns$. □

Remark 6.3. *In this section, the computational efficiency is improved by reducing the original problem to about n/m subproblems whose column dimensions are close to the row dimension m. This is done by successive column basis computations. Note that we can also reduce the column dimension by using successive order basis computations, and only do a column basis computation at the very last step. The computational complexity of using order basis computation to reduce the column dimension would remain the same, but in practice it may be more efficient since order basis computations are simpler.*

7. CONCLUSION

In this paper we have given a fast, deterministic algorithm for the computation of a column basis for $\mathbf{F}$ having complexity $O^{\sim}(n^{\omega}s)$ field operations in $\mathbb{K}$ with s an upper bound for the average column degree of $\mathbf{F}$. Our methods rely on a special factorization of $\mathbf{F}$ into a column basis and a kernel basis. These in turn are computed via fast kernel basis and fast order basis algorithm of [20, 19]. When these computations involve the multiplication of polynomial matrices with unbalanced degrees then they use the fast method for such multiplications given in [20].

In a later publication we will show how this column basis algorithm can be used in efficient deterministic computations of matrix determinant, Hermite form, and computations of column reduced form and Popov form for matrices of any rank and any dimension.

8. REFERENCES

[1] B. Beckermann, H. Cheng, and G. Labahn. Fraction-free row reduction of matrices of Ore polynomials. *Journal of Symbolic Computation*, 41(1):513–543, 2006.

[2] B. Beckermann and G. Labahn. A uniform approach for the fast computation of matrix-type Padé approximants. *SIAM Journal on Matrix Analysis and Applications*, 15(3):804–823, 1994.

[3] B. Beckermann and G. Labahn. Recursiveness in matrix rational interpolation problems. *Journal of Computational and Applied Math*, 5-34, 1997.

[4] B. Beckermann and G. Labahn. Fraction-free computation of matrix rational interpolants and matrix GCDs. *SIAM Journal on Matrix Analysis and Applications*, 22(1):114–144, 2000.

[5] B. Beckermann, G. Labahn, and G. Villard. Shifted normal forms of polynomial matrices. In *Proceedings of ISSAC'99*, pages 189–196, 1999.

[6] B. Beckermann, G. Labahn, and G. Villard. Normal forms for general polynomial matrices. *Journal of Symbolic Computation*, 41(6):708–737, 2006.

[7] Th. G. J. Beelen, G. J. van den Hurk, and C. Praagman. A new method for computing a column reduced polynomial matrix. *System Control Letters*, 10(4):217–224, 1988.

[8] P. Giorgi, C.-P. Jeannerod, and G. Villard. On the complexity of polynomial matrix computations. In *Proceedings of ISSAC'03*, pages 135–142, 2003.

[9] S. Gupta, S. Sarkar, A. Storjohann, and J. Valeriote. Triangular x-basis decompositions and derandomization of linear algebra algorithms over K[x]. *Journal of Symbolic Computation*, 47(4):422–453, 2012.

[10] T. Kailath. *Linear Systems*. Prentice-Hall, 1980.

[11] C. Li. Lattice compression of polynomial matrices. Master's thesis, School of Computer Science, University of Waterloo, 2006.

[12] S. Sarkar. Computing Popov forms of polynomial matrices. Master's thesis, University of Waterloo, 2011.

[13] S. Sarkar and A. Storjohann. Normalization of row reduced matrices. In *Proceedings of ISSAC'11*, pages 297–304, 2011.

[14] A. Storjohann and G. Villard. Computing the rank and a small nullspace basis of a polynomial matrix. In *Proceedings of ISSAC'05*, pages 309–316, 2005.

[15] G. Villard. Computing Popov and Hermite forms of polynomial matrices. In *Proceedings of ISSAC'96*, pages 250–258, 1996.

[16] J. von zur Gathen and J Gerhard. *Modern Computer Algebra*. Cambridge University Press, 2003.

[17] W. Zhou. *Fast Order Basis and Kernel Basis Computation and Related Problems*. PhD thesis, University of Waterloo, 2012.

[18] W. Zhou and G. Labahn. Efficient computation of order bases. In *Proceedings of ISSAC'09*, pages 375–382. ACM, 2009.

[19] W. Zhou and G. Labahn. Efficient algorithms for order basis computation. *Journal of Symbolic Computation*, 47:793–819, 2012.

[20] W. Zhou, G. Labahn, and A. Storjohann. Computing minimal nullspace bases. In *Proceedings of ISSAC'12*, pages 375–382. ACM, 2012.

Author Index

www.ingramcontent.com/pod-product-compliance
Lightning Source LLC
LaVergne TN
LVHW060134070326
832902LV00018B/2785
9781450320597